Avalanche Transit-Time Devices

MODERN FRONTIERS IN APPLIED SCIENCE

Gallium Arsenide Microwave Bulk and Transit-Time Devices
by Lester F. Eastman

Laser Applications
by W. V. Smith

Microwave Engineers Handbook
by Ted Saad

Logarithmic Video Amplifiers
by R. S. Hughes

Parallel Coupled Lines and Directional Couplers
by Dr. Leo Young

Microwave Filters Using Parallel Coupled Lines
by Dr. Leo Young

Data Modem Selection and Evaluation Guide
by V. V. Vilips

Design, Performance and Application of Microwave Semiconductor Control Components
by Kenneth E. Mortenson and Jose M. Borrego

Adaptive Electronics
by Dr. Wolfgang Gaertner

Phased-Array Antennas
by Dr. Arthur A. Oliner and George H. Knittel

Gallium Arsenide Microwave Bulk and Transit-Time Devices
by Lester F. Eastman

Infrared to Millimeter Wavelength Detectors
by Frank R. Arams

ARTECH HOUSE, INC.
affiliated with Horizon House – Microwave, Inc.
610 Washington Street
Dedham, Massachusetts, 02026

Avalanche Transit-Time Devices

George I. Haddad

ARTECH HOUSE, INC.
Dedham, Massachusetts

affiliated with
Horizon House-Microwave, Inc.

Printed and bound in the United States of America.

Library of Congress Catalog card number: 72-77132

FOREWORD

The purpose of this volume is to present, under one cover, a collection of some of the most important articles on Avalanche Transit-Time Devices that have appeared in the literature in order to provide a ready reference for workers active in the field. This has proven to be a difficult task, especially when one considers the number of excellent articles that have been published on the subject. Due to the lack of space, it is impossible to include all of the significant work which has been done by every individual. I sincerely hope that no one will feel slighted if his work has not been included.

The articles in this volume are divided into the following seven sections.

1. General review articles.
2. Characteristics of the IMPATT mode.
3. Harmonic effects in the IMPATT mode.
4. The high-efficiency (TRAPATT) mode.
5. Noise and oscillator stabilization.
6. Various oscillator properties.
7. Other articles of interest.

The avalanche transit-time diode is now established as a useful, reliable, microwave device which can perform many functions including, generation, amplification, frequency conversion and detection of RF power. Its frequency of operation has been extended well into the millimeter-wavelength region and it is now finding increased usage in systems applications. Many manufacturers are able to produce these devices. The fabrication technology has reached such a sophisticated level that volume production at a relatively low cost is feasible. This has provided a great impetus for utilization of these devices in civilian applications, an activity which is growing rapidly.

There are many individuals who deserve credit for the contents of this volume. Perhaps, the greatest credit belongs to B. C. DeLoach and his co-workers at Bell Telephone Laboratories for discovering the device and for making many significant contributions to bring it to its present state of development. I wish to thank the authors of the papers which appear in this volume for their permission to include their work. The help provided by M. S. Gupta and W. E. Schroeder in the selection of papers is gratefully acknowledged and appreciated. Finally, I wish to thank Miss Betty Cummings for her patience, diligence and excellent work in the preparation of this volume.

George I. Haddad

Avalanche Transit-Time Devices

TABLE OF CONTENTS IX

TABLE OF CONTENTS

SECTION V NOISE AND OSCILLATOR STABILIZATION

SECTION VI **VARIOUS OSCILLATOR PROPERTIES**

SECTION VII **OTHER ARTICLES OF INTEREST**

GENERAL REVIEW ARTICLES 1

A Proposed High-Frequency, Negative-Resistance Diode

3

By W. T. READ, JR.

(Manuscript Received March 2, 1957)

This paper describes and analyzes a proposed semiconductor diode designed to operate as an oscillator when mounted in a suitable microwave cavity. The frequency would be in the range extending from 1 to 50 kmc. The negative Q may be as low as 10 and the efficiency as high as 30 per cent.

The diode is biased in reverse so as to establish a depletion, or space-charge, layer of fixed width in a relatively high resistance region, bounded by very low resistance end regions. The electric field has a maximum at one edge of the space-charge region, where hole-electron pairs are generated by internal secondary emission, or avalanche. The holes (or electrons) travel across the space-charge layer with constant velocity, thus producing a current through the diode. Because of the build-up time of the avalanche, and the transit time of the holes across the depletion layer, the alternating current is delayed by approximately one-half cycle relative to the ac voltage. Thus, power is delivered to the ac signal. When the diode is mounted in an inductive microwave cavity tuned to the capacity of the diode, an oscillation will build up. It appears possible to obtain over 20 watts of ac power in continuous operation at 5 kmc.

I. DESCRIPTION

This paper discusses a proposed oscillator consisting of a semiconductor diode biased in reverse and mounted in a microwave cavity. The impedance of the cavity is mainly inductive and is matched to the mainly capacitative impedance of the diode so as to form a resonant system. We shall show that the diode can have a negative ac resistance so that it delivers power from the dc bias to the oscillation. The negative Q may be as low as 10 and the efficiency as high as 30 per cent.

The principle of operation is as follows: a reverse bias is applied to establish a space-charge, or depletion, layer of fixed width in a relatively weakly doped region bounded by highly doped end regions. A possible structure is the n^+-p-i-p^+ structure, where the + denotes high doping and i means intrinsic. This structure is shown in Fig. 1(a). The field distribution under reverse bias is shown in Fig. 1(b). The voltage is always well above the punch-through voltage, so that the space-charge region always extends from the n^+-p junction through the p and i (intrinsic) regions to the i-p^+ junction. The fixed charges in the various regions are shown in Fig. 1(b). A positive charge gives a rising field in going from left to right. A positive field makes holes move to the right. The maximum field, which occurs at the n^+-p junction, is of the order of several hundred kilovolts per cm, so that hole-electron pairs are generated by internal secondary emission (also called multiplication or avalanche). The electrons go immediately into the n^+ region. The holes move to the right across the space-charge region. The field throughout the space-charge region is in the range (above about 5 kilovolts per cm) where carriers move with constant velocity independent of the field. For practical purposes we can forget about the electrons in the space-charge

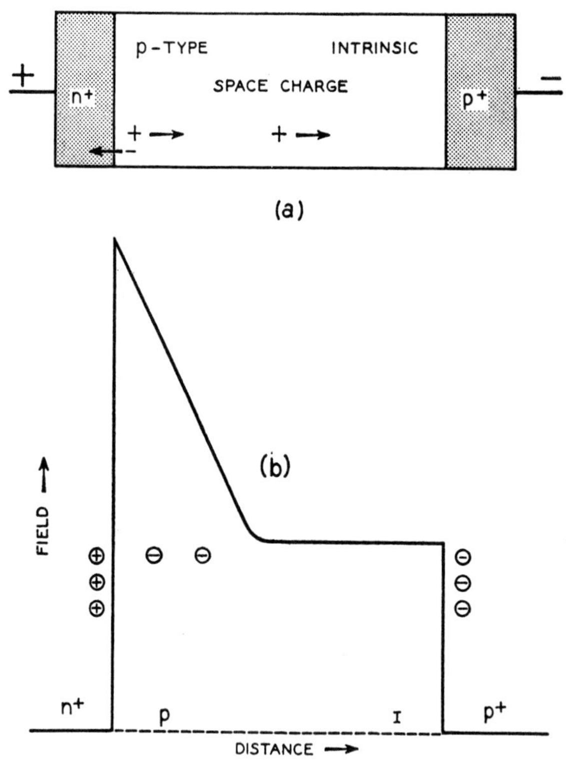

Fig. 1 — The structure (a) and field distribution (b) under reverse bias.

region and consider that current is generated at the n^+-p junction, so that only the holes move in the space-charge region. Thus, the physical picture is as follows: a current of holes $I_0(t)$ is generated at the n^+-p junction. The holes move to the right and traverse the space-charge region, moving with constant velocity, v. In silicon v is about 10^7 cm sec^{-1}. The transit time of a hole across the space-charge region is $\tau = W/v$, where W is the width of the space-charge region. Throughout the discussion we shall illustrate various quantities by giving numerical values for the case $W = 10^{-3}$ cm. Thus τ would be 10^{-10} sec.

The holes moving across the space-charge region produce a current $I_e(t)$ in the external circuit. It can be shown that I_e is equal to the average current in the space-charge region. Since velocity is constant, I_e is simply $v/W = 1/\tau$ times the total charge of the moving holes. (We shall assume throughout that all quantities refer to unit area of junction.) Suppose, for example, that a pulse of holes of charge δQ is suddenly generated at the n^+-p junction. Immediately, a constant current $I_e = \delta Q/\tau$ begins to flow in the external circuit, and continues to flow during the time, τ, that the holes are moving across the space-charge region from the n^+-p junction to the p^+ region. Thus, on the average, the external current $I_e(t)$ due to the moving holes is delayed by $\tau/2$ relative to the current $I_0(t)$ generated at the n^+-p junction. We shall show (in the discussion of multiplication) that the current $I_0(t)$ is delayed by one-quarter of a cycle relative to the ac voltage. Thus, to get a total delay of one-half cycle, we want the delay $\tau/2$ due to transit time to be one-quarter of a cycle. The cavity should therefore be tuned to give a resonant frequency $(2\tau)^{-1}$.

The conductive current $I_e(t)$, which arises from carriers moving through the space-charge layer, should be distinguished from the displacement, or capacitative, current I_c, which charges and discharges the diode regarded as a capacitor. This current, I_c, supplies the variation in charge at the edges of the space-charge region, where the field changes abruptly and, for practical purposes, can be considered discontinuous. Since I_c is 90° out of phase with the voltage it contributes nothing to the power.

1.1 *Multiplication*

Carriers moving in the high field near the n^+-p junction acquire enough energy to knock valence electrons into the conduction band, thus producing hole-electron pairs. The rate of pair production, or multiplication, is a sensitive nonlinear function of the field. By proper doping, the field can be given a relatively sharp peak so that multiplication is confined to a very narrow region at the n^+-p junction. The multiplication rate can be regarded as a function of the peak field, E_0. At a critical field, E_c, breakdown will occur; that is, any current will be self-sustaining; every pair produced will on the average produce one other pair. We shall neglect the thermally generated reverse saturation current, which will be small compared to the generated current $I_0(t)$. Then, when the field is above E_c, the current $I_0(t)$ will be more than self-sustaining and will build up. When the field is below the critical breakdown field E_c, the current is less than self-sustaining and will die down.

In operation, the diode is biased so that the peak field is above E_c during the positive half of the voltage cycle and below E_c during the negative half. Hence the current $I(t)$ builds up during the positive half and dies down during the negative half. Therefore $I_0(t)$ reaches its maximum in the middle of the voltage cycle, or one-quarter of a cycle later than the voltage.

We have assumed so far that the field varies in phase with the voltage. This will be a good approximation if the carrier space-charge can be neglected. This can be seen from Fig. 1(b). If the voltage is above the punch-through voltage, an increase in voltage simply raises the whole field distribution throughout the depletion layer. That is, additional charges simply appear at the edges of the space-charge layer. Hence the field at each point varies in phase with the voltage. The space charge of the carriers will, however, affect the shape of field distribution in the space-charge region. When the current becomes too large, the carrier space-charge cannot be neglected. As we shall see, the effect of carrier space-charge is to reduce the delay between the voltage and the current generated, I_0. In practice this limits the dc bias current.

When the dc bias is small enough so that carrier space charge can be neglected, the operation can be summarized as follows: the peak field varies in phase with the voltage and generates (at one edge of the depletion layer) a current delayed by 90°. This current gives rise to a current through the diode delayed by 90° relative to the current generated and, therefore, by 180° relative to the voltage.

1.2 *Diode and Cavity*

The capacity of the diode and the inductance of the cavity determine the frequency, f, of oscillation. We have seen that the optimum frequency, f, is such that the transit time, τ, is one-half cycle. Hence the angular frequency, ω, should be

$$\omega = 2\pi f = \frac{\pi}{\tau} = \frac{\pi v}{W}. \tag{1}$$

For $W = 10^{-3}$ cm, the frequency, f, would be 5,000 mc.

We have seen that the current into the diode is $I_e + I_c$, where I_e is a conductive current due to holes moving in the space-charge region and I_c is a displacement current required to charge the diode regarded as a capacitor. Thus, the diode acts like a capacity and negative resistance in parallel.

The capacity is related to the width, W, of the depletion layer by

$$\begin{aligned} C &= \frac{\kappa}{4\pi W}, \\ &= \frac{1.06 \times 10^{-12}}{W} \quad \text{for Si,} \end{aligned} \tag{2}$$

where C is in farads per cm^2 and W in cm. The negative conductance of the diode will be small compared to ωC, so the admittance is mainly that of the capacity. Thus for $W = 10^{-3}$ cm the admittance will be about 30 mhos per cm^2.

To make a resonant system we mount the diode in a microwave cavity having a mainly inductive impedance. The inductance, L, of the cavity is chosen so that the cavity and diode together have a resonant frequency $f = \frac{1}{2}\tau$.

Fig. 2 shows a possible cavity with the diode mounted in it. The diode should have narrow end regions so that the depletion layer is as close as possible to the metal base and center post of the cavity. Then the heat generated will flow away rapidly. The cavity is tuned by adjusting the vertical position of the plunger forming the top of the cavity. There must be some break in the cavity so that a dc bias can be applied. This is accomplished in Fig. 2 by insulating the center lead. The break could also be made by insulating the walls from the base. On the right is an outlet through which the ac power is tapped.

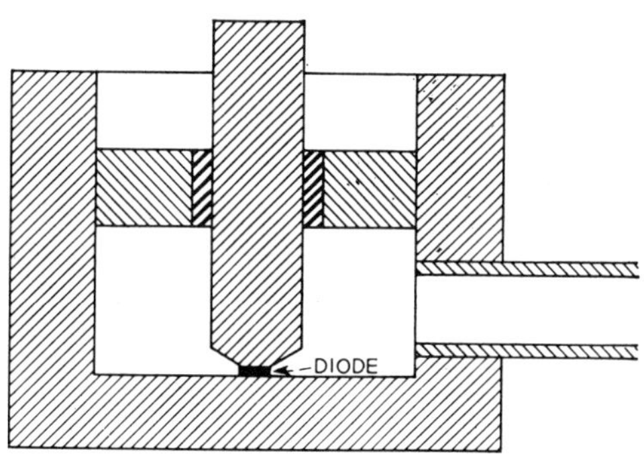

Fig. 2 — Diode mounted in microwave cavity.

The impedance, ωL, of the cavity must be equal to the impedance of the diode. If the diode is circular with radius R, then the junction area is πR^2 and the impedance, $(\pi R^2 \omega C)^{-1}$. Thus the area of the diode is limited by how small the impedance of the cavity can be made. Setting $\omega L = (\pi R^2 \omega C)^{-1}$ and using (1) and (2) we have

$$\begin{aligned} \left(\frac{R}{W}\right)^2 &= \frac{4}{\pi \kappa v \omega L}, \\ &= \frac{9.5 \times 10^3}{\omega L} \text{ for Si,} \end{aligned} \tag{3}$$

where ωL is in ohms. It would be difficult to make ωL less than about 10 ohms. Hence R could not be more than about $30W$.

1.3 *Stable Operation*

We have shown that the diode will deliver power, so an oscillation will build up. In Section III we shall calculate exactly the Q for the small signal case where everything is linear. For reasonable values of the direct current bias, I_d, the Q can be made negative and as low in magnitude as desired. As the oscillation builds up, the behavior becomes highly nonlinear. In the practical range of operation the field will vary by a fraction — probably less than 20 per cent — of its average value but the current will vary by orders of magnitude. The current I_e approaches a square wave, being negligibly small during the positive half of the ac voltage cycle and almost constant during the negative half cycle. Since the direct current I_d is the average conductive current, it follows that the amplitude of variation of I_e is approximately equal to I_d. If V_a is the ac voltage amplitude, the ac power delivered P_r is found to be

$$P_r = \frac{2}{\pi} I_d V_a \tag{4}$$

per unit area of junction. Thus, if the dc bias is applied by a constant direct current generator, the power delivered is proportional to V_a. The energy stored in the capacity is proportional to V_a^2. Hence, $-Q$ is proportional to V_a. In other words, if the amplitude increases, the stored energy, or energy of oscillation, increases faster than the energy delivered per cycle. This is the condition required for a stable oscillation to be possible. At the stable operating point, $-Q$ of the diode is equal to the effective Q of the cavity. If the amplitude increases, the energy delivered by the diode increases less than the energy lost to the cavity. Thus, the amplitude decreases. In the same way a decrease in amplitude is also self-correcting.

The total power output is P_r times the area πR^2 and is therefore proportional to $R^2 I_d V_a$. The maximum R is given by (3). The maximum allowable I_d is limited by the fact that carrier space-charge cannot be too large, as we now show.

1.4 *Space Charge of the Carriers*

The holes moving across the space-charge region will affect the field distribution. As shown in Fig. 1(b) a positive charge gives a rise in field (in going from left to right) and a negative charge, a decrease in field. Thus, the effect of the holes will be to oppose the fixed negative charge in the depletion layer, and hence to flatten the field distribution shown in Fig. 1(b). Therefore, for a given voltage, the effect of the holes is to reduce the peak field at the n^+-p junction. The multiplication rate increases with the peak field. Thus, the current generated has a space charge that tends to reduce the rate of current generation. That is, the current tends to shut itself off. Instead of building up throughout the positive half of the voltage cycle, $I_0(t)$ builds up until the carrier space-charge has reduced the peak field below the sustaining field E_c. Then the current decreases. Thus, I_0 reaches its peak *before* the middle of the cycle. This reduces the delay and hence the power. Increasing the current, therefore, increases the ac power only up to a certain point where carrier space-charge begins to spoil the phase relations.

We have seen that in the practical range of operation the current I_e varies by approximately its dc value I_d. A current I_d corresponds to a total carrier charge τI_d in the space-charge layer (since I_e is the average current in the space-charge layer and all carriers travel with velocity $v = W/\tau$). If the carrier space-charge is to be neglected it must be small compared to the charge CV_a that produces the voltage variation. Thus we want $\tau I_d \ll CV_a$. Actually, as we shall show in Section III, it is sufficient to take $I_d\tau = CV_a/2$. Since the period of oscillation is 2τ, this becomes

$$I_d = \frac{\omega}{2\pi} CV_a. \tag{5}$$

If I_d is no greater than this the current and voltage will be roughly 180° out of phase and we can use (4) for the power per unit area. For larger I_d the increase in power would be more than offset by the effect of carrier space-charge on the phase relations.

1.5 *Power Output*

Combining (3) with (5) we have, for the total ac power output

$$\pi R^2 P_r = \frac{V_a^2}{\pi^2 \omega L}. \tag{6}$$

9

As we have seen, the impedance ωL of the cavity could be as low as 10 ohms. Now consider the maximum V_a. This is limited by the dc voltage V_d. It is seen from Fig. 1(b) that if V_a is too large the field in the intrinsic region will be reduced to zero during the negative voltage cycle. By careful doping, the voltage at punch-through can be made small compared to the dc bias at breakdown. Even allowing for the reduction in field due to the carrier space-charge, we will be safe with $V_a = V_d/2$. This gives an efficiency of about 30 per cent. From (6) the power becomes

$$\text{Power Output} = \pi R^2 P_r = \left(\frac{V_d}{2\pi}\right)^2 \frac{1}{\omega L}. \tag{7}$$

The dc voltage is limited by the following considerations: (a) In the negative half cycle the field must not fall below about 5 kilovolts per cm. Otherwise carrier velocity will depend on field; so the carriers will slow down during the negative half of the ac voltage cycle and thus reduce the power. This can be avoided if V_d is at least 10^4W volts, where W is in cm. (b) To localize the multiplication the field must remain well below E_c except near the n^+-p junction. This will be so if $V_d = 2V_a$ is less than about $(2/5)E_cW$. Then the maximum field in the intrinsic region will always be less than 0.6 E_c. Thus we have $10^4W < V_d < 0.4\, E_cW$.

For W as large as 10^{-3} cm the critical field will be about 350 kilovolts per cm. For W as low as 10^{-4} cm the multiplication would have to be confined to a region no more than 10^{-5} cm in width. This would require a somewhat steeper field gradient and a higher critical field — say about 650 kilovolts per cm. Thus for $W = 10^{-4}$ cm, or 50 kmc, V_d would have to be less than 26 volts. For $\omega L = 10$ ohms, the maximum power output would then be less than 2 watts. At $W = 10^{-3}$ cm, or 5 kmc, the maximum V_d would be 140 volts, which gives a power of 50 watts. At $W = 10^{-2}$, or 500 mc, the maximum voltage would be 1400 volts, and the minimum 100 volts. The power would then be between 25 watts and 5 kw.

1.6 *Zener Current*

Chynoweth and McKay[1] have shown that in sufficiently narrow junctions, where breakdown occurs at around 10 volts or less, the current is generated not by multiplication but by internal field emission. That is, electrons tunnel from the valence to conduction bands. This is known as Zener current. We shall show in Section IV and Appendix E that the device will operate on Zener current, but much less effectively. This, even more than the limited power, may limit the minimum W and hence the maximum frequency that would be practical.

Experiments

An experimental program has been undertaken to construct and test a device operating by avalanche.

II. CURRENT AND SPACE CHARGE

In this section we consider the physics of multiplication, space-charge and carrier flow in more detail and obtain the equations that govern the behavior of the diode.

2.1 *Multiplication*

Electrons moving with velocity v generate hole-electron pairs at a rate of $\alpha v n$, where n is the electron concentration and α is the ionization rate. By definition α is the number of pairs produced on the average by an electron in moving unit distance. Thus α^{-1} is the average distance between ionizations. We shall be dealing with the case where each pair produces roughly one other pair; thus α^{-1} is of the order of the width of the narrow region near the n^+-p junction where the multiplication is localized. For fields below about 600 kilovolts per cm α can be considered a function of field, $\alpha = \alpha(E)$. For larger fields the electrons do not have time to get into equilibrium with the field in the average distance α^{-1} between collisions. We shall be dealing with fields where $\alpha = \alpha(E)$.

McKay[2] has measured α as a function of E for electrons in silicon. Fig. 3 is a plot of the best fit curve to McKay's data, together with the theoretical curve calculated by P. Wolff[3]. A straight line of slope 6 is also shown and is seen to be a good fit to the data below about 500 kilovolts per cm. In the formulas we shall use the general relation $\alpha \approx E^m$ and take $m = 6$ for numerical calculations.

Assumptions

We shall make the simplifying assumption that $\alpha = \alpha(E)$ is the same for both holes and electrons. Actually this is not so. However, α is so sensitive to field that the difference in field for the same α is small. Any difference in the α's will alter both the carrier and field distributions so as to favor the carrier with the lower α. Consequently, it is believed that this assumption will not give misleading results.

12

We also assume that holes and electrons travel with the same velocity in the high fields involved. This will not lead to serious error since the motion of one type of carrier always plays a minor role.

2.2 *DC Case*

We take the x axis normal to the junctions with $x = 0$ at the n^+-p junction and $x = W$ at the i-p^+ junction, as shown in Fig. 4. It can be shown that the direct current I is related to the thermally generated re-

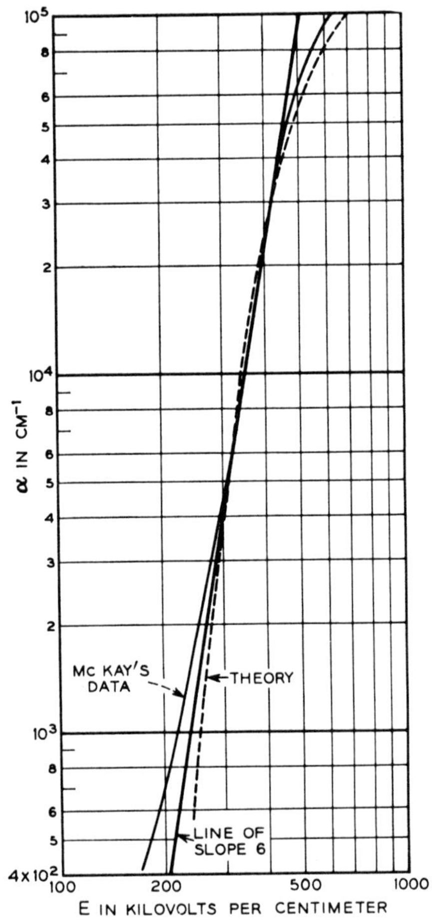

Fig. 3 — Ionization rate vs electric field.

verse saturation current I_s by

$$\frac{I_s}{I} = 1 - \int \alpha(E)\, dx \tag{8}$$

where the integral is taken over the space-charge region. See, for example, McKay[2]. We will be dealing with space-charge regions where most of the multiplication occurs in a narrow region, which we shall call the *multiplication region*, near $x = 0$. Thus we need take the integral in (8) only from $x = 0$ to $x = x_1$, where $x_1 \ll W$ is the width of the narrow multiplication region.

Breakdown occurs, that is, the direct current becomes infinite, when $\int \alpha\, dx = 1$. Physically this means that each hole-electron pair generated in the multiplication regions will, on the average, generate one other pair. We now consider how $\int \alpha\, dx$ depends on the peak field E_0 for various field distributions including the one shown in Fig. 4.

(a) The simplest case is that where the field is a constant E_0 in the narrow multiplication region. This would correspond to a short region of high constant field followed by an abrupt drop to a much lower field.

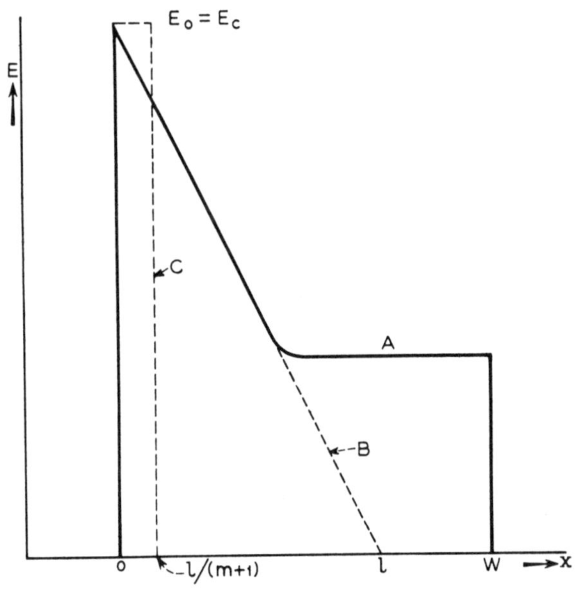

Fig. 4 — Field distribution at breakdown.

Then

$$\int \alpha \, dx = \int_0^{x_1} \alpha(E_0) \, dx = x_1 \alpha(E_0).$$

So if α is proportional to E^m

$$\int \alpha \, dx = \left(\frac{E_0}{E_c}\right)^m,$$

where E_c the critical field for breakdown is given by $\alpha(E_0) = 1/x_1$.

(b) Next consider the linear field distribution shown in Fig. 4. Curve A is the field distribution at breakdown. We can take $E = E_0 - kx$ in the multiplication region. The slope k is proportional to the charge in the p-region. We shall neglect the effect of the carriers on the space charge. Then k will be proportional to the fixed charge in the p-region and will be constant independent of current. In practice k would be well above E_c/W. In order for the carriers to produce a field gradient comparable to this, the current would have to be of the order of $(\kappa v/4\pi)(E_c/W)$, or several thousand amperes/cm^2. Actual currents will be much smaller than this.

In Fig. 4 the flat section of field will contribute negligibly to the multiplication, so we can replace the actual field distribution A by the tangent curve B. Then

$$\int \alpha \, dx = \left(\frac{E_0}{E_c}\right)^{m+1}, \tag{9}$$

where E_c is given by $\alpha(E_c) = (m + 1)/l$ and $l = E_c/k$ is the zero intercept of the tangent curve B at breakdown. Thus the peak field E_c at breakdown is equal to the breakdown field for a constant field region of width $l/(m + 1)$, as shown by curve C in Fig. 4.

(c) For a linear gradient junction the field varies parabolically with x and

$$\int \alpha \, dx = \left(\frac{E_0}{E_c}\right)^{m+1/2}.$$

In what follows we shall use (9), which applies to a linear field, or abrupt junction. This will be perfectly general if m is left arbitrary. The results obtained for $m = 6$ will be a good approximation for a linear gradient junction, since the difference between 7 and 6.5 is within the experimental uncertainty in m.

As an example, consider a step junction with $m = 6$ and $W = 10^{-3}$ and let $l = 0.7W$. Then the critical field E_c is given by $\alpha(E_c) = 7/l = 10^{-4}$ cm. From Fig. 3, $E_c = 350$ kilovolts/cm.

2.3 *Time Dependence*

In the dc case $\int \alpha \, dx$ cannot be greater than unity. That is, we cannot get above breakdown. This is not necessarily so for rapidly varying fields. We now derive a differential equation for the current as a function of time. This will reduce to (8) in the dc case.

Let p and n be the hole and electron densities respectively. The corresponding currents are $I_p = qvp$ and $I_n = -qvn$, where q is the electronic charge. Hole-electron pairs are being generated at a rate $\alpha v(n + p)$. This is so large compared to the rate of thermal generation that the latter can be neglected. The equations of continuity then become 15

$$\frac{\partial p}{\partial t} = -\frac{1}{q}\frac{\partial I_p}{\partial x} + \alpha v(n + p), \tag{10}$$

$$\frac{\partial n}{\partial t} = \frac{1}{q}\frac{\partial I_n}{\partial x} + \alpha v(n + p), \tag{11}$$

where $\alpha = \alpha(E)$. The three variables, p, n and E are determined by (10) (11) and by Poisson's equation and the boundary conditions at the junctions. An exact solution is impractical. Instead, we make an approximation based on the fact that most of the generation is confined to the narrow multiplication region near $x = 0$. This is shown in Fig. 5. The multiplication region extends from $x = 0$ to $x = x_1$. The width x_1 is a small fraction of the width W of the space-charge region.

2.4 *Assumptions*

We shall assume that (a) *all* of the multiplication occurs in the relatively narrow multiplication region and (b) the total current $I(x, t) = I_p(x, t) + I_n(x, t)$ in the multiplication region is a function $I(x, t) = I_0(t)$

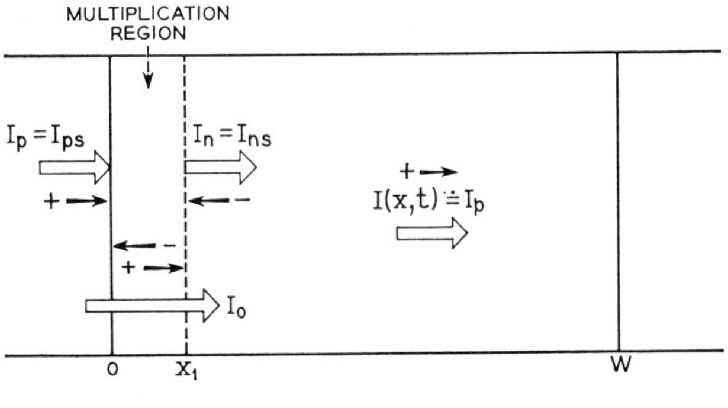

Fig. 5 — Boundary conditions on the multiplication region.

of time only. This, of course, will not be true of I_p and I_n individually in either the ac or dc cases. Assumption (a) will be valid if the net acceptor concentration in the p-region is sufficient to give a sharp drop in field and hence to localize the multiplication. The second assumption will be a good approximation provided the current does not vary by a large percentage in the time $\tau_1 = x_1/v$ required for carriers to cross the multiplication region. This will not be so at large amplitudes. However, the errors for rising and falling currents tend to cancel, so the equation is right on the average. As we shall see, at large amplitudes only the average is involved.

2.5 *Differential Equation for $I_0(t)$*

Adding (10) and (11), using the assumption $I_p + I_n = I_0(t)$ and integrating from $x = 0$ to $x = x_1$ gives

$$\tau_1 \frac{dI_0}{dt} = -[I_p - I_n]_0^{x_1} + 2I_0 \int_0^{x_1} \alpha\, dx, \tag{12}$$

where $\tau_1 = x_1/v$ is the transit time across the multiplication region. The boundary conditions are shown in Fig. 5. The hole current at $x = 0$ consists entirely of the reverse saturation current I_{ps} of holes thermally generated in the n^+-region; these have moved to the n^+p-junction by diffusion. Thus at $x = 0$, $I_p - I_n = 2I_p - I_0 = 2I_{ps} - I_0$. At $x = x_1$ the electron current consists of the reverse saturation current I_{ns} of electrons thermally generated both in the space-charge region and in the p^+-region, so $I_p - I_n = -2I_{ns} + I_0$. With these boundary conditions, (12) becomes

$$\frac{\tau_1}{2}\frac{dI_0}{dt} = I_0\left(\int_0^{x_1} \alpha\, dx - 1\right) + I_s. \tag{13}$$

In the *dc* case I_0 is the direct current I, so this reduces to (8).

The condition for breakdown is $\int_0^{x_1} \alpha\, dx = 1$. If a field that satisfies this is suddenly applied, I_0 will increase linearly at a rate of $2I_s/\tau_1$ and become infinite. If a larger field is applied, I_0 will approach infinity exponentially. For a smaller field I_0 will approach a finite value.

The integral $\int \alpha\, dx$ over the multiplication region depends very little on carrier space-charge. Hence it will be negligibly affected by the small differences in carrier distribution between the ac and dc cases. We can therefore combine (9) and (13) to obtain a differential equation

$$\frac{\tau_1}{2}\frac{d}{dt}\ln I_0 = (E_0/E_c)^{m+1} - 1 + \frac{I_s}{I_0} \tag{14}$$

relating the current $I_0 = I_0(t)$ in the multiplication region to the peak field $E_0 = E_0(t)$.

In most practical cases the current I_0 will be so large compared to I_s that the effect of I_s can be neglected. The correction due to I_s and the detailed formulas for evaluating the effect are given in Appendix D.

2.6 *Example*

17

At low enough amplitudes of oscillation we can expand E_0/E_c in powers of $E_0/E_c - 1$ and retain only the linear term. Then, neglecting I_s, equation (14) becomes

$$\frac{d}{dt} \ln I_0 = \frac{2(m+1)}{\tau_1}\left(\frac{E_0}{E_c} - 1\right). \tag{15}$$

If I_0 is to be periodic, then E_0 must be periodic and the dc bias must be such that the average E_0 is E_c. Suppose we apply a periodic voltage with the proper bias so that $E_0 = E_c + E_a \sin \omega t$, where ω is the optimum frequency π/τ. Then

$$\ln \frac{I_0(t)}{I_0(0)} = \frac{2(m+1)}{\pi} \frac{\tau}{\tau_1} \frac{E_a}{E_c}\left(1 - \cos \frac{\pi t}{\tau}\right). \tag{16}$$

Fig. 6 shows the field and $\ln I_0$ as functions of time. Suppose $\tau/\tau_1 = W/x_1 = 20$ and $m = 6$. Then even if the amplitude E_a of variation of E_0 is as small as 1.3 per cent of E_c, I_0 will vary by a factor of 10 over a cycle. Thus we can have small signals in field and voltage but large signals in current. We shall call this the intermediate range of amplitudes.

From (16) the maximum value of I_0 is seen to occur in the middle of the cycle, where $t = \tau$. Thus, if I_0 varies by a large factor, the current is generated mainly in a pulse in the middle of the ac voltage cycle, as shown in Fig. 6.

Actually the space charge of the current will affect the $E_0(t)$ curve, which will not be exactly sinusoidal for a sinusoidal voltage. However, this does not affect the conclusion that a small field and voltage signal can give a large current signal and that the current I_0 approaches a pulse as the ac amplitude increases.

2.7 *Carrier Space-Charge*

We have now dealt with the multiplication region, and obtained an equation relating I_0 and E_0. It remains to consider the rest of the space-charge region, where current generation is negligible. We shall also neglect the reverse saturation current I_{ns} of electrons, which will be negligible compared to the total current $I(x, t)$. From Poisson's equation we shall derive another relation between I_0 and E_0. This will involve also the voltage $V(t)$ and together with (14) will uniquely determine $I_0(t)$ and $E_0(t)$ for any $V(t)$.

2.8 *Physical Picture*

The physical picture is shown in Fig. 7(a). The width of the narrow multiplication region is small compared to the total width W of the space-charge region. Therefore, in treating the current, space-charge and field distributions throughout the whole space-charge region, we shall assume that the multiplication region has zero width so that all current is generated at $x = 0$. Then $I_0(t)$ is a current of holes flowing out of $x = 0$, and the only carriers in the space-charge region are the holes.

The current $I(x, t)$ at any point x and time t is

$$I(x, t) = I(0, t - x/v) = I_0(t - x/v). \tag{17}$$

Thus the entire hole and current distributions are given in terms of $I_0(t)$.

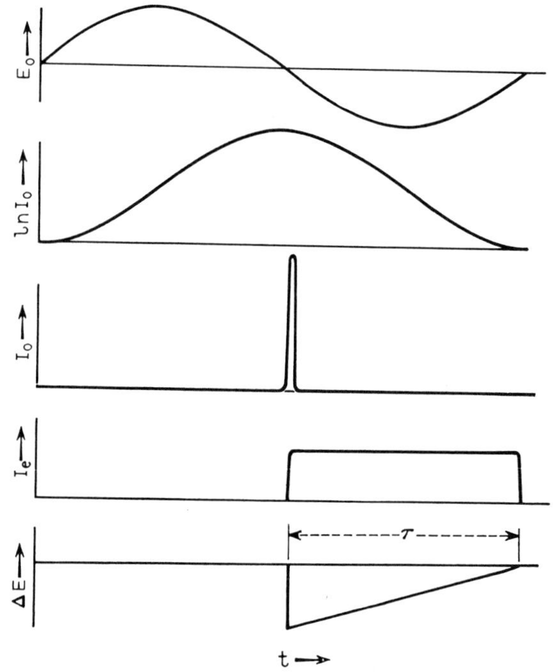

Fig. 6 — Case of a sharp current pulse.

Next consider the effect of the holes on the space charge and the field. The field distribution $E_p(x)$ at punch-through is shown by the dashed curve in Fig. 7(b). If no current flowed the field distribution for any voltage above the punch-through voltage would be simply $E_p(x)$ plus a constant determined by the voltage. The fixed negative charge in the p-material gives a drop in field across the space-charge, or depletion, layer. The positive space charge of the holes opposes that of the acceptors and hence reduces the drop in field. So, for a given voltage, E_0 will decrease as the current flowing in the space-charge region increases and flattens the field distribution. 19

2.9 *External Current*

The holes traversing the space-charge region give rise to a current $I_e = I_e(t)$ in the external circuit. I_e is equal to the average current flow-

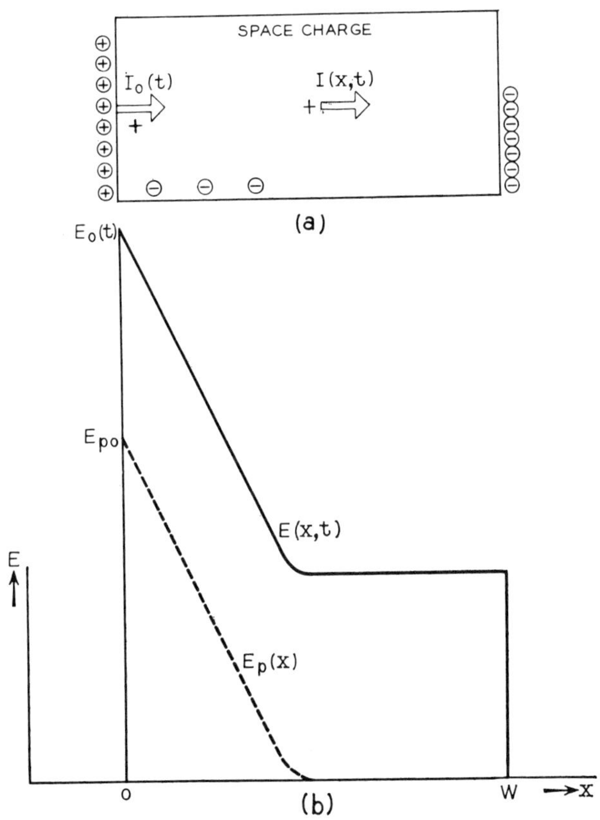

Fig. 7 — Current (a) and field (b) in the space-charge region.

20

ing in the space-charge region. (A rigorous and simple proof of this for the present case of plane parallel geometry is given in Appendix A.) Thus

$$\begin{aligned} I_e(t) &= \frac{1}{W}\int_0^W I(x, t)\, dx \\ &= \frac{1}{W}\int_0^W I_0(t - x/v)\, dx \\ &= \frac{1}{\tau}\int_{t-\tau}^{t} I_0(t')\, dt'. \end{aligned} \tag{18}$$

In other words, the current I_e in the external circuit is the total charge in the space-charge region divided by the transit time τ. Fig. 6 shows I_e for the case discussed at the end of Section III, where I_0 was a sharp pulse in the middle of the cycle. From (18) it follows that the average value of $I_e(t)$ is equal to the average of $I_0(t)$.

In addition to I_e, which arises from carriers moving through the space-charge region, there is a capacitative current

$$I_c = C\frac{dV}{dT} \tag{19}$$

flowing in the external circuit. When the voltage $V(t)$ across the diode is above the punch-through voltage, V_p, the capacity is a constant given by (2). As discussed earlier, I_c is the current required to charge and discharge the diode regarded as a capacitor. It furnishes the variation in charge at the edges of the space-charge region.

2.10 *Effect of Current on Field*

We now show how the space charge of the holes reduces the peak field for a given voltage. The stability of the device comes from the fact that current multiplication increases as E_0 increases but the current carriers give a space charge that reduces E_0.

If there were no current flowing, any increase in V above V_p would simply raise the entire field distribution by an amount $(V - V_p)/W$. Fig. 8(a) is a plot of the difference $E'(x, t) = E(x, t) - E_p(x)$ at a given time. The slope of the $E'(x, t)$ curve is determined entirely by the space charge of the holes; the effect of the fixed charge is already included in $E_p(x)$. The holes give a charge density $I(x, t)/v$. Thus, Poisson's equation is

$$\frac{\partial E'}{\partial x}(x, t) = \frac{4\pi}{\kappa v} I(x, t) = \frac{4\pi}{\kappa v} I_0(t - x/v). \tag{20}$$

The excess of V over V_p is equal to the area under the curve in Fig. 8(a). This is equal to $WE'(0, t) = W[E_0(t) - E_{p0}]$ plus the sum of the areas of a number of horizontal strips like the one shown. The area of such a strip is $(W - x)\, dE'$. So

$$V(t) - V_p = W[E_0(t) - E_{p0}] + \int_0^W (W - x) \frac{\partial E'}{\partial x}\, dx.$$

Substituting (20) into this and setting $t' = t - x/v$ gives

$$\begin{aligned} E_0(t) &= E_{p0} + \frac{V(t) - V_p}{W} + \Delta E(t), \\ \Delta E(t) &= -\frac{4\pi}{\kappa\tau} \int_{t-\tau}^{t} I_0(t')[\tau - t + t']\, dt'. \end{aligned} \tag{21}$$

Here $\Delta E(t)$ is the effect of current on the field. The quantity ΔE is always negative.

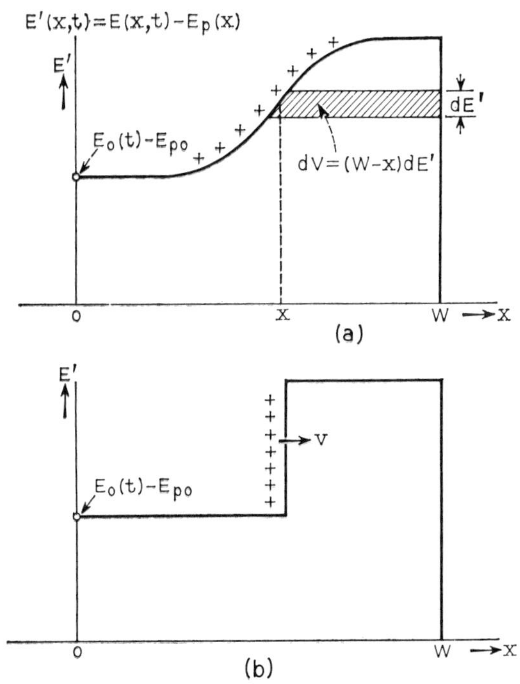

Fig. 8 — Effect of carrier space-charge on field.

2.11 *Effect of a Current Pulse*

To illustrate the physical meaning of (21) we return to the example illustrated in Fig. 6. A sharp pulse of current was generated near the middle of the cycle. The reduction $-\Delta E$ in E_0 due to this pulse is also shown in Fig. 6. If the pulse is instantaneous, $-\Delta E$ jumps at once to its maximum value and then declines to zero linearly in the time τ required for the pulse to cross the space-charge region. The physical reason for this is easily seen by reference to Fig. 8(b), which shows the same thing as Fig. 8(a) except that the carrier space-charge is concentrated at one point, that is, in a pulse. For a given voltage (area under the curve), the reduction in E_0 will decrease from its maximum value to zero as the pulse moves to the right across the space-charge region. From the effect of a single instantaneous pulse of current on E_0, the effect of any arbitrary current pulse can be found by resolving it into a series of instantaneous pulses and superposing the effects. An instantaneous pulse of current of charge δQ gives an instantaneous drop in field of $4\pi\delta Q/\kappa$.

III. ANALYSIS

We now have two equations, (14) and (21), relating the current $I_0(t)$, the field $E_0(t)$ and the voltage $V(t)$. Thus in principle the current can be found for any applied voltage. Actually the exact solution is impractical except in limiting cases. In this section we present (a) an exact solution for the linear small-signal case, (b) an approximate analysis for large amplitudes and (c) a rapidly converging iteration method that yields solutions of any desired accuracy.

3.1 *Voltage*

We shall assume that the voltage varies sinusoidally, $V(t) = V_d + V_a \sin \omega t$. This will certainly be a good assumption in the small-signal range, where the diode is linear. The cavity is linear at all amplitudes. At large amplitudes of oscillation a sinusoidal voltage gives a sinusoidal capacitative current I_c plus a conductive current I_e which approaches a square wave as the amplitude increases. Thus we are assuming that the voltage across the cavity is sinusoidal while the current contains a distribution of higher frequencies. For a cavity like that shown in Fig. 2, this assumption may be a relatively good approximation. If the cavity is tuned to the fundamental it may be almost a short circuit for the higher frequencies.

3.2 *Dimensionless Variables*

It will simplify the discussion to express everything in terms of dimensionless variables, taking as units parameters that characterize the device. For example, W would be unit length, τ, unit time, and E_c, unit field. Then the carriers would travel with unit velocity $v = W/\tau = 1$. As can be seen in Fig. 1, the voltage will be of the order of, but less than, the unit voltage WE_c. It is convenient to choose the unit charge so that $4\pi/\kappa = 1$. Then unit charge produces a unit gradient of field. Since $v = 1$, a unit charge moving in the space-charge region gives unit current. Hence unit hole current in the space-charge region produces unit field gradient, and the average current I_e in the space-charge region is equal to the total drop in field due to the carrier space-charge. The actual current will be small compared to unit current since the curent produces a drop in field that is smaller than the ac field variation, which in turn is small compared to E_c.

From the choice of units it follows that the diode has unit capacity and that optimum frequency is $\omega = \pi$. The following table summarizes the units and gives typical values for a silicon diode with $W = 10^{-3}$ cm. McKay's data,[2] plotted in Fig. 3, gives $E_c = 350$ kilovolts per cm; the effective width of the multiplication region at breakdown is taken to be 10^{-4} cm.

Quantity	Unit	Example
length	W	10^{-3} cm
time	τ	10^{-10} sec
field	E_c	3.5×10^5 volts/cm
voltage	WE_c	350 volts
current density	$\frac{\kappa E_c}{4\pi\tau} = vCE_c$	3.7×10^3 amps/cm^2
power density	$\frac{\kappa v}{4\pi} E_c^2$	1.3×10^6 watts/cm^2

The unit of power is seen to be relatively independent of W, since unit voltage goes as W and unit current as $1/W$. (A given current causes a greater drop in field across a wider space-charge region.)

3.3 *Governing Equations*

Since the peak field $E_0(t)$ varies around E_c it is convenient to define a dimensionless field

$$E(t) = \frac{E_0(t) - E_c}{E_c}. \tag{22}$$

Then, for a sinusoidal voltage variation, equation (21) becomes in dimensionless terms

$$\begin{aligned} E(t) &= E_b + V_a \sin \omega t + \Delta E, \\ \Delta E &= -\int_{t-1}^{t} I_0(t')[1 - t + t']\, dt'. \end{aligned} \tag{23}$$

24

Here ΔE is the effect of current on E and E_b the effect of the dc voltage V_d. Let E_{p0} be the peak field at punch-through. Then, at punch-through, E is $(E_{p0} - E_c)/E_c$. If no current flowed, any increase in dc voltage above the punch-through voltage V_p would simply raise the whole field distribution by $(V_d - V_p)/W$ (see Fig. 7). Thus the dimensionless parameter E_b is given in terms of dimensional quantities by

$$E_b = \frac{E_{p0} - E_c}{E_c} + \frac{V_d - V_p}{WE_c}. \tag{24}$$

The value of E_b will be very small compared to unity.

In equation (14) we neglect I_s and expand the right-hand side in terms of powers of $E = E(t)$. It will be sufficient to stop the expansion at E^2. Then

$$\frac{d}{dt} \ln I_0 = \frac{2(m + 1)}{\tau_1} \left(E + \frac{m}{2} E^2 \right). \tag{25}$$

The current through the diode is the capacitative current

$$I_c = C\, dV/dt = \omega V_a \cos \omega t \tag{26}$$

plus the conductive current I_e, where, from (18),

$$I_e(t) = \int_{t-1}^{t} I_0(t')\, dt'. \tag{27}$$

Average, or dc, Values

We shall let $\langle\ \rangle$ denote time averages and define I_d as the direct current $I_d = \langle I_e + I_c \rangle = \langle I_e \rangle = \langle I_0 \rangle$ where the last step follows from (27). Averaging (23) gives

$$\langle E \rangle = E_b - \frac{I_d}{2} \tag{28}$$

or $\langle \Delta E \rangle = -I_d/2$. Since $I_0(t)$ is periodic, we have, from (25)

$$\langle E \rangle + \frac{m}{2} \langle E^2 \rangle = 0. \tag{29}$$

Thus, in the small-signal limit the average field $\langle E \rangle$ vanishes, and $I_d = 2E_b$, where E_b is given by (24).

3.4 *Linear, Small-Signal Case*

When all quantities vary by small fractions of their average values, then the equations separate into a dc part and a linear ac part which is easily solved. We now derive the impedance as a function of the direct current I_d for the optimum frequency $\omega = \pi$. The ac voltage is $V_a \sin \pi t$ or, in complex form, $V_a e^{i\pi t}$. We write $I_0 = I_d + I_{0a} e^{i\pi t}$ where I_{0a} is complex. Similar expressions can be written for I_e and E. Equations (23), (25) and (27) give the ac relations

$$\begin{aligned} i\pi I_{0a} &= \frac{2(m+1)}{\tau_1} I_d E_a , \\ I_{0a} &= \frac{i\pi}{2} I_{ea} , \qquad (30) \\ i\pi(V_a - E_a) &= I_{0a} - I_{ea} . \end{aligned}$$

From equations (30), $V_a = Z I_{ea}$ where

$$Z = \frac{1}{2}\left[1 - \frac{\pi^2 \tau_1}{2(m+1)I_d}\right] + \frac{i}{\pi} . \qquad (31)$$

The current through the diode has two parts, I_e and I_c, where I_c is a pure capacitative current. Thus the equivalent circuit consists of a unit capacity in parallel with an element of impedance Z, where I_e goes through Z. The impedance Z consists of a fixed reactance and a resistance that varies with the dc bias. A simpler equivalent circuit emerges from considering the admittance Y of the diode. Since the capacity and the impedance Z are in parallel, $Y = i\pi + 1/Z$. Here the Q of the diode is the ratio of the imaginary part Y_i of Y to the real part Y_r. From (31)

$$\begin{aligned} Q &= \frac{\pi}{2}\left[1 - \frac{\pi^2 \tau_1}{2(m+1)I_d}\right] , \\ Y_r &= \frac{\pi Q}{1+Q^2} , \qquad Y_i = \frac{\pi}{1+Q^{-2}} . \end{aligned} \qquad (32)$$

Thus Q varies linearly with τ_1/I_d and is negative for I_d less than $[\pi^2 \tau_1/2(m+1)]$. When I_d is equal to this critical value the diode becomes an open circuit for this frequency. This means that none of the alternating current generated in the multiplication region flows out of the diode. Rather it flows into the edges of the space-charge region and provides the current that charges and discharges the diode regarded as a capacitor. In other words, the unit acts as a capacitor generating its own charge internally. Hence voltage can vary with no external alternating current.

3.5 *Equivalent Circuit*

Equations (32) describe a simple equivalent circuit consisting of a fixed unit capacity in series with a conductance πQ, where Q depends on the dc bias. The equivalent circuit suggests the following practical conclusion: If the cavity is designed to act like an inductance in series with a variable load resistance, then the load resistance can be made equal to the negative resistance of the diode, so that the two resistances cancel, and the equivalent circuit consists of the fixed capacity of the diode and the inductance of the cavity. Hence at small amplitudes the resonant frequency will be independent of the dc bias.

3.6 *Sharp Pulse Approximation*

As the oscillation builds up, the behavior rapidly ceases to be linear, and it is impractical to solve the equations exactly. However, as we have seen, when the amplitude increases the current $I_0(t)$ approaches a sharp pulse and $I_e(t)$ approaches a square wave as shown in Fig. 6 for the optimum frequency $\omega = \pi$. The average current I_d is half the maximum. In the limit of a sharp pulse the problem again becomes simple. We now derive some approximate relations for this case, and show how the oscillation can be stabilized.

The power delivered P_r is

$$P_r = -\frac{V_a}{2}\int_0^2 I_e(t) \sin \pi t \, dt.$$

We can substitute (27) for $I_e(t)$ into this and reduce the double integral to a single integral by integrating by parts. The result is

$$P_r = -\frac{2V_a}{\pi}\langle I_0(t) \cos \pi t\rangle \tag{33}$$

where again the brackets denote the average over a cycle.

Thus, if the current I_0 is generated in a pulse near the middle of the cycle, where $\cos \pi t$ is negative, P_r will be negative and power will be delivered to the ac signal.

We define the Q of the diode as 2π times the ratio of the energy stored in the capacity to the energy lost in a cycle. The stored energy for unit capacity is $V_a^2/2$. The energy lost is the negative of the power delivered, P_r, times the period $2\pi/\omega = 2$. So

$$\frac{1}{Q} = \frac{4}{\pi^2}\frac{\langle I_0(t) \cos \pi t\rangle}{V_a}. \tag{34}$$

Let the pulse of current occur at a time t_1. Then in the limit of an instantaneous pulse (34) becomes

$$\frac{1}{Q} = \frac{4}{\pi^2}\frac{I_d}{V_a} \cos \pi t_1. \tag{35}$$

The phase relations will be ideal for $t_1 = 1$; that is, when the pulse occurs in the middle of the cycle. In this case $-Q$ increases with V_a for constant direct current, so the oscillation is stable.*

We now consider how t_1 depends on V_a and the dc bias. The current pulse becomes sharper as V_a increases and τ_1 decreases, as can be seen from (16). In the limit of vanishing τ_1 the pulse becomes instantaneous and the problem can be solved exactly. This is done in Appendix B. Here we give a simple physical argument that will be a good approximation so long as the duration of the pulse is small compared to a period, as will be the case in the range of practical operation. 27

Fig. 9 shows $E(t) = E_b + V_a \sin \pi t + \Delta E$ for the case where the current pulse occurs at an arbitrary time t_1. As illustrated in Fig. 6, a

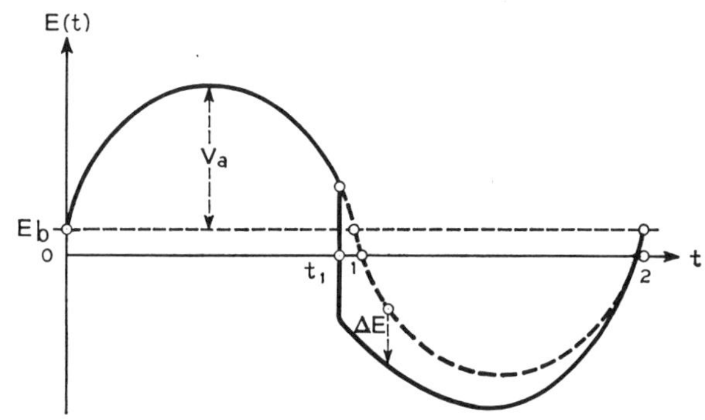

Fig. 9 — Variation of peak field with time for sharp current pulse.

current pulse causes ΔE to drop abruptly and then rise linearly to zero in a transit time. Let E_δ be the abrupt drop in E. We have already seen that the average ΔE is $-I_d/2$. Since ΔE is a triangular wave, lasting half a cycle, it follows that the maximum is four times the average; so $E_\delta = 2I_d$.

Let us consider in some detail the relatively short interval during which current is being generated. The holes can move only a short distance during this time. Hence, as seen from Fig. 8, the field will drop by an amount roughly proportional to the amount of charge generated. It will, therefore, have dropped by $E_\delta/2$ when half the charge has been generated. The pulse will be roughly symmetrical and will have reached its peak when half the charge has been generated. The current is a maximum at $E(t) = 0$ since it builds up for positive E and decreases for negative E. Thus, during the first half of the pulse E drops to zero, the

* I am indebted to J. L. Moll for pointing out the advantage of applying the bias with a constant direct current generator.

drop being $E_\delta/2$. During the second half of the pulse the field continues to drop to $-E_\delta/2$. This is shown in Fig. 9. Before the pulse occurs the field is $E(t) = E_b + \sin \pi t$. Hence in the limit of an instantaneous pulse $E_b + V_a \sin \pi t_1 = E_\delta/2$. Since $E_\delta = 2I_d$

$$I_d = E_b + V_a \sin \pi t_1 \,. \tag{36}$$

Eliminating $\langle E \rangle$ from equations (28) and (29) gives another relation between I_d and E_b :

$$I_d = 2E_b + \left(\frac{m}{2}\right) \langle E^2 \rangle .$$

Since E_b , V_a and $-\Delta E \leqq 2I_d$ are all small compared to unity, the only contribution to $\langle E^2 \rangle$ that cannot be neglected in comparison with I_d and E_b is $\langle V_a^2 \sin^2 \pi t \rangle = V_a^2/2$. So we have

$$I_d = 2E_b + \left(\frac{m}{2}\right) V_a^2. \tag{37}$$

Eliminating E_b between (36) and (37) gives

$$\sin \pi t_1 = \frac{I_d}{2V_a} + \frac{m}{4} V_a . \tag{38}$$

When the right-hand side is larger than unity the current cannot be an instantaneous pulse. What happens then is that the current varies almost sinusoidally and produces a space charge which keeps E small at all times. In this case the phase relation between the voltage and current makes Q positive.

In practice, we begin with a small enough bias current so that $I_d/2V_a$ has become small before the oscillation has built up to the range where the sharp-pulse approximation is valid. The effective Q of the cavity will be chosen so that the oscillation will be stabilized before $(m/4)\ V_a$ becomes comparable to unity. For example, suppose we take $I_d = V_a/2$ as discussed in Section I. Then $-\cos \pi t_1$ will have dropped only to 0.85, even for as large an amplitude as $V_a = 0.2$. For $V_a = 0.1$, $\cos \pi t_1$ would be -0.92.

3.7 *Constant dc Voltage Bias*

If a constant dc voltage bias V_d is applied then E_b is constant. As the oscillation builds up I_d increases in accord with (37).

In terms of E_b and V_a , equations (35) and (38) become

$$\frac{1}{Q} = \frac{8}{\pi^2} \left(\frac{E_b}{V_a} + \frac{m}{4} V_a \right) \cos \pi t_1 , \tag{39}$$

and

$$\sin \pi t_1 = \frac{E_b}{V_a} + \frac{m}{2} V_a. \tag{40}$$

The solid curves in Fig. 10 are plots of $-1/Q$ vs V_a for two values of E_b. In the range shown, $\cos \pi t_1$ is approximately -1. The curves, therefore, have minima at $V_a = 2\sqrt{E_b/m}$. As V_a increases, $-1/Q$ will eventually reach a maximum and begin to decrease because the term $(m/2)V_a$ in (40) will become important, and the pulse will occur too soon in the cycle. This range lies beyond that shown in Fig. 10. 29

At the amplitudes shown in Fig. 10, and especially in the stable range, the main error in the above approximations comes from the fact that the pulse is not sharp. We now turn to a method of obtaining more accurate results at low amplitudes.

3.8 *Iteration Method*

Equation (25) gives $I_0(t)$ when $E(t)$ is known and (23) gives $E(t)$ if $I_0(t)$ is known. Thus we can guess at $E(t)$, find the corresponding $I_0(t)$ from (25) and use it to determine a new $E(t)$ from (23), and so on.

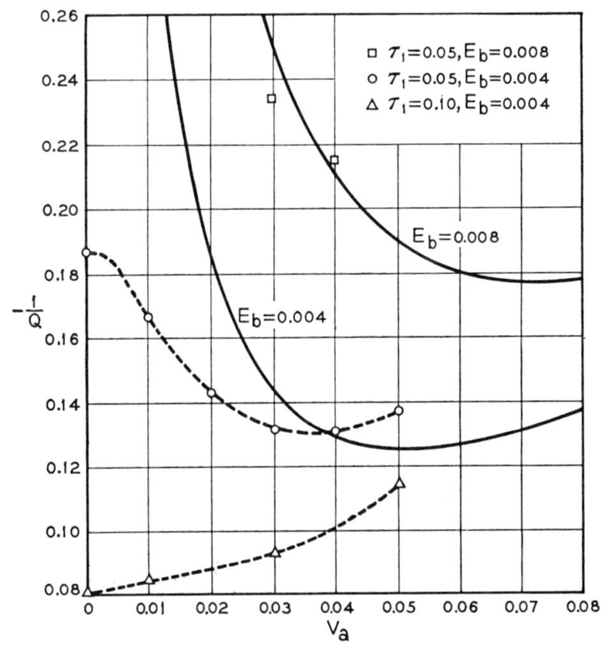

Fig. 10 — Variation of Q with amplitude for several dc voltage biases.

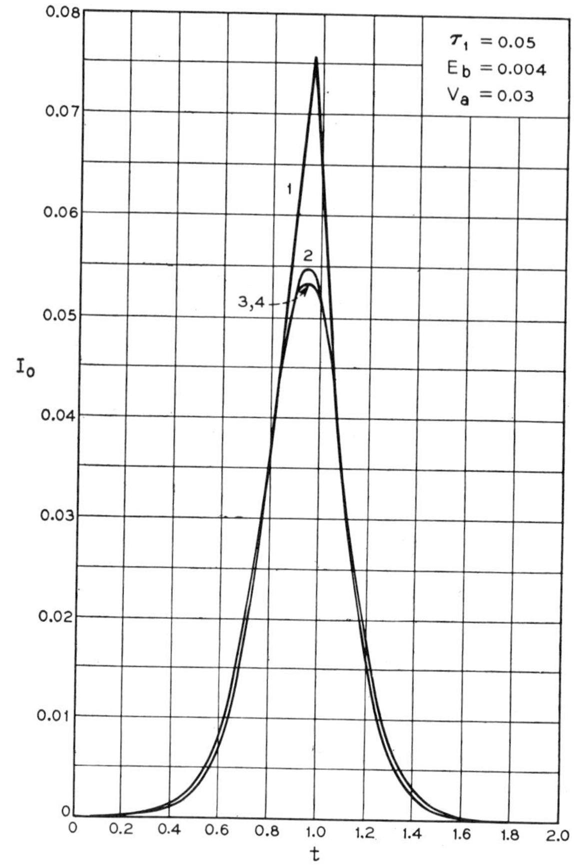

Fig. 11 — Converging solutions for $I_0(t)$.

For the initial $E(t)$ we take the instantaneous pulse solution. The procedure converges rapidly and results in solutions of any desired accuracy. In finding $I_0(t)$ from $E(t)$ in (25) a constant of integration, $I_0(0)$, is involved. What is found directly from $E(t)$ is $I_0(t)/I_0(0)$; $I_0(0)$ is so chosen that the next $E(t)$ will satisfy the condition (29) that the following $I_0(t)$ be periodic. The procedure is discussed in more detail in Appendix C.

Fig. 11 shows plots of $I_0(t)$ for $\tau_1 = 0.05$, $E_b = 0.004$ and $V_a = 0.03$. The various iterations are numbered. The values of $-1/Q$ in successive iterations were (to four places): 0.1416, 0.1330, 0.1321 and 0.1321. Fig. 12 shows $-\Delta E$ vs t for $\tau_1 = 0.05$, $E_b = 0.004$ and various values of

V_a. The corresponding $I_0(t)$ curves are shown in Fig. 13, and $I_\epsilon(t)$ curves are shown in Fig. 14.

Fig. 10 is a plot of $-1/Q$ vs V_a for various values of E_b and τ_1. The points were obtained from the iteration procedure and the solid curves from the sharp-pulse approximation. As expected, this approximation improves as V_a, E_b and $1/\tau_1$ increase. With $E_b = 0.004$ there is a minimum for $\tau_1 = 0.05$ but none for $\tau_1 = 0.1$. In the former case stable oscillations would be possible at amplitudes below about 0.035. 31

3.9 *Intermediate Amplitudes*

We now have solutions for the linear small-signal range and for the large amplitude range, where the sharp-pulse approximation is good. However, there remains an intermediate range where the current variation is too large for the small-signal analysis to apply but not large enough

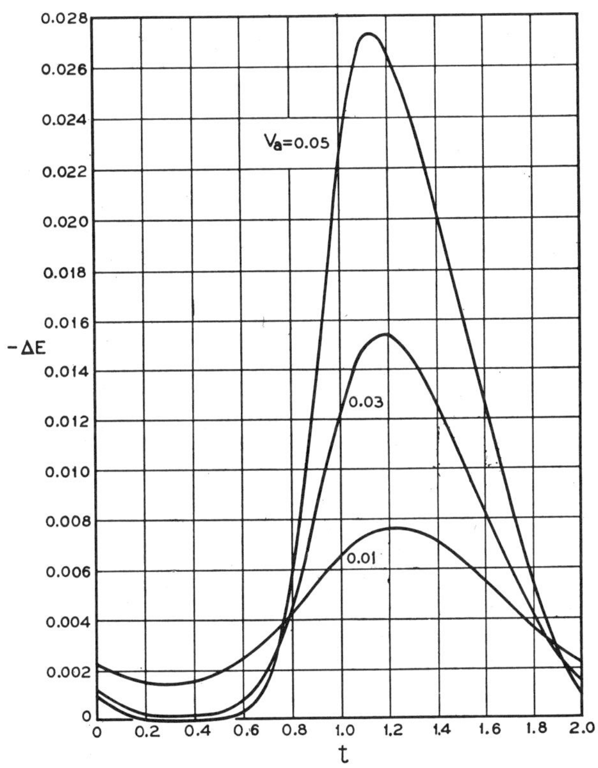

Fig. 12 — Variation of $-\Delta E$ with time for several amplitudes.

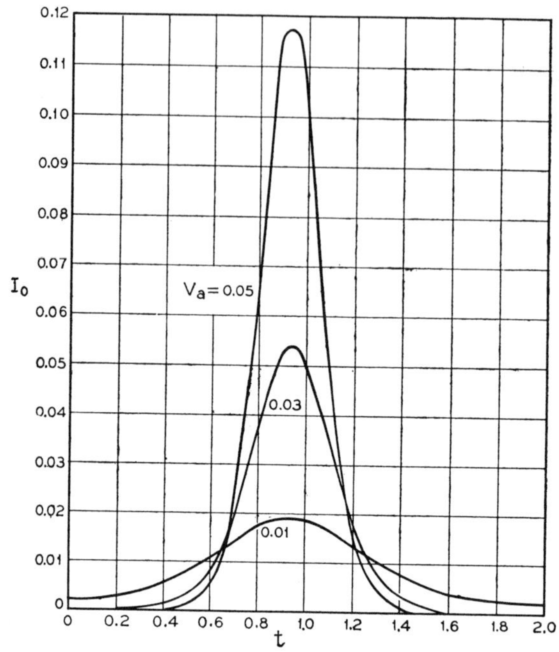

Fig. 13 — Current generated approaches a sharp pulse as amplitude increases.

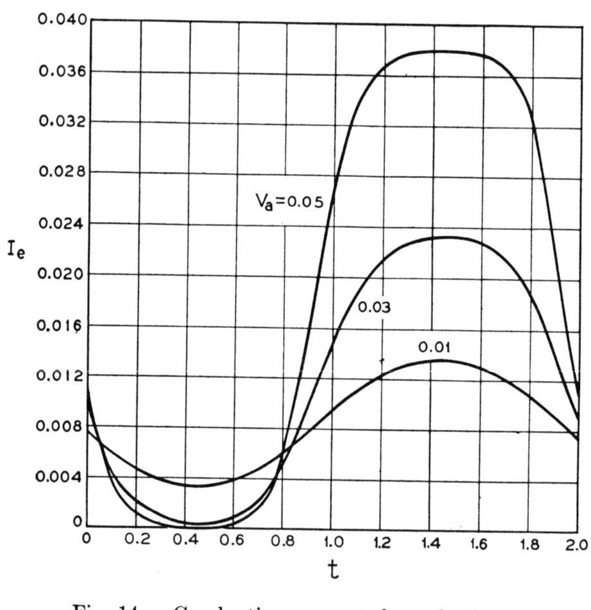

Fig. 14 — Conductive current through diode.

for the sharp-pulse approximation to be valid or even practical as a start for the iteration procedure. This range can be dealt with by an approximation that becomes more accurate as $-Q$ increases and can be within 10 per cent even for $-Q$ as low as 7. The results show how $I_0(t)$ changes from a cosine wave to an increasingly sharp pulse as V_a increases.

If Q^{-2} is small compared to unity then, in the small-signal case, $-Q = \pi V_a/I_{ea}$, where I_{ea} is the amplitude of variation of I_e. The same is true at large amplitudes if $I_e(t)$ is analyzed into a Fourier series and I_{ea} taken as the amplitude of the fundamental. In the linear small-signal range, I_{ea} is proportional to V_a so $-Q$ is constant. If I_d is kept constant as the amplitude builds up, then I_{ea} increases less rapidly than V_a. This can be seen from the fact that V_a can increase without limit while $I_e(t)$ approaches a square wave of amplitude I_d, for which $I_{ea} = (4/\pi)I_d$. Thus Q approaches $-(\pi^2/4)(V_a/I_d)$ as given by (35), for $t_1 = 1$. The phase shift, πt_1, between voltage and I_e is $\cot^{-1} Q$. We have seen that I_e is always 180° out of phase with the peak field. Therefore, if $-Q$ is large enough so that I_e and voltage are approximately 180° out of phase, the peak field is in phase with the voltage and we can use equation (16) with $E_a = V_a$. The current $I_0(t)$ is then

$$I_0(t) = I_0(0)e^{-x(1-\cos\pi t)},$$
$$x = \frac{2(m+1)V_a}{\pi\tau_1}. \tag{41}$$

This is seen to approach a pulse of increasing sharpness as V_a increases, and to reduce to the small-signal results as V_a approaches zero.

The Q can be found from (35) and (41). From (35) $1/Q = (4/\pi^2)(I_d/V_a)f(x)$, where

$$f(x) = -\frac{\langle I_0(t)\cos\pi t\rangle}{\langle I_0(t)\rangle} = -\frac{d}{dx}\int_0^\pi e^{-x\cos\theta}\,d\theta. \tag{42}$$

The function $f(x)$ is the ratio of the first to the zero order Bessels functions of pure imaginary argument. As x decreases, $f(x)$ approaches $x/2$, which means that Q is constant. As x increases, $f(x)$ bends over and approaches unity asymptotically. Values are $f(1) = 0.45$, $f(2) = 0.70$, $f(4) = 0.87$, $f(8) = 0.94$ and $f(16) = 0.97$. When I_d is constant, $Q_0/Q = 2f(x)/x$, where Q_0 is the small-signal Q. Thus the curve of $-1/Q$ vs V_a starts out flat and then decreases and approaches the form $1/V_a$ as V_a increases. This is illustrated by curves 1 and 2 in Fig. 15.

If the $-1/Q$ vs V_a curves for constant I_d are calculated from the sharp-pulse approximation [equations (35) and (38)], they have the form of the solid curves 3 and 4 in Fig. 15. As V_a decreases, the calculated $-1/Q$

goes through a maximum. For I_d below about $4/m$, the maximum would occur at about $V_a = I_d/\sqrt{2}$. However, as V_a decreases, the sharp-pulse approximation will break down. If I_d is small enough so that the small signal Q is negative, then the sharp-pulse approximation must break down before the maximum is reached, since, as we have seen, $-1/Q$ increases monotonically as V_a decreases. So the curves have the form of curves 1 and 2 in Fig. 15. However, if the small-signal Q is negative, then the curves have the form of curves 3 and 4 and the sharp-pulse approximation breaks down in the range of positive Q (shown dotted in Fig. 15). In this range the current varies roughly sinusoidally and produces a space charge that keeps the field variation small.

IV. OPERATION

In this section we consider in more detail some of the practical questions about the design and operation of the diode. In particular, we discuss the stability for both constant current and constant voltage bias, the limitations on both dc bias and ac amplitude, the effects of heating and finally the frequency dependence of the effective admittance.

4.1 *Stability at Constant Direct Current*

Fig. 15 shows the form of the variation of $-1/Q$ with voltage amplitude V_a when the bias is applied with a constant direct current generator. The oscillation will be stable at any point where $-1/Q$ is decreasing with V_a and the Q's of cavity and diode are equal in magnitude. The horizontal dashed line in Fig. 15 represents the effective $1/Q$ of the cavity. Thus the oscillation can be stabilized at the points A, B and C. The curves are numbered in order of increasing direct current I_d. If the direct current is turned on slowly compared to the response time Q_T of the diode, the oscillation will begin when I_d is slightly above the value for curve 1. Thereafter the oscillation will build up and V_a will increase as I_d increases. By slowly varying I_d we can establish a stable oscillation at any desired amplitude.

By raising I_d rapidly enough we could get onto curves 3 or 4 while the amplitude was so low as to be in the unstable range. The oscillation would then start but quickly die out.

Because of the nonlinearity of the diode, the dc voltage will vary as V_a increases at constant I_d. Eliminating E_b between (24) and (37) gives

$$I_d = 2(V_d - V_p + E_{p0} - 1) + \frac{m}{2} V_a^2, \tag{43}$$

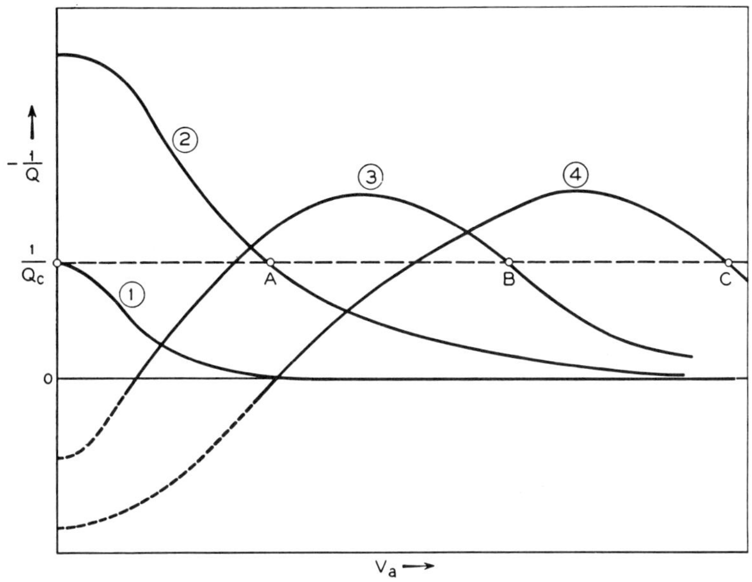

Fig. 15 — Variation of Q with amplitude (schematic) for constant direct current.

where E_{p0} is the peak field at punch through (taking E_c as unit field) and like V_p is a constant of the diode.

4.2 *Stability at Constant dc Voltage*

When the bias is applied by a constant voltage, E_b will be constant as seen from equation (24). Curves for several values of E_b and τ_1 were shown in Fig. 10. The small-signal Q is given by (32) with $I_d = 2E_b$. As E_b is raised, the small-signal Q will become negative and $-1/Q$ will rise. If Q_c, the effective Q of the cavity, is small enough the oscillation will initially be stable. For example, in Fig. 10, if $\tau_1 = 0.05$ and $1/Q_c = 0.16$, the oscillation will begin when E_b is raised to about 0.004. As E_b is further increased, the amplitude V_a will also increase. However, as seen from the figure, the oscillation will not remain stable as E_b increases unless $1/Q_c$ also increases. For example, at $E_b = 0.008$ the minimum $-1/Q$ is almost 0.18. We have seen that at large enough amplitudes $-1/Q$ reaches a maximum and decreases to zero. Thus, if E_b is increased with constant Q_c, the amplitude will suddenly jump from the range shown in Fig. 10 to a much larger value. In the range where the sharp-pulse approximation is good, $-1/Q$ is given by (37) and (40). For example, when V_a^2 is large compared to $(4/m)E_b$, we have, approximately,

$$\frac{1}{Q} = \frac{4m}{\pi^2} V_a \sin 2\pi\tau_1 ,$$
$$\sin \pi t_1 = \frac{m}{2} V_a . \tag{44}$$

36 So there is a maximum in $-1/Q$ at $V_a = \sqrt{2}/m$. The maximum will be at lower amplitudes for larger E_b.

It is seen from (39) and (40) that the oscillation will always be stable when V_a^2 is between two and four times E_b/m, provided E_b is large enough that the sharp-pulse approximation is good in that range. Thus, with constant dc voltage, the oscillation can be stabilized at any amplitude, but to reach the operting point it may be necessary to vary the Q of the cavity as the bias voltage is increased.

In the remainder of this section we shall consider several effects, such as heating, which may cause the power and $-1/Q$ to decrease with increasing amplitude. These may limit the maximum power output (especially at low frequencies) and may also be used to stabilize the oscillation in the case of constant dc voltage.

4.3 *Efficiency*

We define the efficiency, ϵ, as the ratio of the ac power, P_r, to the power P_d delivered by the dc voltage or current source; therefore,

$$P_d = V_d I_d . \tag{45}$$

The power P_h that goes into heat is the difference $P_h = P_d - P_r$. Therefore,

$$P_d = \frac{P_r}{\epsilon} = \frac{P_h}{1 - \epsilon} . \tag{46}$$

Later in this section we consider the temperature rise caused by P_h.

At the optimum frequency, $\omega = \pi$,

$$\epsilon = \frac{P_r}{P_d} = -\frac{2}{\pi}\frac{V_a}{V_d}\frac{\langle I_0(t) \cos \pi t\rangle}{I_d} . \tag{47}$$

At small and intermediate amplitudes this can be evaluated by (41) and (42), which hold when $-Q$ is well above unity. In the sharp-pulse approximation

$$\epsilon = -\frac{2}{\pi}\frac{V_a}{V_d} \cos \pi t_1 . \tag{48}$$

Thus, for $V_a = V_d/2$ and $t_1 = 1$, the efficiency would be over 30 per cent.

4.4 *Limitations on Bias and Amplitude*

For a desired operating amplitude we want V_d to be as small as possible to maximize the efficiency. However, as discussed in Section I, the minimum V_d is limited by the requirement that the field in the intrinsic region must not become negative during the negative half of the voltage cycle, and the maximum V_a is limited by the necessity of keeping the multiplication localized. We now consider these requirements in more detail. 37

In the analysis, we have assumed that the field in the intrinsic region is always high enough so that the carrier velocity remains constant, independent of field. However this will not be so at large enough amplitudes. The field may, in fact, momentarily become negative. This would reduce the negative resistance and eventually destroy it. The effect of field on velocity gives both an upper limit on the allowable amplitude for a given bias and a method of stabilizing the oscillation at any desired amplitude.

The minimum field will occur in the intrinsic region and at the trailing edge of the pulse of holes advancing to the right, as illustrated in Fig. 8. The minimum field will fall below the constant-velocity range only at high enough amplitudes so that we can use the instantaneous pulse approximation. The holes will then be closely bunched as in Fig. 8(b). The pulse causes a drop in field equal to the current I_e, which is $2I_d$ during the half cycle that the pulse is moving. The field E_i immediately behind the pulse is, therefore,

$$E_i(t) = V_d - V_p + V_a \sin \pi t - 2I_d(1 - t + t_1). \tag{49}$$

This has a minimum at a time t_2 where

$$\cos \pi t_2 = -\frac{2}{\pi}\frac{I_d}{V_a}, \qquad \pi < \omega t_2 < 3\pi/2. \tag{50}$$

Equation (38) gives t_1, and (43) gives I_d in terms of $V_d - V_p$ and V_a. Thus for a given diode the minimum $E_i = E_i(t_2)$ is determined by the voltage bias and amplitude.

In general, the average field will be well above the range where velocity depends on field. Hence we can take $E_i(t_2) = 0$ as the critical condition where the drop in velocity in the negative half of the cycle begins to reduce the power appreciably. As V_a increases, the minimum E_i decreases. If the direct current is held constant, so that the oscillation is stable, then we will want $V_d - V_p$ to be large enough to build up V_a to the desired amplitude without reducing E_i below zero and thereby losing power. However, V_d should be made as low as possible to maximize the efficiency. Thus $E_i(t_2) = 0$ is an optimum condition for operation at constant direct current. On the other hand, if we wish to operate with constant dc voltage and in the range which would be unstable for constant velocity, then V_d should be low enough so that the velocity variation will come in at the desired amplitude and stabilize the oscillation; so the condition $E_i(t_2) = 0$ would not only be optimum but necessary.

We have seen that $E_i(t_2)$ depends on $V_d - V_p$, V_a and the peak field E_{p0} at punch-through. Thus for a given diode $E_i(t_2) = 0$ gives a relation between $V_d - V_p$ and V_a. The quantity V_a and the corresponding $V_d - V_p$ could then be chosen to maximize the power. However, an upper limit on the voltages is set by the necessity of localizing the multiplication. The field throughout the intrinsic region must be well below the breakdown field, especially when the current is flowing. From Fig. 8(b) the field at the leading edge of the current pulse is $V_d - V_p + V_a \sin \pi t + 2I_d(t_1 - t)$. This determines how many hole-electron pairs the pulse of holes will produce in moving across the space-charge region. The holes produced join the pulse and add to the power. However, the electrons moving in the opposite direction will disturb the phase relations.

In the arguments above V_d occurs only in the combination $V_d - V_p$. It is therefore desirable to make V_p as small as possible so that V_d is small and ϵ large.

Both the voltage V_p and the peak field E_{p0} at punch-through are determined by the impurity distribution in the p-region. The amplitude at which the diode is designed to oscillate will fix the choice of E_{p0}. The larger the desired amplitude, the smaller E_{p0}. However, if E_{p0} is too small the multiplication will not be localized.

4.5 *Example*

We may illustrate the above discussion by applying it to an actual design. In the discussion at the end of Section I we took $V_d = 2V_a = 4I_d$ as a reasonable operating condition. This would give an efficiency of about 30 per cent. To operate at an amplitude of $V_a = 0.2$ would then require that $V_d - V_p$ be no less than 0.275. For $W = 10^{-3}$ and a unit voltage of 350 volts this would mean $V_a = 70$ volts, $V_d = 140$ volts and, for optimum operation, $V_p = 44$ volts. These parameters are reasonable. In fact, both V_d and V_p could probably be slightly lower so that the efficiency would be higher.

4.6 *Heating*

The power P_h that goes into heat will cause the temperature of the diode to rise above that of the surroundings. We shall assume that the surface of the cavity is kept at constant temperature. If the radius R of the diode is small compared to the radius of the center post of the cavity, as in Fig. 2, then heat can flow away from the diode in almost all directions. The diode can be made thin enough and mounted sufficiently close to the metal so that the temperature drop in the silicon is small compared to that in the metal. Let ΔT be the difference between the diode temperature and the temperature of the surface of the cavity. Then ΔT is related to P_h and R by the formula for spreading resistance

$$\Delta T = \frac{P_h R}{4K}, \tag{51}$$

where K is the thermal conductivity of the metal. For copper, $4K = 16.7$ watts per cm per °C.

The temperature will rise to its equilibrium value in a time of about R^2/D, where D is the coefficient of thermal diffusion, which is about unity for copper.

We may now apply these results to the examples discussed at the end of Section I assuming an efficiency of $\frac{1}{3}$ so that $P_h = 2P_r$. The 50 watts of power output at 5,000 megacycles and $R = 0.03$ cm would produce a temperature rise of about 60°C. At 500 megacycles and $R = 0.3$, however, the 5 kw maximum power would raise the temperature by about 600°C. Thus at low frequencies the maximum power output in continuous operation would be limited by heating rather than by how small the impedance ωL of the cavity can be made. However, the time constant of the temperature rise for $R = 0.3$ cm would be almost a tenth of a second, so the temperature rise in pulse operation would not be serious.

4.7 *Effect of Temperature on Critical Field*

McKay[2] has found that the critical field increases with temperature. For critical fields between 250 and 500 kilovolts per cm in silicon, a change in temperature changes the critical field by 0.05 per cent per °C. Increasing the critical field effectively decreases the dc voltage and current. Thus when the diode begins to get hot it effectively reduces its bias. The heating will therefore stabilize the oscillation.

4.8 *Effect of Reverse Saturation Current*

In silicon the reverse saturation current will become important only at large amplitudes. In the small-signal range it is easily shown that I_s can be neglected if it is small compared to $\pi[\pi\tau_1/2(m+1)]^2$ which will be of the order of 10^{-3}, or a few amperes per cm^2 for $W = 10^{-3}$ cm.

Even at large amplitudes I_s will be small compared to the average current and will have a negligible effect on the space charge. However, as V_a increases, the ratio of the maximum and minimum values of I_0 (which varies exponentially with V_a) will increase much faster than the maximum I_0. Thus the minimum I_0 becomes very small at high amplitudes. It cannot, however, fall below I_s.

If the effect of I_s is included, the equation (25) for current generation becomes

$$\frac{\tau_1}{2}\frac{d}{dt}\ln I_0 = (m+1)\left(E + \frac{m}{2}E^2\right) + \frac{I_s}{I_0}. \qquad (52)$$

As V_a increases and the minimum I_0 decreases, the term I_s/I_0 becomes important near the current minimum, and prevents the current from becoming too low. Thus I_s will be important in the equation only near the current minimum. However, by increasing the minimum I_0, I_s also increases subsequent values of I_0 and hence increases the associated carrier space-charge. The carrier space-charge will therefore shut the current off (by reducing the field) earlier in the cycle. This will reduce the delay between voltage and current and so reduce the power.

The amplitude at which I_s becomes important can be roughly estimated as follows: From (41) the ratio of maximum and minimum values of I_0 is related to V_a by

$$V_a = \frac{\pi\tau_1}{4(m+1)}\ln\frac{I_{max}}{I_{min}}.$$

A formula for I_{max} is derived from the instantaneous-pulse approximation in Appendix B [equation (63)] 11. In practical cases, where I_d will be of the order of 10^{-1}, I_{max} is seen to be of the order of unity. At room temperature the reverse saturation current in silicon can be made less than a microampere per cm^2, which is a dimensionless current of around 10^{-10} in the kmc range. Thus the minimum I_0 will be greater than I_s if V_a is no larger than about 2.6 τ_1. To operate at an amplitude $V_a = 0.2$ we therefore want τ_1 to be no smaller than 0.07.

The effect of I_s is treated in detail in Appendix D and more general relations are found to replace (37) and the equations derived from it.

4.9 *Frequency*

So far we have considered only the optimum frequency $\omega = \pi/\tau = \pi$. The small-signal analysis is easily carried out for arbitrary frequency. The admittance Y of the diode is found to be

$$
\begin{aligned}
Y(\omega) &= \frac{4\omega y[(1 - \cos\omega) + i(2y - \sin\omega)]}{2y^2 - 2y\sin\omega + 2(1 - \cos\omega)}, \\
y &= \frac{\omega}{2}\left[1 - \frac{\omega^2\tau_1^2}{2(m+1)I_d}\right].
\end{aligned}
\tag{53}
$$

This reduces to (32) for $\omega = \pi$. When $-y$ is well above unity, it can be held constant by varying I_d in proportion to ω^3. Then $-1/Q$ will vary only with the phase factor $1 - \cos\omega$. Thus by varying I_d the device can be tuned mechanically over a frequency range extending from $\frac{1}{2}$ to $\frac{3}{2}$ of the optimum frequency.

At large amplitudes the current no longer varies sinusoidally. However, as mentioned in Section III, the cavity may be almost a short circuit for frequencies higher than the fundamental. Hence, to evaluate the diode as an element in the oscillator, we can analyze the current $I_e + I_c$ into a Fourier series and retain only the fundamental. The relation of this to the ac voltage then defines the conductance, G, and capacity, C, of the diode. The results are

$$
\begin{aligned}
G &= \frac{2}{\omega V_a}\left[\langle I_0(t)\sin\omega t\rangle \sin\omega + \langle I_0(t)\cos\omega t\rangle(1 - \cos\omega)\right], \\
C &= 1 + \frac{2}{\omega^2 V_a}\left[\langle I_0(t)\sin\omega t\rangle(\cos\omega - 1) + \langle I_0(t)\cos\omega t\rangle \sin\omega\right]
\end{aligned}
\tag{54}
$$

where again the brackets denote time averages. When current and voltage are approximately 180° out of phase, we can use (42) to evaluate the time averages in (54).

In the sharp-pulse approximation the averages become $I_d \sin\omega t_1$ and $I_d \cos\omega t_1$ respectively. Equation (37), which is independent of frequency, remains valid. Since the drop E_δ in field is $(4\pi/\omega)$ times the average $\langle -\Delta E\rangle = I_d/2$, the general form of (36) becomes

$$
\frac{\pi}{\omega} I_d = E_b + V_a \sin\omega t_1 . \tag{55}
$$

Above the small-signal range the admittance $Y = G + i\omega C$ of the diode depends on voltage amplitude as well as on frequency and direct current. For stable oscillation the admittance $Y = Y(\omega, I_d, V_a)$ of the diode is equal to $-Y_c$, where Y_c is the admittance of the cavity. For a given setting of the plunger (Fig. 2), $Y_c = Y_c(\omega)$ depends only on the frequency. Thus the frequency and amplitude of oscillation are determined in terms of the direct current by $Y(\omega, I_d, V_a) + Y_c(\omega) = 0$. When I_d is varied both V_a and ω will vary. However, it is not certain whether the variation of frequency with I_d could be made large enough for a practical frequency modulation device. Near the optimum frequency and bias the frequency will remain relatively constant as I_d and V_a vary. Thus the frequency is primarily determined mechanically by adjusting the height of the cavity (Fig. 2) and the amplitude of oscillation is determined electrically by the bias.

4.10 *Zener Current*

In sufficiently narrow junctions, where breakdown occurs, the rate of generation is an extremely sensitive function of field. So, as in secondary emission, the generation can be highly localized. The diode could be made to operate by Zener current rather than multiplication if the p region is sufficiently narrow. However, the conditions for negative resistance would be less favorable. The current I_0 generated by field emission is a function of the peak field E_0. Hence I_0 and E_0 are in phase, so all of the delay has to come from the transit time. For ideal phase relations at large amplitudes the bias should be such that the current is generated mainly in a short burst near the voltage peak. Then, if the transit time is $\frac{3}{4}$ of a cycle, the current, I_e, will flow during the last three quarters of the cycle and power will be delivered to the ac signal. However, the Q and efficiency are considerably lower than for secondary emission. The small-signal Q can be varied by varying the bias. A small change in bias will change dI_0/dE_0 drastically (since we are on the knee of the current-voltage curve). For the frequency $0.75/\tau$, the minimum $-Q$ is 25 at small signals and over 100 at large signals. The latter might be improved somewhat by increasing the frequency in relation to $1/\tau$. The efficiency could probably not be raised above about 5 per cent without ruining the Q.

An analysis of the diode operation on Zener current is discussed in Appendix E.

V. ACKNOWLEDGEMENTS

The author wishes to thank J. L. Moll, G. Weinreich, B. Senitzky and R. W. Hamming for helpful discussion and advice. The machine computations leading to Figs. 10 through 14 were programmed by Mrs. G. J. Rowe.

Derivation of I_e

We first give a simple analytical derivation and then a physical argument. A more general proof is given by Shockley (1938).

Subtract (4) from (3) and use Poisson's equation

$$\frac{\partial E}{\partial x} = \frac{4\pi q}{\kappa}(p - n + N)$$

for $p - n$. The result is*

$$\frac{\partial}{\partial x}\left[\frac{\kappa}{4\pi}\frac{\partial E}{\partial t} + I\right] = 0.$$

In other words the quantity in brackets is a function of time only and is the same, at a given time, throughout the length of the diode — not only in the space-charge region, but also in the ends. The current in the ends, or leads, is $I_c + I_e$. Let the ends be of sufficiently low resistance so that the field can be assumed to vanish there. Then

$$\frac{\kappa}{4\pi}\frac{\partial E}{\partial t} + I = I_c + I_e.$$

Averaging over the length W of the space-charge region and using $C = \kappa/4\pi W$ gives

$$C\frac{dV}{dt} + \frac{1}{W}\int_0^W I\,dx = I_c + I_e.$$

The physical argument can be illustrated by Fig. 8(b). Let a small pulse of charge δQ be generated at $x = 0$. It causes a drop in field $\delta E = (4\pi/\kappa)\delta Q$. If the field at the edges remains constant as the pulse moves, the voltage will drop at a rate $v\delta E = (4\pi/\kappa)v\delta Q$. Therefore, if the voltage is to remain constant, a current δI_e must flow in the

* G. Weinreich (private communication) has pointed out that this is a special case of the general three-dimensional result

$$\text{div}\left[\frac{\kappa}{4\pi}\frac{\partial \vec{E}}{\partial t} + \vec{I}\right] = 0$$

which follows from

$$\text{curl } \vec{H} = \frac{\kappa}{c}\frac{\partial \vec{E}}{\partial t} + \frac{4\pi}{c}\vec{I}$$

and div curl $= 0$.

external circuit and increase the voltage by adding charges to the edges of the space-charge region. The rate of voltage increase will be $\delta I_e/C$. Setting this equal to $v\delta E$ gives $\delta I_e = \delta Q/\tau = v\delta Q/W$.

APPENDIX B

Instantaneous Pulse

The equations can be solved exactly for any amplitude and bias in the limiting case of $\tau_1 = 0$. As we have seen, the current I_0 approaches an instantaneous pulse. This solution will give reasonable approximations to the Q, efficiency and power for actual cases if the effective duration of the current pulse is a small fraction of a cycle. We let the frequency ω be arbitrary but less than $2\pi/\omega$, so that only one current pulse is flowing at a time. The solution is completely specified by the time t_1 at which the pulse and the discontinuity in field occur and the magnitude E_δ of the discontinuity. From $\langle \Delta E \rangle = -I_d/2$ and the form of $\Delta E(t)$,

$$E_\delta = \frac{4\pi}{\omega} \langle -\Delta E \rangle = \frac{2\pi}{\omega} I_d. \tag{56}$$

The condition that I_d be periodic is

$$\langle E \rangle + \frac{m}{2} \langle E^2 \rangle = 0 \tag{57}$$

where $E = E_b + V_a \sin \omega t - E_\delta(t - t_1)$ in the internal from t_1 to $t_1 + 1$ and $E = E_b + V_a \sin \omega t$ at all other times. Thus (57) is a quadratic equation for E_δ as a function of E_b, V_a and t_1. Only the smaller of the two roots is meaningful. When E_b and V_a are small compared to unity (57) reduces to (37) and is independent of t_1.

To obtain another relation between t_1 and E_δ we solve (23) and (25) and find the relation between I_0 and E during the pulse. In the limit of an instantaneous pulse, t in equation (23) does not vary during the pulse, so $dE = -I_0\, dt$. We can use this to eliminate dt in (25). Then integrating we have the relation

$$I_{\max} - I_0 = \frac{(m+1)}{\tau_1} E^2 \left[1 + \frac{m}{3} E \right] \tag{58}$$

for $I_0(t)$ as a function of $E(t)$ during the pulse. The values of E at the beginning and end of the pulse are found by setting $I_0 = 0$ in (58). Let E_1 be the field at the beginning of the pulse. Then

$$E_1 = E_b + V_a \sin \omega t_1 , \tag{59}$$

and the maximum I_0 is

$$I_{\max} = \frac{(m+1)E_1^2}{\tau_1}\left(1 + \frac{m}{3}E_1\right). \quad (60)$$

Neglecting terms in E_1^2 in comparison with unity, (58) gives

$$E_\delta = 2E_1\left(1 + \frac{mE_1}{6}\right). \quad (61)$$

Combining this with (57) and (60) we have

$$E_b + V_a \sin \omega t_1 = \frac{\pi}{\omega} I_d\left(1 - \frac{m\pi}{6\omega} I_d\right). \quad (62)$$

In practical cases, the term $(m\pi/6\omega)I_d \approx I_d$ can be neglected in comparison with unity. Then (63) becomes (55). This, together with the quadratic (57), determines E_δ and t_1 for any E_b and V_a.

From (60) and (61),

$$I_{\max} = \frac{(m+1)E_\delta^2}{4\tau_1} = \frac{\pi^2(m+1)}{\omega^2\tau_1} I_d^2. \quad (63)$$

Thus the effective duration of the pulse $\Delta t = I_d/I_{\max}$ is

$$\Delta t = \frac{\omega^2\tau_1}{\pi^2(m+1)I_d}. \quad (64)$$

So the pulse becomes sharper as τ_1 decreases and I_d increases. We can estimate the accuracy of the instantaneous pulse approximation by comparing Δt with the period $2\pi/\omega$.

APPENDIX C

Iteration Method

The procedure in detail is the following: Each iteration goes from an $E(t)$ through an $I_c(t)$ to a new $E(t)$. In the first iteration we have to guess at $E(t)$. The procedure has been to begin with the $E(t)$ corresponding to the instantaneous current pulse and illustrated in Figs. 6 and 9. The magnitude E_δ of the discontinuity is chosen to satisfy

$$\langle E + (m/2)E^2\rangle = 0.$$

The time t_1 when the discontinuity occurs is found from (40). Putting this $E(t)$ into (25) and integrating gives a periodic $I_0(t)/I_0(0)$ from which, using (23), we find a function $F(t) = -\Delta E(t)/I_0(0)$. Putting $E = E_b + V_a \sin \pi t - I_0(0)F(t)$ into $\langle E + (m/2)E^2\rangle = 0$ gives a quadratic equation for $I_0(0)$. Only the smaller of the two roots is meaningful. From the known $F(t)$ and $I_0(0)$ we have a new $E(t)$. Since it satisfies $\langle E + (m/2)E^2 \rangle = 0$, the $I_0(t)/I_0(0)$ calculated from it in the following iteration will be periodic.

To plot $I_0(t)$ in Fig. 11 we have determined both $I_0(t)/I_0(0)$ and $I_0(0)$ from the same $E(t)$. The quantity $I_0(t)/I_0(0)$ is found from (25) and $I_0(0)$ is chosen to satisfy $\langle I_0 \rangle = -2\langle \Delta E \rangle$. This relation is automatically satisfied if $\Delta E(t)$ has been determined from $I_0(t)$ using (23). However, only in the exact solution is it satisfied if $I_0(t)$ was determined from $\Delta E(t)$. To determine $I_0(0)$ from $\Delta E(t)$ we have

$$I_0(0) = \frac{-\,2\langle \Delta E(t) \rangle}{\left\langle \dfrac{I_0(t)}{I_0(0)} \right\rangle}$$

where $I_0(t)/I_0(0)$ is found from $\Delta E(t)$ using (25) with $E = E_b + V_a \sin \pi t + \Delta E$. The value of $I_0(0)$ found in this way, in each iteration, was compared with the value that makes the following $I_0(t)$ periodic. When the two values agreed within a specified amount, usually taken to be about one per cent, the iteration procedure was terminated. The procedure was programmed on an I.B.M. 650 Magnetic Drum Calculator. The machine would give a solution of the required accuracy in about four iterations, or forty minutes, on the average.

In one case, where the current peak occurred at $t_1 = 0.92$, the iteration procedure was repeated starting with $t_1 = 0.85$. The peak in successive iterations moved from 0.85 to 0.92. In other words, the final result was independent of the initial t_1. However, more iterations were required for a poor initial choice.

APPENDIX D

Effect of I_s

It is convenient to consider the solution for a single cycle extending between two current minima. The effect of I_s will be important only near the ends of the cycle where I_0 is small and the term I_s/I_0 cannot be neglected in (52). Let the solution be $f(t)I_0(t)$ where $I_0(t)$ is a solution for $I_s = 0$ and is the correct solution during most of the cycle. Then $f(t)$ will be unity during most of the cycle and will rise sharply at the beginning and end of the cycle. Since the current is always positive, each rise in $f(t)$ must be less than unity. The condition that the current be periodic is $\Delta \ln f + \Delta \ln I_0 = 0$, where Δ denotes the change during a cycle. The change in $\ln I_0$ is found by the same procedure that led to

(37). If Δf is the total change in f in a cycle we have, instead of (37),

$$\ln\left(\frac{2 + \Delta f}{2 - \Delta f}\right) = \frac{4\pi(m + 1)}{\omega\tau_1}\left[E_b - \frac{I_d}{2} + \frac{m}{4}V_a^2\right]. \qquad (65)$$

Replacing $I_0(t)$ by the correct solution $f(t)I_0(t)$ in (52) gives

$$\frac{df}{dt} = \frac{2}{\tau_1}\frac{I_s}{I_0}. \qquad (66)$$

To determine Δf we need to solve this only in the short interval near the current minima where f is changing. In practical cases the current will have stopped flowing slightly before the end of the negative half of the cycle. Hence when f is changing, $E + (m/2)E^2$ will vary as $V_a\omega t$ and

$$\ln\left[\frac{I_0(t)}{I_{\min}}\right] = \frac{(m + 1)V_a\omega t^2}{\tau_1} \qquad (67)$$

where $I_{\min}$ is the minimum of $I_0(t)$ and $t = 0$ is taken at the minimum. Substituting (67) into (66) and integrating over the short interval where f is changing gives

$$\Delta f = \frac{2I_s}{I_{\min}}\sqrt{\frac{\pi}{(m + 1)V_a\omega\tau_1}}. \qquad (68)$$

The ratio of $I_{\min}$ to $I_{\max}$ is found by integrating (25) from the current minima at $E = 0$ to the maximum at $t = t_1$. At the amplitudes where the effect of I_s is important we can use the sharp-pulse approximation, so $E = E_b + V_a \sin \omega t$ during this interval. The quantity $I_{\max}$ is given by (63). Thus Δf is found in terms of E_b, V_a and t_1. Equation (65) now replaces (37) and together with (55) determines t_1 and I_d for a given voltage bias and amplitude.

APPENDIX E

Zener Current

The small-signal case is easily solved. The ac variations in I_0 and E_0 are proportional. At large signals it is probably a good approximation to say that (a) no current is generated until E_0 has risen to a critical value, which we shall define as E_c, and (b) thereafter I_0 will be such as to keep E_0 from rising above E_c. Thus (25) is replaced by $I_0 = 0$ for $E < 0$ and $E = 0$ for $I_0 > 0$. Equations (23) and (26) remain unchanged. The current $I_e(t)$ will flow during the following three intervals:

(A) Beginning at the time t_0 when $E_0 = E_c$, current begins to be generated and continues to be generated until a time t_1 when no more carrier space-charge is required to keep E_0 from rising above E_c, that

47

is, to keep E from becoming positive. The last holes to be generated will have been generated before the first have crossed the space-charge region. During this interval, I_e increases from $I_e(t_0) = 0$ to its maximum $I_e(t_1)$, while I_0 jumps at once to its maximum at t_0 and thereafter declines, reaching zero at t_1. The governing equation is

$$\tau \frac{dIe}{dt} - I_e = C \frac{dV}{dt} = CV_a \omega \cos \omega t. \tag{69}$$

This can be obtained by physical reasoning from Fig. 8 or by differentiating (23) and (26). Equation (69) is easily solved, subject to $I_e(t_0) = 0$. The time t_1 and the maximum current $I_e(t_1)$ are found from $dI_e/dt = 0$ at $t = t_1$.

(B) From the time, t_1, when I_e reaches its maximum and I_0 drops to zero, until the time $t_0 + \tau$, when the first holes to be generated reach the i-p$^+$ junction, I_e remains constant and equal to $I_e(t_1)$. A constant number of holes are moving with constant velocity across the space-charge region.

(C) Between $t_0 + \tau$ and $t_1 + \tau$ the holes generated in the first interval are flowing out at the i-p$^+$ junction. Since the first holes to be generated are the first to flow out, $I_e(t) = I_e(t_1) - I_e(t - \tau)$ during this interval.

Thus, by solving (69) for $I_e(t)$ in the interval $t_0 \leqq t \leqq t_1$ we know $I_e(t)$ throughout the cycle. Except in the first interval, $E \leqq 0$ and $I_0 = 0$.

The equations have been solved for the frequency $0.75/\tau$ or $\omega = 3\pi/2\tau$. The Q is a function of ωt_0, which is determined by the voltage bias and amplitude. The Q is negative for $13° < \omega t_0 < 90°$. The maximum $-1/Q$ is about 0.0095 and occurs at about $\omega t_0 = 40°$. The quantity $[\epsilon/(1 + \epsilon)]$ (V_d/V_a) rises almost linearly with ωt_0 from zero at 13° to 0.21 at 90°. The limitations on V_d and V_a would be about the same as for a diode operating by multiplication.

REFERENCES

1. A. G. Chynoweth and K. G. McKay, Phys. Rev., **106,** (1957), 418.
2. K. G. McKay, Phys. Rev., **94,** (1954), 877.
3. P. Wolff, Phys. Rev., **95,** (1954), 1415.
4. W. Shockley, J. Appl. Phys., **9,** (1938), 635

Reprinted from THE BELL SYSTEM TECHNICAL JOURNAL
Vol. XLIV, No. 2, February, 1965
Printed in U.S.A.

B.S.T.J. BRIEFS

A Silicon Diode Microwave Oscillator

By R. L. JOHNSTON, B. C. DE LOACH, Jr., and B. G. COHEN

(Manuscript received December 28, 1964)

49

Microwave oscillations have been obtained on a pulse basis from silicon diodes. This brief reports fabrication details and performance data. The similarities of this device to that proposed by Read[1] are discussed.

The diodes were made by diffusing boron to a depth of approximately 25 μ into one face of a slice of 0.15 ohm-cm n-type silicon, and then lapping the other face to a final slice thickness of 175 μ. Electroless nickel was then applied to both faces and sintered at 800°C for 5 minutes in N_2. Nickel was then replated and followed by a final plate of gold. The slice was then ultrasonically sectioned into squares 125 $\mu \times$ 125 μ. To remove cutting damage the "wafers" were etched for about 10 seconds in CP-8 (3 parts HNO_3, 1 part HF), and were then incorporated into a microwave encapsulation. A sketch of a wafer before encapsulation is shown in Fig. 1.

Oscillations are observed when a critical reverse voltage is applied to the diode. This voltage has been observed on a variety of samples to correspond to that required to produce enough reverse current to create an electric field on the order of 2 kv/cm in the 150-μ n-type region. A typical reverse *V-I* characteristic obtained on a pulse basis using a sampling oscilloscope is shown in Fig. 1. On samples which were lapped to reduce the drift region length from 150 μ to 75 μ, the required voltage in excess of the avalanche voltage was halved. Voltages in excess of threshold produce more output until a maximum is reached. Some lower frequencies in the 1-2-gc region exhibited several maxima, but the higher frequencies (12 and 24 gc) had but one.

The fact that voltages considerably in excess of the "breakdown" voltage are employed tends to deemphasize the role of microplasmas and nonuniformities in the junction and thus contributes to the ease of fabrication of these devices.

The diodes were tested in either a coaxial system, for the lower frequencies ($f \lesssim 12$ gc), or in a reduced-height waveguide for the higher frequencies (8–24 gc). The mounts incorporated a bypass capacitor which allowed the introduction of a video pulse to power the diode.

Microwave oscillations (when present) were coupled to a spectrum analyzer.

The operation of a particular sample of the geometry shown in Fig. 1 will be described. It was placed in the coaxial circuit and driven with a 2 μsec pulse at a 10-kc repetition rate. When the applied pulse voltage reached a critical value, microwave power output was observed. Power
50 was obtainable over a wide range of frequencies, but some frequencies had particularly high amplitudes. A plot of some of these "high points" is presented in Fig. 2. Included in this figure are two points obtained from a similar sample operated in the waveguide mount. Fig. 2 can also be interpreted as a rough plot of efficiency versus frequency, since pulse powers between 15 and 30 watts were employed for all these points. The 80 mw obtained at 12 gc represents 0.5 per cent efficiency. Similar samples have been operated with duty cycles of 25 per cent (to burnout).

The higher-frequency operation for which the efficiencies are on the order of 0.5 per cent is most likely an oscillation involving primarily the space charge depletion width for the drift space. The extent of this region is of the order of that predicted by Read for this frequency of operation. The requirement of 2 kv/cm in the 150-μ region for this operation most likely assures that fields greater than this exist across

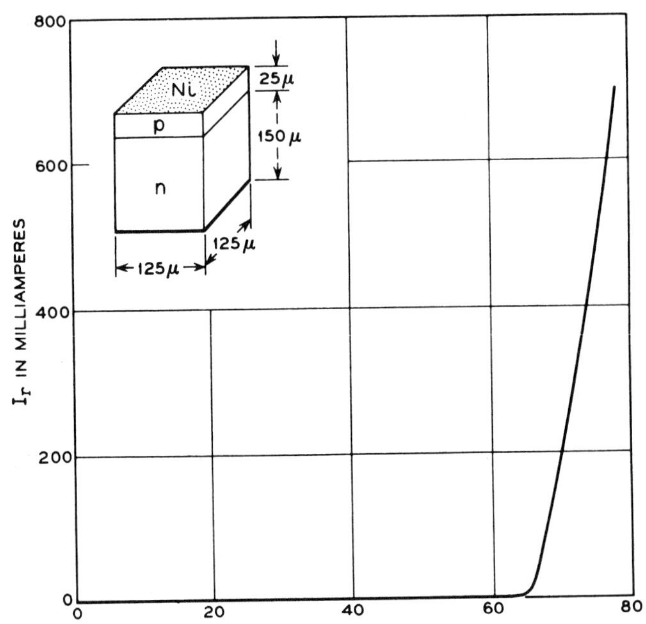

Fig. 1 — A typical sample and its *V-I* curve.

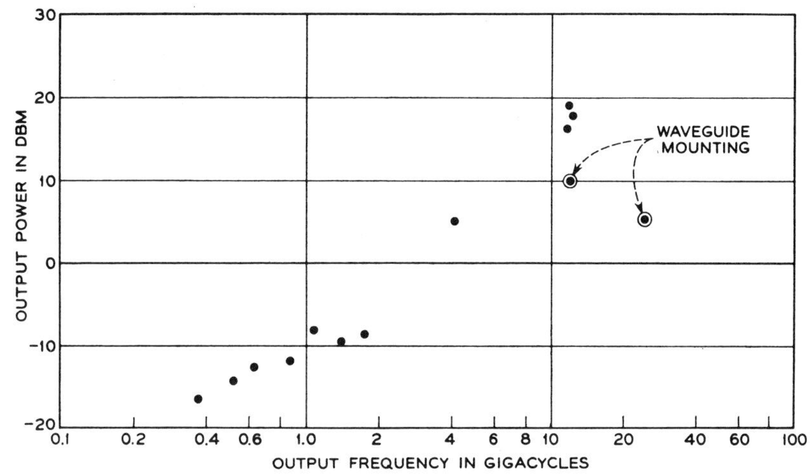

Fig. 2 — Output power vs frequency for a diode in a coaxial circuit and one in a rectangular waveguide circuit.

51

the depletion region. This added length should then be removable without changing the character of these oscillations and indeed should enhance the efficiency of operation. The circuitry employed in obtaining the higher-frequency microwave power was crude, and thus the power obtained in the X and K band regions (12 and 24 gc) should be taken as a poor lower limit of that available.

We are uncertain at present as to the explanation for the lower-frequency points exhibited in Fig. 2. The 2 kv/cm field required is high enough that the mobility has decreased in the 150-μ region. The usual dielectric relaxation time of this material of 10^{-13} sec is thus increased and some bunching of charge is preserved in the region, with charge transport (1×10^7 cm/sec) becoming significant. This "stiffening" of the conductive region could allow it to function as a drift region in the same manner as does the swept region in the above. The approximately linear increase in output power with frequency in this region could then be due to a redistribution of ac field between the space charge depletion layer capacity and the ac impedance of the drift region. That this mechanism is not very effectual can be deduced from the efficiencies of some 1×10^{-3} per cent in the region up to 2 gc.

Many helpful discussions with R. M. Ryder and J. C. Irvin are gratefully acknowledged.

Note Added in Proof:

Subsequent to the pulsed microwave operations described in this

Brief, Lee et al.[2] have obtained low-frequency cw operation in a silicon diode with a npπp structure closely approximating the structures described by Read.[1] Still more recently, continuous microwave oscillations have been obtained by Johnston and De Loach[3] in structures similar to those described herein.

REFERENCES

1. Read, W. T., A Proposed High-Frequency Negative-Resistance Diode, B.S.T.J., *37*, pp. 401–446.
2. Lee, C. A., Batdorf, R. L., Wiegmann, W., and Kaminsky, G., to be published.
3. Johnston, R. L., and De Loach, B. C., to be published.

Reprinted by permission from
IEEE JOURNAL OF SOLID-STATE CIRCUITS
Vol. SC-4, No. 6, December 1969

Modes of Avalanche Diodes and Their Associated Circuits

BERNARD C. DE LOACH, JR., MEMBER, IEEE

Abstract—This paper attempts to put in perspective the modes of operation of avalanche diodes. Although theoretical treatments exist that encompass all known aspects of avalanche diode behavior, the compartmentalization of behavior into well-understood special cases is of enormous help to the device and circuit designer in understanding the peculiarities encountered in experimental work.

The degree of compartmentalization is, of course, an arbitrary choice but the distinction of five modes seems appropriate. These are, respectively, the IMPATT, TRAPATT, and parametric modes, which are potentially useful, and the space-charge feedback and thermal modes, which probably are not. The latter two are important, however, in that they constitute a means of diode self-destruction.

The presentation is tutorial and will hopefully remove some of the confusion introduced by the varied names employed by workers in the field to describe basically similar results. The approach is one of reason rather than legalism, and should this particular compartmentalization stand the test of time, appropriate nomenclature will no doubt be decided by an international committee.

Manuscript received June 10, 1969.
The author is with Bell Telephone Laboratories, Inc., Murray Hill, N. J.

I. Introduction

Avalanche diodes present the greatest challenge to microwave solid-state-circuit design of any semiconductor device yet discovered.

They can combine in particular configurations, the most difficult features of tunnel-diode oscillators with those of multifrequency parametric amplifiers or oscillators with additional boundary conditions on heat flow. The reasons for work on such tricky devices can be summarized by the following.

1) They are at present the most powerful CW solid-

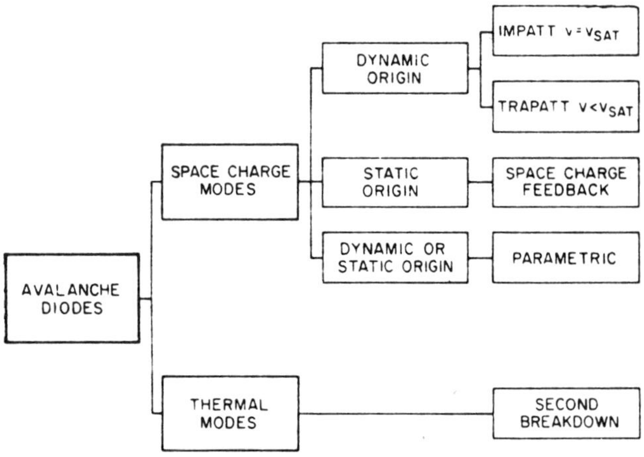

54

Fig. 1. Avalanche diode mode chart.

state sources of high-frequency microwave power (5 W·CW; 14 GHz).
2) They are at present the most efficient sources of high-frequency microwave power (60 percent near 1 GHz and 43 percent at 5 GHz).
3) They can be constructed in Si, Ge, and GaAs and in other semiconductors should the need arise.
4) They have demonstrated long-term reliability capabilities in both Si and Ge and in some cases in GaAs.
5) They offer potentially *cheap,* as well as efficient, long-lived rugged microwave sources.
6) They have operated from 200 MHz to 340 GHz.

Several distinctive "modes of operation" occur within the broad category of avalanche diodes. As might be expected, the boundaries between these modes are somewhat indistinct and indeed different workers in the field separate the plethora of experimental results into different categories or modes.

This paper attempts to describe the distinctive features of the broad class of avalanche diodes and makes an arbitrary decision as to the modal boundaries, which results in five "modes" of operation. These modes are diagramed in Fig. 1.

II. Power-Generation Mechanisms

There are two basically different RF power-generation mechanisms in avalanching junction devices. In the more important of these, giving rise to the space-charge modes, the space charge of mobile carriers changes the electric field within the device, consequently changing the voltage across the device, in such a manner as to produce voltage variations greater than 90° out of phase with device current variations, and consequently leading to the generation of power.

The other mechanism, giving rise to the thermal modes, involves the thermal generation of carriers from the lattice. Hence, the dissipative power from a reverse-biased p-n junction in avalanche causes local heating which in turn changes the local generation rate of mobile carriers. If enough heat is generated, carriers can be thermally generated in sufficient numbers to markedly increase the current, and furthermore, the diode can now supply this current at a drastically reduced voltage since impact avalanche fields no longer need be maintained. (This has been called a "second type of breakdown.") If less heat is dissipated in the high-current low-voltage state, upon cooling, the diode can switch back to the original state. Consequently, an oscillatory condition can result giving rise to RF power generation.

III. Modes Arising from the Two Mechanisms

There exist, within the broad classification of space-charge modes, static, dynamic, and parametric modes (which require the presence of a dynamic or static mode to exist). Historically, theory was first developed for static modes [1]–[6], and then dynamic modes [7]–[9]. As might be expected these are in order of increasing theoretical complexity.

Consider first the simplest case, the static modes. These are defined to be modes in which time variations of electric field and charge density essentially follow dc solutions during oscillation. That is, if one took a snapshot of the charge and field within the diode at some instant of time corresponding to some instantaneous terminal voltage and current, the internal picture would be essentially the same as that obtained if a battery and a resistor were used to achieve the same voltage and current. These modes have consequently been called quasi-static by some authors. The periods of the static modes must obviously be much longer than carrier transit times through the device. Fig. 2 displays the electric field versus distance for a p-i-n diode under high reverse bias, just prior to breakdown. If now the voltage is increased slightly so as to produce holes and electrons (through impact avalanche) uniformly throughout the high field region, it follows that one edge of the depletion layer will have a higher electron density and the other edge a higher hole density, since the electric field pushes holes in one direction (towards the right) and electrons in the other direction (towards the left). The net space charge of holes and electrons necessitates a change in electric field with distance. This is expressed by Poisson's equation, which states, for one-dimensional geometry, that

$$\frac{\partial E}{\partial x} = \frac{\rho}{\epsilon} \tag{1}$$

where E is the electric field, ρ the net charge density, and ϵ the dielectric constant. For static modes, the partial derivative of (1) can be replaced by the total derivative.

Consequently, one might expect an electric field versus distance profile such as shown in Fig. 3. Note that the narrow multiplication regions must now have higher fields in order to produce the required quantity of charge over regions considerably narrower than the total depletion layer width. (This is made possible of course by

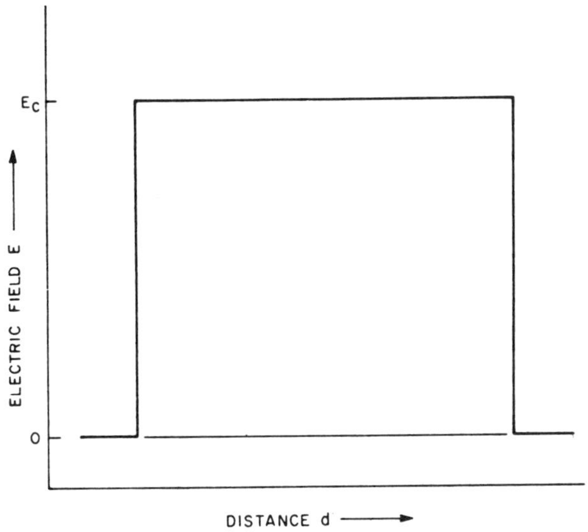

Fig. 2. Electric field versus distance for an idealized p-i-n under high reverse bias voltage, just below breakdown.

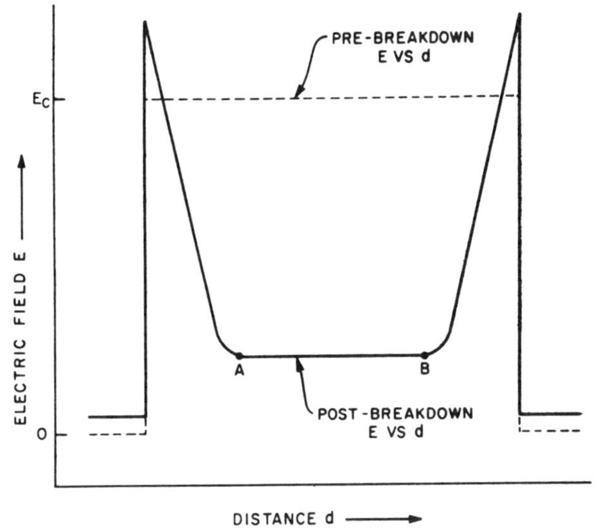

Fig. 3 The p-i-n diode of Fig. 2 displaying both pre- and post-breakdown electric field versus distance profiles.

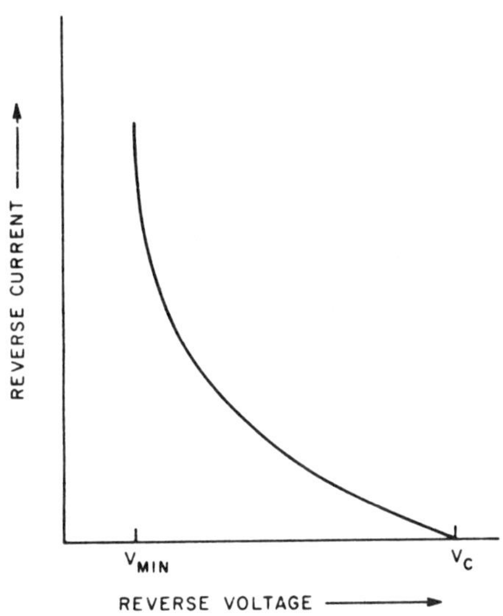

Fig. 4. Qualitative expectation for p-i-n reverse voltage versus reverse current.

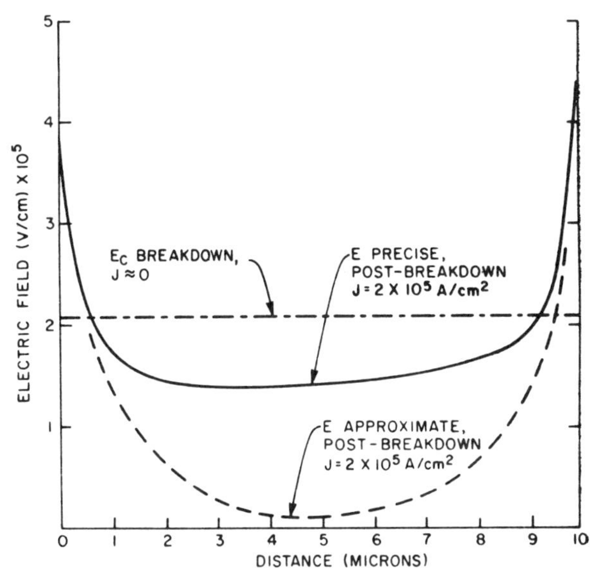

Fig. 5. Precise computer solutions of electric field versus distance for a 10-μm-wide p-i-n having abrupt transitions at the edges. Courtesy D. L. Scharfetter *et al.* [14].

the nonlinearities in the hole and electron ionization coefficients that lead to much higher ionization rates at higher electric fields.)

Should the phenomena indicated in Fig. 3 occur, the expected reverse current versus bias voltage curve (assuming uniform current density) would be as shown in Fig. 4 and obviously gives rise to a negative resistance. For oscillator efficiency to be high, employing such a negative resistance, the minimum voltage V_{min} should be much less than the breakdown voltage V_c.

Previous analytical treatments [3], [10]–[13] have considerably exaggerated the extent to which V_{min} can be reduced. Fig. 5 presents a precise calculation [14] of the electric field versus distance profile for a silicon p-i-n diode in hard breakdown with abrupt steps in the impurity density at the edges of the depletion layer. Experimental values for the hole and electron ionization coefficients were employed. Included are dashed lines for the critical breakdown field, and the field in hard breakdown when an analytic dependence on E is assumed for the ionization coefficients that has previously been employed [3], [10]–[13]. The difference in the precise calculation and approximate calculation is striking. Bowers [15] has taken an approach similar to that of [14].

The difference implied in reverse-bias voltage versus current plots is shown in Fig. 6, which was obtained from a series of such calculations for different current densities. It should be apparent that the approximate calculation is considerably overoptimistic. Efficiency calculations yield 23 percent for the approximate and 9

56

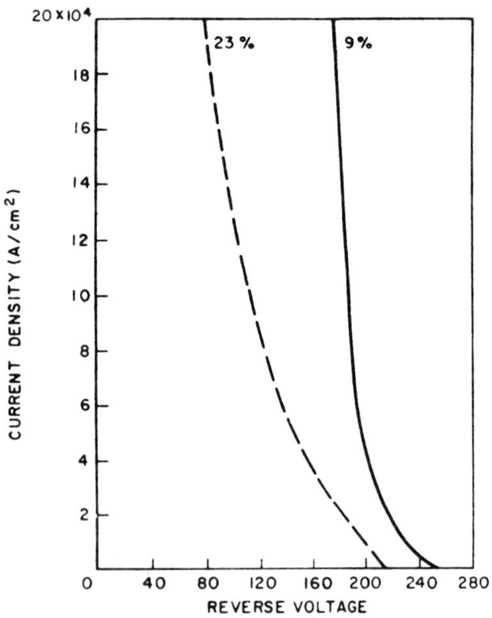

Fig. 6. Precise and "analytic" calculations of reverse current for the structure of Fig. 5. Courtesy D. L. Scharfetter *et al.* [14].

percent for the exact solution, assuming sinusoidal excitation.

The point that should be emphasized [16] is that the lowest value to which the field can drop is that at which multiplication ceases. See points A and B in Fig. 3. Thus, approximate treatments that require larger than realistic values of the ionization coefficients at low values of electric fields are going to be overly optimistic on achievable efficiency. Further, the "precise" efficiency is diminished if realistic doping profiles, external resistances, etc., are introduced. It consequently seems highly unlikely that space-charge feedback negative resistances can explain microwave efficiencies of 60 percent.

In addition to the relatively poor efficiency to be expected in realistic structures, the practical utilization of space-charge feedback negative resistances is highly unlikely due to the filament formation tendencies inherent in this type of operation.

Dynamic Modes

Consider now those modes that utilize dynamic space-charge effects to produce microwave power. By "dynamic," we imply variations of electric field and charge density in times shorter than the transit time of carriers through the device. Consequently in dynamic modes one expects to see space-charge density waves propagating through the diode instead of the smooth uniform growth experienced in the static modes.

The first dynamic analysis was due to Read and appeared in 1958 [7]. Read analyzed a p-n-i-n structure and demonstrated that it possessed a "dynamic" negative resistance of the IMPATT (IMPact Avalanche Transit-Time) type. Following the first report of microwave oscillations [17] in ordinary p-n junctions in 1965, Misawa developed an analysis [8] that showed that these devices also possessed a "dynamic" negative resistance of the IMPATT type. The analysis was quite general and included as a special case the "static" subtransit time space-charge-feedback regime. It should be noted that all IMPATT structures do not possess "static" negative resistances, at least at normal operating current densities. Consequently, the filament forming tendency inherent in the "static" modes can be avoided in IMPATT device design. In practice this has meant the avoidance of p-i-n-type structures.

The first large-signal analysis capable of handling realistic structures was that of Scharfetter and Gummel [9]. It also could treat multifrequency parametric oscillations. Many other authors have contributed to IMPATT theory [18].

The TRAPATT "dynamic" mode was discovered on the computer [19] by Scharfetter and co-workers in seeking explanations for the "anomalously" high experimental efficiencies first reported by Prager, Chang, and Weisbrod [20]. It involves a unique "traveling avalanche zone" and a "trapped plasma." The word TRAPATT is derived from TRApped Plasma Avalanche Triggered Transit.

Some of the properties of these dynamic modes will now be considered.

IMPATT Mode

The details of the dynamic IMPATT mode are more difficult to present than those of the static mode since their origin lies in the dynamics of moving space-charge waves.

IMPATT operation invokes two basic properties of carrier transport in solids at high electric fields. The first of these is that property of a thin avalanche zone that forces a continuing exponential increase of the local carrier population as long as a certain electric field E_c is exceeded. Note that it doesn't matter how much E_c is exceeded. Of course the actual exponential growth rate will be higher the more the actual electric field exceeds E_c. Consequently, if the electric field can be made to exceed E_c for any amount of time, the local particle current will always peak just when the local field finally returns to a value infinitesimally below E_c. Now imagine a thin avalanche zone with a constant dc voltage applied to it to produce a field as shown in Fig. 7(a). (The peak field can be imagined slightly below E_c to prevent current runaway.) If we now superpose a small, sinusoidal voltage as shown in Fig. 7(b), the resulting local current would appear as shown in Fig. 7(c). Note that the particle current peaks 90° later than the applied voltage, as first postulated by Read [7].

The second effect required for IMPATT operation is that of transit-time delay. Contrary to charge motion in a vacuum, in semiconductors carrier velocities rapidly saturate at modest values of electric field, i.e., over the range of approximately 10 000 V/cm to several 100 000

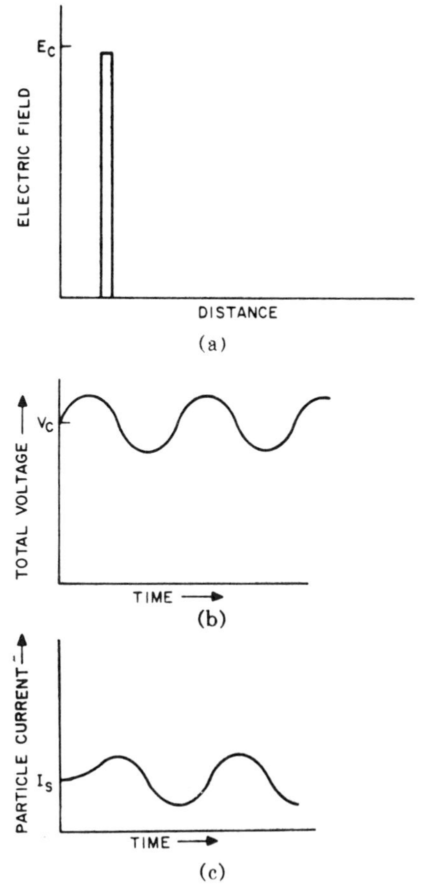

Fig. 7. (a) Electric field versus distance plot for a thin avalanche region. (b) Applied voltage as a function of time. (c) Resulting particle current as a function of time.

V/cm in silicon, carriers move at about 10^7 cm/s. This implies that as long as the fields in an IMPATT device are in this range, a given length of device has a unique transit time. Consequently, by adjusting the length of the drift regions outside the avalanche zone, arbitrary amounts of delay can be obtained.

Finally then, as Read observed, the combination of carriers being delayed 90° from an applied voltage across a thin avalanche region in conjunction with the transit-time delay of carriers drifting through a region of appropriate length, to produce an overall delay of 180° can give rise to an "efficient" negative-resistance device capable of developing watts of power at microwave frequencies (see Fig. 8).

A plot of the two terminal small-signal admittance of a typical germanium (step junction) IMPATT diode, Fig. 9, is presented in Fig. 10. The parameter along the curve is frequency. As expected from our previous discussion of the classical Read structure, there is a peak in the negative conductance where the transit-time delay is optimum, falling off at both higher and lower frequencies. In general, this peak in the negative conductance is a function of operating current density and strictly speaking is not a constant of the device structure. Nevertheless the frequency at which the negative conductance peaks, for a given operating current density, will be called *the* IMPATT frequency of the device. This definition will be utilized in the discussion of TRAPATT and parametric effects.

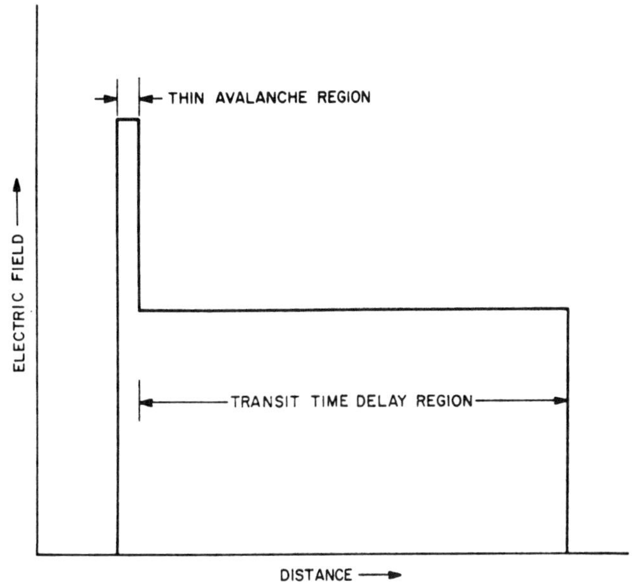

Fig. 8. A Read diode structure employing both a thin avalanche region and a long transit-time delay region.

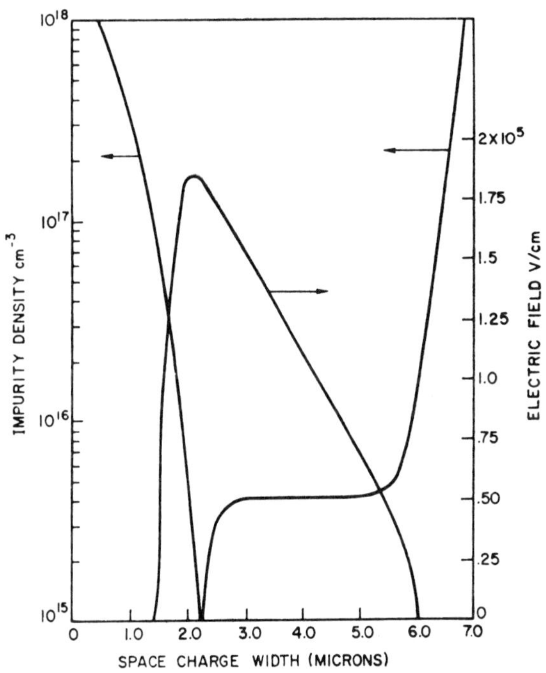

Fig. 9. Impurity profile and electric field profile at breakdown of a typical germanium step-junction IMPATT diode.

It is important to note that no small-signal negative resistance exists below about 7.5 GHz for this structure at the assumed operating current density. Nevertheless, highly efficient oscillations can take place at these low frequencies as will be seen shortly.

57

58

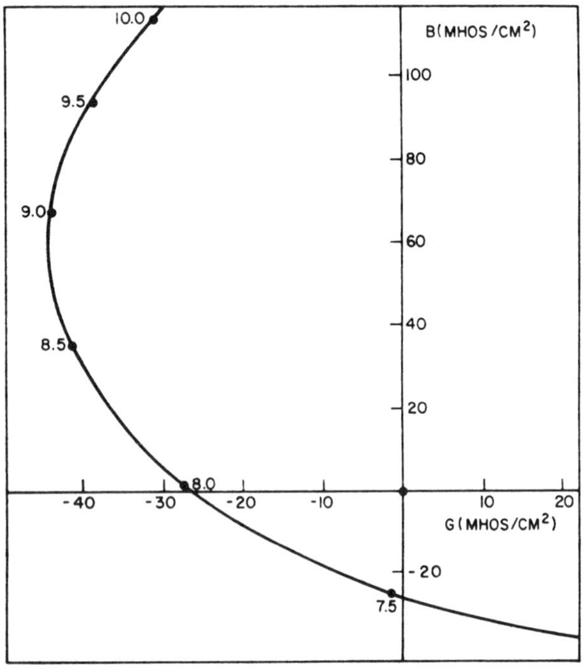

Fig. 10. The two terminal small-signal admittance of the structure of Fig. 9 as a function of frequency at a current density of 1000 A/cm².

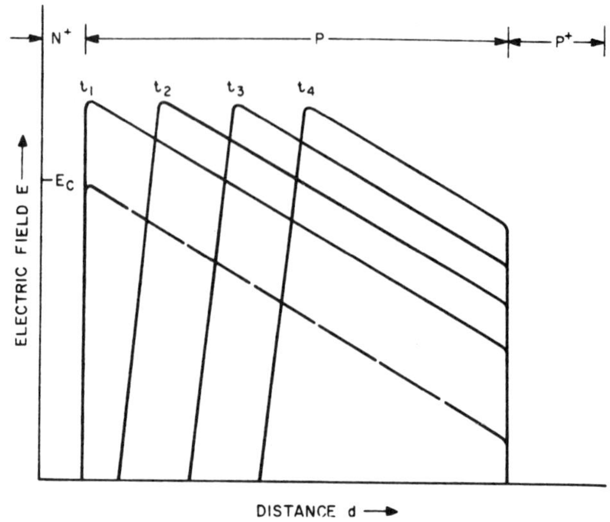

Fig. 11. Sequential field versus distance profiles during the transit of the avalanche zone through an idealized reach-through step junction. The breakdown field is shown dashed.

TRAPATT Mode

It has been experimentally demonstrated [20]–[31] that for certain specific types of IMPATT structures, in complicated multifrequency resonant circuits, remarkably high efficiencies can be obtained at sub-IMPATT frequencies. A computer analysis [19] displays the following sequence of events. First, some number n of IMPATT cycles occurs, i.e., transit-time oscillations in the usual sense. Then when the amplitude of the RF "IMPATT" voltage reaches a critical value on the nth cycle, a traveling avalanche zone is launched through the diode, filling it with a space-charge neutral hole electron plasma, in times less than that required for a carrier to cross the diode moving at saturated velocity.

An idealized time sequence of events as the zone traverses the diode is shown in Fig. 11. The reader is referred to [23] for a precise time sequence. Fig. 11 presupposes a large "launching" voltage, well in excess of breakdown voltage, which to date has been provided by the IMPATT cycle. The zone is then driven through the diode as long as the external circuit can supply the necessary current. Since large numbers of carriers are created at the peak values of electric field, the field is reduced to a very low value in the wake of the zone. The passage of the zone thus reduces the diode voltage to near zero and the dense plasma created allows a high-current low-voltage state. The extraction of this charge now proceeds and the diode recovers to a high-voltage low-current "swept" state. The alternation of these states yields exceptionally high efficiencies.

This type of operation bears a superficial resemblance to the "parametric operation" to be treated in the next section. It can be shown, however, that TRAPATT operation does not, at least in principle, require an IMPATT "pump" to operate [32]. It does appear that the device will not be self-starting in the absence of the IMPATT negative resistance but would require a triggering pulse, but nevertheless, this implies an operation different from what is commonly called parametric.

One still might ask, would the parametric label or classification for TRAPATT operation be helpful, and the answer seems to be no. For one thing an order of magnitude more power is sometimes attainable at the TRAPATT (low) frequency than is obtainable at the IMPATT (pump) frequency. Power frequency scaling laws observed in simple parametric systems seem to be completely violated.

Consequently it seems appropriate to distinguish TRAPATT operation as a separate mode of avalanche diode operation. It is typified by the presence of a traveling zone of avalanche yielding a dense hole electron plasma in which the electric field is reduced below that required for velocity saturation and the latter property gives us a simple test for TRAPATT operation. Thus, if it is a necessary part of the device operation that the field drop below that required for velocity saturation throughout the normal high field region of the device, the mode of operation is distinctive and will be called the TRAPATT or trapped plasma mode.

Parametric Mode

Variable reactance diodes have long been utilized as parametric devices. In typical operation a sinusoidal pump is coupled to the diode and output power is obtained at harmonics or gain is obtained at signal and idler frequencies.

IMPATT diodes have small-signal reactances that are functions of the direct current and voltage, and large-signal reactances that are functions of RF voltage and current amplitudes during oscillations. It should be apparent then that appropriate resonant circuits at frequencies other than the IMPATT frequency can couple to the "self-pumped" variable reactance of an oscillating IMPATT diode. Such effects were first identified [33]–[35] in 1965 and contributed in a large way to the understanding of IMPATT behavior in that an explanation was provided for the out-of-band signals readily observed in most experimental set-ups. These effects were treated more recently [36] in more detail.

The utility of the parametric definition of this type of operation is justified in that it predicts the major features to be expected, i.e., a fall off of power with frequency, less power out at signal and idler frequencies than is available from the pump, three frequency systems having gains predictable from the analysis of Manley and Rowe, etc. Thus, in structures in which it is difficult to launch the traveling avalanche zone, a combination of a new (IMPATT) and an old (parametric) description provides useful insight into experimental observations, and the labeling of a parametric mode of avalanche diode operation seems appropriate.

Parametric operation bears a superficial resemblance to TRAPATT operation, in that both operations require an IMPATT pump at present.

Thermal Modes

The thermal modes are handicapped by two fundamental limitations. 1) They form filaments; and 2) they are limited in operating frequency by thermal time constants. Nevertheless they are important to the avalanche diode designer in that they represent a self-destruction mechanism and can provide low frequency modulation of space-charge mode frequencies and power. No utilization of thermal modes has been reported nor is such use likely. The manifestation of this phenomenon in transistors has been labeled second breakdown. The title seems appropriate and we shall lump thermal modes under this title.

IV. Avalanche Diode Circuits

Having discussed some of the generalities of device physics, now consider the problem of the determination of circuit configurations appropriate for a given mode of operation. It is assumed that problems with thermal modes and static modes have been avoided through appropriate device design.

In general, a device such as that of Fig. 9 is capable of IMPATT, TRAPATT, and parametric operation. The determining factor is consequently the circuit. Note that of these three modes only the IMPATT mode presents a small-signal negative resistance, i.e., both the TRAPATT and parametric modes require large voltage swings across the diode in order to be operative. Consequently, basic IMPATT circuit design is reasonably straightforward.

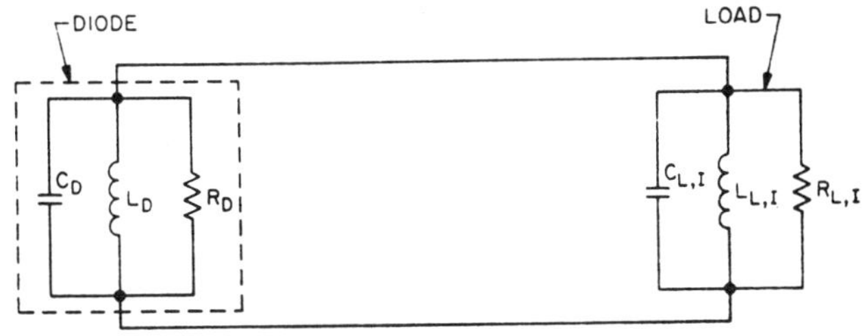

Fig. 12. Required circuit elements for basic IMPATT circuit design. The subscripts L, I imply load at the IMPATT frequency.

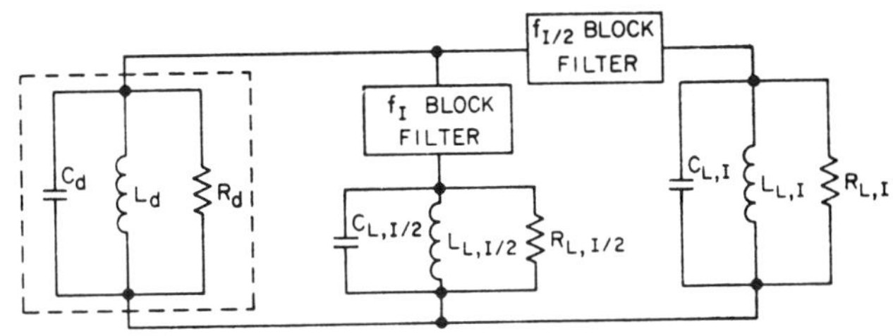

Fig. 13. Required circuit elements for self-pumped degenerate parametric amplifier or oscillator at half the IMPATT frequency. The subscripts L, I imply load at the IMPATT frequency and the subscripts $L, I/2$ imply load at *one-half* the IMPATT frequency.

The necessary circuit elements are shown in Fig. 12. The appropriate circuit reactance must be provided to resonate with the diode reactance at the chosen frequency of operation and the appropriate resistive loading must be supplied to optimize the output power. Estimates for these values can be obtained from the small-signal admittances of Fig. 10. By including a load inductance $L_{L,\,I}$ as well as a load capacitance $C_{L,\,I}$ the Q of the circuit can also be adjusted.

Unfortunately for the circuit designer, the simple considerations considered above do not yield an optimum IMPATT circuit. The difficulty is perhaps best illustrated by considering the effect of resonating the IMPATT diode at twice the "IMPATT" frequency as well as the IMPATT frequency. Such circuits have been shown to remarkably enhance oscillator efficiency and in some cases have almost doubled the achievable efficiency.

Thus, in addition to the circuit specifications of Fig. 12 that apply near the frequency of oscillation, "out-of-band" circuit performance must also be carefully tailored in order to achieve optimum results. Complicated multifrequency waveforms must be carefully studied to predict optimum device-circuit configurations.

The coupling of multifrequency resonant circuits to an IMPATT diode to achieve high-efficiency IMPATT operation is not so remarkably different from coupling multifrequency resonant circuits to achieve TRAPATT or parametric operation. As a case in point consider the design of a self- (IMPATT) pumped degenerate parametric amplifier, i.e., the band of amplification sought is at one-half the normal IMPATT frequency for the device. The required circuit elements are shown in Fig. 13. *Any*

MODE NAMES	USEFUL	ULTIMATE FREQUENCY LIMITATIONS	HIGHEST FREQUENCY	HIGHEST POWER		HIGHEST EFFICIENCY	
				CW	PULSE	CW	PULSE
IMPATT	YES	TUNNELING	340 GHz	5W 14GHz	30W 8.5GHz	15%	10%
TRAPATT	YES	TUNNELING	6GHz	5W 0.5GHz	400 W 1.0GHz	43%	60%
PARAMETRIC	YES	>IMPATTS	UNCERTAIN	--	--	--	--
SPACE CHARGE FEEDBACK	PROBABLY NO	<IMPATTS	UNCERTAIN	--	--	--	--
SECOND BREAKDOWN	PROBABLY NO	THERMAL TIME CONSTANTS <<IMPATTS	UNCERTAIN	--	--	--	--

Fig. 14. Avalanche diode modes with accomplishments and limitations.

60

good IMPATT diode will provide such half-frequency gain (or oscillations) in such a circuit.

Now consider a circuit design for a TRAPATT oscillator at one-half the IMPATT frequency. Our knowledge of the required circuit elements is incomplete but the components of Fig. 13 must be present and apparently more is required of the circuit.

In particular, resonances seem to be required at several multiples of the TRAPATT frequency. Exploratory circuitry work is currently in progress to determine the extent of these extra requirements. It is also apparent that all IMPATT diodes are not TRAPATT diodes. It may turn out, however, that good TRAPATT circuits are almost unavoidably good parametric circuits.

Discussion

Avalanche diodes present many interesting facets to the circuit designer. The IMPATT mode is very well understood in principle and yet no one is able to adequately specify an optimum IMPATT circuit.

Parametric modes appear interesting but the high-noise levels present in IMPATT diodes preclude the usual low-noise parametric applications. Self-pumped power up-converters or harmonic generators seem to be more reasonable parametric applications. The practical importance of parametric applications is yet to be demonstrated.

TRAPATT diode fabrication and circuit work is just beginning. The fascination here is the exceptional efficiency that can be achieved. It appears that TRAPATT's having efficiencies greater than 50 percent at the multiwatt level are foreseeable at least through X band.

The IMPATT, TRAPATT, and parametric modes are the only ones of the five modes of Fig. 1 that seem to have applications potential. Some achievements and limitations of each mode are listed in Fig. 14. The dashed lines represent an inability on the part of the author to obtain conclusive information. The degree of future success in the field of avalanche diodes will depend in large measure on the success of the solid-state-circuit engineer in meeting the challenges of avalanche diode circuit design.

Acknowledgment

The author would like to acknowledge the many helpful discussions with his co-workers at Bell Telephone Laboratories, Inc. D. L. Scharfetter rendered valuable assistance in the section on static modes.

References

[1] K. G. McKay, "Negative resistance semiconductive apparatus," U. S. Patent 2 908 871, October 26, 1954.

[2] J. B. Gunn, "Avalanche injection in semiconductors," *Proc. Phys. Soc.* (London), vol. B69, pp. 781–790, 1956.

[3] ——, "High electric field effects in semiconductors," in *Progress in Semiconductors*, A. F. Gibson, Ed., vol. 2. London: Heywood, 1957, pp. 213–247.

[4] D. J. Rose, "Microplasmas in silicon," *Phys. Rev.*, vol. 105, pp. 413–418, January 1957.

[5] M. A. Leblond and R. Gentner, "Note sur une structure particulière permettant d'obtenir des oscillations de haute fréquence," *Compt. Rend. Acad. Sci.* (Paris), vol. 242, p. 621, 1956.

[6] M. A. Leblond, "Étude théorique de la caractéristique intensité-tension d'une structure semiconductrice p-i-p en tenant compte de l'ionization due aux porteurs," *Compt. Rend. Acad. Sci.* (Paris), vol. 242, pp. 1856–1858, 1956.

[7] W. T. Read, "A proposed high-frequency, negative resistance diode," *Bell Sys. Tech. J.*, vol. 37, p. 401, March 1958.

[8] T. Misawa, "Negative resistance in *p-n* junctions under avalanche breakdown conditions, Part I," *IEEE Trans. Electron Devices*, vol. ED-13, pp. 137–151, January 1966.

[9] D. L. Scharfetter and H. K. Gummel, "Large-signal analysis of a silicon Read diode oscillator," *IEEE Trans. Electron Devices*, vol. ED-16, pp. 64–77, January 1969.

[10] B. Hoefflinger, "High-frequency oscillations on p^{++}-n^{+}-n-n^{++} avalanche diodes below the transit-time cutoff," *IEEE Trans. Electron Devices*, vol. ED-13, pp. 151–158, January 1966.

[11] H. Egawa, "Avalanche characteristics and failure mechanism of high voltage diodes," *IEEE Trans. Electron Devices*, vol. ED-13, pp. 754–758, November 1966.

[12] S. G. Liu and J. J. Risko, "High power punch-through avalanche diode microwave oscillators," *1968 ISSCC Digest Tech. Papers*, p. 146.

[13] M. W. Muller and H. Guckel, "Negative resistance and filamentary currents in avalanching silicon p^{+}-i-n^{+} junctions," *IEEE Trans. Electron Devices*, vol. ED-15, pp. 560–568, August 1968.

[14] D. L. Scharfetter, D. J. Bartelink, and B. C. De Loach, Jr., "Comments on static negative resistance in avalanching-silicon p^{+}-n-n^{+} junctions," (to be published).

[15] H. C. Bowers, "Space charge induced negative resistance in avalanche diodes," *IEEE Trans. Electron Devices*, vol. ED-15, pp. 343–350, June 1968.

[16] D. L. Scharfetter (private communication).

[17] R. L. Johnston, B. C. De Loach, Jr., and B. G. Cohen, "A silicon diode microwave oscillator," *Bell Sys. Tech. J.*, vol. 44, p. 369, February 1965. The interested reader is referred to the following publication appearing in 1966 reporting previously unpublished Russian work in this area. V. M. Val'd-Perlov, A. V. Krasilov, and A. S. Tager, "The avalanching transit-time diode," *Radio Engrg. and Electron. Phys.*, pp. 1764–1777, November 1966.

[18] References to much of this work can be found in *Microwave Semiconductor Devices*, H. A. Watson, Ed. New York: McGraw-Hill, 1969, ch. 15.

[19] D. L. Scharfetter, D. J. Bartelink, H. K. Gummel, and R. L. Johnston, "Computer simulation of low frequency high efficiency oscillations in germanium," presented at the 1968 IEEE Solid-State Device Research Conf., June 17–19, 1968.

[20] H. J. Prager, K. K. N. Chang, and S. Weisbrod, "High-power high-efficiency silicon avalanche diodes at ultrahigh frequencies," *Proc. IEEE* (Letters), vol. 55, pp. 586–587, April 1967.

[21] H. J. Prager, K. K. N. Chang, and J. Weisbrod, "Anomalous silicon avalanche diodes for microwave generation," *Proc. 1967 Cornell Conf. on High Frequency Generation and Amplification* (Ithaca, N. Y., August 29–31, 1967), pp. 266–280.

[22] R. L. Johnston and D. L. Scharfetter, "Low frequency high efficiency oscillations in germanium impatt diodes," presented at the 1968 IEEE Solid State Device Research Conf., June 17–19, 1968 (to be published in *IEEE Trans. Electron. Devices*).

[23] R. L. Johnston, D. L. Scharfetter, and D. J. Bartelink, "High efficiency oscillations in germanium avalanche diodes below

the transit-time frequency," *Proc. IEEE,* vol. 56, pp. 1611–1613, September 1968.
[24] D. E. Iglesias and W. J. Evans, "High efficiency CW impatt operation," *Proc. IEEE* (Letters), vol. 56, p. 1610, September 1968.
[25] S. G. Liu, "High-power punch-through avalanche-diode microwave oscillators," *IEEE J. Solid-State Circuits,* vol. SC-3, pp. 213–217, September 1968.
[26] C. P. Snapp and B. Hoefflinger, "High current density operation and analysis of avalanche transit-time diodes," *Proc. 7th Internatl. Conf. on Microwave and Optical Generation and Amplification* (Hamburg, Germany), September 16–20, 1968.
[27] C. P. Snapp, L. A. Stark, and B. Hoefflinger, "Performance and theory of avalanche resonance pumped impatt oscillators," presented at the 1968 Internatl. Electron Devices Meeting, Washington, D. C., October 23–25, 1968.
[28] C. P. Snapp and B. Hoefflinger, "An efficient multiresonant avalanche diode oscillator in the 1.5 to 11 GHz range," *Proc. IEEE* (Letters), vol. 56, pp. 2054–2055, November 1968.
[29] B. Hoefflinger, C. P. Snapp, and L. A. Stark, "High efficiency avalanche resonance pumped amplification," *1969 ISSCC Digest Tech. Papers,* p. 193; also *Electron. Letters,* vol. 5, pp. 43–45, February 1969.
[30] P. A. Levine and S. G. Liu, "Tunable L-band high-power avalanche diode oscillator," *1969 ISSCC Digest Tech. Papers,* p. 84.
[31] R. S. Ying, R. G. Mankarious, and D. L. English, "High efficiency anomalous mode oscillation from silicon impatt diodes at 6 GHz," *1969 ISSCC Digest Tech. Papers,* p. 86.
[32] B. C. De Loach, Jr., and D. L. Scharfetter, "Device physics of TRAPATT oscillators," (to be published).
[33] B. C. De Loach, Jr., and R. L. Johnson, "Two terminal semiconductor high frequency oscillator," U. S. Patent 3 270 293, February 16, 1965.
[34] R. L. Johnston and B. C. De Loach, Jr., "Continuous microwave oscillations from a reverse-biased silicon diode," Bell Telephone Laboratories, Inc., Microwave Diode Research Rept. 20, Contract DA36-039-sc-89205, pp. 19–22, June 20, 1965.
[35] B. C. De Loach, Jr., and R. L. Johnston, "Avalanche transit-time microwave oscillators and amplifiers," presented at the 1965 Solid-State Device Research Conf., June 1965; also *IEEE Trans. Electron Devices,* vol. ED-13, pp. 181–186, January 1966.
[36] W. J. Evans and G. I. Haddad, "Frequency conversion in IMPATT diodes," *IEEE Trans. Electron Devices,* vol. ED-16, pp. 78–87, January 1969.

Basic Principles and Properties of Avalanche Transit-Time Devices

GEORGE I. HADDAD, SENIOR MEMBER, IEEE, PAUL T. GREILING, MEMBER, IEEE, AND WILLIAM E. ŞCHROEDER, STUDENT MEMBER, IEEE

Invited Paper

Abstract—**The basic principles and characteristics of the various modes of operation of avalanche transit-time devices are presented. Theoretical and experimental results are also presented in order to indicate the present state of development and the kind of performance which has been achieved in these devices.**

I. Introduction

SINCE its inception the avalanche diode, as a microwave device, has received a great deal of attention, and many significant advances have been made concerning power output, efficiency, frequency range, fabrication and evaluation techniques, and other properties.

The purpose of this review article is to discuss the basic principles of the various modes of operation of this device, to present the type of performance which has been achieved, to give some typical data on its characteristics, and to present some information concerning experimental circuits and measurements employed in evaluating the properties of these devices. This article is therefore divided into two main parts. The first deals with the basic principles and the theory, and the second deals with experimental results and measurements.

Avalanche diodes have been fabricated so far in Si, Ge, and GaAs. Due to the basic phenomena involved in the generation of negative conductance in this device, almost any semiconductor material can, in principle, be employed for realizing an avalanche-diode oscillator. This results in a great deal of flexibility in the choice of a material as compared to the Gunn-effect, for example, which is limited to a few materials. The desirable parameters which are important in determining the suitability of a particular material are the following:

1) a high saturated drift velocity of the carriers
2) a high energy gap and low and equal ionization rates
3) a high thermal conductivity
4) availability, reliability, and ease of forming a junction.

Manuscript received June 8, 1970. This work was supported by the Rome Air Development Center under Contract F30602-68-C-0043.

The authors are with the Electron Physics Laboratory, Department of Electrical Engineering, University of Michigan, Ann Arbor, Mich. 48104.

It is therefore possible that other materials may be found which have a combination of properties which are more desirable than the three materials mentioned.

A great deal of flexibility also exists in the formation of a junction. Junction formation can be achieved through 1) diffusion, 2) Schottky barrier, 3) ion implantation, and 4) epitaxial growth.

Diode fabrication and material technology was intentionally excluded from this article because it was felt that this information would be more suitable for other publications.

II. Basic Principles and Characteristics of the IMPATT Mode

Before discussing the physical operation, mathematical analyses, and characteristics of IMPATT diodes, it is informative to derive the output power and efficiency for certain voltage and current waveforms. In the following section it will be shown how currents and voltages of the form shown in Fig. 1 are generated in IMPATT diodes.

The average power dissipated for the waveforms shown in Fig. 1 is

$$P = \frac{1}{2\pi}\int_{y\pi}^{x\pi} I_{max} V_{RF} \sin \omega t \, d\omega t \tag{1}$$

where I_{max}, V_{RF}, and x are illustrated in the figure. Performing the integration, the following result is obtained:

$$P = \frac{V_{RF} I_{max}}{2\pi}(\cos y\pi - \cos x\pi). \tag{2}$$

The maximum current I_{max} is related to the dc current by

$$I_{dc} = \frac{1}{2\pi}\int_{y\pi}^{x\pi} I_{max} \, d\omega t \tag{3}$$

which reduces to

$$I_{dc} = I_{max}\left(\frac{x-y}{2}\right). \tag{4}$$

Therefore,

$$I_{max} = \frac{2I_{dc}}{x-y} \tag{5}$$

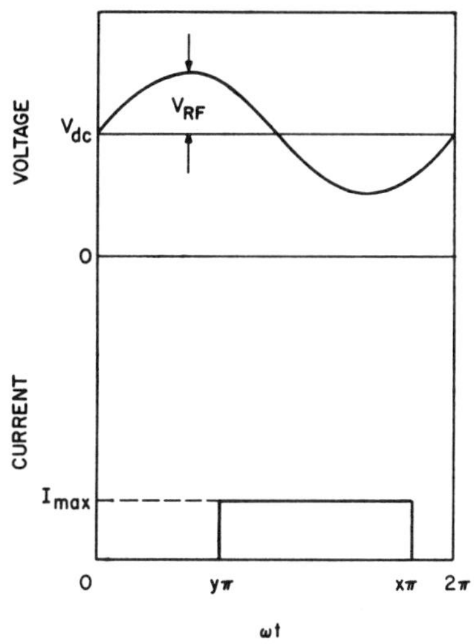

Fig. 1. Voltage and current waveforms.

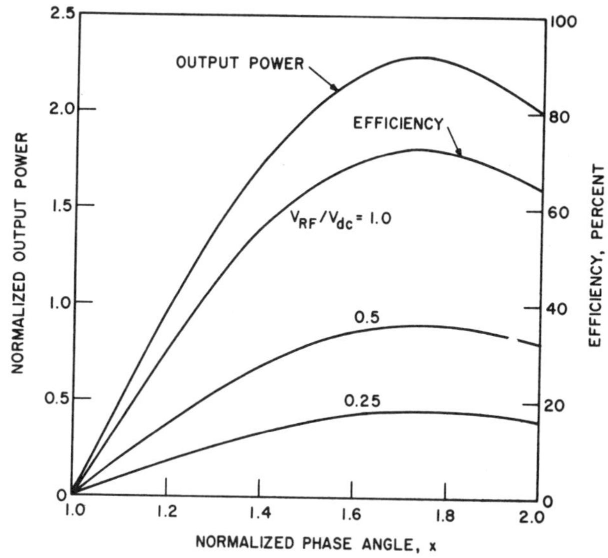

Fig. 2. Normalized output power and efficiency versus normalized phase angle x ($y=1$).

which when substituted into (2) gives

$$P = \frac{V_{RF} I_{dc}}{\pi} \left(\frac{\cos y\pi - \cos x\pi}{x - y} \right). \tag{6}$$

The efficiency η is then given by

$$\eta = \frac{1}{\pi} \frac{V_{RF}}{V_{dc}} \left(\frac{\cos y\pi - \cos x\pi}{x - y} \right). \tag{7}$$

It is obvious from the preceding equations that for $0 \leq y \leq 1$, maximum power output and efficiency will result for $y=1$. For this case (6) and (7) reduce to

$$P = -\frac{V_{RF} I_{dc}}{\pi} \left(\frac{1 + \cos x\pi}{x - 1} \right) \tag{8}$$

and

$$\eta = -\frac{V_{RF}}{\pi V_{dc}} \left(\frac{1 + \cos x\pi}{x - 1} \right). \tag{9}$$

The negative sign in (8) and (9) indicates that power is delivered by the diode.

The normalized power output $[P/(V_{RF} I_{dc}/\pi)]$ and efficiency are plotted in Fig. 2 as functions of x for different values of V_{RF}/V_{dc}. It is seen that maximum power output and efficiency occur at $x \approx 1.742$.

Physical Explanation of IMPATT Diode Operation

An understanding of the dynamic operating characteristics of IMPATT diodes can be obtained best by considering the operation of a Read diode. The structure, field distribution, and current waveforms for a reverse-biased Read diode are shown in Fig. 3. The Read diode consists of two regions: a narrow avalanche region (p-region) in which carrier multiplication by impact ionization occurs, and a drift region (intrinsic region) in which the carriers drift at saturated or field-independent velocities and where no impact ionization occurs.

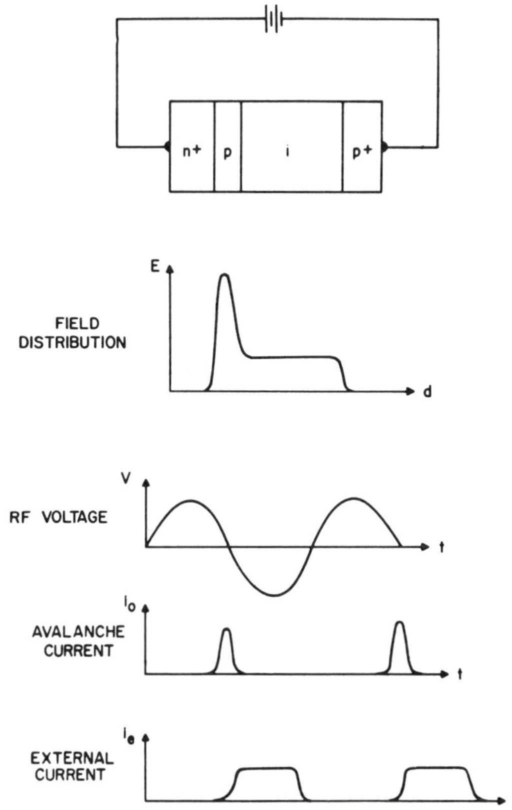

Fig. 3. Read-diode waveforms.

The negative resistance or conductance of a Read diode is attributed to a phase shift between the current through the diode and the voltage across it. This phase shift consists of two components. There is a phase delay of the current caused by the avalanche multiplication process and by the finite transit time of the holes drifting through the drift region. The phase delay caused by the avalanche process results from the condition that the rate of generation of electron-hole pairs in

the avalanche region is proportional to both the electric field and the density of electrons and holes that are already present. If a small RF voltage of sufficiently high frequency is assumed to be superimposed along with a dc voltage near breakdown across the diode, the rate of generation of electron-hole pairs will exceed the rate at which the pairs leave the avalanche region. Thus both the density of carriers and the current will grow exponentially with time as long as the RF and dc voltages add to give a total field above breakdown. When the RF voltage changes sign and subtracts from the dc voltage, causing the field to fall below breakdown, the generation rate starts decreasing, causing the current to decrease. Thus the current generated by the avalanche
64 process will have its maximum when the RF field goes through zero. Thus, as shown in Fig. 3, the avalanche process contributes a 90° inductive phase lag to the current generated in the avalanche region. This current is then injected into the drift region of the diode. The current induced in the external circuit by this charge, drifting through the drift region at a saturated drift velocity, is shown at the bottom of Fig. 3. It is obvious that the fundamental component of the external current is more than 90° out of phase with the RF voltage causing the resistance or conductance of the diode to be negative. In order for the diode to operate as a stable oscillator, the negative conductance of the diode must decrease with increasing RF voltage. The RF voltage across the diode will grow until the admittance of the diode is balanced by the admittance of the diode mount and microwave circuit. The diode will oscillate at a frequency and RF voltage level where the diode admittance is equal and of opposite sign to the circuit admittance.

There are several important effects which must be included in this explanation. First, there is the effect of the space charge of the carriers. As mentioned previously, the current in the diode consists primarily of holes which will cause the electric field to decrease behind them and to increase in front of them as they drift across the drift region. As the holes are generated in the avalanche region and injected into the drift region, they cause the field in the avalanche region to be depressed. The depression of the field by the space charge of the holes causes the field to drop below the breakdown level before the RF voltage does. Therefore, the phase lag of the current due to the avalanche process is decreased. If the RF voltage is held constant and the dc current is increased, which means the space charge in the diode is increased, the field in the avalanche region will drop below the breakdown level earlier in the cycle. Thus, the phase lag of the current will decrease with increasing dc current and constant RF voltage. Conversely, if the dc current is held constant but the RF voltage is increased the phase lag of the current is increased.

The additional phase delay needed for negative conductance is caused by the transit time of the holes. The effect of the transit time on the frequency range of operation can be easily understood by assuming that the current generated in the avalanche region is generated in a sharp pulse. The exact time in the cycle that this pulse is generated will be a function of the dc current and the RF voltage as discussed previously. The phase delay caused by the transit time is proportional to $\omega\tau$, where τ is the transit time in seconds. For a given diode the transit time τ remains constant but the phase delay decreases with decreasing frequency. At a sufficiently low frequency the phase delay of the transit time plus the phase lag of the avalanche process are not sufficient to have the fundamental component of the external current lag the RF voltage by more than 90°. Therefore, below a certain cutoff frequency the conductance of the diode becomes positive. As can be seen from the effect of the RF voltage on the phase lag in the avalanche process, the conductance will be negative at lower frequencies for larger RF voltages. However, for these frequencies the diode will not be a self-starting oscillator since the small-signal conductance is positive. As the dc current is increased, the cutoff frequency is increased due to the effect of the current on the phase lag in the avalanche process.

The Read diode is a convenient structure for analytical purposes and for obtaining a physical understanding of IMPATT diodes, but it is difficult to construct. The p-n junction is easier to fabricate but much more difficult to analyze. In the p-n junction the avalanche region and drift region are not separated, but rather carriers are generated while they are drifting through an appreciable portion of the diode. Also, both electrons and holes take part in the conduction process rather than just one type of carrier. Thus, it is not possible to separate the two phase-delay components as in the Read diode. However the mechanisms causing the phase delay between the voltage and the current in IMPATT diodes are still the transit time and the avalanche process. Another difference in IMPATT diodes is the cutoff frequency. Misawa [1] has shown that the cutoff frequency for a diode with a uniform field in the depletion region (a p-i-n diode) goes to dc. This is due to space-charge effects on the field distribution, which are comparable to those observed in a gas discharge, rather than to transit-time effects. Hoefflinger [2] has shown that for high current densities (large space charge) the cutoff frequency for a Read structure goes to dc due to avalanching in the drift region caused by the large space charge present.

Basic Equations

The equations [3] which govern the operation of an IMPATT diode are Poisson's equation and the continuity equations for holes and electrons. In one dimension, these equations are

$$\frac{\partial E}{\partial x} = \frac{q}{\epsilon}(N_D - N_A + p - n) \tag{10}$$

$$\frac{\partial p}{\partial t} = -\frac{1}{q}\frac{\partial J_p}{\partial x} + g_d + \frac{1}{q}(\alpha_n |J_n| + \alpha_p |J_p|) \tag{11}$$

and

$$\frac{\partial n}{\partial t} = \frac{1}{q}\frac{\partial J_n}{\partial x} + g_d + \frac{1}{q}(\alpha_n|J_n| + \alpha_p|J_p|) \quad (12)$$

where

$$J_p = qp\mu_p E - kT\mu_p \frac{\partial p}{\partial x} \quad (13)$$

$$J_n = qn\mu_n E - kT\mu_n \frac{\partial n}{\partial x} \quad (14)$$

$$g_d = \frac{pn - n_i^2}{\tau_{p0}(n + n_1) + \tau_{n0}(p + p_1)} \quad (15)$$

$$\alpha_n = A_n' \exp\left[-\left(\frac{b_n'}{E}\right)^{m_n'}\right] \quad (16)$$

$$\alpha_p = A_p' \exp\left[-\left(\frac{b_p'}{E}\right)^{m_p'}\right] \quad (17)$$

$$\left(\frac{\mu_{n0}}{\mu_n}\right)^2 = 1 + \left(\frac{N_T}{\frac{N_T}{S_n} + N_n}\right) + \left[\frac{\left(\frac{E}{A_n}\right)^2}{\frac{E}{A_n} + F_n}\right] + \left(\frac{E}{B_n}\right)^2 \quad (18)$$

and

$$\left(\frac{\mu_{p0}}{\mu_p}\right)^2 = 1 + \left(\frac{N_T}{\frac{N_T}{S_p} + N_p}\right) + \left[\frac{\left(\frac{E}{A_p}\right)^2}{\frac{E}{A_p} + F_p}\right] + \left(\frac{E}{B_p}\right)^2 \quad (19)$$

where

- E electric field (V/cm)
- q electronic charge (C)
- ϵ dielectric constant (F/cm)
- N_D, N_A donor and acceptor densities (cm^{-3})
- p, n hole and electron densities (cm^{-3})
- J_p, J_n hole and electron current densities (A/cm^2)
- g_d hole and electron generation recombination rates through a single-level trapping center
- α_p, α_n hole and electron ionization rates (cm^{-1})
- μ_p, μ_n hole and electron mobilities ($cm^2/V \cdot s$)
- k Boltzmann constant (1.38×10^{-23} J/°K)
- T absolute temperature (°K)
- p_1, n_1 hole and electron concentrations in the conduction band when the single-level trap and the Fermi level are equal
- A_p', b_p', m_p' hole ionization rate constants
- A_n', b_n', m_n' electron ionization rate constants
- μ_{p0}, μ_{n0} low-field low doping density hole and electron mobility ($cm^2/V \cdot s$)
- N_T total ionized impurity density (cm^{-3})
- N_p, S_p, A_p, F_p, B_p hole mobility constants
- N_n, S_n, A_n, F_n, B_n electron mobility constants.

Mathematical Analyses

Read [4], in his original paper, made several assumptions in order to simplify the above equations and obtain a closed-form solution for the small-signal admittance of the diode. He assumed equal and field-independent drift velocities for the carriers. Also, the effects of diffusion and carrier generation and recombination through defects were neglected. The ionization rates for holes and electrons were assumed equal and the breakdown condition [5], the integral of the ionization rate, was approximated by a function proportional to a power of the electric field. Read divided the diode into a narrow avalanche region in which the conduction current was assumed independent of position and a drift region in which the carriers drifted at saturated velocities. Besides the small-signal results, qualitative large-signal results were obtained by both a sharp pulse and an intermediate-amplitude approximation.

Misawa [1] showed through a small-signal analysis that negative resistance of the type predicted by Read for the Read diode existed in junction diodes with a wide range of doping profiles. He included the effect of unequal ionization rates and drift velocities for holes and electrons. The effect of nonsaturated drift velocities was also considered. Misawa's computer solutions for the small-signal impedance were compared to those obtained by Read and were shown to be qualitatively similar for a wide range of doping profiles. Gilden and Hines [6] derived a simple equivalent circuit for the Read diode using a closed-form expression for the small-signal impedance and assuming a short transit angle. The electronic tuning effects due to the dc current could be clearly seen from their equivalent circuit. Fisher [7] developed a small-signal analysis in which the avalanche region of the diode, although still assumed to be narrow as in the Read diode, could be located any place in the space-charge region of the diode. Also, several of the assumptions used in other analyses [4], [6], such as equal saturated drift velocities and ionization rates of the carriers, were removed. His closed-form solutions for the impedance were shown to be in agreement with Misawa's results. Gummel and Scharfetter [8] extended Gilden and Hines' small-signal analysis to include diodes in which the avalanche region is not necessarily narrow. They obtained small-signal admittance plots for a Read diode and for more realistic diode structures in which the avalanche region occupies some fraction of the total depletion region. They found that optimum

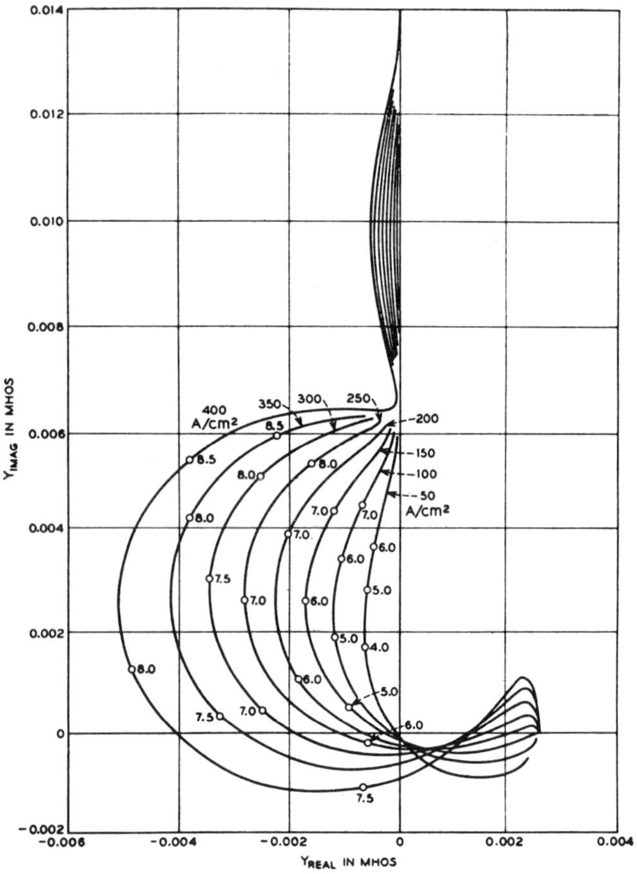

Fig. 4. Small-signal admittance of a Si Read diode as a function of frequency (GHz) and current density (A/cm²), Gummel and Scharfetter [8].

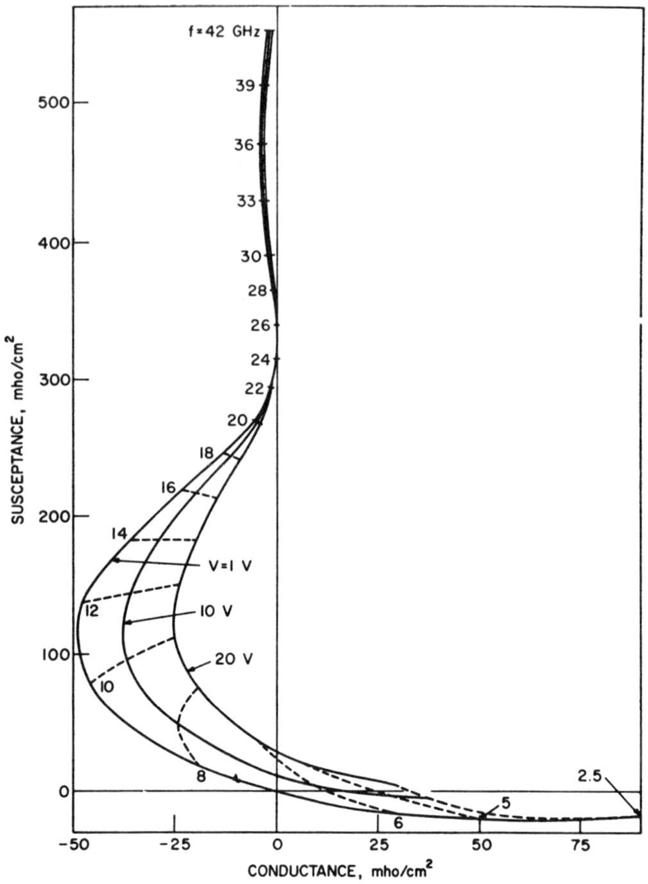

Fig. 5. RF admittance of a Si diode. $V_{RF}=V \sin \omega t$, $I_{dc}=500$ A/cm²; Schroeder and Haddad [13].

performance is achieved when the avalanche region is approximately one third of the drift region. Their small-signal results for a Si Read diode with a 1-μm-wide avalanche region and a 9-μm-wide drift region are shown in Fig. 4 for different bias current densities and frequencies. Two important results are evident. First, there is an optimum frequency of operation for maximum negative conductance. Second, the lower cutoff frequency increases with bias current. This has been shown for a Read analysis (zero avalanche width) to be proportional to the square root of the current [6].

Small-signal results predict the range of current and frequency in which an IMPATT diode will oscillate if placed in the proper resonant circuit. These results also indicate the magnitude of the maximum negative conductance which can be generated by the diode. However, no data on power output, efficiency, and other important effects can be obtained since IMPATT diodes are nonlinear devices and the space-charge distribution under large-signal conditions is much different from the small-signal case.

A large-signal analysis was performed by Evans and Haddad [9] where, in addition to the assumptions made by Read, they assumed a small transit angle in the drift region. With this assumption they were able to obtain closed-form solutions for the equations and thus several important nonlinear properties of the device could be predicted, even though qualitatively, without resorting to extensive computer calculations. This model is also very well suited for studying self-oscillating frequency converters and other nonlinear phenomena which may otherwise be prohibitive to investigate with more exact analyses. Such a model was used by Evans and Haddad [10] in a study of various types of self-oscillating frequency converters. The restriction of a small transit angle was removed in an analysis by Greiling and Haddad [11], and several operating characteristics of a Read-type diode were presented, including the effect of large RF voltage swings where the velocity of the carriers is nonsaturated in the drift region. Blue [12] used a Read-type diode for a large-signal analysis in which the field in the avalanche region was independent of position and the space-charge effects within the avalanche region were neglected. The numerical techniques employed for solving the equations which were obtained by this analysis produced an efficient and inexpensive computer program. Results [13] of an analysis of this type for a Si Read diode in which the avalanche region is 1 μm wide and the drift region is 4 μm wide are shown in Fig. 5. The results show the effect of RF voltage on the admittance and the lower cutoff frequency. Scharfetter and Gummel [3] have developed a computer program for obtaining a numerical solution to the basic equations which govern the operation of the avalanche

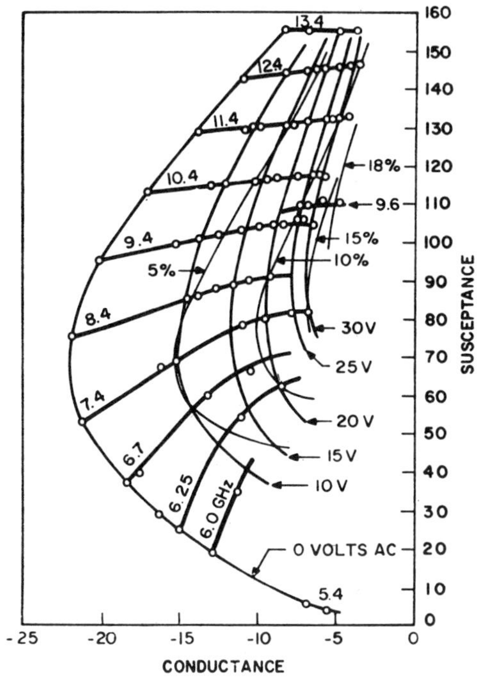

Fig. 6. Admittance and efficiency of a Si Read diode as a function of RF voltage and frequency; $J_{dc} = 200$ A/cm², admittance is in mho/cm²; Scharfetter and Gummel [3].

diode. These are the equations which were presented at the beginning of this section. They have attempted to make their model as exact as possible by including carrier generation and recombination due to impact ionization and defects, respectively, and the variation of mobility as a function of electric field and doping density. They have also assumed that these effects are not equal for both electrons and holes and have included the effect of the microwave circuit on the diode in which it is placed. Large-signal results of their analysis for a Read-type diode with a realistic doping profile are shown in Fig. 6. The results are similar to those shown in Fig. 5. Their program is presumed to give the most accurate and detailed solution for the large-signal operating characteristics of IMPATT diodes. However, the amount of computer time necessary for its execution is prohibitive for any detailed study of the effect of different diode parameters on its operating characteristics. A large-signal analysis was developed by Greiling *et al.* [14], in which IMPATT diodes with any doping profile could be analyzed. Though not all of the physical parameters included by Scharfetter and Gummel were included in this analysis, it does give good physical results for the large-signal operating characteristics of IMPATT diodes. The amount of computer time required for execution is such that it is also possible to use the analysis to study the effect of different diode parameters on the operating characteristics. The results obtained from this analysis and a comparison with experimental results are presented in [15].

Because of the broad-band negative resistance and the nonlinearities, IMPATT diodes operate under multifrequency as well as single-frequency conditions. Although the analyses of Scharfetter and Gummel [3] and Greiling and Haddad [15] can be employed to investigate multifrequency operation, the amount of computer time necessary for execution becomes too costly for a detailed study. Several workers [12], [13], [16]–[18] have investigated the effect of multifrequency operation using simplified analyses. Blue [12] has shown that his analysis can be used quite effectively for studying multifrequency operation. Schroeder and Haddad [13] used this type of analysis and performed an extensive study of the multifrequency large-signal operating characteristics of Read diodes. These results show that the negative conductance of a Read diode can be enhanced by having harmonics or subharmonics of the signal frequency present. This enhancement of the negative conductance is a function of the phase between the two frequencies, the RF voltage amplitudes, and the relative value of the frequencies with respect to the optimum operating frequency (frequency which produces maximum negative conductance). It was shown that for a Read diode (Fig. 5) with a cutoff frequency around 7 GHz and an optimum operating frequency of 11 GHz, the diode could oscillate with significant output power at 5 GHz if a properly phased oscillation at 10 GHz were present. This is due to waveshaping of the RF voltage across the diode which causes an additional phase delay in the generation of the current in the avalanche process. Under single-frequency operation at 5 GHz the total phase delay of the avalanche process and the transit time is not enough for the fundamental current to be more than 90° out of phase with the RF voltage. However, by adding the second harmonic of the signal frequency with the proper phase it is possible to generate a waveform which causes the RF field in the avalanche region to go through zero at a much later time in the cycle and thus increase the phase delay of the avalanche process. This additional phase delay is enough for the conductance to become negative. Therefore, the phase angle between the fundamental and harmonic RF voltages is an important parameter in determining the negative conductance of the diode. It was also shown that it is possible to degrade the performance of the diode with certain phase angles.

Fig. 7 summarizes the results of this analysis for harmonic and subharmonic tuning. The power density is plotted as a function of frequency for two-frequency modes where the harmonic ratio is two and a single-frequency operation curve is plotted for comparison. The phase angle is chosen to provide maximum output power at the designated frequency while maintaining negative conductance at both frequencies. The single-frequency curve is dashed below 7 GHz because the small-signal conductance is positive there.

The curve labeled second-harmonic tuning shows the power output obtainable at the fundamental with an equal amplitude second harmonic present. The improvement is greatest at frequencies below the optimum transit-time frequency and it is possible to obtain power output comparable to and even slightly greater than

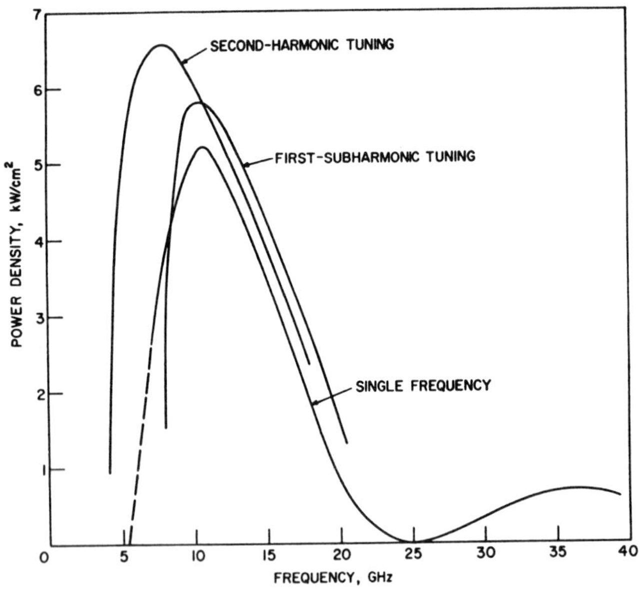

Fig. 7. Optimum power output for a Si Read diode under various operating conditions; $V_{RF} = 20 \sin \omega_1 t + 20 \sin (\omega_2 t + \Psi)$, $J_{dc} = 500$ A/cm²; Schroeder and Haddad [13].

that attained at the optimum single-frequency point. The second-harmonic tuning curve extends below the avalanche frequency, indicating that it is possible to obtain power from the diode at frequencies below the single-frequency operating range. However, for frequencies above the optimum 11 GHz the power improvement is small and for frequencies from 12 to 20 GHz better performance is obtained using first-subharmonic tuning.

The first-subharmonic tuning curve shows that there is little improvement below 11 GHz, but that in the frequency range from 11 to 20 GHz some improvement can be obtained. In particular, at 20 GHz the improvement attainable is nearly 100 percent although the power is not so large as at the optimum single-frequency point. These results were also compared to more realistic diode structures and good agreement was obtained [19]. Also, these theoretical results were found to be in good agreement with experimental results reported by Swan [20].

Basic Principles of Operation and Characteristics of the TRAPATT Mode

In 1967 Prager *et al.* [21] observed experimentally high-efficiency operation of avalanche diodes under pulsed conditions. Due to the low frequency of operation (considerably less than the transit-time mode) and the 60-percent efficiency which was higher than IMPATT theory predicted, they referred to it as the "anomalous mode."

Using the experimental voltage waveforms of an oscillating Ge diode in a computer simulation, Scharfetter *et al.* [22] developed a physical explanation and detailed description of this mode of operation. Their simulation showed that when the large-signal limits of

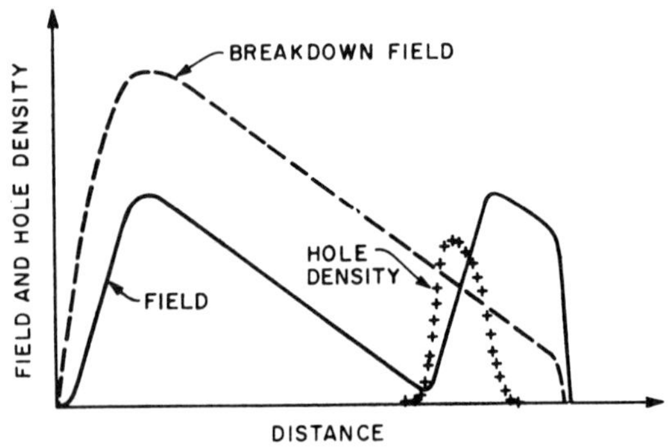

Fig. 8. Large-signal IMPATT field and carrier densities during the drift portion of the RF cycle, Evans [23].

IMPATT operation were exceeded, a low-frequency high-efficiency mode could exist.

In IMPATT operation, as the RF voltage increases, several effects limit the performance of this mode. First, the effect of a large space charge on the field in the drift region increases the field in front of the pulse high enough for impact ionization to occur. Also, for large space charge the field behind the pulse is depressed so that the drift velocity of some of the carriers will drop below saturation and will be trapped in this low-field region for a short time. This effect [23] is shown in Fig. 8. Both the generation and trapping of carriers in the drift region change the transit angle and degrade the performance of the diode. The space charge of the pulse will also depress the field in the avalanche region and decrease the phase delay produced by the avalanche process. This in effect also changes the transit angle and degrades the performance.

Scharfetter *et al.* [22] showed that when these limits were exceeded by having a large RF voltage at the IMPATT frequency and large current densities, the following events occur. A large overvoltage at the IMPATT frequency (approximately twice the breakdown voltage) generates a large space charge in the avalanche region which depresses the field almost to zero behind it and increases it in front of the carriers to a high enough level to cause impact ionization. In Fig. 9 the different waveforms are shown for an instant in time in the cycle. In Fig. 9(a), the hole and electron current densities are large enough to reduce the field almost to zero at the left edge of the diode and increase it on the right to cause avalanching. This avalanche region or zone traverses the diode leaving behind it an electron-hole plasma trapped in the low field. The carriers in the diode which are moving at low-field-dependent velocities are then removed by space-charge-limited flow. In Fig. 9(b) the terminal current and voltage are plotted. The large overvoltage due to the IMPATT oscillations is reduced to a small value as the avalanche zone traverses the diode and fills it with a plasma. The current which is almost all particle current during the low-field state is large, whereas as the plasma is removed the particle current is reduced and the terminal current becomes mostly displacement current. As is shown by the voltage

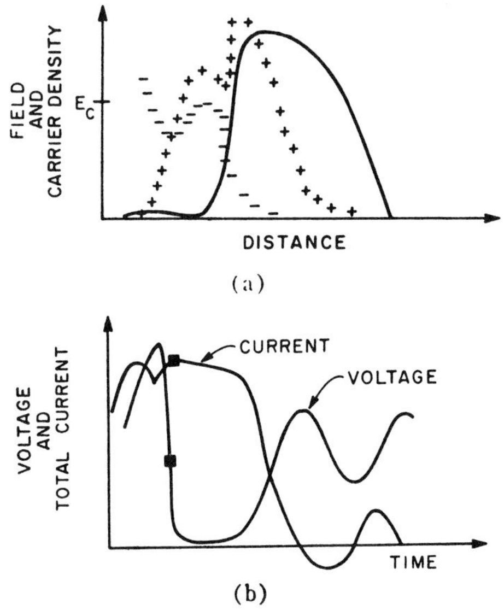

Fig. 9. TRAPATT waveforms. (a) Field and carrier density. (b) Terminal voltage and current, Evans [23].

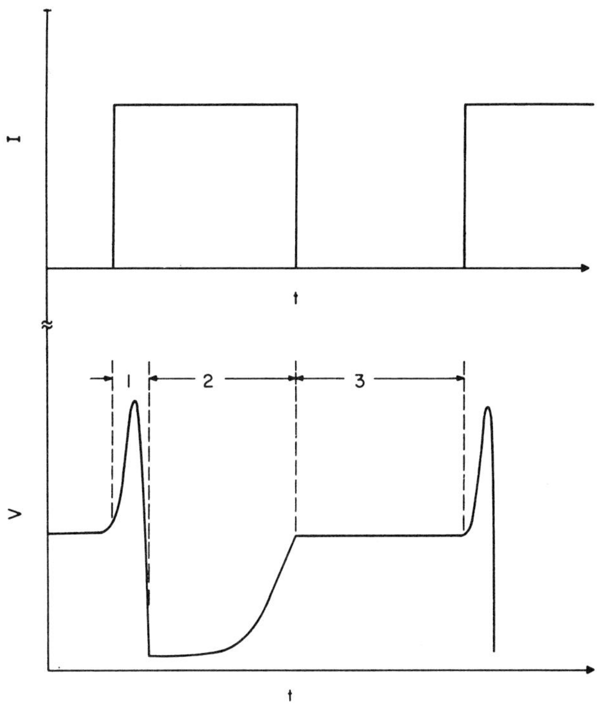

Fig. 10. Current and voltage waveforms for the TRAPATT mode.

and current waveforms, the current is large when the voltage is small and vice versa; thus, this current–voltage ($I-V$) relationship leads to a very efficient mode. For lower current densities it is possible for the avalanche zone not to travel through the entire length of the diode. This depends on the magnitude of the space-charge field and that due to the doping profile. Under these conditions the efficiency can be low (less than 1 percent) even though the diode is in the trapped-plasma avalanche triggered transit (TRAPATT) mode.

Because of the complexity of the complete equations [3] describing this mode it is not feasible to obtain design criteria for the diode or the circuit by a detailed solution of these equations without using a prohibitive amount of computer time. Clorfeine *et al.* [24] and DeLoach and Scharfetter [25] developed a simple and elegant model for this mode of operation which is useful in understanding the physics of this mode and also for obtaining approximate design criteria for optimizing its operation.

By assuming a certain terminal current waveform and neglecting the device-circuit interaction, an approximate design theory for TRAPATT diodes was developed. Justification of this simplified model was done by verification of the results with the more exact analysis [22].

In the simplified model of the TRAPATT mode the cycle is divided into three parts [25]. The first part is a charging period in which an avalanche shock front (ASF) [26] forms and propagates through the diode at a velocity which is greater than the saturated velocities of the carriers. Behind the ASF a high-density electron-hole plasma is trapped due to the low-field region which is created. The second part of the cycle is a charge removal period in which the plasma is removed by a recovery transient. Finally, the RF voltage at the IMPATT frequency builds up again to trigger the ASF. The current and voltage waveforms are shown in Fig. 10. The charging period consists of the large overvoltage and the immediate drop in voltage to almost zero after the ASF has crossed the diode. The recovery period consists of the portion where the voltage builds up from this low value to breakdown again and the final period consists of the high-voltage low-current state. Thus, as shown in Fig. 10, the diode switches from a low-voltage high-current state to a high-voltage low-current state.

The model assumes that the ionization rates for holes and electrons are equal and independent of the field for fields above E_a, and zero for fields below E_a. The velocities of the carriers are equal and saturated for fields above E_s and proportional to the field ($v = \mu E$) for fields below E_s.

Initially a constant current J_0 is applied to the diode. The current in the diode consists of a negligible reverse saturation current and a displacement current. Therefore,

$$J_0 = \epsilon \frac{\partial E(x, t)}{\partial t} \tag{20}$$

where ϵ is the dielectric constant (F/cm). As long as the field remains below the critical value E_a for generation of carriers, it will increase linearly with time throughout the diode as shown in Fig. 11. From (20) and Poisson's equation, the field for $E \leqq E_a$ is given by (an n^+-p-p^+ structure is assumed) [25]

$$E(x, t) = E_a - \frac{qN_a x}{\epsilon} + \frac{J_0 t}{\epsilon} \tag{21}$$

where

q electronic charge (C)
N_a acceptor doping density (cm^{-3}).

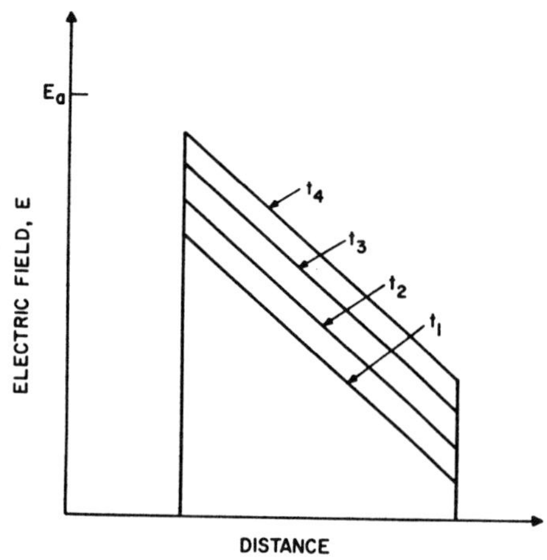

70

Fig. 11. Electric field versus distance at successive time intervals with constant current J_0, DeLoach and Scharfetter [25].

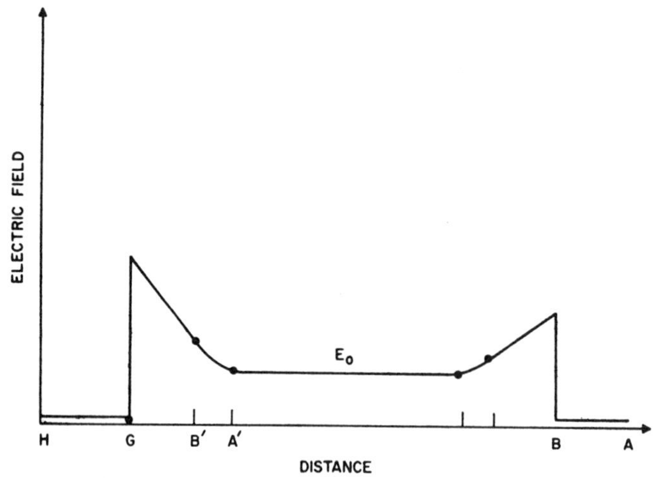

Fig. 13. Electric field versus distance at an instant in time during the recovery portion of the cycle, DeLoach and Scharfetter [25].

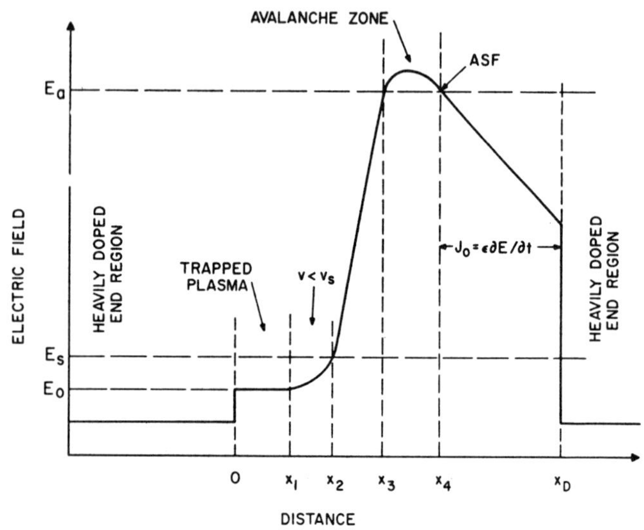

Fig. 12. The electric field versus distance of an avalanche zone in transit.

From this equation the time at which the field reaches E_a for each x is obtained. This yields the velocity v_z of the front of the avalanche region as it progresses through the diode. Thus,

$$v_z \triangleq \frac{dx}{dt} = \frac{J_0}{qN_a} \cdot \tag{22}$$

This velocity v_z is referred to as the avalanche zone velocity. To the right of the avalanche zone, the current is mainly a displacement current and is given by (20). In order for the assumption of (20) to be valid the zone velocity v_z must be larger than the carrier saturated velocity v_s. Therefore, from (22)

$$J_0 > qN_a v_s. \tag{23}$$

After the ASF has formed and progressed some distance, the field in the diode is as shown in Fig. 12. The field behind the ASF decreases below E_a due to the space charge of the carriers generated in the avalanche zone ($x_3 \leq x \leq x_4$).

Since the ASF moves through the diode at a velocity greater than the carriers, the carriers cannot "catch up" to the zone. Thus the field at any point in the diode where $E \geqq E_s$ depends only on the field and carriers to its right. If one changes to a moving reference plane located at x_4 and moving with the velocity of the ASF, the field in the region where $E \geqq E_s$ is time invariant. This assumes a constant reverse saturation current J_s and total current J_0. In this moving reference frame the electric field, hole current, and electron current can be obtained [25].

Behind the avalanche zone the space charge of the carriers decreases the field below the value necessary for saturated drift velocities. The field drops to a value such that the total current is carried by the electron-hole plasma which has formed. The ASF moves through the diode rapidly and leaves the diode filled with a plasma trapped in the low field. (The carriers are moving with a velocity v_p where $v_p = \mu E_0 \ll v_s \ll v_z$.) Thus the voltage across the diode is small but the conduction current is large.

The second portion of the TRAPATT mode is the transient recovery cycle. In order to gain an understanding of this portion a simplified model is used again. The depletion region of the diode is assumed to be filled with a space-charge neutral electron-hole plasma ($p_0 - n_0 - N_a = 0$) with p_0 and n_0 being the hole and electron densities which are constant in the plasma. The current during the recovery cycle is assumed to be constant. Thus, in the plasma the current is

$$J_R = q\mu(p_0 + n_0)E_0 \tag{24}$$

where $E_0 < E_s$. Since carriers are not being generated in the diode, the holes which are moving to the right are depleted from the left edge of the diode and the field increases due to the space charge of the electrons. Similar events occur at the right edge with electrons being depleted. Thus, the field at both edges will increase with time as shown in Fig. 13. The velocity at which these two depleted regions grow is the velocity of the electrons and holes in the plasma v_p. The current in the plasma region is all conduction current, whereas in the two end regions where the field is increasing it consists of both conduction and displacement current.

If the field in the diode exceeds E_a, more carriers will be generated and added to those already being swept out. The minimum time for the recovery will be when the field at the left edge of the diode reaches E_a at the same instant the last carriers are leaving the diode. This minimum recovery time can be divided into two parts: the time for the two depleted regions to grow at a velocity v_p until they meet in the middle, and the time for the remaining carriers to leave the diode at a velocity v_s. Therefore,

$$\tau = \frac{x_D}{2}\left(\frac{1}{v_p} + \frac{1}{v_s}\right). \qquad (25)$$

It can be shown [25] that the recovery current which produces this minimum time is

$$J_R = qv_s(2N_0 - N_a) \qquad (26)$$

where N_0 is the doping density necessary for the field to be equal to zero at x_D and to be equal to E_a at $x=0$. Thus, the recovery current depends only on the depletion-layer width, the field for breakdown, and the doping density. The limits on this current as given by (26) are

$$v_s E_c C_d \leqq I_R \leqq 2v_s E_c C_d \qquad (27)$$

where C_d is the depletion-layer capacitance (F).

After the plasma has been swept out of the diode, the diode will be in a high-voltage low-conduction current state.

The third portion of the TRAPATT cycle consists of the high-voltage low-current state in which the voltage at the IMPATT frequency builds up until the ASF is initiated again. The existence of the IMPATT frequencies to initiate this ASF is a sufficient condition but has not been shown to be a necessary condition.

This simplified model gives some of the physical understanding of TRAPATT operation but does not explain every phase of the mode such as how it is initiated, and the transition from the IMPATT to the TRAPATT mode.

From this simplified model an approximate design theory can be obtained which depends only on the doping density N_a and the width of the depleted region as related to $N_0(N_0=\epsilon E_a/qx_D)$. From computer simulations the current through the diode can be represented by a square wave. Thus if J_0 is related to the recovery current by

$$J_0 = MJ_R \qquad (28)$$

and if the time for the ASF to cross the diode is a fraction of the total TRAPATT period given by T/y, then a relation between the two variables N_a and N_0 can be derived. This relation is

$$N_a = \frac{2N_0^2}{\left(\dfrac{y\epsilon E_a}{Mqv_sT} + N_0\right)}. \qquad (29)$$

The recovery cycle can be taken as a fraction of the TRAPATT period and is given by

$$T \triangleq \frac{1}{f} = \frac{x_D}{2n}\left(\frac{1}{v_p} + \frac{1}{v_s}\right) \qquad (30)$$

where n is the fraction of the period for the recovery cycle. The final parameter is the reverse saturation current of the diode which can be given by 71

$$J_s \approx 1.6 \times 10^2 x_D \qquad (31)$$

for both Ge and Si [25].

Results of these design equations for both Si and Ge are shown in Figs. 14 and 15 as a function of frequency for the case where $M=1$, $y=10$, and $n=1/2$. The results show the effect of different reverse saturation currents and a 50-percent increase in the breakdown field.

The simplified model of the TRAPATT mode which has been presented is useful for the diode design but does not include the interaction with the circuit. A model of a TRAPATT oscillator which is useful for circuit design has been presented by Evans [23]. This model explains the device-circuit interaction and is used to predict the frequency, power output, efficiency, and RF impedance of an oscillator.

Evans presented a simple TRAPATT oscillator circuit in which a lumped capacitor is located next to the diode, a low-pass filter is placed at the end of the transmission line, and a slotted coaxial line tuner is located between the filter and the load. The capacitor is necessary to provide the large pulse of conduction current for the trapped plasma in the diode. This high-current state exists for a small fraction of the TRAPATT period and is provided by the discharging of the lumped capacitor and the transmission line. The length of transmission line between the diode and the low-pass filter determines the TRAPATT frequency. The slotted coaxial line tuner is used to match the diode to the load.

Simple waveforms for the voltage and current were assumed in which the voltage is equal to the breakdown voltage until the trapped plasma is initiated, at which time it is zero for a short time τ_p and then increases linearly to the breakdown value as the capacitor and transmission line are recharged during a time τ_R. The conduction current consists of a large pulse of current of very short duration provided by the lumped capacitor next to the diode and then a smaller but longer pulse of current due to the transmission line being discharged. During the recharging time and the time before the plasma is initiated the current is zero. The fundamental components of these waveforms are a function of the

72

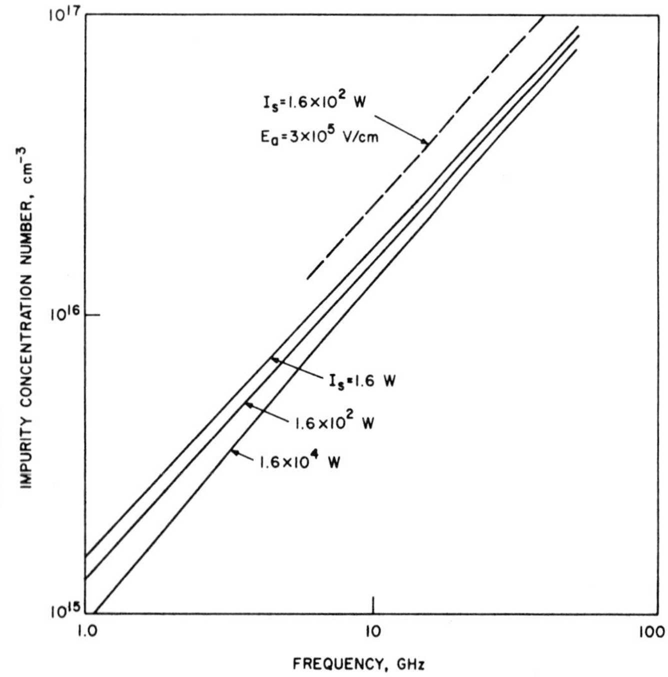

Fig. 14. N_0 as a function of frequency for three assumed values of I_s for $E_a=2\times10^5$ V/cm. The dashed curve shows the effect of increasing E_a to 3×10^5 V/cm for the median I_s, DeLoach and Scharfetter [25].

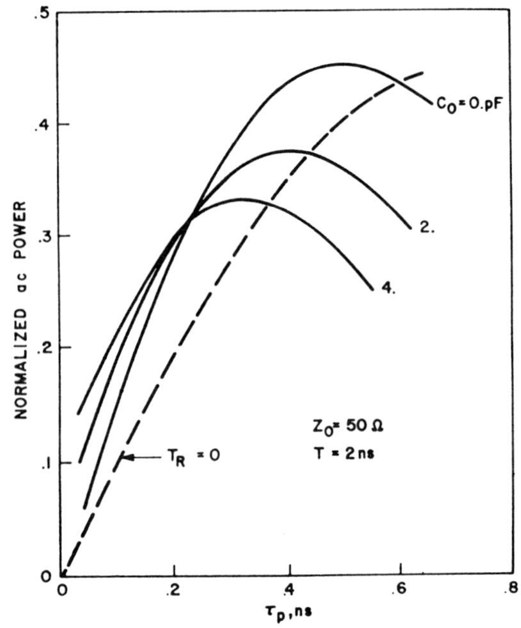

Fig. 16. RF output power versus trapped-plasma duration τ_p for different lumped capacitors C_0, Evans [23].

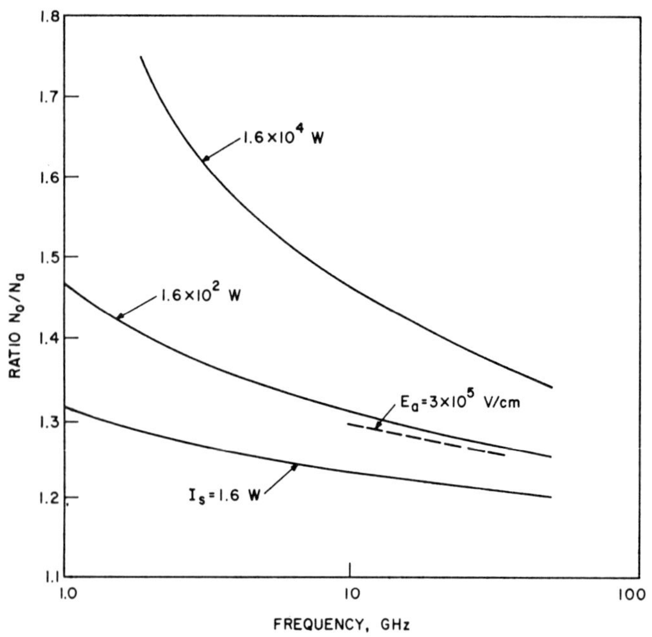

Fig. 15. The ratio of N_0 to N_a as a function of frequency for three assumed values of I_s for $E_a=2\times10^5$ V/cm. The dashed curve shows the effect of increasing E_a to 3×10^5 V/cm for the median I_s, DeLoach and Scharfetter [25].

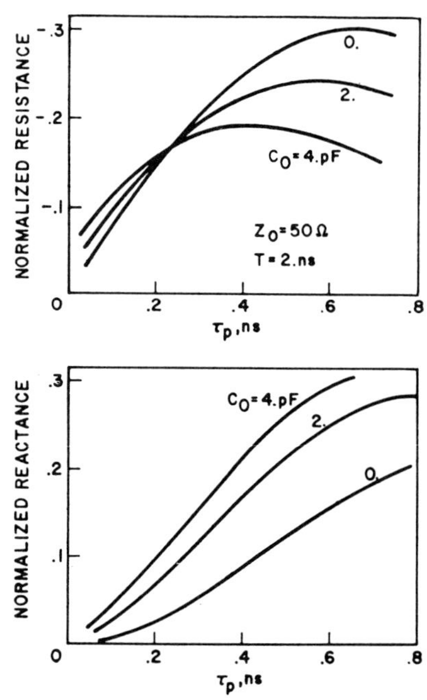

Fig. 17. Diode impedance versus trapped-plasma duration τ_p for different lumped capacitors C_0, Evans [23].

breakdown voltage V_B, the dc current I_{dc}, the characteristic impedance Z_0, the duration of the trapped plasma state τ_p, and the lumped capacitor next to the diode C_0. The output power and impedance can be calculated from these components and are plotted in Figs. 16 and 17 for a 500-MHz oscillator. Fig. 16 shows the normalized (normalized to $I_{dc}V_B$) output power as a function of the duration of the trapped plasma state for different values of C_0. The dashed curve is for the case where the voltage recovers immediately after the trapped plasma and there is no charging or recovery time. Note that for short high-current states (small τ_p) the recovery time improves the output power. The normalized impedance (normalized to V_B/I_{dc}) is plotted

TABLE I
IMPATT-Mode State of the Art

Frequency (GHz)	Power Output	Efficiency (percent)	Duty	Material	Number of Diodes in the Package	References
9.9	540 mW	15	CW	GaAs	1	de Nobel [29]
11.1	1.6 W	15	CW	GaAs	1	Kim and Armstrong [30]
13.3	4.7 W	8	CW	Si	2	Swan [31]
55	650 mW	10	CW	Si	1	Edwards *et al.* [32]
110	75 mW	3.2	CW	Si	1	Edwards *et al.* [32]
6	500 mW	8	CW	Ge	1	Dunn and Olsen [33]
8	30 W	4–5	pulsed	Si	1	Gilden and Moroney [34]
21	8.8 W	4–5	pulsed	Si	1	Gilden and Moroney [34]
341	1 mW	small	pulsed	Si	1	Bowman and Burrus [35]

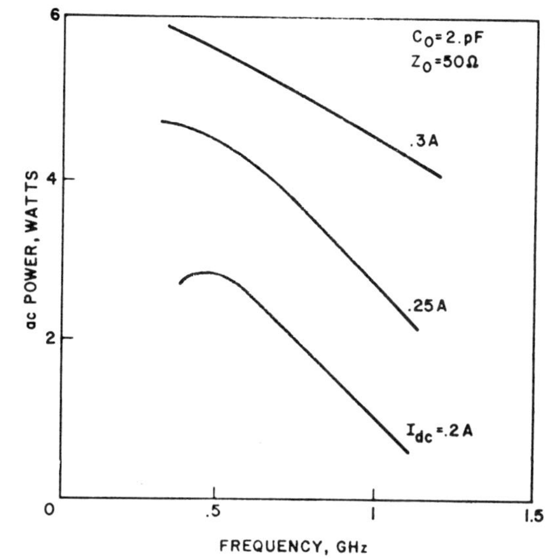

Fig. 18. RF output power versus frequency as a function of dc bias current, Evans [23].

in Fig. 17 as a function of the duration of the trapped plasma. Finally, the output power as calculated by this model is shown to decrease with frequency in Fig. 18.

These results were applied to a TRAPATT oscillator by Evans [23] and were found to be in excellent agreement with his experimental results.

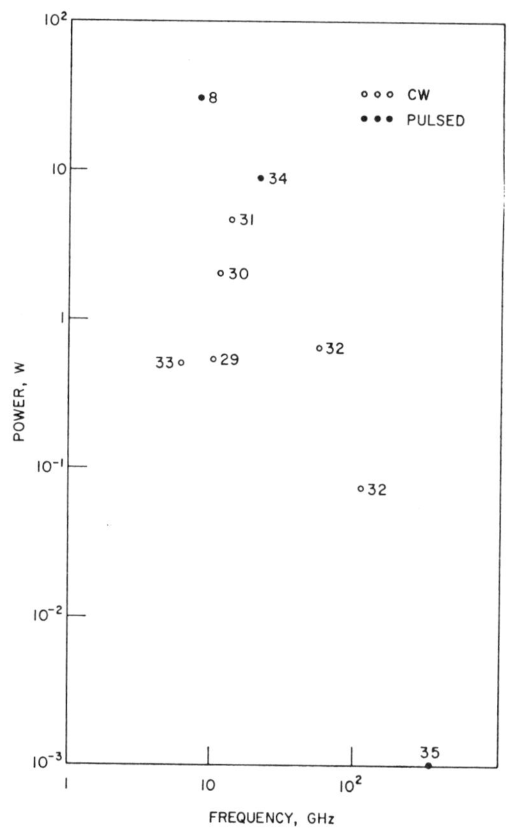

Fig. 19. State of the art in the IMPATT mode. (The numbers next to the diode points refer to the reference numbers.)

III. Experimental Results for Avalanche Diodes in the IMPATT Mode

The first observation of microwave oscillations from avalanche transit-time diodes was reported in 1965 by Johnston *et al.* [27] for a Si p-n junction. At approximately the same time Lee *et al.* [28] reported that they had realized a Read-type diode structure and had observed oscillations, albeit at a much lower frequency. Since these first reports there has been a great deal of effort devoted by many researchers to achieving high output power and efficiency over a wide frequency range. Performance improvements have been achieved very rapidly as better semiconductor wafers became available and as the technology for making and mounting the diodes was improved.

IMPATT oscillators have been fabricated from Ge, Si, and GaAs, although recently most of the research activity has been devoted to the latter two materials. IMPATT oscillators have been operated at frequencies ranging from 1 to 340 GHz. Some results which are representative of the present state of the art are given in Fig. 19 and Table I. It should be pointed out that some of the results of Table I are laboratory results obtained from exceptionally good diodes and operated with elaborate heat-sink arrangements. However, it is indicative of the rapid evolution of the art that there are now commercially available IMPATT-mode oscillators which deliver ≈1 watt at frequencies from 2 to 15 GHz and 100 mW from 15 to 18 GHz with conversion efficiencies in the range from 3 to 8 percent. Further, it appears that millimeter-wave IMPATT oscillators will soon be commercially available.

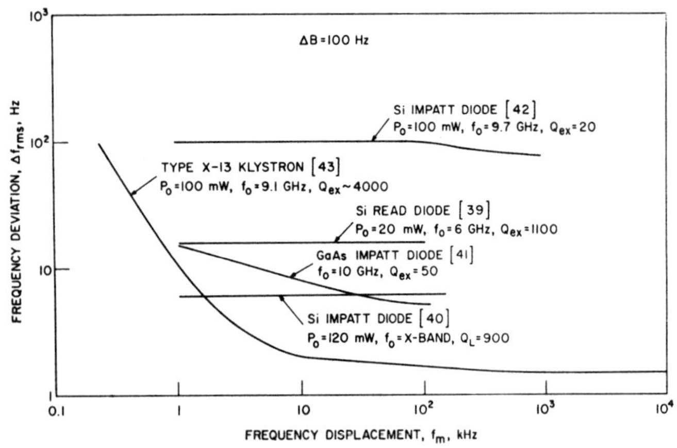

Fig. 20. Comparison of FM noise data for IMPATT-mode oscillators. (The numbers in brackets correspond to the reference numbers.)

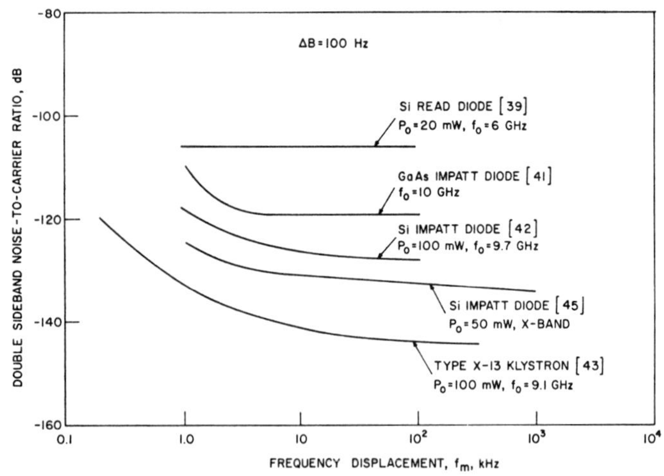

Fig. 21. Comparison of AM noise data for IMPATT-mode oscillators. (The numbers in brackets correspond to the reference numbers.)

Output power and dc to RF conversion efficiency are not the only criteria for microwave oscillators. Noise performance of oscillators is also an important parameter. The output of an ideal oscillator is a sinusoidal signal whose amplitude and frequency do not change with time. All practical oscillators, however, exhibit some random fluctuation of both amplitude and frequency (or phase). Thus it is convenient and useful to describe oscillator noise in terms of amplitude modulation (AM) and frequency modulation (FM) noise. Noise performance data are usually given as the amount of AM or FM noise power which is contained in a specified bandwidth ΔB displaced by a frequency f_m from the oscillator frequency f_0.

Some representative data for FM noise performance of IMPATT oscillators are given in Fig. 20 with that of an X-13 klystron for comparison. All data have been reduced to an equivalent 100-Hz bandwidth. The FM noise has been expressed as an rms frequency deviation, Δf_{rms}, in accordance with conventional practice. The rms frequency deviation is related to the FM noise power for small modulation index by the equation [36]

$$\Delta f_{rms} = f_m \sqrt{\frac{P_{SB}}{P_0}} \tag{32}$$

where

P_{SB} double sideband FM power
P_0 the power of the oscillator.

It should be noted in interpreting Fig. 20 that the loaded Q factor of the oscillator affects noise performance. The following relation for the rms frequency deviation of a negative-conductance oscillator may be deduced from Edson [37]

$$\Delta f_{rms} = \frac{f_0}{Q_L} \sqrt{\frac{kT(\Delta B)}{P}} \tag{33}$$

where

Q_L 2π times the stored energy/total energy dissipated in one cycle
k Boltzmann constant
T effective noise temperature of the diode
P power generated by the diode.

The inverse dependence of rms frequency deviation upon Q has been verified experimentally for IMPATT oscillators by Harth and Ulrich [38]. Bearing in mind the dependence of Δf_{rms} on Q, Fig. 20 shows that GaAs appears to be better in FM noise performance than Si diodes [41]. Ge diodes seem to be better in this respect than Si also [44], although the data are not shown in the figure since detailed results are not available.

Some representative AM noise data for IMPATT oscillators are given in Fig. 21. The reported data have been reduced to double sideband noise-to-carrier ratios in a 100-Hz bandwidth. The effect of Q on AM noise is usually found to be much less important than for FM, especially for frequencies close to the carriers. It has recently been reported [46] that for GaAs diodes operated at high current densities (attained to date only under pulse bias) especially low AM noise-to-carrier ratios are obtained, i.e., more than -145 dB at 1-MHz displacement.

There are circuit techniques for further reducing IMPATT oscillator noise which have been experimentally demonstrated. FM noise may be reduced by the use of a stalo cavity or by injection phase locking to a source which has low FM noise [39], [40], [42], [47], [48]. It has been demonstrated that AM noise performance can be improved by optimizing the bias network impedance [45].

Oscillator Tunability

The IMPATT diode has a broad-band negative conductance and thus has the potential of being tunable over a wide frequency range. Continuous tuning over

TABLE II
TRAPATT-Mode State of the Art

Frequency (GHz)	Power Output	Efficiency (percent)	Duty	Material	Number of Diodes in the Package	References
1.1	1.2 kW	25.6	pulsed	Si	5	Liu and Risko [60]
3	20 W	30	pulsed	Si	1	Grace and Gibbons [61]
6.5	3 W	17	pulsed	Si	1	Ying *et al.* [62]
2.6	≈7 W	40	pulsed	Ge	1	Johnston *et al.* [63]
8.6	10 W	25	pulsed	Si	1	Gibbons and Grace [64]
16.5	8 W	6.9	pulsed	Si	1	Ta [65]
0.45	5.3 W	43	CW	Ge	1	Iglesias and Evans [66]
2.8	2.4 W	20	CW	Si	1	Evans and Johnston [67]

more than an octave range has been achieved with a mechanically tuned circuit [49]. An electronic tuning range of 3 GHz for an X-band oscillator has been reported [50], using YIG as the variable circuit element. Varactor-diode tuning has also been demonstrated [16], [51]. The electronic tuning capability of IMPATT oscillators has been exploited for a millimeter-wave FM system; either bias current or varactor tuning was employed [52], [53].

IMPATT Amplifiers

IMPATT diodes can also be employed in microwave amplifiers. IMPATT amplifiers have been realized in Si, Ge, and GaAs diodes using a circulator-coupled reflection-type amplifier circuit. The saturated power output and dc to RF conversion efficiency at high drive levels are comparable to that obtained from the same diode operated as an oscillator. The noise figures currently reported for IMPATT amplifiers are in the range from 20 to 40 dB. While this type of noise performance may prohibit their use as low-noise preamplifiers, the potential for moderate-noise high-gain, high-power output amplifiers seems very good.

Using Si X-band IMPATTs, Scherer [54] has reported achieving a 13-dB gain over a 1-GHz bandwidth from a single diode. Using three Si diodes in cascade he has achieved a 36-dB gain with a 200-mW power out at 10 GHz. The overall noise figure was measured to be 31 dB. A noise figure of 30 dB has also been reported [45] for Ge X-band amplifiers under high-gain conditions; this is some 6 to 10 dB lower than that usually observed in Si diodes of similar structure.

A GaAs X-band amplifier has been reported by Baranowski *et al.* [41] which gives small-signal gains of 10 to 30 dB with noise figures as low as 20 dB. Kuno *et al.* [55] have observed a noise figure of 17 dB at a 20-dB gain in an X-band GaAs diode. Thus, at present, GaAs diodes used as amplifiers or oscillators provide better noise performance than Si diodes.

It has been demonstrated by Midford and Bowers [56] that an IMPATT diode can be fabricated for use as a traveling-wave amplifier, in contrast to the more conventional lumped negative-conductance reflection-amplifier approach. Several other schemes exist where

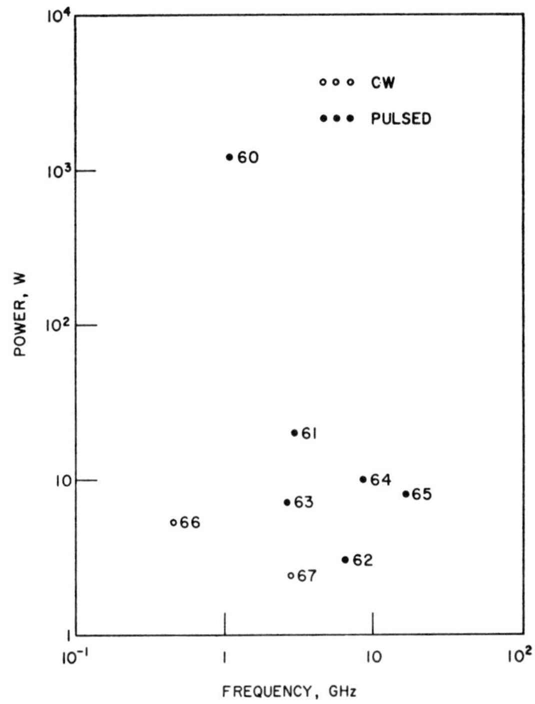

Fig. 22. State of the art in the TRAPATT mode. (The numbers next to the diode points refer to the reference numbers.)

discrete diodes may be employed for traveling-wave amplifier applications. These include coupled cavities [57], a loaded microstrip line on a ferrite substrate [58], and a loaded slow-wave structure or filter network [59].

IV. Experimental Results for Avalanche Diodes in the TRAPATT Mode

Prager *et al.* [21] first observed in 1967 a high-efficiency subtransit-time frequency mode of oscillation in Si avalanche diodes. This TRAPATT mode has subsequently received much attention because of the high dc to RF conversion efficiencies obtainable. Because of the high threshold current densities required to initiate the mode, it has proved more difficult to obtain continuous oscillation in this mode than in the IMPATT mode. A summary of some of the best results reported for TRAPATT-mode oscillators to date is given in Table II and Fig. 22. Although the high efficiency occurs for subtransit-time frequencies, it has been possible to obtain TRAPATT operation at frequencies up to K-band when millimeter-wave transit-time diodes are used. It is significant that no high-efficiency TRAPATT-mode oscillation has yet been reported for GaAs, which has proven to be a very suitable material for the IMPATT mode. It seems likely that the negative differential mobility of electrons in GaAs at low fields is responsible for this difficulty [68].

The circuits which have been found to work well for TRAPATT oscillators are somewhat more complicated than those for IMPATT operation. This is because in all TRAPATT operation which has been reported to date, harmonically related transit-time signals have been
76 present also. This requirement arises from the need to repetitively trigger the avalanche shock front. Consequently, a TRAPATT circuit must have the proper performance at several harmonically related frequencies. The circuit consists, essentially, of a low-pass filter which loads the diode very lightly at the transit-time frequency thus permitting high voltage excursions at the diode, followed by an element for tuning the subharmonic frequency output power [23], [69]. Evans [23] has shown that over a wide frequency range the subharmonic frequency is approximately inversely related to the physical distance between the diode and the high-frequency short plane. Levine and Liu [69] have also observed this relationship and have utilized it to achieve continuous frequency tuning from 1 to 1.5 GHz with variations in power output of only 1 dB. This result indicates the feasibility of electronic tuning of the TRAPATT oscillator.

Hoefflinger *et al.* [70] reported amplification in the TRAPATT mode. They achieved this by tuning the amplifier circuit to establish a transit-time frequency oscillation and suppress all subharmonics; they then observed stable reflection gain at several of the subharmonics. At 1.3 GHz, a high-power gain of 12 dB resulting in a dc to RF conversion efficiency of 25 percent was observed [71].

Prager *et al.* [72] were the first to observe TRAPATT amplification in Si diodes under pulse bias conditions. Gibbons and Grace [64] have recently reported TRAPATT amplification in Si diodes at X-band frequencies. They observed a saturated power output of 19 watts with a 7-dB gain at 29-percent efficiency. The instantaneous bandwidths reported for a 10-dB gain have been 5 percent or less.

Continuous wave (CW) TRAPATT amplification has been obtained in Ge diodes by Evans [73] with approximately a 1-watt output at 500 MHz and 25-percent efficiency.

Detailed noise performance data for TRAPATT mode avalanche diodes have not yet been reported either for oscillators or amplifiers. However, preliminary results indicate that the TRAPATT amplifier is quite noisy [73].

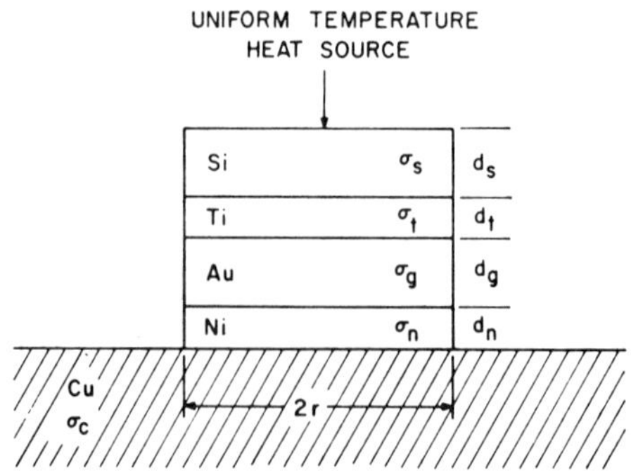

Fig. 23. Simplified diode and heat-sink structure used to determine total heat-flow resistance, Swan *et al.* [74].

V. Technology for Improving Power Output and Efficiency of Avalanche Diodes

One of the major technological problems confronting both IMPATT- and TRAPATT-mode diode fabrication is effective heat removal. In many cases it is heating which limits the power output attainable from CW IMPATT diodes. It is also heating which makes it difficult to obtain TRAPATT mode under CW conditions because of the high current densities required.

In avalanche diodes the dc power which is not converted to RF, and consequently represents heat which must be removed, is generated in the vicinity of the junction. It is therefore advantageous to design the diode with the junction near the semiconductor surface and to mount the diode with the semiconductor surface in contact with a good heat path to the sink. Most avalanche diodes made today are mounted using the so-called inverted chip technology.

A simple analytical model for the heat-flow resistance is illustrated in Fig. 23 [74]. Assuming that the junction acts as a uniform plane heat source and that the copper stud is essentially an infinite half-space heat sink, the total heat-flow resistance (°C/W) is given by

$$\begin{aligned} R_T &= R_s + R_t + R_g + R_n + R_c \\ &= \frac{1}{A}\left(\frac{d_s}{\sigma_s} + \frac{d_t}{\sigma_t} + \frac{d_g}{\sigma_g} + \frac{d_n}{\sigma_n}\right) + \frac{1}{4r\sigma_c} \end{aligned}$$

where the last term represents the spreading resistance in the copper heat sink. For junctions close to the semiconductor surface and typical metallization parameters, R_T is dominated by the spreading resistance. There are two approaches to reducing the spreading resistance. One is to use a heat sink with a thermal conductivity higher than copper, such as type IIa diamond [75]. The other approach is to realize a diode of a given total area with a composite structure consisting of several smaller area diodes connected in parallel [16], [74]. In the latter approach, if the small diodes are spaced electrically close together in a single package they may be regarded as the equivalent of one diode from a circuit point of view. Alternately, one may design the geom-

77

etry of the diode cross section to reduce spreading resistance [76].

Another fabrication-controlled parameter which seriously affects avalanche-diode performance is parasitic series resistance. Series resistance can be introduced by poor metal contacts, by losses in the highly doped end regions of the semiconductor, and by losses in the undepleted epitaxial region. Several experimenters have observed this deleterious effect of the unswept epitaxial layer on device power and efficiency [77], [78]. Proper design can result in diodes which are punched through (or nearly so) at breakdown and thus can eliminate this problem. It has recently been pointed out [79] that at high frequencies skin effect may introduce serious losses even in the highly doped substrate material. This problem can be reduced by thinning the substrate region and/or by using small diameter diodes.

For certain practical applications where high microwave power is desired it is of interest to combine many, perhaps hundreds, of diodes together. This is true notwithstanding the fact that ultimate limits have not yet been established for the power output from a single diode (by single diode, we here include the possibility of many chips in a single package). It was demonstrated early in avalanche-diode device development that several diodes could be combined using waveguide tees or hybrids to yield oscillator power approximately equal to the sum of the individual diode capabilities [80], [81]. Subsequently, various researchers have reported achieving power summation for oscillators and amplifiers utilizing series connection [82] and parallel connection [83] as well as periodically loaded transmission lines [84], [85].

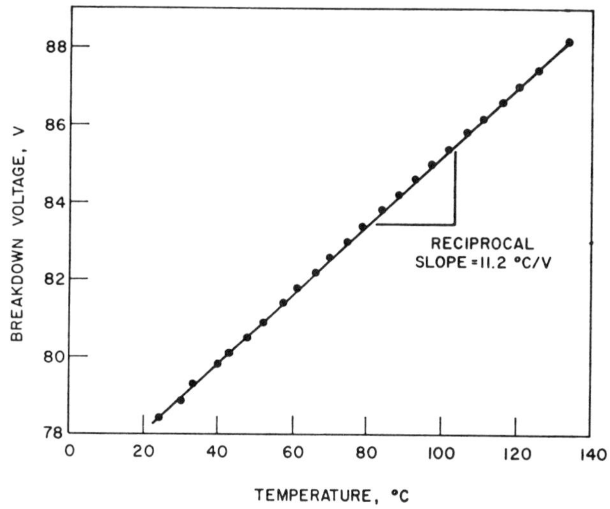

Fig. 24. Temperature dependence of breakdown voltage for a typical Si diode (similar to diode 46, see Fig. 25).

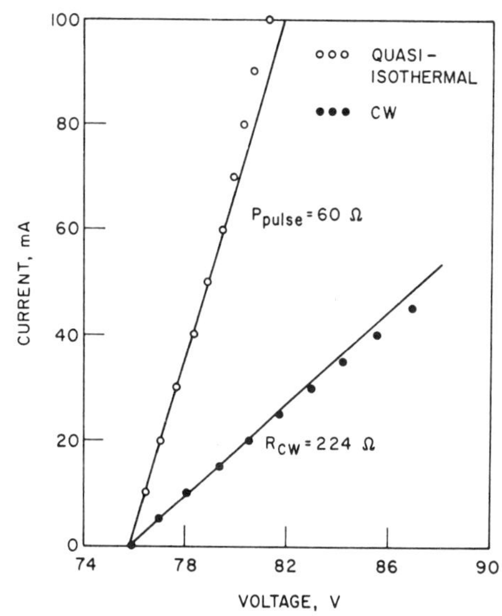

Fig. 25. A comparison of the quasi-isothermal and CW I–V characteristics for a typical Si diode (46).

VI. Experimental Evaluation of Avalanche-Diode Performance

Some experimental results obtained on avalanche-diode oscillators in this laboratory are given here. The diodes used are not one-of-a-kind state-of-the-art devices but are representative of diodes which are readily available. Further, the power output listed for a specific device is not the maximum that could be obtained since the diodes were operated at input power levels well below the burnout point. The intention here is to illustrate what can be readily achieved with a simple circuit and a reasonably good device.

Preliminary Diode Evaluation

1) *Heat-Flow Resistance*

It is desirable to know the heat-flow resistance of an avalanche diode in order to estimate the maximum safe CW input power. The temperature dependence of breakdown voltage may be used to measure heat-flow resistance; the technique to be described is based on a procedure proposed by Haitz *et al.* [86]. First, it is necessary to know the temperature coefficient of breakdown voltage for the diodes at hand; this may be estimated from a knowledge of the diode material and doping profile or it may be measured directly. The result of a direct oven-controlled measurement for an abrupt-junction non-punch-through Si diode is shown in Fig. 24. The I–V characteristic may then be measured in reverse breakdown with CW bias and compared to the characteristic obtained under quasi-isothermal conditions. The quasi-isothermal I–V characteristic is obtained by biasing the diode with short pulses using a low duty cycle. It is important to prevent the diode from oscillating during these measurements. A comparison of the quasi-isothermal and CW characteristics for a typical Si diode is shown in Fig. 25. The electrical resistance R_{pulse} ob-

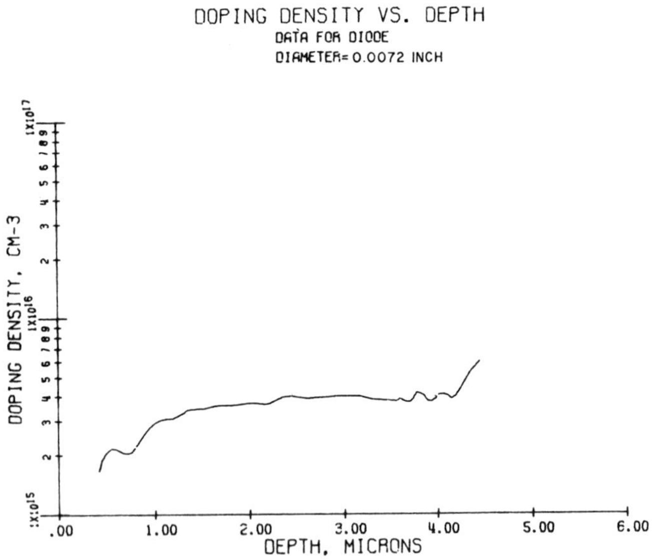

Fig. 26. Doping profile for a Ge diode (X-75). (The depth is measured from the plane of the junction into the relatively lightly doped epitaxial layer.)

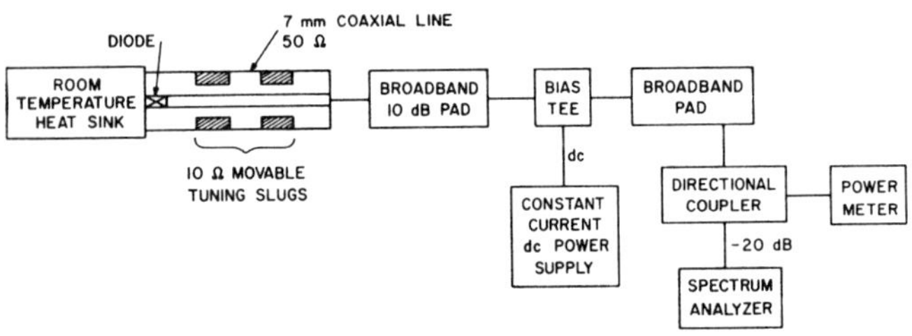

Fig. 27. Block diagram of the IMPATT-diode oscillator circuit.

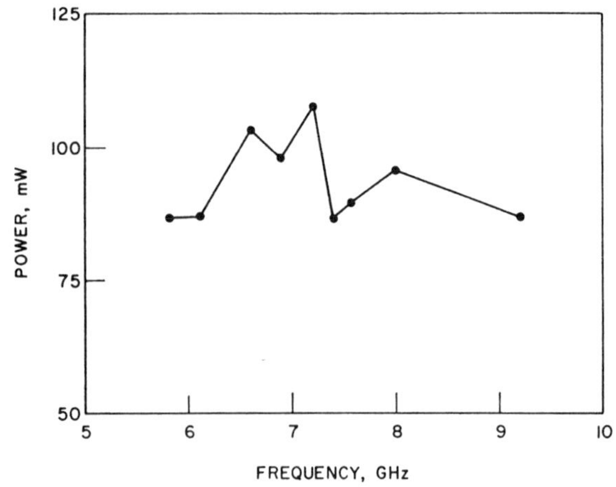

Fig. 28. Power-frequency characteristic for a Ge diode (X-75) operated at constant current. (Operating conditions: CW bias, $V_{dc}=54$ volts, $J_{dc}=380$ A/cm².)

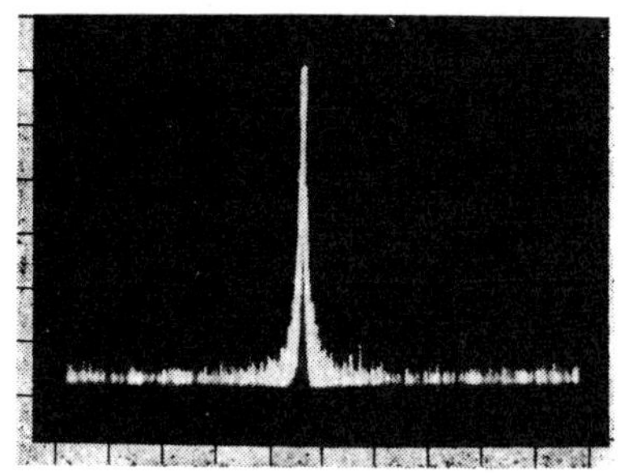

Fig. 29. Output spectrum of a Ge diode (X-75) when delivering 110 mW at 8.02 GHz. (Operating conditions: CW bias, $V_{dc}=54$ volts, $J_{dc}=380$ A/cm². Spectrum analyzer settings: vertical 10 dB/div, horizontal 1 MHz/div, IF bandwidth 10 kHz.)

tained under pulse bias is due primarily to space-charge resistance.

Attributing the difference between the characteristics solely to an increase in the junction temperature in the CW steady state, one finds an expression for heat-flow resistance:

$$r \cong \frac{(R_{CW} - R_{pulse})k}{V_B} \tag{34}$$

where

- R_{CW} electrical resistance under CW bias (Ω)
- R_{pulse} electrical resistance under pulse bias (Ω)
- k temperature coefficient of breakdown voltage (°C/V)
- V_B breakdown voltage at room temperature (V).

For the diode considered in Fig. 25 $r \cong 24.2$°C/W.

2) *Capacitance–Voltage Measurement*

Another measurement which is useful is the junction capacitance as a function of reverse voltage. The results are used to determine the doping profile of the diode in the standard way [87]. A computer program has been developed which uses the capacitance–voltage ($C-V$) measurement as input and produces a plot of the doping profile; a typical result is shown in Fig. 26 for a Ge diode (X-75). To a first approximation, the diode is a one-sided abrupt junction with a doping concentration of 4×10^{15} cm^{-3} in the epitaxial layer. The data clearly show that near the junction the doping concentration drops off, due presumably to a somewhat gradual diffusion in the junction-formation process, and that the depletion region barely extends to the higher doped substrate region at breakdown. Knowledge of the doping profile permits one to estimate the optimum frequency range for transit-time operation of the diode. Also, it has been found that diodes which display the TRAPATT mode at high efficiency are usually strongly punched through at breakdown; this can be determined from a $C-V$ measurement. Capacitance data are useful also to infer junction cross-section area from the corresponding data for a similarly fabricated diode of known area.

3) *IMPATT-Mode Oscillation*

The Ge and Si diodes described were operated as IMPATT-mode oscillators in a coaxial circuit, similar to that of Iglesias [88], which is shown in Fig. 27. The circuit is tuned with two identical low-impedance mov-

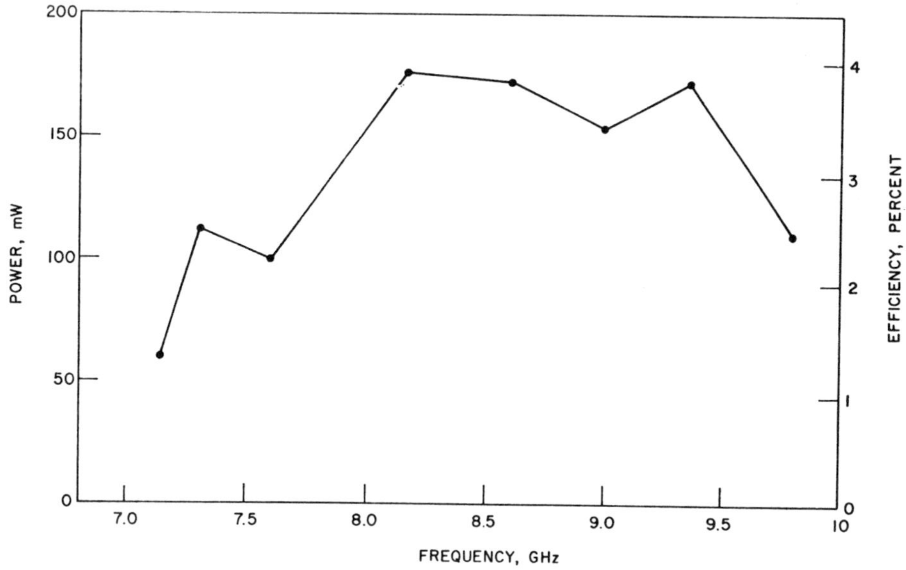

Fig. 30. Power-frequency characteristic for a Si diode (46) operated at constant current. (Operating conditions: CW bias, $V_{dc}=89$ volts, $J_{dc}=500$ A/cm^2.)

able slugs. The main advantage of this circuit is its flexibility. Various size slugs can be interchanged easily and it is possible to tune over a wide frequency range, including several frequencies simultaneously. On the other hand the circuit does have some limitations: it is difficult to tune different frequencies independently and the Q is relatively low.

For the Ge diode X-75 operated at 100 mA (380 A/cm^2) CW, the power output is shown versus frequency in Fig. 28 for single-frequency IMPATT operation. The tuning slugs used were one-quarter wavelength long at 12 GHz. A power output of 100 mW$\pm$0.6 dB was obtained from 5.8 to 9.2 GHz by varying the position of the slugs. This result shows the broad-band nature of the negative conductance. The efficiency was approximately 2 percent. Fig. 29 is a photograph of the output spectrum of the diode when oscillating at 8.02 GHz with a power of 110 mW; the log vertical scale and 1-MHz-per-division dispersion show qualitatively that the output is clean (the flat portion of the baseline corresponds to receiver noise).

Representative data for a Si diode (number 46) taken in the same circuit and using slugs one-quarter wavelength long at 9 GHz are shown in Fig. 30. Here the current was kept constant at 50 mA (500 A/cm^2) and the operating frequency tuned over a range of 3 GHz with power output exceeding 100 mW over the range. Fig. 31 shows the variation of power and efficiency with bias current (the circuit was tuned for best power at each current level). The efficiency is approximately 5 percent and begins to saturate at the higher current levels, although the power is still increasing.

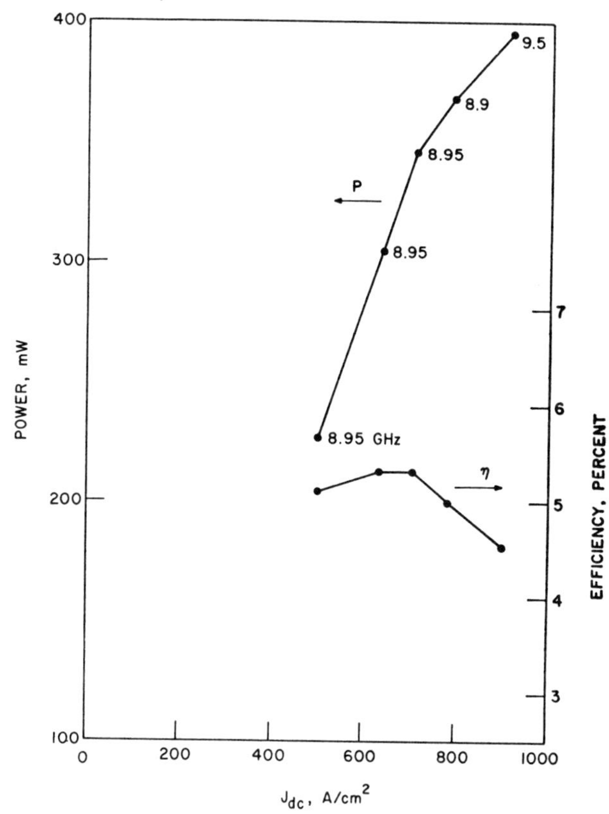

Fig. 31. Power and efficiency as a function of bias current for a Si diode (46).

4) *Multifrequency and TRAPATT Oscillations*

The circuit used to obtain CW oscillation in the TRAPATT mode is similar to that of Fig. 27. The coaxial line in which the diode is mounted is approximately 60 cm long, and five tuning slugs are typically used. The object in tuning is to place one slug close to the diode to tune the IMPATT frequency and use the remaining four slugs to form a low-pass filter located at approximately $\lambda/2$ of the TRAPATT frequency from the diode [66].

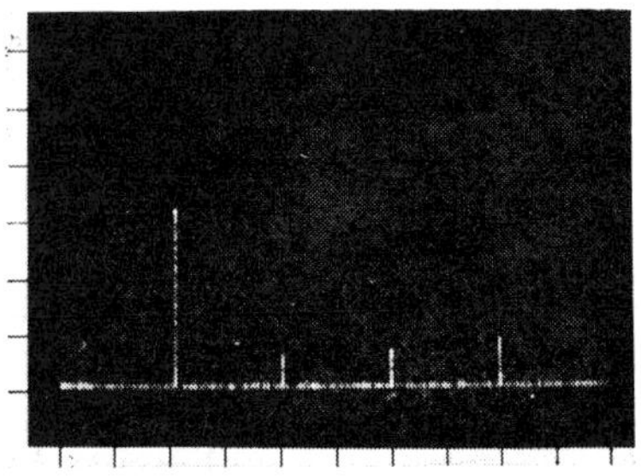

Fig. 32. Representative spectrum analyzer display for the output of a Ge diode (384-12) oscillating in the TRAPATT mode. The frequency of the left-most blip is 402.3 MHz; the three smaller blips are the second, third and fourth harmonics. (Spectrum analyzer settings: vertical 10 dB/div, horizontal 200 MHz/div.)

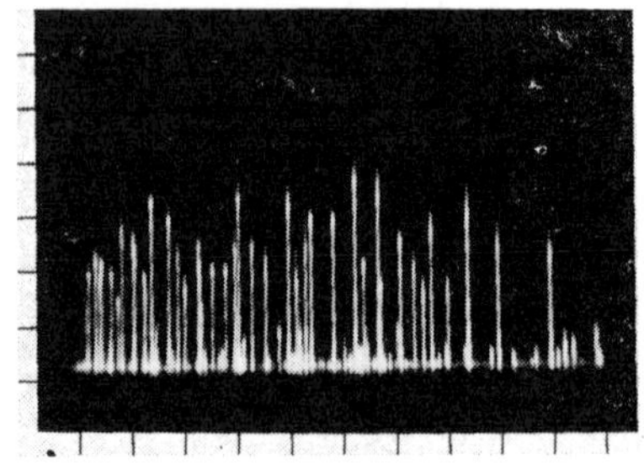

Fig. 33. Spectrum analyzer display for a Si diode oscillating in a multifrequency mode. The signals displayed are the X-band portion of the total spectrum. (Spectrum analyzer settings: vertical set to logarithmic response, horizontal uncalibrated.)

TABLE III

Ge CW TRAPATT Diode 384-12

Fundamental Frequency (MHz)	P_{fund} (watts)	J_{dc} (A/cm²)	V_{dc} (volts)	η (percent)
340	0.440	250	35.0	16.8
340	0.840	330	34.5	24.4
340	1.10	410	35.5	24.8
400	1.05	410	32.9	25.6
440	1.21	410	33.3	29.0
670	0.200	410	34.0	4.7

Several Ge diodes have been operated in the TRAPATT mode in this circuit. Ge diode X-75, described previously, did display the TRAPATT mode but the maximum efficiency observed was 8 percent. Several other Ge diodes which were strongly punched through at breakdown exhibited the TRAPATT mode at much higher efficiencies.

When the TRAPATT mode was obtained, the spectrum analyzer display was similar to that shown in Fig. 32. That is, in addition to the fundamental frequency (402.3 MHz) the second, third, and fourth harmonics were *always* present. The relative magnitude of the harmonics could be made comparable to that of the fundamental by altering the characteristics of the low-pass filter formed by the tuning slugs, but they were always present. In addition, probing the circuit between the diode and the low-pass filter revealed the presence of transit-time range signals which were always harmonically related to the fundamental, for instance the 25th, 26th, and 27th harmonics.

Table III shows some experimental results which were typical of the strongly punched-through Ge diodes in the TRAPATT mode. The breakdown voltage of the diode at 5 mA was 40.9 volts; in the TRAPATT mode the dc voltage decreased 15 to 25 percent. It was noted that, for diodes which displayed the TRAPATT mode, the exact position of the tuning slugs was *not* critical to obtain the TRAPATT mode, although obtaining optimum efficiency required a moderate amount of tuning.

5) *Multifrequency Operation*

Modes of multifrequency operation other than TRAPATT exist which are frequently observed. One such mode involves the presence of harmonic or subharmonic signals in the IMPATT frequency range. Such harmonics are almost unavoidably present in the large-signal operation of an avalanche diode in the IMPATT mode. Depending on the circuit impedance level at the harmonic frequency, the effect of the presence of harmonics on the fundamental frequency performance may be negligible, helpful, or deleterious as just discussed. The effect of harmonic tuning in the IMPATT mode was first observed experimentally by Swan [20] and has subsequently been studied by others [12], [13], [17], [18], [52].

Another kind of multifrequency operation results from the broad-band negative conductance of an IMPATT diode. It is possible, depending on the circuit, to load the diode lightly at several different frequencies within the IMPATT range. When this occurs with small RF amplitudes, two or more IMPATT oscillations may exist essentially independent of each other. For larger RF amplitudes, the nonlinearities of the avalanche diode resulting from the nature of the ionization process and from modulation of the depletion-layer capacitance (for nonpunch-through diode structures) cause frequency mixing to occur. The mixing properties open up a wide range of interesting applications. Avalanche diodes have been operated as self-oscillating frequency converters [10], [89], [90] and parametric amplifiers [91].

An example of the sort of multifrequency oscillations attainable is shown in Fig. 33 of the X-band portion of the output spectrum of a nonpunch-through Si diode (similar to diode 46). The circuit used was the TRAPATT circuit described previously. Output power was measured at 255 MHz corresponding to an efficiency of about two percent. Signals were also present at the second, third, and fourth harmonics and also the ninth, tenth, and eleventh harmonics of 255 MHz. The X-band transit-time signals shown in the photograph were detected with a probe close to the diode followed by a 6.5 GHz high-pass filter. The photograph is somewhat deceptive since each signal frequency is represented several times due to mixing with different harmonics of the analyzer's local oscillator. Nevertheless, the X-band frequencies were carefully measured and found to consist of signals at 11.086 GHz$\pm n \times$255 MHz, where n varies from zero to six. It was concluded that the mode is a complicated mixing phenomena, rather than TRAPATT, since the transit-time signals are not harmonically related to the ultra-high frequency (UHF) signal.

VII. Conclusions

The avalanche diode represents an important and significant addition to the family of microwave devices. It is certain that many applications will be found for this device in the future. The basic principles of operation of this device have been presented, and the present state of development has been discussed. A great deal of work in this area is presently being carried out at various laboratories, and it is certain that the results reported here will be improved further by the time this article appears.

Acknowledgment

The authors wish to thank the authors of the various articles whose work is quoted here, the many individuals at various companies who supplied information and results on their work as well as contributing some diodes, and M. S. Gupta, for his contributions to this article.

References

[1] T. Misawa, "Negative resistance in p-n junctions under avalanche breakdown conditions, pts. I and II," *IEEE Trans. Electron Devices*, vol. ED-13, pp. 137–151, Jaunary 1966.
[2] B. Hoefflinger, "High-frequency oscillations of p^{++}-n^{+}-n-n^{++} avalanche diodes below the transit-time cutoff," *IEEE Trans. Electron Devices*, vol. ED-13, pp. 151–158, January 1966.
[3] D. L. Scharfetter and H. K. Gummel, "Large-signal analysis of a silicon Read diode oscillator," *IEEE Trans. Electron Devices*, vol. ED-16, pp. 64–77, January 1969.
[4] W. T. Read, "A proposed high frequency negative resistance diode," *Bell Syst. Tech. J.*, vol. 37, pp. 401–446, March 1958.
[5] K. G. McKay, "Avalanche breakdown in silicon," *Phys. Rev.*, vol. 94, pp. 877–884, May 1954.
[6] M. Gilden and M. E. Hines, "Electronic tuning effects in the Read microwave avalanche diode," *IEEE Trans. Electron Devices*, vol. ED-13, pp. 169–175, January 1966.
[7] S. T. Fisher, "Small-signal impedance of avalanching junctions with unequal electron and hole ionization rates and drift velocities," *IEEE Trans. Electron Devices*, vol. ED-14, pp. 313–322, June 1967.
[8] H. K. Gummel and D. L. Scharfetter, "Avalanche region of IMPATT diodes," *Bell Syst. Tech. J.*, vol. 45, pp. 1797–1827, December 1966.
[9] W. J. Evans and G. I. Haddad, "A large-signal analysis of IMPATT diodes," *IEEE Trans. Electron Devices*, vol. ED-15, pp. 708–717, October 1968.
[10] W. J. Evans and G. I. Haddad, "Frequency conversion in IMPATT diodes," *IEEE Trans. Electron Devices*, vol. ED-16, pp. 78–87, January 1969.
[11] P. T. Greiling and G. I. Haddad, "Effect of nonsaturated drift velocity on avalanche diode performance," presented at the IEEE Int. Electron Devices Meeting, Washington, D. C., October 1968.
[12] J. L. Blue, "Approximate large signal analysis of IMPATT oscillators," *Bell Syst. Tech. J.*, vol. 48, pp. 383–396, February 1969.
[13] W. E. Schroeder and G. I. Haddad, "Effect of harmonic and subharmonic signals on avalanche-diode oscillator performance," *IEEE Trans. Microwave Theory Tech.* (Correspondence), vol. MTT-18, pp. 327–331, June 1970.
[14] P. T. Greiling, G. I. Haddad, and R. J. Lomax, "Large-signal analysis of avalanche-diode oscillators with various doping profiles and materials," presented at the IEEE Device Res. Conf., Rochester, N. Y., June 1969.
[15] P. T. Greiling and G. I. Haddad, "Large-signal equivalent circuits of avalanche transit-time devices," *IEEE Trans. Microwave Theory Tech.*, this issue, pp. 842–853.
[16] B. Hoefflinger, "Recent development on avalanche diode oscillators," *Microwave J.*, vol. 12, pp. 101–115, March 1969.
[17] C. A. Brackett, "Characterization of second-harmonic effects in IMPATT diodes," presented at the Avalanche Diode Workshop, New York, N. Y., December 1969.
[18] K. Mouthaan, "Two-frequency operation of the avalanche transit-time oscillator," *Proc. IEEE* (Letters), vol. 58, pp. 510–512, March 1970.
[19] W. E. Schroeder, P. T. Greiling, and G. I. Haddad, "Multi-frequency operation of IMPATT diodes," presented at the IEEE Int. Electron Devices Meeting, Washington, D. C., October 1969.
[20] C. B. Swan, "IMPATT oscillator performance improvement with second-harmonic tuning," *Proc. IEEE* (Letters), vol. 56, pp. 1616–1617, September 1968.
[21] H. J. Prager, K. K. N. Chang, and S. Weisbrod, "High-power high-efficiency silicon avalanche diodes at ultra high frequencies," *Proc. IEEE* (Letters), vol. 55, pp. 586–587, April 1967.
[22] D. L. Scharfetter, D. J. Bartelink, H. K. Gummel, and R. L. Johnston, "Computer simulation of low-frequency high-efficiency oscillation in germanium," presented at the IEEE Solid-State Device Res. Conf., Boulder, Colo., June 1968.
[23] W. J. Evans, "Circuits for high-efficiency avalanche-diode oscillators," *IEEE Trans. Microwave Theory Tech.*, vol. MTT-17, pp. 1060–1067, December 1969.
[24] A. S. Clorfeine, R. J. Ikola, and L. S. Napoli, "A theory for the high-efficiency mode of oscillation in avalanche diodes," *RCA Rev.*, vol. 30, pp. 397–422, September 1969.
[25] B. C. DeLoach, Jr., and D. L. Scharfetter, "Device physics of TRAPATT oscillators," *IEEE Trans. Electron Devices*, vol. ED-17, pp. 9–21, January 1970.
[26] D. J. Bartelink and D. L. Scharfetter, "Avalanche shock fronts in p-n junctions," *Appl. Phys. Lett.*, vol. 14, pp. 320–323, May 1969.
[27] R. L. Johnston, B. C. DeLoach, and B. G. Cohen, "A silicon diode microwave oscillator," *Bell Syst. Tech. J.*, vol. 44, pp. 369–372, February 1965.
[28] C. A. Lee, R. L. Batdorf, W. Wiegmann, and G. Kaminsky, "The Read diode—an avalanching, transit-time, negative-resistance oscillator," *Appl. Phys. Lett.*, vol. 6, pp. 89–91, March 1, 1965.
[29] D. de Nobel, "IMPATT diode technology," presented at the Avalanche Diode Workshop, New York, N. Y., December 1969.
[30] C. K. Kim and L. D. Armstrong, private communication.
[31] C. B. Swan, "Improved performance of silicon avalanche oscillators mounted on diamond heat sinks," *Proc. IEEE* (Letters), vol. 55, pp. 1617–1618, September 1967.
[32] R. Edwards, D. F. Ciccolella, T. Misawa, D. E. Iglesias, and V. Decker, "Millimeter-wave silicon IMPATT diodes," presented at the IEEE Int. Electron Devices Meeting, Washington, D. C., October 1969.
[33] C. N. Dunn and H. M. Olsen, "Germanium epitaxial mesa diodes for a 6 GHz 1/2 watt CW avalanche oscillator," presented at the IEEE Int. Electron Devices Meeting, Washington, D. C., October 1967.
[34] M. Gilden and W. Moroney, "High power pulsed avalanche diode oscillators for microwave frequencies," *Proc. IEEE* (Letters), vol. 55, pp. 1227–1228, July 1967.
[35] L. S. Bowman and C. A. Burrus, Jr., "Pulse-driven silicon p-n junction avalanche oscillators for the 0.9- to 20-mm band," *IEEE Trans. Electron Devices*, vol. ED-14, pp. 411–418, August 1967.
[36] M. Schwartz, *Information Transmission, Modulations, and Noise.* New York: McGraw-Hill, 1959.
[37] W. A. Edson, "Noise in oscillators," *Proc. IRE*, vol. 48, pp. 1454–1466, August 1960.
[38] W. Harth and G. Ulrich, "Q dependence of IMPATT-diode FM noise," *Electron. Lett.*, vol. 5, pp. 7–8, January 9, 1967.

82

[39] J. Josenhans, "Noise spectra of Read diode and Gunn oscillators," *Proc. IEEE* (Letters), vol. 54, pp. 1478–1479, October 1966.

[40] J. G. Ondria and J. C. R. Collinet, "Noise-reduction techniques of high-frequency oscillator-high-order multipliers and avalanche-diode-type microwave sources," *IEEE J. Solid-State Circuits*, vol. SC-4, pp. 65–70, April 1969.

[41] J. J. Baranowski, V. J. Higgins, C. K. Kim, and L. D. Armstrong, "Gallium arsenide IMPATT diodes," *Microwave J.*, vol. 12, pp. 71–76, July 1969.

[42] J. R. Ashley and F. M. Palka, "Measured FM noise reduction by injection phase locking," *Proc. IEEE* (Correspondence), vol. 58, pp. 155–157, January 1970.

[43] M. E. Hines and J. G. Ondria, "Theory and measurement of noise in varactor harmonic generator sources," *1966 IEEE Int. Microwave Symp. Dig*, pp. 35–39.

[44] R. L. Rulison, G. Gibbons, and J. G. Josenhans, "Improved performance of IMPATT diodes fabricated from GE," *Proc. IEEE* (Letters), vol. 55, pp. 223–224, February 1967.

[45] E. F. Scherer, "Investigations of the noise spectra of avalanche oscillators," *IEEE Trans. Microwave Theory Tech.*, vol. MTT-16, pp. 781–788, September 1968.

[46] P. A. Levine, "Ultralow noise in pulsed GaAs avalanche diode oscillators," *Proc. IEEE* (Letters), vol. 58, pp. 494–495, March 1970.

[47] J. R. Ashley and C. B. Searles, "Microwave oscillator noise reduction by a transmission stabilizing cavity," *IEEE Trans. Microwave Theory Tech.*, vol. MTT-16, pp. 743–748, September 1968.

[48] K. Kurokawa, "Noise in synchronized oscillators," *IEEE Trans. Microwave Theory Tech.*, vol. MTT-16, pp. 234–240, April 1968.

[49] D. R. Melick and G. A. Foggiato, "Octave mechanically tuned avalanche diode oscillators," *Proc. IEEE* (Letters), vol. 56, pp. 1737–1738, October 1968.

[50] M. I. Grace, "Magnetically tunable transit-time oscillator," *Proc. IEEE* (Letters), vol. 56, pp. 771–773, April 1968.

[51] M. I. Grace, "Varactor-tuned avalanche transit-time oscillator with linear tuning characteristics," *IEEE Trans. Microwave Theory Tech.* (Correspondence), vol. MTT-18, pp. 44–45, January 1970.

[52] T. P. Lee and R. D. Standley, "Frequency modulation of a millimeter-wave IMPATT diode oscillator and related harmonic generation effects," *Bell Syst. Tech. J.*, vol. 48, pp. 143–161, January 1969.

[53] H. E. Brenner, "FM modulation of mm-wave IMPATT oscillators with PCM baseband pulses," *Proc. IEEE* (Letters), vol. 57, pp. 1683–1684, September 1969.

[54] E. F. Scherer, "A multistage high-power avalanche amplifier at *X* band," *IEEE J. Solid-State Circuits*, vol. SC-4, pp. 396–399, December 1969.

[55] H. J. Kuno, J. R. Collard, and A. Gobat, "Microwave amplification with GaAs avalanche diodes," *Electron. Lett.*, vol. 4, pp. 540–542, November 1968.

[56] T. A. Midford and H. C. Bowers, "A two-port IMPATT diode traveling-wave amplifier," *Proc. IEEE* (Correspondence), vol. 56, pp. 1724–1725, October 1968.

[57] T. A. Midford and H. C. Bowers, private communication.

[58] M. E. Hines, private communication.

[59] V. K. Tripathi and G. I. Haddad, "Varactor loaded ladder lines," *Proc. 7th Int. Conf. on Microwave and Optical Generation and Amplification* (Hamburg, Germany, September 1968), pp. 594–599.

[60] S. G. Liu and J. J. Risko, "Stacked kilowatt avalanche-diode microwave oscillators," presented at the IEEE Int. Electron Devices Meeting, Washington, D. C., October 1969.

[61] M. I. Grace and G. Gibbons, "High efficiency microwave oscillations from Si avalanche diodes," *Electron. Lett.*, vol. 4, pp. 564–565, December 13, 1968.

[62] R. S. Ying, R. G. Mankarious, and R. W. Bower, "High-efficiency subharmonic oscillations from silicon diodes at frequencies up to 6 GHz," *IEEE J. Solid-State Circuits*, vol. SC-4, pp. 388–391, December 1969.

[63] R. L. Johnston, D. L. Scharfetter, and D. J. Bartelink, "High-efficiency oscillations in germanium avalanche diodes below the transit-time frequency," *Proc. IEEE* (Letters), vol. 56, pp. 1611–1613, September 1968.

[64] G. Gibbons and M. I. Grace, "High-efficiency avalanche-diode oscillators and amplifiers in *X*-band," *Proc. IEEE* (Letters), vol. 58, pp. 512–513, March 1970.

[65] M. Y. Ta, "High-power 'subharmonic' oscillations in silicon avalanche diodes at K_u band," *Proc. IEEE* (Letters), vol. 57, pp. 2054–2055, November 1969.

[66] D. E. Iglesias and W. J. Evans, "High-efficiency CW IMPATT operation," *Proc. IEEE* (Letters), vol. 56, p. 1610, September 1968.

[67] W. J. Evans and R. L. Johnston, "Improved performance of CW silicon TRAPATT oscillators," (to be published).

[68] M. G. Cottam, "High-efficiency oscillations in gallium-arsenide avalanche diodes," *Electron. Lett.*, vol. 6, pp. 71–73, February 1970.

[69] P. A. Levine and S. G. Liu, "Tunable *L*-band high-power avalanche-diode oscillator circuit," *IEEE J. Solid-State Circuits*, vol. SC-4, pp. 384–388, December 1969.

[70] B. Hoefflinger, C. P. Snapp, and L. A. Stark, "High-efficiency avalanche-resonance pumped amplification," *Electron. Lett.*, vol. 5, pp. 43–45, February 1969.

[71] B. Hoefflinger, C. P. Snapp, and L. A. Stark, "High-efficiency avalanche resonance pumped amplification," *IEEE J. Solid-State Circuits*, vol. SC-4, pp. 391–395, December 1969.

[72] H. J. Prager, K. K. N. Chang, and S. Weisbrod, "Anomalous silicon avalanche diodes for microwave generations," *Proc 1st Cornell Biennial Conf.*, Ithaca, N. Y., August 1967.

[73] W. J. Evans, private communication.

[74] C. B. Swan, T. Misawa, and L. Marinaccio, "Composite avalanche diode structures for increased power capability," *IEEE Trans. Electron Devices*, vol. ED-14, pp. 584–589, September 1967.

[75] C. B. Swan, "The importance of providing a good heat sink for avalanching transit time oscillator diodes," *Proc. IEEE* (Correspondence), vol. 55, pp. 451–452, March 1967.

[76] L. P. Marinaccio, "Ring-geometry IMPATT oscillator diodes," *Proc. IEEE* (Letters), vol. 56, pp. 1588–1589, September 1968.

[77] S. R Kovel and G. Gibbons, "The effect of unswept epitaxial material on the microwave efficiency of IMPATT diodes," *Proc. IEEE* (Letters), vol. 55, pp. 2066–2067, November 1967.

[78] B. B. van Iperen, A. Tjassens, and J. J. Goedbloed, "On the relation between microwave series resistance, capacitance, and output power of IMPATT diodes," *Proc. IEEE* (Letters), vol. 57, pp. 1341–1342, July 1969.

[79] B. C. DeLoach, Jr., "Thin skin IMPATTs," *IEEE Trans. Microwave Theory Tech.* (Correspondence), vol. MTT-18, pp. 72–74, January 1970.

[80] R. J. Socci and R. I. Harrison, "A multidiode avalanche oscillator for increased CW microwave power output," *Proc. IEEE* (Letters), vol. 54, pp. 1006–1007, July 1966.

[81] H. Fukui, "Frequency locking and modulation of microwave silicon avalanche diode oscillators," *Proc. IEEE* (Letters), vol. 54, pp. 1475–1477, October 1966.

[82] F. M. Magalhaes and W. O. Schlosser, "A series connection of IMPATT diodes," *Proc. IEEE* (Letters), vol. 56, pp. 865–866, May 1968.

[83] C. T. Rucker, "A multiple-diode high-average-power avalanche-diode oscillator," *IEEE Trans. Microwave Theory* (Correspondence), vol. MTT-17, pp. 1156–1158, December 1969.

[84] W. O. Schlosser and A. L. Stillwell, "A traveling-wave approach to the high-power solid-state oscillator," *Proc. IEEE* (Letters), vol. 56, p. 1588, September 1968.

[85] F. Ivanek and V. G. K. Reddi, "X-band oscillator and amplifier experiments using avalanche diode periodic structures," *IEEE Int. Solid-State Circuits Conf. Digest* (Philadelphia, Penn., February 1969), pp. 80–81.

[86] R. H. Haitz, H. L. Stover, and N. J. Tolar, "A method for heat flow resistance measurements in avalanche diodes," *IEEE Trans. Electron Devices*, vol. ED-16, pp. 438–444, May 1969.

[87] J. L. Moll, *Physics of Semiconductors*. New York: McGraw-Hill, 1964.

[88] D. E. Iglesias, "Circuit for testing high-efficiency IMPATT diodes," *Proc. IEEE* (Letters), vol. 55, pp. 2065–2066, November 1967.

[89] M. I. Grace, "Down conversion and sideband translation using avalanche transit time oscillators," *Proc. IEEE* (Letters), vol. 54, pp. 1570–1571, November 1966.

[90] Y. Fukatsu and H. Kato, "Frequency conversion with gain through sideband locking of an IMPATT diode oscillation," *Proc. IEEE* (Letters), vol. 57, pp. 342–343, March 1969.

[91] A. S. Clorfeine, "Self-pumped parametric amplification with an avalanching diode," *Proc. IEEE* (Letters), vol. 54, pp. 1956–1957, December 1966.

Microwave Avalanche Diodes

SIMON M. SZE, MEMBER, IEEE, AND ROBERT M. RYDER, FELLOW, IEEE

Abstract—Microwave avalanche diodes of various types (IMPATT, TRAPATT, etc.) can generate power sufficient for microwave receivers and some transmitters. This brief review summarizes mechanisms of operation, power output, efficiency, noise, and some important features of design and fabrication.

I. Introduction

SINCE THE FIRST experimental observation of microwave oscillation from a silicon diode in 1965 by Johnston, De Loach, and Cohen [1], avalanche diodes have become preferred solid-state sources of microwave power. Continuous-wave (CW) power available is sufficient for receivers and some transmitters throughout the microwave frequency range (about 1 GHz to 100 GHz). The power is considerably more than can be obtained from CW Gunn diodes, but at the expense of higher noise and higher operating voltages.

Avalanche diodes can be designed and operated in various modes. The first and the most important mode of operation is called the IMPATT mode. The name IMPATT is an acronym to recall the major mechanisms of operation first proposed by Read [2]—IMPact Avalanche and Transit Time. Ordinarily the IMPATT mode operates with best efficiency near the "transit-time" frequency, and efficiency of dc to RF power conversion is at present limited to about 15 percent.

A second mode of operation was discovered by Prager, Chang, and Weisbrod [3] in 1967. They called it the "anomalous" mode because the operating frequency is substantially lower than the transit-time frequency. The obtainable efficiency is considerably higher (~50 percent) and a substantial change in dc operating voltage occurs when the diode switches into the mode. Subsequent theoretical studies established [37] the mechanism responsible for the new mode. Under certain conditions of large space charge it can happen that the periodic avalanching of the diode begins at the high-field side and sweeps rapidly across the diode [39], leaving it substantially filled by a highly conducting plasma of holes and electrons whose space charge depresses the voltage to very low values. Since the plasma cannot rapidly escape, this mode is called the trapped plasma mode or TRAPATT mode (for TRApped Plasma Avalanche Triggered Transit) [38].

Other modes of operation [4] have been recognized by some authors as occurring under special conditions in some diodes. A parametric mode can occur when pumped variable reactance effects are prominent, and a thermal mode may appear when frequency is low enough so that diode temperature varies significantly during each cycle of oscillation. A space-charge feedback mode has been described, but of course space-charge effects are prominent in all modes.

We shall consider in the next section the two fundamental physical processes pertinent to the device operation, namely the drift velocity and avalanche multiplication. In Section III the operational principle of the IMPATT mode will be considered. The power and efficiency limitations on the IMPATT mode are discussed in Section IV. The noise behavior of IMPATT diodes is considered in Section V. Section VI presents the operational principle of the TRAPATT mode and briefly discusses other modes of operation. Designs and state-of-the-art performances of avalanche diodes as of late 1970 are summarized in Section VII.

Manuscript received December 18, 1970; revised February 8, 1971.
The authors are with the Bell Telephone Laboratories, Inc., Murray Hill, N. J. 07974.

II. Drift Velocity and Avalanche Multiplication

There are two fundamental physical processes most pertinent to the operation of avalanche diodes: 1) the drift velocity at which carriers travel under the influence of an electric field and 2) the avalanche multiplication which occurs when the electric field in a semiconductor is sufficiently high ($3-6\times10^5$ V/cm in Si) that the carriers gain enough energy to excite electron-hole pairs by impact ionization. 83

When the electric field is low, the average drift velocity is directly proportional to the field and the proportionality constant is called the mobility. As the field increases, the carriers interact more strongly with the lattice (especially in emitting optical phonons), so that the average velocity falls much below a linear projection from low-field values. The measured room-temperature drift velocities [5] for high-purity Ge, Si, and GaAs [6], [7] are shown in Fig. 1. We note that for fields above 10^4 V/cm, the drift velocity approaches a limiting value and becomes independent of the field. This value is called the scattering-limited velocity v_s, and is given by [8], [9]

$$v_s \cong \sqrt{\frac{8\,E_{op}\tanh(E_{op}/2kT)}{3\pi m}} \tag{1}$$

where E_{op} is the optical phonon energy, k the Boltzmann constant, T the lattice temperature, and m the mass of the carriers. The values of v_s and E_{op} for the above three semiconductors are listed in Table I along with other parameters. Since all IMPATT diodes are operated at temperatures considerably higher than room temperature (300°K), the corresponding v_s will be smaller. For example, at 500°K the scattering-limited velocities as obtained from (1) are 5×10^6, 9×10^6, and 7×10^6 cm/s for electrons in Ge, Si, and GaAs, respectively.

The region of negative differential mobility for electrons in GaAs, where velocity decreases as field increases, is prominent in the curves. The corresponding negative conductance, due to "valley transfer," is the operating basis for Gunn and LSA oscillators, but has little or no effect on IMPATT diodes because the latter operate mostly at much higher fields of the order of 10^5 V/cm where the velocity is essentially constant at the scattering-limited value v_s.

As the field increases above 10^5 V/cm, impact ionization occurs. The ionization rates are sensitive functions of the field; measured room-temperature results [5] for Ge, Si, GaAs, and GaP are shown in Fig. 2. As we can see, the ionization rate can increase five orders of magnitude for an increase of the field by a factor of only two or three. We also note that the electron ionization rate (α_n) in silicon is substantially larger than that of holes (α_p), while for other semiconductors the electron and hole ionization rates are essentially the same. This fact has special relevance for the noise behavior of IMPATT diodes to be considered in Section V.

As the temperature increases, the mean free path for phonon collisions decreases, causing a reduction of the ionization rate. The temperature dependence of the ionization rate can be obtained from the modified Baraff theory [10], [11].

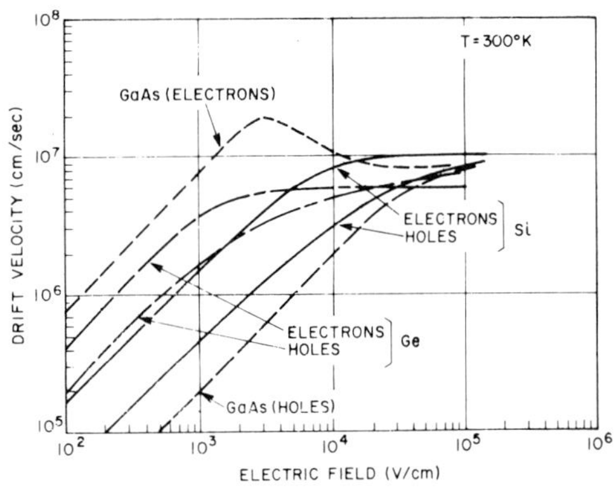

Fig. 1. Drift velocity of charge carriers. At high fields ~ 10^5 V/cm, the scattering limited velocity v_s is nearly independent of field [5]–[7].

84

TABLE I

MATERIAL PARAMETERS

Semiconductor	Ge	Si	GaAs
Energy gap, E_g (300°K)(eV)	0.67	1.12	1.43
Maximum electric field at breakdown,			
E_m (V/cm)	$2\text{–}3 \times 10^5$	$3\text{–}6 \times 10^5$	$3.5\text{–}6.5 \times 10^5$
Dielectric constant ($\varepsilon_s/\varepsilon_0$)	16	11.8	12
Drift mobility, μ (cm²/V·s)			
Electrons (μ_n)	3900	1500	8500
Holes (μ_p)	1900	600	400
Thermal conductivity, κ (W/cm·K)	0.64	1.45	0.45
Optical phonon energy, E_{op} (eV)	0.037	0.063	0.035
Scattering-limited velocity, v_s (cm/s)			
Electrons (300°K)	6×10^6	10^7	9×10^6
Holes (300°K)	8×10^6	10^7	9×10^6

The basic equation which describes avalanche multiplication when it is initiated by electrons is [12]

$$\int_0^W \alpha_n \exp\left[-\int_0^x (\alpha_n - \alpha_p)dx'\right]dx = 1 - \frac{1}{M} \tag{2}$$

where M is the multiplication factor, and W is the depletion width. Self-sustaining avalanche breakdown occurs when M approaches infinity, i.e.,

$$\int_0^W \alpha_n \exp\left[-\int_0^x (\alpha_n - \alpha_p)dx'\right]dx = 1. \tag{3}$$

For equal ionization rates, (3) reduces to

$$\int_0^W \alpha\, dx = 1. \tag{3'}$$

The above equations are used to evaluate the breakdown voltage V_B and the depletion width W at breakdown. The results for one-sided abrupt junctions (and Schottky barrier contacts) are shown [13] in Fig. 3. It is interesting to note that at a given background doping, the breakdown voltage increases with the semiconductor energy bandgap ($V_B \sim E_g^{3/2}$), and for a given semiconductor the value of V_B decreases with increasing doping ($V_B \sim N_B^{-3/4}$). Also shown in Fig. 3 are the maximum electric fields at breakdown. For Si the maximum field is 3×10^5 V/cm at 10^{15} cm^{-3} and increases slowly to 6×10^5 V/cm at 10^{17} cm^{-3}.

III. MECHANISMS OF IMPATT OPERATION

IMPATT diodes are at present the most powerful CW solid-state sources of high-frequency microwave power; up to 1 W at 50 GHz has been realized (see later Fig. 28). The IMPATT diode has covered the complete microwave frequency spectrum from 300 MHz to

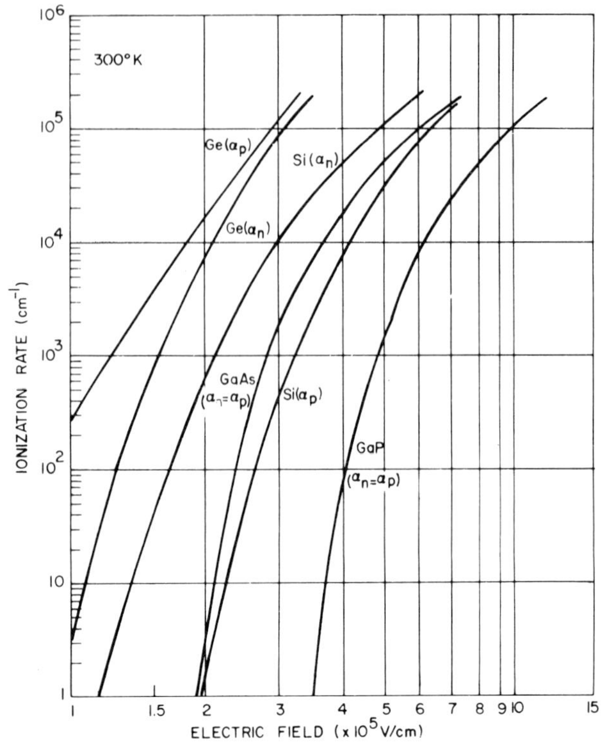

Fig. 2. Ionization rates for avalanche multiplication. For Si, rates for holes and electrons differ drastically; for Ge, GaAs, and GaP, the rates are nearly the same [5].

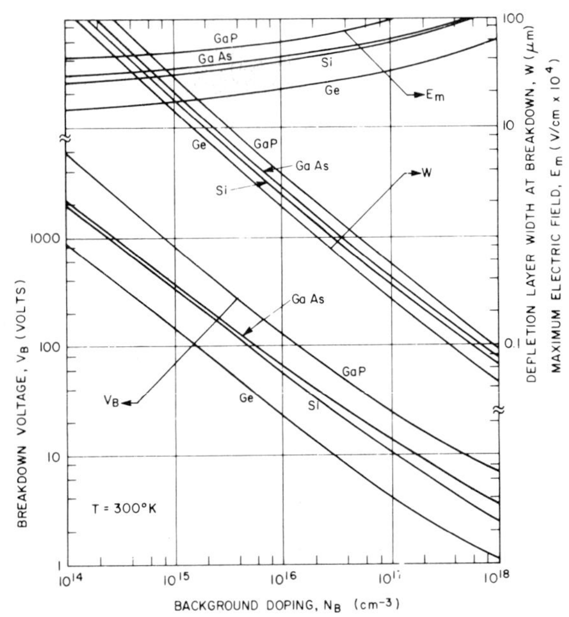

Fig. 3. Breakdown voltage, breakdown field, and depletion-layer width for abrupt p-n junctions [13].

300 GHz (or wavelength range from 1 mm to 1 m). The diodes have been fabricated from Ge, Si, and GaAs and can probably be constructed from other semiconductors should the need arise. IMPATT diodes offer potentially inexpensive, reliable, compact, and moderately efficient microwave sources. We shall now consider the mechanisms of IMPATT operation.

A. Basic Considerations

Many electron-discharge devices may be considered to be analyzable into the following functional parts:

a) a source of carriers;
b) a region where the current of the discharge is modulated by the RF signal;
c) a working region where the carriers produce useful power output by interacting with the RF field;
d) one or more collectors which dispose of the spent carriers.

In IMPATT diodes, the source is the region interior to the diode where impact multiplication is appreciable, called the avalanche region. Holes and electrons are produced in practically equal numbers. Since the multiplication is quite sensitive to the magnitude of field present, the avalanche region is also the place where the carrier transport current is modulated by the RF field. The collectors are the diode contacts, at opposite ends of the interaction space. The working region consists of both the avalanche region itself and the "drift" region between the source and collector, where the field may be too low for impact multiplication. The holes and electrons of course drift in opposite directions. Usually there is only one drift region, so that the opposite carriers disappear into their collector without doing much useful work; but recently some "double-drift" diodes have successfully used both holes and electrons (refer to Section VII).

Ordinarily the resistance of a slab of semiconductor between electrodes is positive; the electric field is approximately in phase with the current; power is dissipated, not generated. But in a number of different ways, the phase of the current may be shifted to make the resistance negative. In the IMPATT oscillators, the principal phase shift mechanisms are the delays due to finite transit time of carriers and due to finite buildup and decay times of the avalanche-produced carriers.

Fig. 4(a) shows the structure of a Read diode [2] which is basically a p^+-n-i-n^+ diode (or its compliment n^+-p-i-p^+). The dc electric field profile under reverse-biased avalanche breakdown condition is shown in Fig. 4(b). There is a high-field avalanche region at the p^+-n junction where electron-hole pairs are generated and a near constant-field drift space in the intrinsic "i" region. Fig. 4(c) shows the energy band diagram under the breakdown condition. The generated holes quickly enter the p^+ region; the generated electrons are injected into the drift space, where they do work which produces external power. As the electric field changes periodically with time around an average value [Fig. 4(d)] the impact ionization rate per carrier follows the field change nearly instantaneously. However, the carrier density does not follow the field change in unison, because carrier generation depends also on the number already present. Even after the field has passed its maximum value, the carrier density keeps increasing because the carrier generation rate is still above the average value. The maximum carrier density is reached approximately when the field has decreased from the peak to the average value. Thus the ac variation of the carrier density lags the ionization rate by about 90°, although the ionization rate is in phase with the ac field. The above situation is illustrated as the "injected" current in Fig. 4(e). The peak value of ac field (or voltage) occurs at $\theta = \pi/2$, but the peak of the injected carrier density occurs at $\theta = \pi$. The injected electrons then enter the drift region which they traverse at scattering-limited velocity. The induced external current is also shown in Fig. 4(e). From comparison of the ac field and the external current it is clear that the diode exhibits a negative resistance at its terminals.

The Read diode thus has a thin avalanche region at one end, providing the modulated avalanche current and about half the needed phase shift; the remaining phase shift is provided by adjoining a drift space about one-half cycle long ($\theta = \pi$ at the desired frequency) in which the field is substantially lower than breakdown values, so that impact ionization is negligible in the drift space.

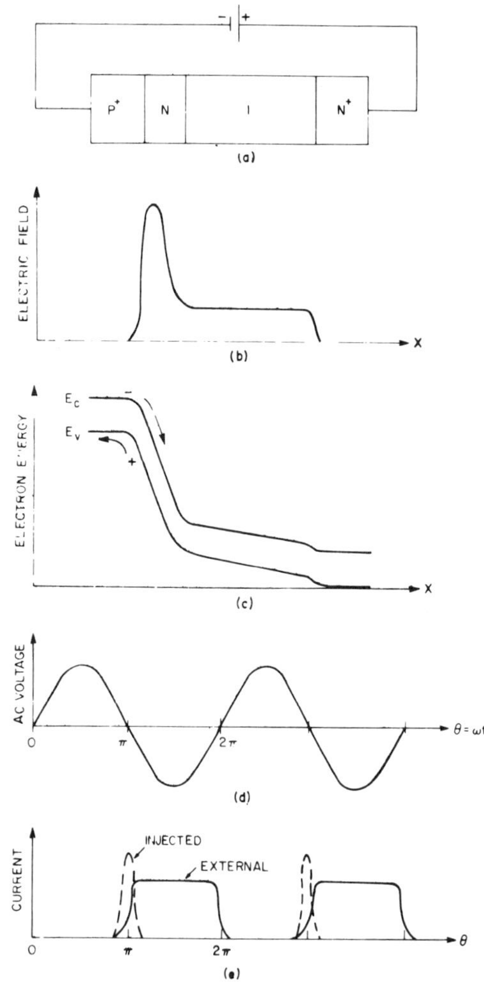

Fig. 4. Read diode. (a) Structure. (b) Field profile showing narrow avalanche region and long drift region. (c) Energy bands. (d) Voltage waveform. (e) Injected current of avalanching carriers compared with induced current in diode electrodes. (After Read, [2].)

Small-signal analysis of a Read diode gives the following expression for the real part of the diode terminal impedance [14]:

$$R \sim \frac{W}{A\varepsilon_s\omega}\,\frac{1}{1-\dfrac{\omega^2}{\omega_r^2}}\left(\frac{1-\cos\theta}{\theta}\right) \tag{4}$$

where W is the depletion width, A the diode area, ε_s the dielectric permittivity, θ the transit angle given by $\omega W/v_s$, and ω_r the avalanche resonant frequency given by

$$\omega_r \equiv \sqrt{\frac{2\alpha' v_s J_0}{\varepsilon_s}} \tag{5}$$

in which the quantity α' is the derivative of the ionization rate with respect to electric field and J_0 is the dc current density. From (5), the resistance can only be negative when $\omega > \omega_r$.

The variation of the negative resistance with the transit angle is plotted in Fig. 5. We note that the peak value occurs near $\theta = \pi$. For transit angles larger than π and approaching $3\pi/2$, the negative resistance of the device decreases rapidly. For practical purposes, although the range of negative resistance may be quite wide, we shall consider that Read-type IMPATT diodes work well only in a frequency range around the π transit angle, i.e.,

$$f \cong \frac{v_s}{2W}. \tag{6}$$

85

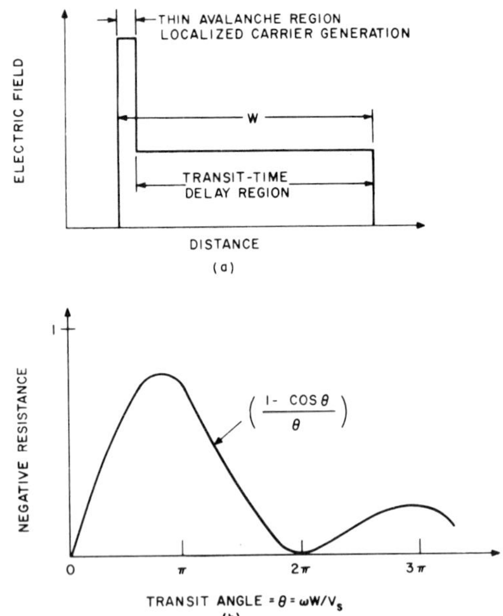

Fig. 5. Read diode. (a) Avalanche and drift regions. (b) Idealized negative resistance versus transit angle.

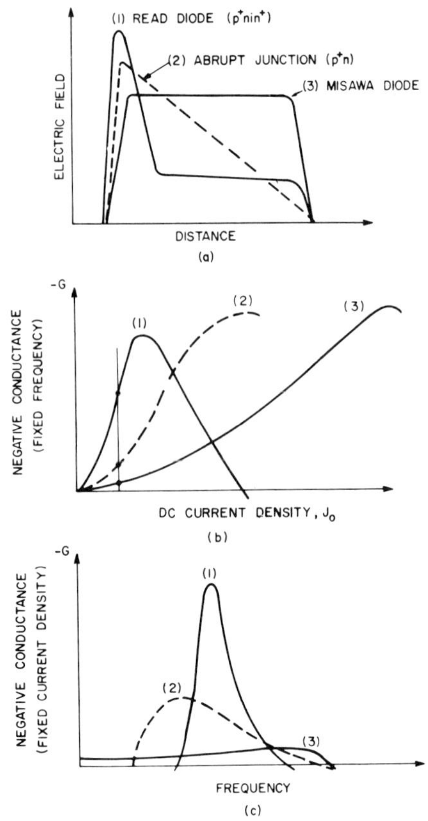

Fig. 6. Effect of doping profile on electric field and conductance for 1) Read, 2) abrupt, 3) Misawa (uniform avalanche) profiles. (a) Electric field at low currents shows narrow avalanche, wide drift for 1) and 2); opposite for 3). (b) Negative conductance versus dc current at a fixed frequency. (c) Read negative conductance is sharply peaked at the "transit-time frequency" where $\theta=\pi$; Misawa negative conductance is broad and flat, and extends to zero frequency.

For example, for a 10-GHz operation in silicon, the depletion width should be about 5 μm.

B. *General Considerations* [15]

Besides the Read diode there are other varieties of diodes which also belong to the IMPATT family. The first reported microwave oscillation was obtained from a simple abrupt p-n junction [1]. Misawa has shown that a negative conductance of IMPATT nature can be possessed by a junction diode with almost any doping profile [16].

To investigate optimization of phase shift between avalanche and transit time, Misawa introduced the concept of "uniform" avalanching and analyzed it by small-signal methods. Misawa showed that a uniformly avalanching region—one in which the rate of carrier production by impact ionization is constant throughout—can exhibit a small negative conductance all by itself without any additional "drift" region. The equivalent circuit shows that for small signals the avalanche can be represented as an electronic admittance roughly proportional to bias current. The electronic admittance, in shunt with the usual depletion layer capacitance, consists of a negative conductance shunted by an inductive susceptance which is an order of magnitude larger, in agreement with Read's suggestion that the phase lag of the current is approximately 90°. The avalanche resonant frequency can thus be defined by the shunt combination of the depletion layer capacitance and the avalanche inductive susceptance; it increases approximately as the square root of bias current [see (5)]. The negative conductance, as a function of frequency, is nearly constant for frequencies up to transit angle $\theta=2\pi$; thereafter it shows transit-time-like variations. Both conductance and susceptance are roughly proportional to bias current. This combination of properties, applying to a uniformly avalanching diode, may be collectively labeled by the term "Misawa diode."

The electric field profiles for the Read diode, p-n junction, and Misawa diode are shown [15] in Fig. 6(a). The abrupt p-n junction can be classified as a Read-type diode, since the effective avalanche region of a p-n junction is also narrow due to the strong dependence of α on the field. The Misawa diode is approximated by a p-i-n diode in which the field is essentially constant.

The negative conductance at a fixed operating frequency for the above three structures is shown schematically in Fig. 6(b) as a function of the biasing dc current density. We note that the negative conductance for a Read diode rises to a maximum value at a relatively low current density, then decreases, and eventually disappears at a current density at which the avalanche resonant frequency [which is proportional to $\sqrt{J_0}$, (5)] equals the operating frequency. Similar behaviors are obtained for the other structures. However, the conductance at low current density is smaller and the peak negative conductance occurs at larger current density, for the abrupt junction. Even larger current density is needed for the Misawa diode.

For a fixed current density, the negative conductance versus operating frequency for the above structures is shown in Fig. 6(c). The Read diode has a relatively narrow peak around the transit-time frequency (6). A wider frequency range is obtainable from the abrupt p-n junction with some reduction of peak conductance value due to nonoptimum phase relation. The reason that the negative conductance disappears at low frequencies (small θ) is that space charge in the drift space can exhibit a large positive resistance at low frequencies [17]. The Misawa diode, however, exhibits a small negative conductance, nearly constant from zero frequency up to the $\theta=2\pi$ frequency.

These features are observable in practical IMPATT diodes provided that space charge is not too large. At large current densities, however, the uniform avalanche model becomes inaccurate. For accurate comparison of practical diodes with theory it is necessary to use numerical computations. An example is shown in Fig. 7 in which both the susceptance and the negative conductance are pre-

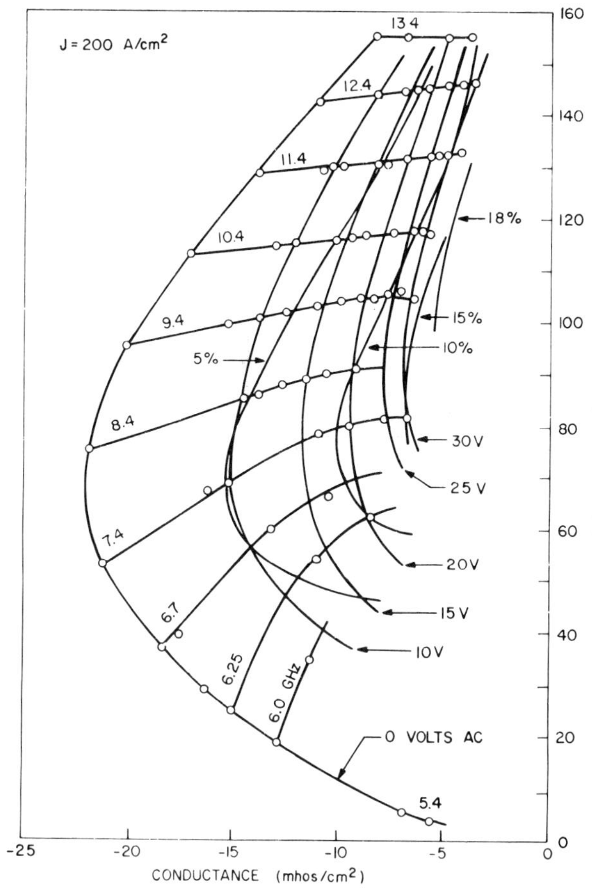

Fig. 7. Numerical calculations of negative conductance and susceptance. Note susceptance is very large and strongly current dependent. (After Scharfetter and Gummel, [18].)

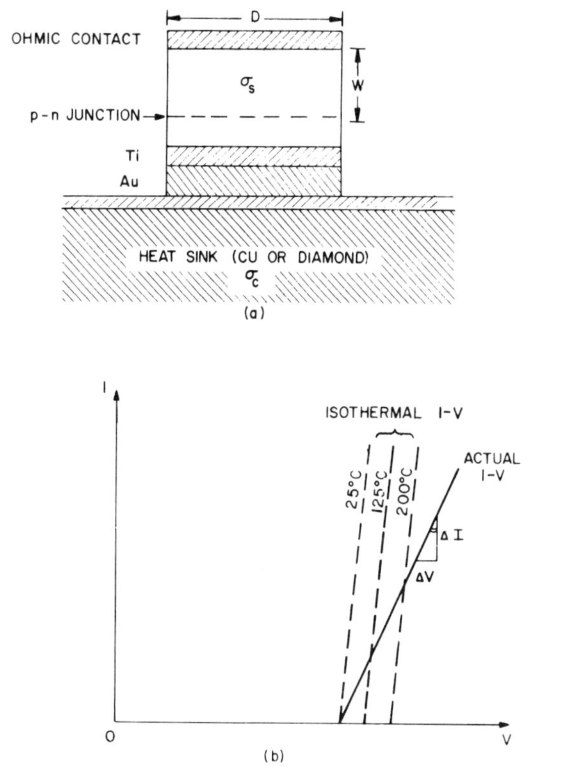

Fig. 8. Thermal considerations. (a) Structure. (b) Strong influence of temperature rise on reverse current voltage characteristic. (After Swan *et al.*, [20] and Gibbons and Misawa, [21].)

87

sented for various ac voltages [18]. Also indicated on the plot are the calculated efficiencies. Note that generally the susceptance increases with increasing ac voltage amplitude whereas the negative conductance decreases. The calculated result shows efficiencies as high as 18 percent. Such high efficiency, however, has not been obtained experimentally from diodes operated in the IMPATT mode. The limitations on efficiency and power output will now be considered.

IV. Power and Efficiency Limitations

A. Power-Frequency Limitation—Electronic

Because of the inherent limitations of semiconductor materials and the attainable impedance levels in microwave circuitry, the maximum output power at a given frequency of a single diode is limited. The limitations on semiconductor materials are 1) the critical electric field E_m at which the avalanche breakdown occurs and 2) the scattering-limited velocity v_s, which is the maximum attainable velocity in the semiconductor.

The maximum voltage that can be applied across a semiconductor sample is limited by the breakdown voltage which, for a uniform avalanche, is given by $V_m = E_m W$ where W is the depletion width. The maximum current that can be carried by the semiconductor sample is also limited by the avalanche breakdown process, because of the fact that the current in the space charge region causes an increase of the electric field (from Poisson's equation). With the maximum field again given by E_m, we obtain $I_m = E_m \varepsilon_s v_s A/W$. Therefore, the upper limit on the power input is given by the product of V_m and I_m.

$$P_m = V_m I_m = E_m^2 \varepsilon_s v_s A. \tag{7}$$

The capacitance associated with a depletion region of W is $C = \varepsilon_s A/W$. Using the transit-time frequency defined in (6), (7) can be rewritten as [19]

$$P_m f^2 = E_m^2 v_s^2 / 8\pi X_c \tag{8}$$

where X_c is the reactance $(2\pi f C)^{-1}$. Assuming we are limited to some minimum circuit impedance, (8) predicts that the maximum power which can be given to the mobile carriers decreases as $1/f^2$. This electronic limit is dominant at high frequencies ($\gtrsim 100$ GHz in Si).

B. Thermal Limitation on Power Output

At lower frequencies, the CW performance of an IMPATT diode is mainly limited by thermal considerations, i.e., by the power which can be dissipated in a semiconductor chip. A typical device mounting arrangement [20] is shown in Fig. 8(a). The cylindrical semiconductor wafer is metallized and attached to a larger metal stud which serves as a heat sink. The heat is generated in the high field region in the semiconductor. The thermal resistance R_T (in units of °C/W), consists of three components: 1) the effective thermal resistance of the semiconductor, approximately $W/2A\sigma_s$, where A is the diode area and σ_s the semiconductor thermal conductivity, 2) the contacting area component which is proportional to $1/A$, and 3) the spreading resistance component of the heat sink $1/2D\sigma_c$ where D is the diameter.

The power P, which can be dissipated in the diode, must be equal to the heat power which can be transmitted to the heat sink, therefore, $P = \Delta T/R_T$ where ΔT is the temperature difference between the junction and the heat sink, and R_T is the total thermal resistance. If the reactance $2\pi fC$ (where $C = A\varepsilon_s/W$) is maintained constant, and the major contribution to the thermal resistance is from the semiconductor, then we obtain for a given temperature increase ΔT:

$$Pf = \left(\frac{\Delta T}{R_T}\right) f \sim \frac{\sigma_s \Delta T}{\varepsilon_s} = \text{constant} \tag{9}$$

so that under the above conditions, the CW power output will decrease as $1/f$.

The thermal resistance can be obtained from dc measurement. Fig. 8(b) shows a typical current–voltage characteristic [21]. Also shown are the isothermal curves at three temperatures. The slope of the isothermal curve is given by the space-charge resistance R_{sc}, which is $(W-W_A)^2/2A\sigma_s v_s$ where $(W-W_A)$ is the drift space width. As the applied voltage increases the current, due to avalanche multi-
88 plication, will increase; this in turn causes the junction temperature to rise. Since the avalanche breakdown voltage increases with increasing temperature, the V–I expression can be written as [22]

$$V = R_{sc} I + V_B(1 + \gamma \Delta T) \tag{10}$$

where V_B is the breakdown voltage at a reference temperature (e.g., 300°K) and γ is the temperature coefficient of breakdown voltage. For small temperature rise $V \approx V_B$, we obtain from the slope:

$$\frac{dV}{dI} = R_{sc} + \gamma R_T V_B^2. \tag{10'}$$

The value of γ can be obtained from theoretical or experimental study; and R_T can be obtained from (10'). For silicon IMPATT diodes with device area of 5×10^{-4} cm^2, the value of R_T generally lies between 5–20°C/W. As the area decreases, the thermal resistance will accordingly increase [20].

The computed results [23] for CW output versus frequency based on the previous considerations are shown in Fig. 9. The maximum temperature rise ΔT is assumed to be 300°C for Si, GaAs, and GaP, and 150°C for Ge. We note that at lower frequencies we have thermal limitation ($P \propto 1/f$) and at higher frequencies we have electronic limitation ($P \propto 1/f^2$). The corner frequency for a given semiconductor depends on the maximum allowed temperature rise and minimum attainable circuit impedance.

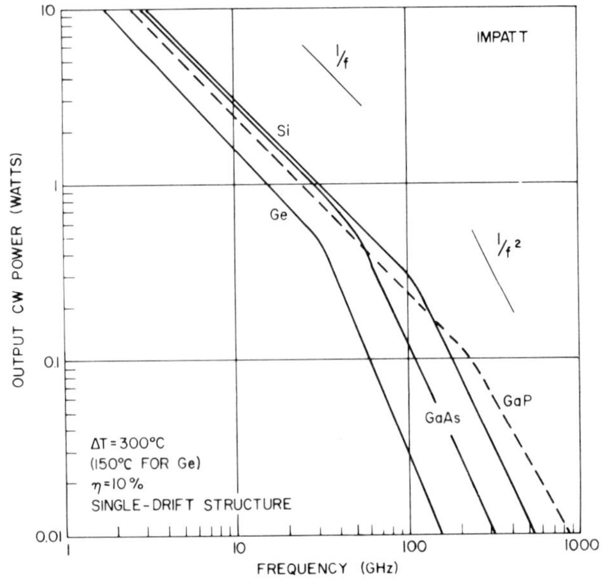

Fig. 9. Calculated power dissipation capability versus frequency. $1/f$ slope indicates thermal limitation; $1/f^2$ slope indicates electronic limitation. For GaP, v_s is assumed to be 10^7 cm/s. If the maximum temperature difference ΔT is limited to 200°C, the curves in the thermal limitation region should be lowered by about 30 percent, (9). (After Scharfetter, [23].)

C. Limitations on Efficiency

Efficiency is the ratio of ac power output to the power delivered by the dc voltage or current source; therefore

$$\eta = \frac{P_{ac}}{P_{dc}} = \left(\frac{V_a}{V_d}\right)\left(\frac{I_a}{I_d}\right). \tag{11}$$

Under the sharp-pulse approximation [2] of the idealized Read IMPATT diode [Fig. 4(e)], we have $(V_a/V_d) \cong 1/2$ and $I_a/I_d \cong 2/\pi$; so that the efficiency would be $1/\pi$ or over 30 percent.

However, for practical IMPATT diodes there are many factors which cause significant reduction of the efficiency. These factors include the space-charge effect, the reverse saturation current [24], the series resistance of the unswept epitaxial layer [25], the minority injection current [26], the high-frequency effect (skin effect) [27], and the high field effect (tunneling [28] and saturation of ionization rates) [29]. We shall briefly consider the above effects on efficiency.

The space-charge effect (Evans [15]) is shown in Fig. 10. The generated holes will cause a depression of the field by the space charge, Fig. 10(a). The reduction in field may turn off the avalanche process prematurely and thus reduces the 90° phase delay provided by the avalanche. As the holes drift to the right [Fig. 10(b)], the space charge may also cause the field to the left of the carrier pulse to drop below that required for velocity saturation. This, in turn, will change the terminal current waveforms and reduce the power generated at the transit-time frequency.

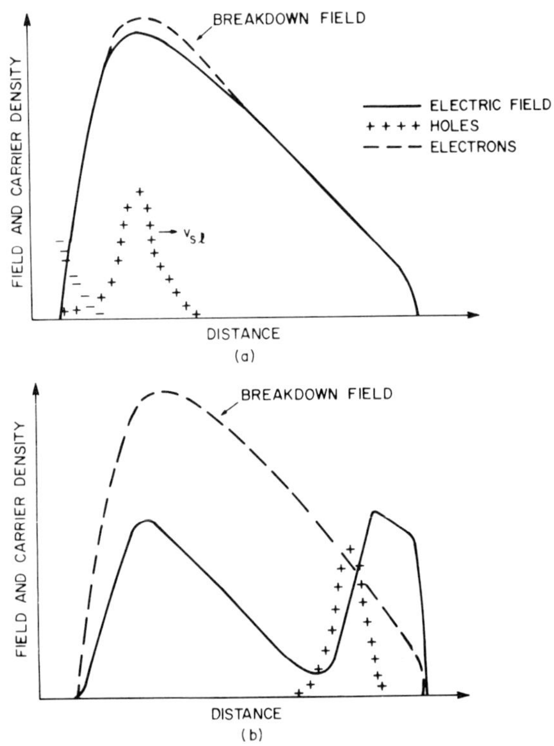

Fig. 10. Instantaneous field and charge distributions in a Read-type diode. (a) Avalanching just completed, charge beginning to move across diode. (b) Charge transit nearly complete. Note strong effect of space charge in depressing electric field. (After Evans, [42].)

The reverse saturation current effect [24] is shown in Fig. 11. As the current increases, the negative conductance decreases and the negative Q (ratio of susceptance to conductance) increases,

89

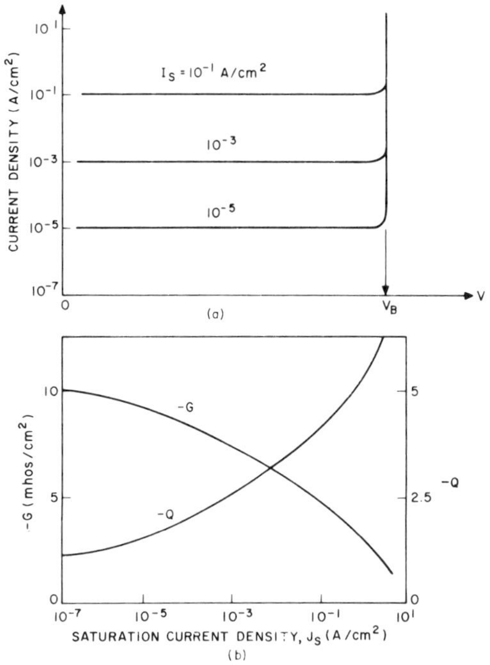

Fig. 11. Tendency of high saturation current to depress oscillation capability. (a) Direct current versus voltage. (b) Negative conductance and Q versus saturation current density. (After Misawa, [24].)

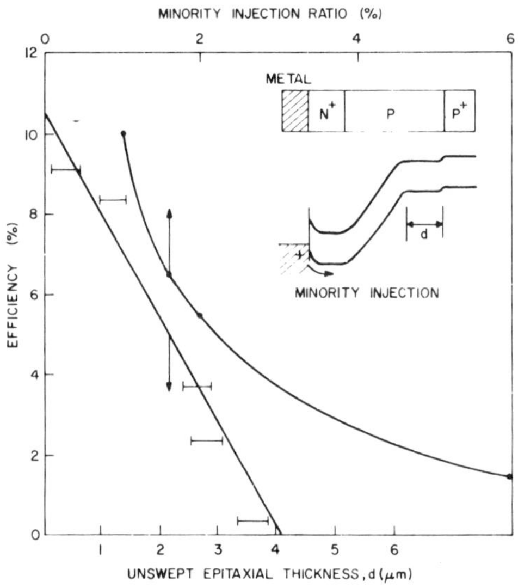

Fig. 12. Effects of unswept epitaxial layer and minority carrier injection (After Kovel and Gibbons, [26] and Decker *et al.*, [25].)

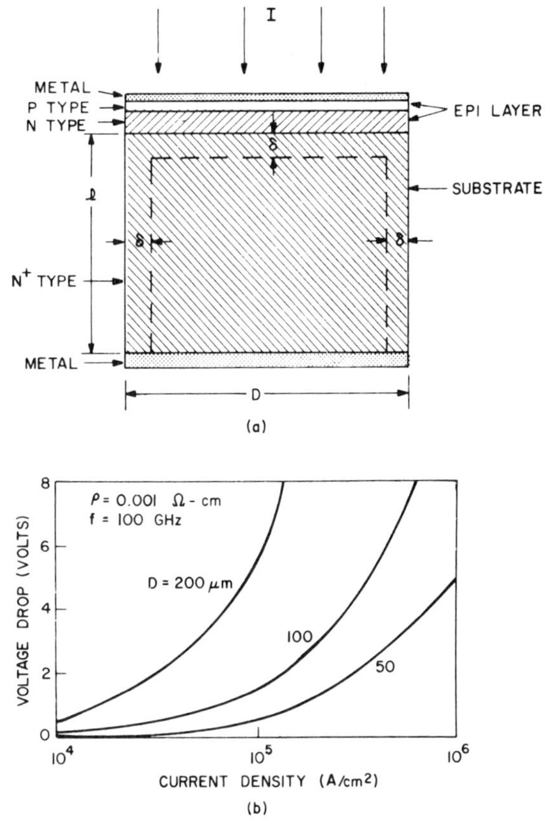

Fig. 13. Skin effect. (a) Current flow confined to a surface lamina of thickness δ causing nonuniformity and resistive loss in the diode. (b) Calculated voltage drop in substrate at 100 GHz for several diode diameters D. (After De Loach, [27].)

causing a reduction of the efficiency. Physically, the difficulty results because a higher saturation current causes the avalanche to build up too soon, reducing the avalanche phase delay again. The minority injection [25] from a poor ohmic contact will also increase the reverse saturation current and thus reduce the efficiency.

The effects of an unswept epitaxial layer [26] and the minority injection current [26] are shown in Fig. 12. The unswept layer of thickness d (shown in the insert) gives rise to a series resistance which reduces the terminal negative resistance. In the example shown the efficiency drops from 10 to 2 percent with 3 μm of unswept layer.

The skin effect [27] is shown in Fig. 13. As the operating frequency of an IMPATT diode is increased into the millimeter wave region, the current will be confined to flow within a skin depth δ of the surface of the substrate. Thus the effective resistance of the substrate is increased and there will be a voltage drop across the radius of the diode as shown in Fig. 13(b). This will cause nonuniform current distribution in the diode and an effective series resistance, both of which reduce efficiency.

For very high frequency operation, the depletion width becomes very narrow [28], and the field required for impact ionization may exceed 10^6 V/cm. There are two major effects at such high fields. The first one is that tunneling current which may be dominant; since it is in phase with the field, the 90° avalanche phase delay is not provided. The second effect is the saturation of the ionization rates, i.e., the rates will vary slowly as the field increases beyond 10^6 V/cm (refer to Fig. 2). This will cause a broadening of the injected current pulse, and change the terminal current waveforms so that the efficiency is reduced [29].

D. *Burnout from Filament Formation*

Burnout may occur not only if the diode is overheated but also, more insidiously, if the carrier current fails to be uniformly distributed over the diode area, but instead concentrates into filaments of locally high intensity. Such untoward behavior can often result when the diode has a dc negative conductance, because then the local region of greatest current density has also the lowest breakdown voltage. For this reason, p-i-n diodes are prone to easy burnout. The moving carrier space charge of drift regions tends to prevent low-frequency negative conductance (10) and therefore helps to prevent filamentary burnout. Some diodes which have positive dc conductance at low currents may develop negative dc conductance and burnout at high currents.

90

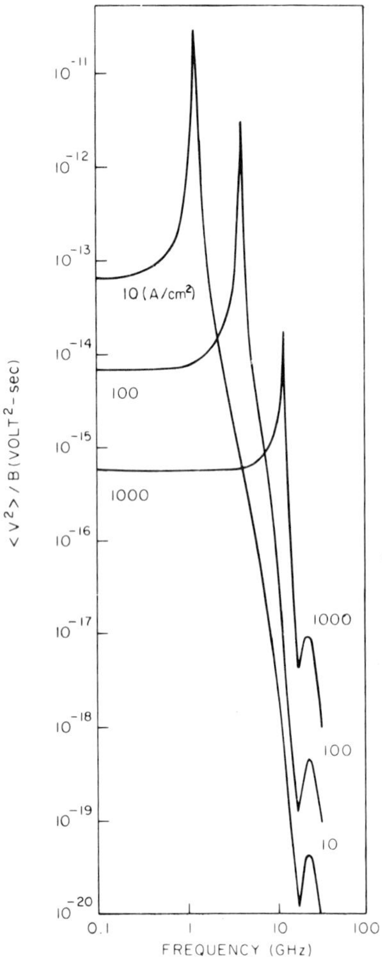

Fig. 14. Calculated equivalent noise voltage versus frequency. The peak at the avalanche frequency and the dip at twice the transit time frequency are really reflections of corresponding changes in diode impedance. The output noise power varies more smoothly with frequency. (After Gummel and Blue, [31].)

V. Noise Behavior

The noise level of IMPATT oscillators is very high, corresponding to amplifier noise figures of 20–40 dB. The reason for such high noise is that an avalanche discharge consists of a large number of carriers which cross the interaction space as a group, but whose time position "jitters" in phase from one cycle to the next because of the random way in which the avalanche grows from a few initial ionizing events. Therefore this "jitter" noise is much greater than shot noise, in which the individual carrier transits are random. The avalanche noise can be greatly reduced by driving the oscillator with a smaller quieter "locking" oscillator which triggers the avalanche more uniformly in successive cycles.

The noise has been quantitatively analyzed in small-signal theories, analytically by Hines [30] and numerically by Gummel and Blue [31]. For a high-gain amplifier under simplified assumptions, the noise figure is [30]

$$\mathrm{NF} = 1 + \frac{qV_A/kT_0}{4m\tau_A^2(\omega^2 - \omega_r^2)} \tag{12}$$

where V_A is voltage and τ_A the transit-time across the avalanche region, and m is an exponent which describes approximately the field dependence of impact ionization, i.e., $\alpha = \alpha_0(E/E_0)^m$. The plot of mean-square noise voltage per unit bandwidth versus frequency for a particular silicon diode [31] is shown in Fig. 14. The two theories agree qualitatively in showing a peak of noise voltage at the avalanche resonant frequency ω_r, falling off rapidly at higher frequencies. Noise can therefore be reduced somewhat by operating well above the avalanche frequency and keeping the current low. These conditions are in conflict with the conditions favoring high power and efficiency so that tradeoffs are necessary to optimize for particular applications.

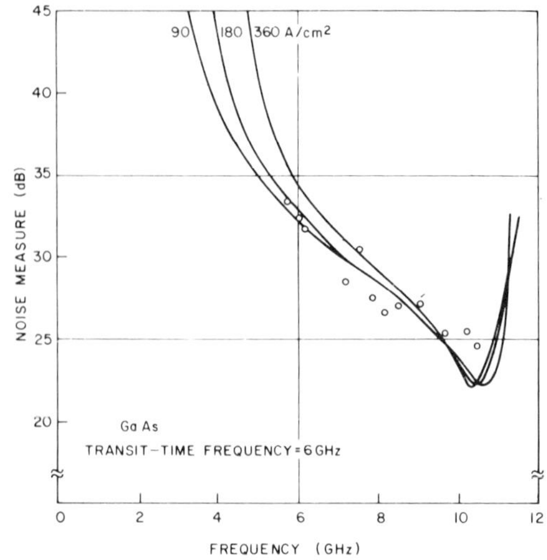

Fig. 15. Noise measure for GaAs IMPATT diodes. (After Blue, [32].)

TABLE II

Noise Measure of IMPATT Diodes

Semiconductor	Ge	Si	GaAs
Small-signal[a] noise measure (dB)	30	40	25
Large-signal[b] oscillator noise measure (dB)	40	55	35

[a] For small-signal amplifier, the noise measure is $M = T/T_0$ where $T_0 = 290$ K.
[b] Oscillator FM noise is given by

$$\Delta f_{rms} = \left(\frac{f_0}{Q_{ext}}\right)\left(\frac{kT_0MB}{P_0}\right)^{1/2}$$

where f_0 = oscillator frequency, Q_{ext} = external circuit Q, M = the large-signal noise measure, B = measurement bandwidth, and P_0 = oscillator power output.

Fig. 15 shows theoretical and experimental results of the noise measure in a GaAs IMPATT diode [32]. At the transit-time frequency (6 GHz) the noise measure is about 32 dB. The minimum noise measure of 22 dB, however, is obtained at about twice the transit-time frequency. One important feature of the GaAs noise measure is that it is substantially lower than that for Si IMPATT diodes. Table II shows some of the state-of-the-art noise measures [33] of Ge, Si, and GaAs IMPATT diodes. The amplifier and oscillator noises in the table are for a lossless circuit at the frequency corresponding to maximum oscillator efficiency, without harmonic tuning.

The main reason for the low-noise behavior in GaAs is that for a given field the electron and hole ionization rates are essentially the same in GaAs while in Si they are quite different. From (2) it can be shown that to obtain a large multiplication factor M the average distance of ionization $1/\langle\alpha\rangle$ is about equal to W_A (the avalanche width) if $\alpha_n = \alpha_p$, but is about equal to $W_A/\ln M$ if $\alpha_n \gg \alpha_p$. So that for a given W_A there have to be considerably more ionization events in Si, which results in higher noise.

The relation between power output and FM noise measure is

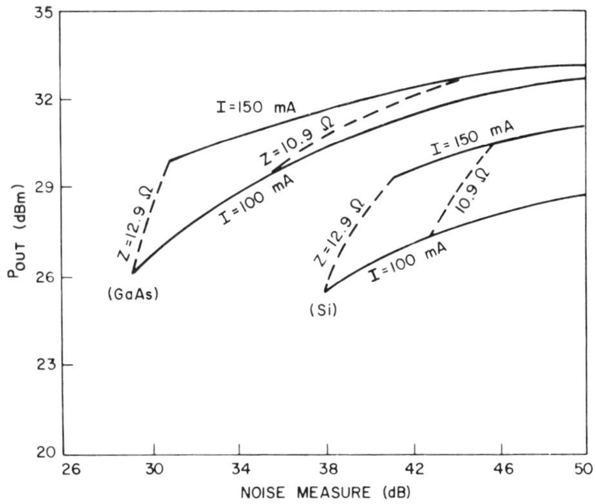

Fig. 16. Noise versus output power for a phase-locked oscillator. Locking power was held constant at +4 dBm. Contours of constant load impedance Z and constant diode current I are shown. (After Irvin *et al.*, [34].)

shown in Fig. 16 for some Si and GaAs 6-GHz IMPATT diodes [34]. The diodes were evaluated in a single tuned coaxial resonator circuit in which the load resistance presented to the resonator was incrementally varied by using interchangeable impedance transformers Z. At maximum power output the noise measure is relatively poor. Lower noise measure can be realized at the expense of a slightly reduced power output. We note again that at a given power level (say 1 W) the GaAs IMPATT diode is about 10 dB quieter than a Si IMPATT diode.

VI. Mechanisms of Trapatt Operation and Other Modes of Operation

A. TRAPATT Mode

The trapped plasma or TRAPATT mode is a high-power high-efficiency mode. To date the highest pulse power of 1.2 kW has been obtained at 1.1 GHz (5 diodes in series) [35] and the highest efficiency [36] of 75 percent has been obtained at 0.6 GHz. However, the TRAPATT operation is a rather complicated manner of oscillation and requires good control of both device and circuit properties. In addition, the TRAPATT mode generally shows considerably higher noise measure than IMPATT mode; and the upper operating frequency seems to be practically limited to below the millimeter wave region.

To understand the initiating of the TRAPATT mode, we consider a current step which is applied at $t=0$ to a n^+-p-p^+ diode, Fig. 17(a). If at $t=0$ the maximum electric field of the diode is smaller than the critical field E_c at breakdown, the field profile will at first move up with time as shown in Fig. 17(b). This is because [37]–[41]

$$J = \varepsilon_s \frac{dE}{dt}. \tag{13}$$

Hence the field can be expressed as

$$E(x, t) = E_m - \frac{qN_A}{\varepsilon_s} x + \frac{Jt}{\varepsilon_s} \tag{14}$$

where N_A is the doping concentration of the p region. Thus the value of t at which the electric field reaches E_m at a given distance x into the depletion region is obtained by setting $E(x, t) = E_m$, yielding

$$\frac{qN_A x}{\varepsilon_s} = Jt \tag{15}$$

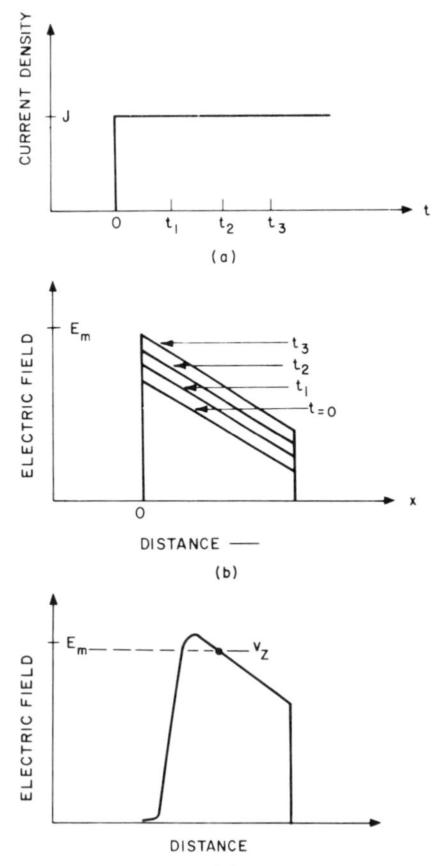

Fig. 17. TRAPATT mode operation (schematic). (a) Current density versus time. (b) Idealized electric field rising with time. (c) Avalanche shock front which sweeps through the diode with velocity v_z greater than the scattering-limited velocity v_s. (After Clarfeine *et al.*, [40], De Loach and Scharfetter, [41].)

91

or upon differentiation

$$v_z \equiv \frac{dx}{dt} = \frac{J}{qN_A} \tag{16}$$

where v_z is the avalanche zone velocity and represents the velocity at which the leading edge of the avalanche region progressed through the diode, Fig. 17(c). For example, for $N_A = 10^{15}$ cm^{-3} and $J = 10^4$ A/cm^2, $v_z = 6 \times 10^7$ cm/s which is much larger than the scattering-limited velocity. Thus the avalanche zone [39] (or avalanche shock front) will quickly sweep across most of the diode, leaving the diode filled by a highly conducting plasma of holes and electrons whose space charge depresses the voltage to low values [37], [38].

The field and carrier density calculated numerically for a particular moment is shown [42] in Fig. 18(a). The corresponding voltage and current wave forms are shown in Fig. 18(b) and (c), respectively. (The dot indicates the situation shown on Fig. 18(a)). We note that the voltage at the beginning of the cycle can become considerably larger than the steady-state breakdown voltage. After the avalanche zone passes through the device the voltage drops to low values, but soon rises again as the field polarizes the space charge. Because of the dependence of the drift velocity on field at low fields (Fig. 1), the electrons and holes will drift at velocities determined by the low-field mobilities, and the time of transit of the carriers can become much longer than W/v_s.

Thus the TRAPATT mode can operate at comparatively low frequencies since the discharge time of the plasma, i.e., the ratio Q/I of its charge to its current, can be substantially greater than the nominal transit time W/v_s of the diode at high fields. Therefore, the

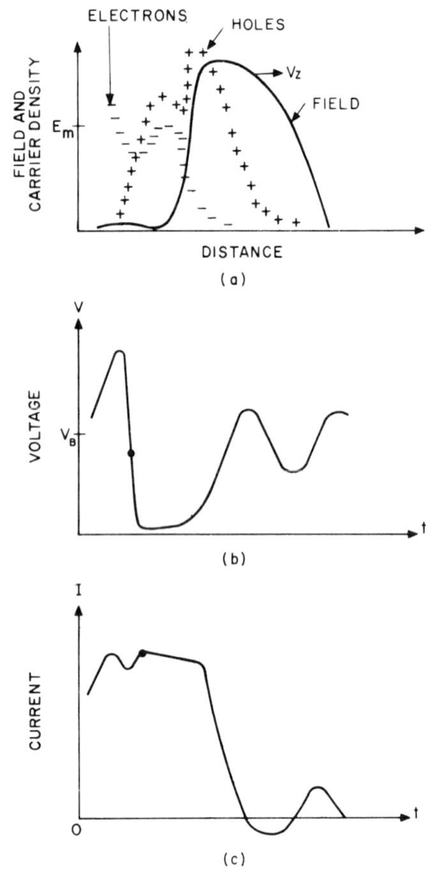

Fig. 18. TRAPATT mode operation. (a) Numerical calculation of instantaneous field and charge. (b) Voltage versus time. (c) Current versus time. Dots on voltage and current curves apply to condition shown in (a). (After Evans, [42].)

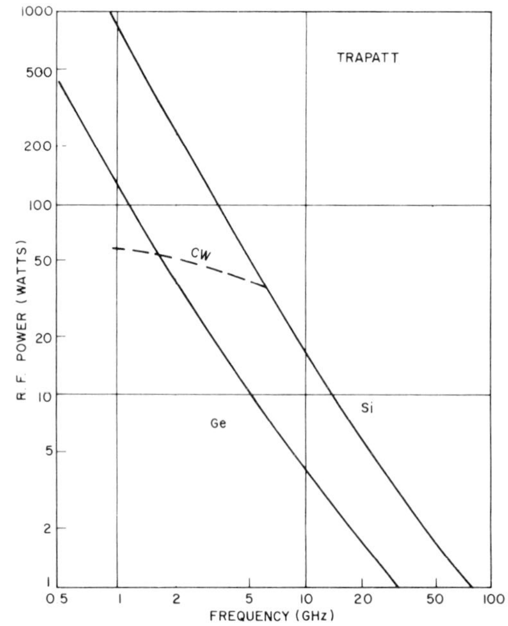

Fig. 19. Theoretical TRAPATT power versus frequency. (After Scharfetter, [43].)

TRAPATT mode is still a transit-time mode in the real sense that the time delay of carriers in transit, i.e., the time between injection and collection, is utilized to obtain current phase shift favorable for oscillation.

The TRAPATT mode also requires a circuit which can support harmonics of the fundamental frequency at high voltage amplitudes. The rich harmonic content is necessary in order to get the requisite phase delay in the current at such low frequencies; the diode must not break down when the fundamental voltage is a maximum, but is required instead to wait until approximately $\frac{1}{4}$ cycle later, when the avalanche is triggered by a peak in the harmonic voltage. Since the terminal voltage can be small when the terminal current is large and vice versa, the TRAPATT mode can have high efficiency.

The numerical Fourier analysis of the resulting voltage waveforms based on a step terminal current can give the oscillator power, efficiency, and impedance. Fig. 19 shows the theoretical power versus frequency behavior that one might expect from this mode of operation [43]. Efficiencies substantially higher than 50 percent have been predicted for some silicon diodes. (75 percent has been obtained in some UHF avalanche diodes [36].)

B. *Other Modes of Operation* [4]

As noted in Section III, an avalanche diode can exhibit large electronic susceptances which are strongly current dependent and have low or even negative loss. Consequently, they can exhibit parametric amplification, in which gain or oscillation can be obtained by "pumping" the low-loss variable reactance at a high frequency. Such "parametric" amplification is really that of a regenerative reactive modulator, operating between a signal frequency S and a lower sideband (usually $(S-b)$ of the pump frequency b). It is also a complicated form of amplification possibly involving harmonics, since finite impedances and finite voltages must be present at the signal frequency S, the "idler" frequency $S-b$, the pump frequency b, and possibly also other frequencies. Oscillations of variable-reactance type were first observed in IMPATT diodes by De Loach *et al.* [4], [44]. Reactance amplifiers have been built and measured by Hoefflinger *et al.* [45] who has noted that at high power levels the various frequencies lock into synchronism and become harmonically related. Under these conditions the parametric amplifier was "self-pumped"—that is, it generated its own pumping frequency.

Another mode of operation is the static mode [46] which may occur in a p-i-n diode. The reverse current–voltage characteristic exhibits a differential dc negative resistance. This mode of operation is believed to be of limited usage, since the lack of space charge resistance (10) has an important consequence of undesirable nature: there is severe danger of easy burnout by current filament formation at high current densities. (See Section IV-D). This is the reason that all practical avalanche diodes are always designed to include some drift space with finite doping so that some space-charge protection against burnout is obtained.

The thermal mode is another mode of operation, in which the phase shift accompanying temperature rise is important to the operation. However, it is handicapped by two fundamental limitations: 1) it may form filaments as does the static mode, and 2) it is limited in operating frequency by the thermal time constant. Nevertheless it is important to avalanche diode design in that it represents a self-destruction mechanism which must be avoided.

VII. Device Design and Performance

A. *Device Fabrication and dc Characteristics*

Based upon the discussion presented in Section IV, IMPATT diodes are usually designed in such a way as to maximize power output and efficiency. Fig. 20 shows some typical high-power IMPATT diode designs.

The configuration of Fig. 20(a) is formed by using a double epitaxial process or by diffusion into an epitaxial layer [47]. The n^+

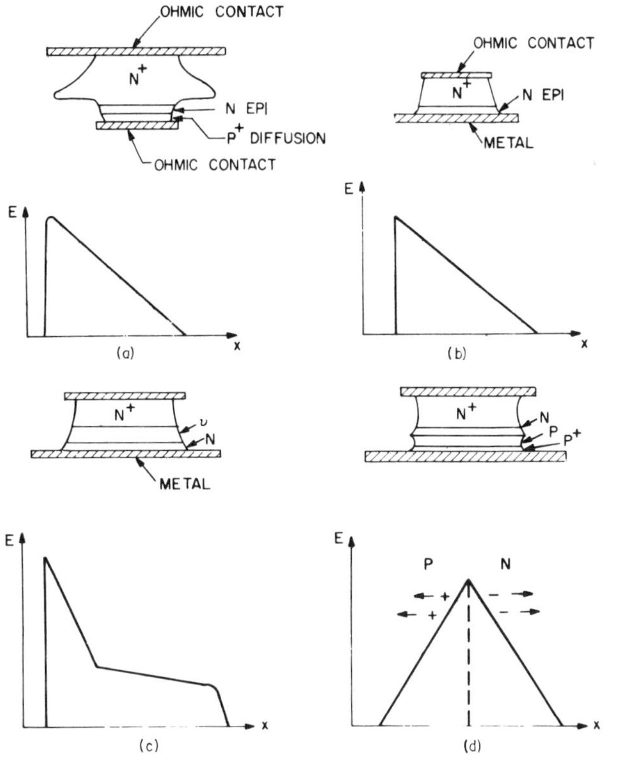

Fig. 20. Outlines of some diode shapes and corresponding field profiles. (a) Diffusion or double epitaxial. (After Misawa, [47].) (b) Schottky barrier. (After Irvin *et al.*, [34].) (c) Improved Read diode. (After Misawa, [51].) (d) Ion implantation. (After Scharfetter *et al.*, [53].)

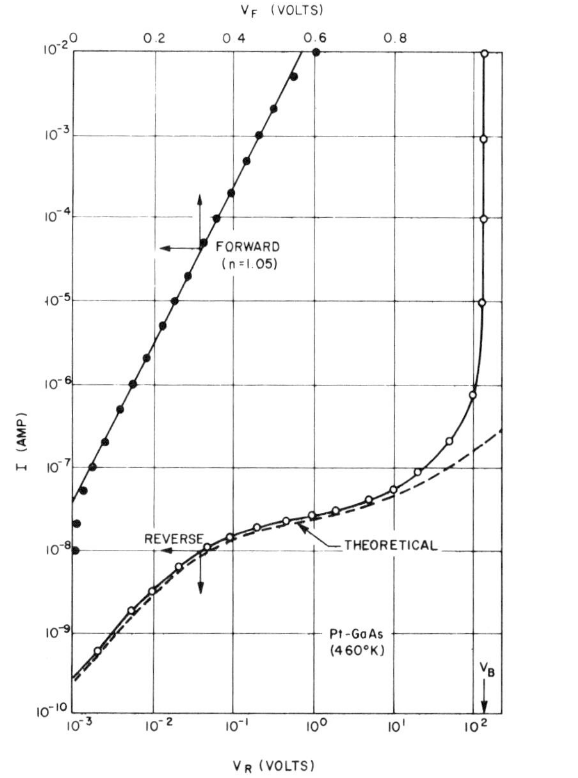

Fig. 21. Current–voltage characteristic of a GaAs Schottky barrier is close to the ideal thermionic-emission theory. (After Coleman *et al.*, [55].)

substrate is used to reduce series resistance; and the epitaxial layer thickness should also be controlled so that at breakdown there is essentially no unswept epitaxial layer left. In other words, the depletion layer width at breakdown should be very close to that shown in Fig. 3. For high-frequency operation even the n^+ substrate should be thinned down [48] to the order of a few micrometers in order to reduce losses and nonuniformity from skin effect.

Fig. 20(b) shows a Schottky barrier IMPATT diode [49], [50]. A Schottky barrier is a rectifying metal–semiconductor contact [5]. Although the field distributions of Fig. 20(a) and (b) are very similar, the Schottky barrier can offer some unique advantages. One is that the maximum field occurs at the metal–semiconductor interface and the generated heat can be readily conducted away from the metal contact. As shown in Fig. 20(b) the device can be formed in a truncated cone shape [34] so that the high-field singularity at the periphery is removed and uniform breakdown can be obtained inside the device. Another advantage of the Schottky barrier approach is that the device can be fabricated at relatively low temperatures such that the original high quality epitaxial layer can be preserved.

The Schottky barrier approach can also be used in modified Read diode structures [51] as shown in Fig. 20(c) by replacing the p^+ layer by the metal contact. Since the Schottky barrier is basically a majority-carrier device, the minority carrier storage effect [52], which occurs in the original Read structure, is eliminated and improved efficiency can be achieved.

Fig. 20(d) shows a new device configuration by using ion implantation technique [53]. For the most common dopants, e.g., B and P, the depth of ion penetration is of the order of 0.5 μm/100 keV, so that 1-μm depth can be readily obtained from ion implantation equipment with a few hundred kiloelectronvolt energy sources. These energy ranges are thus useful in making millimeter wave IMPATT diodes. Of particular importance is the possibility of making "double drift" structures as shown in Fig. 20(d) which is much like two complementary IMPATT diodes in series, using both types of carriers from the same avalanche source. Power output per unit area and impedance per unit area are both approximately doubled. This structure is thus expected to have increased power output with improved efficiency [54].

The dc characteristics of some Schottky barrier diodes correspond closely to theory [55]. Fig. 21 shows the dc current–voltage characteristics of a high-power (3-W CW) 6-GHz GaAs IMPATT diode [34]. The device is fabricated using a Pt Schottky barrier and formed in the shape shown in Fig. 20(b). The device operating temperature as a microwave oscillator is 460°K, and the dc measurement at that temperature shows near-ideal current–voltage characteristics, based on the thermionic emission theory. The breakdown voltage of the same device as a function of heat-sink temperature is shown in Fig. 22 and is also found to be in good agreement with the theoretical result for one-sided abrupt junctions [11], [56].

B. Frequency Scaling and Microwave Performance

From the small-signal theory we can obtain approximate relations for various device parameters as a function of operating frequency. The resistance expression in (4) can be rewritten as

$$-R \sim \frac{W^2}{2A\varepsilon_s v_s}\left(\frac{1}{\dfrac{\omega^2}{\omega_r} - 1}\right)\left(\frac{1-\cos\theta}{\theta^2/2}\right) \tag{17}$$

where θ is the transit angle equal to $\omega\tau$ and $\tau \equiv W/v_s$. For $-R$ to be invariant with ω/ω_r, it is required from (17) that [57]

$$W^2/2A\varepsilon_s v_s = \text{constant} \tag{18a}$$

and

$$\omega_r \tau = \text{constant}. \tag{18b}$$

94

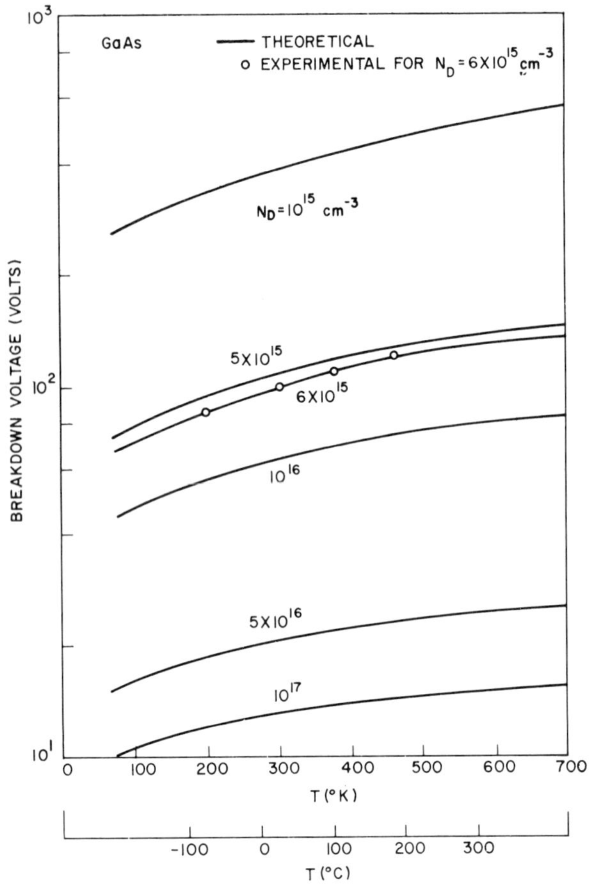

Fig. 22. Breakdown voltage versus temperature follows theory closely. (After Coleman *et al.*, [55].)

TABLE III

APPROXIMATE FREQUENCY SCALING FOR IMPATT DIODES

Parameter	Frequency Dependence
Junction area, A	f^{-2}
Bias current density, J	f
Depletion-layer width, W	f^{-1}
Breakdown voltage, V_B	f^{-1}
Power output, P_{out}	
Thermal limitation	f^{-1}
Electronic limitation	f^{-2}
Efficiency, η	constant

Since the depletion width W is inversely proportional to the operating frequency (6), the device area, which is proportional to W^2, is thus proportional to ω^{-2}. From the avalanche breakdown equation (3'), it can be shown that the ionization rate (α) and its field derivative (α') are inversely proportional to the depletion width. Combination of the relation $\alpha' \sim 1/W$, (5), and 18(b), yields the following result for the dc current density

$$J_0 \sim \frac{\omega_r^2}{\alpha'} \sim \frac{\omega^2}{1/W} \sim \omega. \tag{19}$$

The above frequency scaling relations are summarized in Table III, and are useful as a guide for extrapolating performance and design to new frequencies.

The power output limitations have been considered in Section IV. The efficiency is expected to be only weakly dependent on frequency at low frequencies. However, at millimeter wave regions the operating current density is high ($\sim f$) and the area is small ($\sim f^{-2}$) so that the device operating temperature will be high. This in turn will cause an increase of the reverse saturation current density and a decrease of the efficiency. In addition, the skin effect, the tunneling effect, and other effects associated with high frequency and high field will also degrade the efficiency performance. Hence, as the frequency increases the efficiency is expected to decrease eventually.

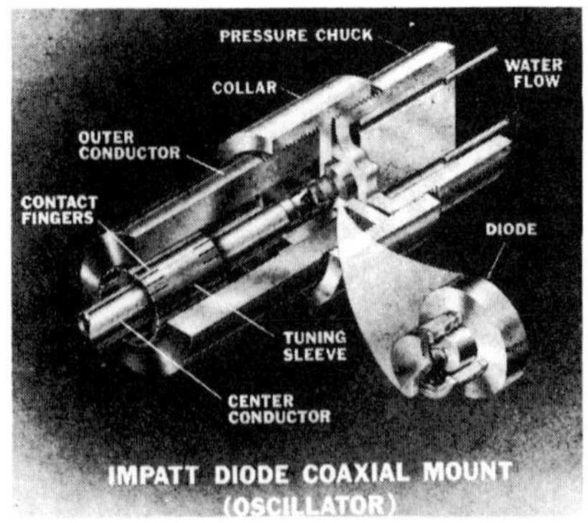

Fig. 23. Drawing of a coaxial oscillation circuit. (After Iglesias, [58].)

TABLE IV

DEVICE PARAMETERS AND PERFORMANCE OF EXPERIMENTAL SILICON IMPATT DIODES

Frequency Range	8–10 GHz	50–60 GHz[a]	100–110 GHz
Junction area (cm^2)	5×10^{-4}	2×10^{-5}	5×10^{-6}
Epitaxial layer thickness (μm)	5	2	0.5
Doping concentration (cm^{-3})	6×10^{15}	6×10^{16}	2×10^{17}
Bias current density (A/cm^2)	~1000	~5000	~50 000
Depletion-layer width (μm)	3.5	0.6	0.25
Breakdown voltage (V)	85	17	9
Power output (W)	3	0.5	0.07
Efficiency	10–12%	10%	3%

[a] Life test shows that the equivalent life time operated at 200°C is longer than 15 000 h.

In the low-frequency range such as C band (5-7 GHz), X band (8–12 GHz), and K band (18–26 GHz), a useful microwave measurement circuit [58] is shown in Fig. 23. Also shown is a device mounted in a microwave package. For the millimeter wave region both the mounting scheme and microwave circuit should be modified [59]. Typical microwave performances of silicon IMPATT diodes are listed in Table IV for three sets of devices [60]. We note that the device parameters follow reasonably close to the frequency scaling listed in Table III. The efficiency in the 100-GHz range is lower presumably due to the aforementioned effects. The relation between the breakdown voltage and frequency is also in good agreement with the computed result shown in Fig. 3.

To demonstrate the importance of the transit-time effect, Fig. 24 shows the breakdown voltage versus the optimum frequency at which the efficiency is maximized [61]. The solid lines are calculated from (6). The agreement is good for all three semiconductors.

Fig. 25 compares the performance [59] of double-drift and single-drift diodes at 50 GHz. The double-drift 50-GHz IMPATT diode mode by ion implantation shows an output CW power of 1 W with a maximum efficiency of 14 percent. These results can be compared with a similar single-drift device which delivers about 0.5 W with an efficiency of about 10 percent.

So far the performance discussion has concerned IMPATT diodes. For a TRAPATT diode the design and performance are more complicated because of the strong device-circuit interaction which dictates most of the device performance (Evans [15]). To reduce the overvoltage, which is the terminal voltage required to initiate the avalanche shock front, a TRAPATT diode is usually designed with a

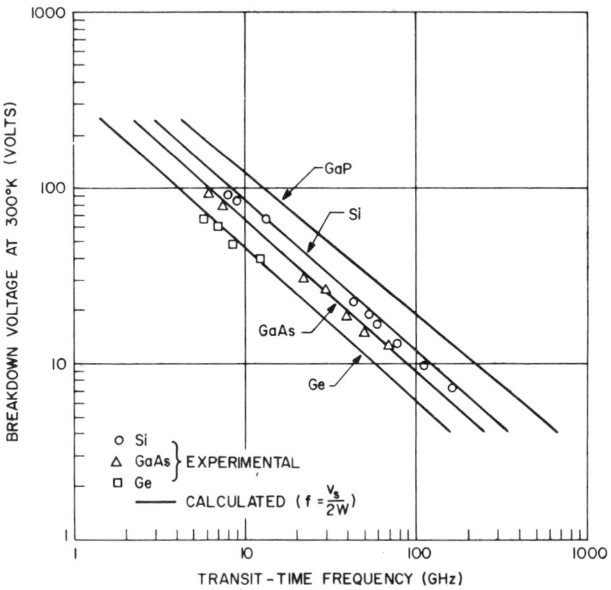

Fig. 24. Calculated breakdown voltage versus frequency for abrupt p-n junction diodes in Ge, Si, GaAs, and GaP. Data points are measured results for diodes operated at optimum frequency at which the efficiency is maximized. (After Dunn, [61].)

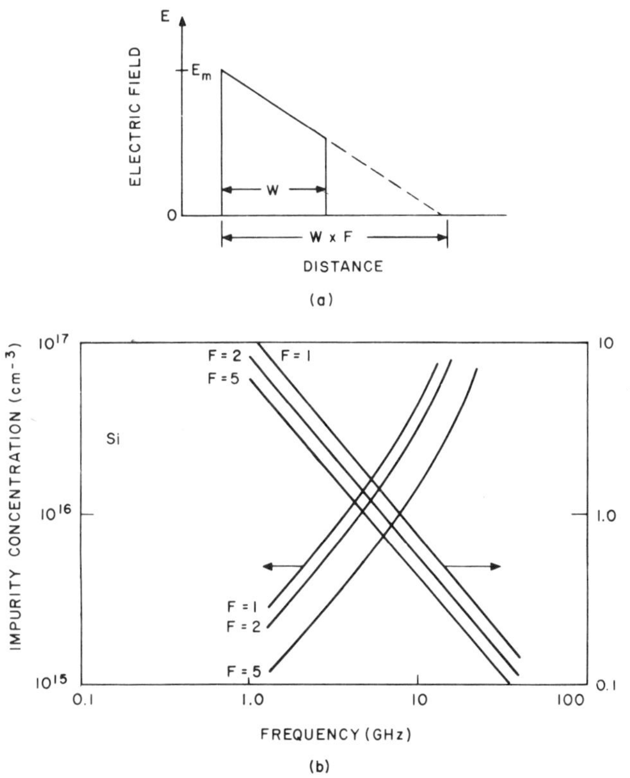

Fig. 26. Design parameters for TRAPATT diodes. The punchthrough factor F is defined in (a). (After Evans, [15].)

large punch-through factor [43]. The punch-through factor F is defined in Fig. 26(a). The larger the factor F, the closer the device resembles a p-i-n diode. The required depletion width and impurity concentration as a function of TRAPATT operating frequency is shown in Fig. 26(b) for three punch-through factors (Evans [15]).

The TRAPATT diode state-of-the-art highest power performance is shown [35] in Fig. 27. The output power of a series connection of five diodes under pulse condition reaches 1.2 kW with an efficiency of 25 percent.

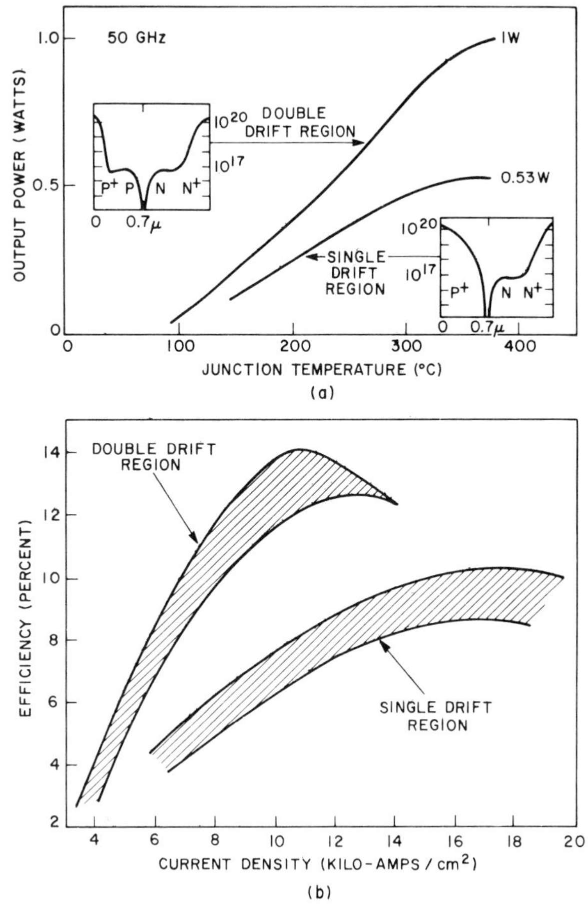

Fig. 25. Power output and efficiency of single-drift versus double-drift IMPATT oscillators at 50 GHz. (a) Power output for two individual diodes. (b) Range of efficiency for four diodes of each type. (After Seidel *et al.*, [59].)

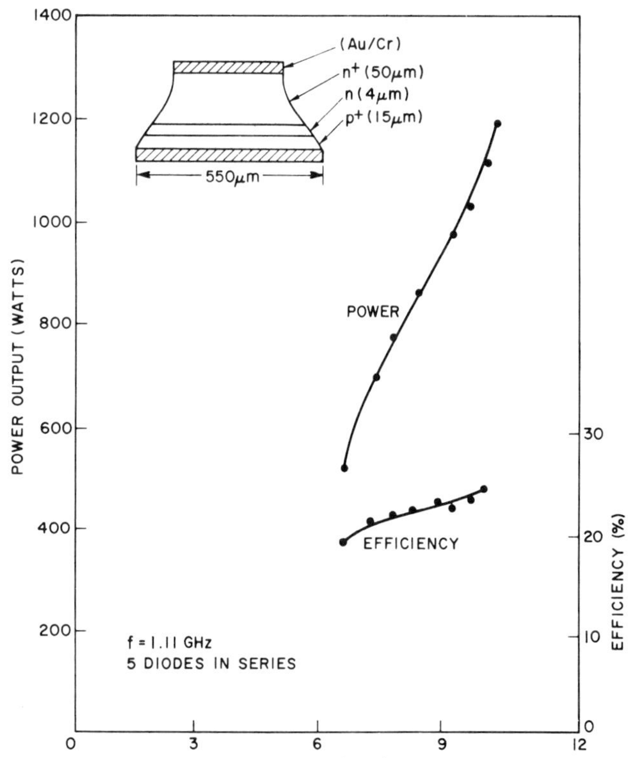

Fig. 27. Power output and efficiency versus current for a stack of 5 diodes at 1.11 GHz. (After Liu and Risko, [35].)

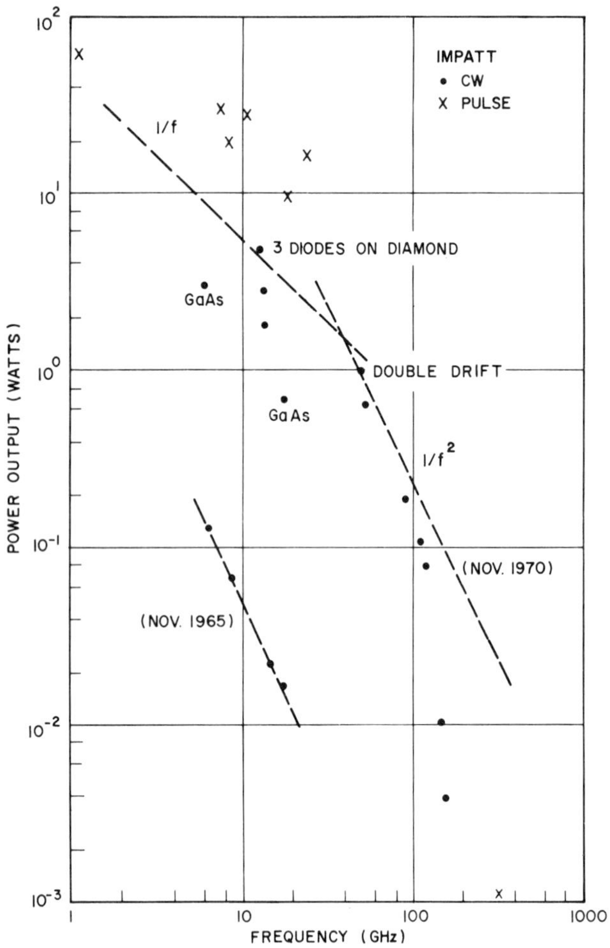

Fig. 28. State-of-the-art performance, power versus frequency, for IMPATT diodes in 1965 and 1970. $1/f$ slope is expected for thermal limitation; $1/f^2$ for electronic limitation ([19], [34], [48], [53], [60], [62], [63]).

96

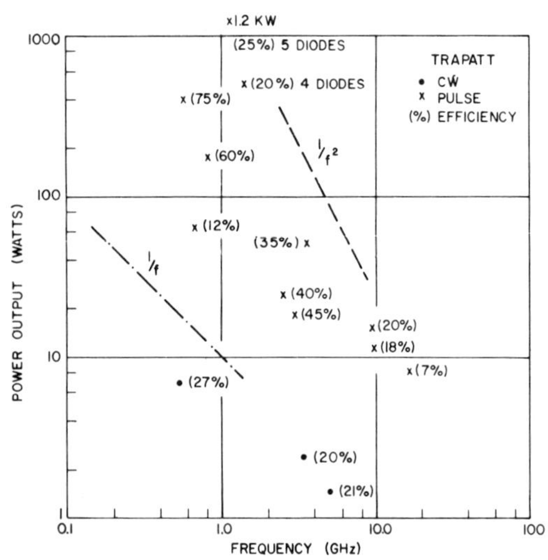

Fig. 29. State-of-the-art power versus frequency, for TRAPATT diodes ([35], [40], [66]–[75]).

C. State of the Art

A summary of the present state of the art is given in Fig. 28 for IMPATT diodes and in Fig. 29 for TRAPATT diodes. Also shown in Fig. 28 are the results obtained [19] in 1965. We note there are two to three orders of magnitude improvement in power output and over one order of magnitude in extending the upper frequency limit. The highest Pf^2 CW result was obtained at 50 GHz with a double-drift structure by Seidel *et al.* [59]. Most of the data are obtained from silicon avalanche diodes. However, the GaAs 6-GHz IMPATT diode, by Irvin *et al.* [34], exhibits to date the highest CW output power from a single chip. The highest power result under pulse conditions, as mentioned previously, was obtained by Liu *et al.* [35].

Acknowledgment

The authors would like to acknowledge the contributions of numerous colleagues at Bell Telephone Laboratories. In particular they wish to thank B. C. DeLoach, Jr., R. Edwards, W. J. Evans, J. C. Irvin, T. Misawa, D. L. Scharfetter, and T. E. Seidel for their published and unpublished work, helpful discussions, and critical reading of the manuscript.

References

[1] R. L. Johnston, B. C. De Loach, Jr., and G. B. Cohen, "A silicon diode microwave oscillator," *Bell Syst. Tech. J.*, vol. 44, Feb. 1965, pp. 369–372.
[2] W. T. Read, "A proposed high-frequency negative resistance diode," *Bell Syst. Tech. J.*, vol. 37, Mar. 1958, pp. 401–446.
[3] H. J. Prager, K. K. N. Chang, and S. Weisbrod, "High-power, high-efficiency silicon avalanche diodes at ultra high frequencies," *Proc. IEEE* (Lett.), vol. 55, Apr. 1967, pp. 586–587.
[4] B. C. De Loach, Jr., "Modes of avalanche diodes and their associated circuits," *IEEE J. Solid-State Circuits*, vol. SC-4, Dec. 1969, pp. 376–384.
[5] S. M. Sze, *Physics of Semiconductor Devices*. New York: Wiley, 1969.
[6] J. G. Ruch and W. Fawcett, "Temperature dependence of the transport properties of gats determined by the Monte Carlo method," to be published.
[7] V. L. Dalal, "Hole velocity in p-gats," *Appl. Phys. Lett.*, vol. 12, June 15, 1970.
[8] E. M. Conwell, *High Field Transport in Semiconductors*. New York: Academic Press, 1967.
[9] C. Y. Duh and J. L. Moll, "Temperature dependence of hot electron drift velocity in silicon at high electric fields," *Solid-State Electron.*, vol. 11, 1968, p. 917.
[10] G. A. Baraff, "Distribution functions and ionization rates for hot electrons in semiconductors," *Phys. Rev.*, vol. 128, 1962, p. 2507.
[11] C. R. Crowell and S. M. Sze, "Temperature dependence of avalanche multiplication in semiconductors," *Appl. Phys. Lett.*, vol. 9, 1966, p. 242.
[12] J. L. Moll, *Physics of Semiconductors*. New York: McGraw-Hill, 1964.
[13] S. M. Sze and G. Gibbons, "Avalanche breakdown voltages of abrupt and linearly graded p-n junctions on Ge, Si, GaAs and GaP," *Appl. Phys. Lett.*, vol. 8, 1966, p. 111.
S. M. Sze, M. P. Lepselter, and R. W. MacDonald, "Metal-semiconductor Impatt diode," *Solid-State Electron.*, vol. 12, 1969, p. 107.
[14] M. Gilden and M. E. Hines, "Electronic tuning effects in the Read microwave avalanche diode," *IEEE Trans. Electron Devices*, vol. ED-13, Jan. 1966, pp. 169–175.
H. K. Gummel and D. L. Scharfetter, "Avalanche region of Impatt diodes," *Bell Syst. Tech. J.*, vol. 45, Dec. 1966, pp. 1797–1828.
[15] For a detailed study of the IMPATT physics, see, for example, [5] and the following references:
J. E. Carroll, *Hot Electron Microwave Generators*. London: Edward Arnold (Publishers) Ltd., 1970, chs. 8–10.
H. A. Watson, Ed., *Microwave Semiconductor Devices and their Circuit Applications*. New York: McGraw-Hill, 1969, ch. 15.
T. Misawa, "Impatt diodes," in *Semiconductor Applications*, vol. 6, *Semiconductors and Semimetals*. New York: Academic Press, 1970.
W. J. Evans, "Avalanche diode oscillators," in *Solid State and Quantum Electronics*, W. D. Hershberger, Ed. New York: Wiley, 1971.
[16] T. Misawa, "Negative resistance in p-n junctions under avalanche breakdown conditions," *IEEE Trans. Electron Devices*, vol. ED-13, Jan. 1966, pp. 137–151.
[17] S. M. Sze and W. Shockley, "Unit cube expression for space-charge resistance," *Bell Syst. Tech. J.*, vol. 46, 1967, p. 837.
[18] D. L. Scharfetter and H. K. Gummel, "Large-signal analysis of a silicon Read diode oscillator," *IEEE Trans. Electron Devices*, vol. ED-16, Jan. 1969, pp. 64–77.
W. J. Evans and G. I. Haddad, "A large-signal analysis of IMPATT diodes," *IEEE Trans. Electron Devices*, vol. ED-15, Oct. 1968, pp. 708–717.
J. L. Blue, "Approximate large-signal analysis of Impatt oscillators," *Bell Syst. Tech. J.*, vol. 48, Feb. 1969, pp. 383–396.
[19] B. C. De Loach, Jr., "Recent advances in solid-state microwave gen-

erators," in *Advances in Microwaves*, vol. II. New York: Academic Press, 1967.

[20] C. B. Swan, T. Misawa, and L. P. Marinaccio, "Composite avalanche diode structures for increased power capability," *IEEE Trans. Electron Devices*, vol. ED-14, Sept. 1967, pp. 584–589.

[21] G. Gibbons and T. Misawa, "Temperature and current distribution in an avalanching p-n junction," *Solid-State Electron.*, vol. 11, 1968, p. 1007.

[22] R. H. Haitz, H. L. Stover, and N. J. Tolar, "A method for heat flow resistance measurements in avalanche diodes," *IEEE Trans. Electron Devices*, vol. ED-16, May 1969, pp. 438–444.

[23] D. L. Scharfetter, "Power-impedance-frequency limitations of IMPATT oscillators calculated from a scaling approximation," *IEEE Trans. Electron Devices*, vol. ED-18, Aug. 1971, pp. 536–543.

[24] T. Misawa, "Saturation current and large signal operation of a Read diode," *Solid-State Electron.*, vol. 13, 1970, p. 1363.

[25] D. R. Decker, C. N. Dunn, and H. B. Frost, "The effect of injecting contacts on avalanche diode performance," to be published.

[26] S. R. Kovel and G. Gibbons, "The effect of unswept epitaxial material on the microwave efficiency of IMPATT diodes," *Proc. IEEE* (Lett.), vol. 55, Nov. 1967, pp. 2066–2067.

[27] B. C. De Loach, Jr., "Thin skin IMPATTS," *IEEE Trans. Microwave Theory Tech.* (Corresp.), vol. MTT-18, Jan. 1970, pp. 72–74.

[28] C. Y. Chang and S. M. Sze, "Carrier transport across metal-semiconductor barriers," *Solid-State Electron.*, vol. 13, 1970, p. 727.

[29] T. Misawa, private communication.

[30] M. E. Hines, "Noise theory for the Read type avalanche diode," *IEEE Trans. Electron Devices*, vol. ED-13, Jan. 1966, p. 158.

[31] H. K. Gummel and J. L. Blue, "A small-signal theory of avalanche noise in IMPATT diodes," *IEEE Trans. Electron Devices*, vol. ED-14, Sept. 1967, pp. 569–580.

[32] J. L. Blue, "Preliminary theoretical results on low noise GaAs Impatt diodes," presented at IEEE Device Research Conf., Seattle, June 1970.
P. A. Levine, "Ultralow noise in pulsed GaAs avalanche diode oscillators," *Proc. IEEE* (Lett.), vol. 58, Mar. 1970, pp. 494–495.

[33] J. G. Josemhans, private communications.

[34] J. C. Irvin, D. J. Coleman, W. A. Johnson, I. Tatsuguchi, D. R. Decker, and C. N. Dunn, "Fabrication and noise performance of high-power GaAs Impatts," paper 8.8 presented at the 1970 IEEE Int. Electron Devices Meeting, Washington, D. C., Oct. 28.

[35] S. G. Liu and J. J. Risko, "Fabrication and performance of kilowatt *L*-band avalanche diodes," *RCA Rev.*, vol. 31, Mar. 1970, p. 3.

[36] D. F. Kostichack, "UHF avalanche diode oscillator providing 400 watts peak power and 75 percent efficiency," *Proc. IEEE* (Lett.), vol. 58, Aug. 1970, pp. 1282–1283.

[37] D. L. Scharfetter, D. J. Bartelink, H. K. Gummel, and R. L. Johnston, "Computer simulation of low-frequency high-efficiency oscillation in germanium IMPATT diodes," *IEEE Trans. Electron Devices* (Solid-State Device Res. Conf.), vol. ED-15, Sept. 1968, p. 691.

[38] R. L. Johnston, D. L. Scharfetter, and D. J. Bartelink, "High-efficiency oscillations in germanium avalanche diodes below the transit-time frequency," *Proc. IEEE* (Lett.), vol. 56, Sept. 1968, pp. 1611–1613.

[39] D. J. Bartelink and D. L. Scharfetter, "Avalanche shock fronts in p-n junctions," *Appl. Phys. Lett.*, May 15, 1969.

[40] A. S. Clorfeine, R. J. Ikola, and L. S. Napoli, "A theory for the high-efficiency mode of oscillation in avalanche diodes," *RCA Rev.*, Sept. 1969, pp. 397–421.

[41] B. C. De Loach, Jr., and D. L. Scharfetter, "Device physics of TRAPATT oscillators," *IEEE Trans. Electron Devices*, vol. ED-17, Jan. 1970, pp. 9–21.

[42] W. J. Evans, "Circuits for high-efficiency avalanche-diode oscillators," *IEEE Trans. Microwave Theory Tech.*, vol. MTT-17, Dec. 1969, pp. 1060–1067.

[43] D. L. Scharfetter, "Power-frequency characteristics of the Trapatt diode mode of high efficiency power generation in germanium and silicon avalanche diodes," *Bell Syst. Tech. J.*, vol. 49, 1970, p. 799.

[44] B. C. De Loach, Jr., and R. L. Johnston, "Avalanche transit-time microwave oscillators and amplifiers," *IEEE Trans. Electron Devices*, vol. ED-13, Jan. 1966, pp. 181–186.

[45] B. Hoefflinger, C. P. Snapp, and L. A. Stark, "High-efficiency avalanche resonance pumped amplification," *IEEE J. Solid-State Circuits*, vol. SC-4, Dec. 1969, pp. 391–395.

[46] D. L. Scharfetter, D. J. Bartelink, and B. C. De Loach, Jr., "Comments on static negative resistance in avalanching silicon p^+-i-n^+ junctions," *IEEE Trans. Electron Devices*, vol. ED-16, Nov. 1969, pp. 970–972.

[47] T. Misawa, "Microwave Si avalanche diode with nearly-abrupt-type junction," *IEEE Trans. Electron Devices*, vol. ED-14, Sept. 1967, pp. 580–584.
——, "CW millimeter-wave IMPATT diodes with nearly abrupt junctions," *Proc. IEEE* (Lett.), vol. 56, Feb. 1968, pp. 234–235.
R. A. Zettler and A. M. Cowley, "Batch fabrication of integrated-heat-sink Impatt diodes," *Electron. Lett.*, vol. 5, Dec. 1969.

[48] R. Edwards, D. F. Ciccolella, T. Misawa, D. E. Iglesias, and V. Decker, "Millimeter-wave silicon Impatt diodes," Int. Electron Devices Meeting, Oct. 29–31, 1969.

[49] D. De Nobel and H. G. Kock, "A silicon Schottky barrier avalanche transit time diode," *Proc. IEEE* (Lett.), vol. 57, Nov. 1969, pp. 2088–2089.

[50] C. K. Kim and L. D. Armstrong, "GaAs Schottky-barrier avalanche diode," *Solid-State Electron.*, vol. 13, Jan. 1970, pp. 53–56.
Y. S. Lee and C. K. Kim, "Two-watt CW GaAs Schottky-barrier Impatt diodes," *Proc. IEEE* (Lett.), vol. 58, July 1970, pp. 1153–1154.

[51] T. Misawa, "Theoretical study of microwave oscillation efficiency in improved Read diodes," to be published.

[52] ——, "Minority carrier storage and oscillation efficiency in Read diodes," *Solid-State Electron.*, vol. 13, 1970, p. 1369.

[53] D. L. Scharfetter, W. J. Evans, and R. L. Johnston, "Double-drift-region (p^+pnn^+) avalanche diode oscillators," *Proc. IEEE* (Lett.), vol. 58, July 1970, pp. 1131–1133.

[54] T. Seidel and P. L. Scharfetter, "High-power millimeter wave IMPATT oscillators with both hole and electron drift space made by ion implantation," *Proc. IEEE* (Lett.), vol. 58, July 1970, pp. 1135–1136.

[55] D. J. Coleman, J. C. Irvin, and S. M. Sze, "GaAs Schottky diodes with near ideal characteristics," to be published.

[56] Y. Y. Chang and S. M. Sze, "Temperature dependence of ionization rates in GaAs," *J. Appl. Phys.*, vol. 40, 1969, p. 5392.

[57] F. A. Blum and N. B. Kramer, "Frequency scaling of IMPATT diodes," *IEEE Trans. Electron Devices*, vol. ED-17, Nov. 1970, pp. 983–986.

[58] D. E. Iglesias, "Circuit for testing high-efficiency Impatt diodes," *Proc. IEEE* (Lett.), vol. 55, Nov. 1967, pp. 2065–2066.

[59] T. E. Seidel, R. E. Davis, and D. E. Iglesias, "Double-drift-region ion-implanted millimeter-wave IMPATT diodes," this issue, pp. 1222–1228.

[60] T. Misawa and L. P. Marinaccio, "100 GHz Si Impatt diodes for CW operation," to be published.
L. P. Marinaccio, "Composite Impatt diodes for 110 GHz operation," to be published.

[61] C. N. Dunn, "Empirical transit-time-frequency-breakdown voltage curves for Impatt diodes," to be published.

[62] S. G. Liu, J. J. Risko, and K. K. N. Chang, "High-power *K*-band silicon avalanche-diode oscillators," *Proc. IEEE* (Lett.), vol. 58, June 1970, pp. 919–920.

[63] C. B. Swan, "Improved performance of silicon avalanche oscillator mounted on diamond heat sinks," *Proc. IEEE* (Lett.), vol. 55, Sept. 1967, pp. 1617–1618.

[64] C. B. Swan, T. Misawa, and C. H. Bricker, "Continuous oscillations at millimeter wavelengths with silicon avalanche diodes," *Proc. IEEE* (Lett.), vol. 55, Oct. 1967, pp. 1747–1748.

[65] L. S. Bowman and C. A. Burrus, Jr., "Pulse-driven silicon p-n junction avalanche oscillators for the 0.9 to 20 mm band," *IEEE Trans. Electron Devices*, vol. ED-14, Aug. 1967, pp. 411–418.

[66] G. Gibbons and M. I. Grace, "High-efficiency avalanche-diode oscillators and amplifiers in *X*-band," *Proc. IEEE* (Lett.), vol. 58, Mar. 1970, pp. 512–513.

[67] R. S. Ying, R. G. Mankariour, and D. L. English, "High-efficiency anomalous mode oscillation from silicon Impatt diodes at 6 GHz," *1969 ISSCC Dig.*, pp. 86–87.

[68] W. J. Evans and D. E. Iglesias, "CW silicon TRAPATT operation," *Proc. IEEE* (Lett.), vol. 58, Feb. 1970, pp. 285–286.

[69] D. E. Iglesias and W. J. Evans, "High efficiency CW Impatt operation," *Proc. IEEE* (Lett.), vol. 56, Sept. 1968, p. 1610.

[70] W. J. Evans and R. L. Johnston, "Improved performance of CW silicon TRAPATT oscillators," *Proc. IEEE* (Lett.), vol. 58, May 1970, pp. 845–846.

[71] W. J. Evans, T. E. Seidel, and D. L. Scharfetter, "A novel TRAPATT oscillator design," *Proc. IEEE* (Lett.), vol. 58, Aug. 1970, pp. 1294–1295.

[72] R. S. Ying and N. B. Kramer, "*X*-band silicon TRAPATT diodes," *Proc. IEEE* (Lett.), vol. 58, Aug. 1970, pp. 1285–1286.

[73] R. J. Chaffin and G. P. Eer Nisse, "A poor man's TRAPATT oscillator," *Proc. IEEE* (Lett.), vol. 58, Jan. 1970, pp. 173–174.

[74] M. Y. Ta, "High-power 'subharmonic' oscillation in silicon avalanche diodes at K_u band," *Proc. IEEE* (Lett.), vol. 57, Nov. 1969, pp. 2054–2055.

[75] S. G. Liu, "Stacked high-power avalanche-diode oscillators," *Proc. IEEE* (Lett.), vol. 57, Apr. 1969, pp. 707–708.

97

GENERAL REVIEW ARTICLES

CHARACTERISTICS OF THE IMPATT MODE II

Electronic Tuning Effects in the Read Microwave Avalanche Diode

M. GILDEN, MEMBER, IEEE, AND M. E. HINES, SENIOR MEMBER, IEEE

Abstract—Read's theory of the negative-resistance avalanche diode has been examined in detail for the small-signal case. The space-charge wave approach has been used in the analysis leading directly to a simple equivalent circuit and a general expression for the small-signal impedance which includes the significant design and operating parameters. The theory indicates that strong tuning effects will occur through variation of the dc avalanche current. This has been verified experimentally.

Introduction

Microwave oscillations from avalanche diodes were predicted a number of years ago by Read [1]. Special doping gradients were recommended to achieve a highly localized avalanche zone at one side of a high-field depletion region in a semiconductor diode. Recently these predictions were realized in diodes closely approximating Read's recommended design [2] and in others using more conventional p-n junction doping gradients [3]. We have also obtained oscillations from p-n junction devices in avalanche, using a variety of different doping profiles and breakdown voltages.

In an attempt to explain the observed behavior, we undertook a more thorough look at the small-signal theory of the device and developed equations which are directly interpretable through a simple equivalent circuit. Read, in his original paper, presented a comprehensive analysis of the device with major emphasis on large-signal nonlinear behavior. He presented an equation for the small-signal admittance[1] which was too complex to be directly interpreted. In this paper, we have simplified and amplified the small-signal theory, presenting charts which illustrate the physical significance of the variation of device parameters.

Space-Charge Wave Analysis

The Model and Plan for Analysis

Figure 1 shows a schematic cross section of one class of p-n junction diode with which we have worked, illustrating the internal fields at avalanche and the drift velocity variation. In this representative case, the diode is formed from uniformly doped n-type material of approximately 0.2 ohm-cm resistivity, with a $p+$ layer diffused into one face. The $p+$ region is thin and highly doped. This provides a highly unsymmetrical p-n junction which is near the left face, as shown in Fig. 1. Under reverse bias on the order of 30–40 volts, a depletion zone forms which is approximately one micron thick. The electric field pattern is approximately triangular as shown, with a peak at the junction near one side of the depletion region. The peak field is estimated to be 5×10^5 V/cm, and is high enough through most of the depletion region to cause saturation in the drift velocity of carriers in transit. Avalanche is confined to the narrow zone where the field is near maximum. We feel that these conditions give a reasonably close approximation to the model which Read proposed, and in our analysis we use his assumptions of saturated drift velocity and a thin avalanche zone at one side of the high-field region. Outside of the depletion zone, the conductivity is high but finite, and this region acts as a simple parasitic resistance. All of the interesting behavior takes place in the depletion zone.

For analysis, we break the diode into three regions, as shown in Fig. 2. These are 1) the *avalanche zone* which is assumed to be thin so that space-charge and signal delay can be neglected; 2) a *drift zone* where no carriers are formed but where space-charge and transit-time effects are significant; and 3) an *inactive zone* which adds undesirable parasitic resistance. For the active zones, using small-signal approximations, we derive expressions for the ac voltages in terms of the *ac circuit current* I and thereby obtain the impedances. The sum of the impedance of the active zones plus that of the passive zone is the total impedance we seek.

The two active zones interact with one another. We treat this by noting that the ac electric field is continuous across the boundary between them and by defining M as the complex fraction relating the avalanche current i_a to the total or circuit current I

$$i_a = MI. \tag{1}$$

The avalanche current is the ac conduction current in the avalanche zone. With our assumption of a

Manuscript received September 13, 1965. The work reported in this paper was supported by the Electronics Technology Division of the Air Force Avionics Laboratory, Wright-Patterson Air Force Base, Dayton, Ohio.

The authors are with Microwave Associates, Inc., Burlington, Mass.

[1] See footnote following our (18).

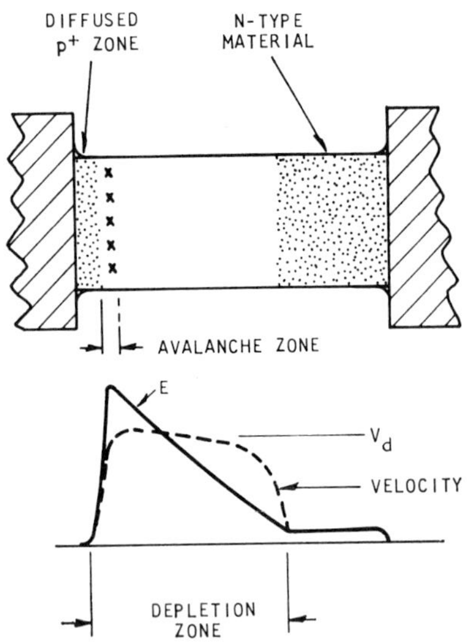

Fig. 1. Typical p-n junction in reverse bias (conditions at avalanche).

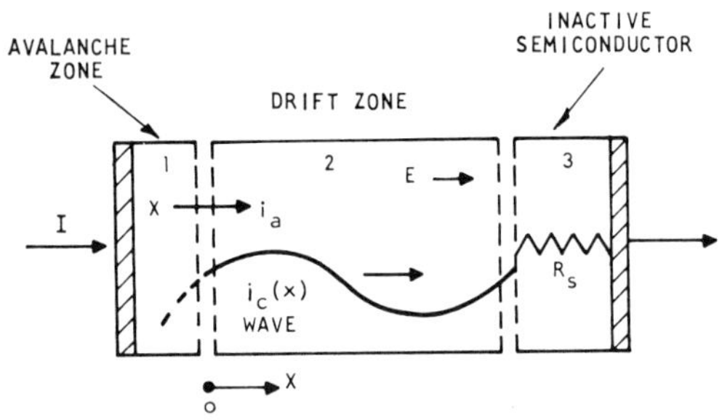

Fig. 2. Model of avalanche diode.

thin avalanche zone, i_a is presumed to enter the drift zone without delay. With the assumption of a saturated drift velocity v_d, the ac conduction current in the drift region propagates as an unattenuated wave at this velocity,

$$\begin{aligned} i_c(x) &= i_a e^{-j\omega x/v_d} \\ &= MIe^{-j\omega x/v_d}. \end{aligned} \tag{2}$$

At any cross section, the total ac circuit current I is equal to the sum of the conduction current i_c and the displacement current i_d, and this sum is a constant, independent of position x.

$$I = i_c(x) + i_d(x) \neq f(x). \tag{3}$$

The displacement current is related to the ac field $\mathcal{E}(x)$ by

$$i_d = j\omega\epsilon A\mathcal{E}(x), \tag{4}$$

where A is the cross-sectional area. These equations combined yield an expression for the ac electric field in the drift zone as a function of x and the total ac current,

$$\mathcal{E}(x) = I\frac{1 - Me^{-j\omega x/v_d}}{j\omega\epsilon A}. \tag{5}$$

Integration of $\mathcal{E}(x)$ gives the voltage across the drift zone in terms of I. The coefficient M is derived in the avalanche analysis below.

In the following sections, we first analyze the avalanche zone, obtaining an expression for its impedance and the fraction M. Then we determine the drift zone impedance and combine these with the passive resistance to obtain the total impedance expression.

The Avalanche Zone

Read showed that the avalanche current can be approximated from the equation

$$\frac{di}{dt} = \frac{2i}{\tau_a}\left[\int_0^{l_a} \alpha\, dx - 1\right] \tag{6}$$

where τ_a is the transit time in the avalanche zone, l_a is its length, i is the total avalanche current (dc and ac) and α is the ionization coefficient. This coefficient, the number of ionizations per cm produced by a single carrier, is a sharply increasing function of the electric field. There is recent work on this dependence on electric field [4]–[7]. A rigorous treatment of the ac properties of the avalanche zone is quite complex, and Read has presented an approximate analysis and discussion of it.

We will simplify (6) by substituting $\bar{\alpha}$ in place of α, where $\bar{\alpha}$ is an average value of α obtained by evaluating the integral over the extent of the avalanche zone. This average is presumed to retain its variability with respect to E. We obtain

$$\frac{di}{dt} = \frac{2i}{\tau_a}(\bar{\alpha}l_a - 1). \tag{7}$$

For steady-state conditions where the time derivative of current is zero, $\bar{\alpha}_a l_a$ is exactly one. The above is a valid large-signal equation.

We now make the customary small-signal assumptions that the current, the E field, and $\bar{\alpha}$ consist of constant dc values plus small ac variations:

$$\begin{aligned} \bar{\alpha} &= \bar{\alpha}_a + \bar{\alpha}_a'\mathcal{E}_a \cdots \\ \bar{\alpha}l_a &= 1 + l_a\bar{\alpha}_a'\mathcal{E}_a \\ I_a &= I_0 + i_a \\ E_a &= E_0 + \mathcal{E}_a, \end{aligned} \tag{8}$$

where $\bar{\alpha}_a'$ is the derivative of $\bar{\alpha}$ with respect to E, E_0 is dc field for steady state, and $\mathcal{E}_a$ and i_a are small-signal quantities. Substitution of the above expressions into (7), neglecting products of small (ac) terms, leads to the expression for the ac component of the avalanche conduction current

$$i_a = \frac{2\bar{\alpha}_a' l_a I_0 \mathcal{E}_a}{j\omega\tau_a}. \tag{9}$$

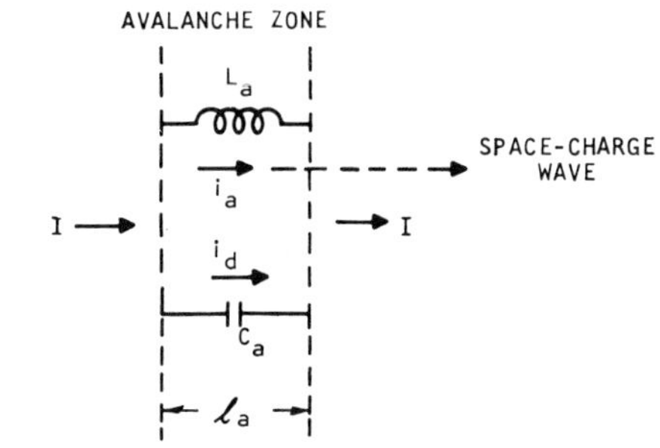

Fig. 3. Equivalent circuit for avalanche zone.

We must also consider the displacement current in the avalanche zone,

$$i_{da} = j\omega\epsilon\mathcal{E}_a A. \tag{10}$$

These are the two components of the total circuit current in the avalanche zone. For a given field, the avalanche current is reactive and varies inversely with ω as in an inductor. The other is also reactive and varies directly with ω as in a capacitor. Thus, the avalanche zone behaves as a shunt combination of an inductor and a capacitor in a parallel resonant circuit.

The equivalent circuit of the avalanche zone is shown in Fig. 3. The equivalent inductance and capacitance are

$$L_a = \frac{\tau_a}{2\bar{\alpha}' I_0} \tag{11}$$

$$C_a = \frac{\epsilon A}{l_a}. \tag{12}$$

The resonant frequency of this combination is an important parameter, significant in itself. This we define as the *avalanche frequency* ω_a,

$$\omega_a^2 = \frac{2\bar{\alpha}_a' l_a I_0}{\epsilon A \tau_a} = \frac{2\bar{\alpha}_a' v_d I_0}{\epsilon A}. \tag{13}$$

The second expression in (13) is interesting in that it does not include the length of the avalanche zone l_a, which is somewhat uncertain. The impedance of the avalanche zone has the simple form

$$Z_a = \frac{l_a}{j\omega\epsilon A}\left[\frac{1}{1 - \frac{\omega_a^2}{\omega^2}}\right] = \frac{1}{j\omega C_a}\left[\frac{1}{1 - \frac{\omega_a^2}{\omega^2}}\right]. \tag{14}$$

The factor M has the simple form

$$M = \frac{i_a}{I} = \frac{1}{1 - \frac{\omega^2}{\omega_a^2}}. \tag{15}$$

A thin avalanche, therefore, behaves as an antiresonant combination with a true resonant frequency which depends upon the fixed geometry, materials, and doping gradients, and upon the *variable* dc current I_0. It is an electronically tunable resonator whose frequency is proportional to the square root of the current.

The Drift Zone

We next substitute our expression for M in (15) into (5) and integrate over the length of the drift zone l_d, obtaining an expression for the voltage across this zone,

$$V_d = \frac{l_d I}{j\omega\epsilon A}\left[1 - \frac{1}{1 - \frac{\omega^2}{\omega_a^2}}\left(\frac{1 - e^{-j\theta}}{j\theta}\right)\right] \tag{16}$$

where θ is the transit angle:

$$\theta = \frac{\omega l_d}{v_d} = \omega\tau_d. \tag{17}$$

We also may identify $\epsilon A/l_d$ with the capacitance of the drift zone and expand (16) obtaining an expression for the impedance[2]

$$Z_d = \frac{1}{j\omega C_d}\left[1 - \frac{1}{1 - \frac{\omega^2}{\omega_a^2}}\frac{\sin\theta}{\theta}\right] + \frac{1}{\omega C_d}\left[\frac{1}{1 - \frac{\omega^2}{\omega_a^2}}\frac{1 - \cos\theta}{\theta}\right]. \tag{18}$$

It is evident that the resistance will be negative for all frequencies above ω_a except for nulls at $\theta = 2n\pi$. The resistance is positive for frequencies below ω_a and approaches a finite value at zero frequency. The low-frequency small-signal resistance is a consequence of space-charge in the finite thickness of the drift zone, an effect usually ignored in the analysis of avalanche diodes.[3]

[2] This equation is the equivalent of Read's (53) which contained errors, apparently typographical. The correct form for the small-signal admittance, using Read's terminology and symbols, is

$$Y(\omega_n) = \frac{2\omega_n y[(1 - \cos\omega_n) + i(2y - \sin\omega_n)]}{4y^2 - 4y\sin\omega_n + 2(1 - \cos\omega_n)}$$

where y has the form

$$y = \frac{\omega_n}{2}\left[1 - \frac{\omega^2}{\omega_a^2}\right].$$

Read used a normalizing scheme such that $\omega_n = \theta$.

[3] This effect is very helpful in obtaining a degree of uniformity in the avalanche current density across the zone. Local regions of excess current will find an effective local resistance in series which limits the local current rise. As the total current increases to high density, eventually all of the avalanche zone can be put into effective use. Thus, at sufficiently high current, the deleterious effects of microplasmas need not be observed.

The Total Impedance

We now have all the information necessary to write an expression for the total impedance, which we take as the sum of the impedances of the avalanche zone (14), the drift zone (18), and the passive resistance of the inactive zone R_S,

$$Z = R_s + \frac{l_d^2}{v_d \epsilon A}\left[\frac{1}{1-\dfrac{\omega^2}{\omega_a^2}}\right]\frac{1-\cos\theta}{\dfrac{\theta^2}{2}}$$

$$+ \frac{1}{j\omega C_d}\left[\left(1-\frac{\sin\theta}{\theta}\right)+\frac{\dfrac{\sin\theta}{\theta}+\dfrac{l_a}{l_d}}{1-\dfrac{\omega_a^2}{\omega^2}}\right]. \tag{19}$$

Equation (19) has been cast in a form which can be simplified directly for the case of small transit angle. We note that $(1 - \cos\theta)2/\theta^2$ and $(\sin\theta)/\theta$ are approximately one for values of θ less than $\pi/4$, yielding a simple equation

$$Z \cong R_s + \frac{l_d^2}{v_d \epsilon A}\cdot\frac{1}{1-\dfrac{\omega^2}{\omega_a^2}} + \frac{1}{j\omega C}\,\frac{1}{1-\dfrac{\omega_a^2}{\omega^2}} \tag{20}$$

where $C = \epsilon A/(l_a + l_d)$, the total depletion zone capacitance, when the diode is biased just below the avalanche voltage so that $I_0 = 0$, but the depletion zone has its full thickness. The second term is the active resistance which becomes negative for $\omega > \omega_a$. The third term is reactive and corresponds to a parallel resonant circuit which includes the diode capacitance and a shunt inductor. This has an equivalent circuit as shown in Fig. 4. It is interesting to note that strong negative-resistance effects occur for transit angles substantially less than π.

It will be instructive to examine the frequency dependence of the total reactance and resistance. For small values of θ, the characteristics are sketched in Fig. 5. The reactance (solid curve) increases with frequency and becomes strongly inductive just below the avalanche frequency. Above the avalanche frequency, the diode becomes capacitive. Well above the avalanche frequency, the reactance finally merges with the reactance that the diode would have in the absence of the avalanche phenomena. The resistance (dashed curve) below the avalanche frequency is positive. Just above the avalanche frequency, the resistance becomes strongly negative.

Conditions for oscillation are illustrated graphically for the case when a simple inductance is added to complete the circuit.[4] The negative of this reactance value is plotted so that the intersection with the avalanche diode reactance curve defines the frequency at which the reactance vanishes, which is approximately the resonant frequency. A vertical projection to the resistance curve shows the value of negative resistance available at

[4] A Nyquist analysis for the conditions and frequencies of oscillation is given by Holmstrom [8].

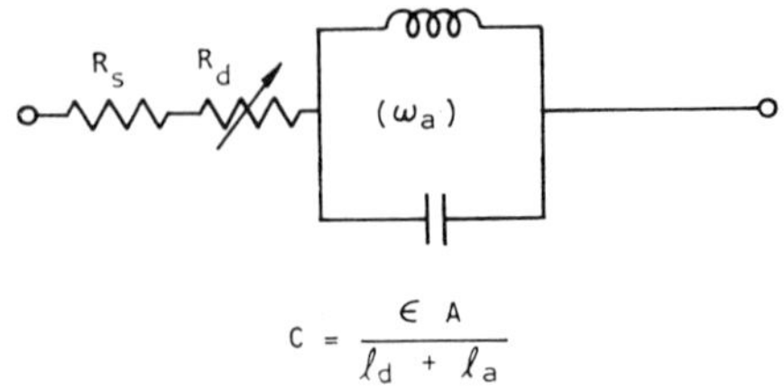

Fig. 4. Equivalent circuit for avalanche diode at small transit angle.

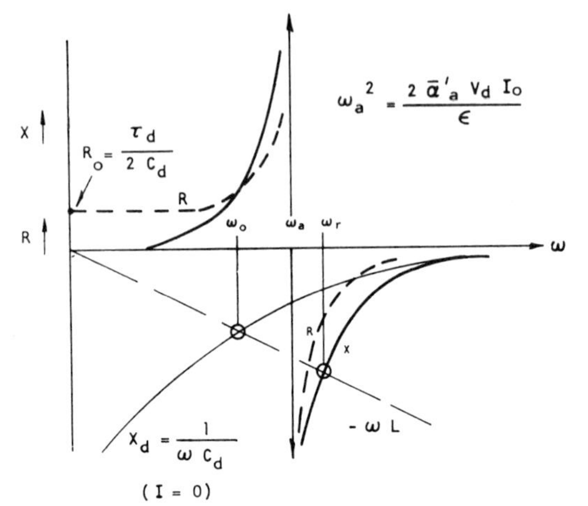

Fig. 5. Frequency dependence of the real and imaginary parts of the diode impedance.

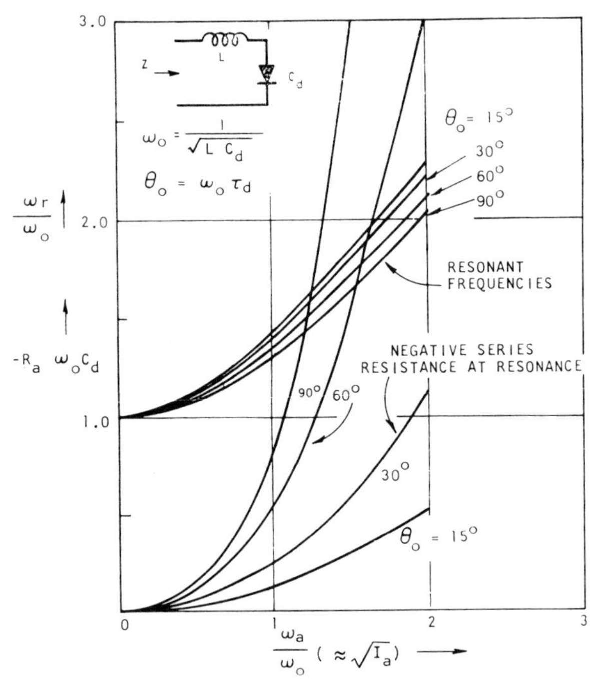

Fig. 6. Electronic tuning characteristics of the avalanche diode oscillator.

104

that frequency. As the value of the series inductance is increased, lower frequencies and larger magnitudes of negative resistance are obtained. The intersection with the curve labeled X_d gives the resonant frequency in the absence of avalanche. Perhaps the most important characteristics are that the frequency of operation is always above ω_a and the frequency is readily varied by controlling direct current which is proportional to ω_a^2. Thus, rapid electronic tuning by means of the diode current is obtainable.

Electronic tuning characteristics for the case of an external series inductor have been calculated and are presented in Fig. 6. It was assumed that the frequency of oscillation would be that for which the total series reactance vanishes. This frequency increases with ω_a/ω_0 and finally becomes a linear function of it for large values. The frequencies are normalized with respect to the ω_0, the resonant frequency of the circuit in the absence of avalanche effects. The range of electonic tuning appears to be limited only by the range of current possible. The normalized negative resistance at resonance increases with the avalanche frequency. It also is proportional to the square of θ_0 where $\theta_0 = \omega_0 \tau_d$. Thus, longer drift regions should result in oscillations starting at lower threshold current.

Experimental Work

The experimental work was carried out with silicon *p-n* diodes in a fundamental mode cavity. The diode breakdown voltages ranged from 20 to 80 volts and the diameters of the junctions were between 1 and 6 mils. The diodes were mounted in a small package, a threaded "pill" type nominally 0.040 inches high and 0.080 inches in diameter, which provided a good heat sink. The details of the fundamental mode cavity are sketched in Fig. 7. The cavity radiates into a waveguide which is connected to the cavity flange. The diameter of the cylindrical cavity is 0.4 inches, and the length was varied from 0–0.100 inches. With this size cavity and a reasonably well-matched output waveguide, the circuit outside of the diode may be approximated as an inductance and resistance in series.

Measurements were made on threshold conditions and electronic tuning. Experimental threshold characteristics for an oscillator using a 40 volt diode are shown in Fig. 8. The transit angles are of the order of one radian. The threshold current and nominal frequency of operation are shown as a function of the height of the cavity h (see Fig. 7); increasing the height of the cavity increases the inductance. The results show that increasing h lowers both the threshold current and the frequency. This is consistent with the theory as discussed in connection with Fig. 5 for small-transit angles. One can readily see this by considering a family of construction lines for various values of load inductance and a constant load resistance. Experimental electronic tuning where frequency is plotted as a function of the square root of current is shown in Fig. 9. The avalanche frequency ω_a is proportional to the square root of current. An estimated value of ω_a appears on the graph based upon the resonance condition

$$\frac{\omega_r^2}{\omega_0^2} = \frac{\omega_a^2}{\omega_0^2} + 1 \tag{21}$$

which can be derived for the case of simple inductive loading and small values of θ. The values of ω_0 and ω_r are taken from the data. Two regions of operation exist, one of passive resonance and the other of oscillation. As the avalanche frequency increases, the passive resonant frequency increases. Then, the threshold for oscillations is reached and it is found that further increases in the avalanche frequency results in further increases in the frequency of oscillation. However, the oscillation frequency does not increase at the rate predicted by theory (the theoretical curve was fitted to the experimental results at the threshold point). There is a second tuning curve for the oscillation region only, for a reduced height cavity. The smaller inductance results in a higher threshold current and a higher range of frequency, again as expected from the theory. An electronic tuning range of 400 Mc/s was achieved here. In other experiments, a ten percent range has been obtained at X band. The departure from theory is most likely due to nonlinear effects which have not been included in the analysis presented in this paper.

Certain measurements were obtained from the avalanche diode oscillator using an output display, Fig. 10, in which the diode bias current is swept at a 60 cycle rate. The horizontal axis is current and the vertical axis in the upper section is detected RF voltage and in the lower section the diode voltage. As the diode is swung into reverse bias, there is an abrupt increase in current indicating the onset of avalanche current. At a value of about 35 milliamperes, the oscillations begin and the oscillation amplitude continues to increase with increasing current. If one uses a wave meter, the frequency is found to increase monotonically with current.

In our experimental work to date, the frequency of oscillation has been in the range from 7 to 20 Gc/s. The higher frequencies, 14 to 20 Gc/s, usually are accompanied by lower frequency outputs. Under pulsed conditions (several microseconds) power levels of approximately 40 mW have been obtained. Under 60 c/s sweeping power levels, up to several milliwatts have been obtained. Under CW condi-

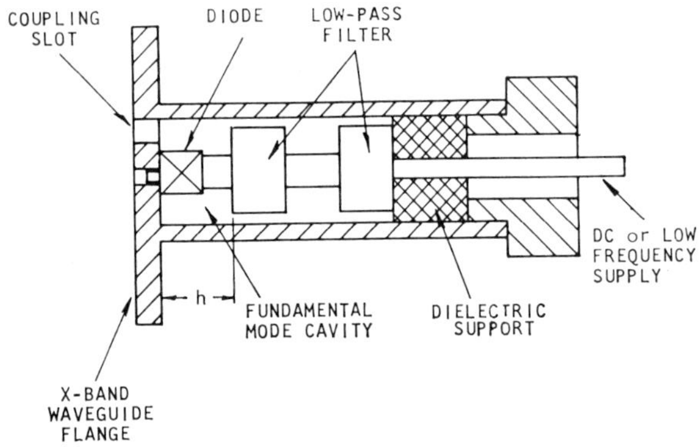

Fig. 7. Fundamental mode cavity and diode mount.

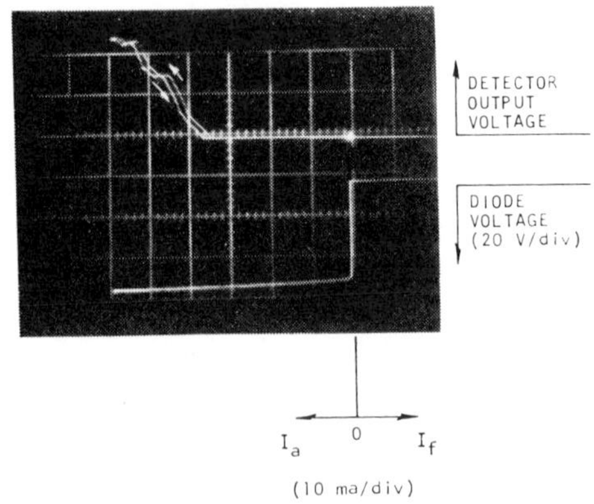

Fig. 10. Typical output display with 60 c/s sweep.

106

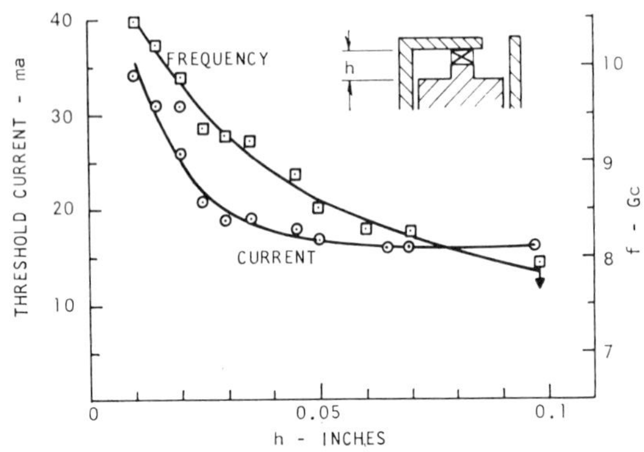

Fig. 8. Experimental threshold conditions.

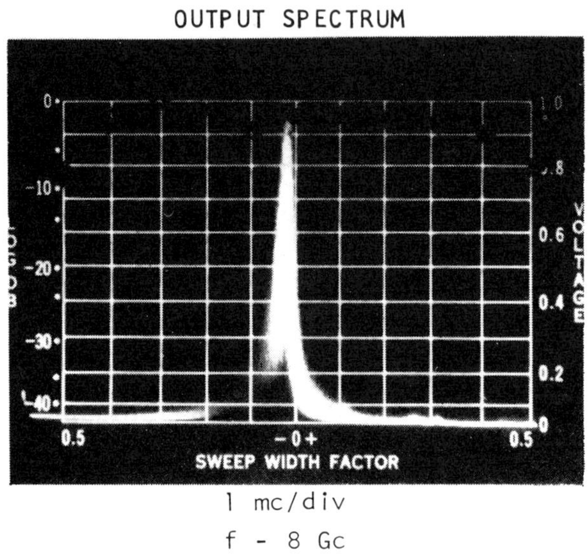

Fig. 11. CW output spectrum.

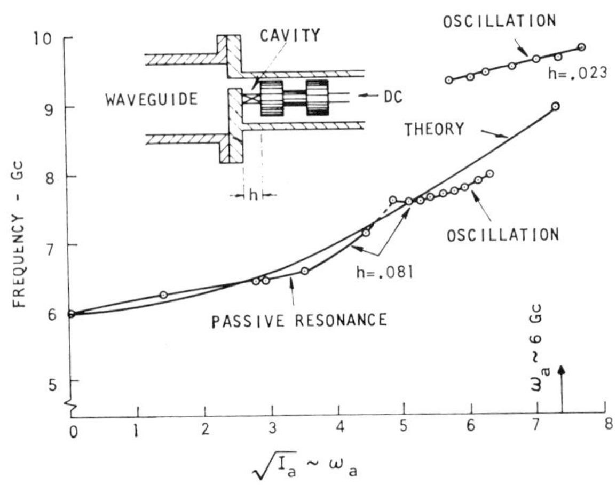

Fig. 9. Experimental electronic tuning.

tions, power levels up to 0.15 mW have been obtained. The diodes giving the above results were approximately 3 mils in diameter; with larger areas and improved heat removal proportionately higher power levels would be obtained. The efficiencies obtained were in the range of 0.1 to 1.0 percent. The decreasing values of power with increasing duty cycle are imposed by thermal considerations. An example of an output spectrum is shown in Fig. 11 for CW operation at 8 Gc/s. The width is approximately 0.5 megacycles. Narrower line widths should be obtainable through interaction with higher Q cavities [9].

Conclusions

A detailed small-signal theory of the avalanche diode oscillator has been given. The central feature of the theory is the concept of an *avalanche frequency* and the parallel-resonant character of the equivalent circuit which is resonant at that frequency. The theory predicts the experimental observations such as wide-band electronic tuning, a useful feature, and it also predicts increasing amplitude with current and lowering of the threshold current with increasing inductance. CW operation was achieved and frequencies up to 20 Gc/s were observed.

Acknowledgment

We would like to acknowledge the helpful discussions with H. Chalifour and W. Moroney who designed and fabricated the avalanche diodes.

REFERENCES

[1] W. T. Read, Jr., "A proposed high-frequency, negative-resistance diode," *Bell Sys. Tech. J.*, vol. 37, pp. 401–446, March 1958.

[2] C. A. Lee, R. L. Batdorf, W. Wiegmann, and G. Kaminsky, "The Read diode—an avalanching, transit-time, negative-resistance oscillator," *Appl. Phys. Letters*, vol. 6, pp. 89–91, March 1, 1965.

[3] R. L. Johnston, B. C. De Loach, Jr., and B. G. Cohen, "A silicon diode microwave oscillator," *Bell Sys. Tech. J.*, *Briefs*, pp. 369–372, December 1964.

[4] G. A. Baraff, "Distribution functions and ionization rates for hot electrons in semiconductors," *Phys. Rev.*, vol. 128, pp. 2507–2517, December 1962.

[5] R. A. Logan, A. G. Chynoweth, and B. G. Cohen, "Avalanche breakdown in gallium arsenide *p-n* junctions," *Phys. Rev.*, vol. 128, pp. 2518–2523, December 1962.

[6] J. L. Moll, and R. Van Overstraeten, "Charge multiplication in silicon *p-n* junctions," *Solid-State Electronics*, vol. 6, pp. 147–157, March-April 1963.

[7] C. A. Lee, R. A. Logan, R. L. Batdorf, J. J. Kleimack, and W. Wiegmann, "Ionization rates of holes and electrons in silicon," *Phys. Rev.*, vol. 134, pp. A761–A773, May 1964.

[8] R. Holmstrom, "Small-signal impedance and oscillation conditions for the Read diode," *this issue*, page 205.

[9] W. A. Edson, "Noise in oscillators," *Proc. IRE*, vol. 48, pp. 1454–1466, August 1960.

Negative Resistance in *p-n* Junctions Under Avalanche Breakdown Conditions, Part I

T. MISAWA

***Abstract*—A one-dimensional, small-signal analysis of the space-charge region of a p-n junction in which avalanche occurs uniformly is presented. The impedance is found to have a negative real part. The impedance is well represented by a parallel connection of the depletion layer capacitance, an inductance, and a negative resistance. The admittance of the latter two is proportional to the bias current. The magnitude of the negative Q is below ten.**

The negative resistance is due to an intrinsic instability in the avalanching electron-hole plasma. A discussion of the instability and a traveling-wave tube-like amplification is given.

108

Introduction

It WILL BE SHOWN that the space-charge region of a *p-n* junction in which avalanche occurs uniformly[1] throughout the space-charge region shows a negative resistance which is due to an intrinsic instability of the avalanching electron-hole plasma.

In the following we discuss first how the dynamics of avalanche multiplication gives rise to the instability and the negative resistance. Then the space-charge region of a *p-n* junction with uniform avalanche all over is analyzed by using simplifying assumptions. The negative resistance will be contrasted with that proposed by Read [1].

In this paper, Part I, many simplifying assumptions are made in order to bring out the essential principles. A more realistic and therefore more complex analysis is given in a later paper, Part II, this issue.

Instability in Avalanching Electron-Hole Plasma and Negative Resistance

When the electric field is high enough, electrons and holes in semiconductors drift at saturated velocities (scattering limited velocities) which are independent of field [2]. Suppose we have an electron density perturbation which varies sinusoidally in space in a field high enough to produce scattering limited velocities. The density perturbation produces the charge and hence, the field perturbation. Since electrons drift at the scattering limited velocity regardless of field, the perturbation does not spread out but propagates without attenuation in the direction of electron drift. The situation is illustrated in Fig. 1. The field wave lags the electron density wave by 90°. We see that the "stiffened" electron-hole plasma can convey a non-attenuating space-charge wave.

Now it will be described how the above space-charge wave grows in the presence of avalanche multiplication. The generation rate of electron-hole pairs is larger both when the electric field is stronger and when there are more carriers [3]. Therefore, in the above case of an electron density perturbation the generation rate peaks somewhere between the place where the field is strongest and the place where the density is largest. This means that the generation rate leads the electron density wave by less than 90°, as shown in Fig. 1. Note that, since in the case of Fig. 1 the dc field is in the negative x direction, the field becomes strongest at its negative peak. This increased generation rate gives rise to an excess electron density which lags the rate by 90°. In the positive half cycle of the rate the excess electron density keeps increasing, and in the negative half cycle the excess density decreases. Now the resultant total electron density gives a current which lags the field by more than 90°. As for holes, although individual holes created by the avalanche drift in the direction opposite to electrons, the hole density wave is dragged by the electron density wave. The hole density wave also lags the field by more than 90°. The situation is shown in Fig. 1. Thus, the ohmic losses due to both electron and hole currents produced by avalanche are negative. This means that the above space-charge wave keeps growing. This is an instability.

One of the easiest ways to produce an avalanching electron-hole plasma is to break down a *p-n* junction. Since we have boundaries in this case, the situation is a little bit different. However, we expect a negative resistance due to the above mentioned instability of the plasma.

The negative-resistance property, or the instability of this confined plasma can be discussed in the following way. The model to be discussed is shown in Fig. 2. Electrons enter from the left side and holes from the right side. We consider the voltage transient after an impulsive current is applied to the avalanching *p-n* junction. First the applied impulsive current produces charge

Manuscript received September 15, 1965.
The author is with Bell Telephone Laboratories, Inc., Murray Hill, N. J.

[1] The uniformity refers to the direction of the current flow.

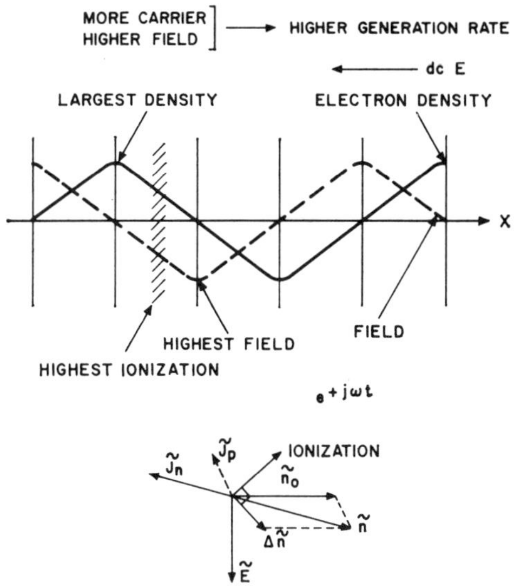

Fig. 1. Instability in the avalanching electron-hole plasma.

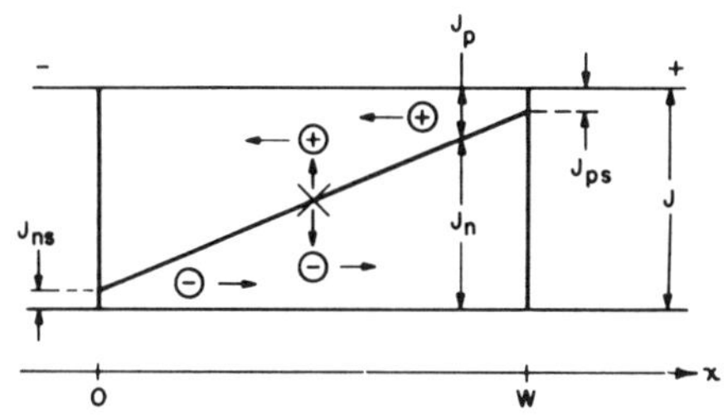

Fig. 2. Space-charge region of *pn* junction under avalanche breakdown conditions.

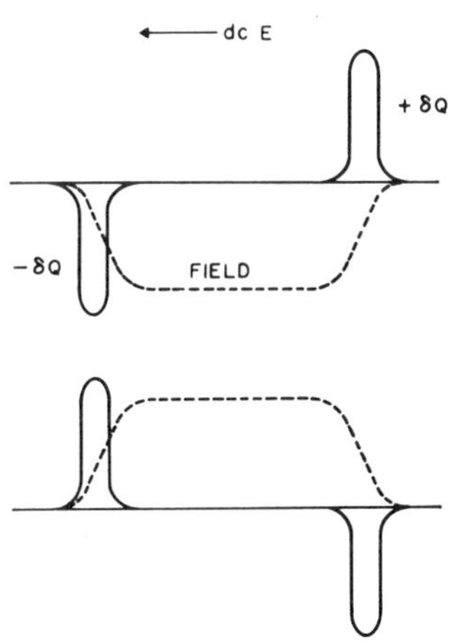

Fig. 3. Two phases of transient after an impulsive current is applied.

spikes at both edges of the space-charge region. This corresponds to a momentary widening of the space-charge region. Since the saturation current of the *p-n* junction is very small, negligible numbers of electrons and holes are emitted. Therefore, almost all the charge in the spikes is not the charge of mobile carriers but the immobile charge of those impurities which were neutralized by carriers. The charge spikes have to wait to be neutralized by carriers generated by avalanche. These charge spikes raise the field in the space-charge region by a constant amount (Fig. 3). This increased field causes more avalanche and generated electrons and holes neutralize the charge spikes. However, the situation is such that we still have a supply of electrons and holes even after the spikes are neutralized. Subsequently, we have too many holes at the left end and too many electrons at the right end. This is a momentary narrowing of the space-charge region. Now these extra charges decrease the field in the space-charge region and we have a situation just opposite to what has just been considered. Since we now have a negative, or decreased, supply, these spikes die away and again the "pendulum" swings beyond the steady state. The situation is repeated over and over again. We again have an instability.

Analysis

Fundamental Equations

The following equations govern the dynamics of the avalanching electron-hole plasma.

Poisson's equation,

$$\frac{\partial E}{\partial x} = \frac{q}{\varepsilon}(N_D - N_A + p - n). \tag{1}$$

Continuity equations for electrons and holes,

$$\frac{\partial n}{\partial t} = \frac{1}{q}\frac{\partial J_n}{\partial x} + g, \tag{2}$$

$$\frac{\partial p}{\partial t} = -\frac{1}{q}\frac{\partial J_p}{\partial x} + g, \tag{3}$$

where the generation rate [3] g is

$$g = \alpha\,|v_n|\,n + \beta\,|v_p|\,p. \tag{4}$$

We neglect diffusion current; then

$$J_n = -qv_n n, \tag{5}$$

$$J_p = qv_p p. \tag{6}$$

Here E is the electric field, q the absolute value of the electronic charge, ε the dielectric constant, N_D and N_A donor and acceptor densities, n and p electron and hole densities, J_n and J_p electron and hole current densities, α and β the ionization rates of electrons and holes, and v_n and v_p the velocities of electrons and holes.

Plane Wave Case

We consider a very simple case in which both electrons and holes have the same ionization rate and the same drift velocity, and the field is so high that the drift velocity v is independent of the field. We divide the solutions of (1) through (6) into the time-dependent part and the constant part. The time-dependent part is assumed to vary like $e^{i\omega t}$ and to be very small compared to the dc part.

When we assume that the dc electric field is independent of x so that the avalanche occurs uniformly all over, the ac solution of the form $e^{j(\omega t - kx)}$ is possible. This is a plane wave with the wave number or the propagation constant k. The dispersion relation for the plane wave is found to be given by

$$k^2 + 2\alpha' J - j2\alpha\omega - \omega^2 = 0, \tag{7}$$

where $\alpha' = d\alpha/dE$ and J is the total dc current, i.e., the sum of electron and hole currents. Although electron and hole currents increase downstream, the sum J is constant. Note that since α is an increasing function of $|E|$, $\alpha' J$ is always positive regardless of the sign of J. Details of the derivation of (7) are shown in the Appendix. We used the various units given in Table I. From (7)

$$j\omega = \alpha \pm \sqrt{\alpha^2 - 2\alpha' J - k^2}. \tag{8}$$

With real k we have a spatially sinusoidally varying perturbation. Equation (8) shows that the perturbation grows exponentially with time. This means that the perturbation is unstable. When the quantity in the square root sign of (8) is negative, the perturbation oscillates with time and the time constant for the exponential growth is $1/\alpha$.

TABLE I

Unit	Expression	Numerical Value	Meaning
Length	w	5μ	width of the region
Time	$\tau = w/v$*	5.88×10^{-11} sec	transit time
Frequency	$\frac{1}{2}\pi\tau$	2.71×10^9 c/s	
Density	N_0	10^{15} cm^{-3}	a reference density
Current	qvN_0	1.36×10^3 A/cm^2	
Field	wqN_0/ϵ	7.54×10^4 volt/cm	
Impedance	$w^2 v/\epsilon$	$2.77 \times 10^{-2}\Omega$ cm^2	impedance of the capacitance with width w at unit frequency
Charge	q	1.6×10^{19} coulomb	

* v is scattering limited velocity, which is assumed as 8.5×10^6 cm/sec.

When (7) is solved for k,

$$k = \pm\sqrt{\omega^2 - 2\alpha' J + j2\omega\alpha}. \tag{9}$$

We see that, since the imaginary part of the quantity in the square root sign in (9) is positive, one of the k's is in the first quadrant and the other in the third quadrant for positive ω. When k is in the first quadrant, the wave propagates in the positive x direction because the real part of k is positive and its amplitude increases toward the positive x axis because the imaginary part is positive. When k is in the third quadrant the wave propagates and grows in the negative x direction. This means that, if it is possible to excite this growing wave by an input probe, an output probe, which is placed in a position distant from the input probe, picks up the amplified wave. This is a traveling-wave tube-type amplification. We are not going to investigate this possibility however, but consider instead the space-charge region of a p-n junction, in which avalanche occurs uniformly.

Space-Charge Region

We assume that the p side region is on the negative side of the x axis as shown in Fig. 2. Electrons enter the space-charge region from the left end and holes from the right end. We designate the currents associated with these primary electrons and holes by J_{ns} and J_{ps}. The Appendix shows that ac field and carrier currents are composed of three terms, i.e., the terms proportional to $\tilde{J}_{ns}$, $\tilde{J}_{ps}$, and $\tilde{J}$. Here $\sim$ sign means the ac part. The ac total current $\tilde{J}$ now contains both carrier currents and displacement current. For example, the ac field is given by

$$E = \rho_1 \tilde{J}_{ns} + \rho_2 \tilde{J}_{ps} + \rho \tilde{J}, \tag{10}$$

where the proportionality constants ρ_1, ρ_2, and ρ are determined by the structure parameters and bias conditions and are dependent on the position x. An important quantity is the voltage $\tilde{V}_a$ across the diode, which is obtained by integrating (10):

$$\tilde{V}_a = Z_t \tilde{J}_{ns} + Z_t \tilde{J}_{ps} + Z\tilde{J}. \tag{11}$$

Since we assumed the same property for both electrons and holes, the integrated coefficients of $\tilde{J}_{ns}$ and $\tilde{J}_{ps}$ are the same.

$\tilde{J}_{ns}$ and $\tilde{J}_{ps}$ are due to thermally generated carriers and depend on the voltages $\tilde{V}_a$. However, the relevant admittances are on the order of the admittance of the reverse biased p-n junction and their product with Z_t is much less than unity. Therefore, the first two terms in (11) are negligible. The impedance of the diode is given by Z.

We consider a specific numerical example. The width of the region is 5 μ. The dielectric constant is 12 as in Si. The ionization rate is taken from Fig. 3 of [1], which is for silicon. Figure 4 shows the impedance as a function of frequency and bias current. Normalization is in the units given in Table I. The three sets of curves are shown there which correspond to three different bias currents, 1, 0.1, and 0.01. The real part shown is all negative. Its magnitude increases rapidly from low frequency, has a maximum value at the resonance frequency, and decreases. The reactance is inductive at lower frequencies and changes to capacitive after the resonance. The resonance frequency increases approximately proportionally to the square root of the bias current.

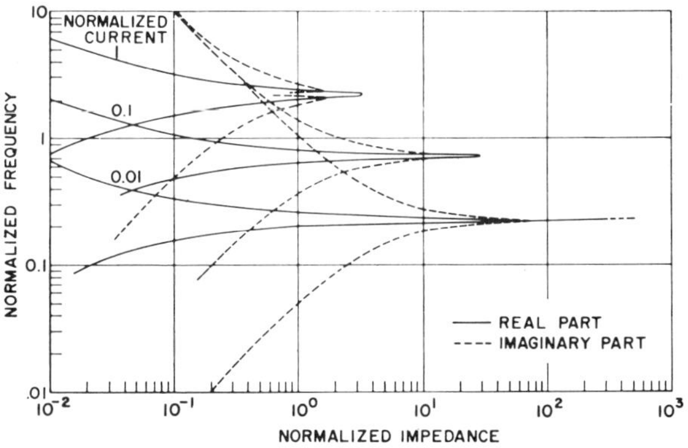

Fig. 4. Small-signal impedance of the structure discussed in the text as a function of frequency and bias current. The real part is negative. The imaginary part is inductive to lower frequencies and capacitive to higher frequencies from the resonance.

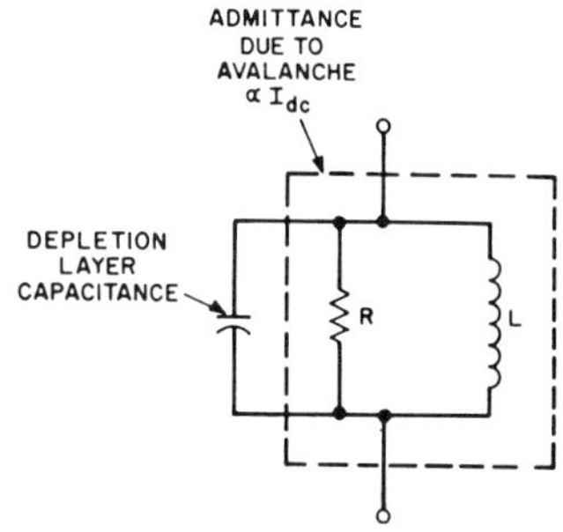

Fig. 5. Equivalent circuit.

This type of behavior of the impedance can be represented by a parallel connection of an inductance, a capacitance, and a negative resistance as shown in Fig. 5. If the capacitance is chosen to be equal to the depletion layer capacitance, the inductance and the negative resistance are fairly frequency independent. Figure 6 shows the admittance of the diode after the admittance of the depletion layer capacitance is subtracted. The negative resistance is fairly constant over a decade of frequency around resonance. The straight line behavior and the slope of the susceptance plot show that the inductance is also fairly frequency independent. It is seen that the admittance is approximately proportional to the bias current. The inductance is inversely proportional to the bias current; hence, it gives, along with the frequency-independent capacitance, a resonance frequency proportional to the square root of the bias current.

The small-signal Q of the device is plotted in Fig. 7. The Q is defined as the angular frequency times the ratio of the average stored energy to the average energy dissipation per unit time. Its calculation is explained in the Appendix. The Q gives information about threshold and the build-up rate of oscillation when the negative resistance is used as an oscillator. A smaller magnitude of negative Q is preferable. From Fig. 7 we see that the Q is better at lower frequencies for a given bias current and at higher currents for a given frequency.

It is illustrative to see how the carrier densities and the electric field behave in the space-charge region. The space-charge wave in the avalanching electron-hole plasma has a wave number given by (9). In the space-charge region this plasma wave propagates back and forth so as to satisfy the appropriate boundary conditions. Although the nature of the wave and the boundary conditions are quite different from those in the ordinary electromagnetic wave in the hollow cavity, we expect, and really

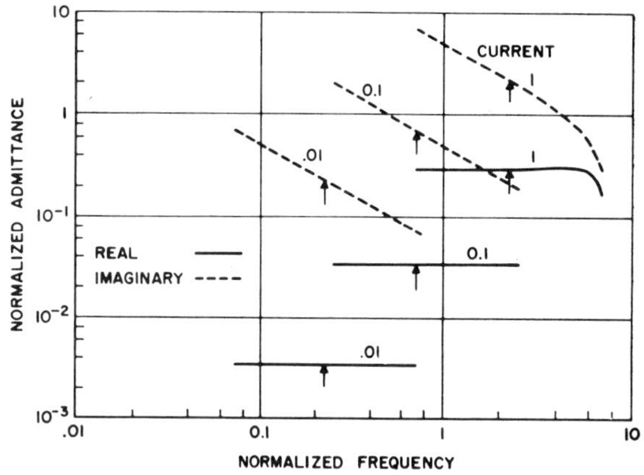

Fig. 6. Small-signal admittance of the structure discussed in the text as a function of frequency and bias current. The contribution from the depletion layer capacitance is subtracted. The real part is negative. The imaginary part is inductive.

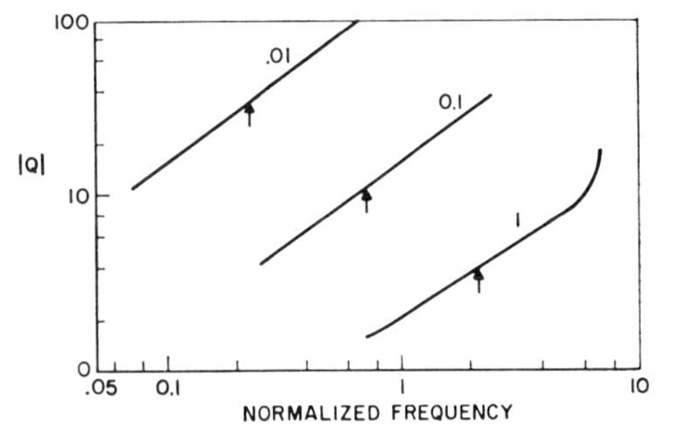

Fig. 7. Small signal Q of the structure discussed in the text as a function of frequency and bias current. The Q is all negative.

have, a resonance phenomenon at a certain frequency as in the latter case. Figure 8 is the phasor or vector diagram[2] of the electric field (a) and the electron current (*b*). Here the phasors of the field and the current at 17 equally spaced points are shown in complex planes for the resonance frequency at a normalized dc current of 1. The phasor of the total current through the space-charge region $\tilde{J}$ is on the positive real axis. The

[2] For general information about the vector diagram, see, for example, Harris A. Thompson, *Alternating-Current and Transient Circuit Analysis*. New York: McGraw-Hill, 1955, p. 79.

112

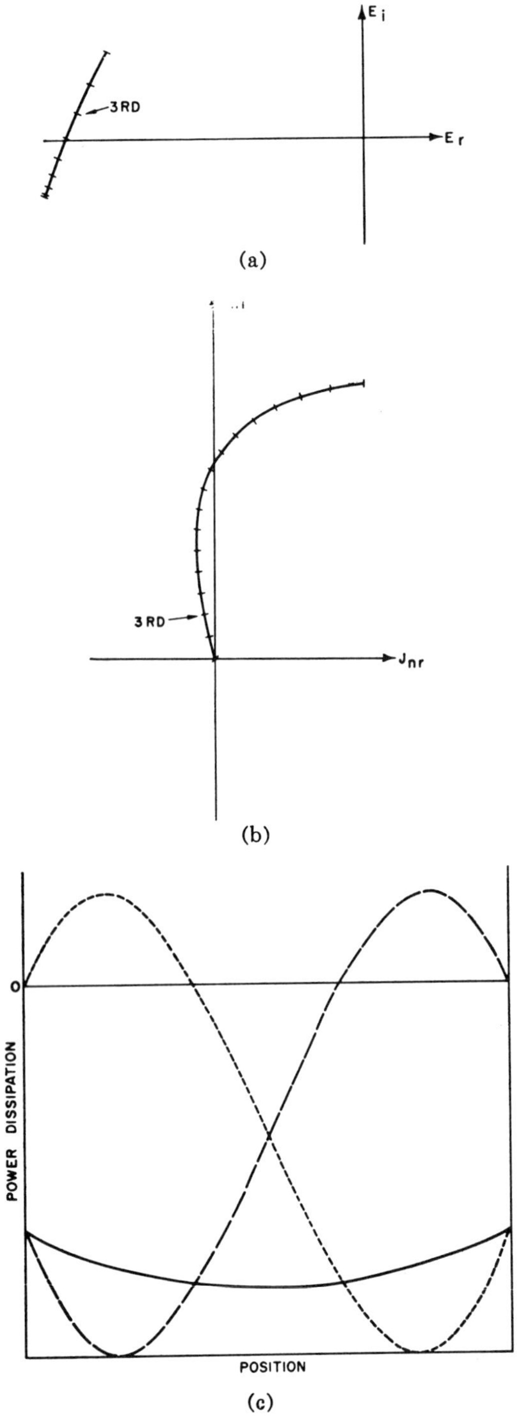

Fig. 8. Phasor diagrams of (a) electric field and (b) electron current. Ticks correspond to 17 equally spaced points and designate end points of phasors. (c) Spatial distribution of power dissipation rate. Dotted line: electron component; dashed line: hole component; solid line: sum. Normalized dc current 1. Near the resonance or normalized frequency 2.19.

electric field is symmetric around the center of the region and is approximately 180° out of phase with the total current. Its phase lags from the center to both ends. This means that the field wave propagates outward from the center. The electron current wave, which is essentially the same as the electron density wave, propagates from left to right although in the left half of the region it looks almost like a standing wave. Its amplitude increases while propagating. The hole current behaves in the same way, but now from right to left. These distributions of field and currents give the power dissipation shown in Fig. 8(c). Electrons dissipate energy in the first one-third of the passage, but their energy dissipation becomes negative after that. Holes behave in the same way and the total energy loss is negative all over the region.

Discussion

We have shown that the space-charge region of a p-n junction in which avalanche occurs uniformly shows a negative resistance, which is due to the intrinsic nature of the avalanching electron-hole plasma. The obtained negative resistance is fairly large in magnitude and the Q of the device is very good.

This negative resistance may be contrasted to another negative resistance utilizing avalanche, which was proposed by Read [1]. The Read diode is understandable by following the same line of reasoning as Shockley's minority carrier delay diode or transit-time mode operation of the transistor structure [4]. Carriers are generated at a "cathode" plane and drift through a "drift space." When the transit time is properly chosen, the phase relation between current and voltage is such that a negative resistance results. Read investigated the dynamics of avalanche multiplication, and proposed using a well-localized avalanche region as the "cathode" and the avalanche-free part of the space-charge region as the "drift space," as Shockley used the emitter of the transistor as a "cathode" and the base region and the collector depletion layer as a "drift space." In Read's structure avalanche is confined to one end of the space-charge region.

Appendix

Fundamental Equations

When equal ionization rate and drift velocity are assumed for both electrons and hole, (1) through (6) in the text become

$$\frac{\partial E}{\partial x} = N_D - N_A + p - n, \tag{12}$$

$$\frac{\partial n}{\partial t} = \frac{\partial J_n}{\partial x} + \alpha(n + p), \tag{13}$$

$$\frac{\partial p}{\partial t} = -\frac{\partial J_p}{\partial x} + \alpha(n + p), \tag{14}$$

$$J_n = -n, \tag{15}$$

$$J_p = -p, \tag{16}$$

in the units given in Table I. The electric field was assumed to be in the negative x direction as in Figs. 1 and 2.

Linearization

The quantities are divided in dc parts and small ac parts. The latter changes as $e^{j\omega t}$, ω being angular frequency. We replace $\partial/\partial t$ by $j\omega$ and retain only the first order terms in the ac part. Then we have

$$\frac{\partial \tilde{E}}{\partial x} = \tilde{J}_n - \tilde{J}_p, \tag{17}$$

$$\frac{\partial \tilde{J}_n}{\partial x} = \alpha' J\tilde{E} + (\alpha - j\omega)\tilde{J}_n + \alpha\tilde{J}_p, \tag{18}$$

$$\frac{\partial \tilde{J}_p}{\partial x} = -\alpha' J\tilde{E} - \alpha\tilde{J}_n - (\alpha - j\omega)\tilde{J}_p, \tag{19}$$

where $\alpha' = \partial\alpha/\partial E$ and J is the total dc current. Since α is an increasing function of $|E|$, $\alpha' J$ is always positive regardless of the sign of J.

Solutions

One of the integrals is

$$\tilde{J} = \tilde{J}_n + \tilde{J}_p + j\omega\tilde{E} \tag{20}$$
$$= \text{constant},$$

in the dimensionless form.

In the case of constant avalanche, α and α' are independent of x. Then the solution of (17) through (19) is given by a superposition of solutions of the form e^{-jkx} with three different values of k. We know, by inserting (20) into (17) through (19), that one of the k's is zero and the other two are the roots of the following equation,

$$k^2 + 2\alpha' J - j2\omega\alpha - \omega^2 = 0, \tag{21}$$

which is (7) in the text.

In the p-n junction case, we have the following boundary conditions:

$$\tilde{J}_n(0) = \tilde{J}_{ns}, \tag{22}$$

$$\tilde{J}_p(1) = \tilde{J}_{ps}. \tag{23}$$

We obtain the following solutions:

$$\tilde{E} = C_1 e^{jkx} + C_2 e^{-jkx} + \frac{2\alpha - j\omega}{k^2}\tilde{J}, \tag{24}$$

$$\tilde{J}_n = \frac{j}{2}(k - \omega)C_1 e^{jkx} - \frac{j}{2}(k + \omega)C_2 e^{-jkx} - \frac{\alpha' J}{k^2}\tilde{J}, \tag{25}$$

$$\tilde{J}_p = -\frac{j}{2}(k + \omega)C_1 e^{jkx} + \frac{j}{2}(k - \omega)C_2 e^{-jkx} - \frac{\alpha' J}{k^2}\tilde{J}, \tag{26}$$

where

$$k = \sqrt{\omega^2 - 2\alpha' J + j2\omega\alpha}, \tag{27}$$

$$C_1 = \frac{1}{\Delta}\begin{vmatrix} J_{ns} + \dfrac{\alpha' J}{k^2} J & -\dfrac{j}{2}(k + \omega) \\ J_{ps} + \dfrac{\alpha' J}{k^2} J & \dfrac{j}{2}(k - \omega)e^{-jk} \end{vmatrix}, \tag{28}$$

$$C_2 = \frac{1}{\Delta}\begin{vmatrix} \dfrac{j}{2}(k - \omega) & \tilde{J}_{ns} + \dfrac{\alpha' J}{k^2}\tilde{J} \\ -\dfrac{j}{2}(k + \omega)e^{jk} & \tilde{J}_{ps} + \dfrac{\alpha' J}{k^2}\tilde{J} \end{vmatrix}, \tag{29}$$

and

$$\Delta = \tfrac{1}{4}\{(k + \omega)^2 e^{jk} - (k - \omega)^2 e^{-jk}\}. \tag{30}$$

$$\tilde{V} = -\int_0^x \tilde{E}\, dx$$
$$= -\frac{C_1}{jk}(e^{jkx} - 1) + \frac{C_2}{jk}(e^{-jkx} - 1) - \frac{2\alpha - j\omega}{k^2}\tilde{J}x, \tag{31}$$

$$\tilde{V}_a = \tilde{V}(1) = Z_t(\tilde{J}_{ns} + \tilde{J}_{ps}) + Z\tilde{J}, \tag{32}$$

where

$$Z_t = -\frac{1}{2\Delta k}\{(k + \omega)(e^{jk} - 1) - (k - \omega)(e^{-jk} - 1)\}, \tag{33}$$

$$Z = \frac{1}{k^2}(Z_t 2\alpha' J - 2\alpha + j\omega). \tag{34}$$

Note that currents are positive in the plus x direction while the potential is measured from the $x = 0$ end. The impedance in the normal sense is obtained by changing the sign of Z. We see that $\tilde{E}$, $\tilde{J}_n$, $\tilde{J}_p$, and $\tilde{V}_a$ are composed of three components proportional to $\tilde{J}_{ns}$, $\tilde{J}_{ps}$, and $\tilde{J}$.

The time average of power dissipation is given by

$$-\left\langle\frac{dw}{dt}\right\rangle = \tfrac{1}{2} Re\,\{(\tilde{J}_n + \tilde{J}_p)\tilde{E}^*\}. \tag{35}$$

The energy is stored as the field energy:

$$\langle w\rangle = \tfrac{1}{2} Re \frac{|\tilde{E}|^2}{2}. \tag{36}$$

The "Q" is given by

$$Q = \frac{\omega\displaystyle\int_0^1 \langle w\rangle\, dx}{-\displaystyle\int_0^1 \left\langle\frac{dw}{dt}\right\rangle dx}. \tag{37}$$

In the numerical example $\alpha = 1$ corresponding to avalanche *breakdown* and $|\alpha'| = 1.672$.

ACKNOWLEDGMENT

The author would like to thank B. C. De Loach, Jr., and R. M. Ryder for valuable discussions, comments, and suggestions. Thanks are also due G. A. Baraff, A. G. Chynoweth, H. K. Gummel, and C. A. Lee for comments and suggestions on an early version of the manuscript.

REFERENCES

[1] W. T. Read, Jr., "A proposed high-frequency, negative-resistance diode," *Bell Sys. Tech. J.*, vol. 37, pp. 401–446, March 1958.
[2] W. W. Gärtner, *Transistors: Principles, Design, and Applications.* New York: Van Nostrand, 1960, p. 48.
[3] Ibid., p. 65.
[4] W. Shockley, "Negative resistance arising from transit time in semiconductor diodes," *Bell Sys. Tech. J.*, vol. 33, pp. 799–826, 1954.

Negative Resistance in *p-n* Junctions Under Avalanche Breakdown Conditions, Part II

T. MISAWA

114

***Abstract*—The small-signal impedance of the space-charge region of p-n junctions under avalanche breakdown conditions is calculated using reasonably realistic dependences of electron and hole ionization rates and drift velocities upon electric field. Two structures are analyzed: one is $p^+\nu n^+$ structure which has a fairly uniform distribution of avalanche multiplication, and the other is a singly diffused junction which is a hybrid of an abrupt and a linear graded junction. Both structures show negative resistance when the transit time of carriers becomes appreciable.**

A computer program was evolved which requires, as input, the impurity profile and field dependences of ionization rates and drift velocities. The program first calculates the dc field and electron and hole currents and then solves the ac small-signal problem. Both the ac small-signal impedance and the Q of the diode are calculated.

INTRODUCTION

A PREVIOUS PAPER [1] showed that the inherent instability of an avalanching electron-hole plasma gives rise to a negative resistance in the space-charge region of a *p-n* junction with uniform distribution of avalanche. The analysis employed several unrealistic assumptions in order to simplify the mathematics. This paper presents results obtained from a computer program designed to include more realistic physical assumptions.

Also, cases intermediate to the above and that treated by Read [2] (in which the avalanche is restricted to one edge of the space-charge depletion region) are calculated. The latter cases are particularly interesting since they can be compared with recently reported experimental results [3].

Manuscript received September 15, 1965.
The author is with Bell Telephone Laboratories, Inc., Murray Hill, N. J.

The program requires as input the impurity profile and the dependence of electron and hole ionization rates and drift velocities on electric field. The program first determines the space-charge region, then calculates the spatial distribution of the dc electric field and electron and hole currents, and then the small-signal ac impedance at a given bias current and frequency. It also calculates the Q of the device which is useful in discussing threshold conditions for oscillation and, hopefully, gives information on large-signal operation.

In this report the results of calculations for two Si structures are reported. One of them, AVX-1, simulates the previously discussed idealized model of uniform avalanche [1]. The other structure, AVX-2, simulates the *p-n* junction which showed CW oscillations in the microwave region [3]. The latter is a typical example intermediate between Read's type and the uniform distribution of avalanche.

It has been found that the first structure shows impedance characteristics quite similar to the previously reported idealized model [1], and the second structure has impedance characteristics which are hybrids of the first type and Read's. It has also been found that the more uniform the avalanche is, the better the Q. This advantage is more pronounced when the carrier drift velocities are dependent on the electric field and the depletion layer capacitance is lossy.

STRUCTURES

The impurity profile, AVX-1, which is intended to be as close as possible to the idealized model previously reported [1], simulates the structure which may be made by a 6 μ thick epitaxial growth of 5 ohm-cm n-type material on a 0.008 ohm-cm As doped substrate followed by a 3.5 μ boron diffusion. A high boron surface concentration of 10^{22} cm^{-3} was assumed in order to take into account the change of diffusion constant at high impurity concentrations [4]. The outdiffusion from the substrate was estimated from the published value of the As diffusion constant and the known thermal history of the sample during epitaxial growth and boron diffusion. The diffusion profiles are approximated by complementary error functions. The nonuniformity of the epitaxial layer resistivity due to transport of substrate impurity through the vapor phase was not taken into account [5]. Figure 1 shows the impurity profile schematically.

The other impurity profile, AVX-2A, which is intended to simulate the intermediate case [3], is shown schematically in Fig. 2, and some details of its space-charge region at breakdown are shown in Fig. 6.

Ionization rates of electrons α and holes β are taken from a recent report by C. A. Lee et al. [6] and are expressed by

$$\alpha \quad \text{or} \quad \beta = a \exp(-b/E), \tag{1}$$

where a and b are constants. They are shown in Fig. 3. β is smaller than α by more than an order of magnitude and β changes more rapidly with field than α. In Fig. 3 and all subsequent figures various quantities are expressed in dimensionless form. The units used are tabulated in Table I.

Two different dependences of carrier drift velocities upon electric field were used in calculation. First, the two structures were analyzed by assuming constant velocities independent of field, 8.5×10^6 cm/sec for electrons and 5×10^6 cm/sec for holes. However, published data of the field dependence of carrier drift velocities in Si [7] indicate that the velocities are still increasing with field at 10^5 volt/cm. Since the field we are interested in is as high as 4×10^5 volt/cm, it is quite important to determine the effects of this nonsaturation of drift velocities. The same two structures were again analyzed by approximating Prior's data on lowest resistivity samples with an E^m type field dependence. Prior's data show that the hole velocity saturates less than the electron velocity. The assumed drift velocities are plotted in Fig. 4.

Analysis and Results

The analysis is essentially the same as that previously reported [1]. However, the space-charge region is now determined from the impurity profile and the given bias current. The ac equations are consequently more complicated, and an analytical solution is no longer possible. Details of the computer analysis are given in the Appendix.

Figure 5 shows spatial distributions of (a), the dc electric field and (b), the electron current in the space-charge region of AVX-1 at four current levels, 0.01, 0.1, 1, and 10. The field-dependent velocities in Fig. 3 were used. Carrier diffusion was neglected in the space-charge region. This is quite reasonable in the region shown in Fig. 5(a) where the electric field is very high.[1] From Fig. 5(a) it is seen that the field distribution is appreciably modified by the space-charge of mobile carriers when the normalized current is greater than 1. We have more net positive charge from holes at the left end and more net negative charge from electrons at the right end. This fact is well reflected in the field distribution. The slope of the electron current curves in Fig. 5(b) is the rate of avalanche multiplication. Since electrons have a larger ionization rate than holes, we have more avalanche multiplication in the n side although the field is smaller there.

Similar curves for AVX-2A are shown in Fig. 6. The field-dependent velocities were used and the assumed current is 1. In this current range the field and current distributions are determined almost solely by the impurity charge due to the high impurity density present. CW microwave oscillations were actually observed with this structure at current densities up to 600 A/cm^2, which corresponds to a normalized current of 0.44. [3] We see that the space-charge region can be divided into three regions of approximately equal width, the center region of roughly uniform avalanche sandwiched between two avalanche-free end regions.

115

The computed space-charge widths and sustaining voltages are tabulated in Table II. Note that AVX-1 with scattering limited velocity shows a slight dc negative resistance while the other three cases do not show negative resistance.

The small-signal impedance of AVX-1 with scattering limited velocities is shown in Fig. 7(a) as a function of frequency at various bias currents. Up to a current of 1 the behavior is quite similar to that of the previously reported case [1]. At $J = 10$ the behavior is complicated because there are two peaks of avalanche. As in the previous idealized case [1] a reasonable equivalent circuit is again a parallel connection of the depletion layer capacitance, an inductance, and a negative resistance. Figure 7(b) shows the admittance of the latter two. The inductance and the negative resistance is fairly constant over the frequency range investigated. Their admittance increases roughly proportionally with the bias current. In

[1] At the edge of the space-charge region there is a region where the field is low and diffusion is important. The present calculation neglects this region.

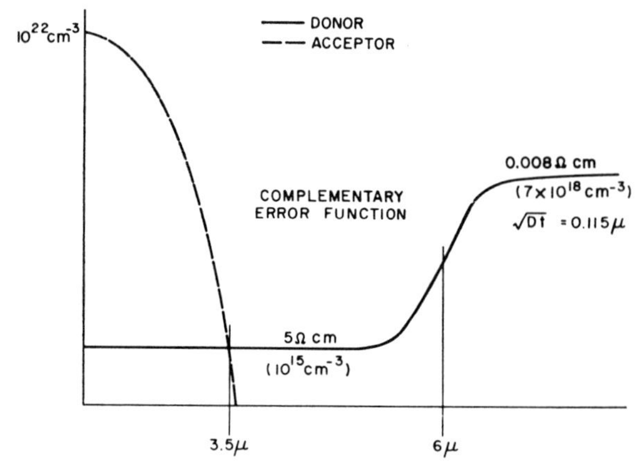

Fig. 1. Impurity profile AVX-1, schematical.

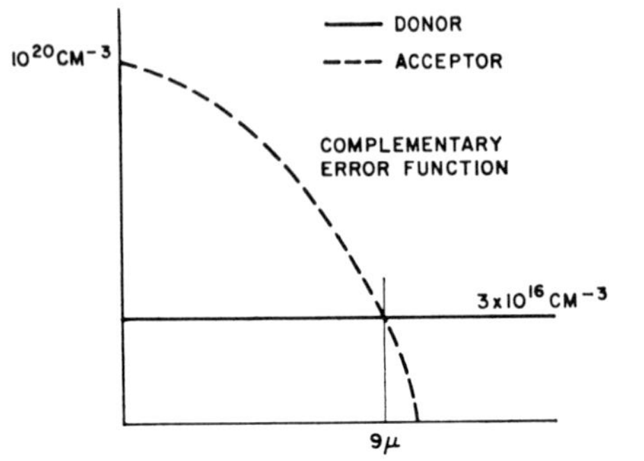

Fig. 2. Impurity profile AVX-2A, schematical.

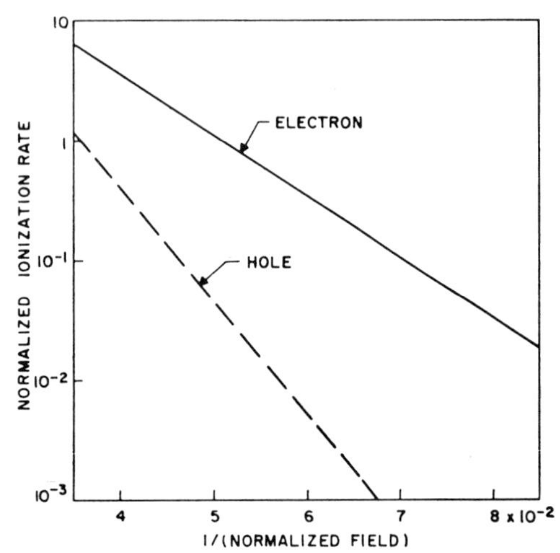

Fig. 3. Ionization rates of electrons and holes.

TABLE I

Unit	Expression	Numerical Value
Length	w	1μ
Velocity	v*	8.5×10^6 cm/sec
Time	$\tau = w/v$	1.177×10^{-11} sec.
Frequency	$1/2\pi\tau$	13.53 Gc/s
Density	N_0	10^{15} cm^{-3}
Current	qvN_0	1.36×10^3 A/cm^2
Field	wqN_0/ϵ	1.507×10^4 volt/cm
Voltage	w^2qN_0/ϵ	1.507 volt
Impedance	w^2v/ϵ	1.107×10^{-3} ohm-cm^2
Charge	q	1.6×10^{-19} coulomb

* v is scattering limited velocity of electrons.

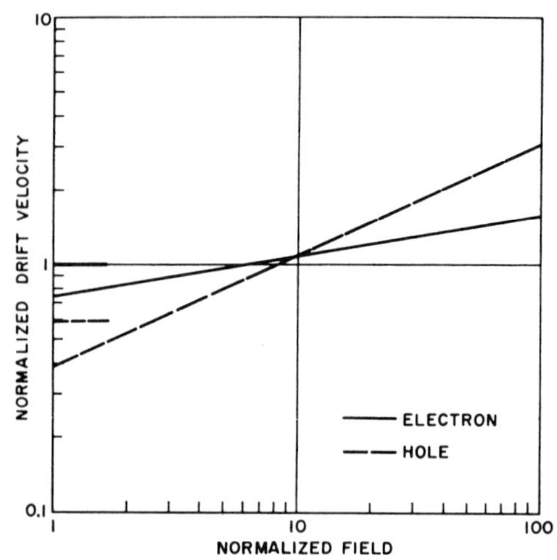

Fig. 4. Assumed relations between drift velocities of carriers and field. Short horizontal lines at the left show used values of scattering limited velocities.

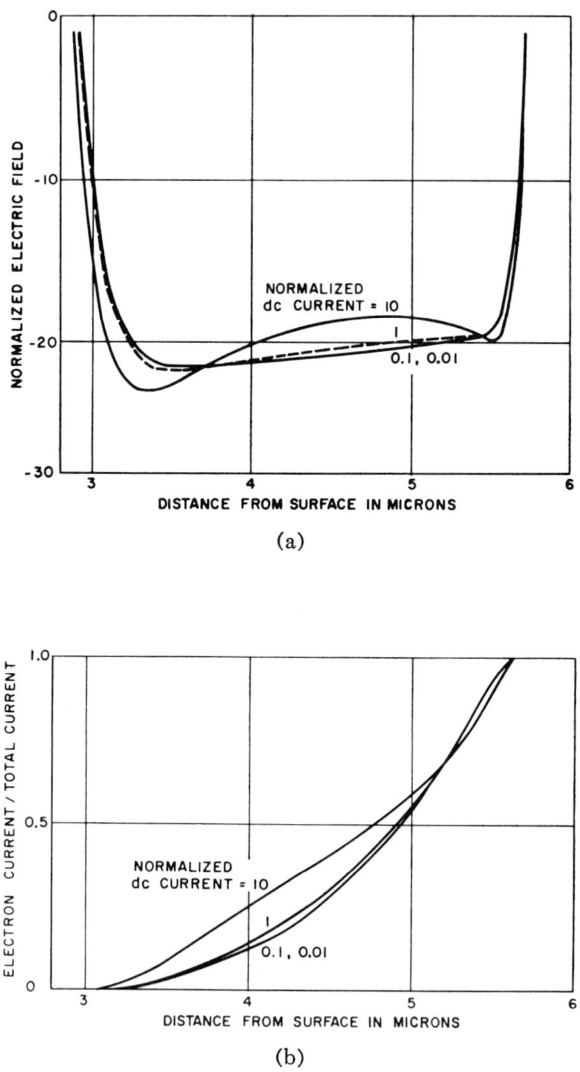

Fig. 5. (a) dc electric field and (b) carrier current for AVX-1 at various dc currents. Field-dependent velocities.

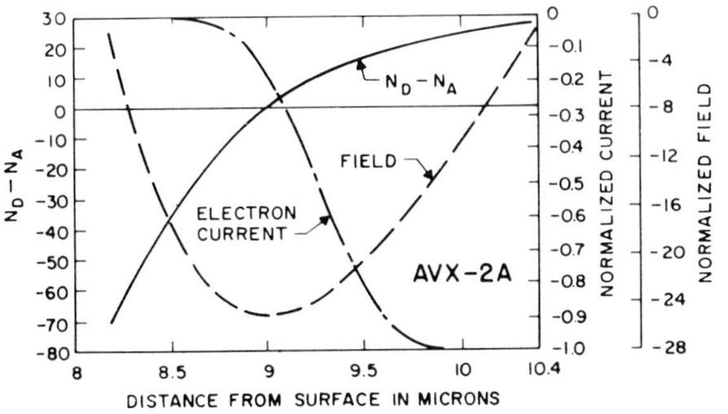

Fig. 6. Impurity distribution, dc electric field, and carrier currents for AVX-2A at normalized dc current = 1. Field-dependent velocities.

TABLE II

	AVX-1		AVX-2A	
J	Scattering Limited Velocities	Field-Dependent Velocities	Scattering Limited Velocities	Field-Dependent Velocities
	space-charge layer width (μ)*			
0.01	2.816694	2.7966826	2.183831	2.183851
0.1	2.8175521	2.7971499	2.187334	2.187186
1	2.8156150	2.8015716	2.222951	2.220915
10	2.8795214	2.83745		
	sustaining voltage (normalized value)			
0.01	54.423283	54.420948	36.485528	36.489729
0.1	54.415021	54.425127	36.53583	36.535982
1	54.283141	54.457478	37.012664	36.958761
10	53.012982	54.495917		

* See Appendix for the definition of the space-charge layer.

Fig. 7(c) the small-signal Q is shown. The behavior is quite similar to the idealized case [1].

Figure 8 shows the ac field, electron and hole currents at 33 equally spaced points in the space-charge region near the resonance frequency (normalized frequency = 0.5703108) for $J = 1$. The electric field is approximately out of phase with the total current whose vector lies on the positive real axis. The phase of the field, however, lags on both sides of a point about one quarter from the n side end of the space-charge region. The lag is more towards the p side end than towards the n side end. We see now that the electron and the hole currents behave quite differently. The electron current, which enters from the left end, keeps increasing its magnitude with little lag in phase, while the hole current suffers appreciable phase delay.

Figure 9 shows the small-signal impedance (a), the admittance (the admittance of the depletion layer capacitance is subtracted) (b), and the small-signal Q of AVX-1 obtained with the field-dependent velocities. As is shown in Fig. 3 the field-dependent velocities are larger than the scattering limited velocities in the range of field present in the space-charge region. Hence, the resonance frequencies have shifted to higher frequencies. The field dependence of carrier velocities results in a

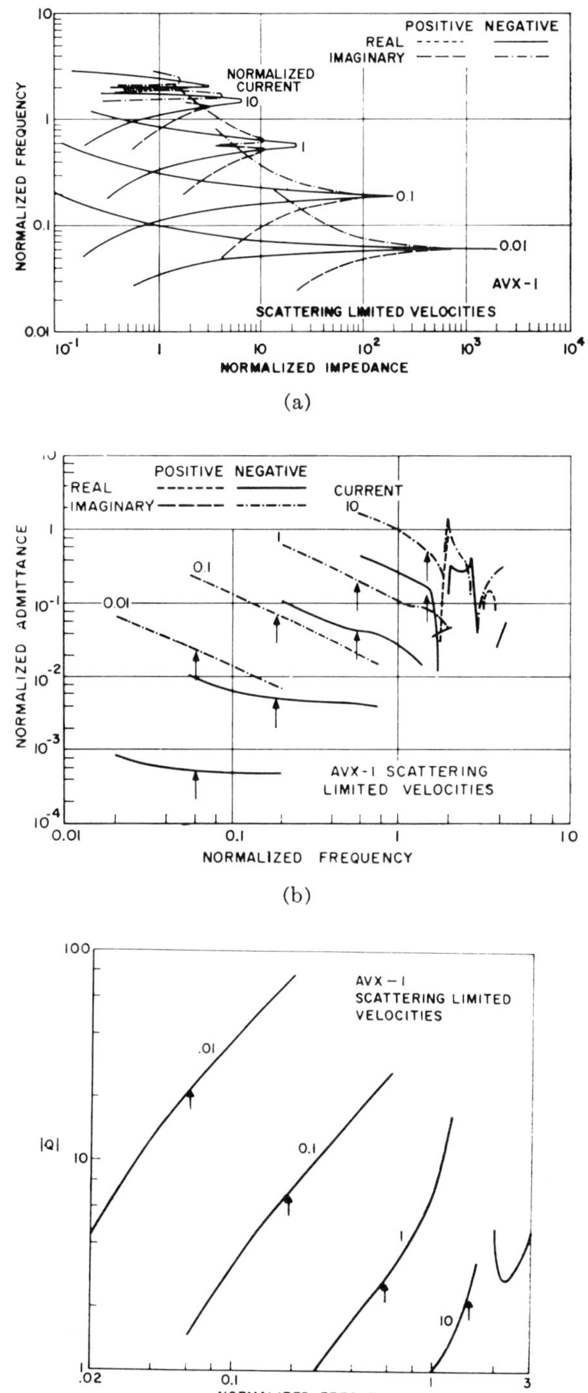

(c)

Fig. 7. (a) Small-signal impedance, (b) admittance (depletion layer capacitance part subtracted), and (c) Q of AVX-1 as a function of frequency for bias currents 0.01, 0.1, 1, and 10. Q is negative. Scattering limited velocities. In (b) and (c) the resonance frequencies are shown by arrows.

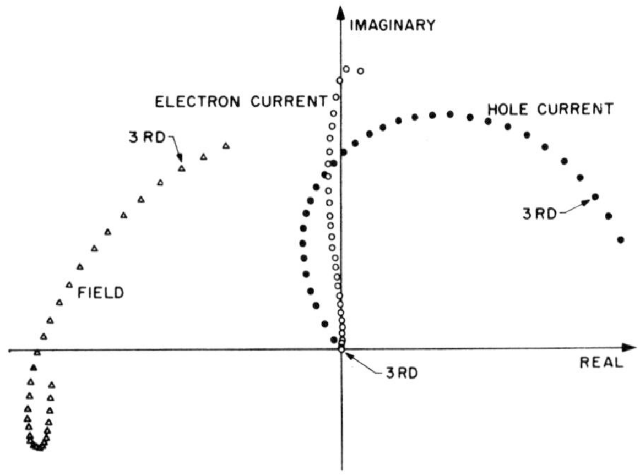

Fig. 8. Phasor diagram of ac electric field, electron current, and hole current at 33 equally spaced points in the space-charge region near the resonance frequency (normalized frequency = 0.5703108) at normalized current 1 for AVX-1. Scattering limited velocities. The total current phasor is on the positive real axis.

nonzero ac conductivity, which produces ohmic loss. The Q has appreciably degraded. Figure 9(b) shows that the conductance and inductance in the equivalent circuit are fairly constant over the frequency range where the impedance is large. How-
118
ever, the conductance goes to zero and becomes positive as frequency decreases from the resonance, corresponding to the dc positive resistance in Table II.

Figures 10 and 11 present the small-signal impedance, the avalanche part of the admittance, and the Q of the structure AVX-2A for the scattering limited velocities and the field-dependent velocities, respectively. The parallel connection of the depletion layer capacitance, an L and a negative R, is still a good equivalent circuit. As Figs. 10(b) and 11(b) show, L and R is nearly proportional to the bias current. The L is fairly frequency-independent. However, the R is less frequency-independent.

The frequency at which the R changes sign has increased and approached the resonance frequency, compared with the preceding case. This shows up in the impedance plots as asymmetry of the curves and positive real part just below the resonance. This behavior is reminiscent of the impedance curve of a Read diode, which is shown in Fig. 12. These curves are for the example discussed in [2] and units for normalization are from [2]. The three current values correspond to our 0.01, 0.1, and 1.[2] In the Read diode there is a certain critical current,

[2] The curves were calculated from Read's (53), which contains the term due to the depletion layer capacitance. Incidentally, there seem to be some misprints there. The actual equation used is

$$Y(\omega) = \frac{2\omega y}{1 - e^{-j\omega} - j2y}$$

$$= \frac{\omega y[1 - \cos\omega + j(2y - \sin\omega]}{1 - \cos\omega + 2(y^2 - y\sin\omega)},$$

$$y = \frac{\omega}{2}\left[1 - \frac{\omega^2\tau_1}{2(m+1)I_d}\right].$$

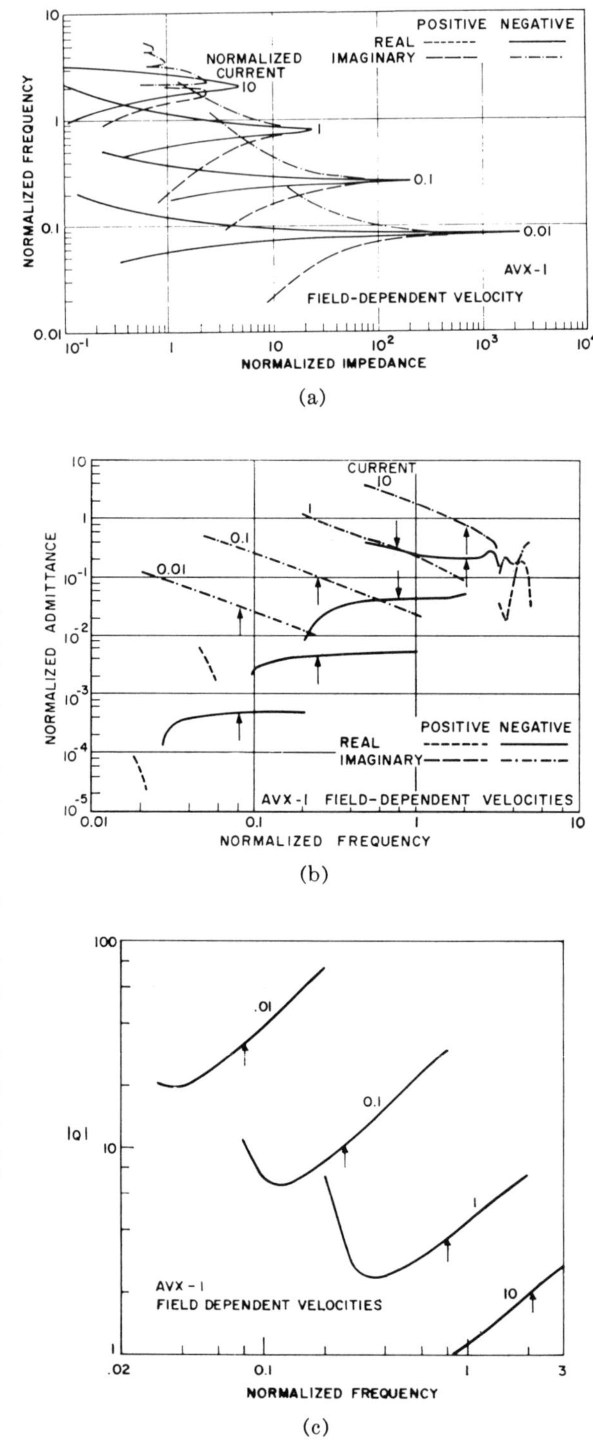

Fig. 9. (a) Small-signal impedance, (b) the admittance (depletion layer capacitance part subtracted), and (c) Q of AVX-1 as a function of frequency for bias currents 0.01, 0.1, 1, and 10. Q is negative. Field-dependent velocities. In (b) and (c) the resonance frequencies are shown by arrows.

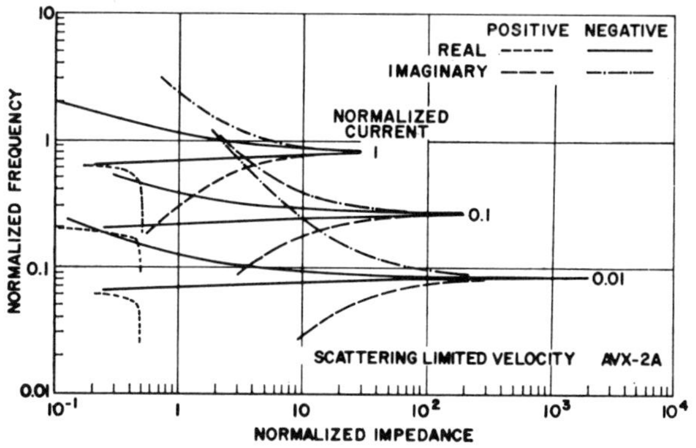

(a)

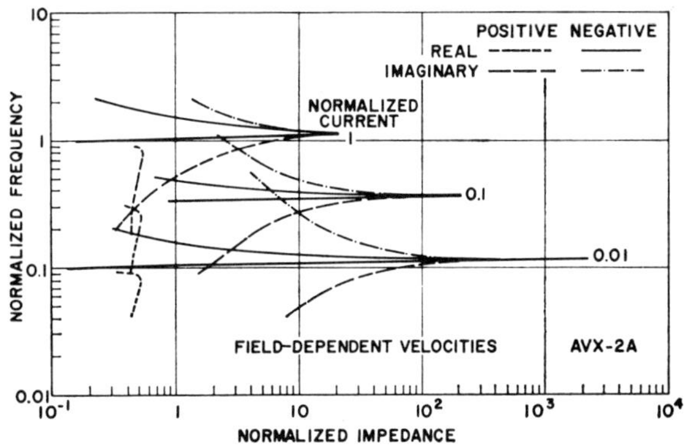

(a)

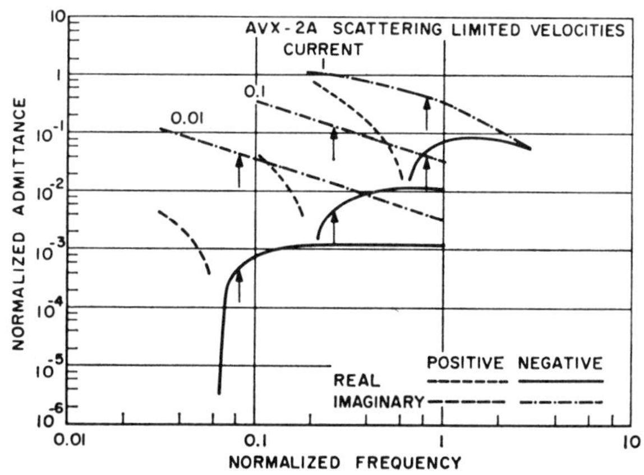

(b)

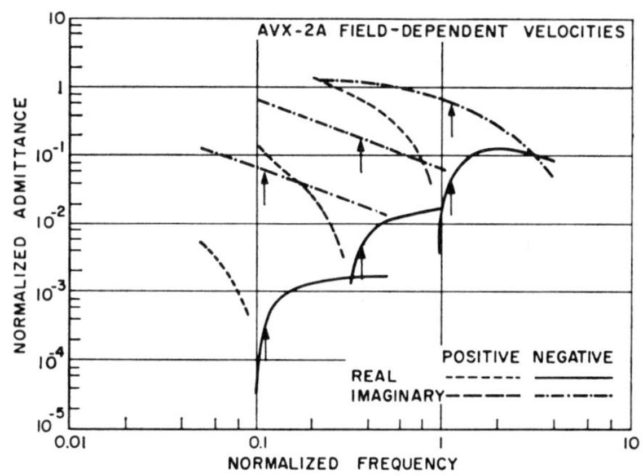

(b)

119

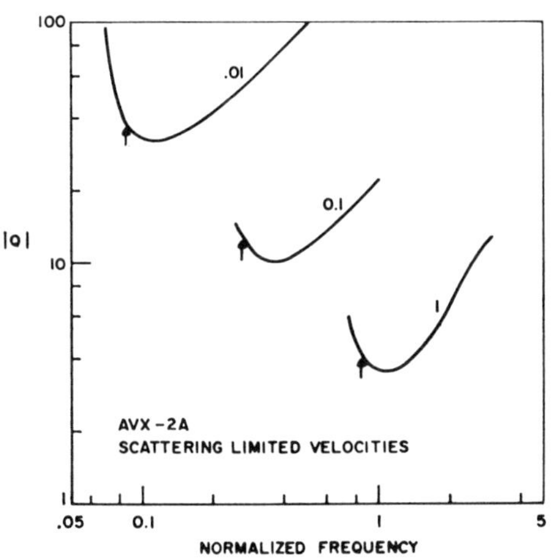

(c)

Fig. 10. (a) Small-signal impedance, (b) the admittance (depletion layer capacitance part subtracted), and (c) Q of AVX-2A as a function of frequency for bias currents 0.01, 0.1, and 1. Q is negative. Scattering limited velocities. In (b) and (c) the resonance frequencies are shown by arrows.

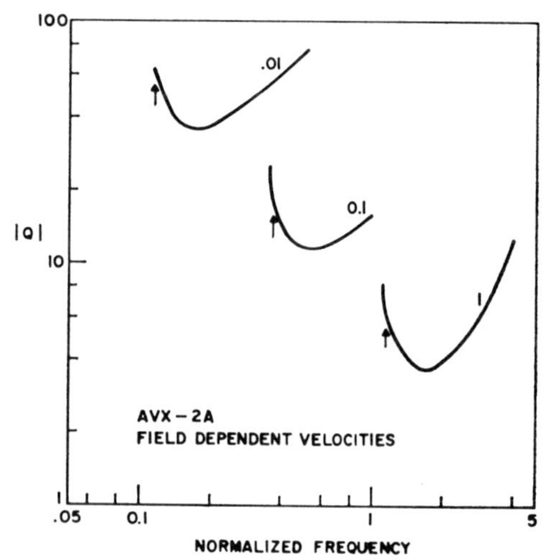

(c)

Fig. 11. (a) Small-signal impedance, (b) the admittance (depletion layer capacitance part subtracted), and (c) Q of AVX-2A as a function of frequency for bias currents 0.01, 0.1, and 1. Q is negative. Field-dependent velocities. In (b) and (c) resonance frequencies are shown by arrows.

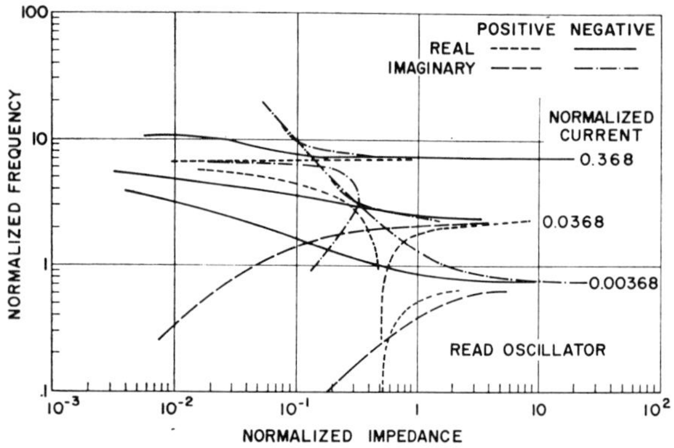

Fig. 12. Small-signal impedance of a Read diode. Units used in normalization are not those in Table I but those in [2]. The three currents correspond to our values of 0.01, 0.1 and 1.

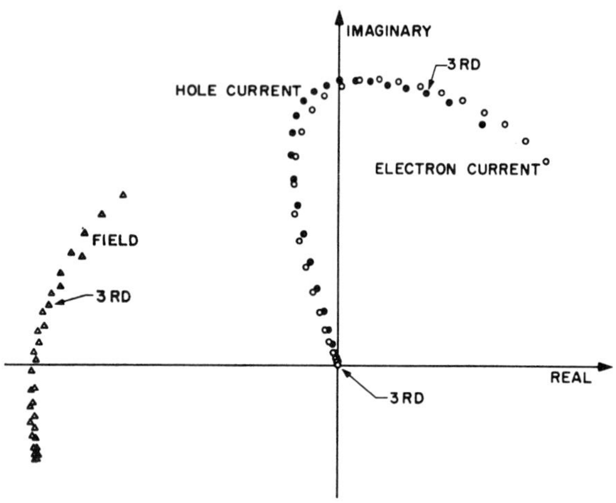

Fig. 13. Phasor diagram of ac electric field, electron current, and hole current at 33 equally spaced points in the space-charge region near the resonance frequency (normalized frequency = 1.1549) at normalized current 1 for AVX-2A, field-dependent velocities. The total current phasor is on the positive real axis.

above which the negative resistance at $\omega = \pi$ or the optimum frequency disappears. In the example shown in Fig. 12 the critical current lies between 0.368 and 0.0368. Below the critical current we have a resonance type curve similar to those obtained in this paper. However, in a Read diode the impedance goes to infinity at resonance and both real and imaginary parts change sign at resonance [8]. Below resonance the impedance is a positive resistance and an inductance, the real part being fairly constant at frequencies well below resonance. In the curve for the intermediate structure AVX-2A, the real part changes sign before resonance, and the impedance remains finite at the resonance where the imaginary part changes sign. The fact that the resonance frequency is proportional to the square root of the dc current is common among the three cases discussed. Figures 10(b) and 11(b) show Q's of AVX-2A for the two different dependencies of velocities. We note that AVX-2A has a poorer Q than AVX-1 in the frequency range where impedance is high. The Q is better above resonance contrary to the case of AVX-1.

Figure 13 shows the spatial distributions of field, electron, and hole currents near resonance (normalized frequency = 1.1549) for current 1 with field-dependent velocities. The total current vector is on the positive real axis. The field, which is approximately 180° out of phase with the total current, lags in phase from the center to both ends. The electron current, which enters from the left end, increases its magnitude in the center region where we have uniform avalanche. In the right avalanche-free region its phase delays without change in magnitude. The hole current behaves quite similarly.

Discussion and Conclusion

It has been found that the structure AVX-1 has characteristics quite similar to those of the idealized uniform avalanche case and AVX-2A exhibits characteristics that are hybrids of the uniform and Read type avalanche. As far as the small-signal Q is concerned, the more uniform avalanche gives better characteristics.

The present small-signal calculation will be useful in discussing small signal amplification by these negative-resistance diodes and also threshold conditions for oscillation when the diodes are used as oscillators.

For example, the fact that the magnitude of Q decreases with current means that there exists a threshold current for the onset of oscillation. The oscillator system is composed of the diode of negative Q and a cavity of positive Q. Spontaneous oscillation is possible only when the total Q of the whole system is negative. Therefore, the threshold current increases with increased loading of the cavity.

Since the device is fairly nonlinear as is seen from the calculated impedance curves, it is difficult to discuss large-signal operation. However, the fact that the negative resistance exists over fairly large-current ranges suggests that reasonable current swing, and therefore reasonable efficiency, is to be expected when the diode is used as an oscillator. Again the improvement of Q at high currents suggests higher efficiency at higher current.

Appendix

Fundamental Equations

The positive direction of the x axis is chosen from the p region to the n region of the p-n junction. Then the fundamental equations read in the dimensionless form

$$J_n = -v_n n, \tag{2}$$

$$J_p = -v_p p, \tag{3}$$

$$\frac{\partial E}{\partial X} = N_D - N_A + p - n, \tag{4}$$

$$\frac{\partial n}{\partial t} = \frac{\partial J_n}{\partial X} + \alpha v_n n + \beta v_p p, \tag{5}$$

$$\frac{\partial p}{\partial t} = -\frac{\partial J_p}{\partial X} + \alpha v_n n + \beta v_p p, \tag{6}$$

$$\frac{\partial V}{\partial X} = -E \tag{7}$$

where v_n and v_p are the absolute values of the electron and holes velocities and all other quantities, which are defined in Table III, can have negative values. We have six unknowns (n, p, J_n, J_p, E, and V) and six equations.

TABLE III

J_n, J_p	—electron and hole current densities.
v_n, v_p	—magnitudes of electron and hole drift velocities, positive quantities.
n, p	—electron and hole densities.
N_D, N_A	—densities of ionized donors and acceptors.
E	—electric field
α, β	—ionization rates of electrons and holes.

Boundary Conditions

The space-charge region is defined as a region where the electric field is so high that the diffusion current is negligible. The field which is referred to as the critical field for the scattering limited velocity will be a good criterion. At the n type end the hole current is given and at the p type end the electron current is given.

DC Solution

Time derivations are zero. The total current J is constant:

$$J = J_n + J_p. \tag{8}$$

By eliminating n, p, and J_p from (2)–(8), we obtain

$$\frac{\partial E}{\partial X} = \left(\frac{1}{v_n} + \frac{1}{v_p}\right)J_n + N_D - N_A - \frac{J}{v_p}, \tag{9}$$

$$\frac{\partial J_n}{\partial X} = (\alpha - \beta)J_n + \beta J. \tag{10}$$

The two simultaneous equations are solved by applying the boundary conditions. Since the boundaries of the space-charge region are not known, the numerical integration is started with a tentative value of the p side end of the space-charge region X_i and given $J_n = J_{ns}$ and $E = E_c$ at $X = X_i$. The integration is terminated when $|E|$ becomes smaller than $|E_c|$. J_p at the terminated point X_1 is generally not equal to the boundary value J_{ps} unless the selected value of X_i fortuitously coincides with the right value. The integration is repeated by changing X_i until J_p satisfies the boundary condition. In actual calculation the successive value of X_i is chosen to be the bisecting point of the last two values of X_i which sandwich the as yet unknown right value of X_i. When the value of X_i approaches the right value, the final value of X_i is obtained by interpolation. Since the relative accuracy of the numerical integration is about 10^{-5}, it is meaningless to specify J_{ps} smaller than $\sim 10^{-5}J$. J_{ps} is set to zero. The dc solution is stored in the machine and is used in the ac solution.

The following values of $|E_c|$ and $|J_{ns}|$ were used.

Scattering Limited Velocities

AVX-1	$\lvert E_c\rvert = 0,\quad \lvert J_{ns}\rvert = 1.44 \times 10^{-14}$
AVX-2A	$\lvert E_c\rvert = 1,\quad \lvert J_{ns}\rvert = 1.4 \times 10^{-14}$

Field-Dependent Velocities

AVX-1	$\lvert E_c\rvert = 1,\quad \lvert J_{ns}\rvert = 1.44 \times 10^{-14}$
AVX-2A	$\lvert E_c\rvert = 1,\quad \lvert J_{ns}\rvert = 1.4 \times 10^{-14}$

AC Solution

Equations for small ac part, which varies as $e^{j\omega t}$ with time, are obtained from (2) through (7) by retaining only the terms linear in the ac parts. The total current, which is again constant, is now composed of electron and hole currents and displacement current

$$\tilde{J} = \tilde{J}_n + \tilde{J}_p + j\omega\tilde{E} \tag{11}$$

in dimensionless form. Eliminating $\tilde{n}$, $\tilde{p}$, and $\tilde{J}_p$ we obtain

$$\frac{\partial \tilde{E}}{\partial X} = \left(j\frac{\omega}{v_p} - \frac{J_n v_n'}{v_n^2} + \frac{J_p v_p'}{v_p^2}\right)\tilde{E} + \left(\frac{1}{v_n} + \frac{1}{v_p}\right)\tilde{J}_n - \frac{\tilde{J}}{v_p}, \tag{12}$$

$$\frac{\partial \tilde{J}_n}{dX} = \left(\alpha' J_n + \beta' J_p - j\omega\beta + j\omega\frac{J_n v_n'}{v_n^2}\right)\tilde{E} + \left(\alpha - \beta - \frac{j\omega}{v_n}\right)\tilde{J}_n + \beta\tilde{J}, \tag{13}$$

where $\sim$ sign means ac part and the quantities without $\sim$ are dc values. The prime means derivative with respect to field. Although the problem is the boundary value problem, it is converted to the "initial value" problem because the equations are linear [9]. The equations are solved in the space-charge region determined in dc solution with the boundary condition, $\tilde{J}_n(X_i) = \tilde{J}_{ns}$, $\tilde{J}_p(X_1) = \tilde{J}_{ps}$. Since the problem is linear, the solution is composed of three terms:

121

$$S = S^{(1)}\tilde{J} + S^{(2)}\tilde{J}_{ns} + S^{(3)}\tilde{J}_{ps}. \quad (14)$$

$S^{(1)}$ is obtained with $\tilde{J} = 1$, $\tilde{J}_{ns} = \tilde{J}_{ps} = 0$, $S^{(2)}$ with $\tilde{J} = 0$, $\tilde{J}_{ns} = 1$, $\tilde{J}_{ps} = 0$, and $S^{(3)}$ with $\tilde{J} = 0$, $\tilde{J}_{ns} = 0$, $\tilde{J}_{ps} = 1$. The voltage across the space-charge region is obtained by integrating $\tilde{E}$:

$$\tilde{V} = \int_{X_l}^{X_s} \tilde{E}\, dX$$

$$= Z\tilde{J} + Z_{tn}\tilde{J}_{ns} + Z_{tp}\tilde{J}_{ps}. \quad (15)$$

$\tilde{J}_{ns}$ and $\tilde{J}_{ps}$ is determined by the applied voltage $\tilde{V}$. Since the relevant admittances are so small that the products with Z_{tn} and Z_{tp} are much less than unity, the second and third terms in (15) are negligible. Z is the impedance of the space-charge region. In the text cnly the $\tilde{J}$ component of various quantities is discussed.

Acknowledgment

The author wishes to express his thanks to B. C. De Loach, D. Kahng, and R. M. Ryder for helpful conversations.

122

References

[1] T. Misawa, "Negative resistance in *p-n* junctions under avalanche breakdown conditions, part I," *this issue*, page 141.
[2] W. T. Read, Jr., "A proposed high-frequency, negative-resistance diode," *Bell Sys. Tech. J.*, vol. 37, pp. 401–446, March 1958.
[3] B. C. De Loach and R. L. Johnston, "Avalanche transit-time microwave oscillators and amplifiers," *this issue*, page 181.
[4] J. J. Chang, "Concentration-dependent diffusion of boron and phosphorus in silicon," *IEEE Trans. on Electron Devices*, vol. ED-10, pp. 357–359, November 1963.
[5] C. O. Thomas, D. Kahng, and R. C. Manz, "Impurity distribution in epitaxial silicon films," *J. Electrochem. Soc.*, vol. 109, pp. 1055–1061, November 1962.
[6] C. A. Lee, R. A. Logan, R. L. Batdorf, J. J. Kleimack, and W. Wiegmann, "Ionization rates of holes and electrons in silicon," *Phys. Rev.*, vol. 134, pp. A761–A773, May 1964.
[7] A. C. Prior, "The field-dependence of carrier mobility in silicon and germanium," *J. Phys. Chem. Solids*, vol. 12, pp. 175–180, January 1960.
[8] C. A. Lee et al. improved Read's analysis and showed that the impedance remains finite. See C. A. Lee, R. L. Batdorf, W. Wiegmann, and G. Kaminsky, "The Read diode—an avalanching, transit-time, negative resistance oscillator," *Appl. Phys. Letters*, vol. 6, pp. 89–91, March 1965.
[9] G. N. Lance, *Numerical Method for High Speed Computers.* London: Iliffe and Sons, Ltd., 1960, p. 63.

A Large-Signal Analysis of IMPATT Diodes

WILLIAM J. EVANS, MEMBER, IEEE, AND GEORGE I. HADDAD, SENIOR MEMBER, IEEE

Abstract—**This paper presents results on RF power output and efficiency of IMPATT oscillators obtained from a large-signal model of these devices. The results are obtained from a closed-form solution of the nonlinear equations describing a Read-type IMPATT diode. The closed-form solution is obtained by assuming a short transit time through the drift region compared to the RF period. The solution is used to obtain the large-signal diode impedance. The analysis shows that the power output of an IMPATT diode depends strongly on the series load resistance presented to the active part of the diode and that the change in diode reactance with increasing bias current also depends on the series resistance. Plots of power output as a function of frequency, bias current, and load resistance are presented. Frequency tuning of the oscillator through current variation is also discussed. Experimental results are presented and compared with the theoretical ones wherever possible. The results lead to an improved understanding of such oscillators and are extremely useful in optimizing their performance and determining their limitations.**

I. INTRODUCTION

SINCE the avalanche diode oscillator was first described by Johnston *et al.* [1] several small-signal analyses [2]–[4] of these devices have provided a considerable insight into their operation. However, several of the important properties of IMPATT oscillators cannot be obtained from a small-signal analysis. Those of particular interest are the power output, efficiency, and frequency tuning of the oscillator and how these quantities depend on the diode parameters, bias current, and the RF circuitry. The purpose of this paper is to describe these important properties of the oscillator. The analysis is similar to that given in Read's [5] original proposal with the additional assumption that the transit time of the carriers through the drift region is small compared to the period of the oscillations. With this assumption it is possible to obtain a closed-form solution for the equations under large-signal conditions and thus eliminate lengthy computer calculations. The large-signal impedance of the diode and the RF power output as well as other important properties are then easily calculated. Experimental measurements on an IMPATT oscillator are presented and compared with the theoretical results.

Manuscript received April 8, 1968; revised May 13, 1968. This work was supported by a grant from Omni Spectra, Inc., Farmington, Mich. It was also partially supported by the Rome Air Development Center under Contract F30602-68-C-0043.

W. J. Evans was with the Electron Physics Laboratory, Department of Electrical Engineering, University of Michigan, Ann Arbor, Mich. He is now with Bell Telephone Laboratories, Inc., Murray Hill, N. J.

G. I. Haddad is with the Electron Physics Laboratory, Department of Electrical Engineering, University of Michigan, Ann Arbor, Mich.

II. BASIC EQUATIONS

The current in the diode and the electric field are related by the continuity equations for holes and electrons and Poisson's equation. The one-dimensional continuity equations with terms added to account for carrier generation by impact ionization are given by

$$\frac{\partial p}{\partial t} = -\frac{1}{q}\frac{\partial J_p}{\partial x} + \alpha_p p v_p + \alpha_n n v_n \quad (1)$$

and

$$\frac{\partial n}{\partial t} = \frac{1}{q}\frac{\partial J_n}{\partial x} + \alpha_p p v_p + \alpha_n n v_n \quad (2)$$

where

p = the hole density (cm^{-3})
n = the electron density (cm^{-3})
J_p = the hole current density (A/cm^2)
J_n = the electron current density (A/cm^2)
q = the electronic charge (coulombs)
α_p = the ionization rate for holes (cm^{-1})
α_n = the ionization rate for electrons (cm^{-1})
v_p = the hole velocity (cm/s)
v_n = the electron velocity (cm/s).

The ionization rates α_p and α_n are dependent on the electric field E. An approximate relation for the ionization rate that can be applied to a number of semiconductor materials may be expressed as [6]

$$\alpha = A' \exp\left[-\left(\frac{b'}{E}\right)^{m'}\right] \quad (3)$$

where A', b', and m' are determined experimentally and depend on the type of carrier and the material. Equation (3) can be used to calculate the breakdown voltage of junction diodes provided that the junction is sufficiently wide to prevent Zener breakdown. In the following analysis, the junction breakdown voltage is assumed high enough so that carrier injection by tunneling may be neglected.

Finally, the field and carrier densities are related through Poisson's equation

$$\frac{\partial E}{\partial x} = \frac{q}{\epsilon}(N_D - N_A + p - n) \quad (4)$$

where

N_D = the donor density (cm^{-3})
N_A = the acceptor density (cm^{-3})
ϵ = the dielectric constant (F/cm).

The quantities N_D and N_A determine the field distribution in the diode before breakdown. In general, the field distribution is such that there exists a region of avalanche breakdown and lower field regions on either side of the avalanche where carriers drift at saturated velocities.

Small-signal computer solutions for (1) through (4) have been obtained by Misawa [2]. These solutions show that IMPATT diodes exhibit a bandpass negative resistance (except the uniform field case, which shows a negative resistance down to dc). They show a considerable similarity between Read-type IMPATTs and a simple $p-n$ junction IMPATT. Therefore, many of the conclusions based on a Read diode can be extended at least qualitatively to IMPATT diodes in general.

In order to obtain a closed-form solution for (1) through (4), a number of simplifying assumptions are made. First the diode is divided into two regions: a narrow end region in which the field is assumed high enough
124 to cause a substantial carrier generation by impact ionization, and a second, larger region, in which the carriers injected from the avalanche drift at saturated or field-independent velocities.

Avalanche Region

In the avalanche region the total carrier current is assumed to be independent of x, i.e.,

$$J_0(t) = J_p(x, t) + J_n(x, t). \tag{5}$$

Adding (1) and (2) and integrating over the avalanche region gives

$$\int_0^{x_1} \frac{\partial}{\partial t}(p + n)dx = \int_0^{x_1} \left(\frac{1}{q} \frac{\partial}{\partial x}(J_n - J_p) + 2(\alpha_p p v_p + \alpha_n n v_n) \right) dx \tag{6}$$

where x_1 is the length of the avalanche region. Now, the additional simplifying assumptions are made that $\alpha_p = \alpha_n = \alpha$ and $v_p = v_n = v$. These assumptions are not valid for silicon devices. However, for the type of diode structure considered here, the avalanche region injects holes only into the drift region and, therefore, the contribution of the electrons to the total current is small. In the case of GaAs the assumption that $\alpha_n = \alpha_p$ is approximately valid. Therefore, because of the feedback of carriers into the high-field region of the diode that occurs when the ionization rates are equal, the avalanche region tends to be more localized in GaAs diodes than it is for the case of nonequal ionization rates. Thus, the assumption of a localized avalanche zone for a Read structure applies approximately to p-n junction GaAs diodes.

The current density is given by

$$J_n = qnv \tag{7}$$

and

$$J_p = qpv \tag{8}$$

where $v = v_n = v_p$. Therefore (6) becomes

$$\frac{x_1}{v} \frac{dJ_0}{dt} = [J_n - J_p]_0^{x_1} + 2J_0 \int_0^{x_1} \alpha(E)dx. \tag{9}$$

At $x = 0$, J_p is equal to the reverse saturation current J_{ps} of thermally generated holes. At $x = x_1$, $J_n = J_{ns}$ the electron reverse saturation current. The transit time of carriers through the avalanche zone is $\tau_1 = x_1/v$. Therefore (9) becomes

$$\frac{\tau_1}{2} \frac{dI_0}{dt} = I_0 \left(\int_0^{x_1} \alpha(E)dx - 1 \right) + I_s \tag{10}$$

where

$I_0 = J_0 A$
$I_s = (J_{ps} + J_{ns}) A$
A = the cross-sectional area of the diode.

In the work that follows, the reverse saturation current in (10) will be neglected. An analysis [7] of dc avalanche breakdown shows that the current multiplication, M, of the avalanche is given by

$$M = \frac{1}{1 - \int_0^{x_1} \alpha(E)dx}. \tag{11}$$

Thus, avalanche breakdown occurs when

$$\int_0^{x_1} \alpha(E)dx = 1 \tag{12}$$

or, in terms of (3),

$$\int_0^{x_1} A' e^{-(b'/E)^{m'}} dx = 1. \tag{13}$$

If E is assumed to be constant in the avalanche region, (13) becomes

$$A' e^{-(b'/E_c)^{m'}} x_1 = 1 \tag{14}$$

where E_c is the critical field for dc breakdown.

The ionization rate is now expanded in a Taylor series about E_c giving

$$\alpha = A' e^{-(b'/E)^{m'}} = A' e^{-(b'/E_c)^{m'}} + A' e^{-(b'/E_c)^{m'}} \cdot \frac{(b')^{m'} m'}{E_c^{m'+1}} \mathcal{E} + \cdots \tag{15}$$

or

$$\alpha \cong A' e^{-(b'/E_c)^{m'}} \left[1 + m' \left(\frac{b'}{E_c} \right)^{m'} \frac{\mathcal{E}}{E_c} \right] \tag{16}$$

where $\mathcal{E} = E - E_c$. Then, from (14) and (16),

$$\int_0^{x_1} \alpha(E)dx = 1 + m' \left(\frac{b'}{E_c} \right)^{m'} \frac{\mathcal{E}}{E_c} \tag{17}$$

$$= 1 + m \frac{\mathcal{E}}{E_c} \tag{18}$$

where m is defined as

$$m \triangleq E_c\alpha'(E_c)/\alpha(E_c). \tag{19}$$

Since $\alpha(E)$ is a very sensitive function of E, small variations in E result in large variations in the current. Therefore, only the first order terms in (15) are retained.

Now, from (10) and (18) the following equation relating the field and current is obtained:

$$\frac{\tau_1}{2} \frac{dI_0(t)}{dt} = mI_0(t) \frac{E(t)}{E_c}. \tag{20}$$

Drift Region

A second relation between the field and the current is obtained from Poisson's equation evaluated in the drift region where $N_D = N_A = 0$ is assumed. For the diode structure considered here the total carrier current in the drift region consists of the hole current $J_p = qpv$. Then, (4) becomes

$$\frac{\partial E}{\partial x} = \frac{I_0(t)}{\epsilon v A}. \tag{21}$$

The voltage drop across the diode consists of the voltage drop across the avalanche region below punch-through, the voltage drop due to the field that extends across the depletion layer above punch-through, and the voltage drop due to the space charge in the drift region. Thus,

$$-V_0(t) = -V_p + l[E_0(t) - E_{p0}] + \frac{I_0}{\epsilon v A} \int_{E_0}^{E} (l - x) dE \tag{22}$$

where

V_p = the dc voltage drop across the avalanche region at punch-through
$E_0(t)$ = the field at $x = 0$
E_{p0} = the field at punch-through at $x = 0$
l = the length of the drift region. Then, from (21)

$$-V_0(t) = -V_p + l[E_0(t) - E_{p0}] + \int_0^l (l - x) \frac{I_0}{\epsilon v A} dx. \tag{23}$$

It is assumed that the field in the drift region is high enough to saturate the drift velocity of the holes. Under these conditions carrier diffusion can be neglected and the current pulse will travel through the drift region without distortion. Thus, the current can be expressed as

$$I_0(x, t) = \mathrm{I}_0(t - x/v). \tag{24}$$

Next, define

$$t' = t - x/v. \tag{25}$$

Equation (23) then becomes

$$-V_0(t) = -V_p + l[E_0(t) - E_{p0}] - \frac{1}{\epsilon v A} \int_t^{t-\tau} v(\tau - t + t') I_0(t') v dt' \tag{26}$$

or

$$-V_0(t) = -V_p + l[E_0(t) - E_{p0}] + \frac{l}{\epsilon v A} \int_{t-\tau}^{t} (\tau - t + t') I_0(t') dt' \tag{27}$$

where $\tau = l/v$ = the carrier transit time through the drift region.

Equation (27) can be rewritten in terms of the field at $x = 0$ as

$$E_0(t) = E_{p0} + \frac{V_p}{l} - \frac{V_0(t)}{l} - \frac{1}{\epsilon \tau A} \int_{t-\tau}^{t} (\tau - t + t') I_0(t') dt'. \tag{28}$$

In (20) the field $E(t)$ is the ac variation about E_c; therefore,

$$E_0(t) = E_c + E(t). \tag{29}$$

Now let

$$V_0(t) = V_d + V(t) \tag{30}$$

where

V_d = the dc bias voltage
$V(t)$ = the ac voltage across the diode.

Then (28) becomes

$$E(t) = E_b - \frac{V(t)}{l} - \frac{1}{\epsilon \tau A} \int_{t-\tau}^{t} (\tau - t + t') I_0(t') dt' \tag{31}$$

where

$$E_b = E_{p0} - E_c - \frac{V_d}{l} + \frac{V_p}{l}. \tag{32}$$

One additional equation is needed to obtain the current flowing in the external circuit. From Maxwell's equation for $\nabla \times \vec{H}$ it follows that

$$\frac{\partial}{\partial x}\left(\epsilon \frac{\partial E}{\partial t} + J_0\right) = 0 \tag{33}$$

that is, the current is constant and independent of x throughout the circuit, as expected. However, E and J_0 are functions of x. Therefore, the external circuit current I is

$$I = A\left(J_0 + \epsilon \frac{\partial E}{\partial t}\right). \tag{34}$$

Integrating over x gives

$$lI = \epsilon A \int_0^l \frac{\partial E}{\partial t} dx + \int_0^l I_0 dx \tag{35}$$

125

or

$$I = -C_d \frac{dV}{dt} + \frac{1}{\tau}\int_{t-\tau}^{t} I_0(t')dt'. \tag{36}$$

Thus the first term is due to the depletion layer capacitance C_d and the second is the current induced in the external circuit by the carrier current I_0 drifting through the depletion layer. For future reference I_e is defined as

$$I_e = \frac{1}{\tau}\int_{t-\tau}^{t} I_0(t')dt'. \tag{37}$$

Equations (20), (31), and (36) result in a unique solution for the ac current and field if the dc bias current and ac voltage are specified.

III. Large-Signal Analysis

The large-signal behavior of the diode can be obtained from (20), (31), and (37). The simplest solution to these equations can be obtained by neglecting the effects of
126 space charge on the field $E(t)$. That is, (31) becomes

$$E(t) = -V(t)/l. \tag{38}$$

However, it is found that the space charge under most operating conditions cannot be neglected. The solution obtained by neglecting the field due to the space charge in the drift region will be referred to in the following work as the intermediate-amplitude case [5].

A more realistic assumption that leads to a manageable solution for the large-signal equations is the short transit time assumption referred to above. This assumption appears to be satisfied for a number of experimental devices. Yamashita *et al.* [8] have reported microwave oscillation from GaAs diodes with a transit angle of $(0.15\sim0.25)\pi$. Irvin [9] and Higgins *et al.* [10] have also reported microwave oscillations at frequencies where the transit angle is small. Much of the experimental work described in Section VII has been conducted with GaAs diodes similar to those used by Higgins. Furthermore, there is a considerable similarity in the small-signal behavior of the Read diodes and diodes for which the short transit time assumption is valid [4]. Therefore, the analysis that follows will provide qualitative information about most IMPATT diodes.

The short transit time assumption is introduced into (31). If $I_0(t)$ is slowly varying over time τ, then

$$E(t) = E_b - \frac{V(t)}{l} - \frac{\tau}{2\epsilon A} I_0(t). \tag{39}$$

Substituting this in (20) yields

$$\frac{\tau_1}{2}\frac{dI_0}{dt} = \frac{m}{E_c} I_0(t)\left(E_b - \frac{V(t)}{l} - \frac{\tau}{2\epsilon A} I_0(t)\right). \tag{40}$$

Equation (40) can be rewritten in a standard form as

$$\frac{dI_0(t)}{dt} = c(t)I_0(t) - aI_0^2(t) \tag{41}$$

where

$$c(t) = \frac{2m}{\tau_1 E_c}\left(E_b - \frac{V(t)}{l}\right) \tag{42}$$

and

$$a = \frac{m\tau}{\tau_1 \epsilon A E_c}\cdot \tag{43}$$

Equation (41) is the Bernoulli equation and has the solution [11]

$$I_0(t) = \frac{e^{-\phi}}{A_0 + a\int e^{-\phi}dt} \tag{44}$$

where A_0 is an arbitrary constant and

$$\phi = -\int c(t)dt. \tag{45}$$

Next, suppose that

$$V(t) = -V \sin \omega t. \tag{46}$$

Then,

$$\phi = -\frac{2m}{\tau_1 E_c}\int\left(E_b + \frac{V \sin \omega t}{l}\right)dt \tag{47}$$

or

$$\phi = -\frac{2m}{\tau_1 E_c}\left(E_b t - \frac{V \cos \omega t}{l\omega}\right)\cdot \tag{48}$$

Thus, the second term in the denominator of (44) grows with time. Therefore, the constant A_1 can be neglected after stable oscillations are established and the solution for $I_0(t)$ is

$$I_0(t) = \frac{e^{-bt-x\cos\omega t}}{a\int e^{-bt-x\cos\omega t}dt} \tag{49}$$

where

$$b = -\frac{2_m}{\tau_1 E_c}E_b = -\frac{m\tau \bar{I}_0}{\tau_1 E_c \epsilon A} \tag{50}$$

and

$$x = \frac{2\ \mathrm{mV}}{\tau_1 E_c l\omega}\cdot \tag{51}$$

The denominator of (49) can be integrated by expanding the exponential in a series of modified Bessel functions [12], $Q_k(x)$, as follows:

$$\Phi = \int e^{c\theta - x\cos\theta}d\theta$$
$$\Phi = \int e^{c\theta}\left(Q_0(x) + 2\sum_{k=1}^{\infty} Q_k(-x)\cos k\theta\right)d\theta \tag{52}$$

and

$$\Phi = e^{c\theta}\left(\frac{Q_0(x)}{c} + 2\sum_{k=1}^{\infty}\frac{Q_k(-x)}{k^2+c^2} \cdot (c\cos k\theta + k\sin k\theta)\right). \tag{53}$$

Then (49) may be expressed as

$$I_0(t) = \frac{\frac{\omega}{a} e^{-x\cos\omega t}}{\frac{Q_0(x)}{c} + 2\sum_{k=1}^{\infty}\frac{Q_k(-x)}{k^2+c^2}(c\cos k\omega t + k\sin k\omega t)} \tag{54}$$

where

$$c = -b/\omega. \tag{55}$$

Thus, the solution for $I_0(t)$ is periodic and the amplitude of the oscillation depends on the normalized voltage amplitude x and the dc bias current, since from (50) and (55)

$$c = \frac{2m\tau \bar{I}_0}{2\epsilon A\tau_1 E_c \omega} \triangleq \frac{\omega\tau}{2}\left(\frac{\omega_a}{\omega}\right)^2 \tag{56}$$

where

$$\omega_a = \left(\frac{2mI_0}{\epsilon A\tau_1 E_c}\right)^{\frac{1}{2}}$$

is the avalanche frequency.

Equation (54) was programmed for evaluation on a digital computer. The solution for several values of c and x is plotted in Figs. 1 and 2. The phase of the RF field and the solution for $i_0(t)$ obtained from Read's "intermediate amplitude" equations are shown for comparison.

IV. Large-Signal Results

The large-signal diode impedance is obtained by expanding the carrier current $I_0(t)$ in a Fourier series, i.e.,

$$I_0(t) = \frac{\omega}{a}\sum_{n=-\infty}^{\infty} R_n e^{jn\omega t} \tag{57}$$

where

$$R_n = \frac{a}{2\pi}\int_{-\pi/\omega}^{\pi/\omega} I_0(t)e^{-jn\omega t}dt. \tag{58}$$

Because of the complexity of (54), the coefficients R_n are found numerically. The external circuit current is then found from (37) and (57).

$$I_e(t) = \frac{\omega}{a\tau}\int_{t-\tau}^{t}\sum R_n e^{jn\omega t}dt. \tag{59}$$

From (59), the dc current is

$$I_d = \frac{\omega}{a}R_0. \tag{60}$$

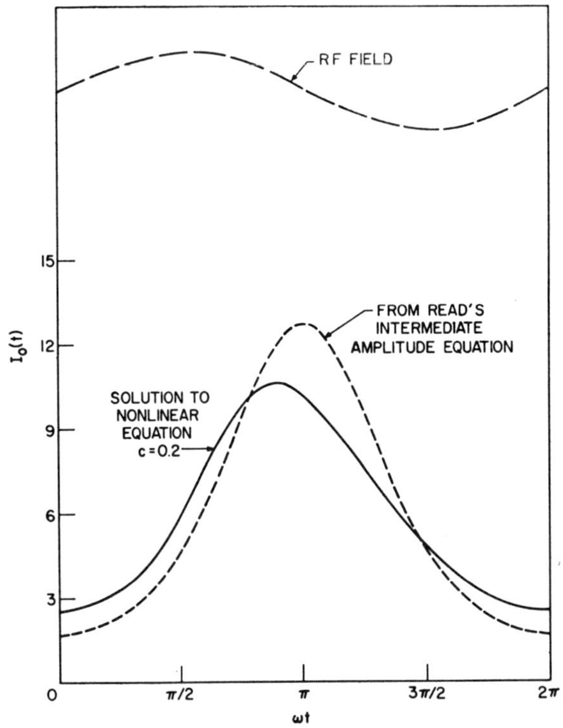

Fig. 1. Normalized particle current versus ωt ($c=0.2$ and $x=1.0$).

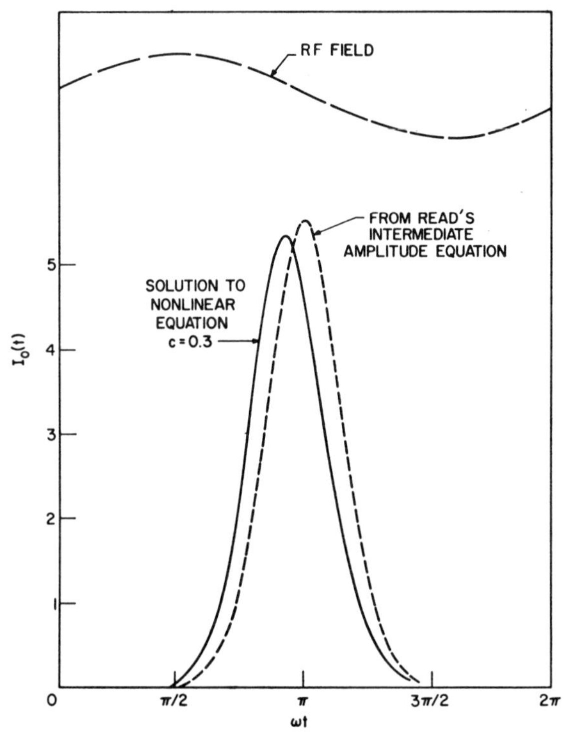

Fig. 2. Normalized particle current versus ωt ($c=0.3$ and $x=5.0$).

However, $(\omega/a)R_0$ is the average value of $I_0(t)$. Therefore, from (43) and (56),

$$R_0 = \frac{a}{\omega}I_d = c. \tag{61}$$

The external circuit current at the frequency ω is

$$i_{e1}(t) = -\frac{jR_1}{\tau a}(1 - e^{-j\omega\tau})e^{j\omega t}. \tag{62}$$

For the current direction assumed, the diode admittance is calculated from (62) and the negative of the voltage given by (46). Thus

$$V(t) = V \sin \omega t = V\left(\frac{e^{j\omega t} - e^{-j\omega t}}{2j}\right) \tag{63}$$

and the diode admittance is given by

$$Y = \frac{i_{e1}}{V} = 2\,\frac{R_1(c, x)}{\tau a V}\,(1 - e^{-j\omega\tau}) \tag{64}$$

or from (51)

$$Y = \frac{2}{\omega\tau}\,g_1\,\frac{R_1(c, x)}{x}\,(1 - e^{-j\omega t}) \tag{65}$$

where $g_1 = 2C_d/\tau$. The susceptance of the depletion layer capacitance must be added to this admittance in order to obtain the total admittance of the diode.

The diode impedance thus depends on the parameters ω, τ, c, and x. The expressions for c and x can be rewritten as

$$c = \frac{l\zeta}{\epsilon v \omega} J \tag{66}$$

and

$$x = \frac{2\zeta}{v\omega} V \tag{67}$$

where

$$\zeta = \frac{m}{\tau_1 E_c}$$

and

$$J = I_d/A. \tag{68}$$

Now from (56)

$$\omega_a^2 = \frac{2\zeta}{\epsilon} J. \tag{69}$$

In Section II it was assumed that the field and the current were independent of x in the avalanche region. Gummel and Blue [13] have shown that this assumption results in an error in the avalanche frequency of $\sqrt{3/2}$. On the other hand, Read [5] has shown that variations of field with distance in the avalanche zone can be accounted for by correcting the value used for m. Therefore, it seems more useful simply to retain ω_a as an independent variable of the analysis that can be estimated either theoretically or experimentally for a given diode.

The diode impedance is shown in Fig. 3 as a function of frequency for several values of the RF voltage amplitude. For this case the avalanche frequency was chosen as 10 GHz at a current density of 1200 A/cm². As expected, the large-signal impedance reduces to that obtained by Gilden and Hines [3] for small RF voltages. Above the avalanche frequency, as the RF voltage is

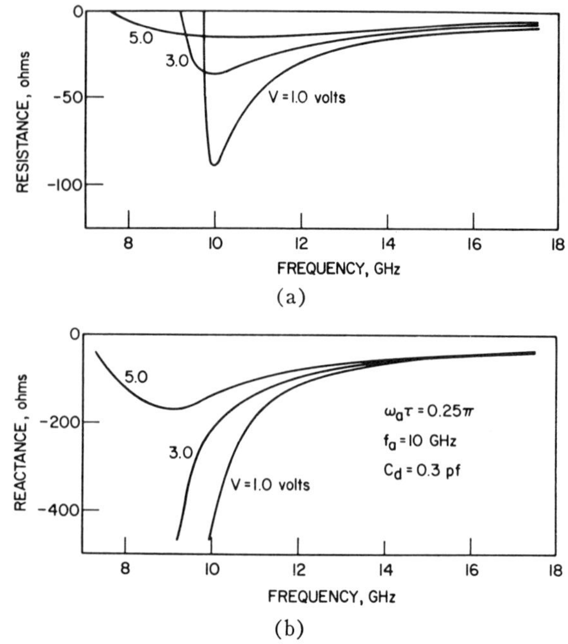

Fig. 3. Diode impedance versus frequency for a bias current density of 1200 A/cm² (I_d=40 mA). (a) Diode resistance. (b) Diode reactance.

increased, the negative resistance of the diode decreases, leading to a condition of stable oscillation for a fixed load resistance. Below the avalanche frequency an unstable region exists in which the negative resistance of the diode increases with increasing RF voltage. However, for sufficiently large voltages the negative resistance at any given frequency does eventually decrease with increasing RF voltage. Small-signal theory predicts that the impedance curves should shift up in frequency as $I_d^{1/2}$. However, when the diode is oscillating, increasing the bias current increases the RF voltage and therefore the impedance does not change as predicted by small-signal theory. Thus, the current tuning of an IMPATT oscillator is somewhat different than that expected from small-signal theory. This point is discussed in more detail below.

Fig. 4(a) shows the variation of the diode resistance with RF voltage. For bias currents above 40 mA the avalanche frequency exceeds the operating frequency and the small-signal diode resistance is positive. Since the diode reactance is much larger than the diode resistance it is generally found that the conductance of the equivalent parallel circuit is also a decreasing function of the RF voltage. The RF power is given as a function of voltage in Fig. 4(b). Although the power output for a given RF voltage does not change greatly with increasing current, it should be noted that the series load resistance required to obtain high power at low current density is much smaller.

The power and efficiency for the diode considered above are shown in Fig. 5 for several values of load resistance. Since the field in the drift region has not been specified (except that it must be high enough to saturate the carrier drift velocity), the voltage across the diode at breakdown is not determined by any of the param-

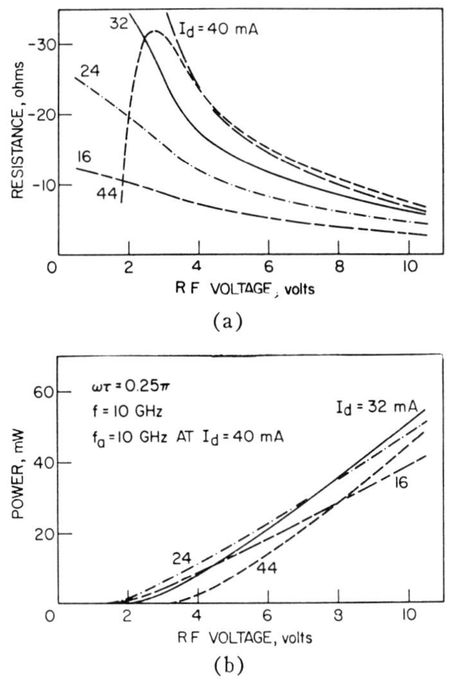

Fig. 4. Resistance and power versus RF voltage. (a) Resistance versus RF voltage. (b) Power versus RF voltage.

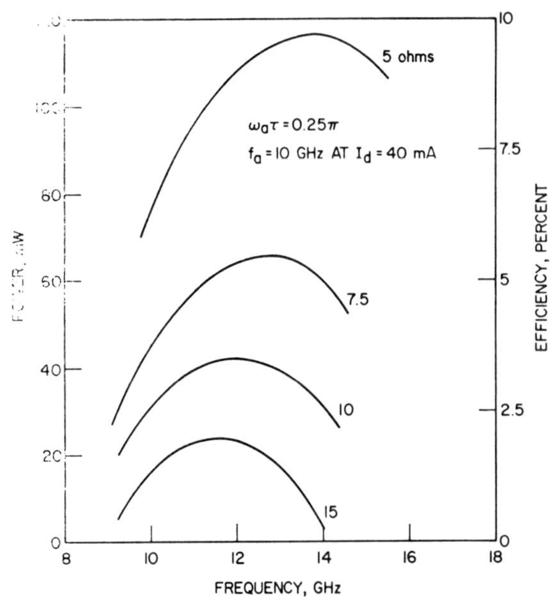

Fig. 5. Power versus frequency for fixed load resistance.

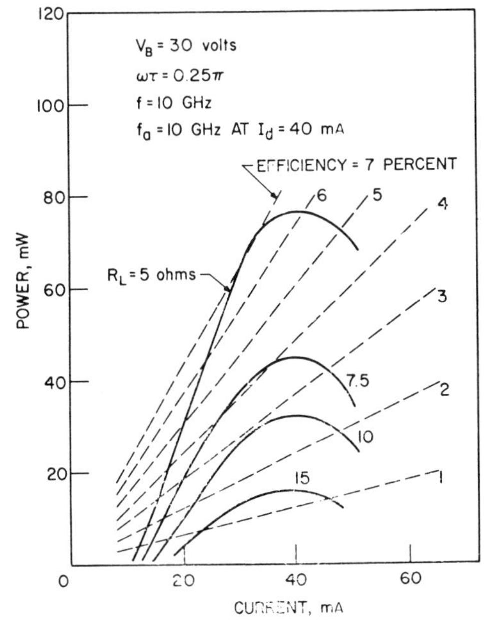

Fig. 6. Power versus bias current for fixed load resistances.

129

eters given above. Therefore, the efficiency in Fig. 5 is calculated for a breakdown voltage of 30 volts, which is a reasonable value for a diode of the given dimensions. Notice that for a fixed bias current the power output is not a maximum at the avalanche frequency. The shift of the maximum to a higher frequency occurs because of the increase in transit angle as the frequency is increased. On the other hand, as the frequency is increased the shunting effect of the depletion layer capacitance reduces the power output. Fig. 6 shows the power output and efficiency as a function of the bias current for several values of load resistance. From these curves the power output is seen to be a maximum, for a given frequency, when the avalanche frequency equals the operating frequency. The maximum efficiency occurs for a bias current about 25 percent below this value.

Figs. 5 and 6 indicate that the power output is strongly dependent on the series load resistance. Fig. 7 shows this relation explicitly for several values of the bias current. These results indicate very high power output at small values of series load resistance. Some caution is required in extrapolating these results to small load resistances because the magnitude of the RF voltage also increases rapidly in this region so that the field in the drift region may drop below that required to maintain a saturated drift velocity. This effect is offset somewhat for a short-transit angle diode because little current is flowing when the field in the drift region reaches a minimum.

The small transit angle approximation was made in obtaining the avalanche injected current, but not in calculating the phase delay of the external circuit current, which is caused by carrier propagation through the drift region. Therefore, the analysis can be used to predict at least qualitatively the behavior of diodes with a longer transit angle. For example, calculations have been made on a diode with a transit angle of 0.5 π radians, a depletion layer capacitance of 1.0 pF, and an avalanche frequency of 5.0 GHz at a current density of 300 A/cm^2. The negative resistance in this diode was found to be somewhat smaller than that given in Fig. 3 because of the larger value of C_d. Under these circumstances a change in the series load resistance of 1 to 2 ohms has a significant effect on the power output of the diode. Thus, the series bulk resistance of the diode can substantially reduce the power output of such an IMPATT diode. Moreover since this bulk resistance is a significant part of the load seen by the active part of the

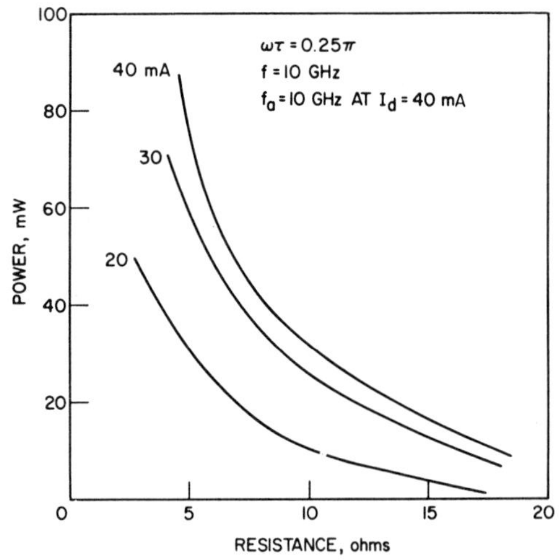

Fig. 7. Power versus load resistance for fixed bias currents.

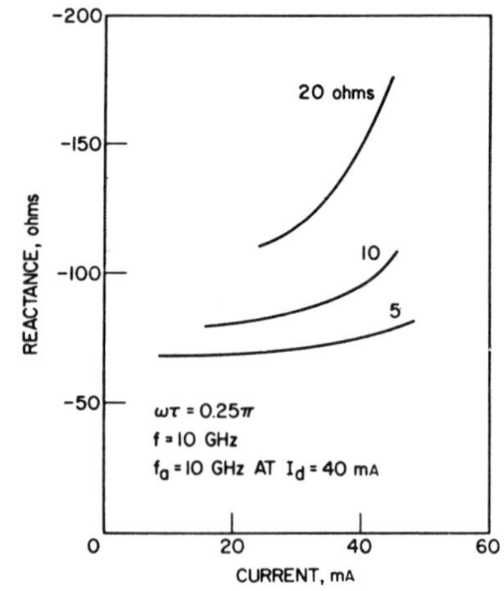

Fig. 8. Reactance versus bias current for fixed load resistance.

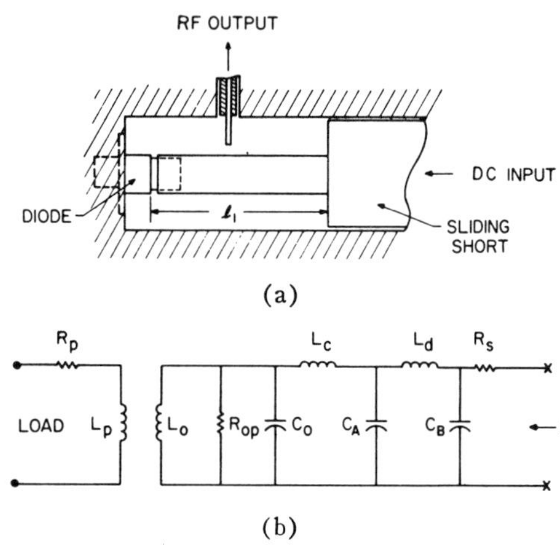

Fig. 9. Oscillator and equivalent circuit. (a) Oscillator. (b) Equivalent circuit.

diode, the available power is further reduced. For the
130 diode considered here with a load resistance of 0.5 ohm, decreasing the parasitic resistance from 2.5 to 0.5 ohms, the available power increases by a factor of 10 to about 1.0 watt. These conclusions are consistent with the experimental results given by Kovel and Gibbons [14].

Another interesting result can be obtained from the previous curves by plotting the diode reactance as a function of the bias current for various values of load resistance. Fig. 8 shows that a large value of series resistance results in a large change in the diode reactance as the bias current is increased. Thus, if the diode is placed in a low Q circuit, low power output and substantial current tuning would be expected. However in a high Q circuit, high power and little frequency tuning would be obtained. Of course, to predict the tuning behavior in a given oscillator the impedance presented to the diode by the circuit must be known.

V. Experimental Results

To confirm some of the results of the large-signal analysis, measurements of power output and current tuning of an IMPATT oscillator were made. The oscillator and its equivalent circuit are shown in Fig. 9. The oscillator consists of a reentrant cavity with the diode mounted at one end and a sliding RF short at the other end. Microwave power is coupled out of the cavity by a simple probe. The equivalent circuit of a packaged diode mounted at the end of a coaxial line has been discussed by Getsinger [15]. To those results are added a circuit for the coupling probe and load and the series bulk resistance (R_s) of the diode. The element values have been determined by measurements and calculations and are discussed in detail elsewhere [16]. From these results it is possible to determine the equivalent series impedance that is seen by the diode and the coupling loss from the active part of the diode to the load.

A number of different types of IMPATT diodes have been operated in this circuit. For comparison with the theoretical results given above an abrupt junction GaAs diode was chosen. Since the ionization rates for holes and electrons are equal in GaAs, the abrupt junction GaAs diode has a somewhat localized avalanche zone and a region where the carriers drift at a saturated velocity. In addition, this diode has a depletion layer width at breakdown of about 1.0 μm, which at 15 GHz gives a transit angle of 0.15 π and thus satisfies the short transit angle assumption.

The avalanche frequency can be estimated experimentally by measurements of the oscillator frequency as a function of the position of the sliding short for small amplitude oscillations. Fig. 10 shows the negative of the reactance that the circuit presents to the diode for several values of the short position. The diode reactance has been matched to the circuit reactance at 15.6 GHz (which is the measured oscillator frequency for a short position of 1.0 cm) by choosing an avalanche frequency of 10.9 GHz at a bias current of 8.0 mA. The intersection of the diode reactance for other short positions then gives the expected tuning curve for the oscillator. Fig. 11 shows that there is good agreement between these values and the measured frequencies of oscillation.

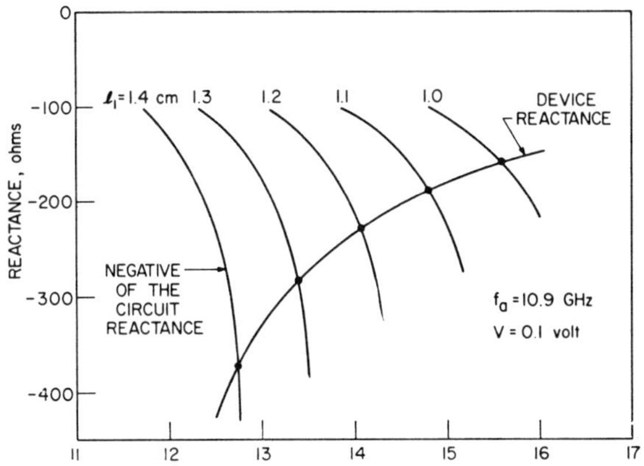

Fig. 10. Device and negative circuit reactance.

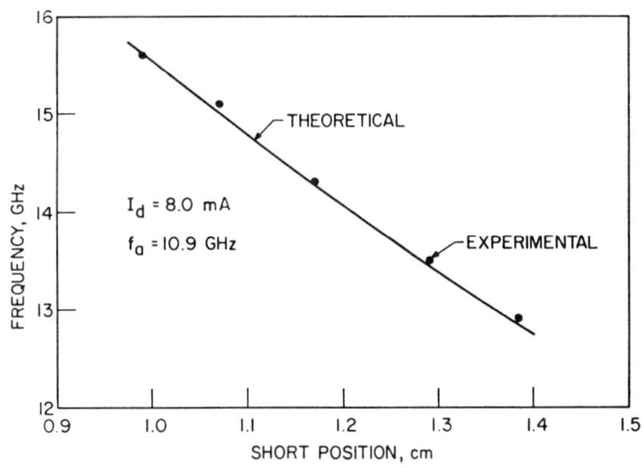

Fig. 11. Frequency versus short position.

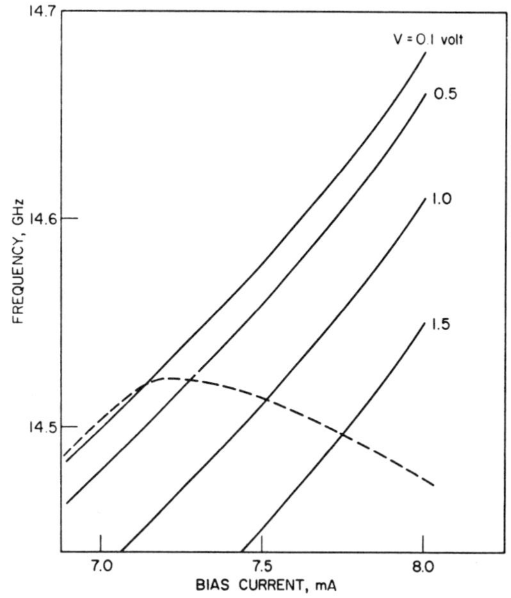

Fig. 12. Predicted frequency versus bias current for experimental oscillator. ($L_1 = 1.15$ cm and $f_a = 10.9$ GHz at 8 mA.) 131

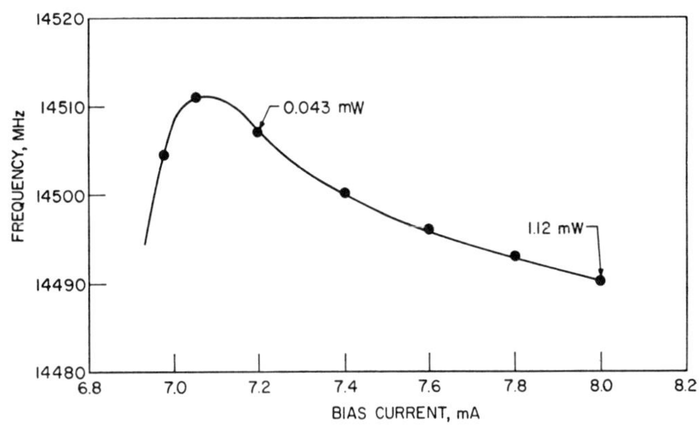

Fig. 13. Frequency versus bias current (experimental).

Having established a value for the avalanche frequency, the results obtained for the diode impedance can be combined with the circuit impedance to predict the oscillator behavior. Fig. 12 shows the oscillation frequency as a function of the bias current for several values of the RF voltage. The curve for $V = 0.1$ volt corresponds approximately to the small-signal tuning to be expected from the diode. Although the circuit causes a significant departure from the $I^{\frac{1}{2}}$ variation predicted for the diode alone, the oscillation frequency does increase rapidly with the bias current. However, the oscillator does not follow a line of constant voltage as the bias current is increased. Using experimental results for power versus bias current and theoretical results for power versus RF voltage, the RF voltage as a function of bias current can be determined. In turn, this result is plotted in Fig. 12 to determine the frequency variation as a function of the bias current. This is shown by the broken line in Fig. 12. Thus, for the particular diode considered here the current tuning behavior differs considerably from that expected from small-signal theory and in fact the oscillation frequency decreases at larger power levels. This predicted behavior is very easily measured and the results are shown in Fig. 13. Considering the number of assumptions that must be made, the results are in good agreement with the theoretical curve.

Other experimental measurements on power output and diode resistance are in good agreement with the theoretical results [16].

VI. Conclusion

The nonlinear equations that characterize the Read-type IMPATT diode have been solved subject to the assumption that the transit time of the carriers through the drift region is small compared to a period of the RF oscillation. This solution is then used to obtain the large-signal impedance of the diode. Several conclusions about the performance of IMPATT oscillators can be drawn from this analysis.

1) The negative resistance of the diode decreases with increasing RF voltage amplitude, thus leading to stable oscillator operation.

2) Oscillator tuning with dc bias current is substantially reduced under large-signal conditions with less tuning for higher efficiency oscillators.

3) The power output of the oscillator is a very sensi-

tive function of the series load resistance presented to the active part of the diode.

4) The power output of the oscillator exhibits saturation with increasing bias current.

References

[1] R. L. Johnston, B. C. DeLoach, and B. G. Cohen, "A silicon diode microwave oscillator," *Bell Sys. Tech. J.*, vol. 44, pp. 369–372, February 1965.

[2] T. Misawa, "Negative resistance in *p-n* junctions under avalanche breakdown conditions," *IEEE Trans. Electron Devices*, vol. ED-13, pp. 137–151, January 1966.

[3] M. Gilden and M. E. Hines, "Electronic tuning effects in the Read microwave avalanche diode," *IEEE Trans. Electron Devices*, vol. ED-13, pp. 169–175, January 1966.

[4] H. K. Gummel and D. L. Scharfetter, "Avalanche region of IMPATT diodes," *Bell Sys. Tech. J.*, vol. 45, pp. 1797–1828, December 1966.

[5] W. T. Read, "A proposed high frequency negative resistance diode," *Bell Sys. Tech. J.*, vol. 37, pp. 401–446, March 1, 1958.

[6] S. M. Sze and G. Gibbons, "Avalanche breakdown voltages of abrupt and linearly graded *p-n* junctions in Ge, Si, GaAs and GaP, *Appl. Phys. Letters*, vol. 8, pp. 111–115, March 1966.

[7] J. L. Moll, *Physics of Semiconductors*. New York: McGraw-Hill, 1964, pp. 212–238.

[8] S. Yamashita *et al.*, "Frequency of microwave oscillations in GaAs *p-n* junctions," *Japan J. Appl. Phys.*, vol. 5, p. 843, September 1966.

[9] J. C. Irvin, "GaAs avalanche microwave oscillators," *IEEE Trans. Electron Devices* (*Correspondence*), vol. ED-13, pp. 208–210, January 1966.

[10] V. J. Higgins, F. A. Brand, and J. J. Baranowski, "Characteristics of varactors biased into avalanche," *IEEE Trans. Electron Devices* (*Correspondence*), vol. ED-13, pp. 210–212, January 1966.

[11] G. M. Murphy, *Ordinary Differential Equations and Their Solutions*. Princeton, N. J.: Van Nostrand, 1960, pp. 23–27.

[12] M. Abramowitz and I. A. Stegun, *Handbook of Mathematical Functions*, Appl. Math. Ser. 55. Washington, D. C.: NBS, 1964, p. 376.

[13] H. K. Gummel and J. L. Blue, "A small-signal theory of avalanche noise in IMPATT diodes," *IEEE Trans. Electron Devices*, vol. ED-14, pp. 569–580, September 1967.

[14] S. R. Kovel and G. Gibbons, "The effect of unswept epitaxial material on the microwave efficiency of IMPATT diodes," *Proc. IEEE* (*Letters*), vol. 55, pp. 2066–2067, November 1967.

[15] W. J. Getsinger, "The packaged and mounted diode as a microwave circuit," *IEEE Trans. Microwave Theory and Techniques*, vol. MTT-14, pp. 58–69, February 1966.

[16] W. J. Evans, "Nonlinear and frequency-conversion characteristics of IMPATT diodes," *Tech. Rept. 104*, Electron Physics Laboratory, Dept. of Electrical Engineering, University of Michigan, Ann Arbor, February 1968.

Reprinted from THE BELL SYSTEM TECHNICAL JOURNAL
Vol. 48, No. 2, February, 1969
Printed in U.S.A.

Approximate Large-Signal Analysis of IMPATT Oscillators

By J. L. BLUE

(Manuscript received August 11, 1968)

133

Insight into the large-signal operating characteristics of IMPATT oscillators has been obtained from detailed numerical calculations of D. L. Scharfetter and H. K. Gummel. However, their calculations are relatively expensive in computer time. Small-signal analyses of IMPATT diodes are much less expensive, but give little reliable information about oscillator performance.

This paper presents a model for large-signal analysis of IMPATT diodes which requires less than 1 percent of the computer time used by Scharfetter and Gummel's method, but still provides a realistic description of IMPATT oscillators in modes of operation which maintain carrier velocity saturation. We show graphical results, based on numerical computations, which provide information about phase relationships in IMPATT oscillators and improve understanding of the two-frequency mode of operation.

I. INTRODUCTION

Considerable insight into the large-signal operating characteristics of IMPATT oscillators has recently been obtained by D. L. Scharfetter and H. K. Gummel through numerical calculations involving a complete modeling of the physical processes taking place in the diode.[1] They solve the one-dimensional partial differential equations for the generation, diffusion, and drift of holes and electrons, as well as Poisson's equation and the differential equations for the circuit in which the diode is embedded. Their analysis, which has led to improved understanding of IMPATT diodes and insight into new modes of oscillation, includes most of the important physical effects. Since their model is so detailed and accurate, their analysis is a large-scale computer project. Even after the computer programs are debugged, analysis of IMPATT diodes is relatively expensive.

Another possible method is small-signal analysis, which is tractable numerically and more economical.[2–4] However, small-signal analysis gives only a limited amount of information about large-signal oscillation.

An intermediate model has been presented by W. J. Evans.[5] However, his simplified large-signal model requires short transit-time in the drift region. He obtained a single first-order differential equation for the diode current, and was able to solve analytically for the (large-signal) diode current and voltage as a function of time. Unfortunately, the restriction to short transit time and his use of other approximations severely limit the applicability of his results.

This paper presents a model for large-signal analysis of IMPATT diodes which incorporates most of the important physical processes in the diodes, but which is more tractable numerically than that of Scharfetter and Gummel. Because the new model requires far fewer computer hours than Scharfetter and Gummel's model, parameter studies are feasible; because the new model contains more physics than Evans' model or the small-signal models, the results are more reliable and more effects can be studied.

II. LARGE-SIGNAL IMPATT MODEL

We consider a one-dimensional diode, neglect diffusion, and assume that the electric field is large enough so that holes and electrons move at scatter-limited velocities. The latter assumption is essential for simplifying the calculations, but means that our model is not applicable to some important oscillator modes, such as analyzed by Scharfetter, Bartelink, and Johnston.[6]

The other important simplification is that in Poisson's equation we neglect the space charge of the ac component of the particle current in the avalanche region. The detailed calculations of Scharfetter and Gummel show that this is a good approximation for the classical Read mode of operation and other modes where the width of the avalanche region remains fairly constant during the oscillation cycle.[7] We include the space charge of the dc component of the particle current in the avalanche region, and both ac and dc components of the particle current in the drift region. This says that the shape of the electric field profile in the avalanche region does not change in time, although the magnitude of the field does; that is,

$$E_{\text{aval}}(x, t) = f(x) + E(t). \tag{1}$$

This is equivalent to the commonly-made assumption that the particle current is constant in the avalanche region.[2, 5, 7] Given a bias current and the doping in the diode, we can calculate $f(x)$ by solving a second-order differential equation by standard methods.[3, 4]

We can treat diodes in which holes and electrons have unequal scatter-limited velocities and unequal ionization coefficients, although the model we have chosen for computation assumes the same values for electrons as for holes. The ionization coefficients may be any given function of the electric field, such as $A\ \exp(-b/E)$.

Besides the simplifications resulting from our diode model, we obtain important simplifications by solving directly for a periodic solution, rather than solving for a transient solution which eventually becomes periodic. We save computation time both because we eliminate solving for the unneeded transient solution, and because we are able to use more advantageous numerical methods. 135

Notice that the essential approximations are constant carrier velocities and equation (1); many combinations of additional approximations may be made, so that we are describing a whole hierarchy of IMPATT models. The aproximation of equation (1) can be overcome by a recursive method of solution, the first step of which is the subject of this paper. However, for most purposes equation (1) is sufficiently accurate.

III. IDEALIZED DIODE

In the rest of this paper, we consider a fairly simple model of an IMPATT diode; we are able to emphasize the essential features of the method without becoming lost in algebra. (We plan to treat the most general case in a subsequent article.)

We consider a diode with a uniform electric field $[f(x) = 1]$ in the avalanche region (of length d_a), and one drift region (of length d_d). We assume equal hole and electron velocities and ionization coefficients, so that the continuity equations for holes and electrons in the avalanche region are

$$\partial p(x, t)/\partial t = v\alpha[E(t)][p(x, t) + n(x, t)] - v\ \partial p(x, t)/\partial x$$

$$\partial n(x, t)/\partial t = v\alpha[E(t)][p(x, t) + n(x, t)] + v\ \partial n(x, t)/\partial x.$$

The importance of the approximations discussed in Section II is that they enable us to avoid dealing with the hole and electron densities as functions of both space and time. We proceed much as did

136

Read[7] and Evans,[5] adding the two continuity equations to obtain an equation involving the particle current density

$$\partial I_p(x, t)/\partial t = 2v\alpha I_p - qv^2\, \partial(p - n)/\partial x. \tag{2}$$

The total current, particle plus displacement, is

$$I_T(t) = I_p(x, t) + \epsilon\, \partial E(x, t)/\partial t. \tag{3}$$

I_T is independent of x; since E in the avalanche region is independent of x, $I_p(x, t)$ is also independent of x. Using this fact, we integrate equation (2) over the avalanche region. Neglecting reverse saturation current, we obtain

$$\partial I_p/\partial t = \frac{2vI_p}{d_a}\{d_a\alpha[E(t)] - 1\}.$$

If we use $\alpha(E) = A \exp(-b/E)$, and define E_c as the "critical" field (that field necessary to maintain the dc current under static conditions) then $\alpha(E_c) = 1/d_a$. We obtain

$$\partial I_p/\partial t = \frac{2vI_p}{d_a}[(Ad_a)^{1-E_c/E(t)} - 1]. \tag{4}$$

(Evans linearized the electric field, and used

$$\partial I_p/\partial t = \frac{2v}{d_a} I_p m\left[\frac{E(t)}{E_c} - 1\right].$$

We retain the more nonlinear version.)

We treat the case in which there is an applied voltage* $V(t)$ in excess of that required to maintain the dc reverse bias current density I_0. The avalanche electric field is reduced by the space charge of the particle current in the drift region; as in Read, we have

$$E(t) = E_c + V(t)/(da + d_d) - \frac{1}{\epsilon}\int_{t-T_d}^{t} I_p(t')\left(1 - \frac{t' - t}{T_d}\right) dt' \tag{5}$$

where $T_d = d_d/v$ is the time for particles to pass through the drift region after leaving the avalanche region. In equation (5) and the remainder of this paper we specialize to the case $d_a \ll d_d$, as Read and Evans did.

Equations (4) and (5), together with suitable initial conditions, may be solved for $E(t)$ and $I_p(t)$. The total current, which is equal

* We could equally well treat the case where there is an applied terminal current $I_T(t)$. It is only slightly more difficult to embed the diode in a circuit; then we use Kirchoff's laws to write $V(t)$ in terms of $I_T(t)$ and the circuit parameters, generally as an integral over I_T.

to the terminal current, may be obtained from equation (3):

$$I_T(t) = \frac{\epsilon}{d_a + d_d} dV(t)/dt + \frac{1}{T_d} \int_{t-T_d}^{t} I_p(t')\, dt'. \tag{6}$$

The first term is the displacement current and the second is the particle-induced terminal current, I_{T_p}.

Instead of solving the initial-value equations for the transient solution, and extending the solution far enough in time for the limiting periodic solution to be obtained, it is preferable to solve directly for only the periodic solution. We assume $V(t)$ to be periodic (not necessarily sinusoidal) with period τ, and find solutions $E(t)$ and $I_p(t)$ which also have period τ. Now we need solve only for t in the interval $(0, \tau)$. It is more convenient to rewrite equation (4) as an integral equation; integrating equation (4) from 0 to t, we obtain ($T_a = d_a/v$)

$$I_p(t) = I_p(0) \exp \left\{ \frac{2}{T_a} \int_0^t [(Ad_a)^{1-E_c/E(t')} - 1]\, dt' \right\}. \tag{7}$$

We substitute equation (5) for $E(t')$, and have a single integral equation for $I_p(t)$, which we shall not write explicitly; since $I_p(t + \tau) = I_p(t)$, we solve only for $0 \leqq t \leqq \tau$. Because of its complexity, it will be solved numerically; for our method to be useful, we must be able to find numerical solutions quickly and reliably. We solve this nonlinear integral equation by a standard iterative method, which converges quickly and stably.

We divide the interval $(0, \tau)$ into M intervals, let $t_n = n\tau/M$, and solve equation (7) approximately for the M values $I_p(t_n)$, $n = 1, 2, \ldots, M$. The integrals are evaluated by, for example, the trapezoidal rule, with integrands evaluated only at times t_n. We obtain a system of M nonlinear equations for the values $I_p(t_n)$, which are easily solved by, for example, the matrix analog of the Newton-Raphson method. More accurate integration rules than the trapezoidal rule may be used. They are more complicated to program, but save execution time because, for a given accuracy, fewer points need be used.

After the $I_p(t_n)$ are found, then the $I_T(t_n)$ may be found from equation (6). Then we Fourier-analyze I_T and V; the ratio of their fundamental components is the large-signal admittance at the frequency $\nu = 1/\tau$. The ac power delivered by the diode also is easily calculated.

The small-signal admittance can be calculated analytically. It is

$$Y(\omega) = \frac{\epsilon}{(d_a + d_d)T_d} \left[\frac{1 - \omega^2/\omega_a^2}{\xi(\omega) + i\omega/\omega_a^2 T_d} \right]$$

137

where

$$\xi(\omega) = \frac{\exp(-i\omega T_d) - 1 + i\omega T_d}{(-i\omega T_d)^2}$$

and

$$\omega_a^2 = \frac{2\alpha'(E_c)vI_0}{\epsilon}.$$

This admittance agrees with that of Gilden and Hines, when their "passive resistance of the inactive zone R_s" is neglected.[2]

IV. RESULTS

We analyze a germanium IMPATT oscillator similar to the one reported by Swan.[8] The results, which are presented graphically, are meant more as an illustration of typical results obtainable than as a comprehensive analysis of Swan's oscillator. In particular, the uniform avalanche region approximation is inadequate. However, even these preliminary results can clarify some aspects of IMPATT oscillator operation.

The diode we analyzed has an avalanche width $d_a = 1.5 \times 10^{-4}$ cm, a drift width $d_d = 3.5 \times 10^{-4}$ cm, and a carrier velocity $v = 5 \times 10^6$ cm per second. The ionization coefficient is $\alpha(E) = 5.9 \times 10^{+6}$ cm^{-1} $\exp[-1.2 \times 10^6 \text{ (V/cm)}/E]$.

Figure 1 shows typical solutions for the ac components of $V(t)$, $I_p(t)$, and $I_{T_p}(t)$. Two cycles are shown; we plot I_{T_p} instead of I_T, because the displacement current component of I_T delivers no power. The frequency is 5 GHz, and the ac voltage amplitude is 10 volts. The avalanche region particle current is extremely nonsinusoidal, and becomes more so with increasing ac voltage amplitude. The fundamental component of $I_p(t)$ lags the ac voltage by about 49.5 degrees; this phase angle also increases with increasing ac voltage, approaching 90 degrees. The phase is closer to 90 degrees for higher frequencies; this may be attributed to the lessened effect of drift region space charge on the avalanche region field. (Read showed that the phase is 90 degrees when the space charge can be neglected, so that the electric field in the avalanche region is in phase with the terminal voltage.) The phase will be discussed further in connection with Fig. 3.

Figure 2 is a complex-plane admittance plot; the dc component of I_T is fixed at 500 A/cm^2, and the ac component of $V(t)$ is a pure sine wave, as it would be if the diode were embedded in a high-Q parallel

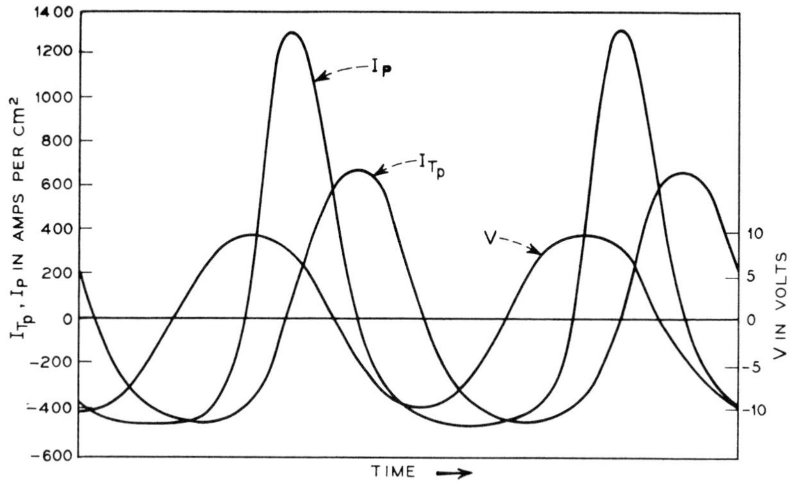

Fig. 1— AC components of V, I_p, I_{T_p} at frequency 5 GHz, current density 500 A/cm^2, ac voltage amplitude 10 V. Two cycles shown.

resistance-impedance-capacitance (RLC) circuit with only one resonant frequency. The solid curves are curves of constant ac voltage amplitude; the dashed lines are curves of constant frequency. For frequencies about 6 GHz and above, the negative conductance decreases with increasing ac voltage amplitude. For lower frequencies, the negative conductance first increases, then decreases with increasing ac voltage amplitude. For some low frequencies, for example 4 GHz, the diode has positive small-signal conductance, but negative large-signal conductance for large enough ac voltage. This type of behavior agrees with that found at low frequencies by the detailed calculations of Scharfetter and Gummel.[9]

Insight into this type of behavior may be obtained from Fig. 3, which shows the amplitude and phase (with respect to the ac voltage) of the fundamental component of $I_p(t)$. Curves of constant ac voltage amplitude are solid and curves of constant frequency are dashed. To transform any point on this plot to a complex admittance plot, one first multiplies the complex number representing the point by

$$[\exp(-i\omega T_d) - 1]/(-i\omega T_d)$$

to give the fundamental component of I_{T_p}. (For small ωT_d, this is approximately the same as rotating the phase plot point of I_p clockwise

140

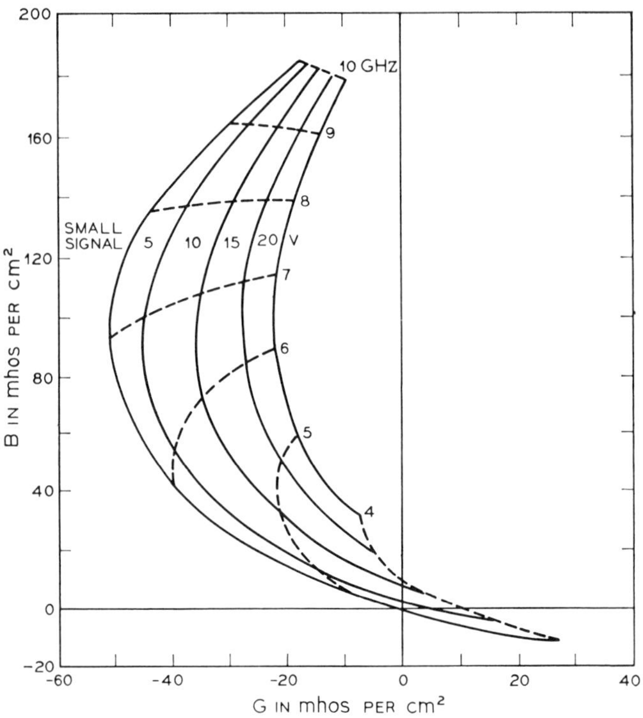

Fig. 2 — Complex-plane plot of diode admittance as a function of frequency and ac voltage amplitude. AC component of $V(t) = A \cos \omega t$. Current density 500 A/cm^2.

by $\frac{1}{2}\omega T_d$.) If the resulting point is in the second or third quadrant, then the diode has negative conductance at that frequency and ac voltage amplitude. Then the amplitude is divided by the magnitude of the fundamental component of the ac voltage amplitude; finally the point is moved directly upwards by $\epsilon\omega/(d_a + d_d)$ to add in the displacement current component.

At low frequencies, increased ac voltage amplitude improves the avalanche current phase considerably; at 4 GHz, where the small-signal conductance is positive, the improvement is sufficient so that the approximately $\frac{1}{2}\omega T_d$ rotation puts the phase plot point into the third quadrant, and negative conductance results for amplitudes above about 12 volts.

Figure 4 shows output power in the fundamental, in watts per square centimeter. Maximum output power is obtained near the fre-

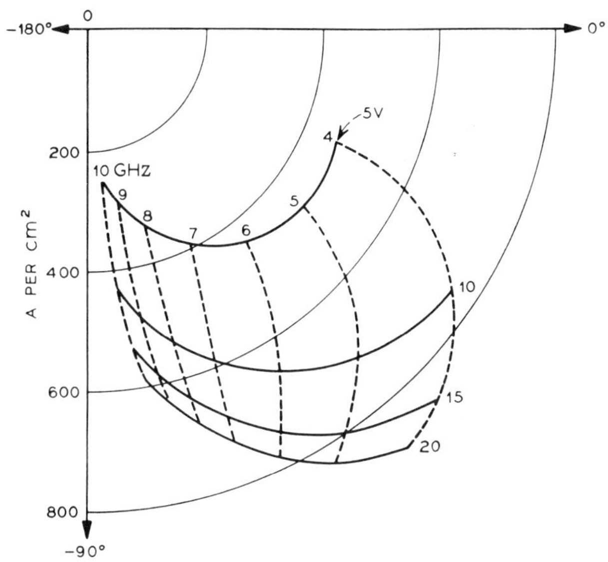

Fig. 3 — Magnitude and phase plot of fundamental component of avalanche particle current, $I_p(t)$, as a function of frequency and ac voltage amplitude. Current density 500 A/cm².

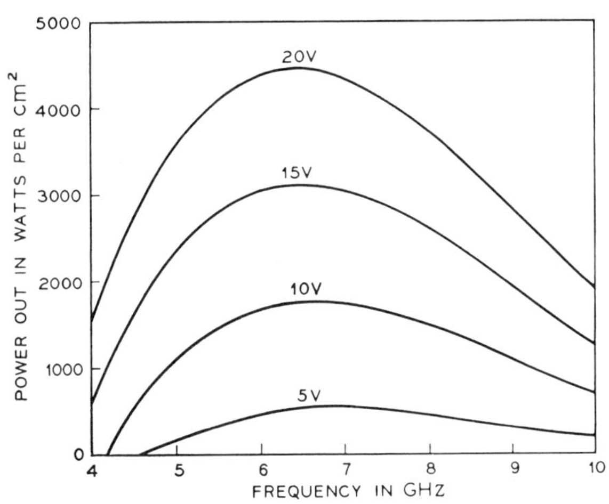

Fig. 4 — Output power vs frequency for various ac voltage amplitudes. Current density 500 A/cm².

141

quency with largest small-signal negative conductance. Efficiency depends upon the dc voltage across the diode, and therefore on the dc electric field in the drift region. The idealized model merely assumes this field large enough to maintain scatter-limited velocity. A better lower bound is obtained by calculating, at each point in the diode, the reduction in field resulting from the space charge of the current pulse drifting away from the avalanche region. Since I_T is independent of x,

$$\epsilon\, dE(x, t)/dt = I_p(t) + \epsilon\, dE_{\text{aval}}(x, t)/dt - I_p(t - x/v).$$

Integrate from 0 to t and subtract the dc component to obtain

$$\Delta E(x, t) = \Delta E_{\text{aval}}(t) + \int_0^t [I_p(t') - I_p(t' - x/v)]\, dt'.$$

Requiring the dc field at x to be as least as large as

$$\max_t\, [-\Delta E(x, t)]$$

gives the minimum field in the drift region allowable, and thus gives the minimum drift voltage and maximum efficiency. The maximum ΔE is also of interest, since in our model it is assumed that no avalanching takes place in the drift region. If E_a is the minimum electric field for which appreciable avalanche ionization can take place, then we must require that

$$\max\, [\Delta E(x, t)] + \max\, [-\Delta E(x, t)] < E_a$$

for all x. This limits the allowable ac voltage amplitude. Numerical results indicate that, for a given current density, the allowable ac voltage amplitude depends only weakly on the frequency of oscillation and the shape of the voltage waveform $V(t)$. For our sample diode, if $E_a \approx 10^5$ volt/cm, 20 volts ac amplitude is, approximately, the limiting voltage for the simple diode model used. A more general model, with a nonuniform avalanche region, could partially include avalanche-broadening, and thus be used to analyze diode modes of operation with larger ac voltage amplitudes.

Swan, Lee, and Standley recently reported marked improvement in the performance of an IMPATT oscillator embedded in a microwave circuit resonant at two frequencies, the fundamental and its second harmonic.[8, 10] As an aid in understanding their results, in the rest of this section we present calculations in which $V(t)$ has the form $V_o\cos \omega t + \frac{1}{2}V_o\sin 2\omega t$. (This particular waveform is not the optimum form, but is illustrative.) "Best" values of the magnitude and phase of the second harmonic depend on the frequency and magnitude of the fundamental, the bias current, the particular diode used, and the standards of performance to be applied to the diode.

Figures 5 to 8 correspond to Figs. 1 to 4; V_o is now chosen so that half the peak-to-peak value of $V(t)$ is 5, 10, 15, or 20 volts. In comparing the oscillations resulting from a two-frequency $V(t)$ with a one-frequency $V(t)$, we have chosen to compare voltage waveforms with equal ac peak-to-peak amplitudes, not with equal amplitudes in the fundamental, because of the ac amplitude limitation discussed earlier.

Figure 5 shows solutions for the ac components of $V(t)$, $I_p(t)$, and $I_{T_p}(t)$; the frequency of the fundamental is 5 GHz, and the maximum ac voltage is 10 volts. The fundamental component of $I_p(t)$ lags the fundamental component of the ac voltage by about 60 degrees instead of 49.5 degrees.

Figure 6 is a complex-plane admittance plot; for clarity, the lines of constant ac voltage are omitted, and dots are placed at 0, 5, 10, 15, and 20 volts. For nonsinusoidal ac voltages, the admittance at the fundamental can move to the left of the small-signal admittance line, resulting in a better negative conductance-to-susceptance ratio. This is especially apparent at 4 GHz; comparison of the 4 GHz, 20

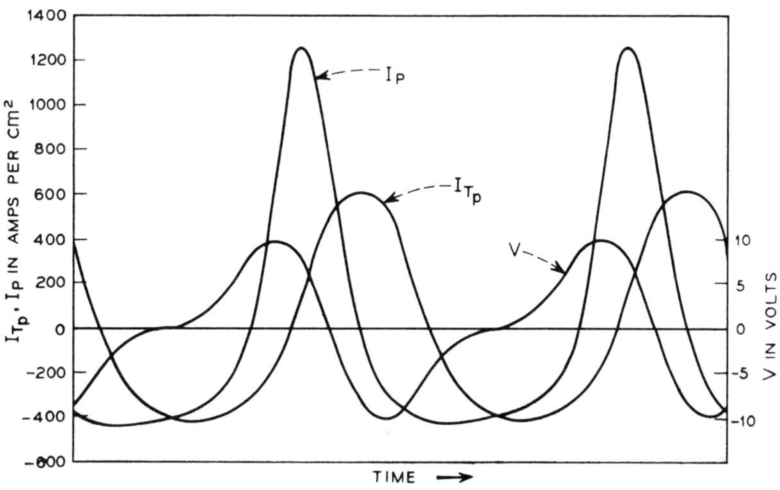

Fig. 5—AC components of V, I_p, I_{T_p} at frequency 5 GHz, current density 500 A/cm², ac voltage amplitude 10 V. Two cycles shown.

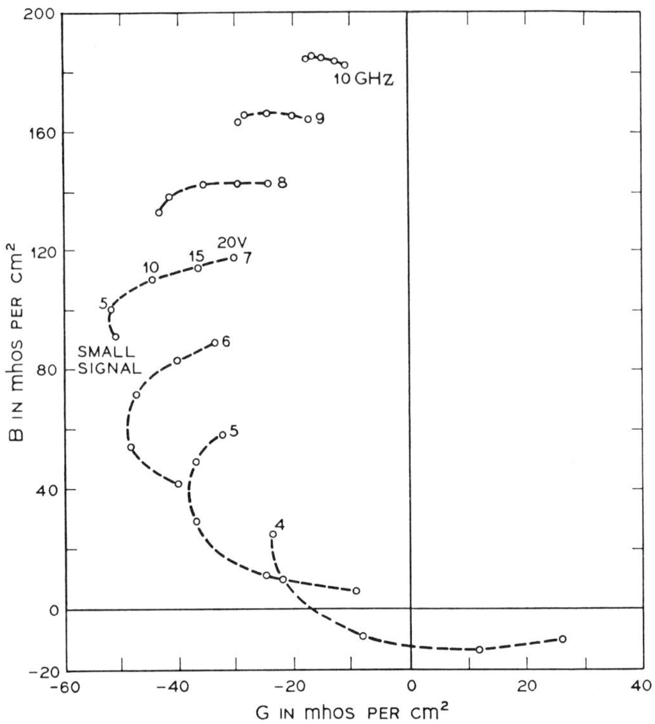

Fig. 6 — Complex-plane plot of diode admittance at fundamental frequency as a function of frequency and maximum ac voltage. AC component of $V(t) = V_o \cos \omega t + \frac{1}{2} V_o \sin 2\omega t$. Current density 500 A/cm².

V points in Figs. 2 and 6 indicates that the addition of a second harmonic component has improved the diode negative conductance by a factor of three, and the negative conductance-to-susceptance ratio by a factor of four.

Figure 7, corresponding to Fig. 3, shows in more detail the phase improvement obtained with addition of the harmonic component to the ac voltage.

Figure 8 shows output power in the fundamental. At higher frequencies, maximum output power is less than that shown in Fig. 4. At 5 GHz, maximum output power is about the same as in Fig. 4, but occurs at an improved negative conductance-to-susceptance ratio. This ratio is important when the diode's parasitic resistance R_p, which we have neglected, is included, since the power is proportional to

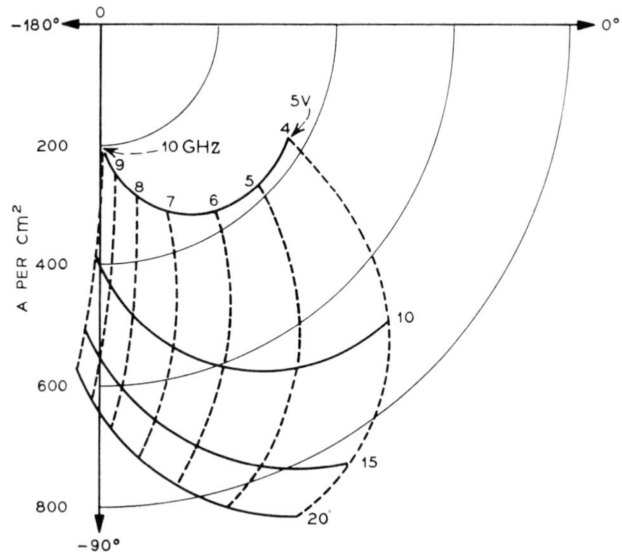

Fig. 7 — Magnitude and phase plot of fundamental component of avalanche particle current, $I_p(t)$, as a function of frequency and ac voltage amplitude. Current density 500 A/cm².

$- G/(G^2 + B^2) - R_p$. At 4 GHz, maximum output power is substantially better than that shown in Fig. 4. The nonsinusoidal form of $V(t)$ improves the phase of $I_T(t)$ sufficiently so that, even though $V(t)$ contains less of the fundamental, the output power in the fundamental remains about the same.

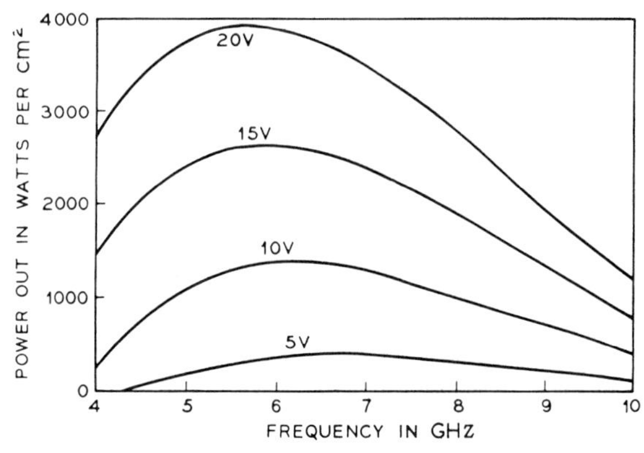

Fig. 8 — Output power in fundamental vs frequency for various maximum ac voltages. Current density 500 A/cm².

V. CONCLUSIONS

We have presented a numerically tractable method for analyzing large-signal IMPATT oscillators. It requires less than 1/100 of the computer time used by Scharfetter and Gummel's method, but still provides a realistic description of IMPATT oscillators in modes of operation which maintain carrier velocity saturation. (One solution takes about one second of computation time on the GE 645 computer, if we use 15 time steps per period.) Numerical calculations based on a simplified version of the model have provided information about the phase relationships in an IMPATT oscillator and improved understanding of the two-frequency mode of operation.

REFERENCES

1. Scharfetter, D. L. and Gummel, H. K., "Large-Signal Analysis of a Silicon Read Diode Oscillator," to be published in IEEE Trans. Elec. Devices, *ED-16,* No. 1 (January 1969).
2. Gilden, M. and Hines, M. B., "Electronic Tuning Effects in the Read Microwave Avalanche Diode," IEEE Trans. Electron Devices, *ED-13,* No. 1 (January 1966), pp. 169–175.
3. Gummel, H. K. and Scharfetter, D. L., "Avalanche Region of IMPATT Diodes," B.S.T.J., *45,* No. 10 (December 1966), pp. 1797–1827.
4. Gummel, H. K. and Blue, J. L., "A Small-Signal Theory of Avalanche Noise in IMPATT Diodes," IEEE Trans. Electron Devices, *ED-14,* No. 9 (September 1967), pp. 569–580.
5. Evans, W. J., "Nonlinear and Frequency Conversion Characteristics of IMPATT Diodes," Technical Report No. 104, Electron Physics Laboratory, Department of Electrical Engineering, The University of Michigan, February 1968. Also see: Evans, W. J., and Haddad, G. I., "A Large-Signal Analysis of IMPATT Diodes," IEEE Trans. Elec. Devices, *ED-15,* No. 10 (October 1968), pp. 708–716.
6. Scharfetter, D. L., Bartelink, D. J., and Johnston, R. L., "High Efficiency Subtransit Time Oscillations in Germanium Avalanche Diodes," Proc. IEEE, *56,* No. 9 (September 1968), pp. 1611–1612.
7. Read, W. T., Jr., "A Proposed High-Frequency, Negative Resistance Diode," B.S.T.J., *37,* No. 3 (March 1958), pp. 401–446.
8. Swan, C. B., "IMPATT Performance Improvement with Second Harmonic Tuning," Proc. IEEE, *56,* No. 9 (September 1968), pp. 1616–1617.
9. Scharfetter, D. L., private communication.
10. Lee, T. P. and Standley, R. D., "Frequency Modulation of a Millimeter-wave IMPATT Diode Oscillator and Related Harmonic Generation Effects," B.S.T.J., *48,* No. 1 (January 1969), pp. 143–161.

Large-Signal Analysis of a Silicon Read Diode Oscillator

D. L. SCHARFETTER, MEMBER, IEEE, AND H. K. GUMMEL, MEMBER, IEEE

***Abstract*—This paper presents theoretical calculations of the large-signal admittance and efficiency achievable in a silicon *p-n-ν-n* Read IMPATT diode. A simplified theory is employed to obtain a starting design. This design is then modified to achieve higher efficiency operation as specific device limitations are reached in large-signal (computer) operation. Self-consistent numerical solutions are obtained for equations describing carrier transport, carrier generation, and space-charge balance. The solutions describe the evolution in time of the diode and its associated resonant circuit. Detailed solutions are presented of the hole and electron concentrations, electric field, and terminal current and voltage at various points in time during a cycle of oscillation. Large-signal values of the diode's negative conductance, susceptance, average voltage, and power-generating efficiency are presented as a function of oscillation amplitude for a fixed average current density. For the structure studied, the largest microwave power-generating efficiency (18 percent at 9.6 GHz) has been obtained at a current density of 200 A/cm², but efficiencies near 10 percent were obtained over a range of current density from 100 to 1000 A/cm².**

I. Introduction

SELF-CONSISTENT large-signal numerical solutions are obtained for equations which describe carrier generation and space-charge balance in a silicon *p-n-ν-n* Read diode [1] microwave oscillator. The solutions describe the evolution in time of the diode and associated circuit (see Fig. 1). The doping profile of the diode (see Fig. 2) was evolved starting with design considerations discussed in [2], and reviewed here, and modified after initial results for improvement of efficiency. From the numerical solutions, the large-signal operating characteristics of the diode and the efficiency at the fundamental frequency (defined as ac power delivered by the diode divided by dc power dissipated) are computed.

Read in his original paper [1] suggested that an efficiency of 30 percent should be obtainable in a silicon diode. This calculation, however, neglected the finite width of the avalanche region, and neglected the difference in hole and electron ionization coefficients in silicon. A modification of Read's calculation, which includes the finite width of the avalanche region and the experimental values of ionization coefficients for silicon [3], is presented in Section II. It is found that the avalanche region must have an appreciable width for the diode to have high power capability and high efficiency. A theoretical efficiency of the order of 15 percent is obtained for conditions which should approach maximum power output.

Manuscript received June 3, 1968.
The authors are with Bell Telephone Laboratories, Inc., Murray Hill, N. J.

The aim of the detailed calculation, in addition to the enhancement of our understanding of the oscillation mechanisms, was to find conditions under which high efficiency and power output can be obtained, and to establish theoretical limits for these quantities. A rigorous optimization, however, was not attempted in this exploratory work. This would have had to start with carefully formulated constraints and would have constituted a very large-scale project. However, the largest efficiency obtained in the detailed numerical calculation reported here is 18 percent, in good agreement with the simplified theoretical estimate.

The large-signal operating characteristics of the diode are calculated in the following two ways:

A. Voltage Driven

To study the performance at a given frequency, but for various bias and load conditions, it is convenient to drive the diode with a sinusoidal voltage applied through a coupling capacitor [see Fig. 1(a)]. From such studies, the admittance at the fundamental frequency is obtained as a function of driving voltage amplitude. Likewise, the ac power delivered by the diode into the voltage generator is obtained.

B. Free-Running Oscillator

For another series of calculations, the diode is imbedded in a resonant circuit [see Fig. 1(b)] and performs as a free-running oscillator. In as much as the ac voltage in the oscillator case is nearly sinusoidal, both calculations should, and indeed do, give comparable results.

In Section III, we present the large-signal operating characteristics and detailed solutions. The operating characteristics, i.e., admittance parameters, average voltage, and power generating efficiency, are shown as a function of diode ac voltage amplitude in Figs. 4 through 9. Fig. 10 is a diode admittance plot (small and large signal) which shows the variation of conductance and susceptance with oscillation amplitude and frequency. Also indicated are the power generating efficiencies. Fig. 10 is a compact presentation of the important results of this paper. Detailed solutions are presented of the hole and electron concentrations, electric field, and terminal current and voltage at various points in time during a cycle of oscillation. These solutions are shown in Fig. 11 ($J_{dc}=200$ A/cm², frequency $=11.4$ GHz) and Fig. 12 ($J_{dc}=1000$ A/cm², frequency $=13.4$ GHz). The dynamics of charge-pulse buildup

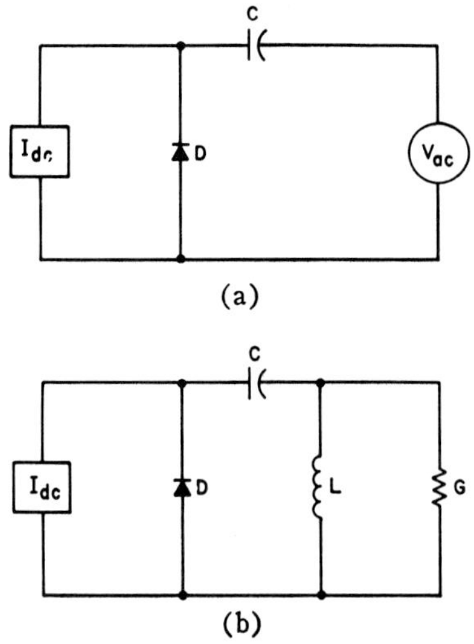

Fig. 1. Diode and associated circuit. (a) Voltage driven. (b) Free-running oscillator.

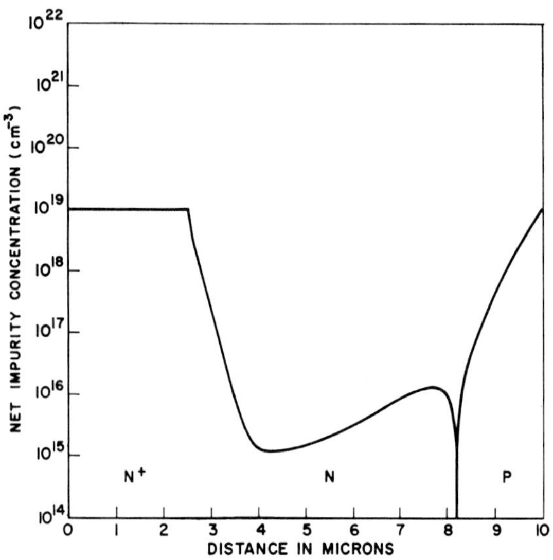

Fig. 2. Net impurity concentration of Read diode 7A versus distance.

in the avalanche region and drift through the drift region is in qualitative agreement with Read's original prediction of operation.

Section IV presents conclusions, and the Appendix describes the time evolution computer program.

II. Simplified Diode Design

In this section, we extend Read's efficiency calculation to include the dc voltage developed across the avalanche region, and to allow for the unequal ionization rates of holes and electrons, as appropriate for silicon [3]. As pointed out by Read, high-power operation requires the generation of as large a charge pulse Q_{max} as possible in the avalanche region, without a reduction of the electric field in the drift region below that required for velocity saturation when this charge moves through the drift region. The motion of Q_{max} through the drift region results in an ac voltage amplitude about one-half the average voltage V_{dcD} developed across the drift region [1], [2]. At the drift frequency (f_D = one-half the inverse transit time of the drift region), the motion of Q_{max} results under favorable conditions in an ac particle current which is ≈ 180 degrees out of phase with the ac voltage across the diode. The average of the particle current is the dc current J_{dc}. The particle current swing is, therefore, at most from zero to twice the dc current. For a square wave of particle current and a sinusoidal variation of drift voltage, both with magnitude and phase as described above, the microwave power-generating efficiency ϵ is

$$\epsilon = \frac{2}{\pi} \frac{\frac{1}{2} V_{dcD}}{(V_{dcD} + V_{dcA})}, \tag{1}$$

where the total dc voltage is divided between the part developed across the avalanche region V_{dcA} and the part developed across the drift region V_{dcD}. Note that we are neglecting any appreciable average ac power contribution from the avalanche region; this is because the avalanche region voltage is inductively reactive relative to the particle current [4]. The displacement current is capacitively reactive relative to the diode voltage and therefore contributes no average ac power. The use of a simple equivalent circuit [4] for the avalanche region, which is derived from small-signal considerations, may of course have no validity under large-signal conditions. However, we will use this approach as a starting point for a simple design theory, and compare detailed numerical calculations with the simple treatment.

In Read's calculation, it was assumed that no dc power loss was associated with the avalanche region since it was considered small in extent compared to the drift region, and that such an avalanche region could produce an ac particle current amplitude equal to the dc current when operated at the drift frequency and operated with ac voltage amplitude equal to one-half the dc voltage [same assumptions as used above in deriving (1)]. Since V_{dcA} was neglected compared to V_{dcD}, Read predicted an efficiency of $1/\pi$ or about 30 percent.

Small-signal calculations [1], [2], [4], [5] show that a resonance frequency f_R is associated with any avalanche region, and that it increases as the square root of dc current. To obtain small-signal negative resistance requires operation above this frequency. However, operation of an avalanche region at a frequency well above its resonance frequency was found to result in very inefficient oscillations in a limited number of exploratory large-signal calculations that were completed on a variety of Read diode structures. Small-signal calculations in the frequency domain [2] showed an optimum growth factor for small-signal oscillations occurring when the operating frequency was about 20 percent higher than the resonance frequency. Therefore, we take as a design criterion that the avalanche frequency of a Read diode should be related to the drift frequency of the diode by the following condition:

$$f_D \simeq 1.2 f_R. \tag{2}$$

The small-signal constraint above, coupled with large-signal conditions imposed on the drift space to be discussed below, results in a tentative design for the Read diode which is analyzed numerically in this paper.

The square of the avalanche region resonance frequency is proportional to the dc current density J_{dc}, the carrier scattering limited velocity v_{sl}, and the partial derivative of the average ionization rate $\overline{B}_A$ for electrons in the avalanche region with respect to the average electric field $\overline{E}_A$. An approximation for the relationship is [1], [2], [4]

$$(2\pi f_R)^2 = 2\overline{B}_A' v_{sl} J_{dc}/\epsilon, \tag{3}$$

where ϵ is the dielectric constant, and $\overline{B}_A'$ the partial derivation of $\overline{B}_A$ with respect to $\overline{E}_A$. The dc voltage developed across the avalanche region is defined as

$$V_{dcA} = \overline{E}_A L_A, \tag{4}$$

and the dc voltage developed across the drift region is similarly defined as

$$V_{dcD} = \overline{E}_D L_D, \tag{5}$$

where L_A and L_D are the effective widths of the avalanche and drift regions. With the definition of drift frequency

$$f_D = v_{sl}/2L_D, \tag{6}$$

V_{dcD} can be expressed as

$$V_{dcD} = \overline{E}_D v_{sl}/2f_D. \tag{7}$$

For 50-percent voltage modulation ($V_{ac} \simeq \frac{1}{2} V_{dcD}$), Q_{max} from Gauss' theorem is about $\simeq \epsilon \overline{E}_D/2$. The ac particle current (which for 100-percent current modulation equals the dc current) equals $Q_{max} f_D$; therefore, the dc bias current density for efficient operation of the drift space is proportional to frequency:

$$J_{dc} = \epsilon \overline{E}_D f_D/2. \tag{8}$$

For ionization rates appropriate for silicon [3], we find, for junctions with breakdown fields between 3 and 4×10^5 V/cm, that the integral of the electron ionization rate with distance is not unity, as is the case for equal rates, but about 3:

$$\overline{B}_A L_A \simeq 3. \tag{9}$$

If we assume for simplicity that $\overline{B}_A$ varies as the sixth power of $\overline{E}_A$, we obtain

$$\overline{B}_A' \simeq 6\overline{B}_A/\overline{E}_A$$

or, using (9) and (4),

$$\overline{B}_A' \simeq 6 \times 3/V_{dcA}. \tag{10}$$

Equation (3), the avalanche resonance relation, when evaluated for the conditions given by (10), (8), and (7), is

$$(2\pi f_R)^2 \simeq 2 \times 6 \times 3 \times V_{dcD} f_D^2/V_{dcA}, \tag{11}$$

and with the condition on f_R and f_D given by (2) results in a relation between V_{dcA} and V_{dcD}:

$$V_{dcA} \simeq 1.1 V_{dcD}. \tag{12}$$

This result is only qualitatively correct, since it is based on a combination of small-signal results for the avalanche region, and large-signal constraints on the drift region. Furthermore, the factor 2 in (3) and (11), the factor 6 in (10) and (11), and the factor 3 in (9) and (11) are all approximate, and depend to some extent upon dc bias, frequency, and oscillation amplitude. However, the qualitative result is important: the avalanche region voltage, for a Read diode designed for efficient high-power oscillator operation, is on the order of the drift voltage and not negligible as suggested by Read. This condition reduces the theoretical efficiency by about a factor of 2 (15 percent instead of 30 percent). Note that for Ge, GaAs, or other materials with nearly equal ionization rates, the factor 3 in (9) becomes a 1 and $V_{dcA} \simeq \frac{1}{3} V_{dcD}$, hence the theoretical efficiency is of the order of 23 percent.

III. Large-Signal Calculation

In this section, we discuss the precise large-signal operating characteristics of a particular Read diode oscillator, as obtained by a numerical calculation performed on a high-speed digital computer. Details of the computer program are discussed in the Appendix. Briefly, the approach is to obtain self-consistent numerical solutions for the equations describing carrier transport, carrier generation, and space-charge balance in a one-dimensional semiconductor structure. The solutions describe the evolution in time of the diode and its associated resonant circuit. Basically, the program solves the following problem at various instances of time during a cycle of oscillation. Given the instantaneous distribution of the hole and electron concentrations and terminal boundary conditions, how will the carriers move in time, i.e., how will the system, diode, and circuit time evolve? Given the carriers, the program determines the electric field by Poisson's equation, and knowing the electric field and carrier concentrations, it obtains the instantaneous hole and electron particle currents (including the dependence of carrier mobilities upon impurity concentration [7] and electric field strength [8], [9]). From the particle currents and net generation rate (including generation and recombination by field-dependent impact ionization [3] and carrier concentration-dependent single-level recombination centers [10], [11]), it obtains the time derivative of the carrier concentrations from the continuity equations. From the time derivatives of the carrier concentrations, the program computes the carrier distributions an instant in time later, and repeats the cycle.

The diode operating characteristics are calculated for the diode imbedded in the circuit shown schematically in Fig. 1(a), and (b). The variation of diode negative conductance, susceptance, average voltage, and microwave power-generating efficiency with frequency, dc current density, and ac voltage amplitude was obtained with the circuit shown in Fig. 1(a). This method made it possible to control directly the frequency and ac voltage amplitude in the calculation and was employed to economize on computer time. Essentially the same results were obtained with the circuit shown in Fig. 1(b), if the admittance of the load inductance L and load conductance G were properly chosen. The coupling capacitor C served only to isolate the dc and ac portions of the circuit so that the dc current density could be directly controlled.

The doping profile for the diode studied in detail is shown in Fig. 2. The corresponding steady-state electric field versus distance for this diode is shown in Fig. 3. To simplify the sign convention on field and current density, the figures represent a n-ν-n-p structure with the substrate on the left and the diffused junction on the right. Note that the donor density in the ν region is not negligible, but causes a continuous fall-off of electric field out to the substrate region. This doping profile evolved in trial runs to eliminate generation by impact ionization as the charge pulse of electrons passed through this region, while still allowing as large an average electric field in the drift region as possible (see Fig. 12(d) which illustrates this effect).

In a given run with fixed dc current density, ac voltage amplitude, and frequency, the program was continued until the limit cycle was obtained. The terminal current and voltage waveforms, during a limit cycle, were analyzed to investigate the variation of diode characteristics at various values of oscillation amplitude. The three fundamental diode characteristics investigated are negative conductance, susceptance, and average voltage. Results are shown in Figs. 4, 5, and 6. The diode's large-signal negative conductance and susceptance values are effective values at the voltage generator frequency obtained by Fourier analysis of the terminal current and voltage waveforms. Detailed results are shown for only one current density because of the great number of lengthy computer runs required to characterize completely the large-signal behavior. However, calculations made at other current densities (100 to 1000 A/cm^2) indicate that the results are qualitatively similar to the detailed results presented for a current density of 200 A/cm^2. The magnitude of negative conductance increases up to about 1000 A/cm^2, but the efficiencies obtained are less than at 200 A/cm^2. The diode's susceptance is predominately due to the diode's space-charge capacitance, and therefore decreased with increasing current due to widening of the space-charge width. The average voltage increases with current density, but because of the rapid increase of ionization rates with electric field, the average voltage decreases with ac voltage amplitude when the average current is held constant. Large-signal results at large currents are more difficult to obtain with the circuit shown in Fig. 1(a) (voltage generator) because parametric [12] effects occur at large dc current densities.

The susceptance shown in Fig. 5 is expressed as a "capacitance" and plotted in Fig. 7. The value is only slightly less than the depletion layer capacitance (1.9×10^{-9} F/cm^2), with the largest deviation at the lower frequencies. The ratio of susceptance to conductance Q is shown in Fig. 8, and the power-generating efficiencies are plotted in Fig. 9.

The conductance and susceptance values are shown as an admittance plot in Fig. 10. Results for frequencies between 6 and 13.4 GHz and ac voltage amplitudes between 0 and 38 volts are shown for a dc current density of 200 A/cm^2. Also shown are constant efficiency contours. The plot indicates, for example, that 10 percent efficiency is obtained at a frequency of 11.4 GHz with a voltage amplitude of 30 volts, while only 20 volts is required at 8.4 GHz. It was found that the efficiency increases with ac voltage until the electric field modulation is so large that, over the negative part of the voltage cycle, the field dropped so low that the carrier velocity was less than the saturated value. This occurs at considerably lower voltage amplitude for the lower frequencies (less than 7 GHz), and made it very difficult to obtain steady-state solutions for these frequencies. Also, note the rapid increase in negative Q (ratio of susceptance to conductance) with ac voltage amplitude. This would cause the efficiency to turn over and decrease with further increases in ac voltage amplitude for devices with series lead and contact resistance, since the diode shunt negative resistance[1] is reduced by a factor of Q^2 in converting from a parallel to a series equivalent circuit.

Detailed "snap shots" of the electric field, and hole and electron concentration as functions of distance during one cycle of steady-state oscillation are shown in Fig. 11. The "snap shots" are shown at approximately one-fourth cycle intervals in Fig. 11(a) through (d). The figures are selected frames from a computer-made movie [6]. A phase plot of the terminal current I and voltage V of the oscillation is included in the figures. Points to note are the following.

1) The generation of pulses of holes and electrons begins where the voltage is a maximum; one-fourth cycle later, the charge pulses are fully formed and begin drifting into their respective drift spaces.

2) The holes disappear quickly from the active region while the electrons drift for approximately one-half cycle and constitute positive particle current while the ac voltage is negative.

[1] The calculation includes all losses that originate between the electrical contacts, i.e., substrate and avalanche space-charge resistances.

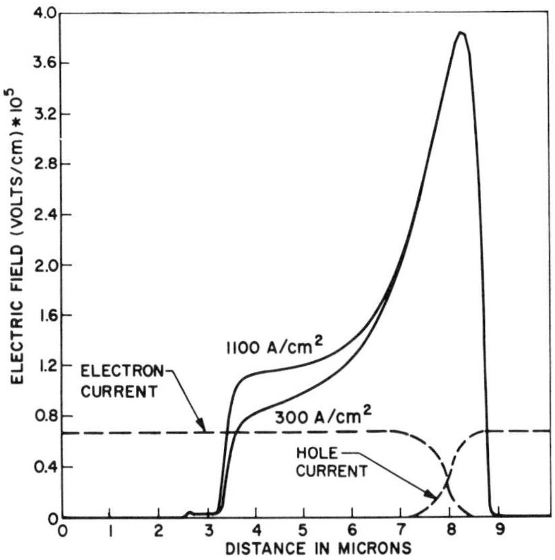

Fig. 3. Static electric field and particle current profile versus distance in dc steady-state Read diode 7A.

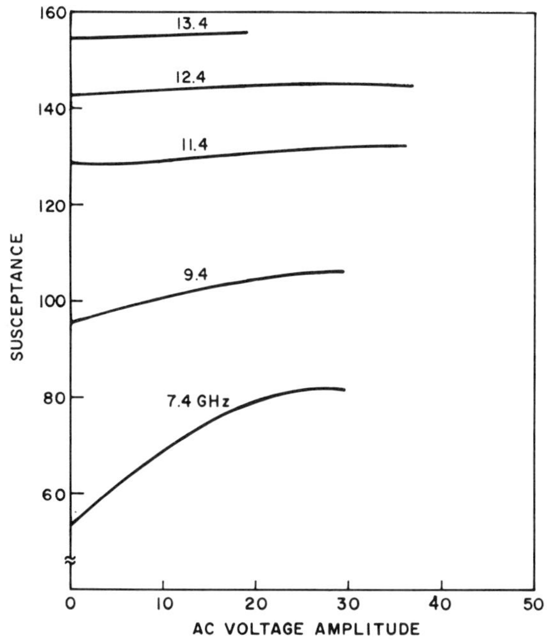

Fig. 5. Diode susceptance (mhos/cm²) as a function of ac voltage amplitude for various fixed frequencies. Current density: 200 A/cm².

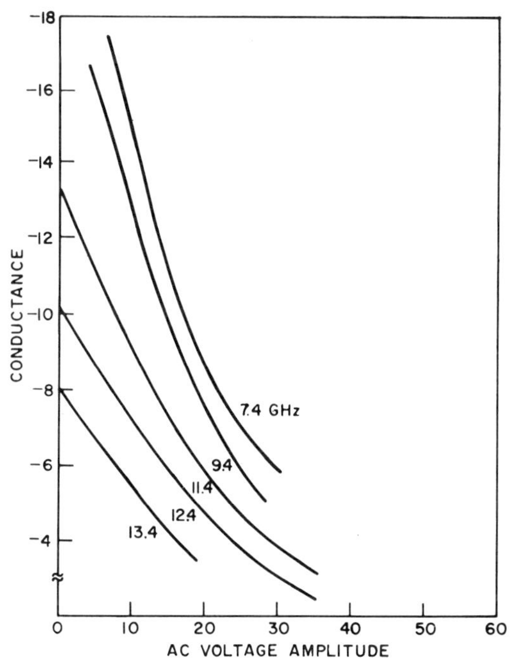

Fig. 4. Diode negative conductance (mhos/cm²) as a function of ac voltage amplitude for various fixed frequencies. Current: density 200 A/cm².

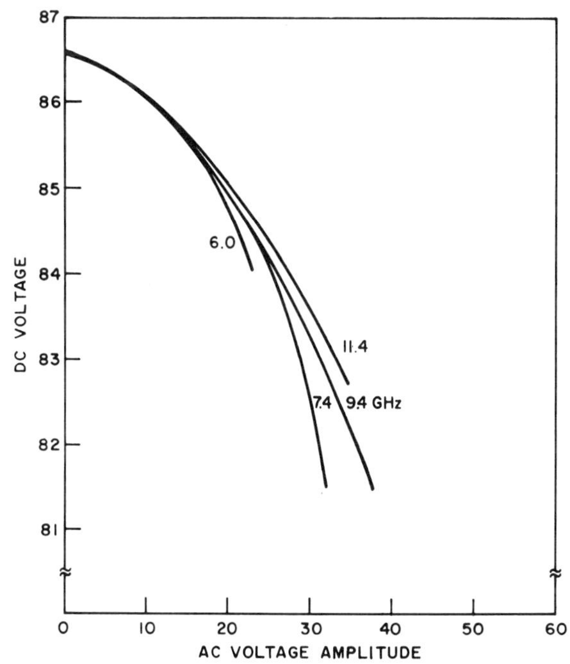

Fig. 6. Diode average voltage as a function of ac voltage amplitude for various fixed frequencies. Current density: 200 A/cm².

151

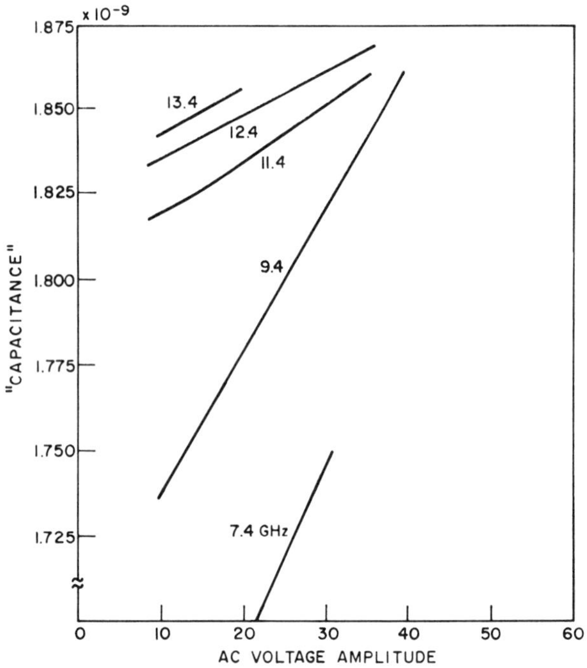

Fig. 7. Diode susceptance expressed as an equivalent capacitance (depletion layer capacitance $=1.9\times10^{-9}$ F/cm^2) versus ac voltage amplitude for various fixed frequencies. Current density: 200 A/cm^2.

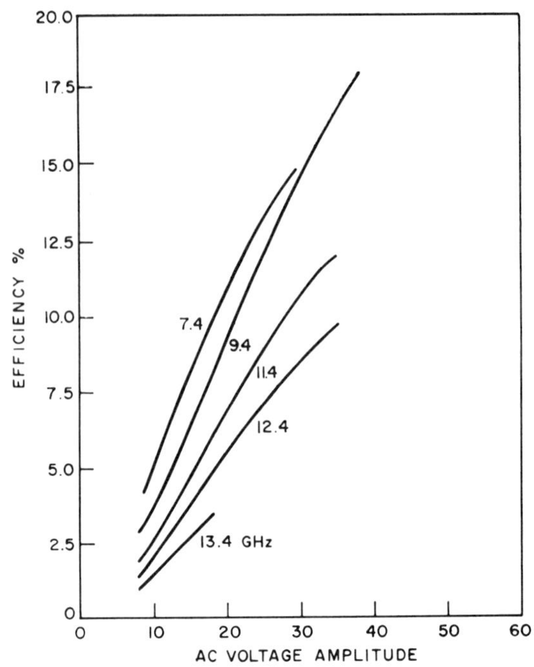

Fig. 9. Power-generating efficiency versus ac voltage amplitude for various fixed frequencies. Current density: 200 A/cm^2.

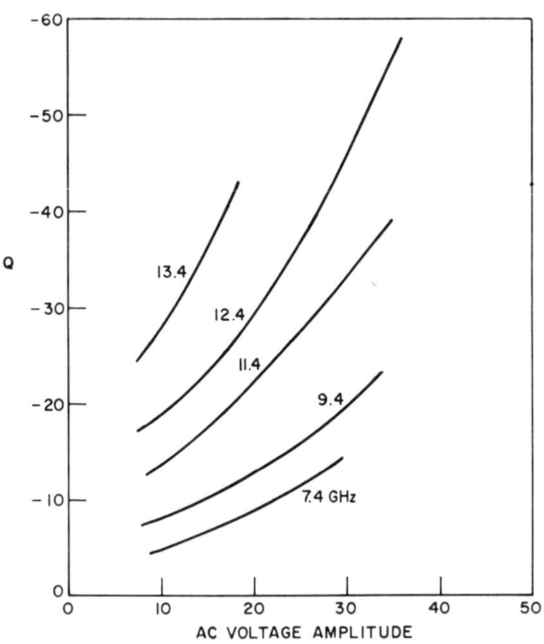

Fig. 8. Diode Q (ratio of conductance to susceptance) versus ac voltage amplitude for various fixed frequencies. Current density: 200 A/cm^2.

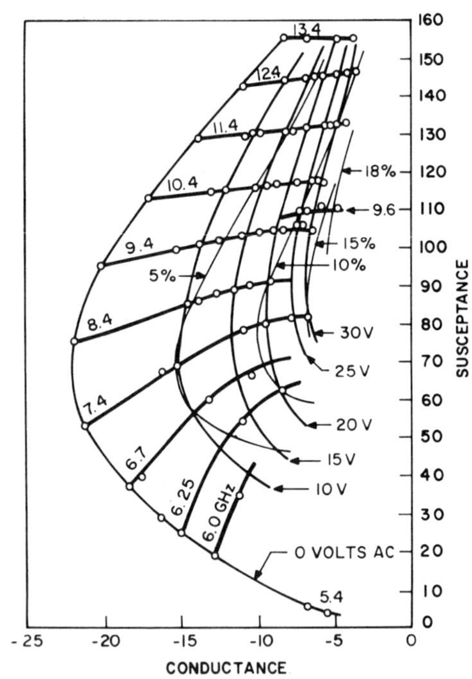

Fig. 10. Diode admittance (susceptance versus conductance) as a function of frequency and ac voltage amplitude, and resultant efficiency indicated. Current density: 200 A/cm^2.

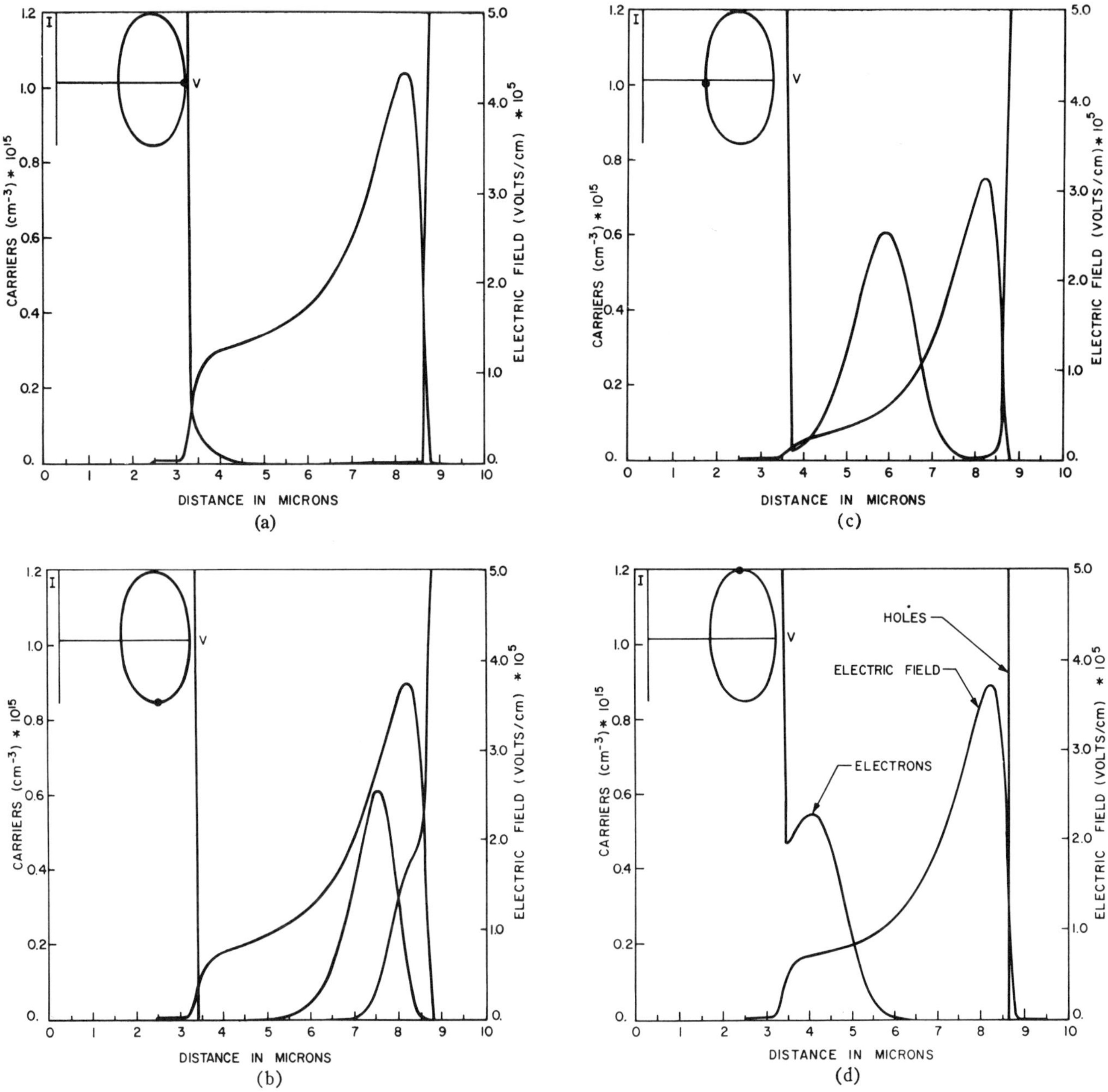

Fig. 11. Solutions of hole and electron concentrations, electric field, and terminal current and voltage (values indicated by ● on phase plot) at various points in time for the diode operating at a frequency of 12.4 GHz, current density of 200 A/cm^2, and efficiency of 12 percent.

3) For the next one-fourth cycle, the remnants of the electron charge pulse are swept out of the picture as the voltage again approaches its maximum value.

4) The displacement current is quite large and has an appreciable swing into the conventional forward direction, while the terminal voltage always remains in the conventional reverse polarity.

5) The behavior is as predicted by Read, except that for silicon diodes an extended avalanche region is required relative to the drift region to obtain the magnitude of charge pulse necessary for sufficient modulation of diode voltage and particle current for efficient oscillations.

Efficient oscillations were also obtained at higher dc current densities by trial and error choices for the load conductance G and load inductance L. "Snap shots" of the oscillation at a dc current density of 1000 A/cm^2 are shown in Fig. 12(a) through (d). The largest efficiency obtained for this bias (9 percent) occurred at an operating frequency of about 13.4 GHz. Note that the carrier concentration scales differ by a factor of 5 between Figs. 11 and 12, as do the dc current densities. However, the large-charge pulse, for the 1000 A/cm^2 case, moving through the drift space results in a significant value of peak field near the substrate just as the electrons approach this region. If this peak field is too large, genera-

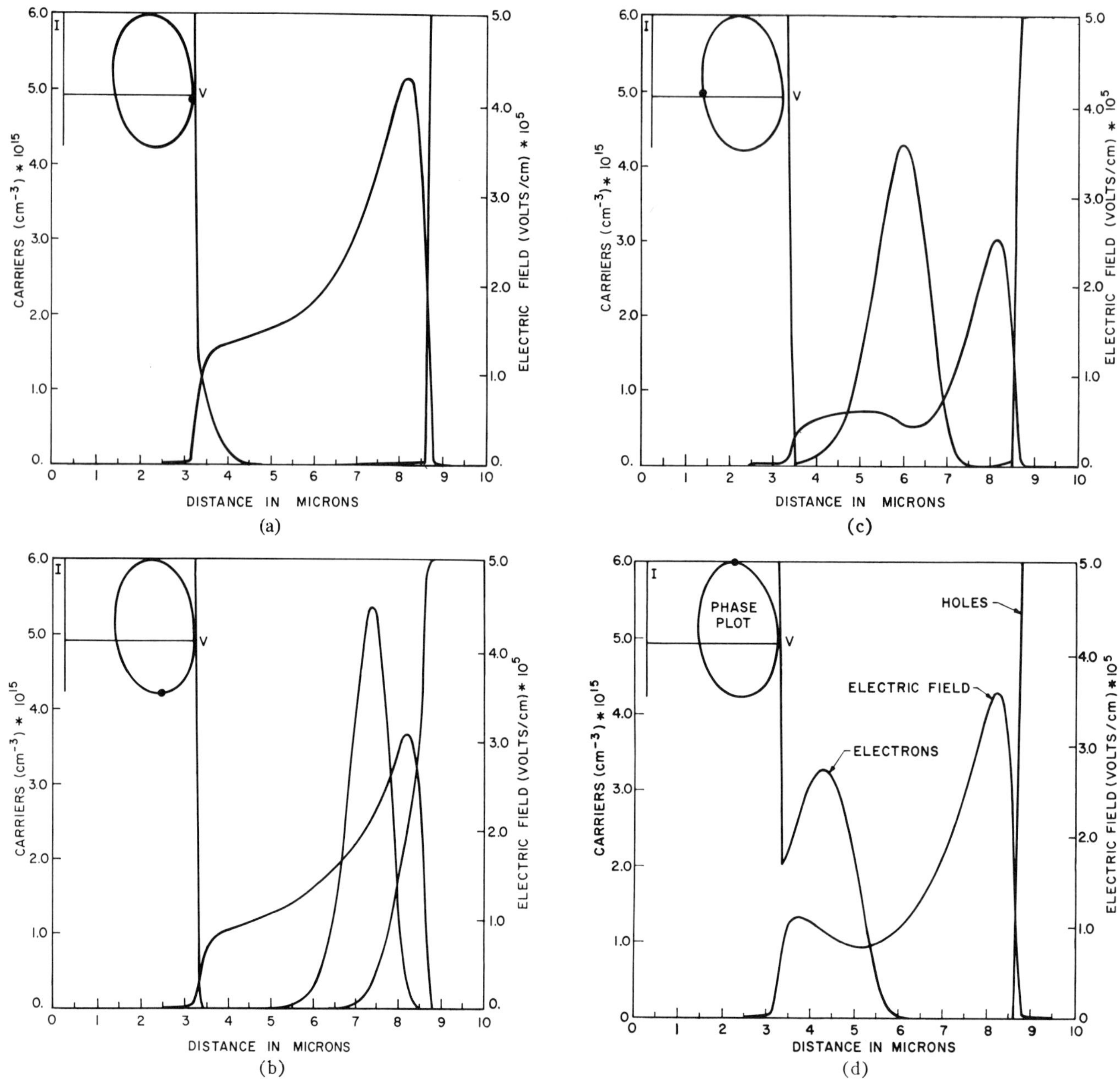

Fig. 12. Solutions of hole and electron concentrations, electric field, and terminal current and voltage (values indicated by ● on phase plot) at various points in time for the diode operating at a frequency of 13.4 GHz, current density of 1000 A/cm^2, and efficiency of 9 percent.

tion by impact ionization occurs, and the subsequent drift of a hole charge pulse toward the p region would constitute positive particle current while the voltage was also positive, a very unfavorable phase relation. This effect was worse in an earlier doping profile (not shown), and was eliminated by altering the profile to provide a continuous fall-off of electric field out to the substrate (see Fig. 3).

The nearly ideal (classical Read [1]) phase relations obtained between diode voltage and particle current is illustrated in Fig. 13. In this figure, we plot the waveforms in time of the particle current and terminal voltage for an ac steady-state solution at 11.4 GHz and dc current density of 200 A/cm^2. The particle current was obtained by subtracting the displacement current from the total current. At the largest oscillation amplitude, the diode "capacitance" changes parametrically [12] and makes an accurate separation of the current components difficult. The result shown in Fig. 13 is for a voltage amplitude of about 17 volts. However, the waveforms for larger amplitudes are similar. The particle current is not a square wave during the half cycle over which the voltage is negative, but has a hump resulting from the extra component of particle current which flows until the holes that were generated along with the electrons are swept into the p layer. For large amplitudes, the hump becomes more pronounced and the particle current bottoms [all carriers are swept out

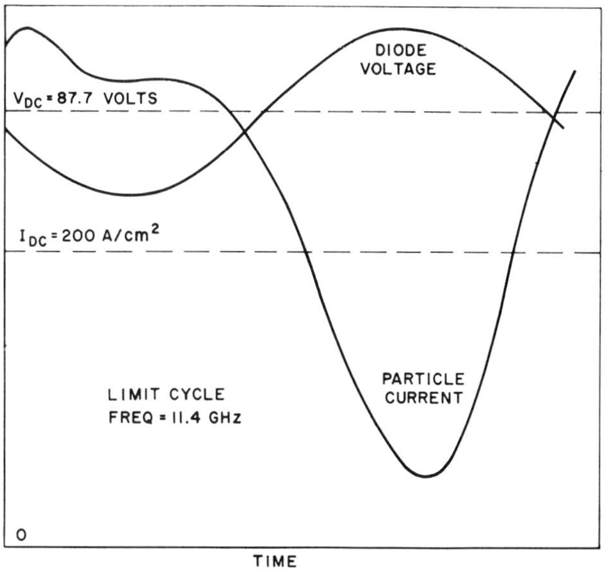

Fig. 13. Diode voltage and particle current versus time for one cycle of steady-state oscillation.

as shown in Fig. 11(a)] over an appreciable part of the cycle.

The results of the calculation presented in this section have shown that a properly designed silicon Read diode oscillator is operable, with efficiencies from 9 to 18 percent, over an order of magnitude variation in dc current. The frequency of operation, amplitudes of voltage and current oscillations, order of magnitude of dc bias, and efficiency are in fair agreement with the prediction of the relatively simple design theory presented in Section II.

IV. Conclusions

Operating characteristics of a silicon Read diode were obtained from large-signal computer calculations of the evolution in time of the diode-resonant-circuit system. The values of ionization rates [3] and carrier velocities [8], [9] appropriate for silicon were used in the calculation. Self-consistent solutions were obtained for the equations describing carrier transport, carrier generation, and space-charge balance within the diode, and which simultaneously satisfied the boundary condition imposed at the diode metallic contacts by the resonant circuit.

Detailed solutions were presented of the hole and electron concentrations, electric field, and terminal current and voltage at various points in time during ac steady-state cycles of oscillation. The largest efficiency obtained (18 percent) is in good agreement with the approximate calculation, which includes the appreciable width of the avalanche region required for silicon devices. Efficient operation (9 to 18 percent) was obtained over a range of dc current density (100 to 1000 A/cm²) and frequency (6.4 to 13.4 GHz).

Appendix

Semiconductor Device Analysis Computer Program

Introduction

This Appendix consists of three sections. Section I describes the device physics included in the large-signal calculation and outlines the solution procedures without detail. Section II describes in detail the time advancement techniques and the mathematical formulation employed in the computer calculation. Section III is a detailed discussion of the matrix inversion subroutine developed specifically for the time evolution computer program.

I. Mathematical Model

A. Charge Transport Equations

The distribution and motion of carriers within a one-dimensional semiconductor device structure can be obtained by solving three basic equations: 1) the continuity equation for holes, 2), the continuity equation for electrons, and 3) Poisson's equation:

$$\frac{\partial p}{\partial t} = g - \frac{1}{q}\frac{\partial J_p}{\partial x}, \tag{13}$$

$$\frac{\partial n}{\partial t} = g + \frac{1}{q}\frac{\partial J_n}{dx}, \tag{14}$$

and

$$\frac{\partial E}{\partial x} = \frac{q}{\epsilon}(p - n + N_D - N_A) \tag{15}$$

where

$$J_p = q\mu_p pE - kT\mu_p\frac{\partial p}{\partial x} \tag{16}$$

$$J_n = q\mu_n nE + kT\mu_n\frac{\partial n}{\partial x}\cdot \tag{17}$$

Boundary conditions are imposed at the contacts by introducing the appropriate restrictions in (13) through (17). For example, current boundary conditions for a *p-n* device are introduced by requiring that (17) is equal to the terminal current density at the *n* contact and that (16) is equal to the terminal current density at the *p* contact. Voltage boundary conditions are imposed by requiring that the integral of $E(x, t)$ over the interval between the two contacts equals the total voltage. In addition, the electric field at the two end metallic contacts is assumed to be zero. Initial values for the hole and electron densities are either furnished by a previous run or are given by the quiescent zero bias solution, i.e., the solution when

$$J_n = J_p = 0 \quad \text{and} \quad \frac{\partial p}{\partial t} = \frac{\partial n}{\partial t} = 0. \tag{18}$$

155

In general, the functions g, μ_p, μ_n vary at each point in the device according to the value of n, p, and E at that point.

B. Carrier Generation-Recombination Expressions

The carrier generation term g is composed of two components: 1) carrier generation and recombination through defects, and 2) impact or avalanche ionization.

Hole electron generation and recombination through defects are represented by a Shockley-Read-Hall (SRH) single-level model which characterizes defects with neutral and single-charge states. The generation-recombination rate through a single-level center is given by

$$g_d = \frac{pn - n_i^2}{\tau_{p0}(n + n_1) + \tau_{n0}(p + p_1)}\cdot \tag{19}$$

The impact ionization rates are strongly dependent upon the electric field intensity and the hole and electron current densities. These generation terms are given by

$$g_I = \frac{1}{q}\left(\alpha_n(E)\,|\,J_n\,| + \alpha_p(E)\,|\,J_p\,|\right) \tag{20}$$

where $\alpha_n(E)$ and $\alpha_p(E)$ are given by the relations

$$\alpha_n(E) = 2.25 \times 10^7 \exp\left(-3.2 \times 10^6/E\right) \tag{21}$$

$$\alpha_n(E) = 3.80 \times 10^6 \exp\left(-1.75 \times 10^6/E\right). \tag{22}$$

C. Mobility Expressions

It is necessary to include in the analysis the variations of mobility (μ) with the electric field (E) and the ionized impurities density (N_D).

The theoretical mobility is approximated by the following expression:

$$(\mu_0/\mu)^2 = 1 + \left(\frac{N_D}{N_D/S + N}\right) + \left(\frac{(E/A)^2}{E/A + F}\right) + (E/B)^2 \tag{23}$$

	μ_0	N	S	A	F	B
Holes	480	4×10^{16}	81	6.1×10^3	1.6	2.5×10^4
Electrons	1400	3×10^{16}	350	3.5×10^3	8.8	7.4×10^3.

(24)

D. Solution Procedures

Because of the nonlinearities in the equations describing the hole, electron, and field distributions, obtaining a transient or even a steady-state dc solution poses a very difficult numerical problem. The structure to be analyzed is first subdivided into a number of small cells. The equations are then normalized to reduce redundant coefficient calculations, and standard difference approximations are used to approximate the spatial derivatives in Poisson's equation and the continuity equations for each increment:

$$\frac{dp}{dt}(N) = g(N) - (J_p(M) - J_p(M-1))/\Delta x \tag{25}$$

$$\frac{dn}{dt}(N) = g(N) + (J_n(M) - J_n(M-1))/\Delta x \tag{26}$$

$$\frac{E(M) - E(M-1)}{\Delta x} = \frac{q}{\epsilon}\left(p(N) - n(N) + N_D(N) - N_A(N)\right) \tag{27}$$

where the Mth mesh point is located midway between the major mesh points $N+1$ and N. It is customary to employ next the standard difference approximations in the current density expressions (16) and (17) and substitute these results in (25) and (26). However, it can be shown that this procedure leads to numerical instability whenever the voltage change between mesh points exceeds $2kT/q$. Rather, (16) and (17) are treated as differential equations in p and n with J_n, J_p, μ_p, μ_n, and E assumed constant between mesh points. The solution of these differential equations then relates J_n and J_p to the other variables:

$$J_p(M) = E(M)\left[\frac{p(N)\mu_p(M)}{(1.0 - \exp(-E(M)\Delta x))} + \frac{p(N+1)\mu_p(M)}{(1.0 - \exp(E(M)\Delta x))}\right] \tag{28}$$

$$J_n(M) = E(M)\,\frac{n(N+1)\mu_n(M)}{(1.0 - \exp(-E(M)\Delta x))} + \frac{n(N)\mu_n(M)}{(1.0 - \exp(E(M)\Delta x))}. \tag{29}$$

These equations provide numerically stable estimates of the current density under all conditions. If the intermesh point voltage is small, these equations approach the standard difference relations; whereas when the voltage change is large, they approach the drift current density at either mesh point N or $N+1$.

E. Definition of Symbols

- p, n hole and electron concentrations, cm^{-3}
- E electric field intensity, V/cm
- q electronic charge, 1.602×10^{-19} coulombs
- J_p, J_n hole and electron current densities, A/cm^2
- g hole and electron generation-recombination rates, $\text{carriers/cm}^3\cdot\text{s}$
- g_I impact ionization generation rate for holes and electrons, $\text{carriers/cm}^3\cdot\text{s}$
- g_d hole and electron generation-recombination rates through a single-level trapping center, $\text{carriers/cm}^3\cdot\text{s}$
- ϵ dielectric constant

k Boltzmann constant, 1.38044×10^{-23} J/°K
T absolute temperature, °K
N_D total ionized impurity concentration, cm^{-3}
N_D, N_A concentration of donor and acceptor impurity atoms, cm^{-3}
μ_p, μ_n hole and electron mobilities, $\text{cm}^2/\text{V}\cdot\text{s}$
μ_{p0} low-field low-doping hole mobility in silicon, (480 $\text{cm}^2/\text{V}\cdot\text{s}$)
μ_{n0} low-field low-doping electron mobility in silicon, (1400 $\text{cm}^2/\text{V}\cdot\text{s}$)
$\alpha_p(E), \alpha_n(E)$ ionization coefficients for holes and electrons, carriers/carrier·cm
p_1, n_1 hole and electron concentrations in the conduction band when the Fermi level coincides with the energy level of the single-level trap.

II. Vector-Matrix Formulation

A. Time Advancement

We represent the hole and electron concentrations by the two-component vector y. The system of equations then reads

$$\mathring{y} = f(y), \tag{30}$$

i.e., the time derivative of y equals some nonlinear functions f of y. We denote y at the beginning of the time interval under study by y_0, and denote the deviations of this value as time increases by $\delta y(t)$:

$$y(t) = y_0 + \delta y(t). \tag{31}$$

For small deviations δy, we expand $f(y)$ as

$$\delta\mathring{y} = f(y_0) + M\delta y. \tag{32}$$

M is the matrix df/dy and represents an integro-differential operator. Now, if M, f, and δy were ordinary numbers, the solution to (32) would be

$$\delta y(t) = (e^{Mt} - 1)f/M \tag{33}$$

or

$$\delta y(t) = f\cdot t(1 + Mt/2 + (Mt)^2/6 + \cdots). \tag{34}$$

Note that if the physical situation is one in which we are near a stable equilibrium state, then the eigenvalues of the matrix M must be negative so that as time goes to infinity, δy reaches a finite value.

For small but finite time steps, we introduce the inverse of the first two terms of (34) to obtain our solution to (30) as

$$\delta y(t) = \frac{ft}{1 - \dfrac{Mt}{2}}. \tag{35}$$

The operator M in the denominator roughly corresponds to integration and as t goes to infinity, we reach an asymptotic value within a factor of 2 of that given by the closed-form expression (33). Written in more conventional form,

$$\left(1 - \frac{Mt}{2}\right)\delta y = f\cdot t. \tag{36}$$

Equation (36) is the vector equation we solve for each increment t in time. When we have solved it, we increment the vector y_0 by the solution δy. To get a better intuitive understanding of what we are doing, let us move the term containing Mt in (36) on the right-hand side:

$$\delta y(t) = \left(f + \frac{M}{2}\delta y\right)t. \tag{37}$$

At time 0, the slope $\mathring{y}$ is $f(y_0)$. At time t, it is approximately $f_0 + M\delta y$. The ratio of the finite differences $\delta y(t)/t$ corresponds to the slope of the straight line through the points 0 and t. What we are doing is propagating the system according to the time derivative at some advanced time $t_a \simeq t/2$, rather than at the initial time. The term in δy on the right-hand side of (37) represents a feedback that gives the system stability. It is worth noting that the expansion of f with respect to y (the derivative matrix M) is correct to first order, but with regard to time it is correct to second order in t. The method of propagation (36) time evolves the system accurately for small time steps t. However, if we are near an equilibrium state and are interested not in the time evolution, but only in the equilibrium solution, time steps t of infinity can be taken with full feedback, i.e.,

$$(1 - Mt)\delta y = f\cdot t.$$

B. The Derivative Matrix

We define the vector y as having components of hole concentration, electron concentration, and electric field:

$$y_N = \begin{Bmatrix} y_1 \\ y_2 \\ y_3 \end{Bmatrix} = \begin{Bmatrix} \Delta\text{HØL}(N) \\ \Delta\text{ELE}(N) \\ \Delta\text{FIELD}(M) \end{Bmatrix} \tag{38}$$

and the right-hand side of the continuity equations as the vector f:

$$f_N = \begin{pmatrix} f_1 \\ f_2 \end{pmatrix} = \begin{pmatrix} \mathring{\text{HØL}}(N) \\ \mathring{\text{ELE}}(N) \end{pmatrix}. \tag{39}$$

The vector equation we solve for each time increment t (36) is

$$\left[\frac{2}{t} - M_N\right](y)_N = 2(f_0)_N + (\Delta f_t) \tag{40}$$

where Δf_t represents changes in f_N at the terminals and where we define the derivative matrix

$$M_N = \frac{\partial f_N}{\partial y}. \tag{41}$$

The matrix elements of M are defined by the expanded form of (40) shown below:

$$[XL]_N * \begin{pmatrix} y_1 \\ y_2 \end{pmatrix}_{N-1} + [XM]_N * \begin{pmatrix} y_1 \\ y_2 \end{pmatrix}_N + [XR]_N * \begin{pmatrix} y_1 \\ y_2 \end{pmatrix}_{N+1} + (FL)_M * y_{3M-1} + (FR)_M * y_{3M} = (RH)_N \qquad (42)$$

where

$$(RH)_N = 2(f_0)_N.$$

Vector equation (42) is solved by the computer program subroutine described in Section III. The 2×2 matrices XL, XM, and XR are obtained by taking the partial derivatives of f_N with respect to HØL and ELE at mesh points $N-1$, N, and $N+1$, respectively, while the two component vectors FL and FR are obtained by taking the partial derivatives of f_N with respect to FIELD at mesh points M and $M-1$, respectively.

III. Matrix Inversion Subroutine

This section describes in detail the method for solving the discretized continuity and space-charge equations for the quantities $\delta H(N)$, $\delta E(N)$, and $\delta F(M)$. The quantities δH and δE are the changes in the hole and electron concentrations, while δF is a normalized electric field; δF is in units of carrier concentration. The boundary conditions have been included in the equations, i.e., the particle currents at the left contact enter into the right-hand side $RH(1)$ of the first continuity equation, and the particle currents at the right contact enter into the right-hand side $RH(LR)$ of the last continuity equation. Since the total current at the contacts is known, continuity equations 1 and LR contain three boundary conditions. The fourth boundary condition is that the electric field at the contact is zero. Since overall space-charge neutrality is built into the equations, a field of zero at one contact forces a field of zero at the other contact. The method for solving the discretized continuity equations will be described for the case of terminal-current boundary conditions. The modifications to this method when boundary conditions are in terms of terminal voltage or a mixture of voltage and current will be described in Section III-C of this Appendix.

A. Input-Output

We describe a computer subroutine which solves the following three equations:

$$\begin{aligned} &XL(1, N) * \delta H(N-1) + XL(2, N) * \delta E(N-1) \\ &\quad + \\ &XM(1, N) * \delta H(N) + XM(2, N) * \delta E(N) \\ &\quad + \\ &XR(1, N) * \delta H(N+1) + XR(2, N) * \delta E(N+1) \\ &\quad + \\ &FL(1, N) * \delta F(M-1) + FR(1, N) * \delta F(M) \\ &\qquad = RH(1, N), \end{aligned} \qquad (43)$$

and

$$\begin{aligned} &XL(3, N) * \delta H(N-1) + XL(4, N) * \delta E(N-1) \\ &\quad + \\ &XM(3, N) * \delta H(N) + XM(4, N) * \delta E(N) \\ &\quad + \\ &XR(3, N) * \delta H(N+1) + XR(4, N) * \delta E(N+1) \\ &\quad + \\ &FL(2, N) * \delta F(M-1) + FR(2, N) * \delta F(M) \\ &\qquad = RH(2, N) \end{aligned} \qquad (44)$$

where N ranges from 1 to LR, and

$$\delta F(M) - \delta F(M-1) = (\delta H(N) - \delta E(N)) * DY(N)$$

where

$$DY(N) = (DX(M) + DX(M-1))/2. \qquad (45)$$

Input quantities and their required range are as follows:

DX_M	$M = 0, LR$	
$[XL]_N$	$N = 2, LR$	
$[XM]_N$	$N = 1, LR$	
$[XR]_N$	$N = 1, LM$	$(LM = LR - 1)$
$(RH)_N$	$N = 1, LR$	
$(FL)_N$	$N = 2, LR$	
$(FR)_N$	$N = 1, LM.$	

Output quantities and their range are as follows:

$\delta H_N = Y(1, N)$	$N = 1, LR$
$\delta E_N = Y(2, N)$	$N = 1, LR$
$\delta F_M = Y(3, M)$	$M = 1, LM$

(note $\delta F(0)$ and $\delta F(LR)$ are zero).

B. Details of Inversion

We represent the output quantities δH and δE by the vector y as

$$(y)_N = \begin{pmatrix} \delta H(N) \\ \delta E(N) \end{pmatrix},$$

and similarly input quantities RH, FL, FR as

$$(RH)_N = \begin{pmatrix} RH(1, N) \\ RH(2, N) \end{pmatrix}$$

$$(FL)_M = \begin{pmatrix} FL(1, M) \\ FL(2, M) \end{pmatrix}$$

$$(FR)_M = \begin{pmatrix} FR(1, M) \\ FR(2, M) \end{pmatrix}.$$

The input quantities XL, XM, XR are represented by matrices $[XL]$, $[XM]$, $[XR]$:

$$[XL]_N = \begin{bmatrix} XL(1, N) & XL(2, N) \\ XL(3, N) & XL(4, N) \end{bmatrix},$$

and similarly for $[XM]_N$ and $[XR]_N$.

With this notation, (43) through (45) can be written as

$$\begin{aligned} &[XL]_N * (y)_{N-1} + [XM]_N * (y)_N + [XR]_N * (y)_{N+1} \\ &\quad + (FL)_M * \delta F(M-1) + (FR)_M * \delta F(M) = (RH)_N. \end{aligned} \tag{46}$$

Equation (45) can be used to eliminate $\delta F(M-1)$ in (46):

$$\begin{aligned} &[XL]_N * (y)_{N+1} + \{[XM]_N - [UFL]_N\} * (y)_N \\ &+ [XR]_N * (y)_{N+1} + \{(FR)_N + (FL)_N\} * \delta F(M) = (RH)_N \end{aligned}$$

where

$$[UFL]_N = \begin{bmatrix} FL(1, N) - FL(1, N) \\ FL(2, N) - FL(2, N) \end{bmatrix} * DY(N). \tag{47}$$

We assume that we can remove the $(y)_{N-1}$ term in (47) and write the resulting equation at mesh point $N-1$ as

$$\begin{aligned} &[UM]_{N-1} * (y)_{N-1} + [UR]_{N-1} * (y)_N \\ &\quad + (UF)_{N-1} * \delta F(M-1) = (Z)_{N-1}. \end{aligned} \tag{48}$$

Recursion relationships for $[UM]$, $[UR]$, (UF), and (Z) are obtained as follows.

Equations (47) and (48) are combined to eliminate $(y)_{N-1}$ terms as

$$\begin{aligned} &\{[XM]_N - [UFL]_N - [T]_N * [UR]_{N-1}\} * (y)_N \\ &\quad + [XR]_N * (y)_{N+1} + \{(FR)_M + (FL)_M\} * \delta F(m) \\ &\quad - [T]_N * (UF)_{N-1} * \delta F(M-1) \\ &= (RH)_N - T_N * (Z)_{N-1} \\ &\qquad [T]_N = [XL]_N * [UM]_{N-1}^{-1}. \end{aligned} \tag{49}$$

The term in $\delta F(m-1)$ can be eliminated in (49) by (45), and therefore by comparing (49) with (48) evaluated at mesh point N, we get:

$$\begin{aligned} [UR]_N &= [XR]_N \\ [UM]_N &= [XM]_N - [UFL]_N - [T]_N * [UR]_{N-1} \\ &\quad + \begin{bmatrix} DY(N) - DY(N) \\ DY(N) - DY(N) \end{bmatrix} * [T]_N * (UF)_{N-1} \\ (Z)_N &= (RH)_N - [T]_N * (Z_{N-1}) \\ (UF)_N &= (FR)_N + (FL)_N - [T]_N * (UF)_{N-1}. \end{aligned} \tag{50}$$

Starting values for $[UM]$, (UF), and (Z) are obtained by comparing (48) (at $N=1$) with (46) (at $N=1$) with the term in $\delta F(1)$ eliminated by (45):

$$\begin{aligned} [UM]_1 &= [XM]_1 + \begin{bmatrix} FR(1, 1) - FR(1, 1) \\ FR(2, 1) - FR(2, 1) \end{bmatrix} * DY(1) \\ (Z)_1 &= (RH)_1 \\ (UF)_1 &= \begin{pmatrix} 0. \\ 0. \end{pmatrix} \end{aligned} \tag{51}$$

Note that (48) can be written as

$$\begin{aligned} \begin{pmatrix} \delta H \\ \delta E \end{pmatrix}_L &= [UM]_L^{-1} * \{(Z)_L - [XR]_L * (y)_{L+1}\} \\ &\quad - (UF)_L * \delta F(L) \end{aligned} \tag{52}$$

and (45) as

$$\delta F(L-1) = \delta F(L) - \{\delta H(L) - \delta E(L)\} * DY(L); \tag{53}$$

therefore, once we have $[UM]$, (UF), and (Z) at mesh points 1 to LR, we can work backwards using (52) and (53) to find

$$\begin{pmatrix} \delta H \\ \delta E \end{pmatrix}_L$$

and δF_{L-1} for $L = LR \cdots 1$.

The computation procedure is as follows.

1) Compute starting values from (51).

2) Compute values at the remaining mesh points $(2, LR)$ using (50).

3) Use (52) and (53), starting at mesh points LR (note $(y)_{LR+1}$ and $\delta F(LR)$ are both zero) and working through mesh point 1 to obtain the values for δH, δE, and δF.

C. General Boundary Conditions

In the discussion thus far, it has been assumed that the desired boundary conditions could be incorporated into the input coefficients. The terminal currents entered into the right-hand side $(RH)_N$ at the contacts $(N=1)$ and $(N=LR)$. More general boundary conditions can be incorporated by using an economical superposition process.

With the time advancement process described in Section II, we want to solve the equations

$$\left[\frac{1}{t} - M/2\right] * (y) = f, \tag{54}$$

where t is the step in time and M is the derivative matrix. If the terminal currents are changing with time, then f is f_0, the right-hand side of the continuity equation at $t=0$, plus one-half of the change in terminal current (the extra term will appear in only two equations, mesh points 1 and LR). Let us multiply (54) by 2 and write it as

$$\left[\frac{2}{t} + M\right](y) = 2 * f_0 + \Delta f, \tag{55}$$

where Δf represents the change in terminal currents in the time t. Equation (55) is identical to (46) and could be solved by the method discussed previously in Section III-A of this Appendix if the change in terminal currents (Δf terms) are known. What is always known at the terminals is one of three possibilities:

1) the terminal current as a function of time is an independent variable;

159

2) the terminal voltage as a function of time is an independent variable; or

3) an equation relating terminal current to terminal voltage is known.

The general solution to (55) for Δf arbitrary can be obtained numerically using superposition. Let us assume that our final solution will be the sum of two solutions:

$$(y) = (y_0) + (y_c). \tag{56}$$

These solutions have the following interpretation:

(y_0) represents the solution for $\Delta f = 0$, that is, for no change in terminal current over the time step t;

(y_c) represents the additional solution for a change in terminal current.

To see this more clearly, let us write (55) as

$$[M]_N(y) = (RH_0)_N + (RH_c)_N \tag{57}$$

where the quantity $2/t + M$ in (55) has been written as $[M]_N$ in (57) and $2f_0$ as $(RH_0)_N$ and Δf as $(RH_c)_N$.

Equation (57) has $(RH_c)_N$ zero everywhere, except at mesh points 1 and LR. Therefore, we solve (57) using the method described in Section III-A of this Appendix with $(RH)_N = (RH_0)_N$. This yields solutions $(y_0)_N$. Next, we solve (57) with $(RH)_N = (RH_c)_N$; this yields solution $(y_c)_N$. Since we do not know the change in terminal current ΔJ, we obtain solutions $(y_c)_N$ for a unit change in terminal current. We then integrate the electric field component of $(y_c)_N$ to obtain the change in terminal voltage per change in terminal current Δv_c. If the integral of the electric field component of $(y_0)_N$ is called ΔV_0, then the change in terminal voltage ΔV is related to the change in terminal current ΔJ by

$$\Delta V = \Delta V_0 + \Delta v_c * \Delta J. \tag{58}$$

References

[1] W. T. Read, "A proposed high-frequency negative-resistance diode," *Bell Sys. Tech. J.*, vol. 37, pp. 401–466, March 1958.

[2] H. K. Gummel and D. L. Scharfetter, "Avalanche region of IMPATT diodes," *Bell Sys. Tech. J.*, vol. 45, pp. 1797–1827, December 1966.

[3] C. A. Lee, R. A. Logan, R. L. Batdorf, J. J. Kleimack, and W. Wiegmann, "Ionization rates of holes and electrons in silicon," *Phys. Rev.*, vol. 137, pp. A761–A773, May 1964.

[4] M. Gilden and M. B. Hines, "Electronic tuning effects in the read microwave avalanche diode," *IEEE Trans. Electron Devices*, vol. ED-13, pp. 169–175, January 1966.

[5] T. Misawa, "Negative resistance in *p–n* junctions under avalanche breakdown conditions, parts I and II," *IEEE Trans. Electron Devices*, vol. ED-13, pp. 137–151, January 1966.

[6] D. L. Scharfetter and H. K. Gummel, "Design of Read diode oscillators," presented at the 1966 IEEE Solid-State Device Research Conf., paper III b-4.

[7] J. C. Irvin, "Resistivity of bulk silicon and of diffused layers in silicon," *Bell Sys. Tech. J.*, vol. 41, pp. 387–410, March 1962.

[8] A. C. Prior, "Field dependence of carrier mobility in silicon and germanium," *J. Phys. Chem. Solids*, vol. 12, pp. 175–180, January 1960.

[9] T. E. Seidel and D. L. Scharfetter, "Dependence of hole velocity upon electric field and hole density for *P*-type silicon," *J. Phys. Chem. Solids*, vol. 28, pp. 2563–2574, 1967.

[10] W. Shockley and W. T. Read, Jr., "Statistics of the recombination of hole and electron," *Phys. Rev.*, vol. 87, pp. 835–842, September 1952.

[11] R. N. Hall, "Electron-hole recombination in germanium," *Phys. Rev.*, vol. 87, p. 387, July 1952.

[12] B. C. DeLoach, Jr., and R. L. Johnston, "Avalanche transit-time microwave oscillators and amplifiers," *IEEE Trans. Electron Devices*, vol. ED-13, pp. 181–186, January 1966.

Nonlinear Analysis of the Avalanche Transit-Time Oscillator

KOENRAAD MOUTHAAN

Abstract—**The nonlinear operating characteristics of the avalanche transit-time oscillator are studied by means of Fourier-series representation. For optimum operation, the oscillator must be designed such that start-oscillation conditions are satisfied simultaneously at the first and the second harmonic of the desired oscillation frequency. Under those conditions the oscillation frequency does not depend on the dc bias current; the signal level increases smoothly with bias current.**

For large signals, the diode exhibits negative resistance for frequencies substantially below the avalanche frequency; the oscillation frequency therefore may be below the avalanche frequency corresponding to the dc bias current required for large-signal operation. A condition for attaining large-signal operation is that the product of drift-zone capacitance and total load resistance must be small compared to the oscillation period; this condition also yields small starting currents.

The output power at the oscillation frequency is obtained explicitly in terms of diode and external circuit parameters. The maximum attainable output power is limited by parasitic series resistance and by permissible RF voltage swing as compared to dc bias voltage. The best power-impedance product is obtained by choosing the transit angle equal to 0.74 π. In practice, it may be advantageous to choose a smaller value for the transit angle, in order that the tuning condition for the second harmonic may be more easily satisfied.

The dc-to-RF conversion efficiency in principle is linearly proportional to the dc current density; the maximum efficiency again is limited by parasitic series resistance and by permissible RF voltage swing.

I. Introduction

The AVALANCHE transit-time oscillator has been analyzed by several authors, primarily from the point of view of diode characteristics in the limit of small RF signals [1]–[6]. The case of large RF signals has received analytical attention [7]–[9] only insofar as diode conduction current response to a single-frequency RF voltage of large amplitude is concerned. The present work is an attempt to analyze the large-signal characteristics of the complete oscillator, i.e., diode plus external circuit, taking into account the generation of higher harmonic voltages.

Due to the strongly nonlinear response of multiplication current to a large-amplitude voltage across the multiplication zone, the conduction current is of high harmonic content. Phase and amplitude relationships amongst currents and voltages are determined by the multiplication process on the one hand and by the external circuit on the other. Coupled equations for the harmonic components result, and solutions may be sought for various levels of oscillation.

Manuscript received January 20, 1969; revised May 26, 1969.
The author is with Philips Research Laboratories, N. V. Philips' Gloeilampenfabrieken, Eindhoven, Netherlands.

In order to arrive at a not unduly complicated analysis, we shall consider the reverse-biased diode to consist of two zones: a multiplication zone and a drift zone. Carrier multiplication takes place in the multiplication zone, where signal delay may be neglected; signal delay takes place in the drift zone, where carrier multiplication may be neglected. Any parasitic series resistance will, at least initially, be considered part of the external circuit. Carrier velocities are assumed to have their field-saturated values throughout both the multiplication and drift zones; the actual values are assumed to be equal for both electrons and holes and will be denoted by v_d. Ionization rates per unit length in the multiplication zone also are assumed to be equal for both electrons and holes and will be denoted by α. Reverse saturation current will be neglected as compared to total conduction current.

II. General Equations

A. *The Multiplication Zone*

With the aforementioned assumptions, the conduction current in the multiplication or avalanche zone equals the avalanche current J_{ca}, and is governed by [7, eq. (13)]:

$$\tau_a \frac{dJ_{ca}}{dt} = 2J_{ca}\left\{\int_0^{l_a} \alpha dx - 1\right\}. \tag{1}$$

Here l_a is the length of the avalanche zone, τ_a is equal to l_a/v_d and the integral is the ionization integral over the extent of the avalanche zone. The solution to (1) is easily obtained in implicit form; we have

$$J_{ca}(t) = J_0 \exp\left[\frac{2}{\tau_a}\int_0^t d\tau \left\{\int_0^{l_a} \alpha dx - 1\right\}\right], \tag{2}$$

where J_0 is the avalanche current at time $t=0$. The equation for the avalanche current is a nonlinear one, as may be seen by considering that the ionization rate α depends on the electric field E in the avalanche zone, whereas E in turn depends on the current J_{ca} through the properties of the remainder of the diode and those of the external circuit. Since (2) in fact represents the nonlinearity of our problem, thus determining the generation of higher-harmonic currents and voltages, we shall consider in some detail the relative importance of the various terms that appear in our subsequent analysis of (2).

The ionization integral may be written in terms of an average ionization rate $\bar{\alpha}$ by means of the definition

$$\int_0^{l_a} \alpha dx = \bar{\alpha} l_a. \tag{3}$$

For a sufficiently thin avalanche zone, the electric field E will be approximately constant over the extent of the avalanche zone. Then $\bar{\alpha}$ in (3) may be replaced by α. We shall consider the electric field E to be the superposition of a nonzero dc part E_c and a time-varying part E_a. Then we have, in the neighborhood of E_c:

$$\alpha l_a = \alpha_c l_a + \alpha_c' V_a + \frac{\alpha_c''}{2l_a} V_a^2 + \cdots. \tag{4}$$

Here the primes indicate differentiation with respect to the electric field and the subscript c indicates evaluation at the field E_c; the voltage $V_a = E_a l_a$ is the difference between the total voltage $V = El_a$ and the dc voltage $V_c = E_c l_a$ across the avalanche zone.

We next set the value of the field E_c equal to the critical value for which $\langle \alpha \rangle l_a = 1$, the notation $\langle\ \rangle$ indicating the time-average value. Then, upon subtracting the time-average value from (4), we obtain (3) in the form

$$\int_0^{l_a} \alpha dx - 1 = \alpha_c' V_a + \frac{\alpha_c''}{2l_a} (V_a^2 - \langle V_a^2 \rangle) + \cdots. \tag{5}$$

For periodic time dependence of $V_a(t)$, higher harmonic generation results primarily from the exponential dependence of the avalanche current on the ionization integral. However, as is seen from (5), for large values of V_a additional higher harmonic generation may result from the nonlinear dependence of the ionization rate on the electric field. In order to make an estimate of the relative importance of the terms on the right-hand side of (5), we shall for the present consider $V_a(t)$ to consist of a first-harmonic part V_1 and a higher harmonic part V_{hh}. For electrons in silicon, we have the empirical result [10] $\alpha = a \exp(-b/E)$ with $a = 1.6 \times 10^6$ cm^{-1} and $b = 1.65 \times 10^6$ $V \cdot$cm^{-1}. Then, for practical values of E_c on the order of 300–500 kV/cm, we find that $a_c' V_a$ is on the order of 3.3–5.5 (V_a/V_c). whereas $\alpha_c'' V_a^2/2l_a$ is on the order of 2–9 $(V_a/V_c)^2$. We may assume that the ratio V_a/V_c remains small of first order as compared to unity so that $\alpha_c'' V_a^2/2l_a$ remains small of first order as compared to $\alpha_c' V_a$. A consequence of this assumption is that $\alpha_c' V_a$ can at most become comparable to unity. Now, in the following, we shall assume that $\alpha_c' V_{hh}$ always remains small compared to unity, whereas the part $\alpha_c' V_1$ will actually be allowed to become comparable to unity. Then, if the latter situation occurs $\alpha_c'' V_1^2/2l_a$ may be comparable to $\alpha_c' V_{hh}$. For that purpose we need to retain the contribution of the second term on the right-hand side of (5) that results from the first harmonic in $V_a(t)$.

We may now explicitly introduce the condition that the oscillation be periodic in time, with period $2\pi/\omega$. All time-dependent quantities can then be written in the form of a real Fourier series. For the voltage $V_a(t)$ we write

$$V_a(t) = \sum_{m=1}^{\infty} V_m \sin(m\omega t + \psi_m). \tag{6}$$

The phase of the first harmonic will be arbitrarily set equal to zero, i.e.,

$$\psi_1 \equiv 0, \tag{7}$$

since only the relative phase of the various harmonics is of consequence. With our estimate of the importance of the second term on the right-hand side of (5), we may then write in (2)

$$\frac{2}{\tau_a} \int_0^t d\tau \left\{ \int_0^{l_a} \alpha dx - 1 \right\} = \sum_{m=1}^{\infty} (v_m + v_m^*) - \sum_{m=1}^{\infty} \{ (v_m - i\sigma v_1^2 \delta_{2m}) e^{im\omega t} + \text{c.c.} \}. \tag{8}$$

Here we have used the complex representation of a real Fourier series; i is the imaginary unit and c.c. denotes the complex conjugate terms. The Kronecker delta δ_{2m} stands for $\delta_{2m} = 1$ if $m = 2$, $\delta_{2m} = 0$ if $m \neq 2$. To obtain (8) we have introduced the dimensionless voltages

$$v_m = \frac{\alpha_c' V_m e^{i\psi_m}}{m\omega\tau_a} \quad (m = 1, 2, 3, \cdots), \tag{9}$$

and the dimensionless coefficient

$$\sigma = \frac{\omega \tau_a \alpha_c''}{8 l_a \alpha_c'^2}. \tag{10}$$

Using our earlier estimates of α_c' and α_c'', we observe that the magnitude of v_m is on the order of $(4/m\omega\tau_a) \cdot (V_m/V_c)$ and that the value of σ is on the order of $\omega\tau_a/16$.

The Fourier expansion of $J_{ca}(t)$ can now be obtained in terms of the voltages v_m by substituting (8) for the argument of the exponential function in (2). Since $J_{ca}(t)$ depends exponentially on the sum of the harmonic voltages, two processes may be distinguished. By the first process, harmonics of each of the frequency components $m\omega$ are generated. By the second process, mixing products of different frequency components $n\omega$ and $m\omega$ are generated. Now, if we assume that

$$\left| \sum_{m=2}^{\infty} \{ (v_m - i\sigma v_1^2 \delta_{2m}) e^{im\omega t} + \text{c.c.} \} \right| \ll 1 \tag{11}$$

for all t, the summation being over $m \geq 2$, the first process leads to harmonics of only the fundamental frequency component of the voltage $V_a(t)$, whereas the second process only leads to mixing products of the higher harmonic components of $V_a(t)$ with the harmonics of the fundamental frequency component generated by the first process.

The condition (11) will generally be satisfied if $|v_m| \ll 1$ for $m \geq 2$ and $\sigma v_1^2 \ll 1$. With our estimates of the magnitude of v_m and that of σ, we find that the condition $|v_m| \ll 1$ for $m \geq 2$ requires that

$$V_m/V_c \ll m\omega\tau_a/4 \quad (m \geq 2),$$

whereas the condition $\sigma v_1^2 \ll 1$ requires that $(V_1/V_c)^2 \ll \omega\tau_a$. In the limit of large-signal operation, we shall also want to have $v_1 \gg 1$, requiring that $4(V_1/V_c) \gg \omega\tau_a$. Thus, in the limit of large-signal operation, we require that

$$(V_1/V_c)^2 \ll \omega\tau_a \ll 4(V_1/V_c).$$

The right-hand inequality is consistent with our earlier assumption of negligible signal delay in the multiplication zone; the left-hand inequality says that if $(V_1/V_c)^2$ is not negligible compared to $\omega\tau_a$, appreciable second-harmonic generation will result from the nonlinear dependence of the ionization rate α on the electric field E.

With the condition (11) it is readily shown (see Appendix I) that the Fourier expansion of $J_{ca}(t)$ is given by

$$J_{ca}(t) = J_{c0}\left\{1 + \sum_{m=1}^{\infty} (\beta_m e^{im\omega t} + \text{c.c.})\right\}. \tag{12}$$

The expansion coefficients β_m are given by

$$\beta_m = b_m/b_0, \tag{13}$$

where

$$b_m = (-1)^m\Big[I_m(2v_1) - \sum_{n=2}^{\infty}(-1)^n\{(v_n - i\sigma v_1^2\delta_{2n})I_{m-n}(2v_1) + (v_n^* + i\sigma v_1^2\delta_{2n})I_{m+n}(2v_1)\}\Big]; \tag{14}$$

and J_{c0} is the dc bias current for nonzero voltages $V_a(t)$:

$$J_{c0} = \langle J_{ca}(t)\rangle = J_0 b_0 \exp\left\{\sum_{m=1}^{\infty}(v_m + v_m^*)\right\}. \tag{15}$$

The functions I_m in (14) represent the modified Bessel functions of the first kind and mth order. The argument $2v_1$ of the modified Bessel functions is real, which is the reason why we have set $\psi_1 \equiv 0$ in (6) for the Fourier expansion of $V_a(t)$. In what follows, we shall be interested in time-varying components only, so we shall omit the dc term from (12).

It may be observed that if the bias is supplied from a constant-voltage dc source, the bias current J_{c0} will depend on the RF signal level. Only if the bias is supplied from a constant-current dc source will J_{c0} be a constant independent of signal level.

To obtain the total current, we must also consider the displacement current $J_{da}(t)$ in the avalanche zone. We have

$$J_{da}(t) = C_a\frac{dV_a(t)}{dt}, \tag{16}$$

where

$$C_a = \epsilon A/l_a \tag{17}$$

is the capacitance of the avalanche zone in terms of the cross-sectional area A and the length l_a. With the definitions

$$L_a = \frac{\tau_a}{2\alpha_c' J_{c0}}$$
$$\omega_a^2 = 1/L_aC_a \tag{18}$$
$$\kappa = \omega^2/\omega_a^2$$

we obtain from (6), (9), and (16)

$$J_{da}(t) = J_{c0}\sum_{m=1}^{\infty}(m^2\kappa v_m e^{im\omega t} + \text{c.c.}). \tag{19}$$

Here the quantity L_a is an inductance associated with the multiplication process; the quantity ω_a will be called the avalanche frequency. It needs to be observed that the quantity L_a, and thus ω_a and κ, depends on the dc bias current J_{c0}. We shall henceforth assume that the bias is supplied from a constant-current dc source, so that J_{c0}, and, hence, L_a, ω_a, and κ are constants independent of the RF signal level.

From (12) and (19) we now obtain the total RF current $J(t)$ through the avalanche zone:

$$J(t) = J_{ca}(t) + J_{da}(t)$$
$$= J_{c0}\sum_{m=1}^{\infty}\{(\beta_m + m^2\kappa v_m)e^{im\omega t} + \text{c.c.}\}. \tag{20}$$

B. The Drift Zone

We next consider the properties of the drift zone. The conduction current generated in the avalanche zone enters the drift zone at $x=0$ without time delay, and subsequently propagates in the direction of the drift velocity as an undispersed wave of velocity v_d, Thus, for the conduction current $J_{cd}(x, t)$ in the drift zone, we have $J_{cd}(x, t) = J_{ca}(t - x/v_d)$; from (12) we obtain

$$J_{cd}(x, t) = J_{ca}(t - x/v_d)$$
$$= J_{c0}\sum_{m=1}^{\infty}(\beta_m e^{-im\omega x/v_d}e^{im\omega t} + \text{c.c.}). \tag{21}$$

The displacement current $J_{dd}(x, t)$ in the drift zone may be obtained in terms ot the time rate of change of the electric field in the drift zone. Writing the time dependence of the electric field $E_d(x, t)$ in the form of a real Fourier series, with harmonic terms $E_{dm}(x)e^{im\omega t}$ +c.c., we obtain

$$J_{dd}(x, t) = \epsilon A \partial E_d(x, t)/\partial t$$
$$= \epsilon A\sum_{m=1}^{\infty}\{im\omega E_{dm}(x)e^{im\omega t} + \text{c.c.}\}. \tag{22}$$

The total current is assumed to be divergenceless, thus independent of x and equal to the total current $J(t)$ entering the drift zone at $x=0$. So we have:

$$J_{cd}(x, t) + J_{dd}(x, t) = J(t), \tag{23}$$

where $J(t)$ is given by (20), $J_{cd}(x, t)$ by (21) and $J_{dd}(x, t)$ by (22). In its explicit form (23) is readily solved for the functions $E_{dm}(x)$. By integrating over the extent of the drift zone, we then obtain the voltage

$$V_d(t) = \int_0^{l_d} E_d(x, t)dx$$

across the drift zone. Writing

$$V_d(t) = \sum_{m=1}^{\infty} (V_{dm}e^{im\omega t} + \text{c.c.}), \tag{24}$$

we find that

$$V_{dm} = \int_0^{l_d} E_{dm}(x)dx = \frac{J_{c0}}{im\omega C_d}\left[m^2\kappa v_m + \beta_m\left\{1 + \frac{e^{-im\theta} - 1}{im\theta}\right\}\right]. \tag{25}$$

Here we have defined the capacitance C_d of the drift zone by

$$C_d = \epsilon A/l_d,$$

and the transit angle θ by

$$\theta = \omega l_d/v_d. \tag{27}$$

C. The Diode Plus External Circuit

We may now include the external circuit in our considerations. We shall assume that the external circuit contains linear elements only; it may therefore be represented by an impedance Z_{em} at the frequency $m\omega$. The sum of the voltages around the circuit must be zero; at the frequency $m\omega$ this condition yields the equation

$$0 = J_m Z_{em} + V_m e^{i\psi_m}/2i + V_{dm}. \tag{28}$$

Here J_m is the total current, $V_m e^{i\psi_m}/2i$ and V_{dm} are the voltages across the avalanche zone and across the drift zone, respectively; all corresponding to the frequency component $e^{im\omega t}$. A little more generality of (28) can be gained by including an external voltage V_{em} on the left-hand side; as we are primarily concerned with the operation of the oscillator, we shall not explicitly include such terms. Taking J_m from (20) and V_{dm} from (25) we now obtain an infinite set of coupled equations which may be written in the following normalized form:

$$\left.\begin{aligned} 0 &= [i(\lambda+1)/m\kappa]j_m z_{em} + v_m + v_{dm} \\ j_m &= \beta_m + m^2\kappa v_m \\ v_{dm} &= \lambda v_m + [i(\lambda+1)/m\kappa]\beta_m z_{dm} \end{aligned}\right\} \quad (m = 1, 2, 3, \cdots). \tag{29}$$

Here the currents have been normalized with respect to the dc bias current J_{c0}:

$$j_m = J_m/J_{c0}; \tag{30}$$

voltages have been normalized as defined in (9):

$$v_{dm} = \frac{2i\alpha_c'}{m\omega\tau_a} V_{dm}; \tag{31}$$

impedances have been normalized with respect to the impedance of the total capacitance $C = C_a C_d/(C_a + C_d)$ of the diode at the frequency ω:

$$z_{em} = \omega C Z_{em}; \tag{32}$$

furthermore, we have introduced the normalized impedances z_{dm} according to

$$z_{dm} = \frac{\lambda}{im(\lambda + 1)}\left\{1 + \frac{e^{-im\theta} - 1}{im\theta}\right\}, \tag{33}$$

and a factor λ measuring the ratio of avalanche-zone capacitance C_a to drift-zone capacitance C_d by

$$\lambda = C_a/C_d = l_d/l_a. \tag{34}$$

If, finally, we eliminate j_m and v_{dm} from (29), we obtain the following set of coupled nonlinear equations governing the combination of diode plus external circuit, i.e., the complete oscillator:

$$0 = m^2\kappa v_m + \beta_m \rho_m \qquad (m = 1, 2, 3, \cdots); \tag{35}$$

where the β_m are determined by (13) and (14) and the coefficients ρ_m are defined by

$$\rho_m = \frac{z_{em} + z_{dm}}{z_{em} - i/m}. \tag{36}$$

III. Solutions to the Oscillator Equations

If we regard (35) as a set of homogeneous equations for the voltages v_m, we may conclude that nonvanishing solutions will exist only if certain conditions on the coefficients ρ_m and κ are satisfied. However, as the β_m are nonlinear in the voltage v_1, no simple determinantal equation results and finding a closed-form solution for the voltages v_m does not appear to be feasible. On the other hand, we may take the point of view that the relevant contributions to the solution depend on the level of the oscillation, in particular on the value of the voltage v_1. Thus, we may seek separate solutions for the case where $v_1 \ll 1$ and the case where $v_1 \gg 1$. This procedure also will provide us with insight into the question of how the oscillation builds up to a particular level.

A. Solution for $v_1 \ll 1$

For sufficiently small values of v_1, the generation of harmonics higher than the second is negligible. Then only the first two of equations (35) need to be considered, whereas in the infinite sum in (14) for the coefficients b_m only the terms corresponding to $n = 2$ need to be retained. Expanding the remaining modified Bessel functions in their respective Taylor series around argument zero, we then find that

$$v_2 = \frac{\rho_2}{6(\rho_2 - 4\kappa)}\{3(1 + 2i\sigma)v_1^2 - 2v_1^4 + \cdots\}, \tag{37}$$

and that

$$0 = (\rho_1 - \kappa)v_1 - \frac{\kappa}{\rho_2 - 4\kappa} \cdot \{\rho_2 - 4\kappa + 2\rho_1(1 + 2i\sigma)\}v_1^3 + \cdots. \tag{38}$$

The v_1^2 term in (37) contributes to the v_1^3 term in (38); the v_1^4 term in (37) first comes into play in (38) in the next higher (i.e., v_1^5) term.

For vanishingly small values of v_1 even second-harmonic generation is negligible. Then (38) reduces to

$$0 = (\rho_1 - \kappa)v_1, \tag{39}$$

yielding a nonzero, though indeterminate, value for v_1 if $\rho_1 - \kappa = 0$. (For infinitesimal values of v_1 and $\rho_1 - \kappa \to 0$, the oscillator may be regarded as a linear noise amplifier of infinite gain, as may be seen by including a noise voltage v_{e1} in series with the oscillator circuit [cf. remark below (28)]. The voltage v_{e1} appears on the left side of (39), showing that v_1 in this limit is proportional to v_{e1} and becomes large as $\rho_1 - \kappa \to 0$.)

For finite values of v_1, (37) and (38) yield

$$v_1^2 = \frac{(\rho_1 - \kappa)(\rho_2 - 4\kappa)}{\kappa\{\rho_2 - 4\kappa + 2\rho_1(1 + 2i\sigma)\}}, \tag{40}$$

$$v_2 = \frac{\rho_2}{2\kappa} \frac{(\rho_1 - \kappa)(1 + 2i\sigma)}{\rho_2 - 4\kappa + 2\rho_1(1 + 2i\sigma)}. \tag{41}$$

Now, v_1^2 depends strongly on the external impedance presented to the diode at the second harmonic of the oscillation frequency. For (40) to be a solution, v_1^2 must be real and positive, imposing conditions on ρ_1 and ρ_2. The (very small) influence of the second derivative of the ionization rate $\alpha(E)$ becomes noticeable through σ in (40) and (41).

B. Solution for $v_1 \gg 1$

For sufficiently large values of v_1, all modified Bessel functions in (14) for the coefficients b_m assume their asymptotic value $e^{2v_1}/\sqrt{4\pi v_1}$, so that $b_m = (-1)^m b_0$ and

$$\beta_m = (-1)^m \quad (m = 1, 2, 3, \cdots; v_1 \to \infty). \tag{42}$$

More precisely, for the mth-order modified Bessel function to be given by the above-mentioned asymptotic value, the argument $2v_1$ must be large compared to unity and large compared to m^2. For instance, for $v_1 = 3$ we have $I_0(2v_1) = 67$, $I_1(2v_1) = 61$ and $I_2(2v_1) = 47$. As a result, the magnitude of β_m in practice will be smaller than unity for the higher values of m.

Upon substitution of (42) in (12) for $J_{ca}(t)$, we obtain

$$J_{ca}(t) = J_{c0}\left\{1 + 2\sum_{m=1}^{\infty}(-1)^m \cos m\omega t\right\} \quad (v_1 \to \infty). \tag{43}$$

Thus, all Fourier components of the current $J_{ca}(t)$ have equal amplitude; that is, we have a sharp pulse of avalanche current occurring at the zero crossings of the first-harmonic voltage (i.e., at $t = (2n+1)\pi/\omega$, where $n = 0, 1, 2, \cdots$). This case corresponds to Read's "sharp-pulse approximation."

With $\beta_m = (-1)^m$, the solution to (35) becomes simple. The infinite set of coupled nonlinear equations reduces to a set of uncoupled linear equations, yielding the result

$$v_m = (-1)^{m+1}\frac{\rho_m}{m^2\kappa} \quad (m = 1, 2, 3, \cdots). \tag{44}$$

This solution is of course subject to the conditions that v_1 must be real and large compared to unity; in addition, the amplitudes v_m of the higher harmonic voltages must remain small compared to unity as expressed by (11). It may be observed that as $|\beta_m|$ in practice is smaller than unity for the larger values of m, also $|v_m|$ in practice will be smaller than indicated by (44) for the larger values of m.

IV. Oscillation Conditions

Having obtained explicit solutions to the oscillator equations (35), we may now consider the conditions that must be satisfied at the various oscillation levels.

A. Impedance of the Diode as a Function of Signal Level

First, we consider the impedance of the diode as a function of signal level in order to find the frequency range over which the diode exhibits negative resistance. We shall define the mth harmonic impedance of the diode as the ratio of the complex voltage and the complex current corresponding to $e^{im\omega t}$ in the respective Fourier expansions. The normalized mth harmonic voltage across the diode is $v_m + v_{dm}$, and the normalized mth harmonic current is $j_m = \beta_m + m^2\kappa v_m$ [cf. (29)]. Taking into account (30), (31) and (32), we then obtain with the third of (29) the following expression for the mth harmonic impedance of the diode normalized with respect to $1/\omega C$:

$$z_m = \frac{\beta_m z_{dm} - im\kappa v_m}{\beta_m + m^2\kappa v_m}. \tag{45}$$

This expression is independent of (28), that is, (45) represents an impedance that can, in principle, be measured in the nonoscillating state by inserting a sinusoidal external voltage of frequency ω in series with the oscillator circuit. The expression (45) is implicit in all of the harmonic voltages v_m because of the presence of β_m. For our purposes, it will suffice to consider the two limiting cases of vanishingly small values of v_1 and of very large values of v_1.

For vanishingly small values of v_1, higher-harmonic generation is negligible and only the first-harmonic impedance has significance. In (14) for the coefficients b_1 and b_0, the infinite sums then can be discarded and the small-argument expansions of the modified Bessel functions $I_0(2v_1)$ and $I_1(2v_1)$ can be used. Then $\beta_1 \to -v_1$ and we find that

$$z_{ss} = \frac{z_{d1} + i\kappa}{1 - \kappa} \qquad (v_1 \rightarrow 0). \tag{46}$$

The subscript (ss) denotes the small-signal case; in order to stress that higher harmonic impedances have no significance in this case, no label for the harmonic number has been added. The result (46) is readily verified to be identical with the result obtained by Gilden and Hines [1]. The real part of z_{d1} is always greater than or equal to zero, so z_{ss} can have a negative real part only for $\kappa > 1$. Thus, at low oscillation levels, the oscillation frequency is necessarily above the avalanche frequency.

For our subsequent considerations (Sections IVB and C), we observe that the small-signal impedance of the diode is a quantity independent of signal level and may be regarded as merely a function of ω. Thus, if $Z_{ss}(\omega) = 1/(\omega C)[z_{d1}(\omega) + i\kappa]/(1 - \kappa)$ is the unnormalized small-signal impedance of the diode at the frequency ω, the expression

$$Z_{ss}(m\omega) = 1/(m\omega C)[z_{d1}(m\omega) + im^2\kappa]/(1 - m^2\kappa)$$

represents the small-signal impedance of the diode at the frequency $m\omega$. From the definition of $z_{dm}(\omega)$, we have $z_{d1}(m\omega) = mz_{dm}(\omega)$. Then, normalizing $Z_{ss}(m\omega)$ with respect to $1/\omega C$, we may write for the normalized small-signal impedance of the diode at the frequency $m\omega$:

$$z_{ss,m\omega} = \frac{z_{dm}(\omega) + im\kappa}{1 - m^2\kappa}. \tag{47}$$

Solving for $z_{dm}(\omega)$ in terms of $z_{ss,m\omega}$ and substituting the result in the expression for ρ_m, we find that

$$\rho_m - m^2\kappa = (1 - m^2\kappa)\frac{z_{em} + z_{ss,m\omega}}{z_{em} - i/m}. \tag{48}$$

For very large values of v_1, we have $\beta_m = (-1)^m$. Then, (45) yields for the normalized mth harmonic large-signal impedance:

$$z_{ls}{}^m = \frac{z_{dm} - (-1)^m im\kappa v_m}{1 + (-1)^m m^2 \kappa v_m}. \tag{49}$$

Again the real part of z_{d1} is always greater than or equal to zero. Therefore, as v_1 is real and positive, the condition for the first-harmonic impedance to have a negative real part now becomes $\kappa v_1 > 1$; that is, κ only needs to be greater than $1/v_1$, which is a small number. We conclude that, for large signals, the diode exhibits negative resistance at frequencies substantially below the avalanche frequency and that the oscillation frequency may be below the avalanche frequency.

If we restrict ourselves to the case $\kappa v_1 \gg 1$, we have

$$z_{ls}{}^1 \rightarrow -i - z_{d1}/\kappa v_1 \qquad (v_1 \gg 1, \kappa v_1 \gg 1), \tag{50}$$

showing that the reactance of the diode approaches the reactance of the diode capacitance C as $\kappa v_1 \rightarrow \infty$. For the large-signal solution (44) to the oscillator equations (35), the expression for the mth harmonic large-signal impedance is readily verified to reduce to $-z_{em}$, as it should.

For large values of v_1 the magnitude of the unnormalized first-harmonic voltage across the diode is given by $|V_{\text{diode}}| = v_1(\lambda + 1)\omega\tau_a/\alpha_c'$, from which we obtain

$$\kappa v_1 = \frac{|V_{\text{diode}}|}{2J_{c0}/\omega C} \qquad (v_1 \gg 1). \tag{51}$$

This expression may be used for numerical evaluation of κv_1 in the practical case.

B. Start-Oscillation Conditions

As is shown by (39), start-oscillation is determined by the condition

$$\kappa = \rho_1. \tag{52}$$

From (48) it is apparent that (52) corresponds to $z_{e1} + z_{ss,\omega} = 0$; that is, vanishing small-signal impedance of the total oscillator circuit at the frequency ω. [It may be noted that, if the system were linear, the conditions $\text{Im}(z_{e1} + z_{ss,\omega}) = 0$ and $\text{Re}(z_{e1} + z_{ss,\omega}) < 0$ would generally assure the oscillation to build up. As the present system is nonlinear, this result does not apply here, and all we can say is that $z_{e1} + z_{ss,\omega} = 0$ is the condition for start-oscillation.] The consequences of the start-oscillation condition are more readily apparent from (52), rather than the impedance representation. For ρ_1 to be real, we must have (see Appendix II):

$$\text{Im } Z_{e1} = \frac{1}{\omega C} + (R_s + \text{Re } Z_{e1}) \cdot \left\{\cot \theta/2 + \frac{1}{\lambda}\frac{\theta/2}{\sin^2 \theta/2}\right\}. \tag{53}$$

Here we have used the unnormalized expression for Z_{e1}; in addition, we have substituted $R_s + \text{Re } Z_{e1}$ for $\text{Re } Z_{e1}$ in order to make the expression explicit in terms of the parasitic series resistance R_s and the useful load resistance $\text{Re } Z_{e1}$. Equation (53) determines the oscillation frequency for start-oscillation. As a supplementary condition, κ must be equal to ρ_1, yielding (cf. Appendix II):

$$\kappa = \kappa_s(\omega) \equiv 1 + \frac{1}{\omega C_d(R_s + \text{Re } Z_{e1})} \frac{\sin^2 \theta/2}{\theta/2}. \tag{54}$$

For any frequency ω satisfying (53), (54) determines the bias current to start the oscillation. The notation $\kappa_s(\omega)$ is introduced to indicate the value of κ required to start the oscillation. Since κ is inversely proportional to the dc bias current, smaller values of $(R_s + \text{Re } Z_{e1})$ lead to smaller starting currents; if $\omega C_d(R_s + \text{Re } Z_{e1})[\theta/2)/\sin^2\theta/2] \ll 1$, we have

$$\left(\frac{J_{c0}}{A}\right)_{\text{start}} = \frac{\omega^2 \epsilon}{2\alpha_c' v_d} \frac{\theta/2}{\sin^2 \theta/2} \omega C_d(R_s + \text{Re } Z_{e1}). \tag{55}$$

The latter result shows that minimum starting-current densities are limited by the product $\omega C_d(R_s + \text{Re } Z_{e1})$.

C. Low-Level Conditions

For values of v_1 that are small compared to unity, v_1^2 is given by (40). The condition that v_1^2 must be real and positive imposes restrictions on ρ_1 and ρ_2.

For start-oscillation ρ_1 must be real, and the bias current such that $\kappa=\kappa_s(\omega)=\rho_1$. If the bias current is increased above the start-oscillation value, κ at constant frequency decreases below $\kappa_s(\omega)=\rho_1$ and $\rho_1-\kappa$ assumes a positive real value. Then, for v_1^2 to attain a finite real value, ρ_2 must be real (note that $2\sigma\ll 1$) and either $\rho_2>4\rho_1$ or $\rho_2<2\rho_1$. If $\rho_2>4\rho_1$ then v_1^2 is positive for all values of κ below $\kappa_s(\omega)$; if $\rho_2<2\rho_1$ then v_1^2 is positive only for values of κ ranging from $\kappa_s(\omega)=\rho_1$ down to $\kappa_s(\omega)/2+\rho_2/4=\rho_1/2+\rho_2/4$. As the latter condition permits the bias current to be increased by at most a factor of 2 above the start-oscillation value, we shall discard the case $\rho_2<2\rho_1$.

If, at start-oscillation, ρ_2 is complex, then upon increasing the bias current above the start-oscillation value v_1^2 may or may not assume a finite value. Certainly, for v_1^2 to assume a finite value, ρ_1 cannot remain real, i.e., the frequency must change from its start-oscillation value. As we shall find below, in the limit of large-signal operation, the frequency must again satisfy (53). We conclude that, if the large-signal limit is to be reached continuously as a function of bias current and possible frequency jumping effects are to be avoided, the external circuit must be designed such that at the desired frequency ρ_2 is real and at least equal to $4\rho_1$. These conditions on ρ_2 take the explicit form (cf. Appendix II):

$$\mathrm{Im}\, Z_{e2} = \frac{1}{2\omega C} + (R_s + \mathrm{Re}\, Z_{e2})\left\{\cot\theta + \frac{1}{\lambda}\frac{\theta}{\sin^2\theta}\right\}, \tag{56}$$

and

$$1 + \frac{1}{2\omega C_d(R_s + \mathrm{Re}\, Z_{e2})}\frac{\sin^2\theta}{\theta} \geq 4\kappa_s(\omega). \tag{57}$$

We observe that $\kappa(2\omega)=4\kappa(\omega)$, and that according to (38), $\rho_2-4\kappa=0$ corresponds to $z_{e2}+z_{ss,2\omega}=0$. Thus (56) and (57) represent start-oscillation conditions for the frequency 2ω, where according to (57) the starting current for the frequency 2ω must be equal to or smaller than that for the frequency ω. It may be noted that satisfying the start-oscillation condition for the frequency 2ω does not mean that the oscillation will also build up at the frequency 2ω as the bias current is increased.

The condition (57) is obtained in explicit form by substituting (54) for $\kappa_s(\omega)$. Then, if both $\omega C_d(R_s+\mathrm{Re}\, Z_{e1})$ and $2\omega C_d(R_s+\mathrm{Re}\, Z_{e2})$ are small compared to unity, we obtain

$$R_s \leq \frac{\mathrm{Re}\, Z_{e1}\cos^2(\theta/2)/4 - \mathrm{Re}\, Z_{e2}}{1-\cos^2(\theta/2)/4}. \tag{58}$$

As R_s is always positive, (58) further requires that

$$\mathrm{Re}\, Z_{e2} \leq \mathrm{Re}\, Z_{e1}\cos^2(\theta/2)/4. \tag{59}$$

According to (58), for any given load resistance $\mathrm{Re}\, Z_{e1}$, higher values of the parasitic series resistance R_s are permissible as $\mathrm{Re}\, Z_{e2}$ is made smaller. In the limit, the second harmonic is not coupled to the load; then $\mathrm{Re}\, Z_{e2}\rightarrow 0$.

If the foregoing conditions on ρ_2 and ρ_1 are satisfied, v_1 increases as the bias current is increased. Then, especially if ρ_1 is large compared to unity, v_1^2 rapidly becomes comparable to unity for increasing bias currents and we enter the range of large-signal operation.

D. High-Level Conditions

For $v_1\gg 1$, the solution to the oscillator equations (35) is given by (44). In particular, we then have

$$v_1 = \rho_1/\kappa, \tag{60}$$

where v_1 must be real and large compared to unity. The condition that v_1 be real requires ρ_1 to be real. This condition on ρ_1 is identical with that for start-oscillation, and requires the oscillation frequency to satisfy (53). We conclude that, for fixed values of $\mathrm{Re}\, Z_{e1}$ and $\mathrm{Im}\, Z_{e1}$, the oscillation frequency in the large-signal limit is the same as that at start-oscillation. Moreover, if the external circuit is designed for simultaneous starting conditions for the second harmonic, the oscillation frequency at intermediate oscillation levels also will satisfy (53). Then large-signal operation can be reached in a continuous fashion, without frequency-jumping effects, by arbitrarily increasing the bias current (of course within the thermal limits of the diode).

The condition $v_1\gg 1$ reduces to the condition that the final bias current must be large compared to the starting current, as may be seen by writing the condition $v_1\gg 1$ in the form

$$v_1 = \rho_1/\kappa \equiv \kappa_s/\kappa \equiv J_{c0}/J_{c0,\mathrm{start}} \gg 1. \tag{61}$$

This condition is more readily satisfied for the smaller values of $J_{c0,\mathrm{start}}$; according to (54) the condition then becomes

$$R_s + \mathrm{Re}\, Z_{e1} \ll 1/\omega C_d. \tag{62}$$

The optimum transit angle for achieving small starting currents at fixed impedance level $1/\omega C_d$ is found by maximizing $(\sin^2\theta/2)/(\theta/2)$, yielding $\theta=0.74\pi$

If $R_s+\mathrm{Re}\, Z_{e1}\ll 1/\omega C_d$ and θ on the order of 0.74π, (53) for the oscillation frequency shows that $\mathrm{Im}\, Z_{e1}$ must be approximately equal to $1/\omega C$; that is, the external circuit must be inductive and nearly in resonance with the diode capacitance.

In practice, it may be advantageous to choose θ smaller than 0.74π, in order for the conditions (58) and (59) for simultaneous starting at the second harmonic to be more easily satisfied. For example, if $\mathrm{Re}\, Z_{e2}\rightarrow 0$ and $\theta=0.74\pi$, then (58) requires R_s to be smaller than $0.04\,\mathrm{Re}\, Z_{e1}$, whereas for $\theta=0.37\pi$ the condition becomes $R_s\leq 0.21\mathrm{Re}\, Z_{e1}$.

Finally, for the present analysis to be valid, the amplitudes $|v_m|$ of the higher harmonic voltages ($m=2, 3, 4, \cdots$) must remain small compared to unity, as is expressed by (11). Generally, considering (44), (11) can be satisfied by minimizing $|\rho_m|$ for $m=2, 3, 4, \cdots$, and by choosing κ not too small (though not necessarily greater than unity). As is shown in Appendix II, $|\rho_m| \to 1$ if Re Z_{em} is not small compared to $1/m\omega C_d$ and Im Z_{em} is preferably large compared to $1/m\omega C_d$. However, as concerns the voltage v_2, it appears that if the oscillator is designed for simultaneous starting at the second harmonic, the voltage v_2 will be on the same order of magnitude as the voltage v_1, thus certainly it will not be small compared to unity. We must conclude that, under these conditions, our results for the large-signal limit have only approximate validity. Fortunately, as remarked below (44), in practice the voltage v_2 will be smaller than indicated by (44), thus improving the situation as concerns our large-signal results.

168

V. Output Power

The power delivered to the external circuit at the mth harmonic of the oscillation frequency is given by

$$P_m = 2j_m j_m^* J_{c0}^2 \text{ Re } Z_{em}, \tag{63}$$

where Re Z_{em} represents the useful load resistance. From the second of (29) combined with (35), we have

$$j_m = -(\rho_m - 1)\beta_m, \tag{64}$$

so that

$$P_m = 2(\rho_m - 1)(\rho_m^* - 1)\beta_m\beta_m^* J_{c0}^2 \text{ Re } Z_{em}. \tag{65}$$

We recall that the coefficients β_m are the coefficients in the Fourier expansion of the avalanche current $J_{ca}(t)$. It may be shown from (13) and (73) that $|\beta_m| \leq 1$; the equality $|\beta_m| = 1$ is attained in the large-signal limit. According to (65), no power will be developed at the higher harmonics if $|\rho_m|$ is made to approach unity. This condition has been met before in specifying that the higher harmonic voltages v_m remain small compared to unity.

The tuning conditions for the fundamental require ρ_1 to be real and greater than unity; for large-signal operation, ρ_1 must actually be real and large compared to unity. Then, if the oscillator is designed for simultaneous starting conditions at the second harmonic, we have $\rho_2 \geq 4\rho_1 \gg 1$ and P_2 in the large-signal limit will have an appreciable value unless Re Z_{e2} is made to vanish. The condition of not coupling the second harmonic to the load, i.e., making Re $Z_{e2} \to 0$, has been met before in specifying the optimum second-harmonic tuning conditions [cf. remark below (59)]. The output power at the fundamental may be obtained explicitly for low and for high oscillation levels. For low oscillation levels, we may write β_1 in terms of v_1 with the aid of (35) and substitute (40) for v_1^2. In the following, we shall restrict ourselves to the large-signal limit. Then $\beta_1\beta_1^* = 1$, so that with ρ_1 from (54) we obtain

$$P_1 = \frac{1}{2}\left[2J_{c0}\frac{\sin^2\theta/2}{\theta/2}Q\right]^2 \text{ Re } Z_{e1}, \tag{66}$$

where $Q = \{\omega C_d(R_s + \text{Re } Z_{e1})\}^{-1}$ must be large compared to unity in order that the condition $\rho_1 \gg 1$ be satisfied. The expression in brackets in (66) represents the RF circuit current. A condition for our analysis to be valid is that the RF voltage swing, i.e., the product of RF circuit current and impedance level $1/\omega C$, must remain small compared to the dc bias voltage. The expression (66) may be maximized as a function of transit angle θ and as a function of load resistance Re Z_{e1}.

As a function of the transit angle, one may consider that C_d is inversely proportional to the length l_d of the drift zone. Upon cancelling the dependence of C_d and θ on l_d, the length of the drift zone appears only in the factor $\sin^4\theta/2$, yielding an optimum transit angle equal to π. In that case, the oscillator certainly cannot be designed for simultaneous starting at the second harmonic as is shown by (57). On the other hand, one may consider the dependence of P_1 on ω. Then we find a $(\sin x/x)^4$ behavior, anomalously yielding an optimum transit angle equal to zero. Rather, we shall maximize P_1 at constant impedance level $1/\omega C_d$, so as to obtain the maximum power-impedance product. Then the optimum transit angle is obtained by maximizing $(\sin^4\theta/2)/(\theta/2)^2$, yielding $\theta = 0.74\pi$. As was mentioned below (62), in practice it may be advantageous to choose θ smaller than 0.74π, in order to more easily satisfy the condition for simultaneous start-oscillation of the second harmonic.

When considering the output power as a function of the load resistance, we observe that P_1 is proportional to $J_{c0}^2Q^2$ Re Z_{e1}. The factor Q^2 Re Z_{e1} is proportional to Re $Z_{e1}/(R_s + \text{Re } Z_{e1})^2$; thus, for given bias current J_{c0} and impedance level $1/\omega C_d$, P_1 is maximized by choosing Re $Z_{e1} = R_s$. Then

$$P_{1,\max} = \frac{1}{2}\frac{J_{c0}^2}{(\omega C_d)^2 R_s}\frac{\sin^4\theta/2}{\theta/2}, \tag{67}$$

a result that has been obtained earlier by Tager [8] and Delagebeaudeuf [9].

According to (67), the maximum output power is inversely proportional to the parasitic series resistance in the oscillator circuit (i.e., diode plus external circuit). We conclude that in order to attain high maximum output power, the parasitic series resistance must be made as small as possible. On the other hand, we observe that if Re $Z_{e1} = R_s$ half of the available RF power is dissipated in the parasitic series resistance, to the detriment of obtaining high dc-to-RF conversion efficiency. Also, if the oscillator is to be designed for simultaneous starting of the second harmonic, R_s must be small compared to Re Z_{e1}, as shown by (58). More importantly, however, for sufficiently large values of $J_{c0}Q$ the output power is limited as the RF voltage swing becomes an appreciable fraction of the dc bias voltage. Denoting the amplitude of the RF voltage swing by

V_{RF} and the dc bias voltage by V_b, we must have $V_{RF} < \gamma V_b$, where the fraction γ is smaller than unity (in our analysis, we have actually assumed $\gamma \ll 1$). With V_{RF} given by the product of RF circuit current as given by the expression in brackets in (66) and impedance level $1/\omega C$, we find a limiting value for the quality factor $Q = \{\omega C_d(R_s + \text{Re } Z_{e1})\}^{-1}$:

$$Q_{\text{lim}} = \gamma\omega C V_b \bigg/ \left(2J_{c0} \frac{\sin^2 \theta/2}{\theta/2}\right). \tag{68}$$

With the corresponding minimum permissible value of $R_s + \text{Re } Z_{e1}$, $(R_s + \text{Re } Z_{e1})_{min} = 1/(\omega C_d Q_{\text{lim}})$, we then find for the limiting output power $P_{1,\text{lim}}$:

$$P_{1,\text{lim}} = \gamma V_b J_{c0} \frac{\sin^2 \theta/2}{\theta/2} \frac{C}{C_d} - \frac{1}{2} (\gamma\omega C V_b)^2 R_s. \tag{69}$$

The second term on the right of (69) represents the power loss in the parasitic series resistance; in order that this power loss be a negligible fraction of the available power, we must have $(\omega C_d R_s)^{-1} \gg Q_{\text{lim}}$. From (69), we also find the limiting dc-to-RF conversion efficiency

$$\eta_{\text{lim}} = \gamma \frac{\sin^2 \theta/2}{\theta/2} \frac{C}{C_d} (\gamma < 1). \tag{70}$$

As the maximum value of $\sin^2(\theta/2)/(\theta/2)$ is 0.725, the mathematical limit on η_{lim} is 72.5 percent. More realistically, we may take $\gamma \approx 0.6$, yielding a limiting efficiency on the order of 40 percent. At values of Q below Q_{lim}, we may obtain the dc-to-RF conversion efficiency from (66). Then the RF output power is proportional to the square of the dc current density. Insofar as the dc bias voltage is independent of current density (note that the junction temperature changes with current density), the input power is linearly proportional to dc current density. Thus, if $Q < Q_{\text{lim}}$, the dc-to-RF conversion efficiency in principle is linearly proportional to the dc current density.

VI. Conclusions

In the foregoing analysis, the nonlinear operating characteristics of the avalanche transit-time oscillator have been studied by means of Fourier-series representation.

At low signal levels, the diode exhibits negative resistance only for frequencies above the avalanche frequency; at high signal levels, negative resistance also occurs for frequencies substantially below the avalanche frequency.

For fixed external impedances, in order to attain the large-signal limit in a continuous fashion as a function of dc bias current, without frequency jumping effects, the oscillator must be designed such that start-oscillation conditions are satisfied simultaneously for the first and the second harmonic of the desired oscillation frequency. The oscillation frequency then becomes nearly independent of bias current, and the oscillation level increases smoothly with the bias current.

A condition for large-signal operation is that the quality factor $Q = \{\omega C_d(R_s + \text{Re } Z_{e1})\}^{-1}$ must be large compared to unity. The output power in the large-signal limit is obtained explicitly in terms of diode and external circuit parameters. The output power in principle is proportional to the square of the bias current; the maximum output power is limited by the parasitic series resistance in the oscillator circuit and by the permissible value of the RF voltage swing as compared to the dc bias voltage. The best power–impedance product is obtained by choosing the transit angle equal to 0.74π; this condition also yields the smallest starting current.

In practice it may be advantageous to choose the transit angle smaller than 0.74π, in order that the oscillator may be more easily designed for simultaneous starting of the fundamental and the second harmonic. Simultaneous starting of the fundamental and the second harmonic further requires that parasitic series resistance be negligible compared to the load resistance at the fundamental, and that the second harmonic not be coupled to the load.

The limitation of output power by parasitic series resistance is borne out by the experiments of Kovel and Gibbons [11]; the importance of tuning the second harmonic is borne out by the experiments of Swan [12]. The high-power results reported by Johnston *et al.* [13] may be partly explained by the aforementioned fact that, in practice it may be advantageous to operate the diode at a transit angle smaller than 0.74π.

The dc-to-RF conversion efficiency in principle is linearly proportional to the dc current density; the maximum dc-to-RF conversion efficiency again is limited by parasitic series resistance and by permissible RF voltage swing.

The output power at the harmonics of the oscillation frequency higher than the second can be made to vanish by suitable choice of the external impedances. The output power at the second harmonic generally cannot be made to vanish if the oscillator is designed for simultaneous starting conditions for the first and the second harmonic.

Appendix I

To find the Fourier expansion of $J_{ca}(t)$, where $J_{ca}(t)$ is given by (2) and (8), let us consider the Fourier expansion of

$$\exp\left\{-\sum_{n=1}^{\infty} (x_n e^{in\omega t} + \text{c.c.}\right\}. \tag{71}$$

If we put

$$\exp\left\{-\sum_{n=1}^{\infty} (x_n e^{in\omega t} + \text{c.c.})\right\} = b_0 + \sum_{m=1}^{\infty} (b_m e^{im\omega t} + \text{c.c.}), \tag{72}$$

then the coefficients b_m are given by

$$b_m = \frac{\omega}{2\pi}\int_0^{2\pi/\omega} e^{-im\omega t}\exp\left\{-\sum_{n=1}^{\infty}(x_n e^{in\omega t} + \text{c.c.})\right\}. \qquad (73)$$

We may write

$$\exp\left\{-\sum_{n=1}^{\infty}(x_n e^{in\omega t} + \text{c.c.})\right\}$$
$$= \exp\{-(x_1 e^{i\omega t} + x_1^* e^{-i\omega t})\} \qquad (74)$$
$$\cdot \exp\left\{-\sum_{n=2}^{\infty}(x_n e^{in\omega t} + \text{c.c.})\right\}.$$

Then if

$$\left|\sum_{n=2}^{\infty}(x_n e^{in\omega t} + \text{c.c.})\right| \ll 1, \qquad (75)$$

we may expand the second exponential function on the right-hand side of (74) and retain only the first two
170 terms; if, in addition, $x_1^* = x_1$ (i.e., x_1 real) we obtain

$$\exp\left\{-\sum_{n=1}^{\infty}(x_n e^{in\omega t} + \text{c.c.})\right\} \qquad (76)$$
$$= e^{-2x_1\cos\omega t}\left\{1 - \sum_{n=2}^{\infty}(x_n e^{in\omega t} + x_n^* e^{-in\omega t})\right\}.$$

Upon substitution of the last expression in (73) we find that

$$b_m = \frac{\omega}{2\pi}\int_0^{2\pi/\omega} dt\left\{e^{-im\omega t} - \sum_{n=2}^{\infty}(x_n e^{-i(m-n)\omega t} + x_n^* e^{-i(m+n)\omega t})\right\} e^{-2x_1\cos\omega t}$$
$$= (-1)^m\left[I_m(2x_1) - \sum_{n=2}^{\infty}(-1)^n \cdot\{x_n I_{m-n}(2x_1) + x_n^* I_{m+n}(2x_1)\}\right]. \qquad (77)$$

Here we have used the integral representation of the modified Bessel function $I_\nu(2x_1)$

$$I_\nu(2x_1) = \frac{\omega}{2\pi}\int_0^{2\pi/\omega} e^{-i\nu\omega t} e^{2x_1\cos\omega t} dt, \qquad (78)$$

and the relationship

$$I_\nu(-2x_1) = (-1)^\nu I_\nu(2x_1). \qquad (79)$$

Letting $x_n = v_n - i\sigma v_1^2\delta_{2n}$ and $\beta_m = b_m/b_0$, we then immediately obtain the Fourier expansion of $J_{ca}(t)$ given in (12).

Appendix II

We want to consider the properties of the coefficients ρ_m defined in (36):

$$\rho_m = \frac{z_{em} + z_{dm}}{z_{em} - i/m}, \qquad (80)$$

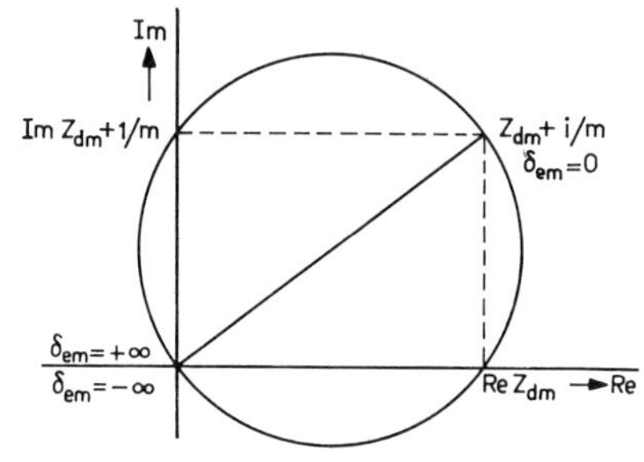

Fig. 1. Locus of $(\rho_m - 1)\text{Re } z_{em}$ in the complex plane as $\delta_{em} = (\text{Im } z_{em} - 1/m)/\text{Re } z_{em}$ ranges from $-\infty$ to $+\infty$.

where z_{em} is the normalized external impedance, and the normalized impedance z_{dm} is defined by (33). If we let

$$\delta_{em} \equiv \frac{\text{Im } z_{em} - 1/m}{\text{Re } z_{em}}, \qquad (81)$$

then we have

$$(\rho_m - 1)\text{ Re } z_{em} = \frac{z_{dm} + i/m}{1 + i\delta_{em}}. \qquad (82)$$

For a passive external circuit, we have $\text{Re } z_{em} > 0$; furthermore, as $\text{Im } z_{em}$ varies from $-\infty$ to $+\infty$, then δ_{em} also varies from $-\infty$ to $+\infty$. As δ_{em} is varied from $-\infty$ to $+\infty$, the factor $1/(1+i\delta_{em})$ describes a circle in the complex plane; the circle passes through the origin, is centered at the real axis, and has unit diameter. Therefore, as δ_{em} is varied from $-\infty$ to $+\infty$, the quantity $(\rho_m - 1)$ Re z_{em} defined in (82) describes a circle through the origin centered at $(z_{dm} + i/m)/2$. This behavior is illustrated in Fig. 1.

We have, from (33),

$$z_{dm} + i/m = \frac{\lambda}{m(\lambda+1)}\left\{\frac{1 - \cos m\theta}{m\theta} + i\left(\frac{\sin m\theta}{m\theta} + \frac{1}{\lambda}\right)\right\}, \qquad (83)$$

so the diameter of the circle is given by

$$|z_{dm} + i/m| = \frac{\lambda}{m(\lambda+1)}\left\{\left(\frac{1 - \cos m\theta}{m\theta}\right)^2 + \left(\frac{\sin m\theta}{m\theta} + \frac{1}{\lambda}\right)^2\right\}^{1/2}. \qquad (84)$$

For purely real values of ρ_m we have

$$\delta_{em} = \frac{\text{Im } z_{dm} + 1/m}{\text{Re } z_{dm}} \quad (\rho_m \text{ real}), \qquad (85)$$

yielding:

$$\frac{\text{Im } z_{em} - 1/m}{\text{Re } z_{em}} = \cot(m\theta/2) + \frac{1}{\lambda}\frac{m\theta/2}{\sin^2(m\theta/2)}, \qquad (86)$$

and

$$(\rho_m - 1)\,\text{Re}\,z_{em} = \text{Re}\,z_{dm} = \frac{\lambda}{\lambda + 1}\,\frac{1 - \cos m\theta}{m^2\theta}\,. \quad (87)$$

From (86) and (87) we obtain (53) and (54) for $m=1$ and (56) and (57) for $m=2$. It may be observed that for purely real values of ρ_m, the quantity (ρ_m-1) Re z_{em} is always greater than or equal to zero.

As concerns the maximum value of $|(\rho_m-1)\text{ Re } z_{em}|$, it is seen from (84) that for large values of m, the maximum value of $|(\rho_m-1)\text{ Re } z_{em}|$ becomes equal to $1/m(\lambda+1)$. Generally, $\lambda = C_a/C_d$ is large compared to unity; if $1/\lambda$ were negligible compared to unity, then $|(\rho_m-1)\text{ Re } z_{em}|$ would vanish as $1/m^2$. Thus it is seen that for not too small values of Re z_{em} (actually Re z_{em} is at least equal to the normalized series resistance r_s), the magnitude of ρ_m rapidly goes to unity as m becomes large.

References

[1] M. Gilden and M. E. Hines, "Electronic tuning effects in the Read microwave avalanche diode," *IEEE Trans. Electron Devices*, vol. ED-13, pp. 169–175, January 1966.

[2] T. Misawa, "Negative resistance in p-n junctions under avalanche breakdown conditions," pts. I and II, *IEEE Trans. Electron Devices*, vol. ED-13, pp. 137–151, January 1966.

[3] S. T. Fisher, "Small-signal impedance of avalanching junctions with unequal electron and hole ionization rates and drift velocities," *IEEE Trans. Electron Devices*, vol. ED-14, pp. 313–322, June 1967.

[4] B. Hoefflinger, "Generalized small-signal analysis of avalanche transit-time diodes," *IEEE Trans. Electron Devices*, vol. ED-14, pp. 563–568, September 1967.

[5] K. M. Johnson, "Small-signal analysis of the Read avalanche diode," *IEEE Trans. Electron Devices*, vol. ED-15, pp. 141–150, March 1968.

[6] F. K. Manasse and J. S. Shapiro, "An improved dispersion relationship for the p-n junction avalanche diode," *IEEE Trans. Electron Devices*, vol. ED-15, pp. 282–289, May 1968.

[7] W. T. Read, Jr., "A proposed high-frequency negative-resistance diode," *Bell Sys. Tech. J.*, vol. 37, pp. 401–446, March 1958.

[8] A. S. Tager, "The avalanche-transit diode and its use in microwaves," *Soviet Phys.—Usp.*, vol. 9, pp. 892–912, May–June 1967.

[9] D. Delagebeaudeuf, "Les diodes à avalanche utilisées comme oscillateurs de puissance aux hyperfréquences. Fonctionnement à haut niveau," *Onde Élec.*, vol. 48, pp. 722–731, July–August 1968.

[10] See, for instance, R. A. Kokosa and R. L. Davies, "Avalanche breakdown of diffused silicon p-n junctions," *IEEE Trans. Electron Devices*, vol. ED-13, pp. 874–881, December 1966.

[11] S. R. Kovel and G. Gibbons, "The effect of unswept epitaxial material on the microwave efficiency of IMPATT diodes," *Proc. IEEE* (*Letters*), vol. 55, pp. 2066–2067, November 1967.

[12] C. B. Swan, "Impatt oscillator performance improvement with second-harmonic tuning," *Proc. IEEE* (*Letters*), vol. 56, pp. 1616–1617, September 1968.

[13] R. L. Johnston, D. L. Scharfetter, and D. J. Bartelink, "High-efficiency oscillations in germanium avalanche diodes below the transit-time frequency," *Proc. IEEE* (*Letters*), vol. 56, pp. 1611–1613, September 1968.

171

Reprinted from IEEE TRANSACTIONS ON *ELECTRON DEVICES*
Volume ED-16, Number 11, November, 1969
pp. 935-945

Computer Calculation of Silicon Avalanche Diodes

MASAKIYO MATSUMURA

Abstract—**The efficiency η and output power P per unit area were computed for silicon p-n-n$^+$ avalanche diodes having an active region of 5 microns. The condition for making both η and P maximum is that the field at the n-n$^+$ interface generated by impurity and bias voltage is about one fourth of that at the junction under oscillation conditions. The maximum η and P were calculated to be about 20.5 per cent and 53 kW/cm^2. Dc current density J_0 for maximum P was about 4200 A/cm^2. On the other hand, J_0 for maximum η was calculated to be about 800 A/cm^2. It was found that a diode having an abrupt type junction is the best.**

The condition under which a diode is represented by use of a diode with abrupt type junction is that the region of a homogeneous field near the junction is less than one fifth of the space-charge region.

172

I. Introduction

SINCE the avalanche diode oscillator was first demonstrated experimentally by Johnston *et al.* [1], small-signal analyses [2]–[4] of these devices have elucidated the important properties of the operation. However, the output power and efficiency cannot be obtained from the small-signal analysis. There are several large-signal analyses [5]–[7] of the diodes. They are based on the simplifying assumptions, such as sufficiently localized avalanche ionization, small transit angle, or negligible space-charge effect of carriers. Computer results indicated that the following three factors play a major role in limiting the output power [8]–[10]: 1) the fact that the avalanche region must have a proper width, 2) the ionization in the drift region, and 3) the decrease of the carrier velocity in the drift region, caused by the low field.

The aim of this paper is to present large-signal computer results of the silicon p-n-n$^+$ diodes and to show how the large-signal properties change with the diode parameters and dc current. Both output power P per unit area, and the efficiency η of the p-n-n$^+$ diodes having an abrupt type junction and active region of 5-micron width, take the maximum values when E_{out}/E_j is about 0.25 under oscillation conditions. E_{out} and E_j are the fields at the n-n$^+$ interface and p-n junction, respectively, induced by the impurity and bias voltage. They were computed to be about 53 kW/cm^2 and 20.5 percent. P did not saturate until the dc current density J_0 became several thousand A/cm^2. However, η became saturated at a J_0 of about one thousand A/cm^2. The maximum P of the p-n-n$^+$ diodes was found by computation to be higher than that of the n-p-p$^+$ diodes. A diode having an abrupt type junction found to have the best microwave properties.

Manuscript received July 7, 1969; revised December 9, 1969.
The author is with Central Research Laboratories, Nippon Electric Co., Ltd., Kawasaki, Japan.

II. Assumptions

The large-signal operating characteristics of the diode can be calculated in two ways. The first is to drive the diode with a sinusoidal ac voltage superimposed on the bias voltage as shown in Fig. 1(a). The second is to embed the diode into a parallel resonant circuit as a free running oscillator as shown in Fig. 1(b). Read [5] stated that under large-signal conditions the induced current takes nearly a square-wave form. The voltage across the terminals of the resonant circuit induced by the diode current becomes a nonsinusoidal wave. But the large-signal negative Q is greater than 10 except for extremely high J_0. If the resonant circuit is tuned to a single frequency, then the values of the higher harmonic components of the ac voltage across the terminals in Fig. 1(b) are estimated to be less than 5 percent of the value of the fundamental component. Thus, both ways of calculation give nearly the same results [11].

To study the performance of the diode at given RF parameters (for various bias and diode parameters), it is convenient to drive the diode with a sinusoidal voltage. Thus, we drive the diode with the sinusoidal voltage $V \sin \phi$ ($\phi = 2\pi ft$) superposed on the bias voltage V_0. P and η are calculated from the equations

$$P = -\frac{1}{2\pi}\int_0^{2\pi} J_{ind}(\phi) V \sin \phi d\phi,$$

and

$$\eta = \frac{P}{J_0 V_0},$$

where J_{ind} is the induced current density defined by the spatial mean value of the electron and hole current densities in the active region of the diode.

Numerical calculation was carried out only in the active region by means of the finite difference method [9]. Fig. 2 shows a diode model. Both edges of the active region contact the highly doped regions. The electric field E does not penetrate into these regions. The thermal generation and recombination of carriers in the space-charge region were not taken into consideration. The diffusion current densities injected from the highly doped regions into the active region were fixed at 1.6 $\times 10^{-7}$ A/cm^2. The field dependence of the carrier velocity was taken into consideration, but not the diffusion of carriers. The saturated velocities for electrons and holes were assumed to be 10^7 cm/s and 7.5×10^6 cm/s, respectively. When the space-charge region is 5 microns, the Read frequencies [5] f_t of electrons and

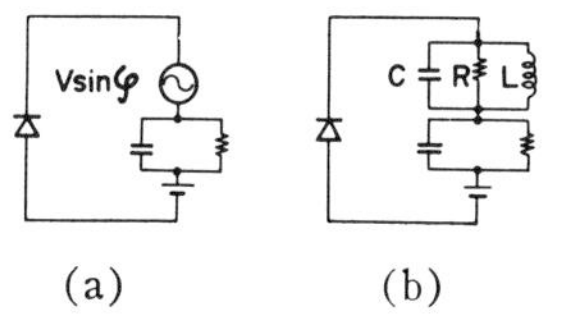

Fig. 1. Method of driving the diode.

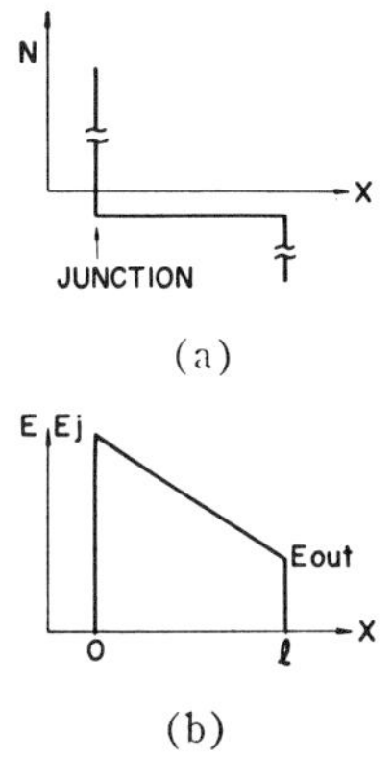

Fig. 2. Diode model. (a) Impurity distribution. (b) Bias field distribution.

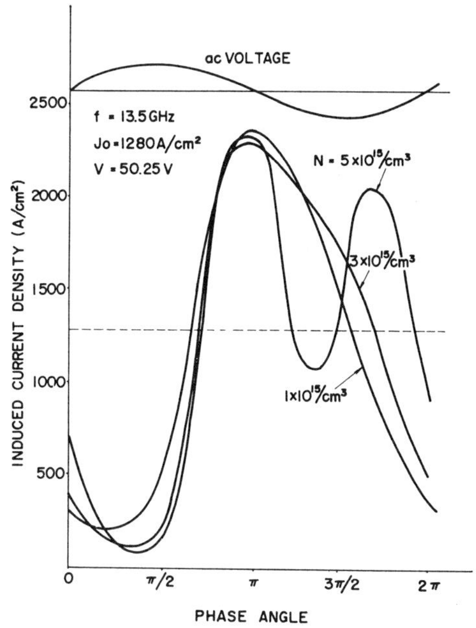

Fig. 3. Induced current density waveforms of the p-n-n^+ diodes as a function of phase angle for various impurity densities.

173

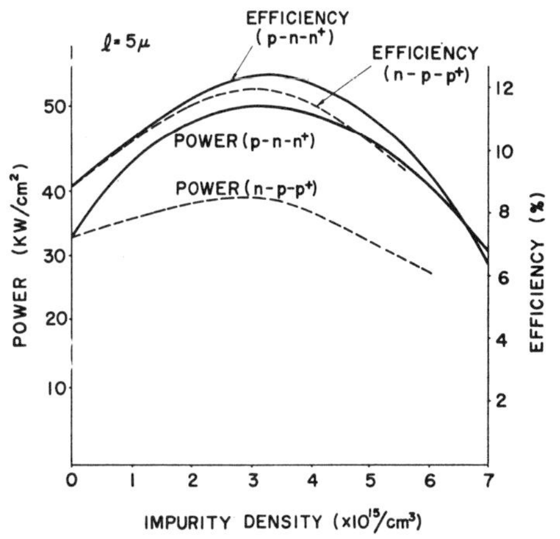

Fig 4. Maximum output per unit area for the p-n-n^+ and n-p-p^+ diodes as a function of impurity density. The ac voltage, dc current density, and oscillation frequency are changed to make the power maximum. Efficiency is a value under the maximum power conditions.

holes become about 10 and 7.5 GHz, respectively. The ionization coefficients were assumed to be proportional to the sixth power of E. These proportionality factors were assumed to be 5.25×10^{-30} cm^5/V^6 for electrons and 2.22×10^{-30} cm^5/V^6 for holes. When the impurity density N of the active region is $5.5 \times 10^{15}/cm^3$, the space-charge region reaches the highly doped region at breakdown condition.

III. Effect of Impurity Density on Output Power

The output power P per unit area was computed as a function of the impurity density N in an active region of 5-micron width. The p-n-n^+ diode is superior to the n-p-p^+ diode.

Fig. 3 shows the waveform of J_{ind} of the p-n-n^+ diodes with an abrupt-type junction as a function of phase angle for various N values. For the case of low N, i.e., large E_{out}, the ionization cannot be localized near the junction, and the conduction current becomes not localized in the active region. Ionization is enhanced at $\phi \doteqdot 3\pi/4$. Consequently, J_{ind} builds up rapidly within range $\phi \doteqdot 3\pi/4$ to $\phi \doteqdot \pi$ and then decreases linearly with increasing ϕ due to the broad waveform of the conduction current. Thus, J_{ind} becomes a sawtooth-like form and takes the maximum value at $\phi \doteqdot \pi$ as seen in the curve of $N = 1 \times 10^{15}/cm^3$ in Fig. 3. The out-of-phase component of J_{ind} with respect to the ac voltage is small. This fact gives rise to low P. The ionization becomes localized with increasing N. The conduction current becomes pulsive and J_{ind} takes a distorted square form. The out-of-phase component of J_{ind} relative to the ac voltage increases and high P can be expected. With further increase of N, E_{out} decreases and the diode becomes similar to a p-n diode. The ionization is further localized and J_{ind} approaches a square waveform. However, J_{ind} decreases at $\phi \doteqdot 3\pi/2$ (where the ac voltage has a maximum negative value) as seen in the curve of $N = 5 \times 10^{15}/cm^3$ in Fig. 3. Dependent upon the values of N and J_0, this effect is caused either by the injection of electrons from the n-n^+ interface or the decrease of the carrier velocity caused by the decrease of the field. As the decrease of J_{ind} near $\phi = 3\pi/2$ gives rise to the small out-of-phase component of J_{ind} with respect to the ac voltage, P decreases with further increase of N. Consequently, there should be an optimum value of N.

Fig. 4 shows peak P and the values of η under peak P conditions as a function of N for the p-n-n^+ and n-p-p^+ diodes having an abrupt-type junction and an active region of 5-micron width. The P of the p-n-n^+ diode becomes maximum at $N \doteq 3.5 \times 10^{15}/\text{cm}^3$ and $J_0 \doteq 4200$ A/cm^2. Maximum P was computed to be about 53 kW/cm^2 (under this condition, η was computed to be about 12.5 percent). P of the p-n-n$^+$ diode was about 60 percent higher than that of the n-p-p$^+$ diode. η of the p-n-n$^+$ diode was computed to be a little higher than that of the n-p-p$^+$ diode. These results are due to the fact that the sustaining field for velocity saturation of electrons is about 10 kV/cm, which is lower than that of holes (about 40 kV/cm). It was concluded that the p-n-n$^+$ diode is superior to the n-p-p$^+$ diode.

At $N = 3.5 \times 10^{15}/\text{cm}^3$, E_j and E_{out} under breakdown conditions were computed to be about 385 and 130 kV/cm, respectively. Thus, E_{out} is estimated to be about one third of E_j. Under constant current density conditions, the bias voltage decreases with increasing amplitude of oscillation. This fact gives rise to $E_{out} \doteq \frac{1}{4}E_j$ under oscillation conditions.

IV. Maximum Efficiency of p-n-n$^+$ Diode

The maximum efficiency η of the p-n-n$^+$ diode having abrupt-type junction and an active region of 5-micron width was computed as a function of N.

Fig. 5 shows η as a function of N for the p-n-n$^+$ diode. In Fig. 5, the value of P corresponding to the maximum value of η is also shown. The maximum value of η was 20.5 percent (P was computed to be about 15 kW/cm^2 under this condition) at $J_0 \doteq 800$ A/cm^2 and $N \doteq 2 \times 10^{15}/\text{cm}^3$ ($E_{out} \doteq \frac{1}{2}E_j$ under breakdown conditions and $E_{out} \doteq \frac{1}{3}E_j$ under oscillation conditions). The decrease of η with decreasing values of N below $N \doteq 2 \times 10^{15}/\text{cm}^3$ arises from the insufficient localization of avalanche ionization. The decrease of η with N increasing from $N \doteq 2 \times 10^{15}/\text{cm}^3$ is caused by the fact that the field-dependent velocity or the injection of electrons from the n-n$^+$ interface appears at lower V.

It is worthwhile to note that P corresponding to maximum η is maximum at $N \doteq 6 \times 10^{15}/\text{cm}^3$, where the diode breaks down just before the space charge region punches through the n region (in this case "unswept region" [12] was about 0.5 microns).

V. Effect of dc Current Density on Power, Efficiency, and Frequency

The output power P per unit area, efficiency η, and frequency f for the p-n-n$^+$ diode with the abrupt-type junction having an active n region of 5-micron width were calculated.

P, η, and f as a function of J_0 for $N = 3.5 \times 10^{15}/\text{cm}^3$ are shown in Fig. 6. It is seen that η is saturated at relatively low J_0 of about 1280 A/cm^2, but P continues to increase with increasing J_0 until J_0 becomes about 4200 A/cm^2. The frequency is about 120 percent of the Read frequency f_t at lower values of J_0 and increases with increasing J_0, becoming saturated at $(3/2)f_t$. These

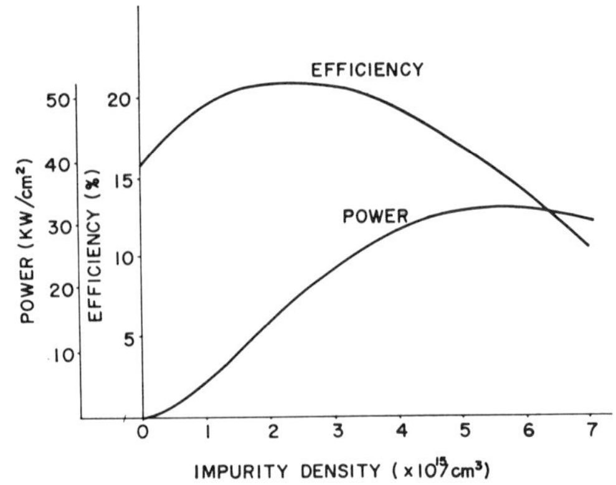

Fig. 5. Maximum efficiency of the p-n-n$^+$ diodes as a function of impurity density. The ac voltage, dc current density, and oscillation frequency are changed to make the efficiency maximum. For $N = 0$, efficiency takes the maximum value of $J_0 \doteq 0$. Power is a value under the maximum efficiency conditions.

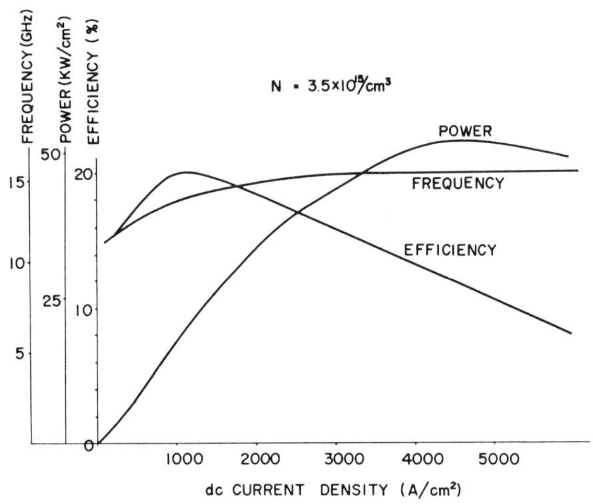

Fig. 6. Power, efficiency, and frequency as a function of dc current density.

results are explained as follows. For low J_0, the field induced by carriers is negligible because the carrier density is much less than N. Due to the negative bias field gradient toward the n$^+$ region, the swing of the ac voltage is apt to make the field at the n-n$^+$ interface negative. Under constant J_0 condition, P and η become saturated at $V = V_{max}$, where the field at the n-n$^+$ interface becomes negative and the injection of electron from the n-n$^+$ interface becomes predominant. As the space-charge effect of carriers is negligible, the field at the junction becomes in-phase with the ac voltage. Between $\phi \doteq \pi/4$ and $\phi \doteq 3\pi/4$, the field near the junction becomes high and the ionization is enhanced. The phase ϕ_0 of the buildup of the electron current becomes near by $3\pi/4$. Therefore, the value of f which makes P maximum becomes about $(2\pi - \phi_0)/\pi f_t \doteq 12$ GHz.

When J_0 increases, the space-charge effect of carriers raises the field at the n-n$^+$ interface. This fact gives rise to the elevation of V_{max} with increasing J_0. At the same time, the electron current pulse width becomes a little narrow with the increase of J_0. These facts give rise to an increase of η and of P as J_0 increases. With increasing J_0, the space-charge effect of carriers makes the field near the junction advance with respect to the ac voltage. Therefore, ϕ_0 advances with increasing J_0. This gives rise to an increase of f and decrease of η. Due to the advancement of ϕ_0, η becomes saturated at $J_0 \doteqdot 1280$ A/cm^2. With further increase of J_0, ϕ_0 approaches $\pi/2$ and f reaches $3f_t/2$. At extremely high J_0, the field near the electron current pulse becomes lower than the sustaining field, and carrier velocity decreases before the field at the n-n$^+$ interface becomes negative. P undergoes restriction by this effect. P is saturated at a value of J_0 of about 4200 A/cm^2, due to the decrease of V_{max}.

When N is 6×10^{15}/cm^3, the diode breaks down before the space-charge region punches through the n region. Due to the residual loss in the unswept region, the diode can not oscillate until J_0 becomes higher than about 400 A/cm^2.

When $N=0$, the diode becomes a p-i-n type. The field at the center of the active region is always the lowest by the space-charge effect of carriers. Consequently, V_{max} is determined by the appearance of the field-dependent velocity and not by the injection of carriers from the p-i or i-n junctions.

It was found that J_0, at efficiency saturation or power saturation, increases with an increase of N. The former lies in a range between 0 and 3000 A/cm^2, and the latter in a range between 3000 and 7000 A/cm^2.

VI. Superiority of Abrupt-Type Junction

As an example of the diodes having a nonabrupt-type junction, we calculated the peak power P per unit area and the efficiency η of the p-i-n-n$^+$ diodes having an active i-n region of 5-micron width, and obtained the condition under which a diode is approximately represented by use of a diode with an abrupt-type junction.

There is no assurance that the abrupt-type junction is the best. As a matter of fact, the junction fabricated by diffusion, epitaxial growth, or ion implantation is not abrupt. The bias field distribution of the actual diode is as shown in Fig. 7(a). The distribution given in Fig. 7(a) is simplified as the p-i-n-n$^+$ diode shown in Fig. 7(b). The i region corresponds to the region having the homogeneous bias field near the junction of the actual diode.

Fig. 8 shows η as a function of the width l_i of the i region for various impurity densities N in the n region. It is seen that η becomes maximum at $N \doteqdot 2 \times 10^{15}$/cm^3 and $l_i \doteqdot 0$ (the abrupt p-n-n$^+$ diode described in Section IV). But η for each N usually has a dip near $l_i=1$ micron, and has a high value at $l_i \doteqdot 0$ and $l_i \doteqdot 2.5$ microns. Let us consider the reason by examining the "snapshot" shown in Figs. 9 and 10, each containing the field E, electron density N_n, and hole density N_p. Each snapshot is shifted one fifth of an oscillation period from $\phi=0$ to $\phi=2\pi$.

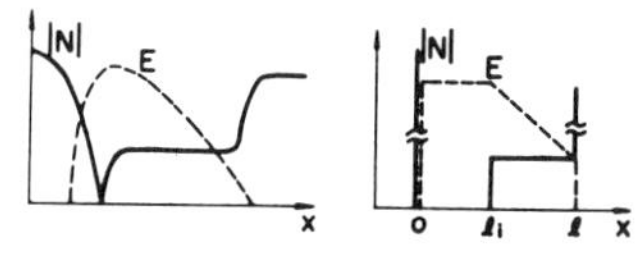

(a) (b)

Fig. 7. Diode model with nonabrupt-type junction. (a) Actual distributions of the impurity and bias field. (b) Simplified distributions of the impurity and bias field.

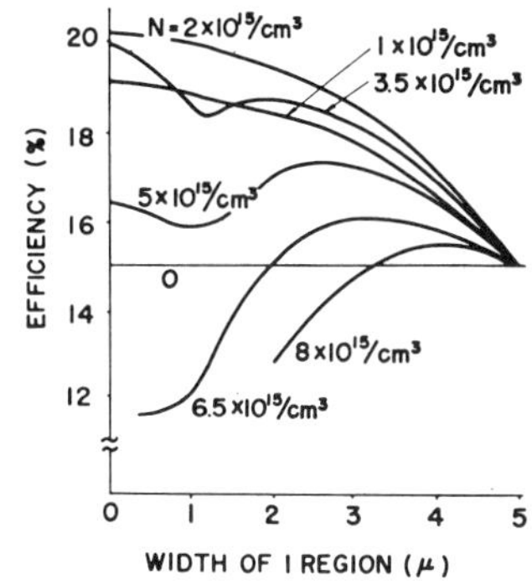

Fig. 8. Efficiency as a function of the width of the i region for various impurity densities. The ac voltage, dc current density, and oscillation frequency are changed to make the efficiency maximum.

For the diode having $l_i=0$, ionization is enhanced near the junction, as shown in Fig. 9. Generated holes drift toward p region and quickly disappear from the active region. Thus, the hole current absorbs little RF power. With increasing l_i, the ionization is enhanced near the i-n interface, as shown in Fig. 10. Generated holes survive in the active region for a relatively long time in the positive half cycle of the ac voltage. Thus, the hole current absorbs the power a great deal. This fact gives rise to the decrease of η with increasing l_i. When l_i becomes about one half of the width of the active region, the region of enhanced ionization moves toward the center of the active region. Generated holes and electrons drift across the different regions, viz., i and n regions, respectively, as shown in Fig. 10. Their transit times become the same. Thus, they act simultaneously to generate power, resulting in the increase of η with increasing l_i. This fact means that the diode can be regarded as a series combination of the p-i-n and p-n-n$^+$ diode, each having the same width of the active region. With further increase of l_i, the diode becomes a p-i-n diode. The ionization cannot be localized and η decreases with increasing l_i. Thus, a dip of near $l_i=1$ micron is caused by the transient state from the abrupt p-n-n$^+$ diode to the diode which can be regarded as a series combination of the p-i-n and p-n-n$^+$ diodes. If l_i is smaller than 1 micron (20 percent of the active region), the diode may be regarded as one with nearly abrupt-type junction. For low N, η decreases monotonically with increasing l_i. This is caused by the decrease of localized ionization which takes place with increasing l_i. For high N, η increases with increasing l_i until l_i be-

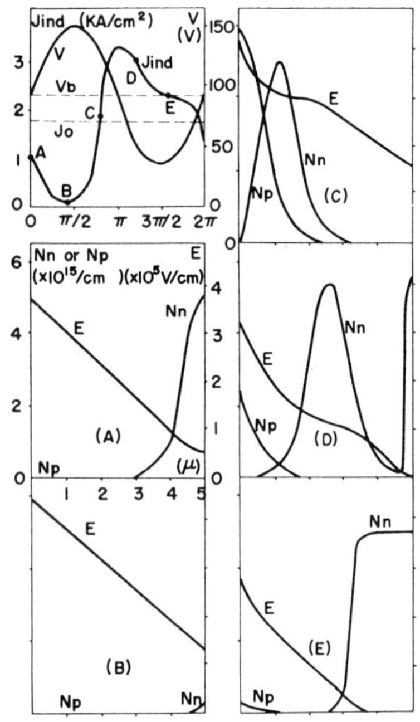

Fig. 9. Field, electrons, and holes in the active region ($l_i=0$).

176

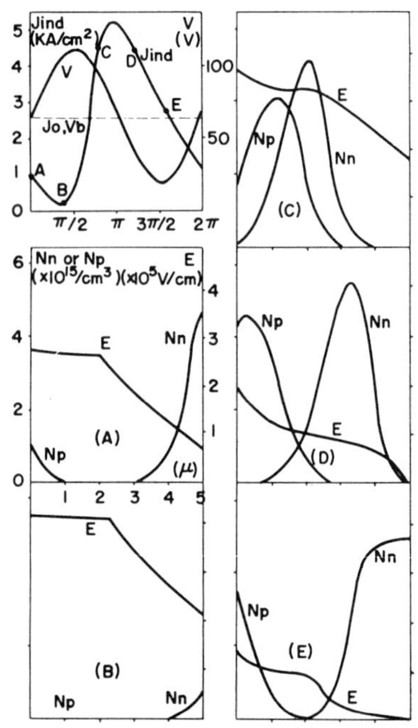

Fig. 10. Field, electrons, and holes in the active region ($l_i=2$ microns).

comes about 2.5 microns. Let us consider the reason. The increase of l_i results in the increase of E_{out}. This makes the increase of V_{max}, resulting in the increase of η. The amount of the increase of η is greater than that of the decrease of η which takes place near $l_i=1$ micron due to a transient state.

Fig. 11 shows maximum P as a function of l_i for various N values. The envelope of P (dashed line in Fig. 11) increases with increasing l_i and becomes maximum at $l_i \doteqdot 2.5$ microns. The maximum P was computed to be

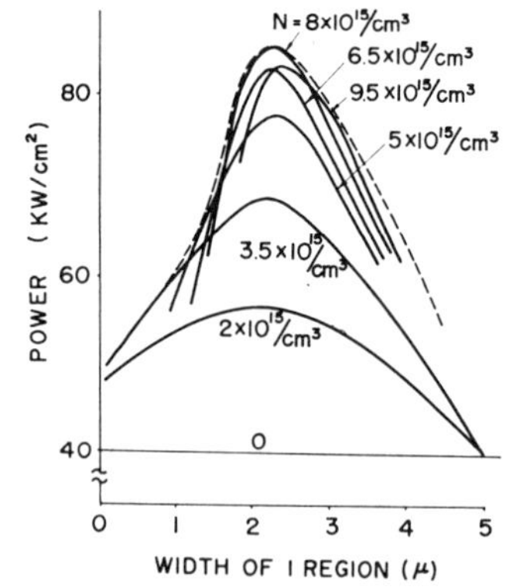

Fig. 11. Power as a function of the width of the i region for various impurity densities. The ac voltage, dc current density, and oscillation frequency are changed to make the power maximum. The broken line represents an envelope of power for various impurity densities.

about 87 kW/cm². J_0, f, and N under the maximum P condition were about 10^4 A/cm², 30 GHz, and 9.5×10^{15}/cm³, respectively. This result is explained by regarding [9] the diode as the series combination of the p-i-n and p-n-n⁺ diodes, each having the same width of active region. The ionization is enhanced near the i-n interface, as shown in Fig. 10. As electrons and holes drift in the different regions, their transit times must be determined by the distance from the i-n interface. Consequently, f_t becomes about 20 GHz corresponding to a drift region of 2.5-micron width. P becomes maximum near $f=3f_t/2$ ($\doteqdot 30$ GHz). J_0 can be estimated to be proportional to fE_j and E_j varies slightly with l_i. Therefore, J_0 becomes about 10^4 A/cm² (twice J_0 of the abrupt p-n-n$^+$ diode). On the other hand, ac voltage V is determined by E_jl. It depends slightly on l_i, if N is not very high. Judging from these facts, P is estimated, and is calculated, to be about twice P of the p-n-n$^+$ diode with the abrupt-type junctions. Since the increase of J_0 with increasing l_i compensates to too great a degree, the decrease of η near $l_i=1$ micron, P has no dip near $l_i=1$ micron.

CW characteristics of the avalanche diodes are restricted by the temperature in the diodes. Thus, it is not desirable that the optimum J_0 is large. For the purpose of good heat removal, heat generation must take place near the heat sink. Thus, the diode with the abrupt type junction is concluded to be the best.

VII. Conclusion

It was concluded that a p-n-n$^+$ silicon diode is superior to a n-p-p$^+$ silicon diode. The maximum power P per unit area of the p-n-n$^+$ diode having an abrupt-type junction and an active region of 5-micron width was computed to be about 53 kW/cm² (at maximum P, the efficiency η was about 12.5 percent) and the maximum η of the p-n-n$^+$ diode about 20.5 percent (at maximum η, P was about 15 kW/cm²). The field E_{out} at the n-n$^+$ interface must be suppressed to one third of the field E_j at the junction under breakdown conditions.

Under oscillation conditions, E_{out} became about one fourth of E_j.

The higher values of power densities compared with Scharfetter and Gummel's results [11] arise not predominantly from the power law approximation of ionization coefficients, but from an intrinsic difference between the p-n-n$^+$ and Read diodes. The equivalent width of the avalanche region of the p-n-n$^+$ diode is wider than that of the Read diode, resulting in the broad electron current waveform. This decreases the space-charge effect of carriers, resulting in the higher dc current densities J_0. For example, when we assume that the electron current varies sinusoidally, the field E_s induced by carrier is given by $J_0/2\pi\epsilon\epsilon_0 f$, and when we assume a sharply pulsed current waveform, E_s increases to $J_0/\epsilon\epsilon_0 f$. As the value of E_s is limited, the waveforms of electron current affects the maximum value of J_0. E_s increases with the increase of the impurity density in an n region.

So far, analysis was carried out without taking diffusion into consideration. No study seems to have been carried out on the diffusion of hot electrons. It seems, however, that the diffusion constant under a high field does not change with field a great deal. As the diffusion current density is proportional to carrier density, the change of current waveform caused by diffusion will be the same for any value of J_0.

The ionization coefficient decreases as the diode temperature T increases, resulting in a higher breakdown voltage. This means that the computer result of $E_{out} \doteqdot (1/3)E_j$ under breakdown condition must be modified. The carrier velocity also decreases with increasing temperature. The saturated velocity at $T = 500°K$ becomes about 80 percent of that at $T = 300°K$. Consequently, the oscillation frequency and the optimum current density at $T = 500°K$ decrease to about 80 percent of those at $T = 300°K$.

If thermal effects are taken into consideration, a diode having an abrupt-type junction is the best. The condition under which a diode is represented by use of a diode with abrupt-type junction is that a region of homogeneous bias field near the junction is less than 20 percent of the active region length.

Under large-signal conditions, the conduction current density is usually much less than the displacement current density. This fact means that the residual loss of the diode affects the microwave performance. Consequently, it is suggested that not only the maximum power and efficiency, but also the optimum bias field distribution will change a great deal with variation of the residual positive resistance. However, these considerations were not taken into account in this paper.

Acknowledgment

The author expresses his thanks to M. Uenohara and I. Uchida for their help in the preparation of this paper.

References

[1] R. L. Johnston, C. B. Deloach, Jr., and B. G. Cohen, "A silicon diode microwave oscillator," *Bell Sys. Tech. J.* (Briefs), vol. 43, pp. 369–372, December 1964.
[2] T. Misawa, "Negative resistance in p-n junctions under avalanche breakdown condition," *IEEE Trans. Electron Devices*, vol. ED-13, pp. 137–151, January 1966.
[3] ——, "Multiple uniform layer approximation in analysis of negative resistance in p-n junction in breakdown," *IEEE Trans. Electron Devices*, vol. ED-14, pp. 795–808, December 1967.
[4] M. Gilden and M. E. Hines, "Electronic tuning effects in the Read microwave avalanche diode," *IEEE Trans. Electron Devices*, vol. ED-13, pp. 169–175, January 1966.
[5] W. T. Read, Jr., "A proposed high-frequency, negative resistance diode," *Bell Sys. Tech. J.*, vol. 37, pp. 401–446, March 1958.
[6] J. L. Blue, "Approximate large-signal analysis of IMPATT oscillators," *Bell. Sys. Tech. J.*, vol. 48, pp. 383–393, February 1969.
[7] W. J. Evans and G. I. Haddad "A large-signal analysis of IMPATT diodes," *IEEE Trans. Electron Devices*, vol. ED-15, pp. 708–717, October 1968.
[8] M. Matsumura, "Computer simulation of silicon Read and p-n-n$^+$ IMPATT diodes under large signal conditions," *Japan J. Appl. Phys.*, vol. 8, pp. 82–84, January 1969.
[9] ——, "Bias field distribution of high power p-n-n$^+$ IMPATT diodes," *Trans. IECE (Japan)*, vol. 53-C, pp. 88–95, February 1970.
[10] ——, "Large signal analysis of Read diodes," *Japan J. Appl. Phys.*, vol. 8, pp. 218–226, February 1969.
[11] D. L. Scharfetter and H. K. Gummel, "Large-signal analysis of a silicon Read diode oscillator," *IEEE Trans. Electron Devices*, vol. ED-16, pp. 64–77, January 1969.
[12] S. R. Kovel and G. Gibbons, "The effect of unswept epitaxial material on the microwave efficiency of IMPATT diodes," *Proc. IEEE* (Letters), vol. 52, pp. 2066–2067, November 1967.

177

Reprinted from IEEE TRANSACTIONS
ON *MICROWAVE THEORY AND TECHNIQUES*
Volume MTT-18, Number 11, November, 1970
pp. 842-853

Large-Signal Equivalent Circuits of Avalanche Transit-Time Devices

PAUL T. GREILING, MEMBER, IEEE, AND GEORGE I. HADDAD, SENIOR MEMBER, IEEE

Abstract—**A large-signal analysis for IMPATT diodes is derived, which allows carrier multiplication by impact ionization to occur at every point in the diode. Therefore, the operating characteristics of IMPATT diodes with a wide range of realistic doping profiles can be investigated. For a given operating frequency, RF voltage, dc bias current, and doping profile, the admittance, power output, efficiency, and bias voltage of a diode can be obtained. An equivalent circuit for the diode package, microwave circuit mount and diode, is obtained experimentally. Using this circuit, the admittance of the diode is measured by a reflection-type circuit and an oscillator circuit as a function of the RF voltage, dc bias current, and frequency.**

I. INTRODUCTION

MOST LARGE-SIGNAL analyses [1], [2] of IMPATT diodes assume a Read-type structure and neglect the divergence of the displacement current in the assumed narrow avalanche region. These analyses can give good qualitative results for certain diode structures but are not applicable to many realistic diodes. As the width of the avalanche region increases, the effect of this assumption becomes appreciable. Therefore large-signal analyses of diode structures in which the avalanching occurs over an appreciable portion of the diode should include the effect of the displacement current in the avalanche region as well as allow for ionization to occur throughout the diode. A large-signal analysis by Scharfetter and Gummel [3] can be used to analyze diodes with any doping profile. The analysis includes the effect of unequal electron and hole mobilities which are a function of the field and the ionized impurity density. Also, the effects of diffusion and carrier generation and recombination due to defects and impact ionization are included. However, in order to study the characteristics of diodes operating in the transit-time mode, it may not be necessary to include all of these effects. Also, in order to investigate many different IMPATT diode structures, the computer time may become prohibitive. The analysis presented in this paper includes only the parameters which are considered important for transit-time operation. Therefore

Manuscript received April 30, 1970. This work was supported by the Air Force Systems Command's Rome Air Development Center under Contract F30602-68-C-0043.

The authors are with the Electron Physics Laboratory, Department of Electrical Engineering, University of Michigan, Ann Arbor, Mich. 48104.

the computer time necessary for investigating many different diode structures is not unreasonable.

A large-signal analysis is presented which allows for impact ionization to occur at every point in the diode. Therefore the operating characteristics of a wide range of diode structures can be investigated. The results of this analysis are compared for a one-sided abrupt-junction diode with different doping densities, and a qualitative discussion of the results of several other diode structures is given. Experimental results of the operating characteristics of abrupt-junction diodes with different doping densities and the dependence of the equivalent circuit of the diode on the doping profile and operating conditions are presented.

II. Large-Signal Analysis

The basic equations governing the operation of an IMPATT diode are Poisson's equation and the one-dimensional continuity equations for electrons and holes with terms included for impact ionization. In order to simplify the equations, several assumptions are made. The velocity of holes and electrons are assumed equal and field independent. Also, the effects of diffusion and recombination are neglected. The ionization rates for holes and electrons are assumed to be equal. The value for the ionization rate is taken to be the average or effective rate of the two carriers. This effective rate is obtained by solving for the dc breakdown voltage of the diode with unequal rates and then obtaining an equal rate for both carriers which gives the same breakdown voltage.

Incorporating these simplifications, a differential equation involving the conduction current density J_c and the hole and electron current densities J_p and J_n can be obtained from the two continuity equations [4]. Thus

$$\frac{1}{v}\frac{\partial J_c}{\partial t} = \frac{\partial}{\partial x}(J_n - J_p) + 2\alpha J_c \qquad (1)$$

where

- v saturated drift velocity (cm/s)
- α ionization rate (cm^{-1}).

Substituting the total current density J_T minus the displacement current density for the conduction current density in (1) and integrating over the entire length of the diode, the following equation is obtained:

$$\tau\frac{d}{dt}J_T - \frac{\epsilon}{v}\int_0^{x_L}\frac{\partial^2 E}{\partial t^2}\,dx$$

$$= (J_n - J_p)\Big|_0^{x_L} + 2J_T\int_0^{x_L}\alpha\,dx - 2\epsilon\int_0^{x_L}\alpha\frac{\partial E}{\partial t}\,dx \qquad (2)$$

where

- τ transit time through the diode (s) $= x_L/v$
- ϵ dielectric constant (F/cm)
- x_L total width of the diode (cm)
- E electric field (V/cm).

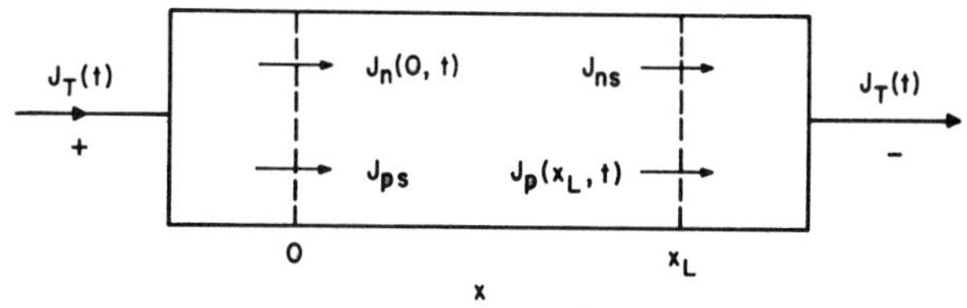

Fig. 1. Boundary conditions for an IMPATT diode.

Note that the equation has been integrated over the entire length of the diode rather than just the avalanche region and therefore impact ionization can occur anywhere in the diode.

The order of integration and differentiation can be interchanged in the second term of (2) to give a term proportional to the second derivative of the voltage across the diode. The sign convention which is assumed is

$$V(t) = \int_0^{x_L} E\,dx. \qquad (3)$$

From the boundary conditions shown in Fig. 1 the term involving the hole and electron currents reduces to

$$(J_n - J_p)\Big|_0^{x_L} = 2J_{\text{sat}} - 2J_T + \epsilon\frac{\partial}{\partial t}[E(0,t) + E(x_L,t)] \qquad (4)$$

where J_{sat} is the total reverse saturation current density (A/cm^2).

The voltage across the diode is assumed to be of the form[1]

$$V(t) = V_B + \sum_{n=1}^{l} V_n \sin(n\omega t + \theta_n) \qquad (5)$$

where

- V_B bias voltage (V)
- n harmonic number of the operating frequency
- l number of harmonics present
- V_n magnitude of the nth harmonic voltage (V)
- θ_n phase angle of the nth harmonic voltage (rad).

Thus (2) reduces to

$$\frac{d}{dt}J_T(t) + A(t)J_T(t) = Y(t) \qquad (6)$$

where

$$A(t) = \frac{2}{\tau}\left(1 - \int_0^{x_L}\alpha\,dx\right)$$

$$Y(t) = \frac{2J_{\text{sat}}}{\tau} + \frac{\epsilon}{\tau}\frac{d}{dt}[E(0,t) + E(x_L,t)] - \frac{2\epsilon}{\tau}\int_0^{x_L}\alpha\frac{\partial E}{\partial t}\,dx - \sum_{n=1}^{l}\frac{n^2\omega^2\epsilon}{x_L}V_n\sin(n\omega t + \theta_n).$$

[1] The validity of this assumption and its significance are discussed in the Appendix.

By assuming an initial electric field which is independent of the total current, the solution to (6) is

$$J_T(t) = \exp\left(-\int_0^t A(\gamma)\,d\gamma\right) \left[J_T(0) + \int_0^t Y(\gamma)\exp\left(\left(\int_0^\gamma A(\beta)d\beta\right)d\gamma\right)\right] \quad (7)$$

which is periodic with a period T when $J_T(0)=J_T(T)$. The conductance and susceptance at each frequency of RF voltage can be found by Fourier analysis of the total current. Therefore for the assumed electric field the correct total current density, conductance, and susceptance are obtained. However the electric field must be modified to include the effect of the space charge of the carriers at each point in the diode and every instant in time. From the total and the displacement current densities the conduction current density can be calculated. Knowing the conduction current, the continuity equations can be solved for the hole and electron current densities to yield [4]

$$J_p(x,t) = J_{ps} + \int_0^x \alpha J_c(x,t)\,dx\,\Big|_{dx=v\,dt} \quad (8)$$

and

$$J_n(x,t) = J_{ns} + \int_x^{x_L} \alpha J_c(x,t)\,dx\,\Big|_{dx=-v\,dt} \quad (9)$$

where J_{ps} and J_{ns} are the respective reverse saturation current densities (A/cm^2). From Poisson's equation a new electric field is calculated which includes the space charge of the carriers. The field is given by

$$E(x,t) = E_D(x) + \int_0^x \frac{J_p(x,t)-J_n(x,t)}{\epsilon v}\,dx + E(0,t) \quad (10)$$

where

$E_D(x)$ electric field due to the doping profile and is independent of time

$E(0,t)$ electric field at $x=0$ and is an arbitrary constant for each t.

The voltage across the diode is given by (5) and is equal to the integral of the field in (10). Thus the field at $x=0$ is given by

$$E(0,t) = \frac{1}{x_L}\left[V_B + \sum_{n=1}^{l} V_n \sin(n\omega t + \theta_n) - \int_0^{x_L} E_D(x)\,dx - \int_0^{x_L}\left(\int_0^x \frac{J_p(\gamma,t)-J_n(\gamma,t)}{\epsilon v}\,d\gamma\right)dx\right] \quad (11)$$

where V_B is selected to give the proper dc total current density for this electric field. For the first iteration the space-charge term in (11) is set equal to zero. For the next iterations the space charge of the previous time is used. When the change in the admittance from one iteration to the next is less than $1\frac{1}{2}$ percent, the solution is assumed to have converged.

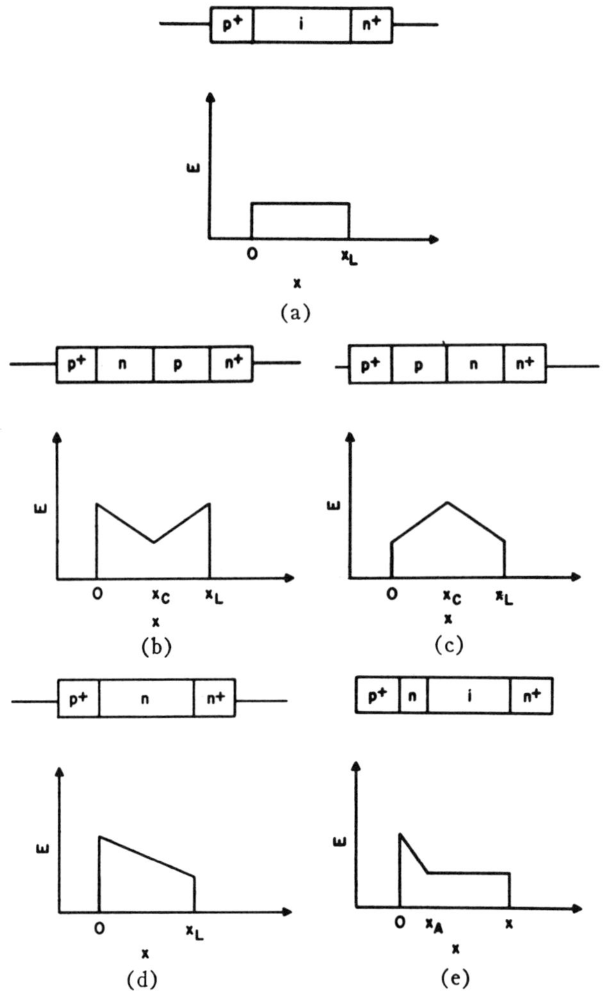

Fig. 2. Diode structures and field profiles. (a) p-i-n diode. (b) Double-injection diode. (c) p-n junction. (d) One-sided abrupt-junction. (e) Read diode.

III. Theoretical Results

The large-signal analysis derived in the previous section has been used to investigate the operating characteristics of IMPATT diodes with different doping profiles. Several of the doping profiles which have been investigated are shown in Fig. 2; however, due to lack of space the only results which are presented are for a Si, one-sided abrupt-junction diode with different doping densities operating in a single-frequency mode. A qualitative discussion of the results of the other diode structures shown in Fig. 2 is given. The results for these other doping profiles [4], for different materials [4] such as Ge and GaAs, and for multifrequency operation [5], which have been obtained by this analysis will not be presented. All of the diodes have a 5-μm-long active region and the assumed doping density in the p^+ and n^+

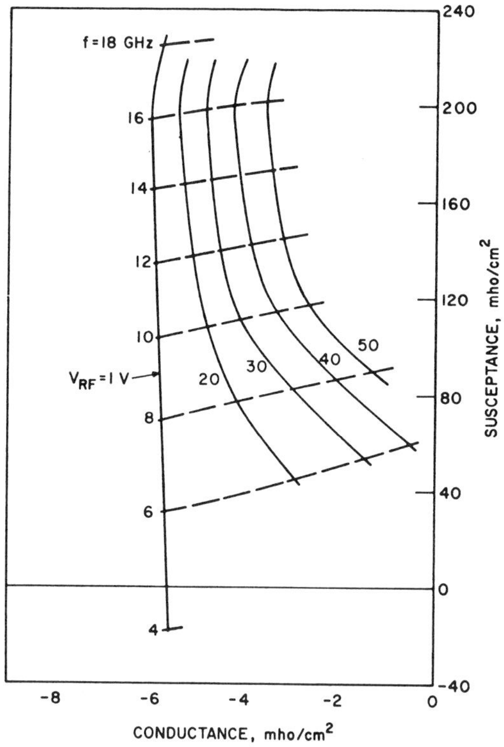

Fig. 3. Admittance for a Si p-i-n diode ($J_{dc}=500$ A/cm²).

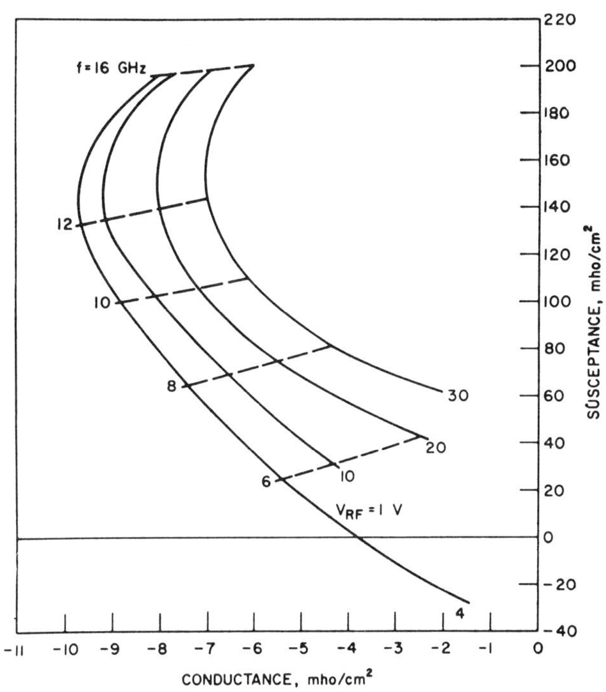

Fig. 4. Admittance for a Si p^+-n-n^+ diode ($J_{dc}=500$ A/cm² and $n=1\times10^{15}$ cm^{-3}).

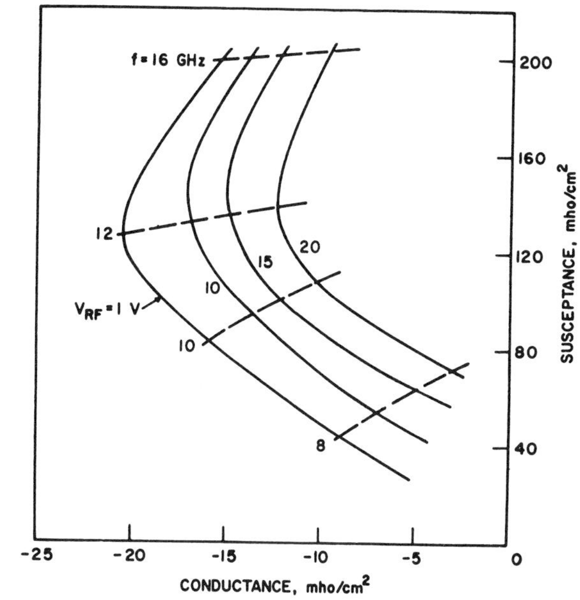

Fig. 5. Admittance for a Si p^+-n-n^+ diode ($J_{dc}=500$ A/cm² and $n=3\times10^{15}$ cm^{-3}).

181

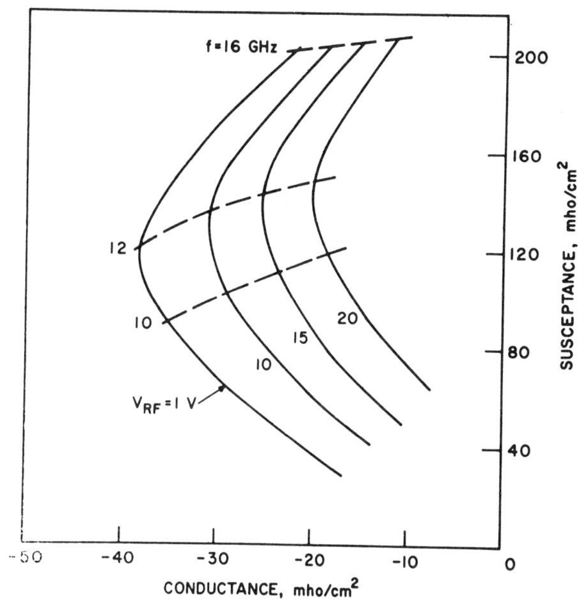

Fig. 6. Admittance for a Si p^+-n-n^+ diode ($J_{dc}=500$ A/cm² and $n=5\times10^{15}$ cm^{-3}).

regions is high enough for the field to be zero at the boundary. Also, for the doping densities considered, the field has "punched through" to the substrate when the diode is biased into avalanche breakdown.

The large-signal single-frequency admittances of the diodes with different doping densities are shown in Figs. 3 through 6. Assuming a single-frequency RF voltage across the diode and obtaining the admittance is the same as solving the problem with the diode in a microwave circuit of this admittance at the operating frequency under steady-state conditions. Inherent in this single-frequency assumption is that all of the harmonics of the RF voltage are shorted by the microwave circuit. This assumption is discussed in the Appendix.

For a current density of 500 A/cm² the dc differential conductance of a p-i-n diode is negative due to space-charge effects. This is shown in Fig. 7 in which the dc current density is plotted as a function of the bias voltage for the different doping densities. Thus for the small-signal case (1 volt RF), the conductance remains negative as the frequency is decreased. For larger RF

182

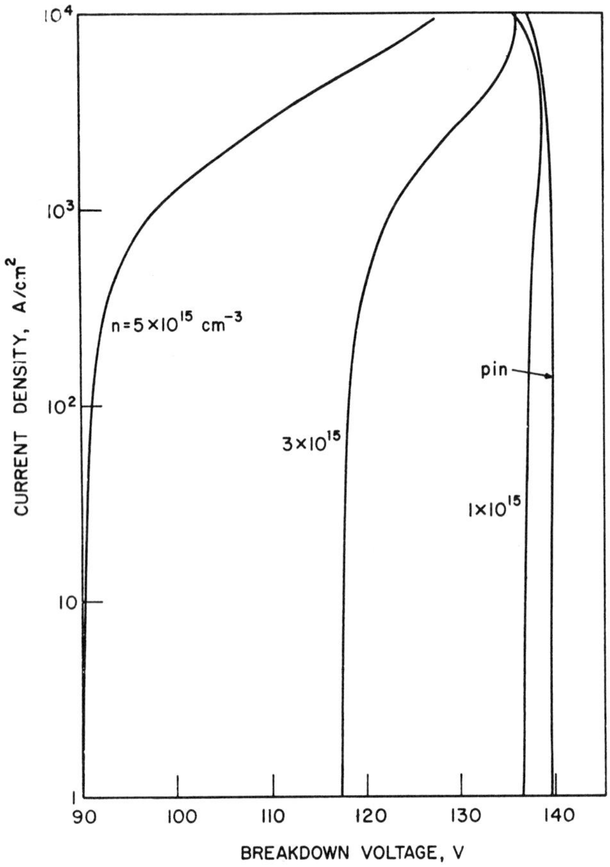

Fig. 7. Static $J-V$ characteristic for a Si one-sided abrupt-junction diode.

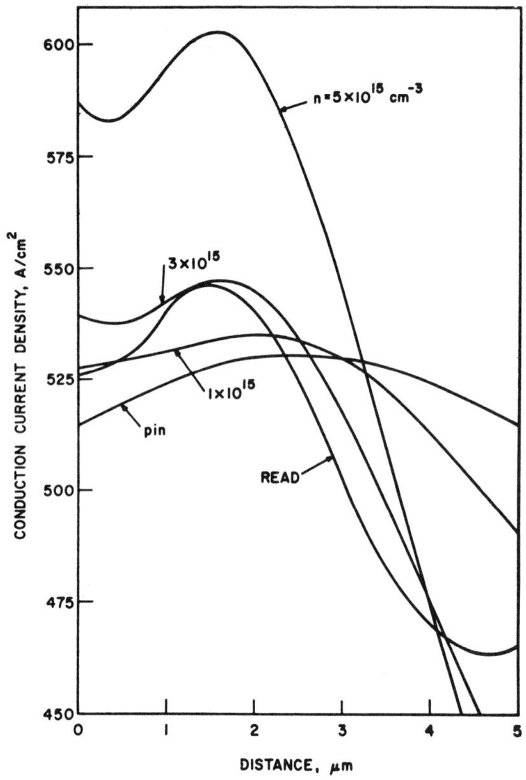

Fig. 9. Conduction current density versus distance for different doping densities ($J_{dc}=500$ A/cm², f = optimum frequency, and $\omega t \cong 180°$).

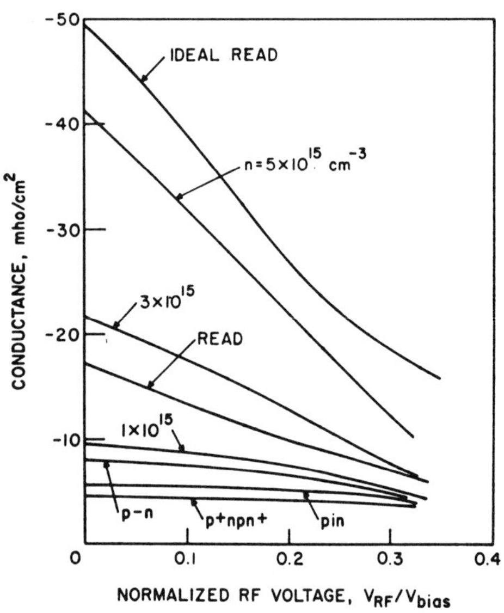

Fig. 8. Conductance versus normalized RF voltage for different Si diode structures at the optimum frequency ($J_{dc}=500$ A/cm²).

voltages the conductance becomes positive as the frequency is decreased. Therefore the small-signal cutoff frequency for the p-i-n diode at this current density goes to dc. Note though that the negative conductance at the low frequencies decreases rapidly with increasing RF voltage. For the other doping densities the dc differential conductance is positive for this current density, as shown in Fig. 7, and thus the small-signal conductance becomes positive for decreasing frequency. In order to compare the diodes with different doping densities, the negative conductance is plotted in Fig. 8 as a function of the normalized RF voltage at the optimum operating frequency. The normalized RF voltage is the RF voltage divided by the corresponding dc voltage. The optimum operating frequency is the frequency for which the negative conductance is maximum. The ideal Read diode results are from an analysis similar to Blue's [2] for a Read diode with a 1-μm avalanche region and a 4-μm drift region. These results are included for reference. The maximum negative conductance for the other structures shown in Fig. 2 is also included. The Read-diode curve in Fig. 8 corresponds to a concentration in the n-layer of 1×10^{16} cm^{-3}, and the concentrations in the p and n regions of the p$^+$n-p-n$^+$ and p-n structures are both 10^{15} cm^{-3}.

The reason for the difference in the magnitude of the negative conductance for the different doping densities can be understood by considering the generation of the conduction current density in the diode. In Fig. 9 the conduction current density is plotted as a function of distance for the diodes with different doping densities. The time in the RF cycle at which this current exists is when the RF voltage goes through zero and begins subtracting from the bias voltage. For the assumed RF voltages in (5) this corresponds to ωt equaling π rad. As can be seen from Fig. 9, as the doping density in the diode is increased the current modulation increases. In the p-i-n diode, due to the large region over which carriers are generated, the transit angle of the carriers varies over a wide range of angles and only a few of them experience the same phase delay. As the doping density is increased in the diode, the avalanche region becomes narrower and the conduction current is generated in a sharper pulse. Thus more of the carriers experience the same transit angle. For the proper frequency, more of the carriers can be made to experience the optimum transit angle for negative conductance as the doping density is increased. Thus the negative conductance of the diode is larger when the distribution of the transit angles of the carriers is concentrated around the optimum transit angle rather than when the distribution is spread over a wide range but has an average equal to the optimum.

Another point which should be explained is the rate of decrease of the negative conductance for the different diodes. This can be understood by considering the effect of large RF voltages on the electric field in the region where the majority of the impact ionization occurs. As the RF voltage is increased in magnitude, the modulation of the conduction current increases, and thus when the carriers are being generated there is more space charge in the diode than for the lower RF voltages. As the RF voltage increases, the field in the avalanche region is flattened or smoothed out over a wider region due to the increased space charge. Thus carriers are generated over a wider region of the diode and the transit angle of the carriers is spread over a wider range. The narrower the avalanche region of the diode (higher doping density) the more effect the spreading out of the avalanche region has on the rate of decrease of the negative conductance. This avalanche widening effect causes each of the diodes to approach the behavior of a p-i-n diode for larger RF voltages. The effect of the different rates of decrease in the negative conductance is apparent in the efficiency of the diodes, which is plotted in Fig. 10. As the avalanche region is spread out, the efficiency of the diodes approaches that of a p-i-n diode.

The other diode structures shown in Fig. 2 behave in a consistent manner with the above discussion. The p-n junction behaves like the one-sided abrupt junction in that the negative conductance increases as the doping density is increased. However, the operating frequency is higher due to the shorter transit time caused by the avalanche region being located in the middle of the diode. The double-injection diode has the least negative conductance since the transit angle of the carriers is spread out even more than in the p-i-n diode by the two avalanche regions. The Read diode which has a doping density of 1×10^{16} cm^{-3} in the 1-μm-wide avalanche region and a 4-μm-wide intrinsic region has less negative conductance than two of the abrupt junctions. This is due to the fact that for this doping profile the field in the intrinsic region at breakdown is large enough for impact ionization to occur during the transit time of the carriers. Thus impact ionization is not confined to the 1-μm region and the transit angle of the carriers is spread out more than the heavily doped abrupt junctions. By doubling the doping density in the avalanche region, the field in the intrinsic region at breakdown was too low for impact ionization to occur. The avalanche region, then, is very narrow and the negative conductance is larger than the ideal Read-diode results [4].

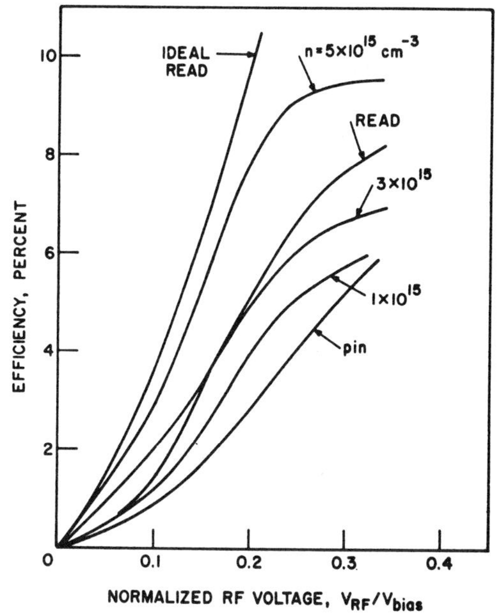

Fig. 10. Efficiency versus normalized RF voltage for different diodes at the optimum operating frequency ($J_{dc}=500$ A/cm^2).

183

IV. Experimental Results

In order to measure the admittance of an IMPATT diode in a microwave circuit, an equivalent circuit for the diode package and mount had to be obtained so that the diode chip could be separated from the rest of the circuit. The equivalent circuit for the diode package and mount is shown in Fig. 11. The lumped elements of the circuit are as follows. The conductance and susceptance of the diode are represented by G and B, respectively. The series resistance due to contact losses, the resistance of the substrate, and the undepleted region of the diode are represented by R_s. The pad capacitance C_{pd} is due to the capacitance between the metal contact on top of the diode and the pedestal on which the diode is mounted. This capacitance, which is a function of the diode geometry, is negligible for the diodes investigated [6]. The lead inductance L_d is the inductance due to the four wire straps between the flange of the diode package and the top of the diode chip. The capacitance of the package and the fringing fields at the end of the coaxial mount are represented by C_{pk}. The series inductance L_s is the inductance caused by the high-impedance shorted transmission

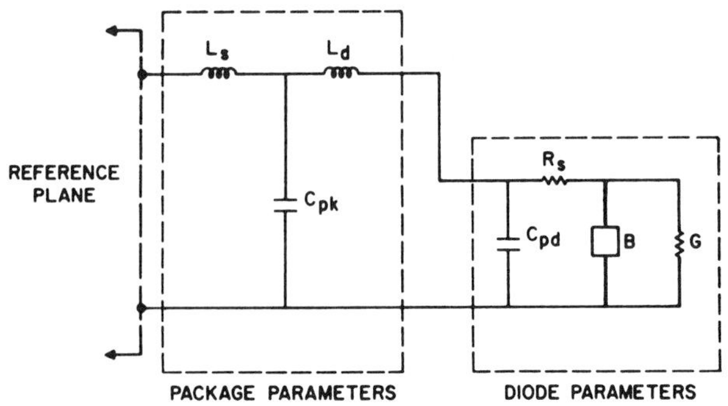

Fig. 11. Equivalent circuit of the diode package and the diode.

line of the pedestal in the diode package [7] and the magnetic energy storage in the cylindrical volume around the diode [8].

The individual element values are measured in the following manner. The series inductance and package
184 capacitance are determined from the measured impedance of an open-circuited package. The lead inductance and a second package capacitance are determined from the impedance of a short-circuited package. The package capacitance for the short-circuited package is different from the open-circuited package capacitance due to the effect on the electric field of the metal straps which connect the flange to the diode chip. The lead inductance is also determined from the impedance of a forward-biased diode since for large currents in the forward direction the equivalent circuit of the diode chip reduces to a small series resistance. The series resistance which is used in the equivalent circuit is obtained from the impedance of the diode when it is biased into reverse breakdown with 1 mA flowing [6]. The element values for the equivalent circuit which were obtained from these measurements are given in the following.

$$L_s = 0.226 \text{ nH}$$
$$C_{pk} = 0.11 \text{ pF}$$
$$L_d = 0.7 \text{ nH}$$
$$C_{pd} = 0 \text{ pF}$$
$$R_s = 1.1\ \Omega.$$

The equivalent circuit for the diode package and mount transform the admittance of the diode from the pedestal in the package to the reference plane in the coaxial mount. Verification that the lumped elements obtained for the equivalent circuit give the correct transformation over the frequency range considered was obtained by measuring the depletion-layer capacitance as a function of the reverse voltage at 1 MHz and at X-band. At 1 MHz the elements in the equivalent circuit are negligible and the depletion-layer capacitance, which is the susceptance of the diode below breakdown, is obtained directly. The impedance of the diode at X-band was measured at frequencies of 8.5, 8.75, 9.0, 9.25, 9.5, 9.75, and 10.0 GHz as a function of the reverse voltage. From this impedance the depletion-layer capacitance and series resistance were obtained by transforming the measured impedance through the equivalent circuit.

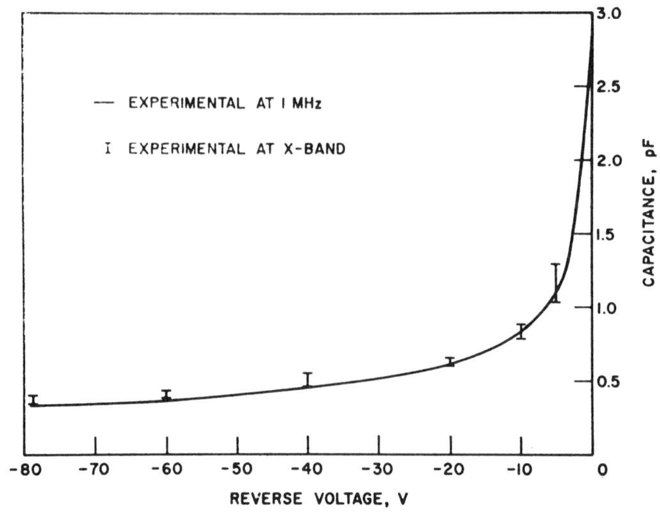

Fig. 12. Capacitance versus reverse voltage for diode TI 40.

The two methods of measuring the depletion-layer capacitance are compared in Fig. 12. The solid curve is for the 1-MHz measurement. The data points indicated by I show the capacitance obtained at X-band using the equivalent circuit element values listed in the table. The spread in the data shows the maximum and minimum capacitance which was obtained at all the frequencies employed. If the depletion-layer capacitance is assumed to be independent of frequency [6], then these results show that the transformation performed by the equivalent circuit element values in the table is correct.

The doping density for a one-sided abrupt-junction diode was determined from the capacitance–voltage (C–V) data obtained above for two diodes. One of the diodes, TI 40, had a doping density of approximately 6×10^{15} cm^{-3} and a 4-μm-wide depletion region at breakdown. The second diode, TI 40B, had a doping density of approximately 3×10^{15} cm^{-3} and a 3.5-μm-wide depletion region with the field punching through to the substrate at breakdown.

The admittance of the diodes was measured by two methods. First a reflection-type amplifier circuit was used to measure the small-signal admittance. A circulator was used to separate the incident signal from the signal generator and the reflected signal from the diode. It was necessary to compare incident and reflected power levels as part of the admittance measurement in order to determine the sign of the voltage standing-wave ratio (VSWR). The admittance was measured as a function of frequency and dc bias current with a slotted line. As the bias current is increased, the negative conductance of the diode increases. For a large enough dc bias current the effective negative resistance is larger than the series resistance and the reflected power is larger than the incident power. The VSWR therefore

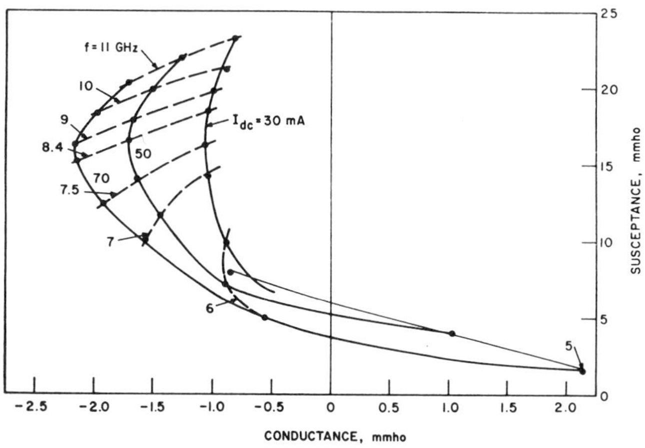

Fig. 13. Admittance for diode TI 40 ($V_{RF} \leqq 1$ V).

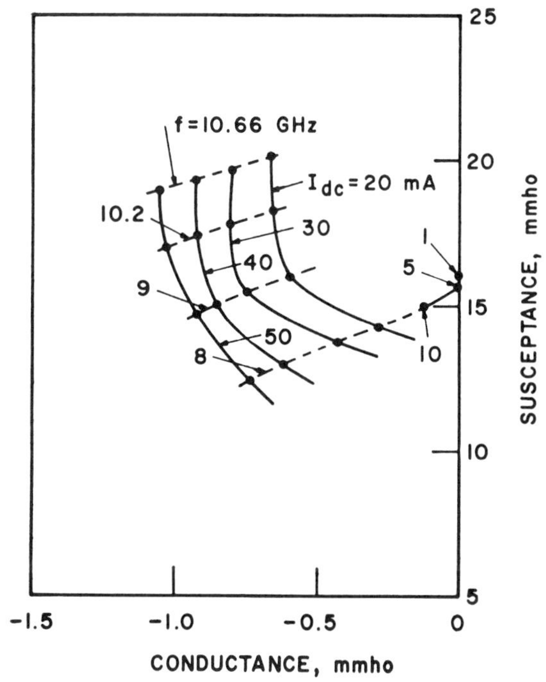

Fig. 14. Admittance for diode TI 40B ($V_{RF} \leqq 1$ V).

increases to a large positive value as the bias current is increased and then when the reflected power is larger than the incident power, the VSWR changes sign and decreases in magnitude. As a check, the diode was mounted in a tunable microwave circuit and was shown to oscillate at the current levels where the VSWR was negative. Because of the high VSWRs which had to be measured, it was necessary to use the double minimum method [9].

In Figs. 13 and 14 the experimentally measured admittances for the two diodes are plotted as functions of frequency and dc bias current. For all of the points shown, the RF voltage across the diode was 1 volt or less. As the incident power was increased, the VSWR and voltage minimum remained constant until the RF voltage was slightly larger than 1 volt. Then the measurements showed the presence of other frequencies which were large enough in magnitude to affect the VSWR and minimum position measurements. This condition exists due to the imperfect connectors and circulators which do not present a 50-ohm load at all frequencies. For the measurements plotted in Fig. 13 two circulators had to be used, one designed for 4- to 8-GHz operation and the other for 8 to 12 GHz. A circulator acts as a reactance rather than as a load out of its designed frequency band and thus the diodes are not necessarily loaded with 50 ohms at all frequencies. Due to the broad-band negative resistance of an IMPATT diode and its nonlinear behavior, the diode can generate power at frequencies other than the impressed frequency when the impressed signal becomes large enough in magnitude and the load at the other frequencies is small. Thus it was necessary for these measurements to keep the RF voltage small. By using better wide-band low VSWR connectors, and a combination of broad-band directional couplers which are lossy out of band rather than the circulator, this problem might be eliminated.

The results plotted in Figs. 13 and 14 do show, however, agreement with the predicted theoretical trends. The negative conductance increases with increasing dc bias current and the susceptance decreases slightly. Also, the cutoff frequency, the frequency below which the conductance is positive, decreases with decreasing dc bias current. Finally, the negative conductance for the abrupt junction with the higher doping density is larger than for the lower doping density. However, in order to compare the admittances of the two diodes at a given bias current, the error involved in the determination of the diode areas should be considered. As shown previously for diode TI 40 the doping density for a 0.005-inch diameter was 6×10^{15} cm^{-3}. Assuming a variation in the doping density of ± 10 percent which is within the range specified by the manufacturer, this yields an error in the determination of the area of ± 5 percent. For diode TI 40B the doping density was 3×10^{15} cm^{-3} for a 0.005-inch diameter. Assuming the same variation in doping density, a ± 5-percent error is also possible. With this tolerance value it is seen that in the worst case the negative conductance for the diode with the higher doping density is still larger than for the one with the lower doping density.

The second method for measuring the admittance of the IMPATT diodes is the indirect method of measuring the admittance of the microwave circuit in which the diode is oscillating. The admittance of the diode package and mount must be the negative of the admittance of the microwave circuit. The measured admittance is then transformed through the equivalent circuit to calculate the admittance of the diode.

The microwave test circuit used for the oscillator measurements consists of a coaxial cavity [10] with a tuning slug connected to an X-band slide screw tuner.

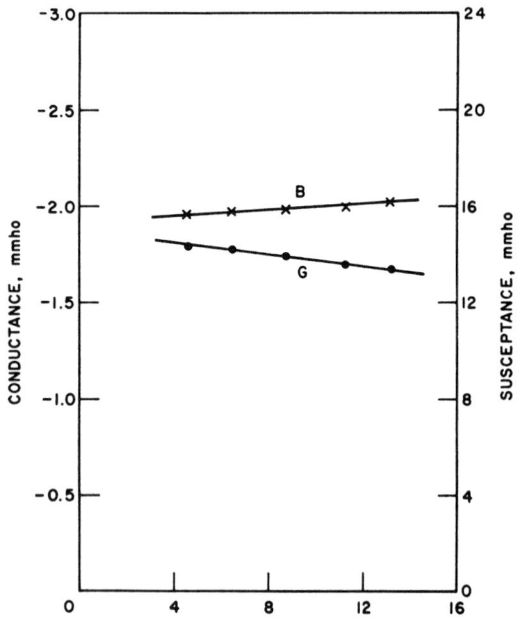

Fig. 15. Conductance and susceptance RF voltage for diode TI 40 ($I_{dc}=50$ mA and $f=8.75$ GHz).

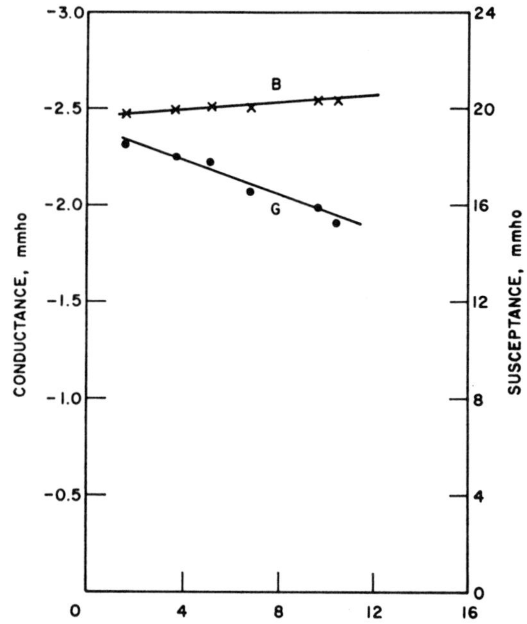

Fig. 16. Conductance and susceptance versus RF voltage for diode TI 40 ($I_{dc}=50$ mA and $f=9.9$ GHz).

a spectrum analyzer, and a power meter. The diode is tuned with the slug in the coaxial cavity to oscillate at a single frequency for a given dc bias current. Single-frequency operation is verified by inserting a probe in the diode cavity above the diode and monitoring the output on the spectrum analyzer. The diode is tuned in such a way that all frequencies except the operating frequency are down in power by at least 30 dB. The frequency is measured by matching the frequency from a signal generator and the oscillating diode on the spectrum analyzer with a 1-MHz/cm spectrum width and then measuring the signal-generator frequency with a wavemeter. The power output or RF voltage level of the diode is varied by changing the insertion of the slide screw tuner. The power level is measured for each insertion of the slide screw tuner. The next step is to remove the diode from the diode cavity and connect a 50-ohm coaxial-line adapter and a slotted line in place of the diode. The admittance of the microwave circuit is measured for each insertion of the slide screw tuner and at the frequency at which the diode was oscillating. The admittance of the microwave circuit is measured with respect to the reference plane located at the flange of the diode and then transformed through the package to the diode. Finally, the input power from the signal generator is measured along with the power which is transmitted through the circuit to the power meter at the end of the circuit. A computer program was used to calculate from the measured data the admittance of the microwave circuit, the admittance of the diode chip, and the RF voltage across the diode chip.

The conductance and susceptance for diode TI 40 are plotted in Figs. 15 through 17 for several different frequencies and a bias current of 50 mA. The results show that the conductance decreases and the susceptance increases for increasing RF voltage as predicted. The decrease in the negative conductance for a change in RF voltage from 1 to 10 volts is approximately 15 percent, whereas the decrease predicted for an abrupt junction with a doping density of 5×10^{15} cm^{-3} is approximately 18 percent. The magnitude of the negative conductance is somewhat less than predicted for a comparable doping

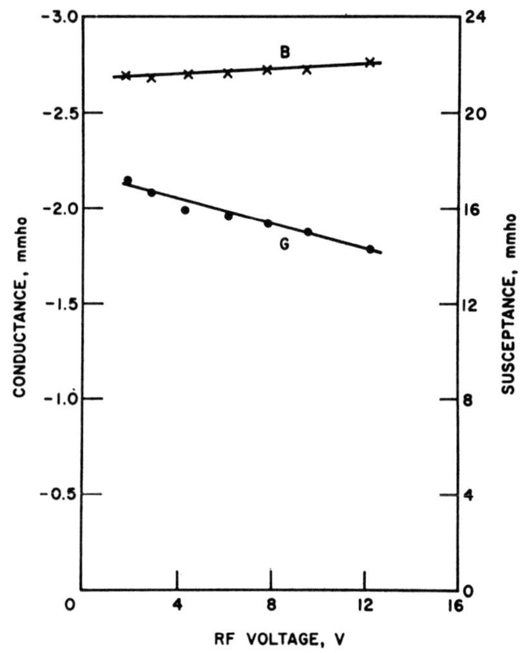

Fig. 17. Conductance and susceptance versus RF voltage for diode TI 40 ($I_{dc}=50$ mA and $f=10.75$ GHz).

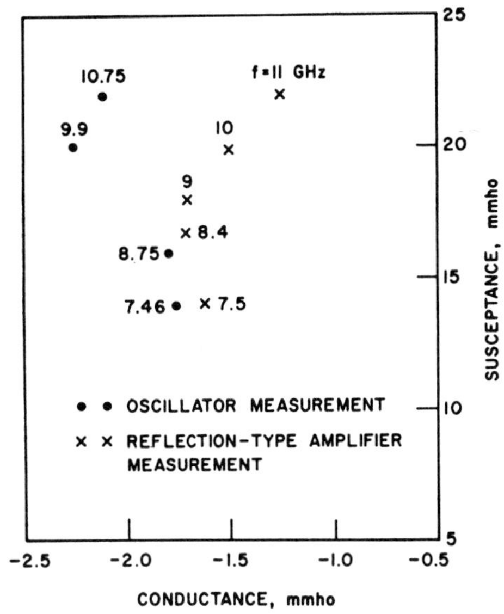

Fig. 18. Admittance of diode TI 40 ($I_{dc}=50$ mA and $V_{RF}\leqq 1$ V).

density. It is difficult to compare the magnitude of the negative conductances for a given current density since the area of the junction is not known exactly for the experimental diode.

Some elaboration must be given to several points about this measurement. The microwave circuit was kept as simple as possible and thus it was not always possible to change the RF voltage with the slide screw tuner while holding the diode in a constant single-frequency mode. Several cases were observed in which the RF voltage remained relatively constant for many different insertions of the slide screw tuner. A second point is that as the slide screw tuner was inserted the spectrum width of the oscillations would change slightly. A point was reached for which the diode would not maintain single-frequency operation. Either several other frequencies would appear or a very noisy spectrum would result. As the insertion was decreased, it was not possible to continuously decrease the RF voltage without having the diode stop oscillating before the voltage reached zero. This is illustrated in the figures since the admittance for 1 volt or less was not obtained. Finally, the frequency of the oscillating diode changed slightly as the slide screw tuner was inserted. The change in frequency was small, 0.25 percent or approximately 20 MHz; however, it was accounted for when the admittance measurements were made. Since the frequency decreased slightly for increasing RF voltage, the susceptance would increase a little faster than measured.

In Fig. 18 a comparison of the two methods of measuring the admittance is shown. The negative conductances and susceptances for the oscillator tests are taken as approximate values for RF voltages less than 1 volt. The dc bias current is 50 mA. The susceptance values are in fairly good agreement. At the lower frequencies where the errors in measurement are less, the conductances agree fairly well, whereas for the higher frequencies there is more discrepancy. However, considering the errors, the agreement is not so bad as indicated. The susceptance is much easier to measure accurately since it is large, whereas the conductance is more difficult to measure since it is only one tenth that of the susceptance. Another point must be made in that the two methods of measuring the admittance do not present the same equivalent circuit to the diode. For the oscillator measurement the circuit is tuned to obtain single-frequency operation, which means the harmonics of the current through the diode are shorted, whereas for the reflection-type amplifier measurement the circuit is not tuned and the impedance seen by the harmonics of the current might be large, as is discussed in the Appendix. Thus it is not apparent that the two methods should yield the same result for all frequencies.

V. Conclusions

The derivation and results of a large-signal analysis for studying the operating characteristics of realistic IMPATT diodes are presented. The analysis allows carrier multiplication by impact ionization to occur at every point in the diode and therefore allows a wide range of doping profiles to be investigated. The admittance, power output, efficiency, and bias voltage can be obtained as a function of the frequency, RF voltage, and dc bias current for different profiles. Several important results are obtained. An optimum frequency exists for maximum negative conductance, and the doping density is an important parameter in determining the admittance.

The operating characteristics are measured experimentally. An equivalent circuit for the diode package and mount is obtained. By C–V measurements the doping density of two different diodes is determined. The admittance is measured by both a reflection-type circuit and an oscillator circuit as a function of RF voltage, dc bias current, and frequency. The experimental results are correlated with the theoretical results.

Appendix

Single-Frequency Operation

The large-signal analysis presented in this paper assumes that the diode oscillates with a single-frequency RF voltage across it. This assumes that all the harmonics of the voltage are shorted by the package and microwave circuit. The experimental results presented were obtained with the diode oscillating essentially in a single-frequency mode with all other frequencies at least 30 dB below the fundamental oscillating frequency. In order to show how the diode could oscillate in a single-frequency mode even though the current through the diode contains many harmonics, the follow-

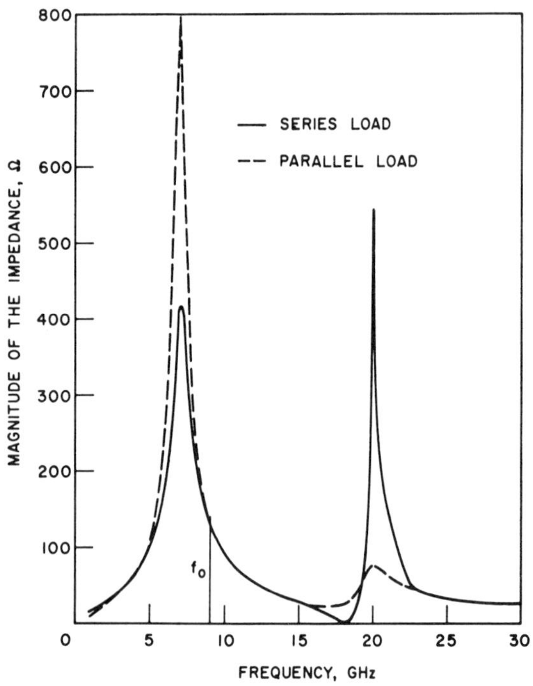

Fig. 19. Magnitude of the impedance versus frequency (f_0=9 GHz).

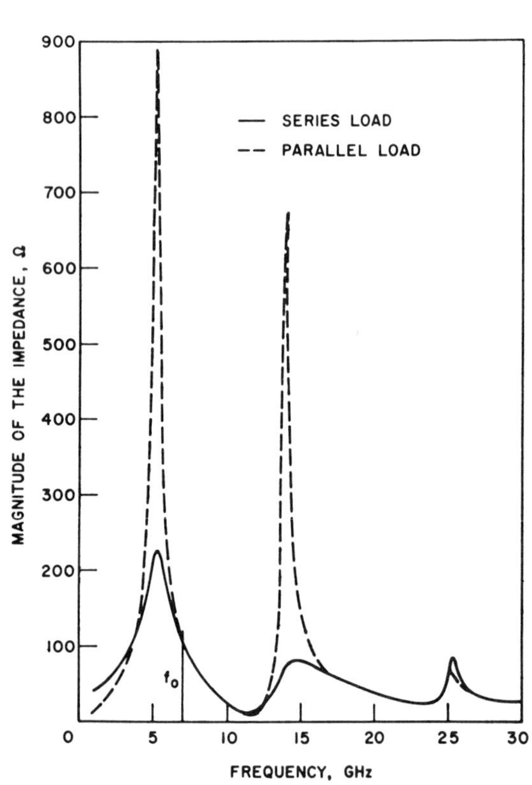

Fig. 20. Magnitude of the impedance versus frequency (f_0=7 GHz).

ing problem was investigated. In order for single-frequency operation to exist, the load impedance seen by the diode must cancel the impedance of the diode at the oscillation frequency and must be small at the harmonics. The microwave circuit was represented by two different loads. One load was a resistance which was independent of frequency in series with a reactance which acted like a sliding short. The second load was a resistance, again independent of frequency, in parallel with a sliding short reactance. One of these loads along with the measured equivalent circuit for the package and mount comprised the load seen by the diode. The susceptance of the diode was divided into its depletion-layer capacitance and an equivalent inductance due to the avalanche transit-time processes. Typical values for the diode conductance and susceptance were selected and the resistance of the microwave circuit and position of the sliding short were obtained for both loads such that the diode would oscillate at the prescribed frequency. Then the impedance seen by the diode was calculated as a function of frequency to investigate the load at the harmonics. The magnitude of this impedance is plotted as a function of frequency for two oscillation frequencies in Figs. 19 and 20.

Even though the impedance of the microwave circuit has been greatly simplified, the results obtained show several important effects. First, for the case where the diode is oscillating at 9 GHz (Fig. 19) the impedance at the oscillation frequency is 133 ohms. The RF voltage at this frequency is equal to this impedance times the magnitude of the current component at this frequency. At the second harmonic, 18 GHz, the magnitude of the impedance seen by the diode for the series load is 1.26 ohms. The magnitude of the second-harmonic current component will be at most one fifth of the fundamental current component, and thus the RF voltage at the second harmonic is at most 1/500 of the RF voltage at the fundamental. This corresponds to essentially shorting the harmonic and having the diode oscillate in a single-frequency mode. The second effect is that for other tuning conditions as shown by the case where the oscillation frequency is 7 GHz (Fig. 20) the opposite condition can exist. The magnitude of the impedance at 7 GHz is 104 ohms, while at 14 GHz the impedance is 670 ohms for the parallel load. Using a similar approximation as before, the RF voltage at the second harmonic would be larger than at the fundamental. Thus the diode would not oscillate in a single-frequency mode. The final effect is the comparison of the parallel and series loads. As can be seen, the frequency response of the impedance seen by the diode is different for the two loads; however, either type of load can be used to make the diode oscillate in a single-frequency mode with the proper choice of circuit parameters. This is due to the fact that the microwave circuit load is only one part of the impedance seen by the diode and that the package parameters which transform this load are the other important part of the impedance.

Acknowledgment

The authors wish to thank Prof. R. J. Lomax, Prof. W. R. Curtice, and W. E. Schroeder for the helpful discussions during the course of this investigation; and Dr. D. D. Khandelwal of Texas Instruments, Inc., for supplying the diodes.

References

[1] W. J. Evans and G. I. Haddad, "A large-signal analysis of IMPATT diodes," *IEEE Trans. Electron Devices*, vol. ED-15, pp. 708–717, October 1968.

[2] J. L. Blue, "Approximate large signal analysis of IMPATT oscillators," *Bell Syst. Tech. J.*, vol. 48, pp. 383–396, February 1969.

[3] D. L. Scharfetter and H. K. Gummel, "Large-signal analysis of a silicon Read diode oscillator," *IEEE Trans. Electron Devices*, vol. ED-16, pp. 64–77, January 1969.

[4] P. T. Greiling, "Large-signal characteristics of avalanche-diode oscillators," Electron Physics Laboratory, University of Michigan, Ann Arbor, Tech. Rep. 117.

[5] W. E. Schroeder, P. T. Greiling, and G. I. Haddad, "Multi-frequency operation of IMPATT diodes," presented at the Int. Electron Devices Meeting, Washington, D. C., October 1969.

[6] J. W. Gewartowski and J. E. Morris, "Active IMPATT diode parameters obtained by computer reduction of experimental data," *IEEE Trans. Microwave Theory Tech.*, vol. MTT-18, pp. 157–161, March 1970.

[7] I. W. Pence, private communication.

[8] W. J. Getsinger, "The packaged and mounted diode as a microwave circuit," *IEEE Trans. Microwave Theory Tech.*, vol. MTT-14, pp. 58–69, February 1966.

[9] E. L. Ginzton, *Microwave Measurements*. New York: McGraw-Hill, 1957, pp. 266–269.

[10] D. E. Iglesias, "Circuit for testing high-efficiency IMPATT diodes," *Proc. IEEE* (Letters), vol. 55, pp. 2065–2066, November 1967.

Large-Signal Silicon and Germanium Avalanche-Diode Characteristics

DAVID R. DECKER, MEMBER, IEEE, CHARLES N. DUNN, MEMBER, IEEE, AND RICHARD L. FRANK

Abstract—A technique for measurement of the large-signal single-frequency microwave amplifier admittance of avalanche diodes is described, and results are presented for silicon and germanium avalanche diodes. Single-frequency amplifier operation can provide a unique characterization of diode-admittance variation with RF drive for diodes operated near the optimum transit angle (the case in which all harmonic voltages are negligibly small compared to the fundamental). Such characterization is useful for predicting diode performance for circuits in which the harmonic voltages are not large enough to have an appreciable effect on the diode admittance at the fundamental frequency.

A process of matching quadratic forms to the above admittance data which may be used for calculation of diode terminal admittance and power output is discussed. The usefulness of the measurement technique is illustrated by the agreement of the calculated maximum power output with the measured power output in a single-transformer coaxial circuit. The corresponding circuit admittance may be used for circuit-design purposes and for evaluating variations in diode-assembly techniques.

The ability to obtain the diode equivalent circuit as a function of incident power allows studies in the design of the associated semiconductor device. For example, one has the capability of obtaining an accurate single-frequency large-signal model near the optimum transit angle, a model which can be studied without building a circuit. With this model it is possible to carry out optimization procedures at considerable savings of time and money.

Introduction

ONE OBJECTIVE of avalanche-diode characterization for oscillator use is to determine the diode output-power capability and the corresponding oscillator-circuit load admittance. As shown by Gewartowski and Morris [1], large-signal single-frequency amplifier admittance characterization closely simulates actual operating conditions for a large class of oscillator circuits. Knowledge of the variation of diode admittance with RF drive level can be used for optimization of diode design and for determination of the circuit-admittance function needed to obtain optimum performance.

To evaluate design changes in both semiconductor processing and in diode assembly, a rapid accurate technique of large-signal admittance characterization is desirable. Computer data-reduction techniques in conjunction with a reflection phase-gain test set accomplish the desired measurement objectives.

Previous techniques described for large-signal characterization of avalanche diodes are best suited for the characterization of small numbers of devices. In one technique the diode is allowed to oscillate in each of several known waveguide circuits [2]. The data thus obtained are subsequently reduced to obtain the diode characteristics. In another technique [1] the diode is placed in a low-impedance coaxial circuit and the wafer reference plane and circuit transformation are obtained by referencing microwave measurements using known wafer properties.

For microwave testing of many devices in a given test band, it is expedient to establish the reference plane with known impedances in configurations which fit the geometry requirements for the particular application. For instance, capacitors which are relatively frequency independent may be constructed in the shape of the microwave package [3], [4].

The following sections present 1) a method of experimentally determining the large-signal single-frequency microwave amplifier admittance of avalanche diodes near the optimum transit angle; 2) measured large-signal admittances of diffused silicon and germanium avalanche diodes having an avalanche frequency of 4 to 4.5 GHz and designed for IMPATT operation in the vicinity of 6 GHz [1], [4]; and 3) reduction of these data to an analytical form which allows the calculation of the required circuit admittance for obtaining the maximum power output.

Circuit Description

A traveling-wave-tube amplifier in conjunction with an admittance test set were used to obtain the large-signal admittance characterization over a range of RF incident power from 0.03 mW to 1.6 watts. Above the 1.6-watt level it became increasingly difficult to prevent the generation of harmonic tones (at various frequencies and significant power levels) that occurred with some of the more efficient diodes, so measurements were restricted to this level.

The spectral content of the reflected measurement signal was investigated at the reflected-power port of the network analyzer. At the second harmonic a voltage transformation ratio of 1:1.4 exists between the diode junction and this measurement port. For all measurements reported in this work, no harmonic or other spurious signal in the band 0 to 12.5 GHz was above the detectable limit of −50 dBm when sampled at the reflected-signal port of the reflection test unit. These results are basically in agreement with the theoretical prediction of Mouthaan [5] for devices operated near the optimum transit angle. The diode series resistance

Manuscript received April 9, 1970; revised September 16, 1970.
The authors are with Bell Telephone Laboratories, Inc., Reading, Pa. 19604

and the external load prevent the buildup of harmonic signals.

Fig. 1 is a block diagram of the large-signal admittance test set. The diode is mounted in an air-cooled holder as the termination of a 50-ohm line. This 20-mmho load conductance across the diode terminals exceeds the diode-negative-terminal conductance and suppresses oscillation. A digital voltmeter in conjunction with the switch panel provides numerical output which is subsequently processed by the computer into the desired data form.

In view of the negative-resistance nature of the device under consideration, it is important to employ components with low standing-wave ratio to assure accurate measurement of the incident power at the diode. The 3-dB broad-band attenuator and the isolator on the generator side of the reflection test unit provide an adequate degree of broad-band match to the diode. A 10-dB directional coupler provides an access port for measuring the test frequency and permits use of a spectrum analyzer for detection of any spurious response that might arise.

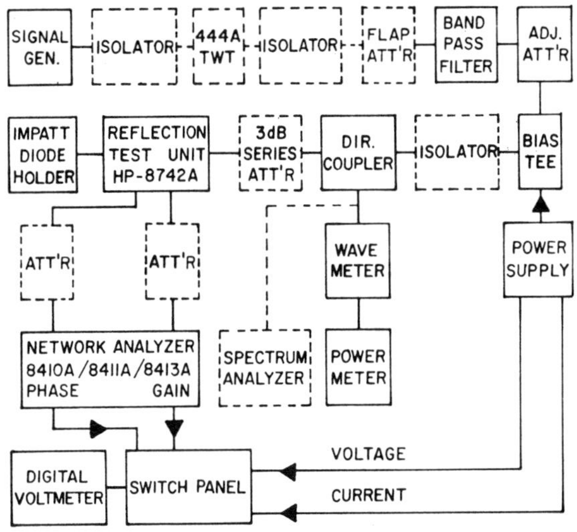

Fig. 1. Block diagram of large-signal test set.

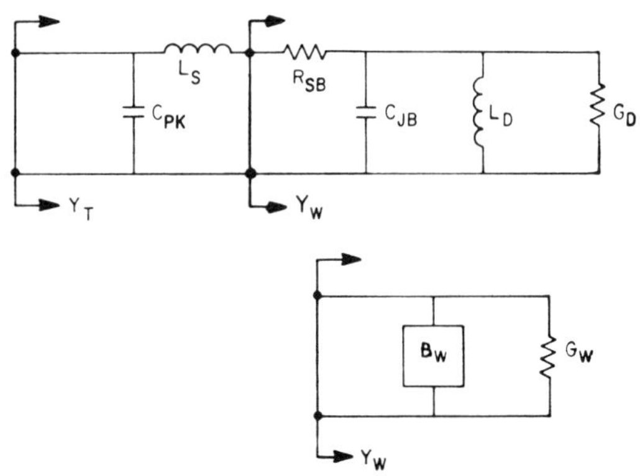

Fig. 2. Avalanche-diode equivalent circuit.

Test Procedure

The principle of the measurements discussed in this paper is based on a set of impedance-transformation equations [6]. The output of a phase-gain test set can be converted from the magnitude (P) and angle (ϕ) of the measured reflection coefficient to the real and imaginary components of impedance by the equations

$$R = Z_0(1 - P^2)/D$$

$$X = 2Z_0(P \sin \phi)/D$$

where

$$Z_0 = 50 \text{ ohms}$$
$$P = 10^{(RG/20)}$$
$$D = 1 + P^2 - 2P \cos \phi$$
$$RG = \text{return gain.}$$

The return gain (RG) is defined as negative return loss in decibels. These equations establish an impedance with respect to some reference plane other than that at the wafer position. To establish the wafer reference plane, it is necessary to determine the transformation of the two-port network connecting the wafer to the measurement system. This transformation is established through the measured values ($Z_{S1,2,3}$) of three known reference impedances [3] ($Z_{R1,2,3}$) by the following equations:

$$Z_{11} = \frac{(Z_{S1} - Z_{S3})(Z_{S1}Z_{R1} - Z_{S2}Z_{R2}) - (Z_{S1} - Z_{S2})(Z_{S1}Z_{R1} - Z_{S3}Z_{R3})}{(Z_{S1} - Z_{S3})(Z_{R1} - Z_{R2}) - (Z_{S1} - Z_{S2})(Z_{R1} - Z_{R3})}$$

$$Z_{22} = \frac{(Z_{S1}Z_{R1} - Z_{S2}Z_{R2}) - Z_{11}(Z_{R1} - Z_{R2})}{(Z_{S2} - Z_{S1})}$$

$$Z_{12}^2 = (Z_{11} - Z_{S1})(Z_{22} + Z_{R1}).$$

With the elements of this Z matrix determined, the diode impedance (Z_R) at the test position is found from the measured impedance (Z_S) by the relation

$$Z_R = \frac{Z_{12}^2}{Z_{11} - Z_S} - Z_{22}.$$

These equations in their complex form are easily processed by computer.

The reference impedances ($Z_{R1,2,3}$) used to calibrate the microwave two-port consisted of 1) a short circuit in the form of a silver package having the same dimensions as the diode package; 2) an empty diode package with capacity 0.224 pF; and 3) a capacitor of 1.061 pF which also maintains the package geometry [4].

Fig. 2 defines the equivalent-circuit elements used in this paper. These elements are determined by using

CHARACTERISTICS OF THE IMPATT MODE

different bias conditions on the test diode. Briefly, the series inductance L_s is measured at high forward bias, and the diode series resistance R_{SB} and junction capacitance C_{jB} are measured at a reverse bias 2 volts less than the breakdown voltage. The package capacitance C_{pk} is measured on a 30-MHz bridge. The wafer admittance YW, which includes the active elements of the avalanche junction, is used for circuit-design and diode-evaluation purposes [4], [7]. From the known values of the incident power, the wafer admittance (GW, BW) and the calculated power output PC as a function of peak RF wafer voltage VW are obtained.

Data Results

It has been found that the large-signal single-frequency wafer conductance and susceptance data as functions of RF wafer voltage can be fitted by least-squares to the parabolic equations

192

$$BW = A_{10} + A_{11}VW + A_{12}(VW)^2$$

$$GW = A_{20} + A_{21}VW + A_{22}(VW)^2$$

where A_{ij} are the least-squares coefficients [8]. Smooth curves for GW, BW, and PC as functions of VW are then generated by computer from these equations. Figs. 3–5 show examples of these curves along with the measured data points for four diodes.

It can readily be seen from Figs. 3 and 4 that increasing the RF voltage at the wafer causes a decrease in negative conductance and an increase in wafer susceptance. Determination of the change in wafer susceptance as a function of RF voltage is primarily important in establishing the circuit admittance necessary for oscillation conditions.

Since the diode output power is given by

$$PC = -(1/2)GW(VW)^2$$

the conductance versus RF voltage curves shown in Fig. 3 can be used to calculate the power curves shown in Fig. 5. The maximum power output from each curve is referred to as the calculated maximum power PA for that diode. Also shown in Fig. 5 are the output powers at 6.2 GHz of these diodes operated as oscillators in a coaxial circuit tuned with a quarter-wave transformer. The second and other harmonic voltages measured under oscillating conditions are no larger than one percent of the fundamental voltage (there is no detectable harmonic power within 40 dB of the fundamental) for this type of circuit. The transformer, which is a quarter wavelength at the fundamental frequency, is essentially transparent to the second harmonic and thus the 50-ohm coaxial line effectively simulates the 20-mmho second-harmonic load present for the amplifier measurements for which no harmonic voltages were detected.

Fig. 6 shows a plot of the measured output power PO versus the calculated maximum power output for a group of 42 germanium and silicon diodes. A linear least-squares fitted line has been drawn through the

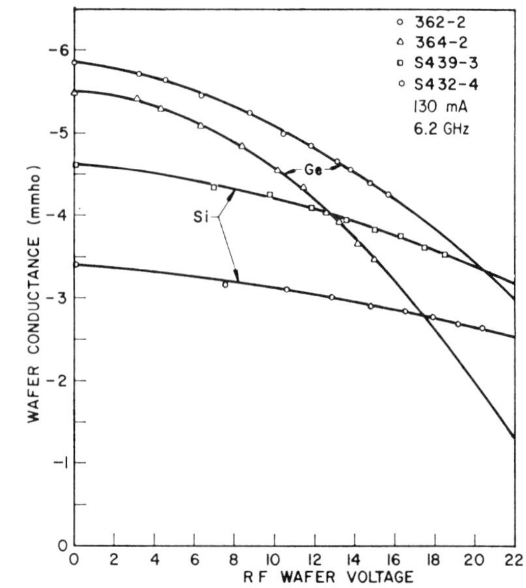

Fig. 3. Wafer conductance versus peak RF wafer voltage.

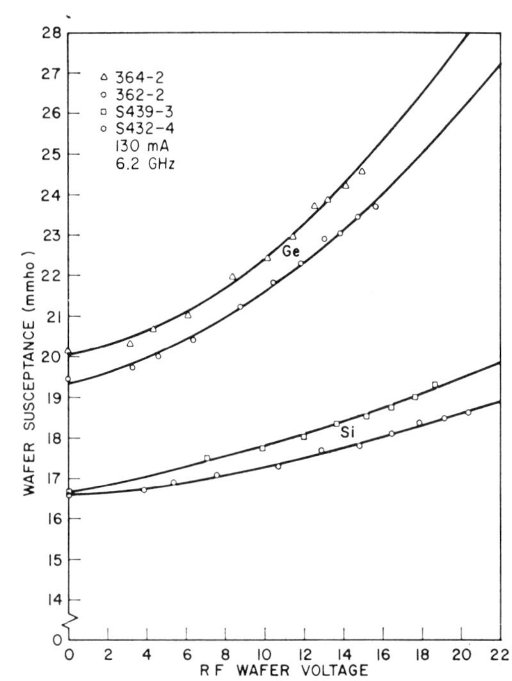

Fig. 4. Wafer susceptance versus peak RF wafer voltage.

data for which the correlation is 0.95. The standard-error-of-measurement lines which are shown indicate that at a given predicted power-output point the measured power output has a one sigma variation of 123 mW from the least-squares fitted line.

In Fig. 7 the large-signal wafer conductance at 6.2 GHz versus RF voltage has been plotted using current density as a parameter. From data like these one can establish the desired circuit admittance and variation of power output as functions of current density. Table I displays the wafer and package terminal characteristics at the maximum-power-output point for one diode at several different current densities. The tape inductance

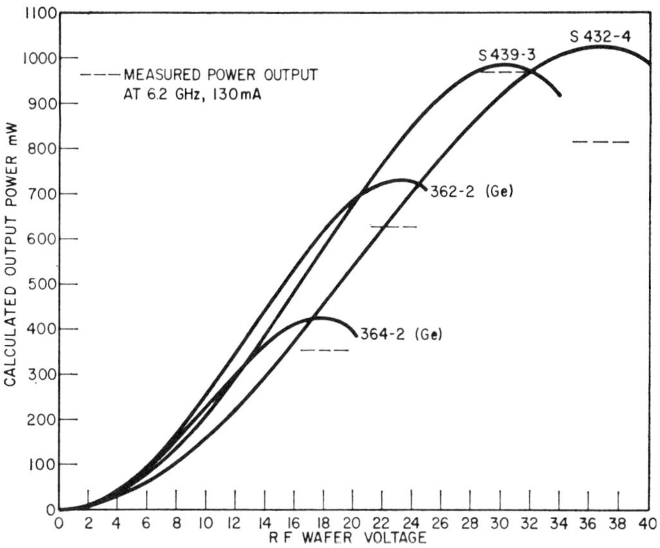

Fig. 5. Calculated power output PC versus peak RF wafer voltage.

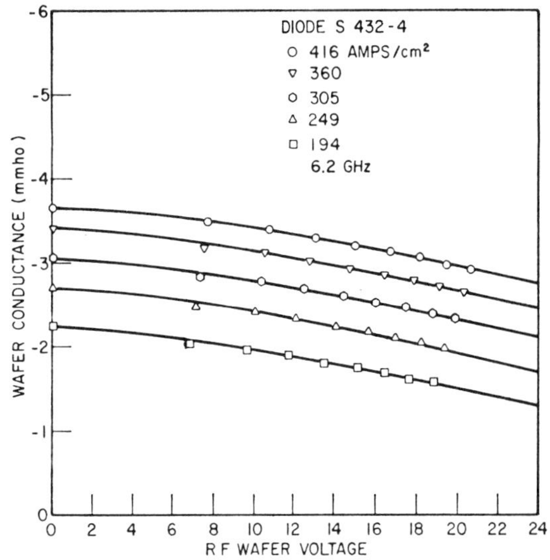

Fig. 7. Wafer conductance versus peak RF wafer voltage; current density as parameter.

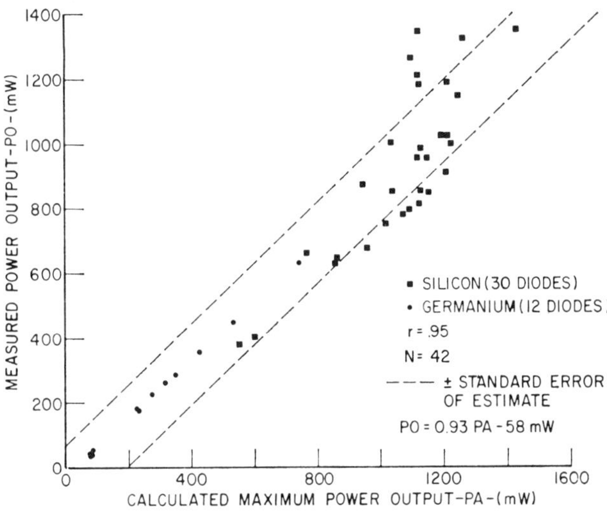

Fig. 6. Measured oscillator power output versus calculated maximum power output.

MATERIAL	SYMBOL	DIODE	L_S (nH)	C_{JBV} (pF)
Si	△	S432-4	.55	.66
Si	•	S439-3	.37	.75
Ge	□	362-2	.48	.88
Ge	○	364-2	.47	.91

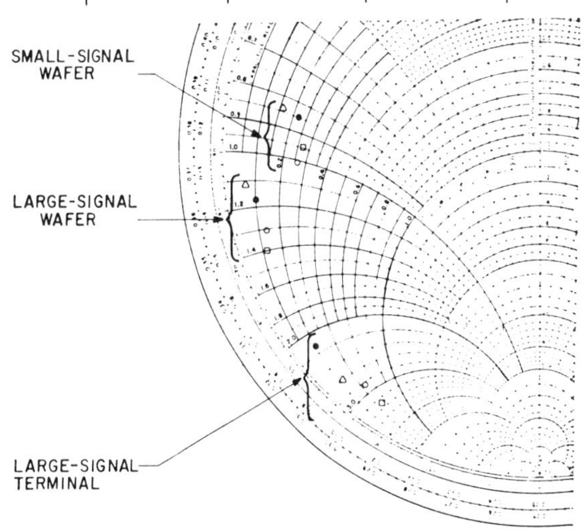

Fig. 8. Admittance plot of circuit to match diode at calculated maximum power output: 130 mA; 6.2 GHz; $Y_{circuit} = -Y_{diode}$; $Y_0 = 20$ mmho.

TABLE I

SILICON-DIODE S 432–4 PROPERTIES AT CALCULATED MAXIMUM OUTPUT POWER VERSUS CURRENT DENSITY

I (mA)	J (A/cm²)	Measured PO (mW)	Calculated PA (mW)	VW_{max} (volts)	GW (mmho)	BW (mmho)	GT (mmho)	BT (mmho)
70	194	68	415	29.4	−.96	22.4	−3.5	51.7
90	249	302	575	31.0	−1.2	21.9	−4.3	49.5
110	305	562	825	34.6	−1.37	22.3	−5.0	51.0
130	360	812	1040	36.9	−1.53	22.5	−5.7	52.1
150	416	955	1250	39.0	−1.67	22.7	−6.3	52.5

L_s of 0.55 nH and package capacitance C_{pk} of 0.224 pF account for the difference between the two admittances. The desired circuit admittance would be the negative of the diode-terminal admittance (GT, BT).

From these data it is seen that the susceptance for this diode is relatively insensitive to current-density changes, whereas the conductance is a rapidly changing function of current density. However, these variations are dependent on the diode structure and, consequently, knowledge of the measured variation is extremely useful in diode and circuit design. Also shown for comparison purposes are the RF voltages at the operating point for the maximum power output, the calculated maximum power output, and the power output measured in the coaxial circuit.

A Smith-chart presentation of the negative of the diode admittance for some of the diodes measured at 6.2 GHz is shown in Fig. 8. The small-signal wafer admittance as well as the large-signal wafer and terminal admittances at the maximum available power-output point of four diodes (two germanium and two silicon) are shown. To illustrate the difference in package assembly, devices with both single (362-2, 364-2, 432-4) and double tapes (439-3) connecting the wafer to the top of the package have been shown. The terminal admittances of the single- and double-tape units are significantly different due to the inductance transformation.

Summary and Conclusions

A method of determining the large-signal single-frequency amplifier characteristics of avalanche diodes operated near the optimum transit angle has been presented. These data provide useful information for establishing the design of both the diode [9] and a class of oscillator circuits. They establish single-frequency oscillator-admittance criteria needed to obtain optimum performance from the diode and provide an evaluation of what performance should be expected. Furthermore, the admittances are more useful as a gauge of the diode

design than are oscillator test results, because the admittances are characteristic of the diode alone rather than of both the diode and the circuit.

Experimentally obtained admittance versus RF wafer voltage measurements on 6-GHz germanium and silicon avalanche diodes are presented. The silicon diodes show a conductance characteristic with wafer voltage that exceeds that of the germanium diodes at large RF voltages, and consequently their calculated maximum power output is greater. The utility of assuming an analytic form for the wafer-conductance variation is demonstrated by the 0.95 correlation coefficient of the predicted and measured oscillator performance with a standard-error-of-measurement one-sigma variation of 123 mW.

For illustrative purposes the circuit admittance of roughly 5.7–52.1 mmho necessary to obtain the maximum-power capability of several diodes is presented and the admittance for different internal tape configurations is shown. The usefulness of the large-signal characterization is shown by the capability of predicting the load susceptance required to tune the diode for operation at the measurement frequency; and the load conductance required to obtain maximum power output at the chosen bias current.

Acknowledgment

The authors wish to thank their many colleagues, especially H. B. Frost; and A. H. Solomon for pointing out the relevance of Mouthaan's analysis to this work.

References

[1] J. W. Gewartowski and J. E. Morris, "Active IMPATT diode parameters obtained by computer reduction of experimental data," *IEEE Trans. Microwave Theory Tech.*, vol. MTT-18, pp. 157–161, March 1970.

[2] N. B. Kramer, "Characterization and modeling of IMPATT oscillators," *IEEE Trans. Electron Devices*, vol. ED-15, pp. 838–846, November 1968.

[3] J. E. Dalley, "Computer-aided microwave impedance measurements," *IEEE Trans. Microwave Theory Tech.*, vol. MTT-17, pp. 572–576, August 1969.

[4] C. N. Dunn and J. E. Dalley, "Computer-aided small-signal characterization of IMPATT diodes," *IEEE Trans. Microwave Theory Tech.*, vol. MTT-17, pp. 691–695, September 1969.

[5] K. Mouthaan, "Nonlinear analysis of the avalanche transit-time oscillator," *IEEE Trans. Electron Devices*, vol. ED-16, pp. 935–945, November 1969.

[6] S. Ramo, J. R. Whinnery, and T. Van Duzer. *Fields and Waves in Communication Electronics*. New York: Wiley, 1965, p. 595.

[7] D. L. Scharfetter and H. K. Gummel, "Large-signal analysis of a silicon Read diode oscillator," *IEEE Trans. Electron Devices*, vol. ED-16, pp. 64–77, January 1969.

[8] J. M. Cameron, *Fundamental Formulas of Physics*, D. H. Menzel, Ed. New York: Dover, 1960, p. 122.

[9] D. R. Decker, C. N. Dunn, and H. B. Frost, "The effect of injecting contacts on avalanche diode performance," to be published in *IEEE Trans. Electron Devices*, vol. 18, March 1971.

MEASUREMENT OF LARGE-SIGNAL IMPEDANCE, OPTIMUM AC VOLTAGE AND EFFICIENCY OF Si pnn+, Ge npp+ AND GaAs SCHOTTKY-BARRIER AVALANCHE TRANSIT TIME DIODES

B.B. van Iperen and H. Tjassens, Philips Research Laboratories
N.V. Philips' Gloeilampenfabrieken, Eindhoven, The Netherlands

Summary

A new accurate method for measuring the large-signal impedance of avalanche transit time diodes is described and results of measurements at X-band on Si pnn^+, Ge npp^+ and n-GaAs Schottky barrier diodes are presented. Optimum values for the ac voltage and the maximum efficiency in the absence of series-and circuit losses are derived from these results. It is found that these quantities are nearly independent of the dc current in the range investigated.
For values of the ac voltage up to one quarter of the dc voltage the large-signal negative ac resistances of the Ge and GaAs diodes show very good agreement with the well-known theory of Tager for the ideal Read structure, whereas for Si diodes a departure from this theory is found. At higher ac voltage all three types show a steep decrease of the negative ac resistance, which is not predicted by the theory.

1. Introduction

In studying avalanche transit time diodes we are interested in quantities that describe the diode properties independent of its encapsulation and external circuitry. In this respect, perhaps the most important quantity is the first harmonic of the hf current flowing in the diode as a function of an applied sinusoidal voltage. The ratio of these two quantities is called the large-signal impedance. Few results of large-signal impedance measurements are available as yet. This might be due to the fact that such measurements are time-consuming and difficult to perform with sufficient accuracy[1,2].
In this paper we report an accurate method of measuring which permits a good comparison with theory. Results of several types of diodes are presented, and optimum values of efficiency and ac diode voltage are obtained. All diodes investigated are of the mesa type, mounted upside down in a normal varactor package, and all measurements were performed at X-band. Details of the diode construction are reported elsewhere[3]. The oscillator circuit used is a simple coaxial line circuit which is matched to a 50Ω coaxial line by one quarter wavelength transformer formed by a slug. The results of CW output measurements are given in Table I.

Table I	Max.values		Typ.values		V_{dc}
	outp.	eff.	outp.	eff.	
Si pnn^+	0.9 W	7%	0.6 W	6%	70 V
Ge npp^+	0.5 W	14%	0.4 W	12%	32 V
GaAs n Schottky	0.7 W	15%	0.5 W	11%	60 V

CW impedance measurements were performed on Si pnn^+, Ge npp^+ and n-GaAs

195

Schottky barrier diodes. The diodes investigated were typical samples in their respective groups.

2. Measurement of large-signal impedance

The measuring procedure starts by measuring the small-signal impedance of the diode as a function of the dc diode current. For this purpose, the diode is placed in a reduced height waveguide mount, and use is made of a waveguide bridge method specially developed for this type of measurement. The method allows quick and easy measurement of the diode impedance with an error smaller than 0.1 Ω in the region of interest.

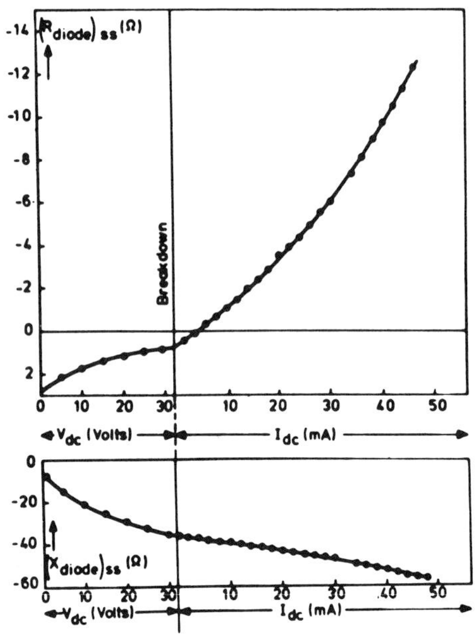

Fig. 1. Real part $(R_{diode})_{ss}$ and imaginary part $(X_{diode})_{ss}$ of the small signal impedance of a Ge npp^+ diode.

Fig. 1 represents the result of such a measurement on a Ge npp^+ diode. After performing the small-signal measurement the diode is mounted in the oscillator circuit and at a fixed frequency the output power P_{load} dissipated in the load resistance is measured for a number of (as yet unknown) values of the load impedance, obtained by inserting transformer slugs of different diameters. For each diameter of the transformer slug the output power is measured as a function of the dc diode current. Assuming that the system is not resonant at any higher harmonic frequency, the large-signal impedance of the diode can be derived now as a function of the ac diode voltage, using the simplified equivalent circuit shown in fig. 2*).

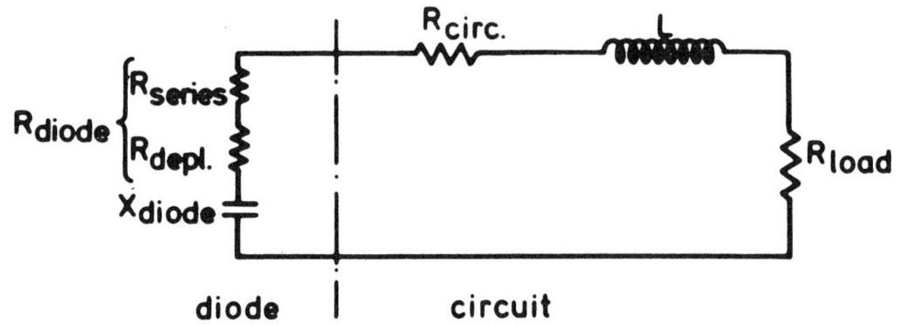

Fig. 2. Equivalent circuit of the oscillator used in discussing the large-signal measurements.

For stable oscillations the conditions $R_{diode}+R_{circ}+R_{load}=0$; $X_{diode}+X_{circ}=0$, where R and X are real and imaginary parts of the respective impedances, are always satisfied. For a given slug, fixed slug position and oscillation frequency, R_{load} and R_{circ} are constant. Hence R_{diode} is also constant and equal to its value at the start-oscillation current which latter value is known from the small-signal impedance measurements. We can thus plot the output power P_{load} as a function of R_{diode} with the dc diode

*) The sole purpose of this equivalent circuit is to explain the method used, and is only correct at a fixed frequency. A more realistic circuit must also contain the coupling network representing the package, the fringing capacitance of the external circuit and the load reactance.

diode current I_{dc} as a parameter. The result obtained in this way for a Ge npp^+ diode is shown in fig. 3.

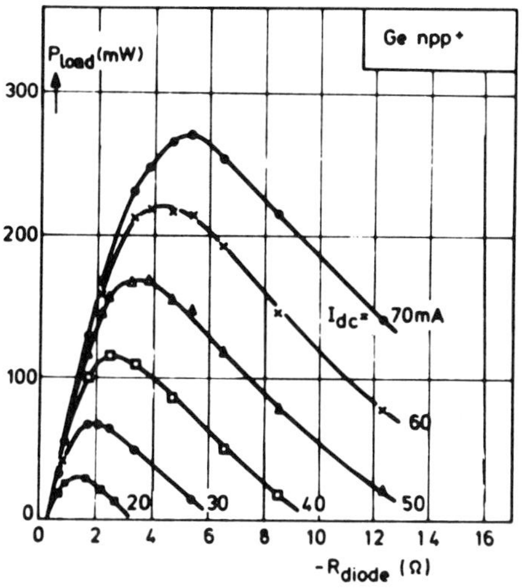

Fig. 3. Output power in load resistance P_{load} as a function of R_{diode}, as measured on the Ge npp^+ diode.

The figure shows that all curves converge to one value R_{diode} (0) for output zero. Since at this point $R_{load}=0$, this enables us to determine R_{circ}, which is simply the opposite of R_{diode} (0). Another consequence is that also R_{load} is now a known quantity for every slug diameter. Furthermore, from the small signal measurements we know the value of the parasitic series resistance R_{series} of the diode, which is the value of R_{diode} at breakdown. Hence we also know the resistance of the active part of the diode (the depleted layer), which is indicated by R_{depl}. Now if I_{ac} is the amplitude of the total ac current, we have the relation

$$P_{load} = \frac{1}{2} I_{ac}^2 R_{load}$$

from which we calculate I_{ac} as a function of P_{load}. If the diode reactance X_{diode} is also known (see below), the amplitude of diode voltage V_{ac} is found from the relation

$$V_{ac} = \left(R_{diode}^2 + X_{diode}^2 \right)^{\frac{1}{2}} I_{ac}$$

From this result and the measured R_{depl} as a function of P_{load} we finally find R_{depl} as a function of V_{ac}.

For a constant oscillation frequency X_{diode} simply is equal to its small signal value. Actually, in varying the dc current the frequency is also varied. This variation, however, is small ($<1\%$) and a corresponding correction on X_{diode} was applied using the complete equivalent circuit of the oscillator. The influence of this effect on R_{depl} was found to be negligible.

Fig. 4 shows the ac resistance of the diode R_{depl} obtained as described above for the Ge diode of fig. 3, together with theoretical curves to be discussed in the next section. Similar graphs are found for Si pnn^+ and n-GaAs Schottky barrier diodes (fig. 8 and 9). 197

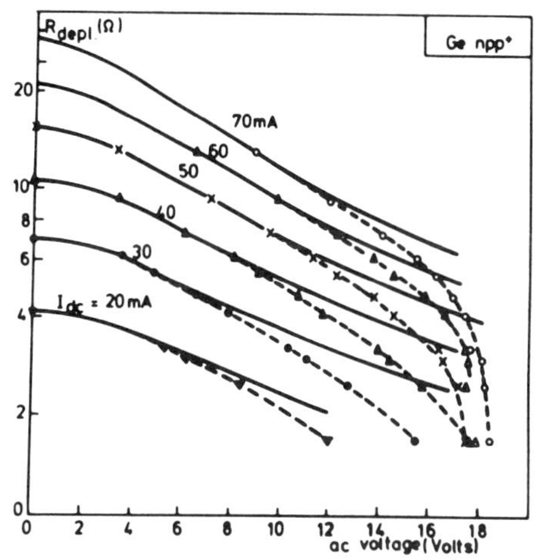

Fig. 4. Measured ac resistance of the active part of the diode R_{depl} as a function of the ac voltage on the diode V_{ac} (same Ge diode as in fig. 3). The full curves are found from the simple large-signal theory as explained in section 3.

The method described above allows for large-signal impedance measurements without a direct determination of the

external load and the circuit resistances presented to the diode. It would be difficult to obtain sufficiently accurate direct measurements of small load resistances such as those occurring here, for instance with a standing wave detector.
The total power generated by the active part of the diode P_{total} (V_{ac}) and the corresponding efficiency η_{total} can now easily be found.

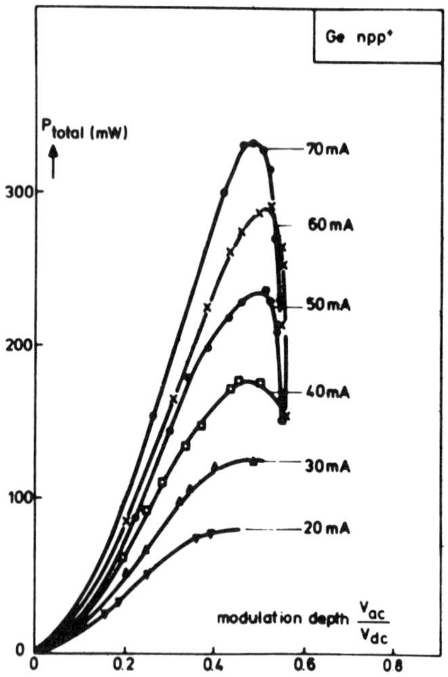

Fig. 5. Total generated ac power of Ge npp$^+$ diode as a function of the ac modulation depth V_{ac}/V_{dc}.

Fig. 5 shows P_{total} as a function of V_{ac}/V_{dc} for the Ge npp$^+$ diode. It can be seen that the position of the maximum is nearly independent of the dc current. This is also true of the other diodes investigated, as is shown in fig. 6. In this figure the maximum total efficiency and the corresponding modulation depth $(V_{ac}/V_{dc})_{max}$ are plotted as a function of the dc current for the three types of diodes investigated. The maximum efficiencies for Si pnn$^+$, Ge npp$^+$ and GaAs Schottky barrier diodes are 7%, 14% and 15% respectively at optimum modulation depths of 0.37, 0.50 and 0.45, independent of the dc current.

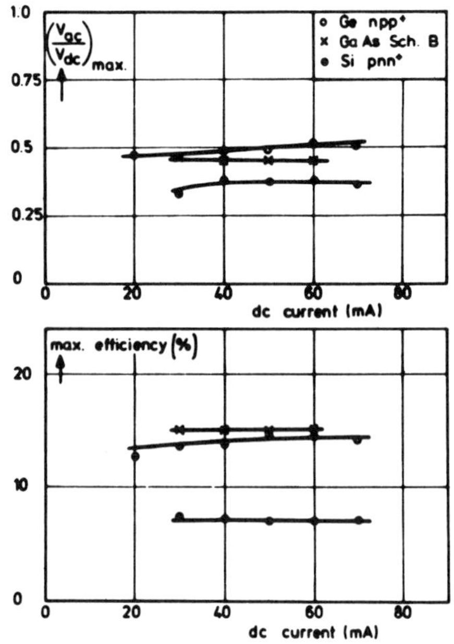

Fig. 6. Max. total efficiency and corresponding optimum modulation depth $(V_{ac}/V_{dc})_{max}$ as a function of dc current I_{dc}.

3. Simple large-signal theory

The measuring results may be compared with the large-signal theory for the ideal Read diode by Tager[4] and Delagebeaudeuf[5]. This theory takes into account the non-linear dependence of the avalanche current on the ac voltage across the avalanche region. It does not predict a limitation to the ac diode voltage caused by other effects, however, and therefore cannot be used to calculate the maximum efficiency. Actually, it predicts a steady increase of P_{total} with increasing V_{ac}. It nevertheless appears to be of interest to find out in how far this theory holds.
According to this theory, the large-signal value of the diode impedance is found simply from Gilden and Hines' expression for the small-signal impedance[6] by multiplying the avalanche frequency ω_a by a reduction factor $\sqrt{k}$ which depends on the diode ac voltage.

$$R_{depl} = \frac{\emptyset/\omega C}{\omega^2/\omega_a^2 k - 1} ,$$

where $\emptyset$ is a transit time function and C the "cold" capacitance of the diode.

In order to calculate R_{depl} we still need to know the quantities $\emptyset/\omega C$ and ω_a. These are found from the diode ac resistance $(R_{depl})_{ss}$, measured at small signal by observing that ω_a^2 is proportional to I_{dc} and inverting the expression for $(R_{depl})_{ss}$ of Gilden and Hines. This yields

$$\frac{1}{(R_{depl})_{ss}} = a\frac{1}{I_{dc}} - b ,$$

where a and b are independent of I_{dc}. As is shown in fig. 7, this linear relation is actually satisfied very well for all diodes investigated.

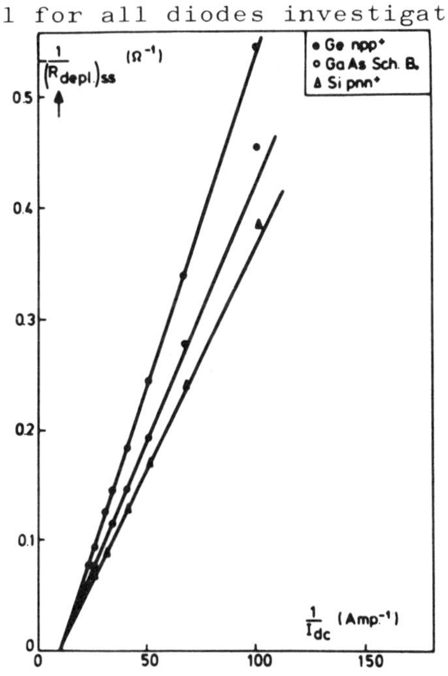

Fig. 7. Experimental relation between small signal diode resistance $(R_{depl})_{ss}$ and dc current I_{dc}.

This enables us to determine a and b and consequently the quantities $\emptyset/\omega C$ and ω_a. Using these values we can calculate R_{depl} from the theory mentioned above without introducing any theoretical constant. In figure 4 the full curves represent R_{depl} for the Ge diode according to this theory. The figure shows that for values of V_{ac} up to 8 V (modulation depth 0.25) this diode follows the theory with an error of less than 0.2 Ω and we may conclude that for this region the ideal Read diode is a very good model for the Ge npp$^+$ diode. At higher ac voltages the measured curve departs downward, as it should do. Fig. 8 shows similar agreement for a GaAs Schottky barrier diode.

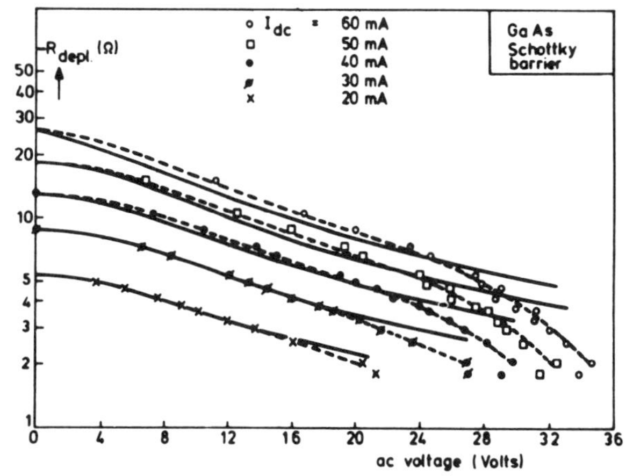

Fig. 8. Same as 4 but for GaAs Schottky barrier diode.

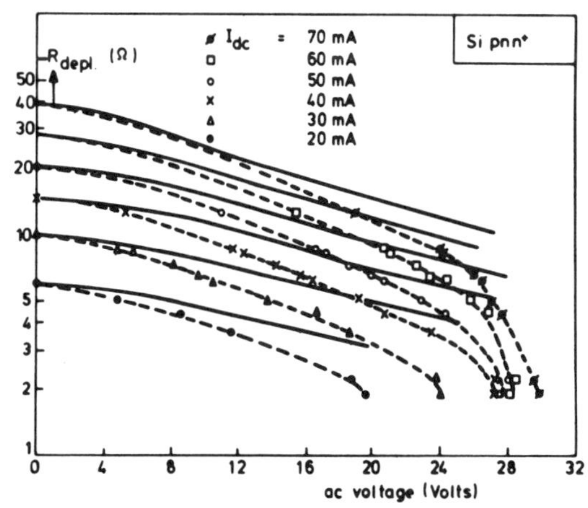

Fig. 9. Same as 4 but for Si pnn$^+$ diode.

Finally, fig. 9 shows the result for a Si pnn$^+$ diode. Here the situation appears to be rather different, since even in the case of small ac voltages the decrease in the measured diode resistance at increasing voltage is stronger than the theoretical prediction. Clearly the ideal Read diode is not a good model for the Si pnn^+ diode. This is probably connected with the fact that in silicon the ionisation coefficient for holes is much smaller than for electrons and as a result the avalanche region is relatively wide and located more towards the middle of the depletion region. For this type of diode therefore a more sophisticated approach seems to be required even to describe the behaviour at small modulation depths.

Acknowledgements

We are indebted to Dr. D. de Nobel and H.G. Kock for providing the diodes, and to J. de Groot and H.J. Knoop for their assistance in preparing the computer programs.

References

1. N.B. Kramer, "Characterization and modeling of IMPATT oscillators", IEEE Trans. ED, 15, 838, 1968.
2. J.W. Gewartowski and J.E. Morris, "Active IMPATT diode parameters obtained by computer reduction of experimental data", IEEE Trans. MTT, 18, 157, 1970.
3. H.G. Kock and D. de Nobel, "Technology of Schottky barrier IMPATT diodes", Proceedings of the conf. on MOGA, Amsterdam, 1970.
4. A.S. Tager, "The avalanche-transit diode and its use in microwaves", Soviet Phys.-USP., 9, 892, 1967.
5. D. Delagebeaudeuf, "Les diodes à avalanche utilisées comme oscillateurs de puissance aux hyperfrequences", Onde élec. 48, 722, 1968.
6. M. Gilden and M.E. Hines, "Electronic tuning effects in the Read microwave diode", IEEE Trans. ED, 13, 169, 1966.

Characterization and Modeling of IMPATT Oscillators

NORMAN B. KRAMER, MEMBER, IEEE

Abstract—**A method has been found for large-signal characterization of IMPATT diodes. Waveguides of different heights are used so that the effect on efficiency of load impedance at a fixed oscillation frequency can be observed. Experimental large-signal admittance plane plots for two *p-n* diodes operating in Ku-band are presented. A nonlinear phenomenological model of an IMPATT oscillator is developed which correctly predicts the experimentally observed dependence of efficiency on load conductance and admittance plane characteristics. In addition to aiding the visualization of oscillator performance, analysis of this parallel circuit model shows that efficiency increases rapidly with small-signal negative conductance.**

Introduction

Microwave oscillations have been observed to occur in a variety of IMPATT (Impact Avalanche Transit-Time) structures. Small-signal theories [1]–[4] for several types of IMPATT devices have been developed which give some hints about the performance of these devices as self-excited oscillators. However, a clear physical correlation of the device and circuit parameters with the frequency and amplitude of stable oscillations does not arise from these linear theories.

In an attempt to improve understanding of oscillator operation, a large-signal characterization technique has been developed. Measurements are made using this technique which allow prediction of frequency and power output for an IMPATT diode operating in a given circuit. A phenomenological nonlinear model which evolves from the small-signal equivalent circuit is shown to predict correctly the dependence of efficiency on load conductance and the admittance plane [5] characteristics of the experimental oscillators. This model aids in the visualization of oscillator performance; in conjunction with the experimentally determined oscillator characteristics, it suggests methods for circuit improvement. Analysis of this model shows that small-signal negative conductance is a useful figure of merit for high oscillator efficiency, provided the device can be represented by a parallel equivalent circuit.

Description of Technique

In order to determine the effect of load admittance on oscillator performance, it is necessary that load conductance and susceptance be capable of independent variation. A test circuit is also required which can be accurately characterized. The diode can be mounted on a cylindrical post in a reduced height waveguide to satisfy the second requirement [6], and use of a set of waveguides of different heights with sliding shorts satisfies the first.

Manuscript received February 26, 1968; revised June 29, 1968.
The author is with Hughes Research Laboratories, Malibu, Calif.

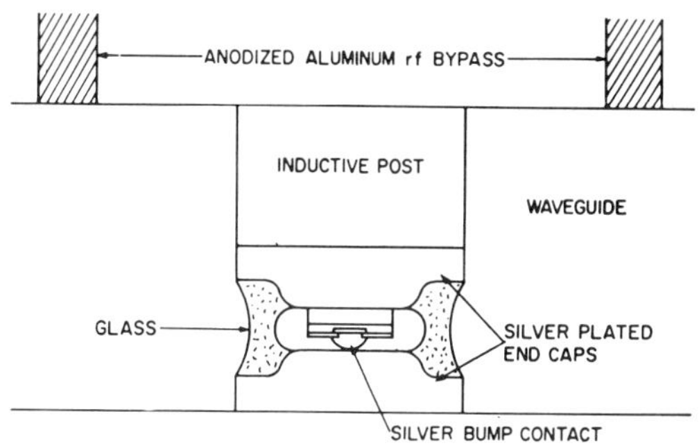

Fig. 1. Pellet packaged diode mounted in reduced-height waveguide circuit. Package is 0.035 inch by 0.060 inch in diameter.

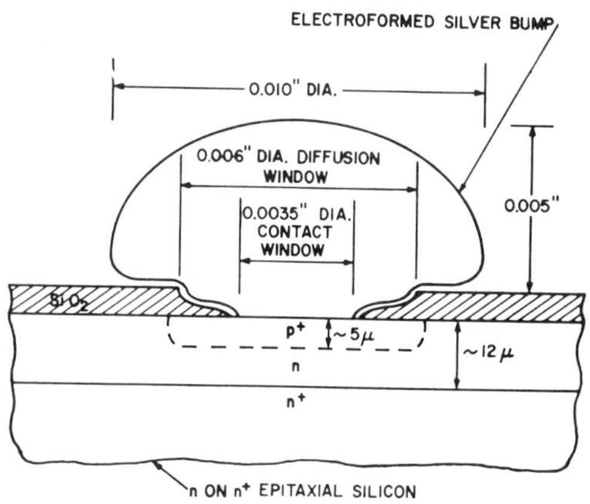

Fig. 2. Junction region of diode.

Measurements were made on planar silicon diodes in a pellet package, as shown in Fig. 1. A detail of the junction region of the diode is shown in Fig. 2. The diodes are Hughes Microglass switching diodes. Epitaxial silicon (n on n^+) which has layer resistivity in the 0.8 to 1.0 $\Omega\cdot$cm range is used; the silicon is subsequently gold-doped. The diodes used in these studies were selected for low breakdown voltage (≈ 60 volts) from a large production lot. (Breakdown voltage is typically between 80 and 120 volts.)

It will be shown later that the passive diode as mounted in Fig. 1 can be represented by a series RLC circuit. R is the series resistance, L is the post inductance, and C is a parallel combination of the depletion layer capacitance and the parasitic capacitance resulting from the electroformed bump. When the values of R, L, and C are known, it is simple to calculate the load admittance seen by the active region of the oscillating diode when it is in the test setup shown in Fig. 3. If Z_a is the active impedance of the depletion layer (caused by avalanche generation and drift of carriers), the experimental oscillator can be represented by the circuit of

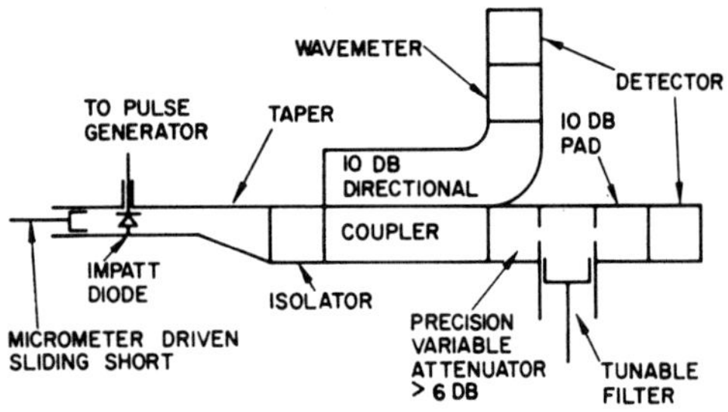

Fig. 3. Experimental arrangement for large-signal measurements.

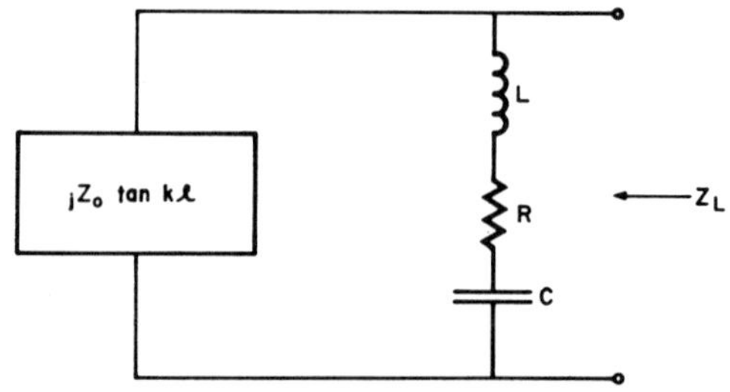

Fig. 5. Load impedance presented to the reflectometer by a combination of a passive diode and a sliding short.

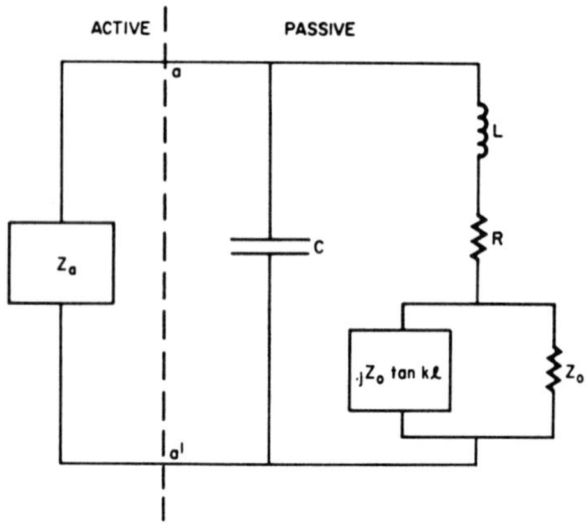

Fig. 4. Equivalent circuit of experimental oscillator.

Fig. 4. Z_0 is the characteristic impedance of the waveguide of height b and width a, and is given by [7]

$$Z_0 = 754 \frac{b}{a}\left[1 - \left(\frac{\lambda}{2a}\right)^2\right]^{-1/2} \tag{1}$$

where λ is the free-space wavelength at the oscillation frequency, k is the rectangular waveguide propagation constant, and l is the distance from the diode to the sliding short. It is convenient to define the active large-signal admittance Y to satisfy the condition that the total circuit admittance vanish:

$$\begin{aligned} Y &\equiv -G - jB \\ &\equiv -j\omega C - [R + j\omega L + (1/Z_0 + 1/jZ_0 \tan kl)^{-1}]^{-1}. \end{aligned} \tag{2}$$

In a given waveguide, Y is determined by a measurement of l and the oscillation frequency. At a fixed frequency, L and Z_0 both increase with b so that the active diode generally will see a different load admittance in each waveguide at that frequency.

Passive Measurements

202 The method used for passive microwave characterization of IMPATT diodes is similar to that of DeLoach [7] for varactor characterization because only relative power measurements are required—no phase information is necessary. The same circuits were used for passive and active measurements, reducing errors caused by unavoidable dimensional variations from circuit to circuit. In a standard Ku-band reflectometer with a klystron source, the short is used to resonate the diode for maximum power absorption. At a fixed frequency, a sharp minimum is observed in the reflection coefficient as the short is moved. A method is now developed for using transmission-line theory to determine the equivalent circuit of the diode mounted as in Fig. 1 from information concerning the reflection coefficient minimum.

If we assume that the mounted diode can be represented by a series RLC circuit, Fig. 5 depicts the load presented to the waveguide transmission line. For convenience, let

$$X_s = Z_0 \tan kl \tag{3}$$

then

$$Z_L = \frac{R + j(\omega L - 1/\omega C)}{1 + (\omega L - 1/\omega C)/X_s - jR/X_s}. \tag{4}$$

The magnitude of the reflection coefficient is given by

$$|\Gamma|^2 = \frac{|Z_L|^2 + Z_0^2 - 2Z_0 \operatorname{Re} Z_L}{|Z_L|^2 + Z_0^2 + 2Z_0 \operatorname{Re} Z_L}. \tag{5}$$

From (5) it can be seen that $|\Gamma|$ will have a minimum approximately where Re Z_L has a maximum. For small values of R, examination of (4) shows that this occurs when

$$-X_s = \left(\omega L - \frac{1}{\omega C}\right). \tag{6}$$

For convenience we can define, for negative values of $\tan kl$,

$$\frac{1}{C_s} = -\omega X_s = -\omega Z_0 \tan kl. \tag{7}$$

Equation (6) can be written

$$\frac{1}{C_s} = \omega^2 L - \frac{1}{C}. \tag{8}$$

If the short position at a fixed klystron frequency is varied until the minimum reflection coefficient is found, the value of $1/C_s$ at this point given by (7) will establish one point on a $1/C_s$ versus ω^2 curve. After this is done for several frequencies, the resulting curve should be a straight line with slope L and intercept $-1/C$ if we have chosen the correct equivalent circuit. Similar considerations for an inductive resonant short show that

$$L_s = 1/\omega^2 C - L \tag{9}$$

where L_s is defined by

$$L_s = Z_0 \tan kl/\omega. \tag{10}$$

This results in a straight line on an L_s versus $1/\omega^2$ plot with slope $1/C$ and intercept $-L$.

At each frequency, $|\Gamma|$ will have some minimum value $|\Gamma|_{\min}$ as a function of l. It can be seen from (5) that $|\Gamma|_{\min}$ will have a minimum as a function of frequency given by

$$|\Gamma|_{\min} = 0 \tag{11}$$

which occurs when

$$Z_L = Z_0. \tag{12}$$

At the resonance defined by (6) we have, from (4),

$$\operatorname{Re} Z_L \approx Z_L \approx X_s^2/R. \tag{13}$$

Therefore, when (11) is satisfied we obtain

$$R \approx X_s^2/Z_0. \tag{14}$$

As a result, the series resistance R is determined by the reactance of the short at the frequency and short position which give maximum absorption. The values of R determined in this way are small compared with the other impedances involved (typically $R<5$ ohms), so that the procedure is consistent.

Five reduced height waveguide circuits were employed in the measurements; they ranged from 0.035- to 0.095-inch high, in 0.015-inch increments. In all five waveguides a straight line behavior of $1/C_s$ versus ω^2 as typified by Fig. 6 was observed for three diodes tested. In most cases, condition (11) was observed to occur within or at the edge of the frequency band over which the measurements were made. For example, condition (11) occurred within Ku-band for diode P-27-64 in all but the 0.095-inch guide, and the values of R determined in the other four guides were all within a few percent of 4.5 ohms. Fig. 7 contains $|\Gamma|_{\min}$ as a function of frequency as measured for diode P-27-64 in the 0.035-inch waveguide. Inductance as a function of waveguide height, as determined for diode P-27-64, is given in Fig. 8. This curve is valid for all diodes of the same package type.

It can be seen from Fig. 6 that the value of C is difficult to determine accurately for these diodes because C is relatively large resulting in a small $1/C_s$ axis intercept. In order to test the validity of using low-frequency (100 kHz) bridge measurements for determining C, an open-circuited pellet diode was evaluated using the resonant short technique. In this case (9) was applicable because C was low (0.34 pF). The resulting data are plotted in Fig. 9; the capacitance is well-defined by the slope of the line and is within 10 percent of its low-frequency value. It is concluded that the low-frequency capaci-

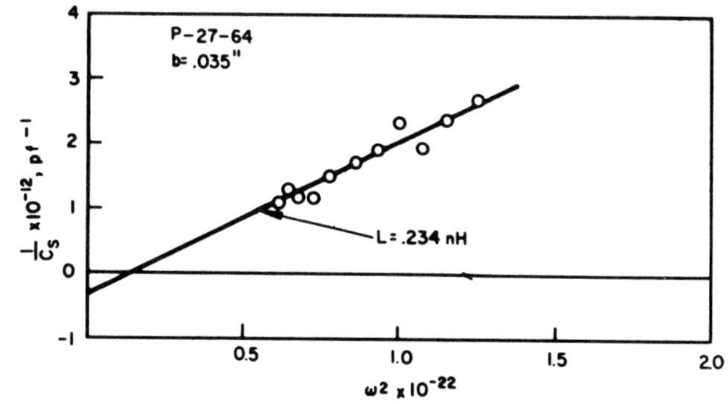

Fig. 6. Curve for determining post inductance.

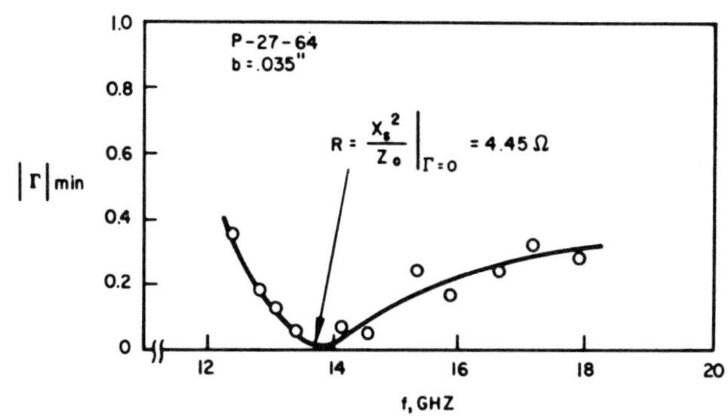

Fig. 7. Curve for determining series resistance.

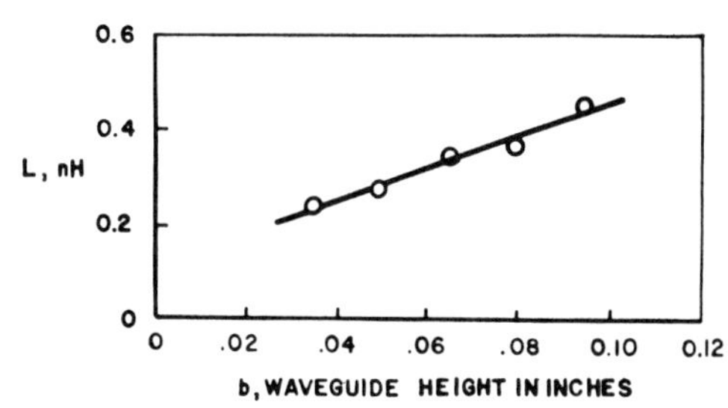

Fig. 8. Post inductance as a function of waveguide height.

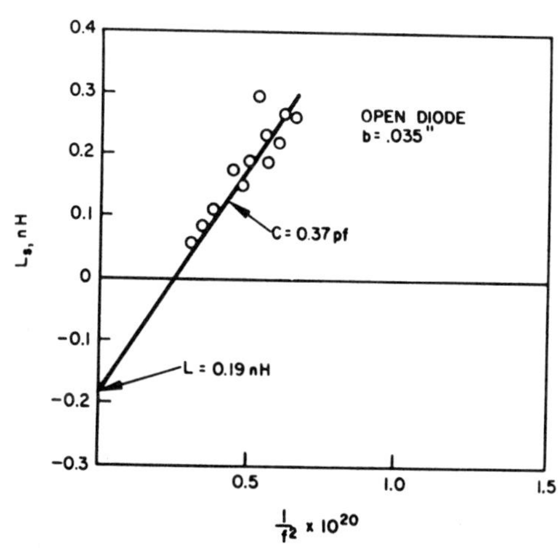

Fig. 9. Curve for determining capacitance of an open diode.

tance value can be used in the RF equivalent circuit. After the measurement the package was opened, and it was observed that the open circuit resulted because there was no contact between the bump and package cap. The somewhat lower inductance measured in this case (intercept in Fig. 9) results from a lack of conduction current through the constricted bump and contact region.

Capacitance–voltage data on the two diodes for which large-signal results are reported below are listed in Table I. The capacitance values of Table I include a package capacitance of 0.2 pF and the bump capacitance. The effect of the bump was determined for several diodes by making $C-V$ measurements before and after the bump was stripped from the diode chip. The bump capacitance varied somewhat from diode to diode because of variations in bump diameter, but it typically decreased from 0.5 pF at 1-volt bias to 0.25 pF at 50-volt bias. The value of C at breakdown was taken as 1.88 pF for diode P-27-64 and 1.6 pF for diode P-30-60.

TABLE I

CAPACITANCE-VOLTAGE CHARACTERISTICS OF EXPERIMENTAL DIODES

Bias Voltage (volts)	Diode Capacitance (pF)	
	P-27-64	P-30-60
1	2.93	2.54
5	2.50	2.22
10	2.32	2.05
20	2.13	1.86
50	1.93	1.64
Breakdown	64 volts	60 volts

ACTIVE MEASUREMENTS

Impedance and power measurements were made in each of the five waveguides employing the experimental arrangement shown in Fig. 3. The loaded Q of the tunable filter cavity was over 700, so that the filter short position could be used to determine frequency with sufficient accuracy. The filter was calibrated for frequency and power over the range of 12.8 to 16.4 GHz. Bias was applied by means of a constant-current pulse generator in the form of short pulses ($\approx$200 ns) at a 5-kHz repetition rate to minimize heating effects.

Data on diode P-27-64 were taken by moving the short behind the diode in 0.010-inch increments and recording the frequency and power output at the maximum amplitude point in the RF pulse. The current pulse amplitude was 600 mA. Total short travel was about 0.7 inches. The reactance of the short was usually capacitive to compensate for the relatively large mounting post inductance L (see Figs. 4 and 8). As the short was moved away from the diode, the frequency decreased from about 16 GHz ($l \lesssim \lambda/2$) to 13 GHz; the frequency then jumped discontinuously to 16 GHz ($l \lesssim \lambda$), and the cycle began again. In the higher waveguides there was usually a range of short positions over which no RF was observed. This appeared as a hole in the RF frequencies, usually toward the high-frequency end.

Using (2), G and B were calculated from the data. Admittance plane curves are contained in Fig. 10. All points plotted to establish a curve for a given frequency were taken within 150 mHz of that frequency. These curves are sufficient to explain the detailed experimental behavior of the IMPATT diode in a given circuit. For example, Fig. 11 contains a calculated admittance plane curve for the 0.035-inch waveguide circuit at the short position where the jump in RF frequency occurred. Intersections with device curves are shown at 13.07 and 16.45 GHz, very close to the frequencies (13.06 and 16.36 GHz) at which the experimental tuning cycle is discontinuous. As another example, the circuit curves for the higher waveguides do not intersect the device curves in the frequency range where no RF output was observed.

The RF power generated by the active region of the diode is a meaningful criterion for the evaluation of diode performance because it becomes the limiting power output with series resistance decreased to a minimum and circuit design optimized. It can be seen from Fig. 4 that the generated power P is given by

$$P = \left(\frac{R}{Z_0 \sin^2 kl} + 1\right) P_L \tag{15}$$

where P_L is the power delivered to the load Z_0. The ideal efficiency will be defined as

$$\eta_i = \frac{P}{VI} \tag{16}$$

where V and I are the bias voltage and current, respectively. The curves shown in Fig. 12 are equal ideal-efficiency contours for a fixed frequency of 14 GHz. These data are replotted from Fig. 10. As shown in Fig. 13, at 14.5 GHz there were also sufficient data points to establish similar efficiency contours.

A different method of taking data was used for diode P-30-60. The short position was set to obtain oscillation at the same nominal frequency in each waveguide, and the RF power, frequency, and bias current were recorded at 20-ns intervals referenced to the leading edge of the current pulse. The peak bias current was 600 mA (in a few cases noise made it necessary to reduce it to 550 mA). In general, the admittance plane characteristics of the experimental diodes were insensitive to peak bias current level; however, the generated RF power increased rapidly with bias current and usually became noisy above 600 mA. Fig. 14 shows the behavior of various quantities during a typical current pulse. Data were taken at 500-mHz increments from 13.5 to 15.5 GHz. Fig. 15 contains admittance plane curves for this diode. The sequence of points for a given pulse is seen to follow

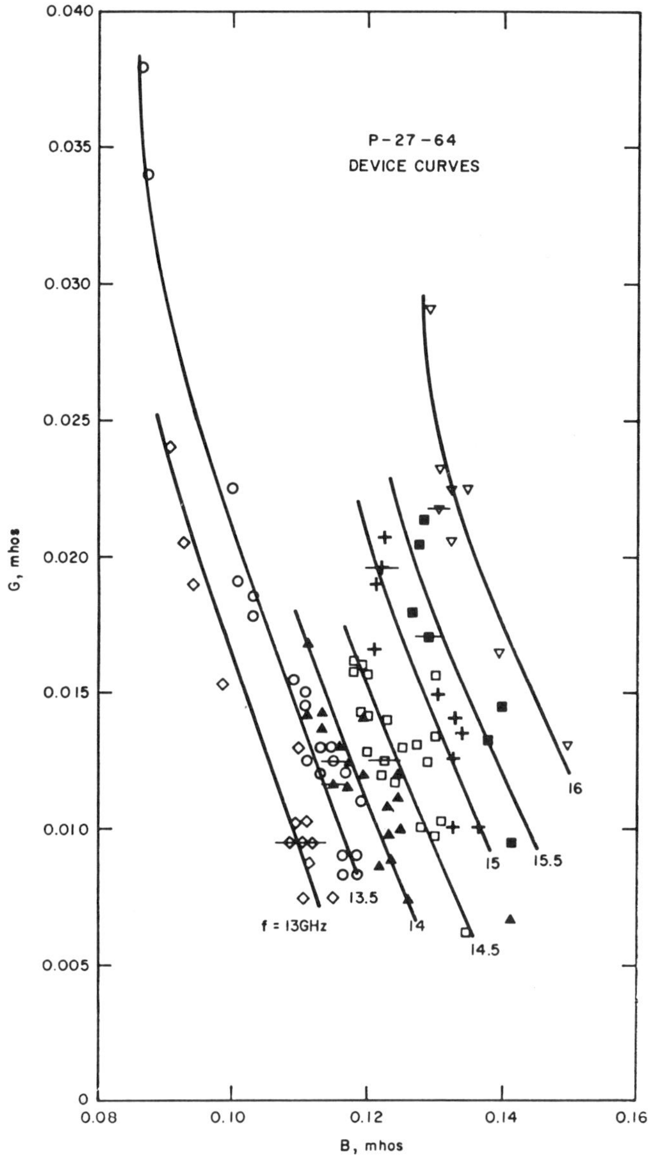

Fig. 10. Admittance plane curves for diode P-27-64 at 600-mA peak bias current. The points of maximum generated power are marked by a horizontal line.

the general trend of the admittance curves. G always increases and B and f always decrease with time from the pulse leading edge. The rise time of the current pulse used was approximately 100 ns. In most cases the RF does not start until about 50 ns after the current pulse has passed through the threshold current level (see Fig. 14). This lag may be due to a thermal effect or a build-up time associated with the growth of oscillations. Using a mercury switch to discharge a coaxial cable gave fast rise time ($\lesssim 1$ ns) high-current (≈ 2 amperes) pulses that were capable of turning the diode on within the 2-ns rise time of the oscilloscope used. However, the RF output of the diodes was noisy at these current levels. It is observed that the admittance plane plots for the two diodes are similar. Maximum power output for diode P-27-64 was 268 mW at 15.79 GHz (correspond-

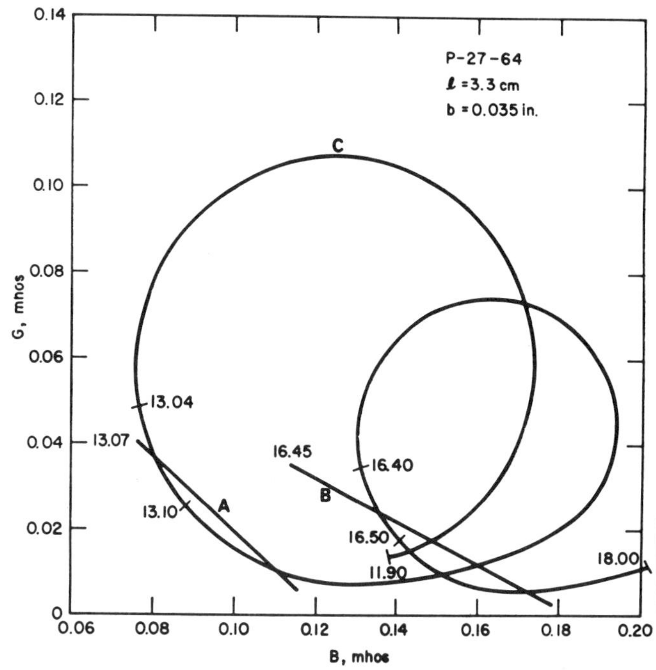

Fig. 11. Admittance plane presentation, showing intersections of device curves A and B with circuit curve C.

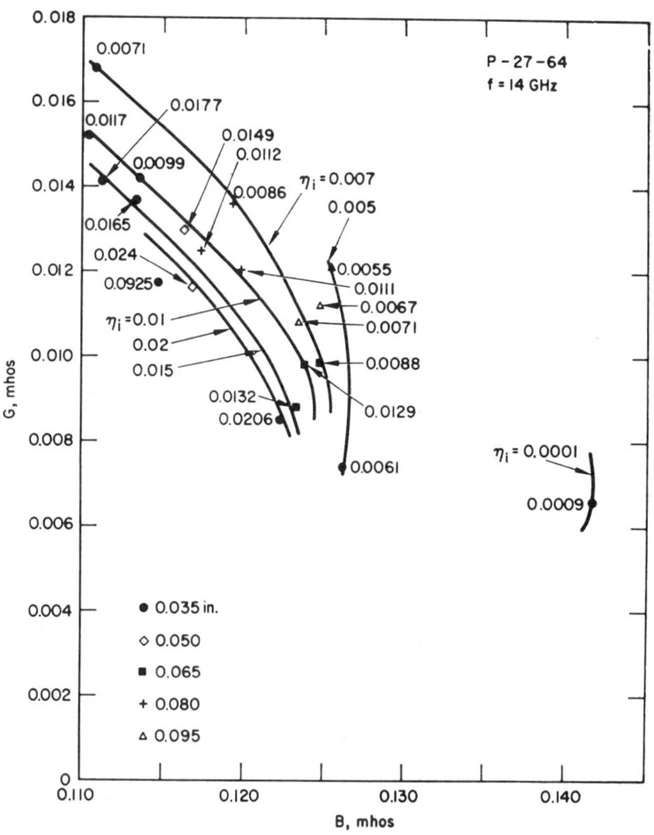

Fig. 12. Efficiency contours in the admittance plane for diode P-27-64 at 14 GHz. Each point is marked with its measured ideal efficiency. Peak bias current = 600 mA.

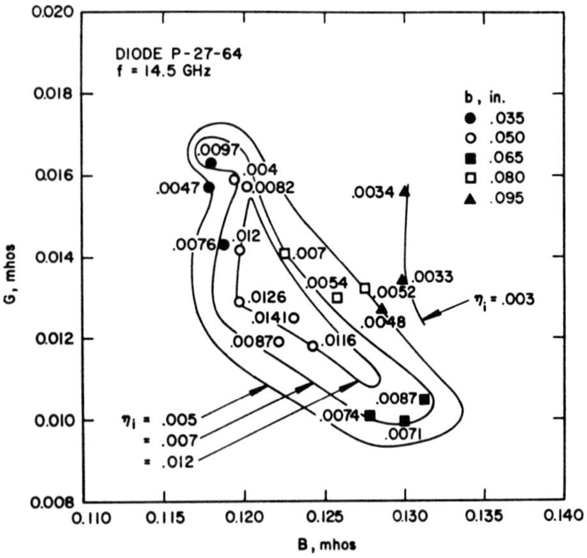

Fig. 13. Efficiency contours in the admittance plane for diode P-27-64 at 600-mA peak bias current. $f = 14.5$ GHz.

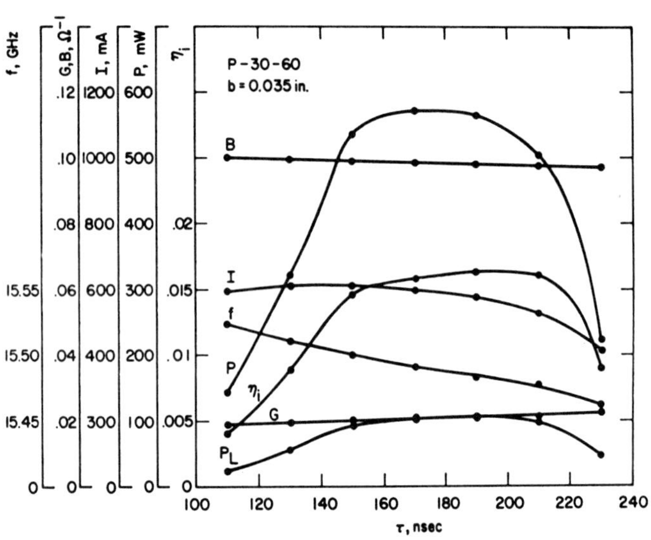

Fig. 14. Bebavior of bias current I, admittance $G+jB$, power generated P, power delivered P_L, frequency f, and ideal efficiency η_i as a function of time τ from current pulse leading edge.

ing to a generated power of 1.33 watts), while maximum power output for diode P-30-60 was 106 mW at 15.48 GHz (corresponding to a generated power of 565 mW).

Nonlinear Model

As pointed out by Read [1], the space charge of carriers generated by the avalanche reduces the peak electric field at a given RF voltage. We will start with a linear equivalent circuit and estimate the change in circuit elements with voltage amplitude required to reflect the effect of space charge. The AVX-2A diode model analyzed by Misawa [3] resembles the structure of the experimental p-n diodes. His calculations indicate that the active mechanism of the diode in the small-signal regime can be well-represented by a parallel combination

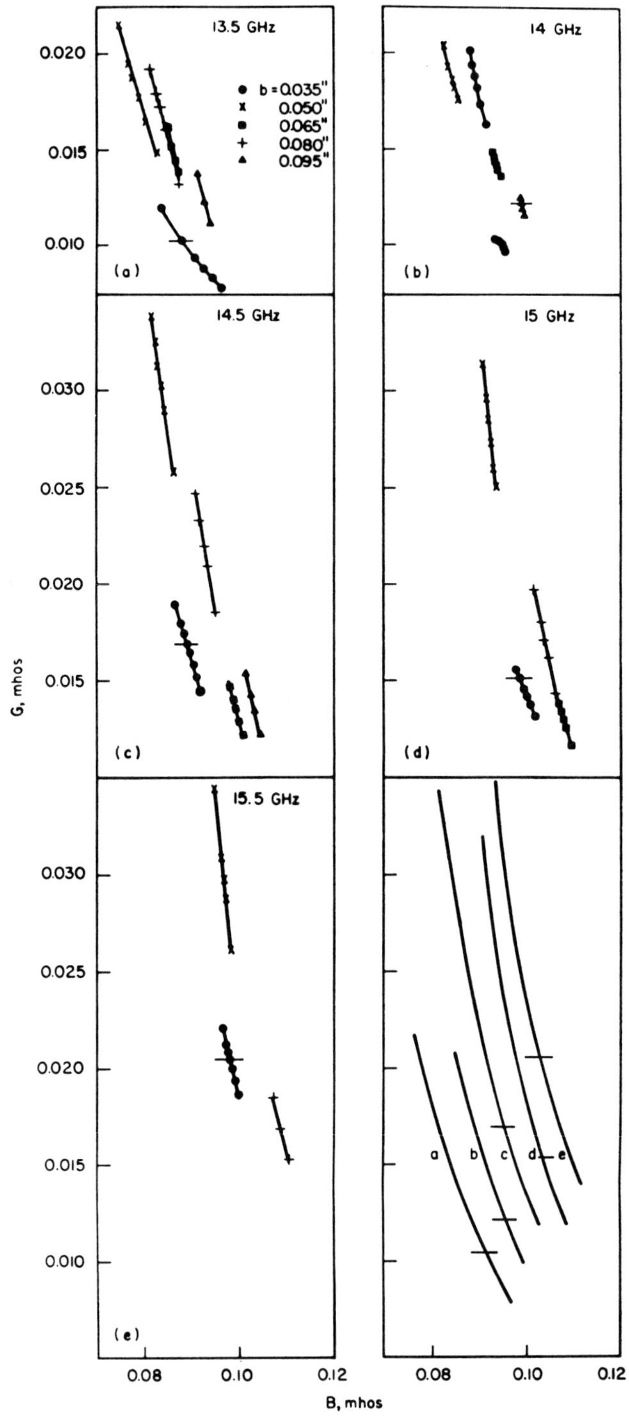

Fig. 15. Admittance plane curves for diode P-30-60 at 600-mA peak bias current. The points of maximum generated power are marked by a horizontal line.

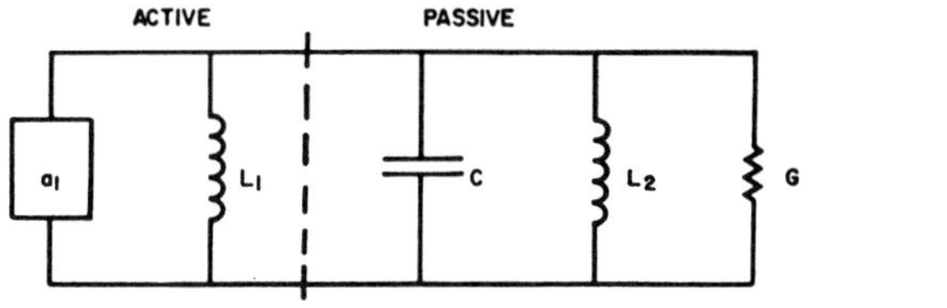

Fig. 16. Model oscillator circuit under small-signal conditions.

of an inductance and negative resistance. Thus we begin with the circuit shown in Fig. 16. L_1 and a_1 represent the small-signal inductance and negative conductance, respectively. C represents the depletion layer capacitance, and L_2 and G are, respectively, the load inductance and load conductance.

As the oscillation builds up, the increasing peak voltage causes the amount of charge generated by the avalanche during the positive half voltage cycle to increase. This increasing space charge causes a reduction in the peak electric field (and subsequent reduction in the charge generation rate) which occurs earlier in the cycle as the oscillation continues to build up, resulting in a decrease in the phase lag between voltage and conduction current and a saturation of the current amplitude. The point of view will be taken that the dominant saturation effect is the degradation in the phase relationship between the voltage and conduction current. In terms of the circuit of Fig. 16, this effect can be represented by a decrease in a_1 and L_1.[1] For convenience we express this saturation effect in terms of symmetrical nonlinear ac characteristics for the conductance

$$i = -a_1 v + a_3 v^3 \tag{17}$$

and for the inductance

$$\frac{di}{dt} = \frac{v}{L_1} + b_3 v^3. \tag{18}$$

With these nonlinear characteristics, the circuit equation for Fig. 16 becomes

$$C\frac{d^2v}{dt^2} + (G - a_1 + 3a_3v^2)\frac{dv}{dt} + \left(\frac{1}{L_1} + \frac{1}{L_2} + b_3v^2\right)v = 0. \tag{19}$$

Consider this equation for the case when $b_3 \to 0$. With the substitutions

$$\omega_0^2 = (1/L_1 + 1/L_2)/C \tag{20}$$

$$\omega_0 t = T \tag{21}$$

$$\epsilon = (a_1 - G)/\omega_0 C \tag{22}$$

$$u = \left(\frac{3a_3}{a_1 - G}\right)^{1/2} v \tag{23}$$

we obtain

$$\frac{d^2u}{dT^2} - \epsilon(1 - u^2)\frac{du}{dT} + u = 0, \tag{24}$$

which is the Van der Pol equation. This equation has a fundamental solution [8] of amplitude nearly 2 for all positive values of ϵ. The frequency depends on ϵ as follows: $\omega = \omega_0$ for $\epsilon \ll 1$, and falls off as ϵ^{-1} for large ϵ. Since $u_{max} = 2$, the power P_0 delivered to the load G at the fundamental frequency will be[2]

$$P_0 = Gv_{max}^2/2 = 2G(a_1 - G)/3a_3. \tag{25}$$

Power output is parabolic as a function of G, peaking when

$$G = a_1/2 \tag{26}$$

and having a maximum value

$$\max P_0 = a_1^2/6a_3. \tag{27}$$

For the estimation of the i–v characteristic of the experimental diode, a_1 and a_3 could be obtained from the measured maximum power and the value of G at which it occurs.

Returning to (19), we consider the case when b_3 is small but nonzero. Instead of attempting to solve this equation directly, we replace the nonlinear negative resistance by a current generator which has the Van der Pol amplitude, as shown in Fig. 17. The circuit equation for this arrangement is

$$C\frac{d^2v}{dt^2} + G\frac{dv}{dt} + \left(\frac{1}{L_1} + \frac{1}{L_1} + b_3v^2\right)v = 2\omega G\left(\frac{a_1 - G}{3a_3}\right)^{1/2} \cos \omega t. \tag{28}$$

With the substitutions (20), (21), and

$$k = G/\omega_0 C \tag{29}$$

$$\omega_1 = \omega/\omega_0 \tag{30}$$

$$\delta = \frac{4b_3G^2\omega_1^2}{\omega_0(\omega_0 C)^3}\left(\frac{a_1 - G}{3a_3}\right) \tag{31}$$

$$v = \frac{2\omega_1 G}{\omega_0 C}\left(\frac{a_1 - G}{3a_3}\right)^{1/2} u \tag{32}$$

we obtain

[1] It is emphasized that this is an assumption about the way in which the saturation occurs. Most certainly the buildup of oscillation occurs in a more complicated way and perhaps in a qualitatively different way. However, as will be seen, this assumption leads to agreement with the experimentally observed behavior.

[2] A value of $\epsilon > 0$ in the Van der Pol equation has the effect of introducing an out of phase component of amplitude u_1 and odd harmonics of amplitudes u_3, u_5, etc. [8]. To second order in ϵ

$$\frac{u_1}{u_{max}} = \frac{u_3}{u_{max}} = \frac{1}{8}\epsilon + O(\epsilon^2)$$

with all higher harmonics being of order ϵ^2. Thus, for values of ϵ typical of the experimental diodes ($\epsilon \lesssim 0.3$), the waveform may be regarded as sinusoidal.

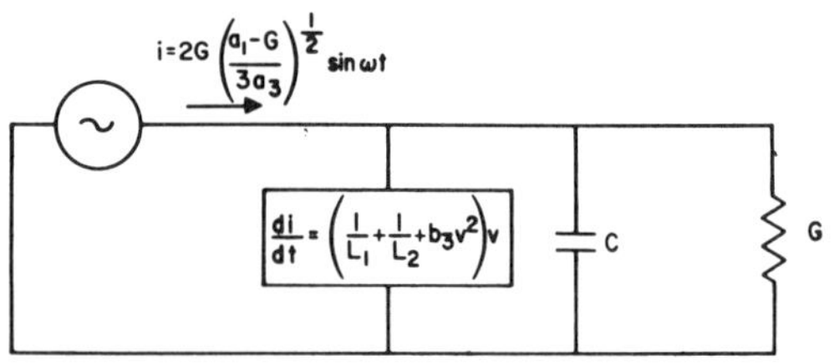

Fig. 17. Driven oscillator substituted for self-excited model oscillator.

$$\frac{d^2u}{dT^2} + k\frac{du}{dT} + (1 + \delta u^2)u = \cos \omega_1 T \qquad (33)$$

which is Duffing's equation with damping. We now assume that the natural frequency of the self-excited oscillator described by (19) will be the frequency of maximum response of the oscillator described by (33). Since δ is small, the method of perturbations can be applied ([8], pp. 354–355), resulting in the following first-order relation between the amplitude of the fundamental β and ω_1

$$[(1 - \omega_1^2 + 3\delta\beta^2/4)^2 + k^2\omega_1^2]\beta^2 = 1. \qquad (34)$$

From (32) we can write

$$v_{\max} = \frac{2\omega_1 G}{\omega_0 C}\left(\frac{a_1 - G}{3a_3}\right)^{1/2}\beta. \qquad (35)$$

The RF power generated is then

$$P = \frac{G v_{\max}{}^2}{2} = (\omega_1 k\beta)^2 P_0 \qquad (36)$$

where P_0 is given by (25). Therefore, in order to determine the natural oscillation frequency, we must find the frequency which maximizes $\omega_1 k\beta$. Defining

$$\delta(\omega = \omega_0) = \delta_0 \qquad (37)$$

we may write (34) in the form

$$(\omega_1 k\beta)^2 = 1 - \beta^2[1 + \omega_1^2(3\delta_0\beta^2/4 - 1)]^2. \qquad (38)$$

The second term on the right-hand side of (38) is always negative, so that the maximum value of $\omega_1 k\beta$ is

$$\omega_1 k\beta = 1, \qquad (39)$$

which occurs when

$$1 + \omega_1^2(3\delta_0\beta^2/4 - 1) = 0. \qquad (40)$$

Substituting (39) into (40) yields

$$\omega_1^2 = 1 + 3\delta_0/4k^2. \qquad (41)$$

Therefore, the natural frequency of (19) is given by (41); and according to (39) and (36), it results in

$$P = P_0. \qquad (42)$$

Thus the inductive nonlinearity raises the frequency of oscillation but does not change the amount of RF power generated.

For comparison with the experimental results we now derive the admittance plane behavior of the model oscillator. From (20), (29), (30), (31), and (41) we have

$$\omega_1^2 = 1 + \frac{b_3}{a_3}\frac{(a_1 - G)}{(1/L_1 + 1/L_2)}. \qquad (43)$$

Defining

$$\equiv b_3(a_1 - G)/a_3 \qquad (44)$$

and

$$x \equiv (1/L_1 + 1/L_2), \qquad (45)$$

we obtain

$$\omega^2 = (x + y)/C \qquad (46)$$

and

$$\omega^2 L_2 C = 1 + L_2(1/L_1 + y). \qquad (47)$$

Thus the active negative susceptance B is given by

$$B = \omega C - 1/\omega L_2 = \frac{1/L_1 + b_3(a_1 - G)/a_3}{\omega}. \qquad (48)$$

In general a_1, L_1, a_3, and b_3 will all depend on ω. We use the small-signal results of Misawa [3] for information about a_1 and L_1. The experimental current density is 2 in Misawa's units, and his admittance unit is 0.2 mho for our diodes. Because of the slightly higher breakdown voltages of the experimental diodes, their intrinsic resonant frequencies should be somewhat lower than that of AVX-2A, which is in Ku-band. Because the load was always inductive in the experiments, we assume that the diode is operating in the frequency range just above the intrinsic resonant frequency; we take this to be 11 GHz because RF power to below 12 GHz was observed for the experimental diodes at the same bias level in a reduced height X-band waveguide. The frequency dependence of a_1 in this range is estimated from an extrapolation of Misawa's curves to a normalized current density of 2:

$$a_1 = (5.1 \times 10^{-2}\omega)^4, \qquad (49)$$

while L_1 is nearly independent of freqeuncy.

At a constant frequency we have, from (48)

$$\left.\frac{\partial B}{\partial G}\right|_{\omega=\text{constant}} = -\, b_3/a_3\omega. \qquad (50)$$

From Fig. 10, a typical experimental value is

$$\left.\frac{\partial B}{\partial G}\right|_{\omega=\text{constant}} = -\,1.73. \qquad (51)$$

Thus we estimate that b_3/a_3 is nearly frequency-independent and that

$$b_3/a_3 = 1.6 \times 10^{11}. \qquad (52)$$

Using (49) and (52) we have, from (48),

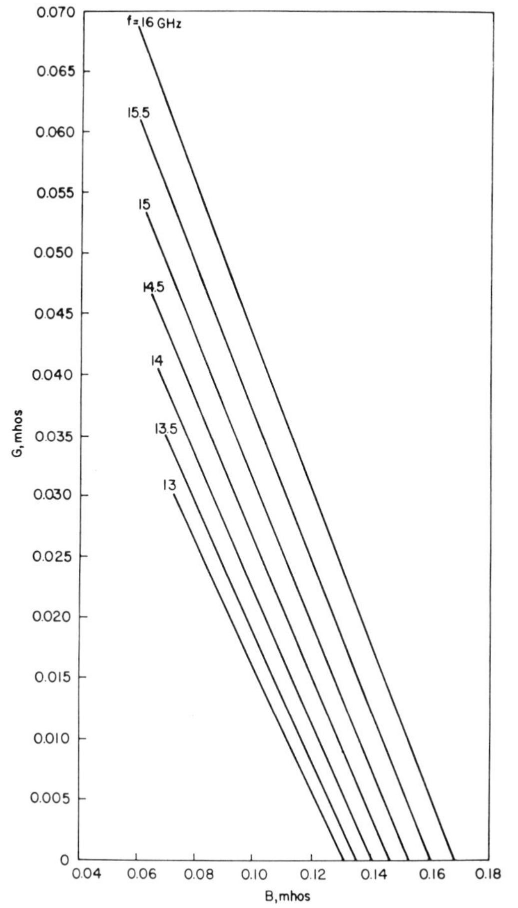

Fig. 18. Admittance plane representation of model oscillator simulating diode P-27-64. Compare with Fig. 10.

$$\left.\frac{\partial B}{\partial \omega}\right|_{G=\text{constant}} = \frac{1}{\omega}\left(\frac{b_3}{a_3}\frac{da_1}{d\omega} - B\right) = 0.22 \times 10^{-11} \quad (53)$$

which is in good agreement with experiment (see Fig. 10).

The experimental admittance plane curves for diode P-27-64 can be duplicated quite well by (48) by using (49) and (52) and choosing $L=0.17$ nH. Equation (48) is plotted in Fig. 18. The maximum value of G for each frequency is determined by $G=a_1$ (oscillation condition). As can be seen from Misawa's [3] Fig. 11(b), the maximum value of G will level off at higher frequencies. From Table I we estimate that the depletion layer capacitance of diode P-27-64 is 1.48 pF (after subtracting the 0.4-pF parasitic capacitance). The intrinsic resonant frequency corresponding to $L_1=0.17$ nH is then 10 GHz, which is in reasonably good agreement with the assumed value of 11 GHz.

Discussion and Conclusions

Experimental measurements typified by Figs. 10 and 15 as well as results obtained for the phenomenological model [see (26) and (42)], suggest that maximum efficiency is obtained for a load conductance equal to approximately one-half the small-signal negative conduc-

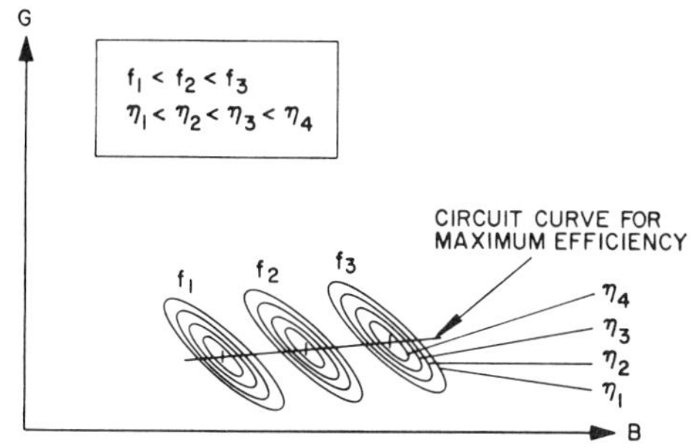

Fig. 19. Conceptual representation of IMPATT diode admittance plane characteristics showing efficiency contours and the circuit tuning curve for maximum efficiency. f_i are frequencies, η_i are efficiencies.

tance. The efficiency contours of Figs. 12 and 13 imply that the diode must see a certain susceptance for maximum power output. In the design of a circuit directed toward optimizing the performance of a particular diode, this has obvious implications as shown conceptually in Fig. 19. Construction of circuit curves in the admittance plane (as shown in Fig. 11 for a reduced-height waveguide circuit) can give valuable information on how various circuit parameters should be changed to increase efficiency at a given frequency and how tuning can best be accomplished over a desired frequency range. Equation (27) indicates that small-signal negative conductance is a useful figure of merit for high efficiency in devices with parallel small-signal equivalent circuits, so that small-signal analyses of various structures (as typified by those of Gilden and Hines [2] and Misawa [3]) should serve as valuable guides for device improvement.

Acknowledgment

The author wishes to thank F. A. Blum for numerous helpful discussions and suggestions and R. E. Johnson for his skillful assistance with the experimental work. H. J. Cohen and F. J. Saia of Hughes Semiconductor Division supplied the diodes for the experiments. The advice and encouragement of D. C. Forster are also gratefully acknowledged.

References

[1] W. T. Read, Jr., "A proposed high-frequency, negative-resistance diode," *Bell Sys. Tech. J.*, vol. 37, pp. 401–446, March 1958.
[2] M. Gilden and M. E. Hines, "Electronic tuning effects in the Read microwave avalanche diode," *IEEE Trans. Electron Devices*, vol. ED-13, pp. 169–175, January 1966.
[3] T. Misawa, "Negative resistance in *p-n* junctions under avalanche breakdown conditions," *IEEE Trans. Electron Devices*, vol. ED-13, pt. 1, pp. 137–143; pt. 2, pp. 143–151, January 1966.
[4] H. K. Gummel and D. L. Scharfetter, "Avalanche region of IMPATT diodes," *Bell Sys. Tech. J.*, vol. 45, pp. 1797–1827, December 1966.
[5] J. C. Slater, *Microwave Electronics*. New York: Van Nostrand, 1950, Ch. 9.
[6] B. C. DeLoach and R. L. Johnston, "Avalanche transit-time microwave oscillators and amplifiers," *IEEE Trans. Electron Devices*, vol. ED-13, pp. 181–186, January 1966.
[7] B. C. DeLoach, "A new microwave measurement technique to characterize diodes and an 800-Gc cutoff frequency varactor at zero volts bias," *IEEE Trans. Microwave Theory and Techniques*, vol. MTT-12, pp. 15–20, January 1964.
[8] D. F. Lawden, *Mathematics of Engineering Systems*, 2nd ed. New York: Wiley, 1959, pp. 328–342.

Noise and Power Saturation in Singly Tuned IMPATT Oscillators

A. MICHAEL COWLEY, MEMBER, IEEE, ZVONKO A. FAZARINC, ROBERT D. HALL, MEMBER, IEEE, STEPHEN A. HAMILTON, CHU-SUN YEN, MEMBER, IEEE, AND ROBERT A. ZETTLER, MEMBER, IEEE

Abstract—The fabrication of *X*-band IMPATT diodes and their evaluation in simple singly tuned circuits are described. Typical performance in *X* band for diodes with junction areas of $\sim 2 \times 10^{-4}$ cm² and breakdown voltages between 65 and 80 volts is 0.7 watt at 7 percent efficiency with junction temperatures <200°C. More than 1.0 watt is easily obtained at 250°C junction temperatures. It is found that oscillator noise and power saturation are related to the existence of a maximum limiting value of ac voltage across the diode. Qualitative agreement of these results with a simple large-signal model is found.

210

I. Introduction

THIS paper presents the important results of an IMPATT diode fabrication and circuit-interaction study. The objectives of the program were 1) to develop a reliable, economical technique for diode fabrication, and 2) to develop simple models for diode–circuit interaction that would allow rational design of useful IMPATT oscillator circuits.

The most difficult technological problem in IMPATT-diode fabrication is the design of an adequate heat-sinking scheme that is compatible with the microwave circuit requirements for operation as an oscillator or amplifier. In the most common approach to this problem, an ordinary mesa diode is inverted and thermal compression bonded to a heat sink, usually copper. This technique was developed at Bell Telephone Laboratories, Inc.; a description of it and references to the relevant literature are given by Sze [1]. It was felt by the present authors that significant improvements in the cost and reproducibility of IMPATT-diode fabrication could be made if the critical portion of the heat flow path could be formed as an integral part of the diode. To this end we have developed a method of electroplating the heat sink directly onto the silicon wafer, before separation of the individual diodes; the details have been presented previously [2] but will be briefly described here for completeness.

Most investigations of IMPATT diodes have utilized microwave circuits with several tuning and loading adjustments, in order to allow trial-and-error optimization of the circuit for a variety of different diodes and frequencies. A good example of this kind of circuit is that designed by Iglesias, as described by Sze [1]. For the present study, the convenience of adjustability was sacrificed in favor of circuit simplicity, in order to allow reasonably straightforward interpretation of the results. The circuits chosen were singly tuned, that is, resonant at only one frequency, and had a fairly low external Q, in the range of 10–30. They were analyzed in terms of the impedance presented to the chip terminals at the frequency of oscillation.

The significant result of this study is the insight into the transit-time mode of operation of avalanche diodes gained through the use of these simple circuits; of particular importance, the noise of the diodes was found to be strongly dependent on the magnitude of the diode RF voltage. Another important feature of IMPATT operation in these circuits is that the diode RF voltage saturates at a peak value of about one-half the breakdown voltage. These results are analyzed in the light of the original large-signal model of Read [3], and the computer simulations of Scharfetter and Gummel [4].

II. Diode Fabrication

The diodes described in this paper were fabricated using double epitaxial p^+-n-n^+ wafers grown from silane at 1050°C. The doping in the n-type layer was uniform, and the thickness was such that at breakdown a small amount ($\sim$0.2 μm) of undepleted material remained. Within a wafer, breakdown voltage variations were typically $\pm$2 volts; diodes were fabricated from wafers with nominal breakdown voltages in the range of 65–80 volts. Typical values of n-type layer thickness and impurity density were 3.8 μm and 7.5×10^{15} cm³; these particular values correspond to a breakdown voltage of 70 volts. The p^+ layers were $\sim$2.5 μm thick, with a resistivity of about 0.003 $\Omega \cdot$cm.

The epitaxial wafer is metallized on the p^+ side by vacuum evaporation of chromium and gold; this side is then electroplated with copper or gold to a thickness of several mils. The silicon wafer is then thinned mechanically and chemically to a thickness of 1–2 mils and then metallized on the n^+ side. The second metallization is then etched into circular pads and these are used to define silicon mesas, which are formed in a second etching step. The finished mesas now stand on a substrate consisting of the electroplated metal heat sink. The individual diodes are separated and the resulting chips are simply soldered into microwave packages or microstrip circuits.

Manuscript received April 13, 1970.
A. M. Cowley, Z. A. Fazarinc, S. A. Hamilton, and C.-S. Yen are with the Hewlett-Packard Company, Palo Alto, Calif. 94304.
R. D. Hall was with the Hewlett-Packard Company, Palo Alto, Calif. He is now with Dean Hall Associates, Los Altos, Calif. 94022.
R. A. Zettler was with the Hewlett-Packard Company, Palo Alto, Calif. He is now with Fairchild Microwave and Optoelectronics Division, Fairchild Semiconductor, Mountain View, Calif.

Fig. 1(b) shows a chip resulting from the above process. The shape of the mesa is such that when the diode is biased into avalanche breakdown, the field at the periphery of the junction is reduced from that in the bulk; this reduces surface leakage and premature breakdown. The mesa shape also lends itself readily to passivation of the diodes while they are still in wafer form, using ordinary dielectric deposition techniques. We have successfully passivated diodes using both evaporated SiO_2 and SiO_2 deposited by the oxidation of silane at ~300°C.

For comparison, Fig. 1(a) shows the commonly used inverted mesa method for heat sinking a diode. In this case the mesas are etched from the p⁺ side, and after separating the individual diode chips, the chips are inverted and bonded junction side down to the heat sink using heat and pressure; no solder is used. This method, under ideal conditions, produces the best heat sinking for an IMPATT diode; measured thermal resistances for diodes fabricated using this technique can be essentially equal to the ideal (calculated) values [5]. This method has the additional advantage that, where the additional expense and trouble are warranted, materials such as diamond can be used for the heat sink; it is well known now that significant improvements in thermal resistance can be obtained in this way [6]. Much of the benefit of using diamond would be lost using our technique, of course, because the major portion of the thermal resistance occurs in the plated heat sink.

Fig. 2 shows schematically the standard varactor-type package used, with the chip and bonding ribbon. The measured values of thermal resistance θ are shown for a wide range of junction areas. For comparison, the calculated value for θ assuming simple spreading into an infinite copper heat sink plus the θ of the silicon itself is shown as the solid line. For the larger areas, the influence of the solder interface and nonideal thermal spreading can be seen as a deviation from the calculated θ. More exact calculations of θ that take these factors into account are within 10 percent of the measured values. Thermal resistance was measured using the dc avalanche resistance method described by Haitz [5]. Most of our X-band diodes had $A \simeq 2.0 \times 10^{-4}$ cm², hence, $\theta \simeq$ 14–17°C/W.

The fabrication technique described above produces diodes whose thermal resistance is comparable to the ideal obtainable using the inverted-mesa technique; the primary benefits it confers are economy and reproducibility. For diodes with breakdown voltages between 65 and 80 volts, and junction areas of the order of 1.7 – 2.0×10^{-4} cm², output powers of about 0.7 watt at 7 percent efficiency were routinely obtained over all of X band, with junction temperatures below 200°C. With higher junction temperatures, typically 230°C, more than 1.0 watt was achieved. Maximum power from these diodes in our circuits was about 1.3 watts; the limit on power was electrical rather than thermal. It has been

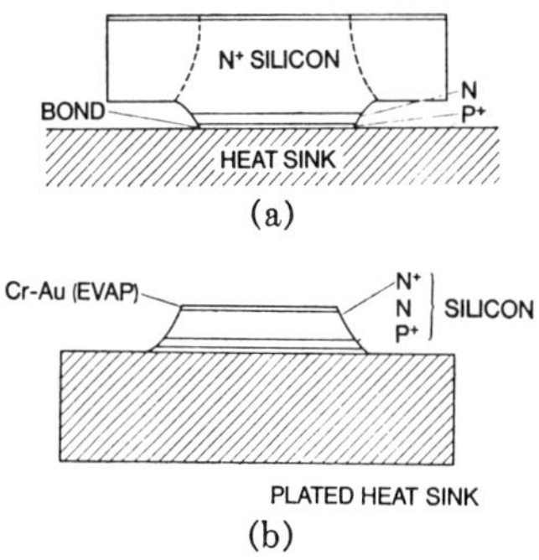

Fig. 1. (a) Conventional inverted-mesa IMPATT-diode heat-sinking technique. The dashed lines indicate the shape of a mesa etched from both sides for improved mechanical stability. (b) Plated heat-sink technique. This chip can be soldered into packages or microstrip circuits using ordinary techniques.

211

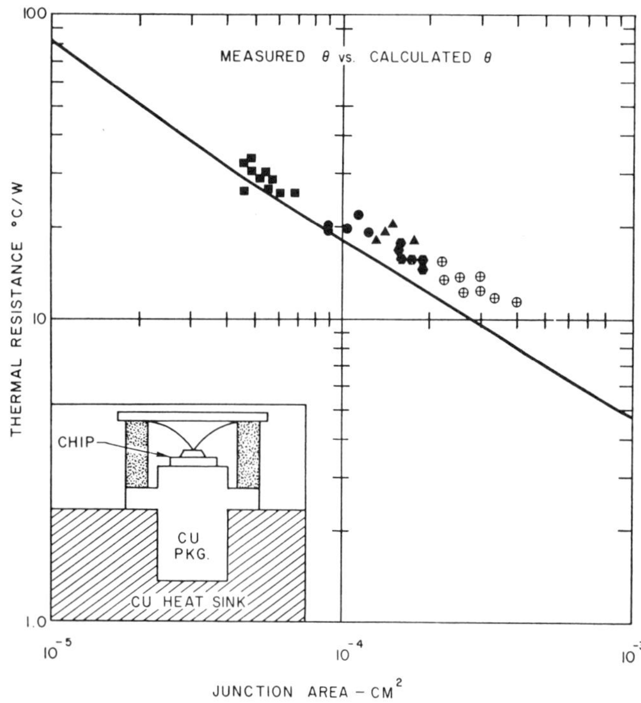

Fig. 2. Measured thermal resistance of plated heat-sink diodes as a function of junction area. The solid line is the calculated thermal resistance of an inverted mesa bonded to a semi-infinite copper heat sink, assuming a perfect bond and 5 μm of silicon between the junction and heat sink.

verified that diode burnout occurs for junction temperatures somewhat above 350°C.

III. Results

The circuit used to obtain all of the results to be reported is shown in Fig. 3. It consists of a length of 7-mm 50-ohm coaxial line with the diode mounted in a clamp at the end for good heat sinking. Tuning and loading are accomplished by adjusting the value and position of a shunt radial capacitor C_L. The value of this capacitor for properly loading in X band typically lies between 1 and 3 pF. The middle picture shows the equivalent circuit, including package parasitics, and the bottom picture shows the equivalent load impedance referred to the chip terminals—we will use the series representation

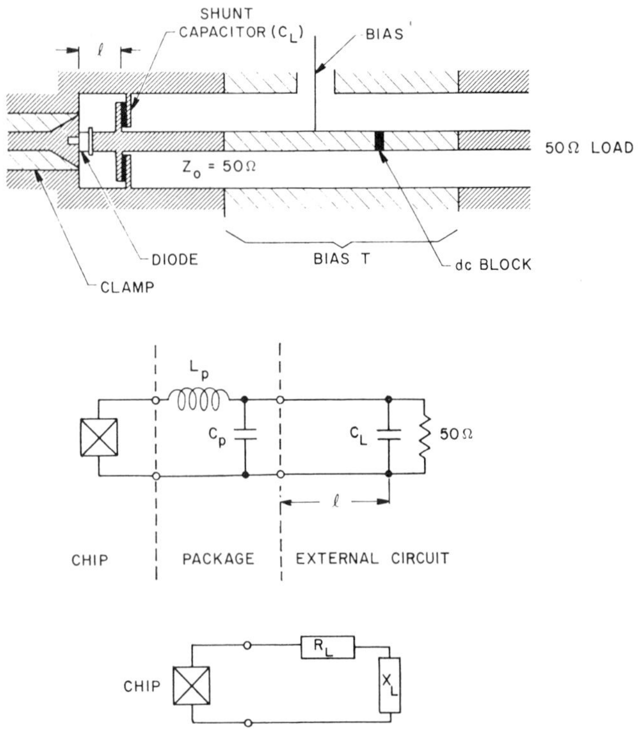

Fig. 3. Schematic illustration of the circuit used in these studies. Package parasitic elements for HPA style 41 package are L_p = 0.6 nH, and C_P = 0.33 pF. Values of C_L for the experiments reported here ranged from 1.0 to 3.0 pF; the length of the transmission line l ranged from 0 to 0.050 inch.

for the diode load. The values of R_L and X_L were determined both by calculation and by means of an HP automatic network analyzer. It was found that R_L and X_L could be calculated quite readily to ± 10 percent accuracy.

For the bias current densities (>200 A/cm^2) used for our studies of diodes in these circuits, the diode reactance, as inferred from the value of X_L and the oscillation frequency, is essentially that of the diode-depletion capacitance at breakdown. The most important circuit parameter in determining diode performance was the value of R_L.

Some results are shown in Figs. 4 and 5. The general behavior shown here is very typical, and is seen in all diodes. As R_L is reduced from some fairly high values, say 3.0 ohms—the power increases slowly, until suddenly a "saturation" effect takes place, and then power drops *linearly* with further reduction in R_L—this linear dependence corresponds to constant ac voltage across the diode. At the same time, the FM noise of the diode increases very rapidly. This trend in behavior is quite reproducible from diode to diode in a wafer, and the value of the "maximum" ac voltage is about the same for similar diodes at the same frequency; for the X-band diodes studied, V_{ac} (max)/V_b is about ⅓ (rms value). This phenomenon seems to be a fundamental one, and has important ramifications in IMPATT-diode circuit design; it indicates that for a given bias current, the series load resistance R_L must not be too small or optimum power will not be achieved, and often more important, noisy oscillations will occur. The noise observed in the RF spectrum when R_L is small is always accompanied by a large increase in noise in the bias circuit, and occasionally by bias circuit oscillations. If R_L is kept above the value for maximum power, the quietest operation is achieved, and the bias circuit has no effect on the stability of the oscillations. In Figs. 4 and 5, the dashed lines to the right of the power maximum correspond to a standard "mismatch" curve, i.e., generator with internal resistance R_S delivering power to a load R_L. The curves were fitted with the measured values of series resistance R_S and the maximum power data point; their method of fitting the curve is somewhat arbitrary, but the reason for trying to fit the data to this curve will become apparent upon reading Section IV and examining Fig. 13. The series resistance was measured in a reduced-height tapered waveguide system using the transmission method [7].

It should be noted that in our simple singly tuned circuits, a *reduction* in R_L causes an *increase* in the value of the external Q, since R_L was varied essentially independently of the reactive elements in the circuit. This fact is somewhat curious, in that we find *worse* noise performance for *higher* Q, other factors remaining the same. This situation is discussed later and is interpreted as a strong dependence of the diode *noise measure* on the value of ac voltage across the diode.

Figs. 6 and 7 show plots of RF power output versus bias current, with R_L as a parameter. The small values of R_L lead to instability (noise) and saturation at relatively low values of I_{dc} and P_o; for larger values of R_L the power at which saturation occurs becomes higher until a maximum is reached. In each case, the rms diode voltage at saturation is approximately the same, ⅓ V_b. It should be emphasized that this is not a thermal effect—it is purely electrical in nature. For all these measurements, T_j was kept below 250°C, and forced air cooling was used on the outside of the cavity. Cavity θ was about 2.5°C/W in addition to ≈16°C/W for the diode itself. Again, the important feature here is that power saturation takes place when the diode ac voltage V_{ac} reaches a certain value—here about 27 volts. When saturation occurs or is imminent, the RF spectrum becomes very noisy, and under certain conditions, bias circuit oscillations can occur. Fig. 7 shows P_o versus I_{dc} for values of R_L large enough such that power does not saturate at safe bias currents. Here, as in Fig. 6, the dependence of P_o on I_{dc} is square law at higher power levels, and much faster near the threshold of oscillation. This is typical behavior, and we will later see that a square-law characteristic conforms to a simple large-signal model for the oscillator.

The effects of power saturation on the RF spectrum are shown in Fig. 8. Photos were taken of the spectrum for R_L = 1.0 ohm, showing the serious degradation of the spectrum as the diode power saturates. Under these RF loading conditions, coherent line spectra accompanied by bias circuit oscillations can also occur if care is

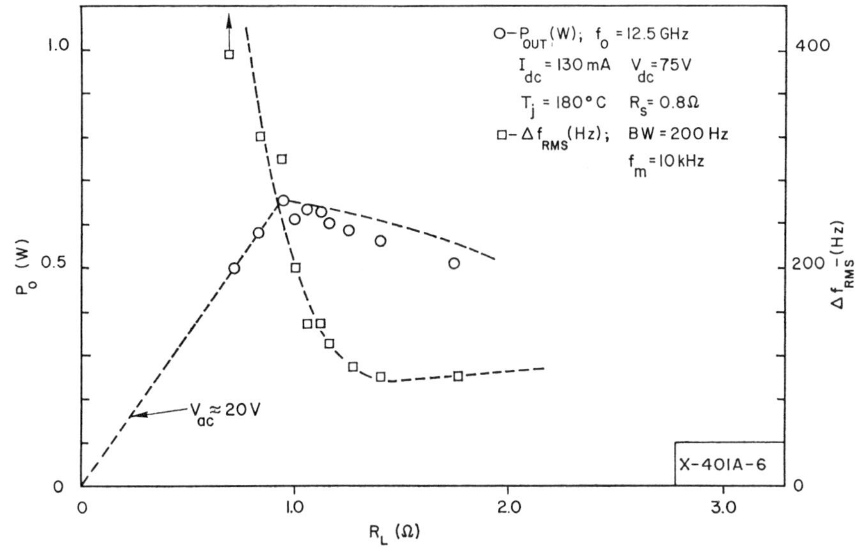

Fig. 4. Typical measurements of output power P_o and FM noise (Δf_{rms} in 200-Hz bandwidth, 10 kHz from the carrier) as a function of series load resistance R_L. Breakdown voltage (V_b) is about 66 volts at 10 μA. Oscillation frequency is 12.5 GHz.

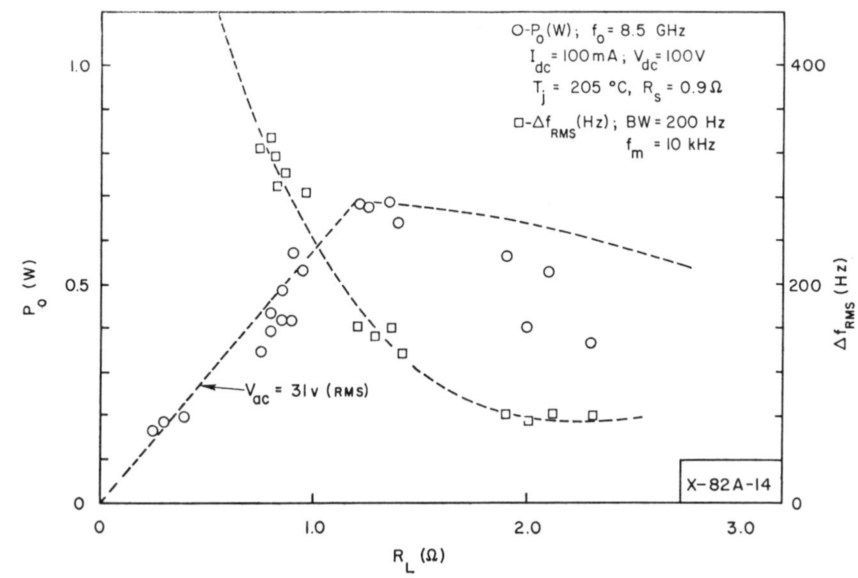

Fig. 5. Data similar to Fig. 4 shown here for a diode with V_b = 80 volts at 10 μA with an oscillation frequency of 8.5 GHz.

not taken to provide a low-Q nonresonant bias circuit. The line spectra are illustrated in the figure, where a series resonance purposely has been introduced; the frequency of the bias circuit oscillation is equal to the separation between lines of the RF spectrum, here about 30 MHz. For loading such that the power is not saturated, the bias circuit is not critical, and bias circuit resonances do not produce oscillations.

The noise performance of an oscillator or amplifier is often stated in terms of the *noise measure* M, which is defined as the ratio of effective diode noise temperature to ambient temperature. We find that M depends mainly on the value of ac voltage across the diode, and in fact seems to be related to the ratio of V_{ac} to V_b, the breakdown voltage; its value is not strongly affected by other parameters, such as bias current. Data are shown in Fig. 9 for several diodes and values of R_L; M varies quite widely with V_{ac}/V_b. Values of M were obtained from measurements of FM noise modulation (Δf_{rms}), and injection locking range δf for a known injection signal power P_i, using the equation[1]

$$M = 4\Delta f_{rms}^2 P_i / \delta f^2 kTB$$

where T is the ambient temperature, k Boltzmann's constant, and B the bandwidth used for measuring Δf_{rms}.

The power saturation phenomenon and its effect on noise measure of the diode are considered to be the most important experimental results found in this study. We will now turn to consideration of possible theoretical explanations for those results.

[1] This equation is derivable from (30) of [8].

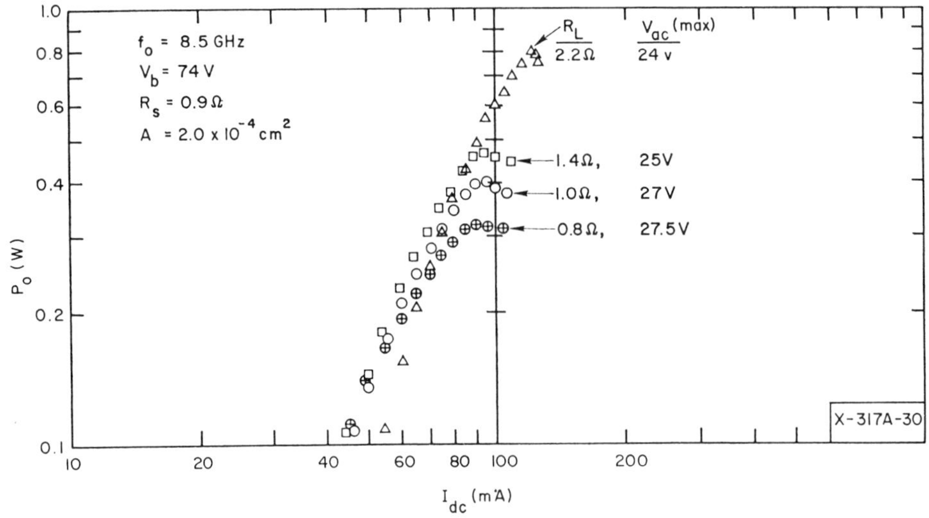

Fig. 6. Power output versus bias current for different values of series load resistance R_L. The power saturates when V_{ac} reaches ~25–27 volts.

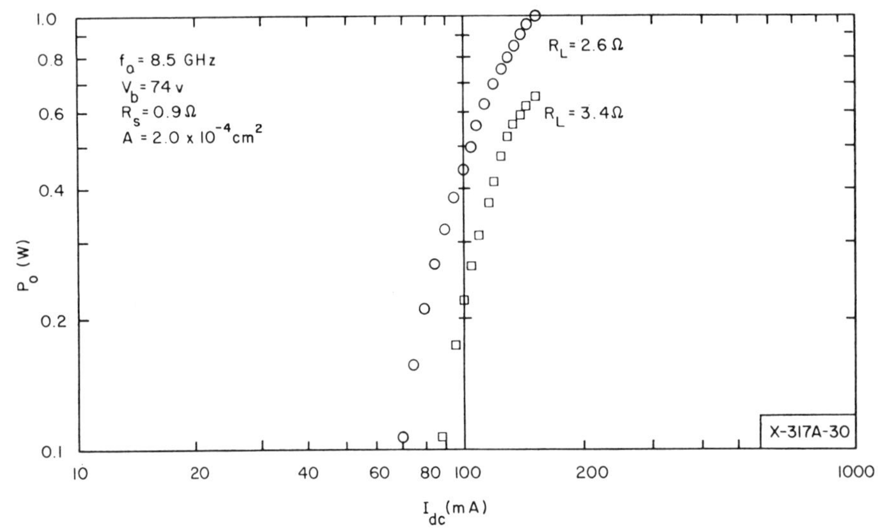

Fig. 7. Power output versus bias current for values of R_L high enough for power saturation not to occur at safe values of bias current.

IV. Analysis

The theoretical framework in which we have tried to understand these results is the large signal model first proposed by Read—his "sharp-pulse" approximation [3]. With certain reservations this can be applied to the abrupt p^+-n-n^+ junction case, where we arbitrarily divide the junction into "avalanche" and "drift" regions shown in Fig. 10. The model is no doubt familiar to most people working in this field; briefly (Fig. 11) we consider the total voltage across the diode to be composed of ac and dc components. The ac voltage swings the field above and below the critical value for breakdown, so that multiplication takes place during the positive half cycle. For large enough ac voltages, the charge generated in the avalanche region approaches a δ function that, due to the inherent inductive property of the avalanche process, is delayed 90° behind the positive ac voltage peak, and occurs when the ac voltage is zero. The charge drifts across the drift space, and induces a current in the external circuit. The time average of the current due to drifting charge is the bias current I_{dc}. The fundamental "in phase" rms component of this current is denoted by I_C and is given by

$$I_C = \sqrt{2} I_{dc} \cdot \frac{1 - \cos(\omega\tau)}{\omega\tau} \tag{1}$$

where τ is the transit time of carriers across the drift region. The current due to particle drift also has a reactive component at the fundamental frequency, and this may be either inductive or capacitive. However, for most practical purposes this component is negligible compared to the displacement current in the diode depletion capacitance I_D, especially for transit angles ($\omega\tau$) near π. In fact, the total current in the diode is mostly displacement current, since I_D is ordinarily much larger than I_C.

The equivalent circuit of Fig. 12(a) is consistent with the above model for the diode; it shows the current I_C, in effect, as the driving source for the diode-depletion

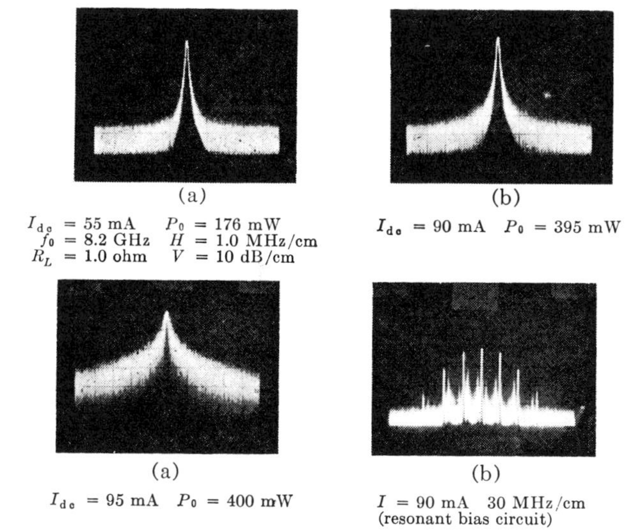

Fig. 8. Behavior of RF spectrum as power saturation is reached. (a) Appearance of the spectrum well below saturation. (b) and (c) Degradation of the spectrum as the power saturates. (c) Effect of connecting a resonant circuit in shunt with the bias port when the power is saturated; bias circuit oscillations occur, which in turn frequency modulate the oscillator, producing sidebands whose spacing is equal to the bias circuit oscillation frequency.

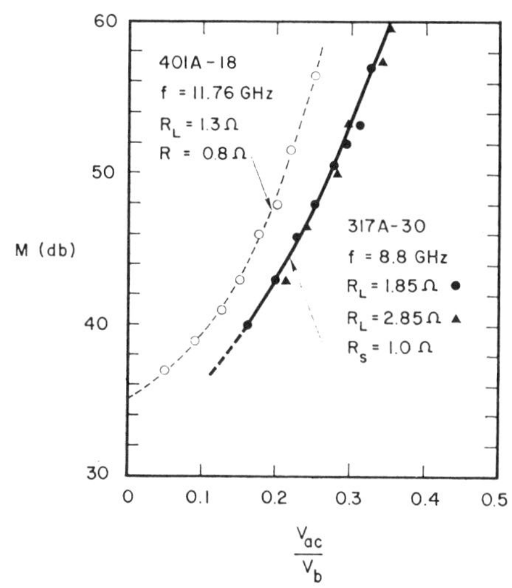

Fig. 9. Noise measure as a function of the rms ac voltage across the diode, V_{ac}.

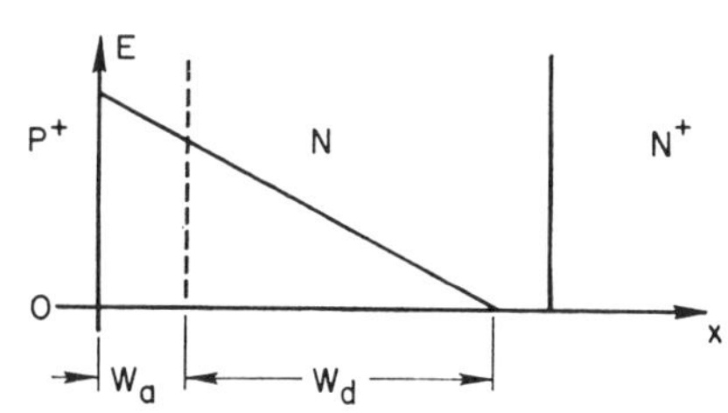

Fig. 10. Electric field versus distance in a reverse-biased p$^+$-n-n$^+$ junction at breakdown. The width of the avalanche region is W_a, and that of the drift region W_d. This E-X diagram corresponds to the case where the diode does not "punch through," i.e., the depletion layer does not reach the n$^+$ substrate at breakdown. In the structures used for our diodes, approximately 0.2 μm of n-type material remains "unswept" at breakdown.

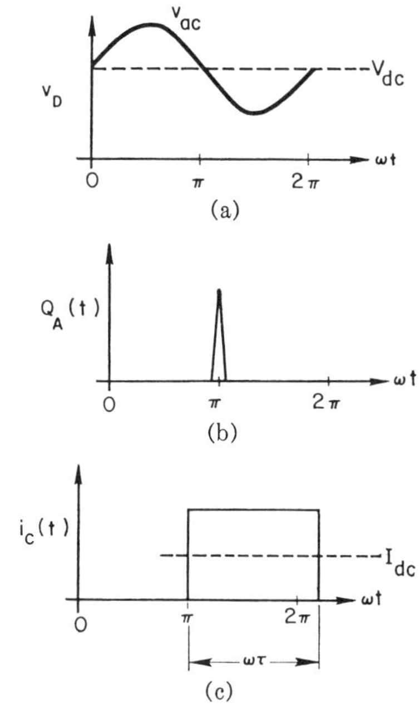

Fig. 11. Read "sharp-pulse" approximation. This model provides qualitative insight into some of our experimental observations. (a) Sinusoidal ac voltage superimposed on the dc voltage, which is assumed to be set equal to V_b. (b) Charge Q_A generated in the avalanche region as a sharply peaked function that *lags* V_{ac} by 90°. (c) Current i_c induced in the external circuit as the Q_A drifts across W_d is shown.

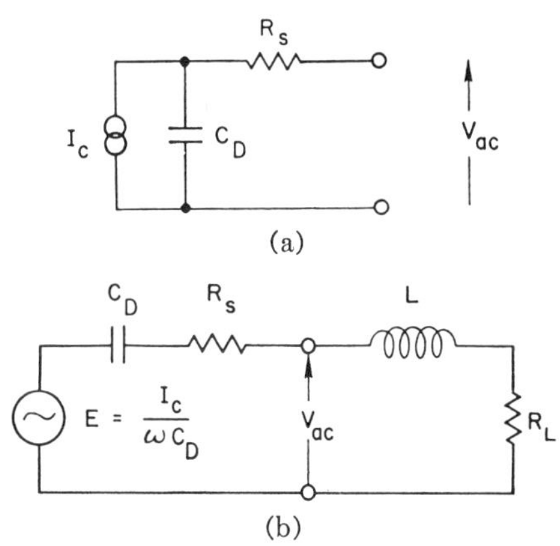

Fig. 12. (a) Equivalent circuit for the diode, based on Read "sharp-pulse" approximation. The in-phase fundamental component of i_c (Fig. 11) is shown driving the diode-depletion capacitance C_D and the external circuit. (b) Circuit transformation allows the power to be calculated practically by inspection.

capacitance in parallel with the external load, where the series resistance is also included. Conversion of the current source I_C into a voltage source [Fig. 12, (b)] allows the power delivered to the external circuit to be derived easily as

$$P_o = \left(\frac{I_C}{\omega C_D(R_S + R_L)}\right)^2 \cdot R_L. \quad (2)$$

When $R_S = R_L$, power is maximum and is

$$P_o(\max) = I_C^2/(4\omega^2 C_D^2 R_S). \quad (3)$$

The above derivations assume implicitly that the diode ac voltage is not limited. If there is a limit on V_{ac} imposed, then in terms of this limit V_{ac} (max), it is easy to verify that when the voltage limit is reached, the power is given by

$$P_o = [\omega C_D V_{ac}(\max)]^2 R_L \tag{4}$$

for a circuit Q greater than 5 or so.

Equations (2) and (4) are shown plotted schematically in Fig. 13. The solid line is (2), and is recognized simply as a "mismatch" curve, which shows the power delivered to a load R_L by a generator with internal resistance R_S; maximum power transfer occurs for $R_L = R_S$. The dashed line, (4), shows the absolute limit on P_o due to the maximum allowed ac voltage V_{ac} (max). The interpretation of this plot is very simple. To the right of the intersection of the two curves, P_o can be calculated from (2), and the power is not saturated, that is, $V_{ac} < V_{ac}$ (max). To the left of the intersection, (4) must be used. Here the power is "saturated," in the sense that further increase in input power (bias current) produces no additional power output, as we saw experimentally in Figs. 6 and 7. By comparison of Fig. 13 and Figs. 4 and 5, it can be seen that there is qualitative agreement between theory and experiment for P_o versus R_L. Also, for the higher powers in Figs. 6 and 7, the power varies roughly as the square of bias current I_{dc} in agreement with (1) and (2).

Using the measured values of R_S in (2), theoretical estimates of the output power were obtained. These were usually 2 or 3 times higher than the measured values of P_o; the reason is probably simply the crudeness of this simple model as a basis for predicting actual IMPATT-diode performance. However, if I_C is adjusted to agree with the measured power at any value of R_L for a fixed value of I_{dc}, the theoretical curve matches the other experimental points very well and the value of R_L at which maximum power is obtained, as well as the value of P_o (max), can be calculated with reasonable accuracy. (The value of I_C typically had to be adjusted to 60 or 70 percent of its theoretical value.) Thus, the equivalent circuits of Fig. 12, and P_o versus R_L curve of Fig. 13 are valid even in the absence of quantitative agreement with the model.

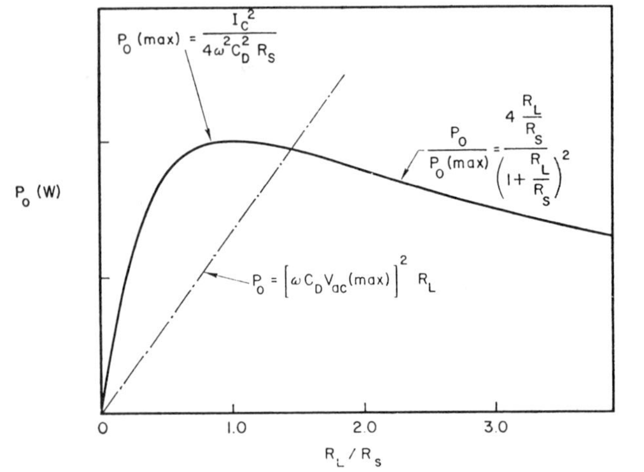

Fig. 13. Predicted P_o versus R_L curves. The solid line is based on the equivalent circuit of the preceding figure, assuming no saturation of V_{ac}. The dashed line is the power delivered to R_L assuming V_{ac} reaches a constant value V_{ac} (max). The interpretation of these curves is that the dashed line is valid to the *left* of the intersection, and the solid line is valid to the *right* of it. There is very good qualitative agreement with Figs. 4 and 5.

The saturation of the diode voltage is apparently fundamental to the IMPATT mode of operation, and similar results have been obtained in computer studies by Scharfetter *et al.* [4], who predict a maximum ac voltage in good agreement with our measured values, viz., rms voltage of about 1/3 the breakdown voltage. The voltage saturation is a self-limiting phenomenon caused by the reduction of the field in the avalanche region by the space charge of the generated carriers.

The measurements reported here were obtained in circuits with relatively low Q, ranging between 10 and 30. The best values of FM noise (Δf_{rms} in 200-Hz bandwidth 10 kHz from the carrier) obtained in these circuits are in the vicinity of 70 Hz, at a power output of 500 mW in X band. With Q values in the range of 100–200, we have obtained $\Delta f_{rms} \approx 10$ Hz with output power of 350 mW. Double sideband AM noise to carrier ratio is typically −125 to −130 dB in a 200-Hz bandwidth, 1.0 kHz–1 MHz from the carrier.

It should be pointed out that in none of this work has any attempt been made to tune higher harmonics of the fundamental oscillation frequency.

V. Discussion

The chief usefulness of the results presented in this paper is that they indicate the circuit conditions necessary to achieve maximum power with reasonable freedom from oscillator noise and instabilities. Basically, for stability and freedom from high noise, the diode must be operated such that its ac voltage is below the "saturated" value, roughly 1/3 V_b for the X-band diodes studied here; in practice this condition is achieved by selecting values of R_L that are somewhat above the value that obtains maximum power. No obvious explanation for the relationship between noise and the onset of saturation has occurred to the authors at the time of writing; further studies are in progress.

VI. Summary and Conclusions

The important results of this paper can be summarized as follows.

1) We have fabricated diodes using a unique new process that provides the heat sink and junction in an integral unit.

2) We have investigated these diodes in simple coaxial circuits, and have observed the P_o versus R_L behavior to be consistent with the existence of a voltage saturation effect in the diodes.

3) Apart from the magnitude of the power in the nonsaturated region, we can account for diode behavior using Read's simple "sharp-pulse" approximation in conjunc-

tion with an assumed "saturation" voltage V_{ac} (max). For the X-band diodes we have studied ($V_b \approx$ 65–80 volts, $A \simeq 2 \times 10^{-4}$ cm^2), the rms value of V_{ac} (max) is roughly ⅓ the breakdown voltage.

4) Noisy diode operation is associated with ac voltage across the diode approaching V_{ac} (max).

The practical consequences of the results presented here are that they allow the circuit designer to predict quite accurately the impedance he must present to a diode at the fundamental frequency for optimum noise and power. The relationship between noise and ac voltage across the diode has been found particularly useful by the authors, and has provided the insight into circuit behavior necessary for design of oscillators with reasonably high powers (>500 mW) and good noise performance with low Q. The ability to obtain low noise with low Q is particularly important in AM and FM telecommunications systems, where IMPATT diodes can be used as injection phase-locked power oscillators in the transmitter. The higher noise of the IMPATT oscillator (compared to varactor multipliers or vacuum tube oscillators with the same power) can be reduced to acceptable level by the locking signal in this case [9]. The noise data for the oscillators built during this study (Δf_{rms} = 70 Hz in 200-Hz bandwidth, $P_o \cong$ 0.5 watt, $Q \simeq 10$) indicate that they would be suitable for most military and commercial telecommunications transmitter applications, when injection phase locked by a quiet locking signal. As a specific example, a typical specification on the maximum allowable value of Δf_{rms} in 200-Hz bandwidth, 10 MHz from the carrier, would be about 5 Hz for a commercial FM telecommunications transmitter; a 10-GHz IMPATT oscillator with the parameters given above, according to [9], could meet this specification when locked by a quiet 20-mW injection signal, with frequency equal to the free-running IMPATT oscillator frequency.

The significance of the technique used to fabricate our diodes is that, from a manufacturing and reliability viewpoint, it will allow IMPATT diodes to compete economically with all of the present-day solid-state devices used to generate microwave power above about 6 GHz, including varactors, step-recovery diodes, transistors, and Gunn effect devices.

The limitations of the analysis given are as follows.

1) It is confined to consideration of voltages and currents at the fundamental frequency of oscillation; harmonics and subharmonics are not considered. This limitation is a reasonable one for single-tuned circuits, but of course will not be suitable for complex circuits having more than one resonance.

2) The model for the analysis, while providing conceptual insight, is not accurate; qnantitative predictions of diode performance in our circuits would require more sophisticated numerical analysis using a computer.

3) The analysis only serves to indicate the *existence* of a relationship between diode noise and ac voltage; a large-signal noise theory is needed to *explain* the relationship.

Acknowledgment

A number of people made significant contributions to this work. C. A. Liechti designed the measurement technique for series resistance. The diodes were fabricated by Mrs. M. Bryan and Mrs. E. Petherbridge. R. C. Patterson, M. R. Gavette, and G. Hom made the measurements and assisted in various aspects of diode fabrication. Discussions with J. Josenhans of Bell Telephone Laboratories were very helpful in understanding the measurement of the noise measure. The authors would also like to thank the reviewers of this paper, who made a number of helpful suggestions.

217

References

[1] S. M. Sze, *Physics of Semiconductor Devices.* New York: Wiley, 1969, ch. 5.

[2] R. A. Zettler and A. M. Cowley, "Batch fabrication of integral-heat-sink IMPATT diodes," *Electron. Lett.*, vol. 5, pp. 693–694, December 27, 1969.

[3] W. T. Read, Jr., "A proposed high-frequency negative-resistance diode," *Bell Syst. Tech. J.*, vol. 37, p. 401, 1958.

[4] D. L. Scharfetter and H. K. Gummel, "Large-signal analysis of a silicon read diode oscillator," presented at the Solid-State Devices Research Conf., Evanston, Ill., June 1966; also *IEEE Trans. Electron Devices*, vol. ED-16, pp. 64–77, January 1969.

[5] R. H. Haitz *et al.*, "A method for heat flow resistance measurements in avalanche diodes," *IEEE Trans. Electron Devices*, vol. ED-16, pp. 438–444, May 1969.

[6] C. B. Swan, "Improved performance of silicon avalanche oscillators mounted on diamond heat sinks," *Proc. IEEE* (Correspondence), vol. 55, pp. 1617–1618, September 1967.

[7] H. A. Watson, Ed., *Microwave Semiconductor Devices and Their Applications.* New York: McGraw-Hill.

[8] K. Kurokawa, "Noise in synchronized oscillators," *IEEE Trans. Microwave Theory Tech.*, vol. MTT-16, pp. 234–240, April 1968.

[9] M. E. Hines, J. C. Collinet, and J. G. Ondria, "FM noise suppression of an injection phase-locked oscillator," *IEEE Trans. Microwave Theory Tech.*, vol. MTT-16, pp. 738–743, September 1968.

Highly Stabilized Half-Watt IMPATT Oscillator

SHIGEMICHI NAGANO AND HIDETOSHI KONDO

Abstract—**An *X*-band IMPATT oscillator having a stabilized output power of over 0.5 watt has been developed. The oscillator consists of a main cavity and a directly coupled reaction-type cavity for stabilization. The oscillator has a frequency stability of 2.6×10^{-5} over a temperature variation ranging from 0° to 50°C and an rms noise deviation of 15 Hz/1-kHz bandwidth at 500 kHz from the carrier. Design considerations have been made concerning the admittance characteristics of the circuit and of the diode to determine preferable circuit conditions for stabilization. The output-power loss due to the stabilization is as small as 0.4 dB. The oscillator is capable of operation in a frequency range of 10.7 to 11.7 GHz.**

I. Introduction

MICROWAVE solid-state oscillators such as IMPATT and Gunn oscillators are coming into practical use. These are very impressive devices but have several practical problems. They include insufficient frequency stability for temperature variations, insufficient output-power level as a transmitter oscillator, and higher noise level. Currently, the output-power level of commercially available IMPATT diodes are improving, and more than several watts is obtainable in the laboratory. If the frequency stability and noise problems are solved without greatly increasing the oscillator size, the IMPATT oscillator can replace conventional oscillators in many applications.

We have developed a compact *X*-band IMPATT oscillator that provides more than a half-watt of stabilized output power at 11 GHz. The stability and noise problems have been solved by employing an oscillator circuit that has a directly coupled reaction-type cavity for stabilization. The oscillator has the frequency stability of 2.6×10^{-5} over a temperature variation ranging from 0° to 50°C and an rms noise deviation of 15 Hz in a 1-kHz bandwidth at 500 kHz from the carrier. The output power is only 0.4 dB lower than that without stabilization. Use of a directly coupled stabilizing cavity makes the oscillator size as small as that of a reflex klystron.

In this paper the authors describe the construction of the oscillator and present design considerations based on a theoretical analysis for the admittance of the oscillator circuit and of the diode. Circuit conditions required for stabilization and prevention of mode jumping are discussed theoretically and experimentally. Problems of mode jumping at power switch-on, and FM noise reduction and output-power loss due to stabilization are also discussed in this paper.

Manuscript received May 9, 1970; revised August 24, 1970.
The authors are with the Central Research Laboratories, Nippon Electric Company, Kawasaki, Japan.

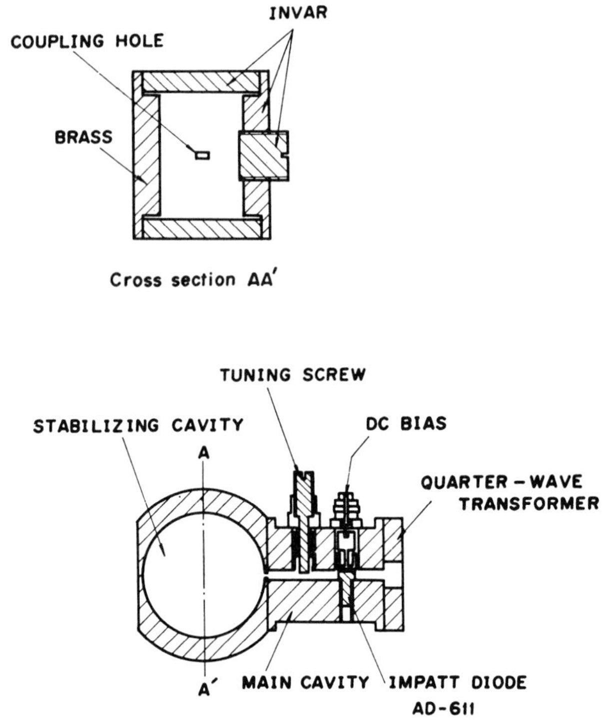

Fig. 1. Drawing of stabilized IMPATT oscillator.

II. Oscillator Configuration

The oscillator consists of a commercially available IMPATT diode (AD-611, NEC), mounted in a rectangular waveguide cavity, and a directly coupled cylindrical cavity for stabilization. The construction of the oscillator is illustrated in Fig. 1. The cylindrical cavity is made of invar and brass, the latter being utilized to compensate dimensional variations due to wide-range changes of temperature. The cavity resonates in the TE_{011} mode and its unloaded Q is about 20 000. The main cavity and the stabilizing cavity are tunable individually from 10.7 to 11.7 GHz by a tuning screw. The two cavities are coupled directly through a coupling hole, and the main cavity is coupled to a standard *X*-band waveguide through a quarter-wave transformer.

III. Design Considerations

To design a highly stabilized practical oscillator, various characteristics other than the frequency stability itself have to be considered. These include mode jumping during operation as well as switching, frequency adjustability, output-power loss, and others.

In a coupled-cavity oscillator there exist more than one possible oscillation mode. To design an oscillator

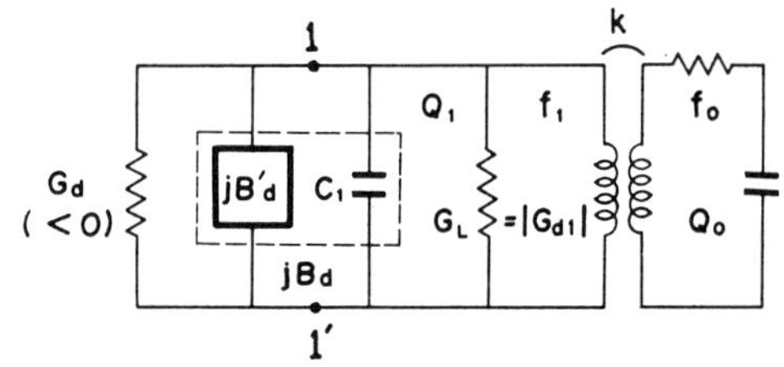

Fig. 2. Equivalent circuit of oscillator.

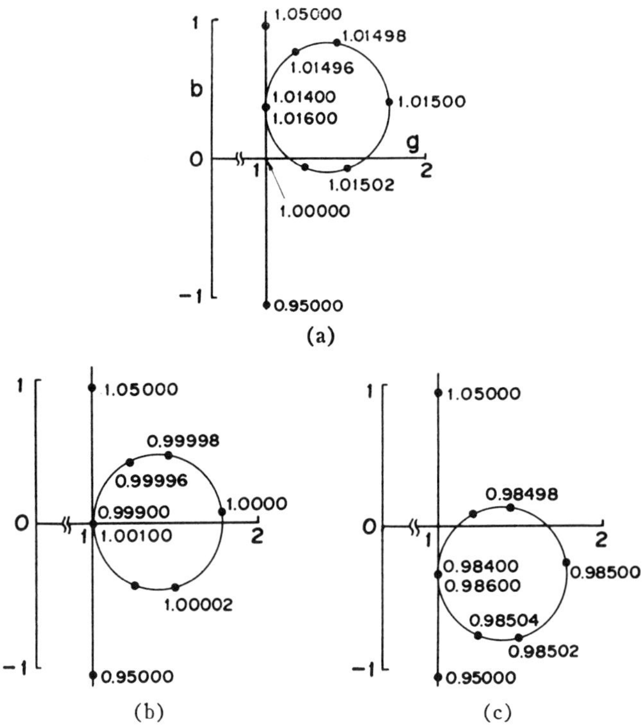

Fig. 3. Normalized admittance loci of circuit. Numerical values written on loci represent f/f_1 values. (a) $f_0/f_1=1.015$. (b) $f_0/f_1=1.0$. (c) $f_0/f_1=0.985$.

circuit as described in Section II, it is necessary to understand the tuning characteristics of the oscillator, or the correlation between the oscillation frequency, and the resonant frequency of the main cavity and of the stabilizing cavity. A theoretical consideration has been made on the tuning characteristics to determine a preferable mode for stabilization and to find necessary circuit conditions.

A simplified equivalent circuit of the oscillator is shown in Fig. 2. In the figure the diode is represented by the conductance G_d and the susceptance B_d (encircled by a dotted-line rectangle); G_L is the load conductance with which maximum power is obtained when $k=0$. The stabilizing cavity, having its own resonant frequency of f_0 and the unloaded Q of Q_0, is coupled to the main cavity with a coupling coefficient k. When $k=0$, the resonant frequency f of the oscillator is equal to f_1 and its external Q is Q_1. The capacitance C_1 represents the equivalent capacitance of the diode for $k=0$, and B_d' is the excursion of the diode susceptance B_d at frequency f from that due to C_1, or

$$B_d' = B_d - 2\pi f C_1. \tag{1}$$

The normalized admittance Y_{in}/G_L looking into the right-hand side at terminal 1–1′ is given as

$$Y_{\text{in}}/G_L = g + jb \tag{2}$$

where

$$g = 1 + \frac{Q_1}{Q_0} \cdot \frac{k^2 f_0 f_1 f^2}{[(1-k^2)f^2 - f_0^2]^2 + f_0^2 f^2/Q_0^2} \tag{3}$$

$$b = Q_1 \frac{f_1}{f} \left\{ \frac{f^2}{f_1^2} - \frac{(f^2-f_0^2)[(1-k^2)f^2-f_0^2] + f_0^2 f^2/Q_0^2}{[(1-k^2)f^2-f_0^2]^2 + f_0^2 f^2/Q_0^2} \right\}. \tag{4}$$

If we take $k^2 \ll 1$ and $k^2 Q_0^2 \gg 1$, then as shown in Fig. 3, the locus of the admittance at frequencies very close to f_0 forms a loop that is almost a circle in the b–g plane. The coordinate of the center of the circle (g_c, b_c) and the radius of the circle r_c are given by

$$(g_c, b_c) = \left[1 + \frac{1}{2} k^2 Q_0 Q_1 \frac{f_1}{f_0},\ 2Q_1 \left(\frac{f_0}{f_1} - 1 \right) \right] \tag{5}$$

$$r_c = \frac{1}{2} k^2 Q_0 Q_1 \frac{f_1}{f_0}. \tag{6}$$

The normalized admittance Y_d/G_L looking into the left-hand side at terminal 1–1′ is expressed as

$$Y_d/G_L = g_d + jb_d \tag{7}$$

where g_d and b_d are equal to G_d/G_L and B_d'/G_L, respectively. As far as a narrow frequency range is concerned, the normalized diode admittances g_d and b_d can be approximated as follows by considering large-signal characteristics of IMPATT diodes [1], [2].

$$-b_d = (-g_d - 1)\tan\theta - a\left(\frac{f}{f_1} - 1 \right) \tag{8}$$

where

$$\theta = \tan^{-1} \left. \frac{dB_d}{dG_d} \right|_{f=f_1, |G_d|=G_L} \tag{9}$$

and a is a positive constant.

The gradient $\tan\theta = (dB_d/dG_d)$ is almost always positive for normal IMPATT operation [1], [2]. The normalized diode admittance following (8) and (9) is schematically shown in Fig. 4.

It may be obvious by superposing Fig. 3 on Fig. 4 that there are three frequencies, f_-, f_u, and f_+ from lower to higher, that satisfy the condition $b+b_d=0$ and $g+g_d=0$. Since f_u corresponds to an unstable mode (3), there exist in general two possible oscillation modes, the f_- mode and f_+ mode. It can also be understood that there is a higher frequency limit for the f_+ mode and a lower frequency limit for the f_- mode. If one tries to change the oscillation frequency f beyond the limits, there occurs a mode jump between the two modes. Graphically, the critical condition for the mode jump is that the straight line for Y_d/G_L is tangent to that for Y_{in}/G_L. The circuit condition for the f_+ and f_- modes free from mode jump-

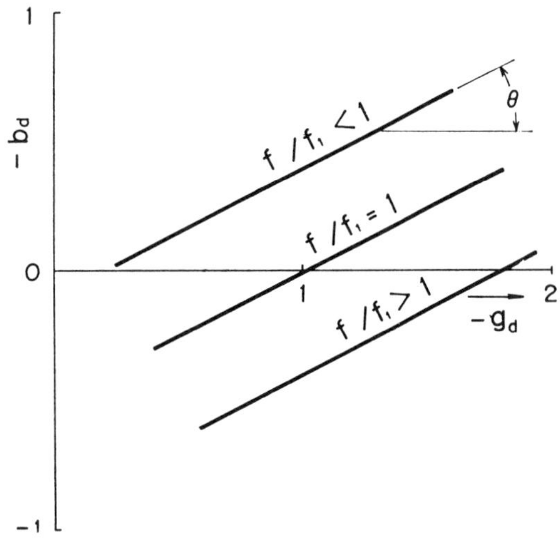

Fig. 4. Normalized diode admittance, $-b_d$ versus $-g_d$ characteristics.

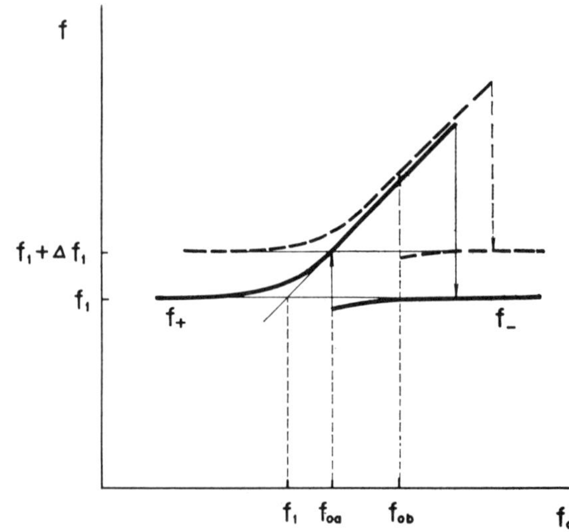

Fig. 5. Tuning characteristic for a fixed f_1.

220 ing is derived analytically by using (5)–(9) as

$$\left(\frac{f_0}{f_1}\right)_\pm \lessgtr 1 \pm \frac{1}{4} \frac{k^2 Q_0}{1 + a/2Q_1} F_\pm(\theta) \tag{10}$$

where

$$F_\pm(\theta) = (1 \pm \sin\theta)/\cos\theta. \tag{11}$$

In the derivation it was assumed $k^2 Q_0 F_\pm(\theta)/(1+a/2Q_1) \ll 1$ and $f \fallingdotseq f_0$. Since $\tan\theta$ is positive for IMPATT diodes, $F_+(\theta)$ is greater than $F_-(\theta)$. The variation of the oscillation frequency f against the variation of f_0 is shown schematically in Fig. 5 by solid lines for a fixed f_1. It can be understood from Fig. 5 by considering that in a situation where f_1 is perturbed while f_0 is fixed a greater frequency stability results with a larger gradient of the tuning curve of Fig. 5 except in the close vicinity of the critical points at which mode jumping occurs. Therefore, in order to obtain good stability, the oscillator should be operated in the f_+ mode.

In practical oscillators f_1 is not constant but varies as the circumferential conditions such as temperature, bias current, phase of load, etc., are varied. On the other hand, when the stabilizing cavity is temperature compensated as described in Section II, the variation of f_0 is negligibly small as compared with that of f_1. In such cases the maximum allowed value of $(f_0 - f_1)$ for the oscillator to operate in the f_+ mode reduces by δf_1, the decrement of f_1. In order to obtain high stability, the value of $(f_0 - f_1)$ should be much greater than δf_1. The value of $(f_0 - f_1)$ required for the f_+ mode is limited by k value, since the radius of the circle in Fig. 3 is proportional to k^2. Once the $(f_0 - f_1)$ value is specified, one can obtain the minimum value of k, $k_{\min}$ from (10) as

$$k_{\min} = 2\sqrt{\frac{\Delta f_{10}}{f_1} \cdot \frac{1 + a/2Q_1}{Q_0} \cdot \frac{1}{F_+(\theta)}} \tag{12}$$

where Δf_{10} represents the value of $(f_0 - f_1)$. This relation is numerically illustrated in Fig. 6.

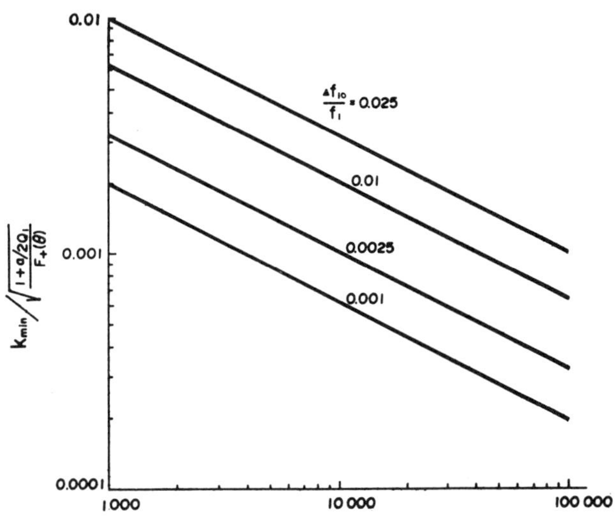

Fig. 6. Minimum value of k as a function of Q_0.

Frequency variation of the stabilized oscillator decreases by a factor S against the value obtained when $k=0$, if δf_0, the variation of f_0, is negligibly small as compared with f_1/S. In this case S is expressed by

$$S = 1 / \frac{df}{df_1}. \tag{13}$$

In ordinary IMPATT oscillators the value of $\delta f_1/f_1$ is about 10^{-3} for 50°C temperature variation. To obtain temperature stability on the order of 10^{-5}, S has to be as large as 100, while $\delta f_0/f_0$ has to be much smaller than 10^{-5}. It can be understood from Fig. 7 that the former condition can be easily obtained under reasonable circuit conditions. The latter condition can be accomplished by the cavity described in Section II.

In order to avoid the operation in the f_- mode invariably when the power supply switch is on, f_0 has to be lower than the critical frequency f_{0a} (refer to Fig. 5). In practical IMPATT oscillators, however, the temperature of the semiconductor element rises gradually to

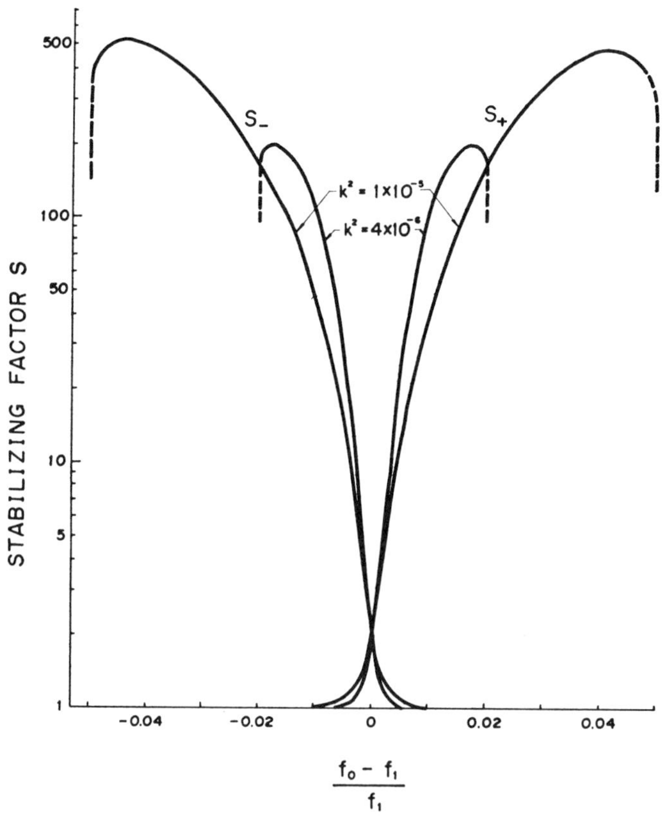

Fig. 7. Stabilizing factor S as a function of $(f_0-f_1)/f_1$. Susceptance B_d' is assumed to be zero; $Q_1=10$, $Q_0=20\,000$.

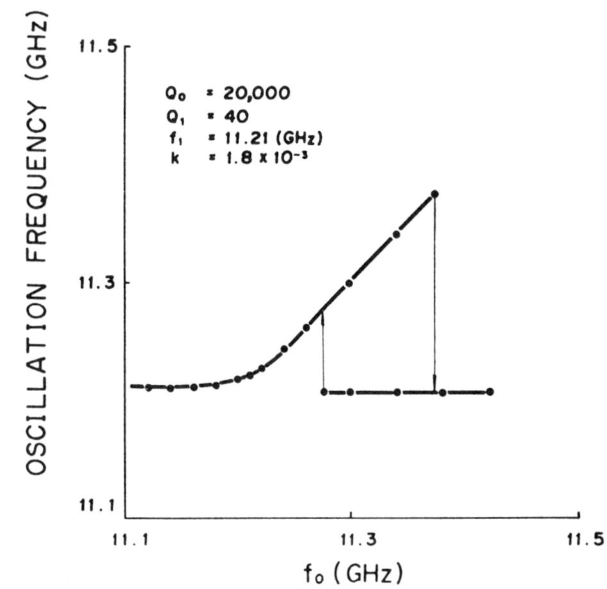

Fig. 8. Experimental tuning characteristic.

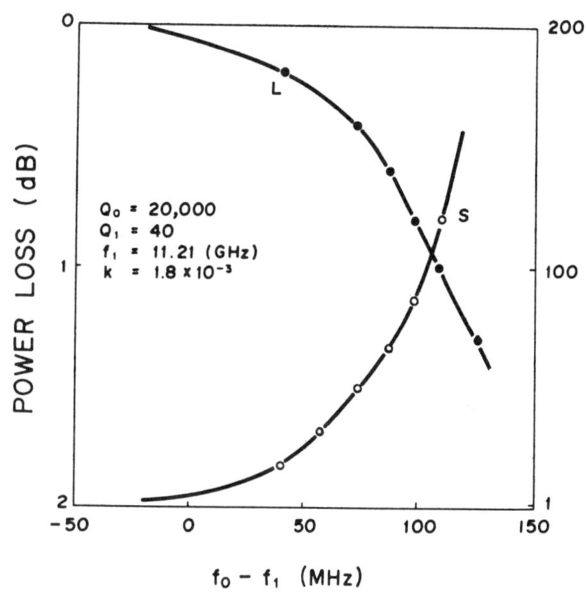

Fig. 9. Stabilizing factor S and output-power loss L as functions of f_0-f_1 when oscillator is operated in f_+ mode.

reach thermal equilibrium after the switch is on. When the temperature is lower, the tuning curves in Fig. 5 shift to the upper right direction as shown by the bold dotted lines in Fig. 5. Therefore, if f_0 is set lower than f_{0b} (refer to Fig. 5), the oscillation of the f_+ mode builds up invariably at the instant of switching on. This implies that we can obtain a larger value of S when we use f_{0b} instead of f_{0a}, since the slope of the tuning curve at f_{0b} is steeper than that at f_{0a}.

IV. Experimental Results

An X-band IMPATT oscillator described in Section II was investigated experimentally. The oscillator was at first optimized by means of a quarter-wave transformer at frequency f_1 by setting the sufficiently detuned stabilizing cavity and subsequently tuning it. The coupling coefficient k was taken as 1.8×10^{-3} at 11.2 GHz. A typical tuning characteristic is shown in Fig. 8. The characteristic is essentially consistent with the one shown in Fig. 5 and verifies the validity of the design consideration described in Section III.

As the resonant frequency f_0 of the stabilizing cavity is tuned to increase the stabilizing factor, the output power decreases because the power is absorbed by the cavity, and the load conductance which the diode sees deviates from the optimum value. The stabilizing factor S and the power loss L of the oscillator operated in the f_+ mode are shown in Fig. 9. The stability factor S was determined by measuring relative values of the pushing figure with respect to the values obtained when the oscillator was unstabilized.

When readjustment for optimum conditions after power switch-on is permitted, (f_0-f_1) can be made higher than $(f_{0b}-f_1)$. In this case (f_0-f_1) is limited only by δf_1 which is influenced by the load mismatch, bias current, and temperature. The critical values of (f_0-f_1) of this oscillator for the variation of these parameters are shown in Table I together with the corresponding stabilizing factor S and power loss L.

When the oscillator is fixed, or when the readjustment after the switch-on is inhibited, f_0 must be lower than f_{0b} which is 73 MHz as given in Table I. The corresponding stabilizing factor and the power loss are 50 and 0.4 dB, respectively.

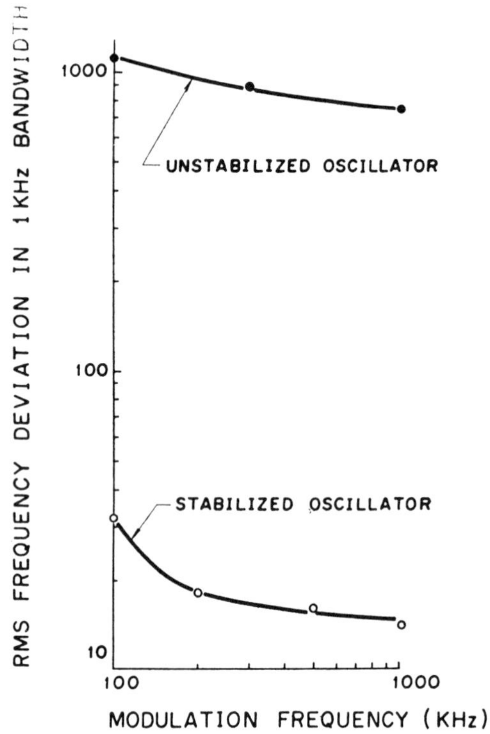

Fig. 10. FM noise of stabilized and unstabilized oscillator.

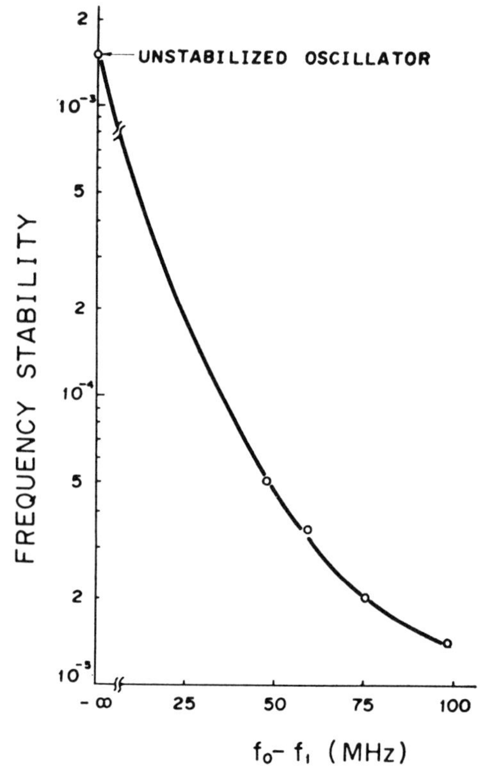

Fig. 11. Frequency stability against temperature variation of 0° to 50° C as function of (f_0-f_1).

TABLE I

CRITICAL VALUES OF THE FREQUENCY DIFFERENCE (f_0-f_1) FOR THE OSCILLATOR TO OPERATE IN f_+ MODE, AND CORRESPONDING STABILIZING FACTOR S AND POWER LOSS L

Conditions	f_0-f_1 (MHz)	S	L (dB)
Load phase (VSWR = 1.5): 0→2π	120	160	1.2
Bias current: 110→109 mA	105	105	0.9
Temperature: 0→50°C	80	57	0.5
Power supply: off→on	73	50	0.4

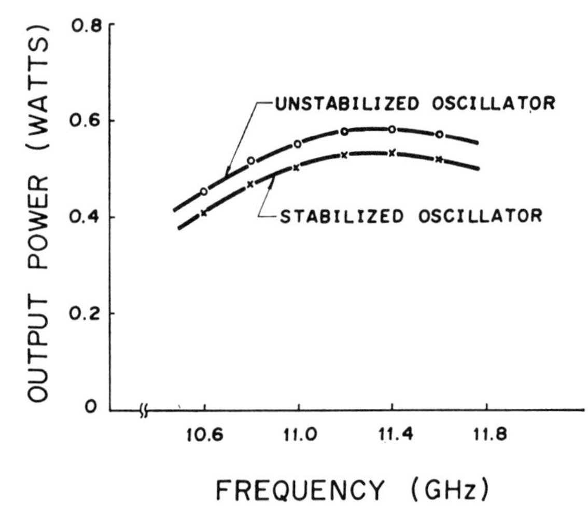

Fig. 12. Output power versus frequency characteristics of stabilized and unstabilized oscillator.

For an oscillator stabilized with a conventional transmission-type cavity, the power loss due to the stabilization is considerably higher, usually about 6 dB to achieve the same S [4]. The obtained power loss of 0.4 dB is remarkably small as compared with that obtained using a transmission-type cavity.

The FM noise of the oscillator is greatly reduced by the addition of the stabilizing cavity. Fig. 10 shows rms frequency deviations of the oscillator under unstabilized $(k=0)$ and stabilized conditions as functions of modulation frequency. The rms noise deviation which is 15 Hz in a 1-kHz bandwidth at 500 kHz from the carrier is improved by a factor of as much as 50 at the sacrifice of only 0.4 dB of power.

The frequency stability against the temperature variation of 0° to 50°C is plotted in Fig. 11 as a function of (f_0-f_1). When (f_0-f_1) was taken to be 98 MHz, a stability of 1.4×10^{-5} was obtained for the 50°C temperature variation. If the oscillator is fixed, however, (f_0-f_1) must be lower than 73 MHz, and in this case the stability was 2.6×10^{-5}.

The oscillator can be tunable from 10.7 to 11.7 GHz by adjusting the resonant frequency f_0 of the stabilizing cavity over the same frequency range and at the same time by varying the resonant frequency f_1 of the main cavity keeping about 70 MHz lower than f_0.

The output-power versus frequency characteristics of the stabilized and unstabilized oscillator are shown in Fig. 12. The power loss due to the stabilization is almost 0.4 dB in the relevant frequency range.

V. Conclusion

A highly stabilized X-band IMPATT oscillator with a half-watt output power has been developed. The oscillator is stabilized by a temperature-compensated reaction-type high-Q cavity that is directly coupled to the main cavity. The oscillator has a frequency stability of 2.6×10^{-5} over the temperature variation of 0° to 50°C and an rms FM noise deviation of 15 Hz/1-kHz bandwidth at 500 kHz from the carrier. The oscillator is capable of operation in a frequency range of 10.7 to 11.7 GHz.

Design considerations have been made concerning admittance characteristics of the circuit and of the diode to determine preferable circuit conditions for stabilization. A stabilizing factor of about 50 was obtained experimentally with output-power loss due to stabilization of only 0.4 dB.

The oscillator can replace conventional oscillators in transmitter oscillators and in many other applications.

Acknowledgment

The authors wish to thank Dr. M. Uenohara and Dr. H. Murakami for their encouragement and guidance, Dr. I. Uchida for the supply of diodes, and Dr. K. Sekido and his colleagues for many fruitful discussions.

References

[1] D. L. Scharfetter and H. K. Gummel, "Large-signal analysis of a silicon Read diode oscillator," *IEEE Trans. Electron Devices*, vol. ED-16, pp. 64–77, January 1969.
[2] M. Matsumura, private communication.
[3] K. Kurokawa, "Some basic characteristics of broadband negative resistance oscillator circuits," *Bell Syst. Tech. J.*, vol. 48, pp. 1937–1955, July–August 1969.
[4] J. R. Ashley and C. B. Searles, "Microwave oscillator noise reduction by a transmission stabilizing cavity," *IEEE Trans. Microwave Theory Tech.*, vol. MTT-16, pp. 743–748, September 1968.

Large-Signal Operation of Avalanche-Diode Amplifiers

ERNST F. SCHERER, MEMBER, IEEE

Abstract—**The effects of large signals on the performance of avalanche amplifiers under various operating and circuit conditions are investigated. Numerical results from a large-signal computer program based on an approximate nonlinear analysis are compared with experiment. It is found that the resulting distortions in the gain response are particularly pronounced in multiple-tuned amplifier circuits. Transfer characteristics of avalanche amplifiers are examined and intermodulation measurements compared with those of TWTs. Preliminary data indicate that avalanche amplifiers compare favorably with TWTs in their distortion characteristics.**

I. INTRODUCTION

224

SOLID-STATE circuits capable of small-signal amplification at microwave frequencies, such as tunnel-diode and parametric amplifiers, have been in existence for many years. It has not been until recently that semiconductor devices with potential high-power amplifying properties, such as the avalanche diode and transferred electron device, have become available. Several interesting circuit concepts have already been realized using avalanche devices that promise to open up a vast new area of systems applications. It is unlikely that the avalanche diode (or any other solid-state device) will simply replace tube devices in the near future. Rather, solid-state amplifier technology will be applied selectively to perform specific systems functions while fully capitalizing on the advantages of the all-solid-state approach. For this reason and to assure successful implementation in new systems concepts, a precise knowledge of the characteristics and limitations of these devices is desirable.

The avalanche diode working in the transit-time mode may be classified, at least phenomenologically, as a negative-resistance device, and some of the circuit design concepts developed for tunnel-diode amplifiers can be readily adopted. However, there are enough differences in the electrical characteristics between a tunnel diode and an avalanche diode to warrant further investigation. Some important features of the avalanche diode are as follows.

1) The avalanche-diode junction does not exhibit a significant negative resistance below a certain transition frequency f_a, which is below the optimum band of operating frequencies as an amplifier.

2) As a result of transit-time effects, the magnitude of the negative resistance varies significantly over the useful range of operation. This is in contrast to the tunnel-diode junction whose negative resistance is practically independent of frequency.

3) Due to the unique negative-resistance mechanism, both the real and imaginary parts of the diode impedance are strongly dependent on bias current and signal level. The bias current itself may depend on the signal level unless supplied from a constant-current source.

The fact that the avalanche diode does not exhibit significant negative resistance below the frequency range of operation simplifies the problem of stabilization compared to that encountered in tunnel-diode amplifiers. Avalanche amplifiers usually do not require out-of-band network stabilization. The implication of feature 2 is that the wide-band performance of avalanche amplifiers is severely degraded by the frequency dependence of the negative resistance unless a suitable form of compensation can be devised. The broad-band design problem has been investigated elsewhere [1], [2].

In this paper consideration is given to the large-signal operation of avalanche amplifiers and the resulting distortions of the gain response. The effect of large signals on the transfer characteristics are investigated, and results of intermodulation-distortion measurements are compared with typical data from TWT amplifiers. A considerable part of the paper is devoted to the description of a large-signal computer program based on an approximate large-signal analysis of the avalanche diode. This program has proven to be very useful in predicting the large-signal performance of avalanche amplifiers and in illuminating some of the basic nonlinear mechanisms peculiar to the avalanche diode. In general, the predicted large-signal behavior has been found to be in close agreement with experimental results.

II. SOME GENERAL CONSIDERATIONS OF LARGE-SIGNAL AMPLIFICATION WITH AVALANCHE DEVICES

In order to realize the full high-power capability of amplifiers, they have to be operated close to output-power saturation and the behavior becomes highly nonlinear. Under these conditions rigorous analysis becomes difficult and it may be necessary to resort to approximate methods and computer simulations such as described in this paper. However, some basic understanding may be gained from studies of a simple model. In its simplest form such a model could consist of a frequency-independent negative conductance whose magnitude is

Manuscript received April 27, 1970; revised September 9, 1970. This work was supported in part by the U.S. Army Electronics Command, Fort Monmouth, N.J.

The author is with Sylvania Electric Products, Inc., Woburn, Mass. 01801.

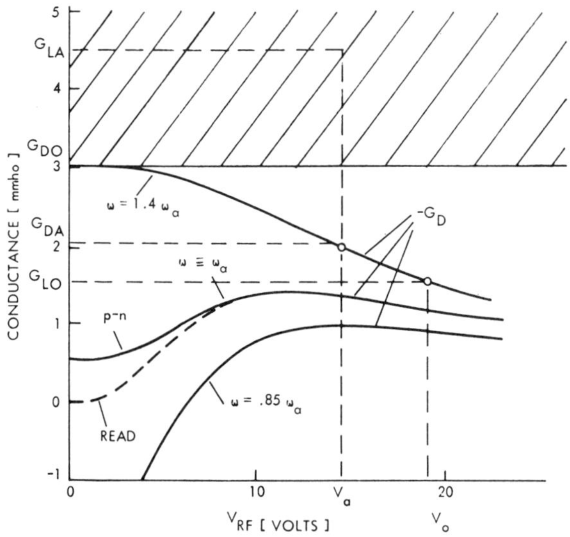

Fig. 1. Dependence of fundamental avalanche-diode conductance component on RF voltage.

some function of the fundamental signal amplitude across it. The parasitic susceptance associated with the device primarily limits the bandwidth and is assumed to be tuned out at the frequency of operation. In Fig. 1 actual plots of the diode conductance G_D as a function of the RF voltage V_{RF} are shown for various modes of operation. In the case where the frequency of operation is above the small-signal avalanche frequency ($\omega = 1.4\ \omega_a$), the G_D curve is a smooth decaying function of increasing signal amplitude. Stable amplification under driving conditions may be obtained as long as the load conductance G_{LA} is greater than the large-signal diode conductance G_D. However, this condition approaches that of the injection-locked oscillator since the amplifier may oscillate with the drive signal removed if the load conductance chosen is smaller than the small-signal diode conductance G_{DO}. The G versus V_{RF} plane can therefore be divided into two regions. In the stable region (dashed area) the load conductance is always larger than the diode conductances and the amplifier is stable under all driving conditions. In the remaining area stable operation is possible only if the locking conditions are satisfied. Even when locked, this mode of operation may not be classified as an amplifier because it does not meet the criteria of unconditional stability and of having an approximately linear input–output characteristic as implied here. This interpretation is a gross oversimplification of the actual situation because possible nonlinear susceptance effects have been neglected. It is approximately true for an amplifier circuit with a single-resonance behavior working sufficiently above the avalanche frequency where the negative real part of the diode is relatively frequency independent. In addition, it is assumed that the circuit is under resonant condition and that the resonance frequency is independent of drive level.

The curves for $\omega = \omega_a$ and $\omega = 0.85\ \omega_a$ in Fig. 1 show typical G_D versus V_{RF} curves when the diode is operated near or below the avalanche frequency. These curves reflect the fact that the small-signal negative conductance of the diode vanishes at $\omega \approx \omega_a$ and becomes positive for $\omega < \omega_a$ and that the effect of increasing signal level is essentially that of lowering the avalanche frequency. In this mode of operation, however, the reactive as well as the real part of the diode admittance are strong functions of the RF level and the assumptions just made are not valid any longer. The conditions for stability are much more complex and would have to be derived from a careful examination of the pole-zero locations of the amplifier network in the complex frequency plane under all possible drive and load conditions. An interesting feature of this mode is the possibility of reducing the ratio of load conductance to diode conductance (higher gain) for a given RF amplitude or, conversely, increasing the output power for a given gain while maintaining unconditional stability of the amplifier. A disadvantage is the resulting highly nonlinear input–output characteristic as a function of RF drive level which basically resembles the respective G_D versus V_{RF} curves shown in the graph. This mode of operation will not be considered here. It may also be seen that as the output-signal level increases, the maximum value of unconditional stable gain decreases because the ratio of G_D/G_{LA} decreases. It is, therefore, important to know the precise nature of the dependence of the average diode negative conductance G_D on the signal level V_{RF} in order to determine the maximum gain and power output. The maximum power delivered by the device under any conditions is that produced by the optimally matched free-running oscillator.

Assuming a static I–V relationship, Hines in a recent paper [3] derived some basic properties of such a nonlinear conductance which show saturation characteristics very similar to those observed in avalanche amplifiers when operating in the mode considered here. It is possible, at least in principle, to derive the coefficients of the power series representing the I–V relationship which would reproduce the actual G_D versus V_{RF} curve of the diode. However, the usefulness of such an approach is questionable since this quasi-static I–V characteristic would itself be a function of various operating and circuit parameters. On the other hand, it may be illustrative to determine the optimum large-signal gain of a reflection-type amplifier as a function of its small-signal gain. The optimum large-signal gain here is defined as the gain where the device delivers its maximum RF power to the output. This relationship is plotted in Fig. 2 for different values of the gain-saturation parameter K, which represents the ratio of the magnitude of the diode conductance at the optimum drive level to the small-signal diode conductance. The curve for $K = 1/2$ corresponds to the case of a cubic nonlinearity. Values for K obtained from amplifier experiments were

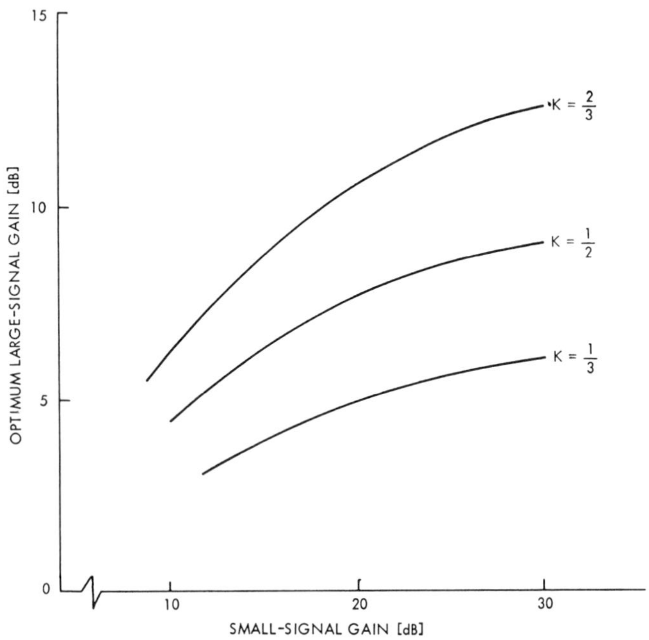

Fig. 2. Optimum large-signal gain as function of small-signal amplifier gain.

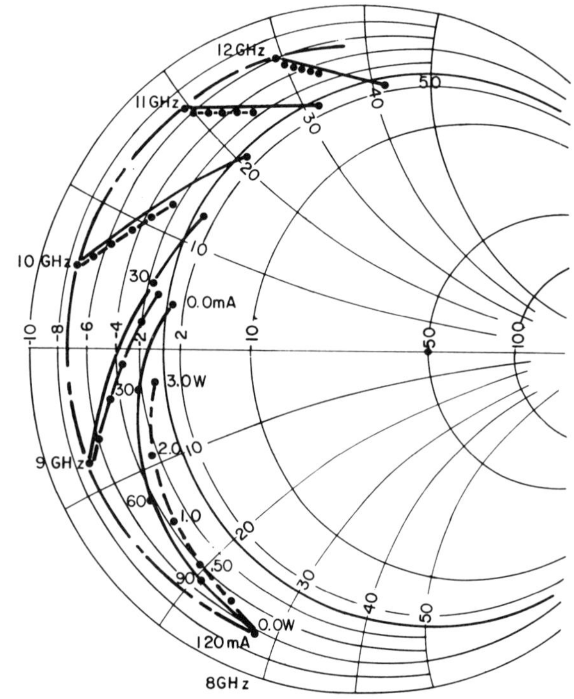

Fig. 3. Measured impedance of packaged avalanche diode as function of bias current and signal level.

found to lie typically in the range of $K=1/2$ to $1/3$. One conclusion to be drawn from this is the fact that the most efficient operation at reasonable gain levels requires a high small-signal gain of the amplifier. Besides creating a possible stability problem, the high-gain condition has a detrimental effect on the transfer characteristics. In practice, stable power amplifiers with 6- to 8-dB gain have been built with output-power levels approaching the maximum oscillating output.

III. Large-Signal Impedance Measurements on Avalanche Diodes

Small-signal measurements of the diode impedance are very useful during the initial stages of device evaluation and for oscillator- and amplifier-circuit design. However, they do not provide any information about maximum power output, saturation characteristics, stability, etc., which are important in oscillators and large-signal amplifiers. A major difficulty in analysis as well as in experimentation arises from the necessity to take into account the effect of higher harmonic impedance terminations. These effects are readily observed in high-power oscillators where independent tuning of the second-harmonic circuit impedance has been found to change the power output by as much a 10 dB depending on the operating conditions. Similar effects were observed in amplifiers but not so pronounced. Theoretical considerations show that a large RF signal in effect lowers the avalanche frequency of the diode and thereby provides a saturation mechanism by reducing the magnitude of the negative resistance. In addition, large changes in the reactance can occur, particularly when the frequency of operation is close to the avalanche frequency of the diode. This is shown in Fig. 3 where the impedance of a packaged diode is plotted at fixed frequencies as a function of bias current and RF power level P_{in}.[1] The solid curves represent loci of small-signal impedances ($P_{in}\approx 0.0$) versus bias current. The dashed curves are impedance-versus-incident RF-power-level plots with the diode biased at the maximum bias current (120 mA). These plots were obtained on a Hewlett-Packard network analyzer with the diode placed in a 50-ohm 7-mm coaxial test holder. One of the interesting features shown by these plots is the result that the effect of a large RF signal on the impedance is essentially the same as that of reducing the bias current. Another observation made during the experiments was the fact that below approximately 8 GHz the fundamental frequency impedance was extremely sensitive to particular impedances presented to the diode junction at the second-harmonic frequency. Above 8 GHz this dependence was greatly reduced and practically disappeared at sufficiently high frequencies. Both results are consistent with theory and the implications of this behavior will be discussed further in Section V. These measurements have to be considered crude, however, since no adequate specifications of the harmonic termination nor a meaningful relationship between RF drive power and signal levels at the junction have been established. Despite these deficiencies, they greatly enhanced the understanding of some of the problems encountered during the large-signal operation of avalanche devices.

[1] The small-signal avalanche frequency of the diode used in this experiment was approximately 6.9 GHz at 120 mA of bias current.

IV. Large-Signal Computer-Analysis Program

Several workers have investigated the large-signal behavior of the avalanche diode either by the derivation of approximate closed-form solutions or by computer-aided methods. These studies were primarily intended for the investigation of various diode structures and of the effects of the diode-parameter variations on the operating characteristics. In general, the methods of analysis are not readily adaptable to amplifier circuits.

Here, a large-signal analysis program which can be executed on a time-share computer and is flexible enough to allow investigation of practically any network configuration of interest is described. The basic idea is to calculate the small-signal diode parameters, which are functions of frequency and bias current, and to modify certain parameters as functions of drive level. It can be shown that under certain restrictions, the gain reduction is entirely due to a lowering of the avalanche frequency with increasing drive level. The problem then reduces to that of finding the appropriate function relating the large-signal avalanche frequency to the external current or voltage amplitude. At sufficiently high drive levels, where the RF-voltage amplitude approaches the dc voltage across the avalanche zone, additional nonlinear effects due to the nonlinearity of the ionization rate and rectification effects will become significant. Most important, the maximum RF-voltage swing cannot exceed the dc bias voltage and therefore imposes an absolute limit on the large-signal operation of the diode. The basic and most important nonlinear effect is that resulting from the restriction on the fundamental particle-current amplitude, which cannot exceed twice the dc bias current. When operated as an amplifier, the convection current emerging from the avalanche zone will increase approximately linearly with the RF-voltage amplitude at low-signal levels. As the drive is increased, the fundamental component of the avalanche current will gradually saturate and eventually reach its large-signal limit. The relationship between the lowering of the avalanche frequency and the particle-current saturation mechanism is readily established by examining the basic current-generation equation of the avalanche region, which can be written as

$$I_a = I_0 \exp\left[\frac{2\alpha_c'}{\tau_a}\int_0^t V_a(\tau)\,d\tau\right]. \tag{1}$$

Noting that the particle emission, as a function of the field, is inductive, one obtains from a Taylor-series expansion the small-signal avalanche inductance

$$L_{a0} = \frac{\tau_a}{2\alpha_c' I_{dc}} \tag{2}$$

where the bias current I_{dc} is the time average of I_a. If the displacement current due to avalanche-zone capacitance C_a is included, the well-known expression for the small-signal avalanche frequency

$$\omega_{a0}^2 = \frac{1}{L_{a0}C_a} = \frac{2\alpha_c' v_d}{\epsilon A} I_{dc} \tag{3}$$

as first defined by Gilden and Hines [4] is obtained. Under the restriction that the phase relationship between avalanche field and particle current remains close to $\pi/2$ (inductive), a large-signal avalanche inductance which is proportional to the fundamental amplitude ratio of the avalanche voltage to the particle current may be defined. It is apparent that the large-signal avalanche inductance increases with increasing RF voltage as a result of the particle-current saturation; e.g., the avalanche frequency is lowered with increasing drive level. In fact, the square of the avalanche frequency is lowered with respect to its small-signal value in the same proportion as the ratio of the avalanche voltage to particle current is increased from its small-signal value. On this basis, the large-signal avalanche inductance can be expressed

$$L_a = L_{a0}\rho(a). \tag{4}$$

The signal-level dependent factor ρ can be determined from the fundamental Fourier coefficient of the particle current. In general, this coefficient is dependent on the amplitudes and phases of all the higher harmonic voltages of $V_a(t)$. In the amplifying mode considered here, $V_a(t)$ can be assumed nearly sinusoidal and higher harmonic voltages may be neglected. Theoretical considerations indicate that the junction voltage at the second-harmonic frequency is only a small fraction of the fundamental voltage amplitude unless a resonance condition exists. This is particularly true if the diode is operating near a transit angle of π for which the harmonic interaction is a minimum. This has also been confirmed by large-signal measurements in the amplifying mode [5], which have shown that the harmonic content of the junction voltage is negligible even at the highest signal levels investigated. Under these conditions, the following simple expression is obtained for the fundamental particle current normalized to the dc component:

$$\beta_1 = \frac{I_{a1}}{I_{dc}} = \frac{I_1(a)}{I_0(a)} \tag{5}$$

where I_0 and I_1 are the modified Bessel functions of argument a and of the zeroth and first order, respectively. The parameter a is related to V_a by

$$a = \frac{2\alpha_c' V_a}{\omega\tau_a} \tag{6}$$

which corresponds to the normalized voltage V_1 of Mouthaan [6]. If the operation of the amplifier is restricted to frequencies sufficiently above the small-signal avalanche frequency where the total diode current

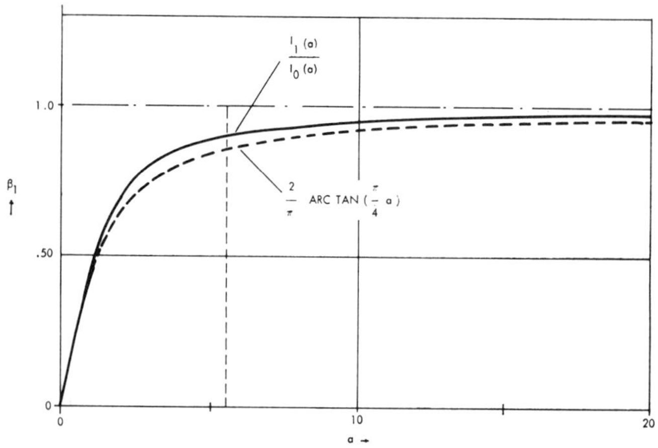

Fig. 4. Fundamental particle-current amplitude normalized to dc component as function of drive level parameter a and simple approximation.

is essentially the displacement current,[2] one can put (6) in the following more convenient form:

$$a = \frac{\omega_{\alpha 0}^2}{\omega^2} \frac{I_1}{I_{dc}} \tag{7}$$

where I_1 is the external sinusoidal RF current. A plot of the fundamental Fourier amplitude β_1 as function of the parameter a and a simple approximation (dashed curve) are shown in Fig. 4. Upon normalization, the desired expression for ρ is now readily obtained:

$$\rho(a) = \frac{a}{2\beta_1(a)}. \tag{8}$$

This function approaches unity for small-drive levels as required and assumes asymptotically a linear relationship between ρ and a with increasing drive level. Using (3), it is possible to define a large-signal avalanche frequency as

$$\omega_\alpha^2 = \frac{1}{C_a L_{a0} \rho(a)} = \frac{\omega_{\alpha 0}^2}{\rho(a)}. \tag{9}$$

Following Read's analysis, an expression for the diode impedance can be written as

$$Z_D = \frac{1}{\omega C_d} \frac{1}{1 - \frac{\omega^2}{\omega_\alpha^2}} \frac{1 - \cos\theta}{\theta} + \frac{1}{j\omega C} \frac{1}{1 - \frac{\omega_\alpha^2}{\omega^2}} \cdot \left[1 - \frac{C}{C_d} \frac{\omega_\alpha^2}{\omega^2}\left(1 - \frac{\sin\theta}{\theta}\right)\right] \tag{10}$$

where C_a and C_d are the avalanche- and drift-region capacitances, respectively, and the total capacitance is given by $C = C_a \| C_d$. Effects of the drift region are invoked through the transit-time angle θ. With the aid of (10) the large-signal behavior of the diode impedance can be examined. For large values of a the inductive component in the equivalent circuit of the diode becomes negligible and the imaginary part of the diode impedance reduces to that of the junction capacitance. At the same time, the real part of the diode impedance is reduced asymptotically to zero. The small-signal diode impedance is obtained from (10) by letting $\omega_a \rightarrow \omega_{a0}$.

[2] This approximation becomes better at high-drive levels since the particle-current amplitude is reduced relative to the displacement current.

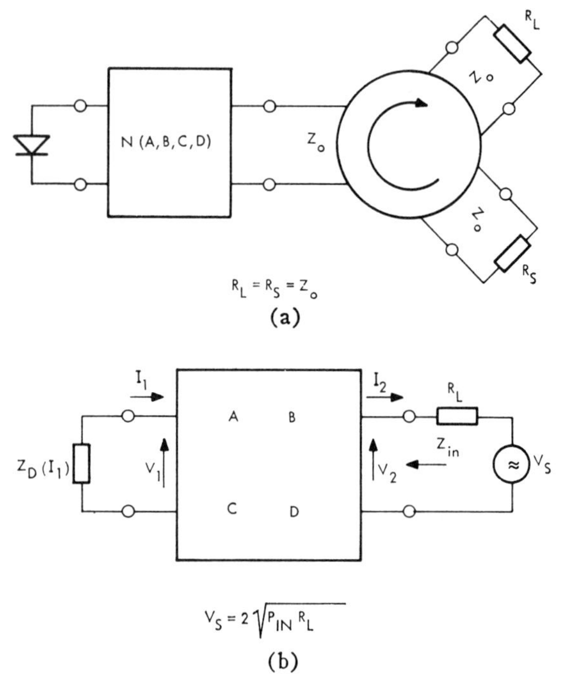

Fig. 5. (a) Schematic of reflection-type avalanche amplifier. (b) Circuit equivalent including driving source and pertinent RF voltages and currents.

With this relation, it is now possible to analyze a reflection-type amplifier circuit as shown in Fig. 5(a). Here an ideal circulator matched at the input and output is assumed, and all the circuit elements external to the diode (and including the package equivalent-circuit elements) are represented by the arbitrary two-port network N. The equivalent network with the pertinent voltages and currents, including the source equivalent, is given in Fig. 5(b). Algebraic manipulation of the four-terminal equations yields for the diode external rms current

$$I_1 = \frac{2\sqrt{P_{in} R_L}}{AR_L + B + Z_D(CR_L + D)} \tag{11}$$

where the matrix elements A, B, C, D are generally complex frequency-dependent quantities. Since the diode impedance Z_D is a function of I_1, (11) is a nonlinear equation of the form $y = f(y)$ and is best solved through an iterative method such as Newton's rule or through successive approximation by specifying that $\Delta y = y - f(y) < \epsilon$. To assure rapid convergence, the variable y has to be chosen so that it remains well bounded and such that $f(y)$ is a single-valued function. The only variable meeting these requirements is found to be the

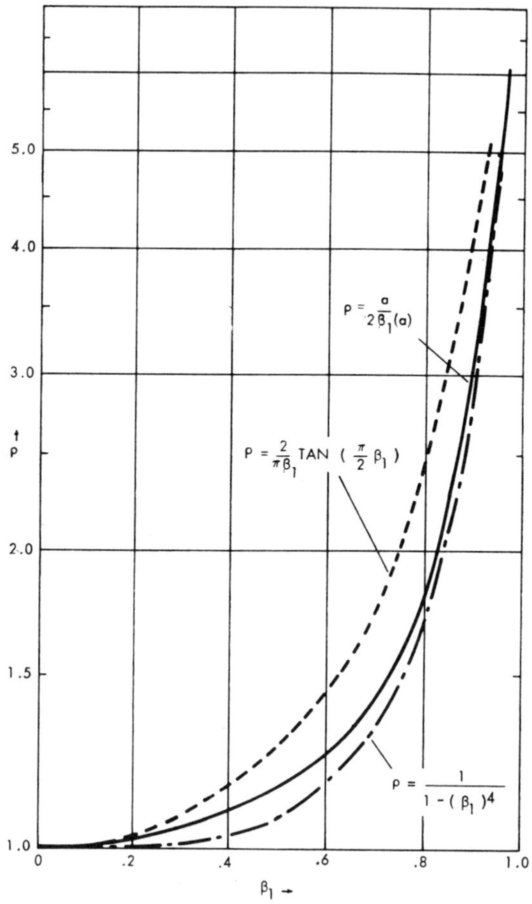

Fig. 6. Relationship between large-signal avalanche-inductance parameter and fundamental Fourier-amplitude β_1 of particle current.

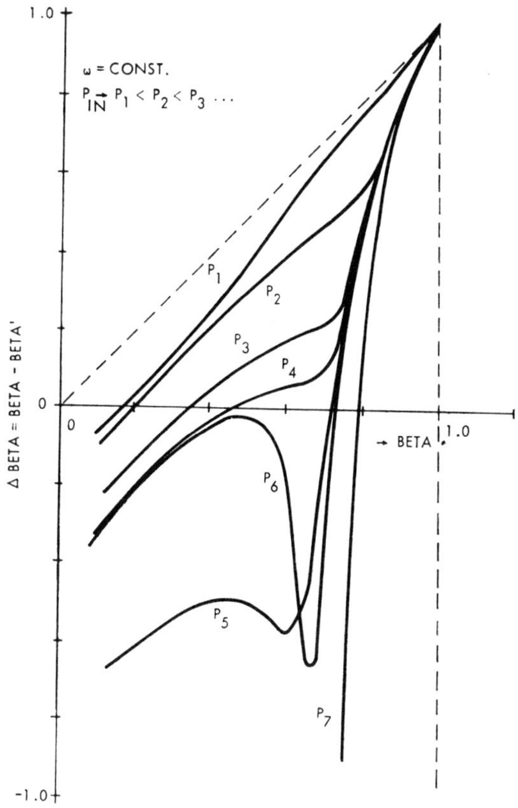

Fig. 7. Functional relationship between β and $\beta - f(\beta)$ for various drive levels.

coefficient β_1, which satisfies the condition $0<\beta_1<1$. The function $\Delta\beta = f(\beta)$ is very involved and is best computed in a sequence of steps as indicated by the following.

1) Set initial β.
2) Compute $L_a = L_{a0}\rho(\beta)$.
3) Compute $Z_D = f(Z_{DO}, L_a, \omega)$.
4) Compute $I_1 = f(A, B, C, D, Z_D, P_{\text{in}}, \omega)$.
5) Compute $f(\beta) + \beta'(I_1)$ using (5) and (7).
6) Find $\Delta\beta = \beta - \beta' = 0$.

The first step requires the computation of $\rho(\beta)$ from (8), which is transcendental in β since β is also implicit in a. A graphical representation of $\rho(\beta)$ is shown in Fig. 6. Two simple approximations are indicated on the same graph. If desired, this function can be solved to the desired degree of accuracy by numerical methods. Typical functional relationships between $\Delta\beta$ and β for different drive levels P_{in} at constant ω are shown in Fig. 7. It is seen that the functions are well behaved and the β corresponding to the solutions $\Delta\beta = 0$ move towards the large-signal limit as the drive level is increased. A similar family of curves is obtained with ω as variable parameter and P_{in} held fixed. In the actual iteration routine of the program, Newton's rule is applied in the vicinity of $\Delta\beta = 0$ where the approximate value of β has been found from stepping β in the interval from zero to unity until $\Delta\beta$ changes sign. It was found that at most a total of 13 iterations are necessary so that $\epsilon < 10^{-4}$. Once the final β and diode impedance Z_D have been found, the gain is obtained from

$$g = \left|\frac{R_L - Z_{\text{in}}}{R_L + Z_{\text{in}}}\right|^2 \tag{12a}$$

and is equal to the magnitude squared of the circuit-reflection coefficient, which also contains the phase information. Z_{in} is the transformed-diode impedance at the input of the network and is given by

$$Z_{\text{in}} = \frac{DZ_D + B}{CZ_D + A}. \tag{12b}$$

No hysteresis effects have been found in numerous computer runs even under extreme drive conditions and arbitrary complex-network configurations. This has also been verified by experimental results and is consistent with theory. It has been shown [7] that for the fundamental resonance mode (zero-harmonic pumping) the stability limit is given by

$$A_0 = \frac{\omega^2}{\omega_{\alpha 0}^2} - \frac{1}{\rho(a)} \geqq 0. \tag{13}$$

According to (4) one can write

$$\rho(a) = \frac{L_a}{L_{a0}} = \frac{\omega_{\alpha 0}^2}{\omega_\alpha^2} \tag{14}$$

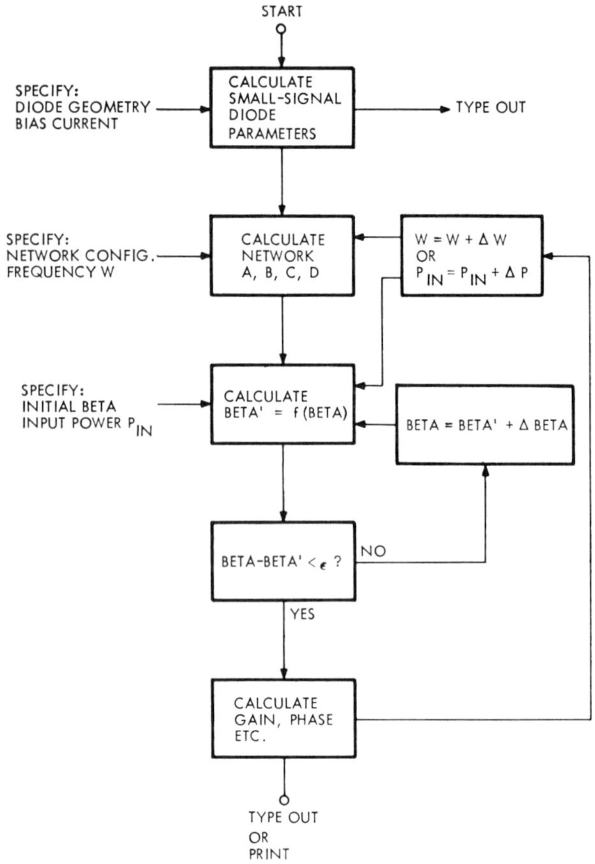

Fig. 8. Simplified flow diagram of large-signal computer-analysis program.

where L_a and ω_a denote the large-signal avalanche inductance and the large-signal avalanche frequency, respectively. Equation (13) can then be rewritten as

$$\frac{\omega^2}{\omega_{\alpha 0}^2} - \frac{\omega_\alpha^2}{\omega_{\alpha 0}^2} \geqq 0 \tag{15}$$

which simply states that the operating frequency has to be larger than the large-signal avalanche frequency for stable operation. A simplified flow-chart diagram of the complete large-signal amplifier program is given in Fig. 8. The actual program also takes into account other nonlinear effects such as field bottoming in the avalanche zone, which was found to be the most important large-signal effect besides the particle-current saturation mechanism described in this section. Executed on a time-share facility, the program requires approximately 1 second of CPU time per data point.

V. Stability and Gain-Response Distortions

Some special problems in the design of avalanche amplifiers and oscillators arise from the fact that both the real and imaginary part of the diode impedance are strong functions of the dc bias current as well as of the RF-signal level. Because the oscillator is usually designed for fixed frequency and amplitude, this does not pose too much of a problem. In an amplifier, however, severe distortion in the gain response and instability under certain conditions can result.

First, the effects of bias-current variations on the amplifier gain response will be investigated because they are directly related to problems arising from large-signal operation. As can be seen from (2), the avalanche inductance changes in inverse proportion to the dc bias current. Therefore, the resonance frequency of an avalanche-diode circuit under small-signal conditions is always lowered if the bias current is reduced. Conversely, an increase in the bias current increases the resonance frequency. In Fig. 9(a), gain versus frequency of a single-resonant amplifier tuned for 10 dB of gain at 10 GHz is plotted for various dc bias-current levels. As would be expected, the resonant frequency and the peak gain decrease smoothly as the bias current is reduced. If the same experiment is performed on a triple-tuned $n=3$ amplifier design, the gain response is altered drastically, even for small changes in the bias current as shown in Fig. 9(b). When the current is further reduced, a very sharp gain peak appears at the lower band edge. The maximum occurs at some critical bias-current level approximately 20 percent below the design operating current for this particular amplifier. It was found that the maximum of the gain peak reaches values of typically 20 to 40 dB, which means that the amplifier is potentially unstable and may go into oscillation under actual conditions. This is a rather discouraging situation but not completely unexpected. Very similar distortions are experienced in the response of passive filters, with the difference, however, that in the case of the amplifier the effects are magnified tremendously because of the gain involved. It might be added that the same phenomena occurs at the high-frequency side if the bias current is increased above its design value, which is, however, of little practical interest. Since the critical current density cannot be avoided in the transition of the bias current from zero to its design operating value or vice versa, one has to look for some means of reducing the gain peak to such a degree that stability is assured under worst-case operating conditions. The frequency and current level at which the gain peak maximizes will depend on various circuit and operating parameters. An analytical solution to the problem of finding the stability criteria in terms of these parameters is not trivial even for an idealized model. Computer simulations have shown, however, that large ratios of operating-to-avalanche frequency, low nominal gain, small ripple, and large initial bandwidth are of primary importance in reducing the problem.

It has been shown earlier that the effect of large RF signals on the diode impedance is very similar to that of reducing the bias current. One would therefore expect distortions in the gain response similar to those experienced in the case of bias-current variations. This is substantiated by the plot in Fig. 10. Here the solid lines represent measured gain curves of a single-resonant amplifier for various input-signal levels. One can see that the gain is continually reduced with increasing drive level while the resonance frequency is shifting to lower frequencies. Because of the resonance effect, the

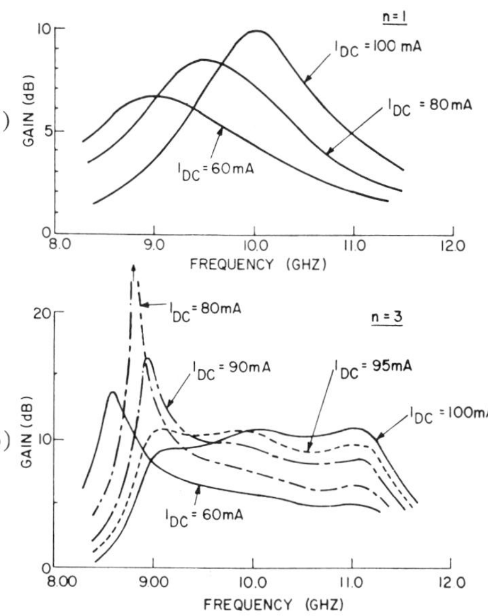

Fig. 9. Distortion of gain response as function of dc-bias-current variations for single-tuned ($n=1$) and triple-tuned ($n=3$) amplifier circuits.

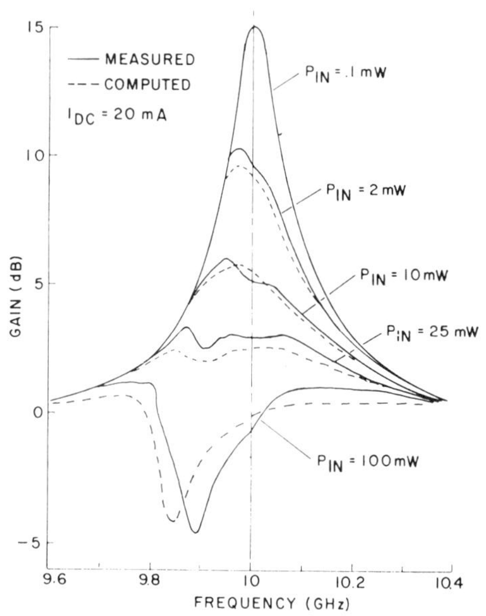

Fig. 10. Measured and computed large-signal gain response of single-tuned avalanche amplifier.

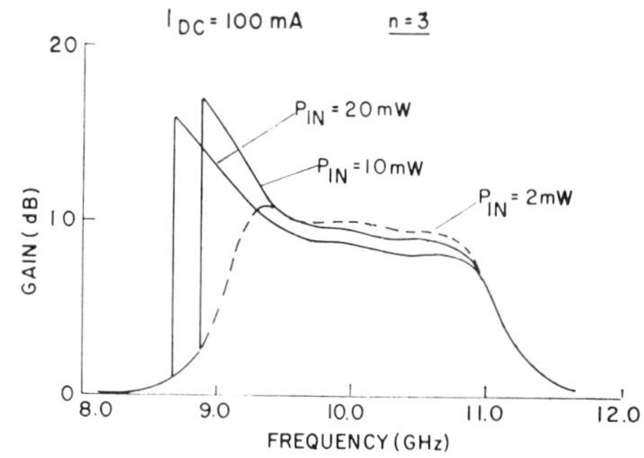

Fig. 11. Computed large-signal gain response of triple-tuned avalanche amplifier.

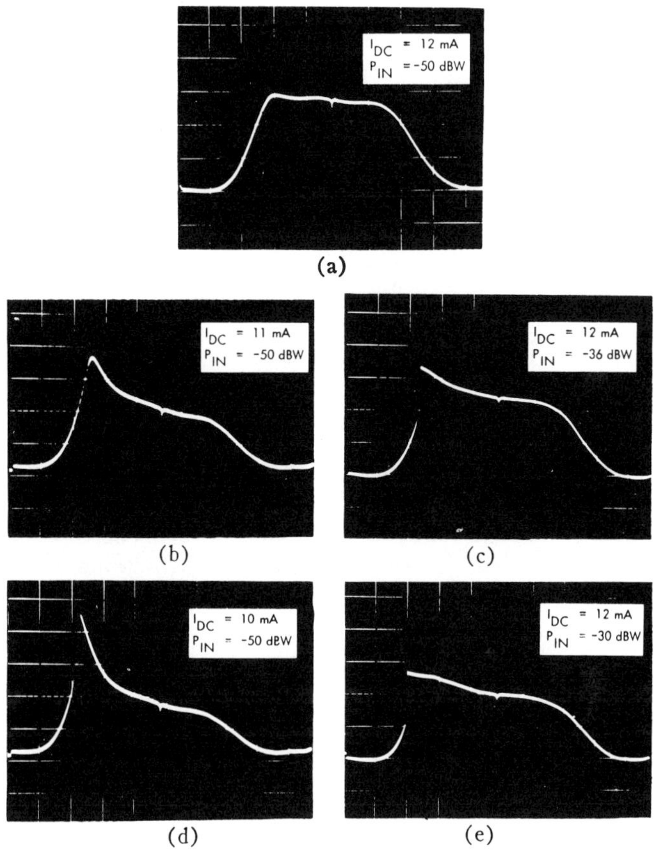

Fig. 12. Experimental gain-response distortions of triple-tuned avalanche amplifier under various operation conditions. Horizontal scale: 50 MHz/div; vertical scale: 5 dB/div.

signal level at the diode is frequency dependent causing the tilting of the resonance curve to the left. As the drive level is increased further, a dip in the gain begins to appear at the resonant frequency and grows to such an extent as eventually to result in negative gain. The dashed curves indicated on the same graph were obtained from a large-signal computer-analysis program described in Section IV. It is interesting to note that the gain actually becomes negative at sufficiently high drive levels; e.g., the average diode conductance becomes positive. This can occur when the drift field drops below the value required to maintain the drifting carriers at saturated velocity. Not only does this reduce the transit angle and therefore the negative resistance, but it also results in undepleted epitaxial material, which will increase the parasitic resistance of the diode. These effects are being accounted for in the computer-analysis program by modifying the drift-region parameters as functions of the RF field. Realizing the highly nonlinear operation of the diode, the match between the measured and computed data may be considered satisfactory. Fig. 11 shows computed gain curves for an $n=3$ amplifier under different driving conditions. Again we recognize some similarity to those obtained previously under variable bias condition, with the difference, however, that the gain peaks do not reach the high values observed previously because of output-power saturation. These distortions have been observed in experimental amplifiers. Fig. 12 shows the distortion of the amplifier response under various operating conditions. Fig. 12(a)

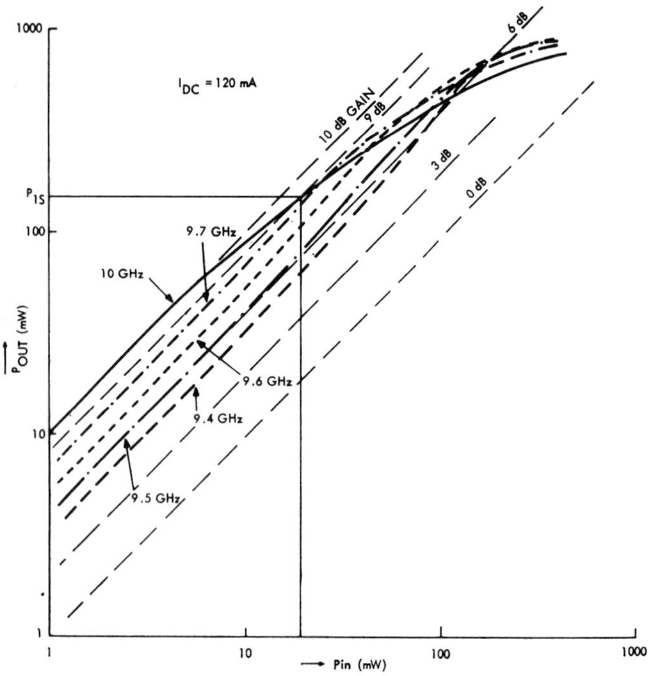

Fig. 13. Saturation characteristics of single-tuned avalanche amplifier as functions of frequency.

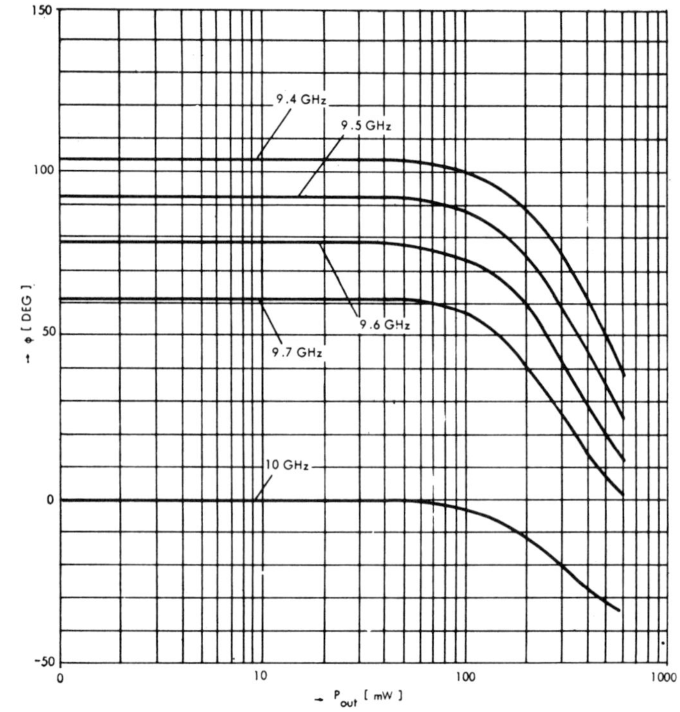

Fig. 14. Transfer phase of same amplifier as function of output power and frequency.

232

shows the triple-tuned response at the design-operating bias current of 12 mA under small-signal conditions. Fig. 12(b) and (d) show the response distortion as the bias current is reduced to 11 and 10 mA, respectively. Fig. 12(c) and (e) show large-signal distortions at the design current of 12 mA with input-power levels of −36 and −30 dBW, respectively.

It is interesting to note that the distortions under large-signal operation in multiple-tuned wide-band amplifier circuits are primarily due to the signal-level-dependent susceptance of the diode. This can be explained from the fact that this distortion always occurs at the low-frequency band edge where the susceptance changes are maximum because of the proximity to the avalanche frequency. In general, multiple-tuned circuits are very sensitive to detuning effects caused by large signals and it may be necessary to devise some methods to counteract these effects. One possibility would be to utilize nonlinear interaction between the fundamental and harmonic signals to achieve the desired effect, a method similar to that proposed by Mouthaan [6] for the avalanche oscillator. Because of the wide-band operation, this poses a formidable circuit problem, and it is questionable whether the method could be applied to the amplifier situation with the same degree of success.

VI. Transfer Characteristics and Signal Distortions

The input–output relationship of an amplifier is usually characterized by a linear region and a saturation region that start typically 4–10 dB below the maximum saturated output. As has been shown, two major saturation mechanisms are responsible for power-output limitations in the avalanche-diode amplifier. First, the amplitude of the fundamental particle current cannot exceed twice the dc bias current. When this limit is reached, the linear input–output relationship becomes approximately a square-root law, e.g., the output power increases approximately linearly with the drive voltage. A second limit is reached when the RF-voltage swing across the avalanche zone becomes comparable to the dc-voltage drop across the avalanche zone. Whereas the first condition causes a gradual saturation, the second condition, which may be called field-bottoming, is responsible for the ultimate limitation of the power output. Fig. 13 shows the input–output characteristics of a typical X-band single-resonant amplifier tuned for a 10-dB (small-signal) gain at a center frequency of 10 GHz. An interesting feature shown by these curves is the gain linearization that can be obtained by operating below the small-signal resonance frequency of the amplifier. This is due to the fact that the resonance frequency is lowered with increasing signal level and therefore partially compensates for the gain reduction under large-signal operation. This effect may be used to advantage in narrow-band applications to reduce distortions. The transfer phase of the same amplifier as a function of the output power is shown in Fig. 14.

In single-carrier FM systems, the power-amplifier stage can be operating in saturation without having a detrimental effect on the modulated signal. In multi-carrier systems like those encountered in satellite communications, electronic countermeasures (ECM),

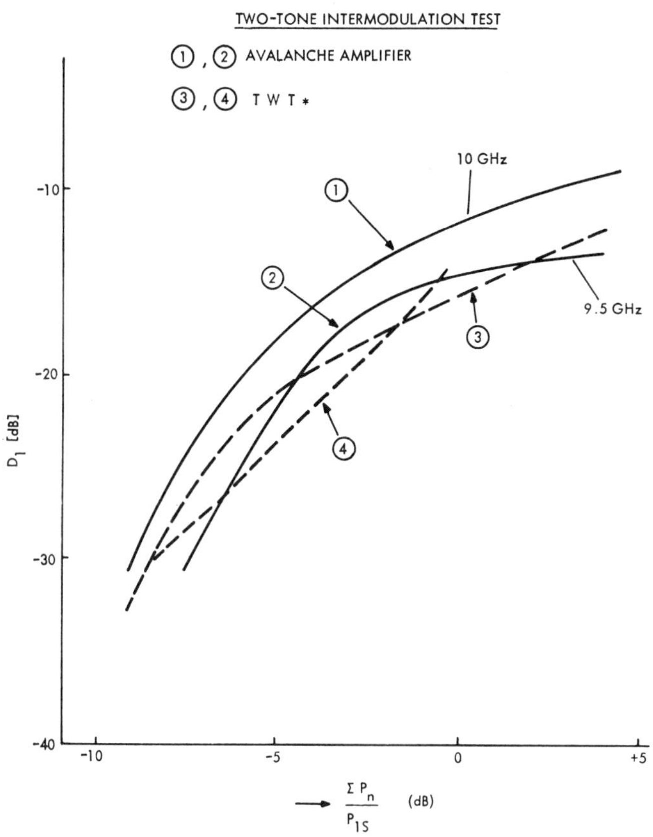

Fig. 15. Results of intermodulation test on avalanche amplifiers and TWTs. *See [8], [9].

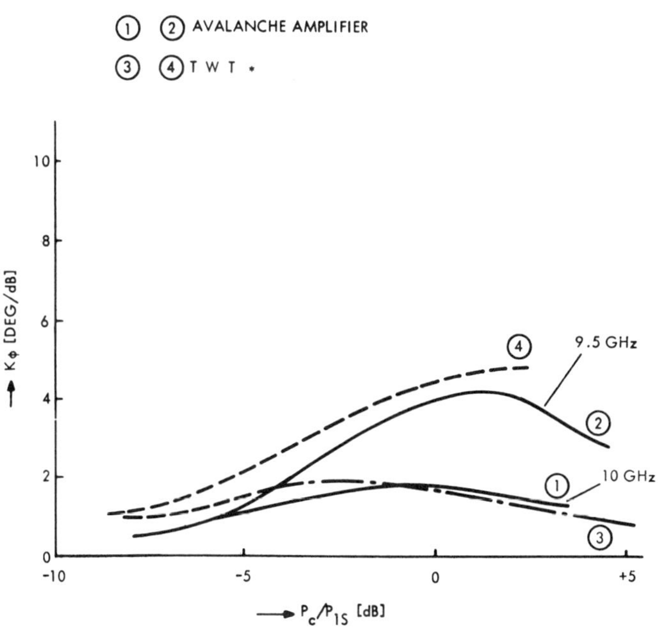

Fig. 16. AM-to-FM conversion in avalanche amplifiers and TWTs. *See [8], [9].

and other systems, the amplifier has to be operated in the quasi-linear region and consequently at reduced output power. A good measure for the linearity of the transfer characteristic is obtained by measuring the intermodulation products in the presence of two or more carriers. A standard method involves the two-tone test, the results of which are shown in Fig. 15. Here the third-order intermodulation product D_1 relative to the carrier signal is plotted as a function of the total carrier power relative to the -1 dB single-carrier output saturation level P_{1S}. The curves 1 and 2 represent data obtained from an avalanche amplifier at 10 GHz (center frequency) and 9.5 GHz, respectively. As expected from the previous graph, the curve taken at 9.5 GHz shows a lower distortion as a result of the more linear transfer characteristic. For comparison, typical results of two different TWTs are included (curves 3 and 4). Another important measure of distortion encountered in communications-systems applications is the AM-to-FM conversion factor K_ϕ. Typical curves for the avalanche amplifier and the TWTs are shown in Fig. 16. Although the data presented are preliminary and incomplete, it shows that the transfer characteristics of avalanche amplifiers compare favorably with those of TWTs.

VII. Conclusion

The effects of large-signal operation on various performance parameters of avalanche amplifiers have been investigated. It has been shown that severe distortions in the gain response and transfer characteristics can occur, which are the result of changes of the diode impedance with signal amplitude. These effects are particularly pronounced in wide-band amplifiers using multiple-tuned matching networks. Theoretical and experimental results indicate that signal-level-dependent distortions are greatest near the avalanche frequency and that for best wide-band performance the ratio of operating-to-avalanche frequency should be as high as is consistent with other circuit requirements. From numerous computer simulations and experiments, the optimum band center frequency was found to be typically 1.5 to 2 times the small-signal avalanche frequency. Signal distortions due to the nonlinear transfer characteristics were found to be similar to those of TWTs when compared relative to the 1-dB compression point. However, the ratio of saturated power output to the 1-dB compression point is somewhat higher, typically 3 to 6 dB.

It is realized that the material presented in this paper is not a comprehensive treatment of the subject. The intention is to illuminate some of the basic problems of large-signal operation and to indicate areas where further research is required. A better understanding is needed in terms of device evaluation and circuit-interaction effects and their relation to the amplifier performance. Evaluation of diodes in various modes of operation have indicated that the requirement of optimum performance as an amplifier or as an oscillator may not be satisfied by the same diode design.

References

[1] E. F. Scherer and W. K. Niblack, "The design of high power and wide-band avalanche amplifiers," *1969 NEREM Rec.*, pp. 78–79.

[2] W. H. Ku and E. F. Scherer, "Gain-bandwidth optimization of avalanche diode amplifiers," this issue, pp. 932–942.

[3] M. E. Hines, "Negative-resistance diode power amplification," *IEEE Trans. Electron Devices*, vol. ED-17, pp. 1–8, January 1970.

[4] M. Gilden and M. E. Hines, "Electronic tuning effects in the Read microwave avalanche diode," *IEEE Trans. Electron Devices*, vol. ED-13, pp. 169–175, January 1966.

[5] J. W. Gewartowski and J. E. Morris, "Active IMPATT diode parameters obtained by computer reduction of experimental data," *IEEE Trans. Microwave Theory Tech.*, vol. MTT-18, pp. 157–161, March 1970.

[6] K. Mouthaan, "Nonlinear analysis of the avalanche transit-time oscillator," *IEEE Trans. Electron Devices*, vol. ED-16, pp. 935–945, November 1969.

[7] W. Harth, "Resonance and parametric effects in IMPATT diodes," *IEEE Trans. Electron Devices*, vol. ED-17, pp. 282–289, April 1970.

[8] R. J. Westcott, "Investigation of multiple f.m./f.d.m. carriers through a satellite TWT operating near saturation," *Proc. Inst. Elec. Eng.*, pp. 726–740, June 1967.

[9] J. Bretting, "Interrelation between efficiency and non-linearities in TWT's for application in communication satellites," *Microwave J.*, pp. 84–89, April 1969.

234

SECTION II

Reprinted from IEEE TRANSACTIONS
ON *MICROWAVE THEORY AND TECHNIQUES*
Volume MTT-18, Number 11, November, 1970
pp. 963-967

A One-Watt CW High-Efficiency X-Band Avalanche-Diode Amplifier

***Abstract*—An *X*-band avalanche-diode amplifier circuit is described which provides accurate and independent in-band impedance control (*R* and $+jx$) and orthogonal second-harmonic reactance control. Low-level gains of 13 dB with 500 MHz of bandwidth have been achieved using this circuit. Power outputs of 1 watt CW with 5 dB of gain and 14.7-percent generation efficiency have also been realized by carefully controlling the in-band and second-harmonic load impedance.**

Efficient solid-state *X*-band power generation can be accomplished by utilizing the negative-impedance characteristics of avalanche diodes in reflection amplifiers. Breadboard 7-GHz circuits have been constructed which exhibit low-level gains of 13 dB with 500 MHz of bandwidth. Power outputs of 1 watt CW, with 5 dB of gain and 14.7-percent generation efficiency have been realized in the same circuit.

The diodes are first characterized by measuring the variation of small-signal negative impedance with current density and frequency.

Manuscript received March 27, 1970; revised September 7, 1970. This work was sponsored by the Department of the Air Force.

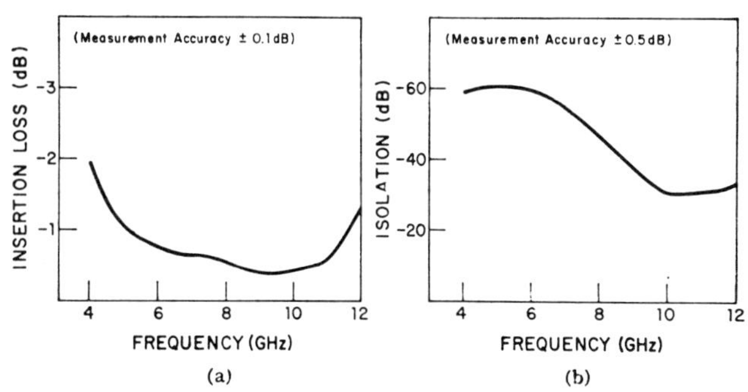

Fig. 1. (a) RF+dc port to RF port bias T insertion loss. (b) RF or RF+dc port to dc port bias T isolation.

The diode-pill-package weld flange is used as an impedance reference. A broad-band DC-T is required for this measurement. One was constructed with the performance shown in Fig. 1(a) and (b). A computer-controlled network analyzer is used to calibrate out the losses associated with the DC-T and a 50-ohm diode holder. The holder consists of a 50-ohm coaxial transmission line linearly tapered down to the dimension of the circuit in which the diode will eventually be placed. Thermal and electrical ground is provided by the same

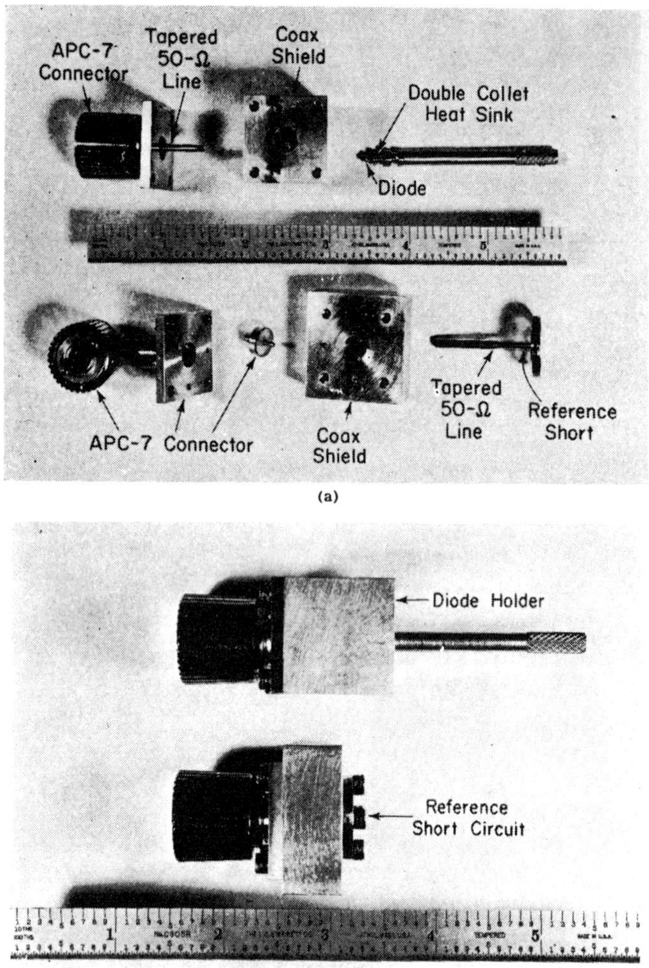

Fig. 2. (a) Disassembled diode holder and short-circuit reference. (b) Assembled diode holder and short-circuit reference.

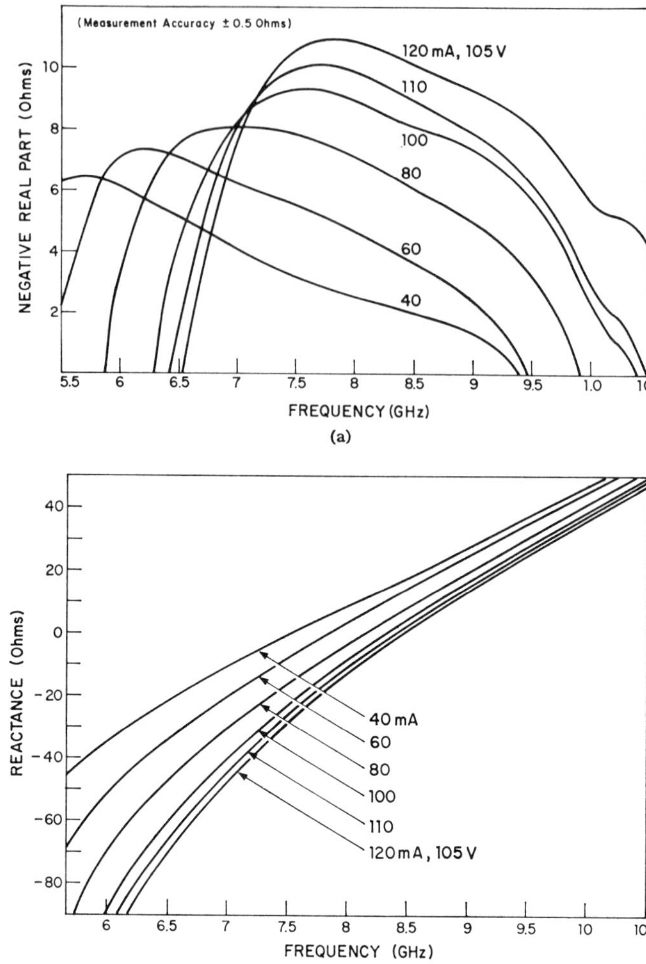

Fig. 3. Small-signal negative impedance for silicon avalanche diode versus bias current and frequency. (a) Real part. (b) Reactance.

236

double-collet heat sink used in the amplifier circuit. A diode-holder reference short circuit was also constructed so that strays associated with the diode mounting would not be included in the network-analyzer calibration. A photograph of the diode holder and reference short circuit is shown in Fig. 2(a) and (b). Typical small-signal impedance characteristics for a silicon avalanche diode, calculated by the network analyzer from measurements of $\rho \angle \phi$, are shown in Fig. 3(a) and (b).

A 10-dB power-gain low-level amplifier was then designed and constructed based upon similar small-signal measurements of a gallium arsenide avalanche diode. The amplifier has independent impedance control (R and $+jX$) at X-band (6–9 GHz). The real part of the presented load impedance is determined by the physical dimensions of a two-section quarter-wave transmission-line impedance transformer. The reactive part is determined by the characteristic impedance of a short-circuited coaxial stub placed in series with the transformer:

$$X_L = Z_0 \tan \beta l$$

where

- X_L inductive reactance of stub ($\beta l < 90°$)
- Z_0 characteristic impedance of stub
- βl electrical stub length.

A set of transmission-line transformers (in 0.5-ohm steps from 1 to 20 ohms) and series short-circuited stubs (in 5-ohm steps from $+j5$ to $+j150$ ohms at 7.2 GHz) was constructed to facilitate low-loss impedance control. Bias is applied through a ferrite-backed single-section low-pass filter. A quarter-wave dielectric-loaded open-circuited stub is incorporated in series with the output to block dc bias. Photographs of the amplifier are shown in Fig. 4(a)–(c).

The amplifier was adjusted such that 10-dB low-level gain occurred with a gallium arsenide avalanche diode biased at 23.3 mA and 68 volts. The load impedance was measured and the theoretical gain [1] was calculated using the equation

$$G = \left| \frac{Z_D - Z_L^*}{Z_D + Z_L} \right|^2$$

where

- Z_D diode small-signal impedance
- Z_L presented circuit impedance
- Z_L^* conjugate of Z_L.

A comparison of the calculated and measured gain response is shown in Fig. 5(a). A tabulation of the measured diode and load impedance is also included in Fig. 5(b) and (c). The maximum output power for this experiment was +5 dBm.

The same circuit was used for measurements of high-power amplifier and oscillator performance. Continuous second-harmonic impedance control, orthogonal to the impedance control at X-band is accomplished by appropriately locating a four-section high-pass filter, terminated in a short-circuited line stretcher, a quarter of a wavelength away from a short-circuit reference. The dc bias line provides the short-circuit reference at the coaxial junction of the bias line and amplifier-output line. This short circuit also serves to short out the out-of-band impedance characteristics of the circulator and external connections at the second harmonic (13.8–15.2 GHz). The phase of the presented load reactance at the second harmonic is con-

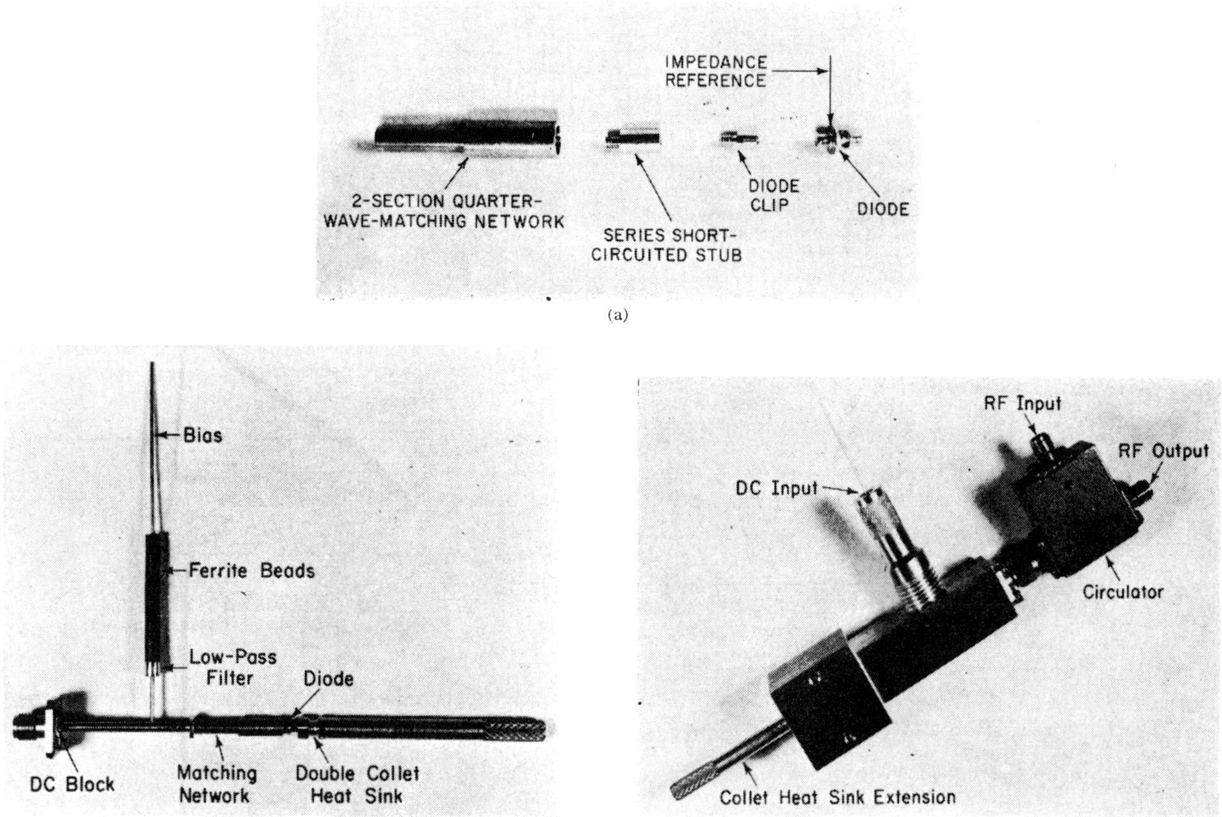

(a)

(b) (c)

Fig. 4. (a) Disassembled *X*-band avalanche-diode-amplifier matching network. (b) Amplifier center-conductor assembly. (c) Assembled low-level amplifier.

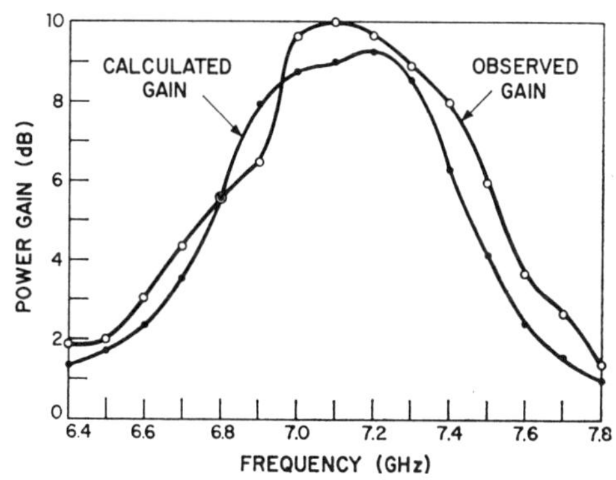

(a)

FREQ	IMPEDANCE REAL	IMPEDANCE IMAG
6000.000	-4.60	-22.71
6100.000	-4.08	-21.58
6200.000	-4.27	-19.34
6300.000	-4.16	-17.41
6400.000	-4.14	-15.21
6500.000	-3.79	-14.67
6600.000	-3.96	-12.89
6700.000	-3.99	-11.57
6800.000	-3.90	-9.61
6900.000	-3.36	-9.23
7000.000	-3.59	-7.51
7100.000	-3.58	-6.08
7200.000	-3.35	-4.17
7300.000	-2.97	-3.59
7400.000	-3.17	-2.19
7500.000	-3.39	-.83
7600.000	-3.25	.77
7700.000	-2.84	1.28
7800.000	-2.94	2.65
7900.000	-3.05	3.96
7999.998	-3.02	5.58

(b)

FREQ	IMPEDANCE REAL	IMPEDANCE IMAG
6000.000	7.26	3.10
6100.000	6.29	2.44
6200.000	5.26	2.87
6300.000	4.80	3.44
6400.000	4.36	3.91
6500.000	4.44	3.94
6600.000	4.30	4.80
6700.000	4.72	5.08
6800.000	5.29	5.33
6900.000	6.09	5.24
7000.000	6.26	5.60
7100.000	6.69	5.09
7200.000	6.54	4.79
7300.000	6.21	4.30
7400.000	5.25	5.05
7500.000	4.83	5.32
7600.000	4.60	5.75
7700.000	4.52	6.35
7800.000	3.76	8.14
7900.000	3.74	8.87
7999.998	3.91	9.94

(c)

Fig. 5. (a) Comparison of measured and calculated gain response for gallium arsenide diode. Bias is 23.3 mA, 68 volts. (b) Gallium arsenide small-signal diode impedance. Bias is 23.3 mA, 68 volts. (c) Circuit impedance of 10-dB low-level amplifier.

237

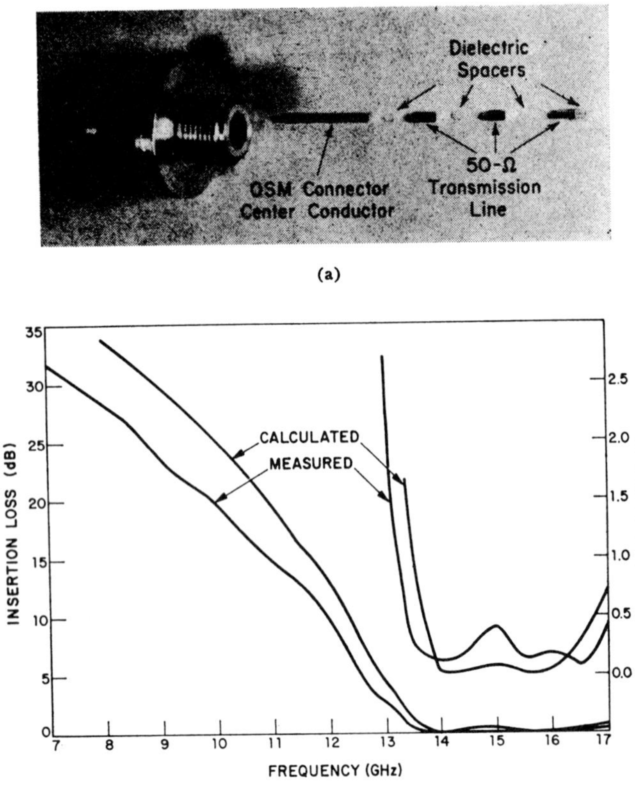

Fig. 6. (a) 14-GHz four-section high-pass filter. (b) Calculated and measured high-pass filter characteristics.

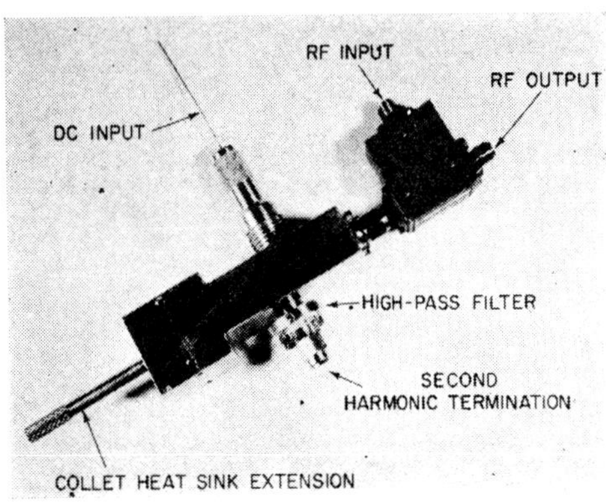

Fig. 7. Assembled high power X-band avalanche-diode amplifier.

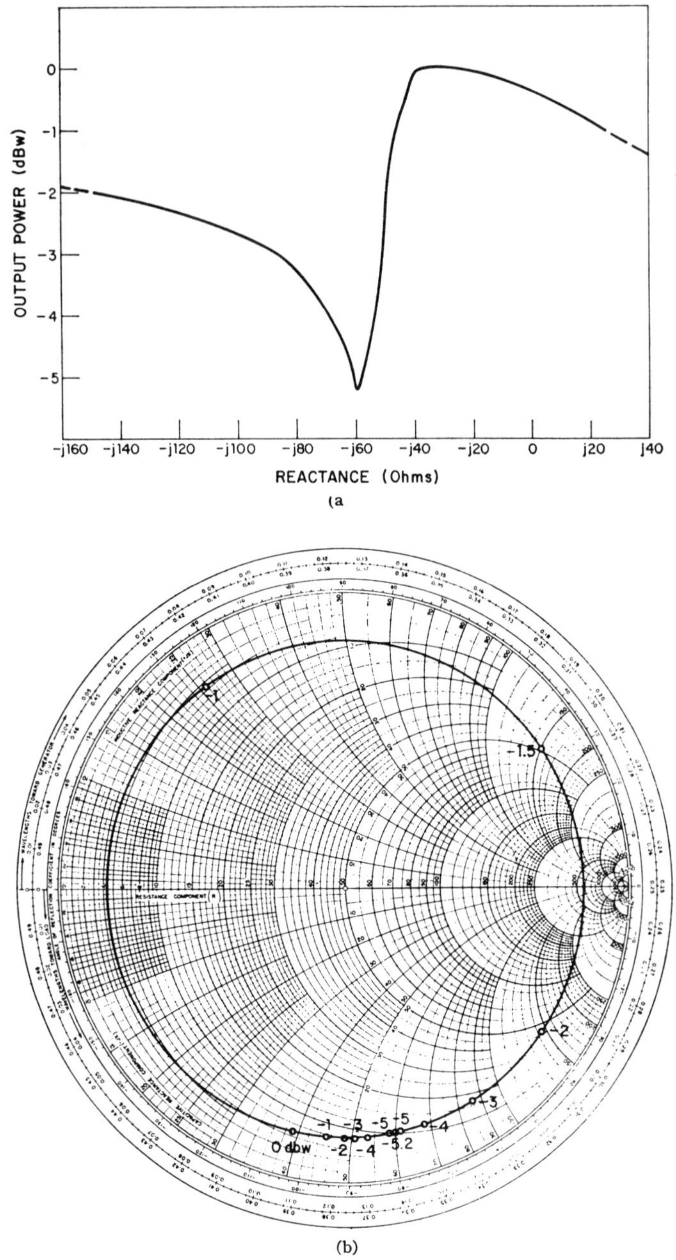

Fig. 8. (a) Variation in saturated output power with second-harmonic (14-GHz) reactance. Silicon avalanche-diode bias is 99.5 mA, 90.1 volts. (b) Smith-chart plot of variation in saturated output power with second-harmonic (14-GHz) reactance.

trolled by varying the length of the line stretcher. A photograph of the disassembled filter is shown in Fig. 6(a). A comparison of measured and calculated filter characteristics is shown in Fig. 6(b). The assembled circuit with independent second-harmonic impedance control is shown in Fig. 7.

Variations in the second-harmonic termination produce reductions in maximum obtainable output power of 1–10 dB when the diode is used either as an oscillator or amplifier. The amount of this variation depends on the internal diode-package configuration. A typical effect of second-harmonic reactance on output power for a silicon avalanche-diode amplifier operating at 7.0 GHz is shown in Fig. 8(a). The amplifier had a maximum output power of 1 watt with a 7-dB power gain and a maximum generation efficiency [2] of 8.8 percent where

$$\eta_g = \text{generation efficiency} = \frac{(\text{RF power out}) - (\text{RF power in})}{(\text{dc power in})}.$$

A Smith-chart plot of the same data is given in Fig. 8(b). The impedance reference used for this measurement was the diode-package weld flange (not the diode chip) so that the particular values of load impedance measured at 14 GHz are relevant only to this circuit.

The best performance measured using this circuit was with a gallium arsenide avalanche diode in an amplifier configuration:

RF output power (7.2 GHz)	1.0 watt
power gain	5 dB
generation efficiency	14.7 percent
3-dB bandwidth	1.1 GHz.

Typical amplifier characteristics are shown in Fig. 9(a) and (b) for a gallium arsenide diode biased at 40 mA and 91.9 volts. The maximum output power was 750 mW at 6.95 GHz with 5-dB power gain and a generation efficiency of 13.9 percent.

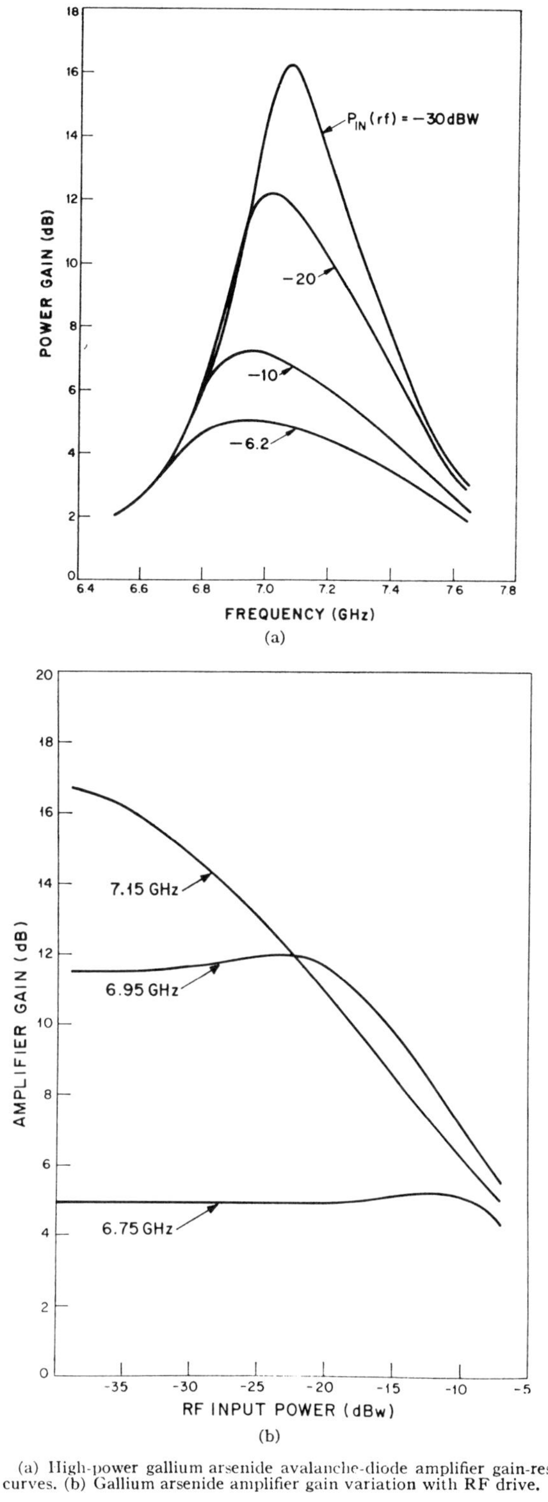

Fig. 9. (a) High-power gallium arsenide avalanche-diode amplifier gain-response curves. (b) Gallium arsenide amplifier gain variation with RF drive.

ACKNOWLEDGMENT

The author wishes to thank R. E. Dolbec and L. P. Corvo for their significant contributions to this paper.

DAVID M. SNIDER
M.I.T. Lincoln Lab.
Lexington, Mass.

REFERENCES

[1] E. Kuh and R. Rohrer, *Theory of Linear Active Networks*. Holden-Day, 1967, pp. 258–265.
[2] M. E. Hines, "Negative-resistance diode power amplification," *IEEE Trans. Electron Devices*, vol. ED-17, pp. 1–8, January 1970.

239

Power-Impedance-Frequency Limitations of IMPATT Oscillators Calculated from a Scaling Approximation

DONALD L. SCHARFETTER, MEMBER, IEEE

Abstract—The obtainable CW power of silicon IMPATT oscillators, as a function of frequency, is calculated by scaling from reference results. The analysis differs from previous treatments in that the microwave circuit impedance limitation, as observed experimentally, is utilized simultaneously with thermal impedance limitations to uniquely determine device diameter, operating currents, and output power. Results are presented for single and multiple (parallel) units on copper and diamond mounting studs, and for both single (p^+-n-n^+) and double-drift-region (p^+-p-n-n^+) structures. Obtainable power falls off essentially as $1/f$ until an ultimate (nonthermal) space-charge-limited current density is reached. Beyond this point the obtainable power varies as $f^{-2.14}$. The calculated results on single-drift-region structures are in agreement with experimental observations over the range of frequencies from 13 to 55 GHz, and the analysis predicts an obtainable power of 300 mW at 110 GHz for a double-drift-region structure with 10 percent conversion efficiency

240

I. Introduction

THE maximum obtainable CW power of an IMPATT oscillator is determined by both circuit and thermal impedance limitations below a certain frequency and only by circuit impedance [1] limitations above this frequency. In this paper, the maximum obtainable power is calculated assuming no limitations due to nonuniform current or temperature distributions across the diode [2], [3]. The approach is to calculate the particular diode area and operating current that simultaneously results in the maximum allowable junction temperatures and minimum diode impedance accepted by the microwave circuitry. This approach is valid until the current density value exceeds the maximum allowable as determined by space-charge considerations of IMPATT theory. This occurs at frequencies well into the millimeter-wave region. Beyond this point, device area and operating current are determined solely by the current density constraint and circuit impedance limitation.

In the design of an IMPATT oscillator, for a particular frequency, two parameters are of first-order importance: diode area and depletion region width. In this paper, design curves are presented for these parameters for the condition that the maximum obtainable CW power is available at burnout. The maximum obtainable able power depends upon electrical circuit and thermal impedance designs. Results presented here pertain to state-of-the-art circuits and devices currently available at Bell Telephone Laboratories. However, the computer code is included so that results for other circuit and device designs can be calculated. For the results presented here, burnout occurs typically at 350°C junction temperature. As stated previously, the design presented here will result in the maximum obtainable output power for a particular frequency at burnout for reliable operation devices would, of course, operate at lower average currents, perhaps by a factor of two, with a corresponding decrease in output power.

In Section II the scaling laws for the thermal and electrical impedance that allow the calculation of the operating conditions for maximum power as a function of frequency, utilizing experimentally observed operating conditions at a particular frequency are developed. In Section III results are presented for silicon single-drift-region (SD) diodes. In Section IV the analysis is extended to predict the maximum obtainable CW power and required operating conditions for silicon double-drift-region (DD) diodes. The agreement with experimental measurements and the implications of the results for IMPATT operation at the highest millimeter-wave frequencies are discussed in Section V.

II. Scaling Equations

In this section simple scaling expressions are developed for the thermal and electrical impedances of an IMPATT diode. The thermal impedance is assumed to be composed of three terms, as illustrated in Fig. 1. The spreading resistance component R_S scales inversely as diode diameter, the contacting area component R_A scales inversely as diode area, and the effective thermal resistance of the silicon R_F scales inversely with both diode area and frequency. The frequency dependence of the latter component occurs due to the increased depletion region width and junction depth with decreasing frequency.[1] In particular, the total thermal resistance R_T for a diameter D at a frequency f is related to the reference values measured for operation at frequency f_0 by

$$R_T = R_{S0}(D_0/D) + R_{A0}(D_0/D)^2 + R_{F0}(D_0/D)^2(W/W_0) \quad (1)$$

Manuscript received August 31, 1970; revised April 5, 1971.
The author is with Bell Telephone Laboratories, Inc., Murray Hill, N. J. 07974.

[1] For the purposes of this analysis, it is assumed that junction depth, like depletion layer width, is inversely proportional to frequency.

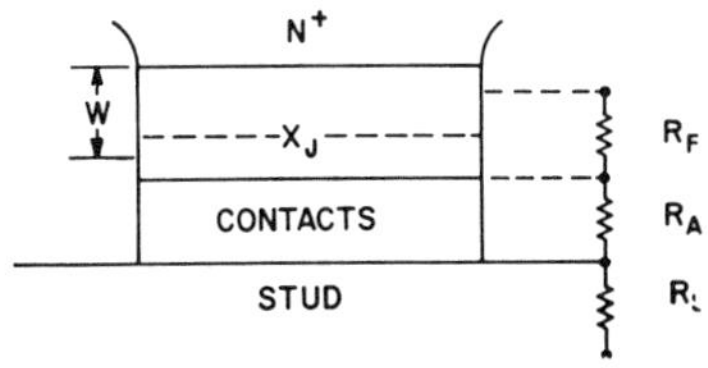

Fig. 1. Thermal resistance components of an IMPATT diode structure. $R_T = R_F + R_A + R_S$; $R_F/R_{F0} = (D_0/D)^2(W/W_0)$; $R_A/R_{A0} = (D_0/D)^2$; $R_S/R_{S0} = (D_0/D)$.

where R_{S0}, R_{A0}, and R_{F0} are the reference thermal resistance components whose total value is R_{T0}, D_0 the reference diameter, and W and W_0 the depletion region width at frequency f and f_0, respectively. This completes the development of a simple scaling law for the diode thermal resistance. A scaling relationship is next developed for the real part of the diode's electrical impedance (7) which, when coupled with a simple scaling relationship for circuit loss (8) and allowable temperature rise (9), yields a relationship for required device diameter (10) for any frequency.

The real part of the diode electrical impedance (negative resistance) is obtained from the simple equivalent circuit picture of an IMPATT diode as a negative conductance G shunted by a large capacitive susceptance S. The negative conductance results from particle current, whose amplitude equals the dc value flowing out of phase with the diode voltage, i.e.,

$$G = \frac{I_{dc}}{\delta V_{dc}}, \tag{2}$$

where δ is an ac voltage modulation factor.[2] For the purpose of the simplified analysis presented in this paper, the negative conductance is assumed to scale according to

$$G/G_0 = (I/I_0)(V_0/V) \tag{3}$$

where the subscript (0) always refers to reference value. The susceptance $S = \omega C$ or

$$S/S_0 = (D/D_0)^2(f/f_0)(W_0/W). \tag{4}$$

The effects of variation of breakdown field with depletion region width [4] are included by scaling the breakdown voltage according to

$$V/V_0 = (W/W_0)^{0.82}, \tag{5}$$

and the consequence of second-harmonic tuning effects [5] by scaling depletion region widths according to

$$W/W_0 = (f_0/f)(q/q_0), \tag{6}$$

where q is defined equal to unity for fundamental mode and less than unity (as determined experimentally ~0.833) for second-harmonic tuning operation. The negative resistance $(G/S)^2$ therefore scales according to

$$R/R_0 = (I/I_0)(D_0/D)^4(f_0/f)^{3.18}(q/q_0)^{1.18}. \tag{7}$$

The minimum real part of the microwave circuit impedance is limited by skin effects [6] and hence the requirement on R will increase as the square root of frequency[3] or

$$R/R_0 = (f/f_0)^{0.5}. \tag{8}$$

The required operating condition is obtained by requiring the temperature rise ΔT to be equal to the reference value ΔT_0 or

$$R_T I V(1 - \mathcal{E}) = R_{T0} I_0 V_0 (1 - \mathcal{E}_0), \tag{9}$$

where $\mathcal{E}$ and $\mathcal{E}_0$ are the efficiency and reference efficiency, respectively.

The analysis can be readily extended to include multiple units N in parallel and (9) reduces to the form 241

$$A(D/D_0)^3 + B(D/D_0)^2 = C \tag{10}$$

where

$$A = R_{S0}$$
$$B = R_{A0} + R_{F0}(q/q_0)(f_0/f)$$
$$C = R_{T0}(N_0/N)\frac{1 - \mathcal{E}_0}{1 - \mathcal{E}}(f_0/f)^{2.86}(q/q_0)^{0.36}. \tag{11}$$

The calculation therefore proceeds as follows. Reference (experimentally determined) values are required for f_0, D_0, I_0, V_0, $\mathcal{E}_0$, q_0, R_{S0}, A_{A0}, and R_{F0}. Equation (10) is solved for D at any other frequency f and the corresponding values of I and V are calculated from the scaling laws and output power calculated from

$$P/P_0 = (I/I_0)(V/V_0)(\mathcal{E}/\mathcal{E}_0). \tag{12}$$

Appendix I extends the treatment to include the ultimate limits on current density. The extension of the treatment to predict results on diamond mounting studs with reference information obtained on a copper stud is presented in Appendix II. Appendix III extends the treatment to include double-drift-region structures and Appendix IV is a listing of the computer program. Results of the analysis for silicon p^+-n-n^+ structures are presented next.

III. Silicon p^+-n-n^+ Structures

A prediction for the power-frequency characteristics of silicon p^+-n-n^+ IMPATT oscillators, based upon the scaling laws presented in the previous section, requires use of the best experimental results available. The

[2] Modulation factor is defined as the ratio of ac (peak) value to dc level. In large-signal operation the particle current amplitude is always nearly equal to the dc value. In addition, it is assumed that the ac voltage modulation factor is also approximately a constant for the condition of maximum output power.

[3] It is recognized that (8) is a simplification of the general impedance limitations. However, more accurate relationships could only be obtained by detailed analysis of individual circuits.

"best" results are those that are not exceeded in predicted power output by any others and, in addition, are accompanied by sufficient and accurate reference values for the thermal design and the operating conditions. To date only one completely specified set of results is available, as reported by Edwards [7] at a frequency of 55 GHz. Nearly as complete are those results reported by Swan [8] for multiple units[4] mounted on a diamond heat sink and operating at 13.3 GHz. It will be shown that Swan's results are in agreement with the scaling laws when referenced to Edward's results.

Presented first are results for single units operating in the fundamental mode with copper heat sinks. Shown in Figs. 2–4 are power output, operating current, and diode diameter versus frequency for various values of obtainable efficiency. Shown in Figs. 5–7 are the corresponding characteristics for multiple units at a fixed efficiency of 8 percent. The limiting characteristic obtained at the highest frequencies occurs from nonthermal space-charge limitations on current density, as
242 discussed in Appendix I.

The results for multiple units on diamond mounting studs are shown in Figs. 8–10. The method for obtaining these results from a reference result on copper is described in Appendix II. Shown in the figures are Swan's experimental values which are in excellent agreement with the predicted results. The good agreement ensures that Edwards' results at 55 GHz can, in turn, be predicted from Swan's results, and this result is discussed more fully later. Required values for the thermal resistance components for all the previous results can be obtained either directly from the tables included in the Appendix or scaled from these according to the previously derived formulas.

The limitation on the power-impedance product imposed by the ultimate space-charge-limited current density for frequencies above approximately 100 GHz may be sufficient to eliminate conventional IMPATT diode sources as candidates for some applications. In addition, the effects of quantum-mechanical tunneling and impact ionization mean-free paths approaching the depletion region thickness are limitations to be considered above 100 GHz. The double-drift-region structures [9] discussed in the next section are expected to considerably improve the power impedance products in the upper millimeter-wave region.

IV. Silicon p^+-p-n-n^+ Structures

An improvement in the power impedance product of a factor of four is predicted [9] for double-drift-region structures over comparable (in frequency) single-drift-region structures. This improvement can be utilized fully only at the ultimate (nonthermal) space-charge-limited current density. As shown in the results, somewhat less improvement is predicted where thermal limitations are dominant. The double-drift-region device should also have improved efficiency [9] over

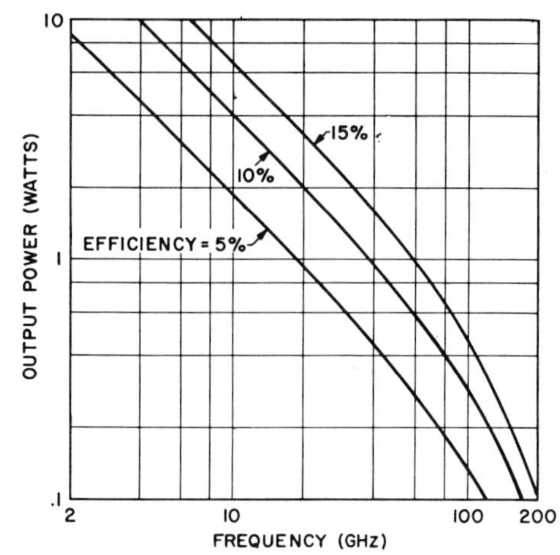

Fig. 2. Power output versus frequency for silicon IMPATT oscillators on copper heat sinks with efficiencies of 5, 10, and 15 percent.

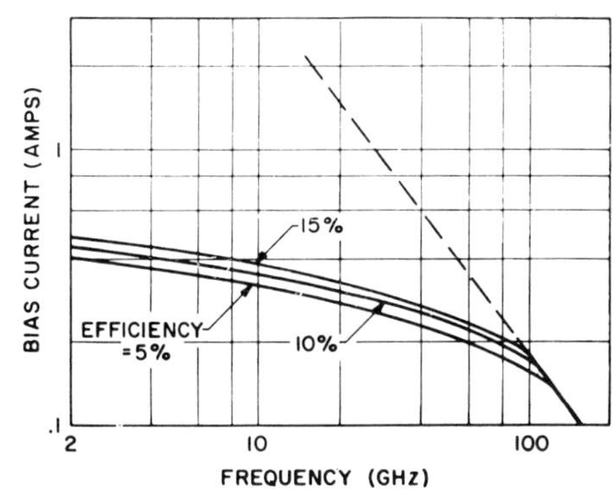

Fig. 3. Operating current versus frequency for silicon IMPATT oscillators on copper heat sinks with efficiencies of 5, 10, and 15 percent.

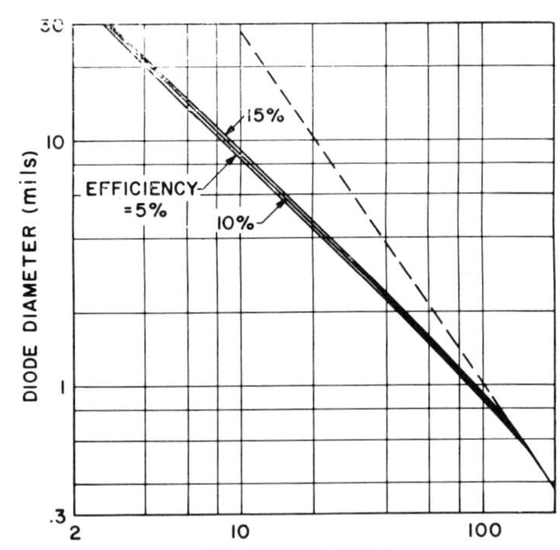

Fig. 4. Diode diameter versus frequency for silicon IMPATT oscillators on copper heat sinks with efficiencies of 5, 10, and 15 percent.

[4] Multiple units refer only to multiply parallel connected units.

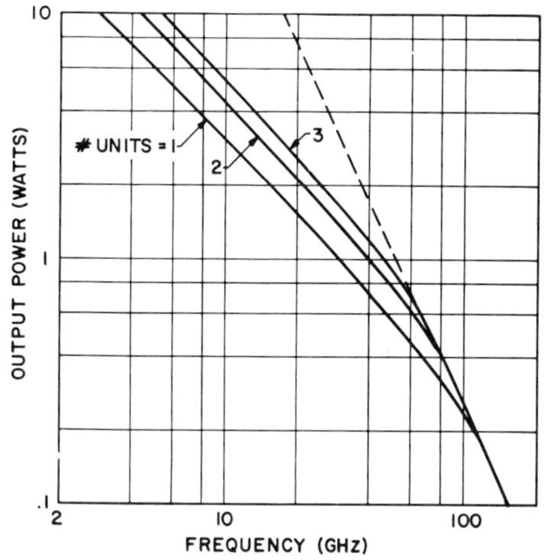

Fig. 5. Power output versus frequency for one, two, and three silicon IMPATT diodes mounted on a coppei heat sink with efficiency of 8 percent.

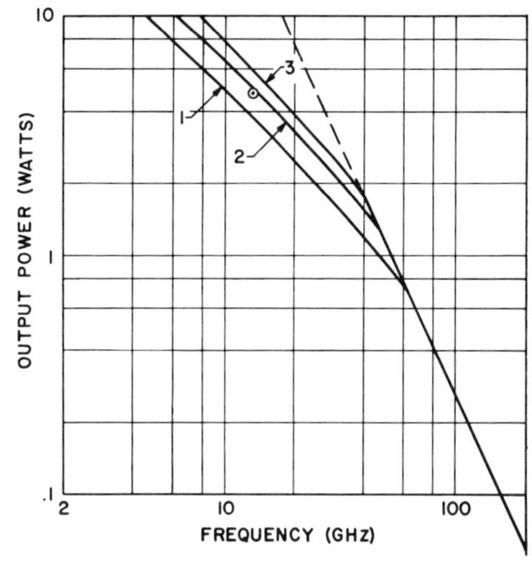

Fig. 8. Output power versus frequency for one, two, and three silicon IMPATT diodes mounted on a diamond heat sink with efficiency of 8 percent.

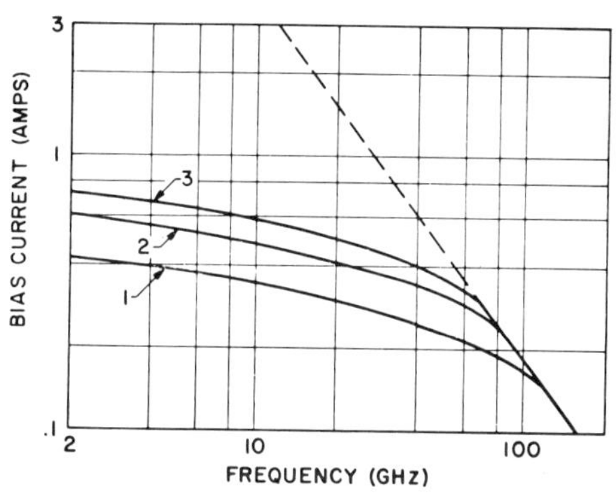

Fig. 6. Operating current versus frequency for one, two, and three silicon IMPATT diodes mounted on a copper heat sink with efficiency of 8 percent.

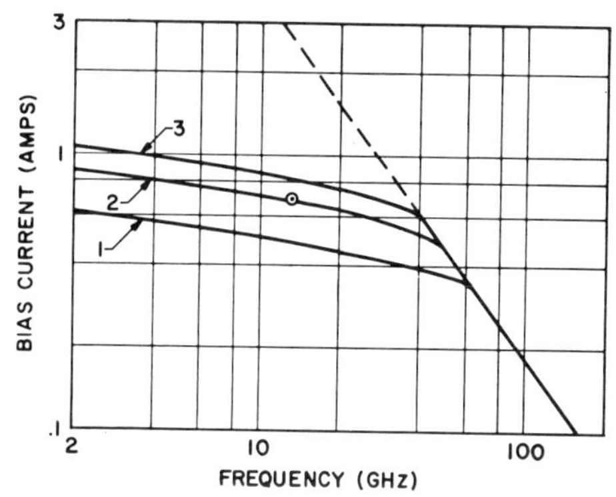

Fig. 9. Operating current versus frequency for one, two, and three silicon IMPATT diodes mounted on a diamond heat sink with efficiency of 8 percent.

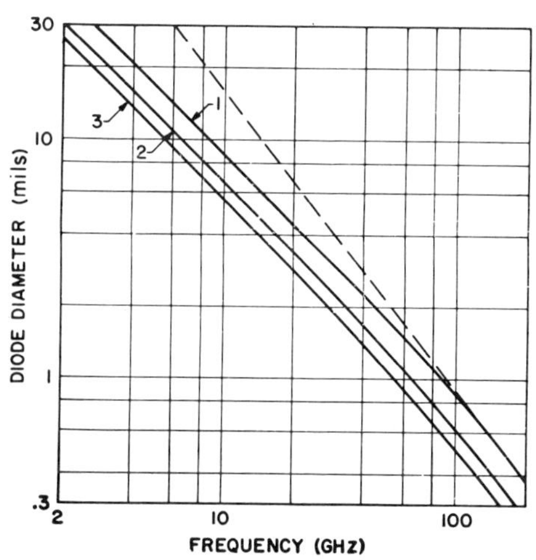

Fig. 7. Diode diameter versus frequency for one, two, and three silicon IMPATT diodes mounted on a copper heat sink with efficiency of 8 percent.

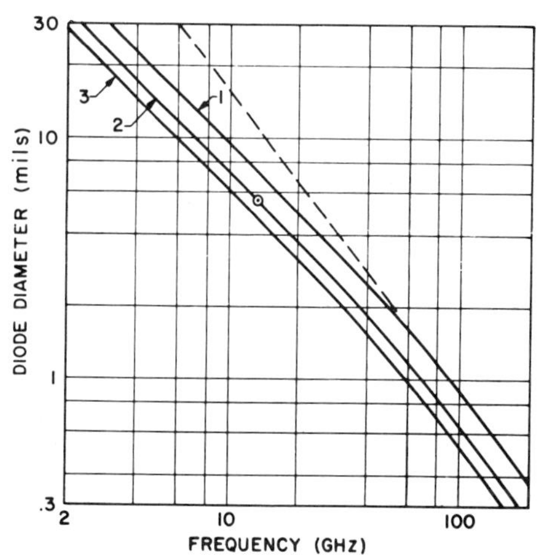

Fig. 10. Diode diameter versus frequency for one, two, and three silicon IMPATT diodes mounted on a diamond heat sink with efficiency of 8 percent.

p^+-n-n^+ devices and results are shown for a 10 percent efficiency in Figs. 11–13. The results are for single and multiple units on a copper heat sink. As predicted earlier, the most significant improvement occurs at the highest frequencies where, for example, 300 mW is predicted at a frequency of 110 GHz. For completeness, Figs. 14–16 show results for multiple double-drift-region units on diamond heat sinks.

V. Discussion

The scaling laws derived in this paper have been checked by comparison of Swan's and Edwards' results. Although this presentation used Edwards' results as reference, because of the detailed thermal data available, concurrent results are obtained by starting with Swan's values. This approach is more dramatic in that Edwards' values at 55-GHz output power, device diameter, and bias current might have been predicted from the much earlier measurements at 13.3 GHz. Inasmuch as the scaling laws predicted observable results up to 55 GHz, they were applied to predict results not yet obtained, i.e., up to 100 GHz. The paper also demonstrated the versatility of the analysis to utilize experimentally determined circuit impedance limitations, as observed with one set of measurements with a particular thermal design (for example, two units on diamond) to predict the expected results with another thermal design (single units on copper). For this reason, a sample calculation and a listing of the program is included in the Appendix. This will enable others with different thermal designs or superior microwave results to scale these results to other frequencies for comparison purposes.

Appendix I

Current Density Limitation

The maximum current density for efficient operation of an IMPATT diode is that value which results in a critical magnitude of generated charge. The critical magnitude of charge is that value which requires (by Gauss's law, see Fig. 17) a field variation in the drift region exceeding the critical value for impact ionization. This limiting value of current density is proportional to frequency and critical field. The critical field scales with frequency to the 0.18 power, hence the limiting current scales as $f^{1.18}$. Information on degradation of efficiencies with operating current density [10] indicates that efficient operation decreases at a current density of approximately 20 000 A/cm² at a frequency of 50 GHz.

The analysis presented prohibits dc bias beyond J_c

$$(J_c/20\,000) = (f/50 \text{ GHz})^{1.18} \tag{13}$$

and evaluates the diameter at this current density to satisfy solely the circuit impedance limitation (7). The dashed characteristics presented in the figures represent operation at this limiting current density. The combination of skin effect and critical field variations with frequency yields a falloff as $f^{-2.14}$ for output power for the limiting characteristic.

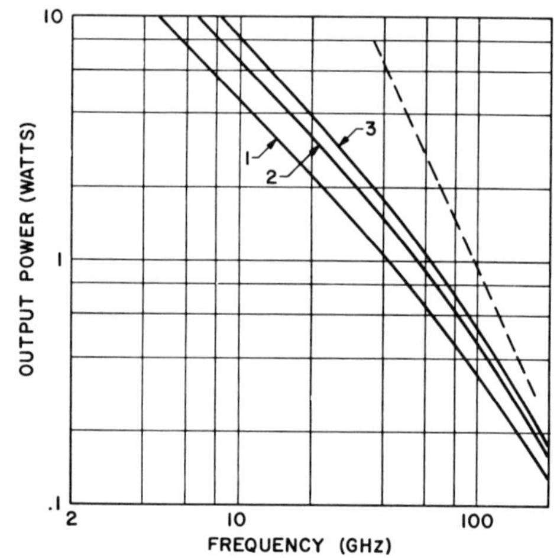

Fig. 11. Output power versus frequency for one, two, and three double-drift-region silicon IMPATT diodes on copper heat sinks with efficiencies of 10 percent.

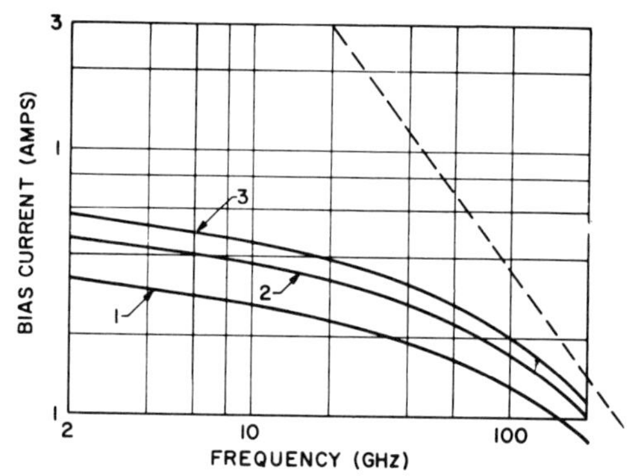

Fig. 12. Operating current versus frequency for one, two, and three double-drift-region silicon IMPATT diodes on copper heat sinks efficiencies of 10 percent.

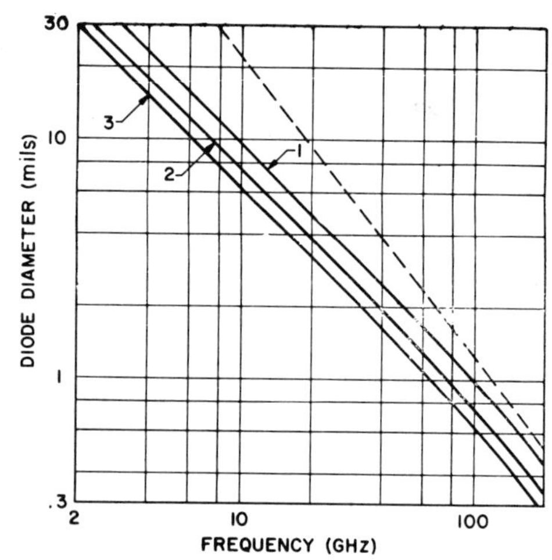

Fig. 13. Diode diameter versus frequency for one, two, and three double-drift-region silicon IMPATT diodes on copper heat sinks with efficiencies of 10 percent.

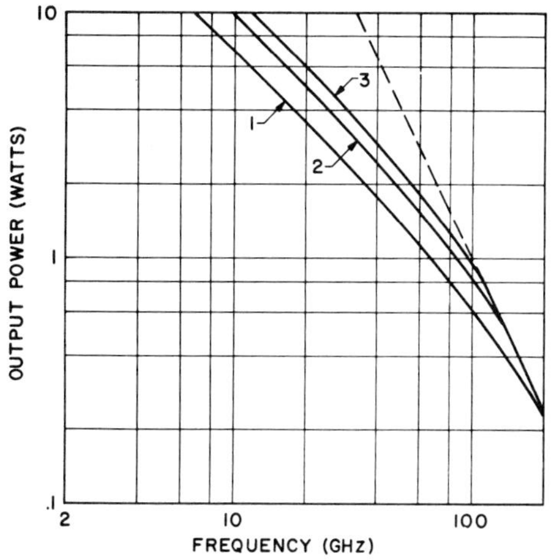

Fig. 14. Output power versus frequency for one, two, and three double-drift-region silicon IMPATT diodes on diamond heat sinks with efficiencies of 10 percent.

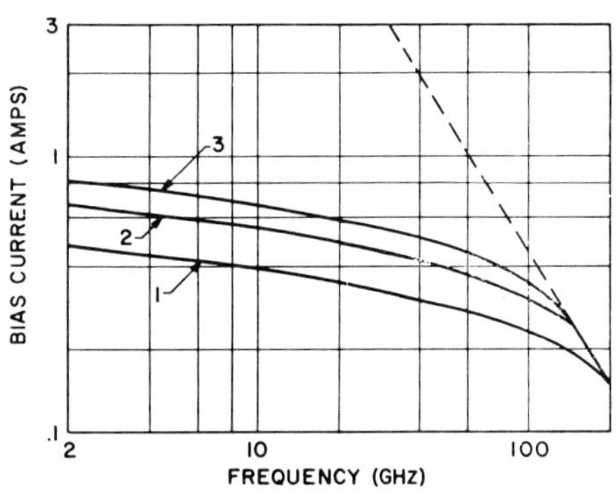

Fig. 15. Operating current versus frequency for one, two, and three double-drift-region silicon IMPATT diodes on diamond heat sinks with efficiencies of 10 percent.

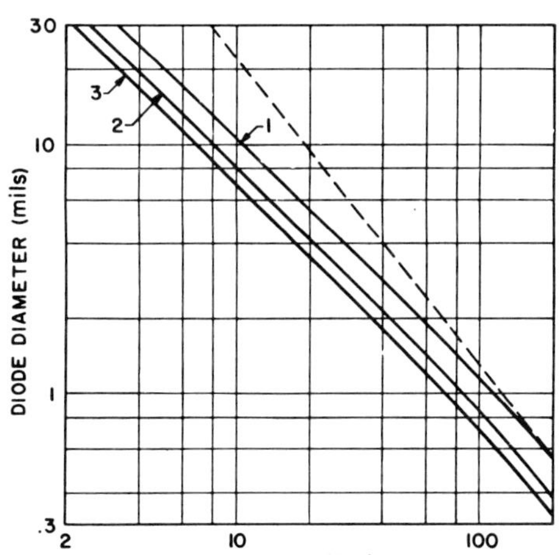

Fig. 16. Diode diameter versus frequency for one, two, and three double-drift-region silicon IMPATT diodes on diamond heat sinks with efficiencies of 10 percent.

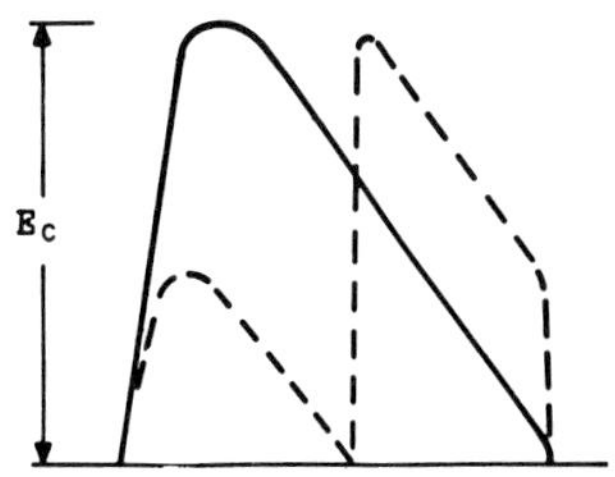

Fig. 17. Schematic of electric field versus distance for the derivation of ultimate limiting bias current density. $E_C = Q_C/\epsilon$; $J_C = Q_C f = \epsilon E_C f$; 20 000 (A/cm²) = 1 (pF/cm) 400 (kV/cm) 50 (GHz); $(J_C/20\,000) = (f/50 \text{ GHz})^{1.18}$.

Appendix II

Variations of Thermal Design

This Appendix presents the modifications necessary to predict output power with a thermal design differing from the reference design.

The approach is to calculate new reference values of P, D, and I consistent with the new thermal design. For the purpose of this Appendix, the new reference values will appear without subscripts. Changes in thermal conductivity, thickness of metallization layers, or junction depth designs are assessed by employing modification factors F_S, F_A, and F_F as given below:

$$R_S = R_{S0}F_S(D_0/D)$$
$$R_A = R_{A0}F_A(D_0/D)^2$$
$$R_F = R_{F0}F_F(D_0/D)^2 \tag{14}$$

and the new reference thermal resistance (total) is

$$R_T = R_S + R_A + R_F.$$

The new operating conditions must satisfy the constraints imposed by temperature rise [see (9)]

$$R_T I = R_{T0} I_0 \tag{15}$$

and negative resistance [see (7)]

$$I/I_0 = (D/D_0)^4. \tag{16}$$

The new reference point is at the same frequency ($f = f_0$, $R = R_0$, and $V = V_0$) and the same mode ($q = q_0$) is assumed. The new reference diameter satisfies the relationship

$$R_{T0}(D_0/D)^4 = R_{S0}F_S(D_0/D) + R_{A0}F_A(D_0/D)^2 + R_{F0}F_F(D_0/D)^2$$

or

$$A(D/D_0)^3 + B(D/D_0)^2 = C$$

where

$$A = R_{S0}F_S$$
$$B = R_{A0}F_A + R_{F0}F_F$$
$$C = R_{T0}. \tag{17}$$

The new reference value of D is obtained as the solution to (17) and determines the new reference value of I

from (16). The new reference value of output power is obtained from

$$P = P_0(I/I_0) \tag{18}$$

and new reference values for the thermal resistance components are obtained from (14).

Appendix III

Variation of Basic IMPATT Structure

This Appendix presents the modification necessary to obtain reference values for a change in the IMPATT structure. For example, double-drift-region structures [9] have an improved negative resistance (per unit area) of a factor of 2 and larger operating voltage by a factor of 1.5. Improved efficiency would also result from the effective sharing of the avalanche region. In addition, provision is made to evaluate results at different operating temperatures and critical limiting current densities. Modification factors are incorporated for negative resistance (F_R), breakdown voltage (F_V), critical current density (F_C), and operating temperature (F_T).

246

Increases in available negative resistance, operating voltage, and operating temperature require a modification to the C terms in (11):

$$C = CF_R/F_VF_T. \tag{19}$$

The resulting calculation for bias current I, voltage V, and critical current density J_c are modified as

$$I = I/F_R,$$

$$V = V * F_V,$$

and

$$J_C = J_C * F_C.$$

A complete listing of the program, which has been run on the Com-Share Time-Sharing System, is contained in Appendix IV.

Appendix IV

```
+C
ØUTPUT:/PØWERB/
ØLD FILE?
ØPTIØNS: CARDS

+RUN
ØPTIØNS:
SPRØG:
XLIBE  JAN 14


 NØ, FØ, PØ, DØ, CØ, VØ, RS, RA, RF, QØ, EØ
1,54.,.464,1.53,.240,18.,30.,10.,10.,.8333,.077


  RFACTØR VFACTØR JCFACTØR TFACTØR
1,1,1,1


FS,FA,FF
.69,.5,1.

 N, Q, E
2,1.,.077

    F         P         D          C        V         RS        RA        RF     J/JC
  2.000    28.837    33.832    0.861  311.841     0.936     0.010     0.663  0.17
  2.500    23.411    27.378    0.839  259.697     1.157     0.016     0.810  0.19
  3.000    19.736    23.027    0.822  223.635     1.375     0.022     0.954  0.21
  4.000    15.059    17.521    0.794  176.640     1.808     0.038     1.235  0.25
  5.000    12.196    14.170    0.772  147.103     2.235     0.058     1.511  0.29
  6.000    10.258    11.911    0.754  126.676     2.659     0.082     1.782  0.32
  8.000     7.791     9.052    0.725  100.056     3.499     0.143     2.314  0.38
 10.000     6.282     7.313    0.702   83.326     4.331     0.219     2.836  0.43
 13.400     4.721     5.523    0.671   65.547     5.734     0.384     3.711  0.51
 16.000     3.962     4.657    0.651   56.676     6.801     0.540     4.372  0.57
 20.000     3.168     3.754    0.625   47.199     8.436     0.830     5.362  0.64
 25.000     2.523     3.024    0.598   39.307    10.475     1.280     6.638  0.73
 30.000     2.087     2.531    0.574   33.848    12.513     1.827     7.893  0.81
 40.000     1.535     1.908    0.535   26.736    16.597     3.214    10.415  0.94
 50.000     1.072     1.487    0.448   22.265    21.296     5.292    13.718  1.00
 55.000     0.874     1.320    0.395   20.591    23.991     6.716    15.827  1.00
 60.000     0.726     1.184    0.352   19.173    26.748     8.348    18.033  1.00
 80.000     0.392     0.826    0.241   15.144    38.323    17.137    27.764  1.00
110.000     0.198     0.555    0.158   11.664    57.061    37.993    44.764  1.00
200.000     0.055     0.263    0.072    7.144   120.471   169.354   109.746  1.00


  RFACTØR VFACTØR JCFACTØR TFACTØR
2.,1.5,1.,1.


FS,FA,FF
1.,1.,1.

 N, Q, E
1,1.,.11

    F         P         D          C        V         RS        RA        RF     J/JC
  2.000    36.339    49.826    0.506  467.762     0.636     0.005     0.306  0.09
  2.500    29.471    40.310    0.493  389.546     0.786     0.007     0.373  0.10
  3.000    24.826    33.899    0.482  335.452     0.934     0.010     0.440  0.11
  4.000    18.925    25.786    0.466  264.960     1.228     0.018     0.570  0.14
  5.000    15.321    20.852    0.453  220.655     1.519     0.027     0.698  0.16
  6.000    12.883    17.528    0.442  190.014     1.807     0.038     0.823  0.17
  8.000     9.787    13.321    0.425  150.084     2.377     0.066     1.069  0.21
 10.000     7.896    10.763    0.412  124.988     2.942     0.101     1.309  0.23
 13.400     5.944     8.133    0.394   98.320     3.894     0.177     1.712  0.28
 16.000     4.995     6.859    0.383   85.014     4.617     0.249     2.015  0.31
 20.000     4.005     5.534    0.369   70.799     5.723     0.382     2.477  0.35
 25.000     3.201     4.461    0.354   58.960     7.100     0.588     3.050  0.40
 30.000     2.658     3.738    0.341   50.773     8.474     0.838     3.620  0.44
 40.000     1.970     2.823    0.320   40.103    11.217     1.468     4.758  0.52
 50.000     1.552     2.268    0.303   33.398    13.966     2.276     5.900  0.58
 55.000     1.399     2.064    0.295   30.887    15.345     2.748     6.475  0.61
 60.000     1.271     1.893    0.288   28.760    16.727     3.265     7.052  0.64
 80.000     0.917     1.421    0.263   22.716    22.295     5.800     9.397  0.74
110.000     0.626     1.028    0.233   17.496    30.797    11.068    13.040  0.86
200.000     0.236     0.526    0.144   10.716    60.236    42.338    27.436  1.00


  RFACTOR VFACTØR JCFACTØR TFACTØR
(SMAINS)30+1
+
-P
?
-CØN
XTR

+QED
*P
```

```
      DIMENSIØN F(20)
      DATA F/2.,2.5,3.,4.,5.,6.,8.,10.,13.4,16.,
     120.,25.,30.,40.,50.,55.,60.,80.,110.,200./
      FS=1.
      AR=1.
      AV=.82
      T2=4.+0.5-AV
      T1=T2-AV
      WRITE(1,3)
3     FØRMAT(/,43H NØ, FØ, PØ, DØ, CØ, VØ, RS, RA, RF, QØ, EØ)
      ACCEPT NØ,FØ,PØ,DØ,CØ,VØ,RSØ,RAØ,RFØ,QØ,EØ
      RTØ=RSØ+RAØ+RFØ
100   CØNTINUE
      WRITE(1,30)
30    FØRMAT(//,S  RFACTØR VFACTØR JCFACTØR TFACTØRS)
      ACCEPT FACR,FACV,FACJ,FACT
      WRITE(1,6)
6     FØRMAT(//,8HFS,FA,FF)
      ACCEPT FS,FA,FF
      IF(FS.EQ.1.) IF(FA.EQ.1.) IF(FF.EQ.1.) GØ TØ 15
      A=RSØ*FS
      B=RAØ*FA+RFØ*FF
      C=RTØ
      X=FUNC(A,B,C)
      DØ=DØ*X
      CØ=CØ*X**4
      PØ=PØ*X**4
      RSØ=RSØ*FS/X
      RAØ=RAØ*FA/X/X
      RFØ=RFØ*FF/X/X
      RTØ=RSØ+RAØ+RFØ
15    CØNTINUE
      WRITE (1,16)
16    FØRMAT(/,8H N, Q, E)
      ACCEPT N,Q,E
      WRITE(1,4)
4     FØRMAT(/,46H    F         P         D          C        V         RS
     123H        RA        RF     J/JC)
      DØ 20 I=1,20
      FFØ=F(I)/FØ
      A=RSØ
      B=RAØ+RFØ/QØ*Q/FFØ**AR
      C=RTØ/N*NØ*(1.-EØ)/(1.-E)/FFØ**T1*(Q/QØ)**(2.-AV)
      C=C*FACR/FACV*FACT
      DDØ=FUNC(A,B,C)
      D=DDØ*DØ
      DØD=1./DDØ
      C=CØ*N*N*(DDØ*FFØ)**4/NØ/FFØ**(AV-.5)/NØ*(QØ/Q)**2
      C=C/FACR

                  V=VØ*(Q/QØ/FFØ)**AV
                  V=V*FACV
                  P=PØ*V/VØ*C/CØ*E/EØ
                  RS=RSØ*DØD
                  RA=RAØ*DØD*DØD
                  RF=RFØ*DØD*DØD*Q/QØ/FFØ
```

```
      CL=2.0E4*(F(I)/50.)**(2.-AV)
      CL=CL*FACJ
      A=3.14159/4.*(D/1000.*2.54)**2
      AL=N*A*A/C*CL
      CA=C/N/A
      S=CA/CL
      IF(S.LT.1.) GØ TØ 20
      S=1.
      D=D*SQRT(AL/A)
      C=N*CL*AL
      P=PØ*V/VO*C/CØ*E/EØ
      DØD=DØ/D
      RS=RSØ*DØD
      RA=RAØ*DØD*DØD
      RF=RFØ*DØD*DØD*Q/QØ/FFØ
 20   WRITE(1,5)F(I),P,D,C,V,RS,RA,RF,S
 5    FØRMAT(8F8.3,F5.2)
      GØ TØ 100
      END
      FUNCTIØN FUNC(A,B,C)
C   THIS FUNCTIØN SØLVES A*X**3+B*X**2=C FØR X
C   FØR A B AND C ALL .GT. 0.
      XØ=.001
      X1=XØ/1.1
 1    X1=X1*1.1
      Z1=A*X1**3
      Y1=C-B*X1**2
      IF(Y1.GT.Z1) GØ TØ 1
      X2=X1/1.1
      DEL1=Z1-Y1
 2    DEL2=A*X2**3+B*X2**2-C
      IF(ABS(DEL2)/C.LT.1.E-7) GØ TØ 3
      X=(X1*DEL2-X2*DEL1)/(DEL2-DEL1)
      DEL1=DEL2
      X1=X2
      X2=X
      GØ TØ 2
 3    FUNC=X2
      RETURN
      END
*XTRAN
+
-LØG
USAGE
CCU:    175
CLT:    0.68  HØURS
P
```

Acknowledgment

The author acknowledges many useful discussions with R. Edwards and R. M. Ryder.

References

[1] B. C. De Loach, Jr., "Recent advances in solid-state microwave generators," in *Advances in Microwaves*, vol. 2, L. Young, Ed. New York: Academic Press, 1967.

[2] R. H. Haitz, "Nonuniform thermal conductance in avalanche microwave oscillators," *IEEE Trans. Electron Devices*, vol. ED-15, June 1968, pp. 350–361.

[3] G. Gibbons and T. Misawa, "Temperature and current distribution in an avalanching *p-n* junction." *Solid-State Electron.*, vol. 11, 1968, pp. 1007–1014.

[4] S. M. Sze and G. Gibbons, "Avalanche breakdown voltage of abrupt and linearly graded p-n junctions in Ge, Si, GaAs, and GaP," *Appl. Phys. Lett.*, vol. 8, 1966, p. 111.

[5] C. B. Swan, "IMPATT oscillator performance improvement with second-harmonic tuning," *Proc. IEEE* (Lett.), vol. 56, Sept. 1968, pp. 1616–1617.

[6] T. Moreno, *Microwave Transmission Design Data.* New York: Dover, 1958.

[7] R. Edwards, D. F. Ciccolella, T. Misawa, and D. E. Iglesias, "Millimeter-wave silicon Impatt diodes," presented at the 1969 IEEE Internat. Electron Devices Meeting, Washington, D. C., Oct. 29–31, 1969, paper 2.3.

[8] C. B. Swan, "Improved performance of silicon avalanche oscillator mounted on diamond heat sinks," *Proc. IEEE* (Lett.,) vol. 55, Sept. 1967, pp. 1617–1618.

[9] D. L. Scharfetter, W. J. Evans, and R. L. Johnston, "Double-drift-region (p^+pnn^+) avalanche diode oscillators," *Proc. IEEE* (Lett.), vol. 58, July 1970, pp. 1131–1133.

[10] R. Edwards, private communication.

Reprinted from IEEE TRANSACTIONS
ON *ELECTRON DEVICES*
Volume ED-18, Number 8, August, 1971
pp. 536-543

HARMONIC EFFECTS IN THE IMPATT MODE III

Reprinted from IEEE TRANSACTIONS
ON *MICROWAVE THEORY AND TECHNIQUES*
Volume MTT-18, Number 6, June, 1970
pp. 327-331

Effect of Harmonic and Subharmonic Signals on Avalanche-Diode Oscillator Performance

Abstract—**The purpose of this correspondence is to show the effects of harmonic and subharmonic signals on the operation of an avalanche-diode oscillator. Generation of harmonic and subharmonic signals is also considered.**

In order to study multifrequency operation of IMPATT diodes it is advantageous to adopt a somewhat idealized model of the physical processes occurring in the diode. This approach provides physical insight and permits an economical examination of a wide range of parameter values. For these reasons the model used is the Read-diode model.[1]

In the Read model the diode depletion region is divided into two adjacent regions. In the avalanche region the electric field is high and impact ionization takes place. In the adjacent drift region it is assumed that no impact ionization takes place and that the avalanche-injected carriers drift at constant velocity. For simplicity the electric field in the avalanche region is assumed to be independent of x. It is assumed that the ionization rates and the saturated drift velocities are equal for electrons and holes. Under these conditions the following equations for the avalanche and drift regions are obtained:[2]

$$\frac{dJ_0}{d\theta} = \left(\frac{2v}{\omega}\right) J_0(\theta)\left(\alpha(E_0) - \frac{1}{x_1}\right) \quad (1)$$

and

$$E_0 = E_c + \frac{\tau}{2\epsilon} J_{dc} - \frac{V(\theta)}{v\tau} - \frac{1}{\epsilon\tau\omega^2}\int_0^{\omega\tau} \theta' J_0(\theta' + \theta - \omega\tau)d\theta', \quad (2)$$

where

J_o = the injected current density (A/cm²),
ω = the radian frequency of the fundamental (rad/s),
v = the saturated drift velocity of the carriers (cm/s),
$\theta = \omega t$,
α = the ionization rate (1/cm),
E_0 = the electric field in the avalanche region (V/cm),
x_1 = the width of the avalanche region (cm),
E_c = the dc field in the avalanche region (V/cm),
τ = the transit time in the drift region (s),
ϵ = the dielectric constant (F/cm),
J_{dc} = the dc component of injected current (A/cm²), and
V = the ac terminal voltage.

Equation (2) may be substituted into (1) to eliminate E_0. This results in a highly nonlinear equation for J_0. In order to obtain a closed-form solution, Evans and Haddad[2] made two further assumptions: they considered the case of diodes with small transit angles to simplify the space-charge integral term in (2) and they linearized the ionization rate. These further assumptions are not made here. Instead, (1) is solved numerically by a digital computer using an iterative technique.

The total current density is given by

$$J = \frac{\epsilon}{l}\frac{dV}{dt} + \frac{1}{\tau}\int_{t-\tau}^{t} J_0(t')dt', \quad (3)$$

i.e., the sum of the displacement current associated with the depletion layer capacitance plus a term representing the current induced by the injected current. The latter term solely determines the real part of the diode admittance.

It is informative to think of the second term on the right-hand side of (3) as repre-

Manuscript received February 13, 1970. This work was supported by the Rome Air Development Center under Contract F30602-68-C-0043.

[1] W. T. Read, "A proposed high-frequency negative-resistance diode," *Bell Sys. Tech. J.*, vol. 37, pp. 401–466, March 1958.

[2] W. J. Evans and G. I. Haddad, "A large-signal analysis of IMPATT diodes," *IEEE Trans. Electron Devices*, vol. ED-15, pp. 708–717, October 1968.

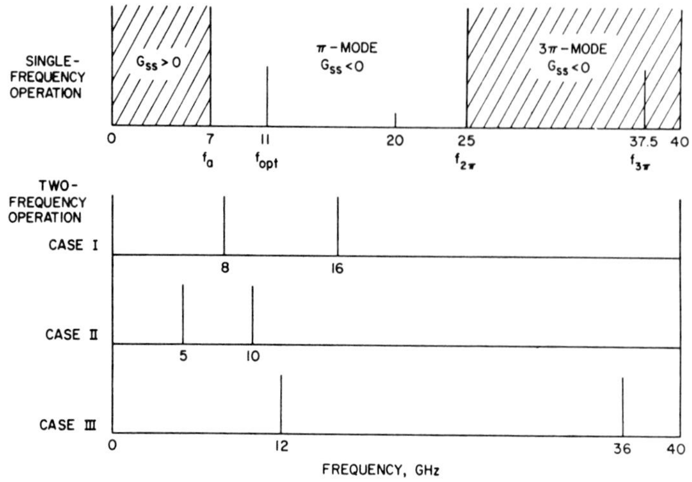

Fig. 1. Comparison of two-frequency cases considered.

senting a linear transformation of the injected current by the drift region. In the frequency domain the transformation is given by

$$\bar{j}_{TP}(\omega) = \bar{c}\cdot\bar{j}_0(\omega), \tag{4}$$

where

$\bar{c} \triangleq \rho\, e^{-i\phi}$,
$\rho = \sin\phi/\phi$,
$\phi = \omega\tau/2$,
$\bar{j}_0(\omega)$ = the complex ω-component of the injected particle current,
$\bar{j}_{TP}(\omega)$ = the complex ω-component of the terminal particle current, and
$\bar{c}$ = a complex constant representing the drift region transformation in the complex plane.

One obvious consequence of this transformation is that for a transit angle of 2π the diode conductance is zero, since $\rho=0$. Thus if a diode is operated in a two-frequency mode, fundamental and second harmonic, where the fundamental is at the near-optimum transit angle π, the second-harmonic conductance is zero. In this configuration the device cannot function as a second-harmonic upconverter. Further, since some loss is inevitably present at the second harmonic, one cannot maintain a second-harmonic voltage unless it is externally supplied.

Another consequence of the transformation is that for transit angles less than π, the drift region gives less than 90 degrees phase delay. Thus, to achieve negative conductance the injected current must have some minimum phase delay. Further, the required phase delay increases with decreasing transit angle.

Equation (1) has been solved for $J_0(\theta)$ and the diode admittance evaluated from (3) for the case of a 5-μm long silicon diode. The parameters used were

$$v = 10^7 \text{ cm/s},$$
$$\alpha(E) = Ae^{-b/E},$$
$$A = 1.315\times10^7 \text{ (1/cm)},$$
$$b = 2.46\times10^6 \text{ (V/cm)},$$
$$x_1 = 1\ \mu\text{m}$$

and

$$J_4 = 500 \text{ A/cm}^2.$$

In single-frequency operation this diode has an avalanche frequency of 7 GHz, an optimum frequency of 11 GHz, and the 2π conductance null occurs at 25 GHz. Several cases of two-frequency operation are considered below; a comparison of the cases with the single-frequency diode characteristic is given in Fig. 1.

Case I: Operation at 8 and 16 GHz

For this case the operating frequencies straddle the optimum single-frequency operation of the diode. For an RF voltage of the form

$$V_{RF} = V_8 \sin\omega t + 10\sin(2\omega t + \psi), \tag{5}$$

Fig. 2 shows the diode admittance for various values of the fundamental voltage V_8 and relative phase angle ψ. Consider the admittance loci for $V_8=20$ volts. When $\psi=135$ degrees the negative conductance at 8 GHz is increased by 50 percent with respect to the single-frequency point; simultaneously, the conductance at 16 GHz is negative. Conversely, the negative conductance at 16 GHz is maximized when $\psi=180$ degrees while the conductance at 8 GHz remains negative; here the increase in negative conductance at 16 GHz is 100 percent. Thus it is possible to increase the negative conductance at either frequency with the proper ψ.

Fig. 3 shows the effect of varying ψ. The figure is a phase-plane plot of the 8 GHz component of injected current. For reference the voltage phasor V_8 is shown and also the injected current phasor for single-frequency operation is indicated. The 8 GHz component of terminal particle current is found by the drift region transformation of (4), which in Fig. 3 corresponds to a clockwise rotation of the injected current locus through $\omega\tau/2$ rad and a radial attenuation corresponding to the value of ρ. Thus the dashed line constructed at an angle of $\omega\tau/2$ in Fig. 3 provides a demarcation for the sign of the conductance. Since in this instance the entire injected-current locus lies above the dashed line, the conductance at 8 GHz is negative for all ψ. The value of ψ which maximizes the negative conductance is found by constructing a line parallel to the dashed line and tangent to the injected-current locus; the point of tangency gives the optimum ψ which, for this case, is approximately 135 degrees.

Case II: Operation at 5 and 10 GHz

This case may be regarded as operation at the optimum frequency and the first subharmonic of the optimum. In single-frequency operation at 5 GHz the diode has positive conductance from small signal to more than 30 volts. However, in the presence of a 10 GHz signal, the solution of (1) shows that both the small- and large-signal conductances at 5 GHz are negative when the proper phase angle is chosen; furthermore, the conductance at 10 GHz remains negative.

Fig. 4(a) shows the RF voltage and injected-current waveforms for

$$V_{RF} = 20\sin\omega t + 20\sin(2\omega t + \psi), \tag{6}$$

where $\psi=270$ degrees, which is the phase angle that produces a maximum overvoltage. Fig. 4(b) shows the 5 GHz components of injected current and terminal particle current for the waveform of Fig. 4(a). Also shown for reference are the same phasors for the case of 20 volt single-frequency operation at 5 GHz. Since $\bar{j}_{TP5}$ lags $\bar{V}_5$ in this case (and the single-frequency case) by less than 90 degrees, it is clear that the conductance at 5 GHz is positive.

On the other hand, if $\psi=135$ degrees the waveforms shown in Fig. 5(a) result. The phasor diagram of Fig. 5(b) shows that $\bar{j}_{05}$ is additionally delayed by this voltage waveform and that the conductance at 5 GHz is large and negative. Thus it is clear that, in order to generate negative conductance at 5 GHz, the voltage waveform must be constructed to delay the injected-current pulse so that $\bar{j}_{05}$ lags $\bar{V}_5$ by at least 54 degrees.

The case of two-frequency operation for higher-order subharmonics of 10 GHz has been examined. We have found that negative conductance at the lower frequencies cannot be obtained in this mode. Specifically, when a signal is present at 10 GHz the conductance at 3.33 GHz is positive, both small and large signal. This can be explained physically as follows.

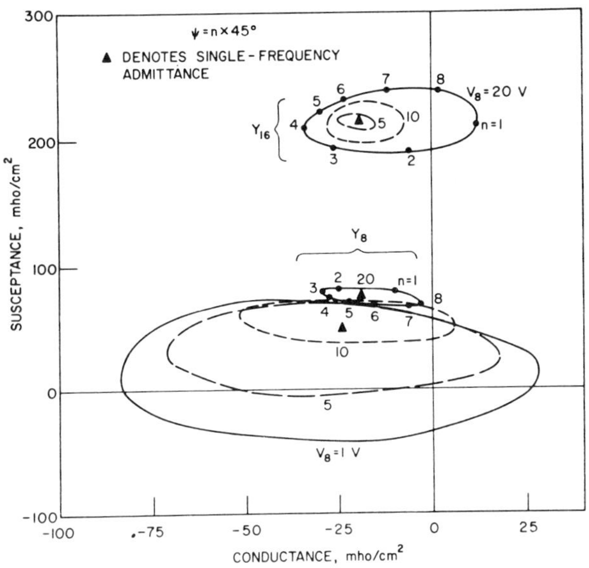

Fig. 2. RF admittance for two-frequency operation. [$V_{RF} = V_8 \sin \omega t + 10 \sin (2\omega t + \psi)$, $f = 8$ GHz, $J_{dc} = 500$ A/cm²]

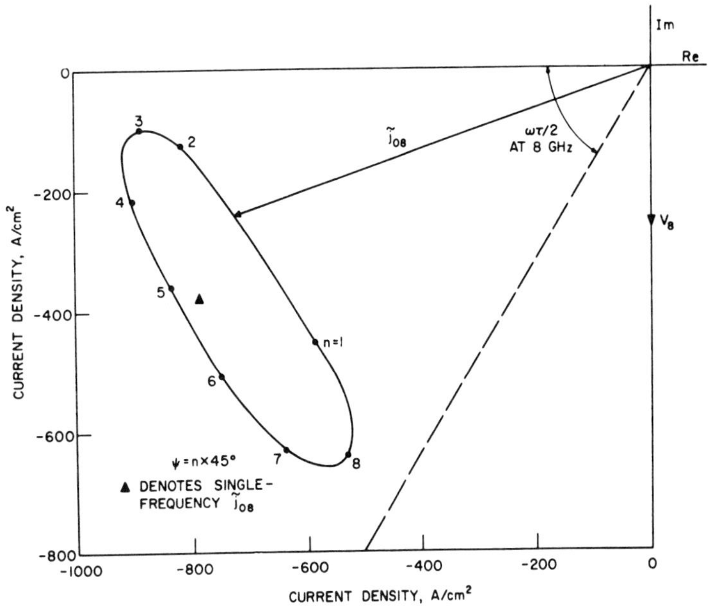

Fig. 3. Phase diagram for fundamental component of injected current. [$V_{RF} = 20 \sin \omega t + 10 \sin(2\omega t + \psi)$, $f = 8$ GHz, $J_{dc} = 500$ A/cm²]

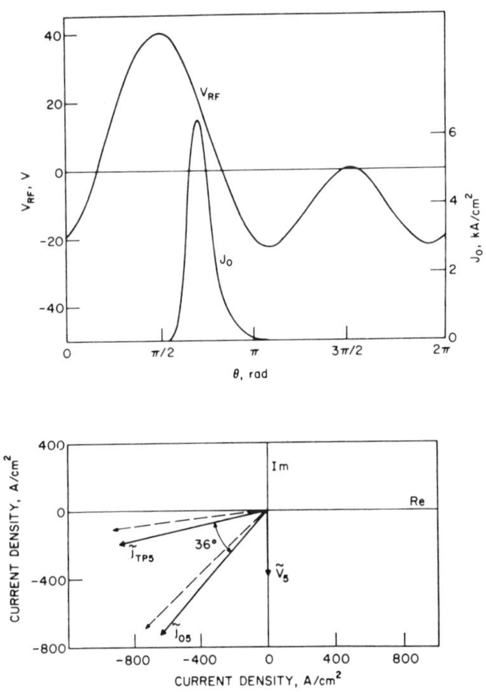

Fig. 4. Voltage and injected-current waveforms. [$V_{RF} = 20 \sin \omega t + 20 \sin (2\omega t + 270$ degrees), $f = 5$ GHz, $J_{dc} = 500$ A/cm²]

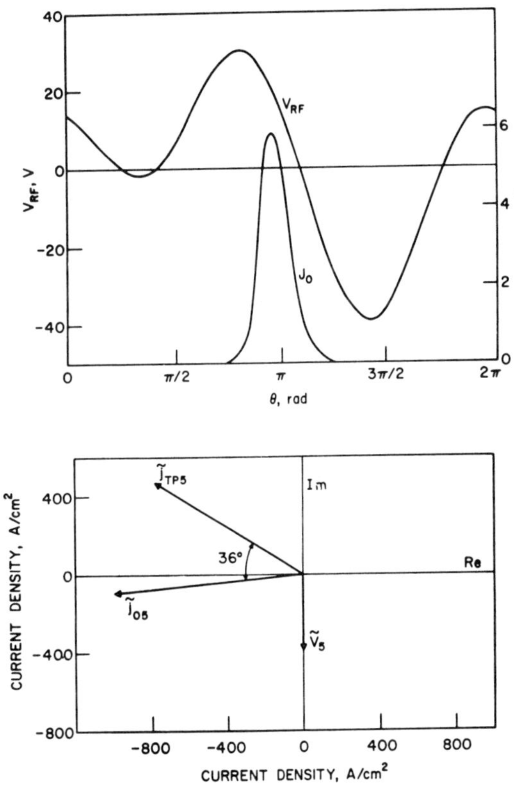

Fig. 5. Voltage and injected-current waveforms. [$V_{RF} = 20 \sin \omega t + 20 \sin(2\omega t + 135$ degrees), $f = 5$ GHz, $J_{dc} = 500$ A/cm²]

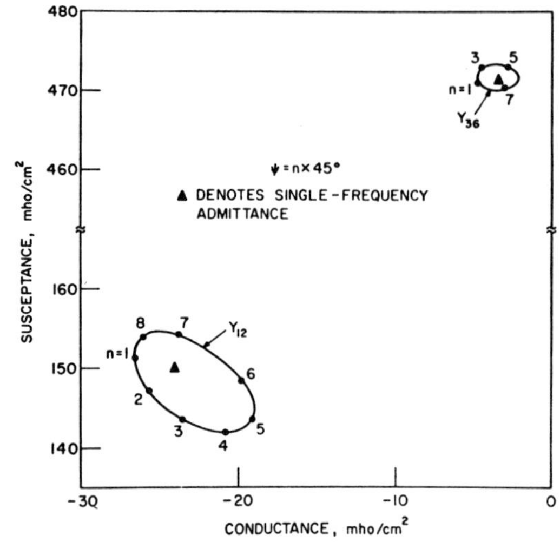

Fig. 6. RF admittance for two-frequency operation. [$V_{RF}=20 \sin \omega t+20 \sin(3\omega t+\psi)$, $f=12$ GHz, $J_{dc}=500$ A/cm²]

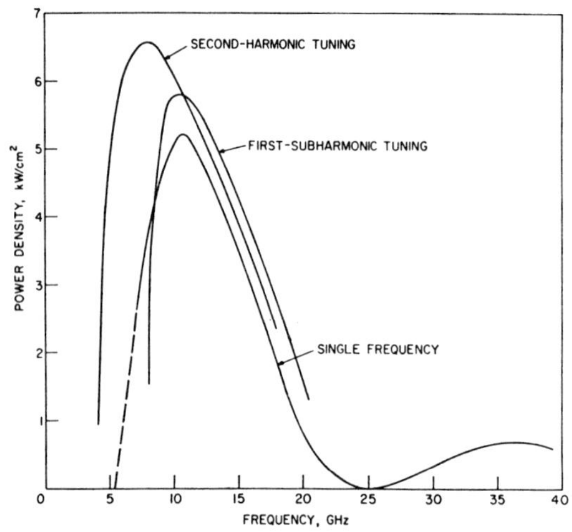

Fig. 7. Optimum power output for various operating conditions. [$V_{RF}=20 \sin \omega t+20 \sin(2\omega t+\psi)$, $J_{dc}=500$ A/cm²]

1) As the frequency is lowered, the drift region supplies less phase delay.

2) As the harmonic ratio of the two-frequency waveform is increased, the maximum phase delay of the injected-current pulse decreases.

Case III: Operation at 12 and 36 GHz

This case may be regarded as operation at the optimum frequency and its third harmonic. (The second-harmonic case is less interesting for the reason discussed previously.) Fig. 6 shows the diode admittance for

$$V_{RF} = 20 \sin \omega t + 20 \sin (3\omega t + \psi). \quad (7)$$

At 12 GHz the negative conductance is at most 15 percent larger than for the 20 volts single-frequency case. Thus third-harmonic tuning does not greatly improve the fundamental frequency performance in this case.

Fig. 6 also shows that the attainable improvement in negative conductance at 36 GHz is, at most, 30 percent. Since single-frequency operation of avalanche diodes in the 3π transit-angle mode has not been found very useful, it is doubtful that the improvement attainable in this two-frequency case is sufficient to make the mode attractive for generating power at 36 GHz. Further, if one considers the PF^2 law[3] for avalanche-diode devices, it can be seen from Fig. 6 that, for constant device reactance, less power is obtained at 36 GHz from the diode operating in this two-frequency mode than could be obtained from a single-frequency π-mode device. It is possible, however, that this two-frequency mode may be useful at very high frequencies where a π-mode diode cannot be fabricated satisfactorily.

[3] B. C. DeLoach, "Recent advances in solid-state microwave generators," *Advances in Microwaves*, vol. 2. New York: Academic Press, Inc., pp. 44–86, 1967.

Conclusions

Fig. 7 shows the power density as a function of frequency for two-frequency modes where the harmonic ratio is two; a single-frequency operation curve is plotted for comparison. For the purpose of the figure it is assumed that the single-frequency voltage is 20 volts and that in the two-frequency modes the signal level is 20 volts at each frequency. The phase angle is chosen to provide maximum output power at the designated frequency while maintaining negative conductance at both frequencies. The single-frequency curve is dashed below 7 GHz because the small-signal conductance is positive there.

The curve labeled second-harmonic tuning shows the power output obtainable at the fundamental with an equal amplitude second harmonic present. The improvement is greatest at frequencies below the optimum transit-time frequency and it is possible to obtain power output comparable to and even slightly greater than that attained at the optimum single-frequency point. The second-harmonic tuning curve extends below the avalanche frequency, indicating that it is possible to obtain power from the diode at frequencies below the single-frequency operating range. However, for frequencies above the optimum 11 GHz the power improvement is small, and for frequencies from 12 to 20 GHz better performance is obtained using first-subharmonic tuning.

The first-subharmonic tuning curve shows that there is little improvement below 11 GHz, but that in the frequency range from 11 to 20 GHz some improvement can be obtained. In particular, at 20 GHz the improvement attainable is nearly 100 percent although the power is not as large as at the optimum single-frequency point.

In summary, the improvement attainable with harmonic or subharmonic tuning depends upon the relationship of the frequency at which power is desired to the optimum operating frequency of the diode. If the desired frequency is equal to the optimum, then the improvement attainable with harmonic tuning is in the range of 10 to 20 percent. If the desired frequency is below the optimum then second-harmonic tuning can provide substantial improvement. However, the power output attainable and the efficiency are still comparable to that attainable at the optimum point. For frequencies above the optimum, first-subharmonic tuning is preferable to second-harmonic tuning and significant improvement can be obtained. It should be emphasized that in this two-frequency mode attainable efficiency does not greatly exceed that attainable for the single-frequency IMPATT mode.

Acknowledgment

The authors gratefully acknowledge discussions with P. T. Greiling concerning this work.

W. E. Schroeder
G. I. Haddad
Electron Physics Lab.
Dept. of Elec. Engrg.
University of Michigan
Ann Arbor, Mich.

Reprinted from the PROCEEDINGS OF THE IEEE
VOL. 56, NO. 9, SEPTEMBER, 1968
pp. 1616-1617

256

Impatt Oscillator Performance Improvement with Second-Harmonic Tuning

***Abstract*—Improvements in power, efficiency, and FM sensitivity are achieved experimentally with Impatt oscillators which are operated so that the second harmonic lies in the frequency range for classical Read operation. The improvement in performance is accomplished by providing a trapped resonance at the second harmonic, in addition to the required tuning at the output frequency.**

Impatt oscillator circuits generally provide a simple fundamental-mode resonance at the desired oscillation frequency which must lie in the frequency range for classical Read [1] operation of the diode. Such a device can be said to be operating in the *normal mode*. A suitable circuit for the normal mode can be achieved with coaxial line using a single interchangeable $\lambda/4$ transformer which creates and couples to the fundamental-mode resonance. It has been found that Impatts tested in such a circuit often generate significantly less RF power than can be achieved with the same diode at the same frequency in the three-slug coaxial line tuner reported by Iglesias [2]. A study of the operation of the Iglesias circuit has revealed that the first tuner (nearest the diode) provides a resonance at the second harmonic. The remaining two tuners are spaced so as to trap or block the second-harmonic power while providing tuning and coupling at the fundamental output frequency.

A comparison of the CW output power versus supply current for a diode operating in the normal mode with that for a tuned second harmonic is given in Fig. 1. The diode is a Ge *n-p* Impatt as reported by Rulison *et al.* [3]. Normal-mode data are shown for six different $\lambda/4$ transformers. In each case the transformer was positioned to give 6 GHz output at 140 mA current. The highest output was about 200 mW and the highest efficiency was less than 4 percent. Tuned harmonic operation with the same diode at the same output frequency produced over 600 mW at about 10 percent efficiency.

Fig. 2 shows experimental CW output power versus frequency for the same diode at a supply current of 140 mA (450 A/cm^2). The data for curve A were obtained with normal-mode operation in a coaxial line circuit. Curve B data were obtained with normal-mode operation in an X-band waveguide circuit where the fundamental-mode resonance was achieved with a small radial line resonator around the diode. The increased output depicted by curve C resulted from providing a trapped second-harmonic resonance with the three-slug coaxial line tuner of Iglesias. Maximum output in the 4–8 GHz band corresponded with efficient trapping of the second-harmonic power. The variation in the output power versus frequency characteristic for the tuned harmonic operation can be attributed to tuning limitations imposed by the particular tuning slugs used. This arises from the difficulty in providing simultaneously resonances at f and at $2f$, while blocking the $2f$ power and optimally loading the oscillator at f. With careful attention to tuning, efficient operation in the tuned harmonic mode has been observed throughout the 4–7 GHz band. The full frequency range for such operation has not been established.

The diode voltage at 140 mA current was approximately 49 volts for the normal-mode operation at 6 GHz, and about 2 volts lower for the tuned harmonic mode at the same output frequency.

It was noted that some batches of diodes which were less efficient than usual with normal-mode operation exhibited no improvement with harmonic tuning.

The electronic tunability of Impatt diodes has been described by Gilden and Hines [4] and by Rulison *et al.* [3]. The modulation sensitivity S is defined by $S=\Delta f/\Delta I$ in MHz/mA. Unfortunately, large modulation sensitivity and low noise are conflicting requirements in an Impatt oscillator. Reducing the inherently large FM noise requires use of a high external Q which decreases the modulation sensitivity. Direct frequency modulation of a device with a low modulation sensitivity involves relatively large dc power variations with the associated undesired amplitude modulation.

A convenient way to compare the FM capabilities for Impatts is to use the ratio of the modulation sensitivity to the FM noise Δf_N, where

$$\Delta f_N = \sqrt{2} f_m \left(\frac{P_{\text{noise}}}{P_{\text{carrier}}}\right)^{1/2}, \quad \text{rms frequency deviation in hertz in a 1 kHz bandwidth}$$

where P_{noise} = single-sideband noise power in a 1 kHz bandwidth at f_m Hz from the carrier. This ratio, $S/\Delta f_N$, is essentially independent of the external Q of the circuit since both numerator and denominator vary approximately as $1/Q_{ex}$ for lossless circuits.

Fig. 2 shows the measured $S/\Delta f_N$ ratios corresponding to the data points on the power-frequency curves A, B, and C. Carrier-to-noise ratios were measured with $f_m = 1$ MHz. It is expected from Josenhans [5] that Δf_N does not vary with f_m. It is seen that the modulation sensitivity is low when operating in the normal mode near the frequency for maximum efficiency. At lower frequencies the modulation sensitivity is improved,

Manuscript received July 1, 1968. The work reported here was presented at the IEEE Electron Device Research Conference, University of Colorado, Boulder, Colo., June 19–21, 1968.

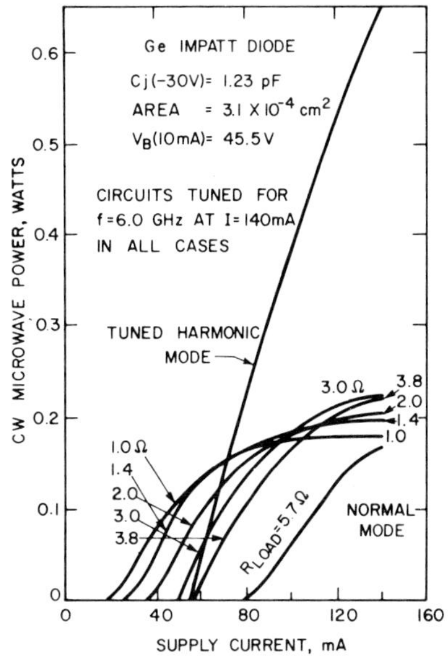

Fig. 1. Normal-mode and tuned-harmonic-mode operation of a Ge *n-p* Impatt. Various transformer impedances were used to determine the most efficient normal-mode operation at 6 GHz. The circuit in the tuned harmonic case was tuned for maximum output at 6 GHz with a supply current of 140 mA.

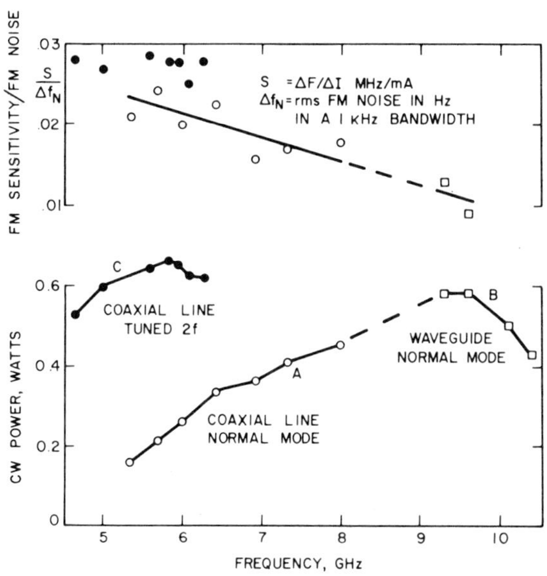

Fig. 2. Experimental CW output power and electronic frequency modulation sensitivity for a Ge Impatt diode operated at 140 mA supply current. Curves A and B are for normal-mode operation in a coaxial line tuner and an *X*-band waveguide tuner, respectively. Curve C is for tuned-second-harmonic operation in a coaxial line tuner. Both coaxial circuits operated into a broad-band attenuator with a 4–8 GHz bandpass filter in front of the power meter.

but the output power and efficiency are seriously reduced. Provision of a tuned second harmonic, in conjunction with operation below the frequency for maximum normal-mode output, permits a relatively high modulation sensitivity to be achieved simultaneously with high efficiency and power.

The tuning requirements for this mode can more readily be met with the circuit of Fig. 3. A simple second-harmonic resonance is provided around the diode by a hat structure. The exact resonance frequency is established by positioning the second-harmonic trap tuner. The tuning and coupling at the output frequency can then be accomplished with a single $\lambda/4$ transformer. By judicious choice of the tuning hat around the diode and of the impedance of the trap section, it has been possible to provide simultaneously the two required resonances, and to optimize the output coupling without requiring a separate $\lambda/4$ output tuning transformer.

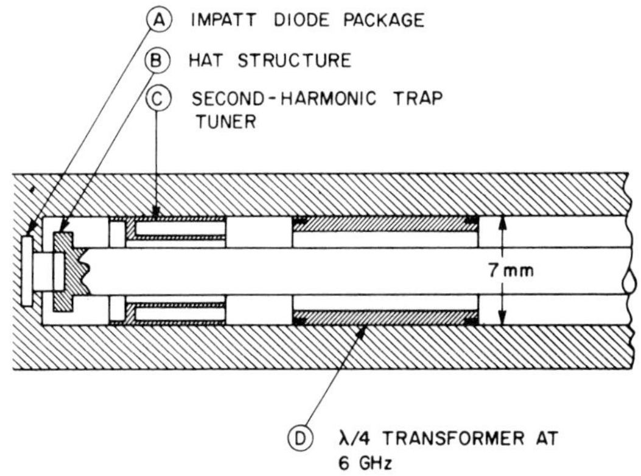

Fig. 3. Coaxial line Impatt oscillator circuit which provides a trapped resonance at the second harmonic in addition to the resonance and coupling required at the output frequency.

An explanation for the improved efficiency of the tuned harmonic mode of operation is proposed in the accompanying letter by Johnston *et al.* [6].

ACKNOWLEDGMENT

The author is grateful to D. E. Iglesias for his cooperation in providing test circuits and test data, and to J. W. Gewartowski, B. C. DeLoach, and D. L. Scharfetter for their many helpful discussions.

C. B. SWAN
Bell Telephone Laboratories, Inc.
Murray Hill, N. J. 07974

REFERENCES

[1] W. T. Read, "A proposed high-frequency negative resistance diode," *Bell Sys. Tech. J.*, pp. 401–446, March 1958.
[2] D. E. Iglesias, "Circuit for testing high-efficiency IMPATT diodes," *Proc. IEEE (Letters)*, vol. 55, pp. 2065–2066, November 1967.
[3] R. L. Rulison, G. Gibbons, and J. G. Josenhans, "Improved performance of IMPATT diodes fabricated from Ge," *Proc. IEEE (Letters)*, vol. 55, pp. 223–224, February 1967.
[4] M. Gliden and M. E. Hines, "Electronic tuning effects in the Read microwave avalanche diode," *IEEE Trans. Electron Devices*, vol. ED-13, pp. 169–175, January 1966.
[5] J. Josenhans, "Noise spectra of Read diode and Gunn oscillators," *Proc. IEEE (Letters)*, vol. 54, pp. 1478–1479, October 1966.
[6] R. L. Johnston, D. L. Scharfetter, and D. J. Bartelink, "High-efficiency oscillations in germanium avalanche diodes below the transit-time frequency," this issue, page 1611.

Reprinted from *Electronics Letters*, Vol. 5, No. 16, 7th August 1969

OCTAVE TUNING AND THE EFFECT OF SECOND-HARMONIC LOADING ON AVALANCHE-DIODE OSCILLATORS

The letter describes a simple coaxial-line 2-bead tuner which allows continuous tuning of an avalanche-diode oscillator over an octave bandwidth. Experiments with such a tuner provide direct evidence that the control of second-harmonic impedance is important if maximum output power is to be achieved; also that large power variations can be obtained with relatively small second-harmonic components in the voltage waveform.

Introduction: It is now widely accepted that pure sinusoidal waveforms do not necessarily give the best performance from avalanche-diode oscillators; recent experiments by Swan[1] and Iglesias[2] suggest that control of the second-harmonic waveform component may be important. Theoretical work by Gittins,[3] Culshaw[4] and Blue[5] indicates that a second-harmonic component of voltage can appreciably affect the fundamental output power. This letter describes experiments which are consistent with these views.

258

Experimental: In experiments at SERL using a 3-bead tuner similar to that described by Iglesias[2] (Fig. 1A), it was noted that, although the maximum power output was high, the bandwidth was low, and it was almost impossible to select or control the exact frequency of oscillation. In an effort to improve the bandwidth and tunability without sacrificing power output, a 2-bead tuner was developed[3] (Fig. 1B) which gave the same maximum power and an octave of bandwidth with continuous tuning (6–12 GHz in a typical case; see Fig. 1C). This 2-bead tuner is quite different from that described by Melick,[6] and had another very interesting and useful property: the power and frequency could be controlled virtually independently. The frequency could be set by positioning the front (quarter-wavelength) bead, and the power could then be maximised at that frequency using the back bead. This feature was considered sufficiently important to merit further investigations.

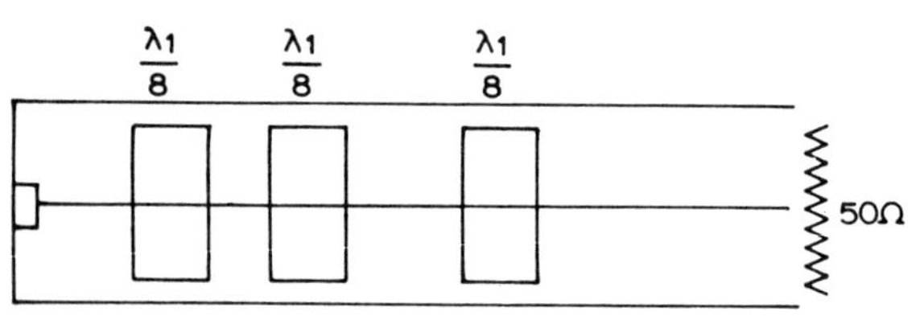

Fig. 1A *Iglesias 3-bead tuner*
λ_1 refers to the fundamental frequency

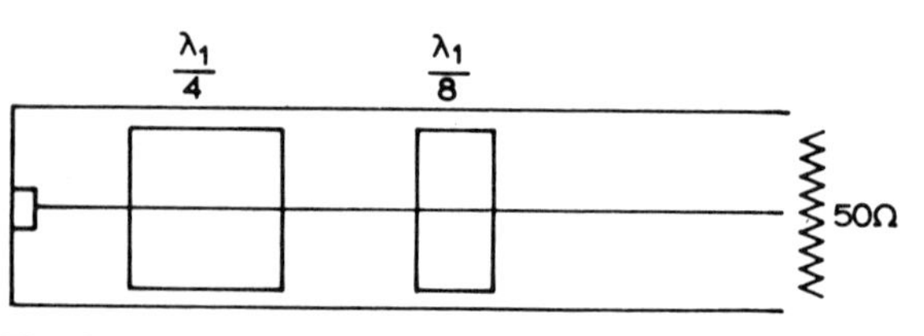

Fig. 1B *SERL 2-bead octave tuner*
Here λ_1 refers to a fundamental frequency near the middle of the octave

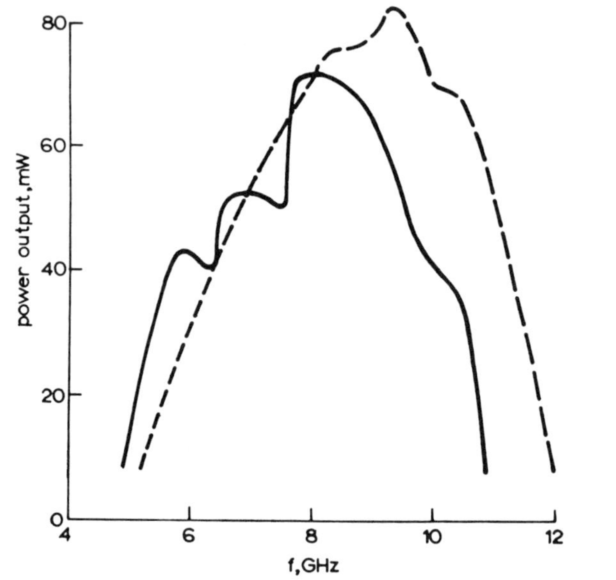

Fig. 1C *Typical performance curves for the octave tuner using SERL diodes*
——— SERL diode DB19A(d)7
- - - - SERL diode DB19A(c)3

The rear ($\lambda/8$) bead in such a 2-bead tuner will, in general, affect the load impedance seen by the diode at both the fundamental and the second-harmonic frequencies, and the relative effect of the two beads on the power output is difficult to assess. A special version of the tuner was therefore made for operation at a single frequency. In this tuner, the front bead was designed to affect only the fundamental load impedance and the back bead only the second harmonic. To achieve this, the back bead was a composite structure (Fig. 3A). The v.s.w.r. performances of the two beads, measured in an otherwise matched coaxial line, are shown in Fig. 2.

The choice of fundamental frequency was determined by two considerations. First, the second harmonic had to be within the range of the coaxial measuring equipment that was being used. Secondly, a frequency of about half the Read value was chosen in order to ensure an appreciable second-harmonic component in the particle-current waveform of the diode. The experiment consisted basically of setting the front bead to achieve the desired frequency, and then moving the back bead away from it, noting the variations of power, frequency and load impedance for a constant direct current. Impedance measurements were made at both fundamental and second-harmonic frequencies. As indicated in Fig. 2, the fundamental frequency was 7·3 GHz. The maximum variation of frequency as a result of moving the back bead during the experiments was ±4%.

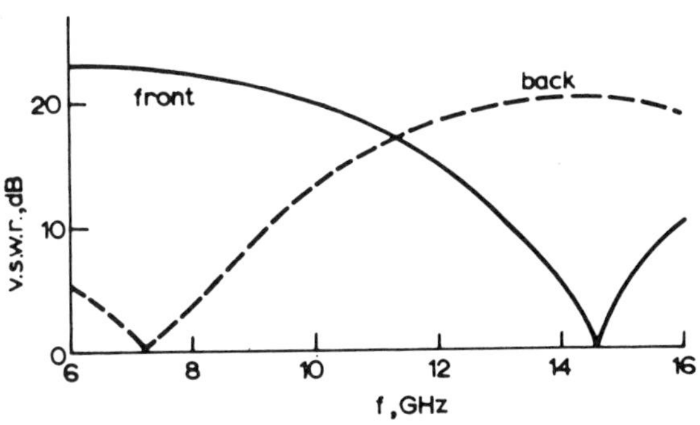

Fig. 2 *V.S.W.R. performance of the front and back beads, measured separately, in an otherwise matched line*
The back bead is transparent at 7·3 GHz and the front bead at 14·6 GHz

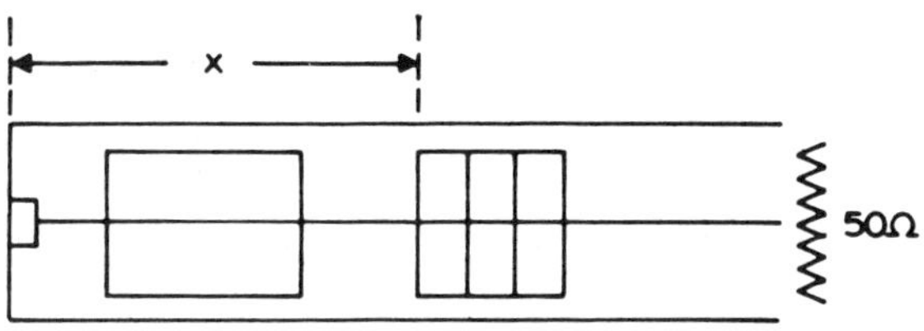

Fig. 3A *Single-frequency version of the 2-bead tuner*
The front bead was fixed in position and the back (composite) bead moved

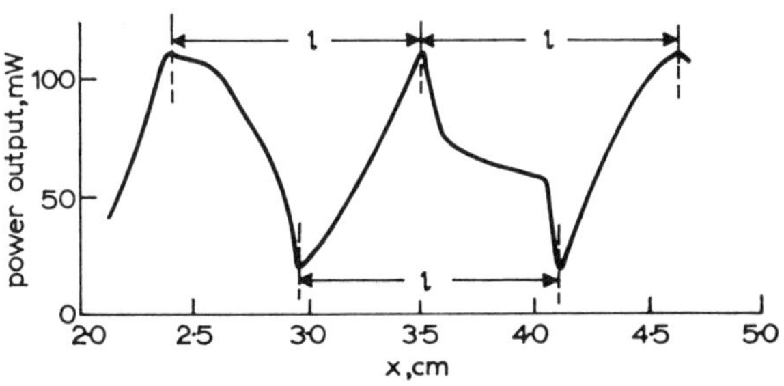

Fig. 3B *Variation of fundamental output power with back-bead position*
The frequency was in the range 7·3 GHz ± 4% throughout

A graph of power against back-bead position (Fig. 3B) shows a clear periodicity with a power variation of about 5:1 between maximum and minimum points. The distance l was found to be $\lambda/2$ at the second harmonic, as expected from the design of the back bead. Direct impedance measurements confirmed that the fundamental load impedance was indeed almost constant, whereas that at the second harmonic varied considerably in magnitude and phase. From the impedance measurements, it was concluded that the second-harmonic voltage component did not exceed 25% of the fundamental amplitude; yet a very large variation in power was observed. The maximum power obtained from the diode with other tuning systems at this frequency was approximately the same as that given in Fig. 3B.

Conclusions: The simple circuit described allows continuous tuning over an octave bandwidth, and the power can be trimmed to a maximum at any frequency in the band. Experiments designed to understand the detailed operation of this tuner have shown that large power variations can be obtained by the addition of only a relatively small second-harmonic component to the voltage waveform. The depletion capacitance of the diode is automatically tuned out at the fundamental frequency, but, in general, it acts as a low-impedance load at the second harmonic. If a large second-harmonic voltage component is to be produced, this depletion capacitance must be tuned out at the second harmonic, necessitating the provision of a second resonant circuit.[1]

This letter is published by permission of the UK Ministry of Defence.

R. A. GIBLIN*
K. G. HAMBLETON
C. A. TEARLE

8th July 1969

Services Electronics Research Laboratory
Baldock, Herts., England

References

1 SWAN, C. B.: 'IMPATT oscillator performance improvement with second-harmonic tuning', *Proc. Inst. Elect. Electron. Engrs.*, 1968, **56**, p. 1616
2 IGLESIAS, D. E.: 'Circuit for testing high efficiency IMPATT diodes', *ibid.*, 1967, **55**, p. 2065
3 GITTINS, J. F.: SERL symposium on avalanche diode oscillators SERL, Baldock, Herts., Dec. 1968
4 CULSHAW, B.: private communication
5 BLUE, J. L.: 'Approximate large-signal analysis of IMPATT oscillators', *Bell Syst. Tech. J.*, 1969, **48**, p. 383
6 MELICK, D. R., and FOGGIATO, G. A.: 'Octave mechanically tuned avalanche diode oscillators', *Proc. Inst. Elect. Electron. Engrs.*, 1968, **56**, p. 1737

* Now at the Department of Electronic & Electrical Engineering, University College London, Torrington Place, London, WC1, England

HARMONIC EFFECTS IN THE IMPATT MODE

Circuit Effects in Second-Harmonic Tuning of IMPATT Diodes

CHARLES A. BRACKETT, MEMBER, IEEE

Abstract—Two-frequency operation of IMPATT diodes is examined in this paper on a conceptual basis. It is shown that a simple interpretation of the "single-frequency" admittances of the diode may be made. An equivalent circuit is presented which shows that the effect of second-harmonic tuning is to introduce a nonlinear resonance into the fundamental input admittance. This occurs when the external circuit admittance is tuned to the neighborhood of the "single-frequency" oscillator admittance of the diode at the second-harmonic. Computer calculations of the tuned-harmonic mode, with the diode embedded in the simplest possible RF circuit, are presented to demonstrate the effect.

I. Introduction

Two-frequency operation of avalanche transit-time diodes, where the frequencies are harmonically related, is a subject that has received considerable attention. Swan [1] demonstrated that fundamental power output and efficiency could be enhanced dramatically (two to three times the output power) by introducing a trapped high Q resonance at the second-harmonic frequency. Several authors [2]–[7] have reported theoretical and experimental investigations into various aspects of this mode of operation. The purpose of this paper is to show that, in spite of the complex nature of the problem, the tuned-harmonic mode may be thought of in terms of a relatively simple circuit model. Computer calculations are presented that demonstrate the behavior of the tuned-harmonic mode in an external RF circuit and that verify the conceptual understanding of the interaction derived from the diode equivalent circuit.

In Section II the equivalent circuit is introduced, and in Section III the results obtained by operating Blue's [2] model of the Read [13] diode in a tuned-harmonic circuit are presented.

II. Equivalent Circuit for Two Frequencies

Whereas in the single-frequency case the diode is completely characterized by the value of a single large-signal admittance, in the two-frequency case a single admittance is not convenient. If we define the fundamental and second-harmonic voltage to be

$$v_1(t) = V_1 \cos(\omega_0 t)$$

and

$$v_2(t) = V_2 \cos(2\omega_0 t + \phi_2)$$

Manuscript received July 26, 1970; revised October 12, 1970.

The author is with Bell Telephone Laboratories, Inc., Murray Hill, N. J. 07974.

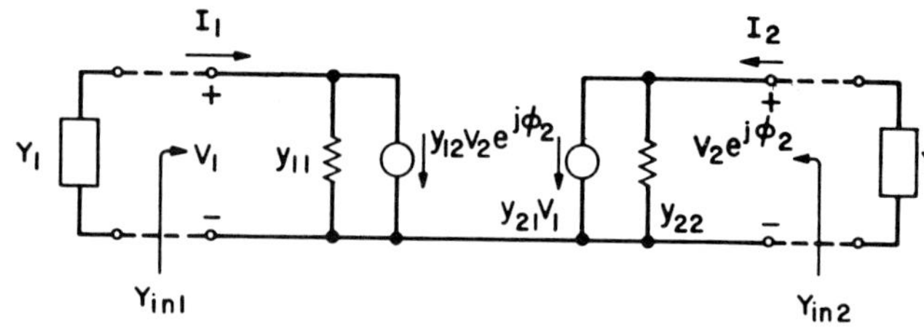

Fig. 1. Equivalent circuit of the IMPATT diode that includes nonzero voltages at two harmonically related frequencies. Port one is the fundamental port and port two is the second-harmonic port; y_{11} and y_{22} are the large-signal single-frequency diode admittances at the fundamental- and second-harmonic frequencies, respectively.

respectively, then we may say that the state of oscillation is uniquely defined by the four independent variables (V_1, ω_0, V_2, ϕ_2). We may then write the input admittances to the diode at the fundamental and second harmonic as $Y_{in1}(V_1, \omega_0, V_2, \phi_2)$ and $Y_{in2}(V_1, \omega_0, V_2, \phi_2)$, respectively. Y_{in1} and Y_{in2} may each be considered to be the sum of two independent parts: the single-frequency admittance (y_{11} and y_{22} at ω_0 and $2\omega_0$, respectively) and a term that provides the coupling, or interaction, between the fundamental and second harmonic. Thus, we may write

$$Y_{in\,1} = y_{11} + y_{12} V_2 e^{j\phi_2}/V_1$$

and

$$Y_{in\,2} = y_{22} + y_{21} V_1/V_2 e^{j\phi_2}. \tag{1}$$

Here, y_{11} and y_{22} are the single-frequency admittances at ω_0 and $2\omega_0$, respectively. Therefore, $y_{11} = y_{11}(V_1, \omega_0)$ and $y_{22} = y_{22}(V_2, 2\omega_0)$ so that y_{11} is independent of V_2 and ϕ_2, and y_{22} is independent of V_1 and ϕ_2. The coupling terms have been written as though they were transfer admittances, so that the standard equivalent circuit of a nonlinear two-port device is obtained as shown in Fig. 1. y_{12} and y_{21} are naturally functions of all four parameters (V_1, ω_0, V_2, ϕ_2) and are also, in general, nonlinear variables. A description of the values of these variables for a 6-GHz Ge oscillator is given elsewhere [8], but it is sufficient to say here that the values are slowly varying functions of V_2 and ϕ_2 and are roughly proportional to V_1. Thus (1) shows the input admittances at the two frequencies for constants V_1 and V_2 to be closed curves that enclose (roughly at their "centers") the points y_{11} and y_{22}. Therefore, the general locations of these curves in the admittance plane are determined by the properties of y_{11} and y_{22}. In addition, if an external circuit admittance Y_2 is connected to the $2\omega_0$-port, the input admittance Y_{in1} may be written

$$Y_{\text{in }1} = y_{11} - \frac{y_{12}y_{21}}{y_{22} + Y_2}. \tag{2}$$

Equation (2) shows that an apparent pole is present in the fundamental input admittance if the condition in the second-harmonic circuit $|y_{22}+Y_2|\to 0$ is approached. This condition may result from changes in the harmonic circuit admittance Y_2, the bias current, or from frequency changes due to tuning at the fundamental. Because of the nonlinear nature of y_{22} it is impossible to operate at this pole since y_{22} moves as V_2 changes. Equation (2) also shows that the argument of the second term is critical when its magnitude is great, since the wrong argument may completely overcome the negative input conductance represented as part of y_{11}. It is therefore seen that conceptually $y_{22}+Y_2=0$ is a resonance point and the desired tuning of Y_2 could be calculated from this equation if the values of y_{12} and y_{21} were known. Complete characterization of a diode for tuned harmonic operation therefore requires the measurement (or calculation) of four (complex) quantities (y_{11}, y_{12}, y_{21}, y_{22}) as functions of the oscillation-state variables (V_1, ω_0, V_2, ϕ_2). However, specification of the admittance y_{22} alone gives a great deal of the information needed to design the proper circuit.

An aspect of this mode of operation not yet mentioned is the circuit conditions required for stable oscillation. This topic is to be treated in a separate paper [8] in a generalization of the work of Kurokawa [9]. The effect of stability considerations on the work presented here is not completely known, and only brief mention of it is made in this paper.

III. Example—A Simple Circuit

Blue's analysis [2] was extended to include an arbitrary circuit model in the numerical treatment so that the oscillation-state variables (V_1, ω_0, V_2, ϕ_2) were chosen, by the same Newton–Raphson process used by Blue, to satisfy the additional constraints posed by the circuit equations

$$Y_1 + Y_{\text{in }1} = 0$$

and

$$Y_2 + Y_{\text{in }2} = 0. \tag{3}$$

Here Y_k is the external circuit admittance (linear) at the frequency $k\omega_0$, and $Y_{\text{in}k}$ is the input admittance (nonlinear) of the diode at $k\omega_0$ as calculated by Blue's analysis. In these calculations no attempt was made to separate $Y_{\text{in}k}$ as in (2).

For the purpose of these calculations Y_1 and Y_2 were taken to be a simple parallel connection of a load conductance (G_1 and G_2) and an inductance (L_1 and L_2). Furthermore, the effect of tuning was investigated by holding G_1, G_2, and L_1 constant and varying L_2. Four different combinations of the variables G_1 and G_2 were taken, resulting in the tuning paths labeled A, B, C, and D in Fig. 2. These four curves are plots of the loci of

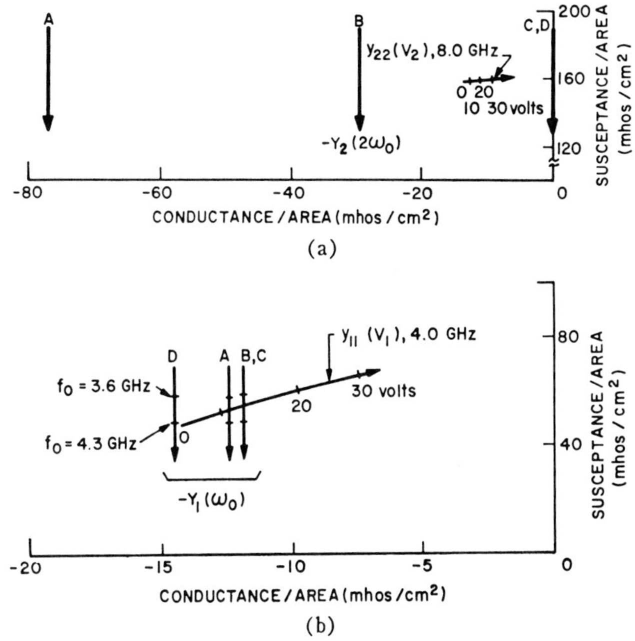

Fig. 2. The complex-plane loci of the circuit and device admittances. (a) Second harmonic. (b) Fundamental. The circuit admittances Y_1 and Y_2 are taken to be constant real conductances G_1 and G_2 plus the susceptance of inductors L_1 and L_2, respectively. Various combinations of G_1 and G_2, with L_1 fixed, produce the tuning paths labeled A, B, C, and D as the second-harmonic inductance L_2 is varied. Paths A, B, C, and D are plots of the negatives of Y_1 and Y_2. Also shown are the single-frequency large-signal device admittances at 4 GHz, y_{11} (V_1), and $y_{22}(V_2)$.

$Y_{\text{in}1}$ and $Y_{\text{in}2}$ ($-Y_1$ and $-Y_2$) as L_2 is varied with G_1, G_2, and L_1 fixed, and the variation with frequency along them is roughly indicated by the two frequencies noted in Fig. 2(b). Also shown in Fig. 2 are the *single-frequency* admittances y_{11} and y_{22} at $\omega_0=4$ GHz. These curves were obtained from Blue's analysis as applied to a 6-GHz Ge IMPATT oscillator described in more detail elsewhere [8]. The inductance L_1 was chosen so that for large detuning of the $2\omega_0$ circuit ($|y_{22}+Y_2|\to\infty$) the oscillation frequency was approximately 4 GHz (see Fig. 4).

The fundamental output power as a function of the second-harmonic circuit susceptance for each of the four tuning paths A, B, C, and D is shown in Fig. 3. Tuning path A corresponds to a heavily loaded second-harmonic circuit and there is only a moderate influence on the fundamental power by the tuning at the second harmonic. P_{OA} is the maximum output power along tuning path A as V_2 goes to zero and is indicated on Fig. 3. Tuning path B is for a less heavily loaded $2\omega_0$ circuit, and the result is a rather dramatic increase in the power. The peak power for path B is a factor of 2.4 greater than the power P_{OB} obtained with zero second-harmonic voltage. Note that P_{OB} is slightly greater than P_{OA} because of a change in the fundamental load conductance.

Tuning path C has the same fundamental load as path B and therefore $P_{OC}=P_{OB}$, but the second-harmonic load conductance for path C is zero giving a completely reactive termination at $2\omega_0$. The power output for path C is much sharper and much higher than

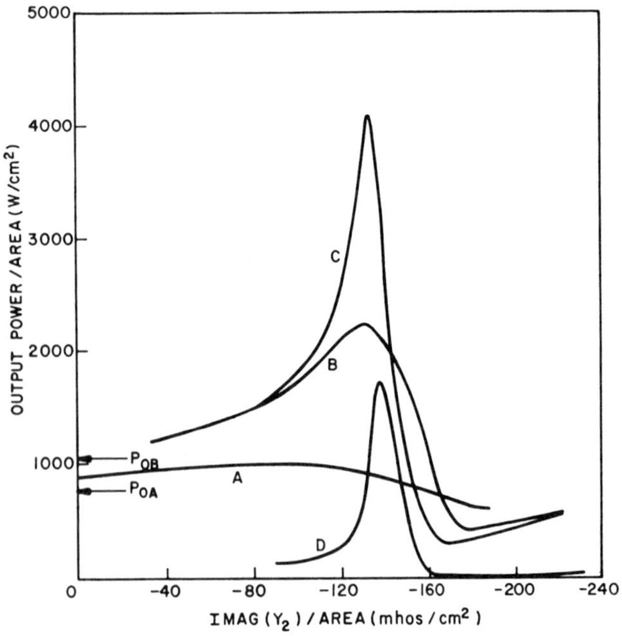

Fig. 3. Fundamental power output versus second-harmonic circuit susceptance for the four tuning paths of Fig. 2.

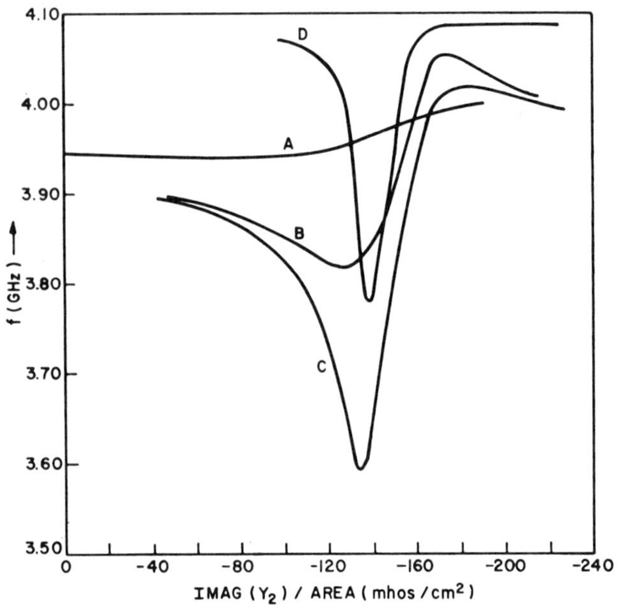

Fig. 4. Fundamental oscillation frequency versus second-harmonic circuit susceptance for the four tuning paths of Fig. 2.

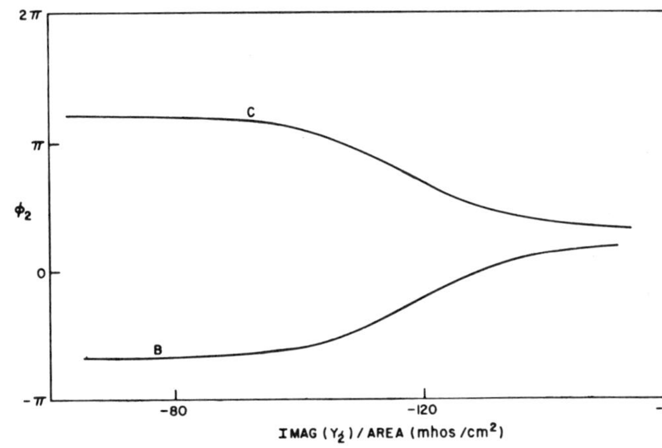

Fig. 5. Relative phase ϕ_2 of the second-harmonic voltage with respect to the fundamental versus second-harmonic circuit susceptance for tuning paths B and C of Fig. 2.

for path B. the peak now being a factor of 4.5 greater than $P_{OC} = P_{OB}$ The reason that path C produces a sharper, higher power resonance is that path C passes closer to the nonlinear admittance y_{22} (V_2) than does path B. In other words, $|y_2 + Y_2|_{B\min} > |y_{22} + Y_2|_{C\min}$. The fact that path C lies on the imaginary axis of the admittance plane is not of particular significance. In fact, if tuning path C had had a slight load conductance, placing it between the imaginary axis and the admittance y_{22}, $|y_{22} + Y_2|$ would have been smaller yet, and the resonance would have been even stronger.

Tuning path D is a case where the single-frequency fundamental oscillator will barely oscillate, but the tuning of Y_2 to a resonance with y_{22} produces a large amount of fundamental output power over a rather narrow range of second-harmonic susceptance.[1]

The variation of oscillation frequency along each of the same tuning paths is shown in Fig. 4. It is seen that there is a large susceptance induced in the fundamental circuit by the second-harmonic tuning, and the minimum in frequency occurs very close to the same tuning condition that produces maximum output power (compare Figs. 3 and 4). Note that a slight increase in frequency also occurs on one side of the resonance. Although in the case shown here, a minimum in frequency is always associated with a maximum in output power, it is not thought that this is necessary. The induced susceptance is a function of the arguments of y_{12} and y_{21}, and these are functions of the device material, field profile, frequency, and bias current density. In a more general circuit, the slopes (dB/dG) of the fundamental and second-harmonic circuit admittance curves will also affect the induced susceptance variation.

Further confirmation of the concept of y_{22} as a resonance point comes from a plot of the relative phase ϕ_2 versus second-harmonic circuit susceptance shown in Fig. 5 for the tuning paths B and C. In Fig. 2 paths B and C pass on opposite sides of the single-frequency admittance $y_{22}(V_2)$. Phase ϕ_2 changes in different directions for these two paths, indicating the presence of a pole between them.

IV. Comments

Due to the approximations made in Blue's analysis [2], it cannot be used to predict the maximum output power, or the optimum circuit for the same, in either single-frequency or two-frequency operation. An attempt to understand harmonic tuning when limited by high RF voltages has been made elsewhere [2], [3] by

[1] Strictly speaking, tuning path D does not indicate any possible single-frequency oscillation with the 4-GHz $y_{11}(V_1)$ shown. Actually, frequencies slightly higher than 4.0 GHz do oscillate (single frequency) because of the frequency dependence of $y_{11}(V_1)$.

restricting the peak-to-peak swing of the voltage. It was found that the most output power was then obtained by operating the diode at its transit-time frequency with as little harmonic interaction as possible. In the results presented here the fundamental and harmonic voltages have been allowed to change in order to meet the circuit constraints, and we have, therefore, assumed that the voltages were low enough that Blue's analysis still applies. The simple Read diode shows an increase in RF output power proportional to V_1 at large V_1 [2], i.e., the negative conductance falls off as $1/V_1$. However, all the experimental indications [10]–[12] are that the negative conductance for real IMPATT diodes saturate with RF voltage more rapidly than $1/V_1$. In this range of operation the Read theory [13] does not apply and the results of this paper may be inaccurate.

As a second comment, the unlikelihood must be pointed out that the suggested tuning paths A, B, C, and D (Fig. 2) could be achieved experimentally with continuous, smooth tuning of the oscillation [8]. This is because of the earlier mentioned requirement of oscillation-state stability. This does not negate the usefulness of the results presented here, however, because it is likely that circuit adjustments can be made that will make almost any small segment of each of the tuning paths stable. Therefore, the results presented here may be considered as the limit of what would happen if stability were achieved.

V. Conclusion

This paper has presented an IMPATT diode equivalent-circuit model, which in theory, would allow the complete characterization of the second-harmonic tuning effect. It appears that actual experimental determination of the elements of this equivalent circuit would be difficult. The equivalent circuit does point out the role that the single-frequency admittances $y_{11}(V_1)$ and $y_{22}(V_2)$, which generally are measurable, play in obtaining an increase in the output power at the fundamental. In particular it was demonstrated that when $|y_{22}+Y_2|$ approaches zero, a pole is induced in the fundamental input admittance that may either enhance or degrade the output power. The second-harmonic single-frequency large-signal diode admittance $y_{22}(V_2)$ is thus seen to be a resonance point as far as interaction with the external admittance $Y_2(2\omega_0)$ is concerned.

Demonstration of these concepts is presented in terms of a computer "experiment" with a simple external circuit connected to the fundamental and second-harmonic ports of the diode.

It is believed that the concepts presented here will aid materially in the practical visualization of the IMPATT diode tuned-harmonic mode, especially as regards the diode-circuit interaction.

Acknowledgment

The author wishes to acknowledge many helpful discussions of this subject with C. B. Swan, J. W. Gewartowski, and K. Kurokawa, and the help of J. L. Blue on the numerical aspect of the problem is gratefully appreciated.

References

[1] C. B. Swan, "IMPATT oscillator performance improvement with second-harmonic tuning," *Proc. IEEE* (Letters), vol. 56, pp. 1616–1617, Sept. 1968.
[2] J. L. Blue, "Approximate large-signal analysis of IMPATT oscillators," *Bell Syst. Tech. J.*, vol. 48, pp. 383–396, Feb. 1969.
[3] C. A. Brackett, "Peak ac voltage limitations in second-harmonically tuned IMPATT Diodes," *IEEE Trans. Microwave Theory Tech.* (Corresp.), vol. MTT-18, Nov. 1970, pp. 992–993.
[4] W. E. Schroeder and G. I. Haddad, "Effect of harmonic and subharmonic signals on avalanche-diode oscillator performance " *IEEE Trans. Microwave Theory Tech.* (Corresp.), vol. MTT-18, June 1970, pp. 327–331.
[5] K. Mouthaan and H. P. M. Rijpert, "Second-harmonic tuning of the avalanche transit-time oscillator," *Proc. IEEE* (Letters), vol. 57, pp. 1449–1450, Aug. 1969.
[6] K. Mouthaan, "Nonlinear analysis of the avalanche transit-time oscillator," *IEEE Trans. Electron Devices*, vol. ED-16, pp. 935–945, Nov. 1969.
[7] K. Mouthaan, "Two-frequency operation of the avalanche transit-time oscillator," *Proc. IEEE* (Letters), vol. 58, pp. 510–512, Mar. 1970.
[8] C. A. Brackett, "Characterization of second-harmonic effects in IMPATT diodes," *Bell Syst. Tech. J.*, vol. 49, Oct. 1970, pp. 1777–1810.
[9] K. Kurokawa, "Noise in synchronized oscillators," *IEEE Trans. Microwave Theory Tech.*, vol. MTT-16, pp. 234–240. Apr. 1968.
[10] J. W. Gewartowski and J. E. Morris, "Active IMPATT diode parameters obtained by computer reduction of experimental data," *IEEE Trans. Microwave Theory Tech.*, vol. MTT-18, pp. 157–161, Mar. 1970.
[11] D. R. Decker, C. N. Dunn, and R. L. Frank, "Large-signal silicon and germanium avalanche diode characterization," *IEEE Trans. Microwave Theory Tech.*, vol. MTT-18, Nov. 1970, pp. 872–876.
[12] D. R. Decker, C. N. Dunn, and H. B. Frost, "The effect of injecting contacts on avalanche diode performance," this issue, pp. 141–146.
[13] W. T. Read, "A proposed high-frequency negative-resistance diode," *Bell Sys. Tech. J.*, vol. 37, pp. 401–446, Mar. 1958.

263

Reprinted from IEEE TRANSACTIONS
ON *ELECTRON DEVICES*
Volume ED-18, Number 3, March, 1971
pp. 147-150

Reprinted from IEEE TRANSACTIONS
ON *MICROWAVE THEORY AND TECHNIQUES*
Volume MTT-18, Number 11, November, 1970
pp. 853-862

Characterization of Nonlinear Interactions in Avalanche Transit-Time Oscillators, Frequency Multipliers, and Frequency Dividers

KOENRAAD MOUTHAAN, MEMBER, IEEE

***Abstract*—Equivalent circuits are introduced for nonlinear operation of the avalanche transit-time diode at a single frequency and at two harmonically related frequencies. Parameters are derived that characterize the nonlinear interaction between the diode and the associated microwave circuit. Particular attention is devoted to the oscillator, the frequency multiplier, and the frequency divider. For the oscillator, parametric interaction with a harmonic of the desired oscillation frequency is shown to give possibilities of enhancing the output power and of significantly improving the noise performance. An experimental two-frequency oscillator is described that has confirmed the theoretical predictions.**

I. INTRODUCTION

THE avalanche transit-time diode, in its capacity of providing a high-frequency negative resistance [1], is finding increasing use in microwave oscillators and amplifiers. Of a more recent origin is the practical utilization of the nonlinear properties of the diode, such as harmonic generation, nonlinear mixing, and parametric interaction, in combination with the negative-resistance properties [2], [3]. Amplifying frequency multipliers and frequency dividers may be mentioned as examples. In the oscillator, parametric interaction between harmonically related frequencies is capable of increasing the output power [4] and of improving the noise performance [5]. Practical utilization of the nonlinear properties of the diode critically depends on the interaction between the diode and the associated microwave circuit. In this paper, a nonlinear equivalent circuit is introduced for the diode, and parameters are derived that characterize the nonlinear interaction between the diode and the microwave circuit. Particular attention will be given to the oscillator; the design parameters for frequency multipliers and frequency dividers will be also indicated.

In order to arrive at a simple equivalent circuit, a number of assumptions will be made that appear to be warranted from the practical point of view. First, the diode will be asssumed to have a Read-type asymmetric multiplication profile. This assumption is approximately valid for the commonly employed asymmetric diode structures (p^+-n-n^+, n^+-p-p^+, Schottky-barrier n-n^+). The avalanche zone is assigned a capacitance C_a; the drift zone is assigned a capacitance C_d and a transit angle θ for the frequency of interest. Second, the particle-current density in the diode will be assumed to remain small compared to the capacitive-current density. The RF electric field in the diode then is approximately in phase with the voltage across the diode. In the Appendix a more detailed equivalent circuit that does not involve the present restriction is derived. Where necessary, the results of the simplified equivalent circuit will be compared with those of the more detailed equivalent circuit. In this context we may observe that the particle current generated by the avalanche has an inductive phase delay with respect to the voltage across the avalanche zone. The avalanche zone therefore is parallel resonant at the frequency where the capacitive current becomes equal to the inductive-particle current. Transit-time negative resistance arises only above the avalanche

Manuscript received April 5, 1970; revised August 24, 1970.
The author is with Philips Research Laboratories, Eindhoven, The Netherlands.

resonance frequency [6]. In practice the diode is usually operated well above the avalanche resonance frequency, which is compatible with our present assumption that the RF current through the diode is mainly capacitive. As a third assumption, the RF voltage across the diode will be taken to remain so small compared to the dc voltage that no significant change in diode behavior results from the nonlinear dependence of the ionization rate on the electric field. In practice, this assumption sets a limit to the RF voltage on the order of 30 percent of the dc voltage. For higher RF-voltage levels the avalanche shifts from inductive toward instantaneous behavior. Also, because of rectification resulting from the nonlinearity of the ionization rate, an RF-voltage-induced negative resistance results that may lead to (parasitic) low-frequency oscillations and premature burnout.

II. Single-Frequency Equivalent Circuit

In terms of the ac voltage, $V_a(t)$, across the avalanche zone, the avalanche-generated particle current, $J_{ca}(t)$, is given by

$$J_{ca}(t) = J_0 \exp\left\{(2\alpha_c'/\tau_a)\int_0^t V_a(\tau)\,d\tau\right\}. \tag{1}$$

Here, α_c' is the derivative of the ionization rate with respect to the electric field as evaluated at the dc bias field E_c, and τ_a is the time required by the carriers to traverse the avalanche zone. In the derivation of (1), both types of charge carrier in the avalanche zone have been assumed to have equal ionization rates and equal saturated drift velocities. The current J_0 is the dc bias current for $V_a(t)\equiv 0$. For $V_a(t)\not\equiv 0$, the dc bias current is given by $J_{c0}=\langle J_{ca}(t)\rangle \neq J_0$. By supplying the bias from a constant-current dc source, the bias current J_{c0} becomes independent of the RF-voltage level and is then independently variable.

The particle current according to (1) is seen to have an inductive phase delay with respect to the voltage. For a single-frequency voltage

$$V_a(t) = V_{a1}\sin\omega t \tag{2}$$

the current $J_{ca}(t)$ may be represented by the Fourier series

$$J_{ca}(t) = J_{c0}\left\{1 + 2\sum_{m=1}^{\infty}(-1)^m\beta_m \cos m\omega t\right\}. \tag{3}$$

The magnitude of the coefficients β_m depends on the magnitude of the voltage V_{a1}. From (1) it may be shown that

$$\beta_m \leq 1 \tag{4}$$

and the upper limit of unity is attained for large values of V_{a1}. In the fundamental-frequency equivalent circuit, of the diode, the avalanche zone may be represented by its capacitance C_a in parallel with an inductive current generator of strength $2\beta_1 J_{c0}$. The complete equivalent circuit will be given in the Appendix. Here, we shall assume the particle current to be small compared to the capacitive current; correspondingly, we shall represent the avalanche zone merely by its capacitance C_a.

The avalanche-generated particle current is injected into the drift zone. There, the particles drift across at field-saturated velocity v_d. From the properties of the drift zone, it may be shown [7] that the total current, $J_{\text{tot}}(t)$, in the drift zone can be written in the form

$$J_{\text{tot}}(t) = \frac{1}{l_d}\int_0^{l_d} J_{ca}(t - x/v_d)\,dx + C_d\frac{dV_d}{dt} \tag{5}$$

where l_d is the length of the drift zone and V_d is the voltage across the drift zone. If we substitute the fundamental-frequency component of $J_{ca}(t)$ according to (3) in (5), we find the fundamental-frequency component of $J_{\text{tot}}(t)$ in the form

$$J_{\text{tot}\,\omega}(t) = -2\beta_1 J_{c0}\frac{\sin^2\theta/2}{\theta/2}\sin\omega t - 2\beta_1 J_{c0}\frac{\sin\theta}{\theta}\cos\omega t + C_d\frac{dV_d}{dt}. \tag{6}$$

The phases of the first two terms on the right of (6) are defined relative to the voltage $V_{a1}\sin\omega t$ across the avalanche zone. In the equivalent circuit of the diode, the drift zone may thus be represented by the capacitance C_d in parallel with two current generators that are in quadrature and of which the phase is determined by the voltage across the avalanche zone. The complete equivalent circuit is given in the Appendix. Here, we shall again assume the particle current to be small compared to the capacitive current. In particular, we assume that the phase of $J_{\text{tot}\,\omega}$ is determined by the fundamental-frequency component of $C_d dV_d/dt$. With the corresponding assumption for the current through the avalanche zone, the voltage across the drift zone is then in phase with that across the avalanche zone. As a result the first term on the right of (6) is in antiphase with the voltage and thus describes the active properties of the diode. Also, we may actually neglect the space-charge-induced reactive current $2\beta_1 J_{c0}\sin\theta/\theta$ in (6) with respect to $C_d dV_d/dt$. In the equivalent circuit for the diode, the drift zone may thus be represented by the capacitance C_d in parallel with a current generator of strength $2\beta_1 J_{c0}\sin^2(\theta/2)/(\theta/2)$ that is in antiphase with the voltage across the drift zone. The equivalent circuit for the diode as a whole further includes a parasitic series resistance R_s and is shown in Fig. 1.

As has been mentioned, the current amplitude β_1 is determined by the RF voltage across the avalanche zone. If we introduce a normalized voltage according to

$$v_1 \equiv \alpha_c' V_{a1}/\omega\tau_a \tag{7}$$

we find that $\beta_1 = I_1(2v_1)/I_0(2v_1)$, where I_0 and I_1 are the modified Bessel functions of the first kind and of order zero and one, respectively. In Fig. 2, β_1 is plotted as a

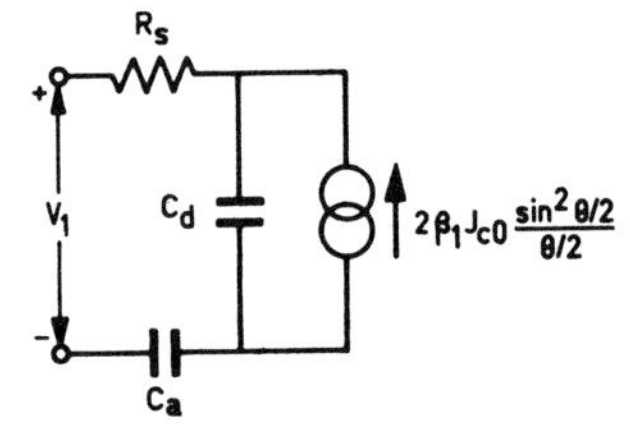

Fig. 1. Diode equivalent circuit for single-frequency operation.

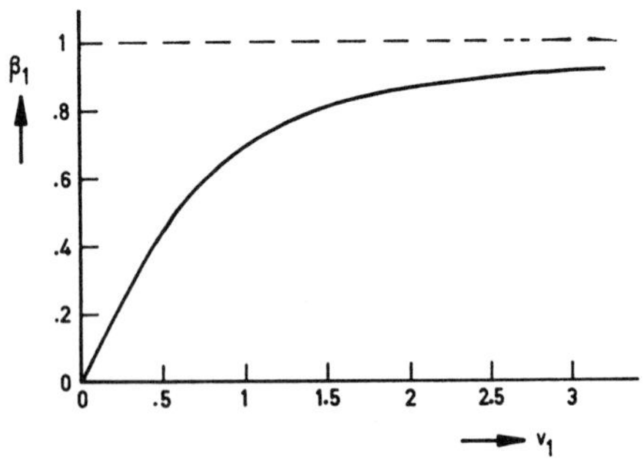

Fig. 2. Current amplitude β_1 in single-frequency equivalent circuit as function of normalized ac voltage across avalanche zone.

function of v_1. For small values of $v_1 (v_1 \ll 1)$, we have $\beta_1 = v_1$. For large values of v_1, we have $\beta_1 \rightarrow 1$. The latter

circumstance makes the use of a current generator in the equivalent circuit attractive. For large values of the RF-voltage swing across the diode, the active current is fixed and limited by the dc current through the diode. An estimate of the magnitude of v_1 relative to the RF voltage V_{a1} in (7) may be obtained by using the data for ionization rates and breakdown voltages given by Sze and Gibbons [8]. For a maximum RF voltage equal to about 30 percent of the dc voltage across the diode and a transit angle of about π for the total width of the depletion layer, we find that v_1 is limited to values of about 2. Thus, for RF voltages that are no higher than about 30 percent of the dc voltage across the diode, v_1 will be no higher than about 2 and β_1 will be no higher than about 0.9.

The equivalent circuit of Fig. 1 can be used for the basic design of an oscillator. The parameters that we shall obtain for the characterization of the single-frequency oscillator will also be found useful for the characterization of the oscillator with parametric interaction between harmonically related frequencies.

In Fig. 1, if an external impedance

$$Z_L = R_L + jX_L \tag{8}$$

is presented to the diode, one condition for oscillation is that the voltage across the diode must be in antiphase with the current generator (as was assumed in the derivation of the equivalent circuit). This condition yields $X_L = 1/\omega C$, where $C = C_a C_d/(C_a + C_d)$. The total current $J_{\text{tot}\,\omega}$ through the circuit then is given by

$$J_{\text{tot}\,\omega} = Q \times 2\beta_1 J_{c0} \sin^2(\theta/2)/(\theta/2) \tag{9}$$

where

$$Q = \{\omega C_d(R_s + R_L)\}^{-1}. \tag{10}$$

In the present picture of the oscillator, the drift zone acts as the negative-resistance element, which is tuned by a reactance $X_L - 1/\omega C_a = 1/\omega C_d$ and loaded by a resistance $R_s + R_L$. The factor Q defined in (10) may therefore be regarded as the loaded quality factor of the oscillator. The reactance tuning condition that is obtained from the complete equivalent circuit given in the Appendix is

$$X_L = \frac{1}{\omega C} + (R_s + R_L)\left\{\cot(\theta/2) + \frac{C_d}{C_a}\,\frac{\theta/2}{\sin^2(\theta/2)}\right\}. \tag{11}$$

As is seen from (11), the error introduced by using the simplified equivalent circuit of Fig. 1 for obtaining the reactance tuning condition $(X_L = 1/\omega C)$ is on the order of $1/Q$, in accordance with our simplifying assumption that the current through the diode is mainly capacitive. With (9) for the total fundamental-frequency current, we immediately obtain the output power of the oscillator. We have

$$P_{\text{out}} = \frac{1}{2} J_{\text{tot}\,\omega}^2 R_L = \frac{1}{2}\left\{2\beta_1 J_{c0}\frac{\sin^2(\theta/2)}{\theta/2} Q\right\}^2 R_L. \tag{12}$$

The output power according to (12) may be maximized with respect to the transit angle θ and with respect to the load resistance R_L. Here, we must observe that β_1 varies with the level of the RF voltage. Only if we assume that β_1 remains approximately constant with varying RF voltage (as is the case for large v_1), can we maximize (12) immediately. The optimum transit angle, assuming a constant impedance level $1/\omega C_d$, then is found to be given by $\theta = 0.74\,\pi$. As a function of the load resistance R_L, the output power varies with $Q^2 R_L$. This yields an optimum for $R_L = R_s$. In that case, however, half of the available power is dissipated in the parasitic series resistance, to the detriment of obtaining high dc-to-RF conversion efficiency. Rather, the power loss in the parasitic series resistance shall remain a negligible fraction of the available power, so we must have $R_s \ll R_L$ for any load resistance R_L that we shall want to use. Then, if $J_{c0}Q$ is increased, either by increasing J_{c0} or by increasing Q, we must observe that the RF voltage across the diode increases according to

$$V_1 = 2\beta_1 J_{c0}\frac{\sin^2(\theta/2)}{\theta/2} Q/\omega C. \tag{13}$$

For any desired operating bias current, (13) sets a limit to the value of Q, since V_1 must remain smaller than a fraction γ (on the order of 30 percent) of the dc voltage. We then find that

$$Q_{\text{lim}} = \gamma\omega C V_{\text{dc}}/\{2\beta_1 J_{c0}\sin^2(\theta/2)/(\theta/2)\} \tag{14}$$

and

$$P_{\lim} = \gamma\beta_1(C/C_d)\sin^2(\theta/2)/(\theta/2)J_{c0}V_{dc} - \tfrac{1}{2}(\gamma\omega CV_{dc})^2R_s. \quad (15)$$

The second term on the right in (15) represents the power loss in the parasitic series resistance. For the case of negligible power loss, (15) also yields the limiting dc-to-RF conversion efficiency:

$$\eta_{\lim} = \gamma\beta_1(C/C_d)\sin^2(\theta/2)/(\theta/2). \quad (16)$$

For $\theta = \theta_{\text{opt}} = 0.74\ \pi$, $C/C_d = 0.75$, $\beta_1 = 0.9$, and $\gamma = 0.3$, (16) shows that $\eta_{\lim} \approx 15$ percent. Slightly higher efficiencies are possible if C/C_d and β_1 can be made more nearly equal to the limiting value of unity. Significant increases in efficiency, however, can result only if the RF-voltage swing may become a larger fraction of the dc voltage across the diode. For example, if $C/C_d \rightarrow 1$, $\beta_1 \rightarrow 1$, and $\gamma = 0.6$, we find that $\eta_{\lim} \approx 40$ percent.

The exact variation of the signal level with the bias current and, in particular, start-oscillation conditions are obtained by equating the quality factor of the drift zone to the Q of the "external" circuit presented to the drift zone. This condition yields

$$\omega C_d V_{d1}/\{2\beta_1 J_{c0}\sin^2(\theta/2)/(\theta/2)\} = Q \quad (17)$$

where Q is given by (10). With $V_{d1} = (C_a/C_d)V_{a1} = (\theta/\alpha_c')v_1$, we may write (17) in the form

$$\beta_1/v_1 = J_{c0,\text{start}}/J_{c0} \quad (18)$$

where $J_{c0,\text{start}} = (\epsilon A/2\alpha_c' v_d)[\omega^2/\{Q\sin^2(\theta/2)/(\theta/2)\}]$. Since for small values of v_1, $\beta_1 = v_1$, (18) shows that $J_{c0,\text{start}}$ is the start-oscillation bias current. From the more detailed equivalent circuit given in the Appendix, we obtain

$$J_{c0,\text{start}} = \left(\frac{\epsilon A}{2\alpha_c' v_d}\right)\frac{\omega^2}{1 + Q\sin^2(\theta/2)/(\theta/2)}. \quad (19)$$

Equation (19) shows that the error introduced by using the simplified equivalent circuit of Fig. 1 is on the order of $1/Q$, as was the case with the reactance-tuning condition.

The factor β_1/v_1 in (18) may be regarded as a normalized value of the negative conductance of the drift zone. If $-Y_r$ represents the negative conductance of the drift zone, then

$$\begin{aligned} -Y_r &\doteq \{2\beta_1 J_{c0}\sin^2(\theta/2)/(\theta/2)\}/V_{d1} \\ &= (\beta_1/v_1)(J_{c0}/J_{c0,\text{start}})(\omega C_d/Q). \end{aligned} \quad (20)$$

The expression (20) for $-Y_r$ may also be used for the design of amplifiers. Dependent on reactance tuning, the current $J_{c0,\text{start}}$ then need not correspond to a start-oscillation current. In Fig. 3, the factor β_1/v_1 is plotted as a function of v_1. The curve for β_1/v_1 is bounded by the hyperbola $1/v_1$. The dashed lines in Fig. 3 correspond to

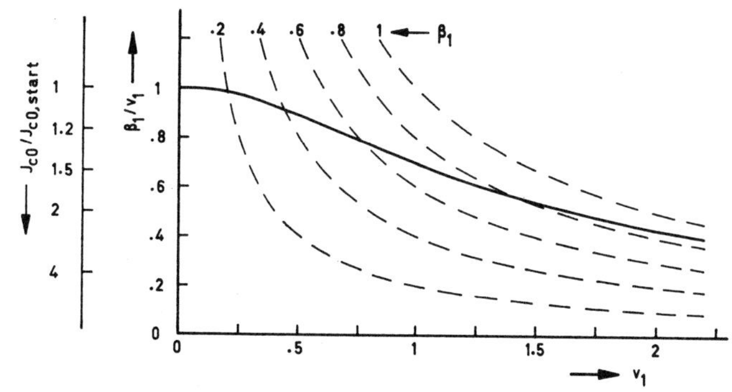

Fig. 3. Normalized negative conductance of drift zone as function of normalized ac voltage across avalanche zone. Dashed lines correspond to given fixed values of β_1. Scale for dc bias current relative to start-oscillation current is shown on left.

given fixed values of β_1. From (18) and Fig. 3 the operating point corresponding to any given value of $J_{c0}/J_{c0,\text{start}}$ can be immediately determined. To that purpose, a scale for $J_{c0}/J_{c0,\text{start}}$ is included on the left of Fig. 3. As an example, for $J_{c0} = 1.5\ J_{c0,\text{start}}$, we find that $v_1 = 1.1$ and $\beta_1 = 0.73$. Conversely, for a given value of v_1, we may determine the corresponding value of $J_{c0}/J_{c0,\text{start}}$. For example, for $v_1 = 2$ we must have $J_{c0}/J_{c0,\text{start}} = 2.3$. For a given diode structure the limit to the RF-voltage swing as a fraction of the dc voltage across the diode determines a maximum value for v_1 (of about 2). From Fig. 3, we may determine the corresponding maximum of $J_{c0}/J_{c0,\text{start}}$. For a given desired operating bias current we shall want to have the highest permissible RF-voltage swing, and thus the highest β_1, in order to obtain maximum output power. Therefore, for maximum output power at a given operating bias current while not incurring too large a voltage swing across the diode, we shall choose a value for $J_{c0}/J_{c0,\text{start}}$ of about 2–2.5.

With respect to the start-oscillation bias current, an important quantity for diode characterization is the start-oscillation current that is obtained under minimum-load conditions. Then,

$$J_{c0,s} = \left(\frac{\epsilon A}{2\alpha_c' v_d}\right)\frac{\omega^2}{1 + Q_s\sin^2(\theta/2)/(\theta/2)} \quad (21)$$

where

$$Q_s = (\omega C_d R_s)^{-1}. \quad (22)$$

The start-oscillation current $J_{c0,s}$ that is obtained for zero useful load is determined by the parasitic series resistance in the diode (and in the measuring circuit). For proper power transfer to the useful load, we must have $R_L \gg R_s$ or, equivalently, $J_{c0,\text{start}} \gg J_{c0,s}$. This condition should be compatible with the condition $J_{c0} \approx 2.5\ J_{c0,\text{start}}$ at the desired operating bias current J_{c0}. Whereas the current $J_{c0,s}$ characterizes the diode, the current $J_{c0,\text{start}}$ relative to $J_{c0,s}$ and relative to the desired or thermally permissible operating bias current J_{c0} characterizes the oscillator as a whole.

III. Two-Frequency Equivalent Circuit

The only nonlinearity that we have considered so far is the decrease of the negative conductance of the diode with increasing RF-voltage level. Essentially, that nonlinearity arises because the ac conduction current is limited by the dc current. The single-frequency equivalent circuit introduced in Section II, therefore, is capable of describing only the limitation mechanism on the oscillation amplitude.

In the actual diode, even for sinusoidal driving voltage $V_a(t)$, strong harmonic conduction currents are generated [cf. (1)]. As a result, if the associated microwave circuit is designed such that a high-quality resonance is obtained at any particular harmonic of the driving frequency, then a significant voltage will develop across the diode at that harmonic. The harmonic voltage in turn influences the current generation. The fundamental-frequency current will contain mixing products of the fundamental-frequency voltage and the harmonic voltage. Moreover, because of the wide-band negative-resistance properties of the diode, the circuit may be designed such that actual oscillation is achieved at a harmonic of the desired operating frequency. The nonlinear interactions between the two frequencies that result in that case may be best described as parametric interactions. If the harmonic voltage across the diode is high enough, the fundamental-frequency current amplitude may not only be enhanced, but, more importantly, it may then be almost entirely controlled by the harmonic voltage. For high-power operation at the fundamental frequency, the loaded Q at that frequency must be low in order to not incur too large an RF-voltage swing. For the single-frequency oscillator this results in poor noise performance. For the two-frequency oscillator, however, the fundamental-frequency current amplitude is controlled by the high-Q harmonic resonance. Thus, a significant improvement in noise performance may be expected, while the low-Q high-power fundamental-frequency circuit is maintained. In order to obtain criteria that characterize such parametric operation of the oscillator, we shall next introduce a two-frequency equivalent circuit for the diode. For convenience, we shall consider the case of frequencies ω and 2ω.

The two-frequency equivalent circuit may be derived along lines similar to those used in the derivation of the single-frequency equivalent circuit. In (1) we now have a fundamental-frequency voltage $V_{a1} \sin \omega t$ and a second-harmonic voltage $V_{a2} \sin (2\omega t+\phi)$. For operation at given frequencies ω and 2ω, we apply as a practical condition that the reactance-tuning conditions shall be independent of the signal level. From (1) it is clear that a signal-level-independent phase relationship between fundamental-frequency current and voltage and, also, between second-harmonic current and voltage is maintained only if the phase ϕ of the second-harmonic voltage is either zero or π. We may therefore work with two voltages $V_{a1} \sin \omega t$ and $V_{a2} \sin 2\omega t$, where V_{a2}/V_{a1} may be either positive or negative. The two-frequency equivalent circuit then is readily found in the form shown in Fig. 4. A more detailed equivalent circuit may be obtained as was done for single-frequency operation and is given in the Appendix. The relative current amplitudes β_1 and β_2 are functions of both V_{a1} and V_{a2}. With the voltage normalization

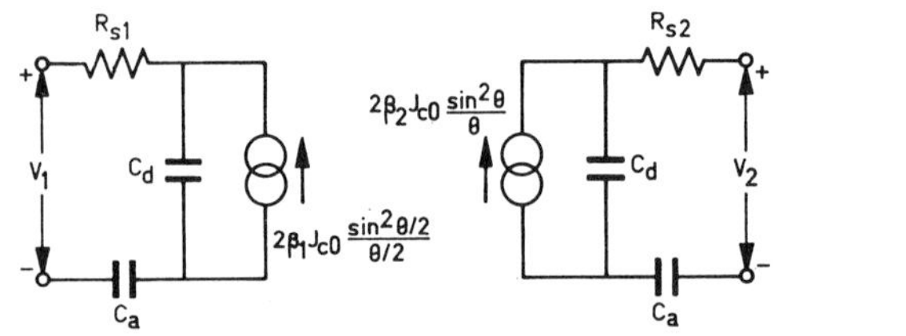

Fig. 4. Diode equivalent circuit for operation at two harmonically related frequencies. Current amplitudes β_1 and β_2 each are functions of both V_1 and V_2.

$$v_m = \alpha_c' V_{am}/m\omega\tau_a \tag{23}$$

it may be shown that with voltages v_1 and v_n at the frequencies ω and $n\omega$, respectively, we have

$$\beta_m = \frac{(-1)^m \sum_{p=-\infty}^{+\infty} (-1)^{p(n+1)} I_p(2v_n) I_{pn-m}(2v_1)}{\sum_{p=-\infty}^{+\infty} (-1)^{p(n+1)} I_p(2v_n) I_{pn}(2v_1)}. \tag{24}$$

Here, the functions $I_\nu(x)$ are the modified Bessel functions of the first kind, the order ν, and the argument x. As in Section II we may consider the negative conductance of the drift zone for the frequencies ω and 2ω. The respective normalized negative conductances are β_1/v_1 and β_2/v_2, which are functions of both v_1 and v_2. A plot of β_1/v_1 as a function of v_1 and with v_2 as a parameter is given in Fig. 5(a); a plot of β_2/v_2 as a function of v_2 and with v_1 as a parameter is given in Fig. 5(b). In Fig. 5(a), lines of constant β_1 are also shown. The curves for β_1/v_1 are bounded by the hyberbola $1/v_1$; the curves for β_2/v_2 are bounded by the hyperbolae $1/v_2$ and $-1/v_2$. From Fig. 5(a), it is seen that a negative voltage v_2 increases the fundamental-frequency current amplitude and thus increases the negative conductance of the drift zone. A positive voltage v_2 decreases the negative conductance of the drift zone. The change in conductance is especially pronounced for the lower values of v_1. From (24) it may be shown that for $v_1 \rightarrow 0$ the negative conductance can be increased by as much as a factor of 2 by high negative values of v_2.

Operation of the diode as a two-frequency oscillator requires tuning of the diode capacitance at two frequencies, ω and 2ω. If

$$Z_{Lm} = R_{Lm} + jX_{Lm} \tag{25}$$

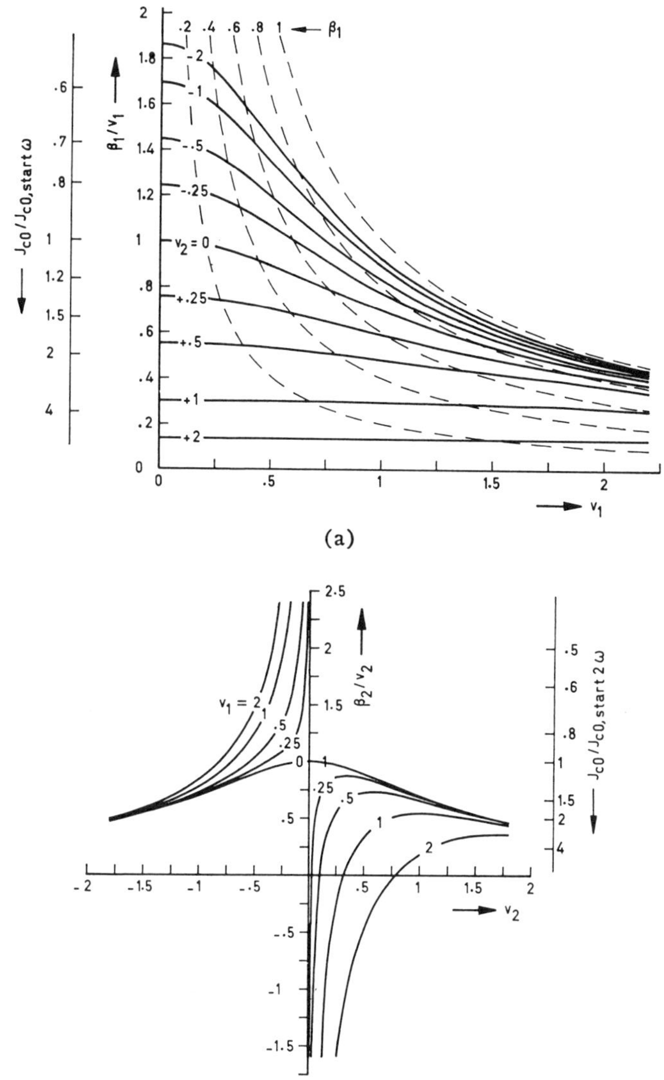

(b)

Fig. 5. (a) Normalized negative conductance of drift zone for frequency ω as function of normalized fundamental-frequency voltage across avalanche zone. Normalized second-harmonic voltage is a parameter. (b) Normalized negative conductance of drift zone for frequency 2ω as function of normalized second-harmonic voltage across avalanche zone. Normalized fundamental-frequency voltage is a parameter.

is the external load impedance presented to the diode at the frequency $m\omega$, then we must have $X_{L1}=1/\omega C$ and $X_{L2}=1/2\omega C$. If the more detailed equivalent circuit given in the Appendix is used, we find that we must have

$$X_{Lm} = \frac{1}{m\omega C} + (R_{sm} + R_{Lm}) \cdot \left\{ \cot(\theta/2) + \frac{C_d}{C_a} \frac{m\theta/2}{\sin^2(m\theta/2)} \right\} \qquad (26)$$

for $m=1, 2$. Here $R_{sm}+R_{Lm}$ is the total load resistance at the frequency $m\omega$. As was found for the single-frequency oscillator, the resistive loading conditions for the frequencies ω and 2ω are best described by the loaded quality factor Q_m of the external circuit that is seen by the drift zone at the frequency $m\omega$ (for $m=1, 2$):

$$Q_m = \{ m\omega C_d(R_{sm} + R_{Lm}) \}^{-1}. \qquad (27)$$

For oscillation the quality factor of the drift zone at the frequency $m\omega$ must be equal to the quality factor Q_m. As in Section II for the single-frequency oscillator, we then find the conditions

$$\beta_1/v_1 = J_{c0,\text{start }\omega}/J_{c0}$$
$$\beta_2/v_2 = J_{c0,\text{start }2\omega}/J_{c0} \qquad (28)$$

where a "start-oscillation current" for the frequency $m\omega$ has been defined according to

$$J_{c0,\text{start }m\omega} = \left(\frac{\epsilon A}{2\alpha_c' v_d} \right) \frac{(m\omega)^2}{1 + Q_m \sin^2(m\theta/2)/(m\theta/2)} \cdot \qquad (29)$$

As in Section II, the simplified equivalent circuit of Fig. 4 leads to a factor $Q_m \sin^2(m\theta/2)/(m\theta/2)$ in the denominator on the right of (29) rather than the factor $1+Q_m \sin^2(m\theta/2)/(m\theta/2)$. The additional constant of unity in the latter expression is obtained if the more detailed equivalent circuit given in the Appendix is used. The current $J_{c0,\text{start }m\omega}$ would be the start-oscillation current in case of single-frequency operation at the frequency $m\omega$. As we shall see below, the start-oscillation bias current for the two-frequency oscillator may differ both from $J_{c0,\text{start }\omega}$ and from $J_{c0,\text{start }2\omega}$.

For the two-frequency oscillator the operating point (β_1, v_1; β_2, v_2) may be determined from Fig. 5(a) and (b) by means of (28). From Fig. 5(a) it is immediately seen that if a negative voltage v_2 is present, then the start-oscillation current is below $J_{c0,\text{start }\omega}$ (by as much as a factor of 2 for large enough values of v_2). If a positive voltage v_2 is present, then the start-oscillation current is above $J_{c0,\text{start }\omega}$. The second-harmonic voltage will be present only if the proper value of $J_{c0}/J_{c0,\text{start }2\omega}$ is achieved simultaneously according to Fig. 5(b). If we take as a condition that for self-starting operation the fundamental-frequency oscillation must start from $v_1=0$, then from Fig. 5(b) we find that $J_{c0,\text{start }2\omega}$ must be below $J_{c0,\text{start }\omega}$. Then, as the bias current is increased from zero, the oscillator will start at the frequency 2ω. As the bias current is increased further, the operating point in Fig. 5(b) moves along the curve for β_2/v_2 with $v_1=0$ until a high enough voltage v_2 is attained relative to the value of $J_{c0}/J_{c0,\text{start }\omega}$ in Fig. 5(a) to start the fundamental-frequency oscillation. The phase relationship between the two frequencies will be such that v_2 is negative as compared to the sign of v_1. It is clear that $J_{c0,\text{start }\omega}$ should not be too large as compared to $J_{c0,\text{start }2\omega}$; otherwise either the maximum permissible bias current or voltage saturation at the frequency 2ω will be reached before the fundamental frequency ω sets in. If, on the other hand, $J_{c0,\text{start }2\omega}$ is above $J_{c0,\text{start }\omega}$, then the oscillator will start at the frequency ω for $J_{c0}=J_{c0,\text{start }\omega}$.

For bias currents greater than both $J_{c0,\text{start }\omega}$ and $J_{c0,\text{start }2\omega}$, a second mode of oscillation satisfying the condition of signal-level-independent reactance tuning may occur, for which $v_2>0$. For $v_2>0$ the fundamental-frequency output power is smaller than for $v_2<0$. Therefore, we shall refer to the case where $v_2>0$ as the low-power mode, as contrasted to the higher power mode for $v_2<0$. Since the start-oscillation bias current for the high-power mode is always below that for the low-power mode, one may expect that moding problems can be avoided in practice.

The fundamental-frequency output power is again given by (12). Now, β_1 must be taken according to (28) and Fig. 5. In the two-frequency high-power mode, β_1 for given v_1 is closer to the limiting value of unity than in the single-frequency case. Correspondingly, the output power will be larger, especially at the lower RF-voltage levels. As we have seen in Section II, v_1 should become no larger than about 2 in order that the diode not be driven into voltage saturation. On the same basis, v_2 should become no larger than about unity.

If the fundamental-frequency oscillation is to start at any particular second-harmonic voltage level, the required ratio of $J_{c0,\text{start }2\omega}$ and $J_{c0,\text{start }\omega}$ may be obtained from Fig. 5(a) and (b). As an example, if we want the fundamental-frequency oscillation to start with $v_2=-0.5$, we find from Fig. 5(a) that at start-oscillation we have $J_{c0}=0.69\ J_{c0,\text{start }\omega}$. From Fig. 5(b) we find that for $v_1=0$, $v_2=-0.5$ we must have $J_{c0}=1.12\ J_{c0,\text{start }2\omega}$. Thus, $J_{c0,\text{start }2\omega}=0.62\ J_{c0,\text{start }\omega}$. From the definition in (29) we find that a condition $J_{c0,\text{start }2\omega}=(1/\delta)J_{c0,\text{start }\omega}$
270 involves a condition on Q_1 and Q_2:

$$\frac{1+Q_2\sin^2\theta/\theta}{1+Q_1\sin^2(\theta/2)/(\theta/2)}=4\delta. \tag{30}$$

For high enough values of both Q_1 and Q_2, (30) may be written in the form $Q_2/Q_1=2\delta/\cos^2(\theta/2)$. As an example, for $\theta=0.74\ \pi$ and $\delta=1.6$, (30) yields $Q_2/Q_1=23+27/Q_1$. The second-harmonic quality factor that is required for two-frequency operation is seen to be large compared to the fundamental-frequency quality factor. Therefore, especially for the lower values of v_1 where β_1 is largely controlled by v_2, we may expect significant improvement in fundamental-frequency noise performance. Although an exact determination would require differentiation of (24) with respect to both v_1 and v_2, we may here make a rough estimate of the equivalent loaded Q of the fundamental frequency oscillation according to $Q_{eq}\approx Q_1Q_2$. For $Q_1=5$ and $Q_2/Q_1=28$, this yields $Q_{eq}\approx 28$, $Q_1^2=700$. As mentioned before, the improvement in noise performance is obtained only as long as β_1 is controlled to a significant extent by v_2. This limits v_1 to about 1, a value that may be compared with a maximum permissible value of v_1 of about 2. Thus, for up to about half the maximum output power, a significant improvement in noise performance of the oscillator may be expected by parametric interaction with the harmonic frequency.

Here, we may observe that the second-harmonic quality factor is limited by the lowest attainable value of the total load resistance for the second harmonic. For highest Q_2, we shall want to make $R_{L2}\rightarrow 0$, that is, no coupling of the second harmonic to the external load. The value of Q_2 then is determined by the parasitic series resistance in the diode and in the microwave circuit. According to (30), a large value of δ can be more easily realized for smaller values of the transit angle. For example, if we wish to have $\delta>1$ and $\theta=\theta_{opt}=0.74\ \pi$, then we must have $Q_2/Q_1>14$. For $\delta>1$ and $\theta=0.37\ \pi$, we only need to have $Q_2/Q_1>3$. On that basis, one may decide to choose the transit angle smaller than the value that gives maximum fundamental-frequency output power. On the other hand, for a given value of Q_2 higher values of δ in (30) can also be obtained by choosing a smaller value for Q_1. ($J_{c0,\text{start }\omega}$ increases with decreasing Q_1.) Since for low-noise performance the value of v_1 must be limited to about unity, we shall in practice choose Q_1 correspondingly lower than would be required for maximum RF-voltage swing at the desired operating bias current. Normalized curves for fundamental-frequency output power as a function of the bias current with the value of 4δ in (30) as a parameter have been reported elsewhere [7].

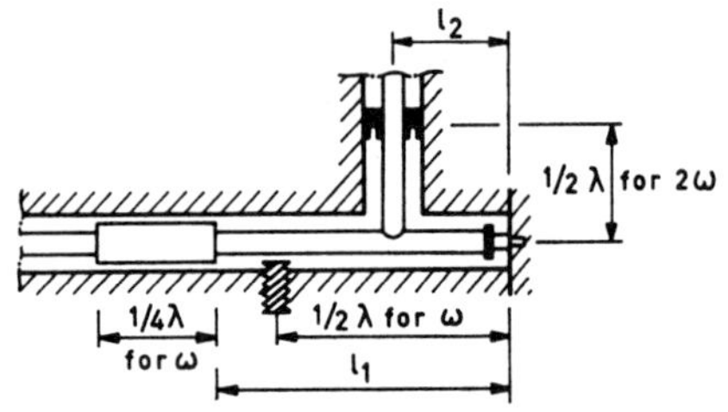

Fig. 6. Experimental coaxial two-frequency oscillator.

In addition to yielding a higher equivalent Q, the interaction with a harmonic of the desired oscillation frequency influences the noise performance of the oscillator in one more aspect. Since the frequency of oscillation in practice varies somewhat with the bias current, low-frequency avalanche-generated noise currents lead to incidental FM in the output of the oscillator. With two-frequency operation, the frequency tends to vary less with varying bias current than in the case of single-frequency operation. Thus, the two-frequency oscillator will also show improved noise performance with respect to incidental FM resulting from low-frequency fluctuations in the bias current.

An experimental oscillator was constructed in order to verify the foregoing ideas. Preliminary results have been reported earlier [5]. The circuit is as shown in Fig. 6. The diode is mounted at the end of a coaxial line. A quarter-wave transformer is employed to obtain the proper load resistance at the desired oscillation frequency, which in this case was 5 GHz. The length of line l_1 between the quarter-wave transformer and the diode provides the proper inductance for oscillation at 5 GHz. A high-Q second-harmonic circuit is obtained by means of a short-circuit stub of length 1/2 λ for 10 GHz. The stub is 1/4 λ for 5 GHz and thus does not influence the tuning for that frequency. The length of line l_2 between the stub and the diode provides the proper inductance for oscillation at 10 GHz. In the experimental circuit, both l_1 and l_2 included an extra one-half wavelength, for 5 GHz and 10 GHz, respectively, and were corrected for diode-package reactance. With a capacitive tuning screw, placed at 1/2 λ for 5 GHz from the diode, the 5- and 10-GHz resonances were brought in exact harmonic relationship. The diodes employed were Si p+-n diffused diodes of 120-volt breakdown voltage ($\theta \approx 0.74\ \pi$ at 5 GHz) and of 200-μ diameter. Start-oscillation bias current was 35–45 mA for 10 GHz; the start-oscillation current for 5 GHz with slightly detuned 10-GHz short-circuit stub was 45–65 mA. Loaded Q was about 7 for 5 GHz and about 100 for 10 GHz ($R_s \approx 0.2$ ohms). Highest CW output power was 1.75 watts, in accordance with what had been obtained earlier [9] with a four-slug coaxial circuit. With single-frequency tuning, the maximum CW output power at 5 GHz was 1.2 watts. Under optimum tuning conditions the present circuit gave an improvement in FM noise performance of more than 20 dB over the case of single-frequency tuning. For instance, at 0.5 watt of output power, the FM noise expressed as single-sideband noise power relative to carrier power and measured in a 3-kHz bandwidth at 100 kHz from the carrier was −46 dB for single-frequency operation and −68 dB for two-frequency operation. The improvement in noise performance could be realized up to nearly 1 watt of output power. For higher output powers, a rapid deterioration of noise performance toward that obtained with single-frequency tuning was observed.

Operation of the oscillator as just described may be called self-pumped subharmonic generation. One may think of an alternative scheme, in which the power is extracted at the upper rather than at the lower frequency. For maximum output power, one would then choose $\theta = 0.74\ \pi$ for the upper frequency. The loaded quality factor for the lower frequency, however, will be small as compared to that for the upper frequency, so that little improvement in noise performance may be expected. In fact, preliminary experiments with a hybrid TEM/waveguide circuit in which only the upper frequency propagated and was coupled to the useful load through the waveguide so far have indicated a deterioration in noise performance as compared to single-frequency operation at the upper frequency.

IV. Frequency Multiplication

If either the lower frequency or the upper frequency is supplied from an external source, possibilities for frequency multiplication or frequency division, respectively, arise. In this section we shall consider the case where the lower frequency is supplied from an external source. [See Fig. 7(a) and (b).] For convenience we shall consider the case of a voltage source ($Z_i = 0$) for the frequency ω. From Fig. 7(a), we see that for frequency multiplication the upper frequency circuit must be tuned according to (26). From Fig. 7(b) it is further clear that $J_{c0,\text{start } 2\omega}$ must be chosen above the desired operating bias current so that no self-oscillation at the frequency 2ω results. By making v_1 high enough, full

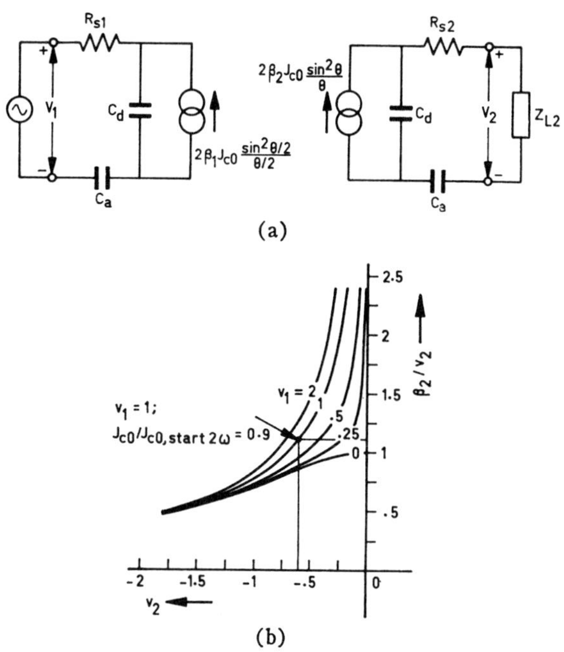

Fig. 7. (a) Circuit representation of frequency multiplier. (b) Operating point of frequency multiplier shown in terms of applied voltage at frequency ω and value of bias current relative to upper frequency start-oscillation current.

current modulation can be obtained at the frequency $2\ \omega$. The upper frequency voltage v_2 may become as high as unity (for J_{c0} just below $J_{c0,\text{start } 2\omega}$). The frequency multiplier can be readily made to have gain. To that purpose, the fundamental-frequency source may also be tuned according to (26), with $J_{c0,\text{start } \omega}$ above J_{c0} in order to prevent self-oscillation at the frequency ω. The transit angle will be chosen equal to $0.74\ \pi$ at the frequency 2ω.

V. Frequency Division

For frequency division, an external voltage will be supplied at the upper frequency. [See Fig. 8(a) and (b).] For convenience, we shall again consider the case where the external source has zero internal impedance. A stable frequency divider is obtained by tuning the lower frequency circuit in Fig. 8(a) according to (26). From Fig. 8(b) we further find that $J_{c0,\text{start } \omega}$ must be chosen above the desired operating bias current J_{c0}. Again, full current modulation can be achieved at the frequency ω by making v_2 high enough; the value of v_1 may become as high as unity (for J_{c0} just below $J_{c0,\text{start } \omega}$). In order to make the frequency divider have gain, the external voltage source may also be tuned according to (26), with $J_{c0,\text{start } 2\omega}$ sufficiently above J_{c0} to prevent self-oscillation at the frequency 2ω. The transit angle will be chosen equal to 0.74π at the frequency ω.

VI. Conclusions

Equivalent circuits have been introduced for operation of the avalanche transit-time diode at a single frequency and at two harmonically related frequencies. The equivalent circuits permit microwave circuit design

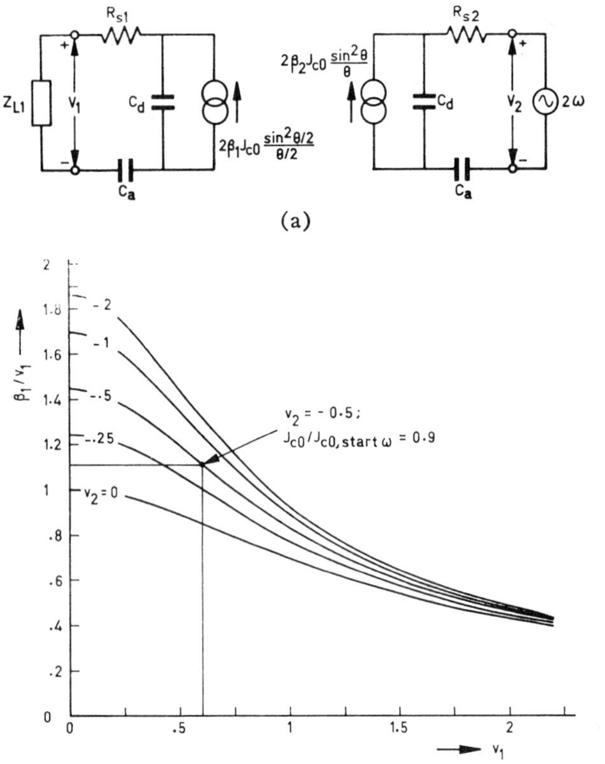

(b)

Fig. 8. (a) Circuit representation of frequency divider. (b) Operating point of frequency divider shown in terms of applied voltage at frequency 2ω and value of bias current relative to lower frequency start-oscillation current.

for optimum nonlinear device-circuit interaction. Parameters have been derived that characterize the diode and the oscillator as a whole. Significant improvement in the noise performance of the oscillator is possible by parametric interaction with a harmonic of the desired oscillation frequency. An experimental oscillator for two-frequency operation has helped in verifying the theoretical predictions. The equivalent-circuit representation puts the design of amplifying frequency multipliers and frequency dividers with avalanche transit-time diodes on a simple basis.

Appendix

The single-frequency equivalent circuit given in Fig. 1 includes neither the effect of the inductive-particle current in the avalanche zone nor the effect of the space-charge-induced reactive current in the drift zone. The corrections that are required to include those effects in the equivalent circuit have already been indicated [following (4) and (6)]. The complete equivalent circuit is given in Fig. 9. The phase of the current generators is given in complex terms and is defined by writing the ac voltage across the avalanche zone in the form

$$V_a(t) = V_{a1} \sin \omega t = V_{a1} \operatorname{Im} e^{j\omega t}.$$

It must be observed that the phase of the current generators is defined *relative to the phase of the voltage across the avalanche zone.* Thus, if one desires to work with a

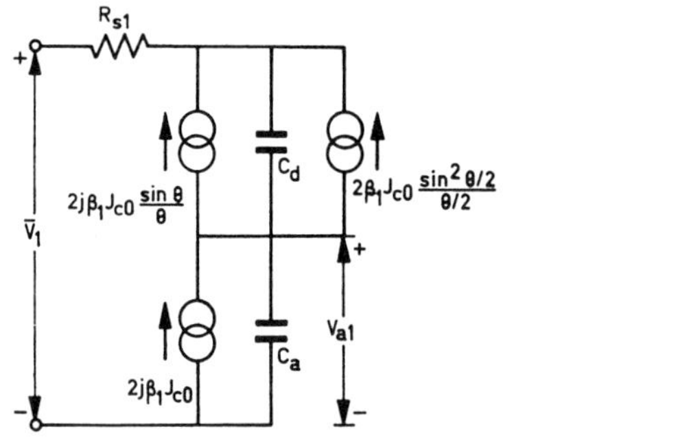

Fig. 9. Complete equivalent circuit for single-frequency operation of diode. Current amplitude β_1 is given by Fig. 2.

complex rather than a real voltage amplitude V_{a1}, the current amplitude β_1 has the same phase factor as V_{a1}. From Fig. 9 the following expression for the impedance of the diode is obtained:

$$Z_{\text{diode}} = \frac{1}{j\omega C} \frac{1}{1 - 2\beta_1 J_{c0}/\omega C_a V_{a1}} + \frac{1}{\omega C_d} \frac{\sin^2 (\theta/2)/(\theta/2) - j(1 - \sin\theta/\theta)}{1 - \omega C_a V_{a1}/2\beta_1 J_{c0}}. \quad (31)$$

The current amplitude β_1 is seen to arise only in the combination β_1/V_{a1}. This expression for the impedance of the diode is identical to that obtained from a more formal analysis [7]. (For comparison of (31) with (16) of [7], it may be observed that in the derivation given in [7] the phase of the fundamental-frequency current relative to the voltage was included in β_1; also, the voltage taken was of the form $\overline{V}_{a1}e^{j\omega t} + \overline{V}_{a1}^* e^{-j\omega t}$, which leads to a difference of a factor of 2 relative to the voltage $V_{a1} \operatorname{Im} e^{j\omega t}$ taken in the present derivation.)

Fig. 9 shows that the avalanche zone exhibits a parallel resonance at the frequency where $\omega C_a V_{a1} = 2\beta_1 J_{c0}$. For small signals we have $\beta_1 = \alpha_c' V_{a1}/\omega\tau_a$, so that then the avalanche resonance frequency ω_a is given by $\omega_a^2 = (2\alpha_c' v_d/\epsilon A) J_{c0}$. On the other hand, for large signals we have $\beta_1 \rightarrow 1$, so that the avalanche resonance frequency then shifts downward with increasing RF voltage according to $\omega_a \approx 2J_{c0}/C_a V_a$. It may be shown [7] that with the oscillator tuning conditions given in Section II, the RF voltage across the avalanche zone grows with the bias current in such a way that the avalanche resonance frequency remains fixed for increasing bias current. [This is equivalent to (18) of Section II.]

From Fig. 9 it may also be shown that, for frequencies above the avalanche resonance frequency, the voltage across the drift zone has a component that is in phase with the voltage across the avalanche zone. With the given direction of the current generator $2\beta_1 J_{c0} \sin^2 (\theta/2)/(\theta/2)$, we thus find that above the avalanche resonance the diode is active. For frequencies below the avalanche resonance frequency, the voltage across the drift zone has a component that is in antiphase with the voltage across the avalanche zone. The diode then is found to be passive.

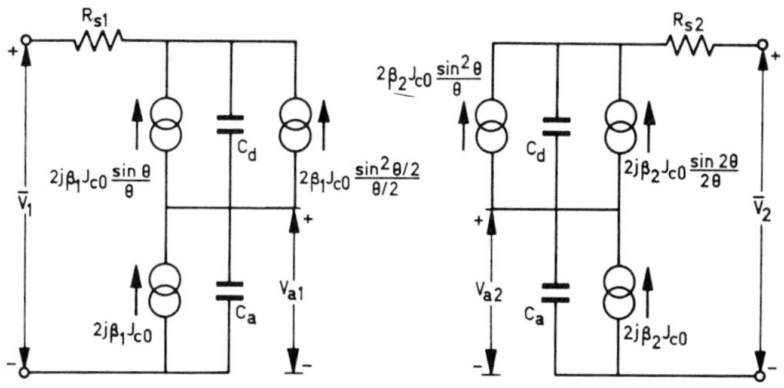

Fig. 10. Complete equivalent circuit for operation of diode at two harmonically related frequencies. Current amplitudes β_1 and β_2 are given by Fig. 5(a) and (b), respectively.

The two-frequency equivalent circuit given in Fig. 4 may be similarly corrected for the effect of the inductive particle current in the avalanche zone and for the effect of the space-charge-induced reactive current in the drift zone. We then obtain the equivalent circuit shown in Fig. 10. The phase of β_1 again is defined relative to that of V_{a1}; the phase of β_2 is defined relative to that of V_{a2}. The phase of the upper frequency voltage is related to that of the lower frequency voltage in such a way that for a voltage $V_{a1} \sin \omega t$ at the frequency ω we have a voltage $V_{a2} \sin 2\omega t$ at the frequency 2ω. The ratio V_{a2}/V_{a1} may be either positive or negative. A more general phase relationship may be assumed such that for a voltage $V_{a1} \sin \omega t$ at the frequency ω we have a voltage $V_{a2} \sin (2\omega t+\phi)$ at the frequency 2ω. In that case, both in the terms in the numerator and in those in the denominator in (24), we must then include a phase factor $e^{jp\phi}$. The phase relationship between β_1 and V_{a1} and that between β_2 and V_{a2} then become dependent on the signal level [cf. remark above (23)].

It may be noted that a negative voltage V_{a2} increases the value of β_1 relative to the value obtained with single-frequency excitation. This gives an upward shift of the avalanche resonance frequency. Similarly, a positive voltage V_{a2} decreases the value of β_1 and we then have a downward shift of the avalanche resonance frequency relative to the case of single-frequency excitation.

References

[1] W. T. Read, Jr., "A proposed high-frequency negative resistance diode," *Bell Syst. Tech. J.*, vol. 37, pp. 401–446, March 1958.

[2] A. S. Clorfeine, "Self-pumped parametric amplification with an avalanching diode," *Proc. IEEE* (Letters), vol. 54, pp. 1956–1957, December 1966.

[3] W. J. Evans and G. I. Haddad, "Frequency conversion in IMPATT diodes," *IEEE Trans. Electron Devices*, vol. ED-16, pp. 78–87, January 1969.

[4] C. B. Swan, "Impatt oscillator performance improvement with second-harmonic tuning," *Proc. IEEE* (Letters), vol. 56, pp. 1616–1617, September 1968.

[5] K. Mouthaan, "Two-frequency operation of the avalanche transit-time oscillator," *Proc. IEEE* (Letters), vol. 58, pp. 510–512, March 1970.

[6] M. Gilden and M. E. Hines, "Electronic tuning effects in the Read microwave avalanche diode," *IEEE Trans. Electron Devices*, vol. ED-13, pp. 169–175, January 1966.

[7] K. Mouthaan, "Nonlinear characteristics and two-frequency operation of the avalanche transit-time oscillator," *Philips Res. Rep.*, vol. 25, pp. 33–68, February 1970.

[8] S. M. Sze and G. Gibbons, "Avalanche breakdown voltages of abrupt and linearly graded p-n junctions in Ge, Si, GaAs and GaP," *Appl. Phys. Lett.*, vol. 8, pp, 111–113, March 1966.

[9] K. Mouthaan and H. P. M. Rijpert, "Second-harmonic tuning of the avalanche transit-time oscillator," *Proc. IEEE* (Letters), vol. 57, pp. 1449–1450, August 1969.

THE HIGH-EFFICIENCY (TRAPATT) MODE IV

High-Power, High-Efficiency Silicon Avalanche Diodes at Ultra High Frequencies

***Abstract*—Silicon avalanche diodes with a pnn^+ mesa structure were made according to suitably chosen design parameters, and carefully fabricated to meet stringent performance specifications. Operating under pulse conditions with a duty factor of approximately 10^{-3}, efficiencies as high as 25 percent, and power outputs up to 435 watts at 425 megahertz, were achieved.**

This letter describes recent tests with new avalanche diodes which have produced oscillations with greatly improved efficiencies at frequencies rarely associated with conventional avalanche diodes. Previous results indicated that, in general, avalanche diodes of the *pn* or *pin* type prefer to operate at a frequency range from several gigahertz on up.[1-3] Only occasional exceptions have been cited, as in the case of $\pi p \pi p$[4] and $p^{++}n^{+}nn^{++}$[5] diodes which were found to operate also at lower frequencies; and certain *pin* diodes[6] with extremely wide depletion regions (150 microns) which were said to have yielded efficiencies of 1×10^{-3} percent at frequencies up to 2 GHz.

The object of our investigation, however, was not to determine the feasibility of fabricating avalanche diodes for use in the high microwave frequency range, but rather in the UHF and *L*-band ranges. Consequently, the diode structure chosen for our tests was quite specifically designed for this lower frequency range. Figure 1 shows the doping profile of our diode as demonstrated by a photomicrograph of a diode mesa cut at a 15-degree angle. A thin substrate to N^+ silicon was prepared with an *N*-type epitaxial layer of 4 to 6 ohm · cm resistivity and a thickness of 8 to 20 microns. Through boron deposition and diffusion a *p*-type region with an abrupt junction was formed. The wafer was then processed by proper masking and etching techniques until a mesa structure with a 26-mil diameter was produced. The successful completion of the final etching step was reached when the monitoring of the reverse breakdown voltage indicated a sharp break in the *I-V* characteristic between 160 and 185 volts. Recognizing the importance of adequate thermal conduction for such high-voltage and high-dissipation devices, a heavy copper pedestal and pin were incorporated in the coaxial encapsulating package. The complete diode assembly exhibited a capacitance of about 10 pF at −6 volts bias, and a corresponding cutoff frequency of approximately 20 GHz.

Fig. 1. Microphotograph of diode mesa.

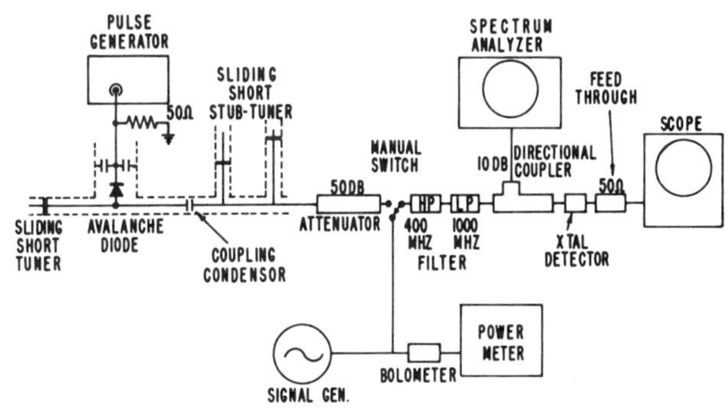

Fig. 2. Block diagram of test set-up.

The performance evaluation of these diodes as RF generators at ultra high or *L*-band frequencies was carried out in the circuit diagrammed in Fig. 2. This circuit can be considered essentially as a coaxial cavity with a sliding short which provided the required cavity length at the resonance frequency. The diode itself was mounted in a special "tee" and driven from a high-voltage pulse generator supplying a reverse-biased voltage pulse of about 1 μs width at a repetition rate of several hundred pulses per second. A 400-MHz high-pass filter was used to eliminate unwanted pulse play-through in the detector circuit; in addition, either a 500- or 1000-MHz

Manuscript received March 2, 1967.

[1] W. T. Read, "A proposed high-frequency negative resistance diode," *Bell Sys. Tech. J.*, vol. 37, March 1958.

[2] B. C. De Loach and R. L. Johnston, "Avalanche transit-time microwave oscillators and amplifiers," *IEEE Trans. on Electron Devices*, vol. ED-13, pp. 181–186, January 1966.

[3] T. Misawa, "Negative resistance in *p-n* junctions under avalanche breakdown conditions, parts 1 and 2," *IEEE Trans. on Electron Devices*, vol. ED-13, pp. 137–151, January 1966.

[4] C. A. Lee, R. L. Batdorf, W. Wiegmann, and G. Kaminsky, "The Read diode—An avalanching transit-time negative-resistance oscillator," *Appl. Phys. Lett.*, vol. 6, March 1965.

[5] B. Hoefflinger, "High-frequency oscillations of p^{++}-n^{+}-n-n^{++} avalanche diodes below the transit-time cutoff," *IEEE Trans. on Electron Devices*, vol. ED-13, pp. 151–158, January 1966.

[6] R. L. Johnston, B. C. De Loach, and B. G. Cohen, "A silicon diode microwave oscillator," *Bell Sys. Tech. J. (Briefs)*, vol. 44, February 1965.

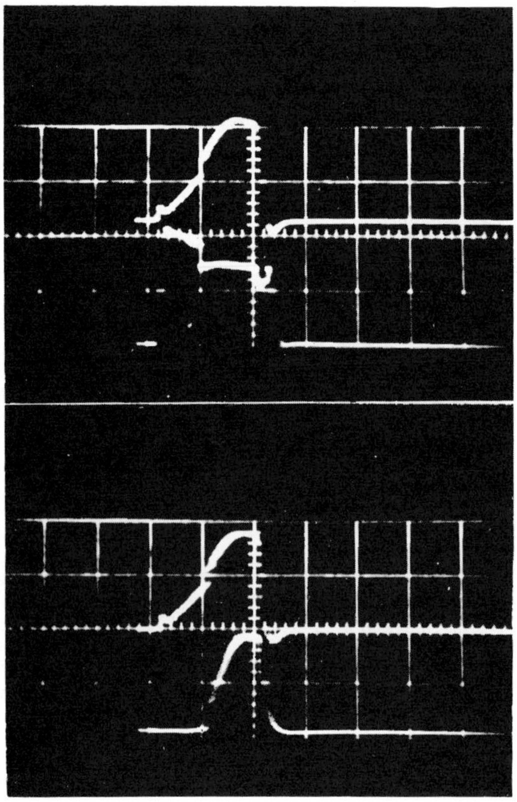

Fig. 3. Oscilloscope tracings. (a) Current pulse = 5 A/div. (b) Voltage pulse = 100 V/div. (c) Current pulse = same as (a). (d) Output of X-tal detector = 150 watts/div. Horizontal time scale 0.5 μs/div.

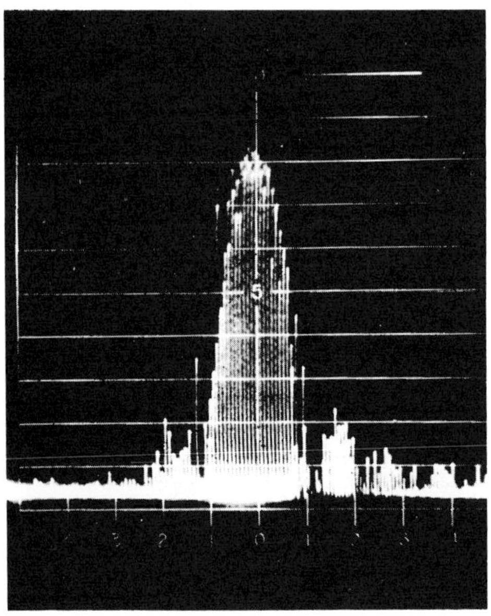

Fig. 4. Oscilloscope tracings of spectrum analyzer. Fundamental resonance frequency = 425 MHz; horizontal scale = 1 MHz/div.

TABLE I
SUMMARY OF DIODE TESTS*

Diode #	C_0 (pF)	C_{-6} (pF)	C_{min} (pF)	V_{BD} (volts)	F_{co-6} (GHz)	P_{in} (watts)	P_{out} (watts)	Eff. (%)	Freq. (MHz)
1	46	11	6.2	172	24.7	1970	435	22	425
9	41	15	5.1	190	14.2	1215	285	23.5	420
23	42	15	5.2	184	13.3	1240	280	22.5	425
68	26	9.2	3.1	186	15.4	1240	250	20	425
70	27	9.5	3.1	184	15.8	800	200	25	420
2A	8.5	3.2	1.9	112	66.9	342	67	19	810
7A	8.2	2.9	1.7	106	66.3	360	84	24	800

* Subsequent to the writing of this letter, another diode, tested in an improved cavity design, yielded 420 watts of output power at 1050 MHz with an efficiency of well over 30 percent.

low-pass filter was used to prevent higher harmonics of the fundamental oscillating frequency from reaching the detector. Several well-calibrated attenuators protected the spectrum analyzer and the diode detector from overloads. Final RF power measurements were obtained in two ways. The first one was through a substitution method in which the RF output of the avalanche diodes was compared with the calibrated output of a signal generator. In the second method, the RF power of the avalanche diode was fed directly into a wattmeter which measured the average power, and the peak power was calculated with the help of the duty factor and the known shape of the actual RF output pulse. Both methods produced results in excellent agreement.

Figure 3 shows oscilloscope traces of the voltage pulse, the current pulse, and the RF power output (as represented by the detector output). In a typical shape of the voltage pulse (when the diode is in its oscillating condition) one observes first a steep rise of voltage in the reverse breakdown voltage, and then an abrupt decrease to a lower level. The pulse current shows first a sharp but relatively small rise corresponding to the onset of reverse breakdown in the avalanche mode, then a gradual rise until another sharp increase occurs simultaneously with the sharp decrease in pulse voltage. When this pulse current bias reaches a certain level, an RF output can be observed which lasts for the duration of the input pulse. When the circuit is detuned, the output power as well as the steep step in the voltage and current pulse disappears. The frequency of the RF power was monitored by a spectrum analyzer, whose oscilloscope tracing is shown in Fig. 4. The highest RF power measured so far was 435 watts, the best power efficiency was 25 percent, at a resonance frequency which was typically around 425 MHz. Slightly different diodes resonated at 800 MHz yielding a maximum of 84 watts with 24 percent efficiency. Further results of power input, power output, efficiency, and various diode characteristics for several representative diodes are summarized in Table I. These experimental results demonstrate that avalanche diodes can indeed be operated efficiently as UHF power generators.

ACKNOWLEDGMENT

The authors wish to express their appreciation for the help and cooperation they received from the members of RCA Electronics Components and Devices Division, Somerville, N. J., where the diodes were fabricated. Special thanks go to A. Pikor of the same division and Dr. H. Kressel, RCA Laboratories, Princeton, N. J., for their development of the exacting process specifications.

H. J. PRAGER
K. K. N. CHANG
S. WEISBROD
RCA Laboratories
Princeton, N. J.

A THEORY FOR THE HIGH-EFFICIENCY MODE OF OSCILLATION IN AVALANCHE DIODES*

BY

A. S. CLORFEINE, R. J. IKOLA, AND L. S. NAPOLI

RCA Laboratories
Princeton, N. J.

Summary—High-efficiency oscillations, previously observed in avalanche diodes and simulated with a computer, are explained by means of an analytic theory in which all the relevant physical processes are clearly displayed and tied together. It is shown how efficiencies of 50-60% can be achieved.

On the basis of the theory, design formulas are derived. Relations involving the choice of width and doping density of the active layer, diode area, bias current density, and fundamental circuit impedance are established. Predictions of the theory are seen to be in good agreement with reported experimental results.

279

I. INTRODUCTION

THIS PAPER presents a theory for the high-efficiency mode of oscillation in avalanche diodes. Since the theory is analytic, all the relevant physical processes are clearly displayed and tied together. It is demonstrated how efficiencies of 50-60% can be achieved and how oscillators can be designed for optimum performance.

The high-efficiency mode was discovered in silicon by Prager, Chang, and Weisbrod.[1] The mode was considered "anomalous" in that, first, the oscillation efficiencies were markedly higher than had been previously observed with avalanche diodes and, second, the oscillation frequencies were considerably lower than could be expected from

* This work was supported by the Air Force Avionics Laboratory, Wright-Patterson Air Force Base, Ohio, under AF Contract No. F33615-68-C-1688, Project No. 4460.

[1] H. J. Prager, K. K. N. Chang, and S. Weisbrod, "High-Power, High-Efficiency Silicon Avalanche Diodes at Ultra-High Frequencies," *Proc. IEEE* (Letters), Vol. 55, p. 586, April 1967.

transit-time theory.[2,3] Subsequently, Johnston, Scharfetter, and Bartelink observed similar oscillations with germanium diodes[4] and were able to simulate these oscillations with efficiencies of up to 26%[5] on a computer. The computer simulations provided insight in two respects. First, they demonstrated that the high-efficiency mode could be achieved without, for example, such effects as filament formation or some fortuitous doping irregularity; also, the simulations revealed the existence of a high-density, slow-moving electron–hole plasma during part of the oscillation cycle.

What was still required, however, was a theory—one that could overcome the limitations of a simulation approach. Such limitations include (a) the need to first establish oscillations in the laboratory so that parameters may be supplied to the computer, (b) the substantial computer time required, (c) the obscuring of the relevant physics by waveform properties that are germane only to the particular nonoptimum diode-circuit combination being simulated, and, finally, (d) the difficulty in applying the simulation procedure to the task of designing and optimizing oscillators.

280

The approach taken in the present paper is, first, to develop a detailed analytic theory of the oscillation physics, and then to consider the question of what diode and circuit parameters are necessary to obtain these oscillations. The answers are seen to follow directly from the analytic work. More specifically, Section II gives a detailed description of the formation of a plasma in an avalanche diode (the plasma is an important ingredient in the theory, in contrast to Hoefflinger's postulated operation[6]). The analytic results obtained for the plasma formation are then used to calculate the response of the diode to certain applied currents (Section III). It is shown that much of the important physics for high-efficiency oscillations follows from a consideration of the diode response to a step current. Next, the criteria for a successful model of the high-efficiency mode are discussed and waveforms consistent with these criteria are introduced and analyzed. The circuit impedances necessary to obtain these waveforms must, of course, be realizable; these can be calculated. Finally, in Sections IV

---

[2] W. T. Read, Jr., "A Proposed High-Frequency Negative-Resistance Diode," *Bell System Tech. Jour.*, Vol. 37, p. 401, March 1958.

[3] H. K. Gummel and D. L. Scharfetter, "Avalanche Region of Impatt Diodes," *Bell System Tech. Jour.*, Vol. 45, p. 1797, Dec. 1966.

[4] R. L. Johnston, D. L. Scharfetter, and D. J. Bartelink, "High-Efficiency Oscillations in Germanium Avalanche Diodes Below the Transit-Time Frequency," *Proc. IEEE* (Letters), Vol. 56, p. 1611, Sept. 1968.

[5] D. L. Scharfetter, "High Efficiency Operation of Impatt Diodes," International Electron Devices Meeting, Washington, D. C., Oct. 1968.

[6] B. Hoefflinger, "Recent Developments on Avalanche Diode Oscillators," *Microwave Jour.* Vol. 12, p. 101, March 1969.

and V, design formulas are presented and predictions of the theory are summarized. These predictions are seen to be in good agreement with reported experimental results.

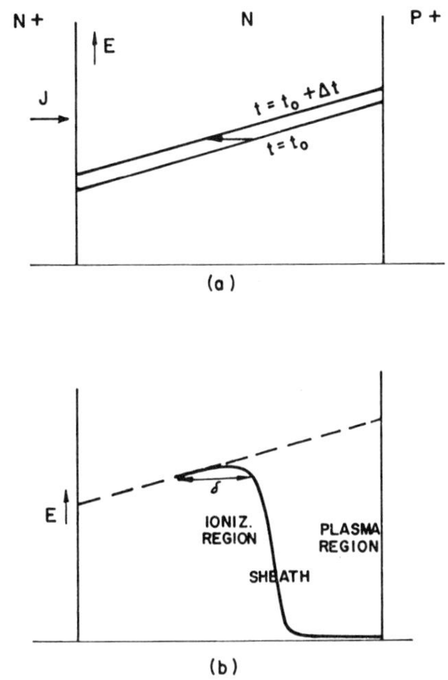

Fig. 1—Plasma formation in avalanche diodes: (a) electric field prior to ionization and (b) electric field subsequent to ionization.

II. Plasma Formation in Avalanche Diodes

A. Wave Propagation

Figure 1(a), in order to be specific, illustrates an abrupt junction p+-n-n+ silicon diode. The results, however, to a large extent, should be applicable to a graded junction as well. In addition, the analysis is appropriate, for example to a germanium p+-p-n+ structure, provided the words hole and electron are everywhere interchanged.

Consider the behavior of the electric field when the diode is under the influence of an external current density, J. At a time, $t = t_o$, the electric field, E, is assumed to have "punched through" the n-n+ boundary; however, it is not yet large enough to produce significant ionization. In the absence of such ionization, there is negligible conduction current, so that the displacement current everywhere is J. At a fixed point in the diode, the electric field will increase in a time interval, Δt, by

$$\Delta E = \frac{J}{\epsilon}\Delta t, \tag{1}$$

where ϵ is the dielectric constant. The slope of the electric field is given by Poisson's equation,

$$\frac{\Delta E}{\Delta z} = \frac{qn_d}{\epsilon}, \tag{2}$$

where $q = 1.6 \times 10^{-19}$ coulomb and n_d is the ionized doping density. The distance that a point of constant electric field moves in Δt is found from Equations (1) and (2) to be $\Delta z = (J/qn_d)\Delta t$. Thus, such a point moves with a wave velocity,

$$v_w = \frac{\Delta z}{\Delta t} = \frac{J}{qn_d}. \tag{3}$$

For sufficiently large J, the wave velocity can exceed the maximum velocity of particles that result from subsequent ionization. It is, in fact, such a condition that leads to the most interesting results and to which attention is now exclusively turned.

B. The Effect of Ionization

The dashed line in Figure 1(b) represents the electric field that would exist at a later time if there were no ionization. When the field exceeds a critical value, E_c, however, electron–hole pairs are created by impact ionization (the concept of a "critical field" is an approximate one, the consequences of which will be examined subsequently). The wave front (represented by the point for which $E = E_c$) moves toward the n-n+ boundary at a wave velocity that exceeds the particle velocities; therefore, the created carriers lag behind the advancing field front. Since the generated electrons drift in the same direction as the wave, while holes drift in the opposite direction, the density of electrons in the immediate wake of the wave front is higher than the density of holes. This creates an electron-rich region that serves to reduce the field. If a sufficient number of electrons are generated to overcome the positive doping density, n_d, the slope of the field profile reverses and the field reaches a very low value. At low fields, the electrons and holes are substantially slowed, leading to the formation of a dense plasma.

For purposes of analysis, three distinct regions are considered: (1) the ionization region, in which electron–hole pairs are created and particle velocities are saturated, (2) the sheath, in which the field is too low for ionization but sufficiently large for velocities to reach saturation, and (3) the plasma, in which the field is sufficiently low that velocities are determined by low-field mobilities. There is a transition region between the sheath and plasma, in which the velocity is not simply expressed; however, this region is so short that ignoring it introduces negligible error. Finally, it is assumed throughout this paper that the defect density is sufficiently low for recombination to be neglected, i.e., relevant times are short compared to recombination times. For carefully made diodes, this approximation is quite good, even in the high-density slow-moving plasma.

C. *Ionization Region*

The continuity equations for electrons and holes are

$$\frac{\partial n}{\partial t} = v_{ns}\frac{\partial n}{\partial z} + \alpha v_{ns} n + \beta v_{ps} p, \tag{4}$$

$$\frac{\partial p}{\partial t} = - v_{ps}\frac{\partial p}{\partial z} + \alpha v_{ns} n + \beta v_{ps} p, \tag{5}$$

where n and p are the densities, v_{ns} and v_{ps} are the saturation velocities, and α and β are the ionization coefficients of electrons and holes, respectively. We look for a solution for which the ionization region (and the sheath) is moving with the wave velocity, v_w; therefore, we define the moving coordinate $x = z + v_w t$ (the positive sign is appropriate, since the wave is moving in the negative z direction). The particle densities are solely functions of x, and so we may reduce the partial differential equations to ordinary differential equations with the transformations

$$\frac{\partial}{\partial t} = v_w \frac{d}{dx}, \quad \frac{\partial}{\partial z} = \frac{d}{dx}. \tag{6}$$

It is also convenient to define a critical current density $J_c = q v_{ns} n_d$, a normalized current density $j = J/J_c = v_w/v_{ns}$, and a saturation velocity ratio $r = v_{ps}/v_{ns}$. J_c is the current density above which the wave travels faster than the generated electrons. When Equations (6) are substituted in Equations (4) and (5) we obtain the reduced continuity relations

$$\frac{dn}{dx} = \frac{\alpha n + r\beta p}{j-1}, \tag{7}$$

$$\frac{dp}{dx} = \frac{\alpha n + r\beta p}{j+r}. \tag{8}$$

Since α and β are functions of electric field, Equations (7) and (8) together with Poisson's equation form a coupled set of three equations. To facilitate their solution, we shall assume (1) that $r\beta p \ll \alpha n$, i.e., ionization by holes can be neglected (for germanium, ionization by electrons is neglected) and (2) that $\alpha = 0$ for $E < E_c$ and α is constant for $E \geqq E_c$.* While the first assumption is not necessary for an analytic solution of Equations (7) and (8), its use is convenient; and, in any case, it is well justified for silicon and germanium, since not only is β significantly smaller than α but $p < n$ everywhere.** Assumption (2) might appear to be rather severe; however, as we shall see, a specific

284

procedure in which the constants α and E_c are chosen leads to results that are in excellent agreement with more rigorous computer computations.

With assumptions (1) and (2), the solutions to Equations (7) and (8) are

$$n = n_o \exp\left\{\frac{\alpha x}{j-1}\right\} \tag{9}$$

and

$$p = \left(\frac{j-1}{j+r}\right) n_o \left[\exp\left\{\frac{\alpha x}{j-1}\right\} - 1\right] + p_o, \tag{10}$$

* A more accurate representation of α is one in which $\alpha = 0$ for $E < E_c$ and α is proportional to $E - E_c$ for $E > E_c$. This form, as pointed out to us by L. S. Nergaard of RCA Laboratories, also results in an equation which is integrable.

** It is interesting to note that for a dc analysis of avalanche breakdown, assumption (1) is not justified, since there is a significant region for which $p \gg n$.

where the left end of the ionization region has been chosen as the origin for x; n_o and p_o are the (extremely small) densities at the origin. The electric field in the ionization region is found by substituting Equations (9) and (10) in Poisson's equation and integrating. The result, after dropping negligibly small terms, is

$$E = E_c + \frac{q}{\epsilon}\left\{ n_d x - \frac{(j-1)(1+r)}{\alpha(j+r)} n_o \left[\exp\left\{\frac{\alpha x}{j-1}\right\} - 1 \right]\right\} \cdot \quad (11)$$

The location of the peak field is found by demanding that the derivative of Equation (11) vanish. Inserting the value of x so determined in Equation (11) shows the peak field to be

$$E_{pk} = E_c + \frac{qn_d}{\epsilon\alpha}(j-1)(\ln\gamma - 1) \quad (12)$$

where

$$\gamma = \left(\frac{j+r}{1+r}\right)\frac{n_d}{n_o}.$$

285

D. Number of Ionizations Required to Turn the Field Around

Referring again to Figure 1(b), δ is defined as the distance for which $E > E_c$. Each electron created at the wavefront produces αl ionizations, where l is the distance traversed by the electron while exposed to an ionizing field. Considering the motion of the electron relative to the ionization wave, it is easily shown that $l = \delta/(j-1)$. Thus, in order to "turn the field around" (i.e., reduce E to a value below E_c), one electron must encounter

$$m = \frac{\alpha\delta}{j-1} \quad (13)$$

ionizing collisions. To find m in terms of more basic parameters, we set $E = E_c$ in Equation (11) at $x = \delta$. The result is the transcendental equation

$$\gamma m = e^m, \quad (14)$$

which, by virtue of the fact that $m \gg 1$, has the approximate solution

$$m \cong \ln [\gamma \ln \gamma]. \qquad (15)$$

For purposes of illustration, we will evaluate m for the sample parameters $j = 2.2$, $n_d = 10^{15}$ cm^{-3}, and $n_o = 10^{10}$ cm^{-3}, leaving a more detailed discussion of these parameters until later. For this case, $m = 14.5$. Hence, the number of ionizations necessary to turn the field around greatly exceeds that required for dc breakdown (for the latter case it is known[7] that if $\beta = \alpha$, $m = 1$; for actual values of β, m is somewhat higher). We may expect, therefore, that peak fields will significantly exceed those required for dc breakdown.

E. The Sheath

The reduced continuity Equations (7) and (8) are also applicable in the sheath, except that, by definition of the sheath, both ionization coefficients are taken as zero. Thus, both particle densities in the sheath are position independent and are those that prevail at $x = \delta$. This substitution in Equations (9) and (10) shows that the sheath particle densities are

$$n_s = n_o e^m \qquad (16)$$

286 and

$$p_s \cong \left(\frac{j-1}{j+r} \right) n_o e^m. \qquad (17)$$

Poisson's equation indicates that the electric field decreases linearly in the sheath as

$$E = E_c - \frac{q}{\epsilon} (n_s - p_s - n_d)(x - \delta). \qquad (18)$$

The sheath thickness, δ_s, is the distance over which the field decreases from E_c to a very small value.

$$\delta_s \cong \frac{E_c}{|dE/dx|} = \frac{\epsilon E_c}{q(n_s - p_s - n_d)}. \qquad (19)$$

[7] K. G. McKay, "Avalanche Breakdown in Silicon," *Phys. Rev.*, Vol. 94, p. 877, May 1954.

F. Choice of α and E_c

The analytic solution for the ionization region was given in terms of two as yet unspecified constants, E_c and α. Here we wish to outline a procedure for choosing these constants. Only by making an effective choice can we hope that our results will be quantitatively accurate.

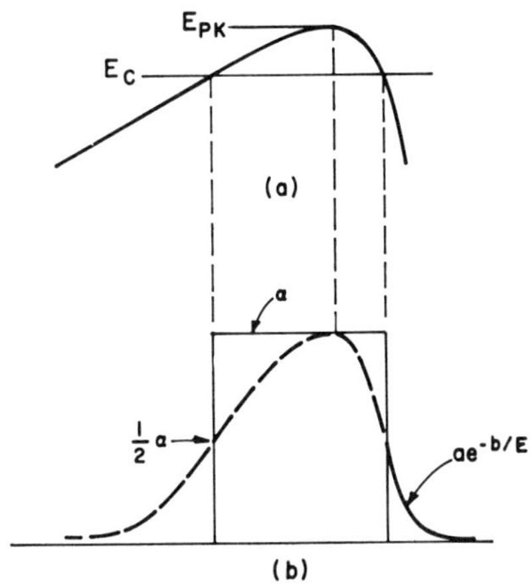

Fig. 2—Determination of ionization coefficient and critical field: (a) electric field versus position and (b) ionization coefficient versus position.

In Figure 2(a), the electric field profile is sketched. The dashed curve of Figure 2(b) represents the actual ionization coefficient, given by the relation $\alpha = ae^{-b/E}$ (a and b are constants of the material) consistent with the field profile of Figure 2(a). The solid curve of Figure 2(b) represents the approximation used in our analysis. Specifically, the criteria for choosing α and E_c are (1) that α be the actual ionization coefficient at the peak field, E_{pk}, and (2) that E_c be the field for which the actual ionization coefficient is half that at the peak. Mathematically, these conditions are written

$$\alpha = a \exp \left\{ \frac{-b}{E_{pk}} \right\} \tag{20}$$

and

$$a \exp \left\{ \frac{-b}{E_c} \right\} = \frac{1}{2} \alpha \tag{21}$$

Along with Equation (12), these restrictions permit a unique determination of α and E_c. In Figure 3, the analytic results so obtained are plotted and compared with computer results for the sample set of parameters indicated. The abscissa, x', is the distance from the peak field. For the computer results, Equations (7), (8), and Poisson's equation were solved with accurate representations for the ionization coefficients of both holes and electrons. The most important results of

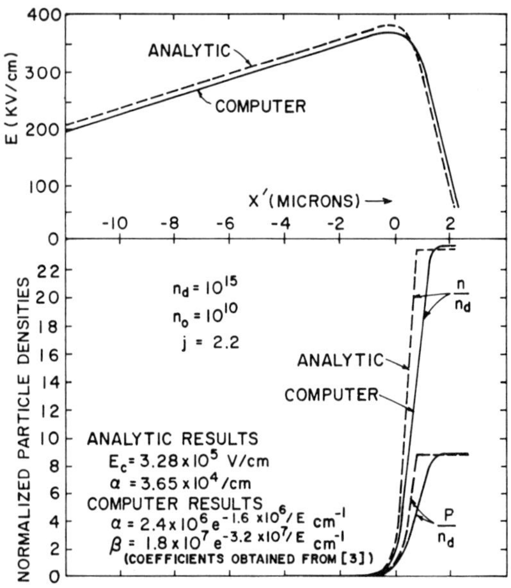

Fig. 3—Comparison of analytic and computer results.

the analytic treatment—those for the peak field and sheath particle densities—are seen to be in particularly good agreement with the computer results. Comparisons for other sample parameters also show excellent agreement.

G. The Plasma

In this section we calculate the electric field and particle densities and velocities in the plasma. The profiles of Figure 4 are stationary relative to the moving coordinate system (although, of course, the carriers are not). Because of this, and since ionization and recombination in the sheath and plasma are negligible, one may equate the electron currents in the two regions, relative to the moving coordinate system. A similar procedure may be carried out for the hole currents. Thus,

$$(v_w - v_{ns})n_s = (v_w - \mu_n E_p) n_p, \tag{22}$$

and

$$(v_w + v_{ps})p_s = (v_w + \mu_p E_p) p_p, \tag{23}$$

where n_p and p_p are the plasma densities and μ_n and μ_p, are the low-field mobilities of electrons and holes, respectively. E_p denotes the electric

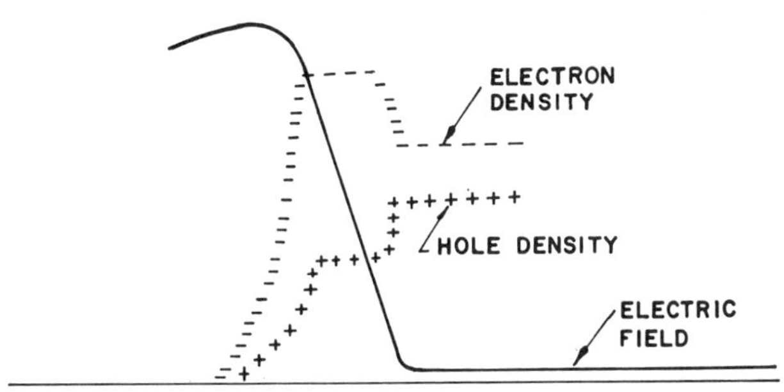

Fig. 4—Particle densities and electric field in sheath and plasma.

field in the plasma. In addition, we invoke plasma neutrality, i.e.,

$$n_p = p_p + n_d. \tag{24}$$

Equations (22) through (24) constitute a set that can be solved for E_p. Dropping negligibly small terms, one obtains

$$E_p = \frac{j^2 v_{ns}}{(j+r)(\mu_p + \mu_n)(p_s/n_d) + j(\mu_n - \mu_p)}. \tag{25}$$

The plasma particle densities may then be found from Equations (22) and (23) in terms of the now known plasma field.

$$n_p = \left(\frac{v_w - v_{ns}}{v_w - \mu_n E_p}\right) n_s, \tag{26}$$

and

$$p_p = \left(\frac{v_w + v_{ps}}{v_w + \mu_p E_p} \right) p_s. \tag{27}$$

Since the slow-moving plasma electrons are actually moving faster, relative to the high-field wave, than are the saturated-velocity sheath electrons, the electron density is lower in the plasma than in the sheath. The same is not true for holes. For the previously used sample set of parameters (and taking $v_{ns} = 1.0 \times 10^7$ cm/sec, $r = 1.0$, $\mu_n = 1220$ cm^2/volt-sec and $\mu_p = 460$ cm^2/volt-sec) we obtain $E_p = 1.04$ kV/cm, $n_p/n_d = 11.7$, and $p_p/n_d = 10.7$. Thus, the plasma field is about 400 times lower than the peak field (note that this low value is consistent with the use of low-field mobilities in the plasma analysis). The plasma electron velocity is $v_{np} = \mu_n E_p = 1.27 \times 10^6$ cm/sec, and the hole velocity, $v_{pp} = \mu_p E_p = 0.48 \times 10^6$ cm/sec. These greatly reduced velocities result in the relatively low frequencies that characterize the high-efficiency mode.

III. Diode Response to Applied Currents

A. Step Response

On the basis of the physics discussed in Section II and additional considerations to be developed here, we shall be able to describe and calculate the diode's response, first, to a step in current, and later, to a more interesting ac current input. Prior to the application of the step current, the diode is assumed to be biased at a voltage V_A that is slightly less than the breakdown value. It is assumed that punch-through has occured, but significant ionization has not. A double integration of Poisson's equation, taking the total charge density as n_d, relates the voltage to the field E_m at the n-p+ boundary,

$$V_A = \left(E_m - \frac{qn_d W}{2\epsilon}\right) W. \tag{28}$$

As discussed in the preceding section, the effect of the current-density step is to raise the field and launch an ionization wave that propagates through the diode leaving a plasma in its wake. This sequence of events is depicted in Figure 5(a). The peak voltage is approximately $V_B = [E_{pk} - qn_d W/(2\epsilon)]W$ and occurs after a charging time $t_{AB} = \epsilon\,(E_{pk} - E_m)/J$. (There are corrections to these terms that can be calculated from the results of the preceding section; however, these corrections are rather small). E_{pk} is given by Equation (12). During the very rapid traversal through the diode by the wave, there is a slight movement of the low-velocity particles in the plasma. Also, as the wave exits the diode, there is a small perturbation in the "feeding" of the plasma at the left end of the diode. Neglecting these effects (which, one can show, can be quite small), a minimum voltage, $V_C = E_p W$, is

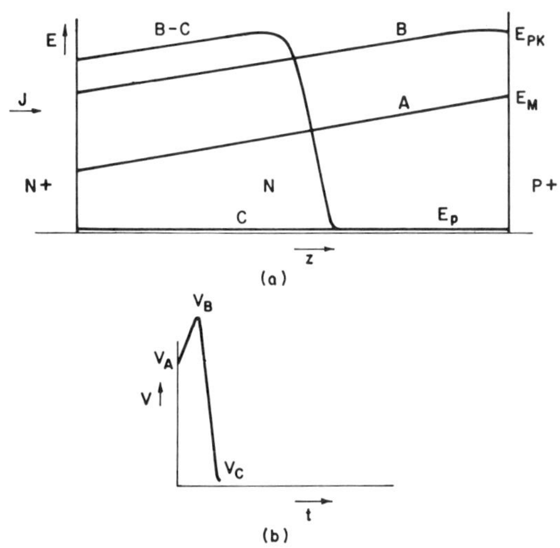

Fig. 5—Diode response to a step current: (a) electric field versus position and (b) voltage versus time.

reached, when the plasma fills the entire diode. The voltage fall time is that required for the peak field and sheath to exit the n region. Since the wave is moving at a velocity, v_w, the fall time is $t_{BC} = (W + \delta_s)/v_w$. The rise and fall of the voltage are shown in Figure 5(b).

We now turn our attention to the extraction of the plasma, with the current still maintained at J. The particle densities and electric field existing when the plasma fills the diode are sketched in Figure 6(a). The situation that prevails at some time during the ensuing extraction period is shown in Figure 6(b). Under the influence of the low plasma field, electrons drift to the left, leaving an electron-deficient region near the n-p+ boundary. Because of the electron deficiency in this region, the field is larger. As a result, the hole velocity in this region is larger, and, therefore, the "residual" hole density, p_r, is lower than in the plasma. The transition between the plasma and the rising field region is illustrated as being abrupt. Actually, particle diffusion and the continuous variation in hole mobility with field results in a rounding effect. Approximate calculations indicate that this rounding effect is slight. In the following treatment, we neglect diffusion and assume particle velocity in the rising-field region to be saturated.

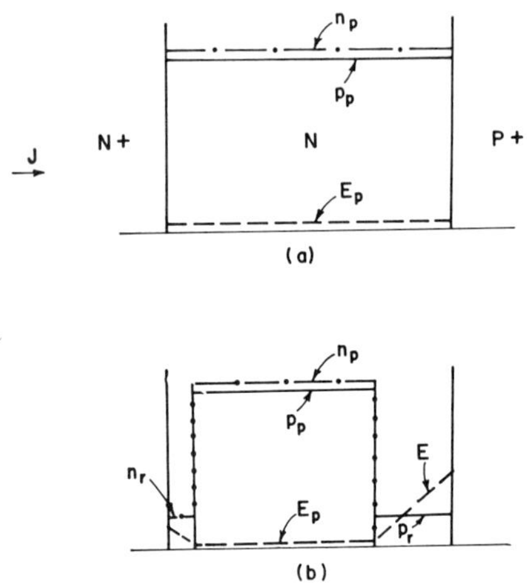

Fig. 6—Particle densities and electric field (a) at the start of the extraction period and (b) during the extraction period.

Since there is no hole accumulation at the moving boundary between the plasma and the rising-field region, we may equate plasma and residual hole currents relative to the moving boundary:

$$p_p(v_{pp} + v_{np}) = p_r(v_{ps} + v_{np}). \tag{29}$$

We can then solve for p_r in terms of quantities already calculated:

$$p_r = \left(\frac{v_{pp} + v_{np}}{v_{ps} + v_{np}}\right) p_p. \tag{30}$$

Similar considerations at the other moving boundary show that the residual electron density is

$$n_r = \left(\frac{v_{np} + v_{pp}}{v_{ns} + v_{pp}}\right) n_p. \tag{31}$$

Having solved for the residual densities and utilizing the fact that the velocities of the right and left boundaries are those of plasma electrons and holes, respectively, we can integrate Poisson's equation twice to obtain the time variation of the voltage during the extraction period. The result, which is included in the voltage-response sketch (Figure 8(a)) is

$$V_{CD}(t) = \frac{q}{2\epsilon}\left[(n_r - n_d) v_{pp}{}^2 + (p_r + n_d) v_{np}{}^2 \right] (t - t_c)^2 + V_C. \quad (32)$$

The plasma width diminishes to zero when the moving boundaries meet. Since the total width swept out by these boundaries is W,

$$v_{np} t_{CD} + v_{pp} t_{CD} = W. \quad (33)$$

Hence, the extraction period is

$$t_{CD} = \frac{W}{v_{np} + v_{pp}}. \quad (34)$$

For a plasma field of 1 kV/cm, the reduced particle velocities result in an extraction time that is six times greater than the transit time based on a saturated velocity.

The situation at the end of the extraction period, when the moving boundaries have coalesced, is illustrated in Figure 7(a). In the ensuing residual extraction period, the residual carriers move apart with saturated velocities (Figure 7(b)) and out of the n region (Figure 7(c)) while the field rises. It is easily shown that residual extraction lasts for a period $t_{DE} = W/(1 + r')v_{ps}$ (where $r' = \mu_p/\mu_n$) after which the voltage is

$$V_E = V_C + \frac{qW}{\epsilon}\left[(n_d + p_r) v_{np} (t_{CD} + t_{DE}) - n_d \frac{W}{2} \right]. \quad (35)$$

We have assumed during this discussion that fields in the extraction and residual extraction periods are not large enough to produce additional ionization. The avoidance of "premature avalanche" must be considered in the design of oscillators and is discussed later.

With the passage of an additional time, $t_{EA} = \epsilon(V_A - V_E)/(WJ)$, the voltage is again at V_A and the cycle repeats itself, as shown in Figure 8(a). Thus, we conclude that the response of the diode to a (sufficiently

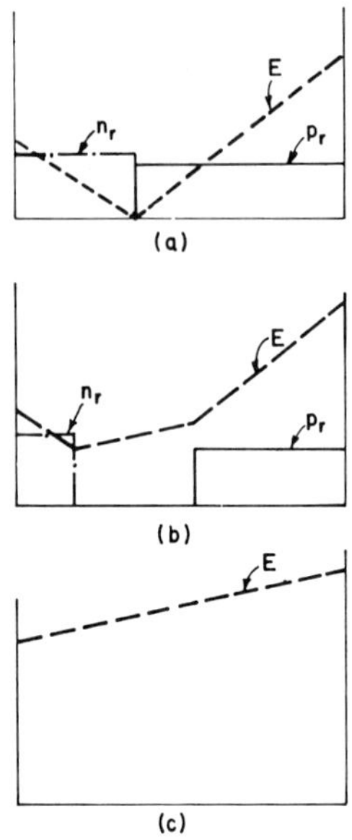

Fig. 7—Particle densities and electric field (a) at the start of residual extraction period, (b) during residual extraction period, and (c) at the end of the residual extraction period.

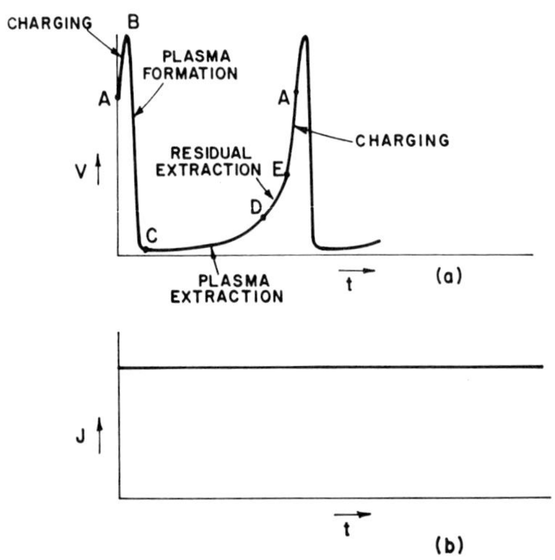

Fig. 8—(a) Oscillatory voltage response to (b) a step-current excitation.

large) step in current density is oscillatory with a large voltage modulation. It is evident, however, that no ac power can be extracted from the device if the circuit current is a constant. We now turn our attention to the consideration of ac waveforms that characterize high-efficiency oscillations.

B. Current Waveform for High-Efficiency Oscillations

A successful model must accomodate a time-varying periodic current that meets the following criteria: (1) the end of the cycle produces a diode field and densities that are the same as those at the beginning

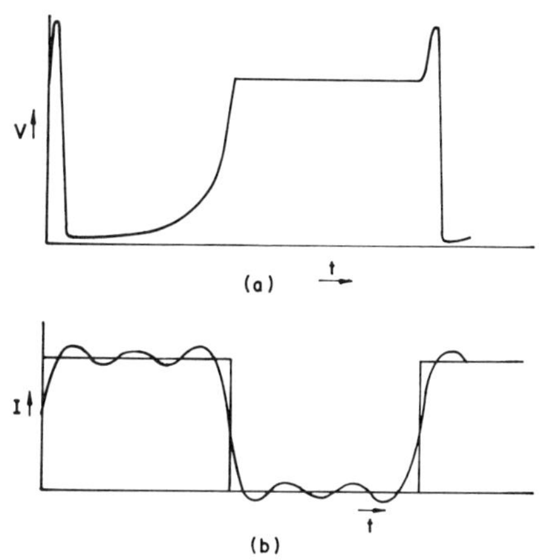

Fig. 9—(a) Voltage response to (b) a square wave of current.

of the cycle (otherwise the steady state is not being described), (2) the current and resultant diode voltage are consistent with passive external circuitry as well as with the diode physics, and (3) predictions of the model are in accord with experiment. In considering the step response of the diode, it was noted that an extended period of low voltage (the plasma-extraction period) accompanied the large current. If, now, a period of high voltage can be coupled with a low current, ac power can be efficiently extracted from the diode.

A simple current waveform, which we now consider, is the square wave of Figure 9(b) (ignore, for the moment, the rippled square wave also shown). The first part of the solution is as described earlier for the step current. Now, however, the current decreases to zero when the voltage returns to V_A. At this point, there is negligible conduction current, and, since the net current is zero, the displacement current is zero. Hence, the voltage remains fixed at V_A, i.e., the voltage is maintained at a high value while the current is zero. The cycle repeats itself when the current again turns on. Thus, it is clear that for a square-wave of external current, the diode will efficiently convert dc power to rf power. The square wave of current, with off and on times equal, can be reasonably well approximated by a fundamental, a third, and a fifth harmonic, as shown in Figure 9(b). A consideration of the finite rise time and the ripples indicates that these should produce no first-order changes in the results.

In order for this current and the resultant diode voltage to be realizable with an external circuit that (except for the bias supply) is both passive and linear, the diode large-signal impedance (i.e., complex ratio of diode voltage to current) at each of the three frequencies must

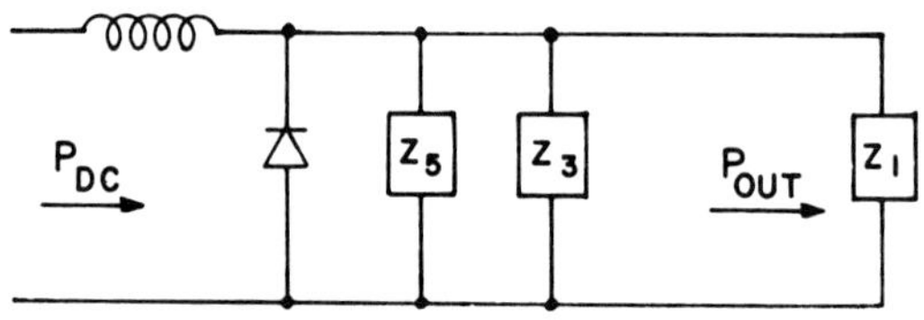

Fig. 10—Equivalent circuit for a high-efficiency oscillator.

not have a positive real part. This being the case, the required waveforms can be supported by the circuit of Figure 10. Z_1, Z_3, and Z_5 are branches with impedances equal to the negative of the diode impedance at the fundamental, third, and fifth harmonics, respectively. Each branch is characterized by a high Q and, therefore, is assumed to be essentially an open circuit at harmonics other than the one to which it is tuned.

The results of a sample calculation for a 0.025-inch-diameter silicon diode the n region of which is 8 microns long and doped with 10^{15} donors per cm^3 are now given. A square wave for which the current density is 3200 A/cm^2 ($j = 2.0$) for one-half cycle and zero for the other, was used to calculate the voltage. This corresponds to a bias current of 5 amperes. A Fourier analysis of the waveforms was carried out and the impedances and efficiency calculated. The above criterion for circuit realizability is found to be satisfied. The dc voltage is noted to have dropped to 63% of the breakdown value. At a fundamental frequency of 840 MHz, the diode impedance is $(-17 - j1)$ ohms, the efficiency is 54%, and the output power is 330 watts. Similar calculations for a wide variety of parameters consistent with the design rules established in the next section show efficiencies from 50 to 60%.

C. *Other Waveforms*

The gross features characterizing the waveforms of Figure 9 are (1) the formation of a low-voltage plasma and its extraction over an extended period of high current and (2) a subsequent extended period of low current in which stored energy maintains the voltage at a high level.

It is evident that these features can prevail, to a large extent, with waveforms differing in detail from those of Figure 9. To the extent that the above features are realized, however, one may expect efficiencies to approach those predicted here. Consider, for example, the waveforms of Figure 11. The minimum current is non-zero, and, conse-

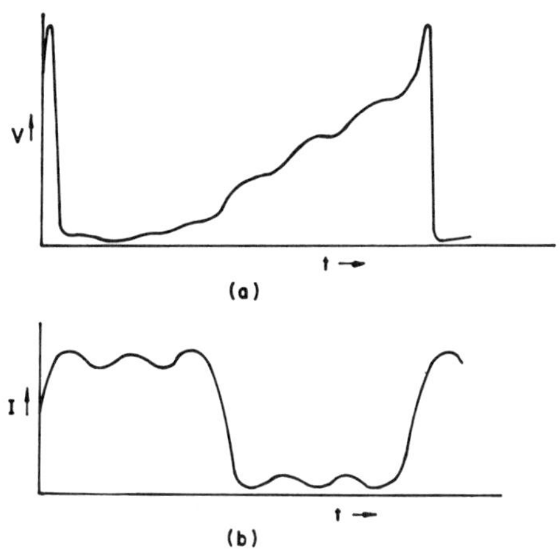

Fig. 11—Another possible set of high-efficiency waveforms: (a) voltage and (b) current.

quently, the corresponding voltage rises in a saw-tooth-like manner. Calculated efficiencies for this situation are slightly lower than those for the previous case considered. The waveforms reported in computer simulations[5] are characterized by the above features to a lesser extent; thus, efficiencies are lower still. Finally, it should be noted that high and low current states need not have equal duration, though, for extensive departures from equality, efficiencies may be expected to suffer.

D. *The Role of Transit-Time Oscillations*

It was reported[8] that conventional transit-time oscillations appear to trigger the high-efficiency oscillatory state. Such would be expected

---

[8] A. S. Clorfeine and R. D. Hughes, "Characterization of Anomalous Avalanche Diodes," IEEE Informal Conference on Active Microwave Effects in Bulk Semiconductors, New York, Jan. 1968.

on the basis of the theory presented here. If a current generator with sufficiently small rise time were available, then no additional starting mechanism would be required. In the absence of such generators, however, it is the transit-time oscillations that provide the rapid current increase necessary to trigger the high-efficiency mode.

As regards the role of transit-time oscillations in the steady state of the high-efficiency mode, we note that at least one important harmonic may lie within what is considered to be the conventional transit-time range for the diode. Thus, variations at a transit-time frequency in the current (and also voltage) waveform, as shown by the rippled curve of Figure 9(b), may be expected. However, it does not seem useful to interpret these variations as a transit-time oscillation that, each cycle, triggers the plasma. Rather, these higher frequencies are simply viewed as very essential harmonics that serve the function of wave shaping.

IV. Design Considerations

298

Realization of near-optimum efficiencies requires proper design of the diode and the circuit and proper bias excitation. In this section, we utilize our theoretical results to develop a number of design formulas. These formulas should be considered first-order, since approximations have been made to keep them simple and, therefore, easily applicable.

A choice of the bias current is discussed first, and then attention is turned to the design of the diode. The oscillation frequency for a given diode is tunable over an appreciable range, the center of which is determined primarily by the width, W, of the high-resistivity region. A properly designed diode must "punch through" prior to breakdown. A more stringent condition regarding the degree of punch-through is developed, however, and results in an optimum n_dW product and, thus, a choice of doping density for a given depletion width. The diode area is determined by the output power required. A general result relating the two is obtained. Finally, a formula for the required fundamental circuit impedance is derived.

A. Bias Current

We have emphasized that the type of operation described here requires that the current density be greater than a critical value $qv_{ns}n_d$ or, equivalently, that the normalized current density j be greater than unity. The detailed calculations indicate that for values of j only slightly greater than unity, there is a marked departure from the waveform of Figure 9(a), and performance may be expected to deteriorate. This effect is not a problem if we choose j in the range 1.5–2.5. As the current is increased further, a point will be reached where premature avalanche occurs (this is discussed in more detail later). By properly choosing other diode parameters, a larger value of j can be accommodated, but there seems to be no particular advantage in this procedure. We shall choose as our nominal design density, $j = 2$ or, equivalently, $J = 2qv_{ns}n_d$. Since the dc current density is half that of the peak value, the desired bias current is

$$I_{DC} \approx qv_{ns}n_dA. \qquad (36)$$

B. *Tunability*

By summing the times required for each portion of the cycle in Figure 9, we obtain a relation for the frequency of oscillation;

$$f \cong \left\{ \frac{(j+r)(j-1)m}{(1+r)j^2} + \frac{4c}{j} \right\}^{-1} \frac{v_{ns}}{2W}, \qquad (37)$$

where c is a function of m and j which, to first order, may be taken as unity. The quantity m has been interpreted as the number of ionizations required to turn the field around, and is given by Equation (15) and, indirectly, by the density ratio n_d/n_o. The quantity n_o is the density of electrons prior to the rapid rise in current. Its minimum value, which is determined by carrier lifetime, can be exceedingly low; however, n_o can be varied over orders of magnitude, by virtue of the multiplication associated with the pre-breakdown voltage, V_A. Thus, a slight variation in V_A can significantly change the frequency of oscillation, since multiplication is a very sensitive function of voltage near the breakdown value. Of course, V_A is only controlled indirectly, namely, by means of the circuitry. Consequently, as the circuit is tuned, i.e., as the proper impedances for the fundamental and relevant harmonics are established at a slightly different set of frequencies, the corresponding changes within the diode are primarily those of the voltage and the electron density in the low-current portion of the cycle. These, in turn, determine the plasma field and extraction time. An inspection of Equation (37) indicates that for $j = 2$, a variation of n_d/n_o from 10^2 to 10^{10} results in f changing from 0.04 v_{ns}/W to 0.12 v_{ns}/W (actually, the range that can be covered while maintaining high efficiency depends somewhat on the diode parameters). Note that a large tuning range has been demonstrated experimentally.[9] Since the conventional transit-time range may extend well below the frequency considered optimum by Read,[2] 0.5 v_{ns}/W, the frequency for high-efficiency operation may be an appreciable fraction of that at which the same diode can oscillate in the transit-time mode.

C. Width of the High-Resistivity Region

We take, as the design frequency (or, in the case of tunable operation, the center of the high-efficiency range over which the diode is to operate), the geometric mean of the above two frequencies, 0.07 v_{ns}/W. Thus, our first design rule pertains to the choice of W,

$$W = 0.07 \frac{v_{ns}}{f}. \tag{38}$$

A 700-MHz diode, for example, should be fabricated with an n region ~10 microns wide.

D. Doping Density

A diode that is tailored so that avalanche occurs at nearly the same voltage at which the electric field punches through to the low-resistivity n+ region gives excellent results in the transit-time mode.[10] From Poisson's equation, it is seen that the above condition occurs when

$$n_d W = \frac{\epsilon E_B}{q}, \tag{39}$$

where E_B is the field at the n-p+ boundary at breakdown. We have assumed in this paper that punch-through occurs, so that Equation (39) represents an upper limit for the $n_d W$ product of the diodes under consideration. We shall see, however, that an even smaller upper limit must be imposed if premature avalanche is to be avoided; thus, a diode tailored to operate in the transit-time mode as described above

---

[9] P. A. Levine and S. G. Liu, "Tunable L-Band High-Power Avalanche-Diode Oscillator," IEEE International Solid-State Circuits Conf., Phila., Pa., Feb. 1969.

[10] T. Misawa, "Microwave Si Avalanche Diode with Nearly-Abrupt-Type Junction," *IEEE Trans. on Electron Devices*, Vol. ED-14, p. 580, Sept. 1967.

cannot be expected to operate optimally in the high-efficiency mode. Such was, indeed, found experimentally by others.[11]

From the results of the preceding section, one can show that the field at the n-p+ boundary at the end of the residual extraction period is

$$E_W \cong \frac{q}{\epsilon}\left[\frac{1+\frac{j}{r}}{1+r'}\right] n_d W. \tag{40}$$

To avoid premature avalanche, E_w must be less than E_B, or, from Equation (40),

$$n_d W < \left[\frac{1+r'}{1+\frac{j}{r}}\right]\frac{\epsilon}{q}E_B. \tag{41}$$

Even for $j = 1$, the coefficient of $\epsilon E_B/q$ (for silicon) is 0.69, or considerably less than the value of unity that characterizes the transit-time diode discussed above. If we permit premature avalanche to commence for $j = 2.5$, it will be safely avoided for our design choice of $j = 2$. Substituting $j = 2.5$ in Equation (41) results in

$$n_d W = 0.4\frac{\epsilon}{q}E_B. \tag{42}$$

Smaller values of $n_d W$ may also be used; however, a diode with a lightly doped n region might be more difficult to fabricate (e.g., it may be less amenable to epitaxial technology). Also, more detailed considerations suggest that for values of $n_d W$ that are too low, a full plasma may not form.

The breakdown field, E_B, which appears in Equation (42), is a weak function of n_d[12]; for silicon, it increases from 3×10^5 V/cm to 4×10^5 V/cm as n_d is increased from 10^{15} cm^{-3} to 10^{16} cm^{-3}. Inserting $E_B = 3.5 \times 10^5$ V/cm in Equation (42) gives $n_d W \cong 9 \times 10^{11}$ cm^{-2}.

[11] R. S. Ying, R. G. Mankarious, and D. L. English, "High-Efficiency Anomalous Mode Oscillations from Silicon Impatt Diodes at 6 GHz," IEEE International Solid-State Circuits Conf., Phila., Pa., Feb. 1969.

[12] S. L. Miller, "Ionization Rates for Holes and Electrons in Silicon," *Phys. Rev.*, Vol. 105, p. 1246 (1957).

301

Thus, we have a design rule that permits a choice for the doping density. By way of example, the 700-MHz 10-micron diode previously mentioned should be doped to a level of slightly less than 10^{15} cm^{-3}.

E. Diode Area

The diode area is determined by the power output required, within, of course, the limits imposed by impedance, thermal, and technological considerations. It is clear from the voltage waveform of Figure 9(a), that the average voltage, V_{DC}, is less than the breakdown value, V_b; calculations indicate that over a wide range of near optimum oscillations, $V_{DC} \cong 0.64\, V_b$. Also, it follows from Equation (42) that the average field $<E>$ at breakdown is 80% of the field at the n-p+ boundary, so that the breakdown voltage is

$$V_b = <E> W = 0.8\, E_B W. \tag{43}$$

Equations (36) and (43) show that the required dc power is

$$P_{dc} = V_{dc} I_{dc} = 0.64\, V_b I_{dc} \cong 0.5\, q v_{ns} E_B (n_d W) A. \tag{44}$$

For a 50% conversion efficiency, the power at the fundamental is
302 $P_1 = 0.5\, P_{dc}$. Using the result obtained from Equation (42) for the $n_d W$ product in Equation (44), we obtain

$$P_1 \cong 0.1\, \epsilon v_{ns} E_B{}^2 A \tag{45}$$

from which the required area can be obtained.

For $E_B = 3.5 \times 10^5$ V/cm, the output available from a properly designed silicon diode is, according to Equation (45), approximately 0.8 watt per square mil.

F. Fundamental Impedance

For the approximate square wave of Figure 9(b), the fundamental current amplitude is $I_1 = (4/\pi) I_{dc}$. The impedance at the fundamental is almost entirely real, and so it follows that for a 50% efficiency oscillation (i.e., $P_{dc} = 2P_1 = V_1 I_1$), $V_1 = (\pi/4) V_{dc}$. Since $V_{dc} \cong 0.64\, V_b$, we find that the required circuit impedance at the fundamental is

$$Z_1 = \frac{V_1}{I_1} = \frac{(\pi/4)(0.64\, V_b)}{(4/\pi) I_{dc}} \cong 0.4 \frac{V_b}{I_{dc}}. \tag{46}$$

For a 0.025-inch-diameter silicon diode with an n-region 8 microns wide and a doping density of 10^{15} cm^{-3}, biased according to Equation (36) and operated optimally, we find that $Z_1 = 15$ ohms.

V. Concluding Remarks

In this paper, we have presented a large-signal analytic theory for high-efficiency oscillations in avalanche diodes. The important periods of the oscillation have been identified, described physically and mathematically, and pieced together. Also, we have seen how the theory leads to a set of design formulas.

Finally, the theory presented here is able to predict features of the high-efficiency mode that can be compared with reported experimental data. Specifically, the theory predicts (a) the magnitude of bias-voltage drop that accompanies mode onset, (b) a bias-current range for optimum performance, (c) the proper doping density for a given width, (d) the width for a given frequency range, and (e) maximum efficiency. Each of these predictions shows good agreement with experimental results.

Device Physics of TRAPATT Oscillators

B. C. DELOACH, JR., MEMBER, IEEE, AND D. L. SCHARFETTER, MEMBER, IEEE

Abstract—This paper utilizes a simplified physical model to describe TRAPATT (TRApped Plasma Avalanche Triggered Transit) operation. By yielding on computational accuracy, a complete high-efficiency device design is generated and the dependence of operation on physical parameters is elucidated. The extreme complexity of the precise differential equations describing TRAPATT operation has made the calculation of a single diode-circuit configuration a tour de force. However, by observing the important features of such a solution, a simplified approach giving realistic answers has been evolved.

A theoretical device design has been evolved. This design provides device width and impurity density as a function of TRAPATT frequency, and indicates a decreasing degree of "reach through" with increasing frequency. In addition, the explicit dependence of width and impurity density on the diode's reverse saturation current has been obtained.

The launching of the avalanche zone through the diode, and, in particular, the limitations implicit in the recovery to a swept-out state, are of broad significance in other types of diodes, particularly p-i-n switches and "snap" diodes.

Introduction

A NEW high-efficiency mode of oscillation in avalanche diodes was first reported [1] in April, 1967, by Prager, Chang, and Weisbrod. The operation was referred to as anomalous because the oscillation frequency was considerably less than the π transit angle frequency (the frequency of operation of the normal IMPATT mode), and the efficiency of power conversion from dc to RF (60 percent) was considerably larger than could be explained by normal IMPATT theory. Initial reports of pulse operation of the silicon units [1], [2] provided only limited information on the characteristics of the new mode.

Scharfetter, Johnston, and co-workers first obtained [3] a physical explanation and detailed description of the high-efficiency mode through computer simulation of the diode-multiresonant circuit system. The simulation required careful characterization of the germanium avalanche diodes employed [4], and precise measurements of the high-efficiency (45 percent) waveforms [3]. This particular mode of operation was labeled TRAPATT for TRApped Plasma Avalanche Triggered Transit. Shortly thereafter the first CW TRAPATT operation [5] was achieved with similar germanium structures.

Alternative explanations have been advanced for the occurrence of the high-efficiency mode [6], [7]. It is hoped that the simplicity of the model presented in this paper will make the more complicated computer solutions of TRAPATT operation, from which it was derived, more plausible.

The "difficult" differential equations describing TRAPATT operation have to date precluded a "thinking" ability in either diode or circuit design. Although a given situation, i.e., diode circuit combination, can be rigorously analyzed, the expenditure of man-hours and computer time is prohibitive on a routine basis. Consequently, very little qualitative information (let alone quantitative) is available as to appropriate diode or circuit design. However, a significant step towards a simplified theory has recently [8] appeared, wherein many of the characteristics of the avalanche zone transit were obtained analytically in a particularly elegant manner.

This paper ignores the complicated device circuit interaction by presupposing certain terminal voltage and current waveforms and the resulting simplification allows the calculation of the essentials of TRAPATT operation. A zero-order design theory results. This device design theory then serves as a starting point for a more complete theory including device circuit interactions. The simplifications employed were justified through the ability to "check" against precise solutions of the differential equations obtained by numerical techniques [3].

TRAPATT operation can be separated into three distinct periods: 1) a transient charging period resulting in a high-current, low-voltage state, followed by 2) a transient charge removal period resulting in a high-voltage, low-current state, and 3) a continuation of the high-voltage, low-current state, in which to date IMPATT voltages build up to again initiate period 1. It is perhaps of some help to visualize period 1 as the closing of a switch and period 2 as the opening of a switch.

The actual device-circuit interactions required to produce these states in a repetitive cycle are complicated. But, as will be seen, the essential device characteristics can be derived by assuming that this cyclic behavior does occur and then determining the necessary device properties.

A considerable effort will first be made to make plausible the propagation of an avalanche "zone" through the entire depletion region of a reverse biased p-n junction under appropriate boundary conditions.

I. The Transient Charging Period

In order to achieve the highest level of understanding, the simplest of approximations will be employed. The degree to which these approximations affect the results

Manuscript received July 1, 1969; revised August 25, 1969.

The authors are with Bell Telephone Laboratories, Inc., Murray Hill, N. J.

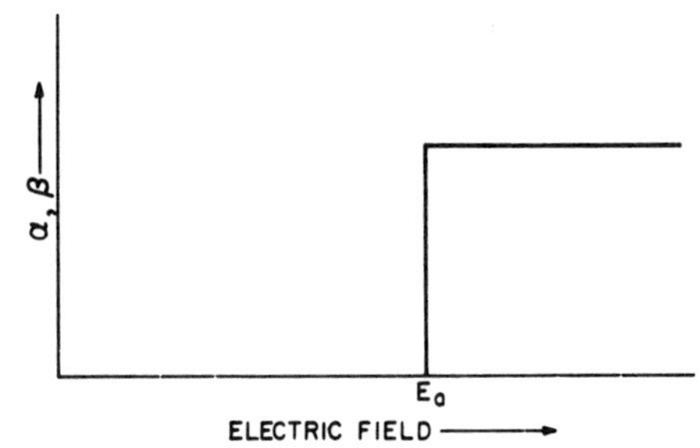

Fig. 1. Assumed dependence of ionization coefficients α and β, for electrons and holes, on electric field.

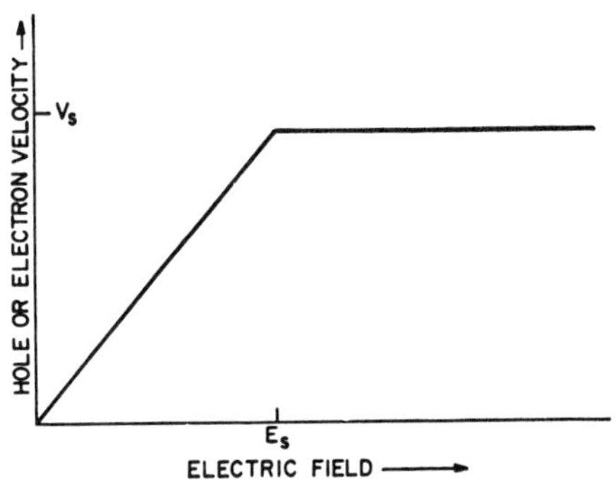

Fig. 2. Assumed dependence of carrier velocities (both holes and electrons) on electric field.

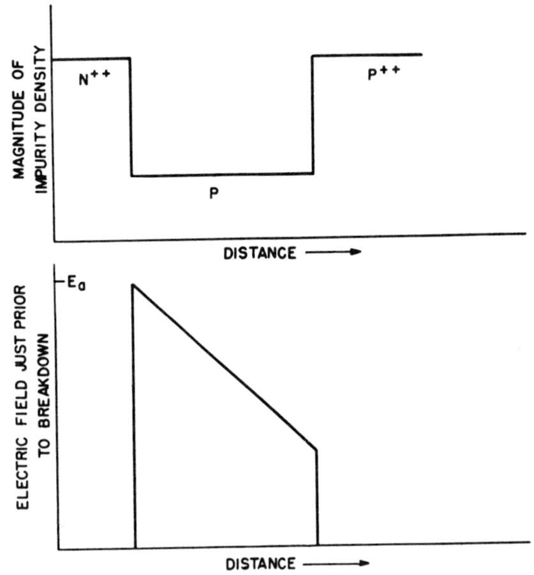

Fig. 3. Impurity profile and corresponding electric field profile assumed for analysis (commonly labeled a "step junction that reaches through"). An n^+-p-p^+ structure is analyzed but results apply to a p^+-n-n^+ structure with the obvious change in notation.

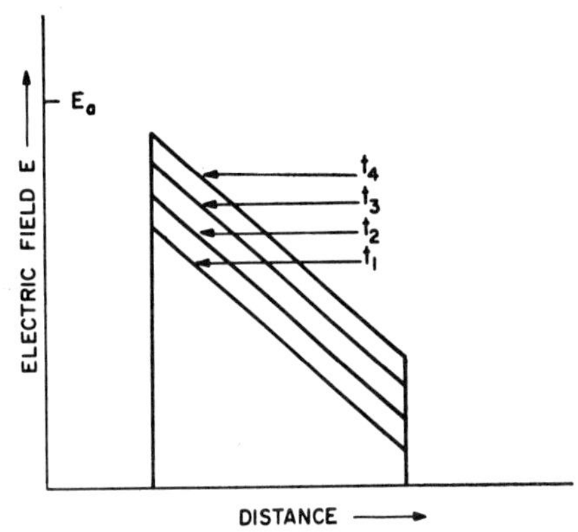

Fig. 4. Snapshots of electric field versus distance at successive time intervals under assumed constant current bias.

is discussed later in the text. The ionization coefficients α and β for electrons and holes are assumed equal, and zero for values below a critical field E_a and constant above E_a (see Fig. 1). Both carrier velocities are assumed to be characterized by the same linear mobility up to a saturation field E_s above which their velocities are equal, constant, and labeled v_s (see Fig. 2). The simplest structure for illustrative purposes is that of Fig. 3, i.e., a step junction that "reaches through" to the substrate at breakdown.

Assume an initial state such that the peak field is below E_a and an exceedingly small but finite I_{sat} is flowing. (Carriers are needed to start the multiplication process.) Now apply a step in current to the diode terminals. The expected events can be illuminated by first considering a well understood system, i.e., a normal parallel plate capacitor with air dielectric and fixed plate separation. A step of current applied to such a capacitor causes displacement current to flow therein, and since this current can be expressed as $\epsilon(\partial E/\partial t)$ (Maxwell's equations), the electric field E between the capacitor plates increases at a constant rate as long as the constant current is maintained. Obviously the air dielectric would break down if this field change continued long enough. Similar events occur in our semiconductor capacitor except that the breakdown occurs preferentially at one edge due to a built-in slope in electric field arising from the net space charge of the ionized impurities. In our model the applied terminal current per unit area I_0 drives the field up uniformly throughout the diode until E reaches E_a. Thus, as long as the peak field is less than E_a, successive time snapshots of equal duration would show electric field versus distance plots, for the structure of Fig. 3, as shown in Fig. 4. Obviously, something different happens, however, when the peak field reaches E_a and charge multiplication begins.

It is at this point that our assumption as to the magnitude of I_0 is crucially important. For relatively low values of I_0 nothing dramatic happens. (This situation has been analyzed in considerable detail and can give rise to space-charge feedback negative resistances which have to date proved operationally uninteresting.) However, something dramatic does happen if I_0 is of such magnitude that the resulting $(\partial E/\partial t)$ in the depletion region drives the point at which E reaches E_a through the diode at a velocity greater than the maximum velocity at which carriers can travel in the semiconductor. It has been shown [3], [8] that under these conditions an avalanching "zone" can propagate through the diode, filling the original depletion region with a dense hole electron plasma and reducing the voltage across the diode to essentially zero.

The current to the right of the position where E reaches E_a is displacement current, plus an extremely small conduction current per unit area, I_s. The velocity at which this "edge" of the avalanching region propagates through the diode is then easily obtainable. For E less than E_a the slope of the electric field with distance is determined from Poisson's law as

$$\frac{dE}{dx} = -\frac{qN_a}{\epsilon} \tag{1}$$

where $-qN_a$ represents the negative charge density of the fixed acceptors and where we have neglected the mobile charge densities of holes and electrons of I_s in comparison with N_a. Letting time begin when the peak field at the left of the diode just reaches E_a, and taking $x=0$ at the left of the original depletion region we can write from (1) that at $t=0$,

$$E(x) = E_a - \frac{qN_a}{\epsilon}x. \tag{2}$$

Due to our assumptions, *to the right of the point at which* $E(x)$ *reaches* E_a,

$$E(x, t) = E(x, 0) + \frac{I_0 t}{\epsilon} \tag{3}$$

where the latter term is obtained from an integration of

$$I_0 = \epsilon\frac{dE}{dt} \tag{4}$$

assuming I_0 a constant. Consequently, from (3), the field maintains its original slope and increases at the same rate for all x (see Fig. 4). Combining (2) and (3) we obtain the field to the right of the point at which E reaches E_a as

$$E(x, t) = E_a - \frac{qN_a}{\epsilon}x + \frac{I_0 t}{\epsilon}\cdot \tag{5}$$

Thus, the value of t at which the electric field reaches E_a at a given distance x into the depletion region is obtained by setting $E(x, t)$ equal to E_a in (5), yielding

$$\frac{qN_a}{\epsilon}x = I_0 t. \tag{6}$$

Or, upon differentiation,

$$\frac{dx}{dt} \equiv v_z = \frac{I_0}{qN_a} \tag{7}$$

where v_z will be labeled the "avalanche zone velocity" and represents the velocity at which the leading edge of the avalanching region progresses through the diode.

It is important to realize that v_z must be greater than v_s in order for our theory to apply, and consequently

$$I_0 > v_s qN_a. \tag{8}$$

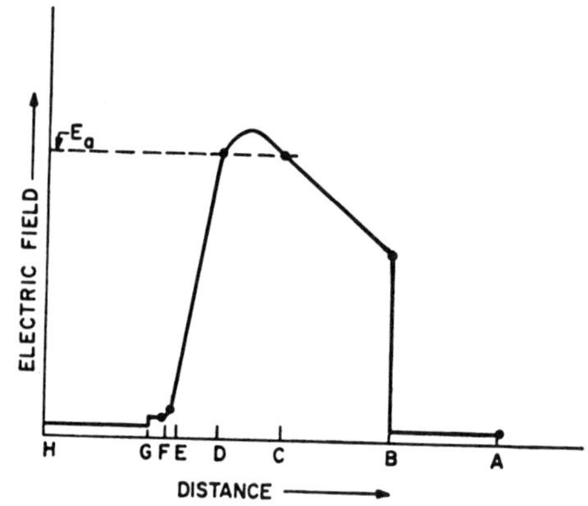

Fig. 5. The electric field versus distance profile of an avalanche zone in transit.

Equation (8) presents the first design information for TRAPATT operation. It specifies the driving current per unit area required to cause the entire depletion region to break down in terms of the impurity density in the depletion region.

Although (8) was derived on the basis of a constant current, it can be shown that avalanche will spread throughout the entire depletion region as long as the driving current always exceeds the value represented by I_0, whatever its time depencence; i.e., it might at times be $2I_0$, $3I_0$, etc.

So far, it has been shown that an applied constant current of sufficient magnitude will cause avalanche to spread throughout an ordinary step junction, which is usually thought of as avalanching only at the end at which the electric field is the highest just prior to breakdown. Still more exotic events occur, under appropriate conditions, when the extent (in space) of the avalanching region is considered.

Since the electric field in the heavily doped end region at the left is very low, i.e., just large enough to carry the total current $I_0 \equiv \mu E q n$ with $n \approx N_d^+$, it is obvious that the electric field must again drop below E_a to the left of the leading edge of our avalanche zone. Thus an avalanching zone of finite extent exists and a snapshot in time must display an electric field versus distance having the general characteristics [3], [8] displayed in Fig. 5. Regions A–B and G–H merely display the very low field end regions where nothing interesting is happening. Plane B represents the right boundary of the original depletion region and in region B–C we have the remaining "depleted region" at this point in time. Hence, to the right of plane C, which represents the onset of avalanche, the current is essentially all displacement current and the electric field has a slope given by (1) and a magnitude given by (5) where x is measured from G, the left-hand boundary of the original depletion region. *Region* C–D *contains all of the diode that is avalanching*, i.e., $E=E_a$ at plane C and at plane D. Region D–E is that region to the left of the avalanche zone in which E remains above E_s, the value of electric field required to keep carrier velocities saturated, and region E–F contains the section of the diode in which E readjusts from E_s to the value of E required to carry the current I_0 in the "plasma" now in the remainder of the depletion region, i.e., region F–G. E drops to a still lower value in region G–H, the heavily doped material at the left of the diode.

Having described completely the phenomena in region B–C by (1) through (7), now consider avalanche region C–D. It is to be noted that the occurrence of events at a given value of electric field within the avalanche zone are only influenced by events to the right of the zone as long as I_0 is flowing, since carriers cannot keep up with the zone [see (8)]. At the leading edge of the avalanche region the particle current I_s is always present as the zone travels through the depletion region. I_s is assumed independent of distance. Also the slope of the electric field at the leading edge of the zone is always the same and is determined by Poisson's equation from the impurity density. Consequently, an observer situated at the leading edge of the avalanche zone, and moving with the same velocity, must see a time invariant E versus distance throughout the avalanche zone. (This statement must be modified if the avalanche zone extends to the edge of the original depletion region as would be the case for start-up.) This follows from our assumption that we have a generator of appropriate design to maintain I_0 through the diode with $I_0 > v_s q N_a$, and a constant I_s initiating the avalanche.

Consider now a moving reference frame attached to plane C and consequently moving with velocity v_z in our original inertial system. To an observer in the moving reference frame electrons appear to move to the left with a velocity $v_s + v_z$ and holes appear to move to the left but with a smaller velocity $(v_z - v_s)$. Thus by the usual conventions, hole and electron currents now appear to flow in opposite directions. Also, the moving observer sees a "particle current" due to the ionized impurities passing by.

The hole and electron charge continuity equations in the moving frame can consequently be written as

$$q(v_s + v_z)\frac{dn}{dz} = q\alpha(n + p)v_s \tag{9}$$

and

$$q(v_z - v_s)\frac{dp}{dz} = q\alpha(n + p)v_s \tag{10}$$

where positive z is measured to the left from plane C and where we have utilized the fact that $(\partial/\partial t)p(z, t) = (\partial/\partial t)n(z, t) = 0$; i.e., in the moving reference frame, quantities are time invariant. Subtracting (9) from (10) and integrating over z yields

$$(n - n_s) = (p - p_s)\frac{v_z - v_s}{v_z + v_s} \tag{11}$$

where n_s and p_s are the electron and hole densities at the leading edge of the avalanche region, i.e., those of I_{sat} where $I_{sat} = qn_s v_s + qp_s v_s$ in the inertial frame. Equation (11) relates the density of avalanche created electrons to avalanche created holes at any given z. It can be made plausible from the following. Imagine an observer at some position z. To his right hole electron pairs are being produced in identical numbers and both electrons and holes move past him at the same rate, i.e., in numbers per second. The electrons, however, are traveling at a higher relative speed $(v_z + v_s)$ than the holes $(v_z - v_s)$ and hence the electrons are on the average spaced farther apart in distance, or their density is less. The ratio of the densities is obviously equal to the inverse of the ratio of the velocities which is precisely what (11) states. It should be noted that within a very short distance into the avalanche zone (11) simplifies to

$$n = p\frac{v_z - v_s}{v_z + v_s} \tag{12}$$

for reasonable values of α.

Utilizing (11) to eliminate n from (10) yields after an integration over z,

$$p = \frac{v_z + v_s}{2v_z}\left[p_s\left(\frac{v_z - v_s}{v_z + v_s}\right) - n_s + (n_s + p_s)e^{Dz}\right] \tag{13}$$

and, similarly,

$$n = \frac{v_z - v_s}{2v_z}\left[n_s\left(\frac{v_z + v_s}{v_z - v_s}\right) - p_s + (n_s + p_s)e^{Dz}\right] \tag{14}$$

where

$$D = \frac{2\alpha v_s v_z}{(v_z + v_s)(v_z - v_s)}.$$

These equations give $p(z)$ and $n(z)$ within the avalanche region. Again, within a very short distance these simplify to

$$p = \frac{v_z + v_s}{2v_z}\cdot(n_s + p_s)e^{Dz} \tag{15}$$

$$n = \frac{v_z - v_s}{2v_z}\cdot(n_s + p_s)e^{Dz}. \tag{16}$$

In order to obtain (13) through (16), α has been assumed constant. $\alpha(E)$ may be included by replacing exponent Dz by

$$\frac{2v_s v_z}{(v_s + v_z)(v_z - v_s)}\int_0^z \alpha(E)dz.$$

As will be seen, this will have surprisingly little effect on our general conclusions.

For a complete delineation of the characteristics of the avalanche region, the extent of the zone and $E(z)$ within the zone must be obtained. First we obtain $E(z)$ through the utilization of Poisson's equation which can be written

$$-\frac{dE}{dz} = \frac{q}{\epsilon}(p - n - N_a) \tag{17}$$

in the moving reference frame. Note the total derivative rather than a partial derivative since quantities are time invariant. The negative sign preceding (dE/dz) is due to the fact that z and x are oppositely directed. Integrating over z

$$E(z) - E_a = -\frac{q}{\epsilon}\int_0^z (p - n - N_a)dz. \tag{18}$$

Utilizing (13) and (14) for $p(z)$ and $n(z)$ yields

$$\begin{aligned} E(z) - E_a = &-\frac{q}{\epsilon}\frac{v_s}{v_z}\frac{n_s + p_s}{D}(e^{Dz} - 1) \\ &-\frac{q}{\epsilon}\left[-\frac{v_s + v_z}{v_z}n_s + \frac{v_z - v_s}{v_z}p_s - N_a\right]z. \end{aligned} \tag{19}$$

Since the first two terms in the last expression are usually (except for a p-i-n) much smaller than N_a,

$$E(z) - E_a \approx \frac{q}{\epsilon}\frac{v_s}{v_z}\frac{n_s + p_s}{D}(1 - e^{Dz}) + \frac{q}{\epsilon}N_a z. \tag{20}$$

Equation (20) then expresses E as a function of z within the avalanche region.

The extent of the avalanche region can be determined from (20). Note that $E(z)$ first rises to a maximum value $((\partial E/\partial z) \equiv 0)$ and then drops to E_a at the left-hand edge of the avalanche region (plane D in Fig. 5). Consequently, setting $E(z) = E_a$ in (20) yields

$$v_s(n_s + p_s)e^{Dz^*} = v_z D N_a z^* \tag{21}$$

where we have neglected 1 compared to e^{Dz} at the left edge of the avalanche region and where z^* represents the width of the avalanche region. This transcendental equation can be solved explicitly for z^* for known values of the parameters. Note that (21) can be rewritten

$$I_s e^{Dz^*} = I_0 D z^*; \tag{22}$$

i.e., the extent of the avalanche region is a function of the reverse saturation current density I_s, the driving current density I_0, the impurity density N_a, and the ionization coefficient α which is contained in D. Again, if α is a function of electric field, (Dz^*) in (22) becomes an integral which does not affect our qualitative interpretation of (22).

In addition to the above, the peak field attained in the avalanche region is of interest. Differentiating (20) with respect to z and setting $(dE/dz) = 0$ yields

$$Dz^+ = \ln\frac{N_a v_z}{v_s(n_s + p_s)} \tag{23}$$

where z^+ is the value of z at which $E(z)$ reaches its peak value. The value of $E(z)$ where $z = z^+$ is given by (20) as

$$\begin{aligned} E(z^+) = E_a &+ \frac{q}{\epsilon}\frac{v_s}{v_z}\frac{n_s + p_s}{D}\left(1 - \frac{N_a v_z}{v_s(n_s + p_s)}\right) \\ &+ \frac{q}{\epsilon}\frac{N_a}{D}\ln\frac{N_a v_z}{v_s(n_s + p_s)}\cdot \end{aligned} \tag{24}$$

Since

$$\frac{N_a v_z}{v_s(n_s + p_s)} \gg 1,$$

$$E(z^+) \approx E_a + \frac{q}{\epsilon}\frac{N_a}{D}\left(\ln\frac{I_0}{I_s} - 1\right)\cdot \tag{25}$$

Thus, the amount the peak field rises above E_a has a rather complicated dependence. For $v_z \gg v_s$ such that $D \approx (v_s\alpha/v_z)$, (25) can be written

$$E(z^+) - E_a = \frac{I_0}{\epsilon\alpha v_s}\left(\ln\frac{I_0}{I_s} - 1\right). \tag{26}$$

Thus the peak field obviously increases as I_0 increases and decreases as α increases. Also it is affected logarithmically by the ratio of I_0 to I_s with larger values of I_s resulting in lower peak values of $E(z)$. This concludes our study of the avalanche zone. It should be remarked that the particle density, (13) and (14), the z dependence of E, (20), the width, (22), and the peak electric field, (24), are all highly dependent on the form of $\alpha(E)$. Thus, these results are qualitative in nature.

The next region to be analyzed is D–E in Fig. 5, i.e., the region to the left of the avalanche zone in which carrier velocities remain saturated. Again, we can invoke time invariance in the frame moving with velocity v_z. This follows from the fact that this region is progressing through the diode in a manner entirely determined by the events to the right of it; i.e., no carriers can catch up to it from the left and hence it is driven at v_z just as the avalanche zone is driven at velocity v_z. (This statement is strictly true only after plane E is distinguishable from plane G.)

The hole and electron densities at plane D can be obtained from (15) and (16) as

$$p(z^*) = \frac{v_z + v_s}{2v_z}(n_s + p_s)e^{Dz^*}$$

and

$$n(z^*) = \frac{v_z - v_s}{2v_z}(n_s + p_s)^{Dz} \tag{27}$$

where z^* is the width of the avalanche zone. These densities are maintained throughout region D–E due to the fact that velocities remain constant in this region and no carriers are created therein. It is interesting that the sum of the hole and electron currents exceed I_0, the driving current in this region. This must be, since $(\partial E/\partial t)$ is negative, and, of course, current continuity must be maintained. An observer in the stationary reference frame would see a total particle current j_c in region D–E given by

$$j_c = qv_s[p(z^*) + n(z^*)] = I_s e^{Dz^*}. \tag{28}$$

This equation can be rewritten to avoid all the mechanics of the avalanche region (i.e., the dependence on α and z^*) through the employment of (22) to yield

$$j_c = I_s \exp\left(\frac{j_c}{I_0}\right). \tag{29}$$

This equation is rather remarkable [8] in that it gives the particle current density to the left of the avalanche zone in terms of I_s and I_0 alone. Note that this is completely independent of the functional dependence of α on E.

The slope of the electric field is constant in region D–E and is obtainable from (17) and (27) as

$$\frac{dE}{dz} = \frac{q}{\epsilon}\left(\frac{v_s(n_s + p_s)e^{Dz^*}}{v_z} - N_a\right) \tag{30}$$

or

$$\frac{dE}{dz} = \frac{qN_a}{\epsilon}\frac{j_c - I_0}{I_0}. \tag{31}$$

Thus, if (29) is employed to obtain j_c as a function of I_0 and I_s, (31) yields (dE/dz) in terms of I_0, I_s, and N_a. Consequently, the particle current and the slope of the electric field with distance are constant throughout region D–E and their values depend only upon I_0, I_s, and N_a, and are independent of the functional form of $\alpha(E)$.

Next consider region E–F. Plane E moves to the right with velocity v_z as previously noted. Within region E–F carrier velocities are given by μE with $E(z)$ determined from Poisson's equation, where z is now measured from plane E. It is perhaps apparent that the electric field versus distance profile, in region E–F, as seen by an observer at plane E moving with velocity v_z is time independent. Again, this is due to the fact that events to the left of plane F do not affect region E–F (as long as I_0 is maintained); i.e., hole velocity to the left of plane F is $\mu E_{\min}$ where as in region E–F it is $\geq \mu E_{\min}$ by definition. Hence, no hole current flows into region E–F from the left. Similarly, electron velocities are higher within region E–F than those to the left of plane F.

Within region E–F we must have continuous hole and electron currents in the moving frame, since E is not a function of time within this frame, i.e.,

$$\begin{aligned} j_h &= (v_z - \mu E)p = q(v_z - v_s)p(z^*) \\ j_e &= q(v_z + \mu E)n = -q(v_z + v_s)n(z^*). \end{aligned} \tag{32}$$

Consequently,

$$p = \frac{v_z - v_s}{v_z - \mu E}p(z^*) = \frac{j_c}{2qv^*(v_z - \mu E)}$$

and

$$n = \frac{v_z + v_s}{v_z + \mu E}n(z^*) = \frac{j_c}{2qv^*(v_z + \mu E)} \tag{33}$$

where

$$v^* \equiv \frac{v_s v_z}{(v_z^2 - v_s^2)}.$$

The particular solutions of (33) at plane F, where E reaches its lowest value, are important because these solutions give the hole and electron density left in the wake of the avalanche zone. Then, when $E = E_{\min}$, and $(dE/dz) = 0$, we must have p-n-$N_a = 0$ or, from (33),

$$\frac{j_c}{2qv^*}\left[\frac{1}{v_z - \mu E_{\min}} - \frac{1}{v_z + \mu E_{\min}}\right] = N_a. \tag{34}$$

This equation can be solved for $\mu E_{\min}$ yielding

$$\mu E_{\min} = v_z\left(\frac{-j_c}{2I_0 v^*} + \sqrt{\left(\frac{j_c}{2I_0 v^*}\right)^2 + 1}\right), \tag{35}$$

and finally (33) and (35) give the hole and electron densities in the wake of the avalanche as

$$p_0 = \frac{j_c}{2qv^* v_z\left[1 + \frac{j_c}{2I_0 v^*} - \sqrt{\left(\frac{j_c}{2I_0 v^*}\right)^2 + 1}\right]}$$

and (36)

$$n_0 = \frac{j_c}{2qv^* v_z\left[1 - \frac{j_c}{2I_0 v^*} + \sqrt{\left(\frac{j_c}{2I_0 v^*}\right)^2 + 1}\right]}.$$

It is to be noted that p_0 and n_0 are independent of the assumed dependence of α on electric field, as was j_c.

The significance of (35) and (36) can be better appreciated by considering the special cases where $\mu E_{\min} \ll v_s \ll v_z$. Then,

$$\mu E_{\min} = \frac{I_0}{j_c} v_s$$

$$p_0 = \frac{j_c\left(1 + \frac{qN_a v_s}{j_c}\right)}{2qv_s}$$

and

$$n_0 = \frac{j_c\left(1 - \frac{qN_a v_s}{j_c}\right)}{2qv_s}. \tag{37}$$

Thus, p_0 and n_0 become comparable (since $(qN_a v_s/j_c) \ll 1$) and essentially proportional to j_c. Since j_c is only a function of I_0 and I_s, p_0 and n_0 are uniquely determined by I_0, I_s, and v_s and, as in the more general case, are independent of the details of the avalanche process.

An expression for $E(z)$ in region E–F can be obtained from the employment of (33) to obtain $p(E)$ and $n(E)$ in Poisson's equation. A partial fraction integration then yields an involved algebraic expression for $E(z)$ which is of little interest and will not be included. The triviality of $E(z)$ in this region is due to the fact that the voltage contributed by $\int^{E_s}_{E_{\min}} E(z)dz$ is a very small fraction of the total diode voltage. Region F–G which eventually occupies the entire depletion region is thus filled with hole electron plasma densities given by (36). The electric field in this region is given by $E_{\min}$, (35), except at the edges of the depletion region where zone formation and zone annihilation occur. Little attention has been paid to the transient in which the zone is actually first formed and nothing has been said about the annihilation transient when the leading edge of the zone arrives at the right-hand boundary (plane B, Fig. 5 of the original depletion layer). However, detailed computer studies [3] indicate that the filling of the depletion region with a constant density hole electron plasma, i.e., ignoring the transients, is a reasonably good approximation.

It has been shown that the traveling avalanche zone first observed in very involved large signal computer solutions can be predicted from the simplest of models. Furthermore, the essential features of this phenomenon for TRAPATT operation, i.e., how fast the zone travels and how much plasma it leaves behind, are independent of the detailed nature of the avalanche process as long as α is sufficiently large that the width of the avalanche zone is a reasonably small fraction of the original depletion layer width.

This completes our study of the traveling avalanche zone. Certainly more accurate modeling of $\alpha(E)$ can be done, but it should be remembered that precise solutions to the differential equations already exist and, as will be seen in the following section, the principal re-
310
sults obtained are actually the plasma densities left in the wake of the avalanche zone and these are independent of the dependence of α on electric field. Consequently, we now turn our attention to the recovery transient.

II. The Transient Recovery Cycle

An interesting sequence of events occurs in TRAPATT operation following the avalanche zone transit. Rather than attempt to include these events in time sequence from the start, a simpler problem will first be treated. As will be seen, this problem contains the essence of the TRAPATT recovery period and with relatively minor additions presents a reasonably accurate picture in comparison with "exact" numerical calculations.

Assume the same punched-through step junction previously discussed (Fig. 3) but having the original depletion layer now filled with holes of uniform density p_0 and electrons n_0 and being space charge neutral, i.e., $p_0 - n_0 - N_a = 0$, and n_0 and p_0 constant throughout the original depletion layer. Further assume that at time zero a *constant* recovery current per unit area I_R is applied to the diode terminals in the same direction as the current which initially drove the avalanche zone through the diode, i.e., the diode continues to be reverse biased in the usual sense.

The ensuing events will now be developed. (This particular problem is also of significance in "recovery" transients in other types of diode operation such as snap diodes and p-i-n diodes and thus is of general importance.) At time zero, our assumed application of I_R causes an electric field to develop within the diode. Two cases are of interest. In the first, the hole electron plasma can initially carry the total current I_R at fields less than E_s, i.e.,

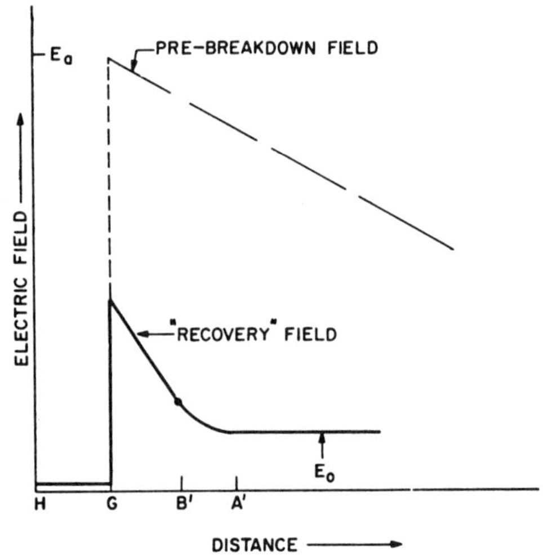

Fig. 6. Time snapshot of electric field versus distance during the recovery transient at the left-hand edge of the original depletion region. The pre-breakdown field is included, dashed, to emphasize the position chosen.

$$I_R = q\mu(p_0 + n_0)E_0 \quad \text{with } E_0 < E_s. \tag{38}$$

In the second, I_R exceeds the limitations of (38) and I_R is in part particle current and in part displacement current. The lower current regime is the more important and will be developed in some detail. As will be seen, the development of the behavior in the high current regime is simply obtained from this treatment in the limit as $E_0 \to E_s$ or, equivalently, as the velocities of the carriers in the plasma become saturated.

Thus, assume that at time zero (38) holds throughout the original depletion layer and now, at a later time t_0, we inquire as to the shape of the electric field versus distance plot at the left-hand edge of the original depletion region, i.e., plane G in Fig. 5. As in the breakdown transient, after an extremely short formation time this region can be characterized by zones (see Fig. 6). Plane A' is defined as that plane at which the electric field first rises above E_0. Thus, immediately to the right of plane A' we have the undisturbed plasma densities n_0 and p_0 where (38) applies and where the electric field is E_0. On the other hand, neglecting diffusion,[1] no holes are to be found to the left of plane A' other than the negligible number appearing due to thermal generation. This results from the fact that there is no source of

[1] The neglect of diffusion is equivalent to assuming that all the holes move with velocity μE rather than simply having the average particle velocity represented by μE, and considerably simplifies the development with very little sacrifice in accuracy.

holes at the left-hand edge of the depletion layer (fields are far below E_a) and the original compliment of holes p_0 in this region has moved to the right with velocity μE_0, which we label v_p, the plasma velocity. Thus, since we have assumed that electron and hole mobilities are identical, (1) can be rewritten

$$I_R = q(n_0 + p_0)v_p. \tag{39}$$

It should thus be apparent that plane A' moves to the right with velocity v_p, and indeed is coincident with the trailing edge of the hole distribution which in turn is moving to the right with velocity v_p.

In our deterministic picture where all carriers respond to electric fields with velocity $v = \mu E$ it is easy to see that holes cannot remain to the left of plane A' since the electric field is higher there by definition and they would "catch up" to plane A'. The inclusion of diffusion would of course place carriers to the left of A' but it is easy to show that this detail is of negligible importance to the argument.

Plane B' is defined by $E = E_s$, i.e., that plane at which E has increased from E_0 to E_s due to the net space charge (in this case electrons and ionized acceptors) to the right preceding plane A'.

Between plane B' and plane G, carrier velocities are saturated and, due to our assumption of constant current, the slope of the electric field is constant.

As in the study of the transiting avalanche zone, it is helpful to utilize a moving coordinate system, and this time it will be fixed at plane A' and move with the "plasma velocity" v_p. It is important to note that to an observer in this frame, the dependence of E on z (distance measured to the left from moving plane A') between planes A' and G is determined solely by events to the right of plane A' and by the driving current I_0. Thus, as long as I_0 is maintained and the hole and electron densities preceding plane A' are n_0 and p_0, E versus z assumes a predictable shape. Consequently, successive time snapshots of E versus z would appear as in Fig. 7.

Since plane A' moves to the right with velocity v_p, and since the electrons to the right of plane A' are moving to the left with velocity v_p, an electron current given by $2qn_0v_p$ must cross plane A'. Since this current is constant, a steady-state field and charge distribution is established in the moving reference frame between planes A' and B' in a time negligibly small in comparison to the transit time of plane A' through the diode. It then follows from the electron current continuity equation that between planes A' and B',

$$qn(v_p + \mu E) = 2qn_0v_p. \tag{40}$$

Also, in the moving reference frame Poisson's equation becomes

$$\frac{dE}{dz} = \frac{q}{\epsilon}(n + N_a). \tag{41}$$

Since z is measured to the left a negative space charge causes an increasing dE/dz to the left. Equations (40) and (41) yield

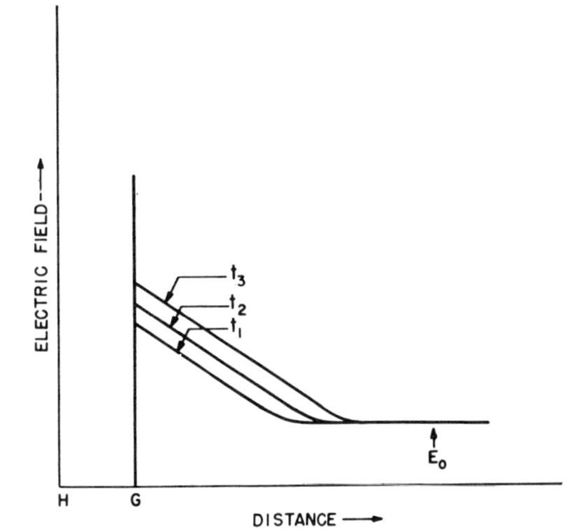

Fig. 7. Successive time snapshots of electric field versus distance during the recovery transient, at the left-hand edge of the depletion region.

$$\frac{dE}{dz} = \frac{q}{\epsilon}\left(\frac{n_0 2v_p}{v_p + \mu E} + N_a\right). \tag{42}$$

This equation can be readily integrated to give $E(z)$ between planes A' and B'. It must be remarked, however, that (42) is markedly affected by our neglect of diffusion. Fortunately, this does not affect the generality of our overall results since the voltage drop from plane A' to plane B' is negligible and consequently the precise shape of E versus z in this region is unimportant.

At plane B', by definition, (40) becomes

$$qn(v_p + v_s) = 2qn_0v_p \tag{43}$$

so that the density of electrons crossing plane B' is given by

$$n = \frac{2n_0v_p}{v_p + v_s}. \tag{44}$$

The shape of the electric field versus distance profile to the left of plane B' is of importance and will now be obtained. Since carrier velocities are saturated in region B'–G, the carrier density n entering at plane B' cannot change with distance. Thus, in this region Poisson's equation becomes

$$\frac{dE}{dz} = \frac{q}{\epsilon}\left(\frac{2n_0v_p}{v_p + v_s} + N_a\right), \tag{45}$$

or the slope of the electric field is a constant in this region, as was indicated in Figs. 6 and 7. Integration of (45) yields

$$E(z) = E_s + \frac{q}{\epsilon}\left(\frac{2n_0v_p}{v_p + v_s} + N_a\right)z \tag{46}$$

where z is measured from plane B' to the left.

This concludes the development of our required

"tools" for the explanation of the recovery transient under our simplifying assumptions.

First, note that a similar treatment can be applied at the right-hand edge of the original depletion region, plane B, Fig. 5; but here the background impurity concentration reduces the net space charge, i.e., the charge on the ionized acceptors is opposite to that of the exiting holes. Thus, the key equations describing the behavior in the vicinity of plane G, i.e., (44) and (46), become in the vicinity of plane B,

$$p = \frac{2p_0 v_p}{v_p + v_s} \tag{47}$$

and

$$E(z) = E_s + \frac{q}{\epsilon}\left(\frac{2p_0 v_p}{v_p + v_s} - N_a\right)z \tag{48}$$

where z in (48) is measured to the right from a plane equivalent to plane A' fixed at the trailing edge of the electron population moving away from plane B, again with velocity v_p. Thus, in general, the entire electric field versus distance profile must appear as in Fig. 8.

The time evolution of the electric field versus distance profile of Fig. 8 is straightforward as long as the electric field is always below E_a. However, $E<E_a$ can be taken as a boundary condition for recovery transients since fields equal to or greater than E_a will produce new carriers which in turn must be swept out.

The time evolution leading to the minimum recovery time, subject to this restriction, is of most interest. Consequently, for a given n_0 and p_0 we seek that value of I_R which will just raise the highest field in the diode (occurring at plane G) to E_a as the last carriers leave. The total resulting elapsed time τ will be labeled the recovery time and the diode cannot be cleared of charge in times less than τ under the assumed conditions of constant recovery current I_R and $E<E_a$.

τ can be divided into two parts, $\tau=\tau_1+\tau_2$, where τ_1 is the time for plane A' to meet its counterpart at the center of the diode and is obviously given by $w/2v_p$ where w is the width of the original depletion region, i.e., the separation of planes B and G in Fig. 5. Thus, at time τ_1 into the transient the electric field versus distance plot is as shown in Fig. 9. τ_2 is the time required for the remaining carriers to leave and, to a good approximation, since the separation of plane A' and B' is extremely small, $\tau_2 \approx (w/2v_s)$. Thus,

$$\tau = \frac{w}{2v_p} + \frac{w}{2v_s}\cdot \tag{49}$$

We now proceed to obtain $E(G, t)$. Then our boundary condition can be imposed by setting $E(G, \tau)=E_a$. $E(G, t)$ is readily obtained since

$$j_c(G) = \frac{2qn_0 v_p v_s}{v_s + v_p} \qquad 0 < t < \tau \tag{50}$$

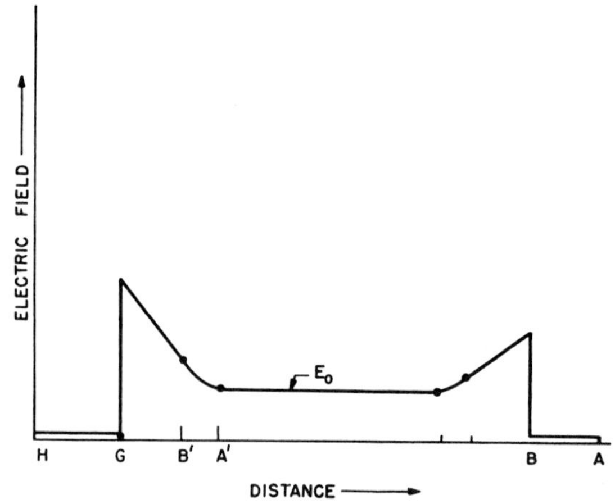

Fig. 8. Time snapshot of entire electric field versus distance plot during recovery transient.

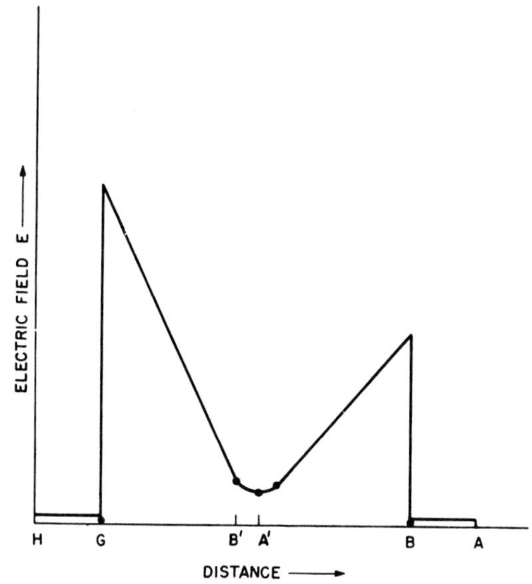

Fig. 9. Snapshot of electric field versus distance at time τ_1 into the recovery transient.

where $j_c(G)$ is the particle current at G. Consequently, since the sum of the conduction and displacement current must everywhere equal I_R,

$$\epsilon \frac{dE}{dt}(G, t) = I_R - j_c(G), \tag{51}$$

and upon integration,

$$E(G, t) = E_0 + \frac{1}{\epsilon}\left(I_R - \frac{2qn_0 v_p v_s}{v_s + v_p}\right)t. \tag{52}$$

Then if $E(G, t)=E_a$ when $t=\tau$, it follows that

$$I_R = qv_s(2N_0 - N_a) \tag{53}$$

when we have neglected E_0 in comparison with E_a and where N_0 is the punch-through impurity concentration, i.e., $N_0 \equiv (\epsilon E_a/qw)$. It is perhaps surprising that the maximum allowable (constant) current for charge removal is independent of n_0 and p_0, the starting plasma density, and instead only depends upon the field E_a at which avalanche begins, the width w of the depletion region, and the impurity density N_a in the depletion region. Thus I_R varies from qv_sN_0 for diodes which are just punched through to $2qv_sN_0$ for p-i-n diodes in which $N_a=0$.

An additional parameter of some interest is the work required to remove the charges represented by n_0 and p_0 from the depletion region. Consider first a unit electronic charge at some distance x to the right of plane G. The work W done on this unit charge is negligible until plane A' reaches it. At this instant in time the field to the left of x is given by

$$E(z) = \frac{q}{\epsilon}\left(N_a + \frac{2n_0v_pv_s}{v_p+v_s}\right)z \tag{54}$$

where we have approximated the field between planes A' and B' by a linear extension of the field in region B'–G. This approximation has a negligible effect on the overall calculation. However, the charge now moves to the left with velocity v_s so that we really require $E(z)$ at a time z/v_s later than the time of arrival of plane A' at x. Having (54) as the initial field, we obtain the field at a given value of z at a later time from an integration of

$$\epsilon\frac{dE}{dt}(z) = I_R - j_c(z), \tag{55}$$

or upon integration,

$$\Delta E(z) = \frac{1}{\epsilon}\left(I_R - \frac{2qn_0v_sv_p}{v_p+v_s}\right)\frac{z}{v_s}\cdot \tag{56}$$

Thus, when in transit, the unit electronic charge sees a field actually given by

$$E(z) = \frac{q}{\epsilon}\left(N_a + \frac{I_0}{qv_s}\right)z \tag{57}$$

and requires energy W given by

$$W = \int_0^x E(z)\,dz = \frac{q}{\epsilon}\left(\frac{I_R}{2v_s} + N_a\right)\frac{x^2}{2}\cdot \tag{58}$$

Similarly, for holes it can be shown that the work per unit charge required for removal is given by

$$W = \frac{q}{\epsilon}\left(\frac{I_R}{qv_s} - N_a\right)\frac{x^2}{2} \tag{59}$$

when x is measured to the left from plane B. Thus for a given p_0 and n_0, the total work required is obtained by integrating (58) and (59) for all charges to obtain

$$W_{\text{total}}\frac{q}{\epsilon}\left(\frac{I_R}{2v_s} + N_a\right)\int_0^{w/2}\frac{x^2}{2}n_0dx + \frac{q}{\epsilon}\left(\frac{I_R}{2v_s} - N_a\right)\int_0^{w/2}\frac{x^2}{2}p_0dx, \tag{60}$$

and upon integration,

$$W_{\text{total}} = \frac{q^2}{\epsilon}\left[2N_0(n_0+p_0) - 2p_0N_a\right]\frac{w^3}{48}\cdot \tag{61}$$

For the important case in which n_0 and $p_0 \gg N_a$, and consequently $n_0+p_0 \approx 2p_0$, it follows that

$$W_{\text{total}} \approx \frac{q^2}{\epsilon}\left[2N_0 - N_a\right]\left[n_0+p_0\right]\frac{w^3}{48} \tag{62}$$

or

$$W_{\text{total}} \approx \frac{Q_TI_R}{\epsilon v_s}\frac{w^2}{48} \tag{63}$$

where $Q_T \equiv q(n_0+p_0)w$ and represents the total removable charge in the diode.

Equation (62) is not necessary for the understanding of TRAPATT operation but it is important in that it contains the principal loss mechanism in the recovery transient other than the usual R_S losses external to the depletion region. The energy dissipated during the avalanche zone transit is discussed in the Appendix. Other loss mechanisms can dominate in p-i-n switches and snap diodes but these losses will always add to that of (62) which cannot be avoided.

The voltage across the diode during the recovery transient can readily be obtained from (46) and (48) and our knowledge of what occurs during the recovery transient. It follows that

$$V(t) = E_0w + \frac{I_Rv_p^2t^2}{\epsilon(v_p+v_s)} \qquad 0 \le t \le \tau_1 \tag{64}$$

and

$$V(t) = \frac{I_R}{\epsilon(v_p+v_s)}\left\{v_p^2t^2 + (v_s^2 - v_p^2)\left(t - \frac{w}{2v_p}\right)^2\right\} + E_0w \tag{65}$$

where the fields in region A'–B', and its counterpart, have been approximated by a linear extension of the high field regions to their left and right, respectively. This has a negligible effect on the accuracy of the calculated $V(t)$.

III. TRAPATT Operation

The mechanisms of zone transit and recovery transient have been examined in some detail for the cases of constant breakdown current and constant recovery current. No attempt has been made to make plausible the cyclic repetition of these two transients. Such a presentation is beyond the scope of this paper. It has been clearly demonstrated, however, that such cyclic operation can be achieved [3]. The starting mechanism to date has arisen from IMPATT voltage swings generated by lightly loading the diode at IMPATT frequencies where a small signal negative resistance exists. When this RF voltage builds up to sufficient amplitude, an avalanche zone can be launched into the depletion region. It is from this operation that the word TRAPATT (TRApped Plasma Avalanche Triggered Transit) originated.

Computer simulations have also indicated that the diode current versus time is reasonably well approximated by a square wave. Consequently, as a start on a design theory we require that the breakdown and the recovery current be equal, and in addition that they be equal to I_R (i.e., we remove the plasma in the minimum time). Consequently,

$$I_0 = I_R = qv_s(2N_0 - N_a). \tag{66}$$

It follows then that

$$v_z = v_s\left(\frac{2N_0}{N_a} - 1\right). \tag{67}$$

By restricting our purview to punched-through step junctions, our TRAPATT design theory has but two variables, N_0 and N_a (i.e., the width of our diode and the actual impurity density contained therein). These two parameters will now be determined for a particular high efficiency TRAPATT design.

Arbitrarily, we require that the time required for the zone to sweep through the diode be one-tenth the desired TRAPATT frequency. This choice keeps short the transition time between the high-voltage, low-current state and the low-voltage, high-current state. This choice is not critical and, in particular, the zone transit time could be chosen to be a somewhat larger fraction of the TRAPATT period T leading to a slightly different design. The tenth period zone transit time does enforce high efficiency, however, and its consequences will be developed. Precisely, we require

$$\frac{w}{v_z} = \frac{T}{10} \tag{68}$$

which, in turn, through (67), requires that

$$N_a = \frac{2N_0^2}{(10\epsilon E_a/qv_sT) + N_0}. \tag{69}$$

Again, arbitrarily, we require that the recovery transient take one-half of the TRAPATT period T, namely,

$$T = \frac{1}{f} = W\left(\frac{1}{v_p} + \frac{1}{v_s}\right). \tag{70}$$

It is perhaps obvious that this design will operate at frequencies both above and below the f of (70). However, our arbitrary assumptions, (68) and (70), enforce high efficiency at frequency f. At frequencies much above f the efficiency will obviously decrease. Thus in a sense, f describes an upper-frequency limit for high-efficiency operation.

TRAPATT operation has thus been divided into three periods: a high-voltage state lasting 4/10 T, a breakdown transient lasting 1/10 T, and a recovery (high-current) transient lasting 1/2 T. Other choices can obviously be made with different optima in mind, but the consequences of this particular design will now be obtained as a function of frequency.

One other "materials parameter" must first be discussed. This is I_s. In general, I_s can be written

$$I_s = \frac{qn_iw}{\tau_e} \tag{71}$$

where τ_e is an effective carrier lifetime in the depletion region, n_i is the intrinsic carrier concentration (at the operating temperature), and w is the width of the depletion region. To be specific, assume $_e = 1\times10^{-7}$ second, and $n_i = 10^{14}$ cm^{-3}, numbers appropriate for both epitaxial Ge and Si at elevated temperatures. Thus, we have

$$I_s \approx 1.6\times10^2 w \tag{72}$$

and (29) becomes

$$j_c = 1.6\times10^2 w \exp\frac{j_c}{qv_s(2N_0 - N_a)} \tag{73}$$

which gives j_c in terms of N_0 and N_a since $N_0 = (\epsilon E_a/qw)$. Also, it follows from (33) that

$$j_c \approx 2qv^*v_z(p_0 + n_0). \tag{74}$$

But, since $I_0 = q(n_0 + p_0)v_p = qv_s(2N_0 - N_a)$, it follows from (68) that

$$j_c = 2qv^*v_zv_s(2N_0 - N_a)\left(\frac{qN_0T}{\epsilon E_a} - \frac{1}{v_s}\right) \tag{75}$$

which again gives j_c as a function of N_0 and N_a and materials constants, but in addition as a function of the TRAPATT period T. We can eliminate j_c from (73) and (75) giving finally a relation[2] for N_0 and an N_a for any given period T. This equation in conjunction with (67) gives N_0 and N_a as a function of T.

Selecting $E_a = 2\times10^5$ V/cm, $v_s = 7\times10^6$ cm/s, and $\epsilon = 1.4\times10^{-12}$ F/cm, numbers roughly appropriate for Ge, one obtains the values shown in Figs. 10 and 11 for $N_0(f)$ and $N_0(f)/N_a(f)$.

Two additional effects are also included in these figures. The first is the variation in the design values as I_s is increased or decreased by two orders of magnitude from our starting assumption. This change is not negligible, and is seen (Fig. 11) to be particularly important

[2] An approximation to the solution of (72) for large I_0/I_s is given by

$$j_c = qv_s(2N_0 - N_a)\left[1.04 \ln\frac{qv_s(2N_0 - N_a)}{1.6\times10^2 W} + 1.6\right].$$

The reader can utilize this form as a rough check of the results of Figs. 10 and 11.

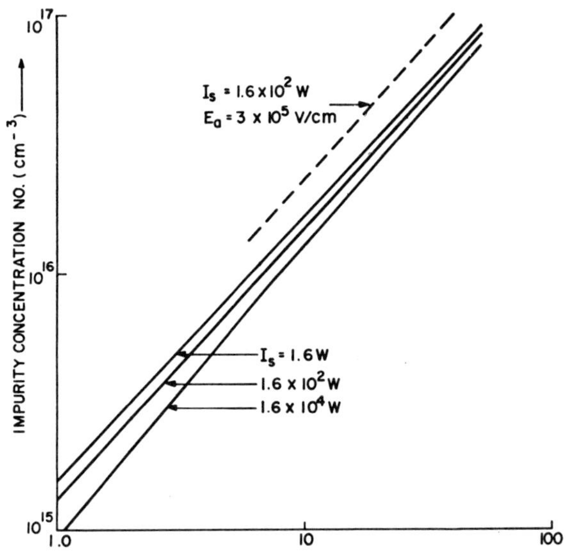

Fig. 10. N_0 as a function of frequency for three assumed values of I_s for $E_a = 2\times10^5$ V/cm. The dashed curve shows the effect of increasing E_a to 3×10^5 V/cm for the median I_s.

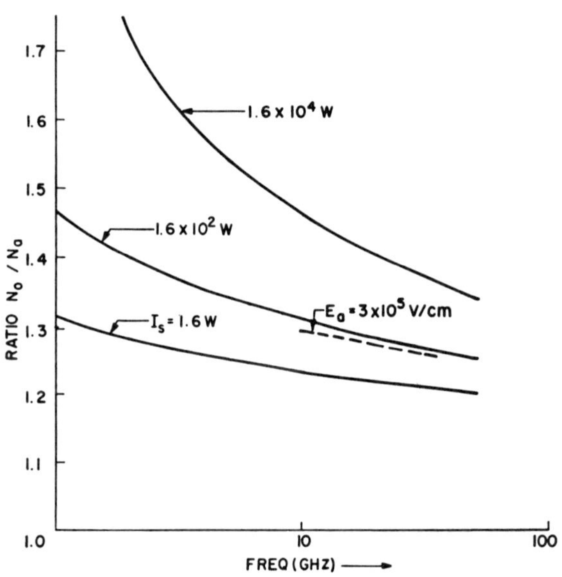

Fig. 11. The ratio of N_0 to N_a as a function of frequency for three assumed values of I_s for $E_a = 2\times10^5$ V/cm. The dashed curve shows the effect of increasing E_a to 3×10^5 V/cm for the median I_s.

at low frequencies. Since I_s is a poorly (if at all) controlled parameter in device fabrication, this may explain some of the difficulties in reproducing results from laboratory to laboratory. The second effect is that incurred by arbitrarily increasing E_a by 50 percent to 3×10^5 V/cm which is in the direction expected for higher frequencies due to the actual nonlinearity in ionization rates. It is seen to have little effect on the ratio N_0/N_a (Fig. 11), but an appreciable effect on N_0, as would be expected.

The values calculated for N_0 and N_a depend on $(\epsilon E_a/v_s)$. It is interesting to note that this parameter is approximately the same for both silicon and germanium, i.e., assuming the values below:

	ϵ(F/cm)	E_a(V/cm)	v_s(cm/s)
Ge	1.4×10^{-12}	2×10^5	7×10^6
Si	1.0×10^{-12}	4×10^5	10×10^6

The curves in Figs. 10 and 11 fit silicon as well as germanium. This does imply a difference in the ratio of the TRAPATT period to the transit time at saturated velocity for the two materials. Further work must be done to establish the relevance of this difference.

This completes our zero-order TRAPATT design. The closeness of this design to functioning [3] Ge structures is to be noted.

The careful reader will observe that our dichotomy of breakdown transient and recovery transient is not complete in TRAPATT operation and indeed that the recovery transient begins at plane G as soon as the high field region associated with the traveling avalanche zone, i.e., plane F in Fig. 5, has passed. Consequently, in the development of Fig. 8 for TRAPATT operation, the field rise at the left-hand edge (plane G) begins at time t' earlier than that at the right-hand edge with t' given by w/v_z. It can be shown that this modifies τ such that

$$\tau = \frac{w}{2v_p} + \frac{w}{2v_s} + \frac{w}{2v_z}\left(1 + \frac{v_p}{v_s}\right) \tag{76}$$

as compared with (49). The correction term will, in general, be small but can be simply included in (52) to give $F(G, \tau)$ and, consequently, a difference in I_R as expressed by (53). The resulting perturbation of the zero-order design theory is then straightforward.

Summary

An attempt has been made to illuminate the device physics of TRAPATT operation. A simplified theory has been evolved that clearly identifies limitations on TRAPATT operation, as well as the frequency dependence of the design parameters.

To date, TRAPATT operation has been observed only through the self-starting feature of an IMPATT negative resistance. This implies a further restriction not considered herein. This requirement can be included in the design at the expense of considerable complication and will be reported elsewhere [9].

The role played by I_s, the reverse saturation current, is intriguing. It is perhaps obvious that I_s can actually be an adjustable materials parameter through careful materials preparation, the intentional insertion of deep traps, etc. Thus, the appropriate magnitude of N_0 and N_a at a given TRAPATT frequency can be markedly affected by the device designer provided I_s can be changed by orders of magnitude. This unsuspected parameter perhaps explains some of the difficulty in reproducing TRAPATT results in different laboratories.

Also, it should be remembered that I_s is a strong function of temperature, another variable in device operation.

The formation of widening depletion layers and the resulting space charge limited currents in recovery transients of p-i-n diodes was first deduced from experimental measurements [10]. This paper contains the first analytical treatment of the limitations imposed by their formation. The form of treatment is readily generalized to time dependent withdrawal currents, but the extensive algebra is beyond the scope of this paper.

Appendix

An approximate form can be readily obtained for the energy dissipated in the diode during the transit of the avalanche zone. First note that carrier growth is exponential with distance in region C–D (Fig. 5), which implies that nearly all the carriers are created in the immediate vicinity of plane D. Consequently, ignore the energy associated with carrier motion in region C–D and instead consider the energy dissipated in region D–E in which we have a high particle current j_c, and also a high electric field.

The work done on a single carrier in traversing this distance is given by

$$W = \int_{E=E_s}^{E=E_a} qE(x)dx \tag{77}$$

where q is the charge of the carrier. Since $E=kx$ in this region, with $k=qN_a/\epsilon\ (j_c-I_0/I_0)$ [see (31)] it follows that

$$W \approx \frac{qE_a^2}{2k} \frac{\text{(joules)}}{\text{(carrier)}}. \tag{78}$$

Since a steady particle current j_c flows in this region, power is being expended at a rate

$$P = \frac{W}{q} j_c \left(\frac{\text{watts}}{\text{unit area}}\right). \tag{79}$$

This power is expended for approximately w/v_z seconds, i.e., the transit time of the zone, yielding the total energy dissipated in one zone transit as

$$W_{\text{total}} = \frac{E_a^2}{2k} \cdot j_c \cdot \frac{w}{v_z} \tag{80}$$

or

$$W_{\text{total}} = \frac{\epsilon E_a^2}{2} \frac{w}{1 - I_0/j_c}, \tag{81}$$

and since in general $j_c \gg I_0$, we have finally

$$W_{\text{total}} \approx \frac{\epsilon E_a^2 w}{2}. \tag{82}$$

This result is rather surprising in that it is independent of I_0, the driving current.

In addition to W_{total} above, the energy required to produce the requisite number of hole electron pairs should properly be included. This term will be on the order of

$$W_{\text{avalanche}} = w \frac{j_c}{2v_s} E_g \tag{83}$$

for dense plasmas and fast zone transits. E_g is the bandgap energy in electron volts.

Equations (82) and (83) contain the principal energy loss terms in the zone transit process. To be precise, they constitute a lower limit on the energy loss.

Acknowledgment

The authors would like to acknowledge many helpful discussions with R. M. Ryder, W. J. Evans, and R. L. Johnston. A device design containing similar features to the one included herein has also been developed by A. S. Clorfeine, L. S. Napoli, and R. J. Ikola, their work to be published in the *RCA Review*.

References

[1] H. J. Prager, K. K. N. Chang, and S. Weisbrod, "High-power, high-efficiency silicon avalanche diodes at ultrahigh frequencies," *Proc. IEEE* (Letters), vol. 55, pp. 586–587, April 1967; also "Anomalous silicon avalanche diodes for microwave generation," *Proc, Cornell Conf. on High Frequency Generation and Amplification* (Ithaca, N. Y., August 29–31, 1967), pp. 266–280.

[2] S. G. Liu and J. J. Risko, "High-power punch-through avalanche diode microwave oscillators," presented at the Internatl. Solid State Circuits Conf., Philadelphia, Pa., February 1968. Also, S. G. Liu, *IEEE J. Solid-State Circuits*, vol. SC-3, pp. 213–217, September 1968.

[3] R. L. Johnston and D. L. Scharfetter, "Low frequency high efficiency oscillations in germanium Impatt diodes," presented at the IEEE Solid State Device Research Conf., June 17–19, 1968; also *IEEE Trans. Electron Devices*, vol. ED-16; pp. 905–911, November 1969.
D. L. Scharfetter, D. J. Bartelink, H. K. Gummel, and R. L. Johnston, "Computer simulation of low frequency high efficiency oscillation in germanium," presented at the IEEE Solid State Device Research Conf., June 17–19, 1968.
R. L. Johnston, D. L. Scharfetter, and D. J. Bartelink, "High-efficiency oscillations in germanium avalanche diodes below the transit-time frequency," *Proc. IEEE* (Letters), vol. 56, pp. 1611–1613, September 1968.
D. L. Scharfetter, "High efficiency operation of Impatt diodes," presented at the Internatl. Electron Devices Meeting, Washington D. C., October 23–25, 1968.

[4] R. L. Rulison, G. Gibbons, and J. G. Josenhans, "Improved performance of Impatt diodes fabricated from Ge," *Proc. IEEE* (Letters), vol. 55, pp. 223–224, February 1967.

[5] D. E. Iglesias and W. J. Evans, "High-efficiency CW Impatt operation," *Proc. IEEE* (Letters), vol. 56, p. 1610, September 1968.

[6] M. W. Muller and H. Guckel, "Negative resistance and filamentary currents in avalanching silicon p^+-i-n^+ junctions," *IEEE Trans. Electron Devices*, vol. ED-15, pp. 560–568, August 1968.

[7] C. P. Snapp and B. Hoefflinger, "An efficient multiresonant avalanche diode oscillator in the 1.5 to 11 GHz range," *Proc. IEEE* (Letters), vol. 56, pp. 2054–2055, November 1968.
C. P. Snapp, L. A. Stark, and B. Hoefflinger, "Performance and theory of avalanche resonance pumped Impatt oscillators," presented at the Internatl. Electron Devices Meeting, Washington, D. C., October 1968.

[8] D. J. Bartelink and D. L. Scharfetter, "Avalanche shock fronts in p-n junctions," *Appl. Phys. Letters*, May 15, 1969.

[9] D. L. Scharfetter, "Power-frequency characteristics of the Trapatt diode mode of high efficiency power generation in germanium and silicon avalanche diodes" (to be published).

[10] R. L. Johnston and B. C. De Loach, Jr., "The transient microwave impedance of PIN diodes," presented at the Internatl. Solid State Circuits Conf., Philadelphia, Pa., February 1965.

Circuits for High-Efficiency Avalanche-Diode Oscillators

W. J. EVANS

Abstract—**This paper describes and analyzes the circuits which have been used successfully for TRAPATT oscillator studies. The results lead to a better understanding of the TRAPATT oscillator and yield a simple model of the oscillator which is useful for circuit design.**

The circuit characteristics of an experimental TRAPATT oscillator are determined from measurements on the circuits and from equivalent circuit model calculations. The following conclusions can be drawn from the analysis. First, the avalanche diode requires sufficient capacitance near the diode to sustain the high-current state required for TRAPATT operation. Secondly, at a distance from the diode corresponding to approximately one half-wavelength at the TRAPATT frequency the transmission line containing the diode should be terminated by a low-pass filter. The function of the filter is to pass the TRAPATT frequency and to provide a shorting plane for the harmonics of that frequency. Finally, on the load side of the filter, tuning for the TRAPATT frequency is required.

The model of the circuit described above suggests a simple explanation of the diode–circuit interaction in a TRAPATT oscillator. Simplified waveforms suggested by the model have been used to calculate power output, efficiency, dc voltage change, and RF impedance for the oscillator. The results agree within a few percent with those obtained for an experimental oscillator. An important conclusion of the analysis is that the high-efficiency operation of avalanche diodes at frequencies in the UHF range can be explained by the TRAPATT theory, even though the trapped-plasma or low-voltage state may last only $\frac{1}{20}$th of the oscillation period.

I. Introduction

A NEW high-efficiency mode of operation using avalanche-diode oscillators has been obtained in silicon diodes by Prager, Chang, and Weisbrod [1] and in germanium by Johnston, Scharfetter, and Bartelink [2]. This mode is characterized by efficient conversion of dc to RF energy (up to 60 percent), operation at frequencies well below the transit-time frequency of operation, and a significant change in the dc operating point when the diode switches into the mode. This paper describes several oscillator circuits which have been used to obtain the high-efficiency or TRAPATT (Trapped Plasma Avalanche Triggered Transit) mode of operation, and suggests a simple model to describe the operation of these oscillators. The next section is a review of the important ideas which describe the TRAPATT mode of operation and were first presented by Sharfetter, Bartelink, Gummel, and Johnston [3]. Section III describes the mechanical and electrical properties of the coaxial circuits used to obtain the TRAPATT mode. The last two sections present a simple model for the oscillator and discuss the conclusions which can be drawn from the model.

Manuscript received July 17, 1969. This paper was presented at the International Microwave Symposium, Dallas, Tex., May 5–7, 1969.

The author is with Bell Telephone Laboratories, Inc., Murray Hill, N. J.

II. The TRAPATT Mode

In order to understand the operation of high-efficiency oscillators, it is necessary to understand what happens when large RF voltage swings exist across an IMPATT diode. Therefore, this section will begin with a review of the large-signal behavior of an IMPATT oscillator.

Fig. 1(a) is a plot of the field profile and carrier densities in an oscillating IMPATT diode at a particular point in time. The profile shown here is for a 6-GHz n^+-p-p^+ epitaxial germanium diode [4]. The plots are made for the particular point in the RF voltage and current waveforms indicated by the squares in Fig. 1(b). The carriers shown have been generated by avalanche breakdown in the high-field region on the left. The holes then drift to the right and the electrons to the left.

One of the saturation mechanisms of this mode of operation is seen in the depression of the field by the space charge of the avalanche-generated carriers. The reduction in field turns off the avalanche prematurely and thus reduces the phase delay provided by the avalanche. This in turn reduces the negative conductance of the oscillator.

Another saturation mechanism is illustrated in Fig. 2. Here the effect of the space charge on the field can be seen after generation has stopped and the holes have drifted into the lower field region of the diode. Again the space charge is depressing the field, so that to the left of the carrier pulse the carrier drift velocity may drop below saturation and some carriers may be trapped for a short time in the low-field region. This change in carrier velocity is not in itself a lossy process as will be seen in the discussion of the TRAPATT mode. However, it does sufficiently change the terminal current waveforms so that the power generated at the transit-time frequency is reduced.

The diode shown in Figs. 1 and 2 will also generate high-efficiency oscillations at frequencies below 3 GHz. However, at the bias current densities used in these oscillators, the diode exhibits no small-signal negative resistance in this frequency range. Therefore in order to start the TRAPATT oscillations, it has been found that very large IMPATT-generated 6-GHz voltage swings must exist across the diode. Thus, in the TRAPATT mode these two saturation mechanisms, i.e., space-charge suppression of the avalanche and carrier trapping, play an important role.

Fig. 3 shows the situation which can occur when the diode shown previously is operated in a high Q circuit where these large RF voltages can develop. The field and carrier densities are shown at a point in time just after the diode voltage has reached a maximum. In this case, the large overdrive in voltage causes the generation of a large amount

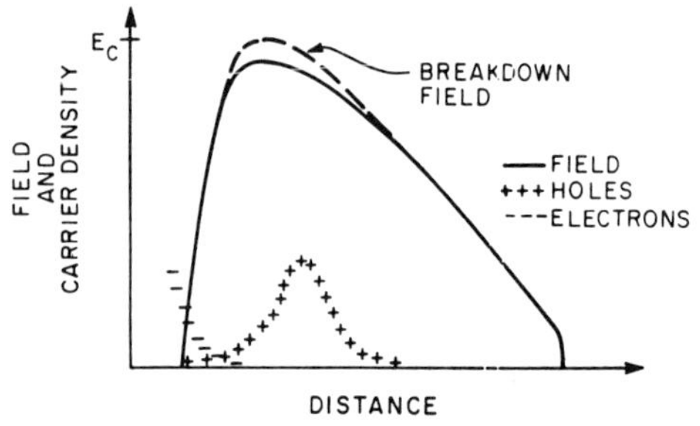

(a)

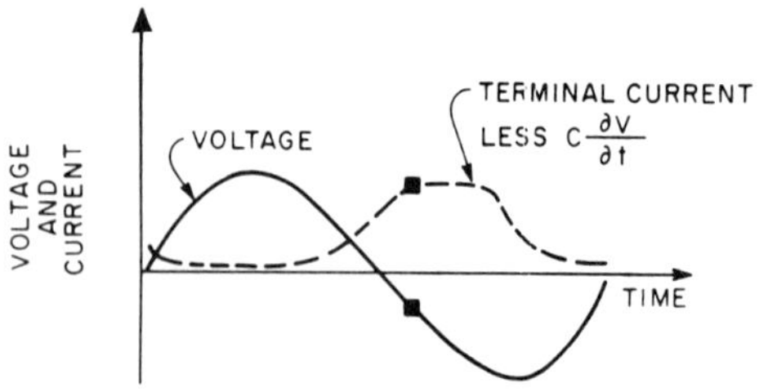

(b)

Fig. 1. (a) Large-signal IMPATT field and carrier densities during the avalanche portion of the RF cycle indicated by the squares in (b).

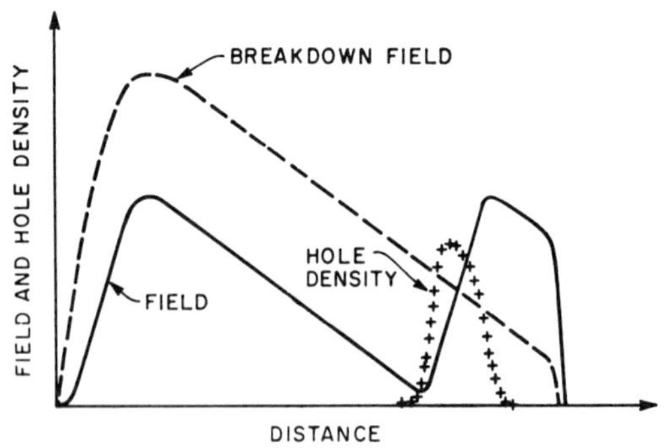

Fig. 2. Large-signal IMPATT field and carrier densities during the drift portion of the RF cycle.

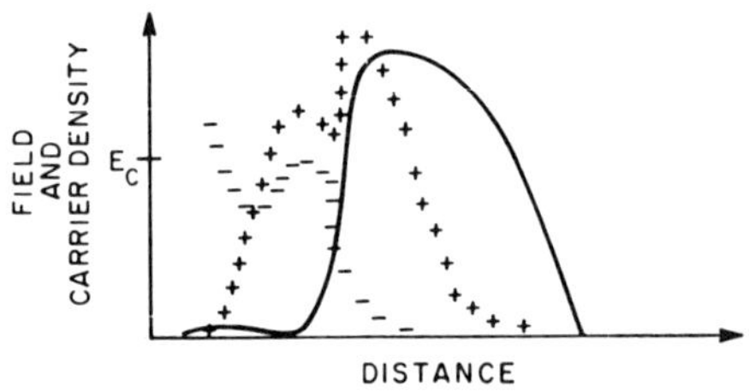

(a)

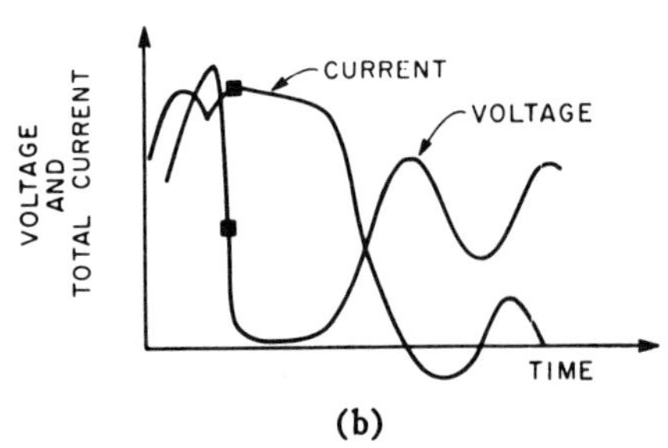

(b)

Fig. 3. TRAPATT waveforms. (a) Field and carrier density. (b) Voltage and current.

of charge which is sufficient to reduce the field to nearly zero on the left. However, the field ahead of these carriers is still above the critical field. Thus, an avalanche zone has been created which travels through the diode leaving the diode filled with a plasma of electrons and holes and nearly zero electric field. Because of the low-field condition the plasma is trapped and is extracted at the boundaries by space-charge-limited flow.

The terminal characteristics of this mode for the full cycle are shown in Fig. 3(b). Initially, there is a current which charges the depletion layer capacitance to a voltage of about twice the breakdown voltage. This initiates the traveling avalanche zone which drops the voltage to almost zero. The particle current is very high during this low-voltage state. After the plasma has been removed from the diode, the current drops to a low average value and the voltage returns to near the breakdown value. Therefore this is a very efficient mode of oscillation, which has been calculated to be capable of efficiencies of over 60 percent in silicon and 50 percent in germanium [5].

A final point to be made is that this mode can exist without generating high-efficiency oscillations. In that case, the peak voltage is slightly less, the traveling avalanche does not travel through the complete diode, and it leaves behind only a small amount of trapped charge. Computer-simulated oscillators have shown trapped-plasma states existing for oscillations with less than 1 percent efficiency.

From this description of the operation of the TRAPATT mode, at least one circuit requirement is clear. Namely, in order to obtain self-starting TRAPATT oscillation, large IMPATT-generated voltage swings must be obtained by trapping the IMPATT oscillation in a high-Q cavity.

III. Circuit Description

In this section a short description and analysis of the circuits used to generate TRAPATT oscillation are presented. The typical circuit used in much of the experimental work is shown in Fig. 4. It consists of a coaxial air line with a number of coaxial tuning sleeves and is similar to the IMPATT circuit described by Iglesias [6]. The particular example shown here is the 500-MHz oscillator which generates 5.5 watts of CW power at 43 percent efficiency. The type of circuit has also been used by Johnston [2] to obtain 3-GHz TRAPATT operation under pulse conditions. The large number of tuners in this circuit would be a handicap for some applications; however, the results of the analysis presented here have led to the construction of a very simple circuit using mainly commercial components. That circuit will be described in Section V.

Measurements on the section of line containing the four tuners to the right of plane A-A can easily be made since the line contains a connector pair at that point. Fig. 5 shows the return loss of the section of line containing the four tuners, looking from A-A toward the load. The general characteristic of the circuit, when it is adjusted for high-efficiency operation of the oscillator, is seen to be that of a low-pass filter. Most frequencies above 3 GHz are reflected by the filter formed by the four tuners. The second point to

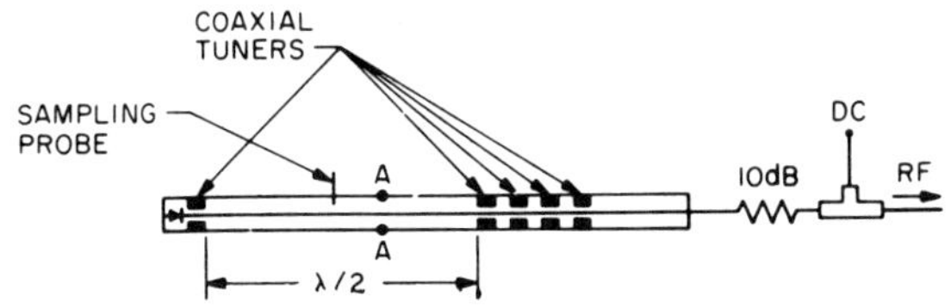

Fig. 4. TRAPATT 500-MHz oscillator circuit.

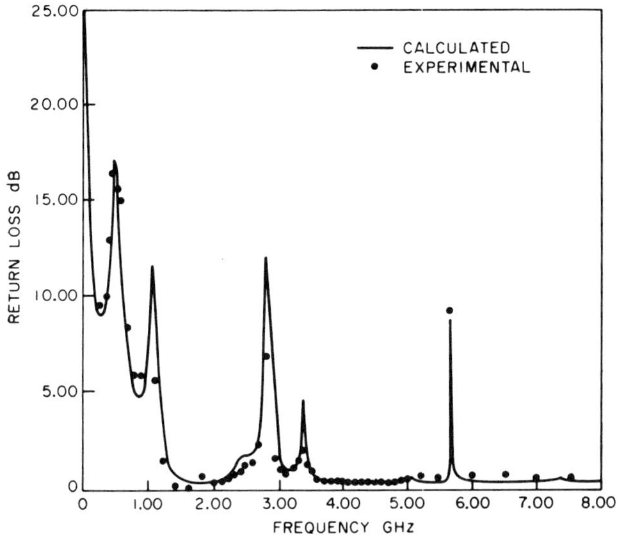

Fig. 5. Return loss versus frequency for the four-slug tuner.

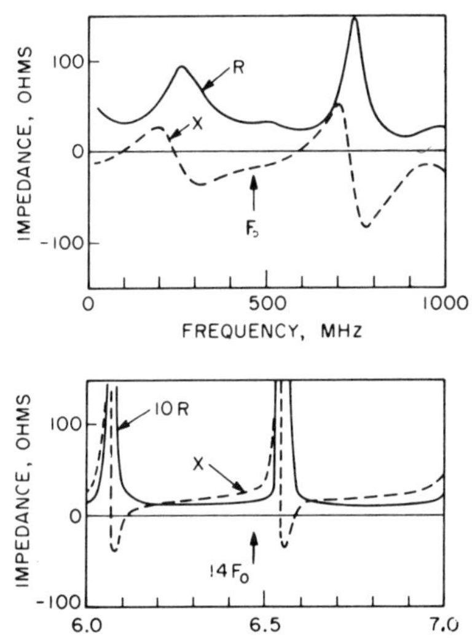

Fig. 6. Circuit impedance at the active terminals of the diode for two frequency ranges.

note is that the circuit matches well to the 50-ohm load at 500 MHz.

Shown also is a calculation of the return loss versus frequency using an equivalent circuit model for the four-slug tuner. The close agreement between these two results indicates that a reasonably accurate calculation of the circuit impedance seen by the active region of the diode can be made by extending the equivalent circuit to the active region of the diode. Fig. 6 shows the results of such a calculation from 0 to 1 GHz and from 6 to 7 GHz. At 460 MHz, the impedance has a substantial real part, about 35 ohms, and a capacitive reactance of 20 ohms. The Q is about 1. In the microwave region, a small part of which is shown, the circuit looks like a shorted transmission line, approximately 30 cm long, which provides a resonance every 500 MHz. At harmonics of the 460-MHz oscillator frequency, the circuit presents a small series resistance (about 2 ohms) to the diode, and a reactance which approximately resonates the diode. The Q at these harmonics is calculated to be about 300 and is consistent with locking range measurements.

IV. Oscillator Model

Next consider the dc operating point for the 500-MHz oscillator. The dc bias current is generally less than 1000 A/cm². On the other hand, the conduction current required [5] for the initiation of the trapped plasma in these diodes is 10 000–20 000 A/cm². Therefore, since the average value of the conduction current is the dc current, the high-current state exists for less than 0.2 ns. On the other hand, the transit time of the carriers through the diode sets a minimum recovery time from the plasma state of about 0.1 ns. Therefore, TRAPATT operation will occur only if the conduction current consists of pulses of current approximately 0.1 ns in duration occurring once during each cycle of the 500-MHz oscillation.

The conclusion suggests the *diode* model shown in Fig. 7(a). Here, it is assumed that the diode can be represented by a current pulse generator and the diode's depletion layer capacitance. Fig. 7(b) shows the sampling probe used to measure the high-frequency waveforms which are confined to the first section of air line. The transfer function for the current i to the probe voltage v is easily calculated for this circuit, and Fig. 8 shows the results of such a calculation. Here a current pulse of 10 000 A/cm², 0.1 ns wide with a period of 2.2 ns has been applied to the circuit. The second plot shows the calculated probe response. The third plot shows a sampling scope photograph of the probe voltage for an operating oscillator. The agreement for such a simple diode model is very good.

These results suggest the following simple interpretation of the operation of the 500-MHz oscillator. The large voltage pulse which passes the probe at A, reaches the diode at B, and triggers the traveling avalanche zone in the diode. This initiates a high-current low-voltage state, from which the diode recovers quickly. The drop in voltage propagates down the line reaching the probe at C. The negative going pulse next reaches the high-frequency short provided by the four-slug tuner where the -1 reflection coefficient causes a positive pulse to be reflected back toward the diode. This pulse passes the probe at A and the process repeats itself.

If this analysis is correct, certain aspects of the oscillator behavior can be predicted. For example, the frequency of operation should be inversely proportional to the length of line between the low-pass filter and the diode. Fig. 9, which gives some measurements on two oscillators, shows this to be the case. The pulsed TRAPATT measurements were

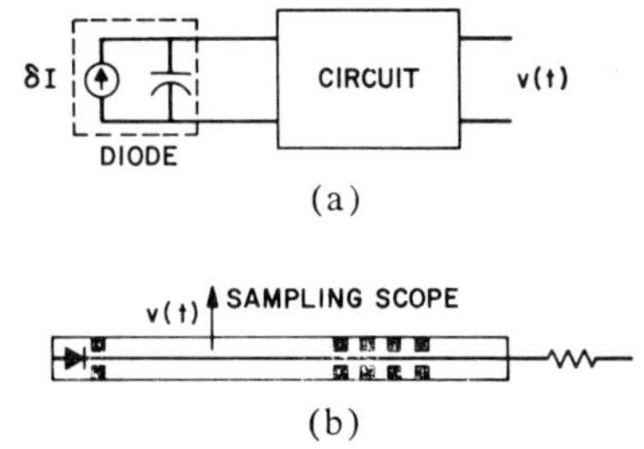

Fig. 7. (a) Simplified diode model. (b) Sampling probe position.

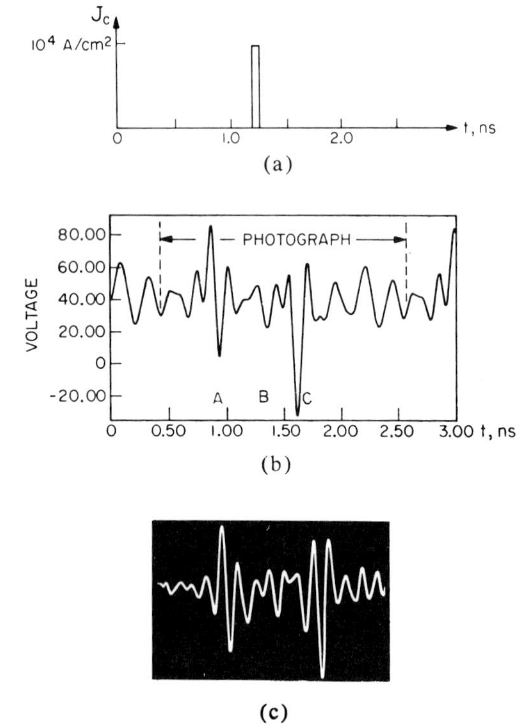

Fig. 8. Probe response. (a) Conduction current versus time. (b) Calculated probe voltage. (c) Measured probe voltage.

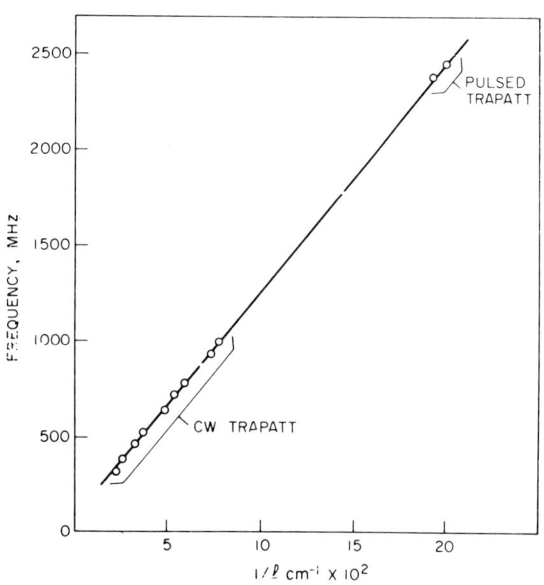

Fig. 9. Oscillator frequency versus filter position.

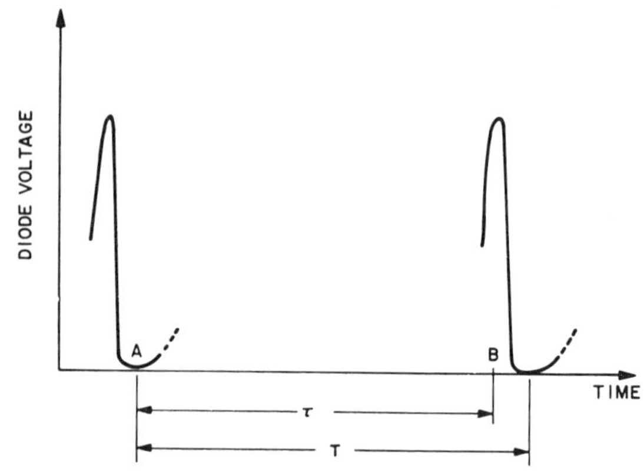

Fig. 10. A portion of the RF voltage waveform for the TRAPATT oscillator.

320

obtained by R. L. Johnston. The CW TRAPATT results were obtained with the oscillator described previously. Note the rather large tuning range for this CW circuit. Over this range the power output decreases approximately linearly with frequency. If the efficiency is assumed to be unsaturated and the dc current fixed, this linear decrease in power output can be predicted from the simple model presented here. These calculations will be discussed in Section V.

The frequency of operation for these oscillators can be very accurately predicted by measuring the distance between the reflecting plane of the filter and the diode. However, the round-trip delay for the pulse in this length of line does not directly give the period of the oscillation. Fig. 10 shows why this is so. Shown there is a very simplified sketch of the voltage across the diode which triggers the traveling avalanche. Because of the -1 reflection coefficient of the filter, it is the drop in voltage at A which produces the over voltage at B. Therefore, the period of the oscillation must include the time required for the diode voltage to drop to zero. This is typically about 0.1 ns for the germanium diodes used in the experiments described here. Since this is a small fraction of the 2-ns period of the 500-MHz oscillator, the typical value is sufficiently accurate for most calculations. For example, with a cavity length of 30.5 cm, the frequency calculated from only the transmission line delay is 491 MHz. However, if 0.1 ns is added to the total period, the frequency is 468 MHz. The measured frequency was 462 MHz.

It is also possible to construct a general *circuit* model for this mode of operation. Fig. 11 shows such a model. The inductance L is the usual diode lead inductance; its size is not critical, but it is helpful in driving the diode voltage to more than twice breakdown to initiate the TRAPATT plasma state. The next element is a capacitance. A calculation of the current required by the diode in the high-current state shows that a 50-ohm line is not able to supply the required current. Therefore, a lumped capacitance is required at this point. Next is a length of transmission line whose length determines the operating frequency. This line is terminated in a low-pass filter which appears to be a short circuit at frequencies above the cutoff. Finally, a tuner is needed to match the diode to the load.

This representation of the coaxial circuit is very simple,

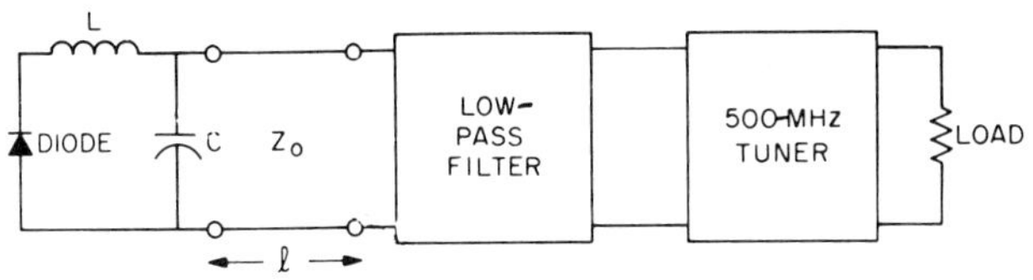

Fig. 11. Simplified TRAPATT circuit model.

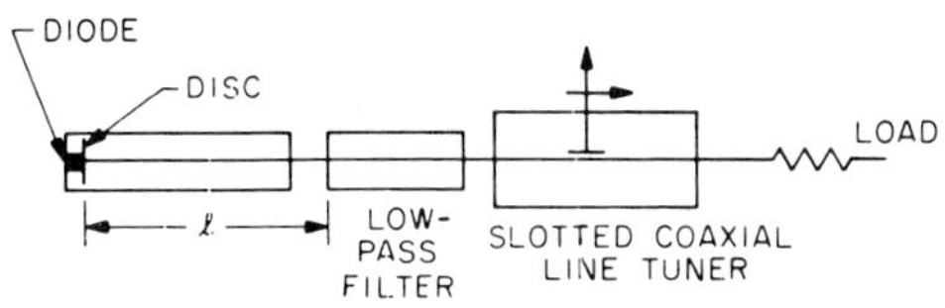

Fig. 12. Simplified TRAPATT circuit.

TABLE I

EXPERIMENTAL DATA FOR CIRCUIT SHOWN IN FIG. 12

Low-Pass Filter Cutoff	Power at 500 MHz (watts)	Power at 1000 MHz (watts)
700 MHz	2.7	—
1000 MHz	1.7	—
1500 MHz	2.5	1.4
3000 MHz	1.3	1.0
4000 MHz	0.9	0.9
4-slug filter	2.6	1.3

$I_{dc} = 200$ mA.

but Fig. 12 shows that this is all that is required. Shown is a simple circuit which is very useful for laboratory measurements. It consists of a coaxial air line with the diode mounted in one end. Contacting the diode package is a 1-pF lumped capacitor formed by a disk on the center conductor. A distance down the line is a commercial low-pass filter. Following the low-pass filter is a commercial slotted coaxial line tuner. The experimental results for such an oscillator are shown in Table I. Notice that generally speaking, as fewer harmonics of the TRAPATT oscillation are confined to the high-Q cavity, the shape of the pulse reflected by the filter is degraded, and the power output decreases.

With a slight retuning of the circuit, 1-GHz oscillations are excited. In that case the cavity is twice as long as it need be, so that two pulses are on the line at the same time. This circuit has also been used to build a high-efficiency 500-MHz CW amplifier by placing a circulator at the load terminals.

V. LOW-FREQUENCY WAVEFORMS

In the last section a simple model of the diode and the circuit for a TRAPATT oscillator were discussed. In this section the model will be expanded so that calculations on oscillator efficiency can be made in terms of some circuit parameters.

It was pointed out above that the conduction current for a 500-MHz oscillator is very large but flows for only a short period of time. Therefore, charge which flows through the diode for this short period of time must come from the

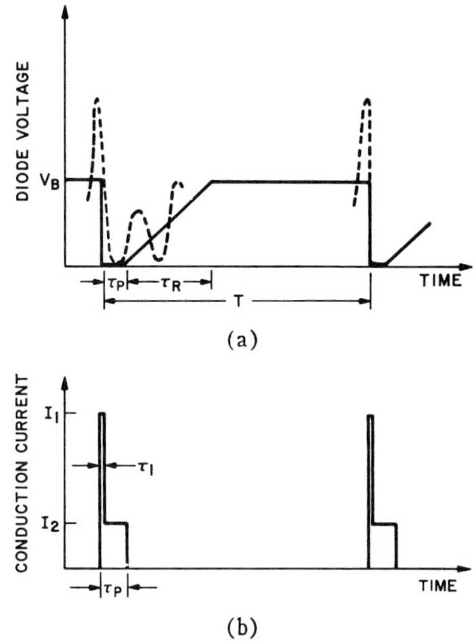

Fig. 13. TRAPATT voltage (a) and current (b) waveforms.

capacitance in the immediate vicinity of the diode.[1] The capacitance which can be discharged during that time consists of the lumped capacitance near the diode and the capacitance of a length of line which depends on the duration of the high-current state.[2] Therefore, the total capacitance can be written as

$$C_T = C_0 + \tau_P / Z_0 \tag{1}$$

where

C_0 = the lumped capacitance near the diode,
τ_p = the length of time that the high-current or trapped-plasma state exists,
Z_0 = the characteristic impedance of the coaxial air line.

After the plasma has been cleared from the diode, the diode voltage is nearly back to its breakdown value. However, the capacitance near the diode can recover only as quickly as the line can recharge it. The diode voltage will thus have a shape similar to that shown by the broken line in Fig. 13(a). The sinusoidal variations shown there are due to ringing of the circuit inductance and capacitance near the diode. To a first approximation, it will be assumed that the average value of the voltage near the diode increases linearly after the high-current state and stops when it reaches the breakdown voltage. This is a reasonable assumption since the capacitance near the diode is recharged by a transmission line. For this model the conduction current will have the shape shown in Fig. 13(b). Initially a large current flows which discharges the lumped capacitance near the diode. This is followed by a lower current which discharges a short length of the transmission line. This type of current waveform has been found in computer simulations of a 500-MHz

[1] In some oscillators this may be present in the form of an RF bypass capacitor in the dc bias connection.

[2] In this regard the oscillator operates very much like the relaxation oscillation analyzed by Ward and Udelson [7].

oscillator. If there is no lumped capacitance near the diode (in this case the diode area is reduced so that the transmission line can supply the required current density) the current waveform is approximately rectangular [5].

Neglecting the initial spike of voltage which initiates the traveling avalanche, the amplitudes of the fundamental components of voltage and current, shown by the solid lines in Fig. 13, are

$$V_{ac} = \frac{V_B}{\pi}\left\{\frac{\cos\omega(\tau_R+\tau_P)-\cos\omega\tau_P}{\omega\tau_R} + j\left(1-\frac{\sin\omega(\tau_R+\tau_P)-\sin\omega\tau_P}{\omega\tau_R}\right)\right\} \quad (2)$$

and since $\omega\tau_1 \ll 1$,

$$I_{ac} \simeq \frac{I_1\omega\tau_1}{\pi} + \frac{I_2}{\pi}\{\sin\omega\tau_P + j(\cos\omega\tau_P - 1)\}. \quad (3)$$

The power and impedance at the fundamental frequency are given by[3]

$$P = \tfrac{1}{2}\,\mathrm{Re}(I_{ac}\cdot V_{ac}) \quad (4)$$

$$Z = V_{ac}/I_{ac}. \quad (5)$$

The recovery time τ_R is estimated by assuming a constant charging current during the recovery and is given by

$$\tau_R \simeq Z_0 C_T. \quad (6)$$

322 The dc current is given by

$$I_{dc} = \frac{Q}{T} = \frac{I_1\tau_1}{T} + \frac{I_2\tau_P}{T} \quad (7)$$

where

$$\omega = 2\pi/T.$$

The charge $I_1\tau_1$ comes from the lumped capacitor C_0 which is charged to approximately $2V_B$ (by the triggering pulse). Therefore,

$$I_1\tau_1 = 2V_B C_0. \quad (8)$$

The charge $I_2\tau_p$ comes from the transmission line which is charged to about V_B. Therefore,

$$I_2\tau_P = \frac{V_B}{Z_0}\tau_P. \quad (9)$$

Equation (7) can therefore be written as

$$I_{ac} = \frac{I_{dc}}{\pi}\left\{\frac{2\pi\gamma}{1+\gamma} + \frac{1}{1+\gamma}\cdot\frac{T}{\tau_P}(\sin\omega\tau_P + j(\cos\omega\tau_P - 1))\right\} \quad (10)$$

where

[3] The actual recovery would be approximately exponential, but the linear approximation is adequate for this discussion.

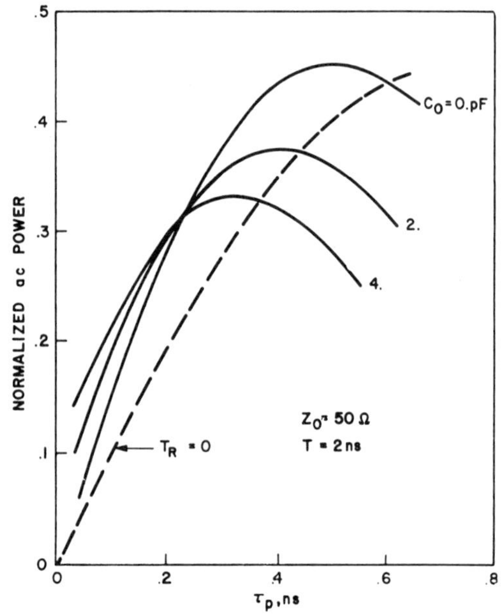

Fig. 14. Ac power versus trapped-plasma duration τ_P for several values of the lumped capacitance C_0.

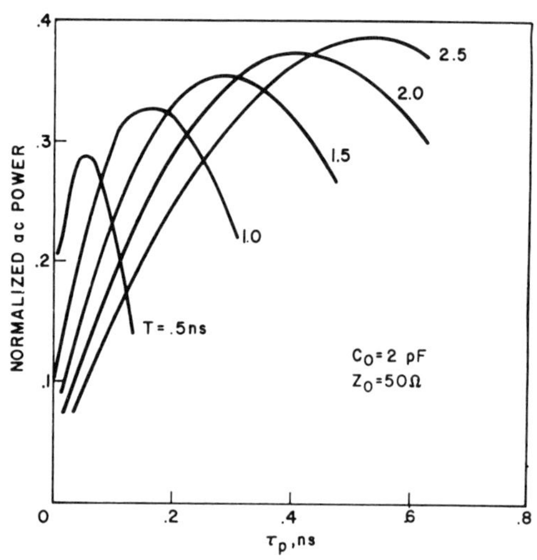

Fig. 15. Ac power versus trapped-plasma duration τ_P for several oscillator frequencies.

$$\gamma = 2C_0Z_0/\tau_P.$$

Although this is a simple model of the oscillator it does yield some interesting results. Fig. 14 shows the ac power output (normalized to $I_{dc}V_B$) for a 500-MHz oscillator with a 50-ohm line and various values of lumped capacitance near the diode. The broken curve is for the case in which the recovery time is neglected, i.e., the voltage recovers instantly after the trapped-plasma state. The figure shows that, for high-current states of short duration, the recharging part of the cycle is instrumental in producing high-efficiency oscillation. For a trapped plasma lasting about 0.1 ns, this increase in ac power is more than a factor of 2 for a lumped capacitance of 4 pF.

Fig. 15 shows the normalized ac power output as a function of τ_P with the oscillator period as a parameter. Using

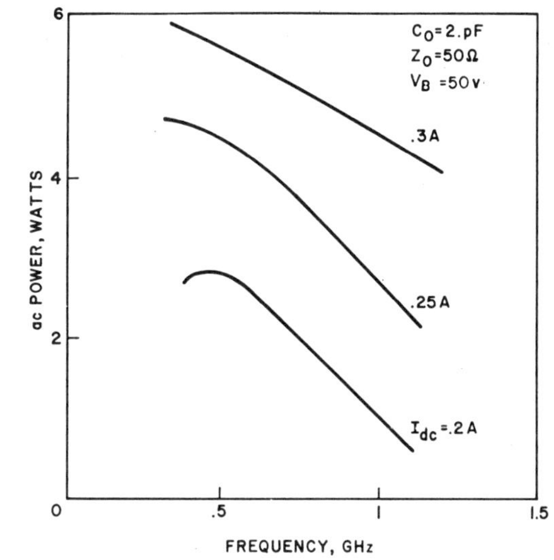

Fig. 16. Ac power versus frequency for values of dc bias current.

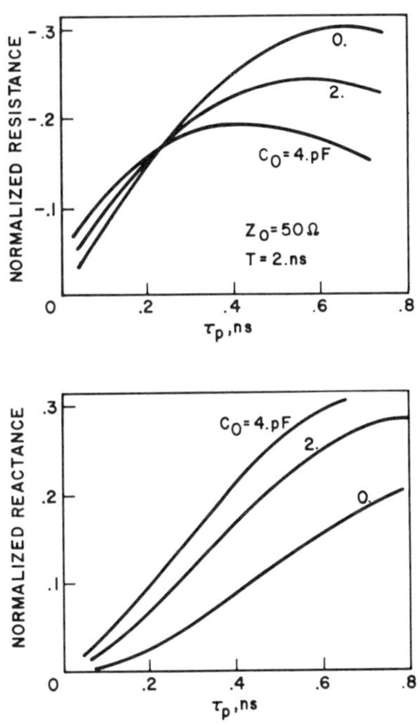

Fig. 17. Diode impedance versus trapped-plasma duration τ_P for several values of the lumped capacitance C_0.

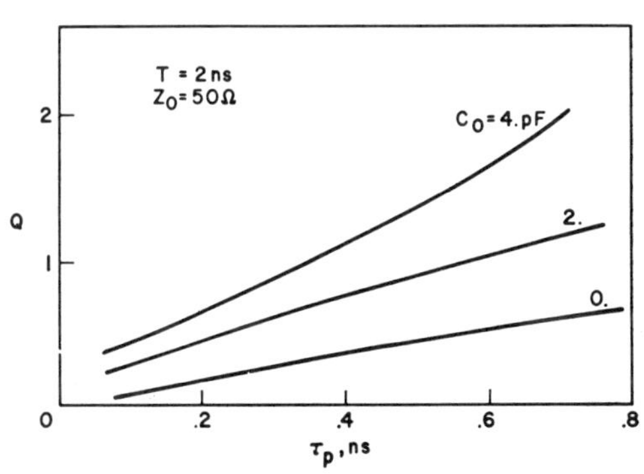

Fig. 18. Diode Q (reactance/resistance) versus trapped plasma duration τ_P.

TABLE II

	Calculation	Experimental
ac power (watts)	2.5	2.6
dc voltage (volts)	41.5	38.9
Efficiency (percent)	30	34
Diode resistance (ohms)	−33.7	−35.5
Diode reactance (ohms)	16.7	19.0

these curves and assuming a diode with a breakdown voltage of 50 volts, the power output as a function of frequency for various bias currents is obtained and is shown in Fig. 16. As mentioned previously, the decrease in power with frequency has been observed experimentally. The normalized diode impedance (normalized to V_B/I_{dc}) is shown in Fig. 17. These results will be used in a specified example given below. The diode Q (the ratio of reactance to resistance) is shown in Fig. 18. The Q is seen to be quite low for most values of τ_P and C_0.

For the 500-MHz oscillator considered in Section IV, the total lumped capacitance near the diode is about 2.8 pF. Using (7)–(9), τ_P is given by

$$\tau_P = \left(\frac{TI_{dc}}{V_B} - 2C_0\right) Z_0. \tag{11}$$

With a diode breakdown voltage of 50 volts and a dc current of 200 mA, τ_P is found to be about 0.17 ns. Using this value of τ_P, the operating point for the oscillator can be found from Figs. 14 and 17. The calculated values are compared with an experimental oscillator in Table II.

The dc voltage is calculated from the voltage waveform shown in Fig. 13. The dc power is the product of this voltage and the dc current. However, the waveforms shown in Fig. 13 do not allow any power dissipation, so that the dc power in excess of that converted to 500-MHz energy must be in the harmonics. However, in the experimental oscillator, the harmonics are reactively terminated so that they do not dissipate power. This situation can be achieved with the waveforms shown by assuming the current pulse begins just after the triggering voltage and requiring the voltage to drop with a finite slope. This is found to make the higher harmonics almost purely reactive while having little effect at the fundamental. The power dissipated in driving the avalanche through a diode is then

$$P_{av} \simeq V_B \cdot I_1 \frac{\tau_1}{T}$$

where τ_1 is the time required for the avalanche to sweep through the diode. This result follows if we assume that, for the transmission-line type of oscillator discussed here, the peak diode voltage is about $2V_B$.

Bartelink and Scharfetter [8] have shown that the velocity with which the avalanche sweeps through the diode is

$$v = \frac{I_1}{qN_a}$$

where N_a is the impurity concentration for a uniformly doped depletion region. Therefore,

$$P_{av} \simeq V_B \frac{qN_a}{T} l$$

where l is the depletion layer width. In a similar way the waveforms could be modified to include the energy required to remove carriers from the diode. For oscillators which have some lumped capacitance near the diode, the current at the end of the trapped-plasma state is small (at the time when the voltage is increasing), and therefore the energy lost during this part of the cycle may be rather small. For a more complete discussion of the energy-loss mechanisms in these oscillators the reader is referred to a paper by DeLoach and Scharfetter [9].

Equation (11) used above to calculate τ_P has some other interesting consequences. Since τ_P must be at least as large as the transit time of carriers through the diode, this puts an upper limit on the lumped capacitance C_0. On the other hand, for large-area diodes, a great deal of terminal current is required initially to drive the avalanche zone through the diode. This in turn requires a large value of C_0. Equation (11) also gives a useful relation between frequency, dc current, and breakdown voltage. In fact, even though the analysis does not appear to contain any material parameters, the equation predicts a substantial difference between germanium and silicon diodes. For example, to obtain results comparable to the germanium oscillator discussed above in a silicon diode with a breakdown voltage of 100 volts required about 0.4 ampere, or twice that used in germanium. (This is four times the power density and requires pulsed operation.) It should be noted also that the impedance of germanium and silicon diodes is the same in this case since the normalizing factor for Z is V_B/I_{dc}. Finally, since the minimum τ_P is fixed for a given diode, the dc current must increase with frequency to maintain the efficiency. If τ_P is greater than the minimum, however, the current need not be strictly inversely proportional to T.

VI. Conclusions

In Section IV the circuit conditions required to initiate the trapped-plasma state and sustain oscillation were discussed. In particular it was found that higher order harmonics must be trapped in order to sustain a voltage pulse which initiates the traveling avalanche. The frequency was found to be controlled by the transit time of this pulse between the diode and the low-pass filter. Finally a simple circuit was presented which was suggested by the analysis and is very useful for laboratory work.

In Section V the low-frequency characteristics of the waveform were examined in order to determine the circuit conditions required for high-efficiency operation. It was found that the time required to recharge the capacitance near the diode after the trapped-plasma state has a significant effect on the oscillator efficiency. In particular, it is found that high-efficiency oscillations are possible even though the trapped-plasma state lasts only a small fraction of the total period (e.g., $\frac{1}{20}$). This occurs because the circuit recovery time is much longer than the diode recovery time. It was also shown that the dc power requirements for silicon and germanium are somewhat different.

The simple model of TRAPATT oscillators which has been presented here is very useful for circuit design and performance calculations. However, very little has been said about the diode design. The analysis assumes that a trapped-plasma state can be excited and that the diode will generate sufficient power in the harmonics to make up for circuit losses, and regenerate the required voltage pulse.

Acknowledgment

The author would like to acknowledge the contributions of D. L. Scharfetter to this paper and helpful discussions with B. C. DeLoach, R. L. Johnston, and D. E. Iglesias.

References

[1] H. J. Prager, K. K. N. Chang, and S. Weisbrod, "High power, high efficiency silicon avalanche diodes at ultra high frequencies," *Proc. IEEE* (Letters), vol. 55, pp. 586–587, April 1967.

[2] R. L. Johnston, D. L. Scharfetter, and D. J. Bartelink, "High-efficiency oscillations in Ge avalanche diodes below the transit-time frequency," *Proc. IEEE* (Letters), vol. 56, pp. 1611–1613, September 1968.

[3] D. L. Scharfetter, D. J. Bartelink, H. K. Gummel, and R. L. Johnston, "Computer simulation of low frequency high efficiency oscillation in germanium," presented at the IEEE Solid State Device Research Conference, June 17–19, 1968.

[4] R. L. Rulison, G. Gibbons, and J. G. Josenhans, "Improved performance of IMPATT diodes fabricated from germanium," *Proc. IEEE* (Letters), vol. 55, pp. 223–224, February 1967.

[5] D. L. Scharfetter, "Power-frequency characteristics of the TRAPATT diode mode of high efficiency power generation in germanium and silicon avalanche diodes," presented at the IEEE Device Research Conference, Rochester, N. Y., June 23–26, 1969.

[6] D. E. Iglesias, "Circuit for testing in high-efficiency IMPATT diodes," *Proc. IEEE* (Letters), vol. 55, pp. 2065–2066, November 1967.

[7] A. L. Ward and B. J. Udelson, "Computer calculation of avalanche-induced relaxation oscillations in silicon diodes," *IEEE Trans. Electron Devices*, vol. ED-15, pp. 847–851, November 1968.

[8] D. J. Bartelink and D. L. Scharfetter, "Avalanche shock fronts in p-n junctions," *Appl. Phys. Letters*, May 15, 1969.

[9] B. C. De Loach, Jr., and D. L. Scharfetter, "Device physics of TRAPATT oscillators," to be published.

Reprinted from IEEE TRANSACTIONS
ON *MICROWAVE THEORY AND TECHNIQUES*
Volume MTT-17, Number 12, December, 1969
pp. 1060-1067

Computer Experiments on TRAPATT Diodes

WILLIAM J. EVANS, MEMBER, IEEE

Abstract—Simplified analytical treatments of the TRAPATT mode of operation in avalanche diodes do not consider in sufficient detail the initial starting conditions for this mode. This paper examines this question in greater detail by analyzing a number of diode structures using a large-scale computer program. The results of these computer experiments have indicated that the requirements for efficient TRAPATT operation and the requirements for IMPATT operation diverge with increasing frequency. Since the IMPATT oscillation is required to start the TRAPATT mode, self-starting TRAPATT oscillators are increasingly difficult to fabricate as the operating frequency is increased. A novel diode structure is proposed which eliminates this problem and is capable of high-frequency CW TRAPATT operation.

Manuscript received June 29, 1970; revised August 28, 1970.
The author is with Bell Telephone Laboratories, Inc., Murray Hill, N. J. 07974.

I. Introduction

In 1968 Scharfetter, Bartelink, Gummel, and Johnston [1] proposed an explanation for the high-efficiency mode of operation of avalanche-diode oscillators that had been observed experimentally [2], [3]. The explanation which they proposed for this high-efficiency or TRAPATT mode was suggested by computer experiments performed with a complex program capable of modeling semiconductor devices rather accurately. Subsequent modeling of a complete diode-circuit system [4] has shown that the original proposal presents an accurate description of the physical phenomena which give rise to the high-efficiency mode of operation. The computer experiments have also suggested where a number of simplifying assumptions can be made, and, as a result, simplified analytical treatments [5]–[7] of the TRAPATT oscillator are available which provide additional understanding of the phenomena involved. Unfortunately, the simplified models do not consider the starting conditions necessary to create the trapped-plasma state. In particular, the limitations imposed by realistic circuit conditions are an important part of the problem. A simple example of such a limitation in the simplified models is that the typical waveforms used in the analyses do not exhibit negative diode resistance at all harmonic frequencies. These solutions therefore could not be sustained in a real circuit. Usually, it is assumed that the phase of the harmonics can be adjusted slightly so that the solution is consistent with the circuit requirements without greatly changing the conclusions of the analysis. In this paper we will identify a more serious limitation and show what effect it has on TRAPATT-oscillator performance.

In Section II of this paper the theory of operation of a TRAPATT oscillator is reviewed and particular emphasis is placed on the circuit aspects of TRAPATT operation. In Section III the procedures and assumptions for a number of computer experiments similar to those originally done by Scharfetter *et al.* [1] are described. In Section IV the results of these experiments are given for a number of typical diodes and for several proposed diode structures. These results are primarily related to the requirements for triggering TRAPATT oscillations; however, they are also useful guides for determining such parameters as oscillator efficiency and frequency. Both germanium and silicon diodes have been considered as well as variations in diode-impurity profile and depletion-layer width. Section V discusses the results in more detail, comparing them to the predictions of the simplified models. The problems of initiating TRAPATT operation are summarized and a new diode structure is proposed for improved TRAPATT operation.

II. Review of TRAPATT Theory

In this section a brief description of TRAPATT operation is given with particular emphasis on the interaction between the diode and the circuit. For a more complete analysis of the TRAPATT oscillator, the reader is referred to the literature [5]–[8]. The TRAPATT mode occurs in avalanche diodes primarily because it is possible in some circuits to rapidly drive the diode to a voltage well in excess of that required for avalanche breakdown. It is the inherent delay in the avalanche process which permits this to occur. However, at some point in time after the application of the excess voltage, the multiplication of carriers begins to occur very rapidly. Because the field is now well above the breakdown value (by 50 to 100 percent), the multiplication greatly exceeds that required to sustain the terminal current of the diode. The excess carriers which

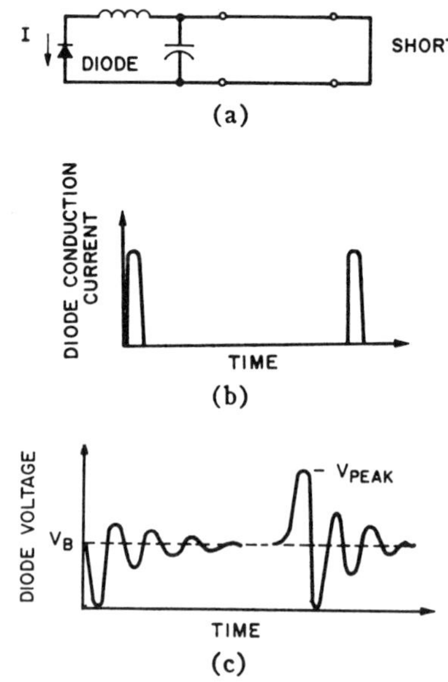

Fig. 1. Simplified circuit analysis. (a) Simplified circuit model. (b) Diode conduction current versus time. (c) Diode voltage versus time.

are generated are therefore sufficient to cause the field in the diode to collapse, leaving the diode filled with a plasma and in a near-zero-voltage state. If the terminal current is maintained, the plasma is removed from the diode and the diode voltage recovers to its breakdown value. The time required to remove the plasma depends on the density of the plasma and the magnitude of the recovery current. However, a recovery time of several times the transit time of carriers moving at the saturated drift velocity is typical.

A simple circuit which can be used for the TRAPATT mode consists of a transmission line with the diode mounted at one end and a low-pass filter at the other end. The low-pass filter is placed approximately one-half wavelength away from the diode and presents a short circuit to all the harmonics of the TRAPATT oscillation frequency. On the load side of the filter some tuning is required to match the diode impedance to the load impedance. Fig. 1(a) shows a simple circuit model of the circuit just described. Since we are primarily interested in the high-frequency components of the TRAPATT waveform, which are responsible for triggering the diode, the low-pass filter has been replaced by a short circuit. The inductance and shunt capacitance are the series lead inductance and the package capacitance of the diode. The capacitance may also include any additional capacitance due to IMPATT tuners or lumped capacitance near the diode.

If the diode current is divided into two parts, the displacement current and the conduction current, it is then possible to calculate the circuit response to an assumed conduction-current waveform. Fig. 1(b) shows a simplified conduction-current waveform for a low-frequency TRAPATT oscillator. The high conduction-current state occurs during the formation of the TRAPATT-plasma state. Fig. 1(c) shows the calculated voltage response to such a current. Note that the diode voltage rapidly drops to a value near zero volts. This negative-going voltage pulse propagates down the line to the short circuit provided by the low-pass filter. The −1 reflection coefficient causes a positive voltage pulse to be reflected back toward the diode. The voltage pulse which reaches the diode therefore has an amplitude of about twice the breakdown voltage of the diode. The energy in this return pulse will overdrive the diode resulting in the field collapse in the diode. Thus a good circuit must be able to transfer this pulse of energy to the diode with no degradation. Thus, for example, low circuit Q or excessive lead inductance would be detrimental to high-efficiency operation. This circuit limitation requires that we examine more carefully the energy required to trigger TRAPATT operation in various diode structures.

III. Computer Experiments

The computer program which has been used in this study is described in a paper by Scharfetter and Gummel [9]. Basically, the program performs a simultaneous time-dependent solution of Poisson's equation, the current equations, and the continuity equations for a one-dimensional structure. Included in the program are accurate expressions for electron and hole mobilities, ionization rates, and thermal generation and recombination. The solution is obtained by subdividing the diode into approximately 100 mesh points and applying specified terminal conditions. The program then obtains the solution for the field and carrier densities at successive time steps.

In this study the terminal current shown in Fig. 2(a) has been applied to the diode. The initial conditions set up in the diode before the application of the step in terminal current are that the diode be reverse-biased to breakdown and carry a dc current of 1 A/cm^2. The step in current has an amplitude of 10^3 to 10^5 A/cm^2.

Fig. 2(b) shows the diode response to the step in current. Initially the diode voltage rises approximately linearly until a time t_0, at which time an avalanche shock front is formed and the field in the diode begins to collapse. Shortly after t_0 the diode is filled with a plasma and the diode voltage is nearly zero. Since the terminal current is maintained, recovery begins immediately and at some later time t the diode voltage has returned to the breakdown value. At this time the current is assumed to be turned off and the diode voltage remains constant at V_B for another interval equal to t, at which time the current is turned on again. This is a very idealistic waveform since, for example, at some of the harmonics of the fundamental period the diode exhibits a positive resistance; i.e., the diode is absorbing energy at those frequencies. However, if the efficiency at the fundamental frequency is calculated, it gives us a very useful figure of merit for the diode, which can be compared from structure to structure. It also is very helpful in that it defines a maximum peak voltage, which is

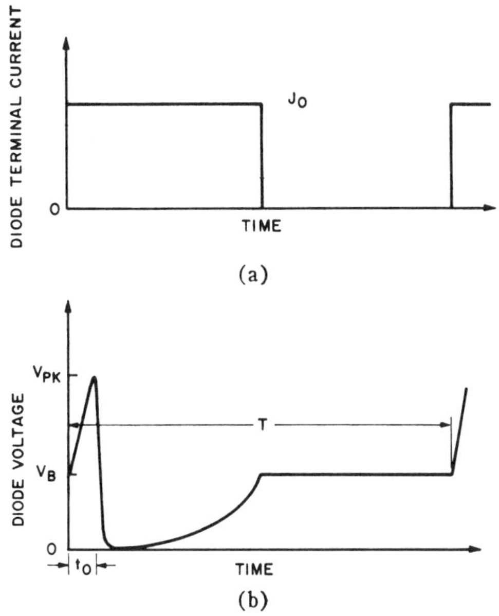

Fig. 2. Diode waveforms for computer experiments. (a) Applied current. (b) Voltage response.

required by each structure for efficient TRAPATT operation.

The results presented in the next section have been obtained in the following way. A solution for the field and carrier densities is obtained for a given diode structure at a reverse-bias current of 1 A/cm^2. A step in terminal current is applied to the diode with the initial conditions determined by the 1 A/cm^2 solution. The solution is allowed to time evolve until the voltage returns to breakdown in the recovery part of the cycle. This point is defined as one-half the oscillator period, and the Fourier coefficients for the dc and fundamental components of the voltage waveform are calculated numerically. The instantaneous I–V product is integrated over one period to find power input to the diode. Thus, for each drive current, the efficiency, frequency, impedance, and peak diode voltage are obtained. Before presenting the results of these calculations, it will be helpful to define several parameters used in the analysis. The first is the punch-through factor F. Fig. 3 (from [7]) shows that this factor is obtained by extrapolating the field versus distance for a punched-through diode of width W (biased to breakdown) to the point of zero field. The ratio of the extrapolated width to the diode width is the punch-through factor F. A second parameter is the overvoltage V_o. This is defined as the maximum diode voltage occurring during the TRAPATT cycle normalized to the diode breakdown voltage. Finally, it is useful to define a current density $J_c = qNv_s$, where N is the depletion-layer impurity concentration and v_s is the saturated drift velocity for electrons or holes. This value of current is, according to the simplified analysis, the minimum current that can form the avalanche shock front that travels through the diode and creates the trapped-plasma state. A number of details which have a small effect on the results and no

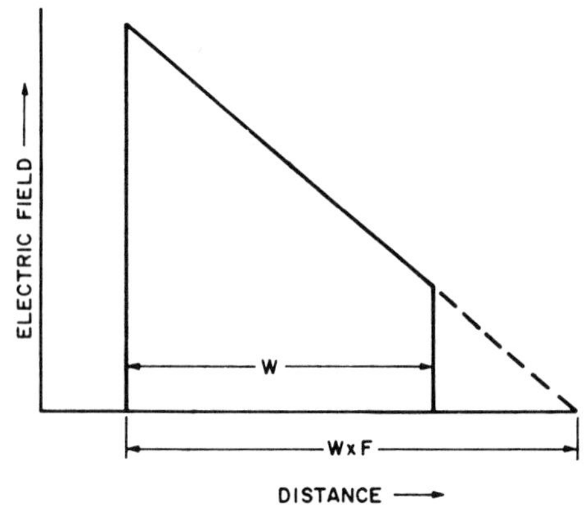

Fig. 3. Electric field versus distance for a typical abrupt-junction diode defining the punch-through factor F.

effect on the conclusions could be mentioned. For example, the doping profiles which have been considered included a realistic (i.e., typical of experimentally measured values) amount of out diffusion from the substrate and a Gaussian tail on the junction-forming diffusion. Ionization rates appropriate for 160°C in silicon and 25°C in germanium were used. A carrier lifetime of 100 ns was used for silicon and 10 ns for germanium. Because the program requires a specific value for each of the many parameters, it is difficult to calculate the overvoltage for optimum efficiency. Therefore, in the results to follow, an attempt has been made to give comparative results for different diode structures. The conclusions drawn from such comparisons are not highly sensitive to the details of the analysis.

IV. Results

Fig. 4(a) shows a plot of the efficiency versus the dc-current density (normalized to J_C) for two germanium diodes. The field profile for these diodes is shown in the inset in the figure. Diode 1 is a typical n^+-p-p^+ 6-GHz germanium IMPATT with a doping of about 4×10^{15} cm^{-3} in the p region. The punch-through factor for this diode is one. Diode 2 is also a germanium diode, but it has been scaled up in frequency by a factor of four. It also has a punch-through factor of one, and a doping density of 2×10^{16} cm^{-1}. (Note that the doping density is more than a factor of four larger than diode 1. This departure from a simple scaling factor of four occurs because the peak field for avalanche breakdown increases as the doping density is increased.) Note that the dc current given in the figure is one-half the drive current J_0 shown in Fig. 2 because the current waveform is assumed to be a square wave. It is found that the TRAPATT frequency does not depend strongly on the dc current since a large value of J_0 creates a larger plasma but also extracts the plasma at a high rate. Therefore in each figure a nominal TRAPATT frequency is given.

An interesting point should be mentioned in regard to Fig. 4(a). Notice that the efficiency of the oscillator does not decrease rapidly when the current falls below 1, i.e., below the critical current J_c required to launch an

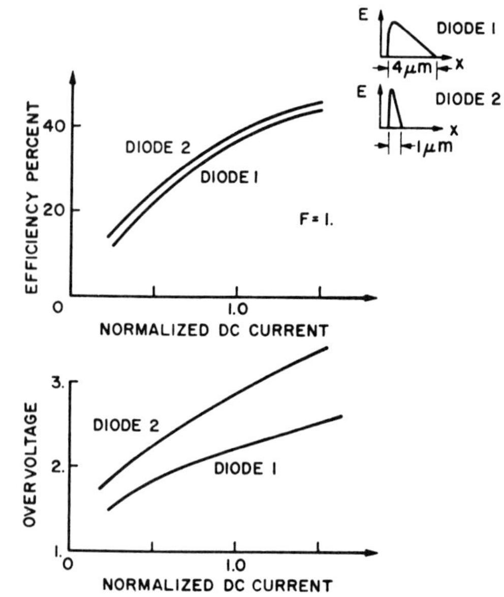

Fig. 4. Germanium TRAPATT efficiency and overvoltage versus dc current (normalized to qNv_s). Diode 1 has breakdown voltage of 50 volts and maximum TRAPATT frequency of ~1.8 GHz. Diode 2 has breakdown voltage of 19 volts and maximum TRAPATT frequency of ~4.5 GHz.

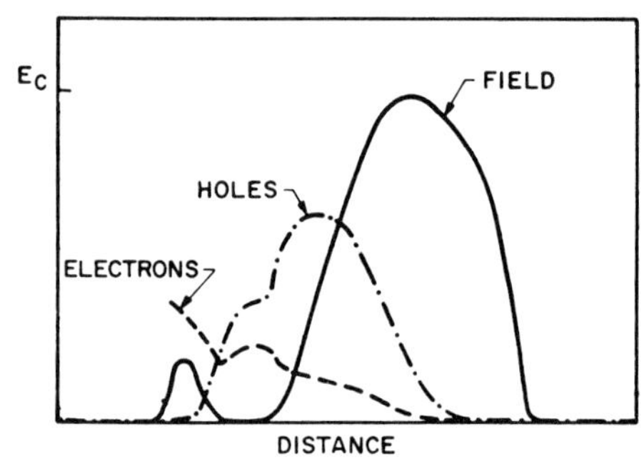

Fig. 5. Germanium diode 1 of Fig. 4 during plasma-formation part of TRAPATT cycle. Shown are electric field and electron and hole densities versus distance. Terminal current is 0.7 J_c.

avalanche shock front. Fig. 5 shows what happens when the current is below J_c for diode 1. This plot shows the electron and hole densities and electric field just after the peak in diode voltage. It can be seen that the field on the left has collapsed as it would if the current were greater than J_c. However, in this case no traveling avalanche zone is created and avalanche multiplication has stopped. Nevertheless, the initial overmultiplication which occurred at the left (in the avalanche zone) has created a large number of excess carriers. The holes in the remaining high-field region will drift to the right at their saturated drift velocity. However their density is so great that the change in field through the bunch (by Poisson's equation) is sufficient to reduce the field to nearly zero for those holes to the left of the peak density. Thus, as these holes drift, the trailing edge of the bunch falls behind and the remaining field in the diode is eliminated. The result of this is that the diode reaches a zero-voltage state and is filled with a trapped plasma even though the terminal current is well below

J_c (by a factor of two or more). This is an interesting conclusion because it says that even at relatively low efficiencies (for TRAPATT, that is) and low currents the circuit requirements are not greatly different from those of "classical" TRAPATT operation. We will also find this effect to be important in very high-frequency structures in which the avalanche zone will not propagate at drive currents even in excess of J_c.

Returning to Fig. 4(b), the overvoltage is plotted as a function of the dc current. It can be seen that there is a significant difference between the two diodes with regard to this parameter. For example, at an efficiency of 35 percent diode 1 requires an overvoltage of about 2.15 times its breakdown voltage, which is consistent with the comments of Section II. However, diode 2 requires an overvoltage of 2.75 times its breakdown voltage. In terms of the energy required to raise these diodes from V_B to V_{PK}, diode 2 requires about twice the normalized energy[1] of diode 1. The expected difference in performance for these two diodes is likely to be even greater in practice since the IMPATT frequency for diode 2 is about 25 GHz and the obtainable circuit Q will be lower. Thus, the higher frequency circuit will develop less overvoltage.

The overvoltage depends on a number of parameters. However, it is possible to obtain a good analytical solution for the voltage as a function of the dc-current density (see Appendix). The overvoltage is found to be approximately proportional to the square root of the dc current and, as indicated by the comparison between diodes 1 and 2, it depends on the width of the diode. These results are discussed in more detail in the next section. Although the analytical model can be used to obtain the overvoltage, it does not permit one to calculate the corresponding efficiency. Thus, it is difficult in some cases to make a good comparison between diodes when only information on overvoltage is available. Therefore, a number of other diode structures will be compared using the computer-generated waveforms.

Fig. 6 shows the results for three silicon diodes similar in geometry to the germanium diodes already considered. The diodes have a punch-through factor of 1.0 and depletion-layer widths of 0.6, 1.0, and 4.0 μm at breakdown. The same increase in overvoltage with decreasing diode width observed in germanium is seen to occur in the silicon diodes, although it is not as severe. The 0.6-μm diode is a particularly interesting case since this diode has the same structure as the 50-GHz millimeter-wave IMPATT reported by Edwards *et al.* [10]. Since these are excellent 50-GHz oscillators (10-percent efficiency), one might expect that they would be a likely candidate for TRAPATT operation at a lower frequency. As Fig. 6 shows, however, they would require an excessive overvoltage to sustain TRAPATT oscillation. In fact, these diodes have been tried unsuccessfully as TRAPATT oscillators in several circuits.

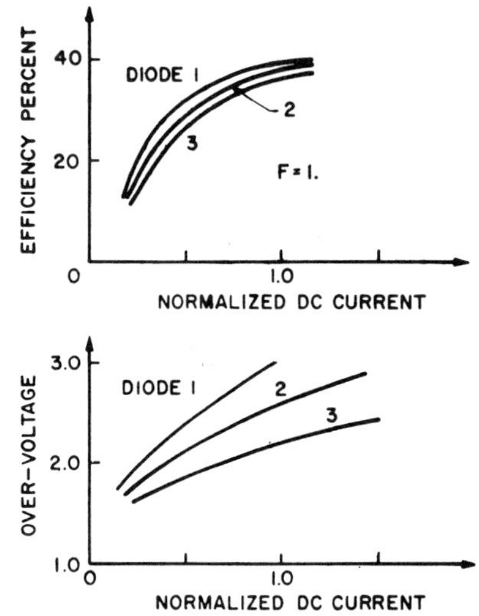

Fig. 6. Silicon TRAPATT efficiency and overvoltage versus dc current (normalized to qNv_s) for diodes with punch-through factor $F=1$, and different depletion widths. For diode 1 width $W=0.6$ μm, breakdown voltage $V_B=20$ volts, and maximum TRAPATT frequency $f_0 \simeq 12$ GHz. For diode 2, $W=1.0$ μm, $V_B=30$ volts, $f_0 \simeq 8$ GHz; for diode 3, $W=4.0$ μm, $V_B=83$ volts, $f_0 \simeq 2.5$ GHz.

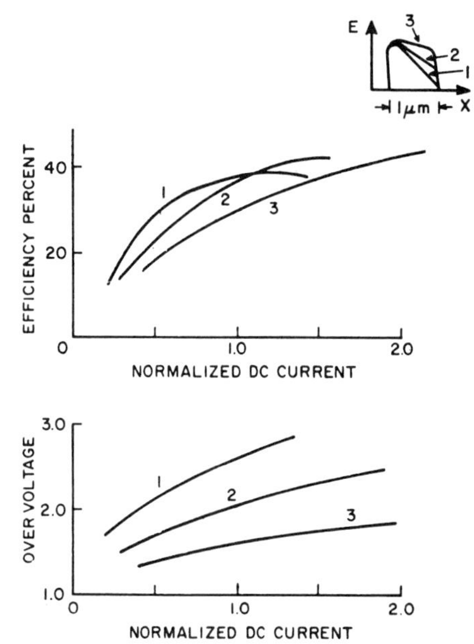

Fig. 7. Silicon TRAPATT efficiency and overvoltage versus dc current (normalized to qNv_s) for diodes with depletion layer of 1 μm and different punch-through factors. For diode 1, $F=1$, $V_B=30$ volts, and $f_0 \simeq 8$ GHz; for diode 2, $F=2$, $V_B=38$ volts, and $f_0 \simeq 6$ GHz; for diode 3, $F=5$, $V_B=45$ volts, and $f_0 \simeq 5$ GHz.

Fig. 7 compares silicon diodes of the same width (1.0 μm) but having different values of F. It is immediately apparent that increasing the punch-through factor results in a significant reduction in the required overvoltage. At 35-percent efficiency, for example, the required overvoltages are 2.4, 2.0, and 1.7 for punch-through factors of 1, 2, and 5, respectively. A further advantage of diodes with high punch-through factors is that a given efficiency can be achieved with less current (recall that the plots are shown for current normalized

[1] Energy normalized to stored energy at breakdown $\frac{1}{2}CV_B^2$.

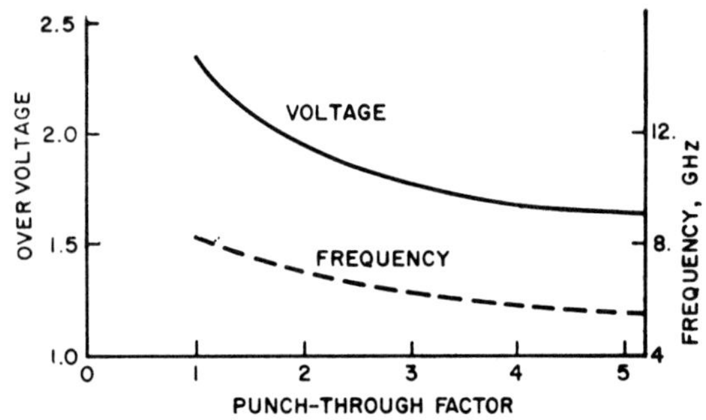

Fig. 8. Overvoltage and TRAPATT frequency versus punch-through factor for diodes of Fig. 7 at 35-percent efficiency.

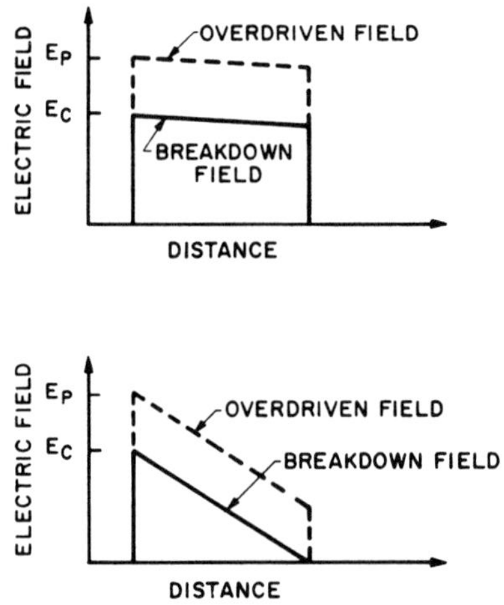

Fig. 9. Comparison of field profiles for punch-through factors of five and one.

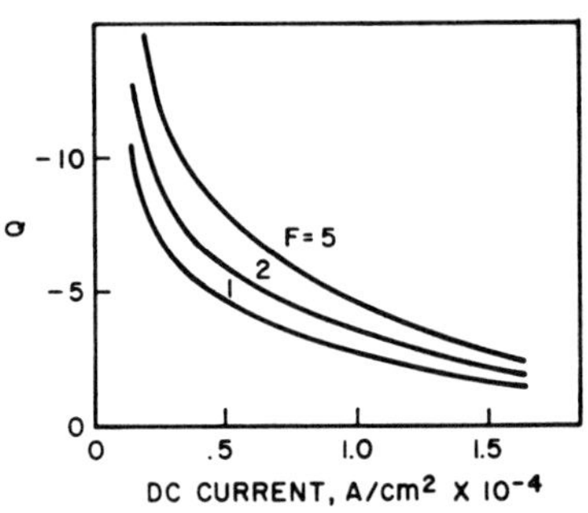

Fig. 10. Q versus dc current for the three diodes studied in Fig. 7. Data are calculated at a frequency for each diode where negative conductance is a maximum.

to qNv_s, and N decreases approximately as $1/F$). For example, at 35-percent efficiency the dc-current densities required are 18 000, 10 000, and 6000 A/cm^2 for $F=1$, 2, and 5, respectively. Finally, the ultimate efficiency with larger values of F is greater. It can be seen from Fig. 7 that for $F=1$ the efficiency saturates at $I_{dc}=1$. This happens because of the premature avalanche breakdown which occurs in the recovery part of the cycle and is discussed elsewhere [6].

There are, however, some disadvantages with larger values of F. The frequency of operation is decreased somewhat with increasing F. Also there is increased difficulty in achieving CW operation. This is discussed in more detail in the next section. The results obtained from Fig. 7 are plotted as a function of F for 35-percent efficiency in Fig. 8.

330

V. Discussion

Some of the results obtained in the preceding section can be understood from simple arguments. The decrease in overvoltage with increasing punch-through factor is probably the easiest to understand. Fig. 9 shows field profiles for a nearly p-i-n ($F=5$) and an abrupt-junction diode ($F=1$). The field E_c is the peak field in the diode required for dc breakdown. If the avalanche region is confined to a small fraction of the diode, then it is approximately the peak field E_p in the diode which determines when the avalanche multiplication will be large enough to start the field collapse. From Fig. 9 it is clear that the same increment of terminal voltage is required for the field to reach E_p (which is typically 1.5 E_c) in the p-i-n diode as in the abrupt-junction case. Thus, the increase relative to the breakdown voltage is much less for larger values of F. This is only approximately true, of course, because the avalanche region widens as the diode approaches the p-i-n case. The increasing overvoltage with decreasing diode width occurs because in a narrow diode the field required for breakdown is higher and the rate of change of the ionization-rate curves with field is smaller. Therefore it takes more excess field to create the same overmultiplication. Thus, the overvoltage is larger (see Appendix).

The results of computer analysis show that the overvoltage problem can be avoided by choosing diodes with a larger punch-through factor and accepting some trade-off in frequency (Fig. 8). However, this conclusion is valid only for pulsed operation. If CW operation is desired, the loss in negative conductance at the transit-time frequency cannot be made up by increased current density. Fig. 10 shows the diode Q (ratio of diode susceptance to conductance) as a function of dc-current density for the three diodes studied in Fig. 7. For example, the diode with a punch-through factor of one has a Q of about -1.5 at 15 000 A/cm^2. To compare this diode fairly for CW operation with a diode having a value of $F=5$, it should be compared at a current density of 10 000 A/cm^2 since the breakdown voltage is 50 percent higher and it is the power-dissipation limitation which is important. At 10 000 A/cm^2 the $F=5$ diode has a Q of -4.5. Thus, approximately a factor of three has been lost. This is very detrimental to large-signal IMPATT operation [9]. Thus, we conclude that high-frequency CW TRAPATT operation is very difficult to achieve with conventional diode and circuit structures. In summary, the problem is simply that the requirements for optimum TRAPATT operation diverge from those for efficient IMPATT operation as frequency is increased.

A solution to this overvoltage problem can be found in modification of the diode structure. Shown in Fig. 11 is a structure which was proposed by Seidel [11].

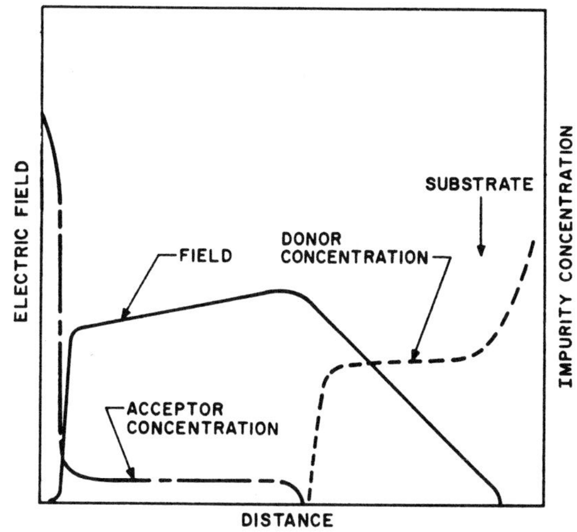

Fig. 11. Asymmetric field profile for high-frequency CW TRAPATT operation.

This structure has been fabricated by ion-implantation techniques and has been successfully operated as a CW TRAPATT oscillator in the 4- to 6-GHz range [12]. In this structure a double-sided design has been employed. The right-hand side has been designed as an efficient IMPATT oscillator whose function is to provide the large IMPATT oscillations which are required to start the TRAPATT mode. The left-hand side is optimized for TRAPATT operation in the 4- to 6-GHz range. This side of the diode has a punch-through factor of about five. Calculation of the diode Q at the frequency of maximum negative conductance (which according to [9] is a good indication of large-signal performance) indicates that this diode is as good at its transit-time frequency as a conventional single-sided IMPATT.

Recalling the discussion in Section IV, it was shown that a zero-voltage state could be reached without the formation of an avalanche shock front. This conclusion is particularly important to the operation of the double-sided diode. Because of the relatively high doping level in the IMPATT side of the diode, the current density required for shock-front formation on the right-hand side is somewhat above the CW operating-current density. However, the total diode field will still collapse by the mechanism discussed in Section IV.

The details of the particular diodes which were fabricated in this manner are discussed in [12]. However, it is interesting to point out some of the tradeoffs in such a design. Using a simplified model of the TRAPATT oscillator, Scharfetter has shown that 50-GHz TRAPATT operation can be obtained assuming the availability of a good 150-GHz IMPATT oscillator [7]. These conclusions are based on an analysis of diodes having a punch-through factor of about 1.1. If the overvoltage problem requires that a punch-through factor of two or more be used, the TRAPATT frequency will drop to around 30 GHz. Now if CW operation is required and the double-sided approach is taken, the frequency will decrease by an additional factor. This occurs because the frequency of the TRAPATT oscillator is primarily determined by the recovery time, and this in turn is a function of the total diode width. However, the tradeoff in the double-sided approach is small because the additional time required to extract the plasma is increased only by a transit time through one side of the diode. In a diode operating in a 5 to 1 mode, this amounts to about a 10-percent reduction in frequency. From this discussion we conclude that efficient (>30 percent) CW operation of TRAPATT oscillators above 25 GHz is unlikely.

VI. Conclusions

In this paper, the mechanism by which the high-efficiency TRAPATT mode of operation is initiated has been studied. It has been shown that the overvoltage required to trigger the TRAPATT mode prohibits some diode structures from operating in the TRAPATT mode. In particular, it has been found that higher frequency abrupt-junction diodes do not make good TRAPATT oscillators. The overvoltage required for TRAPATT increases with decreasing diode width and decreases with increasing punch-through factor. Thus, whereas a 5- to 10-GHz germanium IMPATT may work well as a 1-GHz CW TRAPATT oscillator, a good 50-GHz IMPATT does not operate as an efficient TRAPATT oscillator. The large-signal IMPATT oscillations required for starting the TRAPATT mode and efficient TRAPATT operation are increasingly difficult to obtain in the same diode at higher frequencies.

Since the overvoltage problem is less severe for diodes with a larger punch-through factor, higher frequency TRAPATT operation can be achieved by using diodes which are nearly p-i-n type. However, a tradeoff must be made in operating frequency. A diode with a punch-through factor of five may operate at a frequency of 6 to 10 times below its transit-time frequency, whereas a diode with a punch-through factor of one can operate in a three-to-one mode. A further tradeoff occurs in this type of structure since a favorable negative Q at the IMPATT frequency can only be obtained at current densities requiring pulsed bias [13].

Finally, the analysis has suggested that CW TRAPATT operation can be improved by using a double-sided diode design. The new structure is asymmetric with one side optimized for the IMPATT oscillations and the other side optimized for efficient TRAPATT operation. Even with this improvement, however, it appears that efficient self-starting CW TRAPATT is at best limited to about 25 GHz.

Appendix

An Analysis of the Avalanche Time Delay in TRAPATT Oscillators

In Fig. 2 the voltage response to a step in current was calculated by computer simulation for a number of diode structures. The initial part of the TRAPATT cycle consists of a nearly linear increase in the diode

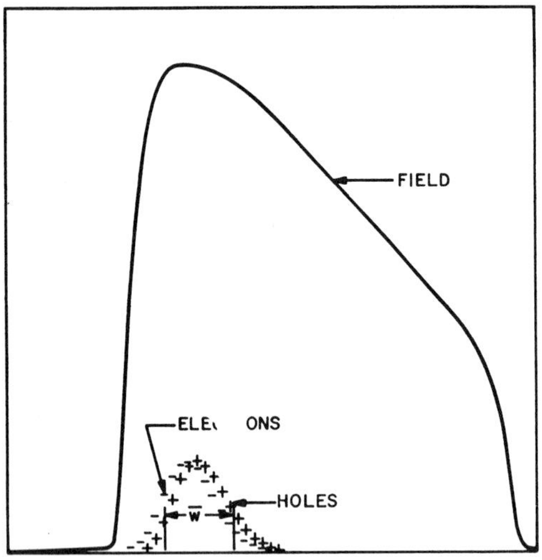

Fig. 12. Field profile and carrier densities for a n^+-p-p^+ 4-μm germanium diode at a point in time corresponding to the maximum diode voltage shown in Fig. 2.

voltage until the creation of an avalanche shock front which causes the voltage to drop rapidly to zero. This Appendix analyzes this part of the TRAPATT waveform.

In order to obtain an expression for the time delay t_0 shown in Fig. 2, a Read model [14] of the avalanche diode has been adopted. Thus, the particle current J_c created in the avalanche region is given by [14, eq. 13]

332

$$\frac{\bar{w}}{2v_s}\frac{dJ_c}{dt} = J_c\left(\int_0^{\bar{w}} \alpha(E)\, dx - 1\right) \tag{1}$$

where

- $\bar{w}$ the avalanche zone width
- v_s the saturated drift velocity
- $\alpha(E)$ the ionization rate (assumed equal for holes and electrons).

Since we are interested in only a small part of the complete TRAPATT cycle, several other assumptions can be made. Fig. 12 shows the carrier densities and field profile for a germanium diode at the time when diode voltage reaches a maximum. This result was obtained by computer and does not make the simplifying assumptions introduced in this Appendix. Note that the field does not change greatly over regions where the electron and hole densities are large. Therefore, (1) can be written as:

$$\frac{\bar{w}}{2v}\frac{dJ_c}{dt} = J_c(\bar{w}\alpha(\bar{E}) - 1) \tag{2}$$

where $\bar{E}$ is the average value of the field over some effective avalanche zone $\bar{w}$. The avalanche width indicated in Fig. 12 is chosen so that the total particle current as obtained by an integral over the carrier densities shown is the same as the peak value times the effective width $\bar{w}$. [The assumption of a distance-independent particle current in the avalanche zone is made in deriving (1).]

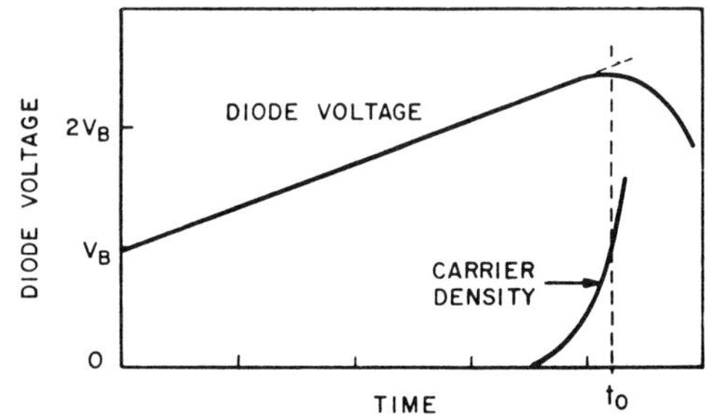

Fig. 13. Expanded scale for the diode voltage shown in Fig. 2 and the generated carrier density versus time.

If initially the diode is assumed to be biased to breakdown and to be carrying a dc current J_{c0}, then a large step in terminal current will typically drive the peak field in the diode to a value about 50 to 100 percent higher than that required to sustain the dc current. Over this range of fields the ionization coefficients are safely approximated by the form

$$\alpha = A_0 e^{\lambda E} \tag{3}$$

provided that A_0 and λ are determined by matching this equation to the experimental values at the two extremes [15].

A good approximation for the time dependence of the field in the diode becomes apparent if we examine another result from the computer simulation. Fig. 13 shows the voltage response from Fig. 2 on an expanded scale. Notice that the response is linear until just before the voltage begins to fall and that the terminal current is principally displacement current up to that time. Therefore, the terminal current J_T is

$$J_T \cong \frac{\epsilon}{w}\frac{dV}{dt}$$

or

$$V = V_B + \frac{wJ_T t}{\epsilon} \tag{4}$$

where V_B is the breakdown voltage. Thus, the field in the avalanche region is

$$\bar{E}(t) = E_c + \frac{J_T t}{\epsilon}$$

where E_c is the critical field. Equations (2) and (3) give

$$\frac{\bar{w}}{2v}\ln J_c/J_{c0} = \int_0^t (\bar{w}A_0 \exp(\lambda(E_c + J_T t/\epsilon)) - 1)\, dt.$$

But the integral of the ionization rate over the avalanche zone $\bar{w}$ at breakdown gives

$$\int_0^{\bar{w}} A_0 e^{\lambda E_c}\, dx = 1. \tag{5}$$

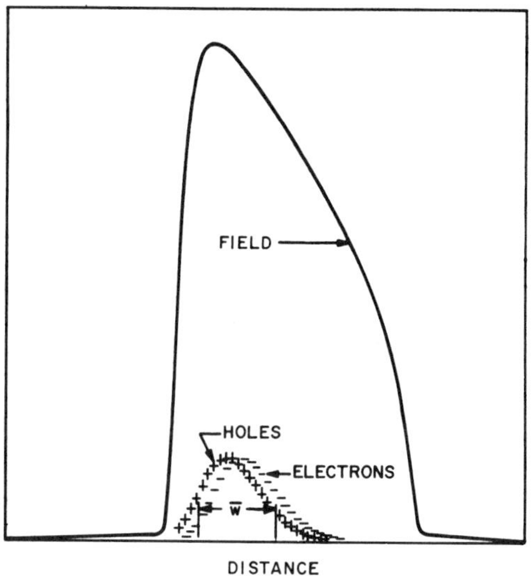

Fig. 14. Field profile and carrier densities for a p+-n-n+ 0.6-μm silicon diode at a point in time corresponding to the maximum diode voltage.

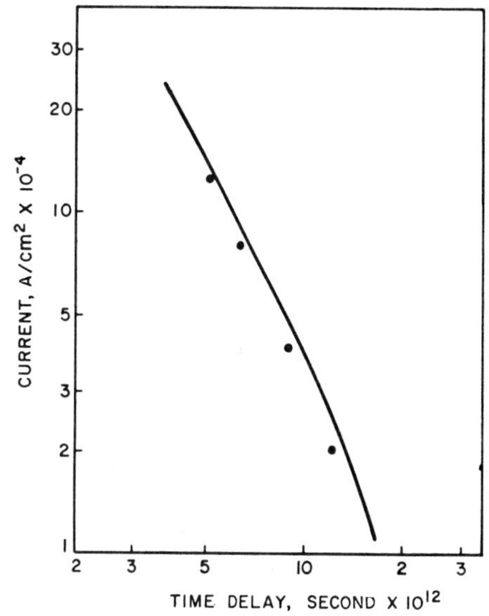

Fig. 15. Time required for diode voltage to reach the maximum value versus amplitude of the applied current step.

Therefore,

$$\frac{\bar{w}}{2v}\ln J_c/J_{c0} = \frac{\epsilon}{J_T\lambda}\left(\exp\left(\lambda J_T t/\epsilon\right) - 1\right) - t.$$

When the conduction current equals the terminal current, the diode voltage has reached the maximum; i.e., the time required for $J_c = J_T$ is approximately

$$t_0 = \frac{1}{\theta}\ln\left[1 + \theta(t_0 + \tau\gamma\ln\theta/\theta_0)\right] \quad (6)$$

where

$$\theta \equiv \lambda J_T/\epsilon$$
$$\theta_0 \equiv \lambda J_{c0}/\epsilon$$
$$\tau \equiv w/2v$$
$$\gamma \equiv \bar{w}/w.$$

Note that this equation is transcendental in t_0 but for most cases it coverages rapidly by successive approximation.

As an example, consider a 50-GHz abrupt-junction silicon IMPATT diode with a depletion width of 0.6 μm. The peak field in such a diode at breakdown is approximately 5×10^5 V/cm. The peak field required to initiate an avalanche shock front may be as high as twice this value; therefore the constants of (3) have been obtained by matching it to $(\alpha_n+\alpha_p)/2$ at 5×10^5 and 10^6 V/cm for silicon at 160°C [16].

The field and carrier densities at the instant of maximum voltage for the 50-GHz diode are shown in Fig. 14. In this case the avalanche region is seen to be relatively somewhat wider than that shown in Fig. 12. Using

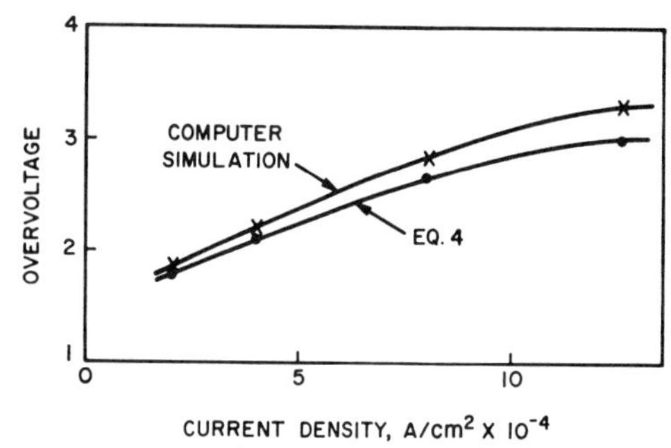

Fig. 16. Maximum diode voltage normalized to the diode breakdown voltage versus amplitude of the applied current step.

a value of $\gamma = 0.3$ in (6), the time delay versus the terminal current density has been calculated for this diode and is shown in Fig. 15. The points shown were obtained by the computer simulation referred to previously. Over the range of interest (i.e., 20 000 to 125 000 A/cm², where high-efficiency TRAPATT operation would occur) the slope of this curve is approximately one half. Using these results and (4), the peak voltage at the time t_0 can be found as a function of current. This is shown in Fig. 16.

In comparing Figs. 12 and 14 it is clear that the effective avalanche width is a smaller fraction of the total diode width for the lower frequency diode. Also, as shown in Fig. 7, in higher frequency diodes the relative delay time t_0 is increased and the overvoltage exceeds twice V_B for the current densities of interest. These effects occur because of the decreasing slope of the ionization rates with increasing field. The decreasing slope requires that the excess field (above E_c) be greater in order to create the same overmultiplication in the diode. This leads to a broader avalanche region and a larger overvoltage.

Fig. 16 shows that the simple calculation given here does not account for all of the overvoltage predicted by the computer simulation. The difference in the two results is due to a broadening of the depletion width with increasing voltage which occurs in the simulated diode. This effect is caused by out diffusion in the simulated structure which is more important in high-frequency diodes. In the example considered here the depletion-layer width increases by approximately 20 percent when the diode voltage is raised from the breakdown voltage V_B to $3V_B$. Thus, in some cases, out diffusion is detrimental to TRAPATT-oscillator performance.

In Section III the initial conditions chosen for comparing a number of diode structures were held constant. In particular, the initial bias current, before the application of a large step in current, was assumed to be 1 A/cm². Equation (6) indicates that a better comparison might have been made if this initial current were scaled with increasing impurity concentration in the high-field region. Thus, in each structure when the terminal current is equal to the minimum value required to create an avalanche shock front ($J=qNv$), the ratio of current to initial current would be the same. The form of (6) indicates, however, that generally this would have a relatively small effect on the overvoltage. Several values of initial current have been used in the computer simulation and they have confirmed the results obtained with (6).

Acknowledgment

The author would like to thank D. L. Scharfetter for assistance in using his semiconductor-device program SEMPAC and for interpretation of the results.

References

[1] D. L. Scharfetter, D. J. Bartelink, H. K. Gummel, and R. L. Johnston, "Computer simulation of low frequency high efficiency oscillation in germanium," presented at the IEEE Solid-State Device Research Conf., June 17–19, 1968.

[2] H. J. Prager, K. K. N. Chang, and S. Weisbrod, "High-power high-efficiency silicon avalanche diodes at ultra high frequencies," *Proc. IEEE* (Letters), vol. 55, pp. 586–587, April 1967.

[3] R. L. Johnston, D. L. Scharfetter, and D. J. Bartelink, "High-efficiency oscillations in germanium avalanche diodes below the transit-time frequency," *Proc. IEEE* (Letters), vol. 56, pp. 1611–1613, September 1968.

[4] D. L. Scharfetter, "High efficiency operation of Impatt diodes," presented at the 1968 Int. Electron Devices Meeting, Washington, D. C.; W. J. Evans and D. L. Scharfetter, "Characterization of avalanche diode TRAPATT oscillators," *IEEE Trans. Electron Devices*, vol. ED-17, pp. 397–494, May 1970.

[5] A. S. Clorfeine, R. J. Ikola, and L. S. Napoli, "A theory for the high-efficiency mode of oscillation in avalanche diodes," *RCA Rev.*, vol. 30, pp. 397–421, September 1969.

[6] B. C. DeLoach, Jr., and D. L. Scharfetter, "Device physics of TRAPATT oscillators," *IEEE Trans. Electron Devices*, vol. ED-17, pp. 9–21, January 1970.

[7] D. L. Scharfetter, "Power-frequency characteristics of the Trapatt diode mode of high efficiency power generation in germanium and silicon avalanche diodes," *Bell Syst. Tech. J.*, vol. 49, pp. 799–826, May–June 1970.

[8] W. J. Evans, "Circuits for high-efficiency avalanche-diode oscillators," *IEEE Trans. Microwave Theory Tech.*, vol. MTT-17, pp. 1060–1067, December 1969.

[9] D. L. Scharfetter and H. K. Gummel, "Large-signal analysis of a silicon Read diode oscillator," *IEEE Trans. Electron Devices*, vol. ED-16, pp. 64–77, January 1969.

[10] R. Edwards, D. F. Ciccolella, T. Misawa, D. E. Iglesias, and V. Decker, "Millimeter-wave silicon Impatt diodes," presented at the 1969 Int. Electron Devices Meeting, Washington, D. C.

[11] T. E. Seidel, "Ion-implantation fabrication techniques and the static properties of double-drift Impatt diodes," presented at the Device Research Conf., Seattle, Wash., June, 1970.

[12] W. J. Evans, T. E. Seidel, and D. L. Scharfetter, "A novel TRAPATT oscillator design," *IEEE Proc.* (Letters), vol. 58, pp. 1294–1295, August 1970.

[13] G. Gibbons and M. I. Grace, "High-efficiency avalanche-diode oscillators and amplifiers in *X*-band," *IEEE Proc.* (Letters), vol. 58, pp. 512–513, March 1970.

[14] W. T. Read, "Proposed high-frequency negative resistance diodes," *Bell Syst. Tech. J.*, vol. 37, pp. 401–446, March 1958.

[15] D. L. Scharfetter, D. J. Bartelink, and B. C. DeLoach, Jr., "Comments on static negative resistance in avalanching-silicon p^+-i-n^+ junctions," *IEEE Trans. Electron Devices* (Correspondence), vol. ED-16, pp. 970–972, November 1969.

[16] C. N. Dunn and D. R. Decker, unpublished work.

THEORY FOR HIGH-EFFICIENCY OSCILLATIONS IN SILICON AVALANCHE DIODES

Indexing terms: Avalanche diodes, Microwave generation, Oscillations

Design and performance criteria have been deduced for high-efficiency oscillations in p^+–n–n^+ and n^+–p–p^+ avalanche diodes. A TRAPATT approach is used with realistic analytic approximations for the carrier velocities and ionisation coefficients. A comparison is made with experimental data.

Introduction: Theories have recently been developed to describe the high-efficiency 'anomalous' mode operation of silicon avalanche diodes at microwave frequencies using a TRAPATT approach,[1, 2] whereby an avalanche zone sweeps rapidly through the diode, leaving behind a trapped plasma of carriers. The cycle is divided into three distinct stages:

(*a*) an avalanche transit period

(*b*) a plasma extraction period

(*c*) a high-voltage/low-current state.

This method has been extended to apply to high-efficiency oscillations in gallium-arsenide avalanche diodes,[3, *] where the effect of negative differential mobility is found to be important.

In this letter, theoretical calculations are reported for the design and performance of silicon p^+–n–n^+ and n^+–p–p^+ diodes in the high-efficiency mode. A TRAPATT approach is employed, but the analysis differs from that of DeLoach and Scharfetter[2] in the following ways:

(i) Realistic velocity/field profiles are used. It is assumed that electrons and holes have mobilities $1220\,cm^2\,V^{-1}\,s^{-1}$ and $460\,cm^2\,V^{-1}\,s^{-1}$, respectively, at low fields, and that the carriers have a common saturation velocity of $9\times10^6\,cm\,s^{-1}$ at higher fields.

(ii) The ionisation coefficient $\alpha(E)$ for electrons is approximated by

$$\alpha(E) = A\exp(\lambda E) \qquad (1)$$

where $A = 7\times10^4\,cm^{-1}$ and $\lambda = 2{\cdot}3\times10^{-6}\,cm\,V^{-1}$. This is in good agreement with experimental data[4] for field values of interest. The ionisation coefficient $\beta(E)$ for holes is much smaller than for electrons, and it is assumed that $\alpha/\beta = 12$. Previous TRAPATT analyses[1, 2] have represented the ionisation coefficients more crudely by step functions.

(iii) No arbitrary assumption relating the avalanche transit time and the plasma extraction time is made, since their connection can be obtained theoretically.

The results of the calculations are now presented, together with an outline of the underlying theory.

Theory and results: Consider a reverse-biased silicon n^+–p–p^+ diode prior to breakdown, with a current density I flowing in the external circuit. The electric field in the diode increases until ionisation occurs at the n^+–p junction, and an ionisation edge moves to the p–p^+ junction with a velocity U given by[2]

$$U = I/qn_d \qquad (2)$$

* COTTAM, M. G.: 'A theory for high-efficiency oscillations in gallium-arsenide avalanche diodes' (to be published)

where n_d is the ionised doping density and $q = 1{\cdot}6 \times 10^{-19}$ C. If the driving current is sufficiently high that U exceeds the hole saturation velocity v_s, an avalanche zone will move through the diode, leaving a dense plasma of electrons and holes.

The analysis of the avalanche transient may be carried out using Poisson's equation and the carrier continuity equations (which are conveniently transformed to a time-independent form by choosing a co-ordinate system stationary with respect to the avalanche zone), following closely the treatment by Cottam[3,*] for the gallium-arsenide case. The hole concentration p may be related to the electric field E by

$$\frac{n_d(c+1)}{2}\log\left\{\frac{p(c+cr-1+r)+n_0(c+1)-p_0(c-1)}{(p_0 r+n_0)(c+1)}\right\}-(p-p_0)$$

$$=\frac{\varepsilon(c+cr-1+r)}{2q(c-1)}\int_{E_I}^{E}\alpha(E)\,dE \quad (3)$$

where n_0, and p_0 are the electron and hole concentrations before breakdown, and c and r denote the ratios U/v_s and β/α, respectively, E_I denotes an electric-field value below which ionisation is negligible, and ε is the relative permittivity. The effect of diffusion is small for the ranges of doping densities and depletion-layer widths considered later, and it has been neglected. Eqn. 3 may be used to deduce an approximate value for the total carrier current density J after avalanche:

$$J = I\log(G\log G) \quad . \quad . \quad . \quad . \quad . \quad . \quad . \quad . \quad . \quad (4)$$

where

$$G = I(c-1+cr+r)/2qv_s c(p_0 r+n_0) \quad . \quad . \quad . \quad . \quad (5)$$

Also, analytic expressions for the avalanche-zone width and for the maximum field E_{max} during avalanche may be obtained using eqn. 3. For example,

$$E_{max} = \frac{1}{\lambda}\log\left\{\exp(\lambda E_I)+\frac{qn_d\,\lambda(c^2-1)}{2A\varepsilon c}(\log G-1)\right\} \quad (6)$$

After the avalanche zone has swept through the diode in a time $\tau_A = w/U$, where w is the depletion-layer width, the electric field falls to a value E_0 below saturation which may be obtained from eqn. 4 and Poisson's equation. The treatment of the plasma-extraction period is similar to that of DeLoach and Scharfetter[2] except that the effect of different low-field mobilities has been taken into account. As the current I continues to flow in the external circuit, the carriers are swept out at the junctions and the electric field rises. By incorporating the condition that, for efficient TRAPATT oscillations, the field at the n^+–p junction has risen to just below breakdown as the last of the carriers are removed, it may be shown that

$$c = \frac{(1+f)\,\varepsilon E_I}{fqwn_d}-\frac{1}{f} \simeq \frac{1{\cdot}38\varepsilon E_I}{qwn_d}-0{\cdot}38 \quad . \quad . \quad . \quad (7)$$

where f denotes the ratio of the electron and hole low-field mobilities. The carriers are extracted in a time τ_E given by

$$\tau_E = fw(v_{oe}^{-1}+v_s^{-1})/(1+f) \quad . \quad . \quad . \quad . \quad . \quad . \quad (8)$$

where v_{oe} is the electron velocity corresponding to the plasma field E_0.

The analysis for a p^+–n–n^+ structure is entirely analogous, with the roles of holes and electrons interchanged, and results similar to eqns. 2–8 may be obtained. The corresponding result to eqn. 7 is

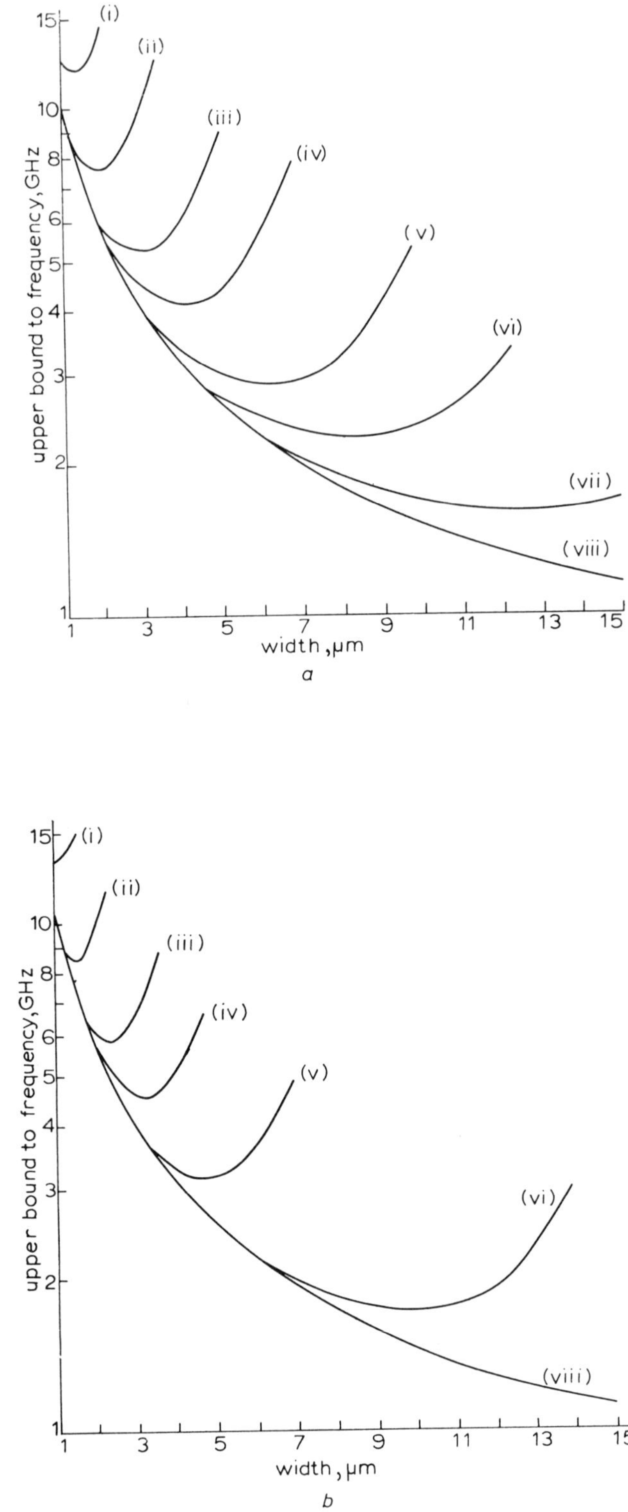

Fig. 1 *Plot of upper bound to frequency against depletion-layer width for different doping densities*

a n^+-p-p^+ diode
b p^+-n-n^+ diode
Doping densities, cm^{-3}
(i) 1×10^{16}
(ii) 6×10^{15}
(iii) 4×10^{15}
(iv) 3×10^{15}
(v) 2×10^{15}
(vi) $1{\cdot}5\times10^{15}$
(vii) 1×10^{15}
(viii) 6×10^{14}

* COTTAM, M. G.: 'A theory for high-efficiency oscillations in gallium-arsenide avalanche diodes' (to be published)

$$c = \frac{(1+f)\,\varepsilon E_I}{fqwn_d} - 1 \simeq \frac{1{\cdot}38\varepsilon E_I}{qwn_d} - 1 \quad . \quad . \quad . \quad . \quad (9)$$

A lower limit to the TRAPATT period, and hence an upper limit to the frequency, can be obtained from the total high-current period $(\tau_A + \tau_E)$, which may be expressed in terms of the parameters n_d and w using the preceding analysis. This has been carried out for doping densities ranging from 6×10^{14} to 1×10^{16} cm^{-3} and depletion layer widths from 1 μm to 15 μm, and the resulting upper bounds to the frequency are plotted in Figs. 1*a* and *b* for n^+–p–p^+ and p^+–n–n^+ structures, respectively. For a particular doping density, eqns. 7 and 9 imply an upper bound to the width w, to satisfy the necessary condition $c > 1$ for TRAPATT oscillations. The curves in Fig. 1 are terminated when this condition is no longer satisfied. The corresponding values for the peak diode-current density I are shown in Figs. 2*a* and *b* for the n^+–p–p^+ and p^+–n–n^+ cases, respectively.

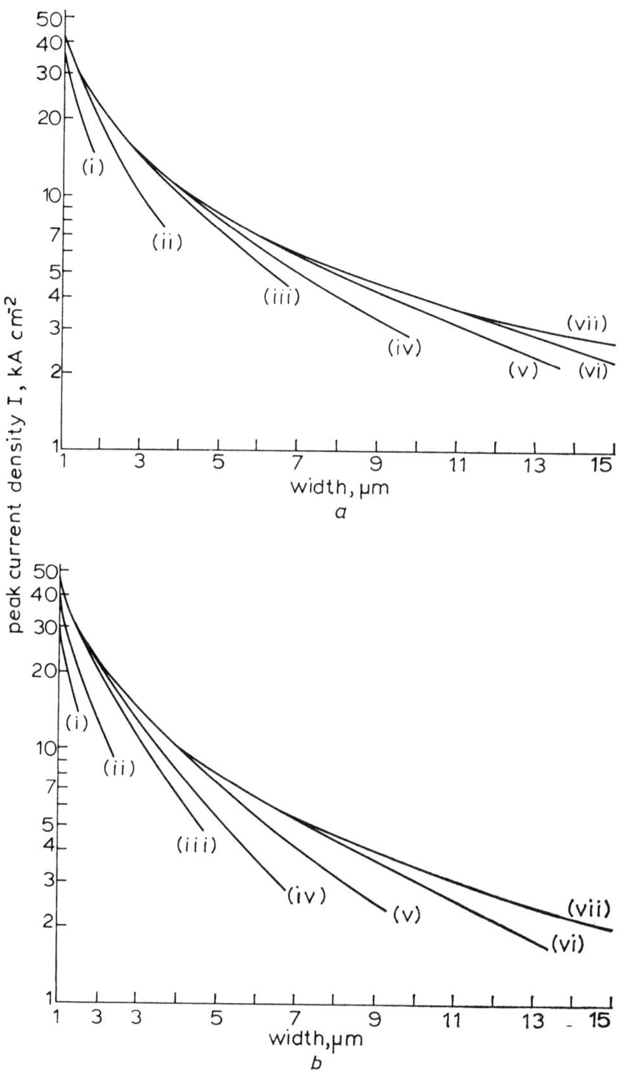

Fig. 2 *Plot of peak diode-current density I against depletion-layer width for different doping densities*

a n^+–p–p^+ diode
b p^+–n–n^+ diode
Doping densities, cm^{-3}
(i) 1×10^{16}
(ii) 6×10^{15}
(iii) 4×10^{15}
(iv) 3×10^{15}
(v) 2×10^{15}
(vi) 1×10^{15}
(vii) 6×10^{14}

Conclusions: Results have been obtained for values of the peak diode-current density and the frequency upper bound for ranges of depletion-layer widths and doping densities. Although TRAPATT oscillations can occur at any frequency below the calculated upper limit, an optimum case for efficiency and power output should be obtained when the frequency is approximately half of the upper bound (corresponding to equal duration of the total high-current state and the high-voltage/low-current state).

The results for the frequency are somewhat dependent on the assumed value for the reverse saturation current, to which the parameter G is related. An estimated value of $160W$ A cm^{-2}, where W is measured in centimetres, has been assumed. The effect of increasing or decreasing this estimate by a factor of 10^2 is correspondingly to increase or decrease the frequency upper bound by up to 25% for wide diodes and up to 15% for narrow diodes. It has been assumed that the same current density I is flowing during the avalanche transit and the plasma-extraction period, so that the diode current has a square-wave variation with time. The effect of relaxing this condition can increase the range of diode widths in Fig. 1 for TRAPATT oscillations, although the efficiency will probably be reduced. It is proposed to investigate this effect in detail and to carry out calculations of the resulting efficiencies.

The predictions of Fig. 1*b* are compared with some experimental results for high-efficiency oscillations in p^+–n–n^+ silicon diodes in Table 1. It can be seen that the general agreement is fairly good.

Table 1 COMPARISON OF THEORY WITH EXPERIMENTAL DATA FOR SILICON p^+–n–n^+ DIODES

Source	Doping density	Width	Operating frequency	Theoretical optimum frequency	Theoretical upper bound to frequency
	cm^{-3}	μm	GHz	GHz	GHz
SERL (Baldock)†	2×10^{15}	5	1·7	1·5	3·1
SERL (Baldock)†	1×10^{15}	1·1	3·3	4·7	9·4
Snapp and Hoefflinger[5]	4×10^{15}	4	3·2	4·0	8·0
Grace and Gibbons[6]	4×10^{15}	3	1–3	2·7	5·4

† CLUNIE, D.: Private communication

The author wishes to thank K. G. Hambleton for a helpful discussion. This work was carried out in part under a CVD contract, and is published by permission of the UK Ministry of Defence (Navy Department) and the directors of the Plessey Co.

M. G. COTTAM
Allen Clarke Research Centre
Plessey Co.
Caswell, Towcester, Northants., England

26th February 1970

References

1 CHLORFEINE, A. S., IKOLA, R. J., and NAPOLI, L. S.: 'A theory for the high-efficiency mode of oscillation in avalanche diodes', *RCA Rev.*, 1969, **30**, pp. 397–421
2 DELOACH, B. C., and SCHARFETTER, D. L.: 'Device physics of TRAPATT oscillators', *Microwave J.*, 1969, **12**, pp. 101–115
3 COTTAM, M. G.: 'High-efficiency oscillations in gallium-arsenide avalanche diodes', *Electron. Lett.*, 1970, **6**, pp. 71–73
4 LEE, C. A., LOGAN, R. A., BATDORF, R. L., KLEIMACK, J. J., and WIEGMANN, W.: 'Ionisation rates of holes and electrons in silicon', *Phys. Rev.*, 1964, **134**, pp. A761–A773
5 SNAPP, C. P., and HOEFFLINGER, B.: 'High current density operation and analysis of avalanche transit-time diodes', Proceedings of the 7th international conference on microwave and optical generation and amplification, Hamburg, Germany, 16th–20th Sept. 1968
6 GRACE, M. I., and GIBBONS, G.: 'High-efficiency microwave oscillations from Si avalanche diodes', *Electron. Lett.*, 1968, **4**, pp. 564–565

THEORY FOR POWER OUTPUT AND EFFICIENCY OF SILICON TRAPATT OSCILLATORS

Indexing terms: Microwave oscillators, Avalanche diodes, Transit-time devices, Semiconductor doping

A TRAPATT analysis is employed to obtain theoretical results for the efficiency and power output of silicon $p^+\text{-}n\text{-}n^+$ avalanche diodes at different frequencies. This is done for a range of doping densities and depletion-layer widths, and enables detailed design curves to be deduced.

In an earlier letter,[1] an analytic TRAPATT theory was employed to describe high-efficiency microwave oscillations in silicon $p^+\text{-}n\text{-}n^+$ and $n^+\text{-}p\text{-}p^+$ avalanche diodes. Results were presented for an upper bound to the frequency, an optimum operating frequency and the peak current density for different doping densities and depletion-layer widths. The conclusions were found to be in fairly good agreement with experimental data.

This treatment is now extended to obtain results for the efficiency and power output of silicon $p^+\text{-}n\text{-}n^+$ diodes for a range of frequencies below their calculated upper bounds, and detailed design curves may be deduced. Initially it will be assumed that a high current density I flows during the avalanche and extraction periods, and there is then a high-voltage/low-current period. By extending the methods of Reference 1, the voltage at any instant is calculated, and the resulting voltage and current waveforms are illustrated in Fig. 1, where τ_A and τ_E are, respectively, the avalanche transit time and the plasma extraction time. The total period T must exceed the high current period $\tau_A + \tau_E$, from which an upper bound to the frequency may be calculated as in Reference 1.

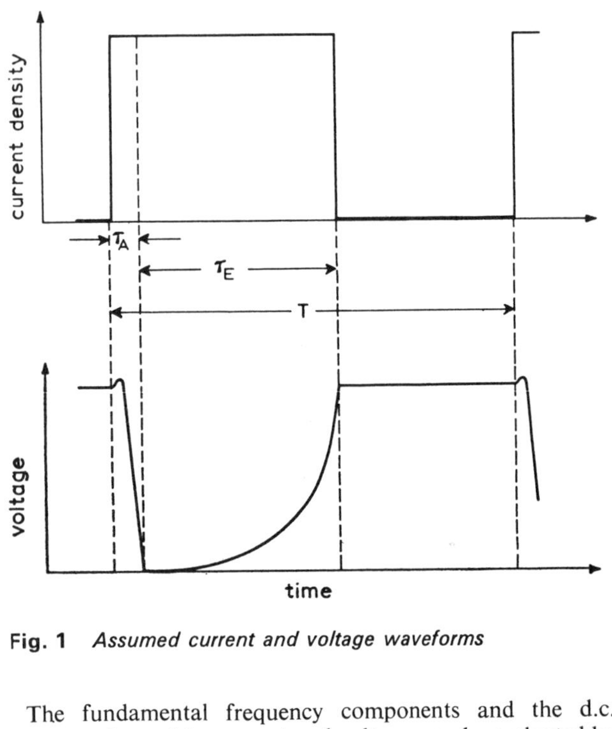

Fig. 1 *Assumed current and voltage waveforms*

The fundamental frequency components and the d.c. average values of the current and voltage may be evaluated by Fourier-analysis techniques, and hence results for the power output per unit diode area and the efficiency may be obtained. This is done for a range of frequencies below the calculated upper bound for each diode, and for doping densities and depletion-layer widths ranging from 6×10^{14} to 1×10^{16} cm^{-3} and 1 to 12 μm, respectively. In each case it is found that the power output and efficiency may be relatively high over a wide range of frequencies, but they have their maximum values at a frequency very close to half the upper-bound frequency.

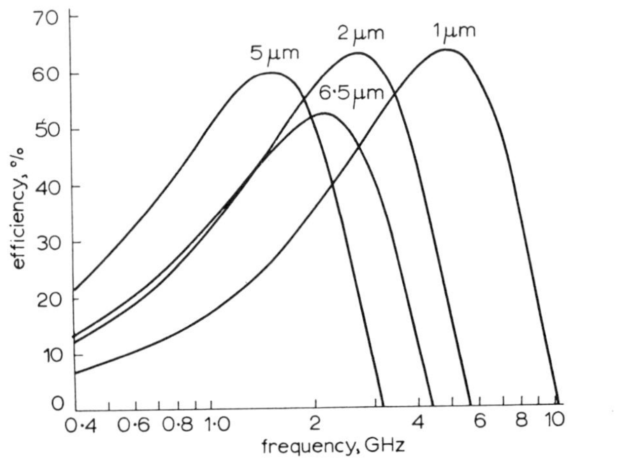

Fig. 2A *Variation of efficiency with frequency for diodes of doping density 2×10^{15} and various widths*

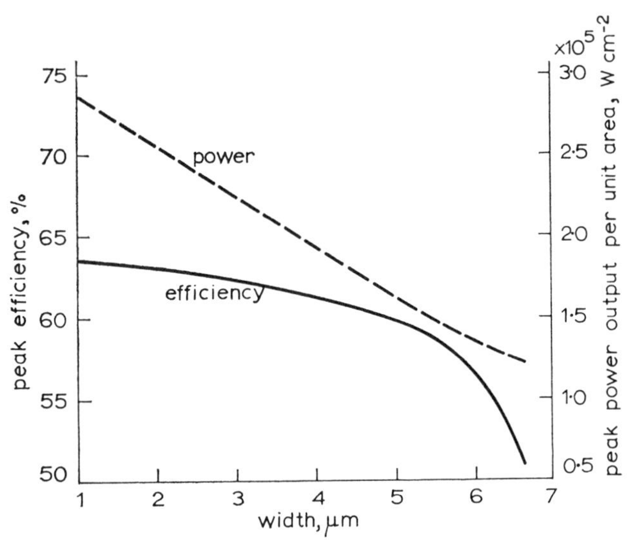

Fig. 2B *Variation of peak efficiency and peak power density with width for doping density 2×10^{15}*

A typical series of results is shown in Fig. 2A, where efficiency is plotted against frequency for silicon $p^+\text{-}n\text{-}n^+$ diodes of doping density 2×10^{15} cm^{-3} and various widths (1, 2, 5 and 6·5 μm). The efficiency falls to zero at the upper-bound frequency, and also tends asymptotically to zero at low frequencies. For example, for the 5 μm curve, the upper bound to the frequency is 3·2 GHz, the efficiency has a maximum of 60% at 1·5 GHz, and the efficiency is above 25% in the frequency range 0·5–2·6 GHz. These results indicate a decrease in the peak efficiency with increasing width for constant doping density, and this is shown more clearly in Fig. 2B for diodes of doping density 2×10^{15} cm^{-3}, together with corresponding results for the peak power output which exhibits an approximately linear dependence on width. The curves are terminated when the avalanche-zone velocity no longer exceeds the electron-saturation velocity. The peak efficiency and power output are also found to decrease as the doping density increases for constant width. This effect is illustrated in Fig. 2C for diodes of 2 μm width.

The results may be summarised as indicating that the highest efficiencies and powers are obtained with narrow diodes of low doping density, and that peak efficiencies and power outputs up to about 60–65% and 3×10^5 W cm^{-2}, respectively, are predicted. Experimentally efficiencies up to 60% and power densities up to about $2{\cdot}5 \times 10^5$ W cm^{-2} have been measured,[2,3] giving close agreement with the theory. More typically the measured efficiencies are somewhat lower presumably because the operating frequency and/or the circuit requirements are not optimised.

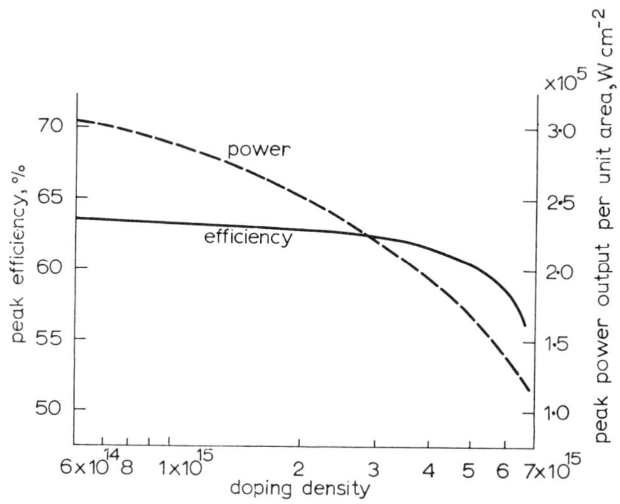

Fig. 2c *Variation of peak efficiency and peak power density with doping density for widths of 2 μm*

The effect of the capacitance of the diode and its package has not as yet been included in this analysis. This capacitance becomes charged during the low-current/high-voltage period and is then discharged during the avalanche transit period, thereby giving an additional current contribution.[4] This effect is estimated to be important only for capacitances of several picofarads or higher, and in these cases it is necessary to generalise the current waveform of Fig. 1, so that the current during the avalanche transit period is greater than during the extraction period. It is a relatively simple matter to modify the theory accordingly, and some results are shown in Table 1 for diodes of doping density 5×10^{15} and width $2\ \mu m$. The peak values of the efficiency and power density are only slightly affected, but the optimum frequency is considerably reduced for a high current ratio. This is because the time taken to extract the carriers is increased. Hence, while the large-capacitance diodes are less likely to 'burn out' since the current need only be large during the avalanche transit period, their optimum operating frequencies may be considerably lower than for the square-wave current case.

Similar calculations are at present being carried out for gallium-arsenide avalanche diodes. This requires a more complicated analysis because the effect of the negative differential mobility is important,[5,6] and in some cases necessitates the inclusion of diffusion.

This work was carried out in part under a CVD contract, and is published by permission of the UK Ministry of Defence (Navy Department) and the directors of the Plessey Co.

Table 1 EFFECT OF DIFFERENT RATIOS OF THE AVALANCHE CURRENT TO THE EXTRACTION CURRENT FOR DIODES OF DOPING DENSITY $5\times10^{15}\ cm^{-3}$ AND WIDTH $2\ \mu m$

Current ratio	Optimum frequency	Peak efficiency	Peak power density
	GHz	%	$W\ cm^{-2}$
1	3·5	60	$1{\cdot}6\times10^5$
2	1·4	62	$1{\cdot}6\times10^5$
5	0·55	63	$1{\cdot}6\times10^5$
10	0·25	64	$1{\cdot}6\times10^5$

M. G. COTTAM* *12th May 1970*

Allen Clark Research Centre
Plessey Co.
Caswell, Towcester, Northants., England

References

1 COTTAM, M. G.: 'Theory for high-efficiency oscillations in silicon avalanche diodes', *Electron. Lett.*, 1970, **6**, pp. 203–205
2 PRAGER, H. J., CHANG, K. K. N., and WEISBROD, S.: 'Anomalous silicon avalanche diodes for microwave generations'. Proceedings of 1st biennial Cornell conference on engineering applications of electronic phenomena, Cornell University, USA, Aug. 1967, pp. 266–280
3 CHANG, K. K. N.: 'Avalanche diodes as UHF and L-band sources', *RCA Rev.*, 1969, **30**, pp. 3–14
4 EVANS, W. J.: 'Circuits for high efficiency avalanche diode oscillators', *IEEE Trans.*, 1969, **MTT-17**, pp. 1060–1067
5 COTTAM, M. G.: 'High-efficiency oscillations in gallium arsenide avalanche diodes', *Electron. Lett.*, 1970, **6**, pp. 71–73
6 COTTAM, M. G.: 'A theory for high-efficiency oscillations in gallium-arsenide avalanche diodes', *J. Phys. D.*, 1970 (to be published)

* Present address: Department of Physics, University of Essex, Colchester, Essex

Subharmonic Generation and the Trapped-Plasma Mode in Avalanching Silicon p^+-n-n^+ Junctions

CRAIG P. SNAPP

Abstract—Experimental evidence is presented which distinguishes two mechanisms whereby an avalanche diode can efficiently generate microwaves at transit angles below those of classical transit-time oscillations.

Measurements of external current waveforms and subsequent calculations of terminal voltage waveforms demonstrated that efficient generation ($\eta \sim 10$ percent) of microwaves was possible at subharmonics of transit-time excitations. This subharmonic generation occurred when various harmonic components interacted such that the peak total voltage was delayed with respect to the fundamental. In such cases of multiharmonic synergetic excitations, the peak displacement currents were less than 0.25 of that required for avalanche shock-front initiation.

The distinguishing features of high-efficiency operation ($\eta \sim 50$ percent) were found to be completely consistent with a trapped-plasma mode interpretation. Experimental external current and dV/dt waveforms have substantiated Evans' circuit analysis [28]. dV/dt waveforms indicated peak displacement currents at breakdown which were on the order of 1.5 times that required for shock-front initiation. Approximate conduction current waveforms exhibited larger currents during the shock-front transit than during the extraction period. General features were consistent with calculations made by Cottam [29].

Circuit characteristics for both subharmonic generation and the trapped-plasma mode were similar. Circuit analysis also indicated the critical importance of each separate harmonic of a trapped-plasma mode fundamental. Harmonic generation utilizing the trapped-plasma mode enabled an extension of the upper bound of the frequency of efficient generation.

I. Introduction

Several explanations have been advanced for the ability of avalanche transit-time devices to generate microwaves at small transit angles with high efficiencies. One of these explanations grew out of large-signal computer simulations made by Scharfetter *et al.* [1]. In this case the physical mechanism would involve the initiation of an avalanche shock front or traveling avalanche zone in the diode's depletion region which results in a high-density compensated plasma that can sustain only a low value of electric field. The carriers in this "trapped-plasma state" would then drift at unsaturated velocities. A diode operated in this trapped-plasma or TRAPATT mode could approach an ideal large-signal oscillator since it could be periodically switched from a poorly conducting state to a highly conducting state in a time considerably less than a transit time.

Another explanation was advanced by Hoefflinger *et al.* [2]. This explanation proposed that certain large-signal interactions caused by a multifrequency excitation would lead directly to high-efficiency operation without the need of a high velocity shock front resulting in a highly conducting state. The excitation frequency of most importance in such operation would be in the avalanche transit-time region and thus near the characteristic avalanche-resonance frequency. This physical mechanism was thus referred to as "avalanche-resonance pumping."

Considerable indirect evidence has been amassed supporting the trapped-plasma mechanism as the correct explanation for truly high-efficiency microwave generation. This paper presents the results of experimental waveform measurements which constitute a direct confirmation of the principal features of the trapped-plasma mode. On the other hand, results based on waveform measurements and amplification studies are also presented which substantiate the basic premise of the avalanche-resonance pumping concept; namely, that efficient generation of microwaves is possible at subharmonics of an avalanche transit-time oscillation under conditions of saturated carrier velocities. In either case, operation was only made possible by achieving a proper multifrequency excitation. The results of calculations which compare and contrast the circuit environments required for the two modes are also presented and the harmonic generating capabilities of the trapped-plasma mode are considered.

Attention will be directed toward two considerably different diode structures whose small transit-angle operating characteristics separately epitomized the two different mechanisms. Only instances of oscillation and amplification where the distinction between the two physical mechanisms was clear are considered.

II. Two Structures: Their Transit-Time and Space-Charge Induced Negative Resistance Characteristics

Before considering the actual oscillator performance resulting from multifrequency excitation, it is informative to detail the structural parameters of the two diodes, their classical transit-time performance, and the effect of space charge on current–voltage characteristics at high current densities. The physical parameters of the two diodes (designated A and B) are listed in Table I. In both diodes the p^+-n junction was formed by a boron diffusion into an n-type silicon epitaxial layer. The shallow p^+ layer diffusion for diode A was approxi-

Manuscript received October 12, 1970; revised December 4, 1970. This work was supported by the Air Force Systems Command, Rome Air Development Center, Griffiss Air Force Base, N. Y.

The author is with the Department of Applied Physics, Cornell University, Ithaca, N. Y. 14850.

TABLE I

	Diode A	Diode B
Breakdown voltage V_B	92 V	88 V
Punchthrough voltage V_p	85 V	20 V
p$^+$ region width W_p	1.5 μm	4.5 μm
n region width W_n	4.7 μm	3.0 μm
n region doping N_d	5×10^{15} cm^{-3}	1.75×10^{15} cm^{-3} (minimum)
Sample area A	1.6×10^{-4} cm^2	1.55×10^{-4} cm^2
Saturation current density (200°C) J_s	$\sim10^{-1}$ A/cm^2	$\sim10^{-2}$ A/cm^2
Effective carrier lifetime (200°C) τ_e	$\sim10^{-7}$ s	$\sim10^{-6}$ s
Threshold conduction current density for static negative resistance J_{th}	6900 A/cm^2	5500 A/cm^2
Critical displacement current density for shock-front initiation ($q\,N_d v_s$) J_c (saturation velocity $v_s=9\times10^6$ cm/s)	7200 A/cm^2	2600 A/cm^2

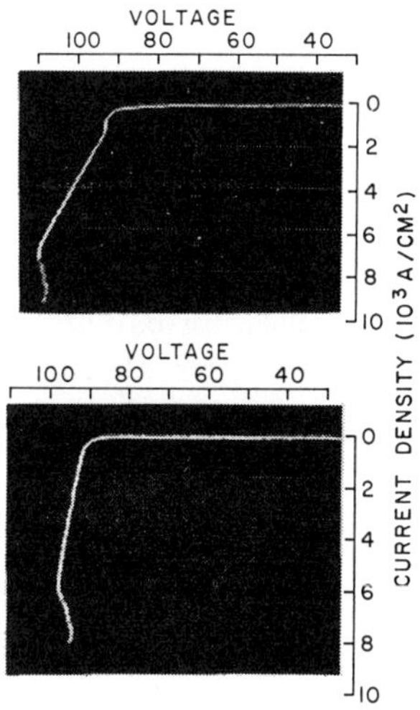

Fig. 1. Oscillographs of static current–voltage characteristics of diode A (top trace) and diode B (bottom trace).

mately 1.5 μ and a very nearly linear relationship existed between the inverse of the square of the depletion layer capacitance and the reverse bias voltage. Diode A thus closely approximated an abrupt junction with a constant n-region doping. The n-n$^+$ interface field of diode A at breakdown was approximately 5 percent of the field at the p$^+$-n junction. Diode B was formed by a rather deep 4.5 μm diffusion which resulted in significant out-diffusion from the substrate and a graded doping profile. The n-n$^+$ interface field of diode B at breakdown was approximately 60 percent of the field at the p$^+$-n junction.

In addition to dimensional and doping differences the two diodes exhibited considerable differences in the effective lifetime of carriers in their respective depletion layers. This information was obtained by measuring the saturation current densities J_s of both diodes and then using the general relationship that $\tau_e=qN_iW/J_s$, where N_i is the intrinsic carrier concentration. This measurement was performed at 200°C in order to minimize the effect of surface currents. As may be seen from Table I, the effective carrier lifetime for diode B was an order of magnitude longer than that for diode A. This difference should be considered in the light of DeLoach's assertion [3] that saturation current should be a critical materials parameter for diodes designed to operate in the trapped-plasma mode.

Bartelink and Scharfetter [4], DeLoach and Scharfetter [3], and Clorfeine *et al.* [5] have shown that an external current density greater than qN_dv_s would be required at breakdown in order for an avalanche shock front to be launched. This current would be almost totally displacement current since the conduction current prior to breakdown would consist only of the saturation current. The values of qN_dv_s for diodes A and B are listed in Table I. It has been pointed out by Clorfeine *et al.* [5] that a fast risetime bias pulse could provide the displacement current required to launch a traveling avalanche zone. Experimental evidence [6], [7] has, however, indicated that transit-time oscillations have so far been the catalyst for high-efficiency oscillations. It is easy to show that for an assumed oscillation with a transit angle of π a voltage amplitude greater than $qN_dW^2/\pi\epsilon$ should provide the necessary displacement current for shock-front initiation. The value $qN_dW^2/\pi\epsilon$ implies voltage amplitudes of 55 V and 8 V for diodes A and B, respectively. The difference in these amplitudes contrasts the difficulty of launching a shock front in diode A as compared to diode B.

The effects of space charge on the current–voltage characteristics of each diode were studied under nonoscillating, minimal heating conditions. Triangular voltage pulses with base widths of 50 ns were applied to the diodes at a repetition rate of 60 Hz. The maximum adiabatic temperature increase during each pulse was estimated to be less than 5°C. Voltage and current were measured with a sampling oscilloscope by utilizing a 20 dB resistive voltage probe and the disk resistor described in Section III. A coaxial directional coupler and tunnel diode microwave detector were used in the circuit to verify that no microwave oscillations occurred during the pulsed current–voltage measurements. It was important that no microwave oscillations be present since nonlinear rectification effects could easily be interpreted as static negative resistances. The oscillographs in Fig. 1 illustrate typical current–voltage characteristics of diodes A and B. The diodes were pulse-biased to current densities above the threshold current density at which avalanche generated space charge induced a static negative resistance. This type of current–voltage characteristic is consistent with the analytical predictions made by Bowers [8] It is coincidental that despite the considerable differences between the two diodes, their thresholds for static negative resistance occurred at approximately the same current density. The threshold current density for diode B was higher than might be expected on the basis of Bowers' calculations due to the graded doping profile of the junction. Although this space-charge induced negative resistance was active at frequencies up to approximately 1 GHz, it was incapable of resulting in efficiencies greater than about 0.5 percent. This low efficiency should be expected considering the small magnitude of the negative resistance exhibited.

The transit-time capabilities of the two diodes were considerably more important from the point of view of subharmonic generation and the trapped-plasma mode. The optimum efficiency for the classical transit-time mode in diode A was approximately 8 percent. This optimum efficiency occurred at a frequency of 9 GHz and a pulsed current density of 800 A/cm². At this current density the range of frequency in which transit-time oscillations were possible was 5–11 GHz. At 2000 A/cm² the maximum efficiency of diode A was 2.5 percent. Diode B, in contrast, did not exhibit good transit-time performance. The optimum efficiency was approximately 2 percent at 11 GHz, at a current density of about 1200 A/cm².

III. Waveform Measurements and Circuit Analysis

The most direct method of obtaining information about the nature of small transit angle oscillations is the accurate determination of sample external current and voltage waveforms. Early efforts by Johnston and Scharfetter [9] in measuring RF voltages close to oscillating samples resulted in degraded performance. The waveforms obtained correctly verified the multifrequency nature of high-efficiency oscillations. Some of these waveforms, however, indicated the existence of a nonphysical forward-biased portion of the RF cycle.
342 This anomaly probably arose because of cartridge inductances and/or significant capacitive coupling to their 10 kΩ resistive voltage probe. The measurements must therefore be considered as only qualitative. The results of current waveform measurements by Snapp *et al.* [7] must likewise be considered as only qualitative since the disk resistor utilized in the coaxial cavity caused deterioration of the efficiency. These measurements, in fact, led to the probably erroneous conclusion that, in certain high-efficiency cases, harmonic components interacted to produce a maximum in overvoltage. These difficulties in the precise determination of current and voltage waveforms were greatly overcome by Torizuka, Yanai, and Yamada [10], [11] through the use of a microstrip line assembly. Unfortunately their assembly did not allow for the direct measurement of microwave power, thus necessitating a Fourier analysis for the determination of efficiency.

The resistive external current and capacitive voltage probes used to obtain the experimental waveforms discussed in this paper were used in conjunction with a standard coaxial resonating cavity and thus allowed the direct measurement of microwave power. Their presence caused no degradation of the pulsed performance of avalanche diodes with areas less than 2×10^{-4} cm². The outputs of the probes were viewed using an RF triggered sampling oscilloscope having a risetime of less than or equal to 50 ps.

A schematic diagram of the thin-film current monitoring resistor assembly is shown at the top of Fig. 2. The resistor had a dc resistance of 0.4 Ω. The resistive annulus width was approximately 0.2 cm and constituted only a small portion of the short-circuiting plane of the GR dimension coaxial cavity. The feedthrough to the precision attenuator and sampling oscilloscope had an impedance of 50 Ω. The attenuator was found to be essential for the elimination of undesirable reflections. The diode samples were mounted on the side of the package which was flush against the center contact of the disk resistor. Measurements of the impedance of the current waveform monitoring assembly have been reported elsewhere [12]. These impedance measurements indicated that only negligible phase error occurred for frequency components up to approximately 9 GHz.

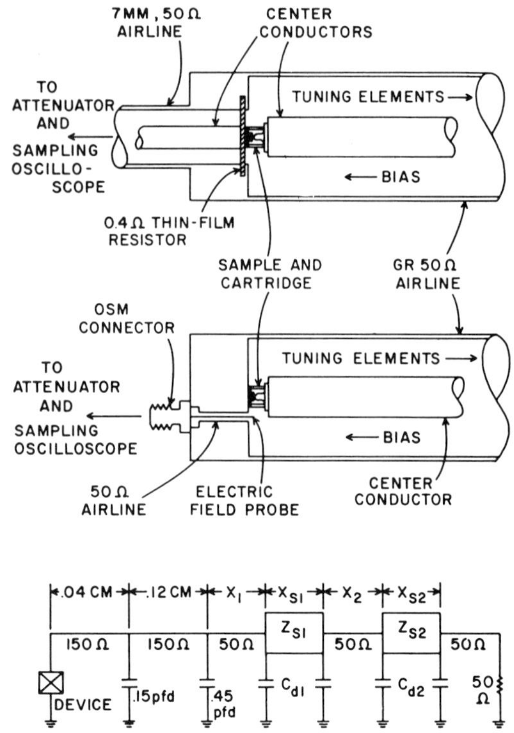

Fig. 2. Schematic diagrams of external current waveform probe, dV/dt waveform probe, and general diagram of the coaxial transmission line circuit including an equivalent circuit of the cartridge.

A typical example of the detail observed in the external current waveforms is illustrated by the oscillograph shown in Fig. 3. In this example, diode B was oscillating at a fundamental frequency of 0.9 GHz with an efficiency of 48 percent. The waveform illustrated has strong components up to about the sixth harmonic.

A schematic diagram of the electric field probe which was used to measure the derivative of the voltage across the cartridge is illustrated at the center of Fig. 2. The probe coupled capacitively to the electric field between the end of the center conductor and the shorting plane. The airline from the OSM connector to the region where the field was monitored had an impedance of approximately 50 Ω and an outer conductor diameter of about 0.1 cm. The measured coupling capacitance of the probe ($\sim$0.01 pF) to the end of the resonator's center con-

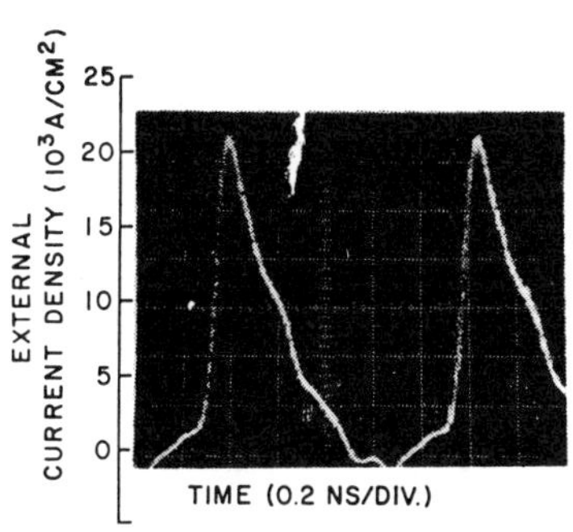

Fig. 3. Example of an experimental external current waveform for diode *B* oscillating at $f_1 = 0.9$ GHz with $\eta_1 = 48$ percent.

ductor was almost two orders of magnitude less than the total capacitance between the inner conductor and the shorting plane. As a result of this low coupling capacitance, the probe produced only a negligible perturbation of the electric field near the sample and cartridge. The voltage output of the probe was normalized to an injected input voltage and found to be linearly related to frequency up to about 7 GHz. This capacitive nature of the probe response resulted in an output which was approximately proportional to the derivative of the voltage waveform across the cartridge. In most cases the two probe assemblies could be interchanged without affecting the performance characteristics of an oscillating diode.

The tuning elements used in the coaxial resonating cavity generally consisted of two or three sleeves covered with 0.003 in low-loss Teflon tape. A generalized schematic of the transmission line equivalent of the circuit is shown at the bottom of Fig. 2. The 0.15 pF and 0.45 pF capacitors at the ends of 150 Ω lines constitute an empirically established equivalent of the cartridge reactance which is valid from approximately 1 to 12 GHz. The discontinuity capacitance at the input and output plane of each slug are also indicated. This simple transmission line equivalent of the coaxial circuit allowed a straightforward calculation of the load impedance seen by the diode chip as a function of frequency. Although the simplifications involved made this circuit analysis less precise than that described by Evans and Scharfetter [13], the principal features of the multiresonant circuits could be determined. The accuracy of the machine calculations were verified for certain test cases by swept-frequency impedance measurements. Certain minor discrepancies involving very weakly coupled lossy resonances were observed, but these were felt to have only a negligible influence on diode operation.

IV. Subharmonic Generation Sans Trapped Plasmas

Attention will now be directed toward those situations where diode *A* generated microwaves efficiently at frequencies below the classical transit-time region and where evidence for the presence of a trapped-plasma state was absent. When this type of operation was first observed [14], the author and others felt that the distinguishing characteristics were in many ways similar to the small transit angle oscillations observed by Prager *et al.* [15] in spite of the fact that efficiencies were much less than 60 percent. This section and the following section will show that this correlation was premature and that two different mechanisms were involved.

A. Circuit Characteristics

The two specific examples which will now be considered were selected because they represented the optimum subharmonic generating capabilities of diode *A* which were observed in the multiple-sleeve coaxial cavity. The impedance for the general circuit illustrated in Fig. 2 with sleeve positions, sleeve impedances, and discontinuity capacitances equal to experimental values was calculated versus frequency from 0 to 10 GHz for both cases. The results for the two cases detailed in Table II are illustrated in Figs. 4 and 5.

For Case A the fundamental frequency was at 2.63 GHz and the efficiency was 6.5 percent. Rectification effects caused a decrease in the bias voltage of approximately 7 percent from the preoscillation value. The fundamental frequency corresponded to a transit angle of approximately 0.28 π and was thus considerably below the 5 to 11 GHz range where classical transit-time oscillations were possible. Both the second and third harmonic were detectable in the vicinity of the diode and also delivered measurable power to the load. No higher harmonics were found to be present. The power observed at each harmonic is listed in Table II. A plot of the calculated load impedance seen by the oscillating sample as a function of frequency is shown in Fig. 4. The position of each of the three frequency components is denoted by a small arrow. As may be seen, the fundamental frequency was in the vicinity of a resonance transmission window where the sample load impedance had a local minimum in reactance. This load impedance had a negative imaginary part. The large-signal sample impedance at this frequency was of course the negative of the load impedance and therefore the reactance of the diode itself was inductive. The imaginary part of the sample impedance at both the second and third harmonic was much greater than the real part. As a result of these relatively large reactances the voltage excitations at these frequencies were significant in spite of the low powers delivered to the load. For the sake of precision the sample load impedance at each harmonic was measured using a slotted line. These impedance values are listed in Table II. Knowledge of the power delivered and the load impedance enabled the calculation of the external current amplitude and terminal voltage amplitude at each frequency in this three-frequency system. These amplitude values are also listed in Table II along with the displacement current amplitude at the nth harmonic, $n\omega_1 V_n C_{diode}$. The maximum possible peak displacement current for this particular example would be the sum of the three components, namely, 280 mA. This would correspond to a current density of 1600

343

TABLE II

Case A: Subharmonic Generation for Load Situation Illustrated in Fig. 4

$I_{bias}=280$ mA $\quad J_{bias}=1750$ A/cm² $\quad \eta_1=6.5$ percent
$V_{bias}=90$ V $\quad \Delta V_{bias}=7$ V
$R_{bias}=330\ \Omega$ $\quad C_{diode}\ (V_B)=0.34$ pF

Frequency (GHz)	Output Power (W)	Measured Load Impedance (Ω)	External Current Amplitude (mA)	Sample Voltage Amplitude (V)	Displacement Current Amplitude (mA)
$f_1=2.63$	1.7	$24-j12$	380	10	60
$f_2=2f_1$	0.2	$8-j26$	220	7	85
$f_3=3f_1$	0.003	$2.5+j140$	50	7	135

Case B: Subharmonic Generation for Load Situation Illustrated in Fig. 5

$I_{bias}=300$ mA $\quad J_{bias}=1880$ A/cm² $\quad \eta_1=9.1$ percent
$V_{bias}=88$ V $\quad \Delta V_{bias}=10$ V
$R_{bias}=290\ \Omega$ $\quad C_{diode}\ (V_B)=0.34$ pF

Frequency (GHz)	Output Power (W)	Measured Load Impedance (Ω)	External Current Amplitude (mA)	Sample Voltage Amplitude (V)	Displacement Current Amplitude (mA)
$f_1=4.25$	2.4	$38-j10$	350	14	125
$f_2=2f_1$	0.01	$1-j55$	200	11	200

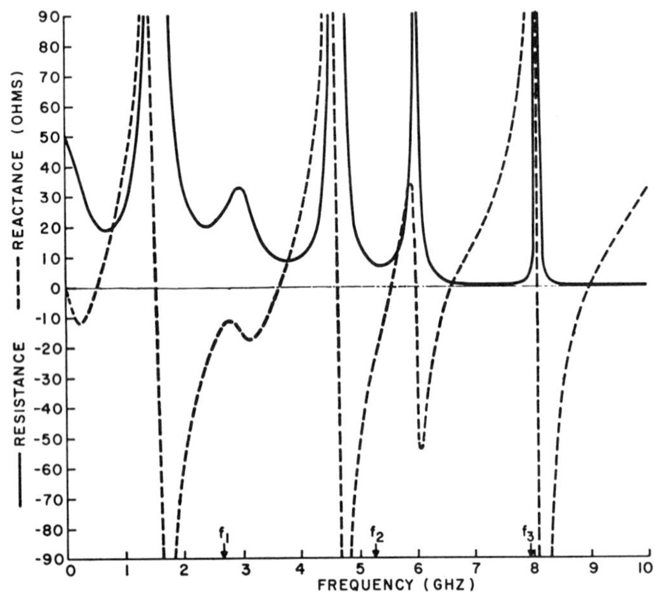

Fig. 4. Load impedance versus frequency for circuit in which diode A generated power at second subharmonic of 7.89 GHz with $\eta_1=6.5$ percent. (Three 10.2 GHz tuning sleeves—$Z_{s1}=Z_{s2}=Z_{s3}=15.5\ \Omega$; $X_1=4.11$ cm, $X_2=0.402$ cm, $X_3=0.723$ cm.)

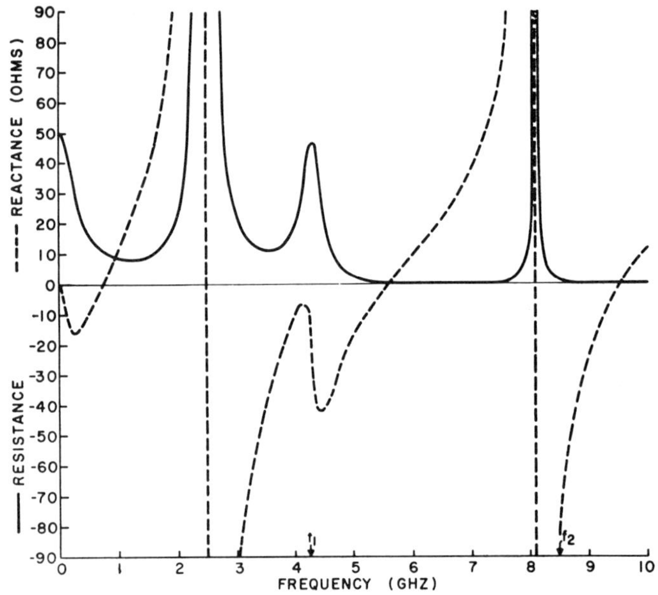

Fig. 5. Load impedance versus frequency for circuit in which diode A generated power at first subharmonic of 8.5 GHz with $\eta_1=9.1$ percent. (Two 10.2 GHz tuning sleeves—$Z_{s1}=Z_{s2}=7.6\ \Omega$; $X_1=2.19$ cm, $X_2=0.363$ cm.)

A/cm². It should be recalled that the displacement current at breakdown for diode A must necessarily be greater than 7200 A/cm² for shock-front initiation. This evidence suggests that this type of multifrequency operation did not involve a trapped-plasma state.

The second case considered was a two-frequency system with a fundamental frequency at 4.25 GHz and a fundamental efficiency of 9.1 percent. Voltage amplitudes were larger than in Case A and as a consequence voltage rectification caused an average voltage reduction of approximately 10 percent. The details of Case B are listed in Table II and a plot of calculated circuit impedance versus frequency is illustrated in Fig. 5. The general circuit characteristics are quite similar to those of Case A. Once again the fundamental frequency was close to a resonance transmission window while the second harmonic was reactively terminated. Although a very weak third harmonic was detectable in the vicinity of the diode, its voltage amplitude was estimated to be negligible compared to the fundamental and second harmonic components. The measured load impedance, external current amplitude, sample voltage amplitude, and displacement current amplitude at the fundamental and second harmonic are listed in Table II. The maximum possible peak displacement current of 325 mA corresponds to a current density of 2100 A/cm². This value is again smaller than that required for shock-front initiation in diode A.

One of the principal distinguishing features of these two examples of multifrequency excitations was the existence of the highest significant harmonic in the region of normal transit-time operation at a transit angle of about 0.8π. Although little power was coupled to the load at these frequencies, the voltage amplitudes were comparable to the fundamental and thus strongly influenced the total voltage waveform.

B. External Current and Voltage Waveforms

The fundamental frequency for the three-frequency system described above was low enough to allow a sampling oscilloscope measurement of the external current waveform. The experimental waveform is illustrated at the top of Fig. 6. This waveform can be very closely approximated by the proper phasing of the three external current amplitudes which are listed in Table II. Since the sample impedance at each of the three harmonics is known, it is straightforward to reconstruct the total terminal voltage waveform. This is illustrated at the center of Fig. 6. It should be noted that the phasing is such that the total voltage did not peak until a later time than the fundamental voltage component. This had the effect of delaying the avalanche current with respect to that which would have occurred if only a single frequency voltage excitation at 2.63 GHz had been present. It should also be noted that the minimum voltage across the diode was approximately 70 V. This is inconsistent with a fully formed trapped-plasma state which would necessarily result in a very low voltage across the diode. The derivative of the total voltage was normalized as displacement current and is illustrated at the bottom of Fig. 6. The peak in the displacement current which just precedes the current buildup was only about 15 percent of that required for shock-front initiation in diode *A*.

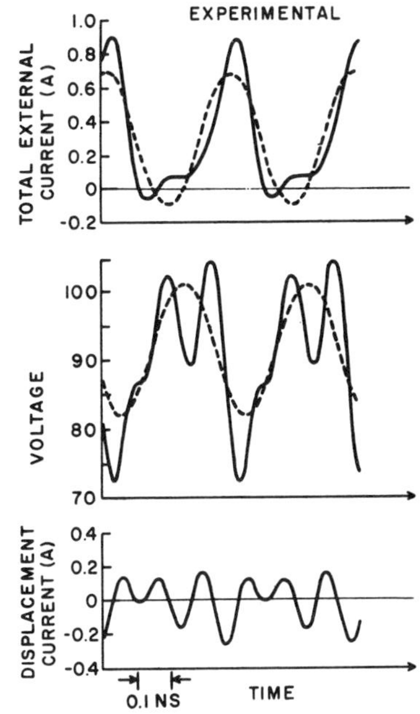

Fig. 6. Experimental external current waveform, computed voltage waveform, and computed displacement current waveform for power generation at second subharmonic of 7.89 GHz with η_1 = 6.5 percent.

C. The Degenerate Amplifier: Rectification Effects

Both the circuit analysis and waveform measurement previously discussed have indicated the essential nature of the harmonic components between the fundamental and diode *A*'s transit-time region. In the transit-time region from 5 to 11 GHz the diode possessed, of course, a small-signal negative resistance. A small-signal negative resistance would not be present for a single frequency excitation at frequencies less than 5 GHz. The presence of an oscillation at a transit-time frequency can, however, apparently result in an active small-signal negative resistance at a subharmonic and the subsequent buildup of an efficient oscillation. If this is a correct interpretation of events, then it should be possible, under proper circuit conditions, for an oscillation in the transit-time region to enable small-signal linear amplification at a subharmonic of that oscillation.

Self-pumped amplification has indeed proved to be possible. The detailed results of such amplification, using diodes similar to diode *A*, are published elsewhere [16], [17]. Amplification is a fortuitous capability of the multifrequency mode being considered because it allows certain observable quantities such as efficiency, gain, and voltage rectification to be studied as a function of signal amplitude. Kuvås [18] has recently analyzed some of these characteristics of the self-pumped degenerate case of amplification in Read diodes. The effect of signal level on the amount of voltage rectification is particularly interesting because it is explainable by an existing large-signal analysis.

For the purposes of this paper it is only necessary to consider the two-frequency degenerate system analogous to oscillator Case B previously discussed. The cases where amplification was possible at smaller transit angles and where more than two frequency components were present are avoided because their higher saturated efficiencies may have been the consequence of a trapped-plasma state.

Degenerate reflection-type amplification [17] was obtained by simply adding a 4–8 GHz circulator to the general circuit shown in Fig. 2. Two 12 Ω 10 GHz quarter-wavelength tuning sleeves were initially positioned so that a free-running transit-time oscillation at 8.72 GHz was established in the cavity. Depending on the input signal strength, quasi-degenerate or degenerate amplification was then possible at frequencies close to the first subharmonic of the 8.72 GHz signal. A linear gain of 16 dB was obtainable at low input signal levels. At maximum power output the gain was approximately 11 dB with an output efficiency of 7 percent at an input signal frequency of 4.385 GHz. As the output power increased from zero, the average bias voltage decreased due to nonlinear rectification. This voltage decrease resulted in an average bias current increase along the bias load line. A 100 Ω load line was utilized because it improved bias circuit stability for this particular example. This current increase was measured using a current probe in the bias circuit and a readout sampling oscilloscope. The resulting data of bias current and bias

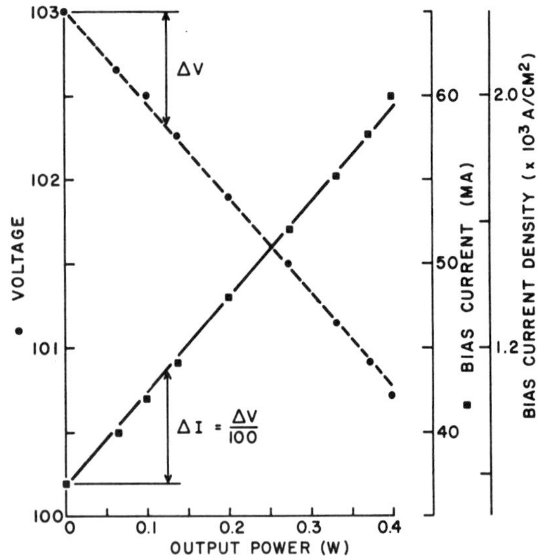

Fig. 7. Rectification effects in self-pumped degenerate amplifier: bias voltage and bias current versus output power at first subharmonic, $f_1 = 4.385$ GHz. Diode area was 0.3×10^{-4} cm^2.

voltage versus signal output power are illustrated in Fig. 7. The relationship of both the bias voltage decrease and the bias current increase as a function of output power was approximately linear. In other words, $|\Delta V_{\text{bias}}| \propto |\Delta I_{\text{bias}}| \propto V_1^2$, where V_1 is the voltage amplitude at the output signal frequency.

This type of behavior is predictable from the analysis by Lee *et al.* [19] of nonlinear processes in avalanching junctions under the assumption of saturated carrier
346
velocities. This analysis showed that voltage rectification should be linearly related to the square of the signal voltage for amplitudes on the order of those present in the degenerate amplifier. The point to be made here is that, although it has not been considered theoretically, the highly nonlinear large-signal processes in a trapped-plasma mode amplifier would probably make higher order rectification terms important and thus result in a nonlinear relationship between the signal amplitude and the amount of bias rectification.

D. Relevance of Existing Theories

The evidence presented above has indicated that the explanation for the operation of diode *A* at small transit angles cannot rest upon the mechanism of a trapped-plasma state resulting from the initiation of a traveling avalanche zone. Another possible explanation comes from the classical parametric theory of Manley and Rowe for simple nonlinear reactances. The weak oscillations which are often observed outside the normal region for transit-time negative resistances in avalanche diodes can be consistent with Manley–Rowe equations. This conclusion is especially true for nonpunchthrough diodes which would consequently have a time-varying depletion layer capacitance. Clorfeine [20] observed degenerate parametric amplification which is also possibly explainable by simple parametric theory and which was superficially similar to the degenerate amplification described in this paper. Unfortunately the Manley–Rowe type of analysis is incapable of accounting for a "scattering" of power to the output signal frequency which is greater than that available from the pump frequency. But greater output power generation at the subharmonics than that available at the transit-time frequency was indeed observed for the two examples which have been considered.

Hoefflinger *et al.* [2] proposed an explanation which would be more applicable to high-efficiency diodes with their approximately p^+-i-n^+ type field profiles and active negative resistances. In this explanation, certain large-signal interactions among multifrequency excitations resulted from the nonlinearity of the avalanche current response and led directly to high efficiency at small transit angles. This parametric interaction was referred to as avalanche resonance pumping because of the importance of the transit-time excitation in the region of the diodes characteristic avalanche resonance frequency. In Hoefflinger's analysis, the successive approximation techniques utilized were not taken beyond third order and, as a consequence, very little was known about program convergence. The lack of knowledge about convergence properties made the quantitative predictions of very high efficiency questionable. The physical explanation given for efficient small transit angle oscillations, however, seems quite plausible. Simply stated, this explanation was that proper phasing of voltage harmonic components could result in a total voltage waveform which peaked at a later time than the fundamental component and thus further delay both the onset of current multiplication and the peak in the fundamental current. This wave-shaping would result in a negative resistance at a transit-time frequency's subharmonic where a single frequency voltage excitation would have only preceded the fundamental current component only by about 90°.

The basic idea of subharmonic generation by avalanche resonance pumping was applied by Kuvås and Lee [21] to Read structures with narrow avalanche zones and well-defined drift regions. This type of structure lent itself to an analytical development which is on relatively firm theoretical ground. The results of this theoretical analysis by Kuvås and Lee indicated that increases in efficiency of as much as 50 percent were possible as the result of multifrequency operation at a reactively terminated transit-time frequency and its first subharmonic. Culshaw [22], Schroeder and Haddad [23], Mouthaan [24], and Harth [25] have also independently considered the two-frequency system where generation would occur at the first subharmonic of a transit-time excitation in Read-type avalanche diodes. Culshaw [22], in addition, showed that systems with more than two harmonically related frequencies could result in high efficiencies at lower frequencies than would be possible in the degenerate two-frequency case. Unfortunately, all of these analyses were restricted to the situation where bias current densities were low and where positive space-charge resistance would not be expected to result in degraded efficiencies [26]. At low current densities it is of course well known that Read diodes should theoretically exhibit efficiencies on the order of 25 percent. It is therefore not surprising that multifrequency excitations can also theoretically lead to high efficiencies at subharmonics of oscillations with transit angles in the vicinity of π. The situation where subharmonic generation and amplification were possible using diode A occurred at current densities where space-charge effects were certainly not negligible. As of this writing no published analysis has dealt specifically with the details of subharmonic generation and amplification such as were observed using diode A. Nevertheless, existing analyses [21]–[23] predict that multiharmonic voltage waveforms with characteristics similar to Case A will result in efficient subharmonic generation. Experimental evidence thus exists for the theoretical expectation that synergetic multiharmonic voltage excitations can provide enough additional current delay to enable efficient subharmonic generation at small transit angles. This physical mechanism does not, however, seem capable of explaining the situations where diode A generated microwaves in the 1 to 2 GHz range with efficiencies as high as 25 percent and at current densities of about 4000 A/cm^2. Such instances gave indications of being consistent with the trapped-plasma mode which will be discussed in the next section with reference to diode B. Possible interactions between simple subharmonic generation and the trapped-plasma mode in diode A were not studied. In certain diode structures it is possible that subharmonic oscillations could act as the catalyst for the trapped-plasma mode. In fact, a distinct transition from a low-efficiency subharmonic mode to high-efficiency output ($\eta \sim 40$ percent) with increasing current density has been observed by Goronkin and Wanuga [27].

V. The Trapped-Plasma Mode

Analytical developments [3], [5], [28] for the trapped-plasma mode have clearly delineated the chain of physical states which should occur during the fundamental period of a high-efficiency oscillation. Initially, the field profile in the diode's depletion layer must be driven above the breakdown condition by a displacement current density (J_t) greater than qN_dv_s. The upshot of the application of this large current to the swept-out diode is the launching of an avalanche shock front or traveling avalanche zone from the p+-n junction. This avalanche zone will then sweep across the depletion layer in a time equal to qN_dW/J_t and leave it filled with a partially compensated electron–hole plasma. In this highly conducting state the voltage across the diode will be small and the velocity of the carriers will no longer be saturated. During this recovery period a large circuit-dependent external current will result in the extraction of the plasma. At the end of the extraction period the diode field profile will again be close to the breakdown

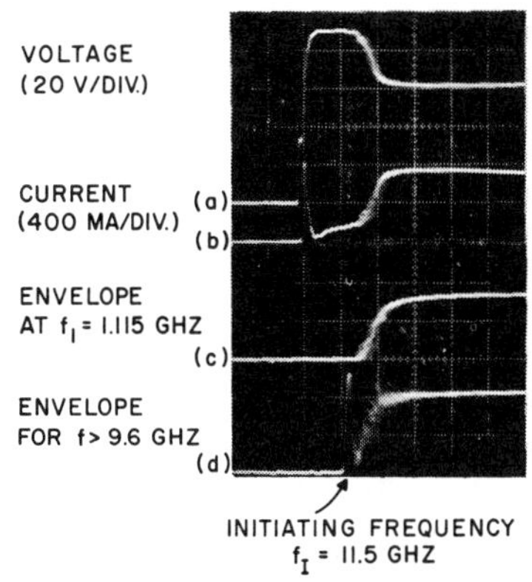

Fig. 8. Pulse turn-on characteristics of trapped-plasma mode in diode B oscillating at $f_1 = 1.115$ GHz with $\eta_1 = 49$ percent: oscillographs of average sample voltage, average sample current, detected fundamental oscillation at $f_1 = 1.115$ GHz, and envelope for detected frequencies greater than 9.6 GHz. Time scale: 50 ns/div.

condition. High-efficiency oscillations will result when the diode–circuit interaction is such that the diode is periodically switched from the poorly conducting prebreakdown state to the highly conducting trapped-plasma state. The operation of diode B in such a trapped-plasma mode is described in this section.

A. General Features of the Trapped-Plasma Mode

The general operating characteristics of the high-efficiency mode were studied using the previously described coaxial circuit whose schematic diagram is illustrated in Fig. 2. Only two or three tuning sleeves were required to obtain optimum results in the 1 to 4 GHz range.

The details of a typical turn-on transient of the high-efficiency mode are illustrated in the oscillographs shown in Fig. 8. The bias pulse length was 400 ns and the repetition rate was 60 Hz. This low-duty cycle insured that only negligible average heating occurred. The oscillographs show the average sample voltage, average sample current, detected RF envelope at the fundamental and the detected RF envelope for frequencies above 9.6 GHz. The fundamental was filtered and detected at the load whereas the frequencies greater than 9.6 GHz were detected with a capacitive probe close to the sample. The bias voltage initially rose very quickly from zero to the breakdown voltage. After breakdown occurred the avalanche current rose slowly for about 60 ns. This slow increase in the current was simply due to a corresponding slow increase in the pulser's charging voltage. When the avalanche current reached 160 mA (1000 A/cm^2) a spike of detector output occurred in trace (d). A tunable bandpass filter was used to determine that this transit-time oscillation was at 11.5 GHz, corresponding to a transit angle of about 0.8 π. This oscillation was followed immediately by a 35 percent decrease in the average voltage, a corresponding increase in the average current to 780 mA, and the onset of a high-efficiency oscillation ($\eta_1 = 49$ percent) at $f_1 = 1.115$ GHz. It thus appears that the transit-time oscillation at 11.5 GHz initiated the high-efficiency oscillation at 1.115 GHz. This type of turn-on behavior is consistent with the waveform measurements made by Johnston *et al.* [6], [9].

It should be noted that the fundamental frequency of the high-efficiency oscillation was not a subharmonic of the initiating frequency. After the high-efficiency oscillation became stable the higher frequency components shown in trace (d) were of course harmonics of the fundamental. Harmonics as high as the fifteenth were detectable near the diode. No justification could be found for placing a hierarchy of importance on the various harmonics. The contention by Clorfeine *et al.* [5] that the harmonics should be viewed as simply necessary components in a high-efficiency waveform seems correct. The designation of any particular harmonic as a transit-time oscillation would be artificial.

The tuning characteristics of the high-efficiency mode in diode B are illustrated in Fig. 9. The bottom curve is a plot of the period of the output fundamental (T) versus the physical distance between the oscillating sample and the input plane of the first tuning sleeve. All of the tuning sleeves were one quarter wavelength long at 10.2 GHz. Two 7.6 Ω sleeves were used for the 0.68 to 1.7 GHz range, three 11.3 Ω sleeves in the 1.9 to 3.5 GHz range, and one 11.3 Ω sleeve followed by two 15.5 Ω sleeves for the point at 4 GHz. The plot shows a linear relationship between the period T and the distance X and an $X=0$ intercept of approximately 80 ps. This intercept resulted from the input plane of the first sleeve being at a position about 1.2 cm closer to the sample than one half the fundamental wavelength. This manner of circuit dependent operation has been observed with high-efficiency Ge diodes by Evans [28] who attributes the $X=0$ intercept to the time required for the voltage across the cartridge to drop from its peak value to its minimum value. In this case the fall time was principally determined by the inductance of the gold ribbon contacting the sample ($L_p \approx 0.8$ nH) and the lumped capacitance near the diode ($C_p \approx 0.6$ pF). Indeed, five diodes that were mounted in identical packages but which had different values of N_d and W all followed the same relationship between T and X. A fall time on the order of 100 ps is also exhibited by the low-frequency voltage waveform which will be described later.

The top and center curves in Fig. 9 illustrate a typical variation in average required bias current and peak efficiency for high-efficiency operation from 0.68 to 4 GHz. The frequency range for optimum operation was between 0.9 to 1.7 GHz where efficiencies of 50 percent were typical. The efficiency began dropping sharply for frequencies above 1.9 GHz. Only 10 percent efficiency was observed at a frequency of 4 GHz. This frequency constituted an upper bound on the fundamental frequency of operation. The curve of required bias current versus X shows a current which was approximately constant at 0.8 A from 0.7 to 1.7 GHz and then increased with frequency to about 1.8 A at 4 GHz.

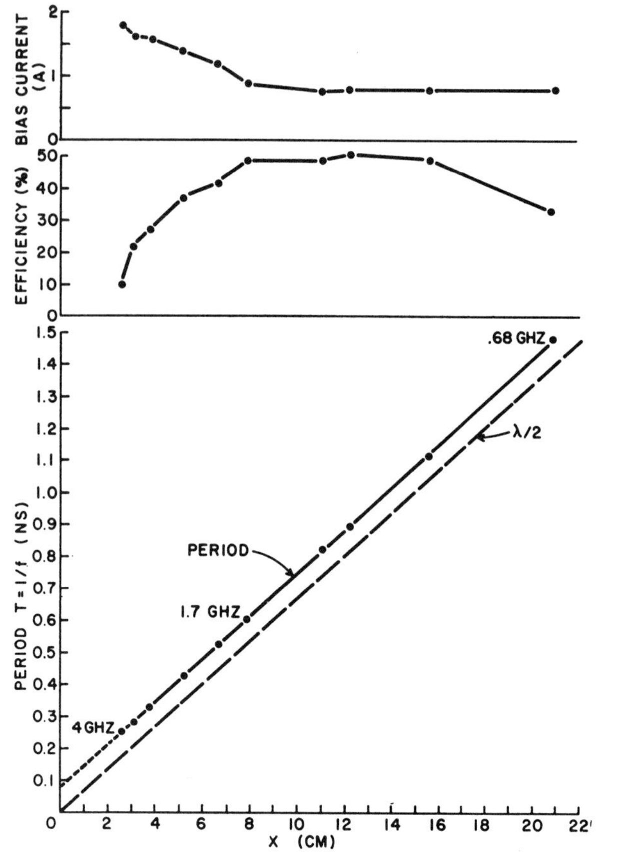

Fig. 9. General trapped-plasma mode characteristics of diode B: bias current, peak fundamental efficiency, and fundamental period versus physical distance between the oscillating sample and the input plane of the first tuning element.

These relationships among frequency, efficiency, and bias current are consistent with an analysis by Cottam [29] of the trapped-plasma mode. Cottam's analysis differs from those by Clorfeine *et al.* [5] and DeLoach and Scharfetter [3] in that realistic carrier velocity versus electric field curves are used, the ionization coefficients are assumed proportional to $\exp(\lambda E)$ instead of step functions, and no arbitrary assumption relating the shock-front transit time and the plasma extraction time is made. This analysis showed that for given values of N_d and W the minimum plasma extraction time (τ_E) is uniquely determined if it is assumed that the current during the extraction period is such that the field at the p^+-n junction is returned to the breakdown value just as the last of the carriers are extracted. The upper bound on the frequency is then equal to $1/(\tau_A+\tau_E)$, where τ_A is the time required for the shock-front transit. τ_E will in general be larger than τ_A since $\tau_A \leq W/v_s$. The 4 GHz upper frequency limit predicted for diode B ($N_d = 1.75 \times 10^{15}$ cm^{-3}, $W = 3$ μ) agrees very well with the observed value. This upper limit corresponds to a frequency of about one third the transit-time frequency. Scharfetter [30] has calculated design curves corresponding to diodes optimized for the 3:1 ratio of transit-time to fundamental frequency. For optimum performance at 4 GHz Scharfetter's design curves suggest a doping of 6×10^{15} cm^{-3} and an n-region width of 4 μ. As might be expected these values of doping and width differ markedly from those of diode B.

Since large conduction currents flow during the extraction period, it is obvious that for an assumed constant conduction current the optimum frequency for diode B would be approximately one half the upper bound frequency. This 2 GHz optimum frequency predicted for diode B may be related to the sharp fall-off of efficiency beginning at 1.9 GHz on the curve illustrated in the center of Fig. 9. Below 1.9 GHz, however, the experimental efficiency remained high until frequencies below 0.9 GHz. This behavior would not be expected if the extraction time remained fixed at its minimum value of 0.25 ns. This apparent discrepancy between experiment and theory can be rectified by allowing for a nonsquare wave of conduction current for the lower frequency cases. Cottam has stated that a large current flowing during the shock-front transit should indeed result in an increased extraction time and a corresponding decrease in the optimum frequency. The increased extraction time would be the result of an increased plasma density and a corresponding decrease in the residual plasma field. The circuit analyses by Evans [28] shows that a large enough lumped capacitance near the diode can result in a nonsquare current waveform having a larger current during the shock-front transit than during the extraction period. Experimental low-frequency waveforms which corroborate Evans' analysis will be considered below.

A comment is in order concerning the amplification properties of diode B at small transit angles. All attempts to obtain reflection-type amplification with small-signal linear gain were unsuccessful. Only "triggered" amplification at large input signals was observed. These characteristics are in sharp contrast to those exhibited by diode A.

B. Waveform Measurements

The details of the low-frequency case for which waveforms were obtained are listed in Table III. This particular case at 0.9 GHz was chosen to insure that as many harmonics as possible would be within the bandwidth of the sampling oscilloscope. The listed values of the fundamental current amplitude (I_1) and fundamental voltage amplitude (V_1) were computed from the measured fundamental output power (P_1) and sample impedance (Z_1). The fundamental current amplitude was approximately 1.7 times the bias current. This amplitude should be compared with the maximum possible Fourier amplitude of 2 I_{bias} which would occur for a train of delta functions with an average bias current of I_{bias}. The bias resistance (R_{bias}) is simply defined as V_{bias}/I_{bias}. This bias resistance is useful for normalization purposes since it can easily be shown that the maximum possible efficiency for a given value of R_1 is simply $2\,R_1/R_{bias}$.

The experimental waveforms of total external current and dV/dt were obtained using the probes described in

TABLE III
PARAMETERS RELEVANT TO TRAPPED-PLASMA MODE WAVEFORMS OF FIG. 10

$f_1 = 0.9$ GHz	$\eta_1 = 48$ percent
$P_1 = 23$ W	$Z_1 = R_1 + jX_1$ $= -26 + 1j\ \Omega$
$I_1 = 1.35$ A	$V_1 = 35$ V
$V_{bias} = 60$ V	$\Delta V_{bias} = 34$
$I_{bias} = 0.8$ A	$R_{bias} = 75\ \Omega$

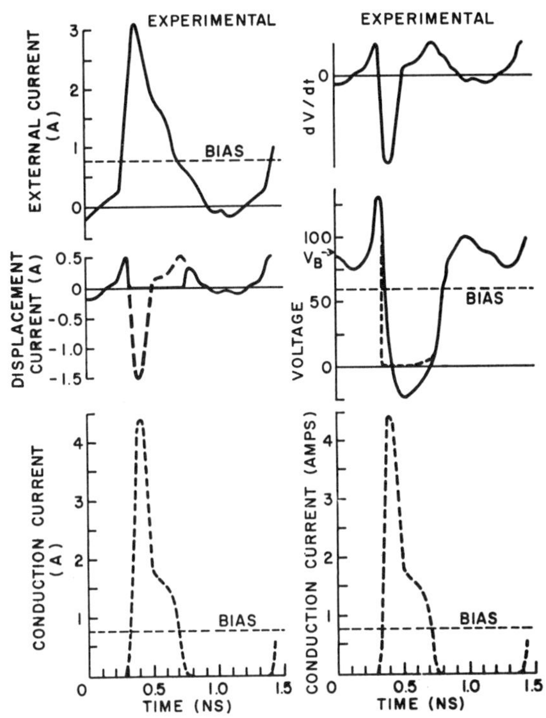

Fig. 10. Trapped-plasma mode waveforms for diode B oscillating at $f_1 = 0.9$ GHz with $\eta_1 = 48$ percent.

Section III. These waveforms are illustrated at the top of Fig. 10. The measured external current consisted of the sum of the conduction current flowing through the diode chip plus displacement current contributions. The output of the capacitive probe was proportional to the derivative of the voltage waveform across the package (V_p).

The right center drawing shows the result of a machine integration of the experimental dV_p/dt waveform. This package voltage waveform was Fourier analyzed and the fundamental amplitude set equal to the voltage amplitude (V_1) listed in Table III. In order to obtain the voltage waveform across the diode chip (V_d) it was necessary to determine the voltage drop across the inductive contacting ribbon. This determination in turn required a knowledge of the current flowing through the sample.

The total sample current was approximated as $I_{ext} - C_p dV_p/dt$, where $C_p \approx 0.6$ pF was the lumped capacitance external to the diode chip. The displacement current waveform ($C_p dV_p/dt$) is depicted by the solid dashed curve illustrated on the center left of Fig. 10. Another current contribution arose from the diode's depletion layer capacitance ($C_d \approx 0.5$ pF at V_B). This capacitance varies in time and space during the sharp voltage drop in V_d associated with a hypothetical shock front. During most of the following extraction period the displacement current arising from C_d is zero and the diode conduction current equal to $I_{ext} - C_p dV_p/dt$. Only negligible error will result if the displacement current arising from C_d is assumed zero during the very short time associated with the shock-front transit. Except when C_d equals zero the displacement current $C_d dV_d/dt$ will approximately equal $C_p dV_p/dt$. This equality is only a coincidence resulting from $C_p \approx C_d(V_B)$. The total $C_d dV_d/dt$ waveform is shown as the solid curve at the center left of Fig. 10. The phasing of the external current waveform and the displacement current waveforms was chosen to result in time-coincident fine structure.

The approximate conduction current through the diode was obtained by subtracting the two displacement current contributions from the total external current. This waveform is shown as the dashed curve at the bottom left and bottom right in Fig. 10. The dashed section of the voltage waveform illustrated at the center right represents the approximate voltage of the diode during the shock-front transit and extraction periods. This voltage differs from the package voltage by an amount approximately equal to L_p times the derivative of the conduction current waveform. In obtaining the chip voltage it was assumed that at no instance did the diode become forward biased.

The current and voltage waveforms which have been obtained possess all of the principal features of the trapped-plasma mode. The sample voltage waveform exhibits a large overvoltage (V_{pk}) of approximately 1.4 V_B. The peak-displacement current density (J_t) at breakdown is on the order of 1.5 times the critical current required for shock-front initiation. Almost immediately after the displacement current peak, the diode voltage drops to approximately zero in a very short time ($\tau_A \approx qN_dW/J_t$). This voltage drop corresponds to the transit of the avalanche zone. Coincident with the sharp voltage drop is a large spike of conduction current. After the shock-front transit the voltage rises slowly from zero until it reaches V_B near the end of the plasma extraction period. The conduction current after the initial spike remains approximately constant until the last of the plasma is removed. The voltage across the swept-out diode then remains in the vicinity of V_B until the next displacement current peak launches another traveling avalanche zone. The limited response of the sampling oscilloscope may have adversely affected the quantitative accuracy of these waveform measurements. In particular the peak voltage may have been greater than 1.4 V_B and undetected transit-time oscillations may have been present during the high-voltage, low-current portion of the fundamental cycle.

The duration of the conduction current (τ_p) as well as its peak value (I_{pk}) and plateau value (I_r) are in good quantitative agreement with the predictions by Evans

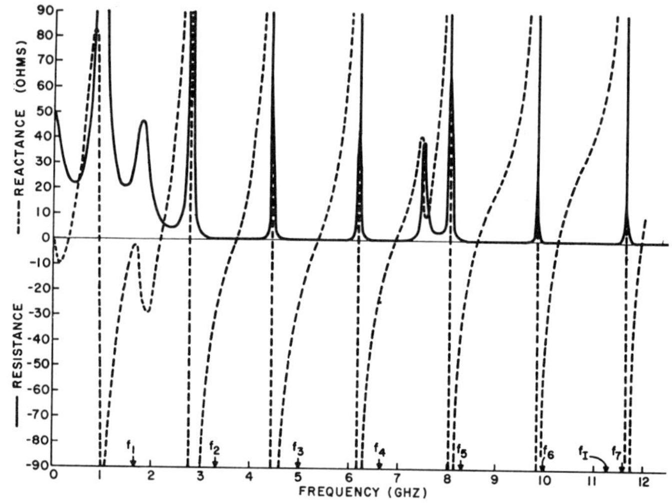

Fig. 11. Load impedance versus frequency for diode B oscillating in trapped-plasma mode at $f_1 = 1.66$ GHz with $\eta_1 = 49$ percent. (Two 10.2 GHz tuning sleeves—$Z_{s1} = Z_{s2} = 7.6\ \Omega$; $X_1 = 7.83$ cm, $X_2 = 2.04$ cm.) f_I indicates frequency of initiating transit-time oscillation.

[28] for similar low-frequency operation. His analysis shows that $\tau_p = [TI_{bias}/V_{pk} - 2(C_d + C_p)]50$, $I_{pk} = (C_d + C_p)V_{pk}/\tau_A$, and $I_r = V_B/50$. The quantity qN_dW/τ_A is better approximated by the peak displacement current density (J_t) than I_{pk}/A as was contended. The predicted values for τ_p, I_{pk}, and I_r are 0.4 ns, 5 A, and 1.8 A, respectively. These quantities agree well with the experimental values: $\tau_p \approx 0.42$ ns, $I_{pk} \approx 4.2$ A, and $I_r \approx 1.5$ A. Inaccuracies resulting from the limited sampling oscilloscope response and the effect of the cartridge inductance can easily account for part of the discrepancy. A set of voltage, current, and displacement current waveforms have thus been obtained which lend considerable validity to the explanation of high-efficiency oscillations based on a traveling avalanche zone and trapped-plasma state.

C. Circuit Essentials

Details of the circuit conditions which existed for typical trapped-plasma mode operation will now be presented. The load impedance seen by a diode chip which oscillated at 1.66 GHz with 49 percent efficiency is illustrated in Fig. 11. The fundamental frequency was in the vicinity of a resonance transmission window where the sample load impedance had a local minimum in reactance. The diode impedance at f_1 was approximately $-33 + j2\ \Omega$ and the corresponding Q factor was almost zero. At each harmonic of the fundamental frequency the imaginary part of the diode impedance was much larger than the real part (e.g., $Z_2 = -1 + j30\ \Omega$). It should be noted that these features of the trapped-plasma mode circuit are similar to the circuit characteristics for subharmonic generation discussed in Section IV. Evans [28] has pointed out the low-pass filter nature of this general type of circuit. Note that all of the higher harmonics of the 1.66 GHz fundamental frequency are above the 3 GHz cutoff of the circuit in

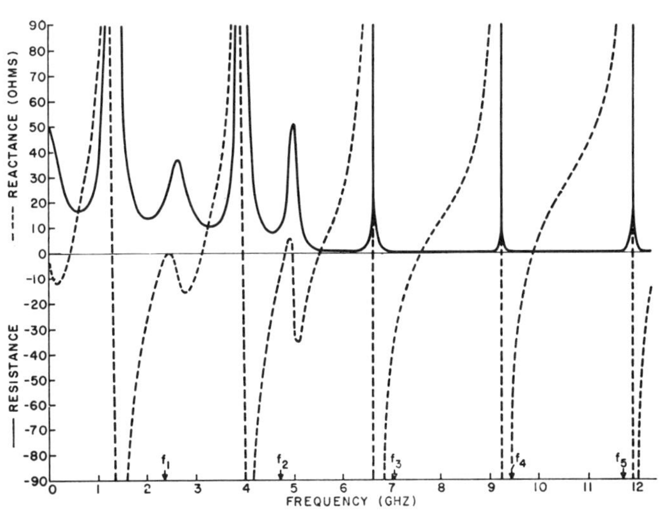

Fig. 12. Load impedance versus frequency for diode B oscillating in trapped-plasma mode at $f_1=2.35$ GHz with $\eta_1=37$ percent. (Three 10.2 GHz tuning sleeves—$Z_{s1}=Z_{s2}=Z_{s3}=11.3$ Ω, $X_1=5.10$ cm, $X_2=0.543$ cm, $X_3=0.642$ cm.)

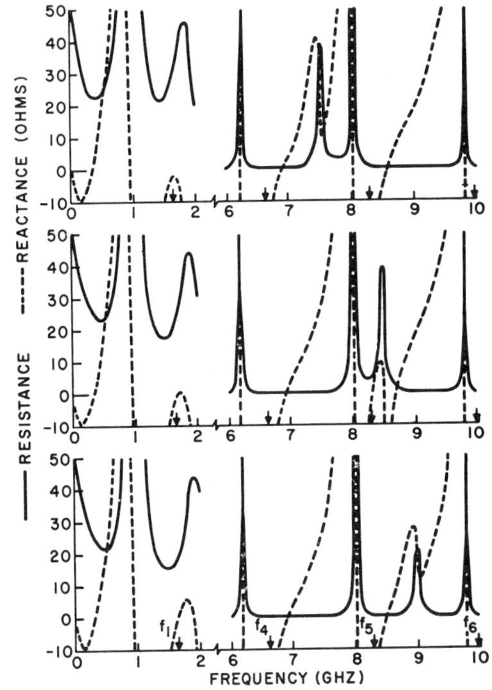

Fig. 13. Load impedance versus frequency for three different settings of the second tuning slug: (upper trace) $f_1=1.66$ GHz, $\eta_1=49$ percent, $P_1=27$ W, $X_2=2.04$ cm; (middle trace) unstable, $X_2=1.80$ cm; (lower trace) $f_1=1.66$ GHz, $\eta_1=50$ percent, $P_1=26.5$ W; $X_2=1.67$ cm.

Fig. 11. All circuit impedance calculations made for high-efficiency situations down to 0.9 GHz showed essentially similar features to those of Fig. 11. In general these two-sleeve tuning conditions resulted in higher efficiency than the low-frequency circuit described by Evans [28].

In order to obtain oscillations at frequencies higher than 1.7 GHz it was necessary to use three tuning sleeves having higher characteristic impedances. The way in which the circuit requirements changed for high-frequency cases is typified by the example shown in Fig. 12. The fundamental frequency was at 2.35 GHz and the efficiency was 37 percent. Once again the fundamental frequency was in the vicinity of a resonance transmission window where a local reactance minimum occurred. The second harmonic, however, was below the low-frequency cutoff at 5 GHz. In fact the diode was terminated with a small reactance and a relatively large resistance at the second harmonic. This implies that the second-harmonic voltage and current amplitudes were relatively small. Weak second-harmonic amplitudes may have resulted in a total current and voltage waveform resembling the square wave situation described by Evans and Scharfetter [13].

A desire to determine the relative importance of the various harmonics involved with the trapped-plasma mode raises interesting experimental problems. Fortuitous circumstances have, however, allowed the shorting out of the fifth harmonic of the 1.66 GHz high-efficiency case illustrated in Fig. 11. This impedance plot shows a resonance at about 7.5 GHz, halfway between the fourth and fifth harmonic. The frequency of this resonance is principally determined by the length of the two tuning sleeves and their separation. The dc to 2 GHz and 6 to 10 GHz range of the Fig. 11 impedance plot is reproduced in the upper trace of Fig. 13. The impedance plot shown in the middle trace is for a tuning situation where the second sleeve was 0.24 cm closer to the first sleeve than for the case in the upper trace. The resonance is now in the vicinity of the fifth harmonic. Under this tuning condition the oscillations were completely unstable and the efficiency was greatly reduced. By moving the second tuning sleeve 0.13 cm closer still to the first sleeve it was possible to again obtain stable high-efficiency oscillations at 1.66 GHz. The impedance plot for this case (the lower trace) has the resonance between the fifth and sixth harmonic. The load impedance below the 3 GHz cutoff and in the vicinity of the fundamental was essentially unchanged for the three different tuning conditions. It is important to stress that the circuit conditions in all three cases were essentially unchanged for frequencies near the initiating transit-time oscillation. Apparently harmonic components up to at least the fifth were essential for high-efficiency operation at 1.66 GHz.

D. Harmonic Generation

Now it seems natural to ask to what degree can energy be coupled out at higher harmonics without affecting oscillator stability. Efficient harmonic generation would obviously require that minimum power be delivered at other harmonics, including the fundamental. Two circuit tuning conditions which met this requirement are illustrated in Figs. 14 and 15. In both cases the fundamental frequency was above the effective cutoff of the circuit and a resonance transmission window allowed power to be delivered at a higher harmonic. The fundamental frequency was determined by the

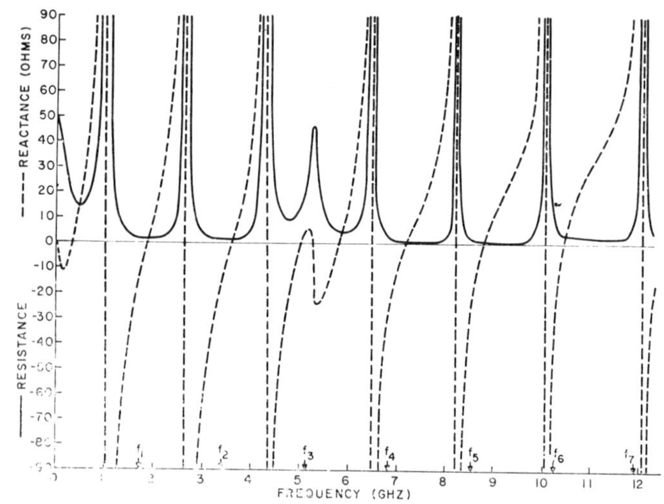

Fig. 14. Load impedance of diode B oscillating at $f_1 = 1.7$ GHz, $\eta_1 = 1.2$ percent, but with principal output at third harmonic $f_3 = 5.1$ GHz, $\eta_3 = 19$ percent. (One 10.2 GHz and one 4.86 GHz tuning sleeve—$Z_{s1} = Z_{s2} = 7.6\ \Omega$; $X_1 = 7.62$ cm, $X_2 = 0.085$ cm.) $I_{bias} = 325$ mA, $V_{bias} = 64$ V.

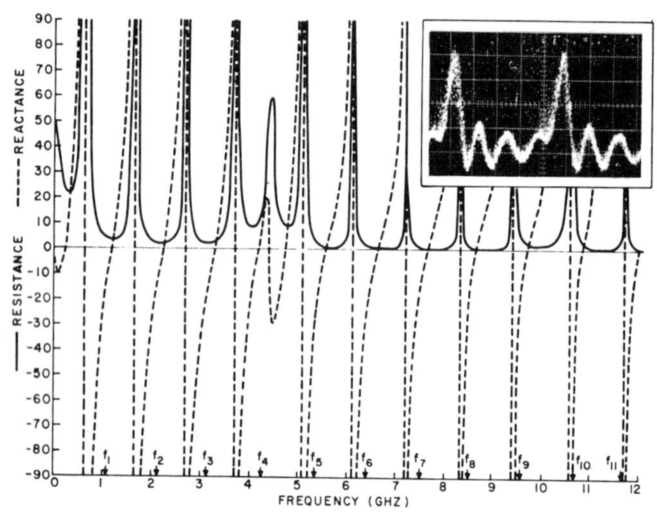

Fig. 15. Load impedance of diode B oscillating at $f_1 = 1.06$ GHz, $\eta_1 = 3.3$ percent but with principal output at fourth harmonic $f_4 = 4.24$ GHz, $\eta_4 = 15$ percent. Oscillograph insert illustrates external current waveform. (One 10.2 GHz and one 4.86 GHz tuning sleeve—$Z_{s1} = Z_{s2} = 7.6\ \Omega$; $X_1 = 12.9$ cm, $X_2 = 0.175$ cm.) $I_{bias} = 375$ mA, $V_{bias} = 78$ V.

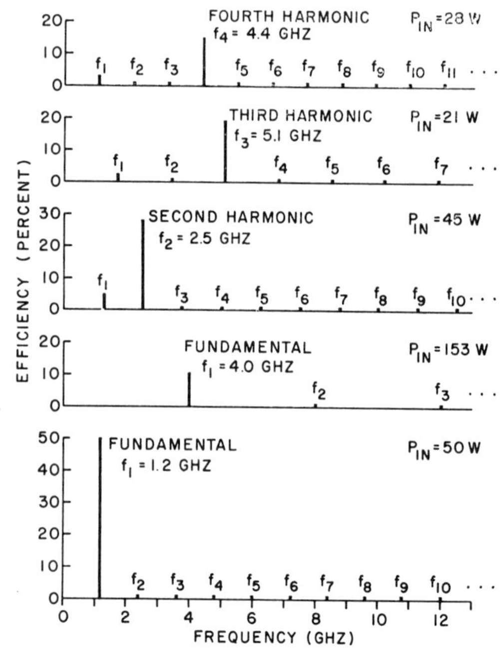

Fig. 16. Bar graphs of efficiency from harmonic generation and fundamental trapped-plasma mode. (Short bars indicate only the presence of the particular harmonic.) $P_{in} = V_{bias} I_{bias}$.

position of the first tuning sleeve, as was also the case with the fundamental trapped-plasma mode. The circuit tuning corresponding to Fig. 14 resulted in 19 percent efficiency at the third harmonic of a 1.7 GHz fundamental. Fifteen percent at the fourth harmonic of a 1.06 GHz fundamental resulted from the circuit tuning of Fig. 15. An oscillograph of the external current waveform is shown at the upper right of Fig. 15. This waveform clearly shows the presence of a strong fourth harmonic. This fourth harmonic contribution was probably displacement current resulting from a fourth-harmonic component in the voltage waveform.

Bar graphs which compare the efficiency capabilities of harmonic generation and the fundamental trapped-plasma mode are shown in Fig. 16. Input powers (P_{in}) are listed for each case. The bar graphs show that harmonic generation resulted in an extension of high-efficiency operation to frequencies above the upper limit for the fundamental trapped-plasma mode. The highest frequency for this type of operation was observed by Stark in a private communication where 2.5 percent efficiency was obtained at the 20th harmonic of a 0.75 GHz fundamental. This efficiency is comparable to that observed by Liu *et al.* [31] for operation at similar transit angles but where apparently only a partial plasma formation occurred.

Unfortunately, precise waveforms were not obtainable for the harmonic generation cases. It is therefore difficult to determine how the shock-front transit or the plasma extraction period differed from normal operation. The low input powers indicate that the peak currents and/or extraction times may have been reduced. In addition the low input powers required for harmonic generation raise the possibility of efficient CW operation at high frequencies.

VI. Summary and Conclusions

Experimental evidence has been presented which distinguishes two physical mechanisms which can result in efficient small transit angle oscillations in p^+-n-n^+ silicon avalanche diodes.

Microwave oscillations which occurred at subharmonics of reactively terminated transit-time oscillations were studied using an avalanche diode with an n-n^+ interface field which was 5 percent of the p^+-n junction field at breakdown. Two examples were considered: a three-frequency synergetic system with an efficiency of 6.5 percent at a fundamental corresponding to a transit angle of approximately 0.28 π and a two-frequency synergetic system with an efficiency of 9.1 percent at a fundamental corresponding to a transit angle of approximately 0.45 π. It was concluded that this latter type of operation was inconsistent with the trapped-plasma mechanism but qualitatively consistent with published analyses of multiharmonic excitations in avalanche diodes. This conclusion was substantiated by the following observations resulting from circuit analysis, measured external current waveforms, and degenerate amplifier characteristics.

1) The phase of the voltage components in the three-harmonic system was such that the peak in the total voltage waveform was delayed with respect to the fundamental voltage component. This delay effectively produced a negative resistance at the second subharmonic of a reactively terminated oscillation which had a transit angle of approximately 0.8 π. The active subharmonic negative resistance effectively ruled out an explanation based on simple Manley–Rowe parametric effects in a pumped nonlinear reactance.

2) The minimum of the RF voltage waveform was approximately 0.75 V_B. This high minimum voltage is inconsistent with a fully formed trapped-plasma state which necessarily results in a low voltage across the sample.

3) The maximum in the computed displacement current waveform was less than 0.25 times the critical current which would be required to launch a shock front or traveling avalanche zone.

4) Rectification effects were studied in a self-pumped degenerate amplifier which exhibited small-signal linear gain. The bias voltage decrease and bias current increase were found to be linearly related to the square of the fundamental voltage excitation. This well-behaved nonlinear rectification is readily explained by existing theoretical considerations of avalanche processes under the condition of saturated carrier velocities and relatively small voltage perturbations.

The practical upper limit of the efficiency of subharmonic generation resulting from multiharmonic excitations was not determined.

Truly high conversion efficiencies of 50 percent were observed using another p^+-n-n^+ avalanche diode with an n-n^+ interface field which was approximately 60 percent of the p^+-n junction field at breakdown. Detailed sampling oscilloscope measurements of external current and dV/dt waveforms for low-frequency cases were found to substantiate Evans' analysis for the interaction between standard coaxial cavities and trapped-plasma mode oscillators. The following principal features of the trapped-plasma mode in a coaxial cavity were observed.

1) Experimental dV/dt waveforms indicated the presence of peak displacement currents prior to breakdown which were on the order of 1.5 times the critical current (qN_dv_s) required to launch a traveling avalanche zone.

2) Approximately determined conduction current waveforms exhibited larger currents during the shock-front transit than during the extraction period. The peak in the conduction current was approximately equal to $(C_d+C_p)\times V_{pk}/\tau_A$ while the current during the extraction period was approximately $V_B/50\ \Omega$. The duration of the large conduction current was approximately $\tau_p \approx [TI_{\text{bias}}/V_B - 2(C_d+C_p)]Z_0$.

3) The sample voltage dropped from a peak value of about 1.4 V_B to approximately zero during the shock-front transit. The sample voltage during the plasma extraction period increased slowly from near zero, indicating the presence of a fully formed trapped-plasma state. At the end of the extraction period the voltage was again near V_B.

The 4 GHz upper bound frequency for the fundamental trapped-plasma mode indicated a minimum plasma extraction time of about 0.25 ns. The frequency range of optimum fundamental mode efficiency was between 0.9 and 1.7 GHz. These features were found to be consistent with the analytical calculation made by Cottam for a diode with doping and width similar to that studied and for nonsquare waveforms similar to those observed.

Detailed calculations were made of the impedance of experimental circuits for both the trapped-plasma mode and subharmonic generation resulting from avalanche-resonance pumping. General circuit characteristics were found to be strikingly similar for both modes. Optimum cases were found to occur when the fundamental frequency was in the vicinity of a resonance transmission window where the load impedance had a minimum in reactance, and when all harmonics up to the transit-time region were reactively terminated. It was concluded that under certain circumstances in identical circuits, different diodes might work in either the trapped-plasma mode or the transit-time subharmonic generation mode. For lower efficiency cases it is thus important to accurately determine current and voltage waveforms before the correct mode of operation can be determined. Circuit analysis also indicated the critical importance of each separate harmonic of a trapped-plasma mode fundamental. For example, it was found that shorting out the fifth harmonic of a fundamental at 1.7 GHz was sufficient to cause bias circuit instability and a marked deterioration of efficiency.

By proper tuning of the coaxial circuit it was possible to reactively terminate the fundamental trapped-plasma mode frequency while placing a resonance transmission window in the vicinity of a higher harmonic. This type of circuit environment enabled an extension of the upper bound of the frequency of efficient generation. An efficiency of 19 percent was possible at the third harmonic of a 1.7 GHz fundamental. The input power required for this example was considerably less than that required for high-frequency generation in the fundamental trapped-plasma mode. It was concluded that harmonic generation can enable efficient operation at frequencies above the upper limit of the fundamental trapped-plasma mode and improve the possibility of efficient CW operation.

Note Added in Proof: External current and dV/dt waveform measurements using a Tektronix 25 ps sampling oscilloscope were made subsequent to the submission of this paper. dV/dt waveforms contained *harmonically* related components up to approximately 25 GHz (twice the normal transit-time frequency). Measured external current waveforms were not radically altered except that the shock-front-launching displacement current peak was in evidence just prior to the major current peak. Aside from these differences the waveforms were similar to those described in the paper.

Acknowledgment

The author is indebted to Prof. C. Ingalls for the loan of his sampling oscilloscope. The author also wishes to thank G. Fitzsimmons of the Boeing Company for his generous donation of the lowly doped silicon avalanche diodes. The contacting and p^+ diffusion of diode A were done at Cornell by R. E. Goldwasser and C. A. Lee, respectively. W. O. Camp was the creator of the program which enabled some machine to calculate the various circuit impedance characteristics. L. A. Stark performed the network analysis which corroborated the machine calculations. Discussions with W. O. Camp, L. A. Stark, B. Culshaw, L. F. Eastman, and C. A. Lee were as enlightening as they were numerous. Thanks are also due to L. A. Stark, Prof. J. Frey, and Prof. C. A. Lee for their reading of the manuscript. This work would not have been possible without the initial inspiration provided by B. Hoefflinger.

References

[1] D. L. Scharfetter, D. J. Bartelink, H. K. Gummel, and R. L. Johnston, "Computer simulation of low-frequency high-efficiency oscillations in germanium," presented at the IEEE Solid-State Device Research Conf., Boulder, Colorado, June 17–19, 1968.

[2] B. Hoefflinger, C. P. Snapp, and L. A. Stark, "High-speed computer program for avalanche diode microwave generators with multiple resonances," in *Proc. 1969 Cornell Conf. Computerized Electronics*, pp. 402–411.

[3] B. C. DeLoach and D. L. Scharfetter, "Device physics of TRAPATT oscillators," *IEEE Trans. Electron Devices*, vol. ED-17, Jan. 1970, pp. 9–21.

[4] D. J. Bartelink and D. L. Scharfetter, "Avalanche shock fronts in p-n junctions," *Appl. Phys. Lett.*, vol. 14, May 15, 1969, pp. 320–323.

——, "Avalanche shock fronts in p-n junctions," *IBM J. Res. Develop.*, vol. 13, Sept. 1969, pp. 596–600.

[5] A. S. Clorfeine, R. J. Ikola, and L. S. Napoli, "A theory for the high-efficiency mode of oscillation in avalanche diodes," *RCA Rev.*, vol. 30, Sept. 1969, pp. 397–421.

[6] R. L. Johnston, D. L. Scharfetter, and D. J. Bartelink, "High-efficiency oscillations in germanium avalanche diodes below the transit-time frequency," *Proc. IEEE* (Lett.), vol. 56, Sept. 1968, pp. 1611–1613.

[7] C. P. Snapp, L. A. Stark, and B. Hoefflinger, "Performance and theory of avalanche-resonance pumped IMPATT oscillators," presented at International Electron Devices Meeting, Washington, D.C., October 23–25, 1968.

——, "Experimental analysis of high-efficiency avalanche-resonance pumped-oscillators," *Electron. Lett.*, vol. 4, Dec. 27, 1968, pp. 595–596.

[8] H. C. Bowers, "Space-charge-induced negative resistance in avalanche diodes," *IEEE Trans. Electron Devices*, vol. ED-15, June 1968, pp. 343–350.

[9] R. L. Johnston and D. L. Scharfetter, "Low-frequency high-efficiency oscillations in germanium IMPATT diodes," *IEEE Trans. Electron Devices*, vol. ED-16, Nov. 1969, pp. 905–911.

[10] H. Torizuko and H. Yanai, "Observation of the current and voltage waveforms of the Si IMPATT diode," *Proc. IEEE* (Lett.), vol. 57, Mar. 1969, pp. 349–350.

[11] H. Yanai, H. Torizuko, and N. Yamada, "Observation of the current and voltage waveforms of the avalanche diode oscillators," *1969 IECE Con. Rec.* (Japan); also, private communication.

[12] C. P. Snapp, "External current-waveform measurements of high efficiency oscillations in silicon avalanche diodes," *Electron. Lett.*, vol. 6, Mar. 5, 1970, pp. 145–147.

[13] W. J. Evans and D. L. Scharfetter, "Characterization of avalanche diode TRAPATT oscillators," *IEEE Trans. Electron Devices*, vol. ED-17, May 1970, pp. 397–404.

[14] C. P. Snapp and B. Hoefflinger, "An efficient multiresonant avalanche diode oscillator in the 1.5 to 11 GHz range," *Proc. IEEE* (Lett.), vol. 56, Nov. 1968, pp. 2054–2055.

[15] H. J. Prager, K. K. N. Chang, and J. Weisbrod, "Anomalous silicon avalanche diodes for microwave generation," *Proc. 1967 Cornell Conf. High Frequency Generation and Amplification*, pp. 266–280.

[16] B. Hoefflinger, C. P. Snapp, and L. A. Stark, "High-efficiency-avalanche resonance pumped amplification," *IEEE J. Solid State Circuits*, vol. SC-4, Dec. 1969, pp. 391–395.

[17] C. P. Snapp and B. Hoefflinger, "Degenerate amplification with semiconductor diodes due to a self-generated avalanche-resonance pump," *Electron. Lett.*, vol. 5, Aug. 21, 1969, pp. 393–395.

[18] R. Kuvås, "Nonlinear study of semiconductor avalanches," Ph.D. thesis, Cornell Univ., Ithaca, N. Y., 1970.

[19] C. A. Lee, R. L. Batdorf, W. Wiegmann, and G. Kaminsky, "Time dependence of avalanche processes in silicon," *J. Appl. Phys.*, vol. 38, June 1967, pp. 2787–2796.

[20] A. S. Clorfeine, "Self-pumped parametric amplification with an avalanching diode," *Proc. IEEE* (Lett.), vol. 54, Dec. 1966, pp. 1956–1957.

[21] R. Kuvås and C. A. Lee, "Nonlinear analysis of multifrequency operation of Read diodes," *J. Appl. Phys.*, vol. 41, Mar. 15, 1970, pp. 1756–1767.

[22] B. Culshaw, "Low-frequency operation of avalanche diodes under large-signal conditions," *Electron. Lett.*, vol. 5, Nov. 13, 1969, pp. 583–585.

——, "High efficiency, low frequency operating modes in avalanche diodes under low current density conditions," *Proc. Inst. Elec. Eng.*, to be published.

[23] W. E. Schroeder and G. I. Haddad, "Effect of harmonic and subharmonic signals on avalanche-diode oscillator performance," *IEEE Trans. Microwave Theory Tech.*, vol. MTT-18, June 1970, pp. 327–331.

[24] K. Mouthaan, "Nonlinear characteristics and two-frequency operation of the avalanche transit-time oscillator," *Philips Res. Rep.*, vol. 25, Dec. 1969, pp. 33–67.

[25] W. Harth, "Resonance and parametric effects in IMPATT diodes," *IEEE Trans. Electron Devices*, vol. ED-17, Apr. 1970, pp. 282–289.

[26] R. Kuvås, "Degradation of the oscillator performance of Read diodes due to space-charge and series resistance," *Electron. Lett.*, vol. 6, Apr. 2, 1970, pp. 216–217.

[27] H. Goronkin and S. Wanuga, "High efficiency S-band silicon avalanche oscillator," presented at the IEEE Int. Solid-State Circuits Conf., Philadelphia, Pa., Feb. 18–20, 1970.

[28] W. J. Evans, "Circuits for high-efficiency avalanche-diode oscillators," *IEEE Trans. Microwave Theory Tech.*, vol. MTT-17, Dec. 1969, pp. 1060–1067.

[29] M. G. Cottam, "Theory for high-efficiency oscillations in silicon avalanche diodes," *Electron. Lett.*, vol. 6, Apr. 2, 1970, pp. 203–205.

——, "Theory for power output and efficiency of silicon TRAPATT oscillators," *Electron. Lett.*, vol. 6, June 11, 1970, pp. 384–385.

[30] D. L. Scharfetter, "Power-frequency characteristic of the TRAPATT diode mode of high efficiency power generation in germanium and silicon avalanche diodes," *Bell. Syst. Tech. J.*, vol. 49, May–June 1970, pp. 799–826.

[31] S. G. Liu, J. J. Risko, and K. K. N. Chang, "High-power K-band silicon avalanche-diode oscillators," *Proc. IEEE* (Lett.), vol. 58, June 1970, pp. 919–920.

Fabrication and Performance of Kilowatt L-Band Avalanche Diodes*†

S. G. Liu and J. J. Risko, RCA Laboratories, Princeton, N. J.

Abstract—A reliable process for the fabrication of high-efficiency high-power avalanche-diode oscillators is described. The key step in the process is a deep diffusion of boron into high-resistivity silicon epitaxial wafers. Such junctions have been found to be consistently free of microplasma effects and capable of handling high avalanche current, which is one of the essential requirements of the high-efficiency high-power avalanche-diode oscillator. Using these diodes in a multiple-stacked configuration, we have reached a 1.2 kW power output at 1.11 GHz with an efficiency of 25.6%. This power level is suitable for such applications as IFF transponder systems.

Combined power output has been obtained from series-connected, as well as from parallel-connected, diodes. The diodes have been operated in both simple coaxial circuits and microstrip circuits. The circuit impedance presented at the position of the diode has been computed for varying circuit parameters.

Introduction

The high-efficiency mode of oscillation in avalanche diodes was discovered at RCA Laboratories.[1] It was found that when a threshold current level is reached in a certain type of reverse-biased avalanche diode, rf oscillation is observed, accompanied by an abrupt decrease in average voltage across the diode and an increase in average current. The frequency of this "anomalous-mode" oscillation is much lower than the corresponding transit-time frequency of the diode; the power output and the efficiency are many times greater than those of the IMPATT diode oscillator.[2] Proposed theories as well as experimental evidence have indicated[3-7] that support of higher-order harmonics, in-

* The research reported in this paper was jointly sponsored by the Air Force Avionics Laboratory, Wright-Patterson Air Force Base, Ohio, under Contract F33615-68-C-1688, the Sandia Corporation, Albuquerque, New Mexico, under Contract 57-6779, and RCA Laboratories, Princeton, N.J.

† A major portion of this paper was presented at the international Electron Device Meeting, Washington, D.C., Oct. 1969.

cluding the transit-time frequency, are necessary for this high-efficiency mode of oscillation.

Combined power output has been obtained from both series-connected and parallel-connected diodes, and series-connected diodes have been operated in both simple coaxial circuits and microstrip circuits. The circuits are capable of matching to the diode over a wide impedance range at the fundamental frequency, and simultaneously supporting the required higher-order harmonic frequencies. The circuit impedance presented at the position of the diode has been computed for varying circuit parameters.

The best duty cycle achieved with the present series-stacked geometry at the kilowatt level is 0.1%. With improved packaging and heat-sinking techniques, operation at 1% duty cycle should be possible. Methods of improving heat dissipation are discussed.

Device Fabrication

High-power high-efficiency avalanche diodes operating at L-band frequencies have the following salient features.[3] First, the n-region of the diode has a low doping density. Secondly, the width of the n-region is narrower than the corresponding depletion width of the n-region at breakdown. Thus, the depletion layer sweeps across the n-region, i.e., the diode "punches through", before reaching the point of avalanche breakdown. Thirdly, the diode must be capable of handling high avalanche current.

356

A reliable fabrication process has been developed that satisfies these requirements and produces diodes capable of power outputs up to the 200-W level. The key element of the diodes is a deeply diffused junction. Such junctions have been found consistently to be free from microplasma effects and capable of handling high avalanche currents. Typical diffusion depths range from 8 to 15 microns. On the other hand, abrupt junctions formed by shallow diffusion from a high-surface-concentration source into high-resistivity materials are found, in general, to display microplasma effects and burn out easily.

Epitaxial Material

The fabrication process starts with silicon epitaxial wafers consisting of an n-layer on heavily doped n-type substrates. The resistivity and thickness of this n-layer are controlled during crystal growth. Epitaxial material is used because the doping density of a thin n-layer can be held uniform across the whole wafer. Silicon epitaxial wafers grown at RCA Electronic Components, Somerville,

New Jersey, have given excellent results in device performance and yield. The epitaxial layer is phosphor doped. Typical resistivity is 4 to 7 ohm-cm, and the layer thickness is 12 to 23 microns. The substrate is antimony doped, and the resistivity is between 0.005 and 0.01 ohm-cm. Both (111) and (100) oriented wafers have been used successfully in diode fabrication.

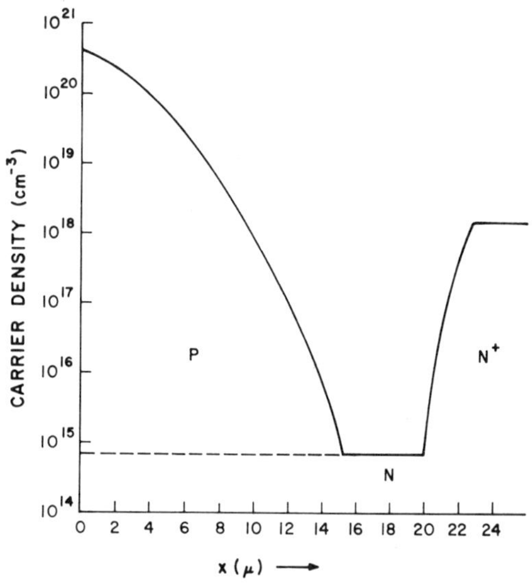

Fig. 1—Diffusion profile of "anomalous-mode" diode with a deep-diffused junction.

Deep Diffusion

A boron silicate glass is deposited onto the silicon wafer at 720°C. Diffusion of boron into the wafer at 1150°C follows. With precise control of the diffusion time, the diffusion depth, and hence the thickness of the n-layer, can be controlled within 0.5 micron. The precise control of the undiffused n-layer is essential to diode performance as a high-power microwave oscillator. An example of the diffusion profile assuming a complementary error-function distribution is shown in Fig. 1. A net undiffused n-region of 4 to 6 microns is desired for diodes operating around 1 GHz. The depletion width at breakdown will extend slightly into the graded p and n+ regions. This depletion width can be calculated from the capacitance measured at breakdown and diode dimensions. The calculated depletion width is about 2 microns wider than the undiffused n-region. It should be noted that the diode has a graded p, as well as a graded n-n+ region; the latter is a result of out-diffusion from the n+-substrate.

The fact that deep-diffused diodes with graded junctions have high current-handling capacity was shown in the following experiment. A silicon epitaxial wafer with a 6 ohm-cm epi-layer was sliced into six sections. Each section was diffused from a high-surface-concentration source for a different length of time at the same temperature. Large-size (0.020-inch diameter) diodes were then made from each section of the diffused wafers, and the *I-V* characteristics of the diodes were examined on a curve tracer. It was found that shallow diffused diodes, up to a depth of about 8 microns, show microplasma effects and burn-out at low current levels. The diodes diffused for longer time showed no microplasma effects and took much higher avalanche current. A comparison of *I-V* characteristics of a 4-hour-diffused diode, and an 8-hour-diffused diode is shown in Fig. 2.

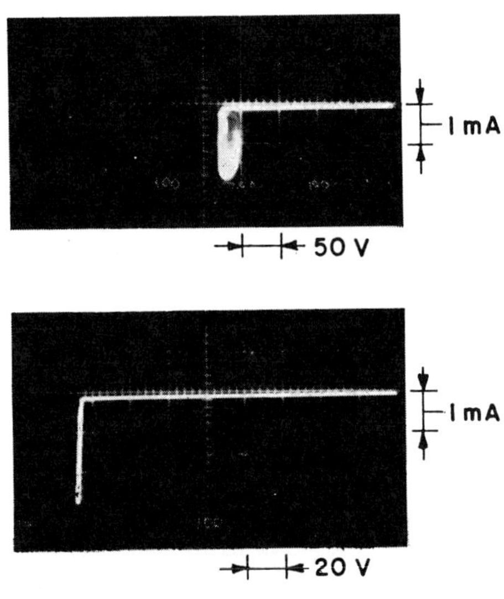

Fig. 2—Reverse-biased *I-V* characteristics of (top) shallow-diffused diodes showing microplasma effect, and (bottom) deep-diffused diodes showing no microplasma effect.

The diffusion process was followed by steps of metalization, masking (with a photoresist), and chemical etching to obtain disk-shaped diode chips. A cross-sectional drawing of the finished disk configuration is shown in Fig. 3.

The disk-shaped chip has a larger diameter on the p-side than on the n-side. This shape is produced by etching from the substrate straight through the wafer. The wafer thickness is reduced to about 3 mils before metalization and etching. The disk configuration has several advantages. The larger diameter on the p-side facilitates handling and reduces possible mechanical damage of the junction when the chips are mounted p-side down for better heat sinking. Furthermore, the increased length at the edge of the junction reduces the electric field strength in this region and thus reduces the possibility of burnout. Diodes made with this process have operated at current levels of 10 A and produced power outputs of 200 W, as described later.

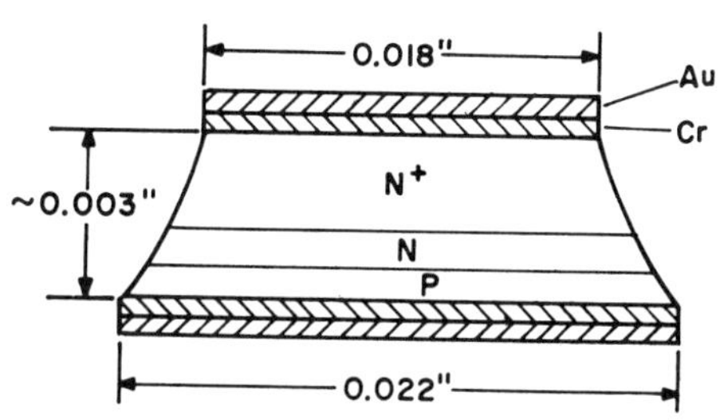

Fig. 3—Disk-shaped diode chip.

Device Testing

The power output from individual diodes using the deep-diffusion process ranges between 100 and 200 W, and efficiencies range from 25% to 40%. Combined power outputs have been obtained from diodes connected in series as a multiple-stacked unit.[8] With this technique, a power level of 1.2 kW has been achieved, the first time that the kilowatt power level has been reached in these devices. Because the diodes are current-controlled, series operation is preferred to parallel operation. In a series string, all pass the same current, and the voltage division between diodes in the string is automatic; in a parallel set, all diodes see the same voltage, and individual currents may vary greatly. Series operation of diodes has been demonstrated in microstrip, as well as in coaxial, circuits.

A block diagram of the experimental apparatus used for power and frequency measurements is shown in Figure 4. Precision calibrated attenuators and a peak power meter were used for accurate power measurement. The precision attenuator has a flat attenuation between dc and 12.4 GHz. A wavemeter connected between the crystal detector output and the oscilloscope is used for frequency measurement. The pulsed rf output is also monitored on a spectrum analyzer, which is not shown in the diagram. The power rating of the attenuators has become important in high-power measurements. The power output from the five-stacked diodes burned out a medium-power attenuator that was rated at 1-kW peak power. Attenuators rated at 2 kW are presently used in power measurement.

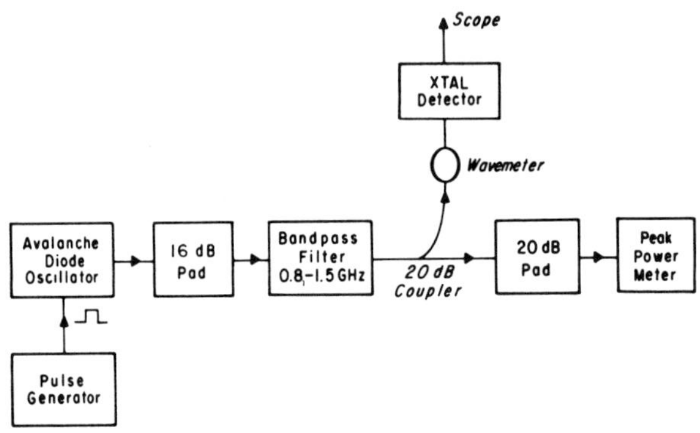

Fig. 4—Experimental setup for measurement of power and frequency.

A filter with a pass band from 0.8 to 1.5 GHz was included in the setup to ensure that only fundamental power was measured. Even without the external filter, the second-harmonic content in the output is usually more than 15 dB below the fundamental for diodes operating in the range of 1.0 to 1.5 GHz. The pulser has an internal impedance of 200 ohms, which is high compared to that of the diodes ($\sim$ 10 ohms). The use of a high-impedance pulse generator minimizes the sometimes self-destructive jump in current as the anomalous mode current threshold is exceeded.

The circuit in which the series-stacked diodes were tested consists of a coaxial line with a movable short-circuit and a low-pass-filter tuning section, as reported by Levine and Liu.[9] A schematic diagram of the circuit is reproduced in Fig. 5. The circuit is capable of matching the diode impedance at the fundamental frequency, while simultaneously supporting the higher harmonics with a reactive termination. The main line has 50 ohms surge impedance, while the tuning section is formed by a length of 125-ohm impedance line and a lumped variable capacitor. The distances from the short circuit to the diode and from the tuning section to the diode determine the oscillation frequency of the diode within its range of oscillation.

Due to the simplicity of the circuit, the impedance presented at the

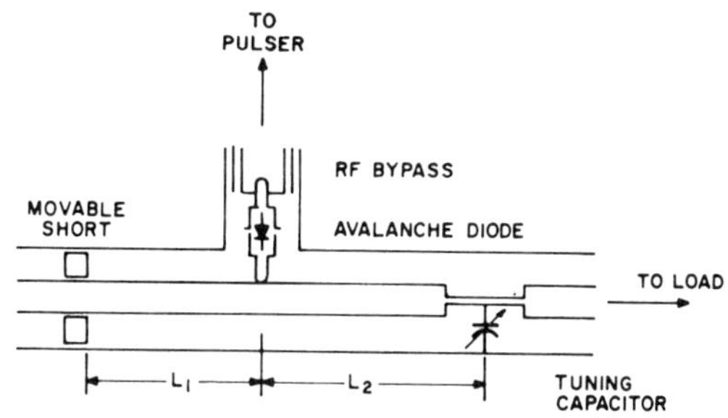

Fig. 5—Schematic diagram of L-band avalanche-diode oscillator.

position of the diode can be easily calculated for varying circuit parameters and frequencies. Such calculated results would provide information on the diode impedance when circuit parameters under optimum power output conditions are known, or vice versa. Fig. 6 shows the calculated impedance variation on a Smith chart, normalized to 50 ohms, for various positions of the short circuit and the tuning section. The frequency is chosen at 1 GHz, and the impedance of the tuning

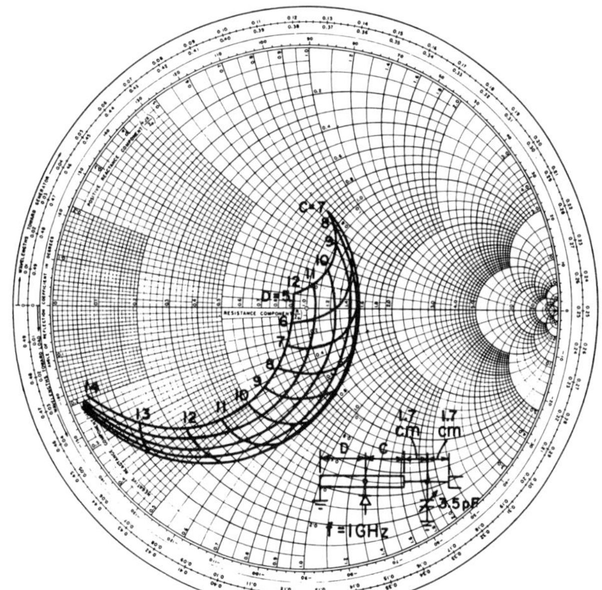

Fig. 6—Impedance calculation with varying positions of short-circuit and tuning section with respect to the diode.

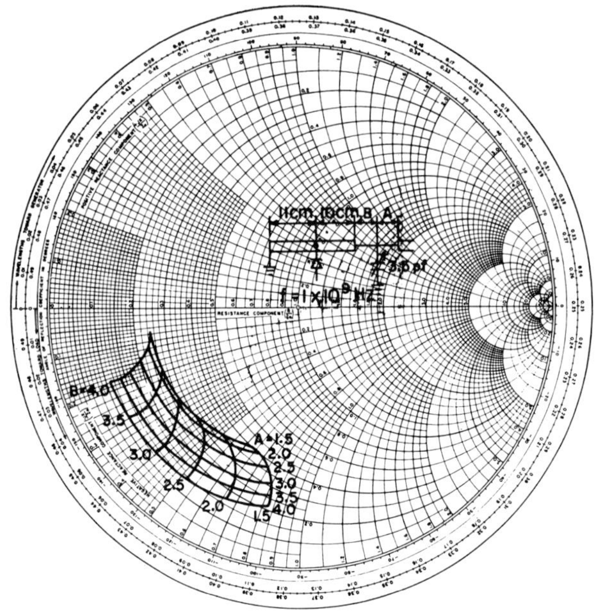

Fig. 7—Impedance calculation with varying length in tuning section.

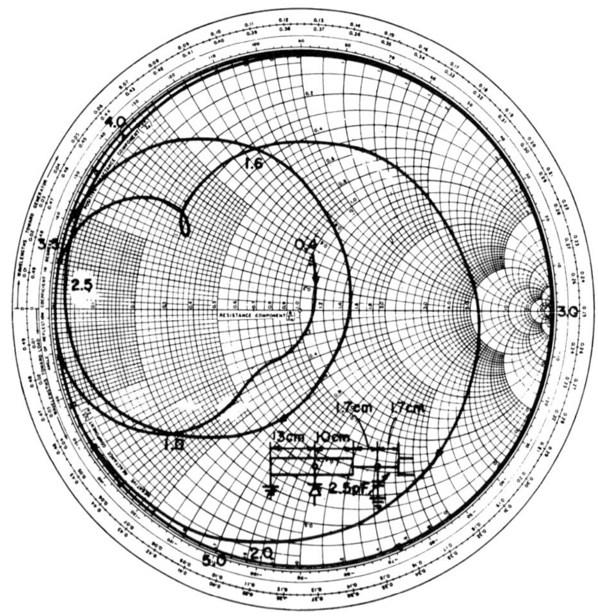

Fig. 8—Impedance calculation with varying frequency.

section is fixed as shown in the figure. The curves show that the circuit is capable of matching to the diode at a wide range of impedance levels. Fig. 7 shows the impedance variation with various lengths of high-impedance line in the tuning section. The frequency and other circuit parameters are fixed as shown. Fig. 8 shows the calculated impedance presented at the position of the diode for a frequency variation from the fundamental through its fifth harmonic. Circuit parameters in this case are fixed and are taken from experimental settings. At harmonic frequencies, the impedances appear to be purely reactive.

Series Configuration

The configuration of a multiple-stacked diode unit is shown in Fig. 9, where five diodes are series-connected in the single package. Individual diode chips are first mounted in pillboxes, which are then inserted into a modified 1N23 rectifier cartridge. The stacked geometry has the advantage that it can be treated as a single unit for the circuit, while each diode can easily be separated for testing or replacement. A schematic drawing of the pillbox with a disk-shaped diode mounted inside is also shown.

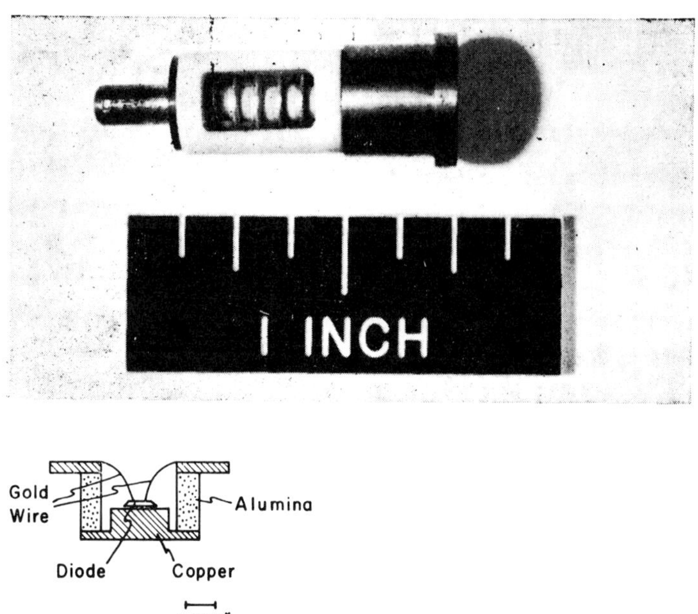

Fig. 9—Photograph of a kilowatt series-stacked five-diode unit and schematic of pill-box with a diode chip mounted inside.

Fig. 10 shows the power output and efficiency of several five-stacked units in the coaxial circuit described above with optimized circuit parameters. The data are plotted against pulsed current passing through the diodes. The maximum power output of all units is above 1 kW, and the efficiency about 25%. The best stacked-unit out of five units tested produced a power output of 1.2 kW with an efficiency of 25.6%. The operating frequency was 1.109 GHz. The upper current

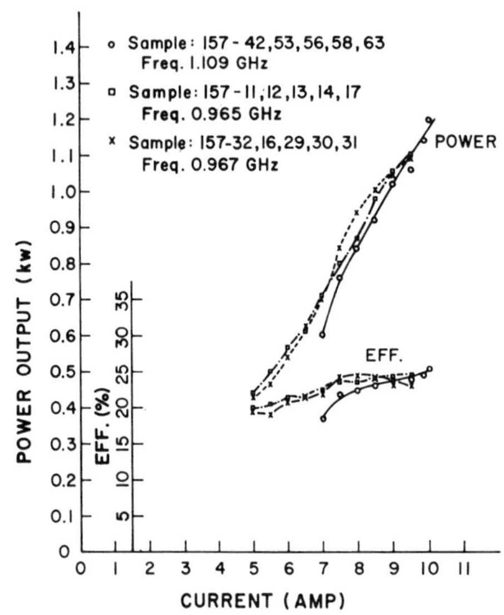

Fig. 10—Power output and efficiency versus current of several series-stacked five-diode units.

level approaches the limit of the pulser. The power output shows no sign of saturation at this current level. Individual diodes can usually take about 12 A before burning out.

The above results were obtained under pulsed conditions with a pulse width of 0.5 microsecond and a repetition rate of 400 Hz. The average voltage across the five series-stacked diodes typically drops from 680 V at breakdown to 460 V during oscillation. The applied pulse current through the diode and the corresponding detected microwave envelope at a power level of 1.2 kW are shown in Fig. 11; the corresponding sin x/x frequency spectrum is shown in Fig. 12.

One feature that seems particularly interesting is that the selection of diodes for series operation is not critical. Table I shows a summary of performance of four five-stacked units. It should be noted that al-

Table I—Summary of Performance of Four Five-Diode Series-stacked Units.

Five-Diode Units	P_{out} (kW)	Frequency (GHz)	Efficiency (%)	Breakdown Voltage Range (V)
1	1.20	1.109	25.6	124 ~ 128
2	1.00	0.965	22.0	132 ~ 140
3	1.09	0.967	23.0	128 ~ 140
4	1.10	0.965	24.6	132 ~ 140

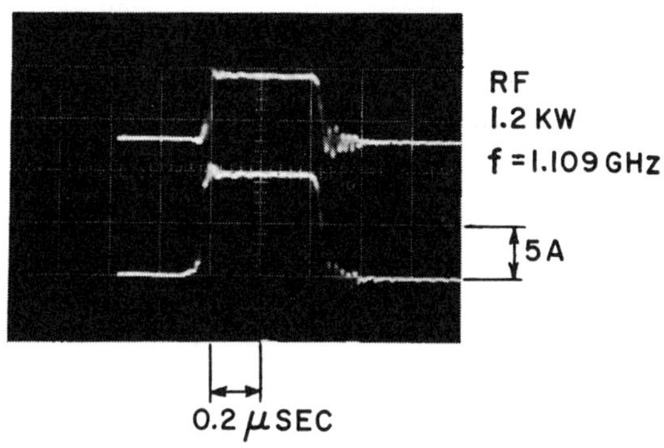

Fig. 11—Waveforms of rf envelope at 1.2 kW, 1.109 GHz (top), and current (bottom) for a series-stacked five-diode unit. Scale: current —5 A/cm; time — 0.2 μs/cm.

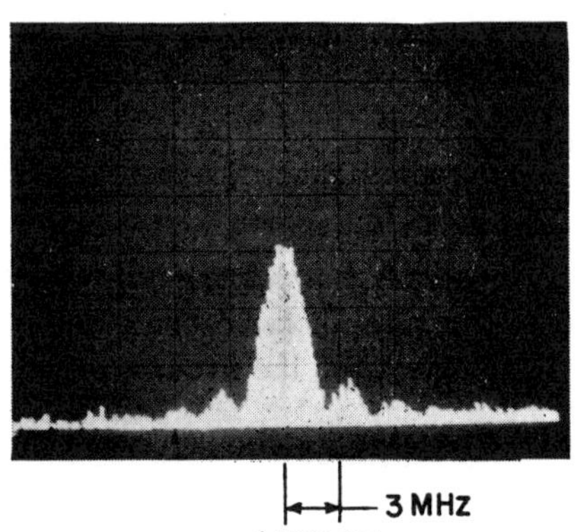

Fig. 12—Frequency spectrum of pulsed power output at 1.1 kW. Center frequency: 1.109 GHz; horizontal scale: 3 MHz/division.

though the breakdown voltages of the diodes varied between 128 and 140 volts, they still operated satisfactorily in series.

Series-connected diode oscillators increase the operating impedance level, a feature that is generally desirable in the circuit design. The circuit impedance seen by the diode and package has been measured on a network analyzer using a probe inserted in place of the diode after the circuit has been tuned for optimum power. Assuming that the circuit presents the negative of the large-signal diode impedance, the impedance of a two-stacked unit including package has a resistance of about −20 ohms, a three-stacked unit −28 ohms, and individual diodes in package −8 to −12 ohms. The susceptances are small at the measuring frequency. It is seen that the increase in circuit impedance is approximately proportional to the number of diodes stacked in series. This measurement was performed on earlier three-stacked units. The diameter of these diodes was 0.018 inch, which is smaller than that used in the five-stacked kW units.

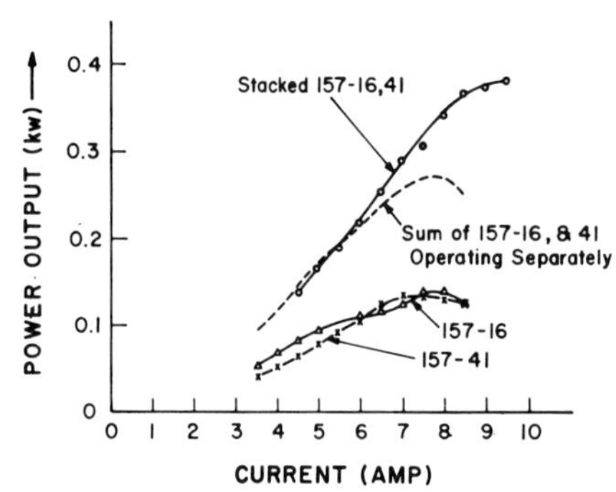

Fig. 13—Power output versus current showing that individual diodes saturate at a lower current level than series-connected diodes.

Another experiment that indicates the importance of circuit impedance level was performed. Individual diodes capable of 200 W were tested in the circuit that is best suited for the five-stacked diodes. All diodes were fabricated from the same wafer, and had a diameter of 0.022 inch. It is noteworthy that, of the four diodes tested, individual diodes showed power saturation at a lower current level than when the diodes were series connected. This result is shown in Fig. 13, where two lower curves show the power output of two diodes measured separately, and the solid curve on the top shows the measured power when two diodes are stacked in series. The curve for the sum of the individual power outputs shows saturation at a current level where the stacked diodes continue to show a power increase.

The capability of frequency tuning of the series-stacked diode units has been studied experimentally. The frequency was varied in steps by varying the positions of the tuning section and the short-circuit with respect to the diode. A unit with five series-stacked diodes was tested at a current level of 8.5 A. Fig. 14 shows the results of the power output versus frequency for a given series-stack. The frequency extends

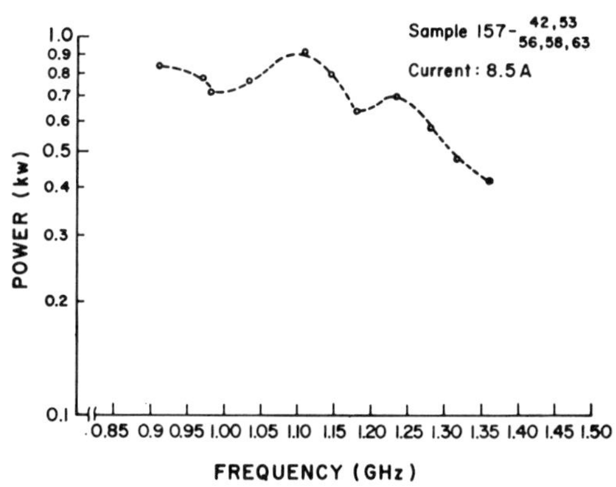

Fig. 14—Power output versus frequency of a series-stacked five-diode unit.

from 0.9 to 1.3 GHz, with a variation in power output of about 2 dB. The upper level of the power is about 800 W to 900 W. The somewhat reduced tuning range compared to the case of a single diode[9] is not surprising, because the diodes are not identical, and do not all have the same preferred frequency.

The circuit and the series-stacked geometry described above work excellently at the kilowatt level. However, the diodes are difficult to heat sink for high-duty cycle operation. The best duty cycle achieved with five-stacked diodes in the present package, operating at the kilowatt level was 0.1%. With improved mounting, packaging, and heat-sinking techniques, it is believed that one can increase the duty cycle to, say, 1%, without much difficulty. Such techniques may require the use of insulating mounting materials of high thermal conductivity, such as berillium oxide. Using such techniques, dissipation of power up to 70 W has been demonstrated in series-connected varactor diodes.[11]

Parallel Configuration

As pointed out previously, parallel operation of the "anomalous-mode" diodes is more difficult to achieve. However, parallel-connected chips have been successfully demonstrated.

Fig. 15 shows four chips connected in parallel in a large-size varactor package. A total power of 300 W was measured at a fre-

Fig. 15—Four parallel-connected diode chips in a large varactor package.

quency of 1.57 GHz and a total current of 16 A. The chips used were mesa structures. The power output of individual chips from the wafer was about 100 W at a 4 A current level. One possible reason for the lower power output obtained with paralleled diodes may be that the lower impedance is difficult for the circuit to match. It appears, however, that with care, a combination of series- and parallel-connected diodes may be used to achieve much higher power level while maintaining a suitable operating impedance level.

Microstrip Circuit

Series operation of the high-efficiency avalanche-diode oscillators has also been achieved on microstrip circuits.[10] The circuit operates on the same principle as the coaxial circuit described above. A schematic diagram of the circuit with a "π" type low-pass-filter tuning section is shown in Fig. 16. The capacitance tuning elements are thin gold-plated steel plates that are held in position by magnetic bars placed underneath the substrate. Details of the circuit are given in Ref. [10].

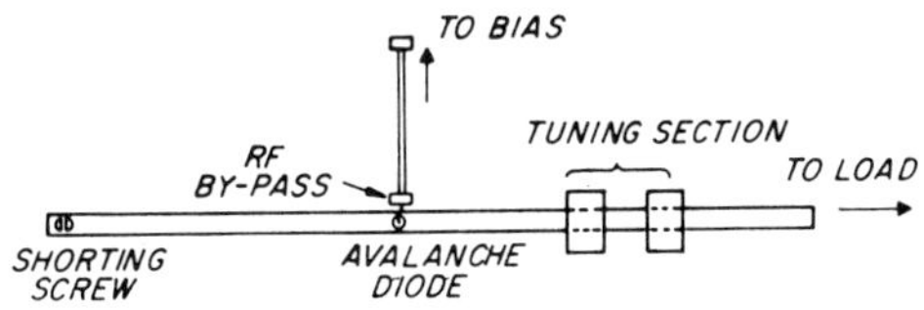

Fig. 16—Schematic diagram of microstrip high-efficiency avalanche-diode oscillator circuit with a "π" tuning section.

The circuit has been modified slightly for mounting more than one diode. Details of two series-connected diodes on the microstrip line are shown in Fig. 17. Each diode chip was first mounted in a pillbox, then connected in series with a gold strip. Diode chips have also been soldered directly onto the line.

Fig. 18 shows results obtained with two series-connected diodes and three series-connected diodes on the microstrip circuit. Power and efficiency are 330 W at 20.2% and 440 W at 23.4%, for the two and three series-connected diodes, respectively. Recently, four series-connected diodes have produced 540 W. The third and fourth diodes were located on top of one of the other diodes.

A comparison has been made between the performance of the microstrip circuit and the coaxial circuit. Power output from the same two series-connected diodes was measured on both circuits. The results are plotted in Fig. 19. It shows that at a current level of 7.5 A, the microstrip circuit performs as well as the coaxial-line circuit. However, the power output becomes jittery and difficult to tune beyond this current level on the microstrip circuit, while the power is still increasing in the coaxial circuit. This may be because fine tuning, such as varying the position of the short-circuit and/or the value of capacitor, is more difficult on the microstrip circuit. 369

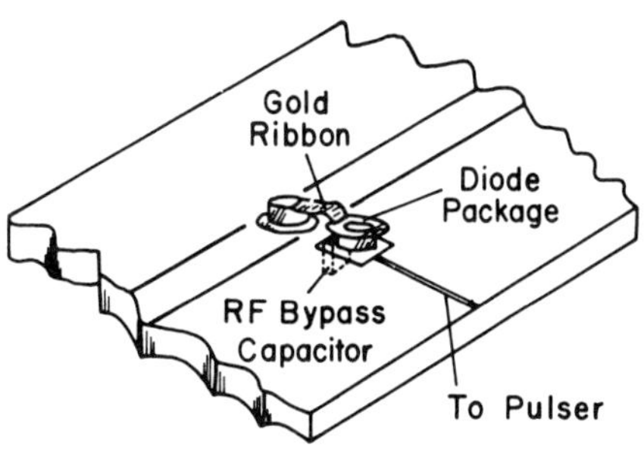

Fig. 17—Details of two series-connected diodes on microstrip circuit.

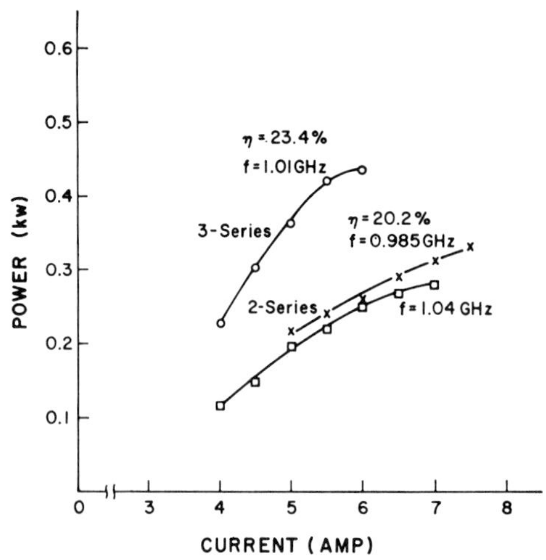

Fig. 18—Power output versus current of series-connected diodes operating on microstrip circuit.

Conclusions

A reliable fabrication process has been developed for high-power high-efficiency avalanche diode oscillators by deep diffusion of boron into high-resistivity n-type epitaxial materials. The large-size deeply

diffused diodes have been shown to be consistently free from micro-

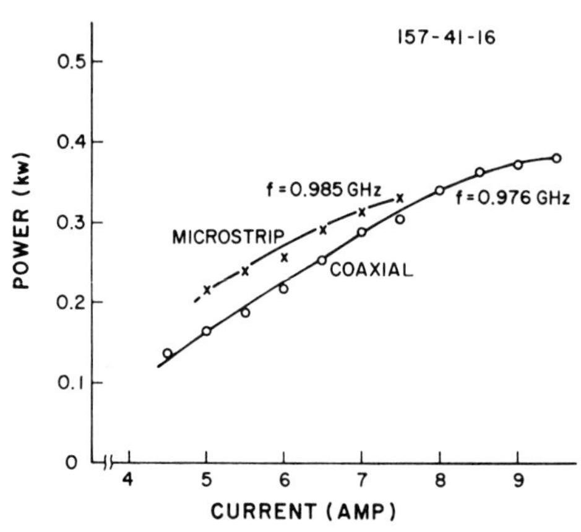

Fig. 19—Comparison of power output of two series-connected diodes on coaxial and microstrip circuits. The coaxial circuit was not re-tuned for optimum power below 8 A.

plasma effects and capable of handling high currents. Diodes are made in disk shape for convenience in flip-chip mounting for improved heat sinking.

Using these diodes we have demonstrated the feasibility of extending the power output from high-efficiency avalanche-diode oscillators to the kilowatt range. Power outputs up to 1.2 kW at 1.11 GHz with an efficiency of 25.6% has been obtained from five series-connected diodes in a multiple-stacked configuration. Best duty cycle achieved at the kilowatt level is 0.1%; with improved packaging and heat sinking techniques, 1% duty cycle at this power level should be feasible. Experiments have also demonstrated combined power output from parallel-connected diodes and from series-connected diodes on a microstrip circuit up to the 500 W level.

Acknowledgment

The authors wish to thank K. K. N. Chang for his constant encouragement, and L. S. Nergaard for critical reading of the manuscript.

References

[1] H. J. Prager, K. K. N. Chang, and S. Weisbrod, "High-Power High-Efficiency Silicon Avalanche Diodes at Ultra High Frequencies," **Proc. IEEE,** Vol. 55, p. 586 (1967).

[2] W. T. Read, "A Proposed High-Frequency Negative-Resistance Diode," **Bell System Tech. Jour.,** Vol. 37, p. 401 (1958).

[3] S. G. Liu and J. J. Risko, "High-Power Punch-Through Avalanche Diode Microwave Oscillators," Presented at the ISSCC in Philadelphia, Feb. 1968; S. G. Liu, **IEEE Jour. Solid State Circuits,** Vol. SC-3, p. 213 (1968).

[4] R. L. Johnston, D. L. Scharfetter, and D. J. Bartelink, "High-Efficiency Oscillations in Germanium Avalanche Diodes Below the Transit-Time Frequency," **Proc. IEEE,** Vol. 56, p. 1611 (1968).

[5] C. P. Snapp, L. A. Stark, and B. Hoefflinger, "Performance and Theory of Avalanche Resonance Pumped Impatt Oscillators," presented at the Inter. Elec. Device Meeting, Washington, D. C., Oct. 1968.

[6] D. E. Iglesias and W. J. Evans, "High-Efficiency CW Impatt Operation," **Proc. IEEE,** Vol. 56, p. 1610 (1968).

[7] A. S. Clorfeine, R. J. Ikola, and L. S. Napoli, "A Theory for the High-Efficiency Mode of Oscillation in Avalanche Diodes," **RCA Review,** Vol. 30, p. 397 (1969).

[8] S. G. Liu, "Stacked High-Power Avalanche-Diode Oscillators," **Proc. IEEE,** Vol. 57, p. 707 (April, 1969).

[9] P. A. Levine and S. G. Liu, "Tunable L-Band High-Power Avalanche-Diode Oscillator," **Digest Tech. Papers, ISSCC,** Vol. XII, p. 84 (Feb. 1969).

[10] S. G. Liu, "Microstrip High-Power L-Band Avalanche-Diode Oscillators," **IEEE-MTT Symposium Digest,** p. 261, May 1969, Dallas, Texas. Also **IEEE Trans. MTT,** Vol. MTT-17 (Dec. 1969).

[11] D. D. Tang, "High Power Varactor Package for Multichip Application," **Proc. IEEE,** Vol. 57, p. 799 (May, 1969).

CW TRAPATT Amplification

Abstract—**High-efficiency CW amplification has been obtained using germanium TRAPATT diodes. Several modes of amplification are described. The first mode gives approximately a 12-dB gain with a maximum efficiency of 25 percent at saturation. The second mode exhibits lower gain, but at a relatively low input power the diode switches into a locked TRAPATT oscillator mode with an efficiency of approximately 35 percent. Each of these modes exhibits a noise figure of approximately 120 dB. A third mode which allows the diode to oscillate in the TRAPATT mode at one frequency and to amplify at a second frequency has also been obtained. The noise figure for this mode is approximately 60 dB.**

Previous authors [1]–[3] have reported high-efficiency amplification with avalanche diodes operating well below their transit-time frequency. In these cases silicon diodes were used and pulsed operation was necessary. In this paper CW amplification using germanium diodes is reported. In addition, several modes of amplification are identified, and it is shown how the noise figure of these amplifiers may be substantially reduced.

Fig. 1 shows the circuit used for the amplifier measurements. The coaxial tuner, low-pass filter, and 30-cm length of line with the diode mounted at one end are the same components used to construct a 500-MHz TRAPATT oscillator [4]. The bias is applied to the diode through a 500-MHz parallel resonant trap. This biasing scheme keeps to a minimum the capacitance that shunts the diode [5]. The amplifier is completed with the usual circulator input. Either amplification or oscillation is obtained by adjustment of the coaxial tuner. This tuner consists of simply a parallel-plate line with a shunting capacitance whose magnitude and position can be controlled (HP Model 872A).

The diode used in these experiments is an epitaxial p^+-p-n^+ germanium diode with a breakdown voltage of about 50 volts and a depletion-layer width of approximately 3.0 μm. The epitaxial p layer has a doping of 5.0×10^{15} cm^{-3}. For the data shown in Figs. 2 and 3, a bias current of 125 mA was used. At this input current room-temperature heat sinking is adequate.

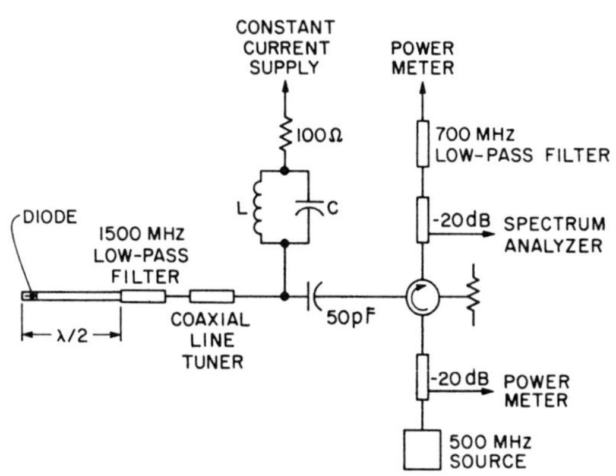

Fig. 1. 500-MHz amplifier circuit.

Fig. 2 shows the power output and efficiency[1] versus the power input at the center of the passband of the amplifier. At this frequency (462 MHz) the amplifier has about a 12-dB gain and saturates at an output power of about 1 watt. The efficiency at saturation reaches about 25 percent. The dc voltage continues to drop with increasing RF input power even after saturation is reached. Below one-half watt output the amplifier exhibits normal passband characteristics with a 3-dB bandwidth of approximately 8 MHz. In this mode of operation the amplifier is quite noisy, having a noise figure of approximately

[1] (RF power out − RF power in)/dc power.

Manuscript received June 12, 1970; revised September 4, 1970. This paper was presented at the 1969 IEEE Electron Devices Avalanche Diode Workshop, New York, N. Y., December 10–12.

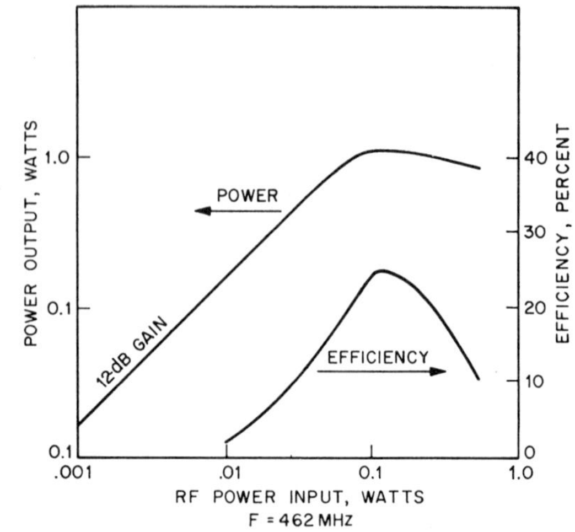

Fig. 2. RF power output and efficiency versus RF power input at 462 MHz.

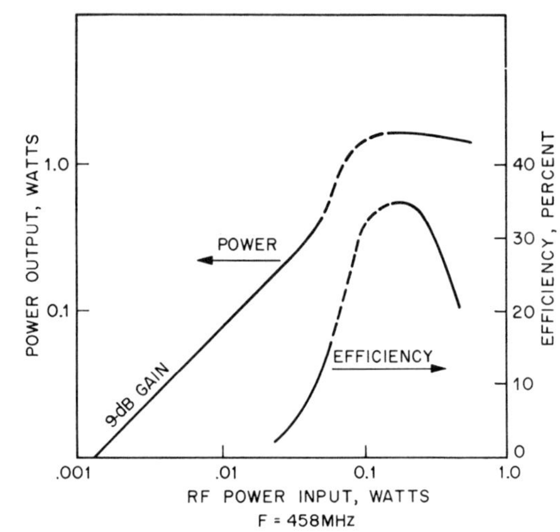

Fig. 3. RF power output and efficiency versu RF power input at 458 MHz.

110 to 120 dB. The noise is particularly high in a CW device because the Q of the high-frequency circuit must be very high in order to operate at low dc currents. Under these conditions large high-frequency RF voltage variations occur in the diode, but because the circuit is detuned at 500 MHz, the output at 500 MHz is not coherent.[2] Probe measurements in the $\lambda/2$-line section indicate that RF power is present at all the harmonics of the input signal up to at least the transit-time frequency of the device. The band of residual oscillation centered at 460 MHz, which accounts for the high-noise figure of the amplifier, is also found to be present in harmonically related bands up to the transit-time frequency. The bandwidth of the amplifier is determined by the Q of the high-frequency circuit since the circuit Q at 460 MHz is only 0.5. Using similar circuits, high-efficiency CW amplification has been obtained at selected frequencies from 450 to 1400 MHz.

If the input signal is tuned about 4 MHz below the center of the amplifier passband, a somewhat different input–output characteristic is obtained. Fig. 3 shows that in this case, when the input power exceeds 70 mW, the amplifier output rapidly increases to about 1.5 watts. This is accompanied by a reduction in the noise near the output signal from about −30 to about −100 dBm/kHz, which is comparable with the noise in the sidebands of a normal TRAPATT oscillator. The efficiency increases to about 35 percent, which is approximately the same as that obtained with the free-running oscillator. Thus, in this mode of operation the diode is operating as a locked oscillator.

[2] The high noise level observed in this case might also be described as residual oscillations [6].

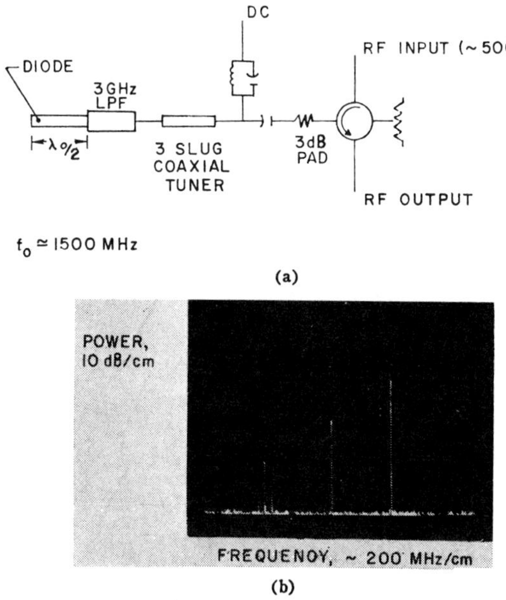

Fig. 4. (a) Circuit of oscillating TRAPATT amplifier. (b) Power spectrum with 1-mW input at 450 MHz for oscillating TRAPATT amplifier.

The very large reduction in noise in the amplifier passband when the diode switches into the oscillating mode suggests that a circuit which supports a TRAPATT oscillation at a frequency outside the passband may provide lower noise amplification. Furthermore, it has been shown experimentally that amplification can be obtained with a diode that is simultaneously oscillating in the high-efficiency mode [7]. Therefore the circuit shown in Fig. 4(a) was constructed. Basically, the circuit consists of a 1400-MHz TRAPATT oscillator. The trap in the bias circuit and the circulator is designed for the 500-MHz amplifier, and a 3-dB pad is required to provide a resistive load for the 1400-MHz oscillator. Fig. 4(b) shows the spectrum for a typical operating point with this circuit. The large signal at the right is a 1400-MHz TRAPATT oscillation. The power output is approximately 2 watts at 20-percent efficiency. The signal at the left is the amplified signal at 450 MHz. In this case amplification of about 10 dB (or 16 dB if the 3-dB pad is excluded) was obtained. Observe that there is a strong conversion of the input signal to the lower sideband of the oscillator. This signal is just above the rather weak second harmonic of the input signal.

As expected, this amplifier is considerably less noisy than the nonoscillating case. A noise figure of approximately 60 dB was measured. In the circuit just described only provision for the 1400-MHz oscillation has been made. Thus, when the input signal reaches about 50 mW, the 1400-MHz oscillation is quenched and no gain is observed at 450 MHz. If the circuit were designed to support 450-MHz TRAPATT operation as well, it is likely that both the relatively low noise and high efficiency could be achieved in the same amplifier. Prager *et al.* [6] have reported on a high-efficiency pulsed amplifier in which a strong signal near the third harmonic was observed. It may be that their amplifier was operating in the manner described here. Proper design of the circuit described in this correspondence requires provisions for the IMPATT, TRAPATT, signal, and idler frequencies. This complexity will likely limit the useful application of the amplifier.

Acknowledgment

The author would like to thank C. N. Dunn for supplying the diodes used in these experiments.

W. J. Evans
Bell Telephone Labs., Inc.
Murray Hill, N. J. 07971

References

[1] H. J. Prager, K. K. N. Chang, and S. Weisbrod, "Anomalous silicon avalanche diodes for microwave generations," presented at Conf. on High Frequency Generation and Amplifications, Cornell University, Ithaca, N. Y., August 1967.
[2] B. Hoefflinger, C. P. Snapp, and L. A. Stark, "High-efficiency avalanche resonance pumped amplification," presented at 1969 Int. Solid-State Circuits Conf.
[3] M. I. Grace, "Parametrically generated high-efficiency low-frequency oscillations in avalanche diodes," 1969 Int. Microwave Symp., Dallas, Tex., May 5–7.
[4] W. J. Evans, "Circuits for high-efficiency avalanche-diode oscillators," *IEEE Trans. Microwave Theory Tech.*, vol. MTT-17, pp. 1060–1067, December 1969.
[5] W. J. Evans, D. L. Scharfetter, and R. L. Johnston, "Tuning initiated failure in TRAPATT oscillators," (to be published).
[6] H. J. Prager, K. K. N. Chang, and S. Weisbrod, "Power amplification with anomalous avalanched diodes," presented at the 1969 Int. Electron Devices Meeting, Washington, D. C., October 29–31.
[7] J. F. Dienst, R. J. D'Aiello, and E. E. Thomas, "Power amplification using an avalanche diode," *Electron. Lett.*, vol. 5, pp. 308–309, July 10, 1969.

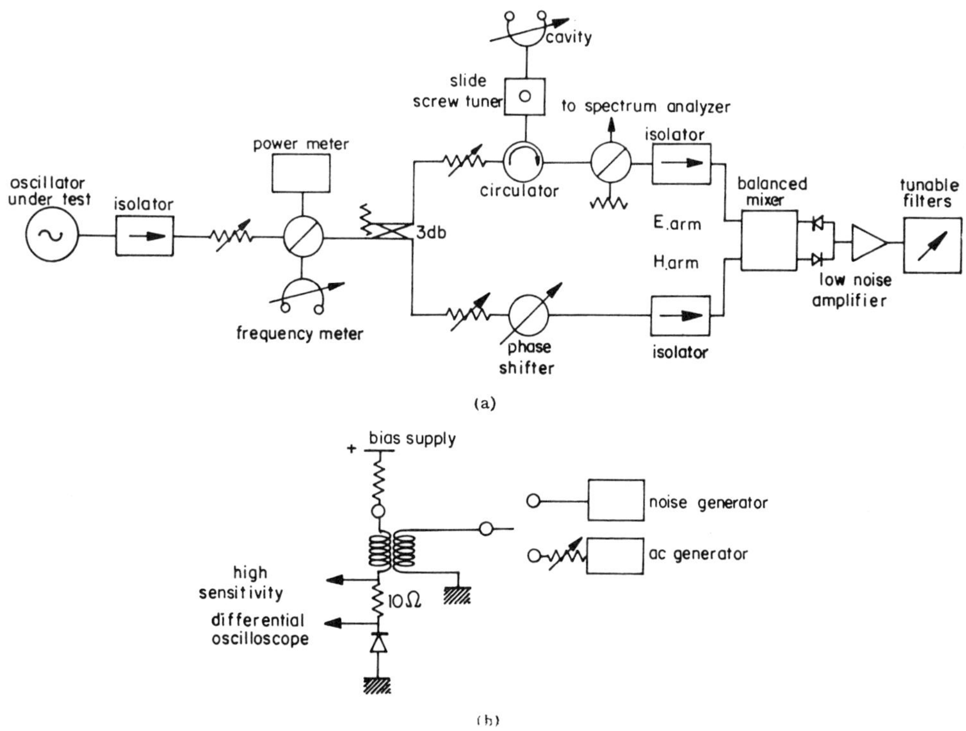

Fig. 1. FM noise measuring equipment.

THE HIGH-EFFICIENCY (TRAPATT) MODE

Power Amplification with Anomalous Avalanche Diodes

H. JOHN PRAGER, MEMBER, IEEE, KERN K. N. CHANG, AND S. WEISBROD

Abstract—The experiment described in this paper is performed for the purpose of obtaining additional information that might help to elucidate the operating principle of the anomalous mode.

The diodes under investigation are of a silicon p^+-n-n^+ mesa structure. The breakdown voltage is 160 volts; the punch-through voltage is 60 volts.

The test setup is similar to a conventional reflection-type amplifier with the diode mounted in a coaxial-cavity-like circuit. The reverse-bias pulse drives the diode to a low avalanche current level at which anomalous operation would normally commence. In this case, however, the tuning and loading of the cavity and its associated stubs is intentionally arranged to minimize free-running oscillations. Any residual RF power is reduced to the level of weak instabilities without any strong harmonics or subharmonics. Only in the presence of an input driving power with a frequency at or near the resonance frequency of the system, does an amplified power of the same frequency appear at the output port.

Starting from the residual level the output power increases proportionally with the input power until at higher drives a saturation level is reached. In this particular case, operating at 410 MHz, saturation is reached at 18.5 watts with a maximum power gain of 12 dB and a bandwidth of 26 MHz. This output is approximately 20 to 30 dB higher in amplitude than any observable peak in the residual broadband frequency spectrum. Operating the diode as a free-running "hard" oscillator at the same current level and frequency, 18.8 watts of RF power is obtained.

The initial and steady-state power levels are substantially the same for both amplifier and oscillator. A plausible evaluation of these results can be obtained following van der Pol's differential equation for a forced negative-resistance oscillator.

Manuscript received April 12, 1970. This work was supported by the Air Force Avionics Laboratory, Wright-Patterson AFB, Ohio, under AF Contract F33615-68-C-1688, Project 4460.

The authors are with RCA Laboratories, Princeton, N. J. 08540.

Introduction

SINCE THE discovery of the anomalous avalanche diode in early 1967 [1], a great deal of effort has been expended by several laboratories [2]–[4] not only to explain the anomaly but also to harness the effect by developing better diodes and better associated circuitry. While most of these activities concentrated on the generation of microwave power, relatively little was done toward the amplification of microwave power. The first cursory results of amplifier operation were reported in August 1967 [5]. This was followed by the paper of Hoefflinger *et al.*, in May 1969 [6], and that of Dienst *et al.* [7] who presented the results of their work on a self-pumped parametric amplifier with the anomalous avalanche diode. The applications for microwave amplification are, however, sufficiently important to warrant still further investigation.

Review of Oscillating Operation of Anomalous Avalanche Diodes

To better highlight the similarities as well as the differences between oscillator and amplifier performance, we shall briefly review some of the pertinent features of the anomalous diodes as oscillators.

The first successful oscillation of the anomalous avalanche diode was accomplished in a coaxial circuit as shown in Fig. 1. The diode was placed in a General Radio (GR) Tee fitted with proper bypass capacitors, while stub tuners were used for tuning to the desired fre-

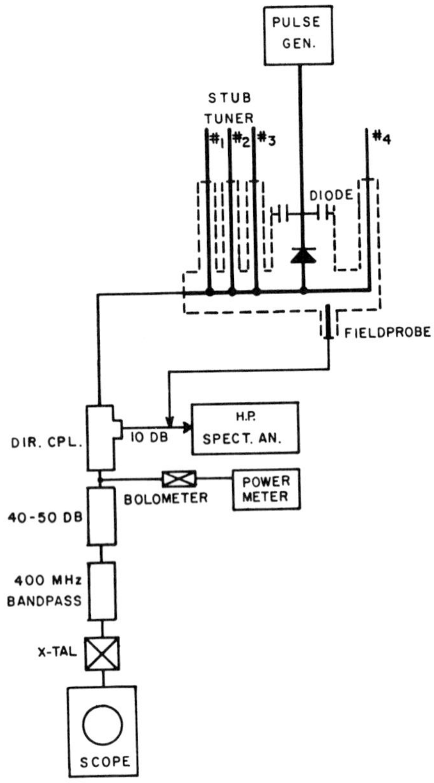

Fig. 1. Test setup for avalanche-diode oscillator.

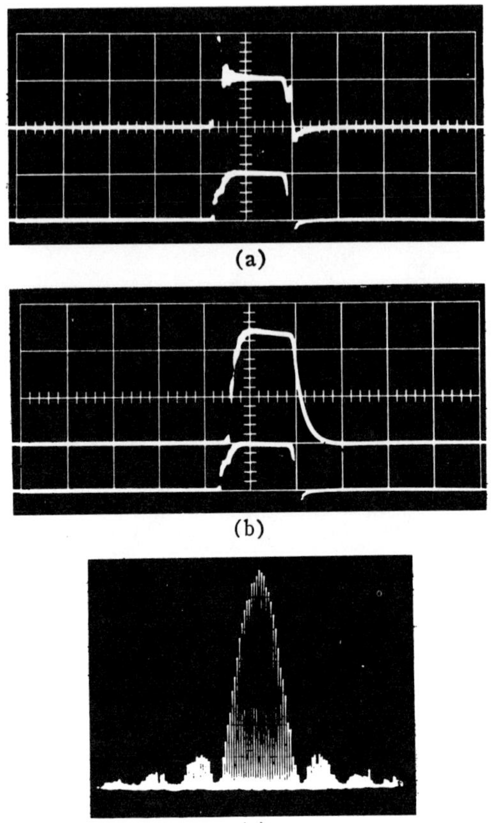

Fig. 2. Pulse waveshapes and frequency spectrum of avalanche-diode oscillator. Scales: (a) Top: voltage, 100 V/div; bottom: current, 5 A/div. (b) Top: RF power, 50 W/div; bottom: current, 5 A/div; time, 0.5 μs/div. (c) Center frequency, 400 MHz; frequency dispersion, 1 MHz/div.

quency as well as for matching to a 50-ohm load. To assure that the power-output measurements were made only at the frequency of interest, a narrow-band pass
376 filter was employed ahead of the crystal detector. For broad-band-frequency studies a field probe was fed directly to a broad-band Hewlett-Packard spectrum analyzer. This probe, consisting of a short section of a 50-ohm OSM cable, was inserted into the GR Tee through a hole in its outer shield. Voltage and current probes were also attached to the diode and brought out to the same scope. The waveshapes of these pulses are shown in Fig. 2. The top trace of Fig. 2(a) is the voltage, and the bottom is the current; the top trace of Fig. 2(b) is the RF power output, while the bottom repeats the trace for the current. The scales are 100 V/cm, 5 A/cm, 50 W/cm, and 0.5 μs/cm. The frequency spectrum is at a center frequency of 400 MHz, with a dispersion of 1 MHz/cm. The dc-to-RF conversion efficiency was 25 percent. These waveshapes are typical for the diodes and the pulse supplies which we have been using. The power supplies provide a tradeoff between constant current and constant voltage. A constant-current source tends to protect the diode from drawing destructively high currents, while a constant-voltage source tends to improve efficiency providing that the diodes can withstand the resultant current levels. These oscilloscope tracings clearly indicate the characteristic increase in current once the critical current level for the anomalous mode has been reached. Corresponding to this jump is the drop in voltage across the diode and the onset of RF power output.

The diodes used in these oscillator and amplifier experiments were silicon diodes with a p^+-n-n^+ profile and a circular mesa structure. The resistivity of the n region was 4.5 Ω·cm; the p region was 8 μm thick, and the n region 10 μm. The mesa had a diameter of 28 mils or an area of 38×10^{-4} cm^2. With a pulse current of 5 amperes under oscillatory conditions, the diode operated with a current density of 1300 A/cm^2. Fig. 3 shows the microphotograph of such a diode after angle lapping and staining.

The capacitance versus voltage characteristic of Fig. 4 reveals a slope of 0.450, which confirms the fact that the diode has a slightly graded junction. The breakdown voltage was approximately 160 volts and the punch-through voltage was 60 volts. The results of a simple computer program relating these capacitance readings to the doping profile of the diode are in good agreement with measurements from the angle lapping.

Amplifier Operation of Anomalous Avalanche Diodes

Having reviewed the results of the oscillator performance, we shall now discuss the results of our amplifier tests. The general test setup as shown in Fig. 5 is that of a reflection-type amplifier. The diode-cavity portion of the circuit is the same as that shown for the

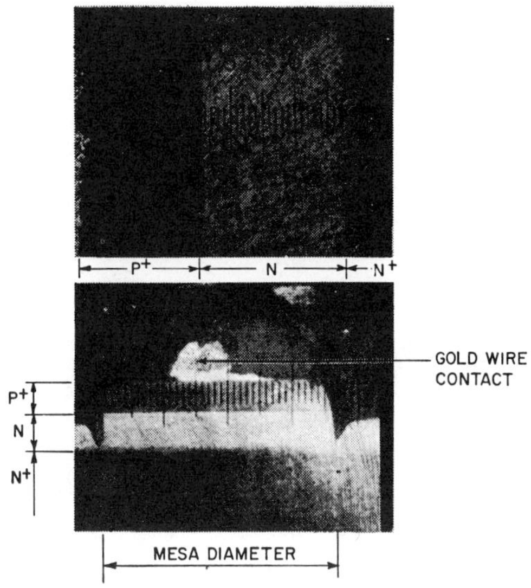

Fig. 3. Microphotograph of angle-lapped junction.

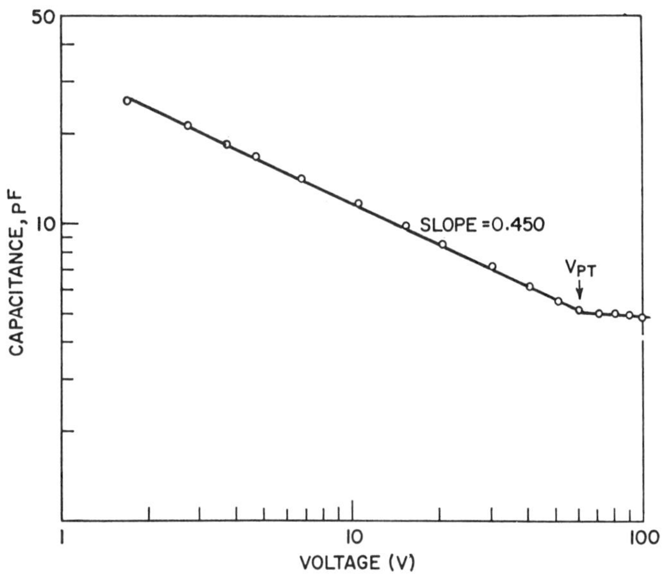

Fig. 4. Capacitance versus reverse voltage.

oscillator, but although the frequency is the same 400 MHz, the stub tuning and matching conditions as well as the pulse-operating conditions are changed to meet the requirements expected from the amplifier. The input signal is provided by a pulse-modulated RF generator operating at the same 400-MHz frequency as the circuit of the diode. The fact that the frequency of the input circuit is exactly the same as the resonance of the diode circuit is an important feature of this amplifier. It clearly differs in this respect from a self-pumped parametric amplifier where there are three separate frequencies for the signal, the pump, and the idler.

The voltage and current waveshapes of the amplifier are shown in Fig. 6. The left-hand side of the figure shows the voltage and current of the diode in the quiescent amplifier condition; i.e., the diode is ON and at operating levels, but the input signal is zero or in the

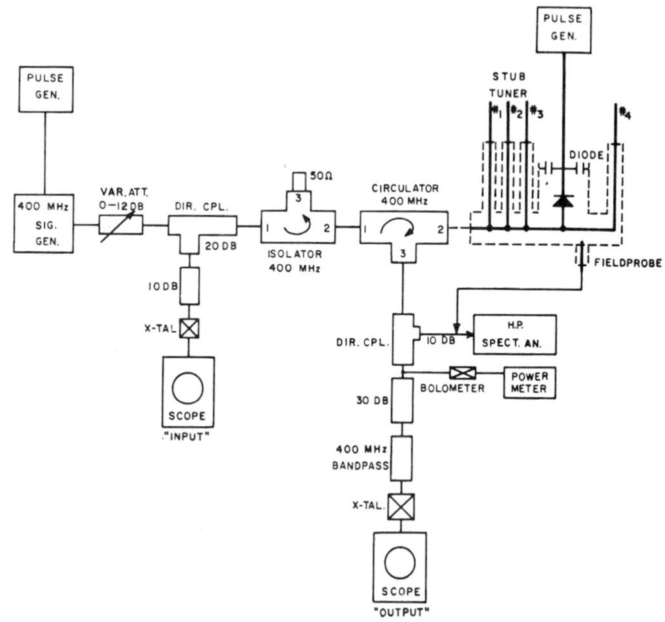

Fig. 5. Test setup for avalanche-diode amplifier.

OFF condition. While considerable diode current is already flowing, the characteristic increase in current is still missing. With the input signal OFF there is as yet no output power present. The right-hand side of Fig. 6 shows the voltage and current when both the diode pulse and the input signal are turned ON. The diode current now shows a small but recognizable increase, and the voltage shows a corresponding dip typical of the anomalous mode. To prove that these changes are not due to a switching phenomenon between two unstable circuit conditions, we gradually varied the input power and measured the gradual change in output power, diode current, and diode voltage. The results have been plotted in Fig. 7. The voltage and current did not change over as wide a range as they did for the oscillator, and accordingly the efficiency of the amplifier is as yet well below the efficiency of the oscillator.

To gain a better insight into the working of the amplifier, we shall now consider the oscilloscope tracings of Fig. 8. These are measurements of RF power as seen by the crystal in the detection circuit following the output of the circulator. In Fig. 8(a) the diode is ON, and the signal is OFF; this means that while the diode was drawing current, as is shown in the left-hand side of Fig. 6, practically no RF power was detectable. Actually with higher scope sensitivity a small RF power is measured, and we shall refer to it as "residual" RF power. The minimizing of this residual RF power during the tuning and adjustment procedure of the amplifier setup is a very important feature of its performance. Although this procedure is empirical, it is guided by a basic plan. Briefly it begins by tuning the diode under self-oscillating conditions to the desired operating frequency of the amplifier. Stub 4 and, to a lesser degree, stub 3 are the most effective controls in determining this desired resonance frequency. With the frequency in

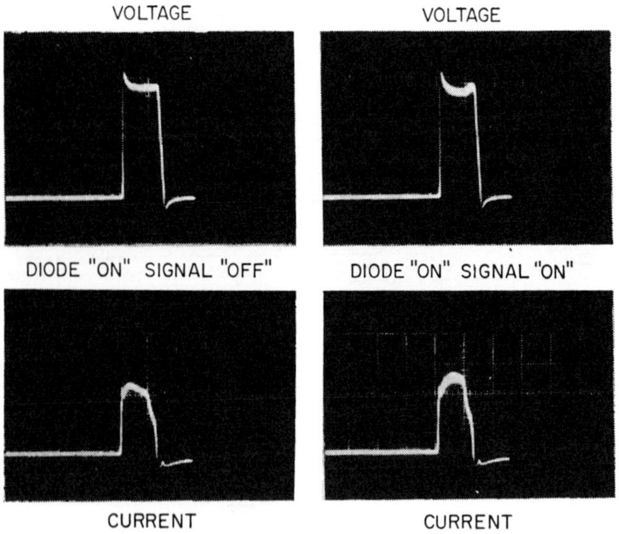

Fig. 6. Voltage and current pulse waveshapes of amplifier for signal ON and OFF conditions. Scales: voltage, 50 V/div; current, 1 A/div; time, 1 μs/div.

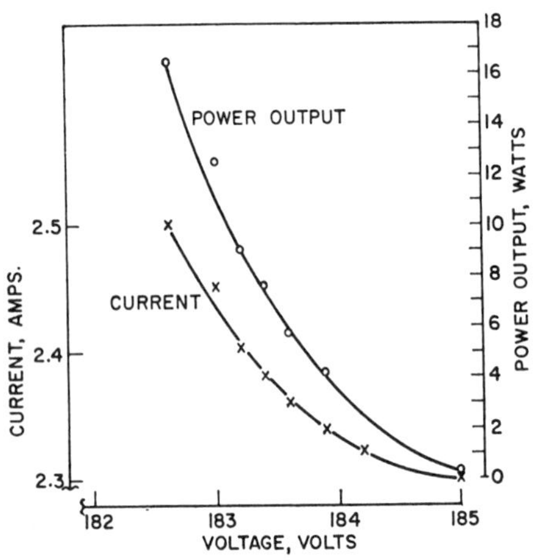

Fig. 7. Pulse current and pulse voltage across diode as function of power output.

the diode circuit thus established, stubs 1, 2, and again to a lesser degree, stub 3, are now adjusted until the oscillatory RF power output is minimized and brought to the lowest attainable residual level. While the first adjustment sets the frequency, the second adjustment changes the dynamic load line until a stable amplifier condition is reached.

Both of these adjustments are required to meet two necessary conditions of the amplifier, but they are not sufficient yet, because a further requirement is that the power output vary gradually (if not linearly) with input. Thus the next step in the adjustment procedure is to verify that this continuity requirement has been met or to modify the stub settings until adequate control has been accomplished.

Turning our attention to Fig. 8(b), we note that it is taken with the diode OFF and the signal ON. Since we are dealing with a reflection-type amplifier, we have of course the reflected portion of the input power appearing

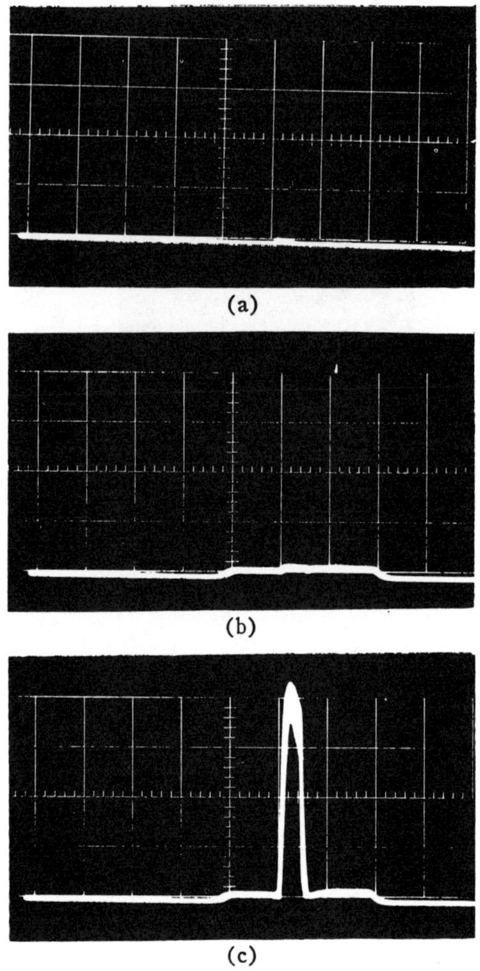

Fig. 8. RF output voltage of amplifier for three distinct operating conditions.

at the output port. (We hasten, however, to point out that for the purpose of measuring the power gain of the amplifier, the input power was measured at the input port and into a 50-ohm load. The picture shown here is only for the purpose of illustrating the qualitative behavior of the amplifier.) In Fig. 8(c) both the diode and the input signal are ON and a strong RF power output now makes its appearance.

Looking at the frequency spectrum under the same three operating conditions, we observe Fig. 9, where the center frequency in all three cases is 400 MHz. In Fig. 9(a) with the diode ON and the signal OFF, there is again very little evidence of any residual oscillation at 400 MHz. Fig. 9(b) shows the frequency spectrum with only the signal ON, and in Fig. 9(c) both the diode and the signal are ON simultaneously. Thus we see again that only in the presence of the signal is the diode driven to give the desired power output.

In Fig. 10 we have plotted this power output and the amplifier gain as a function of input power. It is not a strictly linear amplification. To begin with, at zero input power there is a small amount of residual power output present. Only after very careful adjustment of the stub tuners were we able to reduce this residual output to a bare minimum. But it is of course an essential feature of

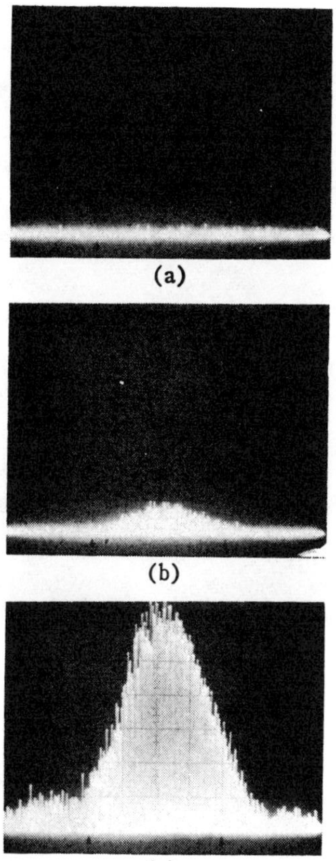

Fig. 9. Frequency spectrum for amplifier for three distinct operating conditions.

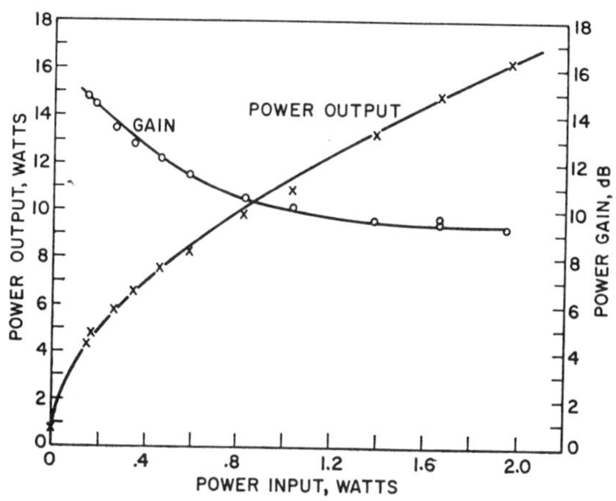

Fig. 10. Power output and gain versus power input.

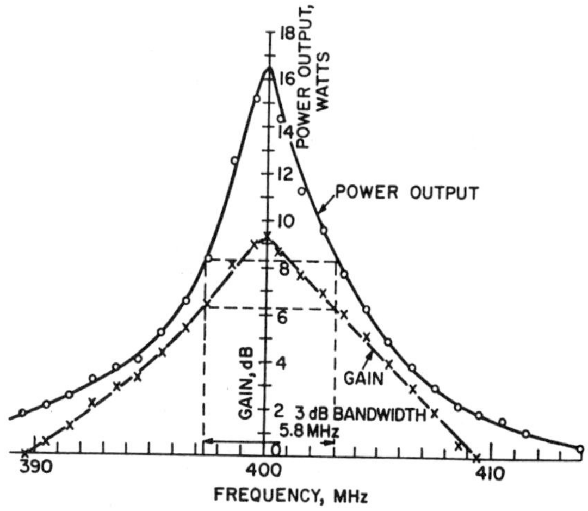

Fig. 11. Frequency response curve of amplifier. Power output (watts) and power gain (dB) versus frequency (MHz).

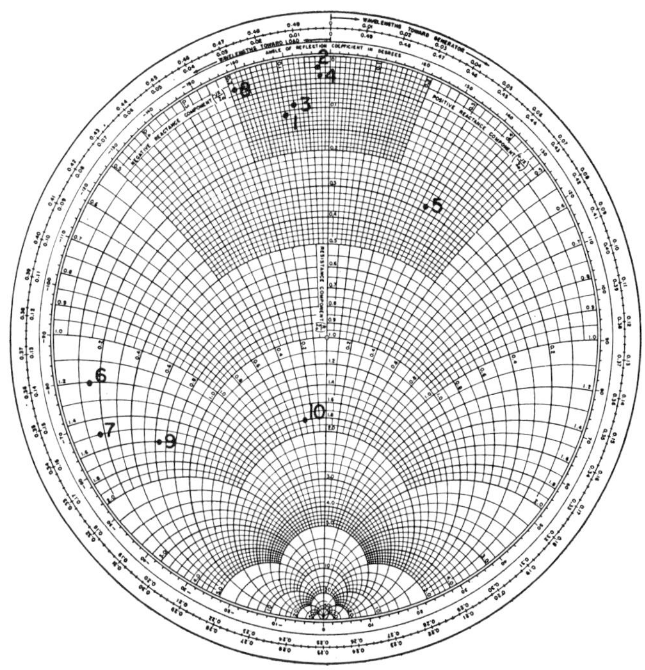

1) 400	MHz	$Z=5.5-j5.0$	5) 2000	MHz	$Z=14.5+j16.5$
2) 800		$1.0-j2.5$	6) 2400		$6.5-j61.5$
3) 1200		$4.5-j4.0$	7) 2800		$8.0-j75$
4) 1600		$1.5-j1.0$	8) 3200		$1.0-j\,9.0$

Fig. 12. Impedence plot for fundamental and harmonic frequencies.

an amplifier that this residual power be as low as possible. When we allowed a slightly higher residual level, we achieved a higher power-output level, but also a more pronounced leveling off at the high end of the curve. Limitations in the available input-power source prevented further exploration of higher saturation levels. (This saturation would not be so apparent if the curves were plotted on log–log paper as is frequently done, and even the residual power at zero input power would then conveniently disappear.) Another amplifier characteristic that is of interest is the bandwidth; this is shown in Fig. 11. Our amplifier circuit utilizes a narrow-band circulator and a triple-stub tuner, which is notoriously narrow banded. It is therefore not surprising that the total bandwidth is only 5.8 MHz or about 1.5 percent.

As part of our experiment we also attempted to measure the circuit impedance at the point at which the diode fits into the circuit. For this purpose we removed the diode from the circuit and replaced it with a 2-inch length of 50-ohm OSM line which was connected to a network analyzer. An exact duplicate of this line with a solid RF short at its end was used to establish and calibrate the reference point. Fig. 12 gives the results of these measurements on a Smith chart. The impedance

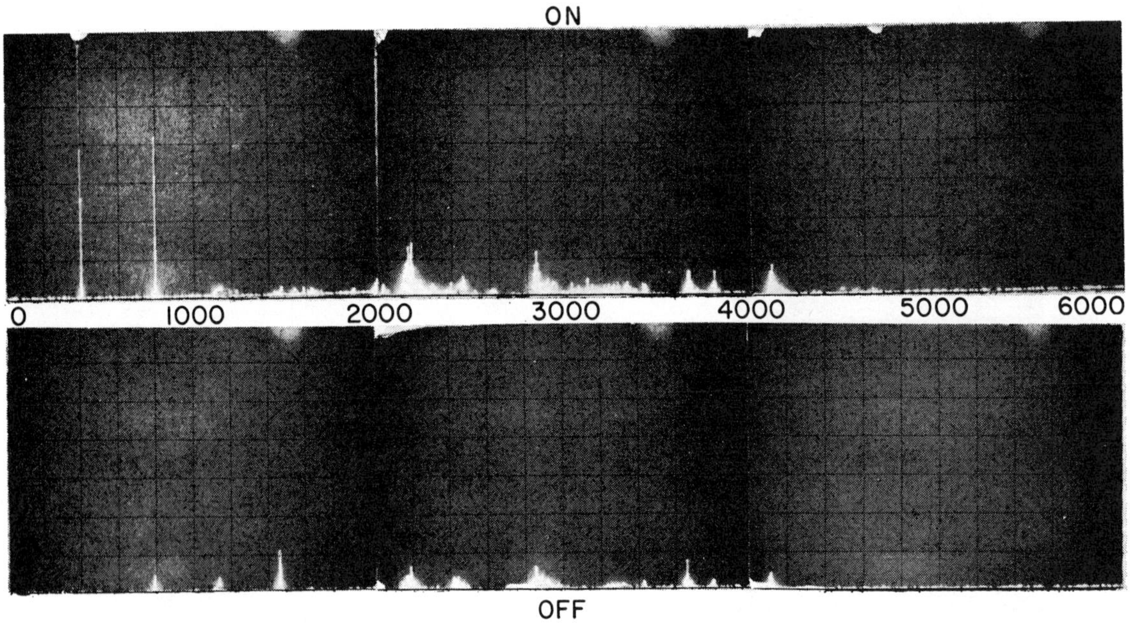

Fig. 13. Wide-band frequency spectrum of amplifier for signal ON and OFF conditions.

for the fundamental frequency does not show a very good match and this might account for the low efficiency of this particular circuit. The second, third, and fourth harmonics are typically clustered around a near short circuit for the real part, and they also show some reactive component. As usual, higher harmonics are much more reactively terminated.

Special Frequency Study of the Amplifier

We had hoped that these amplifier experiments would not only give us a chance for a feasibility study of the amplifier per se, but also a chance to collect further data pertaining to the theoretical model of the anomalous mode. For this purpose we examined the frequency spectrum of the amplifier, employing a field probe near the diode and a wide-band spectrum analyzer. This field probe consisted of a short section of 50-ohm OSM line and was inserted into the GR Tee through a hole in its outer shield. The probe was mounted on a micromanipulator which allowed precise three-dimensional movement of the probe. Fig. 13 is a composite display of the wide-band frequency spectrum from 0 to 6000 MHz, with each division being 200 MHz. The bottom row shows the spectrum when the diode is ON and drawing current and voltage as described before, but with the input signal turned OFF; consequently, there is no 400-MHz trace present. In the top row both the diode and the input signal are ON, and a strong trace at 400 MHz as well as a smaller trace at 800 MHz are now observed. From our previous results we expected to find a strong fundamental and considerable harmonic content. But we also found something else: in the signal OFF condition we observed a small but distinct spectrum line near 1500 MHz and further noticed that this line disappeared in the signal ON condition. This behavior was unique, since there was no other frequency just like it. All other frequency lines show an increase in amplitude when the input signal is ON. We became very curious about this frequency and took a closer look at it.

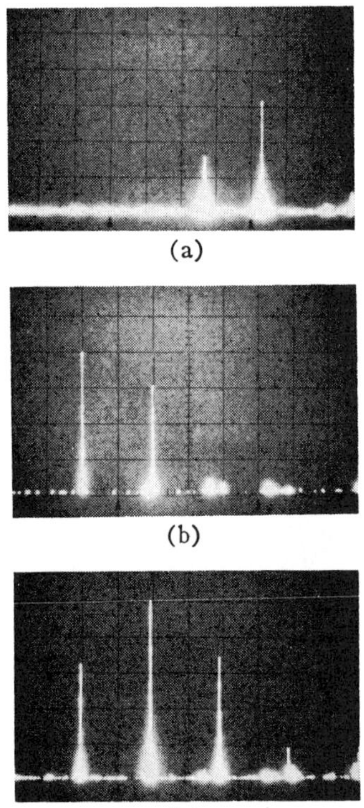

Fig. 14. Free-running frequency. (a) For amplifier OFF. (b) For amplifier ON. (c) For oscillator condition.

Fig. 14(a) shows this 1500-MHz line again in the signal OFF condition; Fig. 14(b) shows the signal ON condition; and finally in Fig. 14(c) the diode has been detuned from its amplifier condition until it became an oscillator at 400 MHz as well as at its harmonics of 800, 1200, and 1600 MHz. And again it can be noted that the

(a)

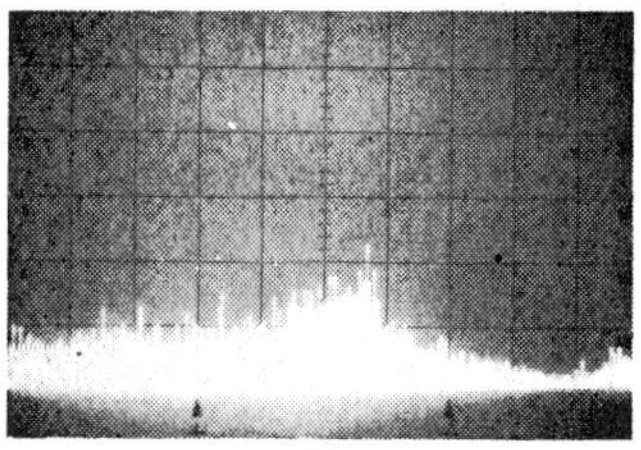

(b)

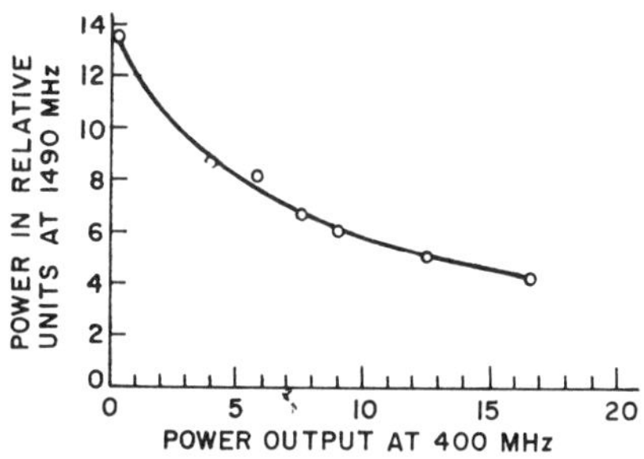

(c)

Fig. 15. Free-running frequency for amplifier at various levels of power output. (a) $f_0 = 1490$ MHz; signal OFF. (b) Signal ON.

1500-MHz line has disappeared not only in the ON condition of the amplifier but also in the oscillator case. Zooming in still closer to this mysterious frequency, in Fig. 15 we found it to be at 1490 MHz. The frequency dispersion in these pictures is now only 10 MHz/cm. Fig. 15(a) depicts the signal OFF condition, Fig. 15(b) shows the signal ON condition, and Fig. 15(c) shows how the amplitude of the 1490-MHz oscillation decreases as the power output of the amplifier increases continuously. The observation of this frequency is quite unexplained because it does not fit into the expected picture. First, although the diodes under test should have a transit-time frequency of approximately 5000 MHz, we could not find any oscillation near that frequency, and 1490 MHz, is too low to agree with any normal transit-time calculation. Furthermore, 1490 MHz is not harmonically related to the 400 MHz of the input and output signal. Second, there is the unexplained cause of the decreasing amplitude of this oscillation with increasing signal input.

Discussion of Frequency Results and Conclusion

In trying to find an explanation for these experimental results, we turned at first to the present theories of the anomalous avalanche diode. Typically both the "plasma" theory [8] and the "avalanche pumped resonance" [6] theory rely on the transit-time frequency

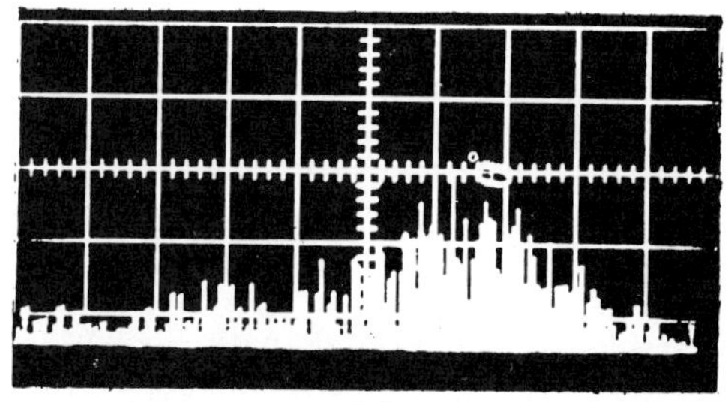

(a)

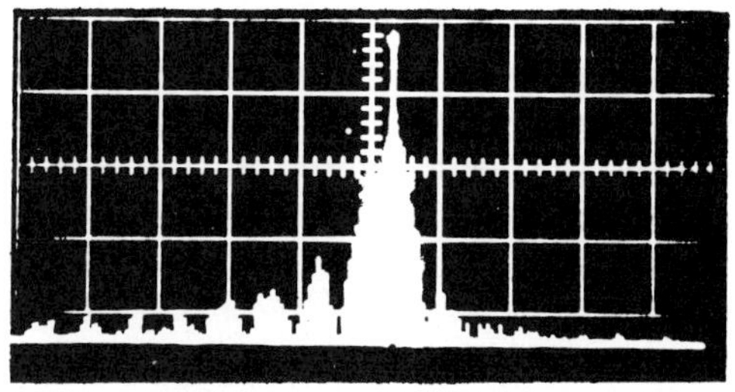

(b)

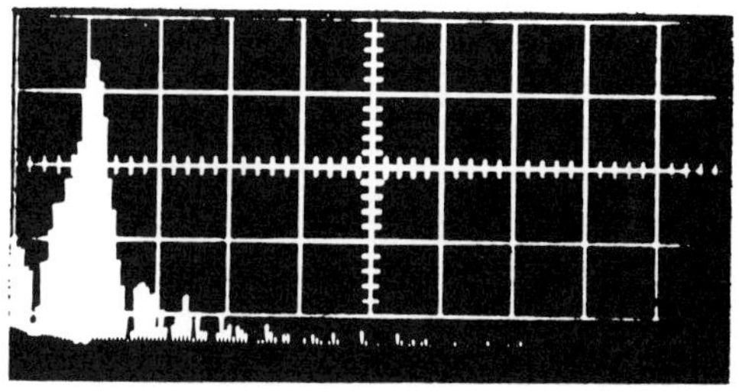

(c)

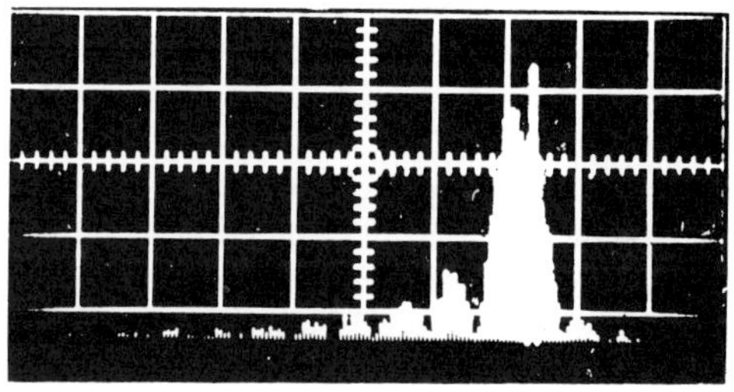

(d)

Fig. 16. Locked oscillator.

oscillation for their starting mechanism and view their operating frequencies in terms of harmonic relation. The nonharmonic relation between our signal frequency and the observed system frequency obviously called for still further explanations. It was for this reason that van der Pol's classical theory of "forced oscillations in a circuit with nonlinear resistance" [9] was given serious consideration. Van der Pol treated both the so-called "free" oscillation of the system and its "forced" oscillation under the influence of an externally applied signal. He described how the buildup of the forced oscillation suppresses the free oscillation and also, significantly for our interpretation, that these two frequencies do not have to be harmonically related.

Still unanswered is the exact origin of the free oscillation of the system. It is however entirely possible that a transit-time oscillation did exist beyond the calculated value of 5000 MHz and also beyond the range of our investigation, i.e., up to 7000 MHz, and that this transit-time frequency in turn caused a weak anomalous oscillation at 1490 MHz.

Locked Oscillator Operation

The type of power amplifier which has been described here is frequently associated with locked-oscillator operation. We have therefore endeavored to demonstrate that type of operation also. This was accomplished with the same type of diodes used in the amplifier experiments, but the circuit was changed to a simplified coaxial structure [10].

Accordingly, the operating frequency was changed to 1090 MHz. The results are illustrated in Fig. 16. Fig. 16(a) shows the frequency spectrum of the oscillation before locking, and Fig. 16(b) after locking. The locked-power output was 25 watts, the locking signal power was 2.5 watts. Fig. 16(c) and (d) show evidence of the frequency locking range with a scale calibration of 1 MHz/cm, thus making the locking range approximately 7 MHz. It should be noted that unlike the amplifier case there is already a strong oscillation present before the locking signal is applied and that the change in power between the unlocked and locked condition is only about 2:1. These two features characterize the two main differences between a locked oscillator and a power amplifier.

Acknowledgment

The authors wish to thank Dr. L. S. Nergaard for suggesting the similarity to van der Pol's theory.

References

[1] H. J. Prager, K. K. N. Chang, and S. Weisbrod, "High-power, high-efficiency silicon avalanche diodes at ultra high frequencies," *Proc. IEEE* (Correspondence), vol. 55, pp. 586–587, April 1967.

[2] R. L. Johnston, D. L. Scharfetter, and D. J. Bartelink, "High-efficiency oscillations in germanium avalanche diodes below the transit-time frequency," *Proc. IEEE* (Correspondence), vol. 56, pp. 1611–1613, September 1968.

[3] C. P. Snapp and B. Hoefflinger, "An efficient multiresonant avalanche diode oscillator in the 1.5 to 11 GHz range," *Proc. IEEE* (Correspondence), vol. 56, pp. 2054–2055, November 1968.

[4] R. S. Ying, R. G. Mankarious, and D. L. English, "High-efficiency anomalous mode oscillation from silicon impatt diodes at 6 GHz," *1969 ISSCC Dig. Tech. Papers*, pp. 86–87.

[5] H. J. Prager, K. K. N. Chang, and S. Weisbrod, "Anomalous silicon avalanche diodes for microwave generation," *Proc. Cornell Conf. High Frequency Generation and Amplification*, pp. 266–280, August 1967.

[6] B. Hoefflinger, C. P. Snapp, and L. A. Stark, "High-efficiency avalanche resonance pumped amplification," *1969 Int. Microwave Symp. Dig.*, pp. 255–260.

[7] J. F. Dienst, R. V. D'Aiello, and E. E. Thomas, "Power amplification using an avalanche diode," *Electron. Lett.*, vol. 5, p. 308, July 1969.

[8] A. S. Clorfeine, R. J. Ikola, and L. S. Napoli, "A theory for the high-efficiency mode of oscillation in avalanche diodes," *RCA Rev.*, vol. 30, pp. 397–421, September 1969.

[9] B. van der Pol, "The nonlinear theory of electric oscillations," *Proc. IRE*, vol. 22, pp. 1051–1086, September 1934.

[10] P. A. Levine and S. G. Liu, "Tunable *L*-band high-power avalanche diode oscillators," *1969 ISSCC Dig. Tech. Papers*, pp. 84–85.

BROADBAND HIGH-EFFICIENCY-MODE AMPLIFIERS AT S BAND*

MARTIN I. GRACE

SPERRY RAND RESEARCH CENTER, SUDBURY, MASS., 01776

The ability to realize broadband high-efficiency-mode (HEM) or TRAPATT amplifiers requires special device and circuit characteristics which are different from those used in conventional TRAPATT oscillators and amplifiers.

The majority of results reported to date for HEM oscillators and amplifiers have been obtained with circuits which utilize time-delayed triggering (TDT) as an oscillation-sustaining mechanism. A description of this excitation mechanism for the HEM has been published by Evans et al.[1]

A fundamental characteristic of all TDT circuits is the microwave filter network described in the circuit model shown in Fig. 1. The network is located an electrical distance $\lambda/2$ from the diode at the output frequency ω_o, where ω_o can be determined as the normal mode frequency of the TDT circuit. Hoefflinger[2] and others[3] have observed that HEM diode circuit combinations can also operate as amplifiers when triggered by injecting an external ac signal. Once the input signal exceeds the threshold value an avalanche shock front is excited. The remainder of the excitation mechanism is similar to Evans's TDT mechanism for self-sustaining oscillations. In the externally excited mode, each rf cycle must be triggered only by the input signal. As might be expected, external excitation of HEM is most easily observed when the frequency of the applied signal is close to ω_o, the normal mode frequency of the TDT circuit. In general, the TDT circuit amplifiers have relatively narrow useful bandwidths (on the order of 2%) due to spurious signal generation.

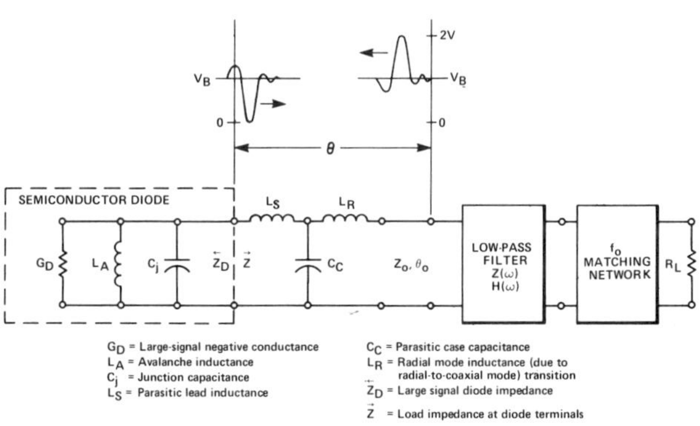

FIG. 1 Large-signal equivalent circuit of broadband amplifier.

The reasons for the narrow useful bandwidth and multiple output frequencies can be attributed to the properties of conventional TDT TRAPATT circuits. When the applied signal frequency ω_s is equal to the normal mode frequency ω_o, the TDT enhances the triggering of an avalanche shock front (ASF). When ω_s is different from ω_o, multiple signal frequencies are observed at the output; in general, the two most important are the applied frequency ω_s and the normal mode frequency ω_o. Observation of the normal mode frequency as well as the applied signal at the output port can be explained as follows. When the applied signal voltage triggers an ASF, the diode voltage drops toward zero and reverses rapidly. Thus, a narrow voltage pulse, whose magnitude is of the order of the breakdown voltage, is generated and travels down the transmission line (as illustrated in Fig. 1). At a distance $0 = \lambda_o/2$, the pulse is reflected from a filter whose high frequency impedance is nearly a short circuit. The pulse is reflected with nearly unity magnitude, but is changed in sign. The reflected pulse then arrives back at the diode with a total time delay of $2\pi/\omega_o = \lambda_o/c$ and instantly drives the voltage at the diode to about twice the diode breakdown voltage, triggering an ASF within the diode. This sequence is just the usual TDT mode, and the frequency of this overvoltage corresponds to the normal mode of TDT frequency ω_o. Obviously, circuits which depend on TDT will not easily be adapted to wide bandwidth operation and should be avoided.

Investigations of TDT amplifiers and oscillators have defined circuit and device parameters necessary for achieving broadband HEM amplification. These parameters are listed below.

(1) The device and circuit must be capable of supporting a diode terminal-current waveform which is consistent with HEM theory (e.g., as shown by Clorfeine et al.).[4] The circuit model developed for the broadband amplification is shown in Fig. 1. The input impedance $Z(\omega)$, as seen by active device, must be such that it satisfies the following criteria:

$\text{Real } Z(\omega_o) = -K \text{ Real } Z_D(\omega_o)$ (required gain value)

$\text{Real } Z(n\omega_o) \approx 0 \quad n = 2,3,4\ldots$ (minimum harmonic power dissipation)

$\text{Imag } Z(n\omega_o) = -\text{Imag } Z_D(n\omega_o) \quad (n = 1,2,3\ldots)$ (conjugate match of harmonics)

$H(n\omega_o) = \begin{cases} 1, & n = 1 \\ 0, & n = 2,3,4\ldots \end{cases}$ (only single frequency output desired)

$Z_D(\omega) = R_D(\omega) + j\,X_D(\omega)$ is the large signal diode impedance

$Z(\omega)$ = Load impedance at the diode terminals.

* Supported in part by the U. S. Air Force for the Rome Air Development Ctr., under Contract No. F30602-70-C-0110.

(2) TDT must be suppressed in the circuit as an oscillation-sustaining mechanism (e.g., the electrical length θ from the active device to the filter network must be much less than $\lambda_o/2$).

(3) The slope reactances ($dX/d\omega$) at all frequencies should be minimized. The tuning element should be located as close as possible to the diode terminals.

A practical microwave circuit which has been used successfully to obtain broad bandwidth amplification is shown in Fig. 2. The HEM diode terminates a 50-Ω coaxial transmission line and a low-pass filter is located at the terminals of the diode. An impedance transformer consisting of 15-Ω slugs, each $\lambda_o/8$ long, is located between the output and the filter network.

The large-signal equivalent circuit of the amplifier is shown in Fig. 1. The HEM device is represented by a parallel circuit comprising a negative conductance G_D, an avalanche inductance L_A (which represents the active portion of the diode), and a capacitance C_j representing the depletion layer capacitance. The inductance L_S and capacitance C_C represent the parasitic lead inductance and capacitance at the diode package. The inductance L_R and C_R represent the discontinuity reactances introduced by the radial-to-coaxial mode transition at the diode-coaxial line interface.

The low-pass filter was chosen so that:

(1) The operating frequency is contained in the passband of the filter.

(2) The stopband of the filter contains the second and third harmonics of the required operating frequency.

(3) The input impedance at the third harmonic is a short circuit.

It was assumed that a square-wave diode-current waveform consistent with HEM theory could, to a first approximation, be satisfied using the fundamental frequency ω_o and its third harmonic $3\omega_o$. At the fundamental frequency, the susceptance of the diode is inductive, and somewhere between the first and third harmonics the diode reactance exhibits an antiresonance. At the third harmonic the diode exhibits a capacitive reactance essentially due to the depletion layer capacitance C_j. The diode depletion-layer capacitance and parasitic lead inductance were chosen such that the diode is series self-resonant at the third harmonic ($3\omega_o$) of the operating frequency. The location of the low-pass filter is such that TDT is suppressed (Fig. 2). The position of the first discontinuity of the filter is of the order of $\lambda_o/5$ away from the diode. This arrangement prohibits TDT and thus eliminates the major problem which arises in conventional TDT circuits.

At the operating frequency the filter is transparent and the two $\lambda_o/8$ slugs are used to resonate with the inductive reactance due to the impact ionization process.

Broadband amplifiers (pulsed-mode operation) exhibiting fractional 3 dB bandwidth in excess of 10% and power gains greater than 6 dB have been observed at 3.5 GHz using the circuit previously described. The bandpass characteristics of a typical amplifier are shown in Fig. 3. The 3 dB bandwidth is 350 MHz for 6 W peak pulse power and the midband gain is 6 dB. The dc-to-rf conversion efficiency is 16%. Peak powers in excess of 10 W at 10 dB gain with efficiencies greater than 20% have been measured.

The HEM amplifiers are large-signal saturated output power amplifiers and do not exhibit net gain until the input signal amplitude exceeds a critical threshold value. For signal levels below the critical threshold value, the HEM diode presents a relatively good impedance match to the circulator and the amplifier exhibits a loss. When the input signal level exceeds the critical threshold, net gain is observed. The amplifier exhibits a small dynamic range (3-5 dB) where it is linear. At higher input power levels the amplifier saturates and the output power remains constant and then decreases. This is illustrated in Fig. 4, which shows the midband input-output power characteristics for a 250 MHz wide amplifier. This type of wideband amplifier design should be capable of yielding HEM amplifiers exhibiting fractional bandwidths as large as 20% at S band and 10% at X band.

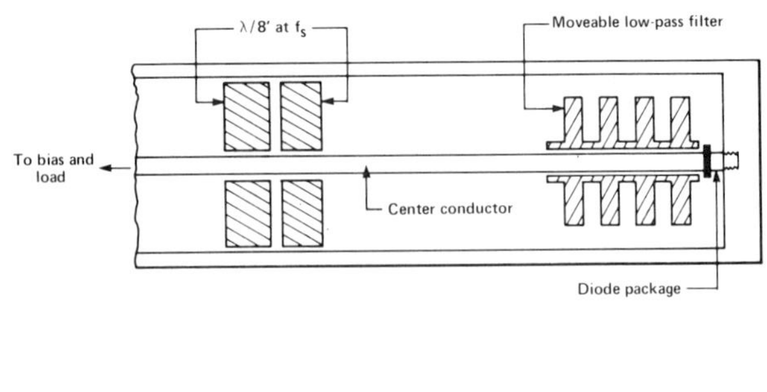

FIG. 2 Cross-section of wideband coaxial circuit with low-pass Tchebycheff filter adjacent to diode.

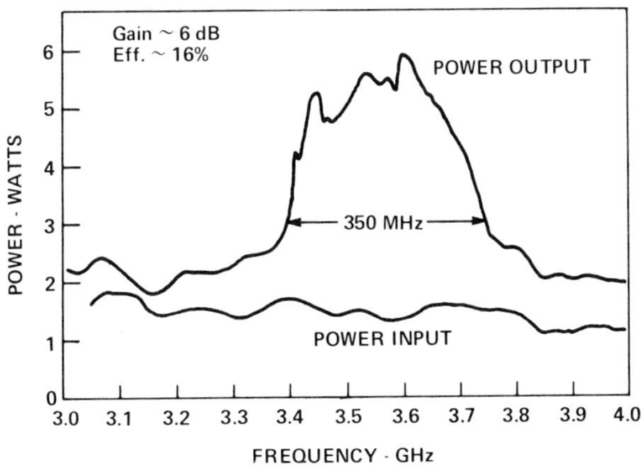

FIG. 3 Bandpass characteristics of a wideband HEM amplifier circuit using Tchebycheff filter.

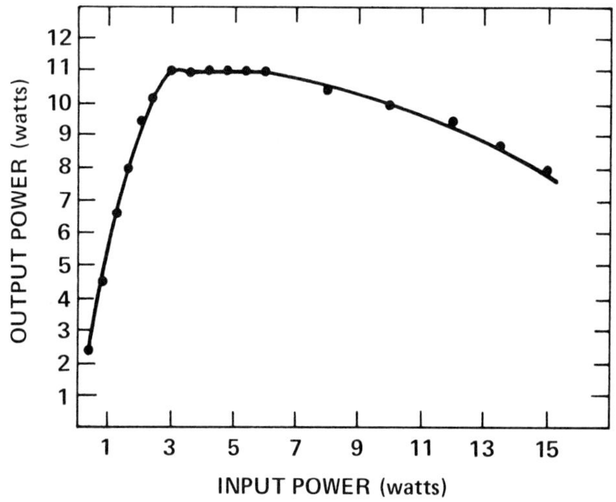

FIG. 4 Input-output power characteristics of wideband HEM amplifier.

REFERENCES:

1. Evans, W. J. and Scharfetter, D. L.: "Characterization of Avalanche Diode TRAPATT Oscillators," IEEE Trans. on Electron Devices, ED-17, May 1970.

2. Hoefflinger, B., Snapp, C. P. and Start, L. A.: "High Efficiency Avalanche Resonance Pumped Amplification," Electronic Letters 5, pp. 43-45, February 1969.

3. Grace, M. I.: "Parametrically Generated High-Efficiency, Low-Frequency Oscillations in Avalanche Diodes," Electronic Letters 5, p. 163, April 1969.

4. Clorfeine, A. S., Ikola, R. J. and Napoli, L. S.: "A Theory for the High Efficiency Mode of Oscillation in Avalanche Diodes," RCA Review 30, p. 397 1969.

NOISE AND OSCILLATOR STABILIZATION V

Noise in Avalanche Transit-Time Devices

389

MADHU-SUDAN GUPTA, STUDENT MEMBER, IEEE

Abstract—The work performed on noise in avalanche transit-time devices is reviewed and presented in detail. Both theoretical and experimental results on the noise mechanisms and performance of these devices when employed as oscillators, amplifiers, and self-oscillating mixers are included. Suggestions for further studies in this area are also given.

I. Introduction

A. Object of the Review

THE INCEPTION of every new electron device triggers a detailed study of the various noise mechanisms in it and ways to reduce the noise. The literature on avalanche transit-time devices, which is now six years old, is no exception and it is worthwhile to make a comprehensive review of the status of understanding of noise in these devices. It is hoped that this review does not only have the advantage of consolidating this knowledge and arranging it systematically so as to put the entire field in perspective, but it also reveals unresolved problems and suggests new areas of investigation.

The importance of noise studies in avalanche transit-time diodes cannot be overstated for two reasons. First, it is generally believed that the noisiness of avalanche diodes may limit their applications and represents their major weakness in comparison with other microwave devices. On the other hand, the large noise generation makes the avalanche diode an attractive noise source for microwave frequencies.

Historically, noise studies in avalanche transit-time diodes began almost as soon as the device became operational. The earliest reported results of measurements of noise in avalanche-diode amplifiers and oscillators were those of DeLoach and Johnston [1] who reported 1) very large amplifier noise figures (50 to 60 dB), 2) reduction of noise figure with an increase in bias current, and 3) very large spectral linewidth (over 100 kHz for

Manuscript received May 6, 1971. This work was supported by the National Aeronautics and Space Administration under Grant NGL 23-005-183.

The author is with the Electron Physics Laboratory, Department of Electrical Engineering, the University of Michigan, Ann Arbor, Mich. 48104.

3-dB width). The first theories of noise in avalanche-diode oscillators and amplifiers were both due to Hines [2], [3]. Substantial improvements have been made since, both in device noise performance and in sophistication of theories of noise in these devices.

B. Sources of Noise in Avalanche Diodes

The noise in an avalanche transit-time diode is primarily made up of three parts which are, in the order of their significance: avalanche noise, frequency-converted noise, and thermal noise. Avalanche noise, which is a type of shot noise, is the noise in carrier current generated by the inherently noisy avalanche multiplication process. Avalanche multiplication acts much like a noisy amplifier, not only multiplying the original signal noise by a factor M (called the multiplication factor), but also by adding an appreciable amount of noise to it. This noise addition is best described in terms of the statistical fluctuations of M. To understand the origin of fluctuations in M, the avalanche mechanism must be considered in detail. Following the discussion of Hines [3], consider an avalanche region with a self-sustaining constant current so that each carrier pair produces, on the average, exactly one carrier pair by impact ionization. Let N be the total number of carrier pairs present in the avalanche region on the average. If the history of one carrier pair in time is followed, it is found that its "offsprings" are generated and swept out of the avalanche region at time intervals of τ_x on the average, where τ_x is the mean time interval between successive ionizations due to a single carrier pair. The chain ionization leads to a pulse-train-like current at the output. The total current through the avalanche region is then given by

$$I = Nq_e/\tau_x \qquad (1)$$

where q_e equals the electronic charge. This current would appear to be a constant if N is large, and τ_x is very small compared to the period of the frequencies of interest. Now the sources of fluctuations in I are looked for.

First of all fluctuations in M should be expected because each carrier pair has a "probability" of ionization; i.e., the ionization process is not deterministic. In terms of the previously mentioned model, this means that not all chains of ionization are interminable and uniform; some of them terminate (when a carrier pair produces no ionizations before being swept out of the avalanche region) and some branch (when a carrier pair produces more than one ionization), even though on the average there is only one impact ionization per carrier pair. This random beginning and ending of pulse trains causes fluctuations in the output current. Again, since τ_x is only the *average* time interval between ionizations in a single chain, the random fluctuation of this interval would give rise to another source of noise in I.

The second source of noise is frequency-converted noise, which is primarily the low-frequency noise upconverted to the neighborhood of the carrier frequency by beating with the carrier. The sources of this low-frequency noise and its nature will be discussed later in Section III-B. Finally, there is thermal noise generated in the diode, which is very much smaller than the avalanche noise, and has therefore been neglected in all noise analyses to date.

It can be seen that generation recombination noise, diffusion noise, and flicker noise are all missing from the preceding list. In practice they would all be present; however, these have not normally been accounted for in avalanche diodes, because their contribution is expected to be unimportant either from theoretical considerations or from experimental evidence. Generation recombination noise is expected to be small because the transit time of the carriers through the space-charge region is of the order of 10^{-10} s, while the carrier lifetime of excess carriers is of the order of 10^{-6} s even in direct gap semiconductors.

TABLE I

APPLICATIONS OF AVALANCHE TRANSIT-TIME DIODES AND CORRESPONDING CHARACTERIZATIONS OF NOISE PERFORMANCE UTILIZED IN THIS PAPER

Type of Network	Application of Avalanche Diode	One Common Characterization of Noise Performance
Passive one-port	noise generator	spectrum of current fluctuations
Active one-port	oscillator	AM and FM noise spectra
Linear two-port	small-signal amplifier	noise figure
	self-oscillating mixer	noise figure
Nonlinear two-port	large-signal amplifier	AM and FM noise of output

Flicker noise, on the other hand, has mostly been neglected until now on experimental grounds. Early work by Josenhans [4] on X-band avalanche diodes showed no $1/f$ effect in noise close to the carrier, leading him to the observation that "no flicker noise was present in Read oscillators." Ondria and Collinet [5] have also reported the absence of $1/f$ behavior in FM noise spectra close to the carrier frequency, indicating that no flicker noise is present in avalanche diodes at low frequencies. More recently, Ashley and Palka [6] have observed a 3-dB/octave slope in noise spectra close to the carrier in some Ku-band avalanche-diode oscillators with a knee below 5 kHz, although the data are not consistent enough to be conclusive.

Obviously, when the avalanche transit-time diode is used as an amplifier or an oscillator the circuit in which it is embedded also contributes to the total noise. Circuit losses, impedance mismatch in the circuit, and noise in the nonreciprocal element (e.g., circulator) are all known to cause additional noise. However, avalanche diodes are very noisy devices, and it is generally believed that the noise contributions from all sources other than the diode can safely be neglected in estimating the noise behavior.

C. Plan of the Paper

The fact that avalanche transit-time diodes have more than one circuit application makes the work of studying noise more difficult. Table I shows the various applications of an avalanche diode and a corresponding commonly employed measure of noise used to describe noise quantitatively. Of course many other measures of noise are also used; e.g., the noise of a passive one-port may also be measured in terms of an equivalent saturated diode current, noise temperature, an equivalent noise resistor, available noise power, noise ratio, or other parameters. The difficulty arises when the noise behavior of the device in one application has to be correlated with its noise behavior in another application. The noise behavior of avalanche diodes in the various applications will be surveyed.

The plan of this paper is as follows. Noise mechanisms in an avalanche diode are discussed in the next section. Early works on random pulse-type noise due to microplasmas in avalanche breakdown are briefly touched upon. Thereafter, two theories of noise, one due to Tager [7] and McIntyre [8], and the other due to Hines [3] and Gummel and Blue [9] are summarized. Experimental work verifying various aspects of these theories is then summarized. In the third section, noise in avalanche-diode oscillators is considered. Theoretical analyses of both a small- and large-signal nature are reviewed. Experimental results under CW and pulsed conditions on unsynchronized as well as phase-locked oscillators, locked with the fundamental or with the subharmonic frequency, are discussed, as are the effects of the circuit on oscillator noise. Section IV is concerned with the theory of noise in small-signal avalanche-diode amplifiers and its comparison with experimental results is reported. Noise in avalanche-diode mixers is briefly discussed in Section V. Finally, in Section VI, some suggestions are made concerning possible areas of investigation.

II. Noise in Avalanche Diodes

A. Noise in Microplasmas

Pearson and Sawyer [10] reported a current instability related to the onset of avalanche breakdown in a reverse-biased p-n junction. McKay [11] showed that the breakdown current consisted of a train of rectangular pulses of equal amplitude but randomly varying pulse length and repetition rate, and explained this phenomenon by predicting the existence of regions of localized ionization in a reverse-biased silicon p-n junction. Such nonuniform carrier multiplication over the junction area was detected by Chynoweth and McKay [12], and was called a microplasma by Rose [13]. Champlin [14] made detailed experimental studies of this breakdown and deduced that the noise is caused by current fluctuations among stable levels. A simple account of a theory based on a random on-off switch model is given by van der Ziel [15]. Haitz [16] later improved the model and characterized the microplasma by four quantities: 1) an extrapolated breakdown voltage; 2) a series resistance; 3) a turnon probability; and 4) a turnoff probability. A summary of this and some later work is given by Mönch [17]. This phenomenon will not be discussed in great detail as the reviews of van der Ziel and Mönch cited previously provide an excellent survey of this field, and also because advances in the technology of preparation of p-n junctions have made it possible to make microplasma-free or "uniform" junctions. Good microwave avalanche diodes are usually fabricated with such uniform junctions.

Prior to the use of avalanche diodes as negative resistance devices at microwave frequencies, they were being used, based on the previously stated current fluctuation mechanism, as RF noise sources (up to several hundred MHz), and were commercially available (Penfield [18]) to replace gas discharge tube noise sources. Haitz [19], [20] has discussed at length the design of uniform avalanche diodes (using guard rings) as noise sources and has described their spectrum, amplitude distribution, and temperature dependence.

B. Theory of Avalanche Noise

Tager [7] presented the first theory of current fluctuations in semiconductors under impact ionization and avalanche breakdown conditions. He made the following assumptions in his investigation of the statistical characteristics of the avalanche process.

1) Avalanche breakdown occurs in a plane layer of the semiconductor.

2) A large uniform electric field is applied across this layer.

3) Carriers of both species move with the same saturated velocity, which is independent of the number of ionizations produced by a given carrier.

4) The width of the avalanching layer of the semiconductor is small enough ($<10^{-2}$ cm in silicon) so that the carrier transit time τ_a through it is much shorter than the carrier lifetime, and therefore thermal generation and recombination in the layer are negligible. At the same time the layer width is much larger ($>10^{-5}$ cm in silicon) than the mean free paths so that only the average motion ("drift") of carriers need be considered.

5) The ionization rates (number of electron-hole pairs produced by impact ionization per unit length per carrier) are the same for electrons and holes. If α_e and α_h, the ionization rates for electrons and holes, respectively, are unequal an equivalent ionization rate may be defined as

$$\bar{\alpha} = \frac{\alpha_h - \alpha_e}{\ln \alpha_h - \ln \alpha_e}. \tag{2}$$

This ionization rate is assumed to be uninfluenced by temperature and is an instantaneous function of the electric field.

6) Fluctuations of the flicker-effect-type and thermal fluctuations are neglected.

7) The primary carrier current I_0 entering the avalanching layer contains full shot noise (the mean square noise current equals $2q_eI_0B$, where q_e is the charge of an electron, I_0 the primary current, and B the bandwidth).

8) The primary current is not rapidly time varying; i.e.,

$$\frac{dI_0}{dt} < \frac{I_0}{\tau_a}. \tag{3}$$

With these assumptions, taking into account 1) the fluctuations in the entering primary current, 2) the fluctuations in the multiplication factor M of the avalanche layer, and 3) the depression of current fluctuations due to the space-charge of the moving carriers, Tager obtained the following results.

1) The mean square fluctuations in the multiplication factor M depend upon the ratio of ionization rates of the two types of carriers. When this ratio is unity (equal ionization rates)

$$\overline{\Delta M^2} = (\overline{M})^2(\overline{M} - 1) \tag{4}$$

while for an infinite ratio (only one species of carrier ionizing)

$$\overline{\Delta M^2} = \overline{M}(\overline{M} - 1). \tag{5}$$

2) The mean square value of M is equal to the cube of the average value of M; $\overline{M^2} = (\overline{M})^3$.

3) The spectrum of carrier current leaving the avalanche layer is

$$S_i(\omega) = \frac{2q_eI_0(\overline{M})^3}{1 + \left(\frac{\omega\overline{M}\tau_a}{2}\right)^2} \tag{6}$$

i.e., at low frequency it is just the shot noise multiplied by $(\overline{M})^3$, while at high frequencies it is inversely proportional to the square of the frequency.

Further developments of this theory were primarily due to an interest in the noise properties of avalanche photodiodes. McIntyre [8] derived a general expression for the spectral density of noise generated in uniformly multiplying p-n junctions for any distribution of injected carriers which is valid at frequencies that are low compared to the inverse of the transit time through the avalanche layer. He used the same set of assumptions as Tager, except that unequal ionization rates for holes and electrons and any field profile are permitted. The expression for the spectral density at low frequencies ($f \ll 1/\tau_a$) is given by

$$S_i(\omega)=2q_e\left\{2\left[I_p(0)M^2(0)+I_n(w)M^2(w)+\int_0^w gM^2(x)dx\right] + I_0\left[2\int_0^w \alpha_e M^2(x)dx - M^2(w)\right]\right\} \quad (7)$$

where w is the width of the avalanche layer, $I_p(0)$ is the primary hole current entering at one end ($x=0$) of the avalanche layer, $I_n(w)$ is the primary electron current entering the other end ($x=w$) of the layer, M is the multiplication factor of the avalanche layer which is a function of x because the electric field is a function of x, and g is the rate of thermal or optical generation of carriers in pairs per unit length. While α_e occurs explicitly in the expression, both α_e and α_h are present through M. Also note that the expression is independent of frequency (which is a direct consequence of omitting all time dependence), resulting in a white noise spectrum at low frequencies.

Under the additional assumption that the electric field is uniform over the entire avalanche region, the previous expression reduces to

$$S_i(\omega) = 2q_eI_0M^3\left[1 + A\left(\frac{M-1}{M}\right)^2\right] \quad (8)$$

where

$$A = \frac{\alpha_e - \alpha_h}{\alpha_h}$$

or

$$A = \frac{\alpha_h - \alpha_e}{\alpha_e}$$

when, respectively, holes or electrons alone are injected into the avalanching layer. For the special case of $\alpha_e=\alpha_h$, this reduces to $S_i(\omega)=2q_eI_0M^3$, which is identical with Tager's result at low frequencies.

On the basis of these results McIntyre concluded that for a given M, noise could be reduced if the ionization were caused primarily by the more ionizing species of carriers. McIntyre also estimated the effect of time-dependent mechanisms which introduce frequency dependence in the noise spectrum at high frequencies. If all multiplications are not instantaneous, but instead there is an average time separation τ_x between ionizations caused by a single pair of carriers the high-frequency end of the spectrum will be attenuated by a factor $(1+\omega^2M^2\tau_x^2)^{-1}$. Also, if C is the diode capacitance and R_p the parallel resistance as seen by the diode, the circuit time constants will introduce an additional frequency-dependent attenuation term $(1+\omega^2R_p^2C^2)^{-1}$.

In an avalanche transit-time diode there is a third source of frequency dependence due to the finite transit time in the drift region which causes a phase shift between the high-frequency voltage and current. This problem was considered by Hines [3] in a small-signal fashion; he assumed the noise to be small compared to the dc bias current on which it is superimposed. The diode discussed by Hines differs from Tager's diode in another very important respect; namely, that the diode current is entirely self-sustaining, requiring no injected carriers. Therefore, the results of the two analyses cannot be compared directly. The method of analysis of Hines is based upon the space-charge wave approach of a small-signal analysis given by Gilden and Hines [21] and involves the following assumptions.

1) The diode is separated into an avalanche region and a drift region.

2) All noise is assumed to originate in the avalanche region.

3) All the assumptions of Gilden and Hines' small-signal theory are carried over; namely, the assumptions of saturated drift velocities which are the same for electrons and holes; a thin avalanche region at one side of a high-field region; equal ionization rates for electrons and holes; constant electric field in the avalanche region; the current, electric field, and ionization rate being expressed as a constant dc value plus a small sinusoidal ac variation; negligible thermally generated reverse saturation current; the drift region current entirely due to one species, the majority carriers; no fixed space charge in the drift region; punch-through operation at all times; negligible series resistance and negligible displacement current so that the total carrier current is independent of distance along the diode length.

4) The ionization rate α varies as a power of the electric field E^m.

5) The noise due to statistical variation of the interval τ_x between ionizations in any given chain of ionizations is negligible.

6) Carrier inertial effects are negligible so that the probability of ionization by a given carrier during its transit is constant and consequently the Poisson distribution can be used.

Under these assumptions, Hines found that the mean square value of total noise current is

$$\overline{I_n^2(\omega)} = \frac{2q_eI_0B}{\omega^2\tau_x^2}\left|\frac{R_d}{Z_t^2}\cdot\frac{1}{1-\dfrac{\omega_a^2}{\omega^2}}\cdot\frac{V_a}{2mI_0}\right| \quad (9)$$

where

$$R_d = \frac{l_d}{\theta\omega\epsilon A}\left[\frac{1-\cos\theta}{1-\dfrac{\omega^2}{\omega_a^2}}\right]$$

I_0 is the dc bias current, l_d is the width of the drift region, θ is the transit angle through the drift region at frequency ω, ϵ is the permittivity of the semiconductor material, A is the area of the cross section of the diode, ω_a is the avalanche frequency of the diode, Z_t is the total impedance around the RF loop containing the diode including the load impedance, and V_a is the dc voltage across the avalanche region. Under the special case of low frequencies ($\omega\ll\omega_a$), the above simplifies to[1]

$$\overline{I_n^2(\omega)} = \frac{q_eBC_d^2V_a^2}{2I_0m^2\tau_x^2} \quad (10)$$

where C_d is the depletion capacitance of the entire depletion zone of the diode, showing that the mean square noise current varies inversely with the dc current I_0. This is due to a decrease in avalanche impedance with the increase of current. At high frequency (9) simplifies to

$$\overline{I_n^2(\omega)} = \frac{q_eB}{\left(1+\dfrac{R_L}{R_d}\right)^2}\cdot\frac{C_d^2V_a^2\theta^2}{m^2\tau_x^2(1-\cos\theta)I_0}. \quad (11)$$

For reasons stated earlier, no comparison with Tager's results is made.

Constant and Semichon [22] have described an analysis similar to Hines', and have found expressions for mean square noise voltage and spectral linewidth.

[1] Equation (34) of Hines [3] has an extra factor of 2.

Gummel and Blue [9] have generalized the analysis of Hines by removing some of the assumptions made, in particular the assumptions of an idealized Read structure, equal saturated velocities for electrons and holes, and equal ionization rates for holes and electrons. The one-dimensional model and small-signal analysis are maintained. In addition it is assumed that the conduction current in the diode has full shot noise; this noise is injected into the avalanche region of the diode, and solution of the carrier transport equations then yields the open-circuit voltage fluctuations caused by the shot noise. A rigorous justification for such an approach has been given by Blue [23] for a stationary quasi-linear system. The mean squared voltage $\langle v^2 \rangle$ is calculated analytically for an idealized Read-type diode structure and numerically for a realistic p-n junction diode with a wide avalanche region. The expression for $\langle v^2 \rangle$ shows the following behavior of the mean square noise voltage.

1) For a fixed bias current this voltage is constant at low frequencies, then it increases and attains a maximum at a resonance frequency, and falls approximately at the fourth power of frequency above resonance.

2) At low frequencies (below resonance) $\langle v^2 \rangle$ is inversely proportional to the bias current, the resonance frequency is proportional to the square root of the bias current density, and in the $1/\omega^4$ domain it is approximately directly proportional to the bias current.

These conclusions agree with the experimental observations of Haitz and Voltmer [24], [25] described in the next section.

C. Experimental Results

Tager's result that the mean square value of fluctuations equals M^3 has been experimentally verified by Naqvi *et al.* [26]. They have measured multiplication noise as a function of the multiplication factor and have plotted the spatial variation of both over the cross section of the diode. Their experimental setup consisted of a sharply focused chopped laser beam, scanning the reverse-biased diode junction, and a phase-sensitive lock-in amplifier to measure the avalanche multiplied photocurrent. They found that: 1) multiplication noise at a multiplication of 88 was as high as 37 dB above the full shot noise in the junction current; and 2) agreement with McIntyre's theory suggests that there is no space-charge correlation effect up to a current density of 100 A/cm^2, which was the maximum used in their experiments.

Baertsch [27], [28] made measurements of noise in a silicon avalanche photodiode current as a function of the multiplication factor and found good agreement with McIntyre's theory.

Constant *et al.* [29] have reported experimental studies of radio frequency and microwave noise in avalanching p-n junctions. Their results on the observed variation of the mean square noise current with bias current and frequency are shown in Fig. 1, and show agreement with Hines' theory.

Haitz and Voltmer [24], [25] and Haitz [30] have reported experimental measurements of the spectral density of a silicon avalanche-diode noise voltage, both at audio frequencies (1 kHz to 11 kHz) and at microwave frequencies (3 GHz to 9 GHz), and have verified Hines' theory of noise in Read diodes as modified for p-n junctions. The main results of this experimental study follow.

1) At low frequencies the spectral density varies inversely as the square root of the dc bias current as expected. There is an excess noise at high frequencies due to thermal effects, and an excess noise at low frequencies due to local variations of the breakdown voltage (as evidenced by the fact that the excess noise increases with area and is not reproducible from diode to diode).

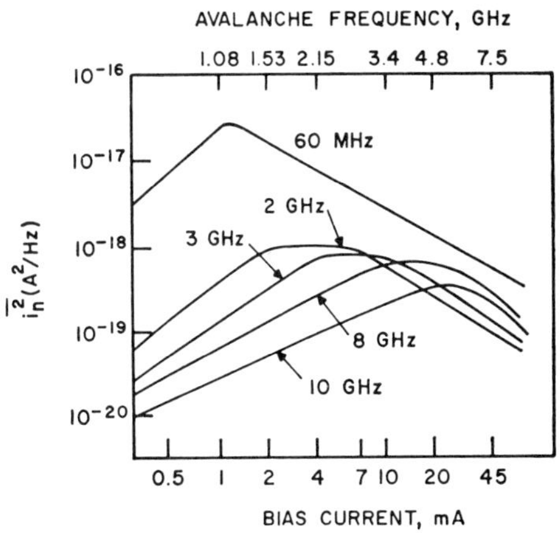

Fig. 1. Mean square noise current of an avalanche diode as a function of bias current and frequency. The avalanche frequency calculated from bias current is also marked on the axis. (Constant *et al.* [29].)

2) The spectral density of the noise voltage is proportional to the breakdown voltage for several different doping profiles of the diode: n$^+$-p, p$^+$-n, and n-p-π-p$^+$.

3) At high frequencies the spectral density varies linearly with bias current for small bias currents (where the avalanche frequency f_a is small), attains a peak, and decreases thereafter as the bias current (and hence f_a) increases.

4) For low bias currents (and hence small f_a), the spectral density of noise shows $1/f^4$ variation with frequency, while at larger bias currents the frequency dependence is much smaller, as expected theoretically.

D. Avalanche-Diode Microwave Noise Generators

Several workers have explored the possibility of using avalanche diodes as microwave noise sources. Such noise sources would have the following advantages over gas-discharge-type noise sources.

1) Much larger noise output, eliminating the need of amplification in many applications.

2) Small size, small weight, low power requirements, and low cost, making it possible to use them for inservice measurement of critical parameters in microwave equipment (Chasek [31]).

3) Possibility of pulse modulation with very narrow pulses (of the order of 1 μs).

The use of flat-topped random pulses of microplasma breakdown in p-n junctions as a source of noise has been mentioned. Minden [32] has carried out some experimental work demonstrating that noise in avalanche diodes is of two distinct types: that associated with microplasma breakdown, and that associated with uniform junction breakdown. He pointed out the following difference between the two. Microplasma breakdown occurs at very low breakdown current (1 to 300 μA), and the noise level as a function of bias current shows several peaks. The uniform junction breakdown noise becomes appreciable at larger reverse bias current and attains a maximum at a value of bias current which is lower than the threshold for coherent microwave oscillations. Minden also found that high-output avalanche-diode oscillators show no microplasma noise, while diodes which generated larger uniform junction noise also generated large microwave output when operated as oscillators.

Val'd Perlov *et al.* [33] described a microwave noise source, verifying some of the predictions of Tager [7]. In particular they found the following.

1) For diodes with breakdown voltages of 10 to 15 V, and at

bias currents of 0.5 to 10 mA, the noise temperature of the source was 10^5 to 10^{6}°K in the centimeter band, and 10^6 to 10^7°K in the meter band.

2) The long-term stability of noise level was very good, with a temperature coefficient of spectral density of noise power being 0.02 dB/°K.

3) The noise source could be modulated by microsecond pulses.

Tager's result (6) shows that at high frequencies the noise level depends only upon frequency and bias current, and not upon the multiplication factor M. Hence the output is independent of reverse saturation current of the diode. At low frequencies, however, it does depend upon M, and consequently upon reverse saturation current and temperature.

Dalman and Eastman [34] reported the following results of noise measurements on diffused p^+-n-n^+ Si avalanche diodes.

1) The avalanche diode itself is an inherently broad-band noise source and the circuit largely determined the bandwidth of the noise source.

2) Noise temperatures from 15 to 25 dB greater than an argon gas discharge tube (30 to 35 dB over 290°K) are available, depending upon the bandwidth of the diode circuit.

Chadelas *et al.* [35] subsequently suggested an automatic radiometer utilizing an avalanche-diode noise generator, and made two observations.

1) The noise temperature of the source (10^5 to 10^6°K) varies linearly with the diode bias current (1 to 10 mA).

2) The noise source would be usable to Q band (35 GHz).

Haitz and Opp [36] have designed an avalanche-diode noise source integrated in a microstrip-line circuit, and reported the following results with the diodes biased at 30 mA (20- to 35-V breakdown voltage).

1) The noise source had a potential range from dc to 18 GHz.

2) The spectral noise power density was in excess of 30 dB above kT_0 from 1 to 11 GHz, with variations within ± 0.7 dB over this frequency range. The variations are further reduced by tuning out reflections from connectors and adaptors.

394 3) The source has a very good long-term stability (± 0.1 dB over a 500-h period).

4) The temperature dependence of the spectral power density (measured at 2.5 GHz) was less than -0.002 dB/°C.

5) The bias voltage dependence of the spectral density was better than -0.005 dB/V.

6) Output variations from diode to diode were small.

Bareysha *et al.* [37] discussed in detail the design of a microwave noise generator using diffused Ge avalanche diodes. They have used a circuit model of the noise generator to study the effect of mismatch between the diode and the load upon noise power output, and the bandwidth of the generator. In addition results of experimental measurements of the dependence of noise spectra on bias current, junction capacitance, breakdown voltage, and temperature are also given. Their main findings follow.

1) Noise-power spectral density at a given frequency attains a broad maximum as a function of diode bias current; the higher the frequency, the larger the bias current required for maximum noise power and the lower the magnitude of this maximum. (See Fig. 1 due to Constant *et al.*)

2) The noise-power spectral density remains practically unchanged as the junction capacitance varies.

3) Noise power is slightly dependent on breakdown voltage for large breakdown voltages where Zener current is negligible.

4) The operation of the noise generator is very unstable if the dynamic resistance of the junction is negative, and its output is strongly dependent upon the load, particularly near the oscillation threshold. Therefore, for high-frequency noise generation the breakdown voltage should be kept low so that the threshold oscillation frequency is high.

5) A change in temperature affects the junction impedance and hence the noise power output. Thermal compensation can be achieved by using a thermistor to control the diode bias current.

III. Noise in Avalanche Transit-Time Diode Oscillators

A. Theory of Oscillator Noise

Hines [2] presented the first noise theory for avalanche transit-time diode oscillators. His analysis, which can only be considered as a first-order approximation, is based upon results found in his small-signal analysis presented in Section II-B. Two sets of assumptions are carried over in this work.

1) The assumptions made by Edson [38] in his analysis of the perturbation of an oscillator by thermal noise.

2) The assumptions made by Hines in his analysis of the noise in the avalanche diode (given in Section II-B).

Thereafter, it is assumed that (9) accurately described noise current in an oscillator and that thermal noise in Edson's theory can be directly replaced by this shot noise. Using these assumptions, Hines found expressions for the AM and FM noise.

Obviously, the assumptions made are quite restrictive. Edson analyzed an oscillator consisting of a simple resonant circuit and a nonlinear conductance only. Therefore, only the device conductance is assumed to be voltage dependent and not the susceptance, and the nonlinearity of the active device conductance is described by using a constant s factor, defined as

$$s = \frac{\Delta G}{G} \tag{12}$$

where ΔG is the small-signal conductance, and G is the total conductance of the device.

Within the limitations of the model, Hines found an AM noise spectrum that is white for low-Q circuits, given by

$$N_{AM}(\omega) = \frac{b^2 q_e I_0 R_L}{s^2 + 4Q_L\left(\dfrac{\omega - \omega_0}{\omega_0}\right)^2} \quad \text{W/Hz}$$

and a constant rms frequency deviation

$$\Delta f_{\text{rms}} = \frac{f_0}{Q_L}\sqrt{\frac{b^2 q_e I_0 B R_L}{2P}} \quad \text{Hz}$$

where

$$b^2 = \frac{\tau_d^2 V_a^2 \omega_0^2 C_d^2 (1 - \cos\theta)}{2m^2 \tau_x^2 I_0^2 \theta^2 \left(1 - \dfrac{\omega^2}{\omega_a^2}\right)^2}$$

τ_d is the transit time in the drift region, and the other symbols have the same meaning as in (9).

Inkson [39] published the first large-signal theory of noise generation in avalanche transit-time diode oscillators. He considered Read's model of an avalanche diode; i.e., he assumed a thin avalanche region, equal drift velocities and ionization rates for electrons and holes, and a power law dependence of ionization rate α on electric field, and he used the sharp pulse approximation also first suggested by Read [40]. The "sharp pulse assumption" maintains that the carrier current consists of one sharp pulse of charge per cycle traveling in the drift region at a saturated velocity with only one pulse being present at any time, and the space charge of the avalanche current being nonzero only when a pulse is present in the avalanche region. It is further assumed that the electrons and holes are always in equilibrium with the field, and hence a Poisson distribution is used for the number of ionizations. Inkson then calculated the jitter in pulse amplitude and interpulse interval due to the random process of ionization, and calculated the AM and FM noise spectra therefrom. The analytical expressions obtained are quite complicated and are not reproduced here; however, the following conclusions can be drawn from Inkson's results.

1) As the bias current of the diode is increased, the output of the oscillator rapidly becomes very noisy.

2) A larger area avalanche diode would make the oscillator less noisy (the diode model being one-dimensional, nonuniformities are assumed to be absent).

3) A wider avalanche region in the diode (i.e., appreciable ionization over a larger part of the depletion layer) leads to a lower noise output from the oscillator (although the analysis is not applicable to such diodes).

4) FM noise (rms frequency deviation) varies inversely as the square root of the loaded Q of the microwave cavity (rather than inversely with Q as in Edson's model), so that the device should be coupled to a high-Q cavity.

Ulrich [41] has also found expressions for AM and FM noise spectra of an avalanche transit-time diode oscillator, but in terms of the nonlinearity of the diode impedance rather than the diode material, structure, and operating point parameters as in the case of earlier works. Ulrich used an approach similar to Kurokawa's general analysis [42], modeling the oscillator by a series circuit containing the device (with RF amplitude-dependent impedance), the impedance as seen by the device (which is frequency dependent), and an all-inclusive noise source independent of RF amplitude or frequency. He found that the AM and FM noise spectra are

$$|\Delta A(f)|^2 = \frac{|e|^2}{2A_0^2\left|\left(\frac{\partial Z_d}{\partial A}\right)_{A_0,\omega_0}\right|^2 \sin^2\phi} \tag{13}$$

and

$$|\Delta\omega(f)|^2 = \frac{|e|^2}{2A_0^2\left|\left(\frac{\partial Z_L}{\partial \omega}\right)_{A_0,\omega_0}\right|^2 \sin^2\phi} \tag{14}$$

and their correlation coefficient is

$$C = -\cos\phi \tag{15}$$

where $|e|^2$ is the mean square noise voltage of the noise source previously mentioned, including all fluctuations produced by the device; A and ω are the current amplitude and frequency of oscillation with mean values A_0 and ω_0, respectively; $Z_L = R_L + jX_L$ is the impedance of the load (including circuit) as seen by the diode; $Z_d = R_d + jX_d$ is the impedance of the diode, and

$$\phi = \tan^{-1}\left(\frac{\partial X_L}{\partial\omega}\bigg/\frac{\partial R_L}{\partial\omega}\right) - \tan^{-1}\left(\frac{\partial X_d}{\partial A}\bigg/\frac{\partial R_d}{\partial A}\right). \tag{16}$$

Two conclusions, both of which have been experimentally verified, can be drawn from the previous analysis concerning the circuit dependence of noise behavior.

1) Only the FM noise depends upon circuit Q, as evidenced by the $\partial Z_L/\partial\omega$ term in (14).

2) A minimum of noise occurs when $\phi = 90°$ which can be obtained by a proper circuit arrangement, and does not require a high-Q cavity.

Experimental results verifying these conclusions are described in a later section.

Vlaardingerbroek [43] found expressions for the output power spectrum and the width of the output spectrum of an avalanche transit-time diode oscillator operating at low and intermediate output levels. He followed the assumptions and method of Hines, discussed in Section II-B, in that the "primary" noise current due to fluctuations in the random avalanche process is related to the noise in the Read diode terminal current; however, nonlinearity is included, as opposed to Hines' small-signal approximations. The output power spectrum is then expressed in terms of the spectrum of the primary noise current.

B. Upconversion of Low-Frequency Noise

The nonlinearity of avalanche diodes, which makes them useful as mixers, introduces another source of noise in avalanche transit-time diode oscillators; namely, upconversion of low-frequency noise to frequencies near the carrier. Experimental as well as theoretical results obtained by Evans and Haddad [44] have shown that upconversion of low-frequency signals to frequencies in the neighborhood of the carrier frequency in an avalanche diode is accompanied by a large gain. This upconversion gain therefore makes low-frequency noise an important contributor to total noise.[2] In addition to thermal noise in the RF circuit, this low-frequency noise could arise from at least two different sources, both of them stronger than the circuit thermal noise.

1) Low-frequency noise (both avalanche and thermal noise with the avalanche noise predominating) generated within the diode.

2) Bias current fluctuations (due to supply fluctuations, thermal noise, or $1/f$ noise) generated in the dc bias circuit.

Goedbloed [45] carried out a theoretical analysis of upconverted noise in an avalanche transit-time diode oscillator using the Read model of the diode (with infinitely thin avalanche region and infinitely high multiplication factor) and a passive single resonant network model for the oscillator circuit. He has shown that if both of the previously mentioned assumptions of the Read model are maintained, there is upconversion of low-frequency noise to AM noise of the oscillator, but not to FM noise. However, if one of these assumptions is discarded, FM noise is enhanced due to upconversion of low-frequency noise. This observation should serve as a warning for caution in model selection before studying noise in such nonlinear devices as avalanche diodes. 395

Scherer [46] has taken the effect of low-frequency noise into account by calculating the components of AM and FM noise due to fluctuations in the bias current. These fluctuations are assumed to be due to avalanche noise alone, and are estimated using a low-frequency noise equivalent circuit of the avalanche diode first given by Haitz [30] and an impedance representing the bias circuit. The low-frequency noise current, which is calculated using this model, modulates the dc bias current, and is transformed into the high-frequency AM and FM noise using ampli-

[2] The reasons why the downconverted high-frequency noise is not considered are twofold. 1) The high-frequency noise can only be generated in the diode and not in the bias circuit, and 2) it suffers a downconversion loss, consequently it is small.

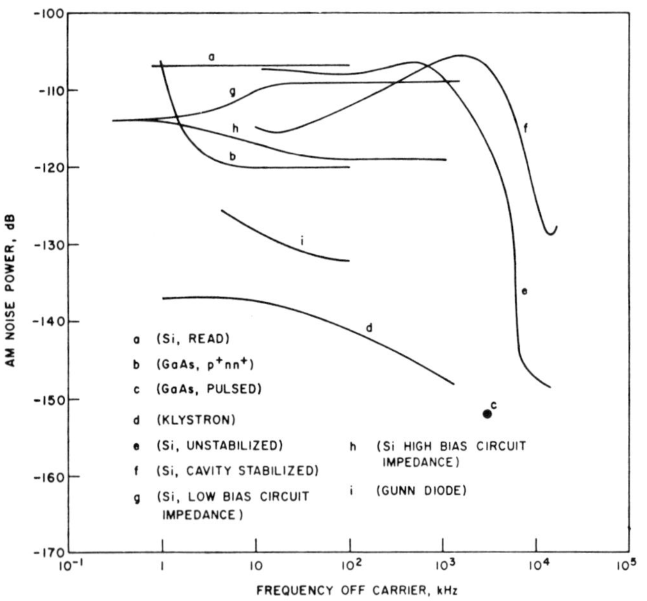

Fig. 2. AM noise spectra of avalanche-diode oscillators. All results have been reduced to double sideband AM noise power to carrier power ratio in a 100-Hz bandwidth. Curves *a* and *b* give noise spectra for a Si Read structure [4] and a GaAs p^+-n-n^+ structure [57], respectively. Point *c* represents AM noise in a pulsed GaAs diode oscillator [59]. Curves *d* and *i* giving AM noise for an X-13 klystron [5] and a Gunn diode [4] are included for comparison. Curves *e* and *f* show AM noise spectra for a Si diode oscillator with and without stabilization using a transmission cavity [63]. Curves *g* and *h* are AM noise spectra for a Si diode oscillator when the dc biasing circuit impedance was very low and very high, respectively, [46]. Output power, bias current density, and frequency of oscillation are different for the various curves.

tude and frequency modulation sensitivities. Using values of avalanche current noise from the computer calculated results of Gummel and Blue and experimentally measured values of modulation sensitivities, he obtained the following.

396 *AM Noise:* noise-to-signal ratio $= -108$ dB;
FM Noise: rms frequency deviation $= 10.6$ Hz

in a 100-Hz bandwidth under a typical (given) set of operating conditions for a Si avalanche diode.

C. Measurements of Oscillator Noise

No attempts are made to discuss the techniques of measurement of AM and FM noise in microwave oscillators; a review on this subject appeared elsewhere [47], and some more recent work will simply be referred to [48], [49]. The noise behavior of "typical" avalanche-diode oscillators is shown in Figs. 2 and 3. Measurements of avalanche-diode oscillator noise have been reported by a rather large number of workers, and no attempt is made to catalog all of them. Instead, only those experimental reports that have studied the dependence of noise behavior of avalanche-diode oscillators on various parameters such as diode material, structure, circuit Q, bias current, power saturation, and bias circuit are included.

The first measurements of AM and FM noise in avalanche transit-time diode oscillators were reported by Josenhans [4]. His results on silicon Read-type diodes follow.

1) The FM noise spectra are symmetrical about the carrier to within ± 1 dB (found using a spectrum analyzer).

2) FM noise measurements up to as close as 400 Hz away from the carrier frequency showed no flicker noise effects.

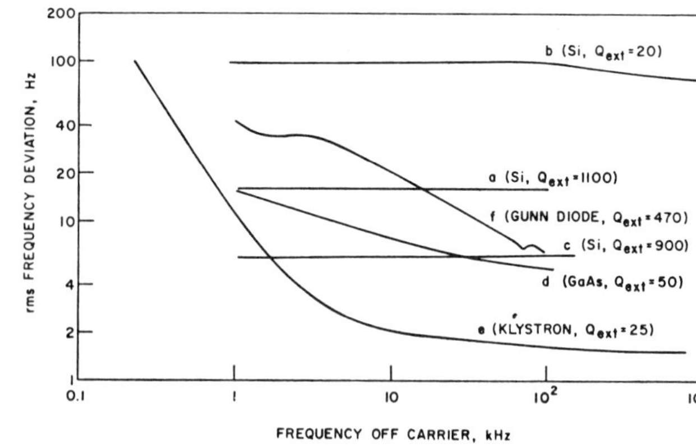

Fig. 3. FM noise of avalanche-diode oscillators. All results have been reduced to double sideband FM noise expressed in rms frequency deviation for a 100-Hz bandwidth. Curve *a* gives the FM noise spectrum for a Si Read diode oscillator with $Q_{ext} = 1100$ [4]. Curves *b* and *c* give these spectra for Si p^+-n-n^+ structures for $Q_{ext} = 20$ [67] and $Q_{ext} = 900$ [5], respectively. Curve *d* is the FM noise spectrum for a GaAs avalanche-diode oscillator for $Q_{ext} = 50$ [57]. Curves *e* and *f* giving FM noise for an X-13 klystron [5] and a Gunn diode [4] are included for comparison.

3) The AM noise spectrum is flat and the rms frequency deviation is constant in the frequency range measured (up to 100 kHz away from the carrier frequency).

4) Noise measurements were performed on some avalanche transit-time diodes operated both as amplifiers and oscillators. The noise behavior of the diode used as an oscillator was also estimated using the results of measurements on the diode used as an amplifier. The estimated oscillator noise was considerably lower than the measured oscillator noise, indicating a large-signal behavior of noise sources in oscillator applications.

Cook [50] reported on avalanche-diode oscillators in which AM noise dropped at 2 dB/octave. Goldwasser *et al.* [51] reported on experimental measurements of FM noise of CW avalanche-diode oscillators close to the carrier frequency (2 to 10 kHz away from the carrier) and made two important observations:

1) The FM noise was strongly dependent upon the dc bias current, and for a given diode and circuit, the current could be optimized to reduce the noise sharply.

2) Upon such optimization, the FM noise in Schottky barrier GaAs diode oscillators was much lower (0.7 Hz rms for a 100-Hz bandwidth) than in diffused p-n-n^+ Si diodes (80 Hz rms for the same bandwidth). An explanation of the superiority of Schottky barrier diodes was also given by Goldwasser.

Cowley [52] experimentally observed the effect of power saturation on avalanche transit-time diode oscillator noise, and concluded that: 1) the occurrence of power saturation or the increase of bias current beyond the saturation level makes the oscillator very noisy. The value of load resistance R_L should therefore be kept somewhat greater than the value that leads to maximum power to achieve better noise behavior; and 2) when saturation occurs, bias current oscillations can arise in the power supply circuit under certain conditions, and would appear as AM and FM modulations in the output, degrading the oscillator spectrum.

In another communication Cowley *et al.* [53] also reported a large increase in noise due to a very small increase in the dc bias current.

The dependence of FM noise of avalanche transit-time diode oscillators on the external Q of the cavity was studied over a wide range by Harth and Ulrich [54]. They also observed the dependence of rms frequency deviation on dc bias current and the correlation coefficient of the AM and FM noise. Their results can be summarized as follows.

1) The rms frequency deviation varies as $1/Q_{ex}\sqrt{P_{out}}$.

2) The rms frequency deviation is constant in the frequency range 100 Hz to 1 MHz away from the carrier frequency, and the contribution of low-frequency upconverted noise to it is negligible in this frequency range as verified by varying the low-frequency impedance in the diode bias circuit.

3) FM noise increases rapidly with an increase of the dc bias current. In the range of bias current between the onset of oscillation and maximum admissible current, the excess noise temperature over T_0 was between 30 and 50 dB, and varied approximately as the fifth power of current density independently of the circuit Q.

4) The correlation coefficient of FM and AM noise was between 0.8 and 1, indicating a common mechanism for their generation.

Scherer [46] has made detailed observations on the effect of the impedance in the dc bias circuit and the bias current of the diode upon the AM noise of the oscillator and reached the following conclusions.

1) The AM noise spectrum at frequencies close to the carrier (less than 1 kHz away from the carrier) is not affected by the impedance of the bias circuit; for large frequency separation AM noise is increased by lowering the bias circuit impedance. The difference in AM noise between the cases of very small and very large bias circuit impedances (compared to diode impedance) increases as the frequency separation increases to 100 kHz and then becomes constant.

2) Some of the diodes show an excess noise with a $1/f$ dependence.

3) For the high bias-circuit impedance case, the measured noise agrees well with the calculated RF noise, indicating that the upconverted component of AM noise is not significant.

4) Noise performance deteriorates with increasing power output, increasing coupling of the load and increasing bias current.

Ulrich [41] observed the effect of RF circuit impedance variation on AM and FM noise and their correlation, and substantiated his theory [(13)–(15)]. All three equations involve the rate of change of circuit impedance with frequency, which was controlled by weakly coupling a high-Q cavity to the oscillator circuit by a magic tee. Tuning this cavity from below to above the oscillation frequency changes the angle ϕ by 360°. Two maxima of both AM and FM noise occurred when ϕ was 0 and 180°, and two minima occurred when ϕ was 90 and 270°. The correlation coefficient was also maximum in magnitude at the noise maxima and minimum at the minima.

FM noise measurements on Ge avalanche transit-time diode oscillators were first reported by Rulison *et al.* [55]. They found that the rms frequency deviation is constant as a function of frequency separation from the carrier, as for Si diodes. However, for a loaded $Q_{ex}=11$, in a 1-kHz bandwidth, the rms frequency deviation was only 300 Hz. In general, Ge avalanche-diode oscillators showed a 10- to 14-dB improvement in noise-to-carrier ratio over Si diodes operating in the same circuit with the same loaded Q_{ex}. Barber [56] also reported on the FM noise in Ge diode oscillators to be 15 dB lower than in Si diode oscillators in cavities of the same Q_{ext} for frequency separations of 1 kHz to 2 MHz from the carrier frequency. GaAs avalanche transit-time diode oscillators have also been reported, by Baranowski *et al.* [57], to be superior to Si diode oscillators in their noise behavior. Their findings may be summarized as follows.

1) The FM noise sideband-to-carrier ratio for a GaAs diode oscillator was at least 10 dB lower when compared to a good Si diode oscillator operating in the same circuit with the same external Q.

2) The FM noise falls off as separation from the carrier frequency is increased.

3) The AM noise sideband-to-carrier ratios were similar for GaAs and Si avalanche-diode oscillators.

Levine [58], [59] reported measurement of AM noise on vapor-grown GaAs avalanche transit-time diodes operated pulsed at high current densities of up to about 2000 A/cm², and obtained AM noise levels lower than in klystrons and Gunn diodes. The best double sideband AM noise-to-signal ratio was −152 dB in a 100-Hz bandwidth 30 MHz away from the carrier frequency (11.1 GHz) for a 25-V breakdown diode operated at 1940 A/cm². In general, AM noise was found to be lower under conditions of high-bias current density, and GaAs diodes had a noise level 20 dB lower than Si diodes.

Avalanche-diode oscillator noise measurements in terms of parameters other than AM and FM noise have also often been reported. Thus DeLoach and Johnston [1] have measured spectral linewidth, Johnston and Josenhans [60], Scherer [46], and Baranowski *et al.* [57] have measured the noise figures of mixers and receivers using avalanche diodes as local oscillators, while Kramer *et al.* [61] have measured the total noise power spectrum. These results will not be discussed in detail.

D. Noise Reduction by Circuit Techniques

In this section some measurements of noise in avalanche transit-time diode oscillators operated with special circuit arrangements intended to reduce noise will be reviewed. It is not the intention here to review the theory or techniques of injection phase locking or other noise reduction methods; such details may be found in references [62]–[67]. The present discussion is confined to the improvement of noise behavior achieved by various methods, such as cavity stabilization and injection locking; these results are shown in Fig. 4.

The simplest method of reducing the noise of an avalanche transit-time diode oscillator is by frequency stabilization using a high-Q transmission cavity. Gilden [62] reported noise measurements on an avalanche transit-time diode oscillator stabilized with a tunable transmission cavity and made the following observations. 397

1) Even though the bandwidth of the cavity was 3 MHz, the width of the oscillator output spectrum was reduced from 300 kHz to 2 kHz as observed on a spectrum analyzer. This indicates that the cavity does not merely act as a filter; instead, the diode noise depends strongly upon the frequency dependence of the external circuit reactance as seen by the diode. This fact was further proved by introducing an isolator between the diode mount and the stabilizing cavity, in which case no noise improvement occurred.

2) Measurement of the AM noise of the oscillator, by a mixer crystal and an IF amplifier, showed a reduction of 15 dB in the noise at the output of the IF amplifier.

Ondria and Collinet [5] on the other hand measured FM noise in cavity-stabilized avalanche-diode oscillators. They used a Stalo cavity and reduced the rms frequency deviation of an avalanche-diode oscillator by a factor of 10. Ashley and Searles [63] observed the effect of cavity stabilization on AM noise of an avalanche-diode oscillator as a function of frequency separation from the carrier frequency. They found that for large separations (above 500 kHz) the AM noise is reduced by cavity stabilization, as expected, by an amount that increases with frequency separation. At 10 MHz, the AM noise reduction is approximately 20 dB. However, for small frequency separations (below 500 kHz), AM noise actually increased when cavity stabilization was used.

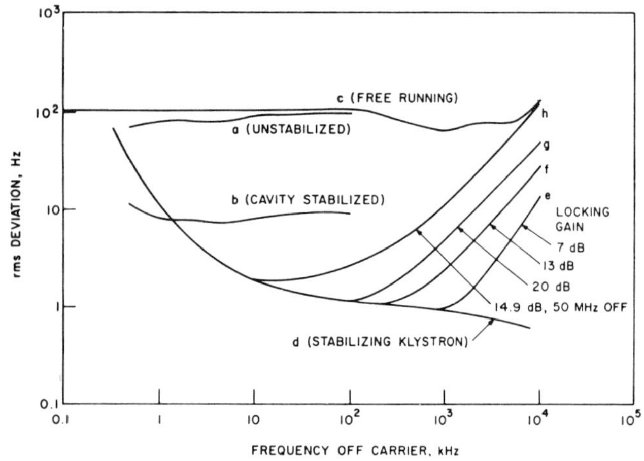

Fig. 4. FM noise reduction by circuit techniques [5]. Curves *a* and *b* are FM noise spectra of an unstabilized and a cavity-stabilized avalanche-diode oscillator. Curve *c* is the FM noise spectrum of another free-running avalanche-diode oscillator which is injection phase locked by a klystron with FM noise spectrum shown by curve *d*. Curves *e*, *f*, and *g* are FM noise spectra of the phase-locked oscillator with locking gains of 7, 13, and 20 dB, respectively, when free-running and klystron frequencies were identical. Curve *h* is the FM noise spectrum for a phase-locking gain of 14.9 dB when the two frequencies were 50 MHz apart. Power output, bias current, and oscillation frequency in cavity-stabilization and phase-locking experiments are different; rms frequency deviation is for one sideband in a 100-Hz bandwidth.

Nagano [64] also measured FM noise in cavity-stabilized oscillators and reported a stabilization factor of 50.

The use of a transmission cavity causes a loss of approximately 6 dB in the signal strength; a method of compensating for this loss has been reported by Ashley and Palka [6]. They used a cavity-stabilized *Ku*-band avalanche-diode oscillator with FM noise of 0.2 Hz rms in a bandwidth of 100 Hz to phase lock another avalanche-diode oscillator, obtaining a locking gain of 10 to 20 dB.

398

Ondria and Collinet [65], [5] have reported the following results on effects of phase locking on the noise spectra of avalanche transit-time diode oscillators. These results include the experimental work which was also reported by Hines *et al.* [66].

1) For phase locking by a small signal (20 dB below carrier), AM noise is reduced by approximately 3 dB; this effect is not evident from the theory of noise in synchronized oscillators and is attributed to the reduction of FM noise. Consequently, the component of AM noise due to FM to AM conversion is reduced.

2) For larger injected signals (10 dB below the carrier), AM noise spectra show degradation of 1 to 2 dB over that of free-running oscillators, a fact predicted theoretically.

3) In an avalanche-diode oscillator phase locked by a klystron, the FM noise level close to the carrier decreased to the same low level as the locking source, but at frequencies farther from the carrier, FM noise increased and asymptotically approached the noise level of free-running avalanche-diode oscillators.

4) The reduction of FM noise extended farther away from the carrier frequency for smaller phase-locking gain (larger injected signal) and for smaller frequency separation between the free-running frequency and the reference frequency.

Ashley and Palka [67] phased locked an *X*-band avalanche transit-time diode oscillator by a stabilized klystron and measured the AM and FM noise of the oscillator. Two major conclusions from their experiments are: 1) no increase of AM noise due to synchronization was detected, as is predicted by the theory of Kurokawa; and 2) the reduction of FM noise by synchronization is less than that predicted by Kurokawa near the carrier frequency, but more than the predicted value far from it.

A noise loading test for an avalanche transit-time diode oscillator, phase locked by an external frequency modulated signal has been reported by Isobe and Tokida [68].

Perichon [69] has reported measurement of FM noise of an avalanche-diode oscillator injection phase locked by a subharmonic signal, using the third and fourth subharmonics as synchronizing signals. The conclusions are in each case identical to those of Ondria and Collinet obtained with fundamental frequency injection. In addition it is noted that the higher the harmonic ratio, the smaller the bandwidth over which FM noise reduction occurs as a result of synchronization. Stated differently, a smaller phase locking gain is available for the same noise reduction.

Scherer [70] has suggested negative feedback methods for reducing oscillator AM and FM noise output. He has pointed out that dc bias current cannot be used for correction of amplitude and frequency, as it would affect both simultaneously. Instead AM noise can be reduced by using a pin modulator for amplitude correction, and FM noise can be reduced by controlling the frequency using varactor tuning. However, no experimental results using these techniques have been reported.

IV. Noise in Avalanche-Diode Amplifiers

A. Theory of Amplifier Noise

The first small-signal noise analysis of a reflection-type avalanche transit-time diode amplifier was presented by Hines [3]. Neglecting the diode series resistance R_s and assuming that fluctuations in the avalanche current are the only source of noise, he expressed the noise figure as

$$F = 1 + \frac{\overline{I_n{}^2} R_L}{GkT_0B} \tag{17}$$

where $I_n{}^2$ given in (9) is the mean square noise current in the external circuit due to fluctuations in the avalanche, R_L is the load resistance as seen by the device, k is the Boltzmann constant, G is the amplifier power gain, B is the bandwidth, and T_0 the reference temperature of 290°K. Since an amplifier power gain is given by

$$G = \left| \frac{R_d - R_L}{R_d + R_L} \right|^2 \tag{18}$$

and under high gain conditions ($R_d \approx -R_L$), the noise figure expression simplifies to

$$F = 1 + \frac{q_e V_a / kT}{4m\omega^2 \tau_x{}^2 \left(1 - \dfrac{\omega_a{}^2}{\omega^2}\right)} . \tag{19}$$

Using typical values of the parameters, a noise figure of approximately 40 dB is predicted by the analysis. Hines also observed that the carrier current becomes very small over part of the cycle and then builds up to the peak value again by avalanche. Since fluctuations become significant when the total current is small, and are multiplied by avalanche, Hines concluded that: 1) the noise output may be large at higher signal levels (when current minima are smaller); and 2) a diode with large leakage current should be quieter at high signal levels than one with a sharp breakdown characteristic, because the leakage current would limit the minimum current.

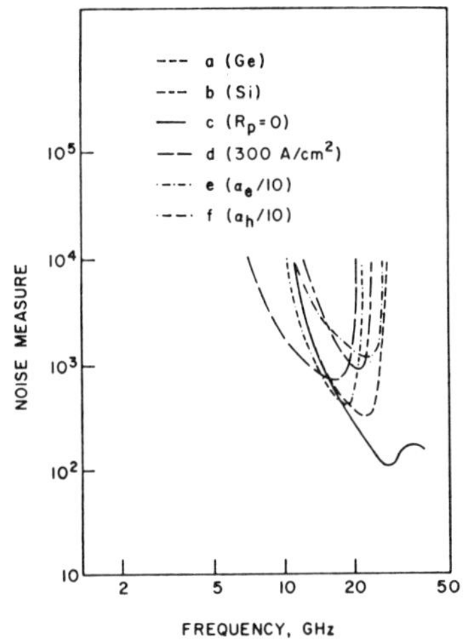

Fig. 5. Optimum noise measure for a p-n avalanche-diode amplifier [9]. Curve *a* gives the noise measure as a function of frequency for a Ge diode with a parasitic resistance of 1 Ω, dc bias current density of 1000 A/cm², and equal electron and hole ionization rates. All other curves are also drawn under the same conditions, except as noted for each curve: *b*, Si diode (unequal ionization rates); *c*, zero parasitic resistance; *d*, 300 A/cm² current density; *e*, electron ionization rate reduced by a factor of 10; and *f*, hole ionization rate reduced by a factor of 10.

Gummel and Blue [9] have used their results of open-circuit mean square noise voltage (described in Section II-B) to calculate the noise measure M of a small-signal avalanche transit-time diode amplifier. The noise measure of a linear two-port amplifier was defined by Haus and Adler [71] as

$$M = \frac{F - 1}{1 - \dfrac{1}{G}} \tag{20}$$

where F is the noise figure of the amplifier and G its gain, to serve as a figure of merit of amplifiers. The optimum value of this noise measure for a negative conductance amplifier has been found by DeLoach to be

$$M_{\text{opt}} = \frac{\langle v^2 \rangle / B}{4kT(-R_d)} \tag{21}$$

where B is the bandwidth, and R_d the real part of the diode impedance. Gummel and Blue have numerically calculated M_{opt} for an avalanche-diode amplifier as a function of operating frequency, parasitic resistance, diode material, and ratio of electron-to-hole ionization rates. Some of their results are shown in Fig. 5. They have reached the following general conclusions.

1) The noise measure shows a sharp minimum as a function of frequency, the frequency of minimum M_{opt} being higher for lower parasitic resistance and for larger bias current density.

2) The lower the parasitic resistance, the smaller the M_{opt}.

3) The larger the bias current density, the smaller the M_{opt}.

4) Ge avalanche diodes have lower M_{opt} than Si avalanche diodes. In general, M_{opt} is lowest for equal ionization rates for holes and electrons, and deteriorates if either α_e or α_h is reduced.

B. Measurements of Amplifier Noise

Most earlier work reported noise figure measurements on Si avalanche transit-time diode amplifiers. Johnston and Josenhans [60] measured a noise figure of 39 dB at 5.5 GHz for a Read-type structure, Josenhans and Misawa [72] reported a noise figure of 45 dB at 7.5 GHz for a p-i-n structure, and DeLoach and Johnston [1] found the lowest noise figure of 49 dB at 11.5 GHz for a p-n structure. These results should not be directly compared because they were obtained at different frequencies, amplifier gains, and bias current densities. DeLoach and Johnston also found that the noise figure decreased with increasing bias current and with decreasing amplifier gain.

Scherer [73] has reported an *X*-band avalanche-diode amplifier with three cascaded stages which has an oveall gain of 36 dB and a noise figure of 31 dB.

Rulison *et al.* [55] measured the noise figure of *X*-band microwave amplifiers using both p-n and n-p-i-p$^+$ (or Read-type) avalanche diodes fabricated with Ge under high gain conditions, and reported noise figures of 30±1 dB, which is an improvement of approximately 10 dB over the results obtained with comparable Si devices reported earlier by Johnston and Josenhans [60].

Kuno *et al.* [74] have measured noise figures of GaAs avalanche transit-time diode reflection-type amplifiers. They reached a lowest noise figure of 17 dB under high gain conditions, which is lower than any reported using Si or Ge diodes. Two general conclusions can be drawn from their work.

1) The noise figure decreases monotonically with increasing dc bias current of the diode for a constant amplifier gain.

2) For a constant bias current, the noise figure of an amplifier slowly increases with increasing power gain, attains a maximum at a gain somewhere between 10 and 15 dB, and then falls for larger gains.

Noise figure measurements on GaAs avalanche-diode reflection-type amplifiers were also reported by Baranowski *et al.* [57] under high-gain conditions (10 to 30 dB) and 10- to 15-dB improvement was found over similar measurements using Si diodes. They also found a considerable reduction of noise figure with increasing diode bias current.

Typical reflection-type amplifier noise figures reported are shown in Fig. 6 as a function of diode bias current.

Although both degenerate [75] and nondegenerate [76] parametric amplifiers using avalanche transit-time diodes have long been reported, no measurements of noise figures in these amplifiers have been reported. On the other hand, noise figures of sideband amplifiers ("amplifiers" in which oscillations are simultaneously occurring at a nearby frequency) have been reported by Evans and Haddad [44], but will not be discussed here. Noisten [77] has measured amplifier noise in terms of AM and FM noise of output; these results are also not discussed here.

V. Noise in Avalanche-Diode Mixers

Avalanche transit-time diodes have been used as self-oscillating frequency converters in two different modes: that in which the signal is uncorrelated to the "local oscillator" output (the usual mixer application), and that in which the signal is derived from the local oscillator output. While the noisiness of the former is adequately described by the noise figure of the mixer, perhaps a more appropriate characterization of noise performance in the second case is the frequency stability or the minimum detectable signal for the oscillator–mixer combination.

No theoretical results on noise in either type of mixer have been reported so far. The mathematical treatment of noise in such a mixer is difficult for two reasons. First, the noise of the local oscillator is large and cannot be neglected, and second, the local oscillator noise is correlated to the fluctuations in the periodically

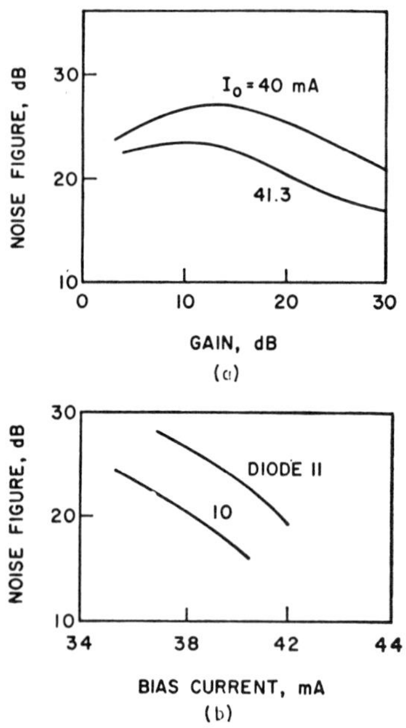

Fig. 6. Single sideband noise figure of GaAs avalanche-diode amplifier at 8.75 GHz [74]. (a) function of amplifier gain. (b) Bias current (for constant gain of 20 dB). Diode diameter is 125 μm.

varying device admittance as they both arise from the same physical mechanism. In addition, a conventional diode mixer noise analysis for a nonlinear conductance or nonlinear reactance type of mixer cannot be simply extended to avalanche-diode mixers, as the conductance and susceptance are both nonlinear and cannot be expressed as functions of voltage independent of frequency (because the *I-V* relationship for the diode is not instantaneous).

Grace [78], who reported on a Si avalanche transit-time diode downconverter (*X* band to 30 MHz) measured a conversion loss of 7 dB, and estimated the noise figure to be approximately 50 dB by measuring the minimum detectable signal power. Evans and Haddad [44] also carried out noise figure measurements on GaAs avalanche diode frequency converters using a noise figure meter, and reported a conversion loss of 6 dB and a noise figure of 55 dB. No later measurements using better diodes, particularly Schottky barrier avalanche diodes, have been reported, and the above figures are poor compared to the Gunn-diode self-oscillating mixers [79].

Measurements of minimum detectable signal on avalanche transit-time diode mixers in doppler radar applications have been carried out [80]. For very short-range radars, reflected signals that are typically 65 dB below the carrier power are detectable 1 Hz away from the carrier frequency.

VI. Conclusions

This exhaustive survey of reported results concerning noise in avalanche transit-time devices has included examination of both theoretical and experimental work on these devices in all of their several applications. Several unexplained experimental results and points of disagreement between experimental work of different workers can be identified. A large number of problems requiring further investigation exist. In particular, experimental and theoretical studies of noise in devices operated in the TRAPATT mode, large-signal noise analysis with a time-dependent multiplication factor, and noise measurements in Schottky-barrier avalanche diodes are some of the possible areas of study. That the work done is not complete is evidenced by the fact that no general set of rules for optimization of the diode structure, material, operating point, and circuit can be stated as yet.

Appendix

Since this paper has been written (in October 1970) several new results in this area have been published. Some of these are summarized below. Naqvi [81] has extended the analysis of McIntyre [8] to higher frequencies by taking into account the finite time required for multiplication. Vlaardingerbroek and Goedbloed [82] have calculated the width of the oscillator output spectrum using a Read model IMPATT diode and taking equal ionization rates for electrons and holes. They reach the conclusion that the p^+-n diode structure leads to a lower noise oscillator than the complementary n^+-p diode structure, which has a smaller avalanche-region width. Sjölund [83] has carried out a numerical large-signal analysis of IMPATT oscillators assuming a rectangular voltage waveform across the diode, and has found that the mean square noise current in the diode is inversely proportional to the ratio I_{min}/I_0, where I_{min} is the minimum conduction current, and I_0 the dc bias current for the diode. Ulrich and co-workers [84], [85] have derived expressions for AM and FM noise spectra of free-running and synchronized IMPATT oscillators by a method similar to the one used for obtaining (13) and (14) and report good agreement between theoretical and experimental results. Josenhans [86] used a series combination of three Si IMPATT diodes in an oscillator and found the oscillator FM noise to be 30 Hz rms in a 1-kHz bandwidth as compared to 90 Hz rms for single diode oscillators under identical conditions, and ascribed this noise reduction to the accompanying power increase. Earlier, Fukui [87] had also reported an improvement over the average FM noise of individual diode oscillators in an oscillator consisting of eight IMPATT diodes mutually synchronized by a hybrid power combiner. Tatsuguchi *et al.* [88] have reported experimental measurements of the noise measure of a phase-locked IMPATT oscillator as a function of cavity load resistance and oscillator output power. Evans [89] has measured the noise figure of Ge TRAPATT diode amplifiers to be about 120 dB in the stable mode, but only about 60 dB when TRAPATT mode oscillations are occurring simultaneously at a different frequency. The FM noise of Ge TRAPATT oscillators at UHF was earlier reported [90] to be between 8 and 10 Hz rms in a 1-kHz bandwidth and for an external Q of 170 to 350. Claassen [91] has presented an approximate expression for the noise figure of an IMPATT diode amplifier in terms of small-signal parameters which can be calculated as a function of the avalanche-region width l_a, and has compared the noise figures of Si, Ge, and GaAs IMPATT diodes to show that in this respect GaAs is the best and Si the worst of the three materials.

Acknowledgment

The author wishes to thank Prof. G. I. Haddad and Prof. R. J. Lomax of the University of Michigan for encouragement, supervision, and helpful discussions.

References

[1] B. C. DeLoach and R. L. Johnston, "Avalanche transit-time microwave oscillators and amplifiers," *IEEE Trans. Electron Devices*, vol. ED-13, Jan. 1966, pp. 181–186.
[2] M. E. Hines, "Potential application and the noise problem in the Read avalanche diode," in *Int. Solid-State Circuits Conf. Dig.* (Philadelphia, Pa.), Feb. 1966, pp. 82–83.
[3] ——, "Noise theory for the Read type avalanche diode," *IEEE*

Trans. Electron Devices, vol. ED-13, Jan. 1966, pp. 158–163.

[4] J. Josenhans, "Noise spectra of Read diode and Gunn oscillators," *Proc. IEEE* (*Special Joint Issue on Optical Electronics with Applied Optics, the Optical Society of America*) (Lett.), vol. 54, Oct. 1966, pp. 1478–1479.

[5] J. G. Ondria and J.-C. R. Collinet, "Noise-reduction techniques of high-frequency oscillator-high-order multipliers and avalanche-diode-type microwave sources," *IEEE J. Solid-State Circuits*, vol. SC-4, Apr. 1969, pp. 65–70.

[6] R. J. Ashley and F. M. Palka, "Noise properties and stabilization of Gunn and avalanche oscillators," in *1970 IEEE G-MTT Int. Microwave Symp. Dig.* (Newport Beach, Calif.), May 1970, pp. 161–164.

[7] A. S. Tager, "Current fluctuations in a semiconductor (dielectric) under the conditions of impact ionization and avalanche breakdown," *Sov. Phys.—Solid State*, vol. 6, Feb. 1965, pp. 1919–1925.

[8] R. J. McIntyre, "Multiplication noise in uniform avalanche diodes," *IEEE Trans. Electron Devices*, vol. ED-13, Jan. 1966, pp. 164–168.

[9] H. K. Gummel and J. L. Blue, "A small-signal theory of avalanche noise in IMPATT diodes," *IEEE Trans. Electron Devices* (*Second Special Issue on Semiconductor Bulk Effect and Transit-Time Devices*), vol. ED-14, Sept. 1967, pp. 569–580.

[10] G. L. Pearson and B. Sawyer, "Silicon *P-N* junction alloy diodes," *Proc. IRE*, vol. 40, Nov. 1952, pp. 1348–1351.

[11] K. G. McKay, "Avalanche breakdown in silicon," *Phys. Rev.*, vol. 94, May 15, 1954, pp. 877–884.

[12] A. G. Chynoweth and K. G. McKay, "Photon emission from avalanche breakdown in silicon," *Phys. Rev.*, vol. 102, Apr. 15, 1956, pp. 369–376.

[13] D. J. Rose, "Microplasma in silicon," *Phys. Rev.*, vol. 105, Jan. 15, 1957, pp. 413–418.

[14] K. S. Champlin, "Microplasma fluctuations in silicon," *J. Appl. Phys.*, vol. 30, July 1959, pp. 1039–1050.

[15] A. van der Ziel, *Fluctuation Phenomena in Semiconductors*. New York: Academic Press, 1959.

[16] R. H. Haitz, "Model for the electrical behavior of a microplasma," *J. Appl. Phys.*, vol. 35, May 1964, pp. 1370–1376.

[17] W. Mönch, "On the physics of avalanche breakdown in semiconductors," *Phys. Status Solidi*, vol. 36, 1969, pp. 9–48.

[18] H. Penfield, "Why not the avalanche diode as an RF noise source," *Electron. Des.*, vol. 12, Apr. 12, 1965, pp. 32–35.

[19] R. H. Haitz, "Controlled noise generation with avalanche diodes I. Low pulse rate design," *IEEE Trans. Electron Devices*, vol. ED-12, Apr. 1965, pp. 198–207.

[20] ——, "Controlled noise generation with avalanche diodes II. High pulse rate design," *IEEE Trans. Electron Devices*, vol. ED-13, Mar. 1966, pp. 342–346.

[21] M. Gilden and M. E. Hines, "Electronic tuning effects in the Read microwave avalanche diode," *IEEE Trans. Electron Devices*, vol. ED-13, Jan. 1966, pp. 169–175.

[22] E. Constant and A. Semichon, "Diodes semiconductrices en regime d'avalanche application aux hyperfrequences," *Onde Elec.*, vol. 48, July–Aug. 1968, pp. 703–721.

[23] J. L. Blue, "Noise calculations in continuous-parameter systems," *J. Appl. Phys.*, vol. 38, Nov. 1967, pp. 4882–4887.

[24] R. H. Haitz and F. W. Voltmer, "Noise studies in uniform avalanche diodes," *Appl. Phys. Lett.*, vol. 9, Nov. 1966, pp. 381–383.

[25] ——, "Noise of a self-sustaining avalanche discharge in silicon: studies at microwave frequencies," *J. Appl. Phys.*, vol. 39, June 1968, pp. 3379–3384.

[26] I. M. Naqvi, C. A. Lee, and G. C. Dalman, "Measurements of multiplication effects on noise in silicon avalanche diodes," *Proc. IEEE* (*Special Issue on Electronic Scanning*) (Lett.), vol. 56, Nov. 1968, pp. 2051–2052.

[27] R. D. Baertsch, "Low-frequency noise measurements in silicon avalanche photodiodes," *IEEE Trans. Electron Devices* (Corresp.), vol. ED-13, Mar. 1966, pp. 383–385.

[28] ——, "Noise and ionization rate measurements in silicon photodiodes," *IEEE Trans. Electron Devices* (Corresp.), vol. ED-13, Dec. 1966, p. 987.

[29] E. Constant, B. Kramer, and L. Raczy, "Sur le bruit presente dans le domaine hertzien et ultrahertzian pas les jonctions semi-conductrices en avalanche," *C. R. Acad. Sci.*, ser. B, vol. 265, Aug. 16, 1967, pp. 385–388.

[30] R. H. Haitz, "Noise of a self-sustaining avalanche discharge in silicon: low frequency noise studies," *J. Appl. Phys.*, vol. 38, June 1967, pp. 2935–2946.

[31] N. Chasek, "Avalanche diodes permit in-service measurements of critical parameters in microwave equipment," *Electronics*, vol. 43, Jan. 19, 1970, pp. 87–91.

[32] H. T. Minden, "Noise behavior of avalanching silicon diodes," *Proc. IEEE* (Lett.), vol. 54, Aug. 1966, pp. 1124–1125.

[33] V. M. Val'd Perlov, A. V. Krasilov, and A. S. Tager, "The avalanching transit-time diode—a new semiconductor microwave device," *Radio Eng. Electron. Phys.* (USSR), vol. 11, Nov. 1966, pp. 1764–1779.

[34] G. C. Dalman and L. F. Eastman, "Avalanche-diode microwave noise generation experiments," presented at the 1967 IEEE Int. Electron Devices Meet., Washington, D. C.

[35] A. Chadelas, E. Constant, and A. Hauducoeur, "Utilisation de diodes a avalanche comme source de bruit a commande electronique," *Onde Elec.*, vol. 48, July–Aug. 1968, pp. 733–734.

[36] R. H. Haitz and F. L. Opp, "A solid state broad-band microwave noise source," presented at the 1968 IEEE Int. Electron Devices Meet., Washington, D. C.

[37] L. I. Bareysha, A. I. Mel'nikov, T. V. Spricheva, A. S. Tager, G. M. Fedorova, and F. M. Shapiro, "Avalanche-transit diode noise generator," *Radio Eng. Electron. Phys.* (USSR), vol. 14, Jan. 1969, pp. 87–92.

[38] W. A. Edson, "Noise in oscillators," *Proc. IRE*, vol. 48, Aug. 1960, pp. 1454–1466.

[39] J. C. Inkson, "Noise generation under large signal conditions in the Read microwave avalanche diode," *Int. J. Electron.*, vol. 25, July 1968, pp. 1–16.

[40] W. T. Read, "A proposed high frequency negative resistance diode," *Bell Syst. Tech. J.*, vol. 37, Mar. 1958, pp. 401–446.

[41] G. Ulrich, "AM noise of IMPATT-diode oscillators," *Electron. Lett.*, vol. 6, Apr. 16, 1970, pp. 247–248.

[42] K. Kurokawa, "Noise in synchronized oscillators," *IEEE Trans. Microwave Theory Tech.*, vol. MTT-16, Apr. 1968, pp. 234–240.

[43] M. T. Vlaardingerbroek, "Output spectrum of IMPATT diode oscillators," *Electron. Lett.*, vol. 5, Oct. 16, 1969, pp. 521–522.

[44] W. J. Evans and G. I. Haddad, "Frequency conversion in IMPATT diodes," *IEEE Trans. Electron Devices*, vol. ED-16, Jan. 1969, pp. 78–87.

[45] J. J. Goedbloed, "On the up-converted noise of an IMPATT-diode oscillator," presented at the 8th Int. Conf. Microwave and Optical Generation and Amplification, Amsterdam, The Netherlands, Sept. 1970.

[46] E. F. Scherer, "Investigations of the noise spectra of avalanche oscillators," *IEEE Trans. Microwave Theory Tech.* (*Special Issue on Noise*), vol. MTT-16, Sept. 1968, pp. 781–788.

[47] B. G. Bosch and W. A. Gambling, "Techniques of microwave noise measurement," *J. Brit. Inst. Radio Eng.*, vol. 21, June 1961, pp. 503–515.

[48] J. R. Ashley, C. B. Searles, and F. M. Palka, "The measurement of oscillator noise at microwave frequencies," *IEEE Trans. Microwave Theory Tech.* (*Special Issue on Noise*), vol. MTT-16, Sept. 1968, pp. 753–760.

[49] J. G. Ondria, "A microwave system for measurements of AM and FM noise spectra," *IEEE Trans. Microwave Theory Tech.* (*Special Issue on Noise*), vol. MTT-16, Sept. 1968, pp. 767–781.

[50] E. J. Cook, "A noise reduction design for IMPATT oscillators," presented at the 1967 IEEE Int. Electron Devices Meet., Washington, D. C.

[51] R. A. Goldwasser, J. Berenz, C. A. Lee, and G. C. Dalman, "IMPATT and Gunn oscillator noise," presented at the 8th Int. Conf. Microwave and Optical Generation and Amplification," Amsterdam, The Netherlands, Sept. 1970.

[52] A. M. Cowley, "Design and application of silicon IMPATT diodes," *Hewlett-Packard J.*, vol. 21, May 1970, pp. 2–13.

[53] A. M. Cowley, Z. Fazarinc, R. Hall, S. Hamilton, C. Yen, and R. Zettler, "The singly-tuned IMPATT oscillator," in *1970 IEEE Int. Solid-State Circuits Conf. Dig.* (Philadelphia, Pa.), Feb. 1970, pp. 10–11.

[54] W. Harth and G. Ulrich, "*Q* dependence of IMPATT diode FM noise," *Electron. Lett.*, vol. 5, Jan. 9, 1969, pp. 7–8.

[55] R. L. Rulison, G. Gibbons, and J. G. Josenhans, "Improved performance of IMPATT diodes fabricated from Ge," *Proc. IEEE* (Lett.), vol. 55, Feb. 1967, pp. 223–224.

[56] M. R. Barber, "New solid state devices for microwave power generation," presented at the 1967 IEEE Int. Convention and Exhibition, New York.

[57] J. J. Baranowski, V. J. Higgins, C. K. Kim, and L. D. Armstrong, "Gallium arsenide IMPATT diodes," *Microwave J.*, vol. 12, July 1969, pp. 71–76.
[58] P. A. Levine, "Measurement of AM noise in pulsed gallium arsenide avalanche diode oscillators," presented at the 1969 IEEE Int. Electron Devices Meet., Washington, D.C.
[59] ——, "Ultralow noise in pulsed GaAs avalanche diode oscillators," *Proc IEEE* (Lett.), vol. 58, Mar. 1970, pp. 494–495.
[60] R. L. Johnston and J. G. Josenhans, "Improved performance of microwave Read diodes," *Proc. IEEE* (Lett.), vol. 54, Mar. 1966, pp. 412–413.
[61] B. Kramer, E. Constant, and B. Boittiaux, "Sur le bruit de modulation des oscillateurs ATT et Gunn," *Onde Elec.*, vol. 48, July–Aug. 1968, pp. 735–736.
[62] M. Gilden, "Efficient high power avalanche oscillators in *X*- and *Ku*-bands," *Micronotes*, vol. 4, Dec. 1966; also *NEREM Reflector*, Nov. 1966.
[63] J. R. Ashley and C. B. Searles, "Microwave oscillator noise reduction by a transmission stabilizing cavity," *IEEE Trans. Microwave Theory Tech.* (*Special Issue on Noise*), vol. MTT-16, Sept. 1968, pp. 743–748.
[64] S. Nagano, "Highly stabilized half-watt IMPATT oscillator," in *1970 IEEE G-MTT Int. Microwave Symp. Dig.* (Newport Beach, Calif.), May 1970, pp. 170–173.
[65] J. G. Ondria and J.-C. R. Collinet, "Investigation of noise reduction techniques of solid-state type microwave sources," in *Int. Solid-State Circuits Conf. Dig.* (Philadelphia, Pa.), Feb. 1968, pp. 82–83.
[66] M. E. Hines, J.-C. R. Collinet, and J. G. Ondria, "FM noise suppression of an injection phase-locked oscillator," *IEEE Trans. Microwave Theory Tech.* (*Special Issue on Noise*), vol. MTT-16, Sept. 1968, pp. 738–742.
[67] J. R. Ashley and F. M. Palka, "Measured FM noise reduction by injection phase locking," *Proc. IEEE* (*Special Issue on Computers in Industrial Process Control*) (Lett.), vol. 58, Jan. 1970, pp. 155–157.
[68] T. Isobe and M. Tokida, "Noise loading performance of a phase-locked IMPATT oscillator for multichannel FM signals," *Proc. IEEE* (Lett.), vol. 56, May 1968, pp. 873–875.
[69] R. Perichon, "Frequency-modulation noise of subharmonically injection phase-locked IMPATT oscillator," *IEEE Trans. Microwave Theory Tech.* (*Special Issue on Microwave Circuit Aspects of Avalanche-Diode and Transferred Electron Devices*) (Corresp.), vol. MTT-18, Nov. 1970, pp. 988–989.
[70] E. F. Scherer, "Circuit techniques for the noise reduction and frequency stabilization of avalanche diode oscillators," in *1968 IEEE G-MTT Int. Microwave Symp. Dig.* (Detroit, Mich.), May 1968, pp. 63–71.
[71] H. A. Haus and R. B. Adler, *Circuit Theory of Linear Noisy Networks.* Cambridge, Mass.: Technology Press; also New York: Wiley, 1959.
[72] J. G. Josenhans and T. Misawa, "Experimental characterization of a negative-resistance avalanche diode," *IEEE Trans. Electron Devices* (Corresp.), vol. ED-13, Jan. 1966, pp. 206–208.
[73] E. F. Scherer, "A multistage high-power avalanche amplifier at *X* band," *IEEE J. Solid-State Circuits* (*Special Issue on Optoelectronic Circuits and Solid-State Microwave Circuits*), vol. SC-4, Dec. 1969, pp. 396–399.
[74] H. J. Kuno, J. R. Collard, and A. R. Gobat, "Low noise epitaxial GaAs avalanche diode amplifiers," presented at the 1968 IEEE Int. Electron Devices Meet., Washington, D. C.
[75] A. S. Clorfeine, "Self-pumped parametric amplification with an avalanching diode," *Proc. IEEE* (*Special Issue on Computers*) (Lett.), vol. 54, Dec. 1966, pp. 1956–1957.
[76] P. Bura, "Self-pumped parametric amplification with avalanche diodes," *Proc. IEEE* (*Special Issue on Radio Measurement Methods and Standards*) (Lett.), vol. 55, June 1967, p. 1076.
[77] J. Noisten, "Avalanche diode power amplifiers for *Ku*-band," in *1970 IEEE Int. Solid-State Circuits Conf. Dig.* (Philadelphia, Pa.), Feb. 1970, pp. 16–17.
[78] M. I. Grace, "Down conversion and sideband translation using avalanche transit time oscillators," *Proc. IEEE* (Lett.), vol. 54, Nov. 1966, pp. 1570–1571.
[79] P. Albrecht and M. Bechteler, "Noise figure and conversion loss of self-excited Gunn-diode mixers," *Electron. Lett.*, vol. 6, May 28, 1970, pp. 321–322.
[80] G. I. Haddad, W. R. Curtice, C. Chao, and J. T. Patterson, "Avalanche-diode and Gunn-effect devices for Doppler radar applications," 1st Ann. Rep., Electron Phys. Lab., the University of Michigan, Ann Arbor, Oct. 1970, pp. 16–37.
[81] I. M. Naqvi, "Avalanche current fluctuations in uniform diodes," Ph.D. dissertation, Cornell University, Ithaca, N. Y., 1969.
[82] M. T. Vlaardingerbroek and J. J. Goedbloed, "On the theory of noise and injection phase locking of IMPATT diode oscillators," *Philips Res. Rep.*, vol. 25, Dec. 1970, pp. 452–471.
[83] A. Sjölund, "Noise in IMPATT oscillators at large RF amplitudes," *Electron. Lett.*, vol. 7, Apr. 8, 1971, pp. 161–162.
[84] G. Ulrich, "Noise of phase locked IMPATT oscillators," in *Proc. 8th Int. Conf. Microwaves and Optical Generation and Amplification* (Amsterdam, The Netherlands), Sept. 1970, pp. 12-41–12-46.
[85] H. J. Thaler, G. Ulrich, and G. Weidmann, "Noise in IMPATT diode amplifiers and oscillators," *IEEE Trans. Microwave Theory Tech.*, vol. MTT-19, Aug. 1971, pp. 692–705.
[86] J. G. Josenhans, "Diamond as an insulating heat sink for a series combination of IMPATT diodes," *Proc. IEEE* (*Special Issue on Transportation*) (Lett.), vol. 56, Apr. 1968, pp. 762–763.
[87] H. Fukui, "Frequency locking and modulation of microwave silicon avalanche diode oscillators," *Proc. IEEE* (*Special Joint Issue on Optical Electronics with Applied Optics, the Optical Society of America*) (Lett.), vol. 54, Oct. 1966, pp. 1475–1477.
[88] I. Tatsuguchi, N. R. Dietrich, and C. B. Swan, "Power-noise characterization of phase-locked IMPATT oscillators," in *1971 IEEE Int. Solid-State Circuits Conf. Dig.* (Philadelphia, Pa.), Feb. 1971, pp. 170–171.
[89] W. J. Evans, "CW TRAPATT amplification," *IEEE Trans. Microwave Theory Tech.* (*Special Issue on Microwave Circuit Aspects of Avalanche-Diode and Transferred Electron Devices*) (Corresp.), vol. MTT-18, Nov. 1970, pp. 986–988.
[90] D. E. Iglesias and W. J. Evans, "High-efficiency CW IMPATT operation," *Proc. IEEE* (*Special Issue on MHD Power Generation*) (Lett.), vol. 56, Sept. 1968, p. 1610.
[91] M. Claassen, "Small-signal noise performance of IMPATT-diodes made from silicon, germanium and gallium arsenide," in *Proc. 8th Int. Conf. Microwaves and Optical Generation and Amplification* (Amsterdam, The Netherlands), Sept. 1970, pp. 12-26–12-29.

Noise Theory for the Read Type Avalanche Diode

M. E. HINES, SENIOR MEMBER, IEEE

Abstract—An analysis is presented for the noise current spectrum of an avalanche diode under assumed conditions of ideal uniform avalanche behavior in a zone which is thin compared with the total high-field depletion zone. The result is applied to the Read diode amplifier. For a typical set of operating parameters, the theory predicts a noise figure on the order of 40 dB. Depending upon particular device parameters, lower noise figures may be possible.

I. Amplifier Configuration and Characteristics

FOR AMPLIFICATION, the Read diode can be used in conjunction with a circulator in much the same way that a tunnel diode is used. The diode is inserted into a resonator which is coupled to a transmission line. This line is coupled to separate input and output lines by means of a circulator as shown in Fig. 1. Because of the negative resistance of the diode, the reflection coefficient at the resonator will have a magnitude greater than one so that the emerging wave will be stronger than the approaching input wave, giving net amplification. The magnitude of the negative resistance and the degree of coupling between the resonator and the line must be chosen so that the net effective resistance is positive in the resonator in order to prevent oscillation. With a close balance between the positive "coupling" or load resistance and the negative diode resistance, the gain may be large.

Figure 2 shows the equivalent circuit upon which our analysis is based. The power gain may be taken as the square of the magnitude of the reflection coefficient in the line, seen at the resonator. Because the reflection coefficient is not altered in magnitude by a matched transformer between two lines, we determine its value by assuming that the input line has a characteristic resistance R_L, and is terminated by the series combination of elements shown in Fig. 2. The reflection coefficient Γ is given by the expression

$$\Gamma = \frac{Z - R_L}{Z + R_L} \tag{1}$$

Manuscript received September 13, 1965. The work reported in this paper was supported by the Electronics Technology Division of the Air Force Avionics Laboratory, Wright-Patterson Air Force Base, Dayton, Ohio.

The author is with Microwave Associates, Inc., Burlington, Mass.

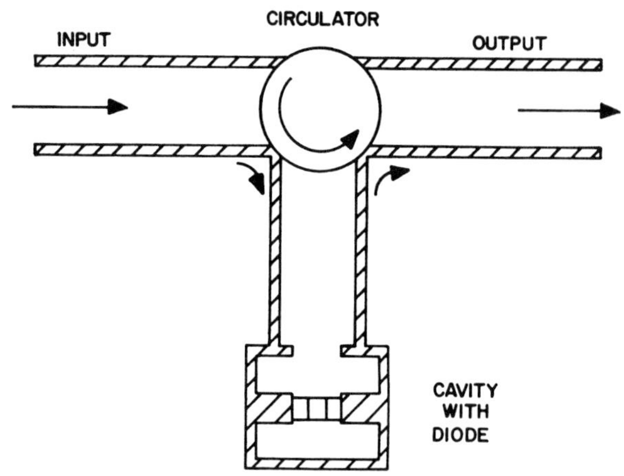

Fig. 1. Reflection technique for obtaining amplification using a negative-resistance diode.

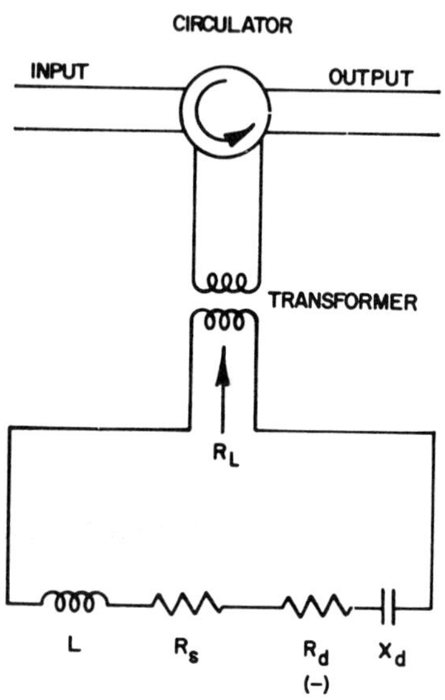

Fig. 2. Equivalent circuit used for analysis of gain and noise figure of a reflection-type amplifier.

where Z is the impedance of all of the elements in the loop except R_L, the load resistance. These include a tuning inductor L, the passive resistance of the diode R_s, the operating reactance of the diode X_d, and the active (negative) resistance of the diode R_d. The gain G is given by

$$G = \left[\frac{-R_L + R_s + R_d + j\omega L + jX_d}{R_L + R_s + R_d + j\omega L + jX_d}\right]^2. \quad (2)$$

At resonance, the reactive terms cancel. We shall also neglect the resistance R_s, leaving

$$G = \left[\frac{R_d - R_L}{R_d + R_L}\right]^2. \quad (3)$$

Gain occurs if R_d is negative. Stability requires $R_d + R_L > 0$.

II. Noise Figure Expression

There are three sources of noise to be considered in this device.

1) Thermal noise from the input line, at T_0.
2) Thermal noise from the passive diode re-resistance R_s at temperature T_d.
3) Fluctuation noise induced by the statistical variations in the avalanche current of the diode.

In order to simplify the expressions and the analysis, we will neglect the series resistance R_s in the analysis which follows. The noise induced by the avalanche is the dominant contribution. The deleterious effects of series resistance can be included in a straightforward manner if the reader desires a more complete expression. A useful noise figure expression is

$$F = 1 + \frac{\begin{pmatrix}\text{Output Noise Power}\\ \text{Arising in the Amplifier}\end{pmatrix}}{[\text{Gain}][kT_0B]}. \quad (4)$$

If $\bar{I}_n^2$ is the mean-square noise current induced in the loop of Fig. 2, caused by the diode, then our noise figure is given by

$$F = 1 + \frac{\bar{I}_n^2 R_L}{GkT_0B}. \quad (5)$$

Our problem is to determine the mean-square circuit current $\bar{I}_n^2$, induced by fluctuations in the avalanche current of the type analyzed in Section III.

III. Noise Current Analysis

In the circuit of Fig. 3, we show our model for analysis. Here, as before, we separate the drift and avalanche zones of the device. The method of analysis is based upon the space-charge wave approach of the previous paper [4]. We assume that the noise originates in the avalanche and this is assumed to take the form of a carrier conduction current driving function i_{n0}. We define i_{n0} as the noise current which would flow in the avalanche if it were possible to maintain the electric field constant at the exact critical value E_0 necessary to maintain a steady avalanche. Thus, i_{n0} is a fictitious, idealized *zero ac field* noise current. An expression for this zero field current spectrum is derived later in this section. At first, we will use the symbol i_{n0} without evaluation.

An avalanche, at high-current density, has a very low impedance, finite for nonzero frequency. In an actual diode, ac noise current flow will induce an ac field in the avalanche zone which will be correlated with the "driving current" i_{n0}. We assume that such a field will induce a correlated ac current i_{ne}, and by linear superposition, the total avalanche conduction current will be the sum of i_{n0} and i_{ne}. The relationship between i_{ne} and the ac field in the avalanche zone was derived

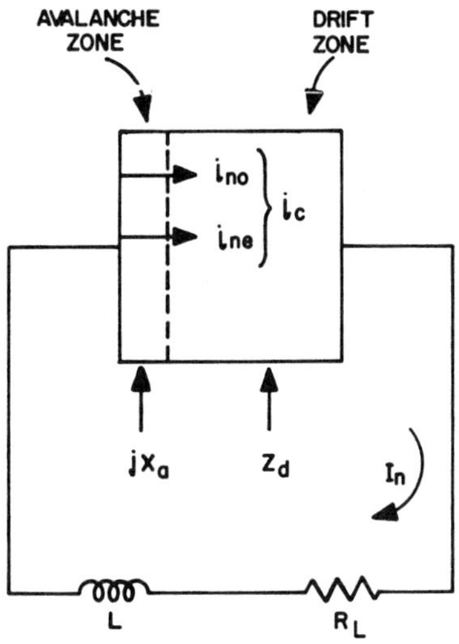

Fig. 3. Circuit used for analysis giving the noise current I_n as a function of the short-circuit avalanche fluctuation current i_{n0}.

in the previous paper [4]. This total avalanche current is injected into the drift zone as a space-charge wave.

There is also a displacement current $j\omega\epsilon A\mathcal{E}_a$ in the avalanche zone. The circuit current I_n will be equal to the sum of these three components

$$I_n = i_{n0} + i_{ne} + j\omega\epsilon A\mathcal{E}_a. \tag{6}$$

As indicated in the previous paper, the circuit current I_n is not a function of x in the drift zone, although the electric field and the drift current i_c both vary with x. In a velocity-limited drift zone,

$$\begin{aligned} i_c &= i_c(o)e^{-j\omega x/v_d} \\ &= {}_{n0} + i_{ne})e^{-j\omega x/v_d}. \end{aligned} \tag{7}$$

Using the relation

$$I_n = j\omega\epsilon A\mathcal{E}(x) + (i_{n0} + i_{ne})e^{-j\omega x/v_d} \tag{8}$$

we obtain an expression for $\mathcal{E}(x)$

$$\mathcal{E}(x) = \frac{I_n}{j\omega\epsilon A}\left(1 - \frac{i_{n0} + i_{ne}}{I_n}e^{-j\omega x/v_d}\right). \tag{9}$$

This is integrated to determine the voltage drop in the drift zone

$$V_d = I_n\frac{l_d}{\omega\epsilon A}\left\{-j + \frac{i_{n0} + i_{ne}}{I_n}\frac{1 - \cos\theta + j\sin\theta}{\theta}\right\} \tag{10}$$

where $\theta = \omega l_d/v_d$, and l_d is the length of the drift region. We now invoke Kirchhoff's law, that the total voltage drop around the loop is zero in Fig. 3, giving the equation

$$I_n\left(j\omega L + R_L + \frac{l_d}{j\omega\epsilon A}\right) + \mathcal{E}_a l_a + (i_{n0} + i_{ne})\frac{l_d}{\omega\epsilon A}\left(\frac{1 - \cos\theta + j\sin\theta}{\theta}\right) = 0. \tag{11}$$

Here $\mathcal{E}_a l_a$ is the voltage drop across the avalanche zone.

From (9) of the previous paper, we obtain

$$i_{ne} = \frac{2\mathcal{E}_a l_a I_0 \bar{\alpha}_a'}{j\omega\tau_a} \tag{12}$$

which may be substituted into (6) to obtain

$$I_n = \mathcal{E}_a\left(j\omega\epsilon A + \frac{2l_a I_0\bar{\alpha}_a'}{j\omega\tau_a}\right) + i_{n0} \tag{13}$$

which yields

$$\mathcal{E}_a l_a = (I_n - i_{n0})\frac{l_a}{j\omega\epsilon A}\frac{1}{1 - \frac{\omega_a^2}{\omega^2}} \tag{14}$$

and

$$i_{ne} = \frac{I_n - i_{n0}}{1 - \frac{\omega^2}{\omega_a^2}} \tag{15}$$

where ω_a is the angular *avalanche frequency*, defined in the previous paper,

$$\omega_a^2 = \frac{2\bar{\alpha}_a' l_a I_0}{\epsilon A\tau_a}. \tag{16}$$

We substitute (14) and (15) into (11), obtaining

$$I_n = i_{n0}\frac{\dfrac{l_d}{\omega\epsilon A}\left[\dfrac{-1}{1 - \frac{\omega_a^2}{\omega^2}}\right]\left(\dfrac{jl_a}{l_d} + \dfrac{1 - \cos\theta + j\sin\theta}{\theta}\right)}{j\omega L + R_L + \dfrac{l_a}{j\omega\epsilon A\left(1 - \frac{\omega_a^2}{\omega^2}\right)} + \dfrac{l_d}{\omega\epsilon A}\left\{-j + \dfrac{1 - \cos\theta + j\sin\theta}{\theta\left(1 - \frac{\omega^2}{\omega_a^2}\right)}\right\}}. \tag{17}$$

This is a complete form of the expression we are seeking at this point. The denominator will be recognized as the total impedance in the loop of Fig. 3, as may be determined from inspection of (14) and (18) of the previous paper. We will designate this total impedance Z_t. Further simplifications become possible by neglecting l_a/l_d as a small number compared with $(\sin\theta)/\theta$,

$$I_n = i_{n0}\frac{\frac{l_d}{\omega\epsilon A}}{Z_t}\left[\frac{-1}{1 - \frac{\omega_a^2}{\omega^2}}\right]\left(\frac{1 - \cos\theta + j\sin\theta}{\theta}\right). \tag{18}$$

We have knowledge only of the mean-square value of i_{n0} and we wish to know the mean-square value of I_n. We determine this relationship by taking the square of the magnitude, yielding

$$\bar{I}_n^2 = \bar{i}_{n0}^2 \left| \frac{1}{Z_t^2} \frac{2l_d^2}{\theta^2\omega^2\epsilon^2} \frac{(1 - \cos\theta)}{\left(1 - \frac{\omega_a^2}{\omega^2}\right)^2} \right| . \tag{19}$$

Using the definition for ω_a^2 and θ, and noting that

$$R_d = \frac{l_d}{\theta\omega\epsilon A} \left[\frac{1 - \cos\theta}{1 - \frac{\omega^2}{\omega_a^2}} \right], \tag{20}$$

we obtain

$$\bar{I}_n^2 = \bar{i}_{n0}^2 \left| \frac{R_d}{Z_t^2} \frac{1}{1 - \frac{\omega_a^2}{\omega^2}} \cdot \frac{\tau_a l_d}{\tau_d l_a} \cdot \frac{1}{2\bar{\alpha}_a' I_0} \right| . \tag{21}$$

In the above, $\bar{\alpha}_a'$ is the derivative of the ionization coefficient with respect to the E field. τ_a and τ_d are the transit times for the avalanche and drift zones, respectively. A simple expression for $\bar{\alpha}_a'$ can be obtained by assuming that $\bar{\alpha}$ varies as the mth power of the E field. Noting that

$$\bar{\alpha}_a l_a = 1, \tag{22}$$

we write

$$\bar{\alpha} l_a = \left(\frac{E_a}{E_0}\right)^m \tag{23}$$

giving

$$\bar{\alpha}' = \frac{m}{E_0 l_a} \left(\frac{E_a}{E_0}\right)^{m-1} \tag{24}$$

and at $E_a = E_0$, we obtain

$$\bar{\alpha}_a' = \frac{m}{E_0 l_a} = \frac{m}{V_a} \tag{25}$$

where V_a is the voltage drop across the avalanche zone. We also may note that a fixed drift velocity implies that

$$\tau_a l_d = \tau_d l_a . \tag{26}$$

Substituting (25) and (26) into (21) yields

$$\bar{I}_n^2 = \bar{i}_{n0}^2 \left| \frac{R_d}{Z_t^2} \cdot \frac{1}{1 - \frac{\omega_a^2}{\omega^2}} \cdot \frac{V_a}{2mI_0} \right| . \tag{27}$$

We now seek an expression for the spectrum of the current fluctuations i_{n0} in a thin avalanche zone under an assumed zero ac electric field. Such a condition will not exist in a practical diode. The analysis is based on a physical picture as follows. The device is biased with a strong dc field such that a free carrier in the zone has a significant probability of causing a lattice collision which will release a new hole-electron pair in the zone before it escapes. One or both of these new carriers is likely to produce another ionization before they escape, etc. There is a critical field value E_0 at which the average number of new hole-electron pairs produced by any one pair is exactly one. This is the condition for a steady avalanche.

Suppose, for illustration, we imagine that there were no fluctuations in this process so that each pair always produced one new pair after a fixed time delay τ_z. Then the new one produced another, etc. Imagine that the zone is initially free of carriers and that one enters the zone from the outside, This would initiate a closely spaced train of electrons leaving one side of the zone spaced at intervals τ_z. Before this event, the current is zero; after it there will be an average current q/τ_z. Normally, τ_z is short compared to a wave period at all frequencies of interest so that the effect would be indistinguishable from that of a simple step-function of current of magnitude

$$\Delta i = \frac{q}{\tau_z} . \tag{28}$$

In a steady, high-current avalanche, there would be many such trains of electrons superimposed, all having been started at some earlier time. *Such an ideal avalanche would be completely free of noise.* In an actual diode, however, noise is present because these trains begin and end in an independent random fashion. Not all new hole-electron pairs produce further new ionizations and some produce two or more new ionizations. Although the average number will be exactly one in a steady avalanche, each deviation from this ideal results in a step-function of current of the type discussed, resulting in a noisy fluctuation of the avalanche.

Another source of noise is the statistical variation of the interval τ_z between ionizations in any given train of electrons. It can be easily shown that any single variation $\Delta\tau_z$ will have the effect of an impulse whose charge is $q\Delta\tau_z/\tau_z$, with a magnitude which will generally be less than q. The combined effect of all such fluctuations will have the same order of magnitude as shot noise and will probably be weaker. This effect is very much weaker than that of the step-functions and will be neglected henceforth.

Our noise, then, consists of a random sequence of step functions. Whenever a pair escapes producing no ionization, we have a negative current step-function, $-q/\tau_z$. When two new ionizations are produced, we have a positive step-function, q/τ_z; when three ionizations occur, the value is $2q/\tau_z$, etc. For the case of one new ionization, as expected, no disturbance occurs. Our problem is to determine the rate at which these various types of disturbances occur and to obtain an expression for the resulting noise current spectrum.

By Fourier analysis, the ac spectrum of a random sequence of equal-magnitude step-functions is given by

$$d[\bar{i}^2] = k\,\frac{2(\Delta i)^2}{\omega^2}\,df \tag{29}$$

where Δi is the magnitude of the steps and k is the average rate of occurrence (number per second).

We wish to determine the spectrum for a high current avalanche in which there are m ionizations per second, producing a dc current $I_0 = mq$. We postulate that $P(n)$ is the probability that any one of these will produce n new ionizations by direct lattice collisions. The number n may be zero, one, two, three, or more giving step-functions of magnitude $\Delta i = (n - 1)q/\tau_x$. The spectrum of the single class producing n new ionizations is, therefore,

$$d[\bar{i}^2]_n = \frac{2mP(n)(n + 1)^2 q^2}{\omega^2 \tau_x^2}\,df \tag{30}$$

and the total spectrum is the sum over all classes. For a small band Δf,

$$\bar{i}^2 = \frac{2qI_0\Delta f}{\omega^2\tau_x^2}\sum_{n=0}^{\infty}(n - 1)^2 P(n). \tag{31}$$

If we assume that carrier inertia effects are negligible so that the probability of ionization by a given carrier during its transit is constant, then the Poisson distribution can be used. If the average number of events is one, then the probability that any number n will occur is

$$P(n) = \frac{1}{n!\,e}. \tag{32}$$

[It is interesting to note that $P(0) = 1/e$, $P(1) = 1/e$, $P(2) = 1/2e$, $P(3) = 1/6e$, etc. Relatively few of the carrier pairs produce the ideal *one* new pair. Disturbances are the rule, not the exception. A typical value of τ_x is 5×10^{-13} seconds, implying that the step-function of a single electron distrubance is 3 microamperes. It is not surprising that the avalanche is noisy.]

For the Poisson distribution, the summation on the right in (31) has a numerical value of one, giving our result

$$\bar{i}_{n0}^2 = \frac{2qI_0B}{\omega^2\tau_x^2} \tag{33}$$

where B is a finite small bandwidth in c/s, ω is the angular frequency of interest, and $I_0 = mq$ is the dc current. The numerator is the familiar shot-noise expression.

There is a hidden inconsistency in the above derivation which can be resolved. Our assumption that the electric field is fixed at the critical value E_0 is not sufficient to specify or determine the dc or average current I_0. Under this assumption, avalanche fluctuations would cause the average current to wander slowly up or down in a "random walk," and there is no theoretical upper limit to this variation. However, in the practical case, space-charge and external resistance provide a current control mechanism when a fixed battery voltage is applied.

It should be emphasized that the above expression does not apply to the circuit noise current of any practical diode, which always includes some effects of resistance and space-charge. The pole at zero frequency is always suppressed by such resistance.

To determine the circuit noise current, we may substitute (33) into (27). The pole at zero frequency will, in general, disappear so that the circuit noise current remains finite. It will also be found that the pole at the avalanche frequency in (27) also disappears when the complete expression for the impedance Z_t is used.

We must also use some caution here to be sure that we are discussing a stable circuit. The presence of oscillation at any frequency will violate the small-signal approximations we have made.

We are interested in two cases, 1) the low-frequency current which the diode will induce when the load resistance R_L is assumed to be zero, and 2) the high-frequency case where the negative resistance is overcompensated by a positive resistance R_L. We can approximate this condition in a practical circuit provided R_L is inductively coupled into the loop so that it is effective only at high frequencies. At low frequencies, we take R_s and R_L to be zero so that $Z_t = R_d$. At high frequencies, we will discuss the resonant condition such that the reactive parts of Z_t vanish, leaving $Z_t = R_d + R_L$ where R_d will be negative and $R_d + R_L > 0$.

Substituting (33) into (27) gives the following results.

A. Low Frequency Case, $\omega \ll \omega_a$

$$\bar{I}_n^2 = 2qI_0B\,\frac{C_d^2V_a^2/I_0^2}{2m^2\tau_x^2}. \tag{34}$$

Here, C_d is the capacitance of the total depletion zone. It is interesting to note that the mean-square noise current varies *inversely* with the dc current.[1] This is the result of a decrease in the impedance of the avalanche as the current rises. The zero-frequency pole in (33) has disappeared because the avalanche impedance is inductive and vanishes at zero frequency.

[1] This relationship is not valid for very small dc currents. We have based this theory on the small-signal properties of a high-current avalanche. At low currents, this theory breaks down and noise fluctuations become "large-signal" compared to the dc values.

B. High Frequency Case, at Resonance

$$\overline{I_n^2} = 2qI_0B \frac{1}{\left(1 + \frac{R_L}{R_d}\right)^2} \cdot \frac{C_d^2 V_a^2 / I_0^2}{m^2 \tau_x^2} \frac{\theta^2/2}{1 - \cos\theta}. \quad (35)$$

This expression is similar to the low-frequency case except for the effects of negative resistance and transit angles included above. Stability requires that $R_d + R_L > 0$.

It should be emphasized that this device differs fundamentally from the case of a photo-detector diode in which a high "multiplication" gain is obtained by biasing near the point where avalanche would occur. We are discussing a full-blown avalanche which is entirely self-sustaining, requiring no injected carriers. In the "multiplying" diode, the internal impedance is relatively high, approaching infinity for no injection. There is no direct method of comparing the theory for these two distinctly different devices.

IV. Amplifier Noise Figure

We now substitute (33), (27), and (3) into (5). This gives

$$F = 1 + \frac{1}{m\omega^2\tau_x^2} \frac{qV_a}{kT} \frac{1}{\left|1 - \frac{\omega_a^2}{\omega^2}\right|} \frac{R_dR_L}{(R_L - R_d)^2} \quad (36)$$

at high gain, $R_d \simeq -R_L$, giving the simple result,

$$F = 1 + \frac{1}{4m\omega^2\tau_x^2} \frac{\frac{qV_a}{kT}}{1 - \frac{\omega_a^2}{\omega^2}}. \quad (37)$$

408

This is worthy of some discussion. The value of V_a cannot be less than the lattice ionization potential, and a value of 3–6 volts is probably minimum for a silicon device. Read has suggested that an appropriate value for m is six. The avalanche frequency ω_a must be less than ω, and we might assume a ratio of 0.5. The delay time τ_x will be on the order of one-half the transit time through the avalanche zone and might typically have a value of 5×10^{-13} seconds. At 10 000 Mc/s, these assumed values will give a predicted noise figure of 11 000 or 40.5 dB. This is a discouraging result.

However, as a high-level amplifier or oscillator, such a noise figure might be acceptable in many applications. It should be pointed out, however, that there are reasons to believe that the noise output may increase at high-signal level. Read has shown that the avalanche current is highly nonsinusoidal at high level, and may almost completely disappear during a large fraction of each cycle. The rate of build-up of an avalanche is strongly dependent upon the initial current. When the currents are small, percentage fluctuations are large. Thus, we may expect, at high level, a large fluctuation in the peak current from cycle to cycle. A diode with a large leakage current should be quieter at high level than one with a clean sharp breakdown characteristic.

V. Conclusions

A theoretical expression has been derived for the noise current spectrum in a semiconductor avalanche. The results have been applied to the Read type avalanche diode amplifier and an expression has been obtained for its noise figure.

The results indicate that this class of device is inherently noisy compared with other types of negative resistance amplifiers.

References

[1] W. T. Read, "A proposed high-frequency negative-resistance diode," *Bell Sys. Tech. J.*, pp. 401–466, March 1958.

[2] C. A. Lee, R. L. Batdorf, W. Wiegmann, and G. Kaminski, "The Read diode—an avalanche, transit-time, negative resistance oscillator," *Appl. Phys. Letters*, pp. 89–91, March 1965.

[3] R. L. Johnston, B. C. De Loach, Jr., and B. G. Cohen, "A silicon diode microwave oscillator," *Bell Sys. Tech. J.*, pp. 369–372, February 1965.

[4] M. E. Hines and M. Gilden, "Electronic tuning effects in the Read microwave avalanche diode," this issue, page 169.

[5] B. C. De Loach, Jr., and R. L. Johnston, "Avalanche transit-time microwave oscillators and amplifiers," this issue, page 181.

[6] A. S. Tager, "Current fluctuations in a semiconductor (dielectric) under the conditions of impact ionization and avalanche breakdown," in translation, *Soviet Physics—Solid State*, vol. 6, p. 1919. February 1965. From *Fizika Tver. Tela*, vol. 6, p. 2418, August 1964.

Reprinted from JOURNAL OF APPLIED PHYSICS, Vol. 38, No. 7, 2935–2946, June 1967

Noise of a Self-Sustaining Avalanche Discharge in Silicon: Low-Frequency Noise Studies

ROLAND H. HAITZ
Texas Instruments Incorporated, Dallas, Texas
(Received 17 October 1966; in final form 3 March 1967)

The noise theory for a self-sustaining avalanche discharge, which was published recently by Hines, is verified experimentally for low frequencies. It was found, in general, that the open-circuit spectral voltage density, $\langle u^2\rangle_{Av}/\Delta f$, arises from three contributions: (1) the inherent avalanche noise calculated by Hines and given by $\langle u_0^2\rangle_{Av}/\Delta f=a^2V_b^2/I$, where $a^2=3.3\times10^{-20}$ A/Hz, V_b=breakdown voltage and I=avalanche current; (2) excess noise at low current densities caused by nonuniformities of avalanche breakdown (even in microplasma-free diodes); and (3) excess noise at high current densities which seems to be caused by thermal effects. The temperature dependence of the noise is entirely due to the two excess-noise contributions. The experimental results are discussed with respect to their application in the design of noise generators, low-noise and low-current avalanche diodes, microwave oscillators, and negative-resistance amplifiers.

INTRODUCTION

Noise studies on stable-burning self-sustaining avalanche discharges have not been of much interest until recently when several new semiconductor devices utilizing avalanche breakdown were conceived and reduced to practice. The classic avalanche diode, which is used quite commonly as a voltage reference or regulator diode, is a large-signal diode and in most applications its noise is uncritical. However, the application of the new generation of avalanche devices as microwave oscillators,[1,2] negative-resistance amplifiers,[3] photomultipliers,[4] or surface-controlled avalanche transistors[5] is severely limited by avalanche noise. It is, therefore, expected that the noise characteristic of these avalanche devices, and consequently their application, should benefit considerably from a thorough investigation of avalanche noise. Of special interest is the noise of an ideal microplasma-free and uniform avalanche discharge.[6]

Earlier studies[7,8] of avalanche noise in commercially-available avalanche diodes revealed large noise variations from diode to diode. The spectral noise density passes through a sequence of maxima and minima with increasing avalanche current. Amplitude and position of these noise peaks and valleys is completely undetermined. Each diode reveals its own noise signature. In addition, the temperature dependence of such a noise characteristic is also undetermined.

Further studies of avalanche noise[8] especially at currents below 100 μA indicated that each of the noise peaks is associated with a microplasma. These microplasmas, which have been studied extensively in the past, are localized regions of avalanche breakdown.[9] If the current through such a microplasma region is small, e.g., of the order of 30 μA, then the avalanche discharge within this microplasma is unstable and switches on and off at random[10] generating a large noise current. With increasing avalanche current through a microplasma region the discharge becomes stable and the on–off switching ceases.[11,12] Such stable conditions are achieved if the microplasma current reaches the order of 50 to 100 μA. During the transition from the unstable to the stable avalanche discharge, the noise current decreases rapidly until the noise level of the stable discharge is reached.

This decrease of the avalanche noise to the noise level of a stable discharge is a rather idealized condition. Usually an avalanche diode contains a large number of microplasmas. At a given bias a fraction of these microplasmas will operate in a stable mode and not contribute any on–off switching noise. Another fraction of the microplasmas will be unstable and generate large amounts of switching noise which will be superimposed onto the noise of the stable microplasmas.

A noise study of a stable avalanche discharge is, therefore, only feasible under rather idealized conditions. Such idealized conditions are obtained either if one can restrict breakdown to a single microplasma or if it is possible to obtain microplasma-free uniform avalanche breakdown.

Melchior *et al.*[8] succeeded in measuring the noise of a stable discharge by studying a diode whose first microplasma broke down at a voltage which was considerably below the breakdown voltage of the second microplasma. The first microplasma became stable at a current of approximately 50 μA. The second microplasma started to break down when the first microplasma was carrying a current of approximately 120 μA. Thus they were able to study the stable discharge of

[1] W. T. Read, Jr., Bell System Tech. J. **37**, 401 (1958).
[2] R. L. Johnston, B. C. DeLoach, Jr., and B. G. Cohen, Bell System Tech. J. **44**, 369 (1965).
[3] B. C. DeLoach and R. L. Johnston, IEEE Trans. Electron Devices **ED-13**, 181 (1966).
[4] K. M. Johnson, IEEE Trans. Electron Devices **ED-12**, 55 (1965).
[5] W. Shockley and W. W. Hooper, IEEE Trans. Electron Devices **ED-11**, 535 (1964).
[6] A. Goetzberger, B. McDonald, R. H. Haitz, and R. M. Scarlett, J. Appl. Phys. **34**, 1591 (1963).
[7] H. J. Stöhr, Rhode & Schwarz-Mitt. **16**, 51 (1961).
[8] H. Melchior and M. J. O. Strutt, Sci. Electrica **9**, 129 (1963).
[9] R. H. Haitz, A. Goetzberger, R. M. Scarlett, and W. Shockley, J. Appl. Phys. **34**, 1581 (1963).
[10] R. J. McIntyre, J. Appl. Phys. **32**, 983 (1961).
[11] R. H. Haitz and A. Goetzberger, Solid-State Electron. **6**, 678 (1963).
[12] R. H. Haitz, J. Appl. Phys. **35**, 1370 (1964).

the first microplasma in the current range from 50 to 120 μA. Over this narrow current range they found that the noise power decreases inversely with the avalanche current and increases slightly with decreasing temperature.

Two approaches have been taken in calculating the noise fluctuations of a self-sustaining avalanche discharge:

(1) In the first approach both Tager[13] and McIntyre[14] calculated the noise current of a photodetector generated by the carrier multiplication process. They obtained expressions for the noise current as a function of the carrier multiplication factor M. Assuming that their equations hold for $M \to \infty$ they calculated the noise of an avalanche discharge.

(2) In the second approach Hines[15] calculated the noise-current fluctuations by taking into account the influence of the space charge of the carriers in the drift zone of the junction on the electric field in the avalanche zone and consequently on the ionization rate.

Both approaches lead to an inverse dependence of the noise power density on the avalanche current I. McIntyre's[14] calculation indicates that the open-circuit spectral voltage density, $\langle u^2\rangle_{Av}/\Delta f$, becomes independent of M for $M \to \infty$:

$$\frac{\langle u^2\rangle_{Av}}{\Delta f}=\frac{2q}{kI(d\delta/dV)^2}, \tag{1}$$

where q denotes the electron charge, and k denotes the ratio of the ionization coefficients of electrons and holes. The term $d\delta/dV$ is a complicated function of voltage, field distribution and ionization coefficients.

The theory of Hines[15] is also quite inaccurate. It neglects the difference between the ionization coefficients of electrons and holes[16]; it uses a power-law approximation for the dependence of the ionization coefficient on field; and it contains several constants which are not accurately known. Hines' theory, however, has two major advantages over McIntyre's theory: (1) it predicts the frequency dependence of the spectral-noise density at frequencies in the range of the reciprocal transit time of carriers through the depletion layer, and (2) it contains only constants which can be estimated within a factor of 2 and no functional terms like $d\delta/dV$ of Eq. (1). Because of these two advantages, Hines' theory will be used in interpreting the experimental results. On the other hand, it is anticipated to obtain more accurate values for the constants in Hines' theory from experimental results.

The noise studies carried out on self-sustaining avalanche discharges are divided into two parts. The remainder of this paper (Part I) will contain the low-frequency noise studies. Part II of this series will deal with noise studies at microwave frequencies.

In Sec. II of Part I we will briefly discuss Hines' theory and introduce a slight modification. Experimental results obtained on a large number of microplasma-free avalanche diodes will be described in Sec. III. These results will then be discussed in Sec. IV. In Sec. V we will discuss a few conclusions relating these noise studies to the design of improved avalanche diodes.

II. THEORETICAL CONSIDERATIONS

In his calculations of the spectral noise density of a self-sustaining avalanche discharge, Hines[15] made several simplifying assumptions:

(1) The ionization coefficients for electrons and holes are assumed to be equal, $\alpha=\beta$.

(2) The field across the avalanche zone is assumed to be independent of the distance x from the edge of the depletion region.

(3) The width of the avalanche region is assumed to be negligible compared with the width of the drift region, $l_a \ll l_d$.

(4) A power law is used to approximate the field dependence of α, $\alpha \propto (E/E_0)^m$, where E_0 denotes a normalizing field.

(5) Spreading resistance, R_{sp}, and thermal resistance, R_{th}, are neglected. These simplifications will be discussed briefly in the next paragraphs.

The first two simplifications can be combined to yield an average ionization coefficient $\bar{\alpha}$ for the case that the field across the avalanche region is not constant. This assumption is rather crude. However, any more realistic assumptions such as the introduction of separate ionization coefficients α and β or a dependence of α on x would complicate the mathematical analysis considerably. Therefore, we will stay with Hines' assumption of an average value $\bar{\alpha}$ independent of x.

The assumption $l_a \ll l_d$ will be a good approximation for Read structures[1] of the type $n^+p\pi p^+$ or $p^+n\nu n^+$. However, for the case of a simple n^+p or p^+n step junction this assumption is no longer valid. The avalanche region extends then over a considerable fraction of the depletion layer and in a good approximation one can assume $l_a=0.3W$, where $W=l_a+l_d$ denotes the total depletion layer width.

Hines assumed a power law for the dependence of $\bar{\alpha}$ on E. A better approximation is given by Lee *et al.*[16] who found the following empirical relationship

$$\alpha=Be^{-b/E}, \tag{2}$$

with different values of the constants B and b for electrons and holes. Similarly the average ionization coefficient $\bar{\alpha}$ can be expressed by Eq. (2) with constants estimated from the respective constants for electrons and holes. The field derivative of $\bar{\alpha}$ with respect to E

[13] A. S. Tager, Fiz. Tverd. Tela **6**, 2418 (1964) [English transl.: Soviet Phys.—Solid State **6**, 1919 (1965)].
[14] R. J. McIntyre, IEEE Trans. Electron Devices **ED-13**, 164 (1964).
[15] M. E. Hines, IEEE Trans. Electron Devices **ED-13**, 158 (1966).
[16] C. A. Lee, R. A. Logan, R. L. Batdorf, J. J. Kleimack, and W. Wiegmann, Phys. Rev. **134**, A761 (1964).

becomes then for $\bar{\alpha}l_a=1$:

$$d\bar{\alpha}/dE=b/EV_a, \tag{3}$$

where V_a denotes the voltage across the avalanche region. The term b/E in Eq. (3) can be calculated from Eq. (2) leading to

$$d\bar{\alpha}/dE=\ln(Bl_a)/V_a\approx m/V_a. \tag{4}$$

The term $\ln(Bl_a)$ in Eq. (4) corresponds to the constant m in Eq. (25) of Hines.[15] With an averaged value of $B=5\times10^6$ cm^{-1}, one obtains values of $\ln(Bl_a)$ of 3.9 and 6.2 for l_a values of 10^{-5} and 10^{-4} cm, respectively. This is only a small correction to Hines' approximation of $m\approx5$.

A considerably larger correction will result from a proper accounting of spreading and thermal resistance. The equivalent circuit of an avalanche diode is shown in Fig. 1. The total series resistance of the diode, R_s, consists of three contributions[10]: (1) space-charge resistance R_{sc},[17] (2) spreading resistance R_{sp}, (3) thermal resistance R_{th}, with

$$R_s=R_{sc}+R_{sp}+R_{th}. \tag{5}$$

At low frequencies R_{sc} is given by[15]

$$R_{sc}=l_d^2/2\epsilon v_d A, \tag{6}$$

where ϵ=dielectric constant, v_d=maximum drift velocity and A=breakdown area. Since the diameter of the breakdown area D is small compared to the thickness of the silicon wafer, both R_{sp} and R_{th} are dominated by current and heat spreading into a hemisphere of diameter D, respectively:

$$R_{sp}=\rho/2D, \tag{7}$$

$$R_{th}\approx(1/2\lambda D)\beta V_b^2, \tag{8}$$

where ρ=specific resistance of the bulk material, λ= thermal conductivity of silicon, and β=temperature coefficient of the breakdown voltage V_b.

The thermal resistance R_{th} is frequency-dependent. R_{th} is caused by an increase of V_b with temperature which results from an increase in power dissipation. The time constant τ_{th} determining the frequency dependence of R_{th} is given by

$$\tau_{th}=rc, \tag{9}$$

where $r=1/2\lambda D$ denotes the thermal resistance in °C/W and c denotes the heat capacity of the breakdown region. For a diode with $D=10^{-3}$ cm one estimates $\tau_{th}\approx0.3$ μsec. This implies that R_{th} disappears for frequencies of the order of 10 MHz or larger.

This frequency dependence of R_{th} will determine the technique for measuring the avalanche noise. To avoid variations of R_{th} that will influence the experimental results one has to measure the open-circuit noise voltage $\langle u^2\rangle_{Av}=\langle i^2\rangle_{Av}R_{sc}^2$. This voltage $\langle u^2\rangle_{Av}$ will be proportional to $\langle i^2\rangle_{Av}$ for all frequencies for which diode and circuit capacitance can be neglected.

[17] W. Shockley, Solid-State Electron. **2**, 35 (1961).

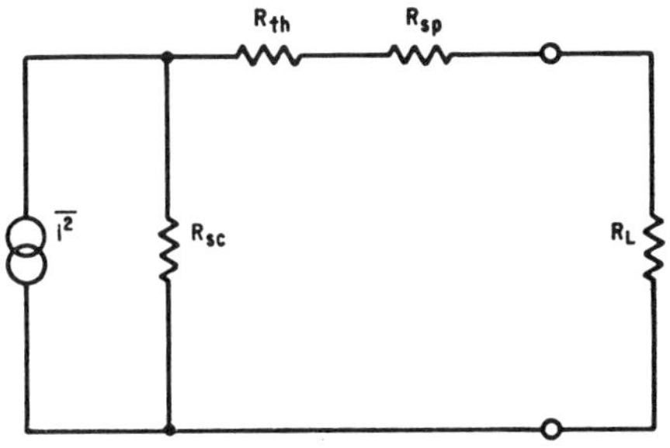

FIG. 1. Noise equivalent circuit of avalanche diode.

At frequencies which are small compared with the avalanche frequency ω_a,[15] the open-circuit noise voltage $\langle u^2\rangle_{Av}$ can be calculated readily from Eq. (34) of Hines,[15] which gives the short-circuit noise current $\langle i^2\rangle_{Av}$ for the case $R_{sp}=0$ and $R_{th}=0$. Multiplying $\langle i^2\rangle_{Av}$ with R_{sc}^2 of Eq. (6) and reintroducing a term $(1+l_a/l_d)^2$, which Hines neglected for $l_a\ll l_d$, one obtains the spectral voltage density:

$$\frac{\langle u^2\rangle_{Av}}{\Delta f}=\frac{q\kappa^2}{2m^2}\frac{\tau_{tr}^2}{\tau_x^2}\frac{V_b^2}{I}. \tag{10}$$

[A numerical error of a factor 2 in the denominator of Eq. (34) of Hines is corrected in Eq. (10).] In Eq. (10), q denotes the electron charge. The avalanche voltage V_a is replaced by the breakdown voltage V_b:

$$V_a=\kappa V_b. \tag{11}$$

For the case of a step junction one finds $\kappa\approx0.5$. The time $\tau_{tr}=(l_a+l_d)/v_d$ denotes the transit time through the depletion layer of width $W=l_a+l_d$ at the saturated drift velocity v_d. The time τ_x is the mean time between two ionization events of a carrier in the avalanche zone.

Equation (10) is of particular interest since it can be reduced to a simple expression, which holds to a good approximation for practically all silicon *pn* junctions. The only two variables in Eq. (10) are V_b and I. The quantity $\kappa\approx0.5$ is constant for all step junctions and only slightly different for linearly graded junctions. The quantity m is practically constant since it varies only from 4 to 6 for a change in l_a from 10^{-5} to 10^{-4} cm. The ratio τ_{tr}/τ_x is also practically constant since $\tau_{tr}\propto(l_a+l_d)=W$ and $\tau_x\propto l_a$. For both step and linearly graded junctions the ratio l_a/W is constant and approximately equal to 0.3. Therefore, one can write

$$\langle u^2\rangle_{Av}/\Delta f=a^2V_b^2/I. \tag{12}$$

Since the errors of κ and m are only of the order of 10% it should be possible to determine the ratio τ_{tr}/τ_x relatively accurately from experimentally determined values of the diode noise constant a^2.

III. EXPERIMENTAL RESULTS

The noise studies to be reported in this section will consist mainly of measurements of the open circuit voltage density, $(\langle u^2\rangle_{Av}/\Delta f)^{1/2}$, as a function of four independent parameters: (1) avalanche current I;

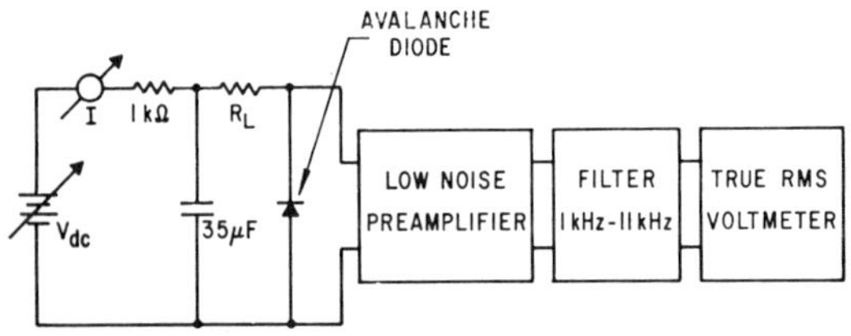

FIG. 2. Circuit and block diagram used for low-frequency noise studies.

(2) ambient temperature T; (3) breakdown voltage V_b; and (4) breakdown diameter D. The experimental methods to measure $(\langle u^2\rangle_{Av}/\Delta f)^{1/2}$ as a function of the four parameters are discussed in the following paragraphs.

Standard low-noise measurement techniques are applied to measure $(\langle u^2\rangle_{Av}/\Delta f)^{1/2}$. The circuit used for the noise measurements is illustrated in Fig. 2. The diode is biased from a dc voltage source through a filter consisting of a 1-kΩ resistor and a 35-μF capacitor. The load resistor is chosen such that $R_L \gg R_s$ and such that the thermal noise of the resistor is negligible compared to the avalanche noise under investigation. Values of R_L are in the range of 300 Ω to 3 kΩ.

The avalanche noise, which is white down to frequencies of less than 100 Hz (Part II), is fed into a low-noise preamplifier and then passed through a band-pass filter with a pass band of 1 to 11 kHz and slopes of 18 dB/octave. The RC filter in the biasing circuit and the lower cutoff at 1 kHz ensure that the 120-Hz ripple from the power supply is not contributing to the measured noise. The pass-band characteristic of the preamplifier–filter combination is measured accurately in order to take both gain variations within the pass band and contributions from the tails in the filter characteristic into account. The output of the filter
412 is measured with a true rms voltmeter. The output of the rms voltmeter is fed into a logarithmic converter and then applied to the vertical input of an X–Y recorder. The diode current is measured with a microvoltmeter as a voltage drop across a 10-Ω resistor, converted to logarithmic units and applied to the horizontal input of the recorder.

The lower end of the current range, over which avalanche noise studies were made, is limited by the stability of the avalanche discharge.[11] Measurements can usually be obtained down to currents of the order of 250 μA. The upper limit of the current range is determined by the power dissipation of the diodes. The higher the breakdown voltage and the smaller the diameter, the smaller is the maximum current the diode can dissipate before going into secondary breakdown due to overheating.[18] Depending on V_b and D the upper limit is in the range of 10 to 50 mA.

Under certain conditions two corrections have to be made to the measured values of $(\langle u^2\rangle_{Av}/\Delta f)^{1/2}$. The first correction has to be made when the thermal noise of R_L and the input noise of the preamplifier are not negligible with respect to the avalanche noise. Such a correction is necessary for diodes with small breakdown voltages at high currents. The second correction has to be made when the diode series resistance R_s is not negligible with respect to R_L. Such a correction is required for small diodes with high breakdown voltages.

The spectral-noise voltage can be studied readily as a function of the first two parameters I and T for any diode of given V_b and D. However, in order to study the dependence of $(\langle u^2\rangle_{Av}/\Delta f)^{1/2}$ on V_b or D different diodes have to be prepared and accurate methods for measuring V_b and D have to be found.

The microplasma-free uniform avalanche diodes were prepared by the well-known guard-ring technique.[6] A series of n^+p and p^+n diodes with breakdown voltages in the range of 5.9 to 46 V were diffused. After diffusion, metalization, and dicing, each chip contained four guard-ring diodes with different diameters of the breakdown region ranging from 6 to 90 μ. Each wafer also contained a planar n^+p or p^+n diode without guard ring for capacitance–voltage measurements, which are required to determine the width W of the depletion layer in the breakdown region. The wafer with all five diodes was mounted in a six-lead TO-5 transistor package. As described earlier,[12] V_b can be determined accurately by extrapolating the V–I characteristic to $I=0$. The diameter D is measured directly by observing the light emission of unmounted diodes under a high-powered microscope.

The measurement of the diode series resistance R_s which is required for an accurate correction of $(\langle u^2\rangle_{Av}/\Delta f)^{1/2}$ is carried out by superimposing a 100-μA, 6-kHz ac current onto the dc current I. The experimental results of R_s are plotted as a function of I in Fig. 3 for four diodes which will be used extensively in the following noise studies. R_s was found to be independent of frequency at least up to 100 kHz.

The open-circuit spectral noise density, $(\langle u^2\rangle_{Av}/\Delta f)^{1/2}$, as a function of I is plotted in Fig. 4 for four 13-V diodes with diameters of 6, 13, 27, and 77 μ. The data are corrected for amplifier noise and diode resistance.

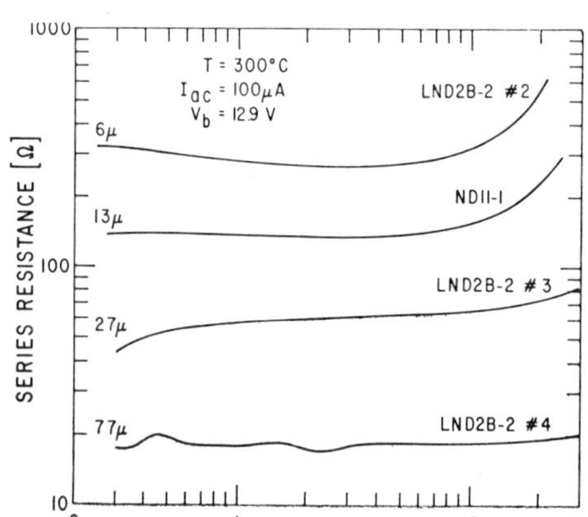

FIG. 3. Diode series resistance as a function of current for four 13-V diodes with breakdown diameter as parameter.

[18] R. H. Haitz, J. Appl. Phys. 36, 3123 (1965).

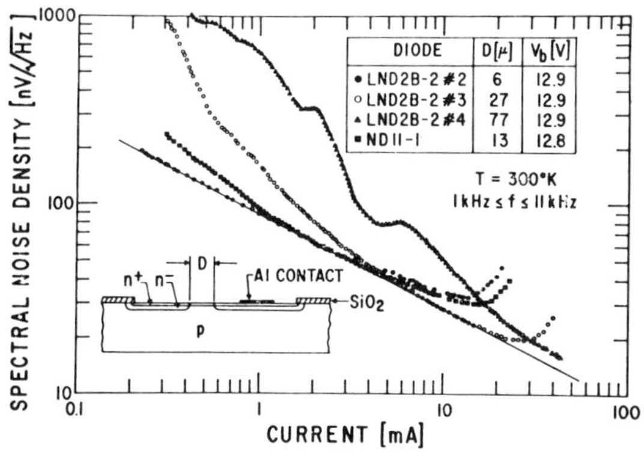

FIG. 4. Spectral-noise density as a function of current for four 13-V diodes with breakdown diameter as parameter. The solid line has a slope of $-\frac{1}{2}$ corresponding to Eq. (10).

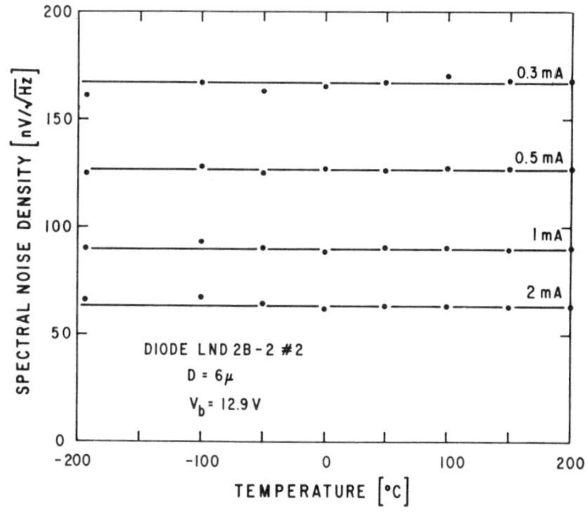

FIG. 6. Spectral-noise density as a function of temperature for a diode following the $I^{-1/2}$ law of noise vs current. Data are taken from the 6-μ diode LND2B-2 #2 at currents between 0.3 and 2 mA.

The solid line has a slope of $-\frac{1}{2}$ as predicted by Eq. (10).

From a first inspection of Fig. 4 one can conclude that the small 6-μ diode follows the predicted $I^{-1/2}$ law at low currents. However, there are deviations at higher currents. With increasing diameter one also observes deviations from the $I^{-1/2}$ law at low currents. However, at intermediate or higher currents the noise density approaches the solid line. These results suggest that the spectral-noise density be described by three contributions: (1) $I^{-1/2}$ noise, (2) excess noise at low current densities, and (3) excess noise at high current densities.

The insertion in Fig. 4 shows a cross section of an avalanche diode. Note that the aluminum contact is located on the guard ring. Such a structure not only allows studies of light emission but also prevents destruction of the diode by alloying of the contact during accidental thermal run-away.

The temperature dependence of a diode exhibiting excess noise at low current densities is illustrated in Fig. 5 for the 27-μ diode LND2B-2 #3. The room temperature curve and the solid line are repeated from Fig. 4. The second curve is measured at 77°K.

The temperature dependence of a diode which obeys the $I^{-1/2}$ law is shown in Fig. 6 for the 6-μ diode LND2B-2 #2. $(\langle u^2\rangle_{Av}/\Delta f)^{1/2}$ is plotted for temperatures between $-196°$ and 200°C. The scatter of the experimental points around the solid lines (average noise density at the corresponding currents) is approximately $\pm 3\%$. This value of $\pm 3\%$ is equal to the estimated experimental accuracy. One can, therefore, conclude that there is no measurable temperature dependence of the $I^{-1/2}$ noise.

The dependence of the open circuit spectral voltage density in case of $I^{-1/2}$ noise on breakdown voltage is illustrated in Fig. 7 for a series of small diodes with diameters up to 20 μ. Both n^+p and p^+n diodes are represented. Also included is a point obtained from an $n^+p\pi p^+$ structure of a Read diode. The solid line has a slope of $+1$ as predicted by Eq. (10).

IV. DISCUSSION

A. Series Resistance

The series resistance of an avalanche diode is, in a first approximation, expected to be independent of

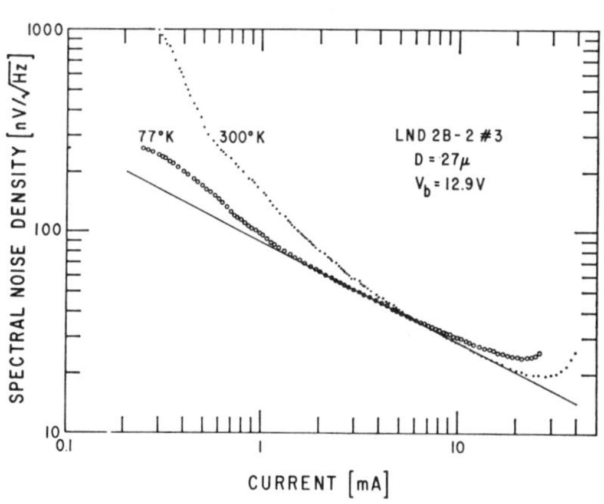

FIG. 5. Spectral-noise density as a function of current at 77° and 300°K for diode LND2B-2 #3 with $D=27$ μ. The room-temperature curve is repeated from Fig. 4.

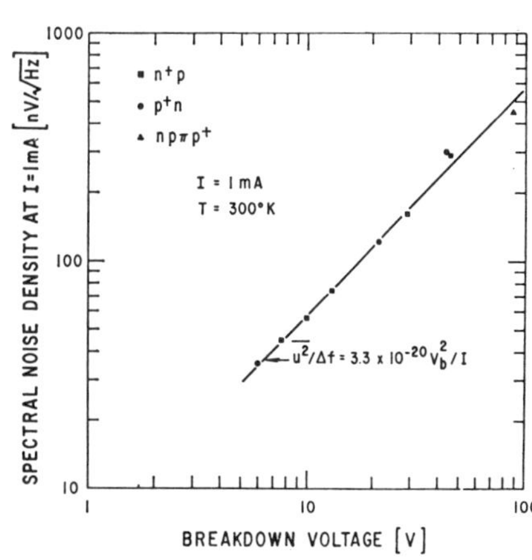

FIG. 7. Spectral-noise density as a function of breakdown voltage at $I=1$ mA for a series of n^+p, p^+n, and $n^+p\pi p^+$ diodes.

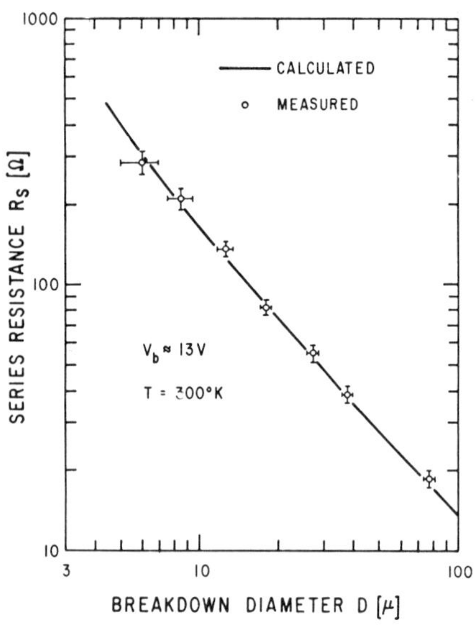

FIG. 8. Diode series resistance as a function of diameter for a series of 13-V diodes. The solid line is calculated from Eqs. (5)–(8).

current. The results of Fig. 3 are in good agreement with such a first approximation. However, there are several small deviations which are discussed in this section.

The two smallest diodes show a slight decrease of R_s at low currents. Such a decrease can easily be caused by a small increase of the breakdown area due to a gradient of the breakdown voltage near the edge of the breakdown area.[19] There is no explanation for the slight increase of R_s at low currents in the case of the two larger diodes.

At high currents all four diodes exhibit an increase of the series resistance with current. This increase is

most pronounced for the two smallest diodes. From Fig. 4 it becomes evident that this increase of R_s with current is accompanied by the earlier-mentioned deviation of the noise density from the $I^{-1/2}$ law. A brief discussion of this effect is given in Sec. IV.C. It is thought that the temperature rise causing this excess noise is also responsible for the resistance rise with current.

It is of interest to obtain an experimental verification of Eq. (5). So far such an experimental verification has not been possible since for large diodes, breakdown is not uniform and R_s usually decreases with current. The values of R_s at a current of $I=3$ mA are plotted in Fig. 8 as a function of D for a series of 13-V diodes including the diodes of Fig. 3. The errors in D are estimated to be the greater of ± 1 μ or $\pm 5\%$. The errors in R_s are estimated from the scattering of R_s for several diodes of equal diameter.

The solid line in Fig. 8 is calculated from Eqs. (5)–(8). The resistivity of the starting material was $\rho=0.12$ $\Omega\cdot$cm. The drift velocity was assumed[20] as $v_d=8.5\times10^6$ cm/sec. The drift length $l_d=0.7W$ was obtained by estimating the space-charge layer width from capacitance measurements to $W=0.35$ μ. For a 13-V diode the temperature coefficient of V_b is equal to

$$\beta=0.7\times10^{-3}/°\text{C}.$$

The agreement between the calculated curve and the experiment points is well within the error limit. From this agreement one can conclude that the accuracy of calculating R_s from Eqs. (5)–(8) is of the order of $\pm 5\%$. Since the calculated curve is based on the assumption that the entire n^+p junction area is breaking down, one can further conclude from the above agreement that the avalanche current is distributed uniformly over the n^+p region even for the large 77-μ diodes.

B. Dependence of Noise on I and V_b

As mentioned in Sec. III, the open-circuit noise density, $(\langle u^2\rangle/\Delta f)^{1/2}$, can be described by three contributions. In this section the noise contribution which decreases inversely with $I^{-1/2}$ is discussed.

The dependence of $(\langle u^2\rangle_{Av}/\Delta f)^{1/2}$ on I and D, which is illustrated in Fig. 4 for the case of diodes with $V_b \approx 13$ V, is typical for all diodes studied during this investigation. The discussion of these n^+p diodes will, therefore, be valid also for p^+n diodes and for diodes with other values of V_b in the investigated range from 5.9 to 46 V.

From Fig. 4 it is evident that the $I^{-1/2}$ contribution to the noise is independent of the diameter D. The small diodes with diameters between 6 and 27 μ approach the solid curve quite well over part of the current range. It is believed that the large 77-μ diode will also approach the $I^{-1/2}$ curve at higher currents.

The results illustrated in Fig. 5 lead to the suggestion that the temperature dependence of $(\langle u^2\rangle_{Av}/\Delta f)^{1/2}$ is entirely determined by the temperature dependence of the excess noise. The measurements reported in Fig. 6 support this hypothesis. One can estimate from the experimental results of Fig. 6 that the temperature coefficient of the $I^{-1/2}$ noise is less than $7\times10^{-5}/°$C.

Equation (10) leads to the suggestion that the $I^{-1/2}$ noise is a linear function of V_b, provided that the ratio τ_{tr}/τ_x is independent of V_b. The constant κ is independent of V_b for all step junctions. The experimental results of Fig. 7 show this expected linear dependence of $(\langle u^2\rangle_{Av}/\Delta f)^{1/2}$ on V_b. One can therefore conclude that τ_{tr}/τ_x is constant.

The constant a^2 of Eq. (12) and the ratio τ_{tr}/τ_x can be determined from the experimental results of Fig. 7.

[19] R. H. Haitz, Second Annual Scientific Report on Contract No. DA 49-186 ORD-1086, Harry Diamond Laboratories (1964), p. 41 (unpublished).

[20] A. C. Prior, J. Phys. Chem. Solids 12, 175 (1960).

From the position of the solid line in Fig. 7 one obtains:

$$a^2=3.3\times10^{-20}\ \mathrm{A/Hz}.^{21} \qquad (13)$$

The error is estimated to $\pm5\%$. The ratio τ_{tr}/τ_x is calculated from Eqs. (10), (12), and (13) to be

$$\tau_{tr}/\tau_x=5.2. \qquad (14)$$

For this calculation it was assumed that $m=4$ and $\kappa=0.5$. Assuming that both constants have an error of $\pm10\%$ one estimates the error of τ_{tr}/τ_x to $\pm15\%$. It should be noted that Eq. (14) and the estimated error of $\pm15\%$ hold only for step junctions.

It is also of interest to estimate the ratio τ_a/τ_x, where $\tau_a=l_a/v_d$ denotes the transit time through the avalanche region. This ratio is equal to the average number of electron hole pairs generated by a carrier which fully traverses the avalanche zone. For the case that the field through the avalanche zone is constant and that the ionization rate of electrons and holes are equal, the ratio τ_a/τ_x should be equal to 1.

For a more realistic case with unequal ionization coefficients α_n and α_p one expects a larger value for τ_a/τ_x with τ_x now denoting the mean time between ionization events of an electron. In a first approximation this ratio can be estimated from the following condition for self-sustaining breakdown[16]:

$$\alpha_n=\alpha_p \exp[(\alpha_n-\alpha_p)l_a]. \qquad (15)$$

The quantities α_n and α_p are the ionization coefficients of electrons and holes, respectively. Equation (15) is solved numerically for various values of α_n/α_p. The product $\alpha_n l_a$, which is equal to τ_a/τ_x, is plotted in Fig. 9 as a function of α_n/α_p.

The ionization coefficients α_n and α_p can be estimated from the experimental curves of α_n and α_p vs $1/E$ measured by Lee *et al.*[16] For this purpose one has to determine the breakdown field for the diodes under investigation. For the 13-V diodes one obtains from capacitance measurements $E_b\approx630$ kV/cm. Similarly, one obtains for the 45-V diodes $E_b\approx420$ kV/cm. The ratios α_n/α_p for these breakdown fields are 1.9 and 6.2, respectively. For these two values one finds the following values for $\alpha_n l_a=\tau_a/\tau_x$ from the curve in Fig. 9:

$$\tau_a/\tau_x=1.4 \quad \text{for } V_b=13\ \mathrm{V},$$
$$\tau_a/\tau_x=2.2 \quad \text{for } V_b=45\ \mathrm{V}. \qquad (16)$$

[21] In earlier publications[22–24] the tails of the band-pass filter characteristic were not properly taken into account. As a result the effective bandwidth was underestimated leading to $a^2=4.8\times10^{-20}$ A/Hz. This value is 1.6 dB larger than the correct value of Eq. (13).

[22] R. H. Haitz, IEEE Trans. Electron Devices **ED-13**, 671 (1966).

[23] R. H. Haitz and F. W. Voltmer, Appl. Phys. Letters **9**, 381 (1966).

[24] R. H. Haitz, *Physics of Failure in Electronics* (Rome Air Development Center, Rome, N.Y.; 1966), series in reliability, Vol. 5.

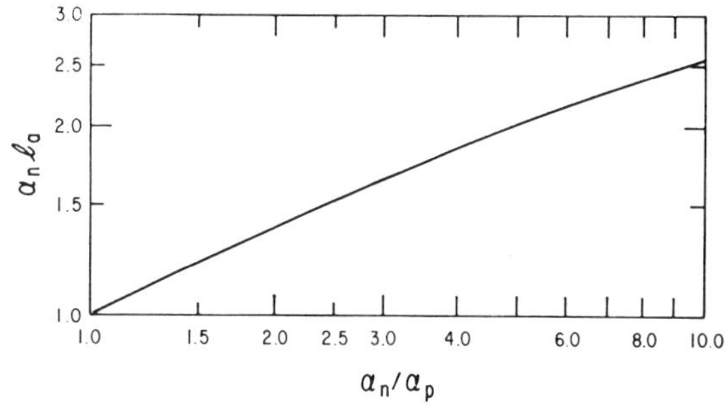

FIG. 9. The function $\alpha_n l_a=\tau_a/\tau_x$ vs α_n/α_p.

The experimental value of τ_a/τ_x can be determined from Eq. (14) by substituting τ_{tr} from

$$\tau_a/\tau_{tr}=l_a/W\approx0.3. \qquad (17)$$

One obtains then for τ_a/τ_x:

$$\tau_a/\tau_x=1.6. \qquad (18)$$

From the good agreement between the estimated values of Eq. (16) and the experimental result of Eq. (18) one can conclude that Hines' theory describes the $I^{-1/2}$ contribution of the noise very well. Even if this good agreement is coincidental due to the crude assumptions in Hines' theory and in calculating the curve of Fig. 9, one is still reasonably safe in assuming that Hines' theory describes the low frequency $I^{-1/2}$ noise within a factor of 2.[25]

C. Excess Noise

In Sec. III it was mentioned that the total avalanche noise can be described by three contributions. The subject of this section is the discussion of the two excess-noise contributions at low and high current densities.

The excess noise at high current densities appears to be associated with thermal effects. This hypothesis is not surprising, since the power dissipation within the breakdown region of the 6-μ diode of Fig. 4 is of the order of 300 mW at a current of 20 mA. This corresponds to a power density of approximately 10^6 W/cm^2. The temperature rise resulting from such a large power dissipation within the 6-μ region is estimated to be of the order of 300°C. Another observation supports the hypothesis of a thermal origin. The excess noise at a given current level can be reduced considerably by improving the heat removal from the diode package.

The excess noise at high current densities was not investigated in detail because the above mentioned preliminary results indicate a thermal origin. Crude

[25] Thanks to an unknown reviewer who estimated the term $d\delta/dV$ of Eq. (1). With this result McIntyre's theory[14] predicts values of a^2 ranging from 6.5×10^{-20} to 10.6×10^{-20} A/Hz for maximum junction fields between 250 and 500 kV/cm. These values are 3 to 5 dB above the experimental result of 3.3×10^{-20} A/Hz.

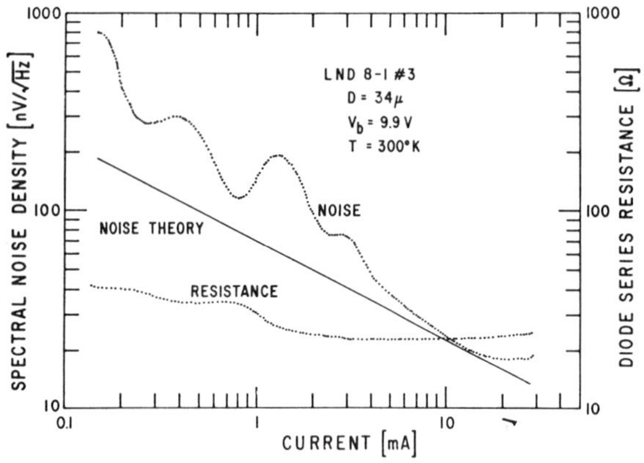

FIG. 10. Spectral noise density and diode series resistance as a function of current for a diode with several excess-noise peaks. The solid line represents Eq. (10).

experimental studies indicate that the spectral density is not white and increases with decreasing frequency.

The excess noise at low current densities seems to be associated with nonuniformities of the avalanche breakdown. Such a hypothesis originates from the observation made in Fig. 4 showing increased excess noise with increasing breakdown area. In case of the largest diodes in Fig. 4 one even observes peaks of excess noise.

Such excess-noise peaks were studied in more detail on diode LND8-1 #3. For this particular diode one has $V_b=9.9$ V and $D=34$ μ. The noise vs current characteristic is shown in Fig. 10. The noise curve exhibits two distinct peaks at 0.4 and 1.3 mA, respectively, and a smaller peak at 2.7 mA. If these peaks are caused by breakdown nonuniformities such as a current dependent expansion of the breakdown area, then this variation of the breakdown area should lead to a corresponding variation of the diode resistance R_s.

Exactly such a variation of R_s with current is observed in the R_s vs I curve of Fig. 10. The decreasing sections of the resistance curve clearly coincide with the rising sections of the noise curves. R_s drops from 40 to 35 Ω at the first noise peak, from 35 to 24 Ω at the second noise peak, and from 24 to 23 Ω at the small third noise peak. For currents above 3 mA the resistance is constant and the excess noise decreases rapidly as evidenced by the rapid approach of the experimental noise curve to the theoretical curve.

The breakdown nonuniformities causing the excess noise peaks in Fig. 10 arise from minor variations of V_b over the breakdown region. The breakdown voltage of the area causing the noise peak at 1.3 mA is only approximately 45 mV above the minimum breakdown voltage of the diode. The corresponding relative difference is less than 0.5%.

The effective breakdown area can be estimated from Eq. (10). The results are summarized in Table I. The effective breakdown area is expressed by a corresponding equivalent breakdown diameter D^*. The total diameter of the breakdown region is estimated from light-emission studies to $D=34\pm1$ μ. Table I indicates that at a small current of 0.2 mA a considerable fraction of the junction is already breaking down. At $I=2$ mA, the breakdown covers practically the entire area. Above 4 mA, the series resistance is constant ($D=D^*$) and the excess noise decreases rapidly, approaching the theoretical noise curve.

From the results shown in Fig. 10 and Table I, one can exclude microplasma breakdown (highly localized) from causing the noise peaks. In case of microplasma breakdown, the initial resistance would be much larger than 40 Ω, e.g., of the order of several kilohms. The nonuniformities causing the noise peaks consist, therefore, of small variations of V_b as discussed by Goetzberger *et al.*[6] which are very likely caused by small local variations of the doping concentration in the starting material.

If the peaks of the excess noise can be contributed to breakdown nonuniformities then one may try to explain the entire excess noise by such nonuniformities. Such a model is discussed in the remainder of this section.

The following discussion of excess noise is based on a *pn* junction model proposed by Shockley.[27] The space-charge layer of the breakdown region is divided into small unit cells of volume W^3, where W denotes the width of the space-charge layer. The breakdown voltage of a given cell k, V_{bk}, is determined by the number of impurities within the cell. Since the number of impurities within the cells fluctuates from one cell to another around an average value, V_{bk} will also fluctuate around an average value $\langle V_b\rangle_{Av}$. Each cell is associated with a series resistance R_{sk} arising from the space charge resistance of Eq. (6). Spreading and thermal resistance are neglected. Since the dimensions of all cells are the same it can be assumed that R_{sk} is constant and equal to

$$R_{sk}=NR_{sc}\approx NR_s, \tag{19}$$

where $N=\pi D^2/4W^2$ denotes the number of cells within the breakdown region. The current through the cell k is then given by

$$I_k=(V_{dc}-V_{bk})/(R_{sk}+R_L), \tag{20}$$

where V_{dc} is the applied dc voltage.

Since Hines' theory holds well for very small diodes we can assume that it also holds for each unit cell. The resulting diode model is shown in Fig. 11. Ac-

TABLE I. Series resistance and effective breakdown diameter as a function of current for the 9.9-V diode of Fig. 10.

I [mA]	R_s [Ω]	D^* [μ]
0.2	40	21
0.5	35	24
2.0	24	33
4.0	23	34

cording to this model the noise contribution from each unit cell can be described by a noise voltage source $\langle u_{ok}^2\rangle_{Av}$ in series with R_{sk}. The open-circuit spectral voltage density for such a unit cell is then obtained from Eq. (12) by replacing I with I_k:

$$\langle u_{ok}^2\rangle_{Av}/\Delta f = a^2 V_{bk}^2/I_k. \tag{21}$$

Each unit cell will contribute to the noise across the external load R_L. The noise-current contribution through R_L from cell k is then given by

$$\langle i_{Lk}^2\rangle_{Av} = \langle u_{ok}^2\rangle_{Av} R_s^2/(R_L R_s + R_s R_{sk} + R_{sk} R_L). \tag{22}$$

Since $R_s \ll R_L$ and $R_s \ll R_{sk}$, one obtains the following expression for the noise voltage across R_L:

$$\langle u_{Lk}^2\rangle_{Av} = \langle u_{ok}^2\rangle_{Av} R_s^2/R_{sk}^2. \tag{23}$$

The total noise voltage across R_L is then obtained by adding the contributions of each cell:

$$\langle u_L^2\rangle_{Av} = \sum_{k=1}^{N} \langle u_{Lk}^2\rangle_{Av}. \tag{24}$$

Replacing the summation by an integration and using Eq. (19) one obtains for the spectral density

$$\frac{\langle u_L^2\rangle_{Av}}{\Delta f} = a^2 V_b^2 \frac{R_s}{R_{sk}} \int_{I_{min}}^{\infty} I_k^{-1} \psi(I_k)\, dI_k, \tag{25}$$

where $\psi(I_k)dI_k$ denotes the probability that the cell k carries a current within the interval I_k to $I_k + dI_k$. Since $1/I_k \to \infty$ for $I_k \to 0$, the lower boundary of the integral is chosen to be $I_{min} = q/\tau_{tr}$, corresponding to the current if only one carrier is present within a cell. Cells which carry no current at all will not contribute to the noise.

It can be shown that Hines' theory represented by Eq. (12) is a special case of Eq. (25). If the dispersion of the distribution function $\psi(I_k)$ is very small, i.e., a very narrow distribution, then one can neglect the variation of $1/I_k$ over the dispersion width of $\psi(I_k)$ and approximate the integral in Eq. (25) by $1/\bar{I}_k$, where $\bar{I}_k = 1/N$. Inserting Eq. (19) into Eq. (25) leads then to the simple form of Eq. (12). No excess noise is present.

A different result will be obtained if the dispersion of $\psi(I_k)$ is so large that one cannot neglect the variation of $1/I_k$ in Eq. (25). It is evident from the integral in Eq. (25) that those cells with small values of I_k, e.g., the cells containing only a few carriers, will contribute excessively to the noise across R_L. This excess noise will

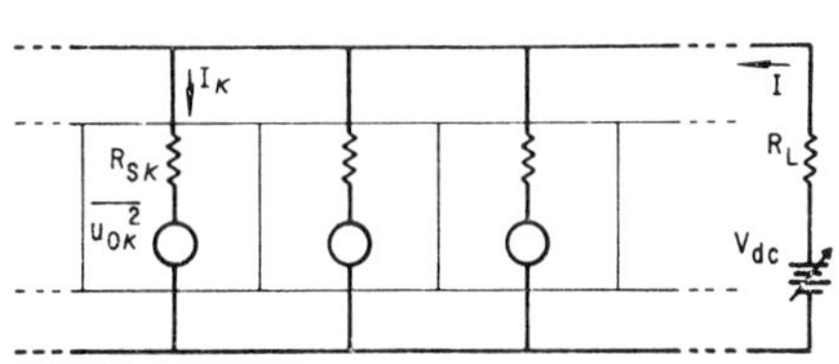

FIG. 11. Shockley's model of a reverse-biased pn junction used in explaining the excess noise at low current densities.

be especially high if most of the diode current I is carried by only a small fraction of the cells resulting in a large number of cells with extremely small currents.

Exactly such a distribution is expected from Shockley's model. To calculate the local fluctuations of V_b Shockley showed that it is reasonable to assume a Poisson distribution for the fluctuation of the number of impurity atoms within the cells. The distribution function $\psi(V_{bk})$ can be calculated to be[26]

$$\psi(V_{bk}) = (2\pi\sigma_b^2)^{-1/2} \exp\left[-\frac{(V_{bk}-\bar{V}_b)^2}{2\sigma_b^2}\right], \tag{26}$$

with the dispersion

$$\sigma_b^2 = b^2 \bar{V}_b^2 / W^3 N_a. \tag{27}$$

The constant $b = 0.425$ is obtained from the experimental relation[26] $V_b = 3.4\times 10^8\, N_a^{-b}$, where N_a denotes the acceptor density in an n^+p step junction. The distribution $\psi(I_k)$ is then calculated from Eqs. (20) and (26) to

$$\psi(I_k) = (2\pi\sigma_I^2)^{-1/2} \exp[-(I_k - \bar{I}_k)^2/2\sigma_I^2] \tag{28}$$

with the dispersion

$$\sigma_I = -\sigma_b/(R_{sk}+R_L) \approx -\sigma_b/R_{sk}. \tag{29}$$

For the 13-V diodes, one has $W = 0.35\ \mu$, $N_a = 3\times 10^{17}/\text{cm}^3$, and $R_{sk} \approx 1/2v_d\epsilon \approx 50$ kΩ. The resulting dispersion becomes then $\sigma_I \simeq 1.0\ \mu$A.

This result has to be compared with the average current per cell, $\bar{I}_k = I/N$. The diode ND11 of Fig. 4 with $D = 13\ \mu$ contains approximately 1100 cells. At $I = 1$ mA, one obtains $\bar{I}_k \approx 0.9\ \mu$A. This value of $\bar{I}_k \approx 0.9\ \mu$A is comparable to $\sigma_I \approx 1.0\ \mu$A. The dispersion of $\psi(I_k)$ is, therefore, very wide and one cannot neglect the variation of I_k in carrying out the integration of Eq. (25). As a consequence one expects large excess noise in diode ND11 at $I = 1$ mA, in contrast to the experimental results in Fig. 4 which show no excess noise.

A solution to this discrepancy between the experimental results and the excess noise model could be found either in a reduction of σ_I or in an increase of $\bar{I}_k$.

A reduction of σ_I could only be achieved if the impurity density fluctuations were overestimated in deriving Eq. (27). The calculation of the dispersion σ_b was based on the assumption that the number of impurities per cell, $W^3 N_a$, fluctuates from one cell to another according to a Poisson distribution with a mean deviation of $(W^3 N_a)^{1/2}$. Shockley[17] has shown that such a Poisson distribution can be expected from the thermodynamics of the diffusion process. Since small diodes, e.g., diode ND11 at $I = 1$ mA, are essentially free of excess noise one can conclude that Eq. (29) overestimates the dispersion σ_I considerably. The above excess-noise model would, therefore, be valid only for

[26] R. H. Haitz and F. M. Smits, IEEE Trans. Nucl. Sci. **NS-13**, 198 (1966).

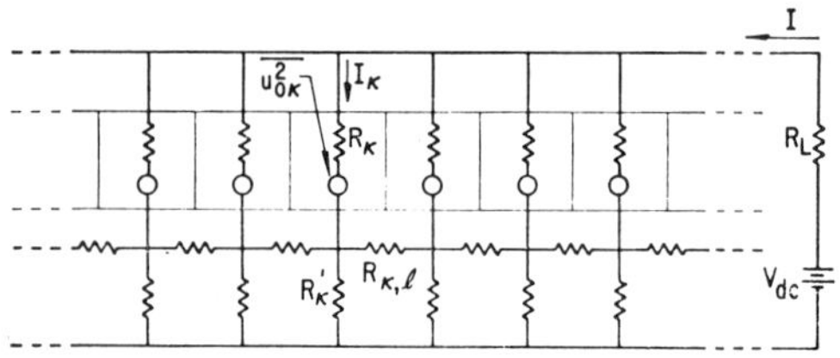

FIG. 12. Refined model of a reverse-biased *pn* junction which includes spreading resistance.

the case that the spatial fluctuation of the impurity density is at least one order of magnitude smaller than the fluctuation expected from a Poisson distribution.

Another model described in the following paragraphs could also account for the observed absence of excess noise in small diodes. This model is based on a splitting of the distribution function $\psi(I_k)$ leading to a small number of cells with relatively large currents I_k and to a large number of cells with no current. The average current of the conducting cells, $\bar{I}_k$, would then be considerably larger than in the preceding model. The dispersion σ_I, however, would be the same as calculated in Eq. (29).

Such a model with a very nonuniform current distribution can be obtained readily if the spreading resistance is introduced into the previous model. A linear model including the spreading resistance is shown in Fig. 12. This model consists of cells with a space-charge resistance R_k and a complex spreading resistance. The spreading resistance is represented by two contributions: (1) the resistance R_k' of a silicon filament with cross section W^2 and approximate length D and (2) the lateral resistance $R_{k,l}'$ with $l=k\pm1$ connecting the cell k with its neighboring cells $k-1$ and $k+1$. Since the ratio of $R_k'/R_{k,l}'\approx D/W$, it is evident that a considerable fraction of the current I_k is distributed laterally and flows to the common ground via the filament resistance of several neighboring cells. Such a current distribution raises the potential at the lower edge of the depletion layer of the neighbor cells to the potential of the conducting cell. Consequently, if the conducting cell is favored for current conduction by a smaller breakdown voltage due to a higher impurity density, then this advantage over its neighbor cells will be sustained even for increased currents. The voltage drop across the spreading resistance is not only reducing the current in the conducting cell but also preventing conduction in the neighbor cells. Such a nonuniform current distribution on a microscopic scale will be sustained until the current density in the conducting cell reaches high values such that the space charge of the mobile carriers offsets the impurity density fluctuations. It is evident that such a mechanism leads to a larger average current $\bar{I}_k$ of the concluding cells without simultaneously increasing the current dispersion σ_I. Such a model may, therefore, be suitable to explain the observed absence of excess noise.

At present it is not possible to decide which of the two models discussed with reference to Figs. 11 and 12 best describes the excess noise at low current densities. From studies of light emission in shallow avalanching *pn* junctions one is tempted to favor the latter model leading to a nonuniform current distribution on a microscopic scale. The light emission of microplasma-free avalanche junctions is never uniform on a microscopic scale but always rather spotty,[6] in agreement with the model of Fig. 12.

V. APPLICATIONS

The noise studies reported in the preceding sections lead to several conclusions regarding the application of avalanche diodes as noise generators, low-noise avalanche diodes, and low current avalanche diodes. The experimental results also allow a more accurate prediction of the minimum noise figure of a negative-resistance amplifier. The importance of low-frequency noise on the noise in avalanche microwave oscillators is also discussed briefly.

A. Noise Generator

The guard-ring avalanche diodes studied in Secs. III and IV have several features which make these diodes ideal noise sources. Their solid-state design guarantees low weight, small space, low power consumption, high reliability, and ease of operation. In the case of small diodes, which are free of excess noise at low currents and follow the $I^{-1/2}$ law of noise vs current, the reproducibility is considerably improved over conventional solid-state noise sources such as the randomly switching avalanche diode[27] or the noise diodes selected from commercial voltage regulator diodes. Since the scattering of the noise output of a batch of diodes can be kept well within $\pm5\%$ a selection of diodes is necessary only for applications requiring extremely narrow tolerances.

The noise characteristics of these avalanche diodes are rather unique. The open circuit spectral noise density is flat within at least ±1 dB from less than 10 Hz to at least 1 GHz.[23] The spectral noise density of diodes with a breakdown voltage of 7 to 20 V can readily reach values between 100 and 400 nV Hz$^{-1/2}$. For diodes with a series resistance of $R_s=100\ \Omega$ the avalanche noise would then be 40 to 50 dB above the thermal noise of R_s. Such a spectral density is still approximately 60 dB below the noise output of a randomly switching avalanche diode.[27] However, the temperature coefficient of the noise density and the upper cutoff frequency are orders of magnitude better than in the case of the randomly switching diode. Another feature of the stable-burning avalanche diode is the possibility of adjusting the noise level by varying the dc current.

It sounds contradictory to choose as noise source those diodes with the lowest noise output at a given

[27] R. H. Haitz, IEEE Trans. Electron Devices **ED-13**, 342 (1966).

breakdown voltage. This decision is determined by the fact that reproducibility and temperature dependence of the excess noise at low currents are rather substantial, making the larger guard-ring diodes inferior to diodes with diameters below 20 to 30 μ.

B. Low-Noise Avalanche Diodes

The same diodes described in the preceding section as ideal noise generators can also be used as low-noise avalanche diodes. As a matter of fact, the $I^{-1/2}$ law of noise vs current [Eq. (10)] describes the minimum noise of an avalanche diode with breakdown voltage V_b at any given current I. It is, therefore, possible to use one and the same diode either as a noise generator (small currents) or a low-noise avalanche diode (large currents).

In most applications of low-noise avalanche diodes the dynamic impedance of the diodes is not critical and impedance levels of the order of 100 Ω can be tolerated. In such cases the small diodes with no excess noise but relatively large impedance are superior to larger diodes.

On the other hand, if the emphasis is on low impedance it will be advantageous to trade impedance against excess noise. In most cases it will be sufficient to increase the diameter to the order of 100 μ in order to meet most impedance specifications of low-power avalanche diodes. The impedance of these intermediate size diodes at low currents, e.g., 250 μA to 1 mA, is far superior to the impedance of large diodes. This behavior is illustrated in Fig. 13 which shows the noise and impedance characteristic of a 9.9-V diode with an 82-μ diameter. At 250 μA, the resistance is approximately 17 Ω, decreasing to 9.5 Ω at 10 mA. A resistance of 9.5 Ω is expected from Eq. (5). For larger diodes the resistance at 250 μ is usually more than an order of magnitude larger while the resistance at currents of the order of 10 mA or more is usually only slightly smaller.

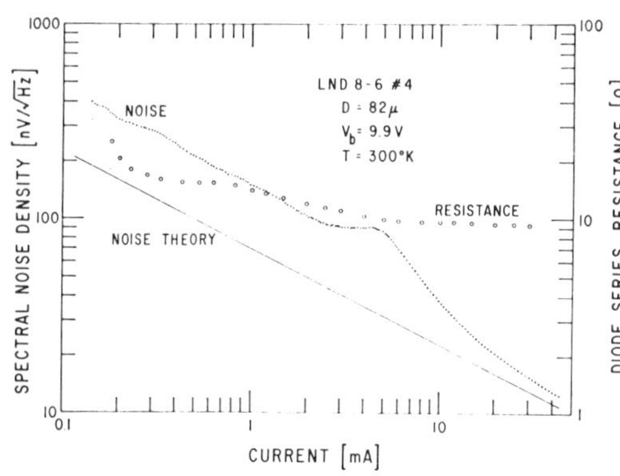

FIG. 13. Spectral-noise density and diode series resistance as a function of current for a 82-μ diode with especially uniform breakdown. The solid line represents Eq. (10).

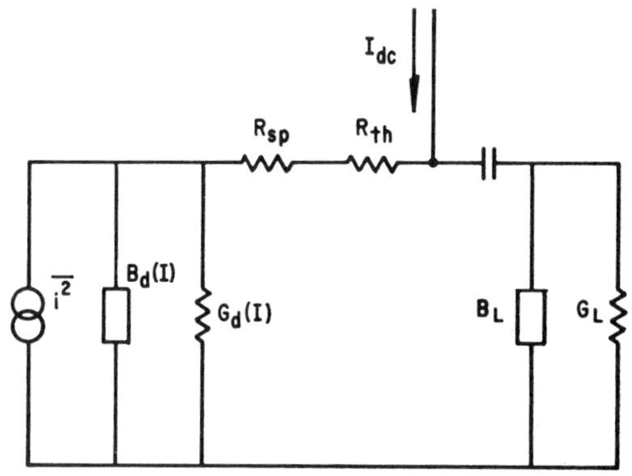

FIG. 14. Noise equivalent circuit of an avalanche transit time diode used as oscillator or negative resistance amplifier. The capacitor symbolizes that only noise at microwave frequencies can propagate directly from diode to load. Low-frequency noise is cut off.

C. Consequences for Oscillators and Amplifiers at Microwave Frequencies

Low-frequency avalanche noise will have serious consequences on the noise of avalanche microwave oscillators. This effect is illustrated in Fig. 14. The equivalent circuit of the oscillating avalanche diode consists of the current dependent conductance $G_d(I)=1/R_{sc}(I)$, which becomes negative above the avalanche frequency ω_a,[15] the admittance $B_d(I)$, the series resistances R_{sp} and R_{th}, and the noise current source $\langle i^2\rangle_{Av}$. The diode is biased by a dc current I_{dc}. The load G_L+jB_L is connected to the diode through a waveguide whose low-frequency cutoff is symbolized by the capacitor in Fig. 14.

The noise source covers the entire frequency range from dc to microwave frequencies. The noise at frequencies above the cutoff frequency of the waveguide can propagate directly to the load. Hines[28] has estimated this contribution. The low-frequency noise current, however, cannot propagate to the load directly but will amplitude and frequency modulate the oscillator through the current dependence of $G_d(I)$ and $B_d(I)$.[22] This amplitude and frequency modulation noise will be the dominating contribution to the oscillator noise. 419

From the above model one has to conclude that it is extremely important to keep the low-frequency noise of the oscillator to a minimum. The oscillator has, therefore, to be designed such that the breakdown is as uniform as possible and the excess noise is minimized. Furthermore, one may attempt to short out the noise current partly through a low pass filter in the bias loop. However, this can be achieved only if R_{sp} and R_{th} are small compared to $R_{sc}=1/G_d$. It is, therefore, required to reduce R_{th} by improving the heat removal from the breakdown region. It should be noted that R_{th} is frequency dependent and decreases rapidly for frequencies above $1/\tau_{th}$, where τ_{th} denotes the thermal time constant of the breakdown region given by Eq. (9).

[28] M. E. Hines, IEEE *Proc. Intern. Solid State Circuits Conf.*, Philadelphia, Pennsylvania, p. 82 (1966).

Hines[15] calculated the minimum noise figure of an avalanche diode negative resistance amplifier to 40.5 dB. However, he underestimated the constant τ_x considerably. Equation (37) of Hines[15] can be rewritten by replacing $\omega=\theta/\tau_{tr}$ where θ denotes the transit angle. For frequencies considerably above the avalanche frequency ω_a one obtains the following expression for the noise figure F:

$$F=1+(\kappa V_b/2m\theta^2)(\tau_{tr}^2/\tau_x^2)(q/kT). \quad (30)$$

[Equation (30) is corrected for a numerical error of a factor 2 in the denominator of Eq. (37) of Hines.[15]] Using the relation $\tau_{tr}/\tau_x=5.2$ found in Sec. IV.B one calculates the minimum noise figure for $V_b=10$ V and $\theta=\pi/2$ to $F_{min}\approx 23$ dB.

ACKNOWLEDGMENTS

The author is indebted to J. C. Lewis and C. H. Douglas for the fabrication of the avalanche diodes.

Reprinted from JOURNAL OF APPLIED PHYSICS, Vol. 39, No. 7, 3379–3384, June 1968

Noise of a Self-Sustaining Avalanche Discharge in Silicon: Studies at Microwave Frequencies

ROLAND H. HAITZ AND FRED W. VOLTMER*

Texas Instruments Incorporated, Dallas, Texas

(Received 25 January 1968)

The studies of avalanche noise reported by Haitz are extended to frequencies up to and above the avalanche frequency ω_a. It is found that the open-circuit spectral voltage density is flat within $\pm 5\%$ from less than 100 Hz up to frequencies approaching ω_a. Near ω_a the open-circuit spectral voltage density increases with frequency, goes through a maximum at ω_a and then decreases rapidly. The spectral power density is similarly flat but fails to exhibit an expected maximum at ω_a. For $\omega \gg \omega_a$ the spectral power density decreases with ω^{-4}. The experimental results are in good agreement with Hines' theory of avalanche noise for $\omega \ll \omega_a$ and $\omega \gg \omega_a$. The discrepancies observed near ω_a are thought to be caused by an oversimplification in Hines' theory which neglects internal diode losses. Compared with conventional noise sources such as temperature-limited diodes or gas discharge tubes, avalanche diodes have several advantages: larger noise output, larger bandwidth, lower $1/f$ noise, low power consumption, small size, and low weight.

I. INTRODUCTION

Recent interest in avalanche devices such as microwave oscillators and avalanche photodetectors has triggered extensive studies of the noise properties of a stable-burning, self-sustaining avalanche discharge. Several papers on the theory of avalanche noise have been published within the last two years. Hines[1] discussed the avalanche noise in terms of a simplified junction model using equal ionization rates for electrons and holes. Gummel and Blue[2] developed a more general theory for arbitrary junction profiles and realistic ionization coefficients $\alpha \neq \beta$. Their theory reduces to Hines' theory for the corresponding simplifying assumptions. In a different approach based on the multiplication of carriers injected into the junction both Tager[3–5] and McIntyre[6] calculated the low-frequency noise of a self-sustaining avalanche discharge by extending the carrier-multiplication factor to infinity.

The noise theory of Hines[1] has several advantages. It yields fairly simple analytical solutions for the spectral noise density at frequencies below and above the avalanche frequency ω_a, while McIntyre[6] gives only a low-frequency solution. The theory of Gummel and Blue[2] yields only a computer solution in its most general case. The analytical solution of Gummel and Blue's theory for a simplified but still more general model than in the case of Hines' theory is still too complicated for most practical purposes. It is, therefore, of considerable interest to determine the validity and accuracy of Hines'[1] theory.

There is also an interesting discrepancy between the theories of Hines[1] and Tager.[3] While Hines' theory predicts that the spectral power density decreases with ω^{-4} for $\omega \gg \omega_a$ Tager's theory predicts only an ω^{-2} cutoff. Furthermore, in Tager's theory the avalanche frequency depends on the carrier-multiplication factor M, which is a sensitive function of the diode leakage current for large values of M. Tager's theory, therefore, predicts a strong temperature dependence of the avalanche frequency, which is expected to be in the range of 10^6–10^8 Hz, while Hines' values for $\omega_a/2\pi$ are in the range of 10^9–10^{10} Hz and practically temperature independent. Avalanche-noise measurements at microwave frequencies are expected to clarify this discrepancy.

The experimental studies of avalanche noise at Texas Instruments were divided into two sections. In the first part[7] we investigated the validity and accuracy of Hines' low-frequency solution of avalanche noise. There it was found that Hines' theory very accurately describes the functional relationship between the open-circuit spectral voltage density and breakdown voltage and avalanche current. In spite of its crude simplifications Hines' low-frequency solution is accurate within better than a factor of two.

It is the subject of this paper to investigate the validity and accuracy of Hines' theory at microwave frequencies, particularly around and above the avalanche frequency ω_a. Preliminary results of this work were reported in an earlier publication.[8]

* Summer Development Program participant from University of California, Berkeley, Calif.

[1] M. E. Hines, IEEE Trans. Electron Devices **ED-13,** 158 (1966).

[2] H. K. Gummel and J. L. Blue, IEEE Trans. Electron Devices **ED-14,** 569 (1967).

[3] A. S. Tager, Fiz. Tverd. Tela **6,** 2418 (1964) [Sov. Phys.—Solid State **6,** 1919 (1965)].

[4] V. M. Val'd-Perlov, A. V. Krasilov, and A. S. Tager, Radiotekhn. i Elektron. **11,** 2008 (1966) [Radio Eng. Electron. (USSR) **11,** 1764 (1966)].

[5] A. S. Tager, Usp. Fiz. Nauk **90,** 631 (1966) [Sov. Phys.—Usp. **9,** 892 (1967)].

[6] R. J. McIntyre, IEEE Trans. Electron Devices **ED-13,** 164 (1966).

[7] R. H. Haitz, J. Appl. Phys. **38,** 2935 (1967).

[8] R. H. Haitz and F. W. Voltmer, Appl. Phys. Letters **9,** 381 (1966).

II. HINES' THEORY OF AVALANCHE NOISE

Hines' theory of avalanche noise leads to a fairly simple expression for the open-circuit spectral voltage density[7,9]

$$\langle u^2\rangle/\Delta f=(q\kappa^2/3m^2)(\tau_{tr}^2/\tau_x^2)(V_b^2/I)[1/(1-\omega^2/\omega_a^2)^2], \quad (1)$$

where the avalanche frequency ω_a is given by[9]

$$\omega_a^2=3mv_dI/\kappa\epsilon AV_b. \quad (2)$$

Here q denotes the electron charge; $m\approx5$ is a constant describing the power dependence of the ionization coefficient on field; $\kappa\approx0.5$ is defined by $V_a=\kappa V_b$, where V_a is the voltage across the avalanche region and V_b the breakdown voltage; $\tau_{tr}=W/v_d$ is the transit time of carriers through the space-charge layer of width W at the saturated drift velocity v_d; τ_x is the mean time between two ionization events of a carrier in the avalanche zone; I is the avalanche current; and A denotes the breakdown area. Since τ_{tr}/τ_x is constant for all practical purposes we can express the first two terms of Eq. (1) by a constant a^2:

$$a^2=(q\kappa^2/3m^2)(\tau_{tr}^2/\tau_x^2). \quad (3)$$

From noise measurements at low frequencies it is found that $a^2=3.3\times10^{-20}$ A/Hz, in good agreement with Eq. (3).

From Eq. (1) it is seen that the open-circuit spectral voltage density is frequency independent up to frequencies approaching ω_a. Near ω_a the open-circuit noise voltage increases rapidly with frequency. For an ideal diode with no internal losses the open-circuit spectral voltage density would reach infinity at ω_a. Above ω_a the noise voltage rapidly decreases with frequency according to the ω^{-4} relation of Eq. (1).

422

The avalanche frequency ω_a is usually located in the microwave range where noise measurements have to be made as power measurements. In a simple circuit containing no reactive elements one obtains the following expression for the spectral power density transferred to the load R_L:

$$w_n(\omega)=a^2(V_b^2/I)[1/(1-\omega^2/\omega_a^2)^2]\times[R_L/(R_{sc}+R_{sp}+R_L)^2]. \quad (4)$$

Here R_{sc} denotes the space-charge resistance which is given by Eq. (5) for small transit angles $\theta\approx\omega\tau_{tr}$:

$$R_{sc}=(0.7W)^2/2\epsilon v_dA(1-\omega^2/\omega_a^2)=R_{sc}^0/(1-\omega^2/\omega_a^2). \quad (5)$$

The term R_{sc}^0 is the low-frequency space-charge resistance. The spreading resistance R_{sp} can be approximated by

$$R_{sp}=\rho/2D, \quad (6)$$

with ρ denoting the resistivity of the lowly doped side of the p–n junction, usually the bulk resistivity, and D the diameter of the breakdown area A.

According to Eq. (4) the spectral power density is independent of frequency for $\omega\ll\omega_a$. A small frequency dependence, however, arises from the thermal resistance of the diode which is neglected in Eq. (4). This frequency dependence is of the order of 1 dB/decade or less.

For frequencies approaching ω_a, R_{sc} is the dominating contribution to the diode resistance. If the external load resistance R_L is tuned for maximum power transfer ($R_L=R_{sc}\gg R_{sp}$) then $w_n(\omega)$ is given by

$$w_n(\omega)=a^2V_b^2/4IR_{sc}^0(1-\omega^2/\omega_a^2). \quad (7)$$

For frequencies $\omega\gg\omega_a$ the spreading resistance R_{sp} is the dominating contribution to the diode resistance. Tuning R_L for maximum power transfer leads to

$$w_n(\omega)=a^2V_b^2/4IR_{sp}(1-\omega^2/\omega_a^2)^2. \quad (8)$$

For $\omega\gg\omega_a$, Eq. (8) predicts an ω^{-4} dependence of noise power on frequency.

III. EXPERIMENTAL RESULTS AND DISCUSSION

A. Diodes

During the studies of low-frequency avalanche noise[7] it became evident that special precautions have to be taken in order to prevent the generation of excessive noise resulting from nonuniform breakdown. The combination of both a guard ring to prevent edge breakdown and a small breakdown area to reduce material nonuniformities have led to satisfactory results.

The small p^+n guard-ring diodes manufactured for the avalanche-noise studies at microwave frequencies are essentially the same as the diodes used for low-frequency noise studies. The only difference is their reduced guard-ring size to cut down the junction capacitance. For a diode with a breakdown region of diameter D (equal to the inner diameter of the guard ring) the external diameter of the guard ring is equal to $D+42\ \mu$. The guard-ring width of 21 μ is partly due to the width of the guard-ring mask (10 μ) and partly due to underdiffusion ($2\times5.5\ \mu$). The diodes are manufactured by a double boron diffusion into n-type Si. The back contact is made by alloying AuSb. The front contact consists of an Al dot of diameter $D+11\ \mu$. The diodes are mounted in a microwave pill package.

Diodes with five different breakdown diameters are obtained from each slice: $D=13$, 36, 63, 89, and 113 μ. The small 13-μ diodes are of no use for these noise studies because of (1) their large diode resistance,

[9] As pointed out by C. A. Lee and R. L. Batdorf at the 1964 Solid State Device Res. Conf., Boulder, Colorado, and by Gummel and Blue[2] the numerical constant in Eq. (2) has to be 3 instead of 2. As a result, the numerical constant in the denominator of Eq. (1) has also to be replaced by 3.

(2) their relatively large uncertainty in breakdown area, and (3) their relatively large guard-ring capacitance. The 36-μ diodes are best suitable for avalanche-noise studies because their series resistance at microwave frequencies is fairly close to the characteristic impedance of a 50-Ω coaxial line. The larger diodes have a correspondingly smaller impedance and the impedance matching required to maximize the noise power transfer leads to correspondingly larger errors. Diodes with breakdown voltages of 9.5 V, 21.5 V, and 44 V were produced. The 21-V diodes were studied most extensively.

B. Measuring Techniques

The open-circuit spectral voltage-density measurements are carried out by biasing the diode through a 3-kΩ bias resistor which is large compared with the internal diode resistance. For measurements at frequencies below 10 kHz the avalanche noise is amplified by a preamplifier with low $1/f$ noise and then measured with a wave analyzer. For measurements between 10 kHz and 18 MHz the avalanche noise is measured directly with another wave analyzer. Between 18 MHz and 200 MHz the avalanche noise is down converted to 1 MHz and then measured with a wave analyzer. The open-circuit spectral voltage density for frequencies above 200 MHz is obtained from noise power and impedance measurements discussed in the remainder of this subsection.

The spectral noise power-density measurements are made with the coaxial system shown in Fig. 1. The diode is biased through a bias tee. The double stub tuner is used to tune for maximum noise power output

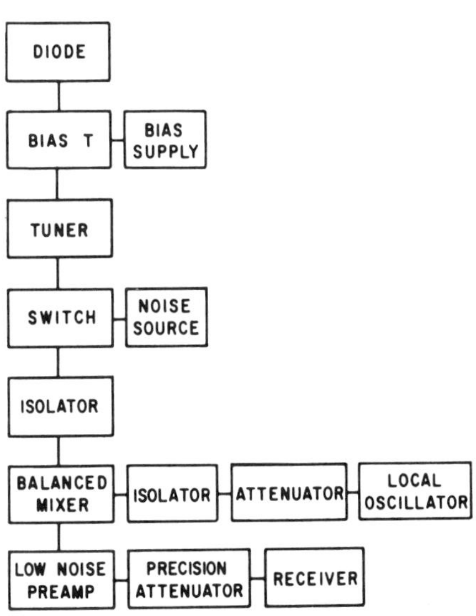

FIG. 1. Block diagram of circuit used to measure spectral noise power density.

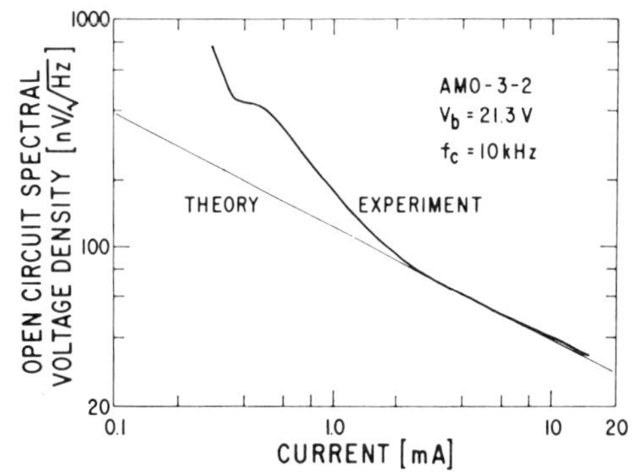

FIG. 2. Open-circuit spectral voltage density at low frequencies as a function of current.

by matching the diode impedance to the load. The mixer converts the avalanche noise to an i.f. frequency of 30 MHz. The mixer output is amplified and measured with a 30-MHz precision receiver. The avalanche noise is compared with the output of a gas discharge noise standard.

The diode impedance is measured using the standard standing-wave technique. A coaxial slotted line is inserted between bias tee and diode. A shorted package is used as reference short. The injected signal is kept low enough so that the peak ac current does not exceed 50% of the dc current.

C. Results and Discussion

The low-frequency open-circuit spectral voltage density shows the typical current dependence of small guard-ring diodes.[7] The results are plotted in Fig. 2. At high currents $(\langle u^2\rangle/\Delta f)^{1/2}$ decreases inversely with $I^{1/2}$ as predicted by Eq. (1). The straight line of Fig. 2 is calculated from Eqs. (1) and (3) using a value of $a^2=3.3\times10^{-20}$ A/Hz. This value of a^2 was found earlier[7] to be characteristic for a large variety of p^+n, n^+p and $n^+p\pi p^+$ diodes with breakdown voltages between 6 V and 90 V. The agreement between these measurements and earlier low-frequency noise studies[7] is excellent.

At currents below 2.5 mA the measured noise is larger than the noise predicted by Eq. (1). This discrepancy, which is typical for avalanche diodes at low-current densities, is not serious. It is caused by extremely small nonuniformities of the breakdown voltage.[10] As for the diode of Fig. 2, the breakdown uniformity of all diodes used in this noise study is checked by a measurement of the low-frequency noise-current characteristic. Only diodes with good agreement between calculated and measured noise at currents in the 3–20 mA range are used.

[10] R. H. Haitz, *Physics of Failure in Electronics*, Vol. V, pp. 447–456 (1967), Rome Air Development Center, USAF.

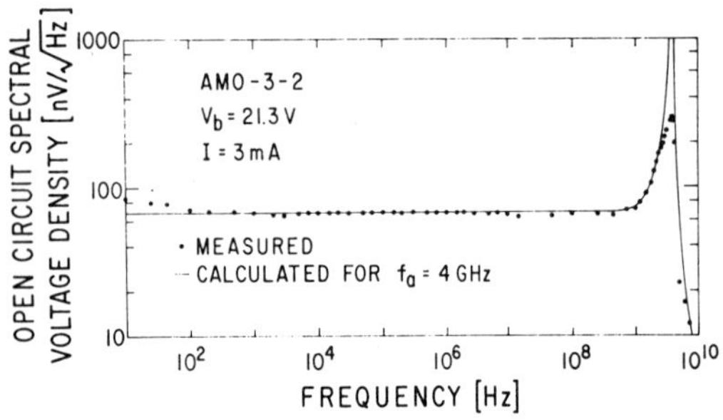

FIG. 3. Open-circuit spectral voltage density at a current of 3 mA as a function of frequency.

The open-circuit spectral voltage density, plotted in Fig. 3 as a function of frequency, exhibits the expected dependence on frequency. For $\omega \ll \omega_a$ the noise voltage is independent of frequency. The variation of $(\langle u^2\rangle/\Delta f)^{1/2}$ between 100 Hz and 1 GHz is less than $\pm 5\%$. Below 100 Hz the noise increases slightly. This increase is believed to be caused either by a $1/f$-type noise contribution superimposed on the avalanche noise or by 60-Hz pickup passing through the filter tails of the wave analyzer. Above 1 GHz the noise voltage increases rapidly with frequency reaching a maximum approximately four times above the low-frequency noise level. The solid line in Fig. 3 is calculated from Eqs. (1) and (3) for $a^2 = 3.3\times 10^{-20}$ A/Hz and $\omega_a/2\pi = f_a = 4.0$ GHz. The discrepancy between experimental and calculated noise near f_a is at least partly caused by simplifications in Hines' theory which completely neglects internal diode losses. It has to be mentioned here that the experimental points above 1 GHz are calculated from admittance and noise-power measurements which will be reported in the remainder of this section.

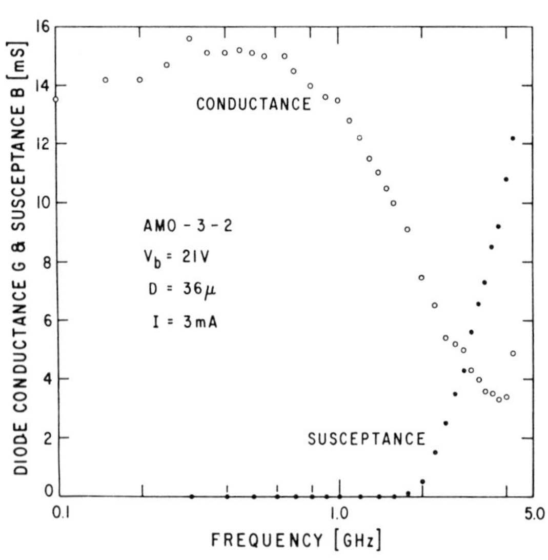

FIG. 4. Diode conductance and susceptance at a current of 3 mA as a function of frequency.

The admittance measurements needed to convert noise-power measurements to open-circuit spectral-voltage density data are plotted in Fig. 4 for diode AMO-3–2. At frequencies below 800 MHz the diode presents a fairly constant conductance of approximately 15 mS. The conductance drops rapidly above 1 GHz reaching a minimum of 3.3 mS near 4.0 GHz. The susceptance is negligibly small up to 2 GHz. Above 2 GHz the capacitive susceptance increases rapidly. The capacitive susceptance at frequencies below the conductance minimum (4.0 GHz) is not expected from Gilden and Hines'[11] avalanche-diode theory, which predicts an inductive admittance for $\omega \leq \omega_a$. Realizing, however, that the guard-ring area, and consequently the guard-ring capacitance, is four times bigger than the breakdown area one finds a quantitative agreement between measured and expected susceptance.

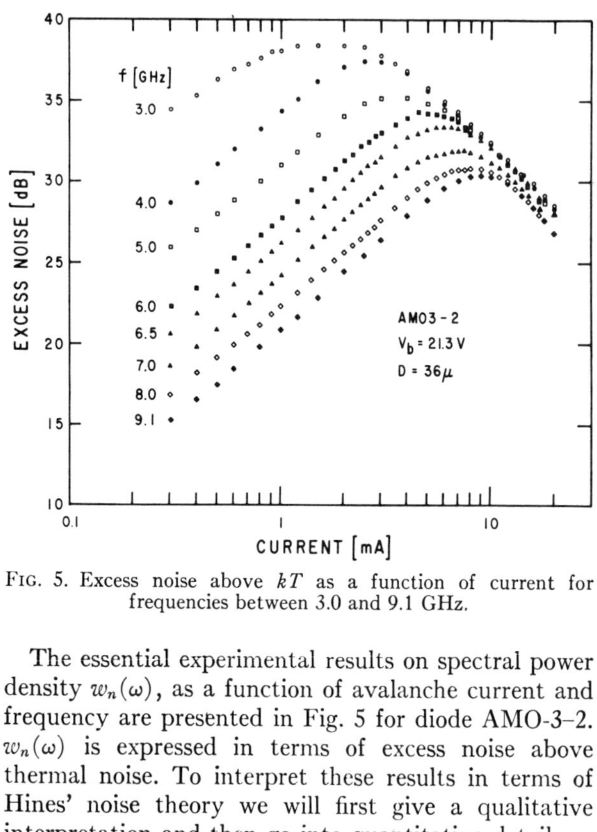

FIG. 5. Excess noise above kT as a function of current for frequencies between 3.0 and 9.1 GHz.

The essential experimental results on spectral power density $w_n(\omega)$, as a function of avalanche current and frequency are presented in Fig. 5 for diode AMO-3–2. $w_n(\omega)$ is expressed in terms of excess noise above thermal noise. To interpret these results in terms of Hines' noise theory we will first give a qualitative interpretation and then go into quantitative details.

From a qualitative inspection of Fig. 5 it is evident that all curves have several characteristic features in common. With increasing current $w_n(\omega)$ first increases approximately linearly, then goes through a maximum and finally decreases linearly with current. Furthermore, with increasing frequency the maximum of the spectral power density decreases and shifts to higher currents. Finally, for currents above the noise maxi-

[11] M. Gilden and M. E. Hines, IEEE Trans. Electron Devices **ED-13**, 169 (1966).

mum, $w_n(\omega)$ is practically frequency-independent while for currents below the noise maximum it is very strongly dependent on frequency.

The qualitative interpretation is continued by comparing the above-mentioned characteristic features with the functional relations predicted by Eqs. (7) and (8). As a first step it is assumed that at the current at which $w_n(\omega)$ reaches a maximum, $I_{\max}$, the avalanche frequency equals the measurement frequency: $\omega_a = \omega(I_{\max})$. Under this assumption, which will find its justification later in the quantitative interpretation, it is evident that ω_a increases with current as predicted by Eq. (2). Furthermore, one can conclude that $\omega \lesseqgtr \omega_a$ for $I \gtrless I_{\max}$, or in other words, to the right of the maxima $\omega < \omega_a$ and to the left $\omega > \omega_a$. For $\omega < \omega_a$ Eq. (7) predicts a frequency dependence in the immediate vicinity of ω_a, but frequency-independent noise for $\omega \ll \omega_a$. While the data of Fig. 5 do not confirm the first prediction they agree very well for $\omega \ll \omega_a$. For $\omega > \omega_a$ Eq. (8) predicts strongly decreasing noise with increasing frequency turning into an ω^{-4} dependence of $w_n(\omega)$ for $\omega \gg \omega_a$. This predicted strong frequency dependence of $w_n(\omega)$ for $\omega > \omega_a$ is obviously met for currents to the left of the noise maxima of Fig. 5.

Summarizing the qualitative interpretation it can be said that Eqs. (7) and (8) together with Eq. (2) fairly well predict the functional relationship between spectral power density and current and frequency.

The quantitative interpretation is discussed with reference to Fig. 6, where $w_n(\omega)$ is plotted as a function of frequency. Noise power-density data from Fig. 5 and data from additional measurements are plotted for three current levels: 0.5, 3, and 15 mA. In keeping with the admittance measurements of Fig. 4 and the noise data of Fig. 6 the avalanche frequency, $f_a = \omega_a/2\pi$, is chosen as 4.0 GHz for $I = 3$ mA. Using the $\omega_a - I$ relationship of Eq. (2) one obtains: $f_a = 1.8$ GHz for $I = 0.5$ mA and $f_a = 9.0$ GHz for $I = 15$ mA. The spectral power density $w_n(\omega)$ is calculated from Eq. (7) for $\omega < \omega_a$ and from Eq. (8) for $\omega > \omega_a$. The resulting curves

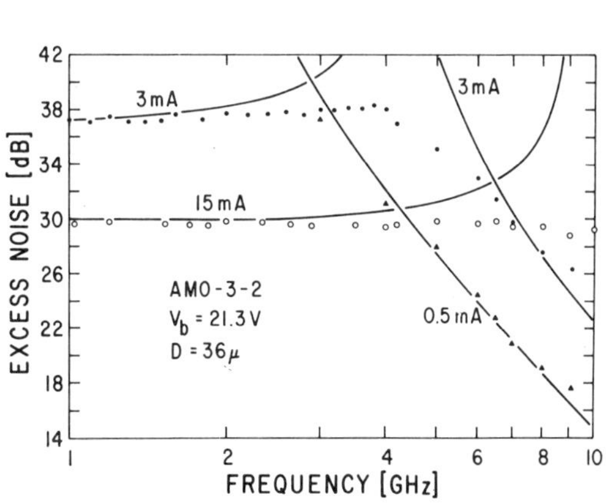

FIG. 6. Excess noise above kT as a function of frequency for currents of 0.5, 3, and 15 mA.

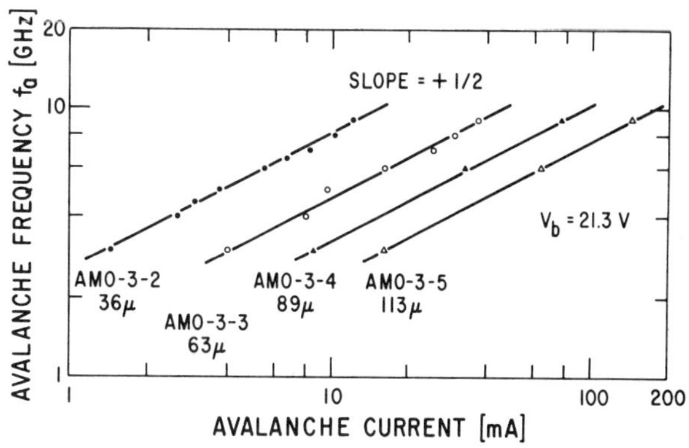

FIG. 7. Experimental values of avalanche frequency f_a as a function of current for diodes with diameters of 36, 63, 89, and 113 μ. The solid lines have a slope of $\frac{1}{2}$.

for $\omega < \omega_a$ fit the experimental data quite well for $\omega \ll \omega_a$ (low-frequency approach). In the vicinity of ω_a the experimental results do not increase as expected. This discrepancy is very likely caused by the fact that Hines' theory does not take internal losses into account. This simplification may be especially critical in the case of guard-ring diodes where the guard ring presents a substantial capacitive load. The spectral power density calculated from Eq. (8) for $\omega > \omega_a$ using $R_{sp} = 20\ \Omega$ is approximately 5 dB above the experimental results. Since the experimental points for $I = 0.5$ mA exhibit the expected ω^{-4} dependence over a range of 20 dB we have used a larger resistance value in order to fit the curves to the experimental points. A value of $R_s = 67\ \Omega$ (equal to the low-frequency diode resistance) is then used to calculate the two curves in Fig. 6 for $\omega > \omega_a$.

To conclude the quantitative interpretation we will investigate the current and area dependence of ω_a predicted by Eq. (2). The avalanche frequency $f_a = \omega_a/2\pi$ is determined experimentally from $w_n(\omega) - I$ plots similar to that of Fig. 6 by extrapolating the linear sections below and above $I_{\max}$. The intersection of the two lines yields a point for the $f_a - I$ curves of Fig. 7. The solid lines used to fit the experimental points have a slope of $\frac{1}{2}$ as predicted by Eq. (2). From these results it can be concluded that Eq. (2) yields the proper current dependence of ω_a.

The dependence of ω_a on area A is investigated by similar $f_a - I$ determinations obtained on larger diodes from the same slice. The three curves for diodes with diameters of 63, 89 and 113 μ are also plotted in Fig. 7. The resulting relationship between f_a and breakdown diameter is plotted in Fig. 8 for $I = 10$ mA. The solid curve in Fig. 8 is calculated from Eq. (2) for $m = 5$, $v_d = 10^7$ cm/sec, $\kappa = 0.5$, $\epsilon = 10^{-12}$ A·sec/V·cm and $V_b = 21.3$ V. For the two larger diodes the agreement between experimental points and calculated curve is quite good. However, with decreasing diameter the experimental points seem to drop further and further

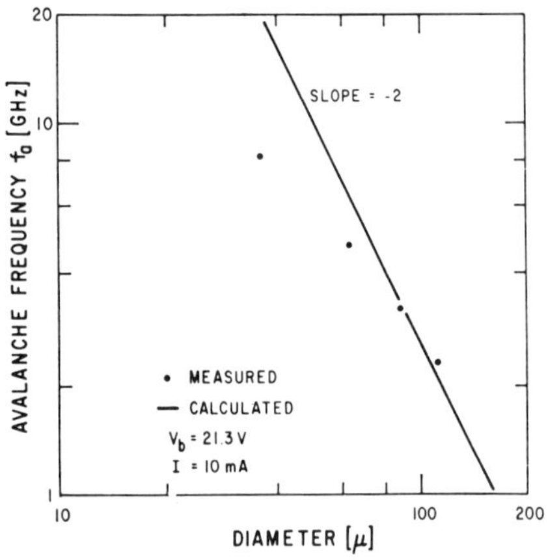

FIG. 8. Avalanche frequency f_a as a function of breakdown diameter. The line with a slope of -2 is calculated from Eq. (2).

below the calculated curve. This discrepancy for diodes with decreasing breakdown diameter is very likely caused by the increasing importance of parasitic capacitance from the guard ring and package.

The 20-dB drop in noise power following an ω^{-4} relationship and the good agreement between measured values of ω_a and the values calculated from Hines' theory strongly indicate that Tager's[3] approach of calculating the noise of a self-sustaining avalanche discharge using a carrier multiplication model is invalid.

IV. APPLICATION AS NOISE SOURCE

The experimental results presented in Figs. 3 and 6 and similar investigations by Constant *et al.*[12] and Dalman *et al.*[13] suggest the application of avalanche diodes as noise sources. Actually, such an avalanche-noise source is expected to have several advantages over conventional noise sources such as a temperature-limited diode or a gas-discharge tube. The following comments hold for a 20-V, 50-Ω diode in a coaxial system:

High Noise Output: Noise figures of 40 dB up to 3 GHz, 30 dB up to 10 GHz, and 25 dB up to 18 GHz can be obtained readily.

Large Bandwidth: A broadband coaxial noise source covering the range from 100 MHz to 12 GHz with a noise figure of 30 dB and an output variation of ±0.5 dB over the entire frequency range can be designed by slightly modifying the AMO-3-2 diode. A broadband coaxial source from 100 MHz to 18 GHz with a noise figure of 25 dB is also feasible.

Low 1/f Noise: Uniform open-circuit spectral voltage density down to frequencies of less than 10 Hz.

Low Power Consumption: An input power of 50 mW is sufficient to drive a diode generating noise 40 dB above kT up to 3 GHz, 300 mW for 30 dB up to 12 GHz and 800 mW for 25 dB up to 18 GHz. The increase in power with frequency is caused by the current dependence of ω_a.

Variable Noise Output: Since the noise output is a linear function of I^{-1} the noise output level can be readily varied thus greatly simplifying a large fraction of conventional noise measurements.

Pulsed Operation: The diodes can be readily pulsed into breakdown with subnanosecond rise and fall times. No voltage spikes are required to trigger the avalanche discharge.

High Reliability: The mean time between failure of such guard-ring avalanche diodes is expected to greatly exceed the failure time of conventional gas discharge tubes or temperature-limited noise diodes.

Large Operating Temperature Range: The avalanche noise diodes are expected to operate satisfactorily over a range from less than −200°C to more than +200°C with only a small temperature dependence of the noise output ($<10^{-2}$/°C).

Small Size and Low Weight: Size and weight of the complete noise source will be determined by coaxial holder and power supply. Holder size can readily be kept below 10 cm³. Compared with a conventional gas-discharge tube the size of the power supply can be greatly reduced because of the small power consumption of an avalanche noise diode.

The above cited advantages of avalanche noise sources make this device the ideal noise source for all applications with critical restrictions in size, weight, power consumption, reliability, output level and bandwidth. It is believed that the frequency range of specially designed avalanche noise diodes can be extended far into the millimeter wave region.

V. CONCLUSIONS

The studies of avalanche noise reported here and in an earlier publication[7] lead to the conclusion that Hines' noise theory,[1] in spite of its simplifications, yields a fairly accurate description of avalanche noise. For frequencies $\omega \ll \omega_a$ Eqs. (1) and (4), which are based on Hines' theory, describe the open-circuit spectral voltage density and the spectral power density quite well. For $\omega \gg \omega_a$ Eq. (4) predicts the proper ω^{-4} dependence of the spectral power density but predicts a value approximately 5 dB above the experimental points. Around ω_a Hines' theory is most inaccurate since it neglects internal diode losses. Part of the inaccuracy near ω_a and above is also thought to be caused by the relatively large guard-ring capacitance of the diodes under investigation. Small mesa diodes recently of interest as avalanche transit time oscillators should be more suitable for accurate noise measurements around and above ω_a.

[12] E. Constant, R. Gabillard, A. Hautducoeur, and A. Chadelas, Compt. Rend. **262B**, 16 (1966).
[13] G. C. Dalman and L. F. Eastman, presented at the 1967 Intern. Electron Device Meeting, Washington, D. C.

426

Noise in IMPATT Diode Amplifiers and Oscillators

HANS-JÖRG THALER, MEMBER, IEEE, GERHARD ULRICH, AND GERHARD WEIDMANN

Abstract—The results of experimental and theoretical investigations of the noise characteristics of IMPATT diode amplifiers and oscillators are presented. The oscillator noise is shown to consist of three different contributions: modulation noise, selectively amplified primary noise, and conversion noise. The influence of the active device nonlinearity and load circuit parameters is discussed in detail. The experimental results are in good agreement with the theoretical predictions. It is especially pointed out that the large correlation between AM and FM fluctuations, usually measured in IMPATT oscillators, indicates nonoptimum AM noise performance. Experimental techniques for achieving optimum AM noise performance are demonstrated (orthogonal noise tuning).

By a simple extension of the model, the noise behavior of an injection phase-locked oscillator can be described. The calculated AM and FM noise power spectra for the synchronized oscillator are also shown to be in good agreement with experimental results.

Finally the signal-to-noise ratio for current modulated IMPATT oscillators is investigated and optimization is demonstrated.

I. Introduction

At THE present, state-of-the-art IMPATT diodes are able to deliver up to 1 W CW at X band. At these power levels IMPATT oscillators are competing with microwave tubes and have advantages as to size, reliability, and costs. A serious disadvantage, however, is the high noise level of IMPATT sources.

The oscillator noise is usually characterized by the amplitude and frequency fluctuations of the microwave signal. These fluctuations may arise from different sources. The correlation existing between AM and FM noise gives evidence to what extent AM and FM noise originate from the same source.

Common noise measuring techniques do not give sufficient information about the different mechanisms contributing to the IMPATT oscillator noise. Moreover, progress in the understanding of the IMPATT oscillator noise is hampered by the fact that the results of a measurement strongly depend on the nonlinear characteristics of the active device and on the properties of the load circuit. That is, the operating point must be specified in order to be able to compare data from different authors.

Manuscript received June 30, 1970; revised January 25, 1971. This work was supported by the Bundesministerium für Bildung und Wissenschaft under Contract RFF 2017/22 (the chapter on IMPATT noise) and by the Deutsche Forschungsgemeinschaft under Contract Mu 247/14 (the chapter on modulation of IMPATT oscillators). The work of H. J. Thaler was supported by Siemens AG, Munich, Germany.

The authors are with the Institut fur Technische Elektronik, the Technical University of Munich, Munich, Germany.

Such a specification is possible on the basis of a general oscillator model applied in this work. If the parameters of the RF and LF oscillator equivalent circuits are fitted to the special requirements of IMPATT oscillators they are able to account for the power and cross-correlation spectra measured for these oscillators.

In the second section the basic features of small and large signal behavior of IMPATT diodes and of the avalanche noise mechanism are reviewed. A discussion of IMPATT amplifier noise follows in Section III. In Section IV the different noise contributions are separated and the noise spectra are calculated by means of perturbation theory. An experimental verification of the expressions derived theoretically is obtained by measuring the power spectra and the cross-correlation coefficient of AM and FM noise.

AM noise and FM noise of the synchronized oscillator are discussed in Section V. The treatment is based on a natural extension of the free-running oscillator model.

Finally, with regard to systems applications, the signal-to-noise ratio of current modulated IMPATT oscillators is calculated in Section VI.

II. Principles of IMPATT Diode Operation

The amplification and generation of a microwave signal with IMPATT diodes is based on the negative resistance presented by the diode in the microwave range.

This negative resistance arises from the transit time delay of carriers injected into a drift space by avalanche multiplication. The physical mechanism was first described by Read [1]. Small signal impedance calculations for simple diode structures were done by Gilden and Hines [2], Misawa [13], and Hoefflinger [4]. The dependence of the large signal impedance on bias current and RF adplitude for a simple Read diode structure is known from large signal calculations by several authors [5], [6], [7]. As the Read model is a reasonable approximation for actual IMPATT diodes, the results of these calculations are also applicable to predict qualitatively the behavior of actual IMPATT diodes.

The carrier multiplication in the avalanche region of an IMPATT diode is an inherently noisy process. A calculation of the mean-squared noise current based on the small signal model described by Gilden and Hines [2] has first been carried out by Hines [8]. An equivalent expression for the mean-squared open-circuit noise voltage $\overline{e^2}$ has been given by Haitz [9]:

$$\frac{\overline{e^2}}{B} = a^2 \frac{V_B^2}{I_0} \cdot \frac{1}{\left[1 - \left(\frac{\omega}{\omega_a}\right)^2\right]^2} \tag{1}$$

where V_B denotes the breakdown voltage, I_0 the diode bias current, B the bandwidth, and $\omega_a = 2\pi f_a$ the avalanche frequency [2]. The calculated frequency dependence of the noise voltage has been confirmed experimentally [9]. From these data the constant a^2 containing technological and geometrical constants is determined to be 3.3×10^{-20} As^{-1}. Below the avalanche frequency the noise voltage apart from thermal effects is white, possesses a pole at ω_a, and decays for $\omega > \omega_a$ with ω^{-4}. For operating points above the avalanche frequency e^2 increases with increasing dc current (see [9]).

III. Noise Figure of a Reflection-Type IMPATT Amplifier

To make use of the diode's negative resistance for RF power amplification, the diode is inserted into a resonator. Incident and reflected waves are separated by means of a circulator as described in [10]. When the diode reactance jX_D is compensated by the load reactance jX_L the power amplification factor G is given by

$$G = \left[\frac{R_D + R_S - R_L}{R_D + R_S + R_L}\right]^2 \tag{2}$$

where R_L is the load resistance as seen by the active element at its terminals and R_S denotes the losses in the diode and the circuit. Amplification is only achieved for
428 $-R_L < R_D + R_S < 0$. Otherwise either oscillations build up when $-R_L > R_D + R_S < 0$ or a gain smaller than 1 results for $R_D + R_S > 0$.

The noise figure F of the amplifier is given by the ratio of the noise power $N_R + N_0$ at the output terminals to the noise power $N_0 = GkT_0B$, which would exist at the output if there were noise free amplification.

The noise power N_R resulting from the avalanche mechanism is given as

$$N_R = \overline{i^2_{\text{no}}} \cdot R_L = \overline{e^2} R_L / (R_L + R_S + R_D)^2. \tag{3}$$

To obtain a convenient expression for the amplifier noise figure F, the small signal diode resistance R_D and the mean square noise current i^2_{no} as given by Hines [8] are combined with (2) yielding the following relation:

$$F = \frac{N_R}{N_0} = \frac{c^2}{\omega^2 - \omega_a^2} \cdot \frac{R_D}{R_D + R_S} \cdot \frac{G - 1}{G}. \tag{4}$$

The noise power N_0, contributed by the input resistance is negligible compared to N_R. The factor of proportionality is

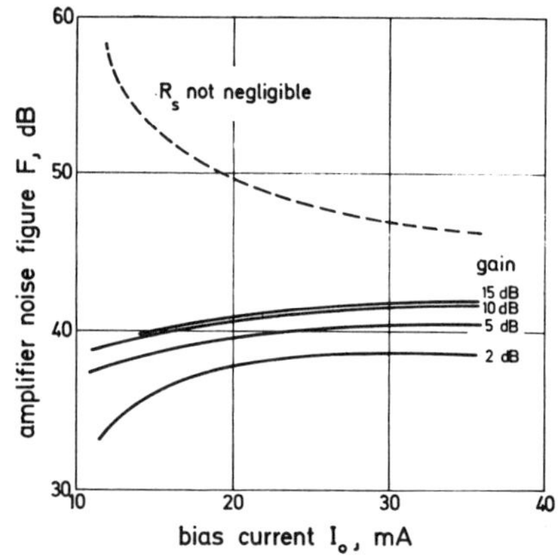

Fig. 1. Noise figure F of an IMPATT diode amplifier as a function of bias current for different gain factors G.

$$c^2 = 2qE_a v_d^2 / kT_0 l_a m \tag{5}$$

where

- q electron charge;
- l_a length of avalanche region;
- E_a electrical field strength in the avalanche region;
- v_d saturated drift velocity;
- $m = \left.\frac{d\alpha}{dE}\right|_{E=E_a} \cdot \frac{E_a}{\alpha}$;
- α ionization coefficient (see [8]).

The dependence of the noise figure on bias current, gain, and frequency has been investigated experimentally. If the loss resistance R_S can be neglected in comparison with the diode resistance R_D, then the noise figure must increase with increasing current, as $\omega_a^2 \sim I_0$. Thus the avalanche frequency, at which the noise maximum occurs, approaches the operating frequency. This behavior is apparent from Fig. 1, where the noise figure F is plotted as a function of current with gain as a parameter. These data were obtained by interpolation from measurements of F at constant currents, when the gain was varied by tuning the amplifier. In agreement with the theoretical expression (3), the noise figure must increase with increasing gain and ranges in magnitude between 36 and 42 dB. Therefrom it is concluded that the avalanche diode amplifier must be driven by small currents and at a low gain, if the noise level is to be kept low.

When R_S cannot be neglected in comparison with R_D the current dependence of the diode resistance must be taken into account. Then the noise figure is not only higher, but may even decrease with increasing current, as illustrated by the broken line in Fig. 1. The latter resulted when large losses were produced by insertion of absorbing material next to the diode. A decrease of the noise figure with increasing current always indicates

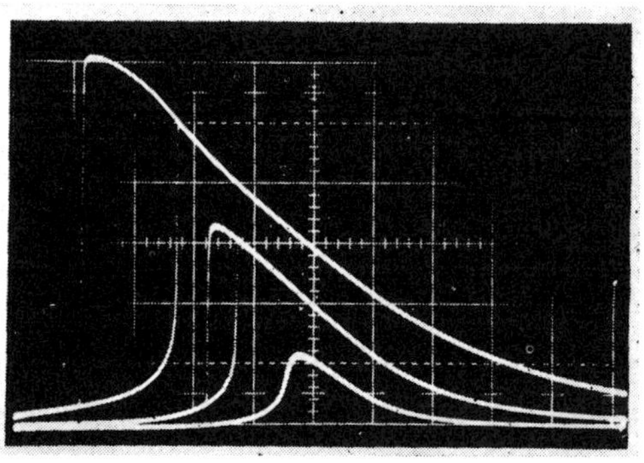

Fig. 2. Shift of the gain versus frequency curve of an IMPATT amplifier at higher power levels. Center frequency: 9.12 GHz. Horizontal scale: 10 MHz/div. Vertical scale: output power, arbitrary units.

that strong losses are present in the diode or amplifier circuit.

The measurement of the noise figure as a function of frequency in the range from 8 to 11 GHz did not show the expected decrease of F (about 5 dB in this frequency range) with increasing frequency according to (3). This is believed to be due to increasing losses within the circuit.

With increasing amplitude of the input signal the amplitude dependence of the diode impedance affects the amplification. As can be seen in Fig. 2, saturation effects occur and the resonance is shifted towards lower frequencies when the power delivered by the amplifier is comparable to the power which would be delivered by the self-oscillating diode. From this performance it can be deduced that the diode impedance becomes more positive with increasing RF amplitude, that is, $\partial R_D/\partial A > 0$ and $\partial X_D/\partial A > 0$, if the load circuit is represented by a simple R–L–C series circuit. This behavior is expected for operating frequencies ω above the avalanche frequency ω_a [10], in agreement with the aforementioned large signal calculations [5], [6], [7].

IV. Oscillator Noise

A negative resistance amplifier becomes an oscillator by means of self-excitation, when the load resistance in the series circuit becomes smaller than the magnitude of the real part of the active impedance. If the latter decreases with increasing RF amplitude, a stable operating point is reached. The investigation of the IMPATT diode amplifier has already shown that both the real and the imaginary part of the active diode impedance depend on the RF amplitude. It is shown below that this amplitude dependence of the diode reactance essentially affects the oscillator noise performance.

The microwave signal emitted by a real oscillator is not exactly sinusoidal but exhibits small amplitude and phase or frequency fluctuations, that is, the oscillator is noisy. It may be expressed by the following equation:

$$x(t) = A(t) \cos [\omega t + \phi(t)] \qquad (6)$$

with

$$\phi(t) = \int^t \Delta\omega(t')dt'.$$

The power spectrum of $x(t)$ therefore only approximately equals the δ-function holding for a purely harmonic signal. Usually the oscillator noise is not characterized by the RF power spectrum, but by the power spectra of the fluctuating quantities $A(t)$ and $\Delta\omega(t)$ or $\phi(t)$, respectively. The fluctuations of $A(t)$ are called AM noise, and those of $\Delta\omega(t)$ or $\phi(t)$ are called FM or phase noise, representing a stochastic amplitude and frequency or phase modulation of the carrier.

A. *Noise Mechanisms*

The deviation of the oscillator signal from the ideal sinusoidal shape arises from two different noise mechanisms. According to Spälti [11] the oscillator signal may be considered as a RF noise signal amplified very selectively; such a RF noise signal is always present in an oscillator and may be caused either by resistance noise or by the noise of the active element. The high selectivity or high Q factor is due to the fact that the effective damping in the oscillator circuit because of the dedamping is very small but always finite.

The second contribution to the oscillator noise arises from the modulation of the oscillator by a noise signal, which is caused by a low-frequency fluctuating signal that is usually produced in the active element itself and modulates the parameters characterizing the active element. Which of the two noise mechanisms is present or predominates in a special oscillator must be investigated in each particular case. An especially suited tool is the investigation of the correlation existing between AM and FM noise. If the modulation noise due to one low-frequency noise mechanism prevails, amplitude and frequency fluctuations of the oscillator signal are entirely correlated. On the other hand, AM and FM noise originating from the mechanism proposed by Spälti [11] or Edson [12] are not correlated. Moreover, the correlation is affected by the nonlinear diode impedance, the influence of which will be pointed out below.

B. *Calculation of the Oscillator Noise*

1) Oscillator Model: The calculation of the oscillator noise is based on the oscillator model shown in Fig. 3(a). The active element is represented by a nonlinear impedance $Z_D(A)$ depending on the RF amplitude. The load resistance is transformed by the usually multiple resonant circuit N into the load impedance $Z_L(\omega)$, which is seen by the active element at its terminals and depends on frequency.

The frequency dependence of the impedance of the active device is negligible in comparison to the frequency dependence of $Z_L(\omega)$. All noise sources in the

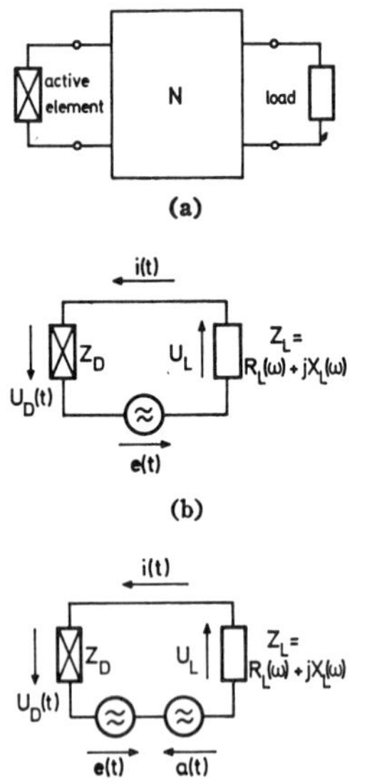

Fig. 3. (a) Oscillator model. (b) Equivalent circuit of the free running oscillator. (c) Equivalent circuit for the injection phase locked oscillator.

oscillator circuit are contained in the noise voltage generator with the voltage $e(t)$, the spectrum of which comprises all frequencies. For IMPATT diode oscillators the noise produced by the avalanche diode by far predominates all other noise sources in the load circuit, which therefore may be neglected.

If the RF amplitude level of the oscillator is sufficiently small, $e(t)$ equals the noise voltage as determined from measurements of the amplifier noise F and the diode and load impedances. In the oscillator equivalent circuit (Fig. 3(b)) the current $i(t)$ corresponds to the oscillator signal $x(t)$. The analysis of the oscillator starts with the application of Kirchhoff's law to the equivalent circuit.

$$V_D(t) + V_L(t) + e(t) = 0. \tag{7}$$

The calculation of the voltages across the lumped elements and insertion into (7) yields a differential equation in the time domain for the oscillator current $i(t)$ with $e(t)$ representing an external force term. A more simple analysis, however, is possible in the frequency domain if $A(t)$ and $\phi(t)$ or $\Delta\omega(t)$, respectively, are slowly varying functions of time, that is if the following conditions hold:

$$\frac{dA(t)}{dt} \ll A\omega_0, \qquad \frac{d}{dt}\,\Delta\omega(t) \ll \omega_0^2.$$

This means a restriction to low modulation frequencies, that is, to noise sidebands near the carrier. In this case it is justified to speak of an instantaneous amplitude and frequency of the oscillator current. The voltages across each element of the equivalent circuit are then calculated to a good approximation in a quasi-stationary manner. Stated otherwise, the voltage across the active element is given as the product of current and active impedance at the instantaneous amplitude, while the voltage across the load impedance is given as the product of current and load impedance at the instantaneous frequency.

In the case of an unperturbed oscillator, that is $e(t)=0$, the voltage balance condition (7) may be directly interpreted as an impedance balance equation. Therefrom the well-known condition for stationary oscillation is deduced:

$$\begin{aligned} Z_D(A_0, I_0) + Z_L(\omega_0) &= 0 \\ -R_D(A_0) = R_L(\omega_0) &= R_0 \\ -X_D(A_0) = X_L(\omega_0) &= X_0. \end{aligned} \tag{8}$$

Equation (8) fixes the oscillating frequency ω_0 and amplitude A_0 in the stationary operating point as a function of the diode bias current I_0.

2) *Representation of the Perturbing Signal $e(t)$ Near the Carrier:* In order to be able to set up a voltage balance equation for the perturbed oscillator in the frequency domain, the primary noise signal must be represented in a form suitable for the frequency domain. Therefore, the perturbing signal $e(t)$ is expressed as sidebands of a carrier at the frequency ω_0 being amplitude and frequency modulated by low-frequency noise signals. In general for a narrow-band noise signal the following representation holds:

$$\begin{aligned} e(t) &= e_c(t)\cos\omega_0 t - e_s(t)\sin\omega_0 t \\ &= \mathrm{Re}\left\{[e_c(t) + je_s(t)]\exp j\omega_0 t\right\} \\ &= \mathrm{Re}\left\{n(t)\exp j\omega_0 t\right\}. \end{aligned} \tag{9}$$

The statistical characteristics (correlation functions and power spectra) of the modulation quantities $e_c(t)$ and $e_s(t)$, which are equivalent to the noise signal $e(t)$, are deduced from the power spectrum $S_{ee}(\omega)$ of $e(t)$. If $e(t)$ is a stationary noise signal then the power spectra $S_{cc}(\Omega)$ and $S_{ss}(\Omega)$ and the cross spectrum $S_{cs}(\Omega)$ of $e_c(t)$ and $e_s(t)$ are given by

$$\begin{aligned} S_{cc}(\Omega) &= S_{ss}(\Omega) = S_{ee}(\omega_0+\Omega) + S_{ee}(\omega_0-\Omega) \\ S_{cs}(\Omega) &= -\,S_{sc}(-\Omega) \\ &= -\,j\{S_{ee}(\omega_0+\Omega) - S_{ee}(\omega_0-\Omega)\} \end{aligned} \tag{10}$$

where Ω denotes the angular frequency deviation from the carrier.

C. Oscillator Perturbation Theory

The following relation is obtained from the voltage balance equation for the perturbed oscillator:

$$\begin{aligned} \{A(t)\exp j[\omega_0 + \Delta\omega(t)]t\}\cdot\{Z_D(A, I) + Z_L(\omega)\} \\ + [e_c(t) + je_s(t)]\exp j\omega_0 t = 0. \end{aligned} \tag{11}$$

The noise signal $e(t)$ effects small statistical deviations $\Delta A(t)$ from A_0 and $\Delta\omega(t)$ from ω_0. The impedances $Z_D(A, I)$ and $Z_L(\omega)$ can be expanded in Taylor series around the operating point; for sufficiently small deviations these expansion may be stopped with the linear terms. According to (8) the zero order terms in the impedance expansions cancel. Then, if one allows also for deviations $\Delta I(t)$ of the diode bias current, (11) leads to the following general perturbation equation:

$$\{[A_0 + \Delta A(t)] \exp j(\omega_0 + \Delta\omega(t))t\} \cdot \left\{ \frac{\partial Z_D}{\partial A}\bigg|_{A=A_0} \cdot \Delta A(t) + \frac{\partial Z_D}{\partial I}\bigg|_{I=I_0} \Delta I(t) + \frac{dZ_L}{d\omega}\bigg|_{\omega=\omega_0} \Delta\omega(t) \right\} + \{e_c(t) + je_s(t)\} \exp j\omega_0 t = 0. \tag{12}$$

In practice one always has $\Delta A(t) \ll A_0$ and $\Delta\omega(t) \ll \omega_0$. With these assumptions and neglecting higher order terms in the small quantities two linear perturbation equations are obtained from the separation of (12) into real and imaginary parts:

$$A_0 \left[\frac{\partial R_D}{\partial A} \Delta A(t) + \frac{\partial R_D}{\partial I} \Delta I(t) + \frac{dR_L}{d\omega} \Delta\omega(t) \right] + e_c(t) = 0 \tag{13a}$$

$$A_0 \left[\frac{\partial X_D}{\partial A} \Delta A(t) + \frac{\partial X_D}{\partial I} \Delta I(t) + \frac{dX_L}{d\omega} \Delta\omega(t) \right] + e_s(t) = 0. \tag{13b}$$

These equations are solved for $\Delta A(t)$ and $\Delta\omega(t)$:

$$\Delta A(t) = -\frac{e_c(t)\dfrac{dX_L}{d\omega} - e_s(t)\dfrac{dR_L}{d\omega} + \Delta I(t) A_0 \left[\dfrac{\partial R_D}{\partial I}\dfrac{dX_L}{d\omega} - \dfrac{\partial X_D}{\partial I}\dfrac{dR_L}{d\omega} \right]}{A_0 \left[\dfrac{\partial R_D}{\partial A}\dfrac{dX_L}{d\omega} - \dfrac{\partial X_D}{\partial A}\dfrac{dR_L}{d\omega} \right]} \tag{14a}$$

$$\Delta\omega(t) = \frac{e_c(t)\dfrac{\partial X_D}{\partial A} - e_s(t)\dfrac{\partial R_D}{\partial A} + \Delta I(t) A_0 \left[\dfrac{\partial R_D}{\partial I}\dfrac{\partial X_D}{\partial A} - \dfrac{\partial X_D}{\partial I}\dfrac{\partial R_D}{\partial A} \right]}{A_0 \left[\dfrac{\partial R_D}{\partial A}\dfrac{dX_L}{d\omega} - \dfrac{\partial X_D}{\partial A}\dfrac{dR_L}{d\omega} \right]}. \tag{14b}$$

This linear relationship between $\Delta A(t)$ and $\Delta\omega(t)$ on one side and $e_c(t)$, $e_s(t)$, and $\Delta I(t)$ on the other is illustrated in a flow graph (Fig. 4). As is seen, $e_c(t)$ produces amplitude fluctuations directly, which by means of a conversion term due to the nonlinear diode reactance also result in frequency fluctuations. In an analogous manner $e_s(t)$ first effects frequency fluctuations, which are then also responsible for amplitude fluctuations because of the FM–AM conversion due to the frequency dependence of the load resistance. The modulating signal $\Delta I(t)$ effects amplitude and frequency fluctuations. The modulation sensitivities are also affected by conversion. These AM–FM and FM–AM conversion terms considerably influence the oscillator noise performance and besides the modulation mechanism are the cause for the correlation existing between amplitude and frequency fluctuations.

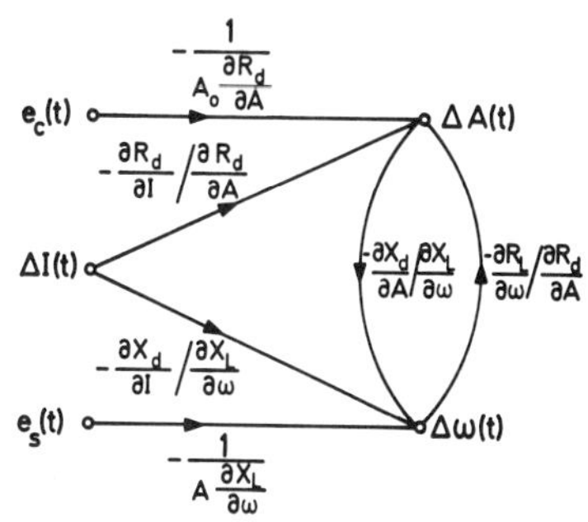

Fig. 4. Flow graph representation of the noise mechanisms in oscillators.

D. Calculation of the Noise Spectra

As is further seen from the flow graph (Fig. 4), the power spectra $S_{AA}(\Omega)$ and $S_{\omega\omega}(\Omega)$ for the oscillator noise quantities $\Delta A(t)$ and $\Delta\omega(t)$ may be taken as the output spectra of a linear system and can be calculated from the power spectra and the cross spectra of the input noise signals and the transfer functions of the system. The input spectra $S_{cc}(\Omega)$, $S_{ss}(\Omega)$, $S_{cs}(\Omega)$, and $S_{sc}(\Omega)$, representing the primary RF noise, are given by (10). $S_{II}(\Omega)$ is the power spectrum of the modulating low-frequency noise current, which is assumed to be uncorrelated with e_c and e_s. In order to arrive at clear formulas for the spectra it is convenient to represent the complex differentials in (12) not by their real and imaginary parts as has been done in the perturbation equation (13), but by their magnitude and phase:

$$\frac{dZ_L}{d\omega} = \frac{dR_L}{d\omega} + j\frac{dX_L}{d\omega} = \left|\frac{dZ_L}{d\omega}\right| \exp(j\theta) \tag{15a}$$

$$\frac{\partial Z_D}{\partial A} = \frac{\partial R_D}{\partial A} + j\frac{\partial X_D}{\partial A} = \left|\frac{\partial Z_D}{\partial A}\right| \exp(j\eta) \tag{15b}$$

$$\frac{\partial Z_D}{\partial I} = \frac{\partial R_D}{\partial I} + j\frac{\partial X_D}{\partial I} = \left|\frac{\partial Z_D}{\partial I}\right| \exp(j\psi). \tag{15c}$$

The angles θ, η, and ψ introduced above are illustrated in the complex plane (Fig. 5). θ denotes the slope angle of the load locus $Z_L(\omega)$ versus the positive real axis, η if the angle between the positive real axis and the locus so

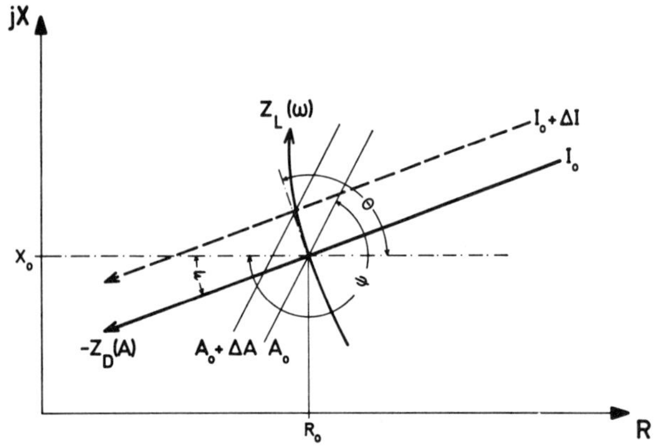

Fig. 5. Complex plane with the loci of the load impedance $Z_L(\omega)$ and the diode impedance $Z_D(A, I)$.

the active impedance as a function of the RF amplitude and for constant bias current, and ψ is the angle included between the positive real axis and the current dependent impedance $Z_D(A, I)$ when the RF amplitude is held constant [13].

The noise spectra are then given by the following.

1) AM noise:

$$S_{AA}(\Omega) = S_{AA}{}^0(\Omega) + K_{AM}{}^2 \cdot S_{II}(\Omega) \quad (16a)$$

where

$$S_{AA}{}^0(\Omega) = S_{cc}(\Omega)/A_0{}^2 \left|\frac{\partial Z_D}{\partial A}\right|^2 \sin^2(\theta - \eta) \quad (16b)$$

and

$$K_{AM} = \left|\frac{\partial Z_D}{\partial I}\right| \sin(\psi - \theta) \Big/ \left|\frac{\partial Z_D}{\partial A}\right| \sin(\theta - \eta) \quad (16c)$$

denotes the amplitude modulation sensitivity.

2) FM noise:

$$S_{\omega\omega}(\Omega) = S_{\omega\omega}{}^0(\Omega) + K_{FM}{}^2 S_{II}(\Omega) \quad (17a)$$

where

$$S_{\omega\omega}{}^0(\Omega) = S_{cc}(\Omega) \Big/ A_0{}^2 \left|\frac{dZ_L}{d\omega}\right|^2 \sin^2(\theta - \eta) \quad (17b)$$

and

$$K_{FM} = \left|\frac{\partial Z_D}{\partial I}\right| \sin(\vartheta - \psi) \Big/ \left|\frac{dZ_L}{d\omega}\right| \sin(\theta - \eta) \quad (17c)$$

denotes the frequency modulation sensitivity.

3) Normalized cross spectrum:

$$\gamma_{A\omega}(\Omega) = \frac{S_{A\omega}(\Omega)}{\{S_{AA}(\Omega) \cdot S_{\omega\omega}(\Omega)\}^{1/2}}$$

$$= \frac{\gamma_{A\omega}{}^0(\Omega) + K_{AM}K_{FM}S_{II}(\Omega)/\{S_{AA}{}^0(\Omega)S_{\omega\omega}{}^0(\Omega)\}^{1/2}}{\left\{\left[1 + K_{AM}{}^2 \dfrac{S_{II}(\Omega)}{S_{AA}{}^0(\Omega)}\right]\left[1 + K_{FM}{}^2 \dfrac{S_{II}(\Omega)}{S_{\omega\omega}{}^0(\Omega)}\right]\right\}^{1/2}} \quad (18a)$$

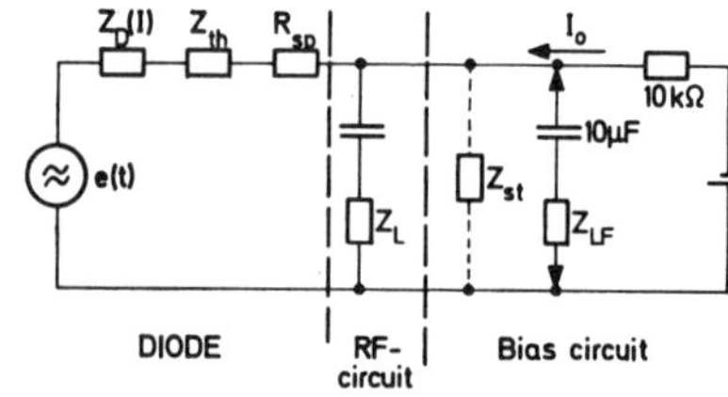

Fig. 6. Low-frequency noise equivalent circuit for an IMPATT diode oscillator.

with

$$\gamma_{A\omega}{}^0(\Omega) = -\cos(\theta - \eta) + \sin(\theta - \eta) \frac{S_{cs}(\Omega)}{S_{cc}(\Omega)}. \quad (18b)$$

The AM and FM noise spectra consist of two terms each. The contributions arising from RF primary noise are $S_{AA}{}^0(\Omega)$ and $S_{\omega\omega}{}^0(\Omega)$. The contributions due to modulation noise are expressed as products of the modulation sensitivities with the spectrum $S_{II}(\Omega)$ of the modulating low-frequency noise current.

The simple quasi-stationary treatment applied in this work imposes a limitation upon the modulation frequency. This limit comes out from an extended analysis [14], which is, however, more involved. Since the AM noise power as calculated by the quasi-stationary method turns out to be white if the noise source is white, and must drop off with 6 dB/octave beyond the quasi-stationary validity range according to [14], the limit frequency can be easily measured.

The cross spectrum $\gamma_{A\omega}$ takes a very complicated form if both noise mechanisms discussed above are taken into account. In the next section the conditions for only one mechanism to be effective are investigated and the simple cross spectrum resulting in this case is discussed.

E. Modulation Noise

The discussion of the modulation contribution to the oscillator noise is based on the low-frequency equivalent circuit of the IMPATT diode oscillator shown in Fig. 6, which is a modification of that proposed by Haitz [9] and Scherer [15]. The noise current source used by these authors was replaced by a voltage source. In both cases the source impedance is given by the electronic diode impedance Z_D. R_{sp} is the parasitic spreading resistance within the diode, Z_{th} the thermally induced diode impedance, which shows a strong frequency dependence [9]. The oscillator RF circuit is schematically represented by a series combination of Z_L and a small capacitor.

From this equivalent circuit it is evident that the modulating low-frequency noise current increases if the external LF impedance Z_{LF} is lowered. The oscillator noise should therefore increase with decreasing Z_{LF}. The oscillator noise for large $Z_{LF}(Z_{LF} \gg Z_D + R_{sp} + Z_{th})$ reduces to the contribution caused by the primary RF noise. The modulation noise contribution has a maximum for $Z_{LF} = 0$ (external low-frequency short). Its

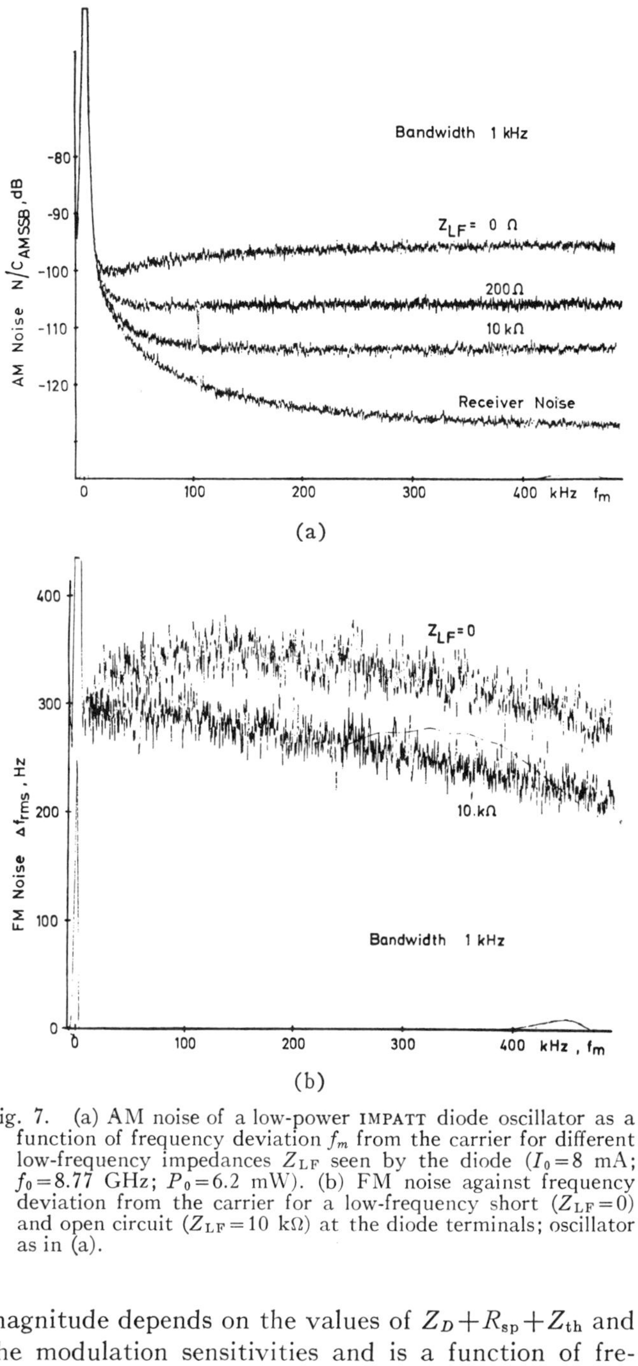

Fig. 7. (a) AM noise of a low-power IMPATT diode oscillator as a function of frequency deviation f_m from the carrier for different low-frequency impedances Z_{LF} seen by the diode ($I_0=8$ mA; $f_0=8.77$ GHz; $P_0=6.2$ mW). (b) FM noise against frequency deviation from the carrier for a low-frequency short ($Z_{LF}=0$) and open circuit ($Z_{LF}=10$ kΩ) at the diode terminals; oscillator as in (a).

magnitude depends on the values of $Z_D+R_{sp}+Z_{th}$ and the modulation sensitivities and is a function of frequency by the frequency dependence of these quantities (especially that of Z_{th}).

Fig. 7(a) and (b) show typical oscillator noise spectra. AM noise is expressed as the ratio of single sideband noise power to carrier power, FM noise as equivalent rms frequency deviation. Both quantities are plotted against frequency deviation from the carrier (equivalent to the modulation frequency). AM and FM noise spectra were measured with a simple direct detection system as described by Mueller [16].

Fig. 7(a) shows a marked increase of the AM noise by more than 10 dB when decreasing the external low-frequency impedance Z_{LF}, whereas the FM noise remains nearly unaffected (Fig. 7(b)). This is in agreement with the observations of Scherer [15] and shows that

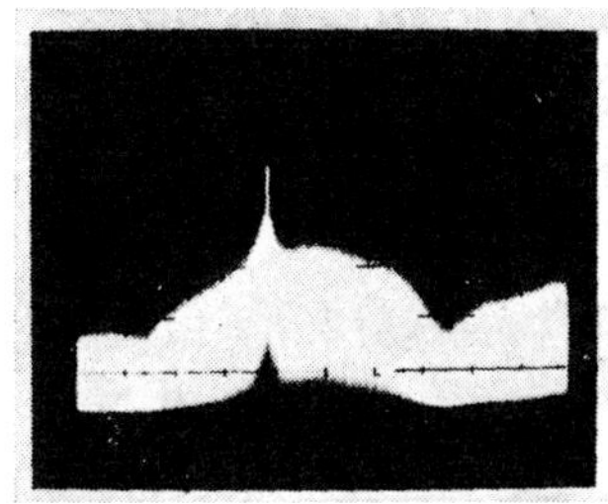

Fig. 8. Asymmetry of the RF power spectrum of an IMPATT diode oscillator (carrier suppressed by 40 dB). Horizontal: frequency, 30 MHz/div. Vertical: power, 10 dB/div.

the modulation contribution to the FM noise of IMPATT oscillators is negligible for all bias impedances. The AM modulation contribution can be made small by the choice of a large Z_{LF}, which requires a current generator as bias source for IMPATT oscillators. With increasing frequency deviation from the carrier, however, modulation noise is difficult to suppress as stray impedances produced by RF bypass capacitors and power supply cables become more and more effective. This has to be kept in mind when applying IMPATT sources as local oscillators. In Fig. 8 the RF spectrum of a noisy oscillator is shown the carrier of which has been suppressed by 40 dB by means of a high Q cavity filter. The striking asymmetry of the spectrum at higher frequency deviations (the upper sidebands are larger than the lower ones) results from the positive AM–FM correlation indicating modulation noise (see also Section IV-F). The power spectrum minima about 100 MHz off the carrier are due to a pole in the external low-frequency impedance produced by a supply cable 60 cm long.

F. Discussion of the Noise Spectra Arising from Primary RF Noise

As pointed out in the previous section the modulation contributions to the noise of IMPATT oscillators can be kept negligibly low at small frequency deviations from the carrier. The AM and FM noise spectra are then given by $S_{AA}{}^0(\Omega)$ (16b) and $S_{\omega\omega}{}^0(\Omega)$ (17b). The normalized AM–FM cross spectrum is simply $\gamma_{A\omega}{}^0(\Omega)$ (18b). The frequency dependence of the AM and FM noise spectra is the same as that of $S_{cc}(\Omega)$. To investigate the frequency dependence of $S_{cc}(\Omega)$, the primary RF noise spectrum $S_{ee}(\omega)$ is expanded into a Taylor series around ω_0. According to (8) the odd terms do not give any frequency dependence at all; the even terms

$$\frac{(\omega-\omega_0)^n}{n!}\left.\frac{d^nS_{ee}(\omega)}{d\omega^n}\right|_{\omega=\omega_0}$$

can be neglected for small $\Omega=\omega-\omega_0$ compared to the zero order term $S_{ee}(\omega_0)$. Then $S_{cc}(\Omega)$ is approximately white. This also results in white AM and FM noise spectra, in agreement with the experimental observations.

Similar arguments apply to $S_{cs}(\Omega)$, so that for small Ω the ratio $S_{cs}(\Omega)/S_{cc}(\Omega)$ is very small compared to unity. Then the normalized AM–FM cross spectrum $\gamma_{A\omega}{}^0(\Omega)$ is simply given by:

$$\gamma_{A\omega}{}^0(\Omega) = -\cos(\theta - \eta) \tag{19}$$

where $\gamma_{A\omega}{}^0(\Omega)$ is real and independent of frequency Ω. Equations (16b) and (17b) demonstrate that the AM noise produced by primary RF noise is determined above all by the nonlinear characteristics of the active element, while the FM noise is affected by the frequency dependence of the load, which is characterized by the oscillator quality factor. The dependence of the FM noise on the oscillator Q factor has been demonstrated experimentally by Ulrich and Harth [17].

Moreover both AM and FM noise depend on the intersection angle $\theta - \eta$, which the locus of the active impedance $Z_D(A)$ makes with the load impedance $Z_L(\omega)$ in the complex plane. The total load dependence of the AM noise and the AM–FM cross spectrum is contained in the load angle θ which is an experimentally accessible quantity. The device angle η is a constant characterizing the active element in its operating point.

For an angle difference $\theta - \eta = 90°$, that is if the two loci intersect each other at right angles, the AM–FM cross spectrum becomes zero and the AM noise reaches a minimum. In Fig. 9 the ratio of the actual AM noise to the minimum AM noise ($\theta - \eta = 90°$) in dB and the AM–FM cross spectrum $\gamma_{A\omega}{}^0$ are plotted as functions of the angle difference $\theta - \eta$. From stability considerations an allowed range $0° < \theta - \eta < 180°$ is deduced [14]. The AM–FM correlation of IMPATT oscillator noise was measured by recording scatter diagrams for amplitude and frequency fluctuations [18]. Fig. 10 shows typical scatter diagrams for three different low-frequency impedances Z_{LF}, together with a diagram obtained for sinusoidal bias current modulation. For $Z_{LF} = 2500\ \Omega$ (approximating a low-frequency open circuit) the sign of the AM–FM cross spectrum $\gamma_{A\omega}{}^0$ is negative and its magnitude approaches unity. For bias current modulation, amplitude and frequency deviations are usually in phase [13] (both amplitude and frequency increase with increasing bias current), resulting in a positive AM–FM correlation for modulation noise contributions. In the particular case shown an external low-frequency short ($Z_{LF} = 0$) produces modulation noise contributions large enough to make the sign of the resulting AM–FM cross spectrum $\gamma_{A\omega}{}^0$ positive. For an intermediate Z_{LF} of about 70 Ω the AM–FM correlation vanishes.

As shown in Section III both $\partial X_D/\partial A$ and $\partial R_D/\partial A$ are typically positive for IMPATT diodes. The device angle η is then in the interval $0° < \eta < 90°$. The load angle θ is always approximately equal to 90° when the oscillator is tuned for maximum power output. Then the AM–FM cross spectrum $\gamma_{A\omega}{}^0$ turns out to be negative, in agreement with the experimental result presented in Fig. 10.

The model presented is further supported by the experimental data shown in Fig. 11. Measured values of the AM–FM cross spectra $\gamma_{A\omega}{}^0$ are plotted against the corresponding values of the angle difference $\theta - \eta$ obtained by a modified pulling technique. The experimental points represent different diodes and oscillation frequencies throughout X band. In all these measurements the oscillators were tuned for approximately maximum power output. The measured AM–FM cross spectra $\gamma_{A\omega}{}^0$ are all in the range of -0.7 to -0.98. This is a characteristic feature of IMPATT diode oscillators.

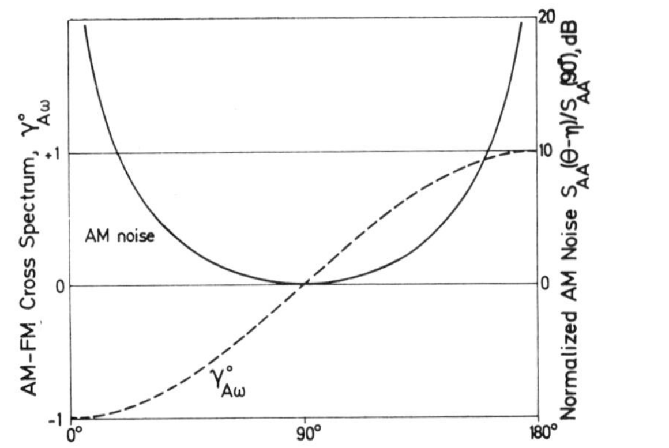

Fig. 9. Normalized AM–FM cross spectrum $\gamma_{A\omega}{}^0$ and ratio of the actual AM noise spectrum to minimum AM noise spectrum as a function of intersection angle $\theta - \eta$.

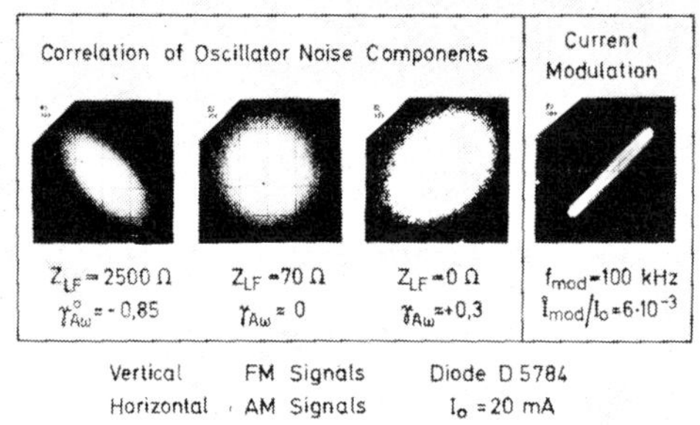

Fig. 10. AM–FM scatter diagrams for oscillator noise at three different low-frequency impedances Z_{LF} at the diode terminals, and for sinusoidal bias current modulation.

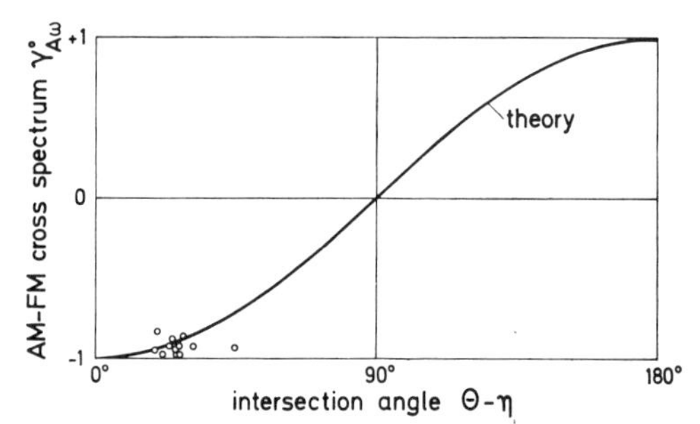

Fig. 11. Experimental AM–FM cross spectra $\gamma_{A\omega}{}^0$ against measured values of the intersection angle $(\theta - \eta)$.

A large AM–FM correlation always indicates that it is possible to improve the AM noise performance. Methods to obtain minimum AM noise with small power loss are described in the following.

One method for obtaining minimum AM noise is orthogonal noise tuning (ONT) which was first demonstrated by the authors [19]. The experimental setup

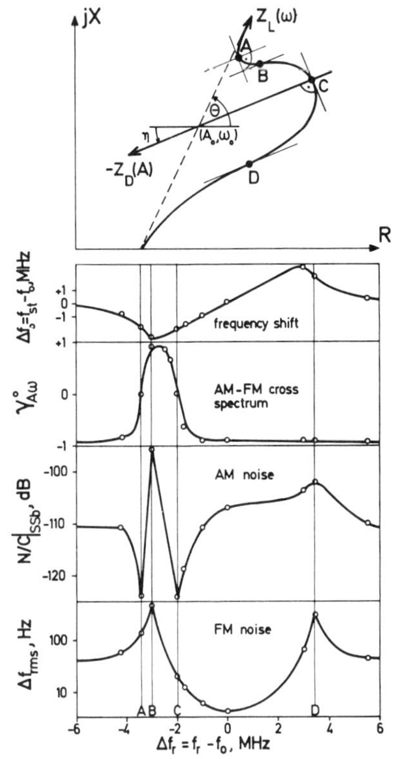

Fig. 12. Shift of the oscillator frequency Δf_0, AM–FM cross spectrum $\gamma_{A\omega}{}^0$, and AM–FM noise (bandwidth: 1 kHz; $f_m = 100$ kHz) versus the resonance frequency of the stabilizing cavity f_T ($f_0 = 9.46$ GHz; $P_0 = 17$ mW).

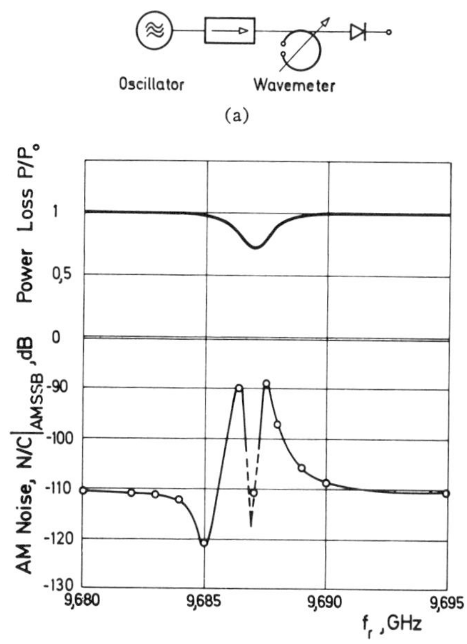

(b)

Fig. 13. (a) Experimental arrangement for AM noise reduction by an external filter. (b) Power loss of the filter and resulting AM noise as a function of filter resonance frequency f_r.

is similar to that commonly used to improve the FM noise by passive stabilization. By means of a high Q resonator weakly coupled to the oscillator, a loop or bump is produced in the locus of the load, which is shifted along the locus of the load when tuning the resonator. Thus the load angle θ and the frequency dependence of the load ($|dZ_L/d\omega|$) are varied. When changing the resonator frequency from lower to higher frequencies, points A to D shown in Fig. 12(a) are reached. In points A and C the loci intersect at right angles and the AM noise minima shown in Fig. 12(b) result. In points B and D the angle difference becomes smallest corresponding to AM noise maxima. Also shown in Fig. 12(b) is the AM–FM cross spectrum $\gamma_{A\omega 0}$ which vanishes when the minimum AM noise adjustment is made, as expected. For intersection angles $\theta - \eta > 90°$ the cross spectrum becomes positive. The FM noise spectrum as given by (17b) is influenced by $|dZ_L/d\omega|$ and θ; the tuning for minimum FM noise does in general not coincide with that for minimum AM noise.

Another technique which is always applicable, especially if only the output port of a black box oscillator is accessible, is to compensate part of the original amplitude fluctuations by adding to the oscillator signal a suitable amount of amplitude fluctuations obtained from the correlated frequency fluctuations by FM–AM conversion. A simple calculation shows that the AM noise spectrum $S_{AA}(\Omega)$ can be reduced by a factor $(1-\gamma_{A\omega}{}^{02})^{-1}$. The resulting amplitude fluctuations are then no longer correlated with the frequency fluctuations which remain practically unaffected.

In the upper part of Fig. 13 a possible experimental arrangement is shown. A simple absorption type cavity wavemeter in the output line of the oscillator is used for FM–AM conversion. The FM–AM conversion slope can be controlled by magnitude and sign by tuning the cavity resonance frequency f_r. In the middle of Fig. 13 the power loss of the filter is plotted as a function of wavemeter tuning, below this the resulting AM noise is shown. By proper tuning the AM noise is reduced by approximately 10 dB, in agreement with the measured value of $\gamma_{A\omega}{}^0$ of about -0.95 in the case shown.

V. Noise in Synchronized Oscillators

A. Synchronization

The injection of a synchronizing signal with amplitude a_0 and frequency ω_s by means of a circulator is accounted for in the oscillator equivalent circuit by insertion of the voltage source $a(t) = a_0 \exp j\omega_s t$ (Fig. 3(c). When the oscillator is locked by the injected signal the ac current $i(t)$ flowing now has the frequency ω_s and exhibits a phase shift ϕ_0 with respect to the synchronizing voltage: $i(t) = A \exp j(\omega_s t + \phi_0)$. As the current flows through the synchronizing source the latter may be represented by an impedance Z_s, assuming $A \approx A_0$

$$Z_s = -\frac{a(t)}{i(t)} = -\frac{a_0}{A_0} \exp(-j\phi_0). \tag{20}$$

In Fig. 14 this impedance is shown in the complex plane together with the loci of the load and the active element. Injection of a synchronizing voltage usually

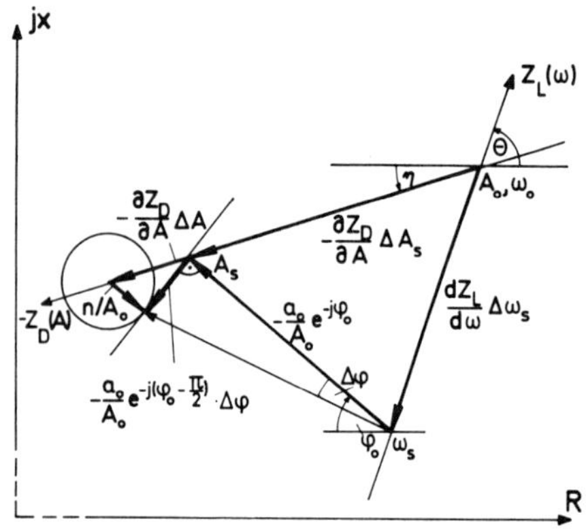

Fig. 14. Representation of the locking source and the noise source as impedances in the complex plane.

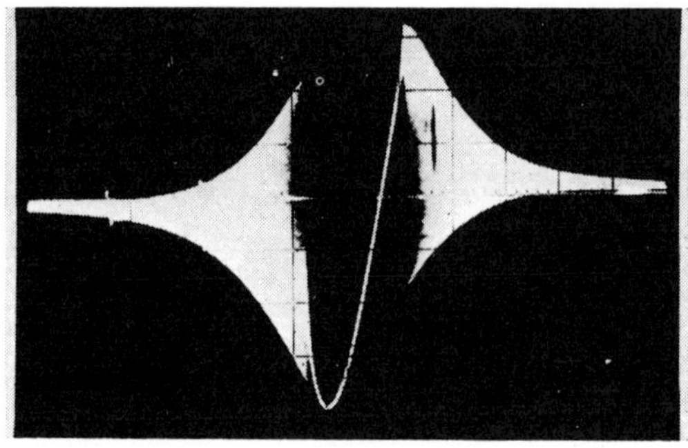

Fig. 15. Output power of the injection-locked oscillator in the locking range: $f_0=9.46$ Hz; $P_0=93$ mW; $P_0/P_s=20$ dB. Horizontal: 5 MHz/div. Vertical: 4.8 mW/div.

effects a change of the oscillator amplitude by $\Delta A_s = A - A_0$ and a change of the free-running oscillation frequency $\Delta\omega_s = \omega_s - \omega_0$. These amplitude and frequency variations in turn effect changes of the active impedance by $(\partial Z_D/\partial A)\Delta A_s$ and of the load impedance by $(dZ_L/d\omega)\Delta\omega_s$. These impedance variations compensate the perturbation of the equilibrium state by the impedance of the synchronizing source. The compensation is expressed as a balance condition for the voltages corresponding to the impedances in the large triangle in Fig. 14.

$$A_0 \frac{dZ_L}{dw} \Delta\omega_s - a_0 \exp(-j\phi_0) + \frac{\partial Z_D}{\partial A} A_0 \Delta A_s = 0. \quad (21)$$

After separating this equation into real and imaginary part the following expressions are obtained for the amplitude and frequency changes, respectively:

$$\Delta A_s = \frac{a_0 \sin(\phi_0 + \theta)}{A_0 \left|\frac{\partial Z_D}{\partial A}\right| \sin(\theta - \eta)} \quad (22a)$$

$$\Delta\omega_s = -\frac{a_0 \sin(\phi_0 + \eta)}{A_0 \left|\frac{dZ_L}{d\omega}\right| \sin(\theta - \eta)} = -\left|\Delta\omega_{s\,\max}\right| \sin(\phi_0 + \eta). \quad (22b)$$

The stable locking range is determined by the condition $\cos(\phi_0+\eta)>0$. Taking into account the further condition $|\sin(\phi_0+\eta)| \leqq 1$ imposed on the locking range, (18b) yields an expression for the total locking bandwidth B_s

$$B_s = 2a_0 \Big/ A_0 \left|\frac{dZ_L}{d\omega}\right| \sin(\theta - \eta). \quad (23)$$

When the nonlinearity of the diode reactance is neglected and a series resonant circuit is assumed for the load, (23) reduces to the well-known Adler formula [20].

From the shift of the operating point by $\Delta R = R - R_0$ and from (22a) the change in power $\Delta P = P - P_0$ is calculated as a function of the frequency $\Delta\omega_s$ within the locking range. An elliptic relationship is obtained:

$$\frac{\Delta P}{P_0} = 2\sqrt{\frac{P_s}{P_0}} \cdot \frac{2 - \frac{\partial R_D}{\partial A}\frac{A_0}{R_0}}{\frac{\partial Z_D}{\partial A}\frac{A_0}{R_0}} \left\{ \sqrt{1 - \frac{\Delta\omega_s^2}{\Delta\omega_{s\,\max}^2}} - \frac{\frac{\Delta\omega_s}{|\Delta\omega_{s\,\max}|}}{\tan(\theta - \eta)} \right\} \quad (24)$$

with

$$P_0 = \frac{A_0^2 R_0}{2}$$

and

$$P_s = \frac{a_0^2}{8R_0}$$

equal to the available power of the locking source. An experimental verification of (24) is shown in Fig. 15, where the frequency of the synchronizer is swept over the locking range and the resulting power change of the oscillator is detected.

B. Noise Spectra

The effect of a noise voltage on the locked oscillator can be represented in the complex plane by an impedance, the magnitude of which is n/A_0. This perturbation results in amplitude and phase fluctuations $\Delta A = A - A_0$ and $\Delta\phi = \phi - \phi_0$, respectively. The condition for the application of the quasi-stationary concept again is, that amplitude and phase only vary slowly, that is $d\Delta A/dt \ll \omega_0 A_0$ and $d\Delta\phi/dt \ll \omega_0$. In a first approximation it is assumed, that no frequency fluctuations occur, that is $d\Delta\phi/dt = 0$ as the frequency is fixed by the noise free locking source. This assumption limits the validity range of the calculated spectra to small frequency deviations from the carrier and to moderately high locking powers.

The exact limiting conditions imposed on the first approximation are then calculated from an extended model which takes into account frequency fluctuations $d\Delta\phi/dt$.

According to Kirchhoff's law the sum of all voltages corresponding to the impedances in the small triangle in Fig. 14 must vanish:

$$-A_0 \frac{\partial Z_D}{\partial A} \Delta A + n + a_0 \exp\left[-j\left(\phi_0 - \frac{\pi}{2}\right)\right] \Delta\phi = 0. \quad (25)$$

Therefrom the noise spectra S_{AA} and $S_{\phi\phi}$ and the correlation coefficient $\gamma_{A\phi}$ are calculated:

$$S_{AA}(\Omega) = S_{cc}(\Omega) \Big/ A_0^2 \left|\frac{\partial Z_D}{\partial A}\right|^2 \cos^2(\phi_0 + \eta) \quad (26a)$$

$$S_{\phi\phi}(\Omega) = S_{cc}(\Omega)/a_0^2 \cos^2(\phi_0 + \eta) \quad (26b)$$

$$\gamma_{A\phi} = -\sin(\phi_0 + \eta). \quad (26c)$$

The basic influence of the locking source is already evident from (26 a, b, c) and is discussed below.

1) Both spectra present a minimum in the center of the locking range ($\phi_0+\eta=0$) and poles at the edge.

2) The AM noise as in the case of the free running oscillator is determined by the nonlinearity of the diode impedance. The minimum takes the same value as has been calculated for the free running oscillator. An increase of the locking power just as AM noise of the locking source does not affect the spectra. FM noise of the locking source, however, enters the expression (26a) via the phase angle ϕ_0.

3) The phase noise spectrum does not depend on the diode and load impedance. Increasing locking power effects a decrease of the phase noise. A lower limit is imposed on this decrease by the phase noise of the locking source, the AM noise of which is of no importance.

4) Due to the nonlinearity of the active impedance AM noise is converted into phase noise. The angle ϕ_0 effects conversion of phase noise into AM noise.

C. Extended Model

In order to extend the validity range of the calculated spectra it is necessary to take into account the velocity of the phase variations ($d\Delta\phi/dt$). This results in an additional perturbation term in (25), which contains the frequency dependence of the load:

$$A_0 \frac{dZ_L}{d\omega} \frac{d\Delta\phi}{dt} - A_0 \frac{\partial Z_D}{\partial A} \Delta A + n + a_0 \exp\left[-j\left(\phi_0 - \frac{\pi}{2}\right)\right] \Delta\phi = 0. \quad (27)$$

Therefrom the more general expressions for the AM noise spectrum S_{AA} and the FM noise spectrum $S_{\dot\phi\dot\phi}$ are deduced:

$$S_{AA}(\Omega) = \frac{S_{cc}(\Omega)}{A_0^2 \left|\frac{\partial Z_D}{\partial A}\right|^2 \cos^2(\phi_0+\eta)} \cdot \frac{\Omega^2 \left|\frac{dZ_L}{d\omega}\right|^2 + \left(\frac{a_0}{A_0}\right)^2}{\Omega^2 \left|\frac{dZ_L}{d\omega}\right|^2 \frac{\sin^2(\theta-\eta)}{\cos^2(\phi_0+\eta)} + \left(\frac{a_0}{A_0}\right)^2} \quad (28a)$$

$$S_{\dot\phi\dot\phi}(\Omega) = \frac{S_{cc}(\Omega)\Omega^2}{a_0^2 \cos^2(\phi_0+\eta)} \cdot \frac{(a_0/A_0)^2}{\Omega^2 \left|\frac{dZ_L}{d\omega}\right|^2 \frac{\sin^2(\theta-\eta)}{\cos^2(\phi_0+\eta)} + \left(\frac{a_0}{A_0}\right)^2}. \quad (28b)$$

From these expressions the conditions for the validity of the first approximation (26 a, b, and c) are evident:

$$\frac{a_0}{A_0} \gg \Omega \left|\frac{dZ_L}{d\omega}\right|$$

and

$$\frac{a_0}{A_0} \gg \Omega \left|\frac{dZ_L}{d\omega}\right| \cdot \frac{\sin(\theta-\eta)}{\cos(\phi_0+\eta)}.$$

The qualitative behavior of the spectra (28 a,b) is shown in Fig. 16, where AM and FM noise are plotted versus frequency deviation from the carrier for two different locking powers. If the latter is increased the frequency range, where noise reduction is effective, gets wider both for AM and FM noise. With increasing frequency deviation the noise approaches the level of the free running oscillator. The deviations from the relationship (28a,b) which are found experimentally (Fig. 17(a) and (b)) are due to the fact that the locking source in practice is always noisy. The spectra are obtained from an IMPATT diode oscillator, which is synchronized by a 2K 25 klystron. The general behavior of the FM noise agrees with (28b). As is expected from (28a), the AM noise could be reduced below the AM noise level of the klystron as the latter does not present a lower limit. The actual minimum AM noise level is believed to be near the 20-dB line.

Within the locking range the noise has also been measured as a function of the difference frequency $\Delta\omega_s$. According to (22b) and (28b) the FM noise is related to $\Delta\omega_s$ by the function $[1-(\Delta\omega_s/|\Delta\omega_{s\,\max}|)^2]^{-1}$. This has been confirmed experimentally.

The AM noise is affected by the FM noise of the klystron. The elliptic relationship Fig. 15 between ΔP_s and $\Delta\omega_s$ demonstrates that the synchronized oscillator acts like a discriminator. Consequently an AM mini- 437

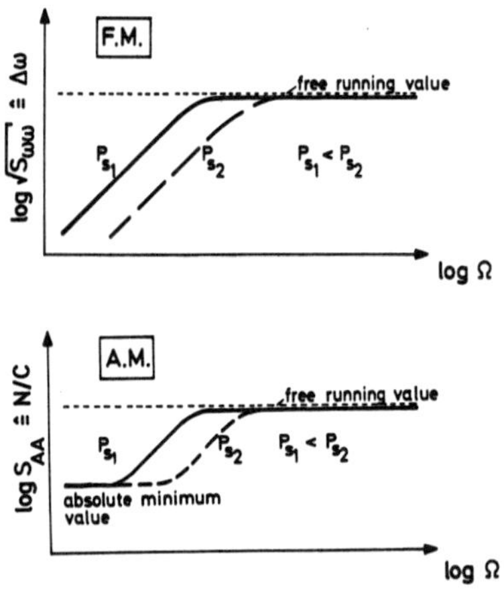

Fig. 16. Qualitative dependence of the AM and FM noise spectra of an injection-locked oscillator on the frequency deviation from the carrier according to (26a, b). (Synchronizer noise neglected.)

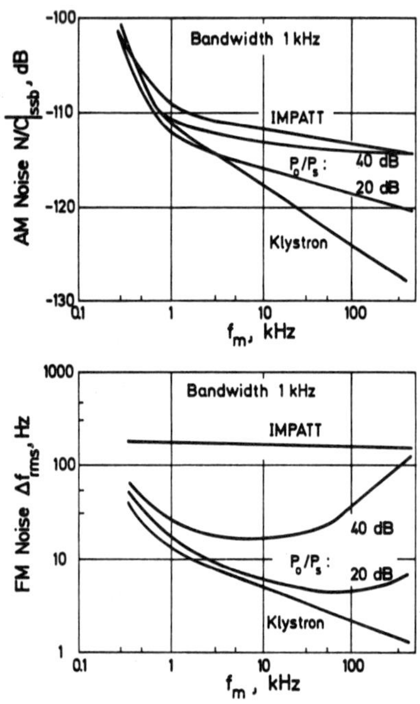

Fig. 17. Measured AM and FM noise spectrum of a free running injection phase-locked IMPATT diode oscillator and of the synchronizing klystron as a function of frequency deviation from the carrier for several locking gains P_0/P_s (f_0=9.46 GHz; P_0=93 mW).

mum was observed in the point where $d\Delta P/d\Delta\omega_s=0$. Outside this extremum the AM noise increased due to FM–AM conversion. As the AM spectra in Fig. 17(a) have been measured in this point of vanishing conversion it is expected that the noise level further reduces when locking is achieved with a source with negligible FM noise.

VI. Signal-to-Noise Ratio

With the use of the expressions obtained above for the noise spectra and modulation sensitivities the signal-to noise-ratio is calculated for a free running IMPATT oscillator the bias current of which is modulated.

Bias current modulation represents the technically most simple method for amplitude and frequency modulation of the microwave signal emitted by the diode. The application of this modulation technique to real systems depends on whether the signal-to-noise ratio meets the specifications of the system under question. The subsequent study gives information about the parameters affecting the signal-to-noise ratio and demonstrates the possibilities for optimization.

A. Definition of the S/N Ratio

1) Amplitude Modulation: The signal-to-noise ratio for AM is defined as the ratio of the AM percentage to the effective AM noise percentage:

$$\left(\frac{S}{N}\right)_{AM} = \frac{K_{AM}\cdot\Delta I}{A_0} \Bigg/ \frac{\sqrt{\int_B S_{AA}(\Omega)d\Omega}}{A_0} \tag{29}$$

where ΔI denotes the modulating current deviation. Assuming white noise within the bandwidth B, the above definition reduces to

$$\left(\frac{S}{N}\right)_{AM} = \frac{K_{AM}\Delta I}{\sqrt{S_{AA}(\Omega)B}}. \tag{30}$$

2) Frequency Modulation: In an analogous manner the signal-to-noise ratio for the FM arising from bias current modulation can be written as:

$$\left(\frac{S}{N}\right)_{FM} = \frac{K_{FM}\Delta I}{\sqrt{\int_B S_{\omega\omega}(\Omega)d\Omega}}. \tag{31}$$

Again the white noise approximation is made:

$$\left(\frac{S}{N}\right)_{FM} = \frac{K_{FM}\Delta I}{\sqrt{S_{\omega\omega}(\Omega)\cdot B}}. \tag{32}$$

B. S/N Ratio

1) Amplitude Modulation: Now the signal-to-noise ratio according to the definition (30) can be calculated from (16) assuming modulation noise to be negligible. As has been pointed out in Section IV-E this condition can always be satisfied by proper biasing of the diode. Then the following expression results:

$$\left(\frac{S}{N}\right)_{AM} = \Delta I\left|\frac{\partial Z_D}{\partial I}\right|\frac{A_0\sin(\psi-\theta)}{\sqrt{S_{cc}(\Omega)B}}. \tag{33}$$

The S/N ratio is seen to depend on the diode characteristics via $\partial Z_D/\partial I$, on the load via θ, and on the modulating current deviation. Of course it is not possible to extend the signal-to-noise ratio to any desired value by means of an increase in ΔI. A limitation is imposed on ΔI by the tolerable distortion of the transmitted signal. As may be seen from Fig. 3, for a fixed oscillator assembly too large a current deviation shifts the locus of the active device outside the linear range of the load impedance.

Moreover higher order terms of the Taylor expansion of Z_D must be taken into account for large ΔI. Yet the latter effect is not so restricting as the curvature of the load locus, has been shown in a recent paper [21].

As is further shown by (33) an optimum signal-to-noise ratio results from a proper adjustment of the load angle θ, that is the sine function becomes a maximum when the locus of the load intersects the line for constant RF amplitude at right angles ($\psi-\theta=90°$). As has already been described, such an optimum adjustment of θ can be achieved experimentally by weakly coupling a resonator to the oscillator circuit. In Fig. 18 the AM signal-to-noise ratio is plotted versus the angle difference $\theta-\eta$. The latter is determined from the correlation coefficient and can be taken as a measure for θ, as η remains constant in this experiment.

In agreement with (33) a sinusoidal dependence of the S/N ratio on θ is found. A simple series resonant load circuit cannot explain this dependence on the load parameter θ. In this case $\theta=90°$ and the S/N is given by:

$$\left(\frac{S}{N}\right)_{AM} = -\frac{A_0 \Delta I}{\sqrt{S_{cc}(\Omega)B}} \frac{\partial R_D}{\partial I}. \tag{34}$$

2) *Frequency Modulation:* For the FM produced by bias current modulation according to the definition (32) and with (17) a similar expression holds for the S/N ratio:

$$\left(\frac{S}{N}\right)_{FM} = \Delta I \left|\frac{\partial Z_D}{\partial I}\right| \frac{A_0 \sin(\eta-\psi)}{\sqrt{S_{cc}(\Omega)B}}. \tag{35}$$

Again, the modulating current deviation ΔI is limited by the onset of nonlinear modulation, which in this case may arise both from the curvature of the load locus and from the nonconstant frequency dependence of the magnitude of the load impedance outside the operating point.

Contrary to the signal-to-noise ratio for AM, the signal-to-noise ratio for FM cannot be optimized by a proper adjustment of the load, as the FM sensitivity and the FM noise depend in the same way on the load parameters. This is confirmed experimentally in Fig. 18 where the measured FM signal-to-noise ratio is plotted versus θ. In agreement with (35) the S/N ratio remains constant.

The angle factor sin $(\eta-\psi)$ appearing in (35) only depends on the parameters η and ψ of the active element. According to large signal theories [5], [6], [7] the RF amplitude dependence of the device resistance and reactance is approximately equal, that is $\eta \sim 45°$. This value is confirmed experimentally by [21]. From the latter work a value for ψ is also deduced ($\psi \approx 240$–$250°$), resulting in an angle factor of about $\frac{1}{4}$. Thus the optimum S/N ratio for AM is about four times higher than the S/N ratio of the FM.

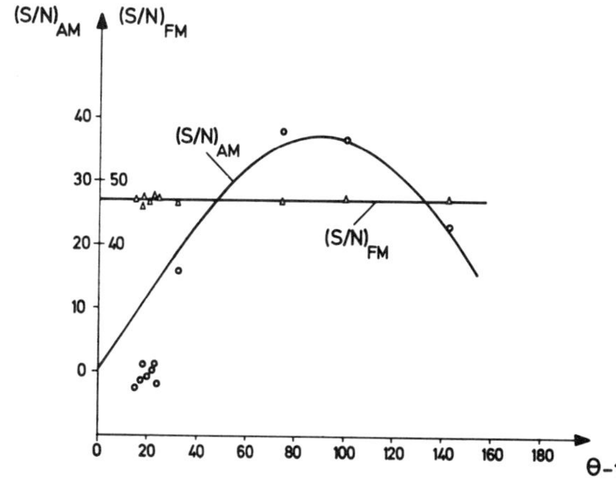

Fig. 18. Signal-to-noise ratio in relative units for amplitude and frequency modulation as a function of the intersection angle $\theta-\eta$.

This advantage is cancelled by the large frequency deviation simultaneously produced with amplitude modulation. On the other hand, the amplitude variations accompanying a reasonable frequency deviation are not serious and can be suppressed by a limiter.

VII. Conclusions

The possible contributions to oscillator noise have been pointed out. For IMPATT oscillators it has been shown that the modulation noise contribution can be made negligible for small frequency deviations from the carrier. The free running oscillator noise arising from RF primary noise is determined by three types of parameters which must be specified in order to characterize the noise performance of the oscillator: the equivalent noise source of the active element, the nonlinearities of real and imaginary part of the active impedance, and the frequency dependence of the load resistance and reactance.

As has been shown, the load dependence of the AM spectrum can be used to minimize AM noise. This minimum condition cannot be satisfied by usual oscillators designed for broadband performance but can only be realized with multiple resonant circuits.

As a further important result with regard to noise reduction, a nonvanishing correlation between AM and FM noise indicates the possibility to compensate the AM noise by an external passive network no matter how the correlation is effected.

By injection phase locking an oscillator to a low-noise reference oscillator both AM and FM noise of the locked oscillator can be reduced. The minimum AM noise is the same as that of the orthogonally tuned free running oscillator, because for a phase-locked oscillator the locus of the load impedance has no influence on the AM noise and there is no FM–AM conversion. The FM noise of the locking source is converted to oscillator AM noise and gives an additive noise component. The minimum FM noise of the locked oscillator is determined by the FM noise of the locking source.

The signal-to-noise ratio of IMPATT diode oscillators is an additional proof that the noise and modulation properties are affected by different parameters, that is, the oscillator noise is not due to a modulation mechanism.

In contrast to the S/N ratio for FM, which only depends on diode parameters, the S/N ratio for AM is sensitive to the oscillator tuning. For optimum load adjustment, a condition that does not coincide with the condition for minimum AM noise, the S/N ratio for AM exceeds the S/N ratio for FM by a factor of 4.

Acknowledgment

The authors wish to thank Prof. R. Müller and M. Bechteler for stimulating discussions.

References

[1] W. T. Read, "A proposed high-frequency negative resistance diode," *Bell Syst. Tech. J.*, vol. 37, Mar. 1958, pp. 401–448.
[2] M. Gilden and M. E. Hines, "Electronic tuning effects in the Read microwave avalanche diode," *IEEE Trans. Electron Devices*, vol. ED-13, Jan. 1966, pp. 169–175.
[3] T. Misawa, "Multiple uniform layer approximation in analysis of negative resistance in p-n junction in breakdown," *IEEE Trans. Electron Devices*, vol. ED-14, Dec. 1967, pp. 795–808.
[4] B. Hoefflinger, "Generalized small-signal analysis of avalanche transit-time diodes," *IEEE Trans. Electron Devices*, vol. ED-14, Sept. 1967, pp. 563–568.
[5] W. J. Evans and G. I. Haddad, "A large-signal analysis of IMPATT diodes," *IEEE Trans. Electron Devices*, vol. ED-15, Oct. 1968, pp. 708–717.
[6] D. L. Scharfetter and H. K. Gummel, "Large-signal analysis of a silicon Read diode oscillator," *IEEE Trans. Electron Devices*, vol. ED-16, Jan. 1969, pp. 64–77.
[7] D. Delagebeaudeuf, "Analyse grand signal du fonctionnement des diodes à avalanche en mode de transit," *Revue Technique Thomson-CSF*, vol. 1, Sept. 1969, pp. 309–337.
[8] M. E. Hines, "Noise theory for the Read type avalanche diode," *IEEE Trans. Electron Devices*, vol. ED-13, Jan. 1966, pp. 158–163.
[9] R. H. Haitz, "Noise of a self-sustaining avalanche discharge in silicon: low frequency studies," *J. Appl. Phys.*, vol. 38, June 1967, pp. 2935–2946.
R. H. Haitz and F. W. Voltmer, "Noise of a self-sustaining avalanche discharge in silicon: studies at microwave frequencies, *J. Appl. Phys.*, vol. 39, June 1968, pp. 3379–3384.
[10] D. Delagebeaudeuf, "Experimental verification of approximate large-signal theory of IMPATT diodes," *Proc. IEEE* (Lett.), vol. 58, July 1970, pp. 1140–1141.
[11] A. Spölti, "Der Einfluß des thermischen Widerstandsrauschens und des Schroteffektes auf die Störmodulation von Oszillatoren," *Bull. Schweiz. Elektrotech. Verein*, vol. 39, June 1948, pp. 419–427.
[12] W. A. Edson, "Noise in oscillators," *Proc. IRE*, vol. 48, Aug. 1960, pp. 1454–1466.
[13] G. Weidmann, "Amplitude and frequency modulation sensitivity of IMPATT diode oscillators," *Nachrichtentech. Z.*, vol. 23, July 1970, pp. 368–371.
[14] K. Kurokawa, "Some basic characteristics of broadband negative-resistance oscillator circuits," *Bell. Syst. Tech. J.*, vol. 48, July–August 1969, pp. 1937–1955.
[15] E. F. Scherer, "Investigations of the noise spectra of avalanche oscillators," *IEEE Trans. Microwave Theory Tech.*, vol. MTT-16, Sept. 1968, pp. 781–788.
[16] R. Mueller, "Noise measurements of microwave local oscillators," *IRE Trans. Electron Devices*, vol. ED-1, Dec. 1954, pp. 43–50.
[17] W. Harth and G. Ulrich, "Q-dependence of IMPATT-diode FM-noise," *Electron. Lett.*, vol. 5, Jan. 1969, pp. 7–8.
[18] G. R. Sugar, "Estimation of correlation coefficients from scatter diagrams," *J. Appl. Phys.*, vol. 25, Mar. 1954, pp. 354–357.
[19] G. Ulrich, "AM-noise of IMPATT diode oscillators," *Electron. Lett.*, vol. 6, Apr. 1970, pp. 247–248.
[20] R. Adler, "A study of locking phenomena in oscillators," *Proc. IRE*, vol. 34, June 1946, pp. 351–357.
[21] J. W. Gewartowski and J. E. Morris, "Active IMPATT diode parameters obtained by computer reduction of experimental data," *IEEE Trans. Microwave Theory Tech.*, vol. MTT-18, Mar. 1970, pp. 157–161.

Investigations of the Noise Spectra of Avalanche Oscillators

ERNST F. SCHERER, MEMBER, IEEE

Abstract—Measured amplitude noise spectra of *X*-band microwave oscillators using silicon avalanche diodes are presented and compared with theoretical calculations. It was found that for the investigated diodes, the up-converted low-frequency noise (modulation noise) is the main contribution to the spectrum. In a frequency range extending from 1 kHz to several hundred MHz, the spectral noise power distribution is a sensitive function of the bias network impedance. Improvements in the noise-to-carrier power ratio of 5 to 15 dB were obtained by optimizing the bias network impedance. The dc current dependence of the amplitude noise spectrum is a complicated function of diode and circuit parameters. In general, the noise-to-signal ratio improves with increasing current. At high current densities, a reversal of this behavior may occur due to excess noise generation in the breakdown region and saturation effects of the signal power output. The equivalent rms deviation of the frequency modulation noise spectrum is typically one order of magnitude below that of the RF phase noise contribution due to RF noise sources. The phase noise spectrum of the oscillator is, therefore, not affected by the bias network impedance. Measurements on recent diodes indicate that the noise characteristics can be improved by careful control of the semiconductor device processing.

Manuscript received March 20, 1968; revised June 3, 1968. This work was supported by the U. S. Army Electronics Command, Fort Monmouth, N. J. under Contract DAAB07-67-C-0671.

The author is with the Semiconductor Division, Sylvania Electric Products, Inc., Woburn, Mass. 01801

I. Introduction

SEVERAL workers [1]–[3] have investigated the noise phenomena associated with a *p-n* junction under avalanche breakdown. Most of the published work relates to the statistical properties and the spectral distribution of the diode current fluctuations. A basic theory for the RF (shot) noise spectrum of oscillators was given by Edson [4] and Mullen [5] and later modified for the case of the avalanche oscillator by Hines [6]. However, little detailed work [7], [8] has been published investigating the significance of the various noise contributions and their dependence on circuit and operating parameters.

The intent of this paper is to present the results of theoretical and experimental work performed on X-band avalanche oscillators under various operating and circuit conditions and to indicate the areas where improvements are possible. Having a more detailed knowledge of the device characteristics, it is then possible to design the external circuitry and specify the operating parameters for optimum noise performance.

II. Calculations

For the discussion of the noise properties of the oscillators, it is convenient to describe the perturbed signal in terms of the phase noise spectrum and the amplitude noise spectrum. The amplitude noise spectrum is commonly measured as a noise-to-signal power ratio (NSR) at some frequency f_m off the carrier frequency f_0 in a specified bandwidth Δf, usually 100 Hz or 1000 Hz. The phase noise spectrum can be converted into an equivalent NSR but is more conveniently expressed as a rms frequency deviation in a given bandwidth.

In order to estimate the effect of the diode noise upon the amplitude noise spectrum of the oscillator, it is necessary to know the magnitude of the mean-square diode current fluctuation and its dependence on dc current, frequency, and external circuitry. The RF noise will propagate directly to the load via signal path, whereas the low-frequency noise modulates the oscillator through the current dependence of the real and imaginary part of the diode impedance. As a result of the nonlinear mixing process, the low-frequency noise components are up-converted into the microwave frequency range and are indistinguishable from the RF noise. The only way these two contributions can be separated is by influencing the low-frequency noise current and examining the effect upon the output spectrum of the oscillator. This was done in the experimental approach.

For the calculations of the noise spectra, the computed
442 values for the noise measure (M) and the short circuit mean-square current fluctuation $\langle i^2 \rangle$ by Gummel and Blue [3] are used. The calculations will be compared with the experimental results of diode 2, whose parameters are very close to those of the model diode used by Gummel and Blue.

A) NSR Due to RF (Shot) Noise

Edson's formula for the amplitude noise spectrum can be rewritten as

$$P_n = \frac{\langle i_n^2 \rangle G_L}{2[s^2 G_L^2 + 4C^2 \omega_m^2]}. \tag{1}$$

Introducing the noise measure

$$M = \frac{\langle i_n^2 \rangle / \Delta f}{4kT_0 |G_D|}, \tag{2}$$

one obtains

$$P_n^{(\Delta f)} = \frac{4kT_0 \Delta f |G_D| G_L M}{2[s^2 G_L^2 + 4C^2 \omega_m^2]}. \tag{3}$$

For frequencies close to the carrier,

$$P_n^{(\Delta f)} = \frac{4kT_0 \Delta f |G_D| G_L M}{2s^2 G_L^2}. \tag{4}$$

Assuming $s=2$ for maximum power output [6] and setting

$$G_L \cong |G_D|$$

(4) reduces to

$$P_n^{(\Delta f)} = \tfrac{1}{2} k T_0 \Delta f M. \tag{5}$$

Finally, the expression for the double sideband NSR becomes

$$(\text{NSR})_{\text{RF}}^{(\Delta f)} = \frac{2kT_0 \Delta f M}{P_0}. \tag{5a}$$

List of Symbols

- P_0 = Oscillator output signal power
- P_n = Noise power
- s = Nonlinearity saturation factor
- G_L = Load conductance
- G_D = Real part of diode impedance
- C = Circuit capacitance
- $\omega_m = 2\pi f_m$
- $(\text{NSR})^{\Delta f}$ = Double sideband noise-to-carrier power ratio in a bandwidth Δf
- $T_0 = 290°\text{K}$.

In Figs. 7 and 8 of Gummel and Blue [3], one finds that for the silicon diode with 1 Ω parasitic resistance at $f_0 = 9.5$ GHz, the noise measure is almost independent of the current (density) and has a value of $\approx 5 \times 10^3$. Using (5a) one obtains for $I_{dc} = 9.5$ mA and $P_0 = 10$ mW:

$$(\text{NSR})_{\text{RF}}^{100\text{Hz}} \triangleq -\underline{124\ \text{dB}};$$

and for $I_{dc} = 19$ mA and $P_0 = 50$ mW:

$$(\text{NSR})_{\text{RF}}^{100\text{Hz}} \triangleq -\underline{131\ \text{dB}}.$$

B. NSR Due to (Up-Converted) Amplitude Modulation Noise

For the calculation of the modulation noise, a detailed knowledge of the diode equivalent circuit at low frequencies and of the diode impedance as a function of frequency is required. The noise equivalent circuit of the diode, first given by Haitz [9], is represented in Fig. 1. The values for the circuit elements were obtained from the impedance versus frequency plot on the same figure. The points indicated by squares represent the noise generator source impedance and were obtained by loading the diode with a calibrated variable resistor until the noise voltage dropped to one-half the open circuit value. This provides a check for the validity of the noise equivalent circuit.

Using the vector representation of an amplitude modulated wave, one can derive the following expression for the noise power contained in the two sidebands of the carrier.

$$(P_n)_{\text{MOD.}} = \frac{(S_a)^2}{4P_0} \langle I_n^2 \rangle, \tag{6}$$

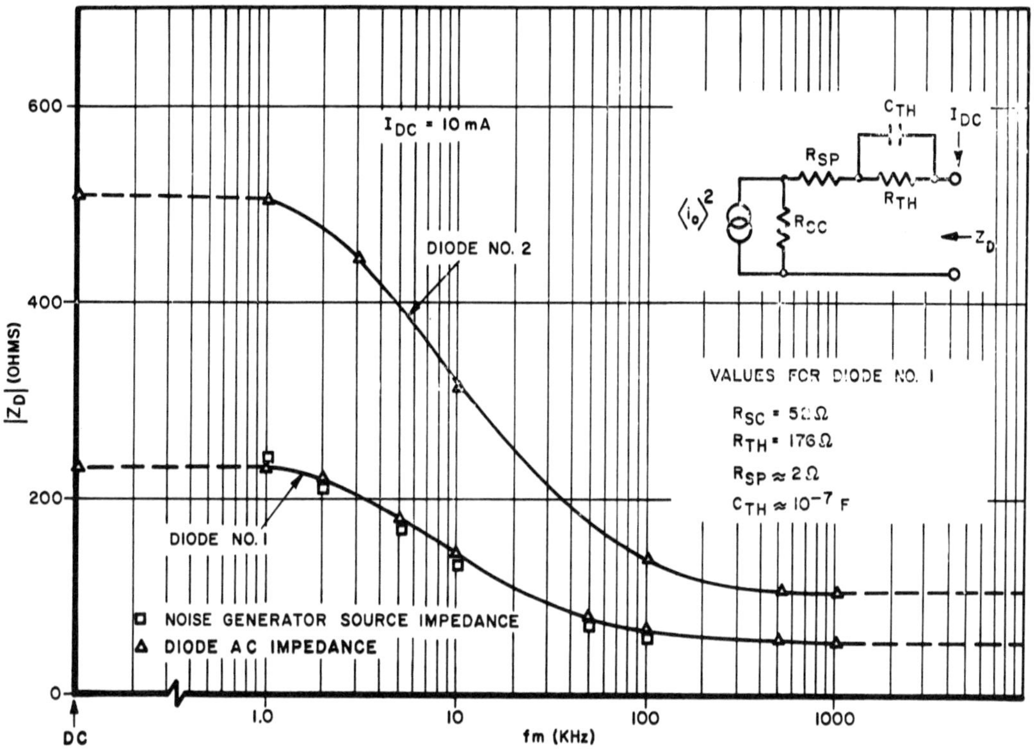

Fig. 1. Impedance of avalanche diodes at low frequencies and diode equivalent circuit.

where

$$S_a = \left.\frac{\Delta P_0}{\Delta I}\right|_{I=I_{dc}}. \tag{7}$$

$\langle I_n{}^2\rangle$ is the mean-square external circuit current superimposed on I_{dc}. Using the noise equivalent circuit of Fig. 1, one can express the effective circuit noise current in terms of the diode noise generator current.

$$\langle i_e{}^2\rangle = \langle i_0{}^2\rangle \frac{R_{sc}{}^2}{|Z_D + Z_B|^2}. \tag{8}$$

Z_B represents the external bias network impedance as seen by the diode. Assuming that an external current source produces the same effect as $\langle i_e{}^2\rangle$ we are able to substitute $\langle i_0{}^2\rangle$ for $\langle I_n{}^2\rangle$ in (6). To make the calculations meaningful, the short circuit case has to be considered.

$$Z_B \ll Z_D.$$

At frequencies where $R_{TH}\rightarrow 0$ and for $R_{SP} \ll R_{SC}$, this leads to the identities

$$\langle i_0{}^2\rangle \cong \langle i_e{}^2\rangle \equiv \langle I_n{}^2\rangle. \tag{9}$$

Using the shot noise ratio (R) of Gummel and Blue, one can write

$$\langle I_n{}^2\rangle \cong 2qI_{dc}R \tag{10}$$

Inserting this expression into (6) one obtains

$$(P_n)_{MOD.} \cong 2qI_{dc}R\frac{(S_a)^2}{4P_0} \tag{11}$$

normalized to a 1-Hz bandwidth.

The double sideband NSR in a bandwidth Δf is

$$(\text{NSR})_{MOD.}^{\Delta f} \cong 2qI_{dc}\Delta f R \frac{(S_a)^2}{4P_0{}^2}. \tag{12}$$

Recalling the $1/I_{dc}$ dependence of $\langle i^2\rangle$, one reads from Fig. 3 of Gummel and Blue for the shot noise ratio (R):

for a current density of 100 A/cm^2 ($I_{dc} \cong 9.5$ mA),

$$R = 350$$

for a current density of 200 A/cm^2 ($I_{dc} \cong 19$ mA),

$$R = 175.$$

The parameter S_a was measured to be 7.5 W/A and 5 W/A at the current levels of 9.5 mA and 19 mA, respectively. Inserting the above values into (12) one calculates for $I_{dc}=9.5$ mA and $P_0=10$ mW:

$$(\text{NSR})_{MOD.}^{100\text{Hz}} \triangleq -\underline{108\,\text{dB}},$$

and for $I_{dc}=19$ mA and $P_0=50$ mW.

$$(\text{NSR})_{MOD.}^{100\text{Hz}} \triangleq -\underline{122\,\text{dB}}.$$

C. RMS Deviation of Frequency Modulation Spectrum

The equivalent rms deviation of the phase noise contribution due to low-frequency noise modulation in a bandwidth Δf is given by

$$(\Delta f_{rms})_{MOD.}^{\Delta f} = S_f\sqrt{2qI_{dc}\Delta f R} \tag{13}$$

where the frequency modulation sensitivity

$$S_f = \left.\frac{\Delta f}{\Delta I}\right|_{I=I\mathrm{dc}} \tag{14}$$

is typically 10^9 Hz/A for an avalanche oscillator with a loaded Q of approximately 100. Inserting the numbers for the shot noise ratio (R) used in the previous section, the rms frequency deviation under short circuit condition of the bias network, when $I_{dc}=9.5$ mA and $P_0=10$ mW, is

$$(\Delta f_{\mathrm{rms}})_{\mathrm{MOD.}}^{100\mathrm{Hz}} = \underline{10.6 \text{ Hz}}$$

and should not vary considerably with current because the effect of the inverse proportional dependence of $\langle i^2\rangle$ on I_{dc} is largely compensated for by the decrease of the loaded Q with increasing current.

This value is considerably lower than what is measured for the RF phase noise (typically 70 Hz) and it is therefore concluded that the phase noise spectrum is not affected by the bias network impedance. This has been substantiated by experiment.

III. Experimental Results

A. AM Spectrum Close to Carrier

Using the techniques described in Section IV, the oscillator noise spectra of two diodes in the frequency range from 0.5 to 500 kHz were measured under two bias network impedance conditions (for convenience they will be referred to simply as open and short circuit conditions) and with the diode current as parameter. From Fig. 2 it is seen that for frequencies above 100 kHz the NSR for open and short circuit conditions differ by approximately 10 dB, the open circuit giving the best NSR. Below 100 kHz the effect of the bias network impedance gradually decreases and disappears for frequencies lower than 1 kHz. This behavior is to be expected from an examination of the equivalent circuit of Fig.1. The frequency dependent R_{TH} is fully effective at low frequencies and prevents a low external impedance from being placed across the junction. The reason for the ascent of the open circuit NSR (solid line) is not obvious. One would expect the open circuit NSR to remain constant for frequencies close to the carrier. It is believed that the rise in the noise sidebands is caused by some type of excess noise with a $1/f$ dependence since a similar behavior of the open circuit noise voltage of the diode was observed. Measuring errors were excluded because the effect was independent of the absolute noise power level and some diodes did not show this dependence at all.

Fig. 3 shows the results of the same measurements on a recently fabricated diode (diode 2). The general characteristics are the same, but the NSR has improved by 10 to 15 dB due to careful control of the semiconductor material processing during diode fabrication.

Indicated on the same graph are the calculated results of Section II Reasonable agreement is obtained if one considers the many simplifying assumptions made in the analysis. It is interesting to note that the measured open circuit NSR is approximately that of the calculated RF noise-to-carrier power ratio $(\mathrm{NSR})_{\mathrm{RF}}$. This leads one to speculate that under open bias circuit conditions the modulation noise is sufficiently suppressed and the RF noise becomes the limiting factor.

B. Noise Bandwidth

The width of the noise spectrum as defined by the 3 dB points is a function of diode current and power level. The widening of the noise spectrum is a consequence of the increasing coupling of the load conductance to the diode resonant circuit. Fig. 4 shows the dependence of the noise bandwidth on the output power level (solid line).

On some diodes, noise spectra of considerably larger bandwidth were observed above a certain power level. A typical curve is shown in the same graph (dotted line). These diodes usually show saturation effects in the signal output which are probably caused by excessive RF losses of the diode.

C. RF Tuning Effects

The dependence of the NSR on the output load coupling at two different power levels is shown in Fig. 5. If the coupling is increased beyond the maximum output power point, the NSR deteriorates rapidly. For a mismatch in the direction of decreasing coupling, an improvement can be obtained. The points indicated by triangles are calculated from the parameter $S_a=\Delta P_0/\Delta I_{dc}$ measured at the -3-dB power points.

D. Dependence of the Amplitude Noise Spectrum on dc Current

In order to estimate the variation of the output spectrum with the current (density), one has to look at the modulation noise and the RF noise separately because their current dependence is different. The noise-to-signal ratio of the modulation noise contribution improves rapidly with increasing current due to the inverse proportional dependence of the mean-square noise current on I_{dc} at low frequencies. At high current densities, a reversal of this behavior can occur due to excess noise generation in the breakdown region and saturation effects on the signal output.

The noise-to-signal ratio due to RF noise is a more complicated function of the current, interdependent on several parameters such as the operating frequency, the magnitude of the negative diode resistance, the avalanche frequency, and the diode loss resistance. As is seen from the expression for the RF noise power (5), the spectral power density P_n/df is a function of the noise measure (M) only. The intensity of the spectrum as a function of current density and operating frequency should therefore closely follow the respective curves given by Gummel and Blue for (M). Because the signal power output increases approximately linearly with current, the same curves could easily be modified to read $(\mathrm{NSR})_{\mathrm{RF}}$. However, the situation is further complicated by the fact that it may be difficult to determine under actual operating conditions which of the two noise contributions is the dominating one. Even without bias network optimization, there may be a crossover point at high current densities where the modulation noise contribution drops below the RF noise level. The current dependence of the amplitude

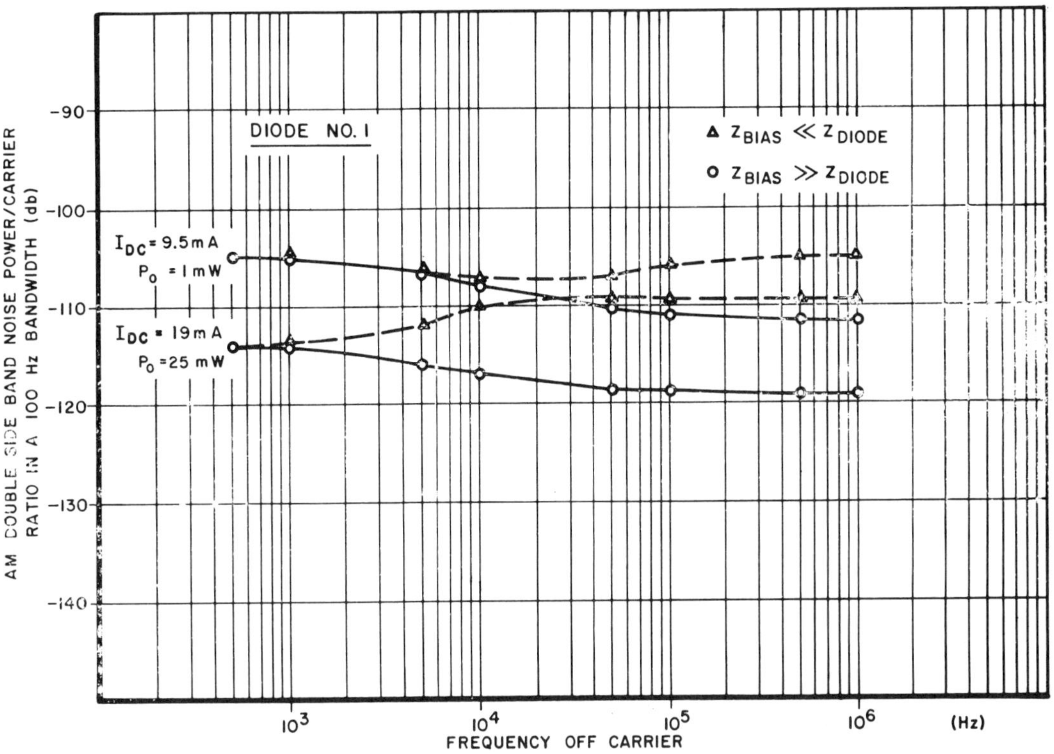

Fig. 2. Amplitude noise spectra of avalanche oscillator close to the carrier as a function of current and bias network impedance (diode 1).

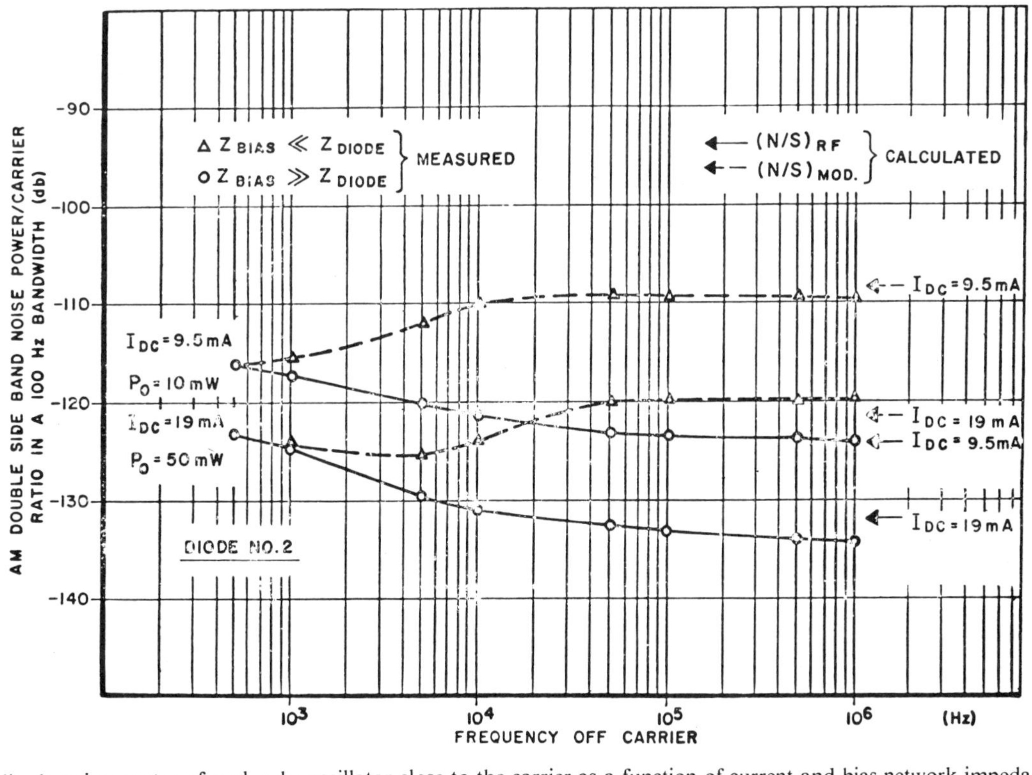

Fig. 3. Amplitude noise spectra of avalanche oscillator close to the carrier as a function of current and bias network impedance (diode 2).

445

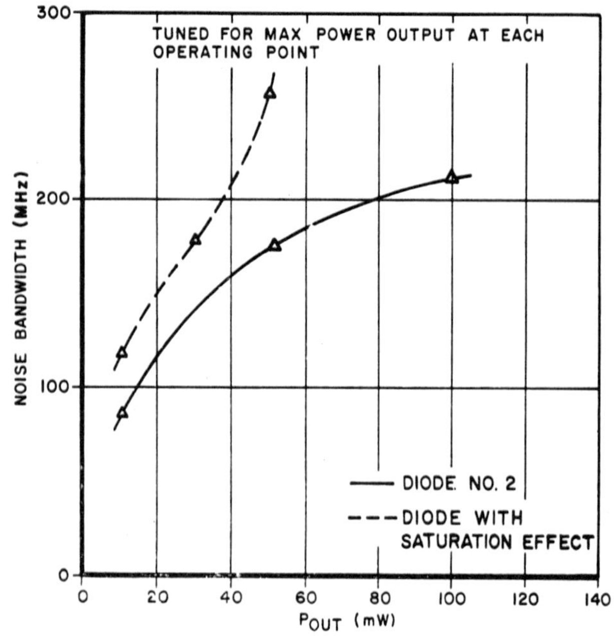

Fig. 4. Bandwidth of oscillator amplitude noise spectrum as a function of power output.

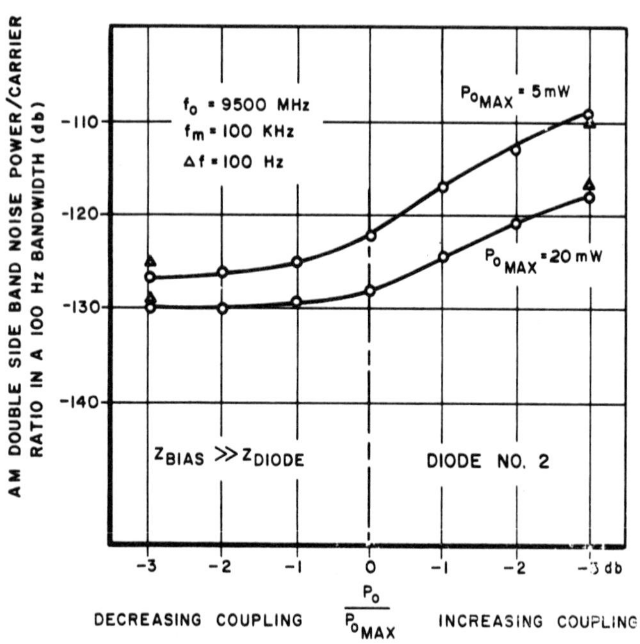

Fig. 5. Dependence of the oscillator amplitude noise spectrum on the RF load coupling.

noise spectrum as predicted from theoretical considerations has been verified experimentally (Figs. 2 and 3). The measured values of NSR check, at least qualitatively, with the calculated numbers.

E. Bias Network Design

At low frequencies ($f_m \leq 1$ MHz) the open-circuit condition was obtained by applying the dc current in series with a resistor of several $k\Omega$ mounted close to the by-pass network. A parallel combination of an electrolytic capacitor and a low-inductance ceramic capacitor was used for the short-

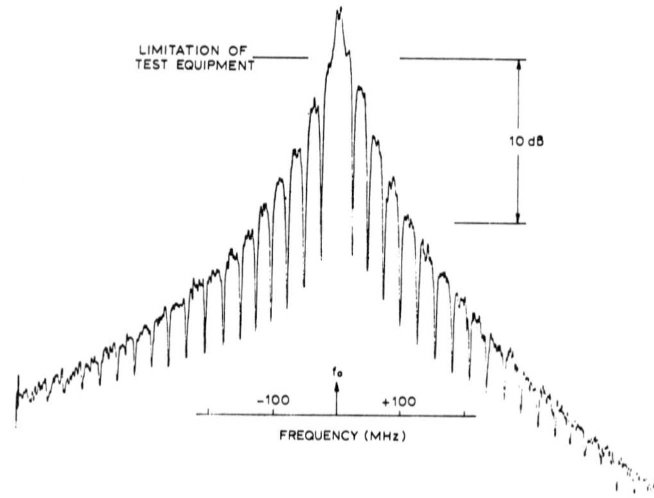

Fig. 6. Measured amplitude noise spectrum of avalanche oscillator with the bias network terminated in a coaxial line with high SWR.

circuit condition. The realization of the open-circuit condition at higher frequencies presents some difficulties because of the low-pass character of the bias network (series $-L$, shunt $-C$). In order to present a high impedance to the diode junction, it is necessary to tune out the shunt reactance. To avoid the problem of high-Q circuits and retuning at each frequency, a length of coaxial cable was connected across the bias output terminal. This provides alternating poles and zeroes of reactance at approximate multiples of 30 MHz. Fig. 6 is a plot obtained on the X–Y recorder showing the upper and lower boundary of the NSR. The diminishing effect of the impedance variations on the spectrum at the high frequencies is due to the residual series inductance of the low-pass filter end section. For practical purposes, a bias network impedance of ten times the low-frequency diode impedance value is sufficient to suppress the noise contribution due to up-converted low-frequency noise components. At higher frequencies it becomes more difficult to maintain the open-circuit condition over a wide frequency range. Some form of series inductor with distributed losses or tuning of the bias network is possible, depending on the application.

F. Complete AM Spectrum

Finally, Fig. 7 shows the complete amplitude noise spectrum typical of the avalanche oscillator. To characterize the spectrum, it is sufficient to evaluate the noise spectrum in the range from approximately 100 Hz to 1 MHz under open- and short-circuit bias network conditons and to specify the noise bandwidth. Implicit in this statement is the assumption that the NSR remains constant in the frequency range from 1 MHz out to frequencies where the filtering effect of the RF resonant circuit begins to affect the noise spectrum. Spot measurements in this frequency range have indicated that the assumption is justified. Beyond the 3-dB points the spectrum falls off theoretically at 6 dB/octave for a simple parallel RLC resonant circuit. Depending on the complexity of the resonant circuit and coupling structure, deviations from the 6 dB/octave law are to be expected. Referring to Fig. 6, it is

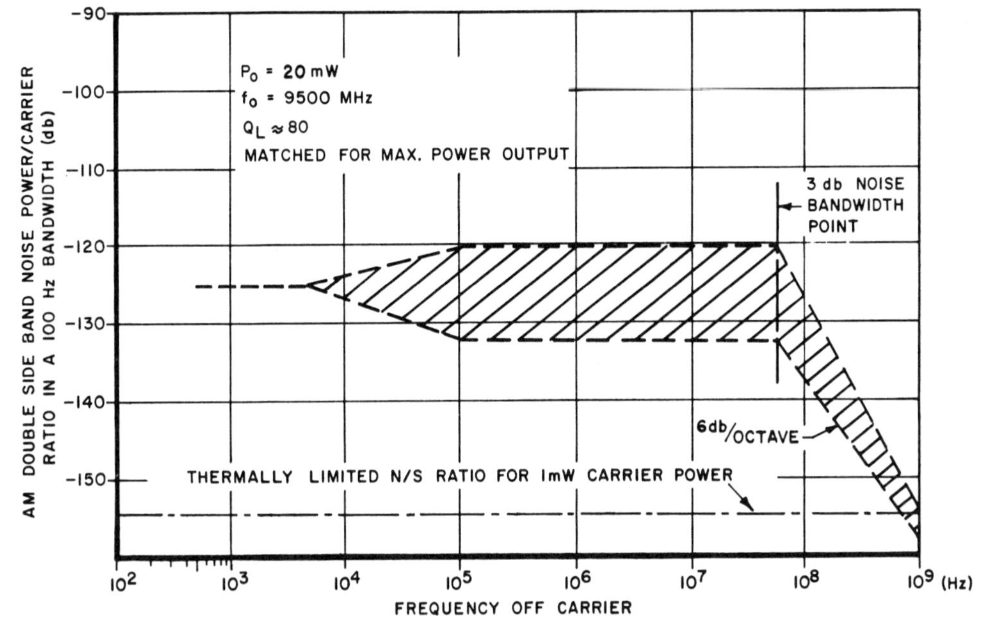

Fig. 7. Typical amplitude noise spectrum of an X-band avalanche oscillator. The shaded area indicates the range where the spectrum is affected by the bias network impedance.

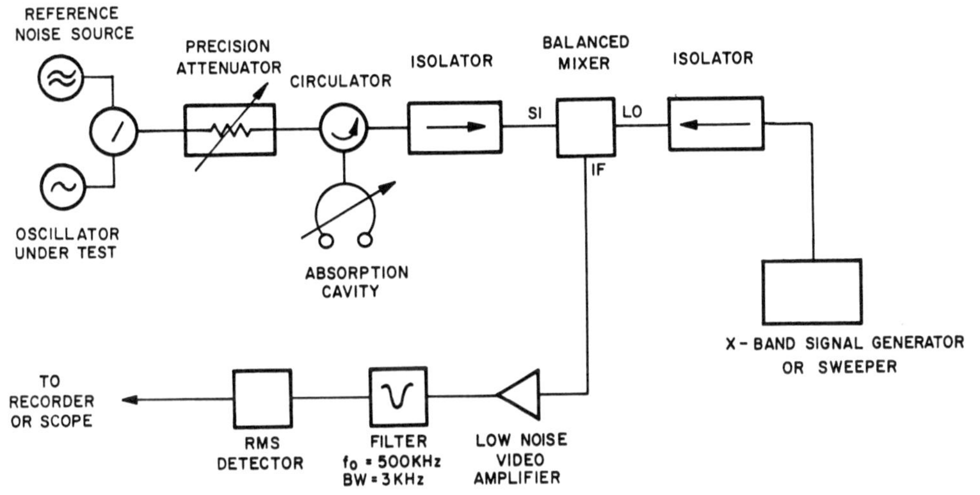

Fig. 8. Block diagram of the test setup for the measurement of amplitude noise spectra far from the carrier.

interesting to note the asymmetry of the noise spectrum. The faster roll-off on the high-frequency side of the spectrum is caused by a pole in the transfer function (above f_0) of the particular network used in this oscillator.

IV. Experimental Techniques

The evaluation of the NSR for frequencies less than 30 MHz off the oscillator signal frequency was performed using the standard methods [10]–[12]. The noisy microwave signal is applied through an isolator and a precision attenuator to a single-ended mixer operating in the direct detection mode. The carrier of the composite signal provides the LO signal, thereby mixing with its own noise side bands. The down-converted noise is then fed to a low-noise video amplifier and measured by a selective rms voltmeter (wave analyzer). The carrier power at the mixer input is held constant at 1 mW. The initial calibration of the system was performed at several frequencies throughout the measuring range. Due to the relatively high noise power levels involved, the residual noise contribution of the system was neglected, resulting in a measuring error of less than 1 dB.

For measurements beyond 30 MHz, the circuit of Fig. 8 shown in block form was employed. Here, a heterodyne system consisting of the balanced mixer and a fixed frequency 1 MHz IF amplifier is used to measure the noise power of the avalanche oscillator. In order to avoid intermodulation problems and to raise the level of the noise side bands, a high Q absorption type cavity ($Q \simeq 4000$) is employed to suppress the carrier of the avalanche oscillator signal. Due to the filtering effect of the cavity and FM to AM conversion, measurements closer than approximately 20 MHz to the carrier cannot be made. The limitation at frequencies far from the carrier is imposed by the system sensitivity and the mixer broad-band performance. Owing to noise figure variations

447

of the mixer, accuracies within 3 dB are anticipated. If the output of the rms detector is amplified in a logarithmic converter, the setup is well suited for swept display on a scope or recorder. Calibration of the system was performed using a broad-band noise source.

V. Conclusions

Below is a summary of the findings.

a) The AM spectrum of the investigated avalanche oscillators is dominated by the modulation noise contribution.
b) The NSR is strongly dependent (up to 20 dB) on the bias network impedance.
c) As long as the modulation noise contribution prevails, the optimum NSR is obtained when the magnitude of the external circuit impedance at f_m presented to the diode is much larger than the low frequency space charge impedance.
d) The NSR is also strongly influenced by the degree of load coupling to the diode circuit. The NSR is improved by under-coupling of the load conductance. If mismatched in the other direction, the noise performance deteriorates rapidly. Matching for maximum power output is a good compromise in terms of noise performance.
e) If the oscillator is operated near the point where saturation of the output signal occurs, a strong dispersion of the output noise spectrum is observed.
f) Optimum noise performance is obtained by operating the oscillator at a signal level as high as possible, but below the point where saturation of the signal output occurs and below the critical current density, at which the avalanche frequency is approaching the frequency of oscillations.

On the basis of analytical and experimental results it is concluded that through optimizing of external circuitry, operating parameters, and improvements in the diode device fabrication, amplitude noise performances of avalanche oscillators approaching those of conventional klystrons and other solid-state sources can be obtained.

Acknowledgment

The author wishes to thank M. E. Hines, R. C. Harrison, W. K. Niblack, and A. H. Solomon for valuable discussions.

References

[1] A. S. Tager, "Current fluctuations in a semiconductor (dielectric) under the condition of impact ionization and avalanche breakdown," *Fiz Tverd. Tela*, vol. 6, p. 2418, August 1964 (transl.: *Soviet Phys.—Solid State*, vol. 6, p. 1919, February 1965).

[2] M. E. Hines, "Noise theory for the read type avalanche diode," *IEEE Trans. Electron Devices*, vol. ED-13, pp. 158–163, January 1966.

[3] H. K. Gummel and J. L. Blue, "A small-signal theory of avalanche noise in IMPATT diodes," *IEEE Trans. Electron Devices*, vol. ED-14, pp. 569–580, September 1967.

[4] W. A. Edson, "Noise in Oscillators," *Proc. IRE*, vol. 48, pp. 1454–1466, August 1960.

[5] J. A. Mullen, "Background noise in nonlinear oscillators," *Proc. IRE*, vol. 48, pp. 1467–1472, August 1960.

[6] M. E. Hines, "Potential applications and the noise problem in the Read avalanche diode," *1966 Internat'l Solid-State Circuit Conf., Digest of Tech. Papers*, pp. 82–83.

[7] J. G. Josenhans, "Noise spectra of Read diode and Gunn oscillators," *Proc. IEEE (Letters)*, vol. 54, pp. 1478–1479, October 1966.

[8] J. G. Ondria and J. C. Collinet, "Investigation of noise reduction techniques of solid-state type microwave sources," *1968 Internat'l Solid-State Circuits Conf., Digest of Tech. Papers*, pp. 82–83.

[9] R. H. Haitz, "Noise of a self-sustaining avalanche discharge in silicon: low-frequency noise studies," *J. Appl. Phys.*, vol. 38, p. 2935, June 1967.

[10] J. G. Ondria, "Study of noise reduction techniques for varactor harmonic generator type microwave sources," U. S. Navy Bureau of Ships, 1st Quart. Report, Contract NObsr.-95009, August–November 1965.

[11] S. Goldman, *Frequency Analysis, Modulation and Noise.* New York: McGraw-Hill, 1948.

[12] A. L. Whitwell and N. Williams, "A new microwave technique for determining noise spectra at frequencies close to the carrier," *Microwave J.*, vol. 2, November 1959.

Reprinted from the PROCEEDINGS OF THE IEEE
VOL. 58, NO. 1, JANUARY, 1970
pp. 155–157

Measured FM Noise Reduction by Injection Phase Locking

Abstract—A 9.72 GHz avalanche transit time oscillator (ATTO) with FM noise of 100 Hz rms/100 Hz was synchronized by a 9.72 GHz stabilized klystron with FM noise of 0.01 Hz rms/100 Hz. The FM noise of the synchronized oscillator was reduced, but not as much as expected for modulation frequencies below 10 kHz. The AM noise was not changed by synchronization.

The purpose of this letter is to report the results of an injection phase locking experiment performed with the intention of making the effect of noise of the synchronizing signal negligible by using a very low noise oscillator.

The block diagram of the experiment is given in Fig. 1. The 9.72 GHz synchronizing signal was derived from a transmission cavity stabilized low noise klystron [1], [2]. The synchronized oscillator was a 100 mW IMPATT diode oscillator [3] using an 85 volt breakdown silicon diode. The external Q of this oscillator was 20. The measurement of FM noise was made with a low threshold discriminator [4] which was calibrated by a known deviation of the klystron.

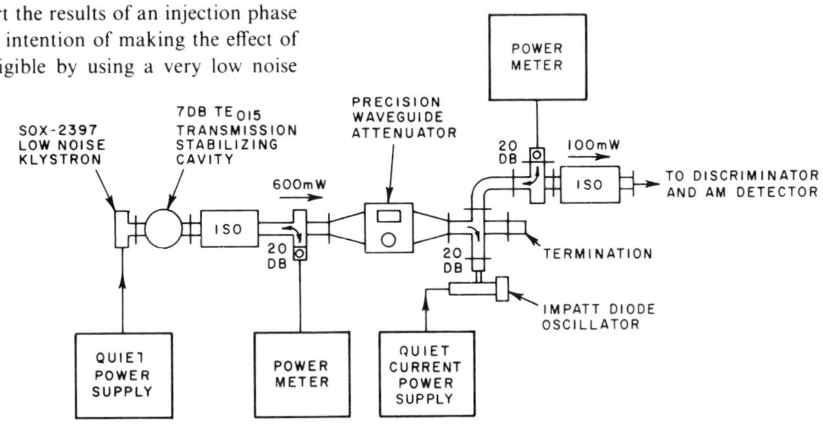

Fig. 1. The injection phase locking experiment.

Manuscript received August 29, 1969; revised October 1, 1969. This work was supported by the U. S. Air Force Avionics Laboratory, Wright-Patterson AFB, Ohio, under Contract F-33-615-68-C-1569.

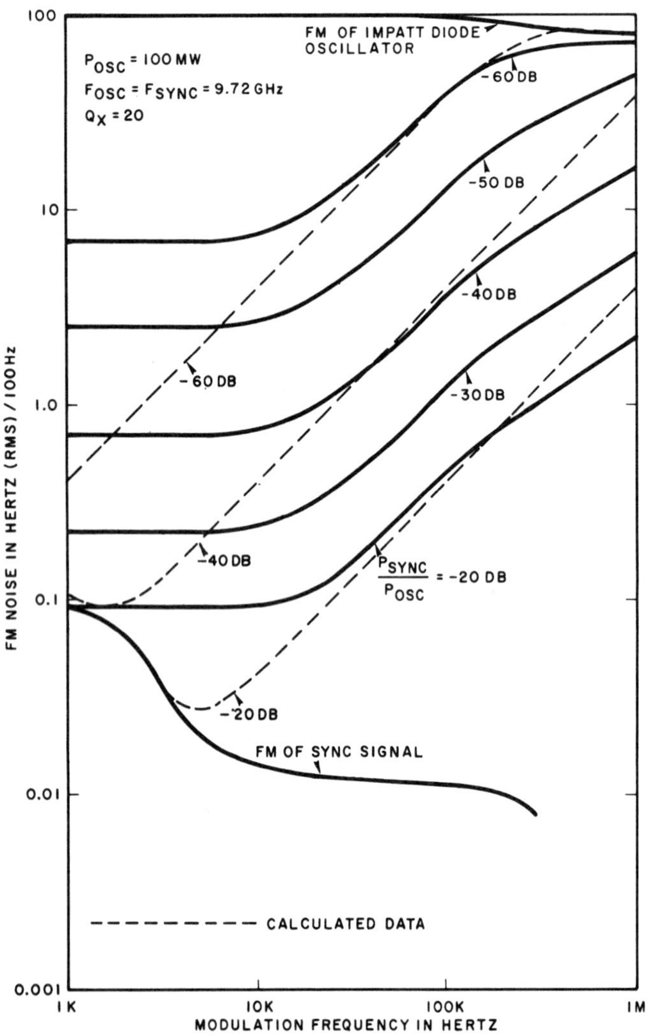

Fig. 2. FM noise in an injection phase synchronized ATTO.

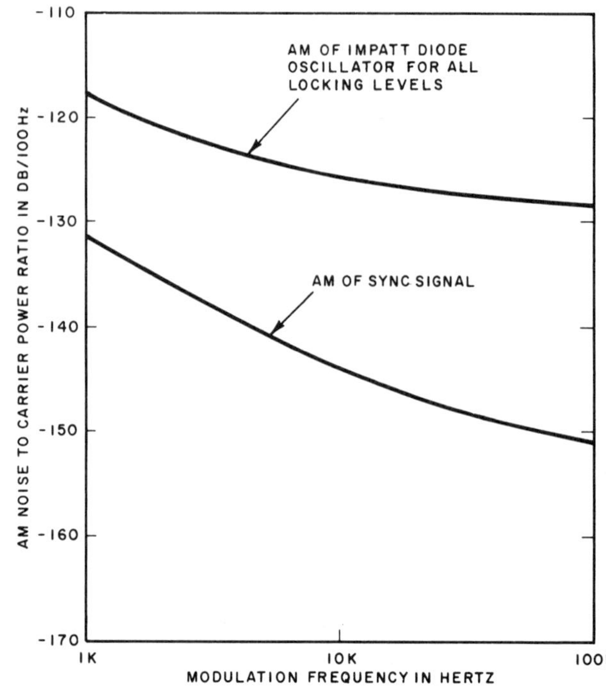

Fig. 3. AM noise of the two oscillators.

The measured FM for various locking levels is shown in Fig. 2. All data were taken with the free-running frequency f_0 set equal to the synchronizing f_{sync}. A check was made for $f_0 - f_{sync} = 5$ MHz and no significant changes from Fig. 2 were observed. Fig. 2 also shows (dotted lines) the FM noise expected of the synchronized oscillator as calculated from Kurokawa's theory as rearranged by Sugiura and Sugimoto [6]. (Equation (2) in [6] was programmed in FORTRAN, and the calculations were performed on a GE time-shared computer terminal.)

The AM noise of the two oscillators was measured (Fig. 3) with Schottky barrier diodes [5]. The AM of the synchronizing signal is within 3 dB of the measurement threshold, while the AM of the ATTO is well above measurement threshold. We searched carefully for the predicted [6] increase in AM under synchronized conditions and, within the 1 dB or so fluctuations of the noise indicator, could not resolve a change in AM with injection locking. At a few modulation frequencies we detected a 1 dB degradation of the AM for the −20 dB locking signal, but this could be explained as the normal range of variation in AM noise.

These FM data were presented at the 1969 IEEE G-MTT International Microwave Symposium [7] and several possible explanations for the data below 10 kHz were advanced by the panelists in later discussions. Some factors brought out in these discussions are worth pointing out.

The first measurements were made in August, 1968, at the Sperry Electronic Tube Division, Gainesville, Fla., on a single diode which operated consistently and quite normally throughout the measurements. Later, this same diode gave similar (unsynchronized) noise data on different power supplies and with the use of another noise measurement setup at Sperry Microwave Electronics Division, Clearwater, Fla. All other performance criteria (power, operating frequency, voltage, current, external Q, etc.) were typical of other diodes from the same lot. After we realized that the low-frequency data might be suspicious, the entire experiment was performed again (May, 1969) on another diode from a later production. The data were in excellent agreement with Fig. 2.

During both experiments, careful checks were made for measurement thresholds, interfering signals, unexpected coupling between oscillators, performance of the synchronizing klystron, etc. No factors were found to make us doubt the data for modulation frequencies below 100 kHz. Ondria and Turlington have also reported this kind of behavior for the FM noise and advanced an explanation [8].

ACKNOWLEDGMENT

The authors wish to acknowledge the assistance of C. B. Searles of Sperry Electronic Division, who built the stabilized klystron used for the synchronizing signal. The associated power supplies and the noise measurement equipment were supplied by Sperry Electronic Tube Division, Gainesville, Fla.

J. ROBERT ASHLEY
University of Colorado
Colorado Springs, Colo. 80907
FRANK M. PALKA
Sperry Microwave Electronics Div.
Sperry Rand Corp.
Clearwater, Fla. 33518

REFERENCES

[1] J. R. Ashley and C. B. Searles, "Microwave oscillator noise reduction by a transmission stabilizing cavity," *IEEE Trans. Microwave Theory and Techniques*, vol. MTT-16, pp. 743–748, September 1968.
[2] C. B. Searles, "An ultra low noise X-band signal source," paper 20.1 presented at the IEEE Internatl. Electron Devices Meeting, Washington, D. C., October 23–25, 1968.
[3] J. W. Amoss and K. E. Gsteiger, "Frequency modulation of avalanche transit time oscillators," *IEEE Trans. Microwave Theory and Techniques*, vol. MTT-15, pp. 742–747, December 1967.
[4] T. Sugiura and S. Sugimoto, "FM noise reduction of Gunn-effect oscillations by injection locking," *Proc. IEEE* (Letters), vol. 57, pp. 77–78, January 1969.
[5] J. R. Ashley, C. B. Searles, and F. M. Palka, "The measurement of oscillator noise at microwave frequencies," *IEEE Trans. Microwave Theory and Techniques*, vol. MTT-16, pp. 753–760, September 1968.
[6] K. Kurokawa, "Noise in synchronized oscillators," *IEEE Trans. Microwave Theory and Techniques*, vol. MTT-16, pp. 234–240, April 1968.
[7] J. R. Ashley, J. G. Josenhans, D. Leeson, D. E. Wunsch, K. E. Gsteiger, T. R. Turlington, J. F. White, and R. A. Campbell, panel discussion on "Noise in solid state devices and systems," IEEE G-MTT Internatl. Microwave Symp., Dallas, Tex., May 5–7, 1969.
[8] J. G. Ondria and T. R. Turlington, "A feedback servo analysis of the injection lock process and noise suppression in Gunn and IMPATT oscillators," Govt. Microcircuit Applications Conf., Washington, D. C., September 18, 1969.

F.M. NOISE OF LOW-LEVEL-OPERATING IMPATT-DIODE OSCILLATORS

Indexing terms: Oscillators, Noise, IMPATT diodes

An analytical relationship for the output-spectrum width of a low-level-operating IMPATT-diode oscillator is verified experimentally by measuring relevant diode and circuit quantities as well as the oscillator noise. Good agreement between theory and experiment is found if the spectrum width is assumed to be predominantly due to f.m. noise.

IMPATT-diode-oscillator noise originates from the random character of the thermal-generation and impact-ionisation processes. As shown by Hines,[1] the noise connected with these processes can be described by a related total noise current flowing through the diode. Two mechanisms can be distinguished by which the total noise current manifests itself in the output spectrum of an oscillating diode.[2] The first mechanism is nonlinear amplification of the high-frequency components of the total noise current, and causes the 'intrinsic' noise. The second mechanism is up-conversion of the low-frequency components of the total noise current into a.m. and f.m. oscillator noise. In this letter, we report an experimental verification, on four types of diodes, of the theory on the intrinsic noise given in Reference 2. For the present discussion, the up-converted noise can be neglected.[3]

For moderately high signal levels, the output signal can be assumed to consist mainly of narrowband linearly amplified noise. The nonlinearity then only enters as the mechanism that limits the magnitude of the output signal. With these assumptions, for a Read model, the width $2\Delta f_0$ of the output spectrum, caused by the intrinsic noise, is given by eqn. 19 of Reference 2* as follows:

$$2\Delta f_0 = \frac{e}{4\pi\tau_i^2}|\Phi|^2 \frac{1}{\left\{\frac{W}{l_d} - \left(1 - \frac{\sin\theta}{\theta}\right)\frac{f_{a,st}^2}{f_0^2}\right\}^2}\frac{I_{dc}R_L}{P_0} \qquad (1)$$

In this relationship, e is the electronic charge, τ_i is the intrinsic response time of the avalanche process, W and l_d are the lengths of the depletion layer and drift region, respectively, θ is the transit angle of the carriers in the drift region, $f_{a,st}$ is the avalanche frequency at the start of the oscillations,[3] f_0 is the oscillation frequency, I_{dc} the d.c. bias, and P_0 is the output power dissipated in the load resistance R_L. The transit-time function Φ is given by

$$\Phi = \frac{l_a}{l_d} + \frac{1 - e^{-j\theta}}{j\theta} \qquad (2)$$

where l_a is the length of the avalanche region.

For the verification of eqn. 1, we performed small-signal impedance and noise measurements on the nonoscillating diode, from which we obtained the values of all diode quantities needed in this equation. The load resistance R_L was found from large-signal impedance measurements.[4] For a given combination of bias current, oscillation frequency and resulting output power, the spectrum width of the oscillator signal can then be predicted from eqn. 1.

Measurements on the nonoscillating diode: The small-signal impedance of the diode was measured as a function of bias current at a fixed frequency in the X band. The values of the parameters W, l_d, θ and b (where b is defined by $f_a^2 = bI_{dc}$, f_a being the avalanche frequency) were found from the measured impedance data, assuming that they could be described by the theoretical expression for this impedance as given by the Read model (eqn. 8 of Reference 2). Noise measurements were carried out in order to obtain τ_i, which governs the internal noise properties of the diode. In the analysis of τ_i, the total noise current of the nonoscillating diode was expressed by the theoretical expression given by the same Read model (eqn. 14 of Reference 2). A full account of the above sketched procedure will be published.† The diode quantities relevant to eqn. 1, as found from the experiments discussed, are listed in Table 1, where V_{BR} is the breakdown voltage, and f is the measuring frequency.

Measurements on the oscillating diode: From large-signal impedance measurements,[4] we found the value of the circuit parameter R_L. Furthermore, information on the signal level of the diode and of the validity of the Read model could be obtained.

If I_{st} is the start-oscillation current and v_a is the amplitude of the alternating voltage across the avalanche region, it can be shown[3] that, for $I_{dc} \geq I_{st}$ and a constant impedance of the external circuit,

$$\frac{I_{st}}{I_{dc}} = \frac{2}{\gamma v_a}\frac{I_1(\gamma v_a)}{I_0(\gamma v_a)} \qquad \gamma = \frac{\bar{\alpha}'}{\omega_0 \tau_i} \qquad (3)$$

Here, $\bar{\alpha}'$ is the field derivative of the average ionisation coefficient, and I_1 and I_0 are modified Bessel functions. Fig. 1 gives an example of the graph $\gamma(v_a)$ for two values of I_{st} as found from the large-signal impedance measurements, also using eqn. 3. In the range where γ is a constant, the diode can be described conveniently by the Read model.[3] It was found that, for all diodes investigated, the point $\gamma v_a = 1$ was within this range.

Oscillator noise: To compare the results of all oscillator-noise measurements, the bias current was always chosen so that $\gamma v_a = 1$; i.e. $I_{dc} = 1{\cdot}12\, I_{st}$ according to eqn. 3. Furthermore, the cavity-loaded Q factor was kept low (<30) in order to study the behaviour of the oscillating diode only.

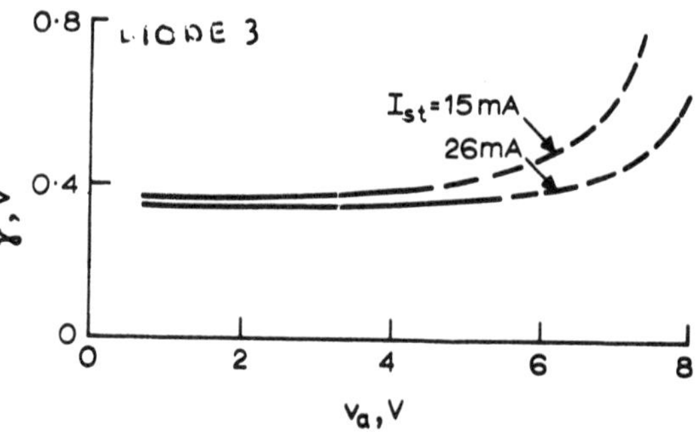

Fig. 1 *Example of graph of γ as a function of v_a for two values of start-oscillation current I_{st}*

The range where γ is a constant determines the range of validity of the Read diode model

If we assume that the intrinsic noise is predominantly f.m. noise, the r.m.s. frequency deviation due to the noise in a strip of bandwidth df, at a distance $f_m \gg 2\Delta f_0$ from the carrier, is given by[2]

$$\Delta f_{rms} = \left(\frac{1}{2\pi} 2\Delta f_0\, df\right)^{\frac{1}{2}} \qquad (4)$$

In Table 2, the values of the calculated and measured Δf_{rms} are given. The calculated Δf_{rms} was found from eqns. 1 and 4 with $df = 100$ Hz, using the measured diode and circuit quantities. The measured Δf_{rms} was found directly from oscillator-noise measurements ($df = 100$ Hz, $f_m = 200$ kHz) and shows very good agreement with the calculated Δf_{rms}. The measured a.m. noise power in two sidebands relative to the carrier power was about −105 dB for all diodes investigated ($df = 100$ Hz, $f_m = 200$ kHz). About −60 dB would be expected if the intrinsic noise were predominantly a.m. noise.

Conclusion: We conclude from Table 2 that, for all diodes investigated, the assumption that the intrinsic oscillator noise is predominantly f.m. noise resulted in a very good agreement between predicted and measured values of Δf_{rms}. Furthermore, we conclude that the theory given in Reference 2

* Eqn. 19 of Reference 2 reads $2\Delta\omega_0 \simeq \omega_0^2 RC$. Eqn. 1 of the present paper is easily found by substituting eqns. 18 and 13 and the imaginary part of eqn. 8 of Reference 2 into eqn. 19. In eqn. 13, we replaced $\tau_a/2$ by τ_i and assumed that $\omega_0 M\tau_i \gg 1$

Reprinted from ELECTRONICS LETTERS, Vol. 7, No. 16

Table 1

Diode number	Type of diode	V_{BR}	f	τ_i	$\|\Phi\|^2$	l_d/W	θ	b
		V	GHz	ps				Hz^2/A
1	Si p^+–n	66	10	3·6	2·2	0·64	1·2	$8{\cdot}1 \times 10^{20}$
2	Si p^+–n	100	10	6·0	1·9	0·59	1·9	$5{\cdot}6 \times 10^{20}$
3	Si n^+–p	110	9	5·4	1·1	0·62	2·7	$5{\cdot}3 \times 10^{20}$
4	Ge n^+–p	32	9	7·4	1·6	0·67	1·6	$9{\cdot}1 \times 10^{20}$

Table 2

Diode number	I_{dc}	R_L	P_0	f_0	Δf_{rms} in 100 Hz (calculated)	Δf_{rms} in 100 Hz (measured)
	mA	Ω	mW	GHz	Hz	Hz
1	35	3·1	9	10	416	423
2	33	1·6	12	10	132	131
3	29	1·6	19	9	86	73
4	39	6·8	15	9	245	300

provides for a good description of the noise behaviour of low-level-operating IMPATT-diode oscillators.

Acknowledgment: The author is indebted to J. H. Verduin for his assistance with the experiments, to H. Tjassens for discussions concerning the microwave circuits, and to H. G. Kock for fabricating the diodes.

J. J. GOEDBLOED

Philips Research Laboratories
NV Philips' Gloeilampenfabrieken
Eindhoven, Netherlands

8th July 1971

† GOEDBLOED, J. J.: 'Determination of the intrinsic response time of semiconductor avalanches from microwave measurements' (to be published)

References

1 HINES, M. E.: 'Noise theory for the Read type avalanche diode', *IEEE Trans.*, 1966, **ED-13**, pp. 158–163

2 VLAARDINGERBROEK, M. T., and GOEDBLOED, J. J.: 'On the theory of noise and injection phase locking of IMPATT diode oscillators', *Philips Res. Rep.*, 1970, **25**, pp. 452–471

3 GOEDBLOED, J. J.: 'On the up-converted noise of IMPATT diode oscillators'. Proceedings of MOGA conference, Amsterdam. Sept. 1970 (Kluwer–Deventer, 1970), pp. 12.36–12.40

4 VAN IPEREN, B. B., and TJASSENS, H.: 'Measurement of large-signal impedance, optimum a.c. voltage and efficiency of Si *pnn*$^+$, Ge *npp*$^+$ and GaAs Schottky barrier avalanche transit-time diodes'. Proceedings of MOGA conference, Amsterdam, Sept. 1970 (Kluwer–Deventer, 1970), pp. 7.27–7.32

Reprinted from the PROCEEDINGS OF THE IEEE
VOL. 59, NO. 8, AUGUST, 1971
pp. 1265–1266

On the Linear Theory of Noise in Read Diodes

Abstract—The linear theory of the noise of Read diodes is extended to the case of different ionizing coefficients for electrons and holes and a finite multiplication factor. It is shown that the noise is smallest when the carriers injected in the avalanche region are of the less strongly ionizing type.

The purpose of this letter is to present an extension of the linear theory of Hines[1] of the noise in Read diodes to the more general case of different ionization coefficients α and β of electrons and holes and a finite M multiplication factor. We deal with the Read model of an avalanche diode—a high field narrow uniform region followed by a drift region; the electrons and holes drift at the same limiting velocity v, independent of the electric field.

In the presence of noise we can write the following Langevin's equation for the electrons and hole currents J_n, J_p.

$$\frac{1}{v}\frac{\partial J_p}{\partial t} + \frac{\partial J_p}{\partial x} = \alpha J_n + \beta J_p + g(x, t)$$

$$\frac{1}{v}\frac{\partial J_n}{\partial t} - \frac{\partial J_n}{\partial x} = \alpha J_n + \beta J_p + g(x, t).$$

The noise source $g(x, t)$ is the sum of two terms $g_1(x, t)$, describing the production of noise in the avalanche region through the generation of electron–hole pairs, and $g_2(x, t)$, the production of noise due to the injected carriers.

Assuming full shot noise we have

$$\langle g(x, t)\rangle = \langle g_1(x, t)\rangle = \langle g_2(x, t)\rangle = 0$$

$$\langle g_1(x, t) g_1(x', t')\rangle = e\langle \alpha J_n + \beta J_p\rangle \delta(t - t')\cdot \delta(x - x')$$

$$\langle g_2(x, t) g_2(x', t')\rangle = e\langle(\alpha J_p(x)\delta(x) + \beta J_n(x)\delta(x - l))\delta(t - t')\rangle.$$

If the transit time $\tau_1 = l\cdot v^{-1}$ across the avalanche region is small, it is straightforward (though perhaps lengthy[2]) to establish the following generalized "Read equation."

$$\frac{2\beta[(\alpha + \beta)l - 2]}{(\alpha - \beta)^2 l}\tau_1 \cdot \frac{dJ}{dt} = J\frac{\beta - \alpha e^{-(\alpha - \beta)l}}{\alpha - \beta} + J_p(0)e^{-(\alpha - \beta)l} + J_n(l) + G(t). \quad (1)$$

The driving current $J(t)$ is defined by

$$J(t) = \frac{1}{l}\int_0^l \langle \alpha J_n + \beta J_p\rangle dx$$

and the noise source (avalanche noise plus injection noise) is characterized by the ensemble averages

$$\langle G(t)\rangle = 0$$

$$\langle G(t), G(t')\rangle = e\delta(t - t')\left[2J_n(l) + 2J_p(0)e^{-2(\alpha - \beta)l} + J\frac{\beta - \alpha e^{-2(\alpha - \beta)l}}{\alpha - \beta}\right]. \quad (2)$$

In the limit of negligible τ_1 (1) and (2) lead to McIntyre's result (his (17)).[3]

From (1) and (2) and the usual relations describing the drift region, we can calculate the open circuit noise voltage of the diode. The noise voltage spectral density at the frequency $f = \omega/2\pi$ is

$$\langle V^2\rangle_f = \frac{\eta^2 + \zeta^2}{C^2\omega^2}\cdot\frac{2eJ_B M^2}{1 + M^2\omega^2\tau^2\left[1 - \frac{\omega_a^2}{\omega^2}\right]^2} \quad (3)$$

where

$$\eta = \frac{C}{C_1}\,\frac{1 - \cos\omega\tau_2}{\omega\tau_2} \qquad \zeta = \frac{C}{C_1} + \frac{C}{C_2}\,\frac{\sin\omega\tau_2}{\omega\tau_2}.$$

C, C_1, and C_2 are the capacitances of the diode and of the avalanche and drift regions, respectively, and $\omega\tau_2$ is the transit angle of the drift region.

$$M = [\alpha - \beta]\left[\beta\exp\left(\frac{(\alpha - \beta)l}{2}\right) - \alpha\exp\left(-\frac{(\alpha - \beta)l}{2}\right)\right]^{-1}$$

$$\tau = \tau_1\cdot\frac{2\sqrt{\alpha\beta}[(\alpha - \beta)l - 2]}{(\alpha - \beta)^2 l}$$

and

$$\omega_a^2 = \frac{J_0}{C_1\tau}\cdot\frac{\alpha^\circ\beta(\alpha l - 1) - \beta^\circ\alpha(\beta l - 1)}{\sqrt{\alpha\beta}(\alpha - \beta)}.$$

J_0 is the dc current and α° and β° are the derivatives of α and β with respect to the electric field.

The equivalent noise current J_B is related to the injected and dc currents through

$$J_B = 2J_n(l)e^{(\alpha - \beta)l} + 2J_p(0)e^{-(\alpha - \beta)l} + J_0\frac{\beta e^{(\alpha - \beta)l} - \alpha e^{-(\alpha - \beta)l}}{\alpha - \beta}.$$

Together with the expression of the impedance of the diode

$$Z = -\frac{j}{C\omega} + \frac{\eta + j\zeta}{C\omega}\,\frac{jM\omega\tau}{\frac{\omega^2}{\omega_a^2} + jM\omega\tau\left(1 - \frac{\omega^2}{\omega_a^2}\right)}$$

is obtained only when the avalanche frequency is negligible (a condition which is met in photodiodes).

2) With McIntyre, we observe that the electrons and holes give different contributions to the noise source. If only holes are injected, the equivalent noise current is

Manuscript received February 4, 1971.

[1] M. E. Hines, "Noise theory for the Read type avalanche diode," *IEEE Trans. Electron Devices*, vol. ED-13, Jan. 1966, pp. 158–163.

[2] More detailed calculations will be given in G. Convert, "Sur la théorie du bruit des diodes à avalanche," to be published in *Rev. Technol. Thomson-CSF*.

$$J_{Bp} = J_0 - \frac{2J_0}{M}\cdot\frac{\alpha - \beta}{\sqrt{\alpha\beta}}. \tag{4a}$$

If electrons are injected

$$J_{Bn} = J_0 + \frac{2J_0}{M}\cdot\frac{\alpha - \beta}{\sqrt{\alpha\beta}}. \tag{4b}$$

The noise source is the *smallest* when the carriers entering the avalanche region are of the *less* strongly ionizing type. In fact there is no contradiction between this result and McIntyre conclusions. This author is interested in the noise-to-signal ratio in photodiodes. This ratio is proportional to $R_1 = M^2 J_{Bp}/J_p$ or $R_2 = M^2 J_{Bn}/J_n$ whether we have holes or electrons; from (4) we obtain

$$R_1 \sim M_p^3 \frac{\alpha}{\beta}\left[1 - \frac{\alpha - \beta}{2M_p\alpha}\right], \qquad R_2 \sim M_n^3 \frac{\beta}{\alpha}\left[1 + \frac{\alpha - \beta}{2M_n\beta}\right]$$

where M_p and M_n are the multiplication factors for holes and electrons. These results coincide with McIntyre's results [his (27) and (28)] (we have neglected M^{-2} terms). We see that, though the noise due to the more ionizing carriers is the highest, it is better to use them in photodiodes because their associated gain is the highest. On the contrary, in an avalanche diode oscillator, within the limit of the Read model, the noise is smaller when the less ionizing carriers are injected. However, the difference between the two types of carriers is not very significant in the limit of the high multiplication factor.

Equation (3) describes completely the behavior of the diode at low level operation, hence we can make the following remarks.

1) The noise power is finite at all frequencies provided M remains finite. At very low frequencies the noise power attains a limit

$$\langle V^2 \rangle_f = \frac{2eJ_B}{C^2\omega_a^4\tau^2}$$

independent of M (which may be assumed to be infinite). The behavior suggested by McIntyre

$$\langle V^2 \rangle_f \sim \frac{M^2}{1 + M^2\omega^2\tau^2}$$

[3] R. J. McIntyre, "Multiplication noise in uniform avalanche diodes," *IEEE Trans. Electron Devices*, vol. ED-13, Jan. 1966, pp. 164–168.

GUY CONVERT
Lab. Central de Recherches
Thomson-CSF, Corbeville
91-Orsay, France

Reprinted from the PROCEEDINGS OF THE IEEE
VOL. 59, NO. 8, AUGUST, 1971
pp. 1266–1267

A Nonlinear Theory of Noise in Read Diodes

Abstract—A nonlinear theory of noise in a Read diode which takes account of the coupling between low and high frequencies has been developed. At high levels, the components of the driving noise voltage of the diode, in phase and out of phase with the signal, have unequal values and are correlated.

The general problem of the noise in nonlinear oscillators has been studied by many authors [1]. Most of these works can be summarized in an equation which gives the perturbation of the current (or voltage) of the oscillator under the action of a driving noise voltage. To be specific, let us consider an active device (a diode) defined by an impedance $Z_d(\omega, J)$, which is a function of the frequency and the amplitude of the current of oscillation $J \exp[j(\omega_0 t + \phi_0)]$. If the diode oscillates on an external circuit with an impedance $Z_e(\omega)$, ω_0 and J are determined, in the absence of noise, by setting the total impedance

$$Z(\omega, J) = Z_d(\omega, J) + Z_e(\omega)$$

equal to zero. If we take account of the noise, the signal is perturbed. It may be written

$$[J + i(t)] \exp [j(\omega_0 t + \phi_0)]$$

where $i(t)$, which is related to the fluctuations of the amplitude $\rho(t)$ and the phase $\phi(t)$ by

$$i(t) = \rho(t) + jJ\phi(t)$$

is assumed to be quasi-stationary within a period of oscillation. The formal expression of the perturbation equation is

$$-j\frac{\partial Z}{\partial \omega}\frac{di}{dt} + J\frac{\partial Z}{\partial J}\cdot\frac{i + i^*}{2} = v(t) \tag{1}$$

where $v(t)$ is defined as the slowly varying amplitude of the driving noise voltage near the ω_0 frequency.

It is commonly assumed [1] that the noise source does not depend on the level of oscillation, so that the description of the noise (for given values of frequency and amplitude) relies only on the two complex quantities $\partial Z/\partial \omega$ and $\partial Z/\partial J$. Such a phenomenological theory indeed provides a good frame for the experimental study of the noise of oscillators. However, to obtain quantitative results, it is necessary to determine explicit expression of $\partial Z/\partial \omega$ and $\partial Z/\partial J$ and of the noise spectra. We have made such a calculation for the particular case of the Read model of an avalanche diode, within the limits of the high level theory developed by Delagebeaudeuf [2].

Manuscript received February 4, 1971.

Because of the mixing of low and high frequencies through the signal, a complete description of the noise necessitates two coupled equations instead of a single one of the type (1). However, if the coupling is small, a condition which can be met by using a high resistance in the supply circuit, the coupled system can be reduced to (1). The expression of $J(\partial Z/\partial J)$ in full form is intricate, but at very high levels of oscillation it becomes simply

$$J\frac{\partial Z}{\partial J} = \frac{\eta + j\zeta}{C\omega}\left|\frac{2J_0}{J}\right|, \qquad Z = R + jX$$

while in the same limit, the diode impedance is

$$Z_d(\omega, J) = -\frac{j}{C\omega} - \frac{\eta + j\zeta}{C\omega}\left|\frac{2J_0}{J}\right|.$$

J_0 is the dc current. Noting C, C_1, and C_2 for the capacitances of the diode and of the avalanche and drift regions, we have

$$\eta = \frac{C}{C_2}\cdot\frac{1 - \cos\omega\tau_2}{\omega\tau_2} \qquad \zeta = \frac{C}{C_1} + \frac{C}{C_2}\,\frac{\sin\omega\tau_2}{\omega\tau_2}$$

with $\omega\tau_2$ the transit angle of the drift region. We see that $J(\partial R/\partial J)$ is equal to the absolute value of the resistance of the diode. With usual design $\partial X/\partial J$ is of the same order or even larger than $\partial R/\partial J$. The noise source $v(t)$ depends on the amplitude of oscillation through two mechanisms.

At high levels, the carriers current in the avalanche region is strongly modulated. Due to this effect, the components of noise in phase and out of phase with the signal are unequal and larger than in low level operation.

As the amplitude of the signal increases, the coupling between low and high frequencies becomes larger introducing new noise components. Calculations, assuming full shot noise, give the following expressions of the spectra of the driving noise components, in phase v_1, and in quadrature v_2 with the signal

$$\langle \overline{v_1^2}\rangle_f = 2eJ_0 d^2 I_0\left[\frac{a^2\eta^2}{2}(I_0 - I_2) + \frac{a^2\zeta^2}{2}(I_0 + I_2) + b^2\eta^2 I_0 - 2ab\eta\zeta I_1\right]$$

$$\langle \overline{v_2^2}\rangle_f = 2eJ_0 d^2 I_0\left[\frac{a^2\eta^2}{2}(I_0 + I_2) + \frac{a^2\zeta^2}{2}(I_0 - I_2) + b^2\zeta^2 I_0 + 2ab\eta\zeta I_1\right] \tag{2}$$

and

$$\langle \overline{v_1 \cdot v_2}\rangle_f = 2eJ_0 d^2 I_0[-a^2\eta\zeta I_2 + b^2\eta\zeta I_0 + ab(\eta^2 - \zeta^2) I_1].$$

The symbol $\langle \cdot \rangle_f$ means ensemble average plus time average of the noise spectrum per hertz of (positive) bandwidth. $I_0 I_1 I_2$ are the modified Bessel functions of the argument U defined implicitly by (ω_a avalanche frequency)

$$J = J_0\left[2\frac{I_1(U)}{I_0(U)} - \frac{\omega^2}{\omega_a^2}U\right].$$

Finally

$$a = 1 + \frac{I_2}{I_0} - \frac{2I_1^2}{I_0^2} \quad \alpha_0 = I_0\left(a - \frac{\omega^2}{\omega_a^2}\right) \quad b = \frac{\omega}{CR_t\omega_a^2}\cdot\frac{I_1}{I_0} \quad d = \frac{1}{C\omega_a^2\tau}\cdot\frac{I_0}{\alpha_0}.$$

R_t is the total resistance of the low-frequency circuit, τ is equal to one-third the transit time of the avalanche region when the ionization coefficients of electrons and holes are equal.

At low levels, v_1 and v_2 are uncorrelated and their spectra are equal. Their value is

$$\langle v_1\rangle_f^2 = \langle v_2^2\rangle_f = \frac{1}{2}\cdot\frac{\eta^2+\zeta^2}{C^2\omega_a^4\tau^2}\cdot\frac{2eJ_0}{\left(1-\frac{\omega^2}{\omega_a^2}\right)^2}$$

which can be obtained from the linear theory of the noise. The correlation observed between v_1 and v_2 at high level is due to the strong modulation of the current of the carriers and the mixing between low and high frequencies.

Because we have assumed that the noise origin is only shot noise, the noise which is generated at low frequencies (b terms) is correlated to the noise generated directly at high frequencies (a terms).

From (2) we can conclude that we cannot minimize at the same time amplitude noise and phase noise.

GUY CONVERT
Lab. Central de Recherches
Thomson-CSF, Corbeville
91-Orsay, France

REFERENCES

[1] W. A. Edison, "Noise in oscillators," *Proc. IRE*, vol. 48, Aug. 1960, pp. 1454–1466; P. Grivet and A. Blaquiere, "Comments on 'Normalized equations of the regenerative oscillator-noise, phase-locking, and pulling'," *Proc. IEEE* (Corresp.), vol. 53, May 1965, pp. 518–519; M. Lax, *Phys. Rev.*, vol. 160, 1967, p. 290; K. Kurokawa, *Bell Syst. Tech. J.*, vol. 48, 1969, p. 1937.
[2] D. Delagebeaudeuf, *Rev. Tech. Thomson-CSF*, vol. 1, no. 3, 1969.
[3] Detailed calculations will be given in G. Convert, "Sur la théorie des diodes à avalanche," to be published in *Rev. Tech. Thomson-CSF*.

INJECTION LOCKING IN IMPATT DIODE OSCILLATORS *

M. S. GUPTA AND R. J. LOMAX
ELECTRON PHYSICS LABORATORY, DEPT. OF ELECTRICAL AND COMPUTER ENGINEERING, THE UNIVERSITY OF MICHIGAN, ANN ARBOR, MICHIGAN, 48104

ABSTRACT

A model for an injection-locked IMPATT diode oscillator is presented and used to calculate the locking behavior of the oscillator at different operating points. The model is compared with earlier theories of injection locking in IMPATT diode oscillators and the calculated results are compared with experimental observations.

INTRODUCTION

A model of an injection phased-locked oscillator is presented which is applicable to an IMPATT diode oscillator. Locking is treated as a forced oscillation phenomenon. The locked oscillator is modeled by a Read-type IMPATT diode, packaged in a Sylvania 023 package and mounted in a coaxial line cavity with sliding slugs for tuning. A large-signal analysis of the diode and a broadband characterization of the circuit are carried out numerically on a digital computer. The dependence of locking bandwidth, phase difference, locking figure of merit and output power change upon the locking gain and the operating point of the IMPATT diode is calculated from the model and the results are compared with previously published and some new experimental results.

INJECTION-LOCKED OSCILLATOR MODEL

In order to clarify the scope and contribution of this paper, a self-explanatory comparison of the various models which have been applied to IMPATT diode oscillators is made in Table I.

The locked oscillator is modeled by a series circuit consisting of the diode, the circuit impedance and the locking source as shown in Fig. 1. The circuit impedance Z_c is that seen at the wafer terminals and is therefore the load impedance transformed by the diode package and the oscillator cavity. As the locking signal also reaches the diode through the two-port representing the package and cavity, the locking signal source e_s shown in Fig. 1 has also been transformed through the two-port. The locking signal and the locked oscillator are both assumed to be noiseless.

The diode is modeled by an n^+pip^+ Read-type IMPATT diode structure with an idealized electric field profile as shown in Fig. 2a. The differential equations governing the terminal voltage and current behavior of the diode, derived under some commonly used simplifying assumptions have been reported earlier[1] and will not be repeated. The resulting diode impedance is calculated by means of a large-signal analysis as a function of the frequency and amplitude of a sinusoidal voltage applied across the semiconductor wafer terminals and is plotted in the impedance plane in Fig. 2b.

The diode is assumed to be packaged in a Sylvania 023 package, shown in Fig. 3a, and it is operated in a coaxial-line cavity with two sliding slugs for circuit tuning. Details of the modeling of this circuit by means of lumped circuit elements and lossless transmission lines have been given.[1] Briefly, the package is represented by a six-element L-C ladder network with experimentally evaluated element values. The cavity is represented by a cascade of lossless transmission lines including the discontinuity capacitances C_s due to tuning slugs. The complete equivalent network seen at the semiconductor wafer terminals is shown in Fig. 3b. The impedance of this equivalent circuit Z_c which has been calculated numerically as a function of frequency is shown in Fig. 4.

ANALYSIS OF LOCKING BEHAVIOR

The diode impedance Z_D is a function of the frequency f and amplitude V of the RF voltage applied across the diode, while the circuit impedance Z_c is a function of frequency alone. In the absence of locking, the condition for oscillation is satisfied if

$$Z_D(V_o,f_o) + Z_c(f_o) = 0 \,, \qquad (1)$$

where V_o and f_o are the amplitude and frequency of the output of the free-running oscillator. (Certain other conditions must also be satisfied if this operating point is to be stable. These are discussed by Kurokawa.[2]) When a locking signal e_s of frequency f_1 and amplitude e_1 is also present, this condition is modified to

$$Z_D(V_1,f_1)\left(1 + \frac{e_1}{V_1}\, e^{j\phi_s}\right) + Z_c(f_1) = 0 \,, \qquad (2)$$

where V_1 is the RF voltage amplitude across the diode when the oscillator is locked at frequency f_1, and ϕ_s is the steady-state phase difference between V_1 and e_s.

In the analyses of Kurokawa[2] and Thaler et al.[3] the impedances $Z_D(V_1,f_1)$ and $Z_c(f_1)$ in Eq. 2 are expressed in terms of $Z_D(V_o,f_o)$, $Z_c(f_o)$ and the in-

*This work was supported by the National Aeronautics and Space Administration under Grant No. NGL 23-005-183.

cremental changes $(V_1 - V_o)$ and $(f_1 - f_o)$ in amplitude and frequency. This requires that impedance sensitivities be defined and the first few terms retained in the Taylor series expansions of the impedances. In the analyses mentioned, Z_D is assumed independent of frequency and both Z_D and Z_c are linearized as functions of amplitude and frequency, respectively. The real and imaginary parts of Eq. 2 are then linear in incremental frequency and amplitude and are solved for the two incremental quantities.

Experiments have shown that the change in the output level of the oscillator upon locking can be very substantial and for coaxial cavity circuits, which have a low selectivity, the locking bandwidths are large. The linearization of the impedance for such a large amplitude or frequency change does not seem justified. This limitation has been removed by using a numerical method of modeling as presented here.

CALCULATION OF LOCKING PARAMETERS

In an experimental situation the diode, circuit and free-running operating point of the diode are usually given and the frequency and power output of the locking signal can be varied. The locking bandwidth, change in output power of the oscillator and locking figure of merit can then be observed. This situation is closely simulated in the present approach. The scheme of calculation is as follows:

1. The diode and circuit impedances are determined at the specified operating point to make certain that the free-running condition of oscillation in Eq. 1 is satisfied. If an experimental setup is being simulated, circuit parameters required for the impedance calculation are already known; for theoretical work the slug positions are used as variables and are chosen to satisfy Eq. 1.

2. The amplitude of the locking signal e_1 is determined from the locking signal power P_S. Usually P_S is measured as the maximum available power from the locking sources; then $P_S = e_1^2/8\,\mathrm{Re}[Z_c(f_1)]$.

3. Given e_1 and the locking signal frequency f_1, Eq. 2 contains the unknowns V_1 and ϕ_S. An iterative numerical method is used to determine their values such that the real and imaginary parts of Eq. 2 are satisfied simultaneously.

4. All other locking parameters such as locking gain and the output power of the locked oscillator can now be calculated using the following expressions:

P_o = output power of the free-running oscillator

$$= \frac{1}{2}\,\frac{V_o^2\,\mathrm{Re}[Z_D(V_o,\omega_o)]}{[Z_D(V_o,\omega_o)]^2},$$

P_ℓ = output power of the locked oscillator

$$= \frac{1}{2}\,\frac{V_1^2\,\mathrm{Re}[Z_D(V_1,\omega_1)]}{[Z_D(V_1,\omega_1)]^2},$$

G = locking gain = P_ℓ/P_S

$$= \frac{1}{4}\left(\frac{V_1}{e_1}\right)^2\left(\frac{\mathrm{Re}[Z_D(V_1,\omega_1)]}{\mathrm{Re}[Z_c(f_1)]}\right)^2$$

5. The locking bandwidth is determined by increasing or decreasing f_1 until ϕ_S cannot be calculated as in Step 3. (Its sine or cosine exceeds one in magnitude.) This gives the highest and the lowest frequencies f_h and f_ℓ for which locking will occur, given a fixed input power level. The following two parameters are then calculated:

B_ℓ = locking bandwidth

$$= f_h - f_\ell,$$

η = locking figure of merit

$$= \frac{B_\ell}{f_o}\sqrt{\frac{P_\ell}{P_S}}.$$

CONCLUSIONS

The locking behavior of an IMPATT diode oscillator has been investigated using the model described above. Computations using this model lead to the following results and conclusions.

1. The calculated variation of locking bandwidth with locking gain is shown in Fig. 5. The variation is linear as predicted by earlier analyses, and the nature of the curve agrees well with the experimental results. However, for large injected power levels (i.e., small locking gain) the locking bandwidth is larger than would be predicted by the linear relationship, so that the curve in Fig. 5 deviates upward.

2. The output power of the locked oscillator varies with frequency separation $(f_1 - f_o)$ in a manner shown in Fig. 6 for a fixed input power level. The nature of this curve agrees very well with the experimental results of Tamai.[4]

3. Fig. 6 also shows that the maximum power output of the locked oscillator occurs at a frequen-

cy different from the free-running frequency of the oscillator.

4. It is found that locking bandwidth is not always centered at the free-running frequency of the oscillator. For locking signal frequencies below the free-running frequency ($f_1 < f_o$), locking occurs up to a larger frequency difference than for $f_1 > f_o$.

5. The manner in which power output varies with locking signal frequency is highly dependent upon the operating point of the diode. Fig. 6 is therefore only valid under the stated operating conditions.

6. For the example given in Fig. 5 the external Q was low (around 10) which gives a high locking figure of merit (0.095 in this case).

REFERENCES

1. Gupta, M. S. and Lomax, R. J., "A Self-Consistent Large-Signal Analysis of a Read-Type IMPATT Diode Oscillator," IEEE Trans. on Electron Devices, vol. ED-18, No. 8, pp. 544-550, August 1971.

2. Kurokawa, K., "Some Basic Characteristics of Broadband Negative Resistance Oscillator Circuits," Bell System Tech. J., vol. 48, No. 6., pp. 1937-1955, July-August 1969.

3. Thaler, H. J., Ulrich, G. and Weidmann, G., "Noise in IMPATT Diode Amplifiers and Oscillators," IEEE Trans. on Microwave Theory and Techniques, vol. MTT-19, No. 8, pp. 692-705, August 1971.

4. Tamai, T., "Phase-Locked Oscillations of Silicon Avalanche Diodes," Japan. J. Appl. Phys., vol. 5, No. 11, p. 1115, November 1966.

5. Adler, R., "A Study of Locking Phenomena in Oscillators," Proc. IRE, vol. 34, No. 6, pp. 351-357, June 1946.

6. Paciorek, L. J., "Injection Locking of Oscillators," Proc. IEEE, vol. 53, No. 11, pp. 1723-1727, November 1965.

7. Vlaardingerbroek, M. T. and Goedbloed, J. J., "On the Theory of Noise and Injection Phase Locking of IMPATT Diode Oscillators," Philips Research Reports, vol. 25, pp. 452-471, December 1970.

8. Jochen, P., "Equivalent Circuit for Injection-Locked Negative Resistance Oscillators," Electronics Letters, vol. 6, No. 3, pp. 61-62, 5 February 1970.

9. Kuno, H. J., Bowman, L. S. and English, D. L., "Millimeter-Wave Silicon IMPATT Power Amplifiers for Phase-Modulated Signals," IEEE Int. Solid-State Circuits Conf. Digest, Philadelphia, Pa., pp. 174-175, 204, February 1971.

10. Huntoon, R. D. and Weiss, A., "Synchronization of Oscillators," Proc. IRE, vol. 35, No. 12, pp. 1415-1423, December 1947.

11. Kurokawa, K., "Noise in Synchronized Oscillators," IEEE Trans. on Microwave Theory and Techniques, vol. MTT-16, No. 4, pp. 234-240, April 1968.

12. Daikoku, K. and Mizushima, Y., "Nonlinear Response of the Avalanche Diode Oscillators," Proc. First Conf. on Solid State Devices, Tokyo, 1969. Also in J. Japan Soc. Appl. Phys., vol. 39, Supplement, pp. 145-150, 1970.

FIG. 1 MODEL OF INJECTION-LOCKED IMPATT DIODE OSCILLATOR.

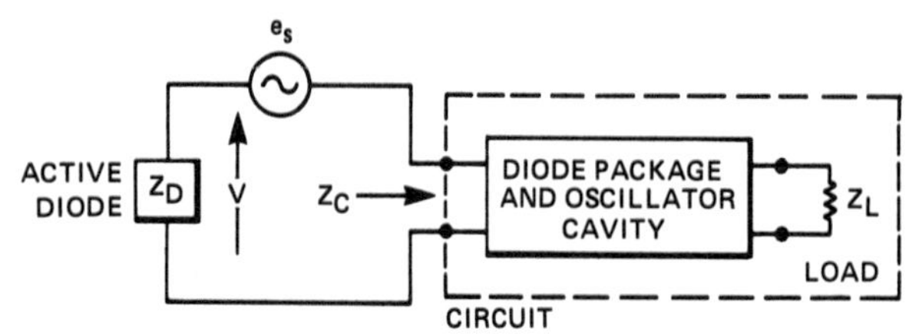

FIG. 2a READ MODEL OF IMPATT DIODE AND IDEALIZED ELECTRIC FIELD PROFILE.

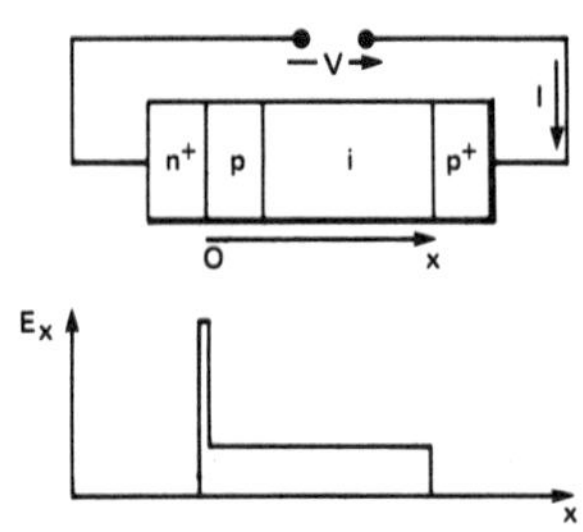

FIG. 2b IMPEDANCE PLANE PLOT OF SINGLE-FREQUENCY LARGE-SIGNAL IMPEDANCE OF READ-TYPE IMPATT DIODE.

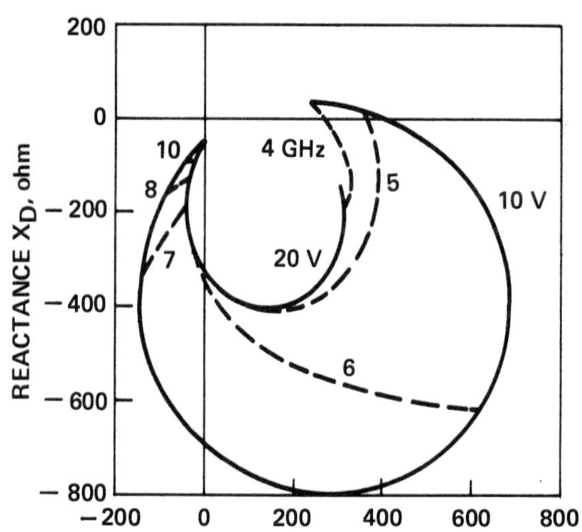

FIG. 3a CONSTRUCTIONAL FEATURES OF DIODE PACKAGE AND OSCILLATOR CAVITY.

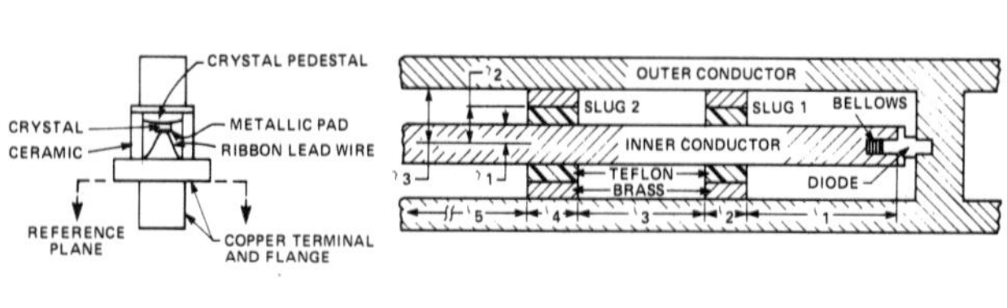

459

FIG. 3b EQUIVALENT NETWORK REPRESENTATION OF THE OSCILLATOR CIRCUIT.

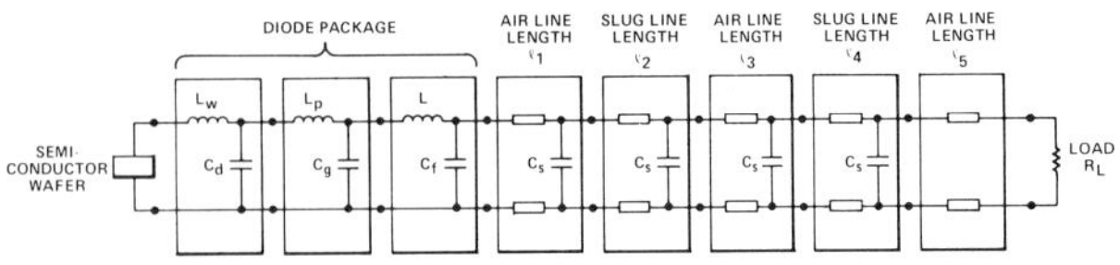

FIG. 4 IMPEDANCE PLANE PLOT OF THE CIRCUIT IMPEDANCE SEEN AT THE DIODE TERMINALS AS A FUNCTION OF FREQUENCY.

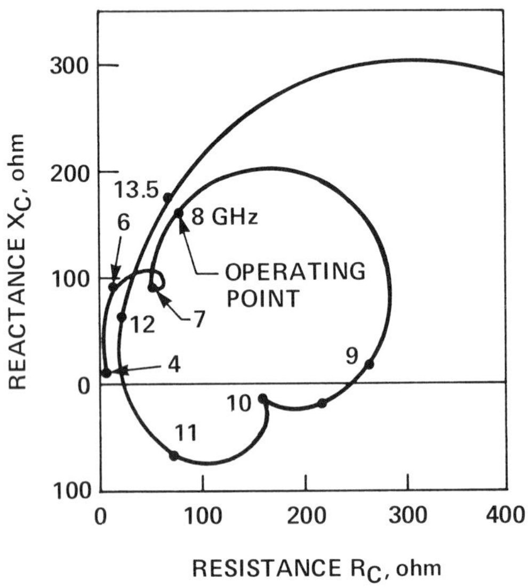

FIG. 5 CALCULATED INJECTION-LOCKING BANDWIDTH OF AN IMPATT DIODE OSCILLATOR IN A COAXIAL CAVITY CIRCUIT. (f_o = 10 GHz, P_o = 150 mW)

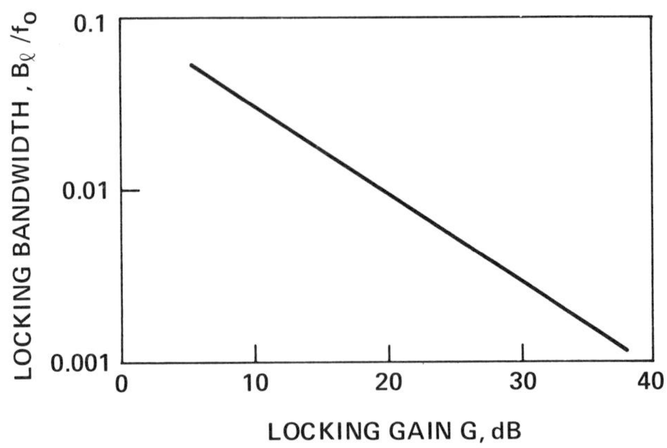

FIG. 6 CALCULATED OUTPUT POWER VARIATION OF AN INJECTION-LOCKED IMPATT DIODE OSCILLATOR AS A FUNCTION OF FREQUENCY DIFFERENCE FOR A CONSTANT INJECTED SIGNAL POWER LEVEL. (f_o = 10 GHz, P_o = 150 mW)

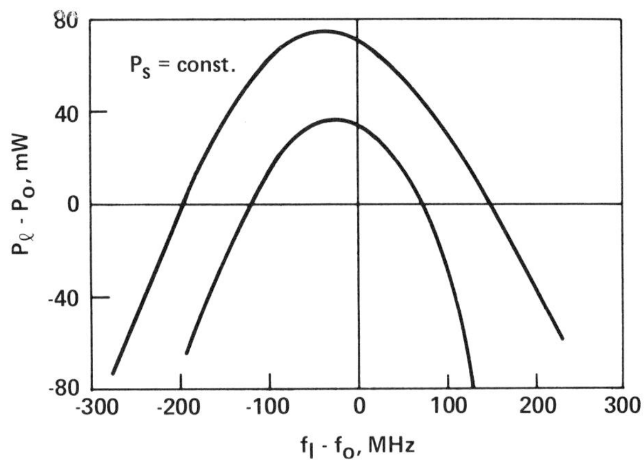

460

Table I

Comparison of Injection-Locked Oscillator Models Applied to IMPATT Diode Oscillators

CW Oscillator Locked by CW Locking Signal

- AFC-Type Feedback Loop Model (Adler,[5] extended by Paciorek[6])
- Reflection-Type Amplifier Model (Vlaardingerbroek and Goedbloed[7])
- Forced Oscillations in a Nonlinear Oscillator Model
  - Oscillator Represented by Instantaneous I-V Characteristic (Jochen[8] and Kuno et al.[9])
  - Oscillator Represented by Load Sensitivity of Output Frequency and Amplitude (Huntoon and Weiss[10])
  - Oscillator Represented by Nonlinear Impedance with Negative Real Part
    - Oscillator Circuit Represented by Single Tuned Resonant Circuit Connected Across Nonlinear Impedance (Kurokawa[11] and Daikoku and Mizushima[12])
    - Oscillator Circuit Represented by a General Network with Arbitrary Frequency-Dependent Resistance and Reactance
      - Diode Impedance Assumed Independent of Frequency and Linear Function of RF Voltage (Kurokawa[2] and Thaler et al.[3])
      - Diode Impedance Nonlinear Function of Frequency and RF Amplitude Determined by Large-Signal Analysis (Gupta and Lomax, Present Analysis)

Reprinted from the PROCEEDINGS OF THE IEEE
VOL. 58, NO. 6, JUNE, 1970
pp. 928–930

Injected Signal Characteristics of High-Power Avalanche Diode Oscillators

Abstract—Efficient injection locking results and other injected signal characteristics of high-power avalanche diode oscillators are presented. An injection locking figure of merit as high as 0.145 has been obtained.

The injection locking of avalanche diode oscillators has been achieved by many investigators [1]–[7] and has been used for such applications as the FM receiver [8]. The purpose of this letter is to report on very efficient injection locking results and other injected signal effects of high-power CW avalanche diode oscillators operated in a two-slug-tuned coaxial circuit at X-band frequencies. The results reported here are for free-running CW oscillator power outputs in excess of 500 mW.

The silicon avalanche diodes used were fabricated using the double epitaxial technique [9]. This process gives hyperabrupt p+n and nn+ junctions for efficient operation of these devices. The antimony doped substrate has a resistivity of 0.008 Ω·cm, and the 2.65 μm thick boron doped p+ layer has a resistivity of 0.0072 Ω·cm. The n layer is 5.33 μm thick with phosphorous doping and has a resistivity of 0.79 Ω-cm and the breakdown voltage of about 78 volts. The double mesa structured devices are flip-chip mounted on a gold plated copper heat sink in package outline [10] S3 using ultrasonic thermocompression bonding. The above parameters represent a non-punch-through operation of these devices. The microwave circuit used consists of a 50 ohm coaxial line terminated in the packaged device at one end and having two teflon coated 18 ohm slugs (0.312 inch long) on the other end.

Fig. 1 shows the schematic of the experimental setup. The locking signal is injected through a circulator and the injected power is measured at the port 2 of the circulator. The oscillator output was displayed on the spectrum analyzer for monitoring locking. The measurements were made by increasing the locking signal power P_{in} for a given frequency separation Δf between the free-running and the locking signal frequencies, until the free-running oscillator was pulled and locked to the input signal. The injection locking results thus obtained are shown in Fig. 2 where the fractional locking bandwidth $2\Delta f/f_0$ is plotted against the locking gain G

Manuscript received February 2, 1970.

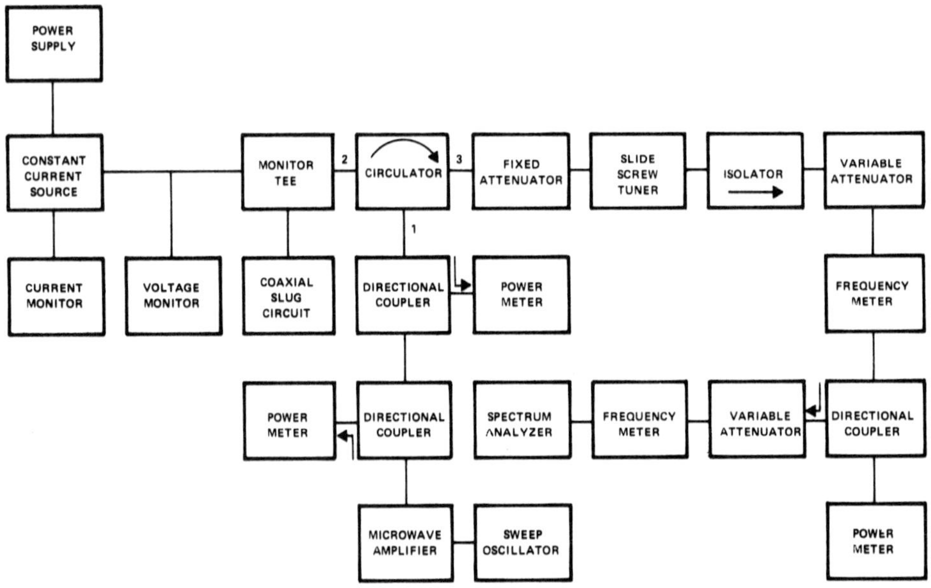

Fig. 1. Experimental setup for the injected signal characteristics measurement.

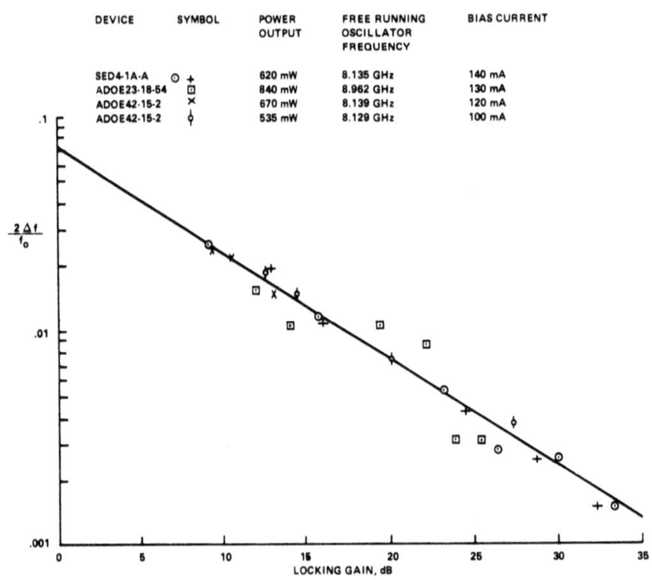

Fig. 2. Injection locking results for different avalanche diode oscillators.

defined as

$$G = 10 \log \frac{P_o}{P_{\text{in}}} \tag{1}$$

where P_o is the free-running oscillator power output. Shown here is the performance obtained on many devices under different bias and circuit conditions. It is observed that most of the points fall on a straight line as expected from the theory [1]:

$$\frac{2\Delta f}{f_o} \cdot \left(\frac{P_o}{P_{\text{in}}}\right)^{1/2} = \eta. \tag{2}$$

The factor η here is a figure of merit for injection locking; and for the results shown in Fig. 2, $\eta = 0.075$, a value higher than those previously reported for CW avalanche diodes. A figure of merit as high as 0.145 was obtained from one device (SED 4-1A-1) when it was biased at a current of 100 mA giving 270 mW at 8.188 GHz.

Under certain bias and circuit conditions, injection locking was obtained only for small frequency separations giving the typical injection locking spectrum with sidebands and frequency pulling effects. For larger frequency separations, the free-running oscillations were quenched and triggered oscillation started at the input frequency.[1] The spectrum changes during the process of obtaining triggered oscillations are shown in Fig. 3 for different levels of input power. It is observed that there is very little frequency pulling and that the sideband generation process is not like that for a locked oscillator. The results, however, have two similarities with the the conventional injection locked oscillator. 1) The triggered oscillator can be swept over a given frequency band. 2) The results when plotted (see Fig. 4) for $2\Delta f/f_o$ versus the locking gain represent an extension of injection locking data. Thus it is believed that the multiple resonant frequency characteristic of the slug-tuned coaxial circuit is causing the triggered oscillations. A verification of this hypothesis requires a detailed evaluation of the microwave circuit, which is being worked out.

Two other observations on the avalanche diode operation in the slug-tuned circuit were made. 1) It was sometimes impossible to injection lock

[1] As used here, the term "triggered oscillation" means the oscillation at the injected signal frequency, in a manner different from conventional injection locking.

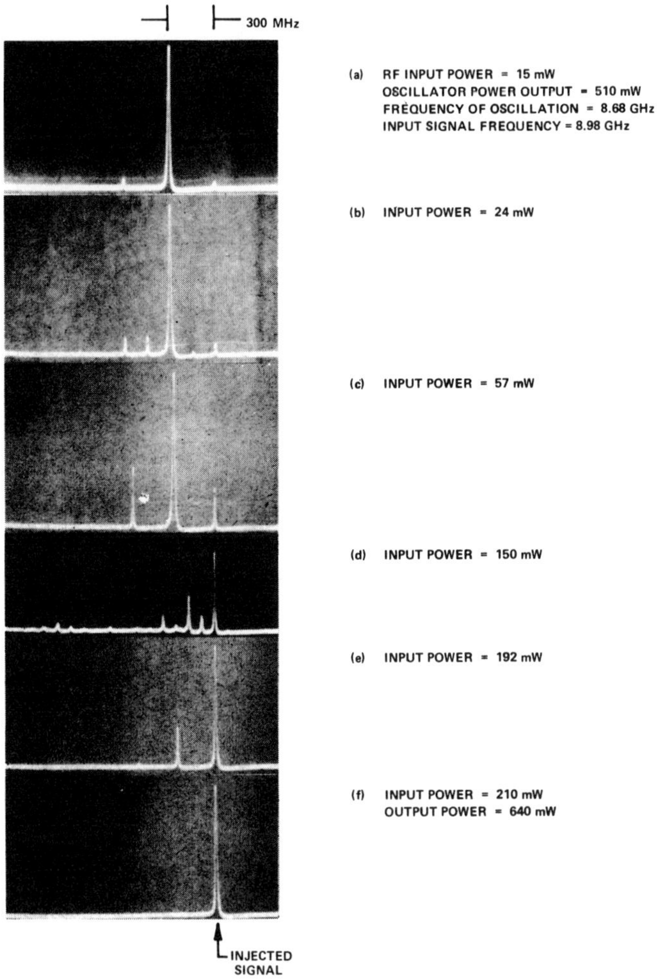

Fig. 3. Spectrum changes during triggered oscillation generation process.

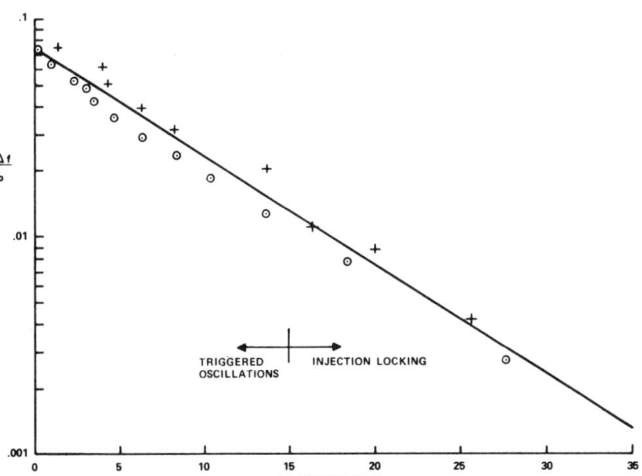

Fig. 4. Injection locked and triggered oscillation characteristics for the device ADOE-23-18-148.

the oscillator even for a negative locking gain for a frequency separation greater than about 50 MHz. Sideband generation did start for higher input power levels. 2) When the input power in an injection locked oscillator was increased, sideband generation sometimes occurred. The sideband frequency was about 135 MHz off from the locked oscillator and had practically no relationship with the frequency separation Δf between the free-running oscillator and injected signal frequencies. The input signal level required for such sideband generation did have a definite relationship with Δf:

$$\frac{2\Delta f}{f_o} \cdot \left(\frac{P_o}{P_{\text{in}}}\right)^{1/n} = K. \tag{3}$$

In one set of measurements, $n=1$ and $K=0.022$ were obtained. These results, however, need further investigation.

In conclusion, many injected signal characteristics of high-power avalanche diodes have been reported including very efficient injection locking and triggered oscillations.

ACKNOWLEDGMENT

The author wishes to thank R. Miller for performing some of the experiments, and N. J. Tolar and H. M. Leedy for the device fabrication.

DEEN D. KHANDELWAL
Texas Instruments, Inc.
Dallas, Tex. 75222

REFERENCES

[1] R. C. Shaw and H. L. Stover, "Phase-locked avalanche diode oscillators," *Proc. IEEE* (Letters), vol. 54, pp. 710–711, April 1966.
[2] T. A. Midford and S. Wanuga, "Phase locking of a silicon avalanche transit time oscillator," *Proc. IEEE* (Letters), vol. 54, pp. 993–994, July 1966.
[3] H. Fukui, "Frequency locking and modulation of microwave silicon avalanche diode oscillators," *Proc. IEEE* (Letters), vol. 54, pp. 1475–1477, October 1966.
[4] T. Tamai, "Phase-locked oscillations of silicon avalanche diodes," *J. Appl. Phys. (Japan)*, vol. 5, p. 1115, November 1966.
[5] M. I. Grace, "Injection locking of pulsed avalanche diode oscillators," *Proc. IEEE* (Letters), vol. 55, pp. 713–714, May 1967.
[6] R. J. Socci and D. A. Fleri, "Relative phase shift in frequency locked avalanche oscillators," *Proc. IEEE* (Letters), vol. 55, p. 1649, September 1967.
[7] F. R. Domer and R. H. Kyle, "Phase-locking an IMPATT device in the pulsed mode," *Proc. IEEE*, vol. 55, pp. 1753–1754, October 1967.
[8] T. L. Osborne and C. H. Elmendorf IV, "Injection-locked avalanche diode oscillator FM receiver," *Proc. IEEE*, vol. 57, pp. 214–215, February 1969.
[9] H. M. Leedy and H. L. Stover, "Double-epitaxial avalanche diode oscillators," presented at the 1968 IEEE Solid-State Device Research Conf., Boulder, Colo.
[10] *Microwave Engineers' Handbook and Buyers' Guide*. Dedham, Mass.: Horizon House-Microwave, Inc., 1966, p. 261.

VARIOUS OSCILLATOR PROPERTIES VI

Frequency Modulation of Avalanche Transit Time Oscillators

JOHN W. AMOSS, MEMBER, IEEE, AND KURT E. GSTEIGER, MEMBER, IEEE

Abstract—**This paper presents experimental data taken to determine the frequency modulation characteristics of avalanche transit time oscillators.**

The active element is a diffused mesa diode with a shallow junction in epitaxial *n-n+* silicon; the details of the construction of the diode are presented and its typical characteristics are discussed.

The basic oscillator consists of the diode mounted in the capacitive portion of a radial mode cavity machined of copper with the outlines of a DO-5 diode header. The frequency of oscillation is dependent upon the diode junction capacitance and is varied between 5 and 8 GHz for the diodes tested. Microwave power levels up to 100 mW have been observed with an efficiency exceeding 3 percent. The frequency drift over the temperature range from -70 to $+100$°C is 2.5×10^{-5} parts/°C.

The frequency modulation characteristics of these oscillators indicate their potential applications in miniature solid-state low-power communications systems.

ELECTRONIC tuning of avalanche transit time oscillators has been treated explicitly by Gilden and Hines.[1] This electronic tuning effect, which is attributed to the variability of avalanche susceptance with current, is evident also in theoretical and experimental investigations of others for both Read and *p-n* type devices.[2],[3]

Manuscript received May 17, 1967; revised September 30, 1967.
The authors are with Sperry Microwave Electronics Division, A Division of Sperry Rand Corp., Clearwater, Fla. 33518.

Read[4] has discussed electronic tuning briefly but was uncertain as to whether the variation of frequency of oscillation with the diode current could be made large enough for a practical frequency-modulation device, particularly when the diode was operating at optimum frequency and bias. For transit angles substantially less than the optimum value, however, Gilden and Hines showed that strong negative resistance effects still occurred and that electronic tuning effects were significant. Recently, Rulison *et al.*[6] reported a germanium *p-n* diode which exhibited a susceptance variation with current between a half and one order of magnitude larger than that of a Read diode under similar operating conditions.

Since moderate CW output power from these devices is now being realized, the ease of electronic tuning suggests their potential usefulness in many practical applications. This paper presents experimental data which were taken to determine the frequency modulation characteristics of avalanche transit time oscillators made at Sperry Microwave Electronics Company in terms of modulation linearity and inherent amplitude modulation.

The active element is a diffused silicon mesa diode, similar to the one described by Misawa.[5] The junction is formed by diffusing boron 2.4 microns deep in epitaxial silicon. The substrate is arsenic doped to a resistivity of 0.008 ohm/cm and the epitaxial layer has a thickness of 7 microns with a nominal resistivity of 1 ohm/cm. It is known that a high boron concentration in silicon leads to an excessive number of lattice defects resulting in a high density of microplasmas. Therefore, a surface concentration lower than customary for the fabrication of varactor diodes was chosen. The uniformity of the junction was further improved by applying a multiple predeposition post-diffusion technique. The resulting impurity profile is represented in Fig. 1. The junctions are of a nearly abrupt type with a sharp breakdown of 80 to 85 volts. Using conventional contacting and masking techniques, mesa diodes with a junction diameter of approximately 100 microns are formed, and after dicing individual diodes are mounted in a microwave pill package with a case capacitance of 0.25 pF and a lead inductance of $\sim$0.3 nH. The diodes exhibit typically a junction capacitance of 0.4$\sim$0.5 pF at breakdown. This fabrication technique led to diodes which gave higher efficiencies for CW operation with no evidence of microplasma breakdown.

The frequency modulation experiments were carried out with the encapsulated diodes mounted in a radial cavity operating in the fundamental mode. A photograph of an oscillator with external dimensions of a DO-5 power diode header is shown in Fig. 2. The basic oscillator consists of the diode mounted in the capacitive portion of the cavity and employs loop coupling with an OSSM output connector. The top cap and base portion of the cavity are separated by a 1 mil mylar spacer and connected together with nylon screws. Direct current bias and the low-frequency modulating signal are applied between top cap and base (Fig. 3). This biasing arrangement provided an effective short at microwave frequencies, yet offered negligible shunting effects at the modulating frequency. Since the cavity is fixed tuned, the oscillating frequency depends somewhat upon the junction capacitance of the diode at breakdown. The frequency of oscillation for typical diodes was between 6.5 and 7.5 GHz. Maximum CW output power for selected diodes was 100 mW at current levels of 40 mA operating at room temperature. In one typical run, 75 percent of the diodes gave between 40 and 60 mW of microwave power at current levels of 30 mA. Other runs consistently produced diodes which gave better than 20 mW at 30 mA.

The equipment which was used to obtain the frequency modulating characteristics of the oscillator is shown in Fig. 4. A low-frequency modulating signal was applied in series with the dc bias voltage using a transformer for isolation. The ac component of diode current was obtained by measuring the voltage across the series resistor. Frequency deviation was measured as a function of applied current either by observing the nulls of the various spectral components or, in the case of high modulation levels at low modulating frequencies, by observing the total deviation directly. A calibrated crystal detector and high-gain oscilloscope were used to measure the amount of inherent amplitude modulation.

In general, the electronic tuning range of an oscillator is inversely proportional to the Q when loaded for optimum output. Thus designing an oscillator cavity to have minimum Q at optimum load has the effect of maximizing the frequency modulating characteristics, i.e. lowering the Q increases the modulation sensitivity and linearity and decreases inherent amplitude modulation. Fig. 5 shows how a change in cavity parameters effects the frequency sensitivity of the oscillator. Frequency deviation is plotted as a function of modulating current for two oscillators which were designed to have significantly different Q values. The internal dimensions of the cavities were proportioned to give different ratios of cavity capacitance to inductance, yet oscillate at essentially the same frequency. A low ratio of cavity capacitance to inductance is conducive to high modulation sensitivity. The data of Fig. 5 were taken by observing the nulls of various frequency components as modulating current was varied. The plots are linear to within measurement accuracy, which is approximately ± 2 percent.

Photographs of the spectrum for various modulation indices for a modulating frequency of 600 kHz is shown in Fig. 6. In the first photograph, the first-order sidebands are zero and in the others the carrier is in its second-, third-, and fifth-order zero. The inherent amplitude modulation deduced from the symmetry and the relative amplitude of the various components agree (within measurement accuracy) with the values obtained with the calibrated detector. The measured amplitude modulation was less than 20 percent for the frequency deviations reported.

Frequency deviation as a function of modulating current and voltage were measured for different modulating frequencies. The modulating current was varied up to 3 mA (rms) yielding a deviation of approximately 35 MHz as shown in Fig. 7. These data, as well as those to follow, were made by observing the deviation directly on the spectrum analyzer display and do not reflect the accuracy of the previous measurements. The plot, however, is essentially linear and independent of modulating frequency. The data plotted as a function of modulating voltage for different modulating frequencies are shown in Fig. 8. For a constant modulating frequency, the deviation is linear with voltage. At a constant modulating voltage, however, the deviation increases with increasing frequency. Thus, for good frequency modulation the oscillator should be modulated with a current source.

The modulating voltage–current relation exhibited an unexpected behavior in the low-frequency impedance of the diode under avalanche conditions, i.e., the ratio of modulating voltage–current decreases with increasing frequency. This implies a circuit time constant three to four orders of magnitude greater than one would expect based on the measured circuit capacitance just prior to avalanche breakdown and the measured dc incremental resistance in breakdown. With the same equipment that was described previously, this behavior was checked for a number of diodes both in and out of the cavity with essentially the same decrease in impedance with frequency. The circuit appears linear at the modulating frequency since no distortion was evident in either the applied voltage or current when observed with the oscilloscope.

A test circuit commonly used to measure the impedance of Zener diodes was constructed (Fig. 9) and confirmed the

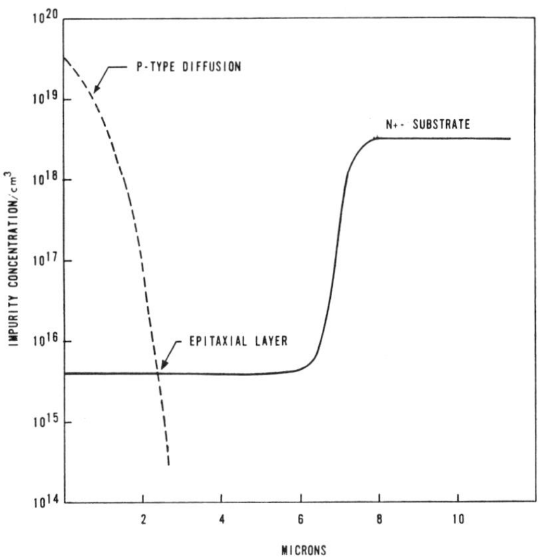

Fig. 1. Impurity profile of oscillator diode.

Fig. 2. Microwave oscillator with DO-5 outline.

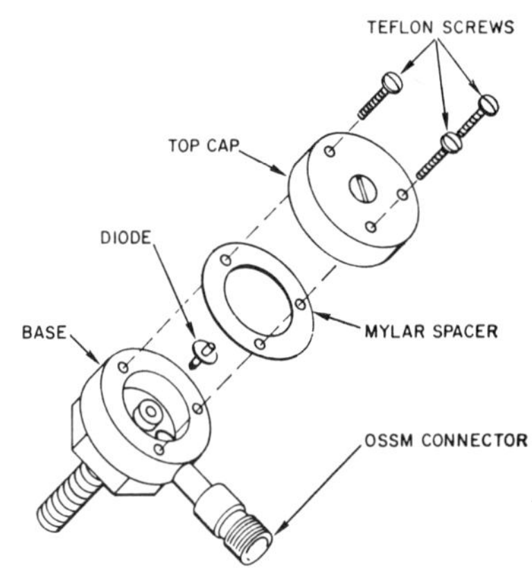

Fig. 3. Sketch of oscillator cavity.

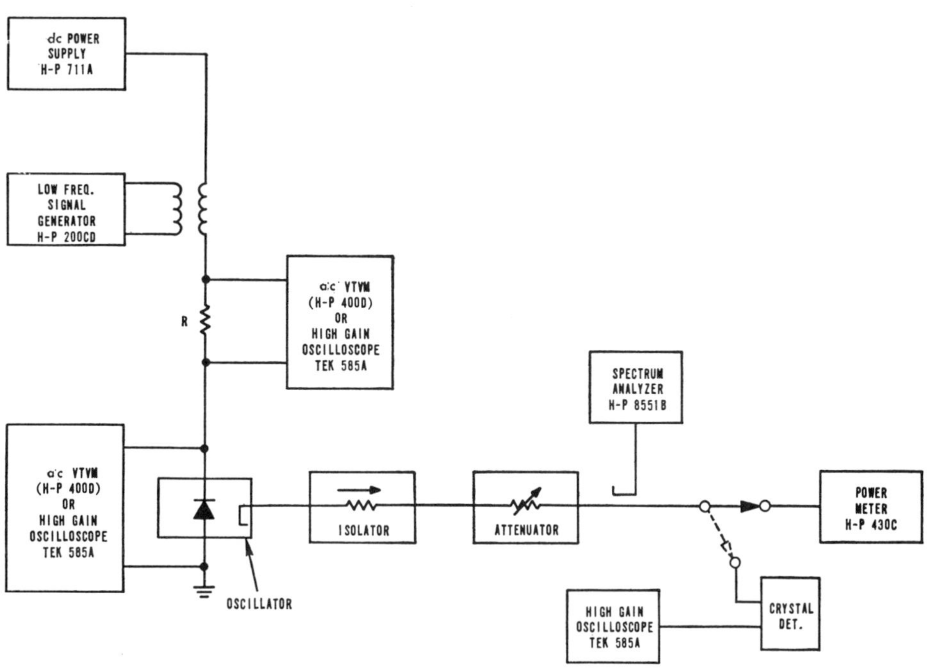

Fig. 4. Equipment used to measure frequency modulation characteristics of oscillator.

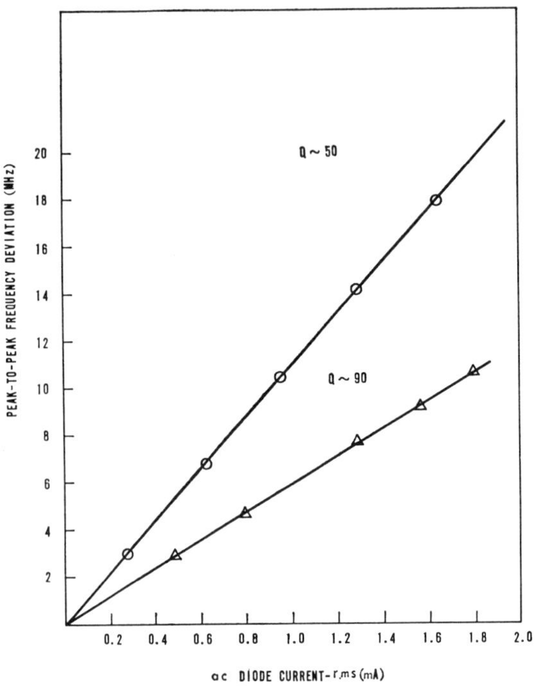

Fig. 5. ΔF versus I_{ac} for two different cavity designs.

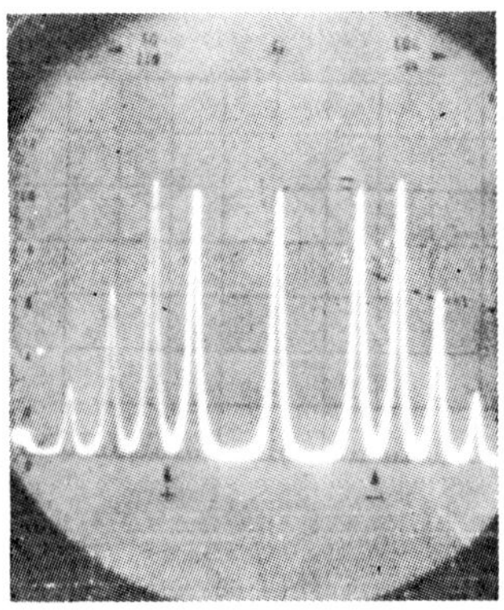

(a) m: 3.8; Horizontal sweep: 1 MHz/div.

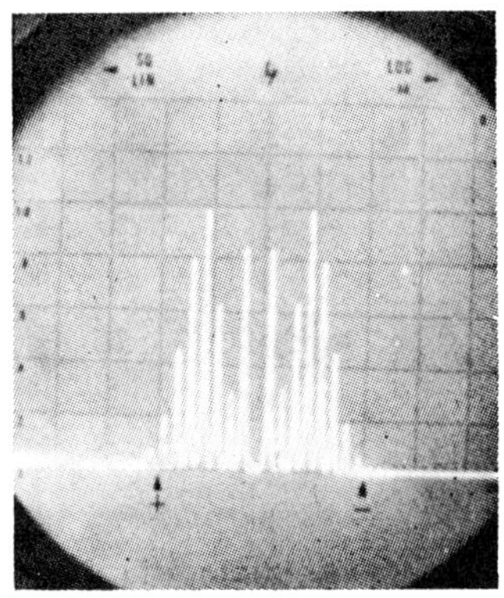

(b) m: 5.5; horizontal sweep: 3 MHz/div.

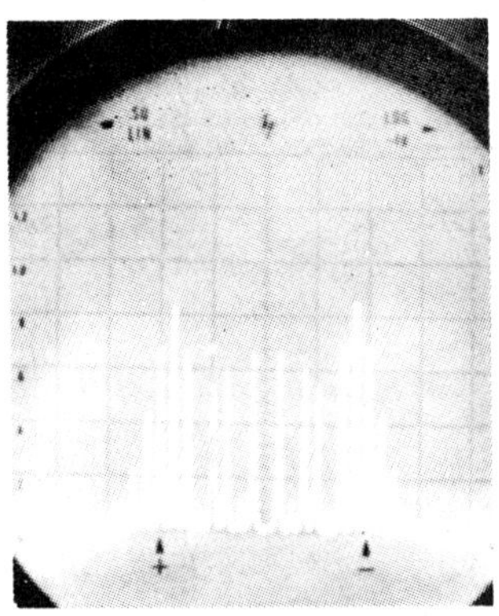

(c) m: 8.7; horizontal sweep = 3 MHz/div.

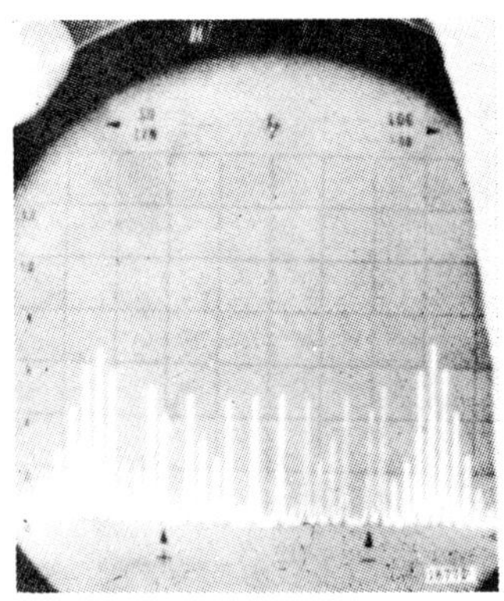

(d) m: 14.9; horizontal sweep = 3 MHz/div.

Fig. 6. Photographs of spectrum analyzer displays for various modulation indices. $f_m = 600$ kHz. $f_c \sim 7$ GHz.

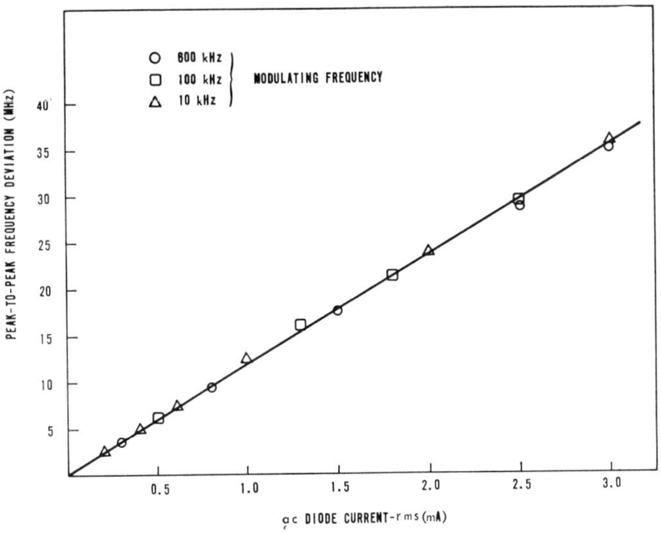

Fig. 7. ΔF versus I_{ac} for different modulating frequencies.

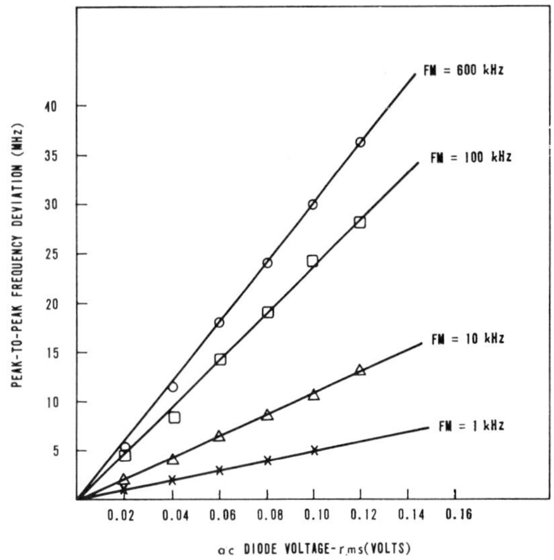

Fig. 8. ΔF versus V_{ac} for different modulating frequencies.

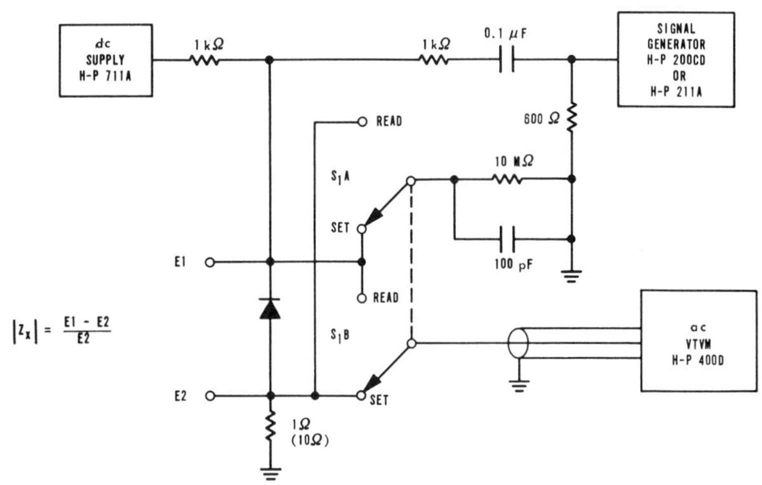

Fig. 9. Low-frequency Zener impedance test circuit.

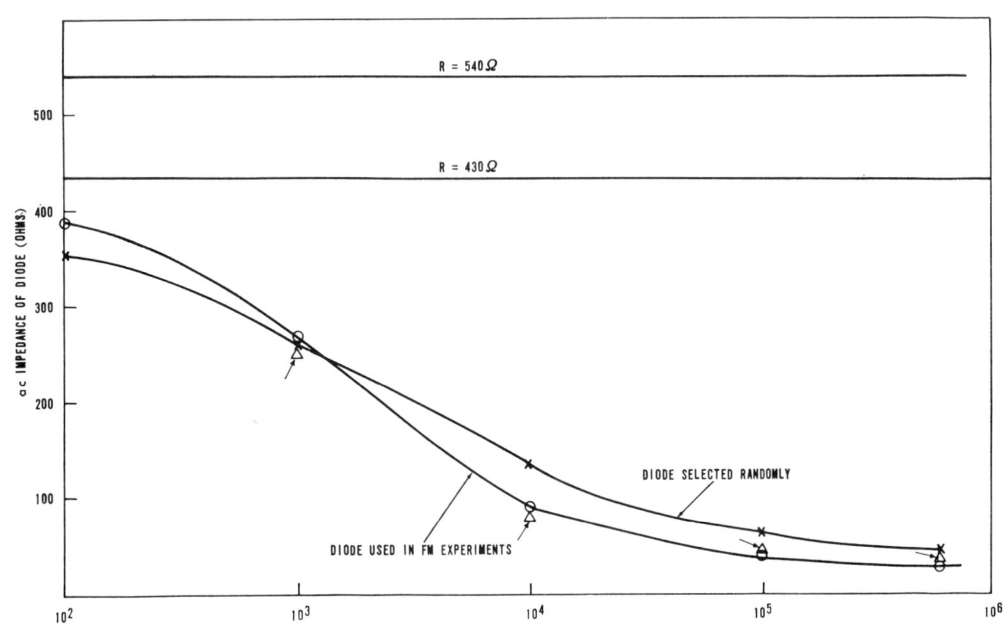

Fig. 10. Small signal ac impedance of diode under avalanche conditions.

471

VARIOUS OSCILLATOR PROPERTIES

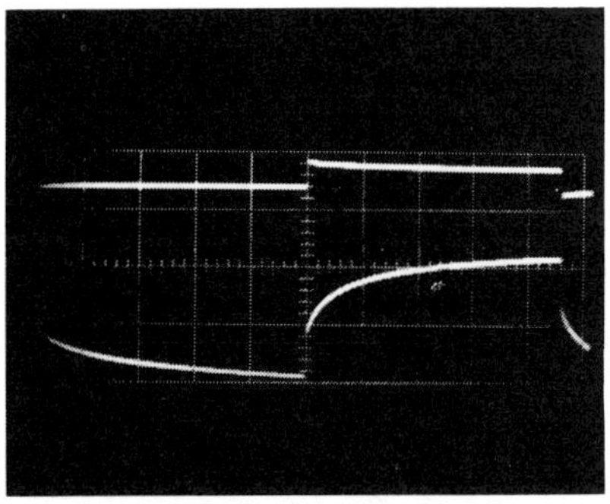

(a) Diode dc biased to 20 mA.

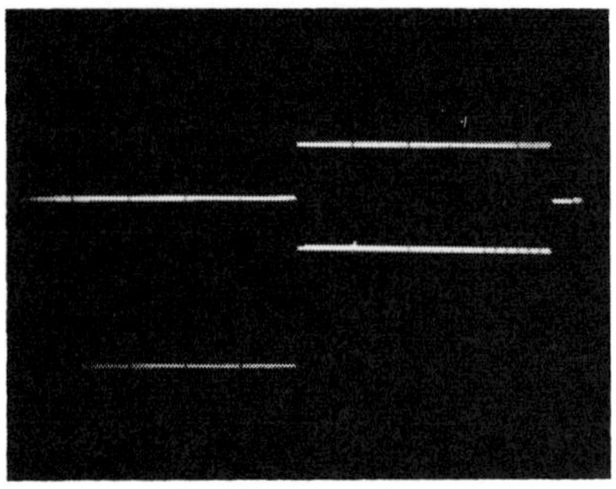

(b) 100 ohm resistor.

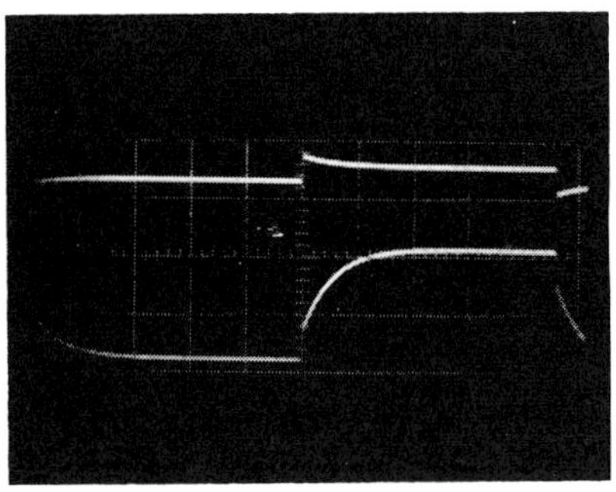

(c) 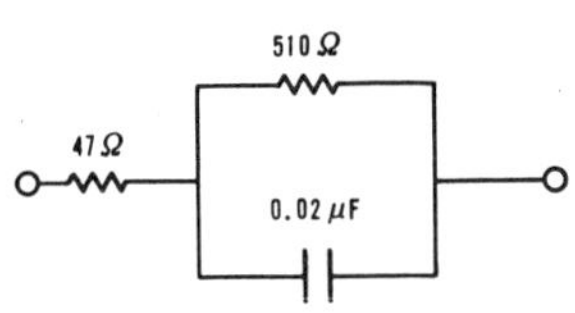

Fig. 11. Current and voltage pulse for 10 kHz square wave using Zener impedance test circuit. The top trace is the voltage across a 10 ohm viewing resistor; vertical scale: 0.05 V/div. The bottom trace is the voltage across the sample; vertical scale: 0.2 V/div. The horizontal scale is 10 μs/div.

previous measurements. With this circuit small signal measurements were made for different dc bias levels between 5 and 20 mA. The ac impedance at a given frequency was slightly dependent upon dc bias level, as would be expected, but retained its characteristic decrease with frequency. Fig. 10 is a plot of the impedance of two typical diodes as a function of frequency. The triangles represent values obtained during the frequency modulation experiments, i.e., the diode was in the cavity and in an oscillating state, and the circles represent values obtained with the diode out of the cavity and were obtained with the Zener impedance test circuit. The impedance curve of another diode selected randomly is also shown. Two resistors of value approximately equal to the incremetal dc resistance of the diodes were measured in the test circuit with no observable decrease in impedance with frequency.

Fig. 11(a) shows the current–voltage waveforms when the square wave current pulse is applied to a typical diode biased at 20 mA in the avalanche region. The top trace is voltage (0.05 V/cm) across the 10 ohm current viewing resistor and the bottom trace is voltage (0.2 V/cm) across the diode sample. The horizontal scale is 10 μs/cm. This characteristic behavior was typical of a large number of diodes, some from two other sources, in which pulse risetimes varied between approximately 5 to 10 μs. For comparison Fig. 11 also shows voltage–current pulses when lumped circuits are substituted into the impedance test circuit. The bottom left trace is for a 100 ohm nonreactive resistor for which the wave shape of the voltage pulse is nearly identical to that of the current pulse. The other trace is for a 47 ohm resistor in series with a 510 ohm resistor and a 0.02 μF capacitor in parallel. Risetime of a 1N978B Zener diode was measured in the same test circuit and was several orders of magnitude less than that for the avalanche transit time diodes.

The phenomena responsible for the long risetime of the avalanche transit time diodes and its effect on their oscillating properties are presently being investigated and will be reported later. C. B. Swan of Bell Telephone Laboratories has suggested that this is a thermal effect, corresponding to a thermal time constant of approximately 5 μs due to an increase in avalanche voltage with junction temperature.[7]

The avalanche transit time oscillator reported in this paper, however, exhibited frequency modulation characteristics comparable to those of a reflex klystron. In the laboratory, these devices have been used to transmit music with good fidelity. Using a battery to supply dc power and a crystal microphone and transformer to achieve current modulation, these devices have been used to transmit voice communication over a distance of $\frac{1}{3}$ mile with approximately 5 mW output power. In both cases, standard waveguide horns with a nominal gain of 17.5$\pm$1.5 dB were used as receiving and transmitting antennas. The receiver consisted of a conventional microwave receiver, slightly detuned to provide FM–AM conversion.

References

[1] M. Gilden and M. E. Hines, "Electronic tuning effects in the Read microwave avalanche diode," *IEEE Trans. Electron Devices*, vol. ED-13, pp. 169–175, January 1966.

[2] T. Misawa, "Negative resistance in *p-n* junctions under avalanche breakdown conditions, pts. 1 and 2," *IEEE Trans. Electron Devices*, vol. ED-13, pp. 137–151, January 1966.

[3] J. G. Josenhans and T. Misawa, "Experimental characterization of a negative-resistance avalanche diode," *IEEE Trans. Electron Devices* (*Correspondence*), vol. ED-13, pp. 206–208, January 1966.

[4] W. T. Read, Jr., "A proposed high-frequency, negative-resistance diode," *Bell Syst. Tech. J.*, vol. 37, pp. 401–446, March 1958.

[5] T. Misawa, "Microwave avalanche diodes with nearly abrupt type junction," presented at the Internat'l Electron Devices Meeting, Washington, D. C., 1966.

[6] R. L. Rulison, G. Gibbons, and J. G. Josenhans, "Improved performance of IMPATT diodes fabricated from Ge," *Proc. IEEE* (*Letters*), vol. 55, pp. 223–224, February 1967.

[7] C. R. Crowell and S. M. Sze, "The temperature dependence of avalanche multiplication in semiconductors," *Appl. Phys. Lett.*, vol. 9, p. 242, 1966.

Reprinted from THE BELL SYSTEM TECHNICAL JOURNAL
Vol. 48, No. 1, January, 1969
Printed in U.S.A.

Frequency Modulation of a Millimeter-wave IMPATT Diode Oscillator and Related Harmonic Generation Effects

By T. P. LEE and R. D. STANDLEY

(Manuscript received May 31, 1968)*

In this paper we report the performance of a continuous wave millimeter-wave IMPATT diode oscillator with a wide-band tunability. The diode is mounted in an iris wafer circuit; its oscillation frequency can be modulated either by a varactor diode or by direct modulation of the IMPATT diode bias current. The oscillator has been successfully used as a millimeter-wave frequency deviator in an experimental pulse code modulation millimeter-wave system.

We also report detailed measurements on subharmonic frequencies in IMPATT diode oscillators. Experimental results show that wide frequency tunability can be obtained with a circuit which provides an "idler" resonance at one-half the fundamental transit-time frequency. The results also show that by providing "idler" resonances at both the transit-time frequency and at one-half of the transit-time, the oscillation at $\frac{3}{2}$ the transit-time frequency is enhanced and yields a useful output power of 2 mW at 86 GHz.

I. INTRODUCTION

A continuous wave (CW) millimeter-wave silicon IMPATT diode oscillator was used satisfactorily as a local oscillator in an experimental pulse code modulation (PCM) millimeter-wave repeater system.[1, 2] That oscillator circuit used a radial-line resonant cavity whose resonant frequency was the primary factor determining the oscillation frequency. The oscillator was difficult to tune either mechanically or electronically. 473

While such characteristics are desirable for fixed frequency local oscillator applications, other applications which demand wideband

* The essential results of this paper were presented at the 26th annual conference on Electron Device Research, Boulder, Colorado, June 19–21, 1968.

performance and tunability are also of interest. For example, a millimeter-wave frequency deviator using an IMPATT diode would be advantageous over the *L*-band deviator and up-converter combination in the PCM repeater system. The IMPATT diode oscillator could deliver power at least one order of magnitude greater than the up-converter.

A previous paper showed that the circuit inductance of the radial-line cavity in shunt with the diode was much smaller than the diode inductance.[1] Therefore, the oscillation frequency could be tuned only a few hundred megahertz by varying the diode bias current.* In addition, the radial-line cavity was loosely coupled to the external waveguide circuit so that a tuning range of only 300 MHz was obtained when a varactor diode was used for frequency deviation. To improve the tunability, it is necessary to have (*i*) the IMPATT diode equivalent inductance dominate that contributed by other circuit elements, and (*ii*) the diode closely coupled to controllable external circuit parameters.

It is important to notice that the tunability can be further improved by providing more than one resonant circuit for the oscillator. The frequencies are harmonically related, and oscillation at each of these frequencies is reactively terminated except the one for the output. Take as an example the two-frequency case. The tunability of the oscillator near the transit-time frequency (say 60 GHz) can be improved if a lossless resonant circuit at half the output frequency (30 GHz) is incorporated in the oscillator circuit. The oscillation at 30 GHz is terminated by a cut-off waveguide and therefore is designated as the "idler," an analog to the idler in harmonic generators using varactor diodes. It is not necessary that the low frequency be used as the idler. Swan has shown that by providing an idler resonance at twice the transit-time frequency, improvements in both the power output and the tunability can be obtained.[3]

474

To achieve wide band tunability, IMPATT oscillators were designed using resonant-iris structures. It can be shown that the diode equivalent inductance is dominating in this circuit compared with that of the radial-line cavity structure. The loaded Q of the iris is about 10 which provides a wide bandwidth for oscillation. The oscillation frequency could be tuned over 9 GHz in the 50 to 60 GHz range by varying the bias current of the IMPATT diode, thus varying the diode

---

* The diode equivalent inductance is inversely proportional to the diode bias current.

equivalent inductance. It could also be tuned over 3 GHz by using a varactor diode closely coupled to the IMPATT diode.

Notice that by using the fundamental transit-time frequency as the idler, the power output at one-half the transit-time frequency is, in general, higher than the power output when the half frequency is used as the idler in our particular circuit. We refer to the frequency which is one-half the transit-time frequency as the subharmonic frequency throughout this paper.

In the sections which follow we describe the circuit structure of the oscillator, the performance as a tunable oscillator, and the results of frequency modulation of the oscillator in an experimental millimeter-wave PCM repeater system. Then measurements of various harmonic frequencies existing in the oscillator circuit, and the identification and effect of the subharmonic oscillation are described in detail.

II. DESCRIPTION OF THE OSCILLATOR

Figure 1a shows the resonant-iris structure. The iris is made in wafer form similar to the Sharpless wafer.[4] The range of the oscillation frequency and the Q of the resonant iris are determined by the size, thickness, and shape of the iris aperture. Refs. 5, 6, and 7 give details of iris characteristics. The wafers used were 0.100 inch thick,

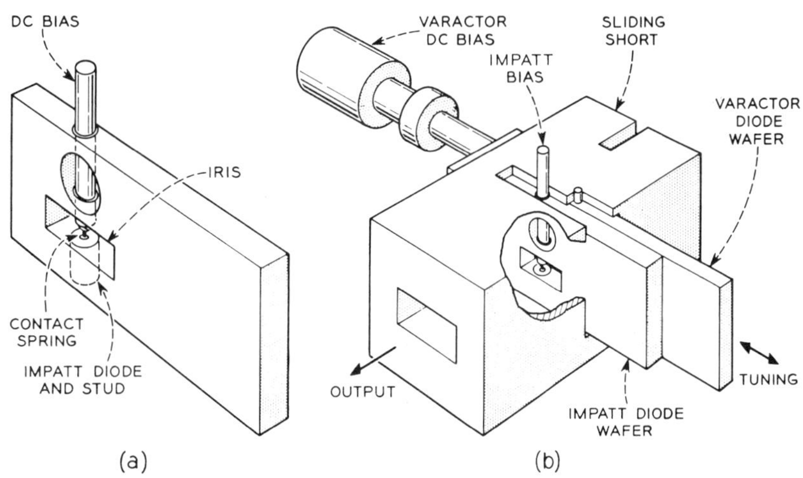

Fig. 1—Artist's view of the oscillator assembly. (a) IMPATT diode wafer. (b) Oscillator mount.

475

and had a rectangular aperture 0.010 or 0.030 inch high and 0.100 to 0.148 inch wide.

The IMPATT diodes had a 0.001 inch diameter mesa structure and nearly abrupt junctions as reported by Misawa.[8] Typical diode characteristics are given in Table I. The diodes were thermal-compression bonded onto gold-plated copper studs 0.063 inch in diameter. After bonding, the diodes were coated with a thin layer of silicone varnish which enchances the mechanical strength of the mesa structure. Electrical contact to the diode was made by a 0.032-inch nickel rod with a welded contact spring, as pictured in Fig. 1a. For the varactor-tuned oscillator, a varactor diode in a Sharpless wafer was mounted adjacent to the iris wafer as shown in Fig. 1b. The relative coupling between the IMPATT diode and the varactor diode (thus the tuning characteristic) was adjusted by varying the relative position of the two diodes in the plane of the cross section of the waveguide.[9] The complete assembly in a RG-98/U waveguide is shown in Fig. 1b.

Table I — Typical Diode Characteristics*

Epitaxial layer thickness†	$2.1\ \mu$
Epitaxial layer doping density‡	6×10^{16} cm^{-3}
Junction depth†	$1.0\ \mu$
Space charge layer width‡	$0.75\ \mu$
Breakdown voltage (at 1 μA)	19V
Capacitance at breakdown	0.19 pF
Junction area A	1.5×10^{-5} cm^2

* All diodes were from the LO 1114 series.
† Measured by staining and interference fringe.
‡ Obtained from voltage dependence of capacitance.

III. OSCILLATOR PERFORMANCE

Figure 2 shows the typical performance of an oscillator without the varactor. (The varactor wafer is replaced by a blank wafer.) The iris used had an aperture of 0.130 by 0.030 inch. The CW ouput power (above 1 mW) and the frequency are plotted as a function of the IMPATT diode current. Oscillation began at a diode current of about 50 mA compared with 100 mA for the cavity structure.[1] The output power was optimized at each current level by adjusting an E-H tuner in front of the diode and a sliding short in back of each

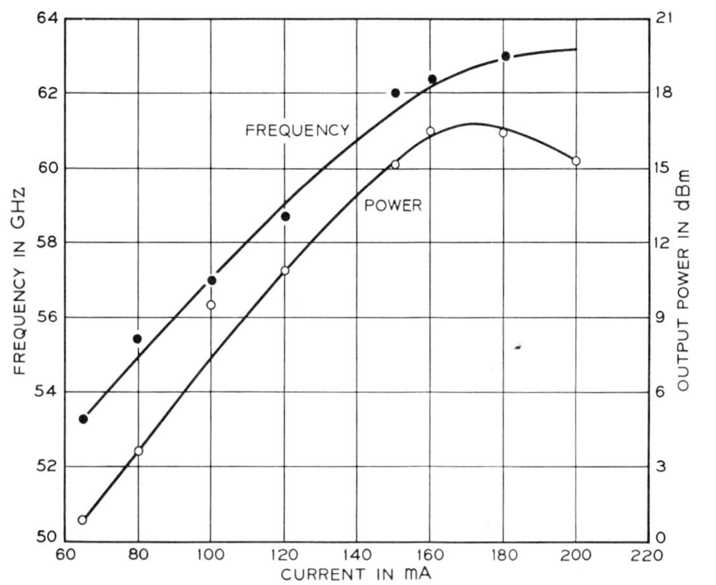

Fig. 2 — Maximum output power and frequency as a function of diode current.

diode. Note that the tunable band (about 10 GHz) is considerably wider than that previously reported for a radial line cavity structure. By varying the bias current with fixed mechanical tuning, the frequency could be tuned over 4 GHz with a 3 dB variation in the output power, and over 9 GHz with a minimum output power of 1 mW.

Table II summarizes the results of various circuits with different iris apertures. Although there are not enough data to draw a meaningful conclusion, it seems that the oscillation frequency was relatively independent of the iris width compared with iris height, which implies that the inductance of the center post (with contacting spring) partially controls the oscillation frequency. This statement is based on the assumption that diodes used in the test are of uniform characteristics since they were made from a single slice and batch processed. The dc characteristics of the diodes showed less than 5 percent variation in capacitance and in breakdown voltage.

IV. FREQUENCY MODULATION

As mentioned in Section I, one goal in this work was to design an IMPATT diode oscillator circuit which could be frequency modulated

TABLE II—RESULTS OF CIRCUITS WITH VARIOUS DIODES AND IRIS APERTURES

Diode No.	Width (in)	Height (in)	C_B‡ (pF)	Tunable¶ bandwidth (GHz)	P_{max} (dBm)
1	148	0.030	0.19	50 – 60	15
2	130	0.030	0.19	53 – 63	16.5
3	120	0.030	0.19	52 – 59	11.5
4	110	0.030	0.18	56 – 63.2	9.4
5	100	0.030	0.19	53 – 61	10.7
6	130	0.010	0.16	64 – 70	11.5
7	148	0.010	0.18	58 – 65	11.1

‡ C_B is the diode capacitance measured at breakdown ($V_B = -19$ volts).
¶ The lower limit on frequency is arbitrarily chosen as the frequency for which the power is 1 mW. The upper frequency limit occurs at a bias current of 160 mA which is well below the burn-out level for the LO 1114 series diodes.

and used as a frequency deviator in a PCM millimeter-wave repeater system.[2] Two approaches were taken. The first used a varactor diode to tune the circuit susceptance outside the diode; the second used the fact that the diode inductance (from avalanching) varies inversely with the bias current.

Both methods have advantages and disadvantages. The power output of the varactor-tuned oscillator remains almost constant over the frequency band so that amplitude modulation is negligible. However, to achieve ultimate performance the circuit is much more complex. The bias-current tuned oscillator has simpler circuitry, but inevitably its power output varies with bias current, which results in AM distortion. However, the AM distortion can be overcome by proper tuning of the circuit if the modulation index is small. (See Section 4.2)

4.1 *The Varactor-Tuned FM Deviator*

The varactor tuned oscillator is shown in Fig. 1b. The varactor diodes used were planar diffused GaAs diodes with a honeycomb structure.[10] The zero-bias capacitance was 0.04 – 0.05 pF and the breakdown voltage was 20 volts. The capacitance varied with voltage approximately as $C = C_0 (1 + V/\varphi)^{-0.4}$, where C_0 is the junction capacitance at zero bias, V is the bias voltage, and φ is "built-in" voltage which is approximately 1 volt for GaAs. The diode was then mounted in a Sharpless wafer which in turn was inserted into the oscillator mount. The varactor diode was about 0.080 inch behind the IMPATT diode. The coupling between the varactor diode and the

IMPATT diode was tuned by sliding the varactor wafer relative to the IMPATT wafer. Additional tuning was provided by a sliding short behind the varactor diode and an E-H tuner in front of the IMPATT diode.

The self-resonant frequency of the varactor diode, as measured by the transmission resonance technique, was used to correlate with the tuning sensitivity of the oscillator. The self-resonant frequency can be varied by changing the length of the contacting wire in the varactor wafer.[11] It was found experimentally that the best tuning sensitivity was obtained when the varactor diode was self-resonant near the "idler" frequency.

A typical frequency tuning characteristic of the oscillator is shown in Fig. 3 where the frequency is plotted as a function of the dc bias voltage (reverse biased) on the varactor. A tunable band of 2.5 GHz was obtained. Notice that the linear region was 1.5 GHz with a sensitivity ot 1 GHz per volt. Also notice that the power output varied ± 1 dB over the tunable band.

Frequency modulation with sinusoidal drive on the varactor was achieved at both the 160 MHz and the 500 MHz modulation fre-

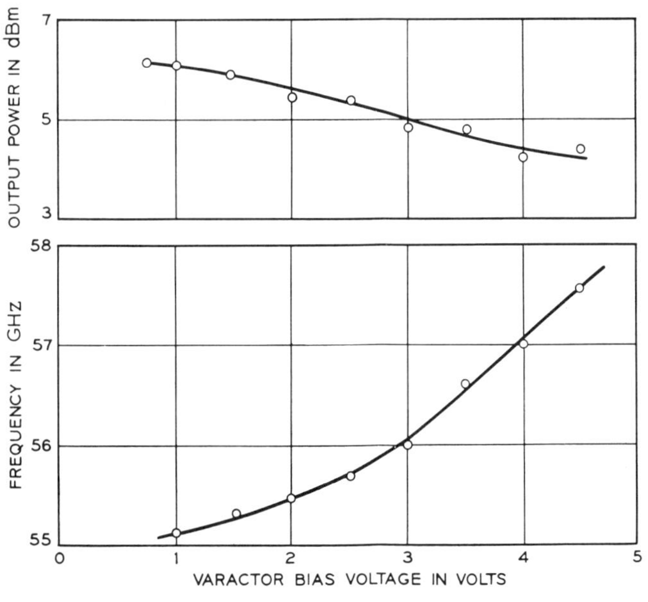

Fig. 3 — Performance of the varactor-tuned oscillator.

quency. Figure 4 shows a series of the FM spectrum at 160 MHz modulation as the modulating power is increased. For the particular diode tested, the minimum modulating power required at 160 MHz for complete carrier suppression was about 4 mW (to a 50 ohm load).

4.2 *The Bias-Current Tuned FM Deviator*

The second method of frequency modulation was achieved by direct modulation of the IMPATT diode bias current (the varactor diode wafer was replaced by a blank wafer). The modulating signal was directly applied to the bias through a coupling capacitor. The dc tuning sensitivity was between 100 to 200 MHz per milliampere at optimum tuning conditions; this resulted in a decrease of 1 dB in output power compared with the maximum power obtained without frequency deviation. For a sinusoidal drive at 160 MHz, complete carrier suppression was achieved at 0.25 mW drive power. Figure 5 shows the spectrum of the modulated signal at various drive levels. When it is tuned properly, no appreciable AM distortion was seen at low modulation index.

4.3 *System Performance with the IMPATT Diode Oscillator-Deviator*

The methods of IMPATT diode oscillator frequency deviation were tested in a two-level PCM millimeter-wave repeater system.[2] The error rate of the system with the IMPATT diode oscillator used as the deviator was measured as a function of the signal-to-noise ratio. Figure 6 is a simplified block-diagram of the test circuit. The IMPATT diode oscillator-deviator is driven by a random-word generator either through the bias of the varactor for varactor-tuned deviation or through the bias of the IMPATT diode for current-tuned deviation.

The random word generator produces 160 megabit pulses per second with a random distribution in polarity. These pulses modulate either the IMPATT diode current or the varactor diode bias to deviate the millimeter-wave output frequency according to the input pulse polarity. The IMPATT diode oscillator-deviator output is down converted to an IF signal in the 1.3 ± 0.25 GHz band. The IF signal is then amplified and injected through a 500 MHz bandwidth filter to a phase-locked oscillator. The function of this oscillator is to act as a limiter thus removing amplitude modulation from the frequency modulated signal. The output of the phase-locked oscillator is further amplified before being injected into the differential phase detector and timing recovery circuit. The latter gives two output voltages. The

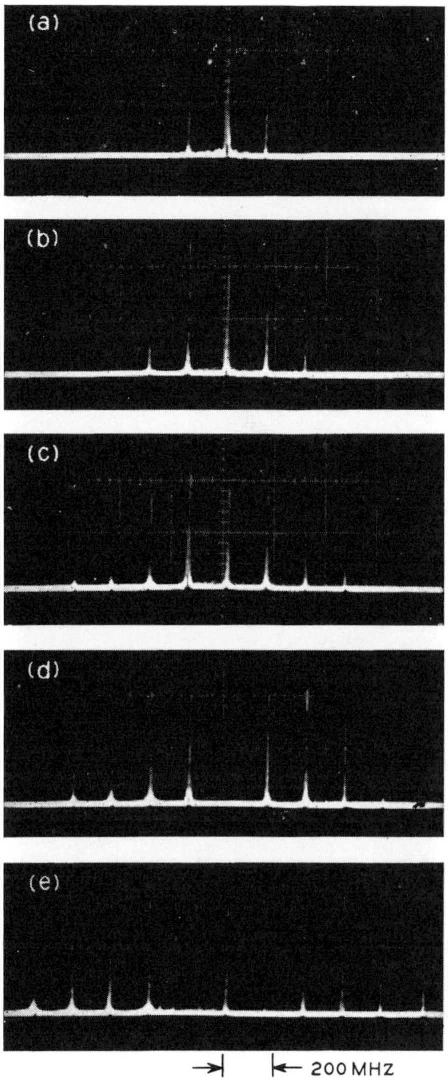

Fig. 4 — FM spectrum at 160 MHz modulation of varactor-tuned oscillator. Complete carrier suppression is shown in (d) with 6 dBm IF power. Others are (a) —9 dBm, (b) —4 dBm, (c) 1 dBm, and (e) 11 dBm (first sideband suppression).

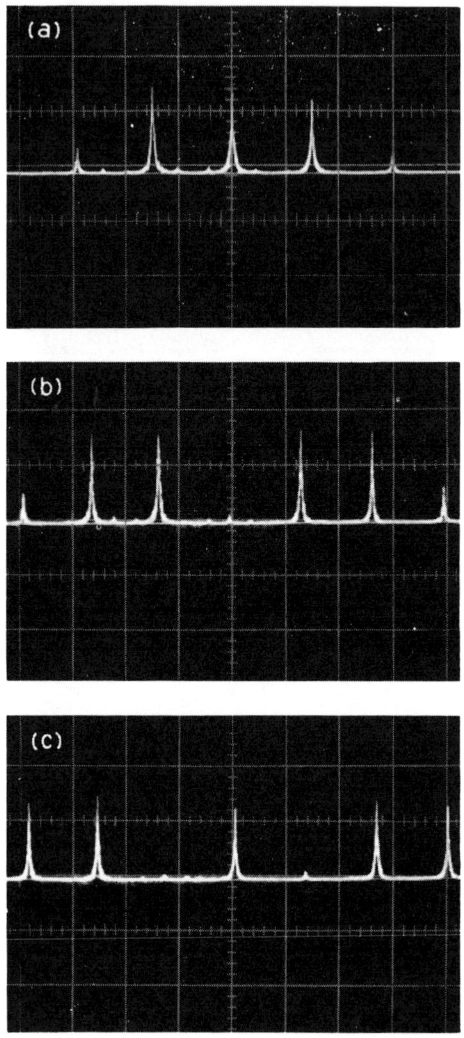

Fig. 5 — FM spectrum at 160 MHz modulation on the bias current. (a) IF power of —11 dBm for equal sidebands and carrier, $\Delta f/f = 1.435$, (b) IF power of —6 dBm for carrier suppression $\Delta f/f = 2.405$, and (c) IF power of —2 dBm for first sideband suppression, $\Delta f/f = 3.832$.

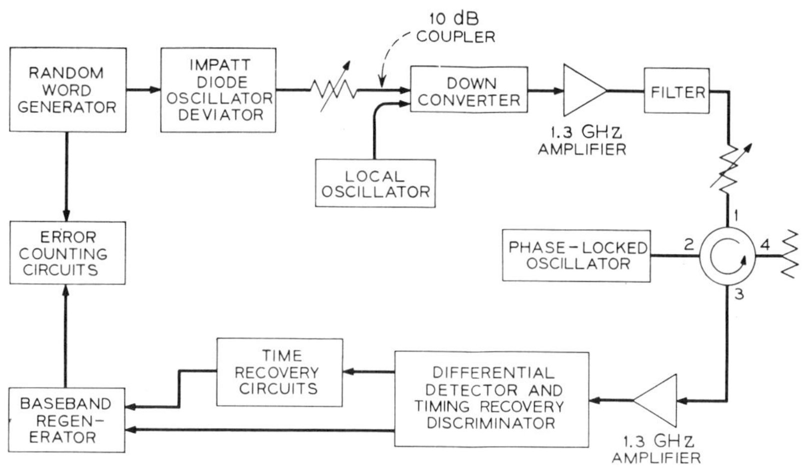

Fig. 6 — Circuit block diagram for the error rate measurement.

first output is ideally a replica of the baseband drive signal from the random word generator. The second output provides a timing signal for the regenerator, which samples the polarity of the first input at the time of arrival of the second input. The regenerator output is then compared with the random word generator output in an error counting circuit. (Reference 12 gives further details of circuit components.)

For the random-word modulation just described, the drive required at the 160 megabit rate was about 0.1 volt peak-to-peak (to a 75Ω load) for the varactor-tuned deviation scheme, and 0.03 volt peak-to-peak for the current-tuned deviation scheme at the optimum error-rate. Figure 7 compares typical eye-diagrams for the two deviation methods tested. For the latter case, the measured error-rate was 10^{-7} for a 16 dB signal-to-noise ratio, which was 2 dB worse than that obtained with a 1.3 GHz tunnel-diode deviator.[2] The major problem appeared to be the FM noise on the IMPATT oscillator output when it was optimally tuned for best deviation. The noise could have been reduced by increasing the circuit Q and consequently increasing the IF drive power.

Error rate improvement could have been obtained by matching the input impedance to the IMPATT bias circuit over the bandwidth of the baseband signal. A narrow band matching results in distortion of the baseband pulses.

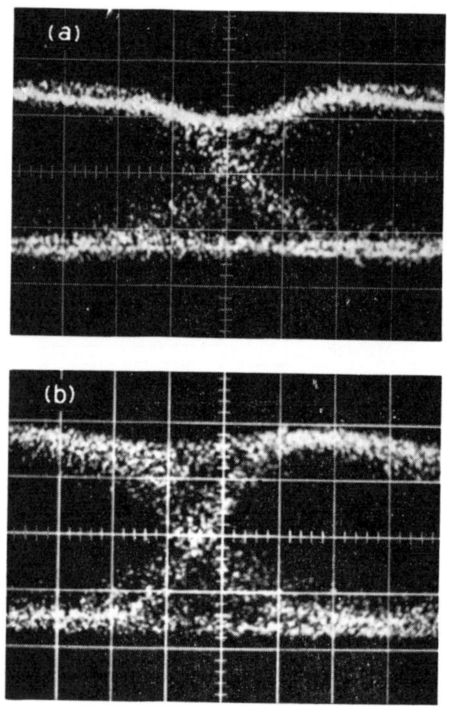

Fig. 7—"Eye-diagrams" of the differentially-coherent-detected random word signal. (a) With the varactor-tuned FM deviator. (b) With the current-tuned FM deviator.

V. MEASUREMENTS OF SUBHARMONIC FREQUENCIES

As mentioned in Section I, the oscillator circuit under study provides at least two resonant circuits for the IMPATT diode, namely, an output circuit which is resonant at the transit-time frequency in M-band (50 to 75 GHz) and an "idler" circuit which is resonant at one-half the transit-time frequency in V-band (26.5 to 40 GHz). The two-frequency arrangement, with the frequencies harmonically related, has improved the tuning sensitivity of the oscillators, as we have shown in previous sections.

Although the existence of such a subharmonic oscillation can be deduced from the experiments described below, it is not surprising in view of the highly nonlinear nature of the impact ionization and avalanche process. Indeed, small-signal theories have predicted that the negative resistance exists over, at least, an octave in frequency.[13–15]

Parametric oscillations and mixing have been observed by DeLoach, Johnston, Evans, and Haddad.[16, 17] High-efficiency, low frequency, subtransit time oscillations have also been reported by Prager, Johnston, Scharfetter, and others,[18–19] and analyzed recently by Scharfetter and others.[20] In Scharfetter's analysis, the low-frequency oscillation is not necessarily harmonically related to the fundamental transit-time frequency. In recent work, Swan[3] showed that improvements in both output power and tuning bandwidth are obtained by providing an idler circuit which resonates at the second harmonic of the transit-time frequency. This result is similar to our findings except that in our case the idler is at one-half the transit-time frequency.

5.1 *Measurements*

Since the idler frequency is below the cutoff frequency of the RG-98/U waveguide, it is impossible to detect its existence directly at the output port of the oscillator as shown in Fig. 1. However, the existence of such a signal at the subharmonic frequency would result in mixing with the fundamental to produce an output at 3/2 the fundamental frequency. It was found that the next higher-order harmonic detected was always 3/2 of the oscillation frequency independent of bias current for all the diodes tested.

For direct detection of the below-cutoff subharmonic oscillation dielectric-filled waveguide tapers which had a cutoff frequency of 18 GHz were used as shown in Fig. 8a. However, a short section of air-filled RG-98/U waveguide with about 30 dB attenuation at 30 GHz remained between the IMPATT diode and the output waveguide taper. This arrangement retained the same subharmonic oscillation circuit conditions. Yet, if enough power exists for the signal at the subharmonic frequency to pass through the short section of air-filled RG-98/U waveguide, both the fundamental and the subharmonic frequencies should be present at the output port. A mixer and a spectrum analyzer were used to detect the subharmonic, while a wavemeter and a diode detector were used for the fundamental frequency. There was indeed an appreciable amount of power (estimated at 9 dBm at the diode) at the subharmonic of 27 GHz; the fundamental was exactly twice the subharmonic frequency, or 54 GHz, within measurement error.

When the short section of the RG-98/U waveguide preceding the diode wafer was also filled with dielectric the power output at the subharmonic was increased appreciably, but the frequency shifted

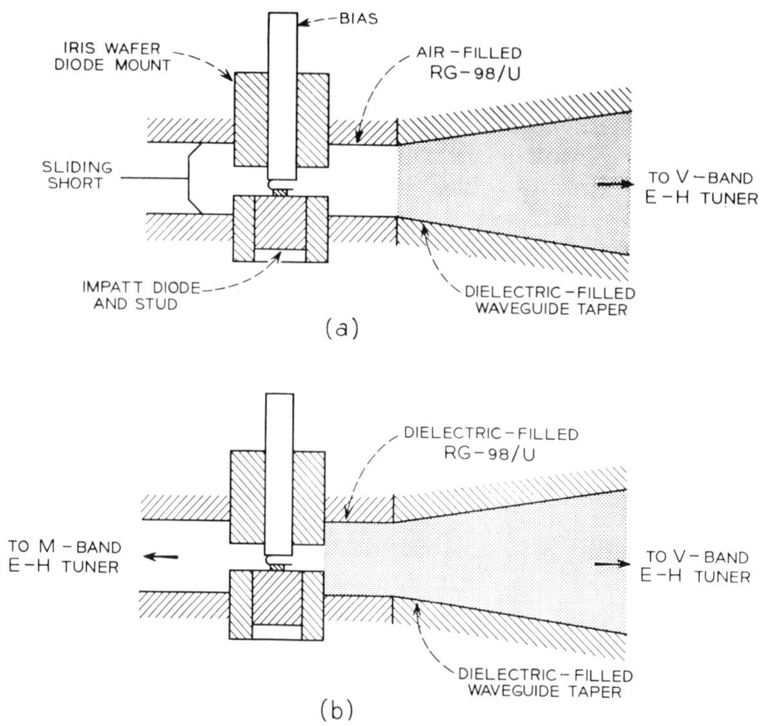

Fig. 8 — Waveguide arrangements for measuring the subharmonic signal.

slightly because of the change in loading impedance. The measurements of the subharmonic power on several diodes are summarized in Table III.

When the sliding short on one side of the oscillator mount was replaced by an E-H tuner followed by a wavemeter and a detector, as shown in Fig. 8b, both the fundamental and the subharmonic could be measured simultaneously. The results again confirmed that the subharmonic frequency was exactly one-half of the fundamental as shown in Fig. 9 for diode No. 3. This harmonic relation held true for any bias current.

The E-H tuner in the M-band circuit serves as an impedance matching device. By mismatching the oscillator to the load in the M-band circuit (thus reducing the 58 GHz power delivered to the load), the power delivered to the V-band load increases, and vice versa. Thus for efficient operation of IMPATT diodes, the subharmonic (and the harmonics)[3] should be reactively terminated. Likewise maximum sub-

Table III—Measurement of Power at One-half the Transit-Time Frequency

Diode No.*	Bias current (mA)	Frequency (GHz)	Power output (dBm)
1	115	28.13	12.05
2	129	28.36	11.2
3	140	29.05	12.0
5	140	29.51	10.2
6	150	34.48	16.9
7	150	32.25	15.2
8	150	32.98	14.7
9	150	31.91	11.1

* The diode number is consistent with that in Table II.

harmonic power can be obtained when the fundamental and all harmonics are reactively terminated. Using this approach, by reactively terminating both the fundamental and the subharmonic, we obtain an output power at 3/2 the transit-time frequency of about 3.3 dBm at 86 GHz.

5.2 *Harmonic Phase-Locking*

The Fig. 8b circuit arrangement also was used for harmonic phase-locking. The locking-signal was injected in the V-band end through

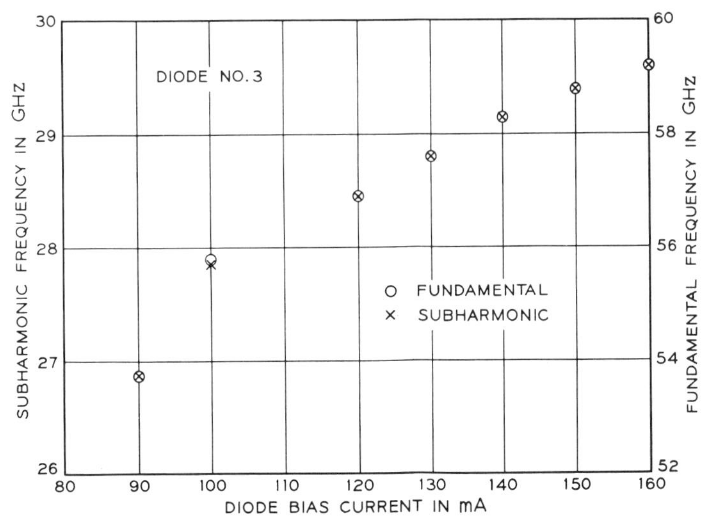

Fig. 9—Fundamental and subharmonic frequencies as a function of diode bias current.

487

a 10 dB directional coupler. Both the locked signals at the V-band and at the M-band were measured. The gain-bandwidth product $(2\Delta f/f_0)$ $(P_0/P_i)^{1/2}$ was about 0.05 to 0.07 for the V-band output.* The M-band output frequency was simultaneously phase-locked with the locked-bandwidth exactly twice that of the V-band. Since the power output in the M-band was about 1 mW compared with about 6 mW at V-band, the apparent gain bandwidth product was much less.

5.3 *V-Band Circuit with Cap Structure*

The experiments of Section 5.1 were conducted on IMPATT diodes in the iris circuit shown in Fig. 1. The same behavior was observed in a different circuit structure. A V-band oscillator was constructed using a resonant cap structure similar to the circuit described in Ref. 1. (see Fig. 10.) Caps were made in various diameters, and could be

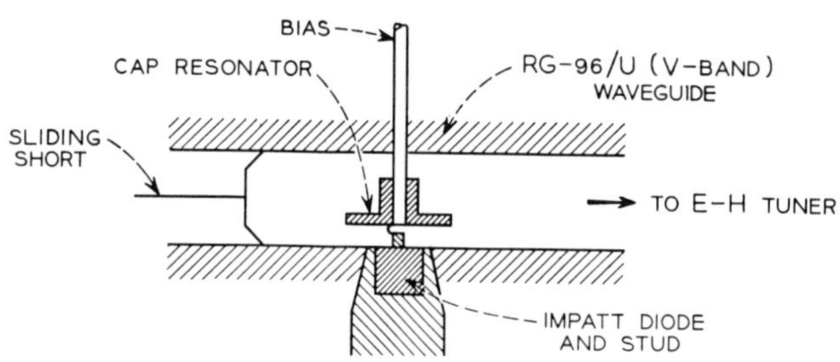

Fig. 10 — A V-band IMPATT oscillator using cap resonator structure.

slid up and down the center bias rod, thus permitting the frequency of oscillation to be varied. A diode with very similar characteristics (breakdown voltage and capacitance) as the ones used before was selected from the same batch (LO 1114). The fundamental oscillation was in the range of 64 to 72 GHz with an output power of 10 dBm at 67 GHz for 150 mA bias. The subharmonic power was also detected.

To optimize the fundamental power, a V-band to M-band taper was used in addition to an E-H tuner so that the subharmonic oscillation was reactively terminated. Table IV shows the results. Notice that when the diode was biased at the same current level, the subharmonic frequency was close to ½ the fundamental frequency. The

* Here we define: f_0 = free running oscillation frequency
P_0 = free running oscillation output power
P_i = injected power at frequency $f_0 \pm \Delta f$
$2\Delta f$ = bandwidth over which the oscillator is phase-locked.

TABLE IV—OSCILLATION IN V-BAND CIRCUIT WITH CAP-RESONATOR STRUCTURE

Subharmonic			Fundamental (Subharmonic reactively terminated)		
Bias (mA)	Frequency (GHz)	Power output (dBm)	Bias (mA)	Frequency (GHz)	Power output (dBm)
80	32.75	7.5	80	64.95	4.0
100	32.87	9.5	105	65.7	5.7
150	33.4	14.0	150	67.0	10.5
160	36.0	14.0	160	72.0	11.1

slight difference resulted from the reactive termination of the subharmonic which required slight retuning for maximum output power at the fundamental frequency.

5.4 *Comparison of the Outputs*

Table V compares the fundamental and subharmonic power outputs in the two different circuits. For the iris-wafer circuit the diode was mounted in the RG-98/U waveguide; dielectric-filled waveguide and tapers were used when the subharmonic frequency power was measured. The dielectric-filled waveguide and tapers had an insertion loss of 1 dB in the frequency range of interest, which was taken into account for the power listed in the fourth column in Table V. For the resonant cap circuit, the diode was mounted in the V-band wave-

TABLE V—COMPARISON OF FUNDAMENTAL AND SUBHARMONIC POWER OUTPUTS

Diode No.	Bias (mA)	Subharmonic Frequency (GHz)	Subharmonic Power (dBm)	Fundamental Frequency (GHz)	Fundamental Power (dBm)
			Iris circuit		
1	115	28.13	12.05	55.02	9.5
3	140	29.05	12.0	57.84	9.9
5	140	29.51	10.2	59.5	7.3
6	150	34.48	16.9	69.1	11.5
7	150	32.25	15.2	63.93	11.1
8	150	32.98	14.7	66.25	10.6
9	150	31.91	11.1	63.7	7.6
			Resonant Cap Circuit		
10	150	33.4	14	67.0	10.5
10	160	36.0	14	72.2	11.1

guide as previously described, so that no dielectric-loading was necessary.

In both circuits, the subharmonic power was maximized by reactively terminating the fundamental power, and vice versa. The subharmonic power is generally greater than that at the fundamental output by about 3 to 6 dB. This is in general agreement with Johnston, Schafetter, and others.[19–20] However, the current density used here is the same as for the diodes operated in the fundamental frequency mode, and the frequencies are harmonically related to each other.

VI. CONCLUSIONS

By using an idler resonance at the subharmonic frequency, one can design oscillators with 20 percent tuning range. Such tunable oscillators and frequency deviators were built and worked satisfactorily in an experimental millimeter-wave PCM repeater system.

Multiple frequency circuits for IMPATT diode oscillators offer an important means of generating millimeter waves with useful power. For example, 2 mW of power at 3/2 the transit-time frequency, or 86 GHz, has been obtained.

VII. ACKNOWLEDGMENT

The authors particularly wish to thank T. Misawa for supplying the IMPATT diodes and for helpful discussions. We would also like to thank F. A. Braun for mounting diodes into various circuits, and G. D. Mandeville for making the error-rate measurements in the PCM system.

REFERENCES

1. Lee, T. P., Standley, R. D., and Misawa, T., "A 50 GHz Silicon IMPATT Diode Oscillator and Amplifier," 1968 International Solid-State Circuit Conference Digest of Technical Papers, *IX* (February 1968), pp. 156–157.
2. Hubbard, W. M., and others, "A Solid-State Regenerative Repeater for Guided Millimeter-Wave Communication Systems," B.S.T.J., *46*, No. 9 (November 1967), pp. 1977–2018.
3. Swan, C. B., "IMPATT Oscillator Performance Improvement With Second Harmonic Tuning," 26th Annual Conference on Electron Device Research, Boulder, Colorado, June 19–21, 1968.
4. W. M. Sharpless, "Wafer-type Millimeter-Wave Rectifiers," B.S.T.J., *35*, No. 6 (November 1956), pp. 1385–1420.
5. Slater, J. C., *Microwave Transmission,* New York: McGraw-Hill Book Company, 1942.
6. Lewin, L., *Advanced Theory of Waveguides,* London: Iliffe & Sons, Ltd., 1951.
7. Tsung-Shan Chen, "Waveguide Resonant Iris Filters with Very Wide Passband and Stopbands," Int. J. Elec. *J(21),* No. 5, (1966), pp. 401–424.

8. Misawa, T., "CW Millimeter-Wave IMPATT Diodes with Nearly Abrupt Junctions," Proc. IEEE, *56,* No. 2 (February 1968), pp. 234–235.
9. Sharpless, W. M., unpublished work. (This tuning method was first used by W. M. Sharpless for an X-band tunnel diode oscillator.)
10. Burrus, C. A., "Planar Diffused Gallium Arsenide Millimeter Wave Varactor Diodes," Proc. IEEE, *55,* No. 6 (June 1967), pp. 1104–1105.
11. Lee, T. P. and Burrus, C. A., "A Millimeter-wave Quadrupler and an Up-converter using Planar Diffused GaAs Varactor Diodes," IEEE Transactions on Microwave Theory and Techniques, *MTT-16,* No. 5 (May 1968), pp. 287–295.
12. Hubbard, W. M. and Mandeville, G. D., "Experimental Verification of the Error-Rate Performance of Two Types of Regenerative Repeaters for Differentially Coherent Phase-shifted-keyed Signals" B.S.T.J., *46,* No. 6 (July 1967), pp. 1173–1202.
13. Read, W. T., Jr., "A Proposed High-Frequency Negative Resistance Diode," B.S.T.J., *37,* No. 2 (March 1958), pp. 401–446.
14. Misawa, T., "Negative Resistance in p-n Junctions under Avalanche Breakdown Conditions," Part I and Part II, IEEE Transactions on Electron Devices, *ED-13,* No. 1 (January 1966), pp. 137–151.
15. Gummel, H. K. and Scharfetter, D. L., "Avalanche Region of IMPATT Diodes," B.S.T.J., *45,* No. 10 (December 1966), pp. 1797–1827.
16. DeLoach, B. C. and Johnston, R. L., "Avalanche Transit-time Microwave Oscillators and Amplifiers," IEEE Transactions on Electron Devices, *ED-13,* No. 1 (January 1966), pp. 181–186.
17. Evans, W. J. and Haddad, G. I., "Frequency Conversion in Read Diodes," Informal Conference on Active Microwave Effects in Bulk Semiconductors, New York, February 2–3, 1967; Also Evans, W. J., "Non-linear and Frequency Conversion characteristics of IMPATT Diodes," Tech. Rep. No. 104, Electron Physics Laboratory, The University of Michigan, Ann Arbor, Michigan, February 1968.
18. Prager, H. J., Chang, K. K. N., and Weisbrod, S., "High Power, High Efficiency Silicon Avalanche Diodes at Ultrahigh Frequencies," Proc. IEEE, *55,* No. 4 (April 1967), pp. 586–587.
19. Johnston, R.L. and Scharfetter, D. L.,"Low Frequency High Efficiency Oscillations in Ge IMPATT Diodes," 26th Annual Conference on Electron Device Research, Boulder, Colorado, June 19–21, 1968.
20. Scharfetter, D. L., Bartelink, D. J., and Johnson, R. L., "Computer Simulation of Low Frequency Oscillation in Ge IMPATT Diodes," 26th Annual Conference on Electron Device Research, Boulder, Colorado, June 19–21, 1968.

Frequency Conversion in IMPATT Diodes

W. J. EVANS, MEMBER, IEEE, AND G. I. HADDAD, SENIOR MEMBER, IEEE

Abstract—**A large-signal model of the Read-type IMPATT diode has been used to analyze the frequency-mixing properties of the oscillating diode. The self-oscillating, two-port frequency converter is described in terms of its short-circuit admittance parameters. It is shown that in the proper circuit, parametric frequency conversion may result in a negative conductance at the input and output ports of the converter. Therefore, high-gain frequency conversion and parametric amplification are possible. Under some conditions, spurious oscillations may occur due to this negative conductance. Experimental circuits have been built which demonstrate conversion gain and parametric amplification and confirm qualitatively the theoretical results. It is also shown experimentally that some of the sideband noise of the IMPATT oscillator is due to low-frequency noise which is up converted from the bias circuit. Some of this noise can be eliminated by proper circuit design.**

I. Introduction

A NUMBER of experimental self-oscillating frequency converters and parametric amplifiers using IMPATT diodes have been reported in the literature [1]–[3]. In such devices an IMPATT diode operates both as the nonlinear mixing element and the local oscillator or pump. In order to describe such a device analytically it is necessary to have a large-signal model of the device which defines the operating point of the oscillating diode in much the same way that the dc equations define the operating point of a conventional mixer diode. A large-signal model of the IMPATT diode which can be conveniently applied to the study of multiple-signal mixing in an oscillating diode has been described in an earlier paper [4]. This paper presents an analysis of frequency conversion in an oscillating IMPATT diode which is based on that large-signal analysis. The results are used to characterize the two-port (three-frequency) converter which for certain loading conditions may also be used as a parametric-type amplifier.

II. Large-Signal Model

The large-signal model of the IMPATT diode which will be used here has been described in detail elsewhere [4]. The assumptions and pertinent results of that analysis are summarized in this section.

The diode structure and field distribution are similar to those analyzed by Read [5]. Starting with the continuity equations for holes and electrons, Poisson's equation, and an expression for the ionization rate, two equations are obtained relating the current injected from a narrow avalanche zone into the drift region and the field. These equations are nonlinear and cannot be solved in closed form. However, if the additional assumption is made that the transit time of the carriers through the drift region is small compared to the period of the RF oscillations, a closed-form solution can be obtained. It is then found that the injected current $I_0(t)$ is given by

$$I_0(t) = \frac{e^{-\phi}}{a\int e^{-\phi}dt}, \qquad (1)$$

where

$$\phi = -\int \frac{2m}{\tau_1 E_c}\left(E_b - \frac{v(t)}{l}\right)dt, \qquad (2)$$

$$a = \frac{m\tau}{\tau_1 \epsilon A E_c}, \qquad (3)$$

where

m = the slope parameter of the ionization rate versus electric field curve
$= E_c\alpha'(E_c)/\alpha(E_c)$

$\alpha(E)$ = the ionization rate for the carriers (assumed the same for holes and electrons)

τ = the transit time of carriers through the drift region

τ_1 = the transit time of carriers through the avalanche region

ϵ = the dielectric constant

A = the diode area

E_c = the critical field for breakdown

$v(t)$ = the ac voltage across the diode

l = the depletion-layer width

and

$$E_b = \frac{\tau}{2\epsilon A} I_d, \qquad (4)$$

where I_d = the dc bias current. The solution given by (1) gives good quantitative results for diodes with a transit angle up to about 0.2π radians. Diodes of this type have been reported by several authors [4], [6], [7]. Good qualitative results are obtained with larger transit-angle diodes since the small transit-angle assumption is made only to calculate $I_0(t)$ and is not used in computing the phase delay of the current propagating

Manuscript received April 25, 1968; revised August 8, 1968. This work was supported by a grant from Omni Spectra, Inc., Farmington, Mich. It was also partially supported by the Rome Air Development Center under Contract F30602-68-C-0043.

W. J. Evans was with the Electron Physics Laboratory, Department of Electrical Engineering, University of Michigan, Ann Arbor, Mich. He is now with Bell Telephone Laboratories, Murray Hill, N. J.

G. I. Haddad is with the Electron Physics Laboratory, Department of Electrical Engineering, University of Michigan, Ann Arbor, Mich.

in the drift region. Finally, the current induced in the external circuit is given by

$$I_e(t) = \frac{1}{\tau}\int_{t-\tau}^{t} I_0(t')dt'. \tag{5}$$

III. Multiple-Signal Analysis

Due to the complexity of the expressions, the multiple-signal analysis is given in the Appendix. The discussion and results presented here will be limited to three-frequency converters.

IV. Three-Frequency Converters

Since the essential features of frequency conversion in an oscillating diode can be obtained from consideration of two signals and the oscillator signal, all V_k are assumed to be zero except the input voltage v_s and the output voltage v_u or v_i. A circuit which satisfies these conditions is shown in Fig. 1. The filters F_s, F_u and F_0 are assumed to be short circuits at all frequencies except f_s, f_u, and f_0, respectively. The two-port representation of this circuit excluding R_s and C_d is shown in Fig. 2.

For the upper-sideband upconverter the three frequencies ω_s, ω_u, and ω_0 are related by

$$\omega_u = \omega_0 + \omega_s. \tag{6}$$

The frequency relation for an upper-sideband downconverter is

$$\omega_s = \omega_0 + \omega_u. \tag{7}$$

Thus, the solution obtained for (6) can be used for the downconverter by simply reversing the definitions of input and output ports. The current at ω_s when $\omega_k = \omega_s$ is found from (47) by requiring $n=0$ in the first term and $m+n=0$ in the second term. Equation (47) then has the form

$$i_{es}(t) = y_{11}{}^{(u)} v_s e^{j\omega_s t}, \tag{8}$$

where

$$y_{11}{}^{(u)} = \frac{\omega_0\alpha}{ja\tau\omega_s} H(0,\omega_s)\left[R_0 - \sum_{n=-\infty}^{\infty}\sum_{l=-\infty}^{\infty} \cdot \frac{(-1)^l\omega_0 R_n Q_l S_{-n-l}}{j(l\omega_0+\omega_s)-b}\right]. \tag{9}$$

The remaining elements of the admittance matrix are obtained in a similar manner and are expressed as follows:

$$y_{12}{}^{(u)} = \frac{\omega_0\alpha}{ja\tau\omega_u} H(-1,\omega_u)\left[R_{-1} - \sum_{n=-\infty}^{\infty}\sum_{l=-\infty}^{\infty} \cdot \frac{(-1)^l\omega_0 R_n Q_l S_{-(n+1)-l}}{j(l\omega_0+\omega_u)-b}\right], \tag{10}$$

$$y_{21}{}^{(u)} = \frac{\omega_0\alpha}{ja\tau\omega_s} H(1,\omega_s)\left[R_{1} - \sum_{n=-\infty}^{\infty}\sum_{l=-\infty}^{\infty} \cdot \frac{(-1)^l\omega_0 R_n Q_l S_{-(n-1)-l}}{j(l\omega_0+\omega_s)-b}\right], \tag{11}$$

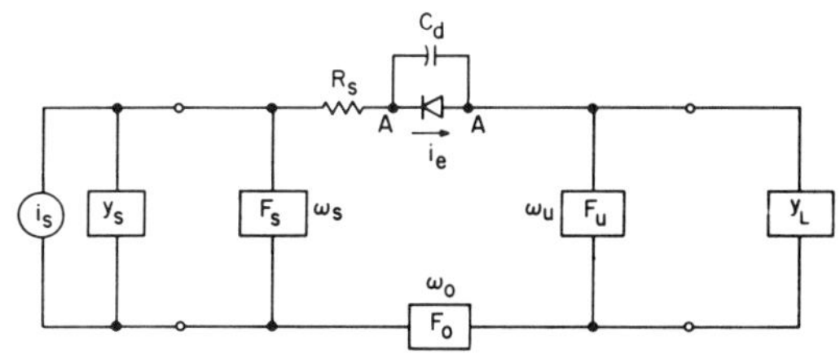

Fig. 1. Converter circuit.

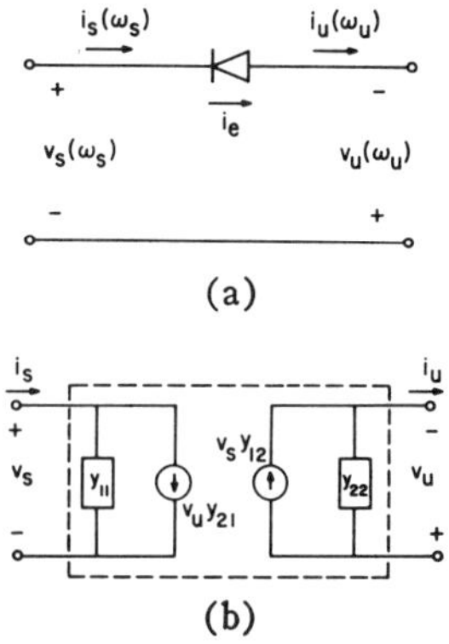

Fig. 2. Circuit representation of nonlinear diode admittance. (a) current and voltage relations. (b) Equivalent circuit.

$$y_{22}{}^{(u)} = \frac{\omega_0\alpha}{ja\tau\omega_u} H(0,\omega_u)\left[R_0 - \sum_{n=-\infty}^{\infty}\sum_{l=-\infty}^{\infty} \cdot \frac{(-1)^l\omega_0 R_n Q_l S_{-n-l}}{j(l\omega_0+\omega_u)-b}\right]. \tag{12}$$

Thus the external-circuit current i_e shown in Fig. 1 is represented by

$$\begin{bmatrix} i_{es} \\ i_{eu} \end{bmatrix} = \begin{bmatrix} y_{11}{}^{(u)} & y_{12}{}^{(u)} \\ y_{21}{}^{(u)} & y_{22}{}^{(u)} \end{bmatrix} \begin{bmatrix} v_s \\ v_u \end{bmatrix}. \tag{13}$$

The admittance matrix for a lower-sideband converter is obtained in a similar fashion. In this case

$$\omega_i = \omega_0 - \omega_s, \tag{14}$$

$$\begin{bmatrix} i_{es} \\ i_{ei}^* \end{bmatrix} = \begin{bmatrix} y_{11}{}^{(L)} & y_{12}{}^{(L)} \\ y_{21}{}^{(L)} & y_{22}{}^{(L)} \end{bmatrix} \begin{bmatrix} v_s \\ v_i^* \end{bmatrix}, \tag{15}$$

where

$$y_{11}{}^{(L)} = \frac{\omega_0\alpha}{ja\tau\omega_s} H(0,\omega_s)\left[R_0 - \sum_{n=-\infty}^{\infty}\sum_{l=-\infty}^{\infty} \cdot \frac{(-1)^l\omega_0 R_n Q_l S_{-n-l}}{j(l\omega_0+\omega_s)-b}\right], \tag{16}$$

$$y_{12}{}^{(L)} = \frac{\omega_0\alpha}{-ja\tau\omega_i} H(1,-\omega_i)\left[R_1 - \sum_{n=-\infty}^{\infty}\sum_{l=-\infty}^{\infty} \cdot \frac{(-1)^l\omega_0 R_u Q_l S_{-n+1-l}}{j(l\omega_0-\omega_i)-b}\right], \tag{17}$$

$$y_{21}{}^{(L)} = \frac{\omega_0\alpha}{ja\tau\omega_s} H(-1,\omega_s)\left[R_{-1} - \sum_{n=-\infty}^{\infty}\sum_{l=-\infty}^{\infty} \cdot \frac{(-1)^l\omega_0 R_n Q_l S_{-n-1-l}}{j(l\omega_0+\omega_s)-b}\right], \tag{18}$$

493

$$y_{22}^{(L)} = \frac{\omega_0 \alpha}{-ja\tau\omega_i} H(0, -\omega_i)\left[R_0 - \sum_{n=-\infty}^{\infty}\sum_{l=-\infty}^{\infty} \cdot \frac{(-1)^l \omega_0 R_n Q_l S_{-n-l}}{j(l\omega_0 - \omega_i) - b}\right]. \tag{19}$$

Equations (9) through (12) and (16) through (19) can now be used to describe the converter performance.

V. Effect of Parasitic Elements

In the last section the admittance matrices associated with the nonlinear part of the diode were derived. However, a diode will always have associated with it a depletion layer capacitance and a series bulk resistance. In this section the converter admittance matrices are transformed to include the effects of these elements. Generally, the performance of a diode at microwave frequencies is strongly affected by additional parasitic elements associated with the particular package and mounting arrangement. These effects may easily be included when desired.

The two-port converter discussed above can be represented as shown in Fig. 3. By representing the currents which are proportional to the transfer admittances y_{12} and y_{21} as current sources, it is possible to separate the input and output circuits as shown. Note that the current direction i_u is reversed from that implied by the analysis of the Appendix. Therefore, the transfer admittances y_{12} and y_{21} obtained there must be multiplied by -1.

To obtain the complete admittance matrix for the converter, the $ABCD$ matrix for the RC coupling network is used. Thus,

$$\begin{bmatrix} v_s \\ i_s \end{bmatrix} = \begin{bmatrix} A_1 & B_1 \\ C_1 & D_1 \end{bmatrix}\begin{bmatrix} v_1 \\ i_1 \end{bmatrix}, \tag{20}$$

where

$$\left.\begin{aligned} A_1 &= 1 \\ B_1 &= -R_s \\ C_1 &= -j\omega_s C_d \\ D_1 &= j\omega_s R_s C_d + 1 \end{aligned}\right\}, \tag{21}$$

and for the output port

$$\begin{bmatrix} v_u \\ i_u \end{bmatrix} = \begin{bmatrix} A_2 & B_2 \\ C_2 & D_2 \end{bmatrix}\begin{bmatrix} v_2 \\ i_2 \end{bmatrix}, \tag{22}$$

where

$$\left.\begin{aligned} A_2 &= 1 \\ B_2 &= -R_s \\ C_2 &= -j\omega_u C_d \\ D_2 &= j\omega_u R_s C_d + 1 \end{aligned}\right\}. \tag{23}$$

These equations are combined with (13) for the upper-sideband converter to obtain the complete admittance matrix

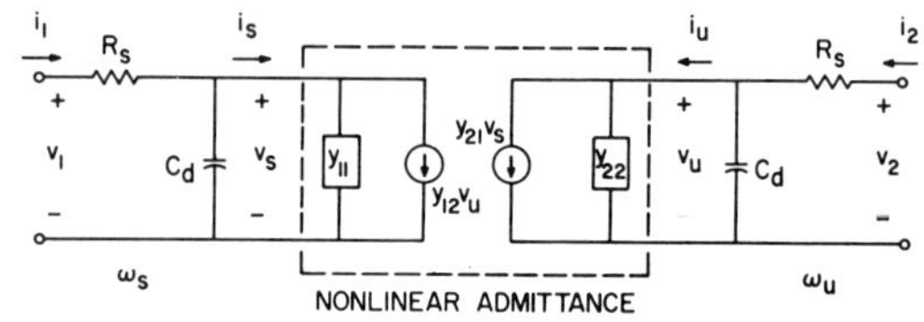

Fig. 3. Complete equivalent circuit including C_d and R_s.

$$\begin{bmatrix} i_1 \\ i_2 \end{bmatrix} = \begin{bmatrix} y_{11}' & y_{12}' \\ y_{21}' & y_{22}' \end{bmatrix}\begin{bmatrix} v_1 \\ v_2 \end{bmatrix}, \tag{24}$$

where

$$y_{11}' = \frac{A - [R_s(j\omega_u C_d + y_{22}) + 1]}{R_s A}, \tag{25}$$

$$y_{12}' = y_{12}/A, \tag{26}$$

$$y_{21}' = y_{21}/A, \tag{27}$$

$$y_{22}' = \frac{A - [R_s(j\omega_s + y_{11}) + 1]}{R_s A}, \tag{28}$$

$$A = [R_s(j\omega_s C_d + y_{11}) + 1][R_s(j\omega_u C_d + y_{22}) + 1] - R_s^2 y_{12} y_{21}. \tag{29}$$

A similar result is obtained for the lower-sideband converter. In that case the output circuit is described in terms of conjugate voltages and currents so that the $ABCD$ matrix for the output circuit of the lower-sideband converter is

$$\left.\begin{aligned} A_2 &= 1 \\ B_2 &= -R_s \\ C_2 &= j\omega_i C_d \\ D_2 &= -j\omega_i R_s C_d + 1 \end{aligned}\right\}. \tag{30}$$

Thus, in the results given in (25) through (28), ω_u is replaced by $-\omega_i$.

VI. Results

In this section the results of the preceding sections are used to calculate some of the characteristics of the self-pumped frequency converter. Parameter values were chosen which correspond to the GaAs diodes used in the experimental work and these are given in Table I. Since there are such a large number of parameters appearing in the analysis, it is not practical to examine the converter behavior as a function of each one of them. As a general rule, however, the better the oscillator performance, the better will be the converter performance. Therefore, many qualitative conclusions on converter performance can be drawn from the results of the oscillator analysis [4]. Since all of the parameters cannot be listed on each figure, the values given in Table I remain constant throughout the figures unless otherwise

TABLE I

GaAs Diode Parameters and Definitions of Certain Quantities

$C_d = 0.12$ pF

$\tau = 5 \times 10^{-12}$ second

$l = 10^{-4}$ cm

$v_d = 2 \times 10^7$ cm/s

$f_a = 12.9$ GHz = avalanche frequency

$A = 1.1 \times 10^{-5}$ cm²

$\epsilon_r = 12.5$

$f_0 = 15$ GHz

$I_d = 8$ mA

$R_s = 2$ ohms

$$b_s = \frac{g_{11}}{2} I_m \left(\frac{y_{12}y_{21}}{g_{11}g_{22}}\right) - b_{11} = \text{optimum source susceptance}$$

$g_s = 0.02$ mho = source conductance

$$b_L = \frac{g_{22}}{2} I_m \left(\frac{y_{12}y_{21}}{g_{11}g_{22}}\right) - b_{22} = \text{optimum load susceptance}$$

$$y_\alpha = y_{11} - \frac{y_{12}y_{21}}{y_{22} - y_L} \quad \text{(input admittance)}$$

$$y_\beta = y_{22} - \frac{y_{12}y_{21}}{y_{11} - y_s} \quad \text{(output admittance)}$$

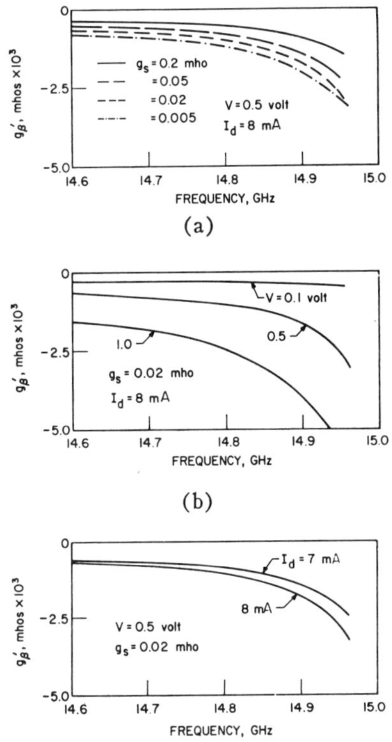

Fig. 4. Output conductance for LSB converter versus frequency. (a) g_s dependence. (b) Voltage dependence (c) Current dependence.

noted on the particular figure. The peak RF voltage amplitude V for the oscillator is given in each figure.

The variations in converter performance caused by various load conditions at the input and output ports are investigated only as a function of the conductance of the termination. The susceptive components of the terminations are assumed to have a value which optimizes the available gain of the network when the input and output conductances are positive [8]. It should be pointed out that these values may not be the optimum value for other considerations (e.g., bandwidth). Note also that all of the following results are computed from (9) through (12), (16) through (19), and (25) through (28); that is, the effects of a series bulk resistance and depletion layer capacitance are included.

Finally, for the magnitudes of voltage and current to be considered, the series expansion appearing in (9) through (12) and (16) through (19) can generally be terminated after a few terms with sufficient accuracy. To check the convergence of these sums the computer program which performs the required calculations lists the value of each term in the sum for $-4 \leq l \leq 4$, $-4 \leq n \leq 4$ or through the 81st term of the sum.

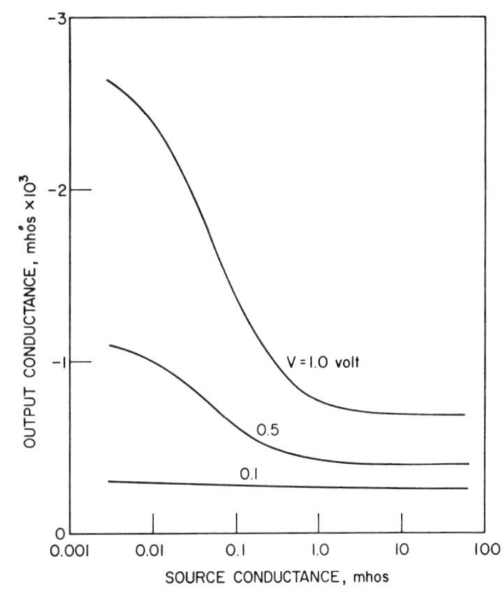

Fig. 5. Output conductance versus source conductance. LSB converter; $f_i = 14.75$ GHz, $f_s = 250$ MHz.

Lower-Sideband Converter

The lower-sideband converter with $\omega_s = \omega_0 - \omega_i$ is considered first. Fig. 4 shows the variation in converter output conductance (including the effects of the depletion layer capacitance and a series resistance) for a number of parameters. First of all, notice that the output conductance near the oscillator frequency (15 GHz) is negative in all cases. This indicates a possible source of instability in the oscillator performance in that spurious sideband oscillations may exist for certain load admittances. Figs. 4 and 5 show the dependence of this conductance on the source or input conductance g_s. It can be seen that by shorting the input port the negative conductance at the converter output is substantially reduced. If the circuit were used as a reflection-type amplifier at, say, 14.7 GHz, then the amplifier gain would be decreased by shorting the "idler" frequency. In a conventional parametric amplifier shorting the idler

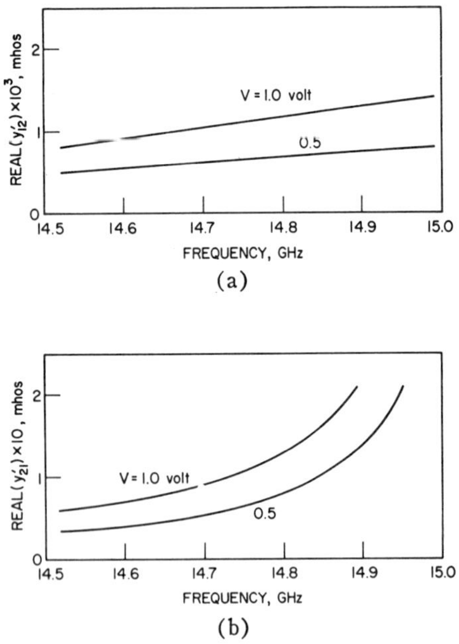

Fig. 6. Transfer admittances for LSB converter versus output frequency. (a) Real part of y_{12}' versus frequency. (b) Real part of y_{21}' versus frequency.

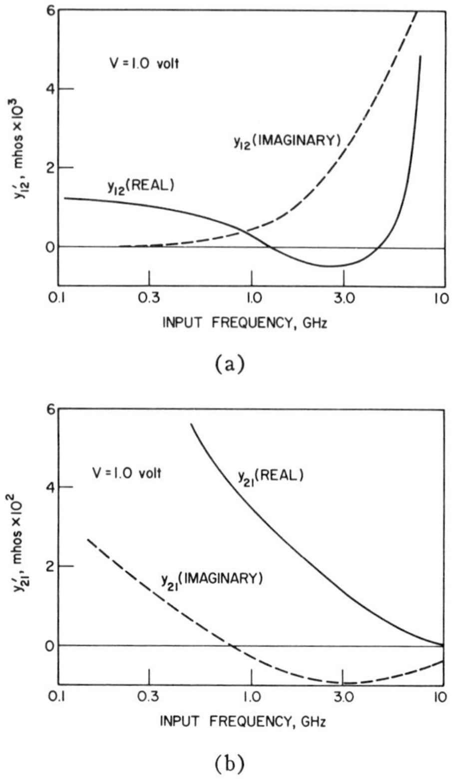

Fig. 7. Transfer admittance for LSB converter versus input frequency. $V = 1.0$ volt. (a) y_{12}' versus frequency. (b) y_{21}' versus frequency.

frequency would destroy the negative conductance seen at the input port.

In the case considered here this does not happen because there are two sources of negative conductance for the lower-sideband converter. The first is the usual negative conductance which occurs in a lower-sideband converter due to regenerative feedback from idler to input port for any nonlinear time-varying admittance. A second contribution to the negative conductance occurs in the case of an oscillating IMPATT diode due to the intrinsic broad-band negative conductance of the diode itself which appears around the oscillation frequency. It is interesting to note that the negative conductance predicted in Fig. 4 is approximately ten times larger than the negative conductance predicted for the oscillator. Thus, the regeneration due to the nonlinear admittance is dominant for the LSB converter.

Fig. 4 shows the increase in output conductance with increasing RF oscillator voltage amplitude. This increase is due to the increased coupling between the input and output circuits for larger oscillator voltages. Finally, Fig. 4 shows the dependence on dc bias current.

Fig. 6 demonstrates the voltage dependence more clearly. Shown is a plot of the real part of the transfer admittances y_{12}' and y_{21}'. As expected, they increase with the RF voltage thus producing more coupling between the input and output circuits. For frequencies near ω_0, the imaginary parts of the transfer admittances are much smaller than their corresponding real parts (see Fig. 7). This indicates that a nonlinear reactive mixing process occurs in this frequency range. On the other hand, at frequencies in the microwave range, the transfer admittances are dominated by their imaginary components indicating a nonlinear resistive mixing process. As expected, the sign of the imaginary parts of y_{11}' and y_{22}' in this range corresponds to a time-varying inductive susceptance.

The output conductance of the LSB converter is shown in Fig. 8 for input, or idler, frequencies extending into the microwave range. (Note the frequency scale is plotted in terms of the input frequency, $\omega_s = \omega_0 - \omega_i$.) The output conductance remains negative well into the microwave region, thus suggesting the possibility of parametric oscillations or amplification in this range.

Although the output circuit of the converter can be stabilized by choosing a load conductance larger than the negative output conductance, the input circuit may be unstable. Fig. 9 shows the input conductance for a source conductance of 0.02 mho, a matched load susceptance and a load conductance of $(1.01)\,|g_\beta|$. This choice of load conductance stabilizes the output circuit of the converter but is small enough to provide a very high regenerative gain in the output circuit.

The result of such a choice is that above 7 GHz, where the output conductance is positive, the converter input conductance is negative but smaller in magnitude than the source conductance of 0.02 mho. However, below this frequency the input conductance may be negative and larger in magnitude than the source conductance. Thus, the converter would be unstable. By making use of this regenerative feedback from the output circuit, a reflection-type parametric amplifier can be built. The high but finite value of negative conductance at approximately 4 GHz is due to the diode resonance near

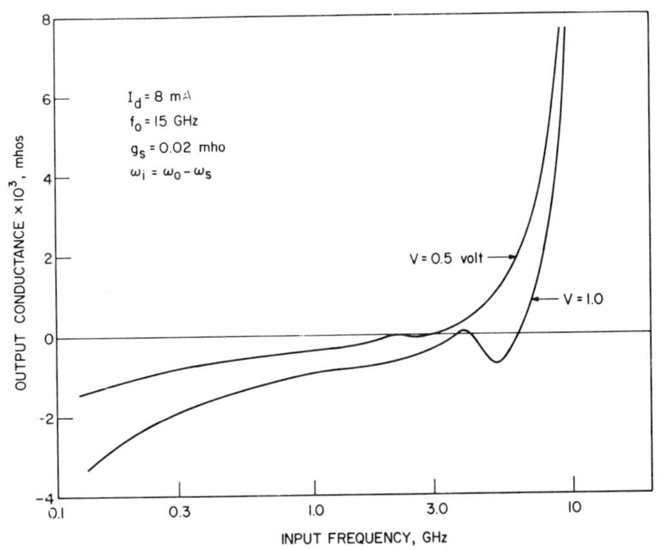

Fig. 8. Output conductance of LSB converter versus input frequency f_s.

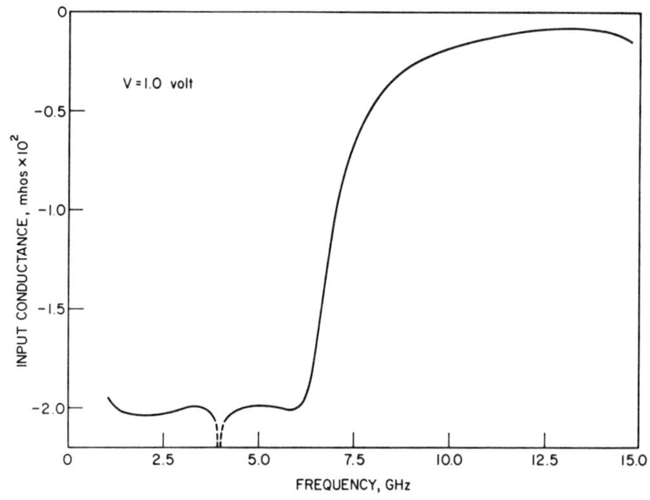

Fig. 9. Input conductance versus input frequency. LSB converter; $g_L = 1.01|g_\beta|$, $V = 1.0$ volt.

the avalanche frequency. (This occurs below f_a because the RF voltage amplitude of 1 volt affects the diode resonance.)

Upper-Sideband Converter

Results similar to those obtained for the LSB converter are shown in Fig. 10 for the USB converter (with $\omega_u = \omega_0 + \omega_s$). The output conductance is approximately ten times smaller than that calculated for the LSB converter. This would be expected from the comments of the previous section on the origin of the negative conductance in the LSB converter. For upper-sideband conversion in passive nonlinear elements no instability occurs, that is, the output conductance is positive. Thus, the negative conductance shown in Fig. 10 is due to the intrinsic negative conductance of the IMPATT diode at the microwave frequency. Thus the output conductance, although showing the same kind of dependence on the source conductance, is affected to a much lesser degree by the input circuit.

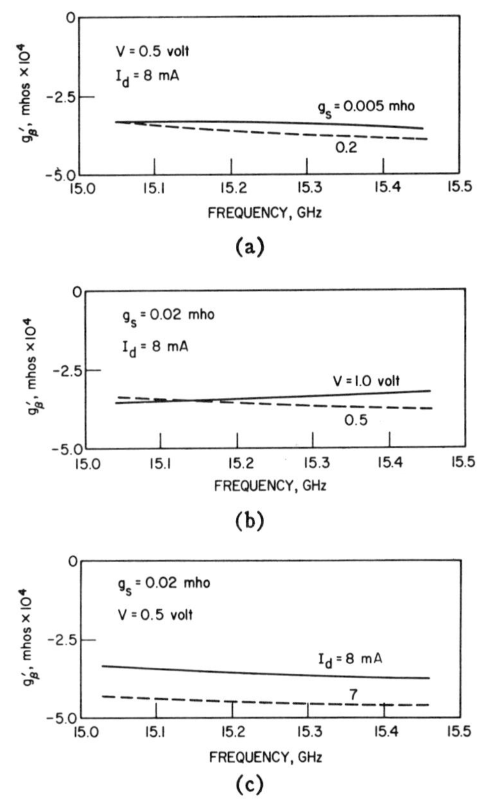

Fig. 10. Output conductance for USB converter versus output frequency. (a) g_s dependence. (b) Voltage dependence. (c) Current dependence.

VII. Converter Measurements

Circuit

The frequency converters analyzed in the preceding section have been built and operated using GaAs and silicon IMPATT diodes. The circuit consists of a reentrant cavity with the diode mounted at one end and a sliding short at the other end. The short is noncontacting to the outer wall of the cavity so that bias and VHF power can be applied to the diode. The oscillator output and upconverted signals are coupled out of the cavity by a probe and coaxial line. The diode mount is then placed in the circuit shown in Fig. 11. In addition to the components shown in the figure, isolators, attenuators and directional couplers are added to control and monitor the circuit operation. The TWT provides a high level noise source which can be calibrated against the noise tube. This allows measurement of noise figures as high as 50 dB by the Y factor method [9]. For higher values, CW noise figure measurements are made with the signal generator. The slide screw tuner in the diode arm of the circulator provides an additional degree of freedom in tuning the diode oscillator/converter circuit. Finally, the bandpass filter is required in the microwave receiver arm of the circulator to prevent the output of the diode oscillator from overloading the receiver mixer.

The circuit is set up so that several modes of operation

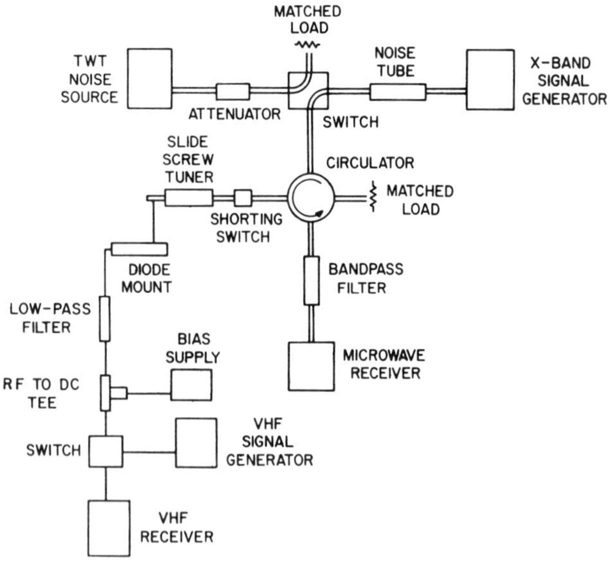

Fig. 11. Simplified experimental setup

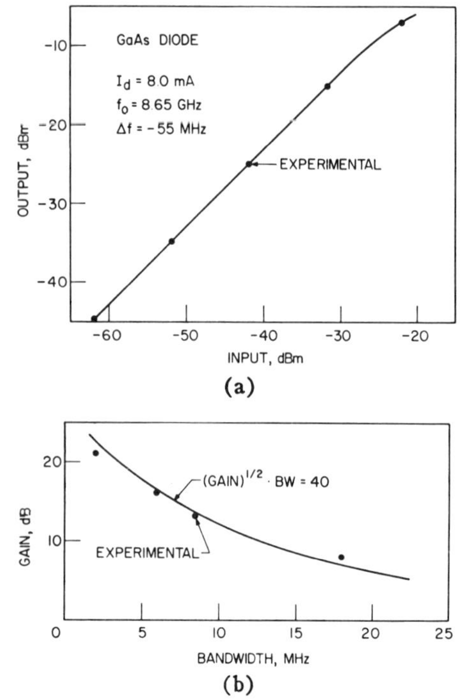

Fig. 12. Saturation and gain bandwidth for lower-sideband reflection amplifier. (a) Output versus input. (b) Gain versus bandwidth.

can be studied without disturbing the diode mount or the slide screw tuner adjustments. Thus it is possible, for example, to measure the power output from the oscillating diode, the reflection gain at a sideband frequency, the upconversion or downconversion gain and noise figure without disturbing the diode oscillator. Of course, with this simple circuit, it is not possible to short all of the unwanted frequency components as was assumed analytically. However, the circuit does demonstrate qualitatively a number of the conclusions reached in the preceding section.

Amplification

In Section VI it was shown that the oscillating diode exhibits a negative conductance which is particularly large at frequencies near the oscillator frequency. The existence of this negative conductance was demonstrated experimentally by reflection-type amplification measurements at frequencies near (within 200 MHz) the oscillator frequency.

This mode of operation as well as the frequency converters discussed below will be described in terms of gain-bandwidth product, saturation, and noise figure.

Fig. 12 shows the saturation characteristics and the gain-bandwidth (at the 3 dB points) for a lower-sideband reflection-type amplifier using a GaAs diode. The sideband amplifier generally does not show the usual saturation behavior but becomes unstable for signals comparable to the oscillator power. The $(\text{gain})^{1/2}\cdot$bandwidth product is approximately constant as expected for negative resistance amplification. Relatively narrow bandwidth is obtained in this mode of operation compared to that possible with a conventional (nonoscillating) IMPATT amplifier due to the fact that only one tuned circuit is available for matching. By using the slide screw tuner and the sliding short in the diode mount, a $(\text{gain})^{1/2}\cdot$bandwidth product of about 250 was obtained for the conventional amplifier. The noise figure of the sideband amplifier is also higher because of the low-frequency noise which is upconverted by the oscillating diode.

For some conditions of tuning and bias current the negative conductance at frequencies near the oscillator frequency is sufficient to cause spurious sideband oscillations. The theoretical results given in Fig. 5 indicate that these oscillations can be suppressed by proper termination of the difference frequency. These conclusions have been verified experimentally [10].

Frequency Conversion

The results of measurements on the various types of frequency converters are shown in Table II.

As indicated by the results of the preceding section, the negative conductance at the upper sideband is somewhat smaller in magnitude than it is at the lower sideband. This is found to be the case experimentally. Generally, it was found to be very difficult to adjust the circuit so that sufficient gain at frequencies above f_0 could be obtained to excite spurious upper-sideband oscillations while maintaining a reasonable power output at f_0. Therefore the following results were obtained with an oscillator output of only 0.2 mW (as compared to about 2 mW for the lower sideband amplifier). The measured gain at a bandwidth of 5 MHz was 7 dB. Saturation occurred about 5 dB below 0.2 mW or at -12 dBm. The oscillator spectrum was observed to be quite noisy for this type of amplifier and this is reflected in a measured amplifier noise figure in excess of 60 dB. It should be pointed out that the high noise in this case is not an inherent property of the upper-sideband amplifier but

TABLE II

TYPICAL CHARACTERISTICS OF GaAs DEVICES

	$(\text{Gain})^{1/2}$ ·Bandwidth (MHz)	Saturation (dBm)	Noise Figure (dB)
Lower-Sideband*			
Upconverter	70	0	34 (25)‡
Downconverter	2	−17†	55 (−6)
Amplifier	40	− 2	46.5
Upper-Sideband**			
Upconverter	16	−15	61 (11)
Downconverter	0.7	−25†	77 (−17)
Amplifier	11	−12	>60 (8)

* Oscillator power level for LSB devices was 2 mW.
** Oscillator power level for USB devices was 0.2 mW.
† Input power.
‡ Number in parentheses indicates device gain in dB for the noise measurement.

is the result of not being able to take advantage of the tuner in the microwave circuit as a stabilizing circuit for the oscillator.

Most of the conclusions drawn from the upper- and lower-sideband amplifiers are also true for the converters; the lower-sideband devices show higher gain, and because of the higher power output in this particular circuit they are less noisy and have a greater dynamic range. Note that in the case of the downconverter, saturation occurs for a relatively low input power. This happens because of the amplification which occurs at the microwave frequency. If these input powers are corrected for the reflection gain in each case, the saturation is found to occur at approximately the same RF power that results in saturation in the other devices. The gain bandwidth products given in Table II are taken from measured data with no correction for coupling loss. Therefore, in a circuit designed to maximize the coupling and bandwidth, a considerable improvement in performance could be expected. Even so, the results presented here show that high upconversion gain (e.g., for the lower-sideband upconverter at 3 MHz bandwidth, 27 dB) and downconversion with some gain (e.g., for the lower-sideband downconverter at 1 MHz bandwidth, 6 dB) can be obtained. Similar measurements have been made on a silicon diode operating in some of the circuits discussed above. The only difference is a small increase in the noise (3 to 5 dB) and an increase in the saturation level due to a higher oscillator power output.

The measured noise figure for the GaAs amplifier was 46.5 dB, and for the silicon amplifier, 46 dB. However, the noise figure for this mode of operation (as well as the gain) is very dependent on the termination at the difference frequency (about 55 MHz in this case). Fig. 13 shows the excess noise power at frequencies above and below the oscillator frequency f_0. The excess noise was measured for three terminations at each difference frequency. Because of the circuitry (low-pass filter, dc to RF tee, and a length of transmission line) which must be connected between the VHF termination and the

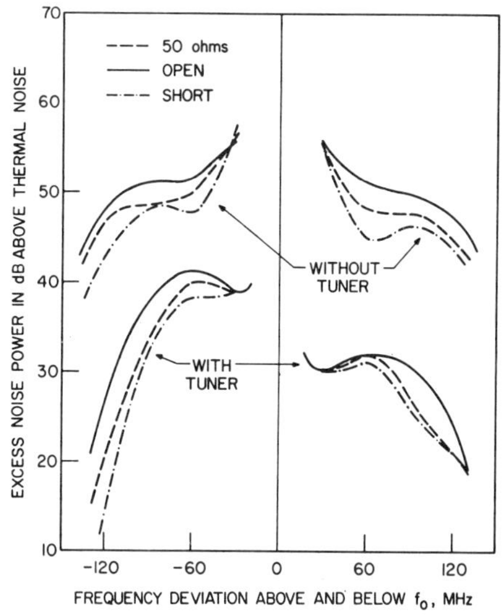

Fig. 13. Excess noise power in oscillator sidebands.

diode, the designation of open, short, and 50 ohms are only the approximate load conditions seen by the diode and therefore are intended only to give a general idea of variation in excess noise with the termination. First, notice that the excess noise is quite large when the slide screw tuner is removed from the microwave circuit. The tuner reduces the sideband noise in the oscillator because it acts as a stabilizing cavity for the oscillator [11], [12]. With the tuner in the circuit and tuned below the oscillator frequency, the noise at the lower sideband is relatively high compared to the upper sideband. In particular it was observed that a small increase in the bias current resulted in spurious sideband oscillations at approximately 70 MHz below the oscillator frequency. These results indicate that much of the noise at the sideband frequencies is due to noise which is upconverted and amplified from the low-frequency part of the spectrum. Shorting the diode at the VHF frequencies will reduce this noise but it also reduces the gain of the sideband amplifier.

VIII. CONCLUSIONS

Starting with the solution for the injected current in a Read-type oscillator, the analysis presented here has shown that these oscillators can also be employed as self-oscillating frequency converters. In particular it was shown that the lower-sideband converter exhibits a large negative conductance over a significant range of frequencies. This negative conductance is a result of the large variation in the diode reactance which occurs because of the current oscillations in the diode. The result is that the operation of the self-oscillating converter is very much like that of a conventional parametric frequency converter. Thus, as in the conventional parametric devices, reflection-type amplification and frequency conversion with gain are possible. As might be expected, the input conductance at one port of such a device depends on the terminations of the various idler frequencies. The degree of this dependence is a function of the amount of coupling between the input and idler ports. In the case of a three-frequency converter it was shown that this coupling depends on the amplitude of the diode oscillations.

Experimental devices have been built which demonstrate the feasibility of self-oscillating frequency converters. It was found that the lower-sideband converter will operate as a high-gain converter. In fact, for some terminations the gain is high enough to cause spurious oscillations. The high upconversion gain and the relatively high noise output at low frequencies suggests that a major source of noise at frequencies near the oscillator frequency is noise which is upconverted from the bias circuit. Experimental measurements of noise near the oscillator show it to be dependent on the low-frequency termination.

Appendix

If multiple signals are present, the voltage across the diode is assumed to be

$$-v(t) = V \sin \omega_0 t + \sum_k V_k e^{j\omega_k t}, \tag{31}$$

where the sum k is over all small-signal voltages appearing across the diode. Then, for steady-state oscillations, (1), (2), and (31) give

$$I_0(t) = \frac{\exp\left(-bt - x \cos \omega_0 t + \gamma \sum_k \frac{V_k}{j\omega_k} e^{j\omega_k t}\right)}{a \int \exp\left(-bt - x \cos \omega_0 t + \gamma \sum_k \frac{V_k}{j\omega_k} e^{j\omega_k t}\right) dt}, \tag{32}$$

where

$$x = \frac{2mV}{\tau_1 E_c l \omega}, \tag{33}$$

$$b = -\frac{m\tau I_d}{\tau_1 E_c \epsilon A}, \tag{34}$$

$$\gamma = \frac{2m}{\tau_1 E_c l}, \tag{35}$$

and $\omega_0 =$ the oscillator frequency.

In many of the applications where such converters will be useful, the voltages V_k will be much smaller than the oscillator voltage V_0. This is assumed in the following analysis. With this assumption the current $I_0(t)$ can be expanded in a Taylor series about $V_k = 0$ as follows:

$$I_0(t, V_k) = I_0(t, 0) + \sum_k \left.\frac{\partial I_0(t, V_k)}{\partial V_k}\right|_{V_k=0} \cdot V_k + \cdots. \tag{36}$$

The small-signal currents flowing in response to the voltages V_k are, neglecting higher order terms in (36),

$$i_0(t) = \sum_k \left.\frac{\partial I_0(t, V_k)}{\partial V_k}\right|_{V_k=0} \cdot V_k. \tag{37}$$

From (32),

$$\sum_k \frac{\partial I_0}{\partial V_k} = \frac{\exp\left(-bt - x \cos \omega_0 t + \gamma \sum_k \frac{V_k}{j\omega_k} e^{j\omega_k t}\right)}{a \int \exp\left(-bt - x \cos \omega_0 t + \gamma \sum_k \frac{V_k}{j\omega_k} e^{j\omega_k t}\right) dt} \cdot \left[\sum_k \frac{\gamma e^{j\omega_k t}}{j\omega_k} - \frac{\int \exp\left(-bt - x \cos \omega_0 t + \gamma \sum_k \frac{V_k}{j\omega_k} e^{j\omega_k t}\right) \gamma \sum_k \frac{e^{j\omega_k t}}{j\omega_k} dt}{\int \exp\left(-bt - x \cos \omega_0 t + \gamma \sum_k \frac{V_k}{j\omega_k} e^{j\omega_k t}\right) dt}\right]. \tag{38}$$

Combining (5), (39), and (38), the small-signal external circuit current is obtained:

$$i_e(t) = \frac{\gamma}{a\tau} \int_{t-\tau}^{t} \frac{e^{f(t)}}{\int e^{f(t')} dt'} \cdot \left[\sum_k \frac{V_k}{j\omega_k} e^{j\omega_k t} - \frac{\int e^{f(t')} \sum_k \frac{V_k}{j\omega_k} e^{j\omega_k t} dt'}{\int e^{f(t')} dt'}\right] dt, \tag{39}$$

where

$$f(t) = -bt - x \cos \omega_0 t. \tag{40}$$

The coefficients R_n are defined by

$$\frac{e^{f(t)}}{\int e^{f(t)} dt} = \omega \sum_{n=-\infty}^{\infty} R_n e^{jn\omega_0 t}. \tag{41}$$

Next define the Fourier coefficients S_l by

$$\frac{e^{-bt}}{\int e^{f(t)} dt} = \omega \sum_{l=-\infty}^{\infty} S_l e^{jl\omega_0 t}. \tag{42}$$

Finally, it can be shown that [13]

$$e^{-x\cos\omega_0 t} = \sum_{m=-\infty}^{\infty} Q_m(-x)e^{jm\omega_0 t}, \tag{43}$$

where the Q_m are the modified Bessel functions of order m.

Equations (39) through (43) can be combined to give

$$\begin{aligned} i_e(t) = \frac{\gamma}{a\tau}\sum_k \frac{V_k}{j\omega_k}\int_{t-\tau}^{t}\Bigg[\omega \sum_{n=-\infty}^{\infty} R_n e^{jn\omega_0 t} \\ \cdot\left(e^{j\omega_k t} - \int \sum_{m=-\infty}^{\infty} Q_m(-x)e^{j(m\omega_0+\omega_k)t'-bt'}dt'\right. \\ \left.\cdot e^{bt}\omega \sum_{l=-\infty}^{\infty} S_l e^{jl\omega_0 t}\right)\Bigg]dt. \end{aligned} \tag{44}$$

The integral over t' in (44) can now be carried out giving the following expression for $i_e(t)$:

$$\begin{aligned} i_e(t) = \frac{\omega\gamma}{a\tau}\sum_k \frac{V_k}{j\omega_k}\int_{t-\tau}^{t}\sum_{n=-\infty}^{\infty} R_n e^{jn\omega_0 t} \\ \cdot\left(e^{j\omega_k t} - \sum_{m=-\infty}^{\infty}\frac{\omega_0 Q_m(-x)e^{j(m\omega_0+\omega_k)t}}{j(m\omega_0+\omega_k)-b}\right. \\ \left.\cdot\sum_{l=-\infty}^{\infty} S_l e^{jl\omega_0 t}\right)dt. \end{aligned} \tag{45}$$

The doubly infinite product over m and l can be rewritten as

$$P = \sum_{m=-\infty}^{\infty}\left(\sum_{l=-\infty}^{\infty}\frac{\omega_0 Q_l(-x)S_{m-l}}{j(l\omega_0+\omega_k)-b}\right)e^{j(m\omega_0+\omega_k)t}. \tag{46}$$

Then (45) becomes

$$\begin{aligned} i_e(t) = \frac{\omega\gamma}{a\tau}\sum_k \frac{V_k}{j\omega_k}\sum_{n=-\infty}^{\infty} R_n\Bigg(H(n,\omega_k)e^{j(n\omega_0+\omega_k)t} \\ - \sum_{m=-\infty}^{\infty}\sum_{l=-\infty}^{\infty}\frac{\omega_0 Q_l(-x)S_{m-l}}{j(l\omega_0+\omega_k)-b} \\ \cdot H(m+n,\omega_k)e^{j[(m+n)\omega_0+\omega_k]t}\Bigg), \end{aligned} \tag{47}$$

where

$$H(n,\omega_k) = \frac{1-e^{-j(n\omega_0+\omega_k)\tau}}{j(n\omega_0+\omega_k)}. \tag{48}$$

The small-signal current at a particular frequency ω_x is obtained from (47) by setting

$$n\omega_0 + \omega_k = \omega_x \tag{49}$$

in the first term and

$$(m+n)\omega_0 + \omega_k = \omega_x \tag{50}$$

in the second term. Equation (47) is in a form which allows the calculation of the small-signal current for any number of idler or image frequencies.

References

[1] M. I. Grace, "Down conversion and sideband translation using avalanche transit-time oscillators," *Proc. IEEE (Letters)*, vol. 54, pp. 1570–1571, November 1966.

[2] A. S. Clorfeine, "Self-pumped parametric amplification with an avalanching diode," *Proc. IEEE (Letters)*, vol. 54, pp. 1956–1957, December 1966.

[3] C. C. Shen *et al.*, "Avalanche diode oscillators: power, efficiency, impedance, and mixing studies," *Proc. 1st Biennial Conf. on Engineering Applications of Electronic Phenomena* (Cornell University, Ithaca, N. Y., August 1967), pp. 253–265.

[4] W. J. Evans and G. I. Haddad, "A large signal analysis of IMPATT diodes," *IEEE Trans. Electron Devices (Correspondence)*, vol. ED-15, October 1968.

[5] W. T. Read, "A proposed high frequency negative resistance diode," *Bell Sys. Tech. J.*, vol. 44, pp. 401–446, March 1958.

[6] V. J. Higgins *et al.*, "Characteristics of varactors biased into avalanche," *IEEE Trans. Electron Devices*, vol. ED-13, pp. 210–212, January 1966.

[7] S. Yamashita *et al.*, "Frequency of microwave oscillation in GaAs p-n junctions," *Japan J. Appl. Phys.*, vol. 5, p. 843, September 1966.

[8] H. C. Torrey and C. A. Whitmer, *Crystal Rectifiers*. New York: McGraw-Hill, 1948.

[9] M. Sucher and J. Fox, *Handbook of Microwave Measurements*, vol. 3. New York:Wiley, 1963, p. 874.

[10] W. J. Evans and G. I. Haddad, "Sideband generation in avalanche transit-time diodes," *Proc. 1st Biennial Conf. on Engineering Applications of Electronic Phenomena* (Cornell University, Ithaca, N. Y., August 1967), pp. 281–300.

[11] E. F. Scherer, "Large-signal amplification with avalanche devices," *Proc. IEEE (Letters)*, vol. 55, pp. 464–465, March 1967.

[12] R. N. Haitz and F. W. Voltmer, "Noise studies in uniform avalanche diodes," *Appl. Phys. Lett.*, vol. 9, pp. 381–383, November 15, 1966.

[13] S. A. Schelkunoff, *Electromagnetic Waves*. Princeton, N. J.: Van Nostrand, 1943, p. 43.

CHARACTERISTICS OF THE AVALANCHE DIODE IN A SELF-MIXING DOPPLER RADAR SYSTEM★

C. CHAO AND G. I. HADDAD

ELECTRON PHYSICS LABORATORY, DEPT. OF ELECTRICAL AND COMPUTER ENGINEERING, THE UNIVERSITY OF MICHIGAN, ANN ARBOR, MICHIGAN, 48104

Experimental Setup

The block diagram of the experimental setup of the doppler-type radar system using an avalanche diode simultaneously as the transmitting oscillator and mixer is shown in Fig. 1. The avalanche diode is placed in a coaxial cavity and biased through a bias tee by a battery. The RF signal generated in the cavity by the diode is coupled to the antenna which radiates the signal toward the moving target. The reflected signal whose frequency is shifted in proportion to the velocity of the moving target ($f_d = 2v_of_o/c$) falls on the oscillating diode and mixes with the oscillator signal and thus a difference frequency (f_d) results. The frequency shift (f_d), or the doppler frequency, is coupled out through the bias circuit through a low-pass filter and can be displayed on an oscilloscope or measured by other means.

In order to perform detailed measurements on the diode properties, several components are inserted in the main signal path. These include:

1. A tuner to facilitate changing the oscillation frequency and power level of the diode.
2. Two directional couplers; one to check the frequency and power level of the oscillator and the other to measure the reflected power level.
3. A precision variable attenuator to simulate the amplitude of the reflected signal from a target at various distances from the antenna. For a prototype doppler radar system using the avalanche diode operated at a fixed frequency and power level the only components which are required are a battery, a cavity with an appropriate biasing network and an oscilloscope or another instrument for measuring the doppler frequency and amplitude of the reflected signal.

In order to compare the performance of the self-mixing avalanche-diode oscillator with an external mixer using the avalanche diode as a local oscillator, the experimental setup in Fig. 1 is modified and the setup in Fig. 2 is used. In Fig. 2 the main addition is a hybrid tee junction with one of its opposite arms connected to a detector mount which houses the crystal and the other to a moving load, and the E- and H-arms of the hybrid tee are coupled to the reflected power and local oscillator power, respectively. The external mixer is biased at the optimum operating point by applying the proper amount of RF power to it. The doppler signal is observed for both the self-mixing avalanche-diode oscillator and the external mixer (with the avalanche diode as a local oscillator) on an oscilloscope (Tektronix 502A).

Three types of moving targets were used for simulation purposes. The first is a rotating wheel which has a radius of 10 inches and 12 teeth located at 30-degree intervals. A one-square-inch metal plate was attached to each tooth to increase the reflected power. The antenna is aimed at these metal plates and can be located at various distances from the wheel. The rotating wheel has a variable speed and the metal plates moving past the center line of the antenna are capable of simulating a linear speed of 1.2 mph to 35 mph. The second type of moving target is a 4-inch by 4-inch metal plate geared to move slowly along the center line of the antenna at a constant speed which is variable between 2.54 cm/s to 12.3 cm/s. The third type of moving target is a man 6 feet tall and 135 lb walking across the center line of the antenna at various distances from the antenna.

Experimental Results

Experimental measurements have been performed using the setups shown in Figs. 1 and 2 for an avalanche diode operating in the X-band frequency range.

1. The Effect of the Operating Power Level of the Diode on the Mixing Property. The signal amplitude at the doppler frequency has been measured as a function of the power reflected from the target for different target velocities and at different oscillator power levels. These are shown in Figs. 3 through 5. In these figures, the doppler frequency signal amplitude is plotted as a function of the ratio of the reflected power to the oscillator power in dB. In this set of curves, the rotating wheel was used as the moving target and is placed at a distance of 18 inches from the antenna. Since the power level, the bias current and voltage, and the frequency of the diode are interdependent this set of curves was obtained at different power levels with all other parameters kept approximately the same. These curves indicate that the diode will mix well at an operating power level of 5 mW as well as at 60 mW. The minimum detectable power for the diode operated under these conditions is on the order of 60-70 dB below the output power of the diode. For the range of power levels measured here, the diode behaves essentially as a square-law detector. It is worth noting that the doppler frequency signal voltage was measured on a Tektronix 502A oscilloscope and a 20 kHz low-pass filter. Improved performance may be achieved by using narrow-band low-noise amplifiers

* This work was supported by the Highway Safety Research Institute, The University of Michigan.

and detectors in the bias circuit.

2. The Effect of the Operating Point of the Diode on the Mixing Property. The doppler frequency signal amplitude as a function of the reflected power from the target for different target velocities and for the diode operated at nearly the same power level but at a different frequency, bias current and voltage, with the fan located at a distance of 12 inches from the antenna are shown in Fig. 6. The observed doppler signal amplitude for the diode operated at 9 mW is larger by a factor of four on the average than that of the diode operated at 10 mW. The minimum detectable power is 70 dB for the former and 60 dB for the latter. This implies that the operating point of the diode has a large effect on the mixing properties. Under certain operating conditions, a negative conductance may exist at the doppler frequency which may lead to conversion gain. This depends on the RF and IF impedances presented to the diode and has been demonstrated and examined in detail by Evans and Haddad.[1]

The doppler signal amplitude as a function of the reflected power for different velocities of the rotating wheel placed at distances of 12 inches and 18 inches from the antenna are plotted in Fig. 7 for the avalanche-diode oscillator at a power level of 25 mW and a frequency of 9.18 GHz. The experimental data all fall approximately on a single line. This is what is expected since they are plotted as a function of normalized reflected power. For a given oscillator condition, the observed signal will be the same for a given value of normalized reflected power, regardless of the position of the moving target.

3. Comparison Between the Performance of a Self-Mixing Avalanche-Diode Oscillator and That of an External Mixer Using the Avalanche Diode as the Local Oscillator. The doppler signal amplitude as a function of the reflected power for an external mixer (a 1N23C crystal) and the self-mixing avalanche-diode oscillator with the moving target located at a distance of 12 inches and 18 inches are plotted in Figs. 8 and 9, respectively. The external mixer is biased at the optimum operating point by applying the proper amount of RF power from the local oscillator. The doppler signal amplitudes from the external mixer are higher by a factor of 10 to 15 than that of the self-mixing oscillator. The minimum detectable power is approximately 20 dB lower in the external mixer system.

4. Experimental Results on a Moving Flat Plate. In order to find the minimum speed and the minimum detectable power that the diode is capable of, a second kind of mechanical moving device was used for simulating a moving target. A 4-inch by 4-inch metal plate geared to move slowly along the center line of the antenna was used in place of the rotating wheel. The metal plate moves away from the antenna at a distance of 18 inches to approximately 48 inches. The signal amplitude as a function of the reflected power at different velocities is plotted for two different levels of oscillator power in Fig. 10. The diode is able to detect the simulated target moving as slow as 2.54 cm/s (or 0.06 mph) at a return power level of 70 dB below the operating power (30 mW) of the diode. In other words, the diode can detect a return signal on the order of 70 dB below the oscillator power at an IF as small as 1.2 Hz. Again it is worth mentioning that the Tektronix 502A oscilloscope was employed for measuring the doppler signal level.

5. Experimental Results on a Walking Man as a Moving Target. The doppler signal amplitude as a function of distance of a walking man for various oscillator power levels is plotted in Fig. 11. The man is 6-feet tall and weighs 135 lb and is walking across the center line of the antenna. The doppler signal was observed for a moving man as far as 25 feet when the oscillator power is 63 mW.

6. The Observed Signal. Some typical observed signal waveforms for both the rotating wheel and the linearly moving plate employed as moving targets are presented in Figs. 12 and 13.

In Fig. 12 the observed signal waveform for the diode operating at an output power level of 10 mW, a frequency of 9.5 GHz, a bias current of 36 mA and a bias voltage of 81 V with the wheel rotating at 20 rpm (34 Hz) and at a distance of 12 inches from the antenna is presented. The time scale for all the photographs in Fig. 12 is 0.1 s/cm. The doppler frequency observed is 34 Hz which is the same value as that of the calculated one. The reflected power in reference to the operating power of the oscillator in dB is designated as ΔP and is shown with the value of the vertical scale and time scale for each photograph. The periodic fluctuation in magnitude of these waveforms is due to the 12 teeth of the rotating wheel which sweep past the center line of the antenna. The reflected power is largest when the incident power is normal to the teeth. The teeth can be identified from this picture.

In Fig. 13, the observed signal waveform for the diode operating at an output power level of 30 mW, a frequency of 9.57 GHz, a bias current of 48.5 mA and a bias voltage of 84 V with a 4-inch by 4-inch metal plate moving away linearly along the center line of the antenna at a constant speed of 8.5 cm/s (5.4 Hz) and beginning at a distance of 18 inches from the antenna is shown. The time scale for all the photographs is 1 s/cm. The observed doppler frequency is 4.8 Hz. It would be expected that the reflected power will be inversely proportional to the fourth power of the distance of the moving target from the antenna. Therefore, for the detector employed here the doppler signal amplitude will be inversely proportional to the square of the distance of the moving target from the antenna. This indeed holds true for all the data shown in Figs. 10 and 13. For example in Fig. 13a the observed doppler signal amplitude is approximately 25 mV at 18 inches and 8 mV at 32 inches. These give a voltage ratio of 3.13 and a square of the distance ratio of 3.16.

REFERENCE

1. Evans, W. J. and Haddad, G. I., "Frequency Conversion in IMPATT Diodes," *IEEE Trans. on Electron Devices*, vol. ED-16, No. 1, pp. 78-87, January 1969.

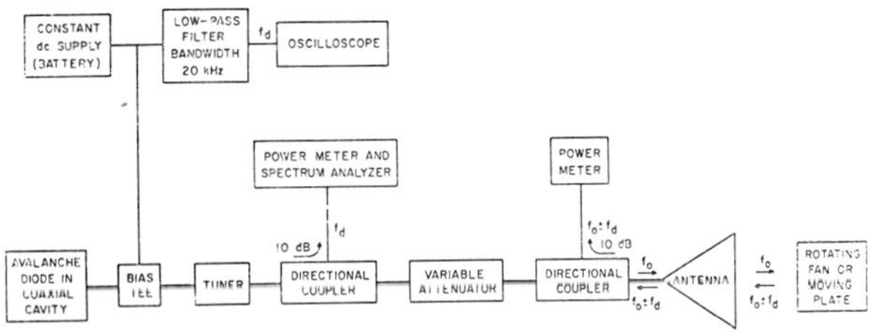

Fig. 1 Experimental Setup of the Doppler-Type Radar System Using an Avalanche Diode as the Oscillator and Mixer.

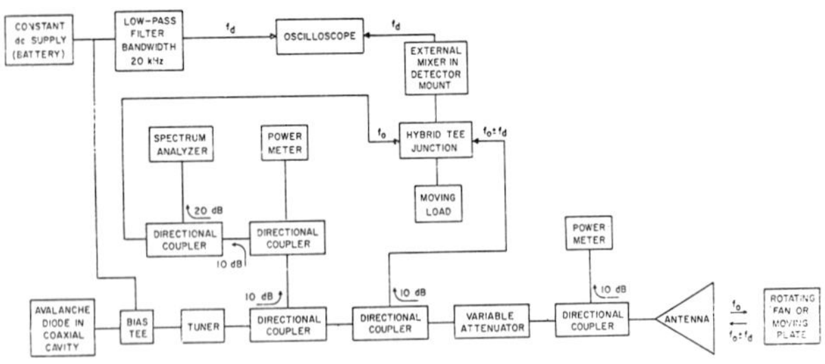

Fig. 2 Experimental Setup for Comparing the Performance of the Self-Mixing Avalanche Oscillator with the External Mixer Using the Avalanche Diode as the Local Oscillator.

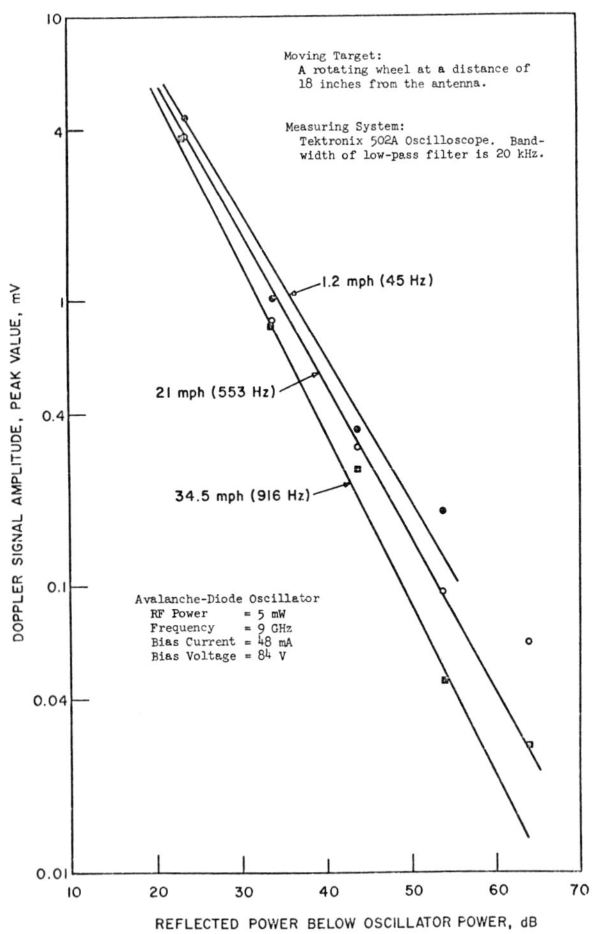

Fig. 3 Doppler Signal Amplitude vs. Reflected Power at Different Target Velocities.

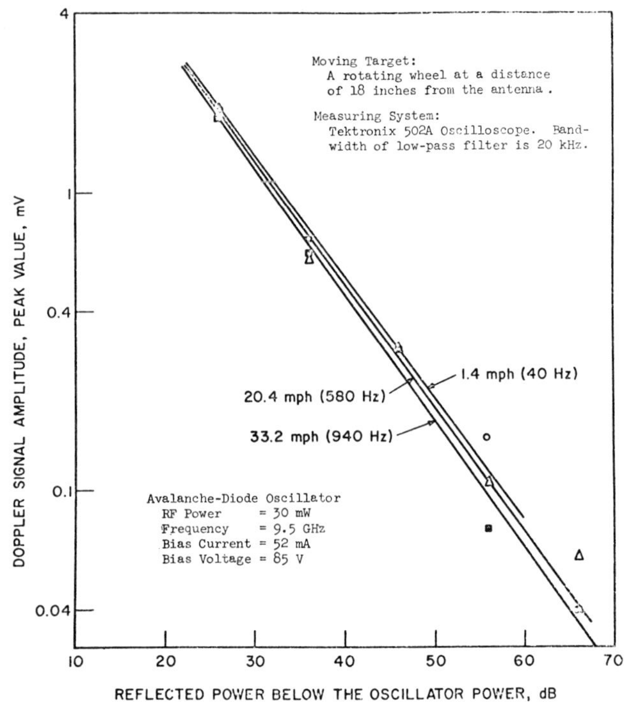

Fig. 4 Doppler Signal Amplitude vs. Reflected Power at Different Target Velocities.

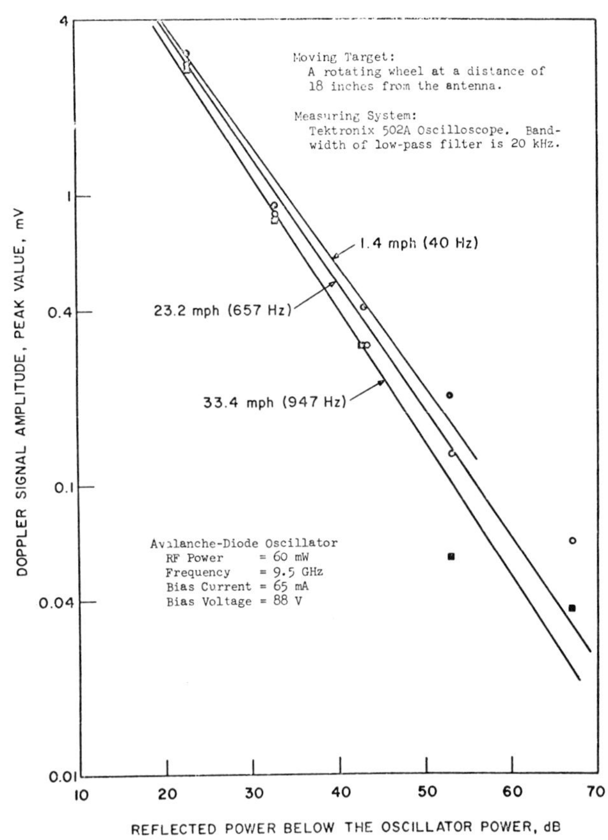

Fig. 5 Doppler Signal Amplitude vs. Reflected Power at Different Target Velocities.

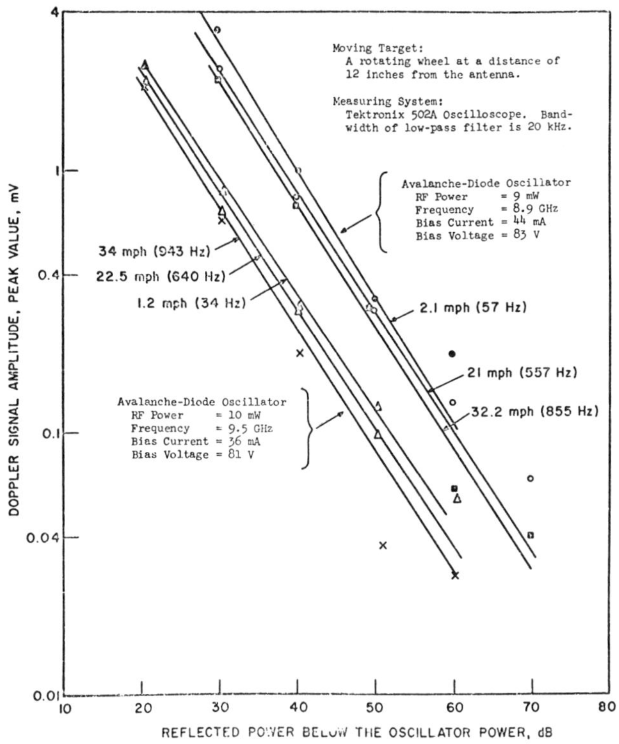

Fig. 6 Doppler Signal Amplitude vs. Reflected Power at Different Target Velocities.

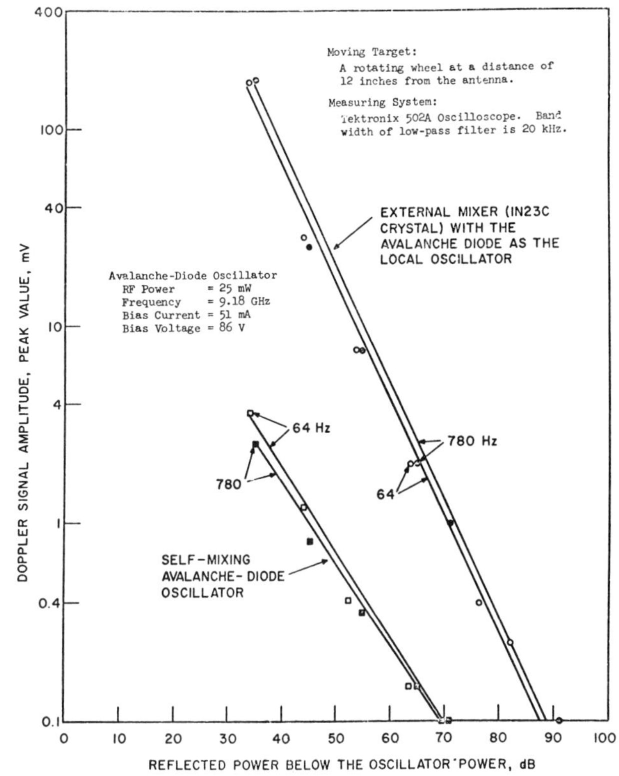

Fig. 8 Doppler Signal Amplitude vs. Reflected Power for the External Mixer and Self-Mixing Avalanche Diode at Different Target Velocities.

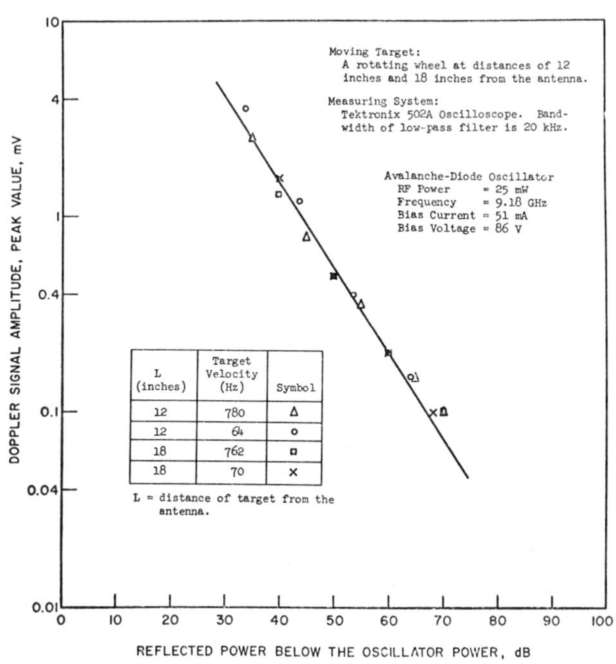

Fig. 7 Doppler Signal Amplitude vs. Reflected Power at Different Target Velocities.

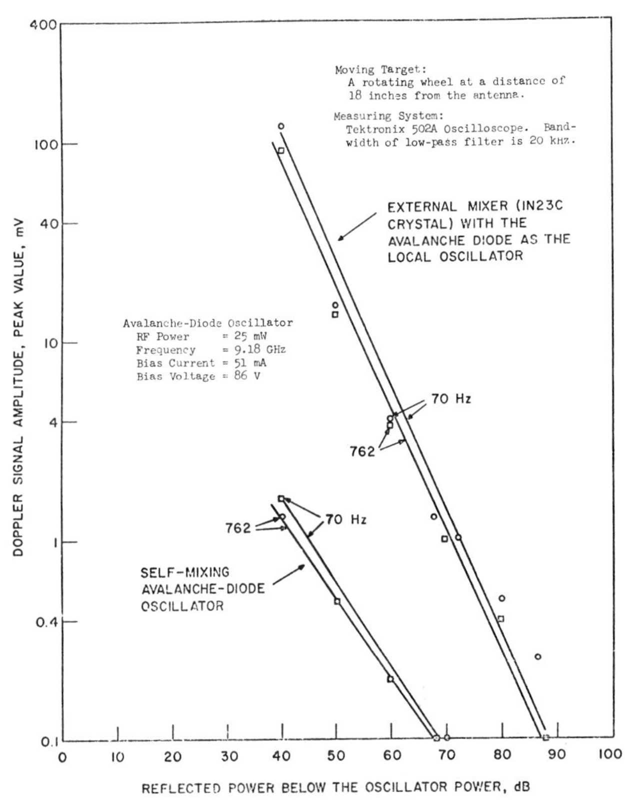

Fig. 9 Doppler Signal Amplitude vs. Reflected Power for the External Mixer and Self-Mixing Avalanche Diode at Different Target Velocities.

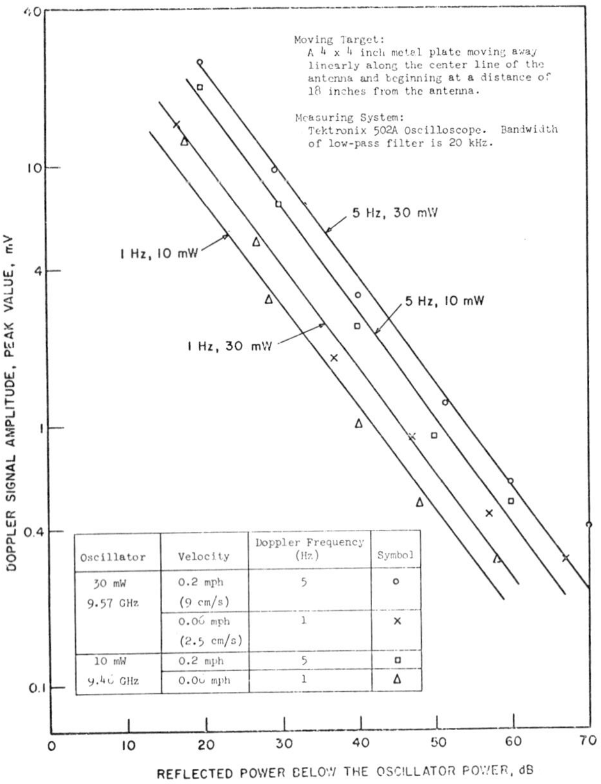

Oscillator	Velocity	Doppler Frequency (Hz)	Symbol
30 mW 9.57 GHz	0.2 mph (9 cm/s)	5	o
	0.06 mph (2.5 cm/s)	1	×
10 mW 9.46 GHz	0.2 mph	5	□
	0.06 mph	1	Δ

Fig. 10 Doppler Signal Amplitude vs. Return Power at Different Target Velocities.

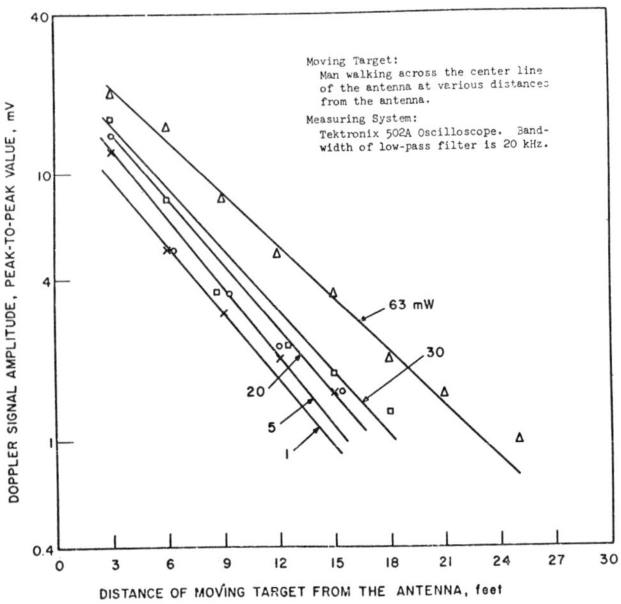

Fig. 11 Doppler Signal Amplitude vs. Distance of Moving Target from the Antenna for Various Oscillator Powers.

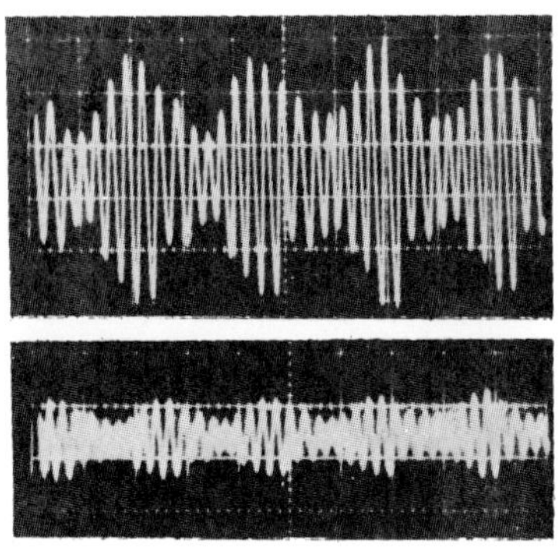

(c) Δp = 18.5 dB, 1 mV/cm, 0.1 s/cm

(b) Δp = 28.5 dB, 1 mV/cm, 0.1 s/cm

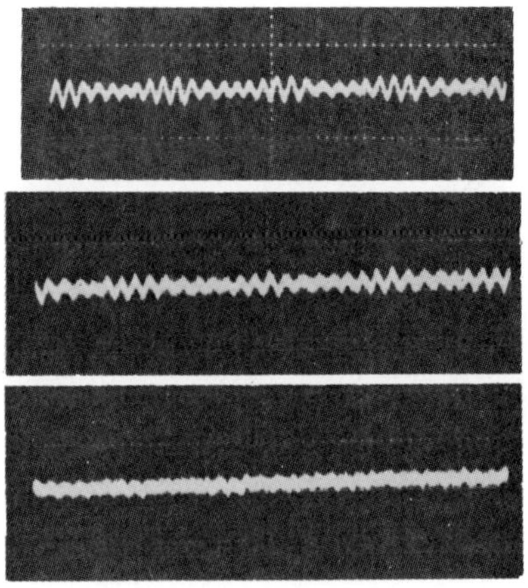

(c) Δp = 38.5 dB, 1 mV/cm, 0.1 s/cm

(d) Δp = 48.5 dB, 0.5 mV/cm, 0.1 s/cm

(e) Δp = 56.5 dB, 0.5 mV/cm, 0.1 s/cm

Fig. 12a-e Observed signal waveforms as a function of the reflected power for the wheel rotating at 20 rpm (34 Hz) and at a distance of 12 inches from the antenna. (Measuring system: Tektronix 502A Oscilloscope. Bandwidth of the low-pass filter is 20 kHz.)

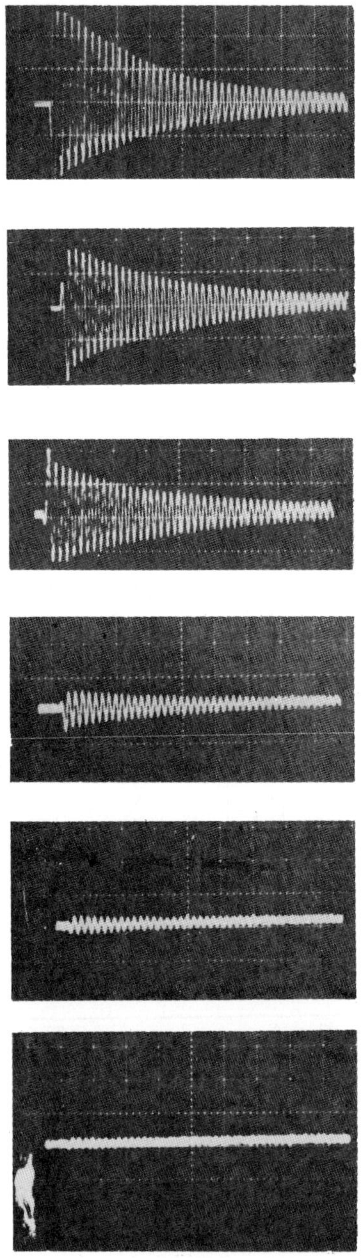

(a) Δp = 20 dB, 10 mV/cm, 1 s/cm

(b) Δp = 30 dB, 5 mV/cm, 1 s/cm

(c) Δp = 40 dB, 2 mV/cm, 1 s/cm

(d) Δp = 50 dB, 2 mV/cm, 1 s/cm

(e) Δp = 60 dB, 2 mV/cm, 1 s/cm

(f) Δp = 70 dB, 2 mV/cm, 1 s/cm

Fig. 13a-f The observed signal waveform as a function of the reflected power for a linearly moving metal plate moving away at a speed of 9 cm/s and starting at a distance of 18 inches from the antenna. (Measuring system: Tektronix 502A oscilloscope. Bandwidth of the low-pass filter is 20 kHz.)

OTHER ARTICLES OF INTEREST VII

100 GHz Si IMPATT DIODES FOR cw OPERATION

T. Misawa and L. P. Marinaccio
Bell Telephone Laboratories, Inc., Murray Hill, NJ

Si Impatt diodes with nearly-abrupt junctions have given efficiencies of 3.2% in the frequency range from 100 to 110 GHz in cw operation. An output power of 74 mW was obtained at 111 GHz with a single-die diode. A composite diode produced 110 mW. At 151 GHz an output power of 3.5 mW was observed.

Diodes for the 100 GHz region have a space-charge-layer width ranging from $\frac{1}{5}$ to $\frac{1}{3}$ μ and a breakdown voltage from 9 to 12 volts. They are made by a single boron or phosphorous diffusion into an n-type or p-type epitaxial film with a uniform doping of about 2×10^{17} cm^{-3} and a thickness ranging from 0.5 to 1 μ. The diodes have typical diameters of 20 μ with a corresponding space-charge-layer capacitance on the order of 0.1 pF. High conversion efficiency is obtained at a bias current above 50 kA/cm^2 which is near the upper limit thermally possible.

The diode is mounted on the broad wall of standard WR-8 waveguide. A metal cap, brought down from the opposite wall, forms a conical antenna around the diode to facilitate matching between the low diode impedance and the high wave-guide impedance.

Results of dc and small-signal analyses of representative structures are presented for temperatures ranging from room temperature to the highest operating temperature. It is shown that the field profile and the optimum oscillation frequency change appreciably with junction temperature. The analyses indicate that a current density higher than 50 kA/cm^2 is required for optimum operation of the diode.

I. INTRODUCTION

Previously we reported on Si Impatt diodes with nearly-abrupt junctions for cw operation in X band [1] and the lower mm wave region [2, 3]. This same type of diode was scaled down further and has proved to be usable cw Impatt oscillators in the 100 GHz range.

Diodes with a space charge region width ranging from $\frac{1}{5}$ to $\frac{1}{3}$ μ and a breakdown voltage from 9 to 12 volts oscillated well between 100 GHz and 120 GHz. The highest efficiency is always observed just prior to burnout, where the diode voltage and hence the space charge region width are appreciably larger than the room temperature values mentioned above. The bias current density is of the order of 10^5 A/cm^2, just prior to burnout.

The fabrication techniques were essentially those used at lower frequencies. However in almost every respect the techniques had to be pushed beyond the limit of common usage. The required thickness of $\frac{1}{2}$ μ and accurate control of donor concentration in the range of 10^{17} cm^{-3} for the epitaxial layer were barely possible with the $SiCl_4$ method. Epitaxial growth using the pyrolysis of silane

Presented at the Symposium on Submillimeter Waves
Polytechnic Institute of Brooklyn, March 31, April 1 and 2, 1970.

at lower temperatures, which became available in the later part of the study, indicated the possibility of easing this problem considerably [4]. For junction formation the conventional open tube diffusion utilizing boron or phosphorus halides turned out to be satisfactory far beyond original expectations. In the case of boron, the junction depths in the range of $\frac{1}{10}\ \mu$ or less were reliably reproduced.

Our main effort was directed towards making a slice with a suitable impurity profile and testing properties of different profiles. As for the geometrical structure of the diode, we chose the simplest fabrication method. A diode was simply etched out from the slice and subsequently etched down to a desired size after being bonded on a stud. It is believed that optimization of the after-slice processing, especially by taking advantage of the thin slice technique developed by Edwards et al. [5], will improve the oscillator performance further. Results with a moderately thin slice already indicate this possibility.

The "cap tuning," developed in the 50 GHz range [3], was used for the diode evaluation. It was found that diodes could be incorporated in this circuit and the low impedance handled satisfactorily without calling for an unusual degree of mechanical precision.

In the following, computed dc and small-signal characteristics are presented in Section II. Section III describes thermal considerations. Diode fabrication (Section IV), measurement (Section V), and characteristics (Section IV) are subsequently described.

II. THEORETICAL dc AND SMALL-SIGNAL CHARACTERISTICS

Typical structures were analyzed with the small-signal computer program previously developed [6]. Revised expressions for carrier velocities and ionization rates were used as explained in the Appendix. Because the space charge layer is very narrow and the breakdown voltage is very low, the validity of the fundamental equations used for the analysis are questionable. Nonetheless the computed results are not far from those observed experimentally and they served well as a semiqualitative guideline in the diode design.

We shall discuss the results obtained for two structures which are typical of the structures actually made. One has a very abrupt junction and the other has a field profile with a rounded top more closely resembling a pin structure. The abrupt junction diode is made by a $0.3\ \mu$ deep acceptor diffusion into an epitaxial layer with a donor concentration of $1.6 \times 10^{17}\ \text{cm}^{-3}$. These parameters apply to the slice L01733 to be described later. In the pin type junction, AVX-147, the donor concentration increases exponentially towards the substrate and the acceptor diffusion has a much lower surface concentration than in the abrupt junction diode.

The field profile of the abrupt junction diode is triangular at low bias current and the space charge region terminates before reaching the substrate, as depicted by profile A in Figure 1. As the bias current increases the space charge region widens because of the mobile carrier space charge. At a current density of 100 kA/cm^2 the space charge region penetrates well into the substrate (profile C). The corresponding current voltage relation is shown in Figure 2. It

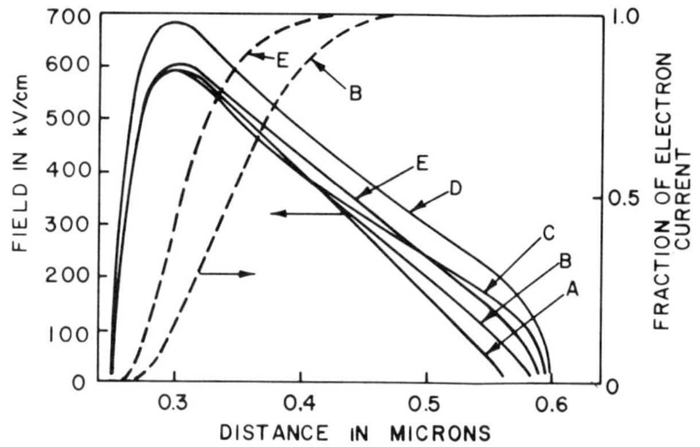

Fig. 1. Computed field profiles and current distributions in an abrupt junction diode, L01733, at several bias currents and temperatures. Results for both p on n and n on p structures are shown.

A	5 kA/cm^2	300 °K	p on n
B	50 kA/cm^2	300 °K	p on n
C	100 kA/cm^2	300 °K	p on n
D	50 kA/cm^2	600 °K	p on n
E	50 kA/cm^2	300 °K	n on p

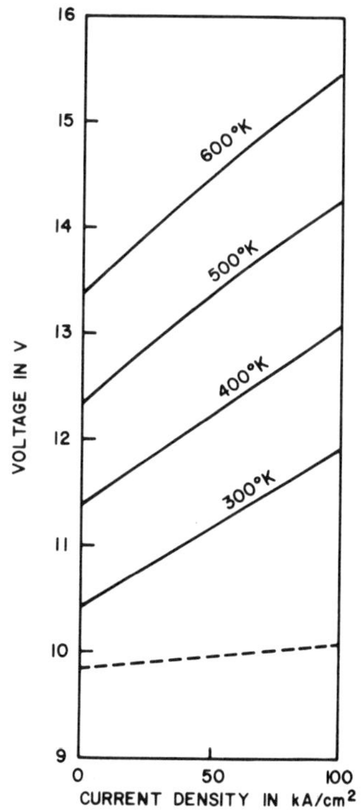

Fig. 2. Computed current-voltage relation for the p on n abrupt junction diode at several temperatures. The dotted line is for a pin type diode of Figure 4.

indicates a more or less constant space charge resistance over the current density range of interest.

As the junction temperature rises, the ionization rates decrease [7] and the scattering-limited velocities also decrease [8]. Both effects tend to increase the diode voltage and computed results are shown in Figure 2. The reduced ionization rate has an important effect on the field profile. Since a higher field is required to maintain the same bias current, the space charge region widens and penetrates into the substrate. Therefore the parasitic series resistance is appreciably reduced from what is expected from the profile at breakdown. This change occurs naturally with actual cw operation of a diode, since the junction temperature rises due to heat generation.

Since the ionization rate of electrons is four to five times higher than that of holes the avalanche region extends toward the substrate and occupies about one-third of the space charge region, as the electron current distribution shown in Fig. 1 indicates. When the polarity of the structure is reversed, i.e., in the n on p structure rather than the above p on n structure, the avalanche region shrinks appreciably (profile E). We know that this improves the quality of the diode negative resistance at low currents, [9] as will be discussed below.

The computed small-signal impedance is plotted in Fig. 3 together with the quality factor, Q. We know that it is desirable to achieve the best Q, [9] which requires a bias current of 50 kA/cm^2. Since the diode typically has an area of 4×10^{-6} cm^2 the diode resistance is about $-50\ \Omega$ at the resonance which occurs at 150 GHz. The best Q is obtained at a higher frequency where the resistance is considerably lower. The real diode always has some parasitic series resistance. This tends to favor the operation at a frequency closer to resonance where the diode has a larger resistance. Therefore the best overall quality factor, including the effect of the parasitic resistance, occurs between the frequency of peak impedance and the frequency of the best Q of the active diode region alone. In the 100 GHz range the diode impedance is generally low and hence this effect should be appreciable.

The characteristics at 600 °K are also shown in Fig. 3 (dot-dash line) for a bias current of 50 kA/cm^2. Because of the widening of the space charge layer and decrease of the scattering-limited velocity discussed before, the frequencies for the peak impedance and the best Q are lower than at room temperature and closer to experimentally observed values for this structure.

The behavior of the Q is also plotted in Fig. 3 for the diode with the reverse polarity or the n on p structure. Here the holes play the major role. As expected from the narrower avalanche region explained above the quality factor at lower bias current improves. A bias current of 20 kA/cm^2 is almost as good as 50 kA/cm^2 of the p on n structure.

The room temperature characteristics of the "pin" type diode are shown in Figure 4. In accordance with the previous study the Q keeps improving up to a higher bias current than in the abrupt junction unit [9]. Another feature is that the best Q is obtained very close to the impedance peak. As discussed before this can be an advantage at our high frequency.

In summary the analysis indicated that the impedance level is not prohibitively low at the 100 GHz range. The next problem is to see if the desired current density can be achieved in cw operation.

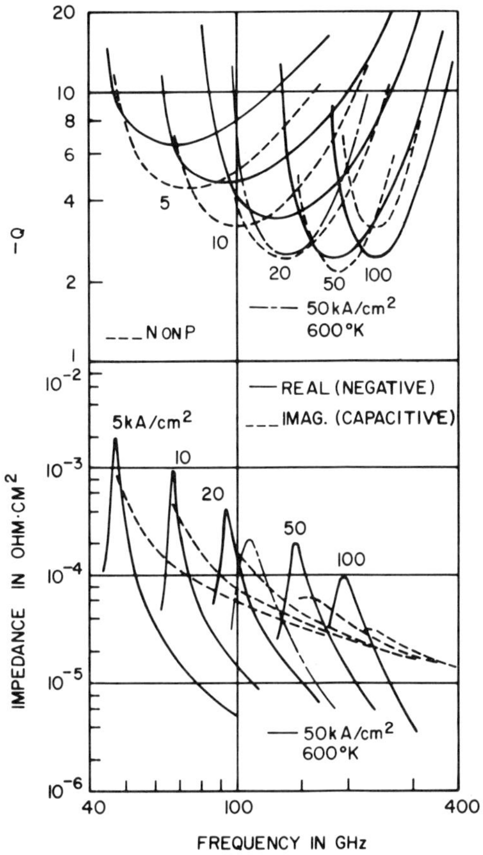

Fig. 3. The small-signal impedance and quality factor, Q, of the abrupt junction diode at 300 °K. Results at 600 °K are shown by dot-dash lines and dotted lines are for the n on p structure.

III. THERMAL CONSIDERATIONS

A great deal of heat is generated in the diode and it has to be removed to keep the temperature down. Heat generation is proportional to the electric field; therefore most of the heat is generated in the space charge region. The generated heat is removed through the semiconductor and the metal contact, and disperses into the heat sink.

First let us consider the thermal spreading resistance in the heat sink. Figure 5 shows the permissible input power for a given temperature rise as a function of the diode diameter. The heat sink is made of copper whose thermal conductivity is 3.85 W cm/°C. Uniform heat flow over the contact area between the diode and the heat sink is assumed and the temperature rise refers to the maximum temperature [10]. For example, for the 50 kA/cm² operation of the abrupt junction diode discussed in the preceding section an input power density of

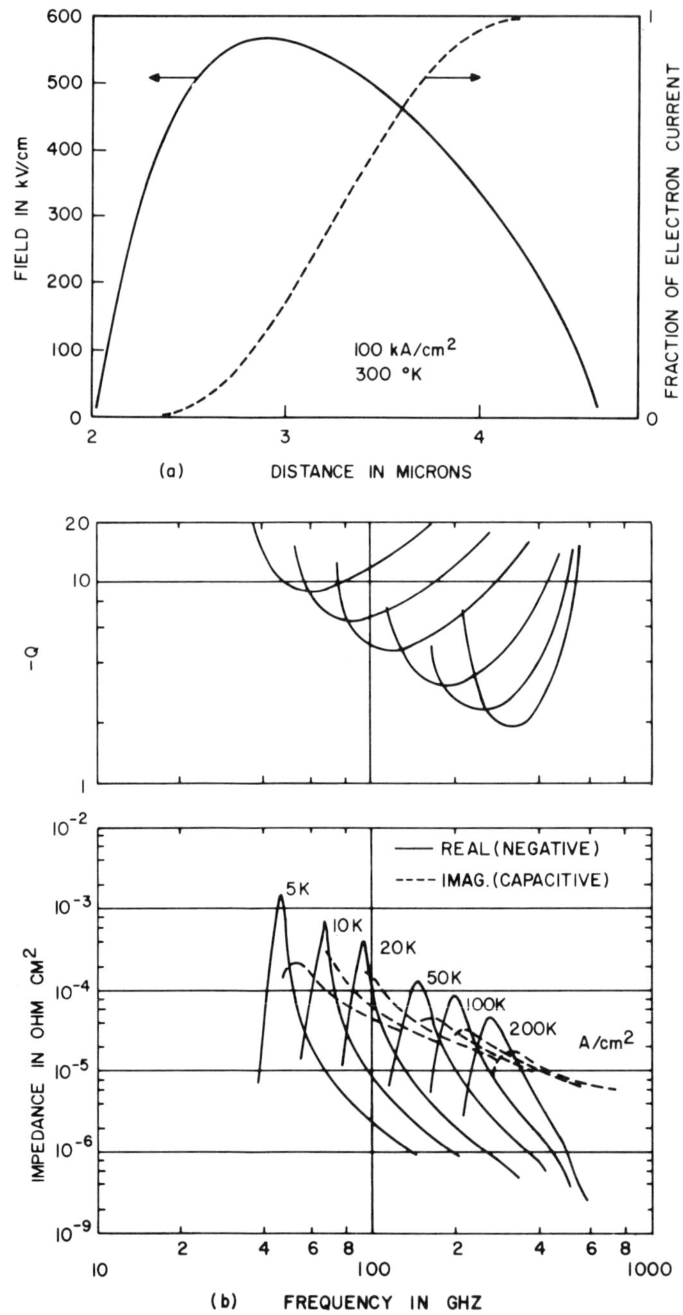

Fig. 4. (a) Computed field and current profiles and (b) small-signal characteristics of a pin type diode.

514

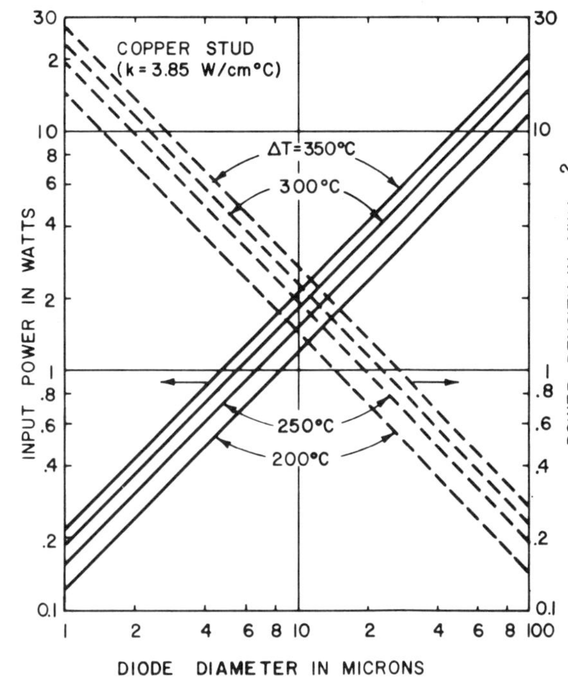

Fig. 5. Permissible input power and input power density for several given temperature rises as functions of junction diameter. For temperature rise in the infinite copper heat sink only.

0.6 ~ 0.7 MW/cm^2 is required. This is achieved in a 25 μ diameter diode with a temperature rise of 200 to 250 °C.

Previously it was estimated that diodes would go into a thermal runaway condition and burn out when junction temperatures approached 300 °C above room temperature [1]. More recent studies by Ciccolella have indicated a still higher burn-out temperature [11]. In a typical diode with a junction diameter of 24 μ burn-out occured at an input power of 3 W. The thermal resistance due to the silicon material itself is estimated as 20 °C/W which causes another 60 °C temperature rise beyond that due to the copper heat sink. It seems that our diode is not very far from the theoretical limit.

It is known that the bias current does not flow uniformly over the junction area because it is cooler at the edge than in the center [12, 13]. The degree of nonuniformity can be estimated from the previous numerical study for X band diodes by using the theoretical dc characteristics shown in Figure 2. A critical parameter in the problem is $B = GV\beta a/k^{14}$, where G is the isothermal differential conductance, or the slope of the I-V curve, V the bias voltage, β the temperature coefficient of breakdown voltage, a the junction radius, and k the thermal conductivity of heat sink material. For our abrupt junction diode B is computed as 3.4 from data given in Figure 2. According tc the previous study [13] the current at the edge of the diode is 1.6 or 2.1 times of that at the center for B = 2.6 or

5.1 respectively. Since our B is in a similar range we expect similar nonuniformity in current density.

IV. FABRICATION

From the preceding discussions one can see the problems encountered in fabrication of the diode. Since a space charge layer width of $\frac{1}{3}\ \mu$ is required we need an epitaxial layer thickness not much more than $\frac{1}{2}\ \mu$ and a junction depth not more than $\frac{1}{3}\ \mu$. The junction diameter should be below 25 μ.

Epitaxial slices were either grown in house or bought from an outside supplier.* It turned out that the major problem was the control of the impurity density in the epitaxial layer. The desired thickness was achieved through the dexterity of the operator in the case of the high-temperature $SiCl_4$ furnace. Later development of the low temperature silane furnace [4] improved the situation.

The pn junction was formed by a standard diffusion process. Diffusion depth was used ranging mostly from 0.3 μ to less than 0.1 μ in the case of boron and 0.3 to 0.6 μ in the case of phosphorus. The depth is tailored in conjunction with the epitaxial thickness to achieve the proper field profile in breakdown, as discussed previously [1].

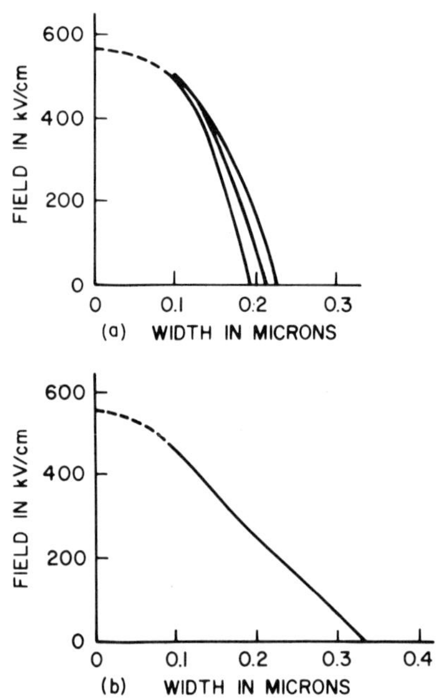

Fig. 6. Measured "field profiles" of (a) L01320 pin type and (b) L01733 abrupt junction type.

*Material for the slice "Indiana" to be discussed in Section VI was purchased from Komatsu Electronics Metals Co., Ltd., in Japan.

Two typical field profiles, obtained from measured capacitance versus voltage relations [15], are shown in Figure 6. Slice L01320 is of the "pin" type and L01733 is of the abrupt junction type. The plot is not exactly the profile itself but shows the width of the portion in which the field is larger than a value given on the *Y* axis. This type of plot gives enough information for our purposes. The structure L01320 in Fig. 6 (a) was actually made on the "out diffusion" tail from the substrate. It belongs to the pin type discussed in Section II.

The diffused slices were metalized by evaporating Ti-Au and pelletized by etching. The wafer was bonded junction side down on a gold-plated copper stud and was etched to the desired diameter.* The three versions of the diode thus made are shown in Figure 7. The structure of Fig. 7 (a) was primarily used. The

(a) (b) (c)

Fig. 7. Three structures of the Impatt diode.

height of the structure is the thickness of the original slice which was usually about 60 μ for convenience of handling. The "dimple" diode of Fig. 7 (b) was made to reduce the parasitic series resistance and to improve the parasitic reactance. A "dimple" was etched from the back side of the slice and the cup-shaped metalization was formed. When Si is etched away to obtain a desired diameter the structure shown results. The last structure of Fig. 7 (c) utilizes the thin slice which was developed by Edwards et al. [5] The gold button on top of the wafer is for simulating the structure of Figure 7 (a). It was found that this facilitates the coupling between the diode and the outside world.** This structure was used in the slice "Patty" which will be discussed later.

V. MICROWAVE MEASUREMENT

The output power was measured in the circuit shown in Figure 8. The diode is mounted on the broad wall of the WR(8) waveguide. A conical antenna is formed with a metal cap, which also acts as a bias lead. The cap-diode assembly is then turned to match the load with the movable short and the E-H tuner. 517

The tuning cap facilitates matching of the low-impedance diode to the high

*The process is explained in more detail in [14].

**This way of assembling the diode made from the thinned slice was devised by R. Edwards and D. F. Ciccolella, private communication.

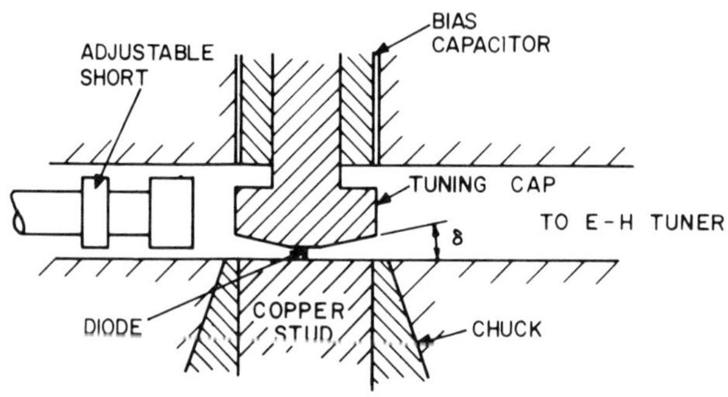

Fig. 8. Oscillator circuit with a tuning cap.

impedance of the full-height waveguide. The characteristic impedance of the conical section is

$$Z_0\ [\text{ohm}] = 1.05\,\delta\ [\text{degrees}]\ , \tag{1}$$

when the opening angle, δ, is small [16]. The diode with a space charge layer capacitance on the order of 0.1 pF was operated with the opening angle around 10°. The reactance of 0.1 pF is 16 Ω at 100 GHz and the characteristic impedance of 10° line is 10.5 Ω. The radius of the cap is approximately a quarter wavelength. From measurements of composite diodes [17] it has been learned that when the diode capacitance is greater then 0.3 pF the loss associated with the impedance transformation becomes appreciably worse.

The actual tuning by the movable short and the E–H tuner was critical and touchy, as had been the case with the lower frequency circuits. A hysteresis in tuning was usually observed at high output power. Namely the pistons in the short and tuner had to be moved in certain directions (inwards or outwards) to obtain the high output power. Once the optimum positions were passed the output power dropped suddenly and it was recovered, not by moving the pistons back, but only by starting the tuning process all over again.

The output power was normally measured with a TRG calorimeter. However, some of the readings to be reported in the next section were obtained with a tunable thermistor mount and then calibrated against the calorimeter.

VI. CHARACTERISTICS

A wide variety of Impatt structures have been fabricated with regard to the configuration of avalanche and drift regions in the space charge region. Some representative structures are listed in Table I. They are classified into three groups: p on n pin-type, n on p pin-type, and p on n abrupt junction type. "p on n" refers to the structure with the n type epitaxial layer and the diffused p type layer, in which electrons play more important roles than holes. "n on p" is the complementary structure in which holes play a more important role. In addition to the differences in properties of electrons and holes, such as discussed in

TABLE I.

WAFER	L01320	GPH658	INDIANA	GE3535-2-60	L01733	OKLAHOMA	PATTY
Epitaxial layer							
Thickness (μ)*	0.9	1.0	1.0	0.9*	0.6*	0.4 ~ 0.5*	0.56***
Conductivity type	N	P	P	N	N	N	N
Impurity density (cm^{-3})**	$\sim 3 \times 10^{17}$	$\sim 3 \times 10^{17}$	$\sim 3 \times 10^{17}$	3.5×10^{17}	1.6×10^{17}	1.5×10^{17}	2×10^{17}
$SiCl_4$ or Silane	$SiCl_4$	$SiCl_4$	Silane	$SiCl_4$	$SiCl_4$	$SiCl_4$	Silane
Junction depth (μ)*	0.9	0.6	0.45	0.3*	~0.4	$\lesssim$0.1*	<0.1*
Breakdown voltage (V)	8.7 ~ 9.8	9.9 ~ 10.4	11.0 ~ 11.2	8.3	10.2	11.8 ~ 12.0	10.0
Space-charge layer width (μ)**	0.21	0.24	0.24	0.19	0.34	0.34	0.27
Type of field profile**	PIN TYPE	PIN TYPE	PIN TYPE	← ABRUPT JUNCTION TYPE →			

*Measured by angle lapping and staining.
**Obtained from capacitance voltage relation.
***Infrared interference technique.

TABLE II.

DIODE	L01320 No. 7	GPH658 No. 14	INDIANA No. 10	GE3535-2-60 No. 4	L01733 No. 4	OKLAHOMA No. 5	PATTY No. 13
Capacitance (pF)	0.11	0.09	0.09	0.085	0.12	0.08	0.10
Junction diameter (μ)	17	16	17	14	24	18	18
Operating Condition							
Current (mA)	175	145	160	160	215	140	175
Voltage (V)	12.2	14.4	15.0	11.8	14.1	16.1	13.2
Input power (W)	2.1	2.1	2.4	1.9	3.0	2.3	2.3
Current density (A/cm^2)	8.0×10^4	6.9×10^4	7.3×10^4	1.0×10^5	5.6×10^4	5.4×10^4	6.8×10^4
Output power (dBm)	15.7	17.9	16.8	5.5	18.5	17.4	18.7
Frequency (GHz)	106	108	109	151	106	105	111
Efficiency (%)	1.6	3.0	2.0	0.19	2.3	2.4	3.2
Remarks	POWER	EFFICIENCY	POWER	FREQUENCY	POWER	POWER	POWER AND EFFICIENCY

connection with Figs. 1 and 3, differences in metallurgy of fabrication were considered to have some consequence.

Observed results of oscillator performance of diodes are listed in Table II. In the case of the pin type junction higher efficiency was observed with the n on p unit than with the p on n unit. A junction capacitance of about 0.1 pF usually gave the best efficiency. The junction diameter for this capacitance ranged from 14 to 24 μ depending upon the space charge layer width. The small diameter is desirable because high current density is required as discussed in Section III. In addition nonuniformities in bias current due to heating [12, 13] and in current due to the skin effect are less in a smaller diode. The former effect was estimated in Section III and the latter was discussed by DeLoach [18]. The skin depth at 100 GHz in the 0.001 Ω cm substrate is 5 μ; therefore with diameters actually used the major portion of our junction area lies within a skin depth of the surface.

The output power increased with bias current up to the burn-out point. However, the rate of increase became small near burn-out in most cases and sometimes completely flattened. Actually this slow-down was a reliable indicator of the diode's approaching burn-out. Output power is plotted as a function of the bias current density (averaged over the entire junction area) in Fig. 9 for some representative diodes. The curves were obtained by reading the highest output

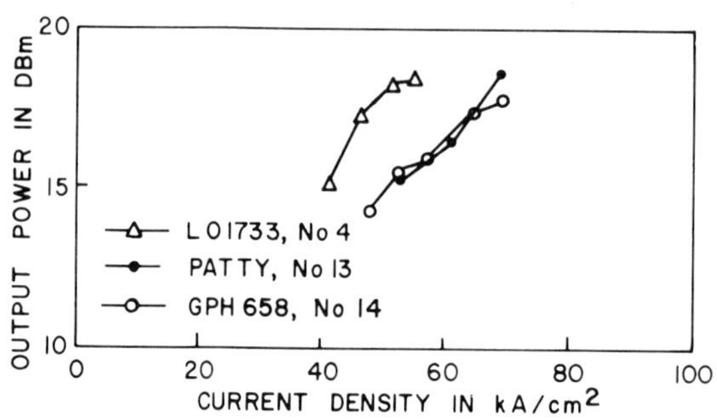

Fig. 9. Output power versus bias current.

power at a given bias current. Although no extensive adjustment was done at lower bias currents, the plot gives a rough idea of how the output power decreases as bias current is reduced.

The operating condition given in Table II corresponds to the highest input power achieved. Since the diode usually burns out due to a tuning transient, the input power listed may not exactly correspond to the thermal capability of the diode. The bias voltage observed during oscillation is lower than the quiescent value by $\frac{1}{2}$ to 1 volt. The largest output power does not correspond to the lowest voltage, which indicates that as more rf power is extracted the rf voltage amplitude decreases to some extent. The current densities observed are close to what we wanted to have from the theoretical small-signal analysis. The pin type L01320 and the narrow GE 3535-2-60 required larger current density. The wider, abrupt junction type L01733 and Oklahoma had the smallest current density, as expected from small-signal theory.

Output powers from ~ 16 to ~ 19 dBm were achieved in the frequency range of 105 ~ 110 GHz. The highest output power was always observed in this rather narrow frequency range despite the fact that the oscillation frequency could be adjusted over a much wider range. It is believed that this is due to narrow-banded nature of the circuit including the commercially purchased components rather than due to diode properties. The narrow GE 3535-2-60 gave an output power of 5.5 dBm at 151 GHz. Indications are that a substantially larger power can be achieved in this high frequency range when an appropriate circuit is used. Performance of this slice in the lower frequency range was fair.

The transit angle in the pin type diodes is about $\pi/2$, which is more or less what is expected from the theoretical small-signal analysis. The abrupt junction type diode had a slightly larger transit angle, again in line with the small-signal analysis. The theory indicates actually that a quite wide range of frequency of operation is achievable with a given diode. It seems difficult to anticipate the frequency of the highest observable output power because it is a result of the interaction of many factors, such as the properties of the space charge layer itself, the parasitic series resistance, and circuit losses.

An efficiency of 2 to 3% was achieved. Two slices which gave the highest efficiencies are worthy of special attention. Slice GPH658 is an n on p "pin" type diode and gave an efficiency of 3%. Considering the lower efficiency of the complementary L01320 which has a similar field profile, the polarity of n on p must have helped increase efficiency. The Patty diode is different from L01733 and Oklahoma in two respects although their field profiles are very similar to each other. First it was made on a silane epitaxial layer and secondly the diode has the structure shown in Fig. 7(c) while diodes from the other slices have the structure of Figure 7(a). Theoretically the silane epitaxial layer, which is grown at lower temperatures, has a sharper transition between the epitaxial layer and

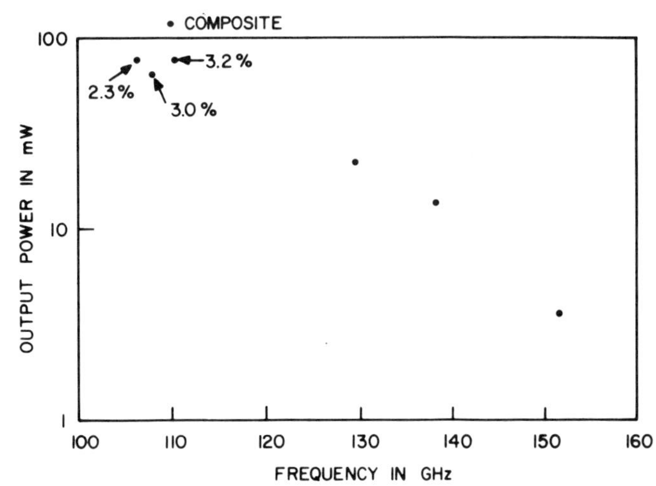

Fig. 10. Output power and efficiency of Impatt diodes.

521

the substrate. In the structure of Fig. 7(c) about half of the substrate silicon is replaced by metal. Both factors should contribute to lowering the series resistance. Especially with thinning down the substrate one can avoid harmful skin effect as discussed by DeLoach [18].

Finally we plot the best output powers obtained, in Figure 10. The highest output power was obtained with a composite Patty with two dies, the details of which will be reported separately by Marinaccio [17].

VII. CONCLUSION

The capabilities of Si Impatt diodes in the vicinity of 100 GHz were investigated. A cw efficiency up to 3.2% was observed and an output power of about 70 mW was obtained with a single chip diode. An output power of 110 mW was achieved with a composite diode with two chips.

The major problem was the control of the impurity density in the epitaxial layer. This may be because the epitaxial reactor used for this study was for general use. The reactor specialized for a narrow range of resistivity should ease the problem.

APPENDIX

The carrier velocity, v, changes with field, E, as follows

$$v = \frac{\mu_0 E}{1 + \mu_0 E / v_s}\,, \tag{A1}$$

where μ_0 is the low field mobility, and v_s the scattering limited velocity. For electrons $\mu_0 = 1{,}400\ \mathrm{cm^2/V\ sec}$ and $v_s = 10^7$ cm/sec and for holes 480 cm^2/V sec and 1.05×10^7 cm/sec [19] were used. The electron mobility changes with temperature [20], T, as $T^{-2.6}$; whereas the hole mobility as $T^{-2.3}$. The saturation velocites change linearly with T. They reduce by 20% from 300 °K to 500 °K [9].

For the ionization rate Crowell-Sze's expression [8] based on Baraff's theory [21] was used. It contains the temperature dependence.

ACKNOWLEDGEMENTS

The authors would like to thank B. C. De Loach, Jr., R. Edwards, and B. T. Murphy for helpful discussions. He also wishes to thank R. C. Manz for specially adjusting the epitaxial reactor for this study, A. C. Adams and G. Skelly for supplying silane epitaxial films and V. Decker for thinning slices. The dexterity of L. J. Oster and R. H. Hey in epitaxial growth and G. D. Peiffer, A. A. Todd, and K. Schuetz in diode assembly was essential and is acknowledged with thanks.

REFERENCES

[1] T. Misawa, *IEEE Trans. Electron Devices*, **ED-14**, 580 (1967).

[2] C. B. Swan, T. Misawa, and C. H. Bricker, *Proc. IEEE*, **55**, 1747 (1967).

[3] T. Misawa, *Proc. IEEE*, **56**, 234 (1968).

[4] C. A. Adams and J. Simpson, "The Growth of Submicron Epitaxial Si Layers by Silane Pyrolysis," meeting of Electro Chem. Soc., New York (May, 1969).

[5] R. Edwards, D. F. Ciccolella, T. Misawa, D. E. Iglesias, and V. Decker, "Millimeterwave Silicon Impatt Diodes," IEEE 1969 Electron Devices Meeting.

[6] T. Misawa, *IEEE Trans. Electron Devices*, **ED-13**, 143 (1966).

[7] C. R. Crowell and S. M. Sze, *Appl. Phys. Lett.*, **9**, 242 (1966).

[8] C. Y. Duh and J. L. Moll, *IEEE Trans. Electron Devices*, **ED-14**, 46 (1967).

[9] T. Misawa, *IEEE Trans. Electron Devices*, **ED-14**, 796 (1967).

[10] H. S. Carslaw and J. C. Jaegor, *Conduction of Heat in Solids*, 216 (Oxford at the Clarendon Press, 1959).

[11] D. F. Ciccolella, private communication.

[12] R. H. Haitz, *IEEE Trans. Electron Devices*, **ED-15**, 350 (1968).

[13] G. Gibbons and T. Misawa, *Solid State Electr.*, **11**, 1007 (1968).

[14] T. Misawa, "Impatt Diodes," *Semiconductors and Semimetals*, 7 (R. K. Willardson and A. C. Beer, eds., Academic Press, to be published).

[15] T. Misawa, "Silicon Avalanche Transit-Time Diodes," *IEEE Trans. Electron Devices*, **ED-17**, 299 (1970).

[16] S. A. Schelkunoff, *Electromagnetic Waves*, 450 (New York: D. Van Nostrand Co., Inc., 1943).

[17] L. P. Marinaccio (to be published).

[18] B. C. DeLoach, "Thin Skin Impatts," *IEEE Trans. MTT*, **MTT-18**, 72 (1970).

[19] T. E. Seidel, private communication.

[20] F. J. Morin and J. P. Maita, *Phys. Rev.*, **96**, 28 (1954).

[21] G. A. Baraff, *Phys. Rev.*, **128**, 2507 (1962).

Composite Avalanche Diode Structures for Increased Power Capability

C. BURKE SWAN, MEMBER, IEEE, TOSHIO MISAWA, MEMBER, IEEE, AND LOUIS MARINACCIO

Abstract—The possibility of operating several avalanche oscillator wafers in parallel to obtain higher power and/or higher efficiency CW operation is explored analytically and experimentally. Experiments show that over a wide range the efficiency is roughly proportional to the power density in the semiconductor. The power densities required for good efficiency are very high and cannot be achieved in large area junctions without an excessive temperature rise caused by the thermal spreading resistance of the heat-sink material. The scheme delineated herein considers small area wafers spaced sufficiently close electrically that they operate as a single avalanche oscillator whereas their physical separation permits essentially independent heat sinking. It has been found that, as expected, the efficiency for CW operation improves approximately inversely with the diode diameter whereas the power capability for a given size wafer increases directly with the number of such wafers employed The relative merits of mounting diodes on copper and on diamond are discussed. Experimental work indicates that the present approach is capable of producing 10 to 15 watts CW at 14 GHz in a single oscillator with available silicon diodes.

Manuscript received June 9, 1967. This work was supported in part by the U. S. Army Electronics Command, Fort Monmouth, N. J., under Contract DA 28-043 AMC-01445(E).

The authors are with Bell Telephone Laboratories, Inc., Murray Hill, N. J.

I. Introduction

An abrupt-junction *pn* silicon diode which gave 1.1 watts of CW power at 12 GHz with 7.7 percent efficiency was described recently by Misawa.[1] When cooled with liquid nitrogen, 2.7 watts of CW power was produced by one of these diodes at 13.8 GHz.[2] The efficiency was 10.2 percent. This improvement demonstrated the value of achieving a high-current density without incurring an excessive diode temperature. The use of small area junctions to take advantage of the inverse dependence of the thermal spreading resistance on the radius of the heat source was suggested.

The present work considers the area dependence of the power and efficiency for continuous operation of a single diode. The advantages of using a parallel connected array of small area junctions to obtain higher power with good efficiency is explored experimentally. The array of junctions is used as a composite avalanche oscillator diode with a single tuning circuit. The goal is to maximize the power that can be achieved on a continuous basis from an individual oscillator. A further power increase can always be achieved by combining the outputs of a number of such oscillators in the manner described by Fukui.[3],[4] Such techniques are not considered in this work.

II. Calculated Effect of Diode Area on Power Capability

Avalanche oscillator diodes generally dissipate several times as much power in the form of heat as is converted to microwave power. Unusually high-current densities are required to obtain good efficiency in present diodes. This can only be achieved on a continuous basis in structures which can adequately dissipate the associated high power densities without excessive diode heating.[2] Reliability considerations may well restrict the operating temperature more severely than do efficiency and stability considerations.

In a practical application there will be some specified maximum allowed temperature, and there will be some minimum current density that is required to achieve the desired efficiency. Because of the heat dissipation requirements, these specified conditions for continuous operation cannot be met with arbitrarily large diode areas. The following thermal analysis illustrates the restrictions thus imposed on the area and on the maximum power capability for a single diode.

We shall consider a typical silicon diode, suitable for K_u band (12–18 GHz) operation, which has been mounted by thermal compression bonding on a copper heat sink as shown in Fig. 1. For the analysis we shall assume an ideal thermal contact between all the metallic layers and that negligible spreading of the heat flux occurs in the thin layers of gold and nickel on the copper. This permits the simplified picture shown in Fig. 2 to be used. We shall further assume that all the heat flux passes through the bottom contact to the infinite half-space heat sink.[1] The total thermal resistance for a circular heat source of radius r at a depth d_s in the silicon is then given by:

$$R_T = R_s + R_t + R_g + R_n + R_c \ {}^\circ\text{C/W} \qquad (1a)$$

$$= \frac{1}{A}\left(\frac{d_s}{\sigma_s} + \frac{d_t}{\sigma_t} + \frac{d_g}{\sigma_g} + \frac{d_n}{\sigma_n}\right) + \frac{1}{4r\sigma_c}. \qquad (1b)$$

The symbols are defined in Fig. 2. The final term gives the thermal spreading resistance for the infinite half-space heat sink[5] and assumes that the temperature is uniform over the contacting area.[2] The various components of the thermal resistance are summarized in Table I which gives the thermal conductivities and the approximate material thicknesses for each layer in a typical K_u band diode. The various components of the thermal resistance are plotted in Fig. 3 as a function of area for a circular diode. It is observed that the very thin layers of Ti and Ni, although relatively poor thermal conductors, contribute negligible thermal resistance to the structure.

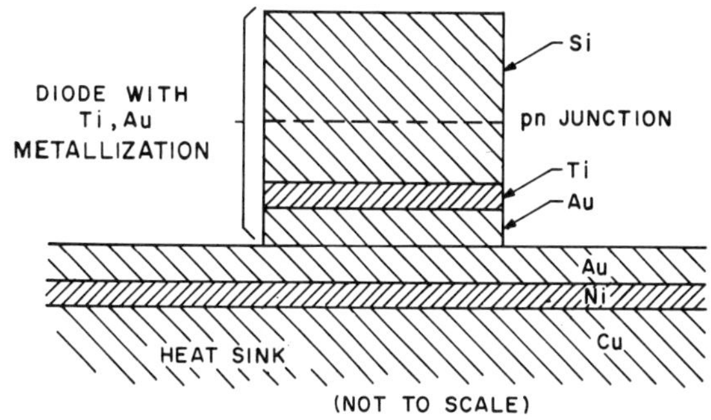

Fig. 1. Diode mounted on copper showing the metallizations used for a gold to gold thermal compression bond.

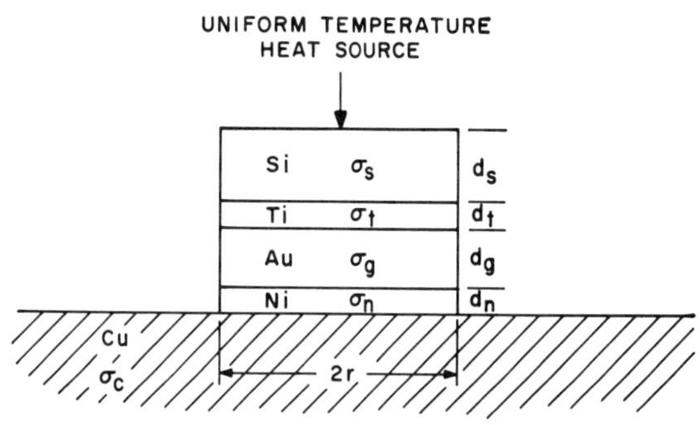

Fig. 2. Simplified diode and heat-sink structure used in the thermal analysis.

The dashed curves in Fig. 3 show the thermal spreading resistance R_D for a diamond heat sink and the corresponding total thermal resistance R_T. The thermal conductivity of the diamond is assumed to be three times that of copper.[3]

Fig. 4 shows power densities and the total dissipation power levels that can be achieved on a copper heat sink with temperature rises of 150°C and 200°C. Fig. 5 compares the power capabilities on copper and on diamond heat sinks for a diode temperature rise of 200°C. It is seen that a significantly higher power can be handled with diamond. This can be better appreciated by con-

[1] Diode wafers could advantageously be designed to dissipate some heat through the top contact. Such structures are not considered here.

[2] This assumption is only a rough approximation. It has some justification for avalanche devices, however, since any tendency of a spot to have a higher temperature is countered by an increased local breakdown voltage which diminishes the electrical current density and thus the power density in this region.

[3] The use of diamond as a heat sink in this application was suggested in Swan.[2] The thermal conductivity of Type IIa diamond has been reported to be about five times that of copper at room temperature.[7] Its thermal conductivity, however, falls off as $1/T$ as the temperature is raised. Since the temperature in the vicinity of the diode will generally be higher than room temperature, we assume a thermal conductivity of three times that of copper. In this application the diamond sample need only be a few times larger than the diode to be effective.

TABLE I

THERMAL CONDUCTIVITY AND LAYER THICKNESS FOR MATERIALS IN A K_u BAND DIODE AS SHOWN IN FIG. 2

Material	Thermal Conductivity* Watts/cm °C	Thickness d cm	d/σ Watts^{-1} cm^2 °C
Silicon	$\sigma_s=0.80$	3×10^{-4}	3.8×10^{-4}
Titanium	$\sigma_t=0.16$	2×10^{-6}	0.13×10^{-4}
Gold	$\sigma_g=3.0$	1.25×10^{-3}	4.2×10^{-4}
Nickel	$\sigma_n=0.71$	2×10^{-6}	0.28×10^{-4}
Copper	$\sigma_c=3.9$	—	—

* The thermal conductivity for silicon is obtained from Maycock[6] for a temperature of 500°K which is assumed as a maximum operating temperature. Room temperature values are used for the other materials in this table since these are slowly-varying functions of temperature.

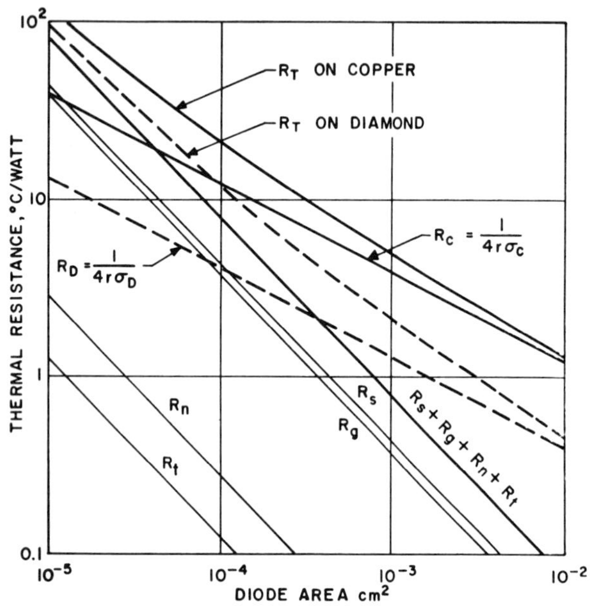

Fig. 3. Thermal resistance for the silicon diode, the interface metallic layers, and for the heat sink. The diode has a circular area and is mounted, as shown in Fig. 2.

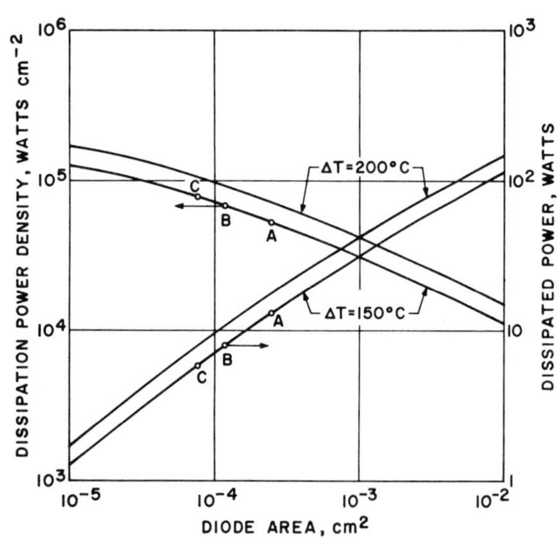

Fig. 4. Area dependence of the dissipated power density and of the total dissipated power for total temperature rises of 150°C and 200°C on a copper heat sink, as shown in Fig. 2.

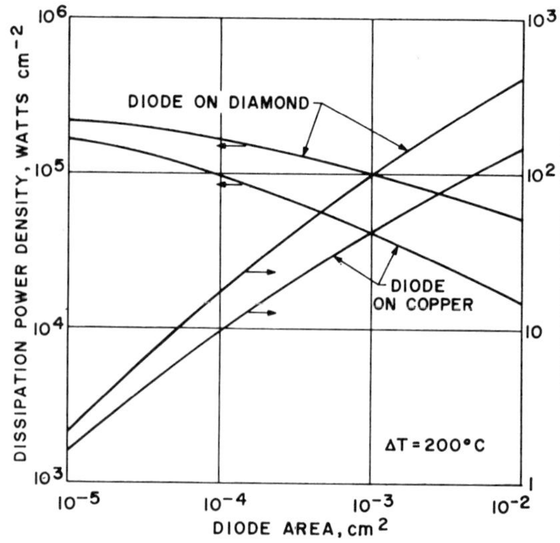

Fig. 5. Comparison of copper and diamond heat sinks showing the power density and the total dissipated power for 200°C temperature rise.

sidering the dependence of oscillator efficiency on the power density. With available silicon diodes it has been found experimentally that a power density of about 10^5 W/cm^2 is required to achieve 10 percent efficiency at K_u band. To achieve this power density with an allowed temperature rise of 200°C, it is seen from Fig. 5 that a copper heat sink can dissipate a maximum of about 9 watts with an area of 9×10^{-5} cm^2 whereas a diamond heat sink can dissipate about 90 watts with an area of 9×10^{-4} cm^2. Thus, for a predetermined efficiency and temperature, it is expected that the single diode maximum power capability is enhanced ten times by using diamond. This assumes that the diode impedance level can be accommodated by the circuit in both cases.

The power capability is also strongly dependent on the temperature rise that can be tolerated. If the allowed temperature rise is only 150°C and we still want 10 percent efficiency, it is seen from Fig. 4 that the dissipation limit on copper is only 3 watts and the area is only 3×10^{-5} cm^2. This is a reduction of three times from the values for a 200°C temperature rise. A temperature rise of less than 150°C on copper for continuous operation appears to be practical only for low power or low efficiency applications.

For diode areas less than 4×10^{-4} cm^2 on copper, the dominant temperature rise occurs in the silicon and gold and not in the spreading resistance of the heat sink. With diamond this occurs for areas less than 4×10^{-3} cm^2. Thus, with practical diode areas, the full benefit of diamond will not be realized without further reducing the thermal resistances arising in the semiconductor and in the contact region. The silicon thermal resistance can possibly be reduced by making shallower junctions (i.e., by reducing d_s). The Schottky barrier diode would be expected to have a significantly smaller effective d_s.

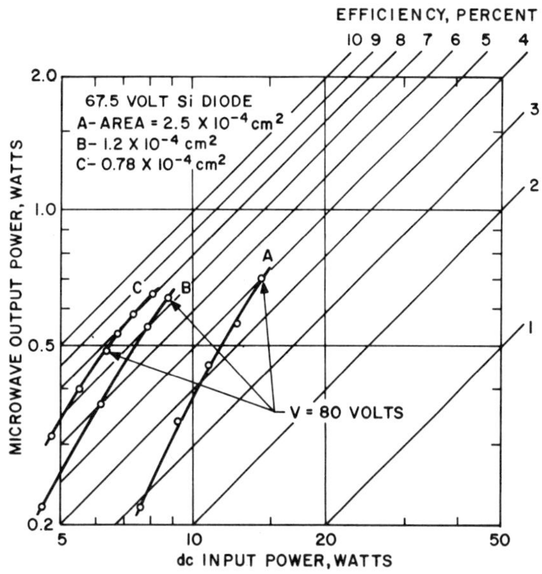

Fig. 6. Performance data for a single silicon avalanche diode mounted on copper. The diode has been successively etched to give smaller areas. Test frequency was 12 to 13 GHz.

TABLE II

DATA FOR A 67.5 VOLT BREAKDOWN SILICON DIODE OPERATED AT 12 TO 13 GHZ WITH 80 VOLTS APPLIED. THE VARIOUS JUNCTION AREAS WERE OBTAINED BY SUCCESSIVE ETCHING OF THE SAME DIODE

Curve Fig. 6	C_B pF	Area cm^2	I at 80 volts amps	de Power watts	Microwave Power W.	Dissipated Power W.	Dissipated W/cm^2	Efficiency percent
A	0.80	2.5×10^{-4}	0.178	14.25	0.71	13.54	5.4×10^4	4.9
B	0.38	1.2×10^{-4}	0.112	8.97	0.65	8.32	6.9×10^4	7.3
C	0.25	0.78×10^{-4}	0.080	6.40	0.49	5.91	7.5×10^4	7.7

The thickness of the gold layer can possibly be reduced to 1 μm or less by using polished surfaces on both the diode and on the heat sink. Good thermal compression bonds have been achieved on diamond with about 1 μm of gold on the diamond and silicon surfaces.

III. EXPERIMENTAL EVALUATION OF A DIODE WITH DIFFERENT JUNCTION AREAS

Fig. 6 shows experimentally that the efficiency can be improved by using smaller area diodes which permit larger current densities. The diode used is an abrupt junction silicon diode with breakdown at 67.5 volts. The diode is mounted as shown in Fig. 1. Curve A is for a circular junction whose area is 2.5×10^{-4} cm^2 and which has a capacitance at breakdown $C_B=0.80$ pF. Curve B is for the same diode after etching down the junction area to 1.2×10^{-4} cm^2 ($C_B=0.38$ pF). The efficiency has increased from 4.9 percent to 7.3 percent. Further etching to 0.78×10^{-4} cm^2 ($C_B=0.25$ pF) increased the efficiency to 7.7 percent. All measurements were made with the diode tuned in the frequency range 12 to 13 GHz where the efficiency was essentially independent of the frequency. In each case the input power was increased until the diode reached approximately the same temperature as indicated by the increased breakdown voltage.[8]

Comparative measurements at the 80-volt operating points, which are indicated in Fig. 6, are tabulated in Table II. It is seen that the dissipated power densities are significantly increased for the etched down structures and, furthermore, the output power has decreased by much less than the ratio of the diode areas. Using the power density and the diode area data from Table II, we can enter the 80-volt operating points on the curve of Fig. 4. This shows that the calculated mean temperature rise, which assumes an ideal bond to the heat sink, is about 150°C in each case. The data for this diode indicates that it is somewhat futile to attempt to achieve higher output power on copper by increasing the diode area. Section IV considers the operation of parallel connected small area diodes which are sufficiently separated that their heat dissipation occurs essentially independently.

IV. PARALLEL CONNECTED *pn* AVALANCHE DIODES

It is evident from Fig. 6 that increased power capability cannot best be achieved by increasing the diode area. Figs. 7 to 9 show experimental results for parallel

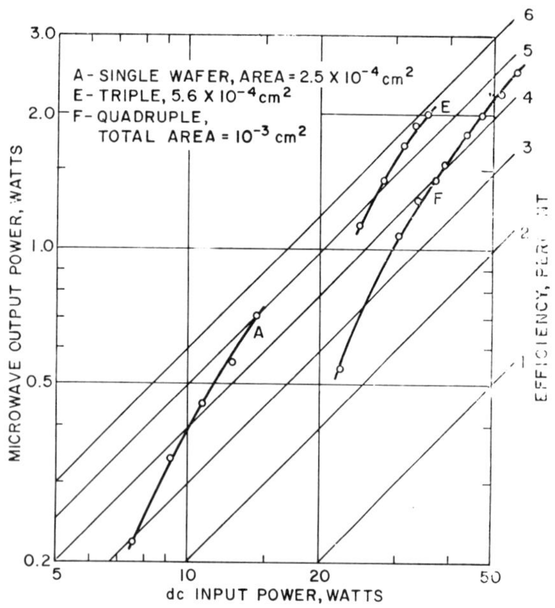

Fig. 7. Performance data for parallel connected Si avalanche diodes. A is the single unit in Fig. 6. F is for four such units in parallel. E is for three such units etched down to 0.75 times their original area. Test frequency was 12 to 13 GHz.

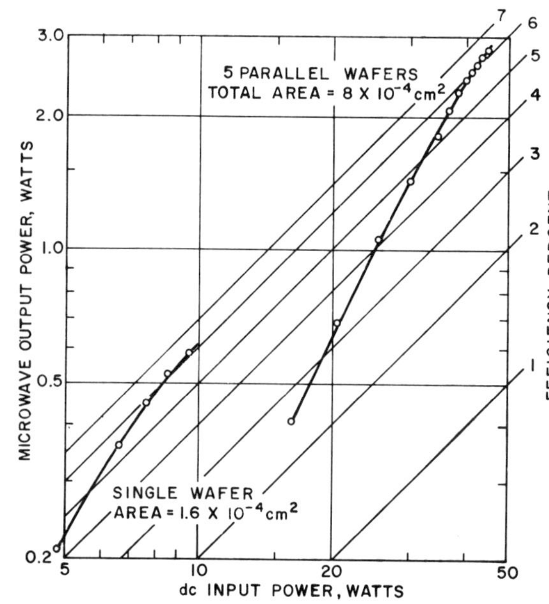

Fig. 9. Performance for a single and a quintuple Si avalanche diode tested at ~12 GHz. These wafers are from a Si slice that had lower efficiency and lower power capability than reported in the previous figures.

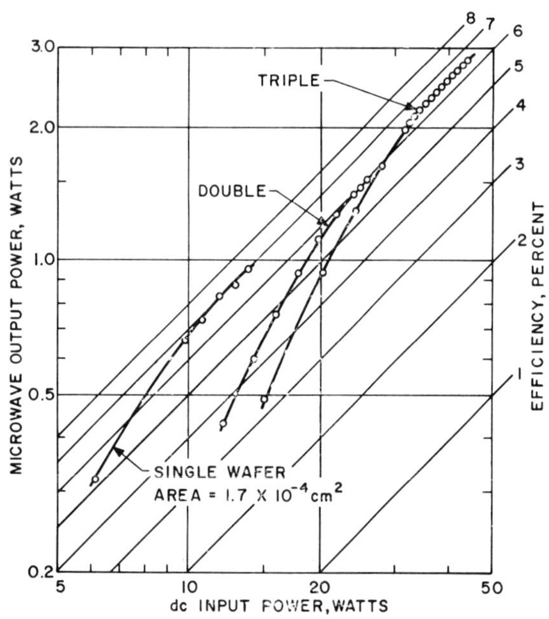

Fig. 8. Performance for a single, a double, and a triple Si avalanche diode tested at 11 to 13 GHz. Wafer areas are $\sim 1.7 \times 10^{-4}$ cm² and are from the same slice used in Figs. 6 and 7.

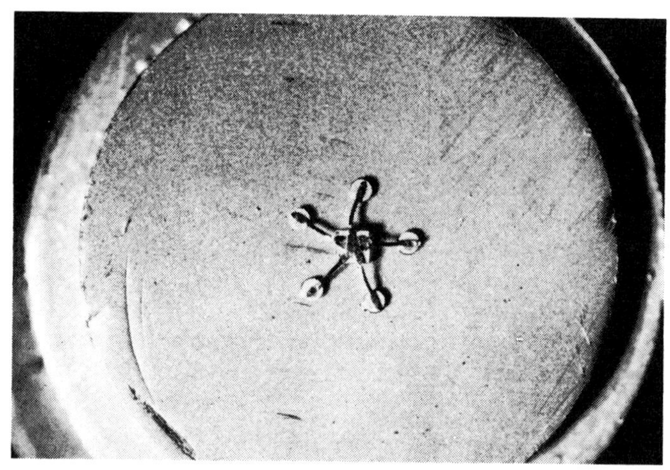

Fig. 10. A composite avalanche diode employing five parallel connected wafers on a copper stud. The face of the copper stud has a diameter of 0.25 inch (0.635 cm), each wafer has a diameter of 0.010 inch (0.025 cm), and the gold sticthing wire has a diameter of 0.002 inch (0.005 cm).

528

connected diodes tested in a single oscillator circuit. Fig. 7 is for the 2.5×10^{-4} cm² units of the same slice used for Fig. 6. Curve A in Fig. 7 repeats the curve A in Fig. 6. Curve E is for a triple unit in which the diode capacitances have been etched down to about 0.75 times their original size. A significant improvement in both power and efficiency was achieved. Curve F is for a quadruple unit with full-sized wafers. The efficiency is only slightly less than that of the single unit, curve A, but the input and output powers are roughly four times higher.

Higher efficiency operation requires the use of higher efficiency individual units. Fig. 8 shows how this was achieved by using diode wafers from the same slice of silicon but with areas $\sim 1.7 \times 10^{-4}$ cm². Data are given for a single, a double, and a triple unit. Again the expected improvement in power without a significant change in efficiency was obtained by using the parallel connected diode wafers. The output was 2.8 watts CW for the triple unit.

Fig. 9 shows a single unit and five parallel units from a different slice which had both a lower efficiency and a somewhat lower power capability than those of Fig. 8. The efficiency is essentially the same in both cases, and the output power of 2.8 watts is five times higher in the composite diode than for a single diode. In all these tests the diodes are not encapsulated. The wafers are bonded junction side down on a 0.375 inch diameter copper stud and are stitched with short 0.002 inch diameter gold leads to a small quartz standoff as shown in Fig. 10.

V. Conclusions Regarding the Continuous Power Capability

It was shown in the previous sections that the CW power for a *single* diode oscillator at any specified efficiency is limited by the maximum area that can be used. This limitation arises from the temperature rise associated with the power density requirements in the semiconductor. The power for the *composite* diode is limited by the maximum diode capacitance that can be accommodated in the oscillator circuit. This accommodation implies that the composite diode can be tuned at the desired frequency and that its power can be efficiently coupled to the load.

The composite diode used in curve F of Fig. 7 had a breakdown capacitance $C_B = 3.2$ pF and was tuned at 11.75 GHz. A higher capacitance or a higher frequency could have been used for these measurements. In another experiment a total C_B of 3.0 pF was readily accommodated at 14 GHz. Although a higher capacitance can undoubtedly be accommodated with circuit refinement, we shall use this experimental value to estimate the power capability of the present techniques.

A typical 14 GHz silicon diode has a capacitance $\sim 3 \times 10^3$ pF/cm^2. Thus a 3 pF diode has an area $\sim 10^{-3}$ cm^2. Operation at about 10 percent efficiency is possible with a power density $\sim 10^5$ W/cm^2. Assuming a maximum temperature rise of $\Delta T = 200°$C, Fig. 5 shows that 10^5 W/cm^2 can be achieved with a diode area of 10^{-4} cm^2 on copper. Ten such diodes in parallel for a total $C_B = 3$ pF could dissipate ~ 90 watts with 10 watts RF output. A similar result could be achieved on diamond with a single diode whose area is 10^{-3} cm^2 provided that there are no bonding problems with the larger area diodes.

If we specify an increased power density on diamond of 1.5×10^5 W/cm^2, the maximum individual diode area is $\sim 2 \times 10^{-4}$ cm^2. Five such units could be accommodated at 14 GHz for a total dissipated power ~ 150 watts and an expected CW output exceeding 15 watts. These per unit calculated results have been obtained in experiments with single and double silicon units on diamond.[9]

Modification of the diode structure, to generate the heat closer to the surface (as in a Schottky barrier), and improvement in the diode junction geometry to achieve a better heat distribution over the diode area, can be expected to yield a further increase in the power density with a corresponding improvement in efficiency. We cannot anticipate any really significant increase in the attainable power density, however, and further power increases for an individual oscillator will require either higher efficiency diodes or developing circuits to accommodate larger capacitance composite diodes. A further step would be to combine the outputs of synchronized oscillators. Fukui[3] has demonstrated that a power enhancement of essentially eight times can be attained with eight individual oscillators connected with an eight-way hybrid combiner.

Assuming that the power-impedance product is inversely proportional to the square of the frequency,[10] we can conclude that the techniques discussed herein are capable of continuously producing 15 watts at 14 GHz, about 30 watts at 10 GHz, and about 1.2 watts at 50 GHz. Preliminary experimental results[11] at mm wavelengths have indicated that the $1/f^2$ relationship is roughly applicable and these power values appear quite realistic.

Acknowledgment

The authors are grateful to J. Hill for the mechanical design of the circuits, to A. Todd for painstakingly assembling the composite diode structures and to B. C. DeLoach for his continuing encouragement.

References

[1] T. Misawa, "Microwave Si avalanche diodes with nearly abrupt type junction," presented at the Internat'l Flectron Device Meeting (Washington, D. C., 1966).

[2] C. B. Swan, "The importance of providing a good heat sink for avalanching transit time oscillator diodes," *Proc. IEEE* (*Letters*), vol. 55, pp. 451–452, March 1967.

[3] H. Fukui, "Frequency locking and modulation of microwave silicon avalanche diode oscillators," *Proc. IEEE* (*Letters*), vol. 54, pp. 1475–1477, October 1966.

[4] ——, "A multiple silicon avalanche diode oscillator," presented at the Internat'l Electron Device Meeting (Washington, D. C., 1966).

[5] Carslaw and Jaeger, *Conduction of Heat in Solids*, 2nd ed. London: Oxford University Press, 1959, p. 217.

[6] P. D. Maycock, "Thermal conductivity of silicon, germanium, III-V compounds and III-V alloys," *Solid-State Electronics*, vol. 10, pp. 161–168, March 1967.

[7] R. Berman, *Physical Properties of Diamond*. Oxford: Clarendon Press, 1965.

[8] C. R. Crowell and S. M. Sze, "Temperature dependence of avalanche multiplication in semiconductors," *Appl. Phys. Lett.*, vol. 9, pp. 242–244, September 15, 1966.

[9] C. B. Swan, "Improved performance of silicon avalanche oscillators mounted on diamond heat sinks," *Proc. IEEE* (*Letters*), vol. 55, pp. 1617–1618, September 1967.

[10] B. C. DeLoach, *Advances in Microwaves*, L. Young, Ed. New York: Academic (to be published).

[11] C. B. Swan, T. Misawa, and C. H. Bricker, "Continuous oscillations at millimeter wavelengths with silicon avalanche diodes," *Proc. IEEE* (*Letters*), vol. 55, p. 1747, October 1967.

OTHER ARTICLES OF INTEREST

Reprinted from IEEE TRANSACTIONS
ON ELECTRON DEVICES
Volume ED-15, Number 6, June, 1968
pp. 350-361

Nonuniform Thermal Conductance in Avalanche Microwave Oscillators

ROLAND H. HAITZ, MEMBER, IEEE

***Abstract*—The local dc current density $j(x)$ in avalanche microwave oscillators is a function of the local breakdown voltage $V_b(x)$, which in turn depends on local temperature $T(x)$ and field $E(x)$. Recent progress in junction fabrication has led to very uniform field distributions such that local temperature variations of the order of a few degrees are no longer negligible. Calculations of $T(x)$ for various diode configurations show that the spherical heat flow near the diode edge causes considerable variations of $T(x)$. The resulting variations in $j(x)$ are so significant even in the case of a reasonably ideal mounting (flip-chip mesa on a copper substrate) that they will affect the oscillator performance with respect to efficiency and oscillator noise. The dynamic thermal properties are discussed in terms of a simplified oscillator model consisting of two parallel subdiodes with different but uniform current densities. Nonuniform breakdown combined with an internal current redistribution resulting from nonuniform internal heating causes AM and FM modulation and is also responsible for the often observed delayed onset of microwave oscillations under pulsed conditions.**

I. INTRODUCTION

IN PREVIOUS publications on avalanche microwave oscillators[1]–[5] the diodes have always been treated as well-behaved junctions carrying a uniform current density across their entire junction area. In reality, however, one finds considerable deviations from such a uniform current density model. It is the object of this paper to point out the origin of these current nonuniformities, to estimate their magnitude, and to discuss their influence on the performance of avalanche microwave oscillators.[6]

The local current density $j(x)$ in avalanche diodes is a function of the local breakdown voltage $V_b(x)$ which in turn depends on local field $E(x)$ and local temperature $T(x)$. In recent years large and successful efforts in reducing local field variations led to junctions with fairly uniform breakdown.[7] Since then further progress in reducing local field variations was so great that one now is very often faced with local temperature variations as the dominating source of current nonuniformities.

The first experimental evidence for thermally originated local current-density variations can be found in Fig. 5(a) of a publication on structurally perfect junctions by Goetzberger *et al.*[8] There it was observed that the light emission from shallow guard ring diodes is generally brighter near the edge of the breakdown area. At that time the importance of thermal effects was not realized and it was thought that the larger current density near the edge was caused by a slightly lower breakdown voltage. However, there was no convincing explanation for the cause of such a lowering of V_b near the edge.

A thermal origin for local current-density variations can readily be understood if one realizes that the heat flow between junction and heat sink is not uniform. This effect is illustrated in Fig. 1. A heat source of uniform power dissipation density is located at the top of a plain slice of material of constant thickness. The lower surface of this slice is kept at a constant temperature T_0 by connecting it with a heat sink. The heat flow is illustrated by flow lines. In the center part of the heat source the flow lines are parallel, resulting in a constant temperature. In the edge region, however, radial heat flow leads to a larger thermal conductance per unit area and, consequently, to a lower temperature denoted by $T(x)$. It is shown in Section II that, in the case of avalanche diodes, the variation of $T(x)$ is large enough to cause significant variations in the current density $j(x)$.

In pulsed avalanche diodes the time dependence of the junction temperature $T(x, t)$ will cause an internal redistribution between initial and steady-state current distribution. This redistribution and its consequences on oscillator properties are discussed in Section III.

Manuscript received September 22, 1967; revised March 4, 1968. Part of this work was supported by Contract F33-615-67-C-1909, U. S. Air Force Avionics Laboratory, Wright-Patterson AFB, Dayton, Ohio.

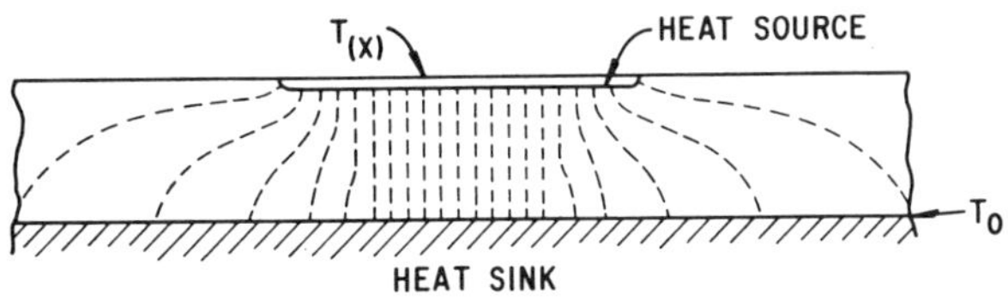

Fig. 1. Nonuniform heat flow near the edge of a flat heat source.

II. Steady-State Thermal Properties and Effects

A. Diode Model

The effects of internal heating of an avalanche diode are fairly easy to analyze in the case of uniform thermal conductance. Suppose a heat source of cross section A and uniform power density is located at one end of a slab of material of length L and of the same cross section A (Fig. 2). Heat flow across the side walls shall be negligible. The other end of the slab is held at constant temperature T_0. Under these conditions the temperature across the heat source will be uniform.

If the heat source is a p-n junction in avalanche breakdown then the current-density–voltage characteristic is represented by a single line intersecting the voltage axis at $V_b(T_0)$. The slope of this line is determined by three contributions to the resistance per unit area R_s: 1) spreading resistance R_{sp}; 2) space charge resistance R_{sc}[9]; and 3) thermal resistance R_{th}.[10] The first two resistance contributions are conventional electrical resistances. The third contribution arises from the temperature dependence of the breakdown voltage. For the example of Fig. 2, R_{th} is given by

$$R_{th} = (L/\lambda)\beta V_{b0}^2, \tag{1}$$

with λ denoting the thermal conductivity of the material, and β the temperature coefficient of the breakdown voltage. The operating point (j_0, V_0) is determined by the intersection of load line and j–V characteristic. For given operating conditions the breakdown voltage $V_b(T)$ is related to the operating point via a $-V$ characteristic with resistance $R_{sp}+R_{sc}$, with R_{sp} usually being negligible.

The temperature across the heat source will become nonuniform if the cross section of the heat source is smaller than the cross section of the heat-conducting material such as in the structure of Fig. 1. As a result V_b and j will become functions of position denoted by local breakdown voltage $V_b[T(x)] = V_b(x)$ and local current density $j[V_b(x)] = j(x)$. The local j–V characteristic no longer is represented by a single line but by an infinite number of lines with slopes determined by the effective local resistance $R_s(x)$. In the case of diodes with highly doped regions on both sides of the depletion layer one can neglect the ohmic spreading resistance and describe $R_s(x)$ by the two remaining contributions

$$R_s(x) = R_{sc} + R_{th}(x). \tag{2}$$

For plane junctions R_{sc} is independent of position.

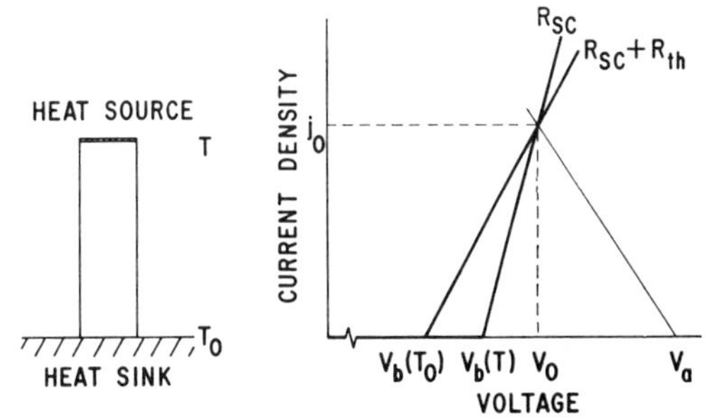

Fig. 2. Structure and current-density–voltage characteristic of a diode with uniform temperature across avalanche region.

The effective local thermal resistance $R_{th}(x)$ is given by the equation

$$R_{th}(x) = \beta V_{b0}^2/g^*(x) \tag{3}$$

where $g^*(x)$ denotes the effective local thermal conductance (in W/cm² °C). $g^*(x)$ is defined by

$$1/g^*(x) = r^*(x) = [T(x) - T_0]/p \tag{4}$$

with p denoting the dissipated power density.

It should be noted that $g^*(x)$ is not the actual thermal conductance between area element Δx and heat sink. As defined by (4), $g^*(x)$ denotes a lumped effective thermal conductance taking into account both the actual thermal conductance between an area element and heat sink *and* mutual heating from neighboring area elements.

The dependence of $T(x)$ on x is the result of a nonuniform heat flow from source to sink. This nonuniformity can be visualized by considering each area element as a point heat source with spherical heat propagation. An area element near the center of the heat source will propagate exactly as much energy to the neighboring elements as it receives from them. As a result one obtains a uniform heat flow towards the sink in the center region of the source. An area element near the edge will propagate heat uniformly into all directions of the hemisphere. However, it will not receive a corresponding heat flow from area elements outside of the heat source. As a result, these area elements near the edge will reach a lower temperature than the area elements near the center, thus causing a nonuniform temperature distribution $T(x)$.

As mentioned earlier, the $j(x)$–V characteristic of a nonuniform diode consists of an infinite number of branches, two of which are shown in Fig. 3. The branch labeled j_c represents the j–V characteristic of an area element at the center of the breakdown region. The branch labeled $j(x)$ represents an area element near the edge. The characteristic labeled $\langle j\rangle$ is the average j–V characteristic of the diode.

The operating point of the diode is determined by the intersection of the $\langle j\rangle$–V characteristic with the load line. If V_0 denotes the voltage at the operating point then the local current density is given by the value of the $j(x)$–V characteristics at the voltage V_0. Furthermore,

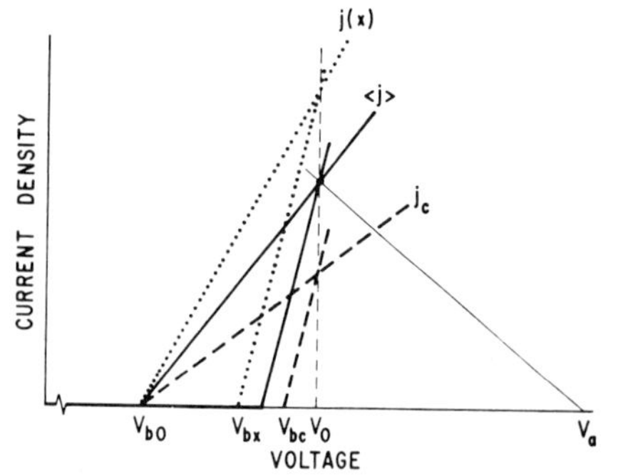

Fig. 3. Current-density–voltage characteristic of a diode with nonuniform temperature across avalanche region.

the breakdown voltage $V_b(x) = V_{bx}$ is given by the intersection of the voltage axis with a line of slope R_{sc} through the operating point $j(x)$, V_0.

The calculation of $j(x)$ is carried out in two steps. In Section II-B the temperature distribution $T(x)$ is calculated. In Section II-C $T(x)$ is used to calculate $V_b[T(x)]$ and $j(x)$.

B. Temperature Distribution

The calculation of $T(x)$ will be carried out for two methods of mounting avalanche diodes. The diode model of Fig. 4(a) represents a good approximation for the thermal properties of guard ring, planar, or shallow mesa diodes with conventional face-up mounting. The planar approach, however, is of little interest because of nonuniform breakdown at the edge due to the depletion layer curvature.[11] The diode model of Fig. 4(b) represents a good approximation for the thermal properties of flip-chip mounted diodes. Because of capacitance problems, flip-chip mounting is only of interest for mesa diodes leading to a configuration as shown schematically in Fig. 4(c).

Further simplifications are made in order to use existing computer programs for the calculations. A rectangular geometry is chosen with $2a \times 2a$ and $2b \times 2b \times c$ being the dimensions of heat source and diode chip, respectively. The heat source of zero thickness is assumed to be located at the surface of the silicon chip. The metal representing the diode package has the dimensions $2b \times 2b \times d$. No heat flow across the side walls of the silicon chip and metal block is assumed. The bottom surface of the metal block is kept at constant temperature T_0. The temperature profile is calculated for a line through the center of the heat source parallel to the chip edge. This line is located at the surface of the chip for Fig. 4(a) and at the interface for Fig. 4(b). The thermal resistance of the silicon–metal interface is neglected.

The calculation of $T(x)$ is carried out numerically by using an existing computer program. This program neglects the temperature dependence of the thermal conductivity in order to obtain a solution which is linear in power. The following dimensions were chosen: $2a = 100\ \mu$, $2b = 1000\ \mu$, $c = 100\ \mu$, and $d = 1000\ \mu$. Copper is chosen as the metal substrate.

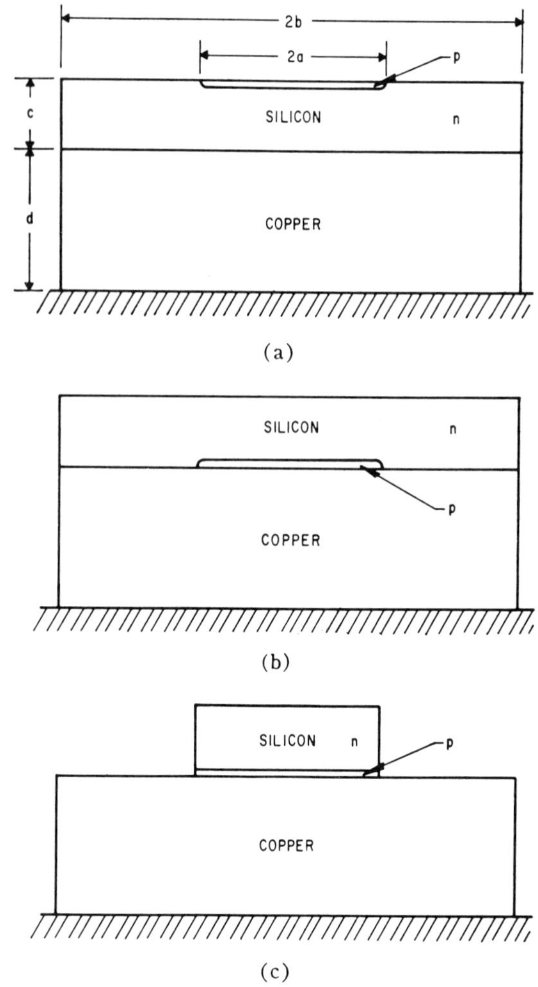

Fig. 4. Three possible mounting configurations for avalanche diodes. (a) Conventional face-up mounting. (b) Flip-chip mounting of a planar diode. (c) Flip-chip mounting of a mesa diode.

The results of $T(x) - T_0$ normalized to a power dissipation density of 1 W/cm^2 are plotted in Fig. 5. As expected one finds a uniform temperature within the center region of the heat source and a decreasing temperature towards the diode edge. The temperature decreases further between heat source edge and chip edge. The temperature difference between diode center and heat sink is approximately three times higher for the face-up mounted diode compared with the flip-chip mounted diode. This ratio is roughly equal to the ratio of the thermal conductivities of copper and silicon.

The difference of the two curves clearly illustrates the superiority of flip-chip mounting silicon diodes onto a substrate of high thermal conductivity. A further advantage of this flip-chip mounting technique will become evident in the next subsection.

C. Current Density Variation

The temperature variation between center and edge leads to a corresponding variation of V_b:

$$V_b(x) = V_b(T_0)\{1 + \beta[T(x) - T_0]\}. \tag{5}$$

Equation (5) takes only those variations of $V_b(x)$ into

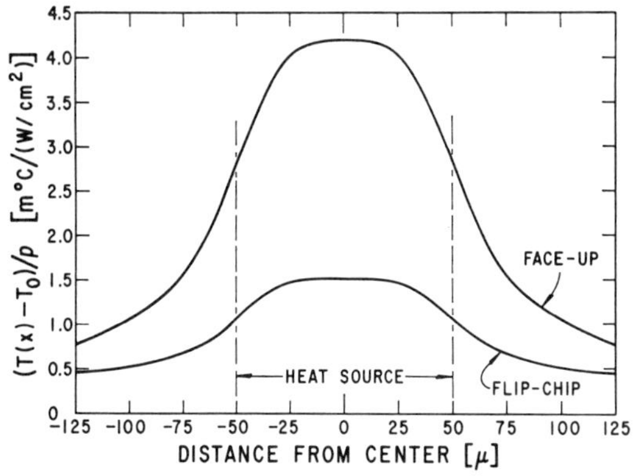

Fig. 5. Temperature distribution along copper surface through diode center for face-up and flip-chip mounting. The temperature difference $T(x)-T_0$ is normalized to a power dissipation density of 1 W/cm².

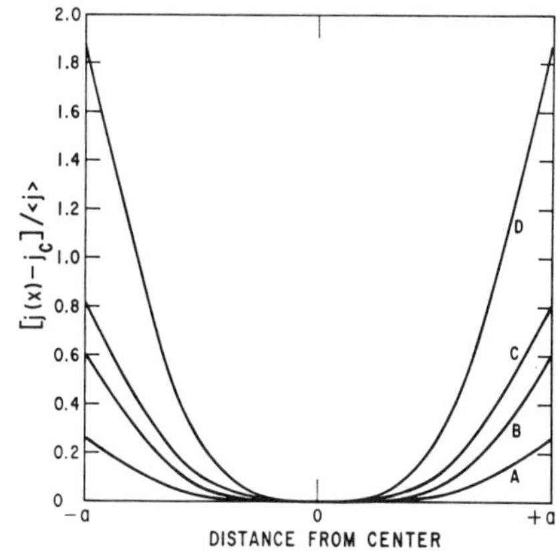

Fig. 6. Current-density variation normalized to average current density $\langle j\rangle$. (a) Flip-chip mounted p^+-n diode. (b) Flip-chip p-i-n. (c) Face-up p^+-n. (d) Face-up p-i-n.

account which result from a nonuniform thermal conductance. Variations of V_b resulting from nonuniform impurity concentrations or from curvatures in the depletion layer are not considered in this section.

The temperature difference $T(x)-T_0$ of (5) is proportional to the power dissipated per unit area, $\langle j\rangle V_{b0}$, and is given by

$$T(x) - T_0 = \langle j\rangle V_{b0} r^*(x). \tag{6}$$

Here $\langle j\rangle$ denotes the average current density, and $r^*(x)$, defined by (4), is obtained from the numerical calculation plotted in Fig. 5. Inserting (6) into (5), one obtains

$$V_{bx} = V_{b0} + \langle j\rangle \beta V_{b0}^2 r^*(x). \tag{7}$$

According to Fig. 3 the current density is then simply given by

$$j(x) = (V_0 - V_{bx})/R_{sc}. \tag{8}$$

Here it is important to realize that only the space charge resistance R_{sc} is controlling the current-density variation. As mentioned earlier, spreading resistance R_{sp} is neglected. The thermal resistance, already included in V_{bx} of (7), has only a negligible effect on the internal distribution of the current density.

It is of interest to calculate the relative deviation of the local current density $j(x)$ resulting from the nonuniform thermal conductance. The j–V characteristic of Fig. 3 is very helpful in visualizing this calculation. Inserting V_{bx} of (7) into (8), one can write

$$j(x) = [(V_0 - V_{bc}) + (V_{bc} - V_{b0}) - \langle j\rangle \beta V_{b0}^2 r^*(x)]/R_{sc}. \tag{9}$$

Since $V_{bc} - V_{b0} = \langle j\rangle \beta V_{b0}^2 r_c^*$ and $V_0 - V_{bc} = j_c R_{sc}$ one obtains from (9) the following expression:

$$\frac{j(x) - j_c}{\langle j\rangle} = 2\epsilon v_d \beta (V_{b0}/W)^2 [r_c^* - r^*(x)]. \tag{10}$$

In (10) we have replaced the space-charge resistance per unit area by

$$R_{sc} = W^2/2\epsilon v_d \tag{11}$$

with ϵ denoting the dielectric constant of silicon, v_d the saturated drift velocity, and W the space-charge layer width.[1]

The normalized current-density variation of (10) is calculated for two diode configurations: 1) an abrupt p-i-n diode with a constant field in the depletion layer, and 2) an abrupt p^+-n junction with a field decreasing linearly to zero. Both diodes are assumed to have a breakdown voltage of $V_{b0}=100$ volts. The ratio V_{b0}/W is obtained from the Appendix to 282 kV/cm for the p-i-n junction and 185 kV/cm for the p^+-n junction. The remaining constants are $\epsilon=1.06$ pF/cm, $v_d=8.5\times10^6$ cm/s,[12] and $\beta=9.4\times10^{-4}$/°C. The effective inverse local thermal conductance at the center is $r_c^*=4.20$ m°C/(W/cm²) for a face-up mounted silicon chip on a copper substrate and $r_c^*=1.52$ m°C/(W/cm²) for flip-chip mounting on a copper substrate. Inserting $r^*(x)$ from Fig. 5 into (10), one obtains the four curves plotted in Fig. 6.

The accuracy of the curves in Fig. 6 is affected by several approximations. For the calculation of $r^*(x)$ we have assumed a constant power density. However, as Fig. 6 indicates, this assumption is not correct because of the higher current density near the edge. This effect should partly offset the nonuniform temperature distribution. A first iteration using a power density distribution varying in three steps between edge and center led only to a small correction in the corresponding $[j(x)-j_c]/\langle j\rangle$ curve. Furthermore, the variation of the thermal conductivity with temperature, which was neglected in order to obtain a solution linear in power, will add to the first error. For the case of a flip-chip mounted diode it was assumed that the heat flow takes place entirely within the copper substrate. Heat flow within the silicon

[1] As pointed out by one of the reviewers, (11) does not correctly describe the space-charge resistance of a p-i-n diode. However, it is quite useful in estimating its magnitude for carrier densities n, p which are small compared with the net impurity density $|N_a - N_d|$ of the i-layer. Since (11) always overestimates R_{sc} of p-i-n diodes, it should be noted that the current-density variations in p-i-n diodes presented in the remainder of this subsection represent only lower limits.

chip from center to edge, which will partly compensate the first two errors, is neglected. The error resulting from all the above approximations is estimated to be less than 30 percent.

The magnitude of the current-density variation shown in Fig. 6 is quite surprising. The current density of an ideally flip-chip mounted *p-i-n* diode will vary by 60 percent from center to edge. This is an appreciable variation considering the strong dependence of oscillator properties on current density. Fig. 6 predicts even stronger variations for face-up mounted diodes: 190 per cent for *p-i-n* diodes and 90 percent for step junctions. Such strong variations are in agreement with earlier studies of light emission from shallow guard ring avalanche diodes.[8] During these earlier studies a considerably brighter light emission was always observed near the edge of the breakdown region (see Fig. 5(a) of Goetzberger[8]). It is now believed that the enhanced light emission near the edge is caused by nonuniform internal heating.

The results of Fig. 6 also show the superiority of the flip-chip mounting technique. Compared with face-up mounting, this technique not only reduces the average temperature (Fig. 5) but also leads to a considerable improvement of the current-density distribution. The current-density variation is reduced by a factor which is approximately equal to the ratio of the thermal conductivity of copper and silicon. Considerable further improvements could be obtained by flip-chip mounting the diode on a substrate of larger thermal conductivity such as diamond.[13]

Fig. 6 also shows that p^+-n diodes have a smaller current-density variation than *p-i-n* diodes with the same breakdown voltage. Such behavior is expected from (10) since the ratio V_{b0}/W is approximately 50 percent larger for a *p-i-n* diode compared with a p^+-n diode. Further improvement in current-density variation can be expected for p^+-n-i-n^+ or n^+-p-i-p^+ Read diodes because of their lower V_{b0}/W ratio.

D. Effects on Oscillator Properties

Local current-density variations will mainly affect three properties of CW avalanche oscillators: efficiency, frequency, and noise. A discussion of these effects will be given in this subsection.

The negative resistance of an avalanche diode is a sensitive function of the current density. As a result the oscillator frequency and efficiency become current-density dependent. This result holds for an oscillator of uniform current density. An oscillator of nonuniform current density can be considered as a number of parallel subdiodes each operating at a different current density. Very likely all of these subdiodes will be phase-locked together and oscillate at a single frequency. However, only a small fraction of these subdiodes will oscillate under optimum conditions with maximum efficiency. The remaining subdiodes will either oscillate at reduced efficiency or only contribute to the internal losses. Consequently, the efficiency of avalanche oscillators will be severely affected by a nonuniform current-density distribution.

In the preceding subsection it was found that the local current-density variations are reduced by going from *p-i-n* to p^+-n and p^+-n-i-n diodes. However, as shown by Gummel and Scharfetter,[4] the current-density dependence of the negative resistance increases from *p-i-n* towards p^+-n and p^+-n-i-n diodes. At present it is not known which of the two effects is stronger. Comparing the published efficiency data of p^+-n and p^+-n-i-n diodes,[14],[15] which indicate higher efficiencies for p^+-n diodes, one is inclined to believe that the second effect is stronger. Then, however, current-density variations could be responsible for the discrepancy between predicted and observed efficiency.

In order to see how a nonuniform current-density distribution affects the oscillator noise, one may utilize the results of earlier avalanche noise studies. There it was concluded that the oscillator noise is mainly caused by the low-frequency avalanche noise current.[16],[17] This low-frequency noise current modulates the amplitude and frequency of the oscillator via the current dependence of diode conductance and susceptance. The low-frequency noise current, however, is a very sensitive function of the uniformity of the dc avalanche current.[17],[18]

The spectral noise current density $\langle i^2\rangle/\Delta f$ is related to the readily measurable open circuit spectral voltage density $\langle u^2\rangle/\Delta f$:

$$\langle i^2\rangle/\Delta f = (\langle u^2\rangle/\Delta f)(A/R_{sc})^2. \tag{12}$$

Here R_{sc} is defined by (11), and A is the breakdown area. From studies of avalanche noise it is known that the measured values of $\langle u^2\rangle/\Delta f$ are often several orders of magnitude above the theoretical minimum.[17],[19] As discussed in detail elsewhere, such large contributions of excess noise are caused by nonuniform current distributions.[19]

In case of frequency modulation noise it is easy to estimate the rms frequency deviation Δf_{rms} if the variation of oscillator frequency f_0 with avalanche current I is known:

$$\Delta f_{\text{rms}} = (df_0/dI)\langle i^2\rangle^{1/2}. \tag{13}$$

Experimental values of df_0/dI are usually of the order of 10^{10} Hz/A and depend inversely on the external quality factor Q_{ex}. The minimum noise current $\langle i^2\rangle$ within a 1-kHz bandwidth is typically of the order of 10^{-8} amperes, leading to values of Δf_{rms} of the order of 100 Hz per 1-kHz bandwidth for Q_{ex} of the order of 100. Any low-frequency excess noise on top of the theoretical minimum noise current will increase the oscillator frequency modulation proportionately.

III. Dynamic Thermal Properties and Effects

A. Diode Model

In Section II we have only considered the effects of thermal nonuniformities during steady-state conditions of CW avalanche oscillators. For pulsed microwave oscillators, however, additional complications arise during the transition period between the onset of breakdown and steady-state avalanche conductance. It will be shown in this section that thermal nonuniformities are responsible for some of the unexplained effects observed in pulsed oscillators.

In this section another simplification of Section II is dropped. In Section II it has been assumed that the breakdown voltage at constant temperature, $V_b(T_0)$, is independent of position. Under this assumption the diode has been treated as a junction with an ideally uniform field distribution. Unfortunately, the field distribution in actual junctions is usually slightly nonuniform because of variations in the depletion layer width. Such variations in width are usually the result of small variations in the impurity density from one edge of the diode to the other edge. In crystals grown from the melt such impurity variations are quite common, leading to periodic impurity density variations across the diode.[8] In epitaxial material a concentration gradient across the slice is also quite common. Furthermore, if *p-i-n* diodes are manufactured from epitaxial material then a thickness variation of the epitaxial layer will also lead to a corresponding variation of the depletion layer width. Since concentration variations of the order of 1 percent will lead to variations of the breakdown voltage which are large compared with the variations caused by the thermal nonuniformities of Section II, we will include the discussion of slightly nonuniform breakdown.

The discussion of dynamic properties of avalanche diodes will include a second independent variable, time, in addition to the independent variable x used in Section II. A computation of $j(x, t)$ will be very complicated. Since an actual diode will deviate substantially from the ideal model treated in Section II because of the breakdown voltage variations discussed above, and because of local variations of the thermal resistance of the silicon–metal interface, so far neglected, we will treat only a highly simplified qualitative model. This simplified model neglects the continuous variation of the variable x and considers the current density at only two points, x_1 and x_2. Looking at this simplified model in another way one can describe it as a diode model consisting of two lumped subdiodes with uniform electrical and thermal properties. This second way of describing $j(x, t)$ has also the advantage of clarity by providing simple relations between thermal and dc electrical properties of the diode and the properties of the diode as a microwave oscillator.

The simple, two-subdiode model is illustrated in Fig. 7.

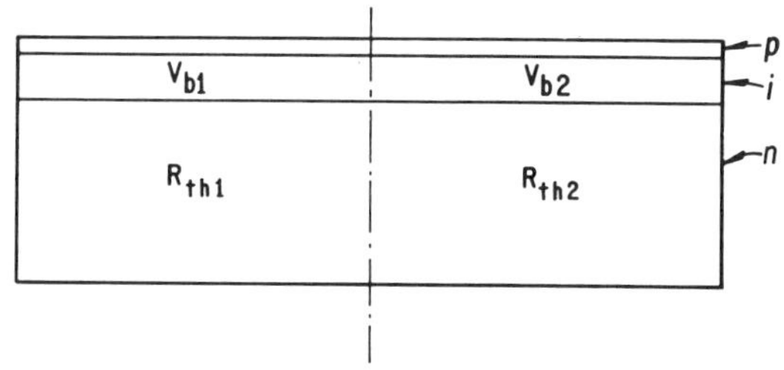

Fig. 7. Simplified two-subdiode model used to discuss the effects of time-dependent current-density variations caused by internal heating.

The two subdiodes are each characterized by two parameters V_{b1} and R_{th1} and by V_{b2} and R_{th2}, respectively. It should be noted that V_{b1} and V_{b2} denote the breakdown voltages of the two subdiodes right at the onset of avalanche breakdown, before internal heating raises the junction temperature above T_0. The increase of V_b with temperature is characterized by R_{th} and need not be treated explicitly. In the above model the only continuous variable is time t.

The dynamic properties of the entire diode can now be treated by superposition of the dynamic properties of the subdiodes taking their mutual thermal interaction into account. This thermal interaction will be considered for two extreme cases: 1) ideal thermal isolation (coupling coefficient $\theta = 0$) and 2) ideal thermal coupling ($\theta = 1$). The second case is only of interest for $V_{b1} \neq V_{b2}$.

B. Current-Density Variations

The dynamic properties of diode and subdiodes will be discussed in a sequence of six cases of increasing complexity. The conditions for these cases are listed in Table I. The conditions for the first four cases are clear. The condition $V_{b2} - V_{b1} > R_{sc} j_0$ of Case V denotes the case of substantial differences in breakdown voltage.

Case I—$V_{b1} = V_{b2}$, $R_{th1} = R_{th2}$: This case represents an ideal diode with a uniform current density throughout the entire junction area. As the calculations in Section II have shown, this case is unrealistic. However, the simplicity of this case will yield an easily understandable introduction to the dynamic problems of avalanche diodes.

The j–V characteristic of avalanche diodes is not independent of time. Only under steady-state conditions the time dependence disappears. For the transition period between onset of breakdown to steady-state conditions the diode can be represented by an infinite number of j–V characteristics, which all intersect with the voltage axis at V_b. Here V_b shall denote the breakdown voltage at the initial junction temperature T_{j0}, which is also the temperature of the heat sink. This infinite number of j–V characteristics is limited by two extreme characteristics: 1) instantaneous characteristic, and 2) steady-state characteristic, both of which are shown in Fig. 8.

TABLE I

CONDITIONS FOR THE SIX CASES FOR WHICH THE CURRENT DENSITY–TIME RELATIONSHIP IS DISCUSSED

Case	V_b	R_{th}	θ	$V_{b2}-V_{b1}$
I	$V_{b1}=V_{b2}$	$R_{th1}=R_{th2}$	0	
II	$V_{b1}=V_{b2}$	$R_{th1}\neq R_{th2}$	0	
III	$V_{b1}\neq V_{b2}$	$R_{th1}=R_{th2}$	0	$V_{b2}-V_{b1}<R_{sc}j_{10}$
IV	$V_{b1}\neq V_{b2}$	$R_{th1}=R_{th2}$	1	$V_{b2}-V_{b1}<R_{sc}j_{10}$
V	$V_{b1}\neq V_{b2}$	$R_{th1}=R_{th2}$	0	$V_{b2}-V_{b1}>R_{sc}j_{10}$
VI	$V_{b1}\neq V_{b2}$	$R_{th1}\neq R_{th2}$	$0<\theta<1$	

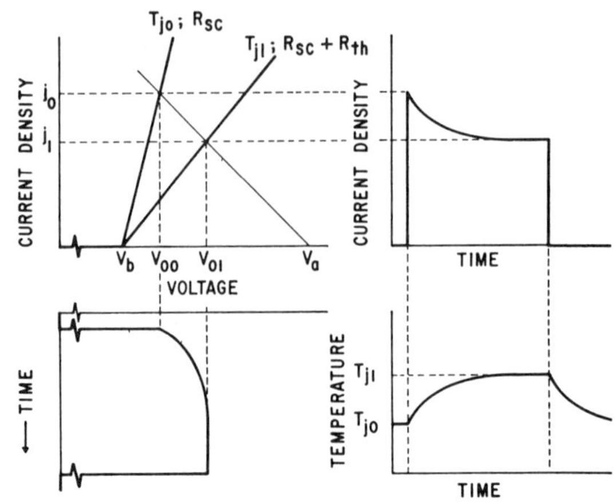

Fig. 8. Current-density–voltage characteristic and time dependence of current density, diode voltage, and junction temperature for an ideally uniform diode.

The instantaneous characteristic, which holds at constant junction temperature $T_{j0}=T_0$, describes the j–V relationship at the beginning of a pulse before internal heating starts to raise the temperature. Since it does not take internal heating into account, its slope is determined by $R_{sc}+R_{sp}\approx R_{sc}$, with R_{sp} usually being negligible.

The steady-state characteristic represents the j–V relationship at temperatures determined by internal heating. As stated by its name it is applicable after internal heating has led to a stable junction temperature, which shall be denoted by T_{j1}. The slope of the steady-state characteristic is determined by the total diode series resistance $R_s=R_{sc}+R_{sp}+R_{th}\approx R_{sc}+R_{th}$.

Load line and j–V characteristics determine the time-dependent operating point. At the beginning of a pulse the operating point is given by (V_{00}, j_0). After internal heating has led to new steady-state conditions, the operating point is given by (V_{01}, j_1). During the transition period the operating point moves along the load line from (V_{00}, j_0) to (V_{01}, j_1). The resulting voltage rise and current drop are illustrated schematically in the lower left and upper right sections of Fig. 8. The lower right section of Fig. 8 indicates the temperature rise from the initial temperature T_{j0} to the steady-state temperature T_{j1}.

At this point it is appropriate to discuss the influence of the load line on the oscillator properties. Since the oscillator frequency is strongly dependent on current density it is of advantage to bias the diode from a source with a large internal impedance. Biasing the diode from a current source would be ideal in stabilizing the diode current. However, in fast pulse circuits this condition is not realizable. It nevertheless should be tried to bias the diode from a source with an internal resistance which is large compared with the thermal resistance of the diode.

To discuss the remaining five cases we will make extensive use of j–V and j-time plots as shown in the upper section of Fig. 8. To preserve clarity we will letter only those parameters which deviate from the ideal case of Fig. 8.

Case II—$V_{b1}=V_{b2}$, $R_{th1}\neq R_{th2}$: This case represents a diode with an ideally uniform breakdown voltage but nonuniform thermal resistance R_{th1} and R_{th2}. This case, illustrated in Fig. 9, is the simplified version of the diode treated in Section II.

The instantaneous characteristic with slope R_{sc} is the same for both subdiodes and consequently coincides with the average instantaneous characteristic. The steady-state characteristic is split into two branches with slopes $R_{sc}+R_{th1}$ and $R_{sc}+R_{th2}$, respectively. The characteristic describing the $\langle j\rangle$–V relationship (heavy line) obviously lies between the two subdiode characteristics. Instantaneous and steady-state operating points are determined by load line and $\langle j\rangle$–V characteristic. The current densities of the individual subdiodes are given by the values of the subdiode characteristic at a voltage corresponding to the operating point of the average characteristic.

The j-time dependence is illustrated in the right section of Fig. 9. The current density is initially the same in both subdiodes. Unequal heating represented by different values of R_{th1} and R_{th2} lead to different values of the subdiode current densities j_1 and j_2.

The j-time characteristic of Fig. 9 represents the case of a fairly small bias source resistance. Operating the diode from a source with large internal resistance would prevent the average current density from dropping. However, it would not prevent the buildup of a difference in subdiode current density. In this case j_1 would increase above its initial value j_{10} while j_2 would decrease below j_{20}.

Case III—$V_{b1}\neq V_{b2}$, $R_{th1}=R_{th2}$, $\theta=0$: Case III and the two cases that follow represent diodes with uniform thermal conductance but nonuniform breakdown voltage. As shown in Section II, the case of uniform thermal resistance is unrealistic. However, these three cases will help to understand the complex but realistic Case VI.

In Case III it is assumed that both subdiodes are thermally isolated from each other. This assumption implies that the temperature of each subdiode is determined by its own power dissipation and is not influenced by heat flow from the other subdiode. Consequently, the slope of the steady-state characteristic is given by space charge *and* thermal resistance. The heavy lines in Fig. 10 as well as in Figs. 11 and 12 always represent the $\langle j\rangle$–V characteristics.

SECTION VII

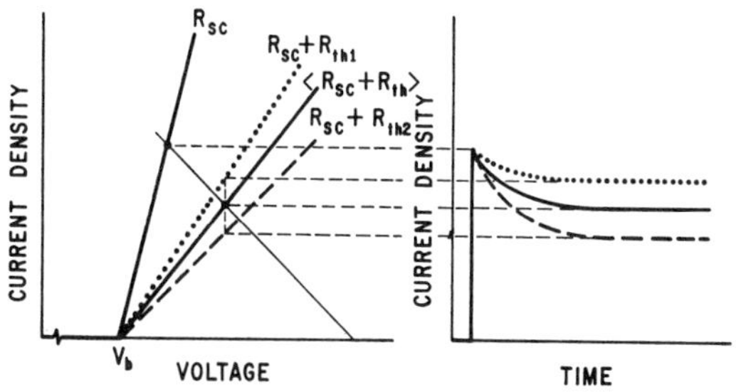

Fig. 9. j–V and j–time plots of a diode with uniform breakdown but nonuniform thermal conductance.

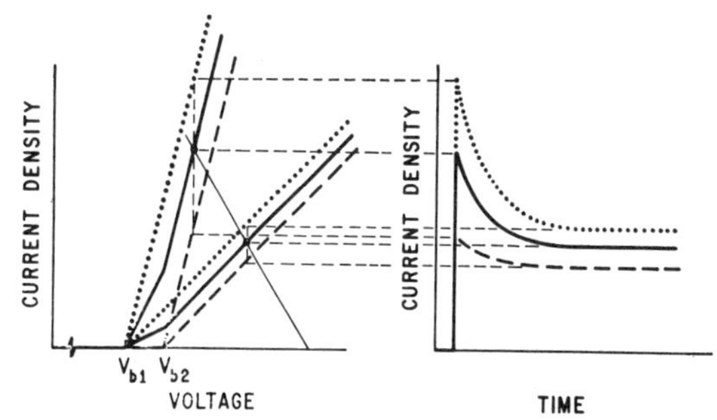

Fig. 10. j–V and j–time plots of a diode with nonuniform breakdown but uniform thermal conductance; no thermal coupling between subdiodes.

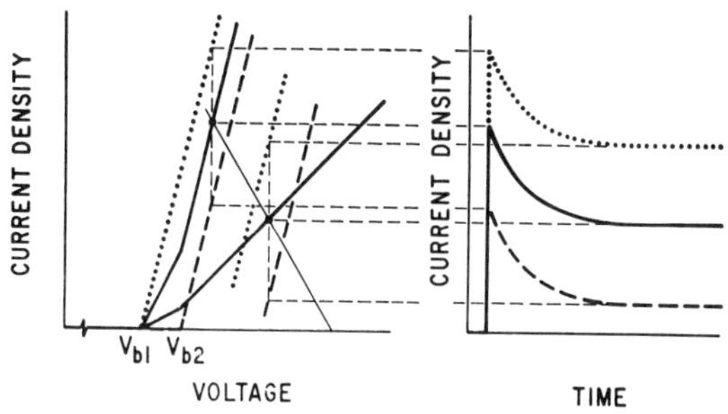

Fig. 11. j–V and j–time plots of a diode with nonuniform breakdown but uniform thermal conductance; good thermal coupling between subdiodes to achieve uniform temperature.

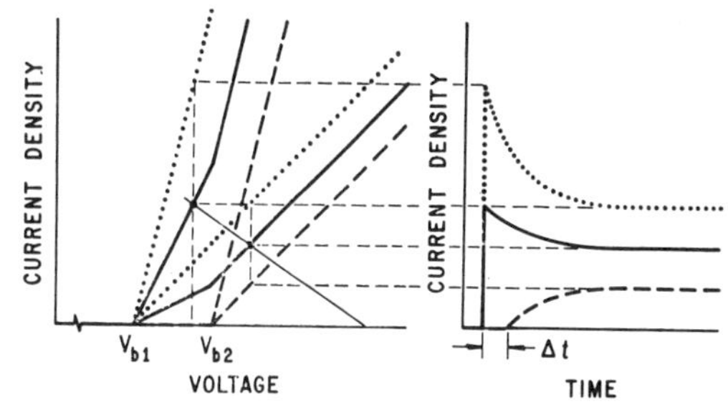

Fig. 12. j–V and j–time plots of a diode with extremely nonuniform breakdown but uniform thermal conductance; no thermal coupling between subdiodes.

In Fig. 10 instantaneous as well as steady-state characteristics are split because $V_{b1} \neq V_{b2}$. The instantaneous and steady-state operating points of both subdiodes are obtained in the same way already described for Case II. The initial current densities are different but with increasing time the absolute and relative difference between j_1 and j_2 decreases. Initially this difference is given by

$$j_{10} - j_{20} = (V_{b2} - V_{b1})/R_{sc}. \quad (14)$$

Under steady-state condition the difference is

$$j_{11} - j_{21} = (V_{b2} - V_{b1})/(R_{sc} + R_{th}) \quad (15)$$

where $R_{th} = R_{th1} = R_{th2}$. From (14) and (15) it is evident that the thermal resistance helps to offset current-density variations resulting from breakdown nonuniformities.

Case IV—$V_{b1} \neq Vb_2$, $R_{th1} = R_{th2}$, $\theta = 1$; Case IV represents a diode similar to Case III but with good thermal coupling between the two subdiodes. Such a coupling implies that both subdiodes will be at the same temperature even if their individual power dissipation density is different.

The essential difference from Case III is in the steady-state characteristics (Fig. 11). The slope of the steady-state $\langle j \rangle$–V characteristic is given by $R_{sp} + R_{th}$. This characteristic determines the average operating point. In contrast to Case III, the operating points of the subdiodes are here determined by characteristics of slope R_{sc}. Because of the uniform junction temperature the initial current-density difference is maintained inspite of the different power dissipation density:

$$j_{10} - j_{20} = j_{11} - j_{21} = (V_{b2} - V_{b1})/R_{sc}. \quad (16)$$

The thermal resistance does not help to reduce current-density differences.

Case V—$V_{b1} \neq V_{b2}$, $R_{th1} = R_{th2}$, $V_{b2} - V_{b1} > R_{sc} j_{10}$: Case V represents a case with substantially nonuniform breakdown as may be found in planar diodes where the junction curvature at the diode edge causes a considerable reduction of the breakdown voltage (Fig. 12). Thermal isolation between the two subdiodes is assumed, which makes this case similar to Case III.

The essential difference from the previous cases arises from the premise $V_{b2} - V_{b1} > R_{sc} j_{10}$ which states that the voltage at the initial operating point, $V_{b1} + R_{sc} j_{10}$, is smaller than V_{b2}. Under such a condition the initial current is carried entirely by the first subdiode. Internal heating of this subdiode raises the voltage above V_{b2}, thus initiating breakdown in the second subdiode at a time Δt after onset of breakdown in the first subdiode.

Case VI—$V_{b1} \neq V_{b2}$, $R_{th1} \neq R_{th2}$—Case VI represents the most general case of nonuniform breakdown and nonuniform heating with light thermal coupling between the subdiodes ($0 < \theta < 1$). The $j - V$ characteristics can be described by a combination of $j - V$ characteristics of the preceding cases. The instantaneous characteristics will be similar to Case V. The steady-state characteristics will have different slopes because $R_{th1} \neq R_{th2}$ and they will not intersect at V_{b1} and V_{b2} because $\theta > 0$ (Case IV).

The current-density time dependence will be similar to Case V with $j_{10} \neq j_{20}$, $j_{11} \neq j_{21}$, and $j_{10} - j_{20} \neq j_{11} - j_{21}$. For a sufficiently large difference $V_{b2} - V_{b1}$ conduction in the second subdiode will be delayed and j_2 will increase with time while j_1 decreases.

C. *Effects on Oscillator Properties*

The current-density variations discussed in the last subsection will affect the properties of pulsed avalanche oscillators to a much larger extent than the variations discussed in Section II affected the properties of CW oscillators. In Section II it was shown that a steady-state but nonuniform current-density distribution resulting from internal heating will reduce the efficiency and increase the oscillator noise. In the case of pulsed oscillators the time dependence of the current density will introduce additional unwanted effects.

1) *Amplitude and Frequency Modulation:* From investigations of CW oscillators it is known that oscillator amplitude and frequency are sensitive functions of the current density. Consequently, every variation of current density with time will cause corresponding amplitude and frequency modulations.

The variations of current density with time generating these modulations can be contributed to two separate effects which are both caused by internal heating:

1) reduction of the average current density,
2) internal redistribution of current.

The first effect arises from the overall temperature rise within the diode between initial and steady-state conditions. The second effect is caused by nonuniform internal heating.

Measures to minimize both effects will first be discussed for diodes with uniform breakdown voltage but nonuniform thermal resistance (Case II). The overall current-density variation can be minimized by driving the diode from a bias source of sufficiently large internal resistance, thus stabilizing the diode current. However, as mentioned earlier, this stabilization method is difficult to apply in the case of fast pulse circuits because of large RC time constants. This current-density variation can also be reduced by minimizing the thermal resistance (flip-chip mounting).

The internal redistribution of current cannot be reduced by external circuit means. Only the local thermal resistance controls this redistribution. Flip-chip mounting obviously offers the best solution.

538

The internal current redistribution is a very serious threat to the application of pulsed avalanche oscillators. Since the steady-state current distribution is nonuniform due to a basic thermal conductivity problem there is also no solution to eliminate the time dependence of the current between initial uniform and final nonuniform conductance.

Nonuniform breakdown voltage complicates the situation even more. A large thermal resistance together with a good thermal isolation between subdiodes will tend to offset the current-density variations caused by nonuniform breakdown (Case III). However, a large thermal resistance will lead to a large overall reduction in diode current (see above). Because a low thermal resistance has other considerable advantages such as larger possible power dissipation, it is thought that flip-chip mounting is of advantage even for diodes with nonuniform breakdown.

Similarly to the case of frequency-modulation noise discussed in Section II-D, the frequency modulation resulting from current variations can be reduced by increasing Q_{ext} (undercoupling of the oscillator). However, such a reduction in FM modulation is usually accompanied by a considerable increase in AM modulation.

In case of a large thermal resistance and a small internal resistance of the bias source the current decrease can be substantial, very often causing the oscillator to jump into different modes. Similarly, a substantial internal current redistribution can also cause such a jumping. A special case of such jumping modes is the often observed delayed start of microwave oscillation which will be discussed in the next subsection.

2) *Delayed Start of Oscillations:* The delayed start of pulsed avalanche oscillations has been repeatedly observed since the early days of avalanche oscillators.[20]–[22] Such a delay poses serious difficulties for the application of pulsed oscillators. In this subsection we will try to explain this effect in terms of a model based on the dynamic thermal properties discussed earlier in this section.

There are two different mechanisms which can cause such delays. They are most readily distinguished by their dependence of delay time on drive current. In one case delay time decreases with increasing drive current while in the other case it increases. In most diodes both effects are present simultaneously. The two effects are illustrated on two diodes in which either one of the effects dominates.

The first of the two effects causing a delayed start of microwave oscillations is illustrated in Fig. 13 for a diode with $V_b = 120$ volts. Diode current and microwave power output at X band are plotted as a function of time for six different levels of applied bias. The current decreases with time, as expected from internal heating. The start of microwave oscillations is delayed with respect to the start of the current pulse. The delay time, which is of the order of 10^{-6} seconds, decreases with increasing drive current. The broken line indicates the starting time on the current versus time plot. The increase in diode current observed right after the onset of oscillations is thought to be caused by the additional cooling mechanism presented by the microwave power generation and propagation.

With the models discussed in Section III-B in mind one is tempted to explain the delayed start of microwave oscillations as a thermal effect. Indeed, a diode as discussed in Case V (Fig. 12) seems especially suitable for such an explanation in terms of the simple subdiode model. Suppose we have a current-density distribution as illustrated schematically in Fig. 12. The first sub-

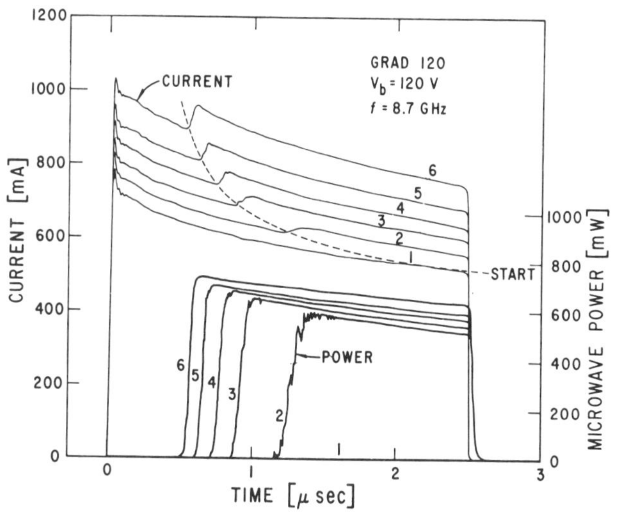

Fig. 13. Delayed start of microwave oscillations with respect to beginning of current pulse. Delay time decreases with increasing drive current.

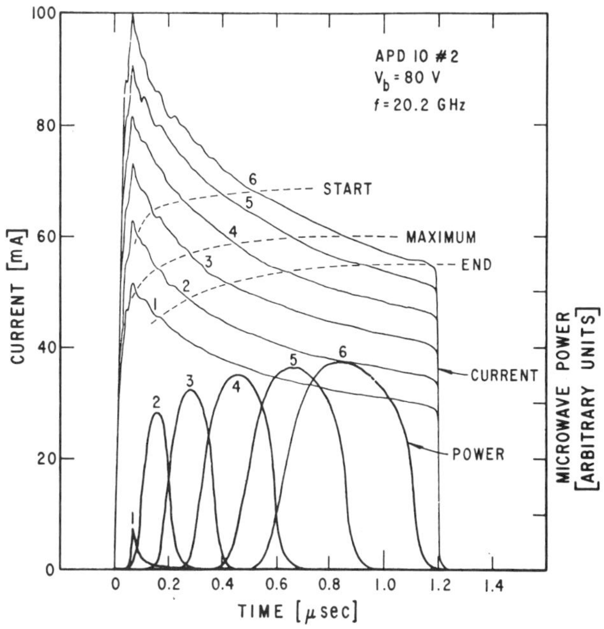

Fig. 14. Delayed start of microwave oscillations with respect to beginning of current pulse. Delay time increases with increasing drive current.

diode shall carry a current density which would be sufficient to drive it into oscillations without the presence of the second subdiode. At the beginning of the pulse the second subdiode is not conducting and acts as a capacitive load to the first subdiode. After conduction starts in the second subdiode it presents an additional conductive load. With increasing current density the conductance of the second subdiode will decrease and finally become negative. Somewhere during this transition the gain in the first subdiode will exceed the losses to second subdiode and circuit and the diode will start oscillating.

This model also predicts the proper dependence of delay time on drive current. With increasing drive current the rate of temperature rise increases and conduction in the second subdiode starts earlier. Consequently, the delay time is expected to decrease with increasing drive current. This conclusion agrees with the observations of Fig. 13.

The second mechanism, which can account for delayed oscillations, is illustrated in Fig. 14. Diode current and output power are plotted as a function of time. In this set of data the delay-time increases with drive current. Furthermore, the oscillations cease after the current drops below a certain sustaining current. Both the delayed start and the interruption of the oscillations can be explained by a strong current dependence of the negative resistance for this particular oscillation mode. The diode will oscillate in this mode only within a small current interval indicated by the broken lines labeled "start" and "end" in Fig. 14. For the mechanism just described one expects that delay-time and oscillation duration increase with drive current. This prediction agrees with the experimental results of Fig. 14.

One expects that the position of the "start" and "end" lines depends only on diode and circuit constants. However, since it depends also on time (shift towards higher currents in Fig. 14), one concludes that internal current redistributions (first mechanism discussed above) are also present.

From the above discussion one can conclude that there are two mechanisms which may account for the often observed delayed start of microwave oscillations under pulsed conditions. The first mechanism is based on a redistribution of the avalanche current within the diode caused by internal heating. The second mechanism is based on the current dependence of the negative resistance, for which the current variation is also caused by internal heating. In most diodes both mechanisms occur simultaneously, making an accurate analysis quite complex.

It should be noted that the second mechanism can be reduced considerably or even circumvented completely by means of biasing the diode from a high impedance source thus stabilizing the diode current. The first mechanism, however, cannot be circumvented at all by external means since it is based on an internal current redistribution.

Both mechanisms together may also be used to explain the observation that pulsed avalanche diodes jump into another oscillation mode. This effect could also be described as delayed starts of the second, third, etc., oscillation modes.

539

The planar diode represents a special case of nonuniform breakdown. Due to the cylindrical field distribution at the edge of the space charge layer the breakdown voltage is reduced considerably below the breakdown voltage of the center region.[11] Thus at the beginning of a pulse the entire current is conducted by the edge region. Rapid internal heating of this edge region increases the breakdown voltage to the breakdown voltage of the center region. Only then will the diodes start oscillating. From the earlier discussions it is also obvious that such diodes will operate at a considerably reduced efficiency. Taking further into account that planar diodes cannot be flip-chip mounted, one arrives at the conclusion that as microwave oscillators planar diodes are far inferior to mesa diodes.

Parallel operation of several avalanche diodes can also be described by the subdiode model. If breakdown voltage and thermal resistance of these diodes are not very accurately matched, a situation resembling that of Case VI arises leading to delayed oscillation start, mode jumping, AM and FM modulation, and reduced efficiency. Applying the calculations of Section II to the subdiode model, it is estimated that breakdown voltage has to be matched within the order of 1 percent and thermal resistance within the order of 10 percent.

IV. Conclusions

The discussion of nonuniform thermal conductance and its effects on the properties of avalanche microwave oscillators has led to a series of interesting conclusions. All of these conclusions are directly related to the fact that optimum oscillator performance can be achieved only for a uniform distribution of the avalanche current. Two parameters affect the local current density: local breakdown voltage and local temperature.

The calculations in Section II have shown that a nonuniform temperature resulting from a nonuniform thermal resistance will lead to significant variations of the local current density. To minimize this current-density variation one has to reduce the overall thermal resistance by flip-chip mounting the avalanche diodes onto substrates of good thermal conductance such as copper or, even better, diamond.[13]

It is interesting to note that flip-chip mounting is already a widely accepted technique in mounting avalanche diodes. Fip-chip mounting reduces the overall thermal resistance and, therefore, was applied to increase the power dissipation limit of avalanche diodes. As discussed in Section II, it also helps to improve efficiency and oscillator noise.

The flip-chip mounted mesa structure seems to be the ultimate oscillator design. Such a design may appear to be a step backwards from planar and guard ring designs, but guard ring or planar diodes cannot be flip-chip mounted without significantly increasing the diode capacitance. Furthermore, the reduced breakdown voltage at the edge of planar diodes really prohibits a planar design.

Low thermal resistance is also of great advantage in pulsed operation since it helps to reduce 1) the overall current variation during the transition period and 2) the internal redistribution of current caused by nonuniform heating. These two effects cause AM and FM modulation of the oscillator and, together with breakdown nonuniformities, are responsible for the often observed delayed start of microwave oscillations.

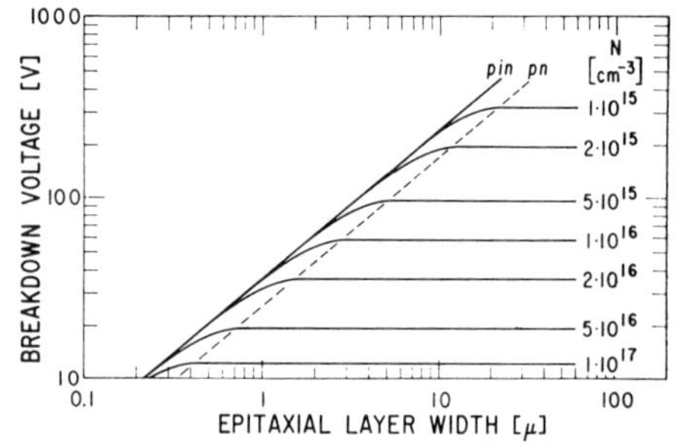

Fig. 15. Breakdown voltage of p^+-n^--n^+ junctions as a function of epitaxial layer width W, for various doping concentrations of the n^- region.

Appendix

The dependence of breakdown voltage on impurity concentration has been calculated for abrupt as well as for linearly graded junctions by Sze and Gibbons.[23] For avalanche microwave oscillators the case of p-i-n junctions is of greatest interest. In this Appendix we will present the results of similar numerical calculations on p-i-n diodes.

The calculations are carried out for a p^+-n^--n^+ junction with the metallurgical p^+-n^- junction located at $x=0$ and the n^--n^+ junction located at $x=W_1$. The case of an abrupt p^+-n^- junction treated by Sze and Gibbons[23] is included in these calculations as that limiting case for which $W \leq W_1$ with W denoting the width of the depletion layer at breakdown. The condition for avalanche breakdown is

$$\int_0^W \alpha_n \exp\left[-\int_0^x (\alpha_n - \alpha_p)dx'\right]dx = 1, \qquad (17)$$

with α_n and α_p denoting the ionization coefficients of electrons and holes, respectively. For silicon, α_n and α_p are approximated by equations of the form

$$\alpha = A \exp(-b/E) \qquad (18)$$

with $A_n = 2.6 \times 10^6$ cm^{-1} and $b_n = 1.65 \times 10^6$ V/cm for electrons, and $A_p = 1.5 \times 10^7$ cm^{-1} and $b_p = 3.13 \times 10^6$ V/cm for holes. The field is a function of voltage V, impurity concentration of the n^- layer N, width W, and position x:

$$E(x, V, N, W) = \frac{V}{W} + \frac{qN}{2\epsilon}(W - 2x). \qquad (19)$$

Inserting (18) and (19) into (17) and solving the resulting integral equation numerically, one obtains the relation between V_b and W_1 plotted in Fig. 15 for various values of N. The dashed line agrees with the earlier calculation of Sze and Gibbons for the abrupt p^+-n junction.

The two limiting cases, p-i-n and p^+-n junctions, display slopes of their V_b–W characteristic which are smaller than 1. This dependence implies that the maximum junction field decreases with increasing breakdown

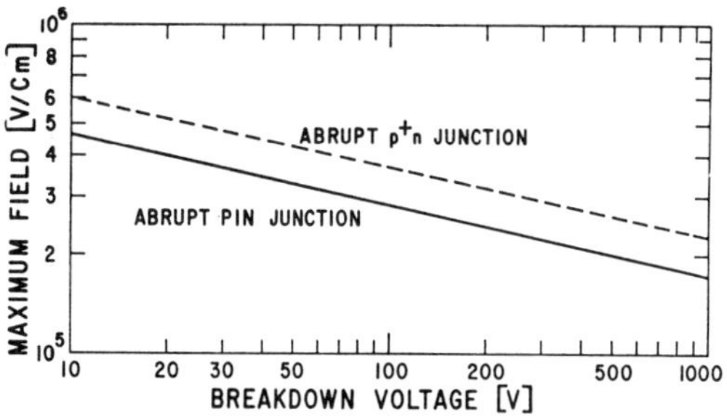

Fig. 16. Variation of maximum field in the depletion layer of p-i-n or p^+-n junctions as a function of breakdown voltage.

voltage. For p-i-n junctions one finds $E_{max}=\langle E\rangle = V_b/W$, for p^+-n junctions one finds $E_{max}=2\langle E\rangle = 2V_b/W$. The relation between E_{max} and V_b is plotted in Fig. 16 for the two limiting cases.

Acknowledgment

The author is indebted to Dr. H. R. Plumlee for the numerical calculation of the temperature distribution of Fig. 5, and to D. R. Powell for the computer solution of (17).

References

[1] W. T. Read, "A proposed high-frequency negative-resistance diode," *Bell Sys. Tech. J.*, vol. 37, pp. 401–446, March 1958.

[2] T. Misawa "Negative resistance in p-n junctions under avalanche breakdown conditions," Pts. I and II, *IEEE Trans. Electron Devices*, vol. ED-13, pp. 137–151, January 1966.

[3] M. Gilden and M. E. Hines, "Electronic tuning effects in the Read microwave avalanche diode," *IEEE Trans. Electron Devices*, vol. ED-13, pp. 169–175, January 1966.

[4] H. K. Gummel and D. L. Scharfetter, "Avalanche region of Impatt diodes," *Bell Sys. Tech. J.*, vol. 45, pp. 1797–1827, December 1966.

[5] S. T. Fisher, "Small signal impedance of avalanching junctions with unequal electron and hole ionization rates and drift velocities," *IEEE Trans. Electron Devices*, vol. ED-14, pp. 313–322, June 1967.

[6] A short version of this paper, entitled "Thermal and noise problems in avalanche microwave oscillators," was presented at the 1967 Solid-State Device Research Conference, Santa Barbara, Calif., June 20, 1967.

[7] R. L. Batdorf, A. G. Chynoweth, G. C. Dacey, and P. W. Foy, "Uniform silicon p-n junctions. I. Broad area breakdown," *J. Appl. Phys.*, vol. 31, pp. 1153–1160, July 1960.

[8] A. Goetzberger, B. McDonald, R. H. Haitz, and R. M. Scarlett, "Avalanche effects in silicon p-n junctions. II. Structurally perfect junctions," *J. Appl. Phys.*, vol. 34, pp. 1591–1600, June 1963.

[9] W. Shockley, "Problems related to p-n junctions in silicon," *Solid-State Electronics*, vol. 2, pp. 35–67, January 1961.

[10] R. J. McIntyre, "Theory of microplasma instability in silicon," *J. Appl. Phys.*, vol. 32, pp. 983–995, June 1961.

[11] K. H. Zschauer, "Einfluss von Randkrümmungen auf die Durchbruchsspannung von Silicium-Dioden," *Z. Naturforsch.*, vol. 19a, pp. 653–655, 1964.

[12] A. C. Prior, "The field dependence of carrier mobilities in silicon and germanium," *J. Phys. Chem. Solids*, vol. 12, pp. 175–180, January 1960.

[13] C. B. Swan, "The importance of providing a good heat sink for avalanching transit time oscillator diodes," *Proc. IEEE*, vol. 55, pp. 451–452, March 1967.

[14] T. Misawa, "Microwave silicon avalanche diodes with nearly-abrupt-type junctions," presented at the Internat'l Electron Device Meeting, October 1966.

[15] R. L. Johnston and J. G. Josenhans, "Improved performance of microwave Read diodes," *Proc. IEEE*, vol. 54, pp. 412–413, March 1966.

[16] R. H. Haitz, "Noise in avalanche diodes," presented at the 1966 Solid State Device Research Conference, Evanston, Ill., June 1966; *IEEE Trans. Electron Devices*, vol. ED-13, p. 671A, August 1966; "Frequency modulation noise in avalanche and Gunn oscillators," presented at ICAMEBS, New York, N. Y., January 1968.

[17] ——, "Noise of a self-sustaining avalanche discharge in silicon: Low-frequency noise studies," *J. Appl. Phys.*, vol. 38, pp. 2935–2946, June 1967.

[18] R. H. Haitz and F. W. Voltmer, "Noise studies in uniform avalanche diodes," *Appl. Phys. Letters*, vol. 9, pp. 381–383, November 1966.

[19] R. H. Haitz, "Studies of the uniformity of avalanche breakdown by means of low-frequency noise measurements," *Phys. Failure in Electronics*, vol. 5, pp. 447–456, June 1967.

[20] L. S. Napoli and R. J. Ikola, "An avalanching silicon diode microwave amplifier," *Proc. IEEE*, vol. 53, pp. 1231–1232, August 1965.

[21] F. A. Brand, V. J. Higgins, J. J. Baranowski, and M. A. Druesne, "Microwave generation from avalanching varactor diodes," *Proc. IEEE*, vol. 53, pp. 1276–1277, September 1965.

[22] F. K. Manasse and J. S. Shapiro, "Current dependent modes in pulsed avalanche diodes," *Proc. IEEE*, vol. 55, pp. 702–703, May 1967.

[23] S. M. Sze and G. Gibbons, "Avalanche breakdown voltages of abrupt and linearly graded p-n junctions in Ge, Si, GaAs and GaP," *Appl. Phys. Letters*, vol. 8, pp. 111–113, March 1966.

OTHER ARTICLES OF INTEREST

A Method for Heat Flow Resistance Measurements in Avalanche Diodes

ROLAND H. HAITZ, MEMBER, IEEE, HARRY L. STOVER, AND NEAL J. TOLAR

Abstract—Avalanche transit time oscillators are operating at power densities approaching 10^6 W/cm², unprecedented in semiconductor device history. At such power densities, heat flow resistance problems at the interface between the flip-chip mounted silicon chips and the metal substrate, as well as between the package and the heat sink, are extremely critical. This paper describes a new, nondestructive and accurate method of measuring the heat flow resistance between junction and heat sink by utilizing the temperature dependent breakdown voltage $V_b(T)$ as a conveniently built-in temperature sensor. Variations in junction temperature ΔT with power $\Delta P = V_b \, \Delta I$ are, therefore, related to variations in breakdown voltage ΔV_b with current ΔI resulting in a contribution to the electrical small signal resistance of the diode. This thermal resistance contribution R_{th} can be separated readily from spreading and space charge resistance R_{sp} and R_{sc} because of the frequency dependence of $R_{th}(\omega)$. Furthermore, the frequency dependence of $R_{th}(\omega)$ allows the separation of heat flow resistance contributions originating in the immediate vicinity of the junction (Si–metal interface) from contributions originating at a poor thermal contact between package and heat sink. In keeping with calculations on simplified geometrical configurations, for which analytical solutions of the frequency dependent heat flow in a distributed circuit could be obtained, experimental results are presented which indicate that both heat flow resistance contributions can be extracted and separated with sufficient accuracy from as few as three electrical resistance measurements, e.g., at dc, 100 Hz, and 1 MHz. The simplicity of such measurements and their evaluation make this technique ideal for in-line testing of production devices.

I. Introduction

THE LATEST generation of avalanche transit-time oscillators is operating at unprecedented power densities of nearly 10^6 W/cm². At such extreme power densities the heat flow between junction and package and between package and heat sink is one of the most crucial problems in the design of avalanche microwave oscillators.

Several techniques presently used in heat flow resistance measurements are either time consuming, inaccurate, or require open diode packages. Most accurate are temperature measurements with an infrared radiometer or a thermocouple. Both techniques are impractical for routine measurements of production devices because they require open packages and either the attachment of a thermocouple to a silicon chip of 50–200 μ in diameter or the coating of such a small chip with a paint of known emissivity. A third technique which utilizes the temperature dependence of the forward characteristic as a means to measure the junction temperature is fairly inaccurate because the spatial distribution of the power dissipation in the forward direction is different from the reverse or avalanche case. Furthermore, in this technique the temperature is commonly measured by continuously switching the diode from the power dissipation state into the temperature measuring state. The resulting large transients prevent temperature measurements directly after switching, and the required oscilloscope readout is not very accurate.

The technique discussed in this paper utilizes the temperature dependence of the reverse characteristic as a means to measure the junction temperature. The temperature dependent breakdown voltage $V_b(T)$ presents a conveniently built-in temperature sensor with a sensitivity of the order of 10–100 mV/°C. Since power dissipation and temperature measurement coincide in space and time, this technique is basically very accurate.[1] Details of this technique are discussed in Section II.

II. Measurement Technique

There are two ways of measuring the heat flow resistance of an avalanche diode by means of the temperature dependence of V_b. Both methods are discussed with reference to Fig. 1.

The first method is based on a measurement of the incremental current-voltage characteristic. The heavy line of Fig. 1 represents a schematic I–V characteristic of an avalanche diode intersecting the voltage axis at $V_b(T_0)$. The finite slope of this characteristic is to a large extent caused by the increase in V_b with junction temperature T_j to values above the sink temperature T_0. For a given operating point (I, V) the temperature increase caused by internal heating is given by

$$T_j - T_0 = rIV \approx rIV_b(T_j) \tag{1}$$

with r denoting the heat flow resistance in °C/W between junction and sink. The temperature difference T_j–T_0 can be calculated from the well-known relation

$$V_b(T_j) = V_b(T_0)[1 + \beta(T_j - T_0)] \tag{2}$$

where we have substituted $V \approx V_b(T_j)$. The constant β is the temperature coefficient of V_b and is defined by

Manuscript received September 19, 1968; revised December 16, 1968. This work was supported in part by the Air Force Avionics Laboratory, Wright-Patterson Air Force Base under Contract F-33-615-67-C-1909.

R. H. Haitz and N. J. Tolar are with Texas Instruments Incorporated, Dallas, Tex.

H. L. Stover was with Texas Instruments Incorporated, Dallas, Tex. He is now with the University of Southern California, Los Angeles, Calif.

[1] This technique for heat flow resistance measurements was first reported by R. H. Haitz at the 1968 Solid State Device Research Conference, Boulder, Colo.

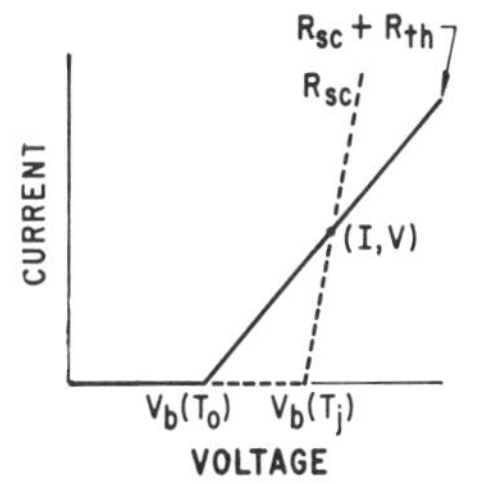

Fig. 1. Current–voltage characteristic of an avalanche diode at dc (heavy line) and high frequencies (≥ 1 MHz, broken line). The slope of the dc characteristic is determined by $R_{sc}+R_{th}$ while the slope of the high frequency ac characteristic is determined by R_{sc} only.

$\beta=(1/V_b)(dV_b/dT)$. By simplifying $V_b(T_j)\approx V_b(T_0)$ $\approx V_b$ in all terms which do not contain the difference $V_b(T_j)-V_b(T_0)$, one finds

$$r \approx [V_b(T_j) - V_b(T_0)]/\beta I V_b^2. \tag{3}$$

All quantities in (3) can be measured readily or can be obtained from voltage regulator data sheets. Consequently, (3) can be used to estimate r from measurements of the dc I–V characteristic. Since the I–V characteristic is linear for uniform breakdown and small internal temperature rises ($<50°$C), it is sufficient to measure the characteristic at two points.

The second and more accurate method is based on the fact that the temperature dependence of V_b results in a contribution to the small signal ac resistance of avalanche diodes. This contribution, called the thermal resistance R_{th}, can be derived by inserting (1) into (2), solving (2) for $V_b(T_j)$ and finally differentiating $V_b(T_j)$ with respect to I. With the simplification $V_b(T_0)=V_b$ and by neglecting higher order and mixed terms of $r\beta V_b I \ll 1$ and $R_{sc}I/V_b \ll 1$, one obtains the following expression for R_{th}:

$$R_{th} = r\beta V_b^2(1 + 2r\beta V_b I + 2R_{sc}I/V_b). \tag{4}$$

In most cases the approximation

$$R_{th} \approx r\beta V_b^2 \tag{5}$$

derived by McIntyre [1] for microplasma breakdown is sufficiently accurate.

At this point it should be mentioned that throughout this paper we will use the expression *thermal resistance* for the contribution to the electrical small signal ac resistance given by R_{th} of (4). The ratio $r=(T_j-T_0)/V_b I$ will be called *heat flow resistance*.

There are two additional contributions to the total small signal series resistance R_s of an avalanche diode: the spreading resistance R_{sp} and the space charge resistance R_{sc}. In avalanche microwave oscillators R_{sp} is usually negligible. For frequencies $\omega \ll \omega_a$, with ω_a denoting the avalanche frequency [2], R_{sc} is given by Shockley [3] as

$$R_{sc} = (1/2\epsilon v_d)(l_d^2/A). \tag{6}$$

Here ϵ is the dielectric constant, v_d the maximum drift velocity of electrons, l_d the drift length, and A the breakdown area.

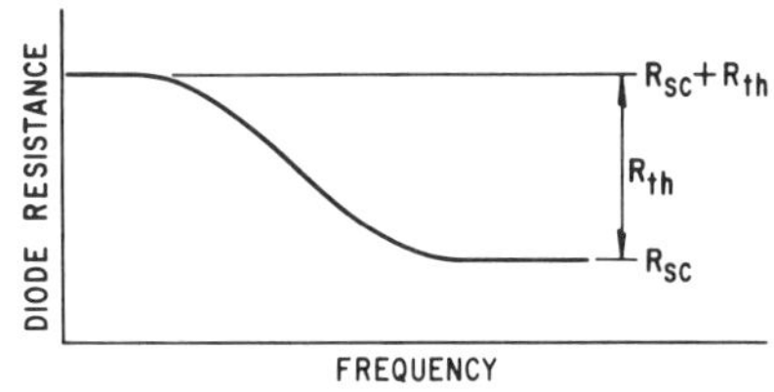

Fig. 2. Schematic dependence of total diode resistance R_{sc} on frequency. The frequency dependence of the thermal resistance R_{th} (ω) yields a convenient technique to separate $R_{th}(\omega)$ from R_{sc} by means of two measurements at low and high frequencies.

The measurement of the heat flow resistance r is now reduced to the measurement of resistance R_{th}, breakdown voltage V_b, and temperature coefficient β. For most avalanche oscillators V_b is known and β can be estimated readily from Section IV, thus reducing the heat flow resistance measurement to a thermal resistance measurement.

In order to measure the thermal resistance R_{th}, one has to separate the contributions to the total diode resistance R_s.[2] The separation is relatively easy because R_{th} is strongly frequency dependent while R_{sp} and R_{sc} are frequency independent for $\omega \ll \omega_a$. Fig. 1 illustrates this effect. At dc or at frequencies below 100 Hz the slope of the I–V characteristic is determined by R_{sc} and R_{th} (heavy line). R_{sp} is neglected. At sufficiently high frequencies the slope at the operating point (I, V) is determined by R_{sc} alone and the characteristic extrapolates to $V_b(T_j)$. This frequency dependence of the diode resistance is shown schematically in Fig. 2. The thermal resistance is simply obtained as the difference between the low and high frequency values of R_s [4].

III. Frequency Dependence of $R_{th}(\omega)$

In order to use thermal resistance measurements as a convenient and accurate analytical tool for heat flow resistance studies, it is necessary to discuss the frequency dependence of $R_{th}(\omega)$ and, consequently, of $r(\omega)$.

Accurate calculations of $r(\omega)$ can be carried out for a simplified geometrical configuration shown in Fig. 3(a) consisting of a heat source of strength i located at the end of a slab of material of constant cross section A, length L, thermal conductivity λ, specific heat c^*, and density ρ. The other end is connected to a heat sink of temperature T_0. The frequency dependence of the heat flow resistance $r(\omega)$ can be calculated readily from the distributed heat flow circuit model of Fig. 3(b). The distributed circuit elements are heat flow resistance per unit length $r'=1/\lambda A$ and heat capacitance per unit length $c'=c^*\rho A$, respectively. The solution of the resulting heat flow equations is found in standard text-

[2] It should be pointed out that it is more convenient and more accurate to determine R_{sc} experimentally rather than calculating it from (6). As pointed out in an earlier publication [4], (6) overestimates R_{sc} for p-i-n and p^+-n-n^+ diodes.

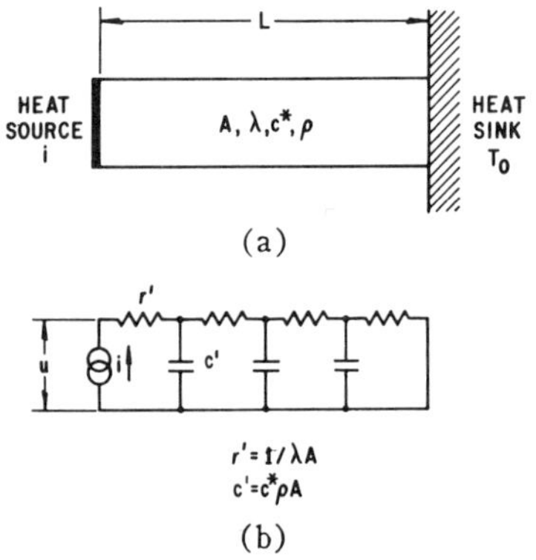

Fig. 3. Idealized sample geometry (a) to calculate frequency-dependent heat flow impedance by means of a distributed heat flow resistance–capacitance circuit (b).

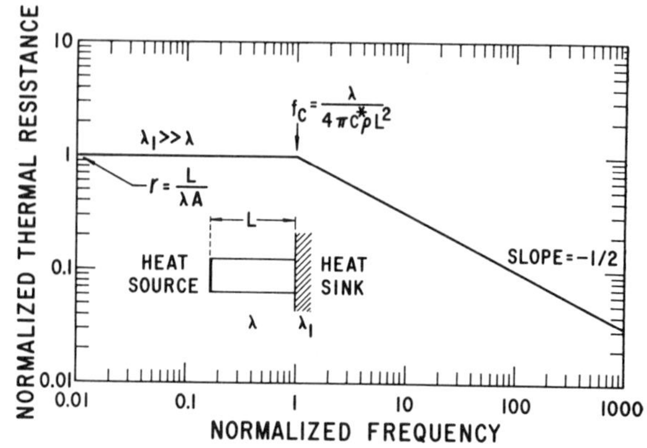

Fig. 4. Normalized thermal resistance versus frequency dependence for the idealized sample geometry of Fig. 3.

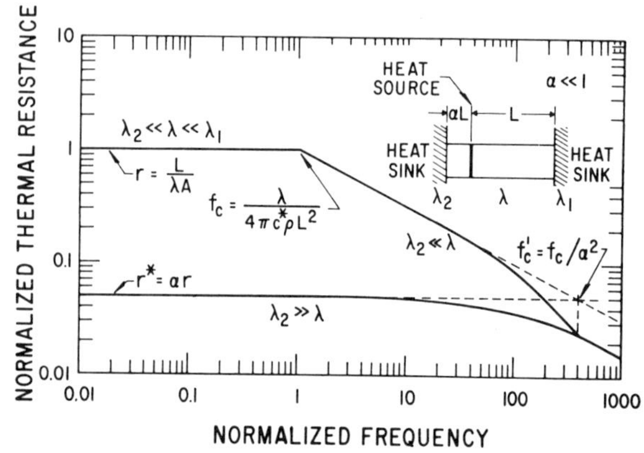

Fig. 5. Normalized thermal resistance versus frequency dependence for a more realistic sample geometry shown in the insert.

books [5] and conveniently expressed by the complex heat flow impedance z given by

$$z = \sqrt{\frac{r'}{2\omega c'}} \cdot \frac{\sinh 2a + \sin 2a + j(-\sinh 2a + \sin 2a)}{\cosh 2a + \cos 2a} \qquad (7)$$

with

$$a = \sqrt{\omega c r/2} = L\sqrt{\omega c^* \rho / 2\lambda}. \qquad (8)$$

Here $r = L/\lambda A$ and $c = c^* LA$ denote total heat flow resistance and heat capacity of the slab of material shown in Fig. 3(a).

Of particular interest are the solutions at low and high frequencies with $2a \ll 1$, and $2a \gg 1$, respectively. From (7) one obtains

$$z = r(1 - j\omega/6\omega_c) \qquad \text{for } 2a \ll 1 \qquad (9)$$

and

$$z = \sqrt{r'/2\omega c'}\,(1 - j) \qquad \text{for } 2a \gg 1. \qquad (10)$$

The quantity ω_c is the cutoff frequency and is defined by $2a = 1$:

$$\omega_c = \lambda / 2c^* \rho L^2. \qquad (11)$$

The cutoff frequency is independent of cross section A.

For $\omega \ll \omega_c$ the imaginary term in (9) disappears and $z = r$. For $\omega \gg \omega_c$ real and imaginary terms of (10) are equal causing a 45° phase delay between temperature and power dissipation.

The heat flow resistance for a more complex but more realistic structure can be estimated from the solution for the simplified structure given in (9) and (10). The real part of the two asymptotic solutions for $\omega \ll \omega_c$ and $\omega \gg \omega_c$ of (9) and (10) are plotted in Fig. 4. The more complex structure is illustrated in the insert of Fig. 5. In this case the heat source is located a distance L from one heat sink of thermal conductivity λ_1 and a distance αL from a second heat sink of conductivity λ_2. Since λ_1 and λ_2 have finite values which are not very large compared with λ, it is of little use to seek an accurate solution for $z(\omega)$ for this case. The estimated solution for this structure is shown in Fig. 5. The upper curve for $\lambda_2 \ll \lambda$, $\lambda_1 \gg \lambda$ is identical to the solution in Fig. 4 up to frequencies approaching the cutoff frequency determined by the length αL. This cutoff frequency f_{c2} is related to f_{c1} by

$$f_{c2} = f_{c1}/\alpha^2. \qquad (12)$$

Beyond f_{c2} the heat flow resistance is half the value of Fig. 4. The lower curve represents the case $\lambda_2 \gg \lambda$. This solution is fairly accurate if the spreading resistance in the heat sink $r_{sp2} = 1/2\lambda_2 D$ is small compared with $r_2 = \alpha L/\lambda A$. Usually $r_{sp2} \geqq r_2$ leading to a heat flow resistance versus frequency curve in between the two curves of Fig. 5.

From the above discussion of the frequency dependence, one expects that the heat flow resistance of avalanche diodes is frequency independent at low frequencies and decreases approximately with $1/\sqrt{\omega}$ at sufficiently high frequencies.

IV. Dependence of β on Impurity Profile

The temperature coefficient β is not only a function of V_b but also a function of the field distribution within the depletion layer and, therefore, a function of the impurity profile.

The impurity profile with the best known temperature

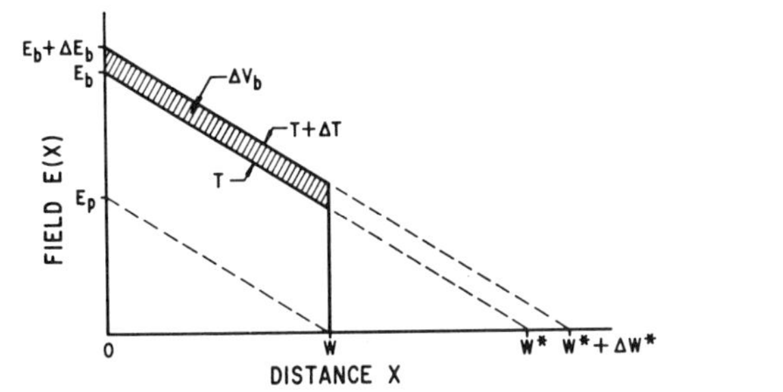

Fig. 6. Field distribution within a p^+-n-n^+ junction with punch-through below breakdown. The temperature increase ΔT causes a corresponding ΔE_b of the breakdown field.

coefficient is the familiar step junction for which reasonably accurate values of $\beta(V_b)$ are found in voltage regulator data sheets. Instead of calculating β analytically as a function of scattering and ionization coefficients of electrons and holes, field distribution, etc., we will normalize β to the empirical values of a p-$^+$n step junction $\beta_0(V_b)$. We will carry out this normalization for two impurity profiles which occur frequently in the design of avalanche microwave oscillators:

1) β^* for a step junction with chopped-off tail as found in p^+-n^--n^+ structures with punch-through before breakdown;
2) β' for a linearly graded junction.

The case of a p^+-n^--n^+ structure is illustrated in Fig. 6. The coordinate $x=0$ represents the position of the p^+-n^- junction and $x=W$ the position of the n^--n^+ junction, respectively. The voltage drop in the highly doped p^+ and n^+ layers is neglected. The field $E_p=2V_p/W$ is the maximum field at the punch-through voltage V_p. The field E_b is the maximum field within the depletion layer at breakdown. The case of a p^+-n step junction without punch-through is obtained by extrapolating the $E(x)$ curve to the intersection with the x-axis at $x=W^*$. The breakdown voltage V_b is related to E_b and W by

$$V_b = \int_0^W E(x)dx = E_b \int_0^W (1 - x/W^*)dx. \tag{13}$$

The temperature dependence of V_b can be expressed in terms of the temperature dependence of the breakdown field E_b. Since the avalanche region of a p^+-n junction is typically 1/3 of the total width W^* one concludes that for a p^+-n^--n^+ junction with $W \geq 0.5\ W^*$ the breakdown field E_b and its temperature dependence is independent of W and, therefore, equal to the case of a p^+-n step junction. For this case of a p^+-n step junction it can readily be shown that

$$\frac{1}{E_b}\left(\frac{\partial E_b}{\partial T}\right)_W = \frac{1}{2V_b}\frac{dV_b}{dT} = \beta_0/2. \tag{14}$$

Expressing V_b in terms of E_p, E_b, and W one finds from (13) or from Fig. 6

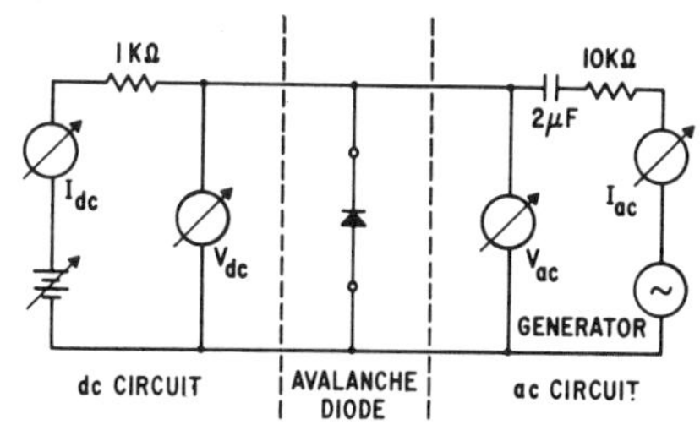

Fig. 7. Test circuit to measure incremental dc and ac resistance of avalanche diodes as a function of frequency and dc bias current.

$$V_b = E_bW - E_pW/2. \tag{15}$$

Realizing that in (15) only E_b is temperature dependent one finds $dV_b/dT = W(dE_b/dT)$. The temperature coefficient β^* of a p^+-n^--n^+ junction with breakdown voltage V_b is then readily expressed in terms of V_b, V_p, and β_0, where it should be remembered that β_0 refers to an ordinary p^+-n step junction with the same breakdown voltage V_b:

$$\beta^* = \frac{\beta_0}{2}\left(1 + \frac{V_p}{V_b}\right). \tag{16}$$

The punch-through voltage V_p is given in terms of the impurity concentration N within the n^- layer by:

$$V_p = qNW^2/2\epsilon \tag{17}$$

with q denoting the electron charge and ϵ the dielectric constant of the semiconductor material. For the case that W^* is known β^* can be expressed as

$$\beta^* = \beta_0/[2 - W/W^*]. \tag{18}$$

A similar analysis of the temperature coefficient β' of a linearly graded junction in terms of β_0 of the step junction yields

$$\beta' = 0.75\beta_0. \tag{19}$$

The preceding analysis indicates that the temperature coefficients of p-n junctions with impurity distributions frequently used in the design of avalanche microwave oscillators deviate significantly from the step junction case without punch-through. To obtain reasonably accurate heat flow resistance measurements the dependence of β on impurity profile has to be taken into account.

V. Experimental Results and Discussion

A. Measurement Procedure and Precautions

The circuit used in these studies to measure the electrical small-signal series resistance makes a compromise between versatility, simplicity, and accuracy. The circuit is shown in Fig. 7. The left half of the circuit is the dc bias circuit. The dc current voltage characteristic can be measured with good accuracy by means of the current meter I_{dc} and the digital voltmeter V_{dc}. The 1 $k\Omega$ resistor provides sufficient isolation during the ac measurements. The right half of the circuit is the ac bias and resistance measuring circuit. A small ac cur-

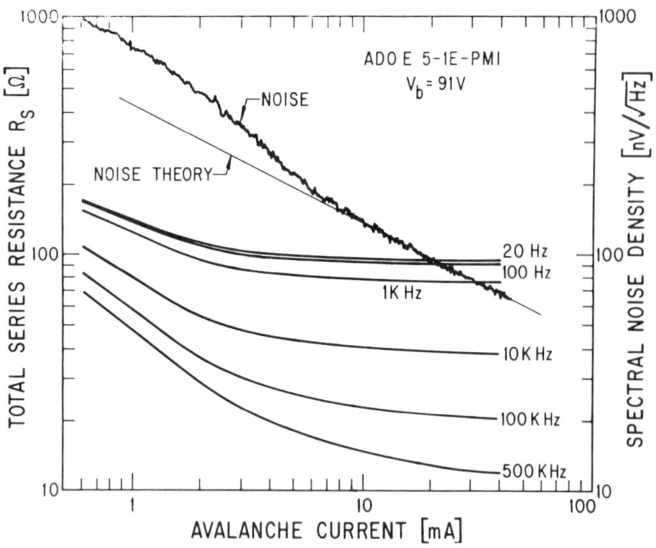

Fig. 8. Total series resistance and open circuit spectral voltage density as a function of bias current for diode ADOE5-1E-PM1. A flat resistance characteristic and a slope of $-\frac{1}{2}$ of the noise characteristic indicate uniform breakdown.

rent of the order of 100 μA to 1 mA is superimposed onto the dc current through a 10 kΩ resistor and a 2 μF blocking capacitor. The small-signal resistance is measured as the ratio of V_{ac}/I_{ac}. For sufficient accuracy the blocking capacitor has to meet the condition $1/\omega C \ll 10$ kΩ. The influence of the 1 kΩ resistor in the dc bias circuit can readily be corrected.

Since avalanche breakdown is very often highly nonuniform across the breakdown region, [6], [7] certain precautions have to be taken. Unless the entire junction area is participating in the breakdown, the small-signal resistance R_s and heat flow resistance r will be larger than expected. Usually, the breakdown region will expand with increasing bias current until it covers the entire junction area [7]. Then R_s becomes independent of current. To insure proper bias conditions for thermal resistance measurements we usually record R_s as a function of current at frequencies between 20 Hz and 500 kHz (Fig. 8). For the diode of Fig. 8 sufficient accuracy is reached for currents above 7 mA.

Another precaution to be taken results from avalanche noise. Nonuniform breakdown is usually the source of large amounts of low frequency noise. This noise integrated over the bandwidth of the ac voltmeter has to be small compared with the ac voltage at the test frequency. Either a check of the spectral noise density such as the upper curve of Fig. 8 or a narrow band pass filter in series with the ac voltmeter is advisable. A noise versus current characteristic such as Fig. 8 is also a convenient check on breakdown uniformity. At currents below 7 mA the measured noise curve deviates significantly from the $1/\sqrt{I}$ law predicted by theory [6]. This deviation indicating nonuniform breakdown coincides with values of R_s that are significantly higher than the R_s minimum at high currents.

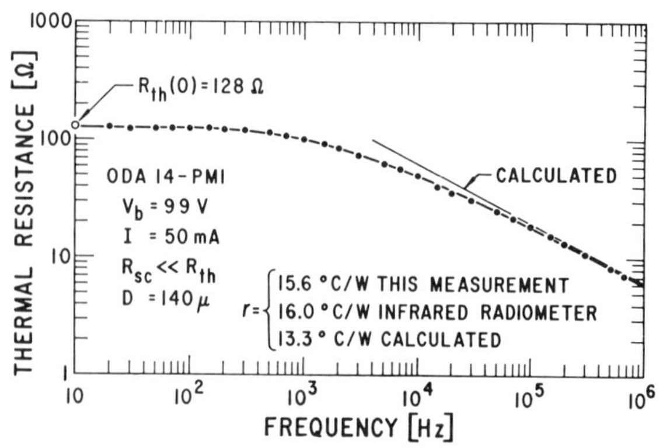

Fig. 9. Thermal resistance as a function of frequency for the flip-chip mounted p-i-n diode ODA14-PM1. Since R_{sc} of this p-i-n diode is negligible, no curve fitting is required to obtain a slope of $-\frac{1}{2}$ at high frequencies.

B. Results

In the case of p-i-n diodes with negligible space charge resistance, the thermal resistance R_{th} is the dominating contribution to R_s. For such a diode the frequency dependence of $R_{th} = R_s$ is shown in Fig. 9. As predicted in Section III, R_{th} is frequency independent at low frequencies and decreases with $\sqrt{f}$ at high frequencies.

Good agreement is found between the low frequency resistance of 125 ohms and the dc incremental resistance of 128 ohms. This agreement implies that there is only negligible heat flow resistance between the microwave pill package and ambient. Any heat flow resistance between the package and ambient would have a cutoff frequency below 20 Hz and result in a discrepancy between the thermal resistance values at 20 Hz and dc.

The asymptote at high frequencies is calculated by multiplying (10) with $\sqrt{2}\,\beta V_b^2$. The $\sqrt{2}$ factor is necessary because the ac voltmeter measures the magnitude of z and not only the real part. Measured values are used for β, V_b, and $A = \pi D/4$. Since the cutoff frequency f_{c2} defined by (12) is of the order of 300 kHz for the length $\alpha L = 5\mu$ we have to use (10) rather than half of it as in Fig. 5. The agreement between calculated and experimental results in Fig. 9 is quite good, supporting the assumption that $R_{sc} \ll R_{th}$.

The thermal resistance for a diode for which R_{sc} is not negligible is shown in Fig. 10. Here we have to subtract an estimated space charge resistance of $R_{sc} = 5.0$ ohms in order to obtain a slope of $-1/2$ at high frequencies. Since the space charge layer in these diodes reaches the n^+ epitaxial substrate before breakdown, it is difficult to calculate R_{sc} accurately from (6). It is, therefore, preferred to determine R_{sc} empirically by fitting $R_{th} = R_s - R_{sc}$ to the calculated asymptote.

The thermal resistance of a diode mounted faceup is shown in Fig. 11. For this diode one also finds a frequency independent thermal resistance at low frequencies and a decrease with $\sqrt{f}$ over two decades at high frequencies. For this diode $R_{sc} \ll R_{th}$. The magnitude of the cutoff frequency agrees well with (11).

The results of Fig. 12 represent a diode which is

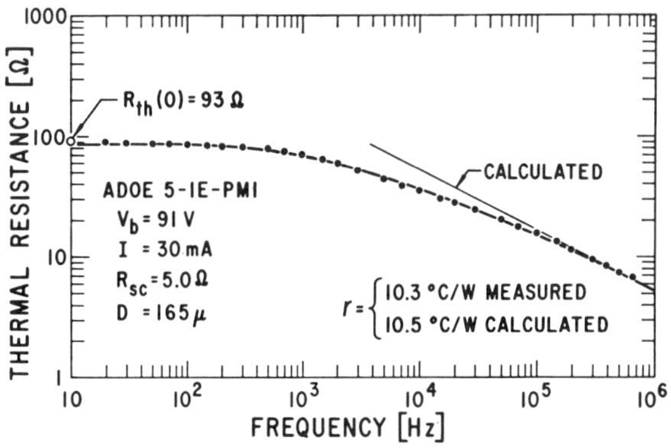

Fig. 10. Thermal resistance as a function of frequency for the flip-chip mounted diode ADOE5-1E-PM1. A slope of $-\frac{1}{2}$ is obtained at high frequencies after subtracting $R_{sc}=5.0$ ohms from the measured values of $R_s=R_{th}(\omega)+R_{sc}$.

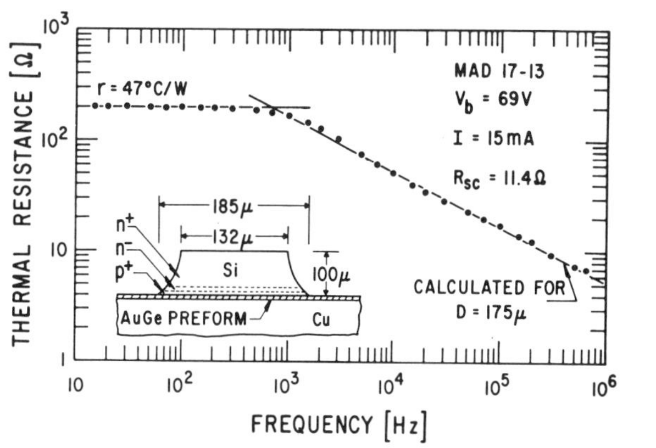

Fig. 12. Thermal resistance as a function of frequency for diode MAD 17-13 flip-chip mounted by means of a Au–Ge preform. The calculated value of $R_{sc}=11.4$ ohms is subtracted from R_s to obtain a slope of $-\frac{1}{2}$ at high frequencies.

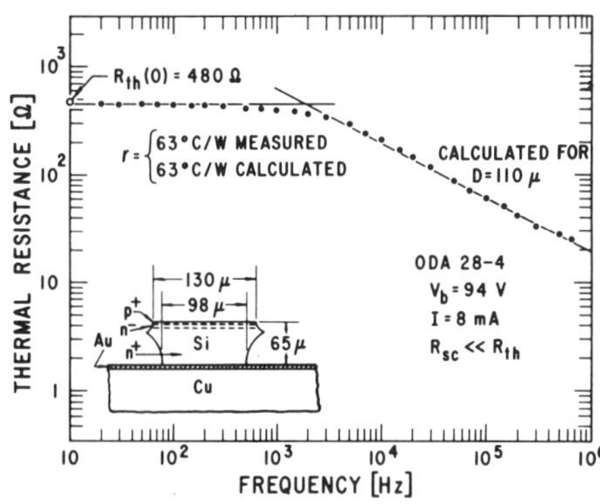

Fig. 11. Thermal resistance as a function of frequency for the face-up mounted p-i-n diode ODA28-4. Since $R_{sc} \ll R_{th}(\omega)$ for this p-i-n diode no curve fitting is required to obtain a slope of $-\frac{1}{2}$ at high frequencies.

flip-chip mounted by means of a Au–Ge preform. This Au–Ge alloy between the silicon chip and the copper package has a poor thermal conductivity and thus presents a large heat flow resistance. As a result the low frequency thermal resistance is very large. At high frequencies the thermal resistance is determined by heat flow into the distributed heat flow resistance–heat capacitance network as shown in Fig. 3 resulting in a decrease of R_{th} with $\sqrt{f}$ over nearly three decades. Since R_{sc} is not negligible in this diode with an undepleted n layer, it was necessary to subtract $R_{sc}=11.4$ ohm from the measured values of R_s. This value of R_{sc} was calculated from (6).

The experimental data from the diodes of Figs. 9 through 12 together with data from four additional diodes are summarized in Table I. In general, a good agreement between R_{th} at dc and 100 Hz is observed, implying that the heat flow resistance is determined by the silicon chip and heat spreading into the copper package. In the case of p-i-n diodes, R_{sc} is negligible. The values of β are measured values which agree quite well with the estimated corrections discussed in Section IV. The values of V_b were measured by extrapolating the linear part of the I–V characteristic to zero current. The current levels at which R_{th} was measured were chosen from plots of R_s and $\langle u^2\rangle/\Delta f$ versus I such as Fig. 8. The diameter of the silicon chips was measured either by microscope or by cross sectioning of the diodes. The distance L_{si} between junction and package was either known from the slice preparation or measured directly in the case of face-up mounting. For these diodes with known chip size we obtained fairly good agreement between the measured and calculated heat flow resistance values. From this agreement we can conclude that the heat flow resistance across the silicon–copper interface is negligible compared with heat flow resistance within the silicon and copper unless soldering preforms are used.

C. Error Discussion

So far two sources of errors have been mentioned: temperature coefficient β and nonuniform breakdown. Other errors result from uncertainties in estimating R_{sc}, from a slight frequency dependence of R_s at low frequencies, from the inaccuracy in measuring V_{ac} and I_{ac}, and from the approximation made in deriving (5) by neglecting the two small terms in (4).

The main error results from the uncertainty of β. If β is measured for the diode under test the error can readily be kept below 5 percent. If β is estimated from punch-through voltage or depletion layer width in homogeneous material then it will be difficult to reduce the error much below 10 percent.

As mentioned in Section V-A the errors resulting from nonuniform breakdown can readily be eliminated by taking proper precautions. The error resulting from inaccurate estimates of R_{sc} can be reduced if R_{sc} is determined by fitting the R_{th} versus frequency curve to an asymptote with a slope of $-1/2$ (see Fig. 10). Otherwise, an error of approximately 3 percent in R_{th} is introduced if one assumes that $R_{sc}=R_s$ (1 MHz). This 3 percent estimate is due to the fact that the cutoff frequency is usually of the order of 1 kHz. At 1 MHz R_{th} is then reduced to 3 percent of its low frequency value.

The slight frequency dependence of R_{th} at low fre-

TABLE I

Diode Number	R_{th} (dc) (ohms)	R_{th} (100Hz) (ohms)	R_{sc} (ohms)	β (10^{-4}/°C)	V_b (volts)	I (mA)	D (μ)	L_{si} (μ)	r_{meas} (°C/W)	r_{calc} (°C/W)	Mounting
ODA14-PM1	128	125	—	7.5	99	50	140	5	15.6	13.3	Flip-chip
ODA14-X1	121	108	—	7.4	93	40		5	12.5		Flip-chip
ADOE5-133	290	250	70	10.0	96	30		3	25.0		Flip-chip
ADOE5-1E-PM1	93	87	5.0	10.0	91	30	165	3	10.3	10.5	Flip-chip
ADOE5-1U	900	750	90.0	10.0	96	8	80	49	76	75	Face-up
ADOE9-1U	740	690	—		101	12	117	62	65	60	Face-up
ODA28-4U	480	450	—	7.4	94	8	110	60	63	63	Face-up
MAD17-13		200	11.4	9.0	69	15	175	5	47		Flip-chip Au–Ge Preform

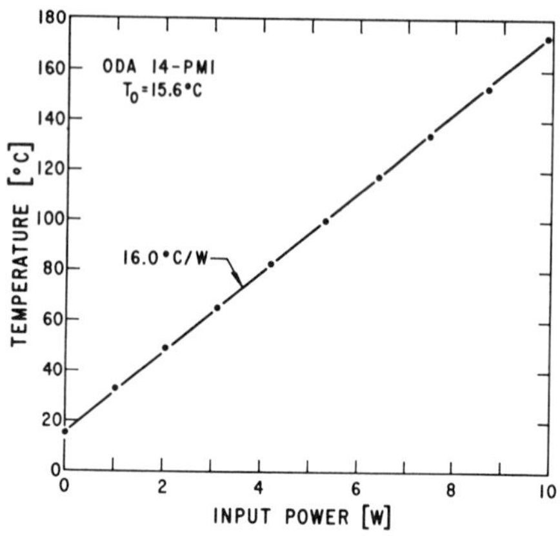

Fig. 13. Surface temperature as measured by an infrared radiometer as a function of input power for diode ODA 14-PM1. The line corresponds to a constant heat flow resistance of 16.0°C/W.

quencies is usually very small and from our experience it is concluded that this error is also less than 3 percent. The errors in measuring V and I are also estimated to be less than 3 percent each.

Another error may result from the approximation given in (5). This error is reduced considerably if V_b in (5) is replaced by the terminal voltage V measured at the bias current I. Then the error is reduced to the error of the digital voltmeter V_{dc} in Fig. 7.

Considering the frequency dependence of R_{th} and taking the above errors into account, it can be concluded that the heat flow resistance of an avalanche diode can be determined with an accuracy of better than 12 percent from two small signal resistance measurements at 100 Hz and 1 MHz. This error can be reduced to less than 8 percent by additional measurements of β.

To substantiate this conclusion, we have measured the temperature at the top of diode ODA-14-1PM with an infrared radiometer. The accuracy of this radiometer was better than ± 1°C. The temperature plotted as a function of input power in Fig. 13 increases linearly as expected with a slope corresponding to $r = 16.0$°C/W. This value is only 2.5 percent above the 15.6°C/W determined by our new method.

Acknowledgment

The authors gratefully acknowledge the assistance of H. M. Leedy and W. A. Little in preparing the avalanche diodes. They are also indebted to H. R. Plumlee for the infrared radiometer measurements.

References

[1] R. J. McIntyre, "Theory of microplasma instability in silicon," *J. Appl. Phys.*, vol. 32, pp. 983–995, June 1961.
[2] M. Gilden and M. E. Hines, "Electronic tuning effects in the Read microwave avalanche diode," *IEEE Trans. Electron Devices*, vol. ED-13, pp. 169–175, January 1966.
[3] W. Shockley, "Problems related to p-n junctions in silicon," *Solid-State Electronics*, vol. 2, pp. 35–67, January 1961.
[4] R. H. Haitz, "Nonuniform thermal conductance in avalanche microwave oscillators," *IEEE Trans. Electron Devices*, vol. ED-15, pp. 350–361, June 1968.
[5] K. Küpfmüller, *Einführung in die theoretische Elektrotechnik.* Berlin: Springer, 1957, p. 326.
[6] R. H. Haitz, "Noise of a self-sustaining avalanche discharge in silicon: low-frequency noise studies," *J. Appl. Phys.*, vol. 38, pp. 2935–2946, June 1967.
[7] R. H. Haitz, "Studies of the uniformity of avalanche breakdown by means of low-frequency noise measurements," *Phys. Failure in Electronics*, vol. 5, pp. 447–456, June 1967.

Solid-State Electronics Pergamon Press 1970. Vol. 13, pp. 53–56. Printed in Great Britain

GaAs SCHOTTKY-BARRIER AVALANCHE DIODES

C. K. KIM and L. D. ARMSTRONG

Micro State Electronics, Operation of Raytheon Company, Murray Hill, N.J., U.S.A.

(*Received* 7 *April* 1969)

Abstract—Efficient microwave power generation in a mesa type GaAs Schottky-barrier avalanche diode is reported. An output power of 600 mW (c.w.) at 8.2 GHz with an efficiency over 10 per cent was obtained when the diodes were mounted with the metal–semiconductor interface adjacent to the heat sink. The diodes were fabricated using nickel on n-type epitaxial GaAs substrate.

It is observed that for the same doping profile, the diodes performed similarly to (zinc) diffused abrupt p^+–n junction avalanche diodes, and the initial result indicates that GaAs Schottky-barrier avalanche diodes are comparable to ordinary p–n junction avalanche diodes in their microwave performance.

Résumé—On reporte la génération de puissance à micro-ondes dans une diode à avalanche de type messa en GaAs à barrière Schottky. Une puissance de sortie de 600 mW (cw) à 8,2GHz ayant un rendement supérieur à 10 pour cent a été obtenue quand les diodes ont été montées avec l'intervalle métal–semiconducteur adjacent à la trappe de chaleur. Les diodes ont été fabriquées en employant le nickel sur la couche épitaxiée en AsGa de type n.

On observe que pour le même profil de dope les diodes ont aussi bien fonctionné que les diodes à avalanche à jonction p^+–n abrupte diffusée (zinc) et le résultat initial indique que les diodes à avalanche en GaAs à barrière Schottky sont comparables aux diodes ordinaires à avalanche à jonction p^+–n dans leur fonctionnement aux micro-ondes.

Zusammenfassung—Über eine leistungsfähige Mikrowellenerzeugung bei einer GaAs-Schottky-Diode mit Mesastruktur wird berichtet. Als Ausgangsleistung wurden im Dauerbetrieb 600 mW bei 8,2 GHz mit einem Wirkungsgrad von mehr als 10 prozent erzielt. Die Dioden waren so aufgebaut, dass der Halbleiter-Metallübergang direkt an die Wärmesenke anschloss, wobei als Kontaktmetall Nickel auf epitaktischem n-Typ Galliumarsenid verwendet wurde. Es wurde beobachtet, dass die Dioden bei gleichem Dotierungsprofil ähnlich arbeiteten, wie mit Zink diffundierte Lawinendioden mit abruptem p^+–n-Übergang. Dieses erste Ergebnis zeigt, dass GaAs Lawinendioden mit Schottkybarriere in ihrem Mikrowellenverhalten den normalen Lawinendioden mit pn-Übergang vergleichbar sind.

It is demonstrated for the first time in this paper that GaAs Schottky barrier diodes can be operated as high power and high efficiency avalanche diodes. Ever since the avalanche diode showed a promise as one of the most powerful solid state microwave sources, various diode structures have been attempted for better heat dissipation. An efficient microwave operation of avalanche diodes requires a high d.c. input power density, and the output power is mainly limited by the power handling capability of the diode.

In ordinary (p^+–n) junction avalanche diodes, mounted with the junction side on a heat sink, the heat generated in the avalanche region has to pass through the thermally poor (p-type) semiconductor and the ohmic contacting metals which are typically 1–2 and 0.1–0.5 μ respectively. For this type of diode structure, the thermal resistance contribution from the semiconductor layer is appreciable and one possible solution to this problem is to make a 'metal–semiconductor diode' where the (p-type) semiconductor layer is entirely eliminated. One major difficulty in high power Schottky-barrier avalanche diode fabrication has been the mounting of diodes with the metal–semiconductor interface adjacent to a heat sink without deteriorating the diode breakdown characteristics. Attempts have been made to passivate the diodes with guard ring diffusions to avoid this diode degradation.

An early attempt in constructing GaAs Schottky-barrier avalanche diodes resulted in a limited pulse power output.[1] However, feasibility of c.w. output power from a Schottky-barrier diode was first shown by WHEELER,[2] who developed in early 1967 a passivated GaAs Schottky-barrier avalanche diode which gave 5 mW (c.w.) output power at X-band. Recently, SZE *et al.*[3] developed a passivated Si Schottky-barrier avalanche diode with a diffused guard ring structure. The maximum output power of their guard ring devices was 13.5 mW (c.w.) at 8 GHz with the unexplained low efficiency of 0.35 per cent. More recently, in late 1968, GaAs guard ring Schottky-barrier diodes were developed by HAMILTON,[4] which gave up to 48 mW peak power output at 15 GHz, but were limited in their c.w. outputs.

In this paper, the feasibility of a high power and high efficiency Schottky-barrier avalanche diode is demonstrated. Figure 1 shows the cross section of the mesa type GaAs Schottky-barrier avalanche diode with no guard ring structure. The diodes were fabricated using nickel on *n*-type GaAs as the metal–semiconductor contacts with areas of about 10^{-4} cm^2. After the contacts were formed, the GaAs wafers were lapped and metallized for mounting. Pellets were formed by diamond scribing and then bonded with the metal–semiconductor interface down to the heat sink in a coax

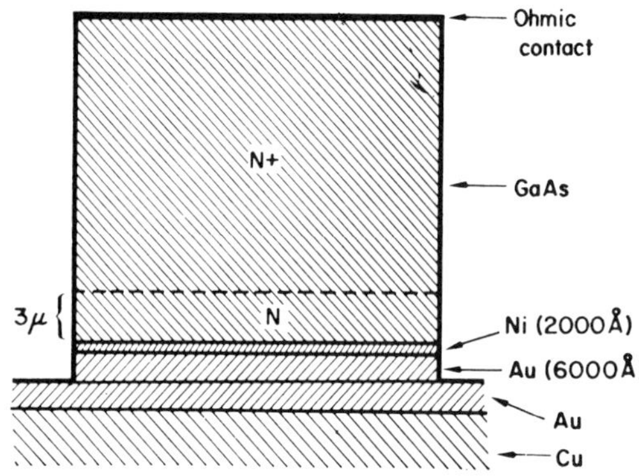

FIG. 1. Cross sectional view of a mesa-type GaAs Schottky-barrier avalanche diode.

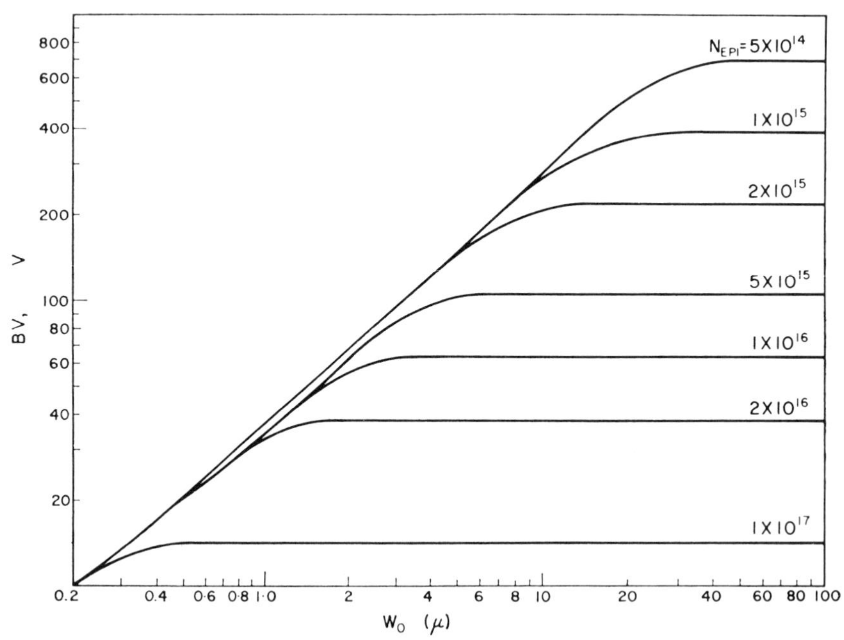

FIG. 2. Breakdown voltage vs. impurity density and epi-layer thickness for abrupt GaAs p^+–n junction.

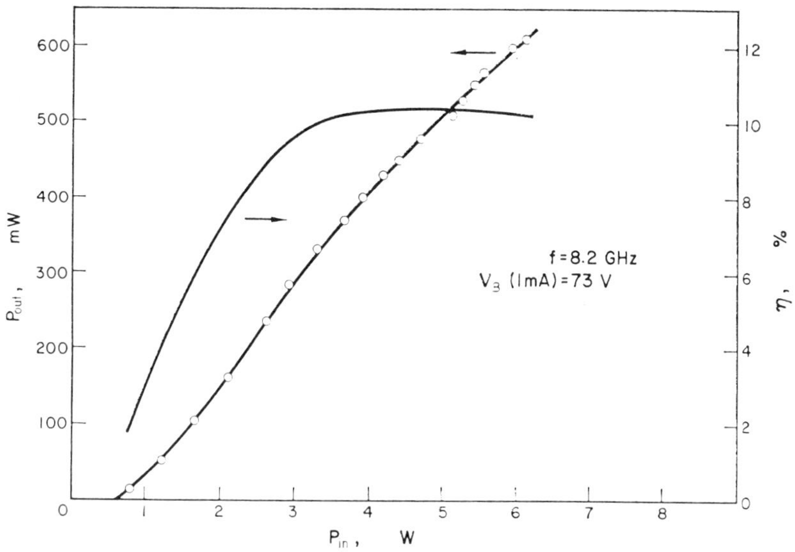

FIG. 3. Microwave c.w. power output vs. d.c. input power of a mesa type GaAs Schottky-barrier avalanche diode.

pin type microwave package (Micro State Type 8), using heat and ultra-sonic energy. Diode breakdown voltages were typically 70 V with the epitaxial doping density and thickness of about $6 \times 10^{15}\ \text{cm}^{-3}$ and $3\ \mu$ respectively. This is in agreement with the theoretically calculated breakdown voltages shown in Fig. 2. The calculation was based on the ionization rates[5] by LOGAN and SZE. In Fig. 3, curves of output power (c.w.), and efficiency vs. input power are presented and it is shown that 600 mW with over 10 per cent efficiency was obtained. A similar diode gave 700 mW with slightly less efficiency, 9.3 per cent. The circuit shown in Fig. 4 was employed for

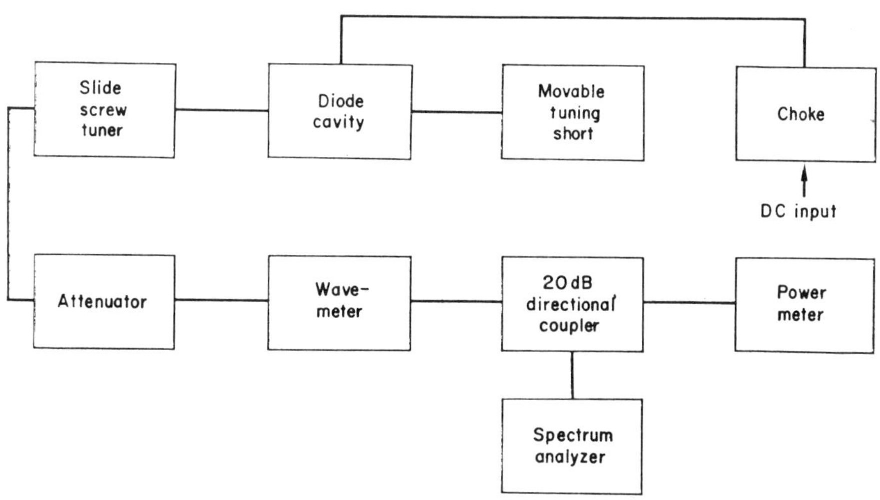

FIG. 4. X-band waveguide test circuit.

measuring the output power with the microwave tuning achieved by using both the sliding short and slide screw tuner.

An analysis[6] had been developed for abrupt p^+–n GaAs avalanche diodes and was used for the diode design. The Schottky-barrier avalanche diodes were observed to be similar to zinc diffused abrupt p^+–n GaAs avalanche diodes in their microwave performance, and the result is comparable to that of the zinc diffused p^+–n abrupt junction GaAs diodes,[7] where 500 mW (c.w.) at 11.1 GHz with an efficiency over 11.5 per cent was obtained.

Acknowledgement—The authors wish to thank W. Matthei for his helpful discussion.

REFERENCES

1. J. C. Irvin, *IEEE Trans. Electron Devices* **ED-13,** 208 (1966).
2. A. J. Wheeler, private communication (micro state report No. IR-706, May 1967).
3. S. M. Sze, M. P. Lepsilter and R. W. McDonald, *Solid-St. Electron.* **12,** 107 (1969).
4. V. Hamilton, monthly status report No. 5, USAECOM Contract No. DAAB07-68-C-0287, Nov. 6 (1968).
5. S. M. Sze and G. Gibbons, *Appl. Phys. Lett.* **8,** 111, (1966).
6. C. K. Kim, *IEEE Trans. Electron Devices* **ED-16** (1969).
7. C. K. Kim and L. D. Armstrong, *Appl. Phys. Lett.* May 1 (1969).

A Multiple-Diode High-Average-Power Avalanche-Diode Oscillator

Abstract—**This correspondence presents a simple circuit which allows direct combining of the power obtainable from several avalanche diodes. The circuit does not require extremely close matching of dc or RF diode characteristics, and no particular isolating networks such as hybrid combiners are necessary. CW power output exceeding 4 watts at 7 GHz and 3 watts at 9 GHz has been demonstrated in a device employing five diodes. The number of diodes which can be combined using this technique is limited by geometric and heat sink considerations.**

Descriptions of typical single-diode and multiple-diode oscillators are given along with equivalent circuits. The circuits employ resistors or resistor networks to suppress low-frequency oscillations and undesired resonances which occur when several oscillators are coupled together. Negligible insertion loss is incurred at the design output frequency.

Measured performance is given on similar single- and multiple-diode oscillators. Data include frequency, power, AM noise, FM noise, temperature, and loaded Q.

Introduction

In 1958, Read [1] proposed a semiconductor avalanche-diode microwave oscillator. The impurity profile of the diode consists of a hyperabrupt p-n junction followed by a high-resistivity region from which all free carriers are swept by the time avalanche breakdown occurs. Avalanche is limited to the narrow zone in the vicinity of the junction. The carriers which are generated drift through the high-resistivity region. Read showed the avalanche process is inductive, and that the current lags the voltage by a phase angle of 90°. The high-resistivity drift region of the structure introduces an additional 90° phase lag, so that at a given frequency the total phase lag is 180° and the device is a pure negative resistance at that frequency.

Further work by Misawa [2] revealed that negative resistance can occur even when the avalanche and drift regions coincide. Misawa and others have made detailed calculations which differentiate between the operation of

Manuscript received June 11, 1969. This work was presented at the International Microwave Symposium, Dallas, Tex., May 5–7, 1969.

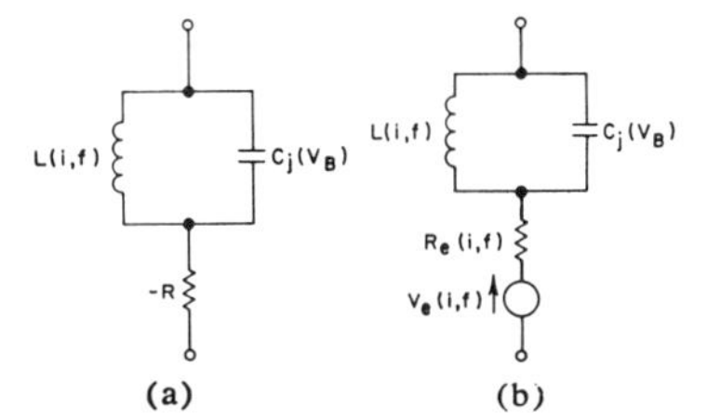

Fig. 1. Junction equivalent circuits for avalanche diode.

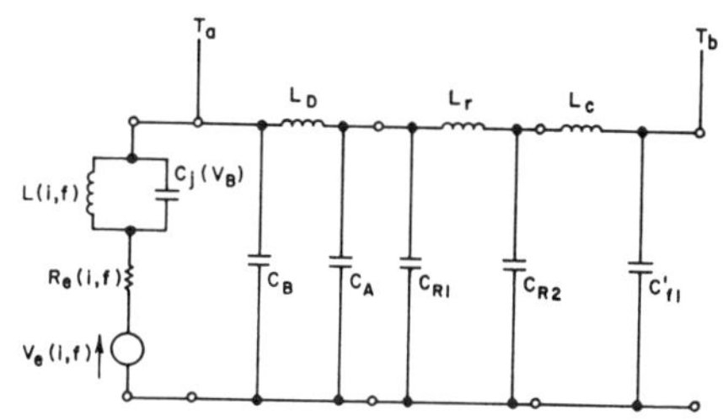

Fig. 2. Overall equivalent of mounted avalanche diode

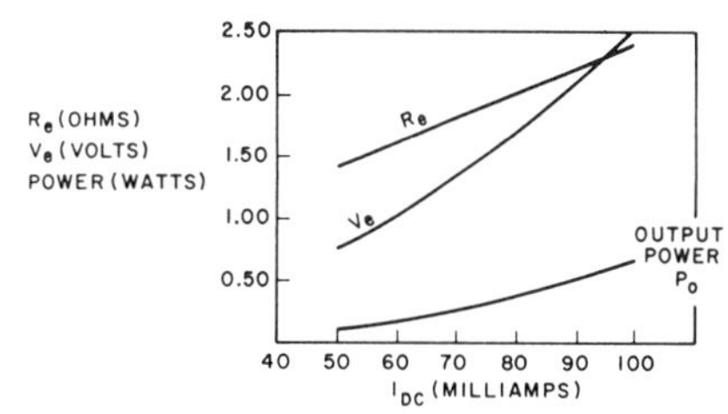

Fig. 3. Effective source voltage and resistance for optimum load.

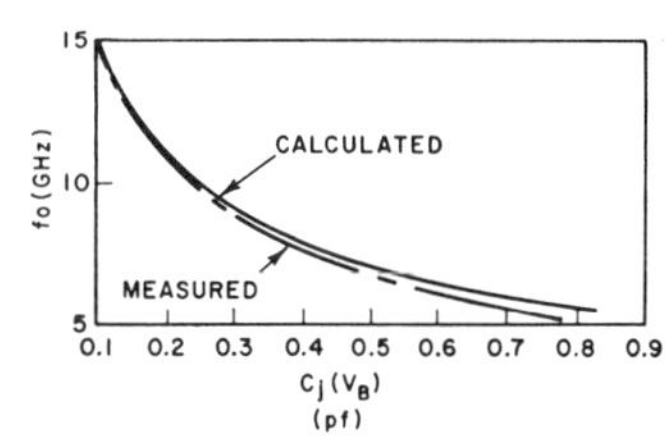

Fig. 4. Self-resonant frequency.

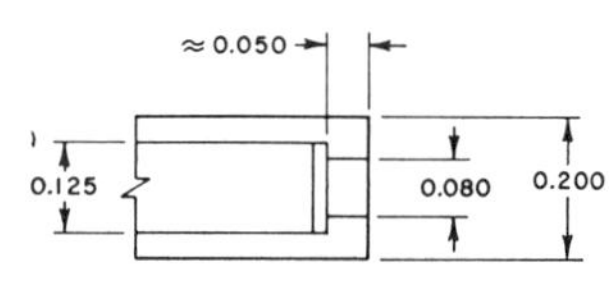

Fig. 5. Pill diode and mount.

Read structures, p-γ-n structures, and various types of simple p-n junctions. All the calculations expressly include the fact that the width of the avalanche region is commensurate with that of the drift region and that an appreciable fraction of the transit length occurs in the avalanche region. The theories have shown, however, that any type of avalanching junction will show negative resistance effects.

Evans and Haddad [3] have presented the results of a large-signal computer analysis of a Read-type diode. These results are quantita-

553

tive only for the specific model analyzed. They are also very useful to a qualitative understanding of other similar oscillators.

Equivalent Circuits

The avalanche diode junction can be represented as shown in Fig. 1(a); a qualitative understanding of the behavior of the circuit can then be obtained by reference to various convenient graphs provided by Evans and Haddad.

For quantitative purposes, the junction equivalent circuit of Fig. 1(b) is convenient. In this circuit the negative resistance has been replaced by an appropriate voltage source V_e and effective generator resistance R_e, both of which vary with frequency and current. If such a junction is packaged and then mounted at the end of an appropriate coaxial mount, the overall equivalent circuit of Fig. 2 results. The notation is according to Getsinger [4]. All the equivalent circuit parameters between reference terminals T_a and T_b can be determined from Getsinger. We have performed measurements to determine the junction parameters to the left of terminal T_a. These parameters are valid for matched load conditions. Measured values for V_e and R_e are shown in Fig. 3. The values were determined by measuring power output as a function of load impedance for various output power levels.

It is convenient to chose the junction capacitance $C_j(V_B)$ to provide self-resonance of the packaged, mounted diode at terminals T_b. Fig. 4 shows this self-resonant frequency as a function of junction capacitance for a standard pill package diode mounted at the end of the coaxial line of Fig. 5.

The graph of Fig. 4 assumes constant $L(i, f)$. In our experience, this assumption results in relatively small errors of a few hundred megahertz at X band. These errors are easily corrected by minor circuit changes or by selecting a slightly different junction capacitance.

The conclusion from the foregoing equivalent circuit considerations is that the packaged-mounted avalanche diode can be treated as a self-resonant RF source having frequency, voltage, and internal impedance which can be readily determined. Such a source can be attached to various transformers or circuits to obtain explicit oscillator types.

The remainder of this correspondence describes performance obtained with a unique multiple-diode oscillator circuit and a similarly made single-diode oscillator.

Specific Oscillators

A typical single diode-coaxial avalanche-diode oscillator is shown in Fig. 6 along with its equivalent circuit, Fig. 7. The "diode" in this case is assumed to be self-resonant and to include all the package and mount reactances discussed in the preceding paragraphs. The load impedance R_b is optimized by adjustment of capacitor C_c where

$$R_b \approx Z_0^2(\omega C_c)^2 R_L \sin^2 \theta.$$

Any attempt to parallel or otherwise connect two or more oscillators at terminals T_0 will usually result in unstable operation. The instabilities can be caused by parametric effects, imperfect synchronous tuning, or multiple resonances occurring in the aggregate circuit. These problems have been overcome in the five-diode oscillator shown in Fig. 8. This oscillator consists of five diode/transformer sections arranged radially about a common bias network and a common output coupling capacitor.

A simple resistive network (which occupies no extra volume) is incorporated into the coaxial center conductors which are attached to each of the five diodes. The resistor network is arranged in a manner such that no RF current flows in the resistors at the desired output frequency. Instabilities at half frequency or other spurious effects are eliminated by the resistive losses. Any inequality of RF output voltages (at the desired output frequency) is also resistively terminated.

Any combination of RF signal sources must provide a means of phase locking the individual sources. This is accomplished by the capacitive output coupling disc which provides capacitive cross coupling between the individual oscillators in addition to coupling the combined output to the load.

The equivalent circuit of a two-diode oscillator is shown in Fig. 9. Additional oscillator transformer assemblies are simply connected at terminals T_0 to achieve higher power output.

The phase-locked resistively stabilized mode of operation has provided some unexpected benefits. When operated in this fashion, each of the five diodes may be operated at significantly higher input power levels than would be permissible in a single-diode oscillator using the same diode. This observed characteristic is not fully understood at present.

Fig. 6. Typical single-diode oscillator.

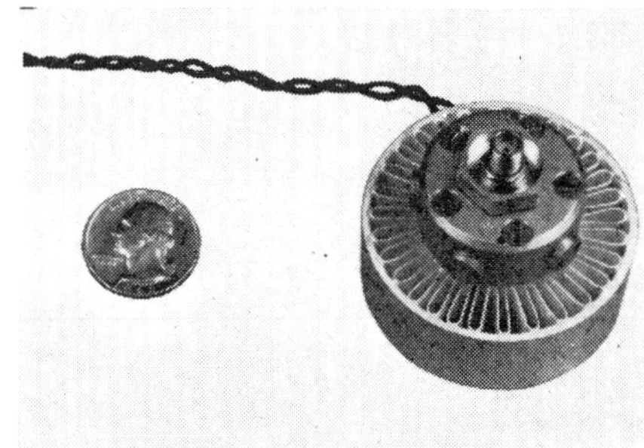

Fig. 8. Five-diode oscillator.

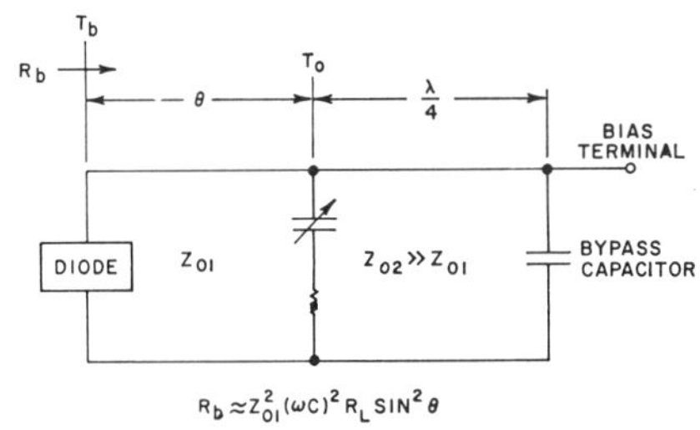

Fig. 7. Equivalent circuit of single-diode oscillator.

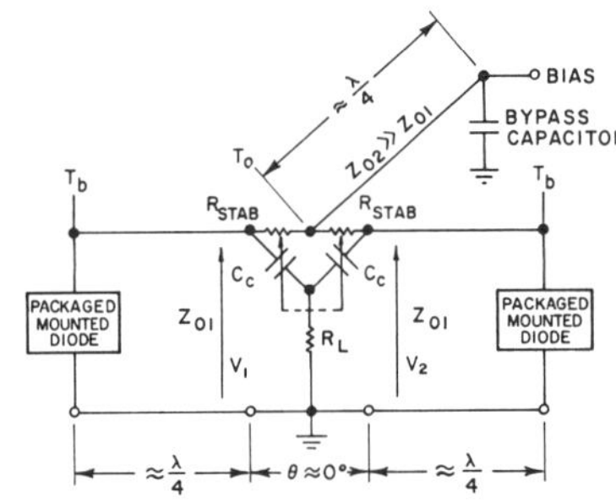

Fig. 9. Equivalent circuit of two-diode oscillator.

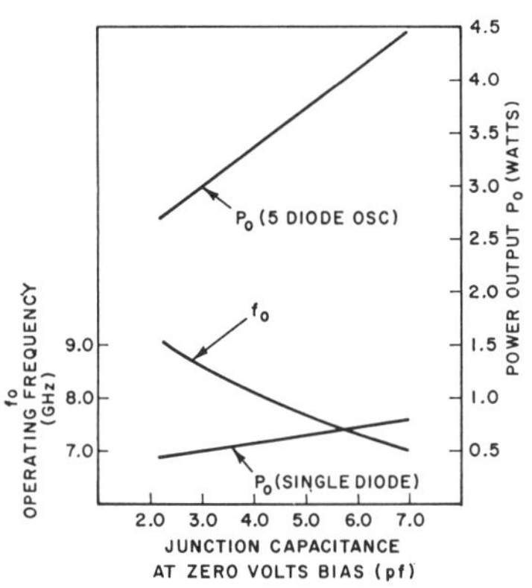

Fig. 10. Power and frequency.

Oscillator Performance

The following paragraphs summarize measured performance obtained on the five-diode oscillator. Where applicable, single-diode oscillator performance is shown for comparison.

Frequency, Power Output, and Efficiency

The operating frequency of the five-diode oscillator is fixed by selection of the diode junction capacitance. The existing oscillator functions at frequencies between 7.0 and 9.0 GHz as shown in Fig. 10. Power output is also shown along with typical single-diode oscillator output. Approximate efficiency for both single- and five-diode oscillators is given by Fig. 11. The slight dropoff in efficiency at the highest power is probably due to junction heating. Higher efficiencies approaching 10 percent have been achieved with numerous diodes in single-diode oscillators. The curve of Fig. 11 is typical of performance which is readily obtainable.

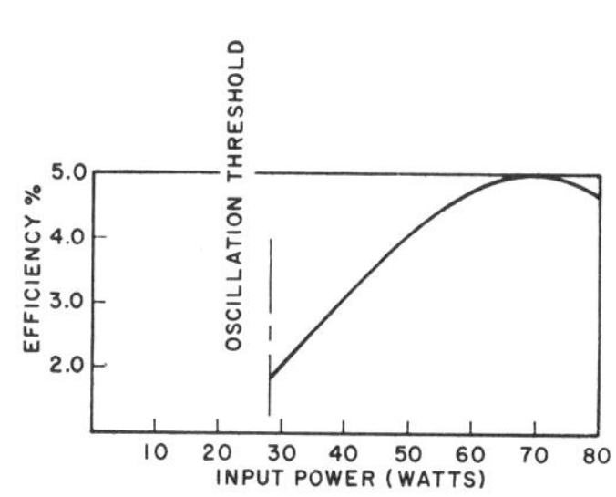

Fig. 11. Oscillator efficiency.

External Q and Noise

The five-diode oscillator exhibits external Q_x of about 18 which is somewhat lower than that of a similarly constructed single-diode oscillator using the same diodes and having $Q_x = 35$.

Measured FM and AM noise characteristics are given by Figs. 12 and 13, respectively. The FM noise for the five-diode oscillator is insignificantly different from that of the single-diode oscillator. The AM noise is improved by about 8 dB in the five-diode case as would be expected to result from parallel operation.

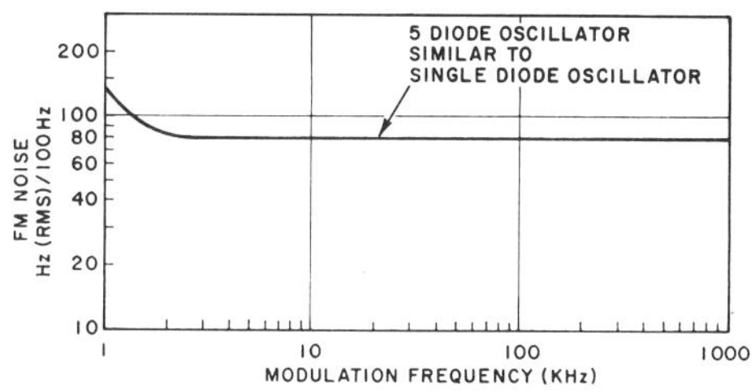

Fig. 12. FM noise performance.

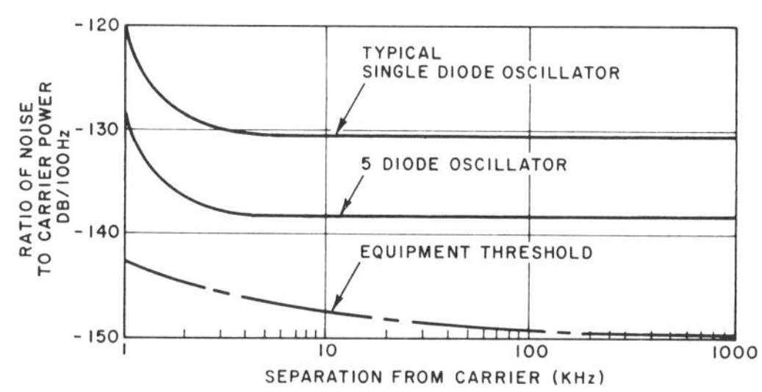

Fig. 13. AM noise performance.

Temperature

Extensive temperature tests have not been made on the five-diode oscillator. Initial tests indicate a relative well-behaved frequency and power dependence on temperature. Initial results are plotted in Fig. 14.

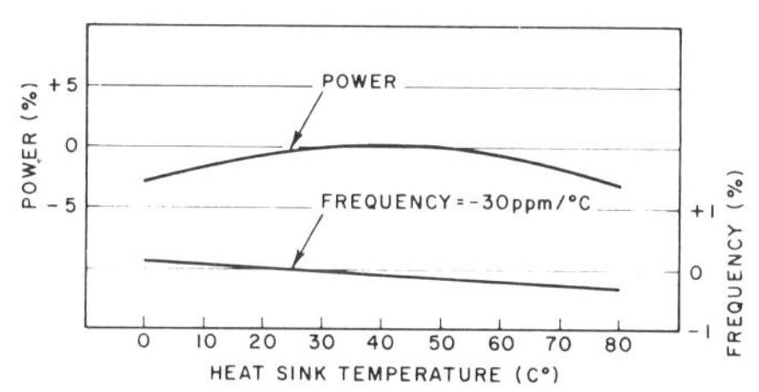

Fig. 14. Frequency and power; five-diode oscillator operated at 1-watt output level.

Conclusions

A novel circuit has been developed that combines the outputs of several (at least five) avalanche-diode oscillators. Performance obtained approaches and sometimes improves on that obtainable with a similar single-diode oscillator. The circuit is made practical by a resistive network which suppresses instabilities that would otherwise occur.

C. T. Rucker
Sperry Microwave
Electronics Div.
Sperry Rand Corp.
Clearwater, Fla. 33518

References

[1] W. T. Read, "A proposed high frequency negative resistance diode," *Bell Sys. Tech. J.*, vol. 37, pp. 401–446, March 1958.
[2] T. Misawa, "Negative resistance in p-n junctions under avalanche breakdown conditions," pt. I and II, *IEEE Trans. Electron Devices*, vol. 13, pp. 137–151, January 1965.
[3] W. J. Evans and G. I. Haddad, "A large signal analysis of IMPATT diodes," *IEEE Trans. Electron Devices*, vol. ED-15, pp. 708–717, October 1968.
[4] W. J. Getsinger, "The packaged and mounted diode as a microwave circuit," *IEEE Trans. Microwave Theory and Techniques*, vol. MTT-14, pp. 58–59, February 1966.

OTHER ARTICLES OF INTEREST

The Single-Cavity Multiple-Device Oscillator

KANEYUKI KUROKAWA, MEMBER, IEEE

Abstract—**The output power from 12 IMPATT diodes has been combined in a single-cavity multiple-device oscillator. The oscillator is free from the well-known moding problem of multiple-device oscillators. The objective of this paper is to present the oscillator-circuit theory, which clearly indicates why this particular circuit configuration can give a stable operation free from moding problems. To handle the formidable equations necessary for the analysis, the eigenfunction approach is extensively used. The condition for stable operation, the noise performance, and the locking behavior of the oscillator are all discussed. The noise performance is similar and the locking behavior identical to those of single-device oscillators.**

I. Introduction

GREAT progress has been made in the output power of microwave solid-state devices such as IMPATT and Gunn diodes. For maximization of their output power, past studies have been aimed primarily at improvement of the device impedance level and more efficient heat removal. The series connection of several chips in a package [1] and the use of the double-drift IMPATT structure [2], [3] both contribute to increasing the terminal impedance of the device. Heat removal is expedited by mounting diodes junction side down [4], [5] on diamond dice [6].

Independent of the improvement of the device itself, however, it is always interesting to explore a complementary method for higher power, namely, the combination of the output power from several devices. A multiple-device oscillator can be used to obtain either higher output power than a single device can provide, or a longer system life by operating each device conservatively, or both.

In the early days of microwave solid-state oscillator development, individual oscillators were first built and then their output power was combined by means of hybrid circuits [7] or transmission lines. The combined oscillator of this configuration is in a near-degenerate state and hence inherently difficult to adjust and sensitive to small changes in the load condition, in the bias supply, and even in the environmental temperature. It can oscillate in several different modes, and if the oscillation mode changes due to the external conditions, both frequency and amplitude will change giving unacceptable performance. This is the well-known moding problem of multiple-device oscillators [8]. To get around this difficulty, an ingenious method was proposed by Rucker [9] and the mechanism of suppression of unwanted modes was clarified by the present author [10].

More recently, a single-cavity multiple-device oscillator with similar stability was described by Magalhaes and the author [11]. In this oscillator each diode is mounted at one end of a coaxial line which is coupled to the sidewall of a common waveguide cavity. The other end of the coaxial line is terminated by a tapered absorber. Twelve IMPATT diodes were combined in this way and 10.5 W of CW power was obtained at 9.1 GHz with 6.2-percent efficiency. No spurious oscillations were observed during the circuit adjustment. The adjustment was easy and the spectrum was clean. Its excellent stability may well qualify this single-cavity multiple-device oscillator circuit as one of the most promising power-combining schemes in the micro- and millimeter-wave regions.

The purpose of this paper is to present the theory of the single-cavity multiple-device oscillator.[1] The theory clarifies the condition under which undesired modes are suppressed and stable operation is obtained. The noise and locking behavior are also discussed.

II. Equivalent Circuit

Fig. 1 shows the essential features of the oscillator. The coaxial lines are coupled to the sidewalls of the waveguide cavity at the positions where the magnetic field and hence the coupling become maximum when the cavity resonates in the desired mode. Fig. 2 shows the equivalent circuit of the oscillator near each resonant frequency of the cavity, assuming that the resonant mode is not a degenerate one.[2] The reference planes of the N-port network (N is the number of devices) are physically located along the coaxial lines approximately an integer multiple of a half-wavelength from the center of the waveguide height. The quantities L, C, and G_c represent the waveguide cavity, G_L the load, and G_0 the tapered termination, and the transformers represent the coupling between the coaxial line and the cavity. The device admittance is indicated by $-\overline{Y}_j$, which is really the admittance looking toward the diode at the reference plane in the jth coaxial line.[3] The device admittance is a function of the amplitude of the RF voltage at the

Manuscript received February 8, 1971; revised April 28, 1971.
The author is with Bell Telephone Laboratories, Inc., Murray Hill, N. J. 07974.

[1] For a simple explanation of the one-diode case, the reader is referred to [17].

[2] To obtain the equivalent circuit, the author first considered the one-diode case using symmetric and antisymmetric excitations of the coaxial line together with a dual form of the equivalent circuit of one-port resonators derived in [14, ch. 4]. Then, he extended the result to the N-diode case following the method illustrated in [14] for two-port resonators.

[3] The existence of a well-defined device admittance is assumed. This is probably the most important assumption for our discussion to be valid. For the physical meaning the reader is referred to [15, ref. 1, 5–8].

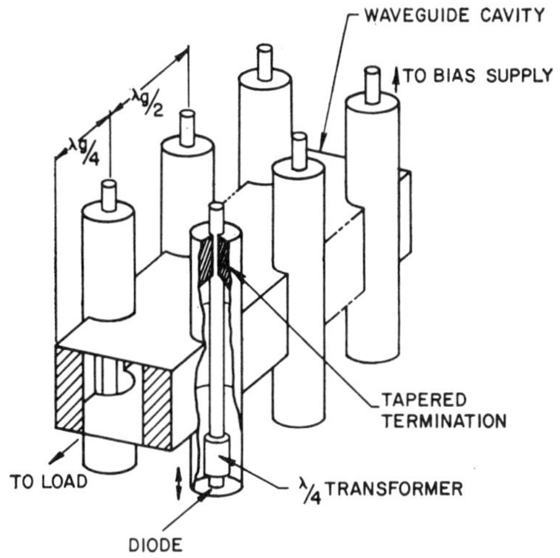

Fig. 1. Single-cavity multiple-device oscillator.

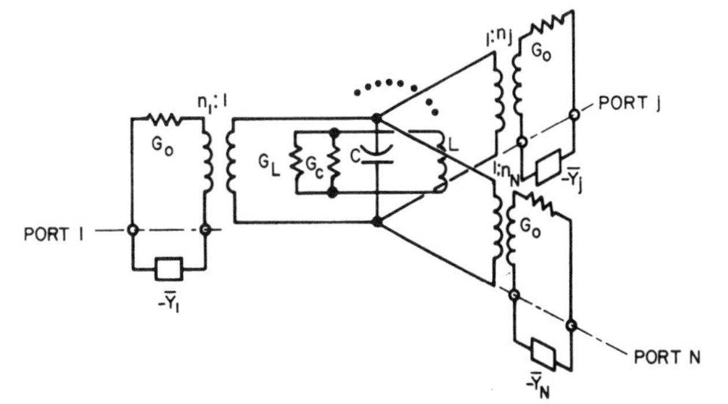

Fig. 2. Equivalent circuit of the oscillator shown in Fig. 1 near each resonant frequency of the cavity. Dots signify the existence of N branches, although only three are shown.

reference plane. Unless otherwise stated, it will be assumed that all the devices have an identical characteristic $-\overline{Y}(A)$ as a function of the amplitude A of the RF voltage. However, the RF voltage and hence the value of the device admittance may be different from one port to another. While considering each resonant mode of the cavity, it will be further assumed that the device admittance is independent of frequency. This assumption simplifies the analysis and is justifiable since the device admittance varies slowly with frequency compared to the resonant admittance of the cavity.

The effects of the discontinuities introduced in the coaxial line and the waveguide cavity by the coupling between them are expressible by a series reactance in the coaxial line and a parallel susceptance in the waveguide. The former can be incorporated into the device admittance and the latter into the LC resonator representing the cavity. With this understanding, the discontinuity effects are taken into account in the following analysis. On the other hand, the direct coupling between any pair of coaxial lines is neglected, compared with the coupling through the resonant field in the cavity.

III. Free-Running Condition

Corresponding to the condition that the total admittance be equal to zero, a condition necessary for free-running oscillation[4] is given by

$$\boldsymbol{Y}\boldsymbol{v} - \overline{\boldsymbol{Y}}\boldsymbol{v} = 0 \tag{1}$$

where $\boldsymbol{v}$ is the voltage vector with its component the RF voltage at each port, $\boldsymbol{Y}$ the admittance matrix representing the N-port network, and $\overline{\boldsymbol{Y}}$ a diagonal matrix with its diagonal components the negative of the device admittances, $\overline{Y}_j$ $(j=1, 2, \cdots, N)$. Equation (1) is a formidable equation to handle. The ij component of $\boldsymbol{Y}$ has a complicated expression. Fortunately, the ij component of its reciprocal $\boldsymbol{Z}=\boldsymbol{Y}^{-1}$ has a relatively simple form. It is given by

$$Z_{ij} = \frac{\delta_{ij}}{G_0} + \frac{n_i n_j}{Y_a} \tag{2}$$

where δ_{ij} indicates the Kronecker delta, n_i and n_j are transformer ratios, and

$$Y_a = \sqrt{\frac{C}{L}}\left\{j\left(\frac{\omega}{\omega_a} - \frac{\omega_a}{\omega}\right) + \frac{1}{Q_a} + \frac{1}{Q_{ext}}\right\}$$

$$\omega_a = \frac{1}{\sqrt{LC}}, \qquad \frac{1}{Q_a} = \frac{G_c}{\omega_a C}, \qquad \frac{1}{Q_{ext}} = \frac{G_L}{\omega_a C}. \tag{3}$$

Let us, therefore, investigate the eigenvectors of $\boldsymbol{Z}$. They will enable us to handle (1) systematically.

The eigenvectors of $\boldsymbol{Z}$ are given by

$$\mathbf{x}_0 = \begin{bmatrix} n_1 \\ n_2 \\ \cdot \\ \cdot \\ n_N \end{bmatrix}, \qquad \mathbf{x}_m = \begin{bmatrix} 1/n_1 \\ \exp(jma)/n_2 \\ \cdot \\ \cdot \\ \exp\{jm(N-1)a\}/n_N \end{bmatrix} \tag{4}$$

and the corresponding eigenvalues by

$$\lambda_0 = \frac{1}{G_0} + \frac{c}{Y_a}, \qquad \lambda_m = \frac{1}{G_0}, \qquad m = 1, 2, \cdots, N-1 \tag{5}$$

where $a=2\pi/N$ and

$$c = n_1^2 + n_2^2 + \cdots + n_N^2. \tag{6}$$

557

Altogether, N eigenvectors have been obtained. From these eigenvectors a set of row vectors $\tilde{\mathbf{x}}_i$ can be constructed to satisfy

$$\tilde{\mathbf{x}}_i \cdot \mathbf{x}_j = \delta_{ij} \tag{7}$$

where i, j run from 0 to $N-1$. The procedure is as follows. First, we notice that $\mathbf{x}_0$ is perpendicular to all the $\mathbf{x}_m$. This can be easily verified by calculating $\mathbf{x}_0^+ \cdot \mathbf{x}_m$ where $+$ indicates the adjoint vector (or matrix). Next, let row vectors $\boldsymbol{u}_m$ be defined by

$$\boldsymbol{u}_m = [n_1 \; n_2 e^{-jma} \; \cdots \; n_N e^{-jm(N-1)a}] \tag{8}$$

[4] Whether a particular mode of oscillation does or does not grow from a vanishingly small amplitude will be discussed elsewhere.

and let us suppose that

$$\alpha_1\mathbf{x}_1 + \alpha_2\mathbf{x}_2 + \cdots + \alpha_{n-1}\mathbf{x}_{N-1} = 0. \quad (9)$$

Multiplying by $\boldsymbol{u}_m$, we obtain

$$\alpha_m \boldsymbol{u}_m \cdot \mathbf{x}_m = 0.$$

Since $\boldsymbol{u}_m \cdot \mathbf{x}_m = N$, α_m must be equal to zero. This means that if (9) is satisfied, all the coefficients must be equal to zero. In other words, $\mathbf{x}_1, \mathbf{x}_2, \cdots, \mathbf{x}_{N-1}$ are linearly independent. Hence, in the space spanned by the $\mathbf{x}_m$ there is a nonzero vector which is perpendicular to all but one vector $\mathbf{x}_k$ $(1 \le k \le N-1)$. Such a vector can be obtained by Gram–Schmidt orthogonalization [12], [13]. This vector is, of course, perpendicular to $\mathbf{x}_0$. Therefore, properly adjusting the magnitude and taking the adjoint vector $\tilde{\mathbf{x}}_k$, we obtain

$$\tilde{\mathbf{x}}_k \cdot \mathbf{x}_j = \delta_{kj}$$

where k runs from 1 to $N-1$ and j from 0 to $N-1$. Joining $\tilde{\mathbf{x}}_0$ defined by

$$\tilde{\mathbf{x}}_0 = \frac{1}{c}[n_1\ n_2 \cdots n_N] \quad (10)$$

with the above $\tilde{\mathbf{x}}_k$, we obtain the desired set of row vectors.

The spectral representation of $\boldsymbol{Z}$ is given by [14]

$$\boldsymbol{Z} = \sum_{k=0}^{N-1} \lambda_k \mathbf{x}_k \tilde{\mathbf{x}}_k$$

and hence its reciprocal $\boldsymbol{Y}$ by

$$\boldsymbol{Y} = \sum_{k=0}^{N-1} \lambda_k^{-1} \mathbf{x}_k \tilde{\mathbf{x}}_k. \quad (11)$$

Thus the first term in (1) becomes

$$\boldsymbol{Y}\boldsymbol{v} = \sum_{k=0}^{N-1} \lambda_k^{-1} \mathbf{x}_k (\tilde{\mathbf{x}}_k \cdot \boldsymbol{v}). \quad (12)$$

Similarly, expanding $\overline{\boldsymbol{Y}}\boldsymbol{v}$ in terms of $\mathbf{x}_k$, we have

$$\overline{\boldsymbol{Y}}\boldsymbol{v} = \sum_{k=0}^{N-1} \mathbf{x}_k (\tilde{\mathbf{x}}_k \cdot \overline{\boldsymbol{Y}}\boldsymbol{v}). \quad (13)$$

Substituting (12) and (13) into (1) gives

$$\sum (\tilde{\mathbf{x}}_k \cdot \boldsymbol{v})\lambda_k^{-1}\mathbf{x}_k - \sum (\tilde{\mathbf{x}}_k \cdot \overline{\boldsymbol{Y}}\boldsymbol{v})\mathbf{x}_k = 0$$

which is equivalent to

$$\tilde{\mathbf{x}}_k \cdot (\lambda_k^{-1}\boldsymbol{I} - \overline{\boldsymbol{Y}})\boldsymbol{v} = 0, \qquad k = 0, 1, 2, \cdots, N-1. \quad (14)$$

From the $(N-1)$ equations with $k = 1, 2, \cdots, N-1$, we conclude that

$$(G_0\boldsymbol{I} - \overline{\boldsymbol{Y}})\boldsymbol{v} = \alpha\mathbf{x}_0 \quad (15)$$

where α is a constant. Otherwise the $(N-1)$ equations cannot be satisfied simultaneously. Let us now assume that $G_0 - \overline{Y}_j$ does not vanish for all the j's $(1 \le j \le N)$; then $\alpha \ne 0$ for nonzero $\boldsymbol{v}$ and (15) becomes

$$\boldsymbol{v} = (G_0\boldsymbol{I} - \overline{\boldsymbol{Y}})^{-1}\alpha\mathbf{x}_0.$$

Substituting into (14) with $k=0$ gives

$$\tilde{\mathbf{x}}_0 \cdot (\lambda_0^{-1}\boldsymbol{I} - \overline{\boldsymbol{Y}})(G_0\boldsymbol{I} - \overline{\boldsymbol{Y}})^{-1}\alpha\mathbf{x}_0 = 0$$

which is equivalent to

$$Y_a = \frac{n_1^2 G_0 \overline{Y}_1}{G_0 - \overline{Y}_1} + \frac{n_2^2 G_0 \overline{Y}_2}{G_0 - \overline{Y}_2} + \cdots + \frac{n_N^2 G_0 \overline{Y}_N}{G_0 - \overline{Y}_N}. \quad (16)$$

Equation (16) is a necessary condition for free-running oscillation,[5] but it is not sufficient. For the oscillation to be realizable, another condition has to be satisfied as we shall see later (Section VI).

IV. Desired Mode of Oscillation

For the desired mode of oscillation, all the coaxial lines see the same field configuration near the coupling aperture except for possible difference in the sign. The amplitudes of the RF voltage are the same for all the ports. Thus we have

$$n_1^2 = n_2^2 = \cdots = n_N^2 \equiv n^2$$

$$\overline{Y}_1 = \overline{Y}_2 = \cdots = \overline{Y}_N \equiv \overline{Y}(A_0)$$

and

$$\boldsymbol{v}_0 = A_0 \frac{1}{n} \mathbf{x}_0 \exp(j\varphi_0) \quad (17)$$

where A_0 and φ_0 are the amplitude and phase of the RF voltage. Equation (16) reduces to

$$Y_a = \frac{n^2 N G_0 \overline{Y}(A_0)}{G_0 - \overline{Y}(A_0)}$$

or equivalently

$$\overline{Y}(A_0) = \frac{1}{(1/G_0) + (n^2 N / Y_a)}. \quad (18)$$

On the Smith chart (normalizing admittance: G_0), the right-hand side of (18) represents a circle passing through the center as shown in Fig. 3. To support the desired mode of oscillation, therefore, the device line $\overline{Y}(A)$ has to intersect the circle, and the intersection would determine the oscillation frequency ω_0 and amplitude A_0. In practice, the actual diode admittance, which is generally large compared to G_0, is transformed by the associated quarter-wavelength transformer shown in

[5] One might think that (16) is obvious from the equivalent circuit, indicating zero total admittance across the resonator terminals. However, it is not obvious that (15) and (16) are equivalent to (1) provided $G_0 - \overline{Y}_j \ne 0$ for all the j's. Note that a separate consideration of one mode at a time in (4) is not allowed, since the principle of superposition does not hold.

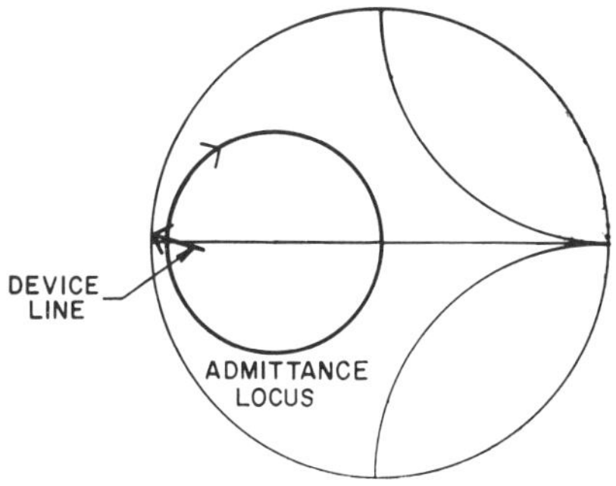

Fig. 3. Admittance locus $[(1/G_0)+(n^2N/Y_a)]^{-1}$ and device line $\overline{Y}(A)$.

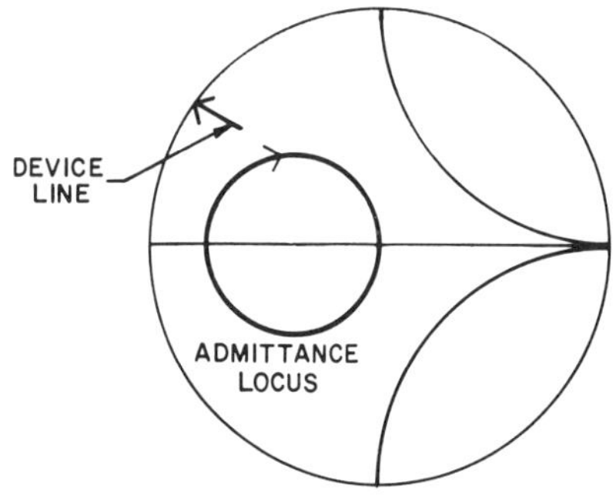

Fig. 4. Admittance locus $[(1/G_0)+(c/Y_a)]^{-1}$ and corresponding device line $\overline{Y}(A)$.

Fig. 1 and brought to the desired position on the Smith chart as illustrated in Fig. 3.

In the derivation of (16), we have assumed that $G_0-\overline{Y}_j\neq 0$. If the device line is as illustrated in Fig. 3, this condition is certainly satisfied, which is the case with ordinary IMPATT diodes.

For the desired mode, we have also assumed that the RF voltages have the same amplitude for all the ports. If the RF amplitudes are different, the device shows different admittances. Under this condition (15) and (16) might still be satisfied and lead to another mode of oscillation. Such a possibility will be eliminated as follows. We first notice that

$$|\lambda_m^{-1}| = G_0 \gg |\overline{Y}_j|, \qquad 1 \leq m \leq N-1, \quad 1 \leq j \leq N.$$

Then following a discussion dual to the one presented in [10], we immediately see that the free-running condition (1) is not satisfied unless v is proportional to $\mathbf{x}_0$ in "almost every case." Thus, the possibility of getting another mode of oscillation with the same resonant mode of the cavity is practically eliminated. If we further follow the discussion given in [10], the oscillation condition can be obtained for the case in which the device characteristics are slightly different from each other.

V. Suppression of Undesired Modes

Let us next consider a different resonant mode of the cavity, which will occur at a considerably different frequency. Since $\overline{Y}(A)$ is the admittance transformed by a certain length of transmission line (including the $\lambda/4$ transformer), $\overline{Y}(A)$ is now located at a different position on the Smith chart. For example, if the cavity mode is the next higher one, $\overline{Y}(A)$ may become something like the one shown in Fig. 4.

We assert that if the device line $\overline{Y}(A)$ is located outside the circle corresponding to $[(1/G_0)+(c/Y_a)]^{-1}$, then (16) is not satisfied. The proof is as follows.

By hypothesis, $\overline{Y}$ is outside the circle $[(1/G_0)+(c/Y_a)]^{-1}$; therefore $(1/\overline{Y})$ is outside the circle $(1/G_0)+(c/Y_a)$, $(1/\overline{Y})-(1/G_0)$ is outside the circle (c/Y_a), and $c[(1/\overline{Y})-(1/G_0)]^{-1}$ is outside the circle Y_a. However, since Y_a is a constant conductance circle, this means that

$$\text{Re } (\Sigma n_j^2) \frac{G_0\overline{Y}}{G_0-\overline{Y}} < \text{Re } Y_a$$

regardless of the value of A. Then, since

$$\text{Re } \Sigma \frac{n_j^2 G_0 \overline{Y}_j}{G_0-\overline{Y}_j} \leq \max \text{ Re } (\Sigma n_j^2) \frac{G_0\overline{Y}}{G_0-\overline{Y}} < \text{Re } Y_a$$

(16) cannot be satisfied.

Thus, as long as the device line is outside the circle $[(1/G_0)+(c/Y_a)]^{-1}$ as shown in Fig. 4, (16) is not satisfied and the corresponding mode of oscillation is suppressed.

For those resonant modes of the cavity other than the desired one, the coaxial lines are no longer at the maximum magnetic field positions and some of the couplings become considerably smaller than the original coupling n^2. This means that the circle $[(1/G_0)+(c/Y_a)]^{-1}$ is now smaller than the one shown in Fig. 3. Also, $\overline{Y}(A)$ is displaced from the position most favorable for the intersection. For these reasons it is not difficult with IMPATT diodes to arrange the circuit so as to place the device line $\overline{Y}(A)$ outside the circle $[(1/G_0)+(c/Y_a)]^{-1}$ for all the resonant modes of the cavity but the desired one. Furthermore, the desired cavity mode permits only one mode of oscillation, as discussed in Section IV. This is the reason why the oscillator shown in Fig. 1 can be made free from moding problems. If the terminations G_0 are replaced by short circuits, the circle touches the perimeter at off-resonant frequencies: it is then obviously much more difficult to avoid intersections of the device line with the circle and, therefore, to ensure the suppression of undesired modes.

VI. Stability of the Desired Mode

Suppose condition (16) is satisfied for the desired mode. We have still to investigate whether or not the corresponding intersection represents a stable operating point. To do so, we need dynamic equations for the amplitudes and phases of the RF voltages. Assuming that these are slowly varying functions of time and following the procedure developed for single-device oscillators [15], we obtain

$$\sum_j \left\{ Y_{ij} + Y_{ij}'\left(\frac{d\varphi_j}{dt} - j\frac{1}{A_j}\frac{dA_j}{dt}\right)\right\} A_j \exp(j\varphi_j) = \overline{Y}(A_i)A_i \exp(j\varphi_i) \quad (19)$$

where Y_{ij} indicates the ij component of the admittance matrix and the prime indicates the derivative with respect to ω. A_j and φ_j are the amplitude and phase of the RF voltage at the reference plane in the jth coaxial line. Essentially the same expression was previously obtained by Okabe and Okamura [16]. In the steady state (19) reduces to (1); therefore, from (17), the steady-state solution is given by

$$A_j \exp(j\varphi_j) = A_0 \frac{n_j}{n} \exp(j\varphi_0). \quad (20)$$

For small perturbations from the steady-state solution, we write

$$A_j \exp(j\varphi_j) = (A_0 + \delta A_j)\frac{n_j}{n} \exp(j\varphi_0)\exp(j\delta\varphi_j) \quad (21)$$

where δA_j and $\delta\varphi_j$ indicate the perturbations and φ_0 is now a function of time. We impose an additional condition to the $\delta\varphi_j$; i.e.,

$$\sum_j \delta\varphi_j = 0 \quad (22)$$

which is necessary since φ_0 is otherwise arbitrary. Substituting (21) into (19), we obtain

$$\begin{aligned}\sum_j Y_{ij}(\delta A_j + jA_0\delta\varphi_j)\frac{n_j}{n} &+ Y_{ij}'\left(A_0\frac{d\varphi_0}{dt} + A_0\frac{d\delta\varphi_j}{dt} - j\frac{d\delta A_j}{dt}\right)\frac{n_j}{n} \\ &= \overline{Y}(A_0)(\delta A_i + jA_0\delta\varphi_i)\frac{n_i}{n} + \frac{\partial\overline{Y}}{\partial A}A_0\delta A_i\frac{n_i}{n} \quad (23)\end{aligned}$$

where we have used the fact that (20) is the steady-state solution. Using (11), the above equation can be rewritten in the form

$$\begin{aligned}\sum_k \lambda_k^{-1}\mathbf{x}_k\tilde{\mathbf{x}}_k\cdot(\boldsymbol{\delta A} + jA_0\boldsymbol{\delta\varphi}) &+ \sum_k (\lambda_k^{-1})'\mathbf{x}_k\tilde{\mathbf{x}}_k\cdot\left(A_0\frac{d\varphi_0}{dt}\frac{1}{n}\mathbf{x}_0 + A_0\frac{d\boldsymbol{\delta\varphi}}{dt} - j\frac{d\boldsymbol{\delta A}}{dt}\right) \\ &= \overline{Y}(A_0)\sum_k \mathbf{x}_k\tilde{\mathbf{x}}_k\cdot(\boldsymbol{\delta A} + jA_0\boldsymbol{\delta\varphi}) + \frac{\partial\overline{Y}}{\partial A}A_0\sum_k \mathbf{x}_k\tilde{\mathbf{x}}_k\cdot\boldsymbol{\delta A} \quad (24)\end{aligned}$$

where

$$\boldsymbol{\delta A} = \begin{bmatrix}\frac{n_1}{n}\delta A_1 \\ \frac{n_2}{n}\delta A_2 \\ \vdots \\ \frac{n_N}{n}\delta A_N\end{bmatrix} \qquad \boldsymbol{\delta\varphi} = \begin{bmatrix}\frac{n_1}{n}\delta\varphi_1 \\ \frac{n_2}{n}\delta\varphi_2 \\ \vdots \\ \frac{n_N}{n}\delta\varphi_N\end{bmatrix}. \quad (25)$$

Equating coefficients of $\mathbf{x}_k$ $(k\neq 0)$ in (24), we have

$$G_0\tilde{\mathbf{x}}_k\cdot(\boldsymbol{\delta A} + jA_0\boldsymbol{\delta\varphi}) = (\overline{G} + j\overline{B})\tilde{\mathbf{x}}_k\cdot(\boldsymbol{\delta A} + jA_0\boldsymbol{\delta\varphi}) + \left(\frac{\partial\overline{G}}{\partial A} + j\frac{\partial\overline{B}}{\partial A}\right)A_0(\tilde{\mathbf{x}}_k\cdot\boldsymbol{\delta A}) \quad (26)$$

where $\overline{Y}$ is written in the form

$$\overline{Y} = \overline{G} + j\overline{B}. \quad (27)$$

Rewriting (26) in the form

$$\tilde{\mathbf{x}}_k\cdot\left[\left(G_0\boldsymbol{\delta A} - \overline{G}\boldsymbol{\delta A} + \overline{B}A_0\boldsymbol{\delta\varphi} - \frac{\partial\overline{G}}{\partial A}A_0\boldsymbol{\delta A}\right) + j\left(G_0A_0\boldsymbol{\delta\varphi} - \overline{G}A_0\boldsymbol{\delta\varphi} - \overline{B}\boldsymbol{\delta A} - \frac{\partial\overline{B}}{\partial A}A_0\boldsymbol{\delta A}\right)\right] = 0$$

we can immediately see that the real and imaginary parts in the bracket are each proportional to $\mathbf{x}_0$. Thus we have

$$\begin{aligned}\left(G_0 - \overline{G} - \frac{\partial\overline{G}}{\partial A}A_0\right)\tilde{\mathbf{x}}_k\cdot\boldsymbol{\delta A} + \overline{B}\tilde{\mathbf{x}}_k\cdot A_0\boldsymbol{\delta\varphi} &= 0 \\ -\left(\overline{B} + \frac{\partial\overline{B}}{\partial A}A_0\right)\tilde{\mathbf{x}}_k\cdot\boldsymbol{\delta A} + (G_0 - \overline{G})\tilde{\mathbf{x}}_k\cdot A_0\boldsymbol{\delta\varphi} &= 0. \quad (28)\end{aligned}$$

The determinant of the coefficients is given by

$$|G_0 - \overline{Y}|^2 + A_0\frac{1}{2}\frac{\partial}{\partial A}|G_0 - \overline{Y}|^2.$$

If, as is usually the case, $|G_0 - \overline{Y}|$ increases with A, this determinant is positive, and the only possible solution of (28) is a trivial one:

$$\tilde{\mathbf{x}}_k\cdot\boldsymbol{\delta A} = 0 \qquad \tilde{\mathbf{x}}_k\cdot A_0\boldsymbol{\delta\varphi} = 0, \qquad k\neq 0. \quad (29)$$

This means that as soon as the source of the perturbations is removed, the $\mathbf{x}_k$ $(k\neq 0)$ components of $\boldsymbol{\delta A}$ and $A_0\boldsymbol{\delta\varphi}$ become zero. Considering the fact that these components see G_0 as the load, which is free from stored energy, the result is understandable.

For $k=0$, (24) becomes

$$(g + jb)\tilde{\mathbf{x}}_0\cdot\boldsymbol{\delta A} + (g' + jb')\tilde{\mathbf{x}}_0\cdot\left(A_0\frac{d\varphi_0}{dt}\frac{1}{n}\tilde{\mathbf{x}}_0 - j\frac{d\boldsymbol{\delta A}}{dt}\right) = (\overline{G} + j\overline{B})\tilde{\mathbf{x}}_0\cdot\boldsymbol{\delta A} + \left(\frac{\partial\overline{G}}{\partial A} + j\frac{\partial\overline{B}}{\partial A}\right)A_0\tilde{\mathbf{x}}_0\cdot\boldsymbol{\delta A} \quad (30)$$

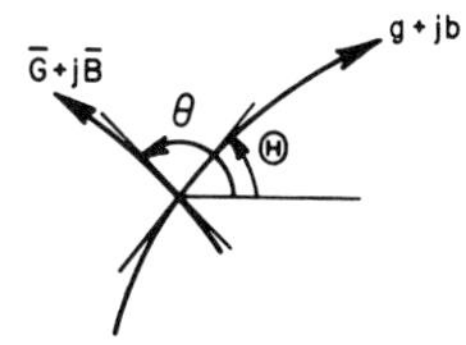

Fig. 5. Explanation of stability criterion. Arrow on the admittance locus indicates the direction of increasing ω; that on the device line the direction of increasing A.

where λ_0^{-1} is written in the form

$$\lambda_0^{-1} = g + jb = y \tag{31}$$

and use is made of (22) which is equivalent to

$$\tilde{\mathbf{x}}_0 \cdot \delta\boldsymbol{\varphi} = 0. \tag{32}$$

Since $g=\bar{G}$ and $b=\bar{B}$ at the operating point, the first terms on both sides of (30) cancel each other. Equating the real and imaginary parts separately, (30) gives

$$b' \frac{d}{dt} \tilde{\mathbf{x}}_0 \cdot \delta\mathbf{A} - \frac{d\bar{G}}{dA} A_0 \tilde{\mathbf{x}}_0 \cdot \delta\mathbf{A} + g' \frac{1}{n} A_0 \frac{d\varphi_0}{dt} = 0$$

$$-g' \frac{d}{dt} \tilde{\mathbf{x}}_0 \cdot \delta\mathbf{A} - \frac{d\bar{B}}{dA} A_0 \tilde{\mathbf{x}}_0 \cdot \delta\mathbf{A} + b' \frac{1}{n} A_0 \frac{d\varphi_0}{dt} = 0. \tag{33}$$

Eliminating $d\varphi_0/dt$ from (33), we have

$$(b'^2 + g'^2) \frac{d}{dt} \tilde{\mathbf{x}}_0 \cdot \delta\mathbf{A} - \left(b' \frac{d\bar{G}}{dA} - g' \frac{d\bar{B}}{dA}\right) A_0 \tilde{\mathbf{x}}_0 \cdot \delta\mathbf{A} = 0.$$

For stable operation, $\tilde{\mathbf{x}}_0 \cdot \delta\mathbf{A}$ has to decrease with time. This requires

$$b' \frac{d\bar{G}}{dA} - g' \frac{d\bar{B}}{dA} < 0. \tag{34}$$

Conversely, if (34) is satisfied, both $\tilde{\mathbf{x}}_0 \cdot \delta\mathbf{A}$ and $d\varphi_0/dt$ decrease with time. Taking angles Θ and θ as shown in Fig. 5, we have

$$b' \frac{d\bar{G}}{dA} - g' \frac{d\bar{B}}{dA} = -\left|y'\right| \left|\frac{d\bar{Y}}{dA}\right| \sin(\theta - \Theta).$$

Thus, if the intersecting angle measured counterclockwise from the admittance locus to the device line is less than π, $\tilde{\mathbf{x}}_0 \cdot \delta\mathbf{A}$ and $d\varphi_0/dt$ decrease with time, and vice versa.

From the above discussion we see that the intersection shown in Fig. 3 represents a stable operating point, since the intersecting angle is less than π and $|G_0 - \bar{Y}|$ increases with A, leading to (29) for all nonzero k's.

VII. Circuit Efficiency

We now calculate the output power P_o of the oscillator. For simplicity we assume that the oscillation frequency coincides with the resonant frequency of the cavity. Referring to Fig. 6, we have from the current relation

$$Y_a V_o = \Sigma n_j I_j = \Sigma n_j (\mathbf{Y}\mathbf{v}_0)_j. \tag{35}$$

Since

$$Y_a = G_c + G_L \qquad (\mathbf{Y}\mathbf{v}_0)_j = g A_0 \frac{n_j}{n} \exp(j\varphi_0)$$

$$g = \frac{1}{(1/G_0) + n^2 N/(G_c + G_L)} = \bar{G} \tag{36}$$

at the operating point, (35) gives

$$V_o = \frac{g A_0 n N}{G_c + G_L} \exp(j\varphi_0) = \frac{\bar{G} A_0 n N}{G_c + G_L} \exp(j\varphi_0).$$

The output power is given by

$$\begin{aligned} P_o &= \frac{1}{2} G_L V_o V_o^* = \frac{1}{2} \bar{G} A_0^2 N \frac{G_L}{G_c + G_L} \frac{n^2 N g}{G_c + G_L} \\ &= P_d N \frac{1}{1 + \dfrac{Q_{\text{ext}}}{Q_a}} \left(1 - \frac{\bar{G}}{G_0}\right) \\ &= P_d N \left(1 - \frac{Q_{\text{ext}}}{Q_{\text{ext}} + Q_a}\right)\left(1 - \frac{\bar{G}}{G_0}\right) \end{aligned} \tag{37}$$

where P_d is the power from an individual device and use is made of (36). Thus the circuit efficiency can be defined by

$$\eta = \left(1 - \frac{Q_{\text{ext}}}{Q_{\text{ext}} + Q_a}\right)\left(1 - \frac{\bar{G}}{G_0}\right). \tag{38}$$

For good efficiency, $\bar{G}$ must be small compared with G_0 and Q_a must be large compared to Q_{ext}. If the terminations in the coaxial lines are eliminated ($G_0 \to \infty$), the second factor disappears improving the circuit efficiency. However, without the terminations the suppression of undesired modes may become quite difficult, as explained in Section V.

VIII. Noise Performance

The output voltage of the oscillator is given by

$$\begin{aligned} V &= \sum_j \sum_i n_j Y_a^{-1} Y_{ji} (A_0 + \delta A_i + j A_0 \delta\varphi_i) \frac{n_i}{n} \exp(j\varphi_0) \\ &\quad + \sum_j \sum_i n_j (Y_a^{-1} Y_{ji})' \left(A_0 \frac{d\varphi_0}{dt} + A_0 \frac{d\delta\varphi_i}{dt} - j \frac{d\delta A_i}{dt}\right) \frac{n_i}{n} \exp(j\varphi_0) \\ &= \mathbf{c}\tilde{\mathbf{x}}_0 \cdot Y_a^{-1} \mathbf{Y} \left(\mathbf{x}_0 \frac{A_0}{n} + \delta\mathbf{A} + j A_0 \delta\boldsymbol{\varphi}\right) \exp(j\varphi_0) \\ &\quad + c\tilde{\mathbf{x}}_0 \cdot (Y_a^{-1}\mathbf{Y})' \left(A_0 \frac{d\varphi_0}{dt} \frac{\mathbf{x}_0}{n} + A_0 \frac{d\delta\boldsymbol{\varphi}}{dt} - j \frac{d\delta\mathbf{A}}{dt}\right) \exp(j\varphi_0) \end{aligned} \tag{39}$$

where $\delta\mathbf{A}$ and $\delta\boldsymbol{\varphi}$ are the small perturbations due to

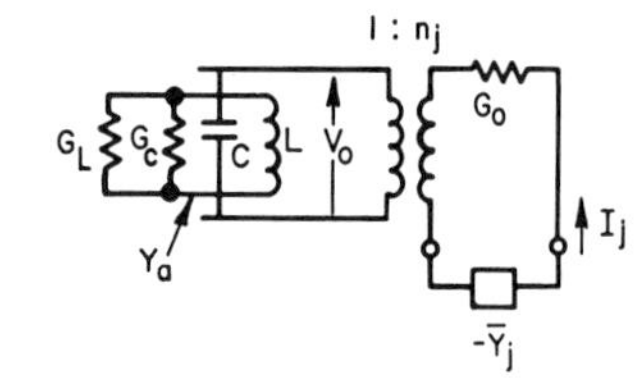

Fig. 6. Calculation of the output voltage V_o.

noise. Since

$$cY_a^{-1}\boldsymbol{Y} = \sum_k n^2 N Y_a^{-1}\lambda_k^{-1}\mathbf{x}_k\tilde{\mathbf{x}}_k\cdot$$

$$= \left\{1 - \frac{1}{G_0}(g + jb)\right\}\mathbf{x}_0\tilde{\mathbf{x}}_0\cdot + \sum_{k\neq 0} G_0(cY_a^{-1})\mathbf{x}_k\tilde{\mathbf{x}}_k\cdot$$

we have

$$c\tilde{\mathbf{x}}_0\cdot Y_a^{-1}\boldsymbol{Y} = \left\{1 - \frac{1}{G_0}(g + jb)\right\}\tilde{\mathbf{x}}_0\cdot$$

$$c\tilde{\mathbf{x}}_0\cdot(Y_a^{-1}\boldsymbol{Y})' = -\frac{1}{G_0}(g' + jb')\tilde{\mathbf{x}}_0\cdot.$$

Substituting into (39) gives

$$V = \left\{1 - \frac{1}{G_0}(g + jb)\right\}\left(\frac{A_0}{n} + \tilde{\mathbf{x}}_0\cdot\delta\boldsymbol{A}\right)\exp(j\varphi_0)$$
$$-\frac{1}{G_0}(g' + jb')\left\{\frac{A_0}{n}\frac{d\varphi_0}{dt} - j\frac{d}{dt}(\tilde{\mathbf{x}}_0\cdot\delta\boldsymbol{A})\right\}\exp(j\varphi_0). \quad (40)$$

To evaluate (40), we need $(d\varphi_0/dt)$ and $\tilde{\mathbf{x}}_0\cdot\delta\boldsymbol{A}$. They are obtainable from a dynamic equation similar to (19). We neglect the contribution of the noise originating in the load G_L, in the cavity loss G_c, and in the terminations G_0, compared with that originating in the devices. The noise contribution of each device can be represented by a noise-current source connected in parallel. Thus, corresponding to $\boldsymbol{Y}\boldsymbol{v} - \overline{\boldsymbol{Y}}\boldsymbol{v} = \boldsymbol{i}$, we have

$$\sum_k \lambda_k^{-1}\mathbf{x}_k\tilde{\mathbf{x}}_k\cdot(\delta\boldsymbol{A} + jA_0\delta\varphi) + \sum_k(\lambda_k^{-1})'\mathbf{x}_k\tilde{\mathbf{x}}_k$$
$$\cdot\left(A_0\frac{d\varphi_0}{dt}\frac{\tilde{\mathbf{x}}_0}{n} + A_0\frac{d\delta\varphi}{dt} - j\frac{d\delta\boldsymbol{A}}{dt}\right)$$
$$= \overline{Y}\sum_k \mathbf{x}_k\tilde{\mathbf{x}}_k\cdot(\delta\boldsymbol{A} + jA_0 d\varphi) + \frac{d\overline{Y}}{dA}A_0\sum_k \mathbf{x}_k\tilde{\mathbf{x}}_k\cdot\delta\boldsymbol{A}$$
$$+ \sum_k \mathbf{x}_k\tilde{\mathbf{x}}_k\cdot\boldsymbol{i}\exp(-j\varphi_0) \quad (41)$$

where $\boldsymbol{i}$ represents the noise sources. When $\boldsymbol{i}=0$, (41) reduces to (24). Equating the real and imaginary parts of the $\mathbf{x}_0$ component of (41) separately, we have

$$b'\frac{d}{dt}\tilde{\mathbf{x}}_0\cdot\delta\boldsymbol{A} - \frac{d\overline{G}}{dA}A_0\tilde{\mathbf{x}}_0\cdot\delta\boldsymbol{A} + g'\frac{1}{n}A_0\frac{d\varphi_0}{dt} = \tilde{\mathbf{x}}_0\cdot\boldsymbol{i}_c$$
$$-g'\frac{d}{dt}\tilde{\mathbf{x}}_0\cdot\delta\boldsymbol{A} - \frac{d\overline{B}}{dA}A_0\tilde{\mathbf{x}}_0\cdot\delta\boldsymbol{A} + b'\frac{1}{n}A_0\frac{d\varphi_0}{dt} = \tilde{\mathbf{x}}_0\cdot\boldsymbol{i}_s \quad (42)$$

where

$$\boldsymbol{i}_c = \mathrm{Re}\left\{\boldsymbol{i}\exp(-j\varphi_0)\right\} \qquad \boldsymbol{i}_s = \mathrm{Im}\left\{\boldsymbol{i}\exp(-j\varphi_0)\right\}.$$

Equation (42) reduces to (33) when $\boldsymbol{i}_c = \boldsymbol{i}_s = 0$. Solving (42) and substituting the result into (40), we get the desired expression for the output voltage. For simplicity, however, we again assume that the oscillation frequency coincides with the resonant frequency of the cavity. Then, we have

$$b = 0 \qquad g' = 0. \quad (43)$$

Furthermore, we assume that

$$\frac{d\overline{B}}{dA} \simeq 0. \quad (44)$$

The two equations in (42) now become independent of each other and we can easily obtain

$$\overline{(\tilde{\mathbf{x}}_0\cdot\delta\boldsymbol{A})^2} = \frac{\overline{(\tilde{\mathbf{x}}_0\cdot\boldsymbol{i}_c)^2}}{\omega^2 b'^2 + \left(\frac{d\overline{G}}{dA}A_0\right)^2} \quad (45)$$

$$\overline{\left(\frac{1}{n}A_0\frac{d\varphi_0}{dt}\right)^2} = \frac{\overline{(\tilde{\mathbf{x}}_0\cdot\boldsymbol{i}_s)^2}}{b'^2}. \quad (46)$$

At the operating point we have

$$\frac{\omega_0 b'}{2\overline{G}} = \eta Q_{\mathrm{ext}}. \quad (47)$$

We define a saturation factor s by

$$s \equiv \frac{1}{\overline{G}}\frac{d\overline{G}}{dA}A_0.$$

The saturation factor is approximately two for the maximum output power condition. Consequently, the first term in the denominator of (45) can be neglected compared with the second term, provided that $Q_{\mathrm{ext}}(\omega/\omega_0)$ is small compared with 1. Under the same assumption, the second term on the right-hand side of (40) can be neglected compared to the first term. To the first-order approximation, frequency fluctuation arises from $\overline{(d\varphi_0/dt)^2}$ only. Thus, a little manipulation shows that

$$\frac{\overline{(\Delta V)^2}}{V_o^2} \simeq \frac{kTB}{NP_d(s/2)^2} = \frac{\eta kTB}{P_o(s/2)^2} \quad (48)$$

$$\frac{\overline{(\Delta f)^2}}{f_0^2} \simeq \frac{kTB}{NP_d Q_{\mathrm{ext}}^2\eta^2} = \frac{kTB}{P_o Q_{\mathrm{ext}}^2\eta} \quad (49)$$

where

$$kTB = \frac{\overline{(i_{cj})^2}}{4\overline{G}} = \frac{\overline{(i_{sj})^2}}{4\overline{G}}$$

k is Boltzmann's constant, B the bandwidth, and T the noise temperature of the device. In (48) and (49), $\overline{(\Delta V)^2}$ and $\overline{(\Delta f)^2}$ are the mean square values of the amplitude and frequency fluctuations. They are both inversely proportional to the number of devices N. Thus, as the number of devices increases, the oscillator becomes less noisy.

IX. Injection Locking

Suppose that the oscillator is injection locked by an external signal. The signal can be represented by a current source i_0 parallel to the load G_L. In the steady state we have

$$\boldsymbol{Y}v - \overline{\boldsymbol{Y}}v + \frac{1}{n^2N}\left\{1 - \frac{1}{G_0}(g+jb)\right\} i_0\mathbf{x}_0 = 0. \quad (50)$$

The locking frequency ω may not be the same as the free-running frequency ω_0. So, rewriting $\boldsymbol{Y}$ in the form $\{\boldsymbol{Y} - \boldsymbol{Y}(\omega_0)\} + \boldsymbol{Y}(\omega_0)$, (50) becomes

$$\begin{aligned}\sum_k \{\lambda_k^{-1} - \lambda_k^{-1}(\omega_0)\}\mathbf{x}_k\tilde{\mathbf{x}}_k \cdot \frac{1}{n} A_0\mathbf{x}_0 \\ + \sum_k \lambda_k^{-1}(\omega_0)\mathbf{x}_k\tilde{\mathbf{x}}_k\cdot(\Delta\boldsymbol{A} + jA_0\Delta\varphi) \\ = \overline{Y}(A_0)\sum_k \mathbf{x}_k\tilde{\mathbf{x}}_k\cdot(\Delta\boldsymbol{A} + jA_0j\Delta\varphi) \\ + \frac{\partial\overline{Y}}{\partial A}A_0\sum_k \mathbf{x}_k\tilde{\mathbf{x}}_k\cdot\Delta\boldsymbol{A} \\ - \frac{1}{n^2N}\left\{1 - \frac{1}{G_0}(g+jb)\right\} i_0\mathbf{x}_0\exp(-j\varphi_0). \quad (51)\end{aligned}$$

The $\mathbf{x}_k$ $(k \neq 0)$ component of (51) gives only trivial solutions. For $k=0$, (51) becomes

$$\begin{aligned}\frac{d\overline{G}}{dA}A_0(\tilde{\mathbf{x}}_0\cdot\Delta\boldsymbol{A}) &= \Delta g\frac{A_0}{n} + \frac{i_0}{n^2N}\{(1 - G_0^{-1}g) \\ &\quad \cdot\cos\varphi_0 - G_0^{-1}b\sin\varphi_0\} \\ \frac{d\overline{B}}{dA}A_0(\tilde{\mathbf{x}}_0\cdot\Delta\boldsymbol{A}) &= \Delta b\frac{A_0}{n} + \frac{i_0}{n^2N}\{-(1 - G_0^{-1}g) \\ &\quad \cdot\sin\varphi_0 - G_0^{-1}b\cos\varphi_0\} \quad (52)\end{aligned}$$

where

$$\Delta g = g(\omega) - g(\omega_0), \qquad \Delta b = b(\omega) - b(\omega_0). \quad (53)$$

Eliminating $(\tilde{\mathbf{x}}_0\cdot\Delta A)$ from (52), we obtain

$$\sin(\varphi_0 + \psi) = \frac{\dfrac{nNA_0}{i_0}\left(\dfrac{d\overline{G}}{dA}\Delta b - \dfrac{d\overline{B}}{dA}\Delta g\right)}{\left|\dfrac{\partial\overline{Y}}{\partial A}\right|\left|1 - G_0^{-1}g + jG_0^{-1}b\right|} \quad (54)$$

where

$$\tan\psi = \frac{\dfrac{d\overline{B}}{dA}(1 - G_0^{-1}g) + \dfrac{d\overline{G}}{dA}G_0^{-1}b}{\dfrac{d\overline{G}}{dA}(1 - G_0^{-1}g) - \dfrac{d\overline{B}}{dA}G_0^{-1}b}. \quad (55)$$

Since the magnitude of sine is less than 1, a locking formula can be derived from (54). For simplicity, we will assume that the free-running frequency coincides with the resonant frequency of the cavity. Then $\Delta g = 0$, $b=0$, and (54) reduces to

$$\sin(\varphi_0 + \psi) = \sqrt{\frac{P_o}{P_i}}\,Q_{\text{ext}}\frac{\Delta\omega}{\omega_0}\frac{\dfrac{d\overline{G}}{dA}}{\left|\dfrac{d\overline{Y}}{dA}\right|}$$

where P_i is the injection signal power and $\Delta\omega = \omega - \omega_0$. The corresponding locking range is given by

$$\left|\frac{\Delta\omega}{\omega_0}\right| < \sqrt{\frac{P_i}{P_o}}\frac{1}{Q_{\text{ext}}}\frac{\left|\dfrac{d\overline{Y}}{dA}\right|}{\left|\dfrac{d\overline{G}}{dA}\right|}. \quad (56)$$

This is identical to the formula for single-device oscillators. The dynamic equation for injection locking can be derived from (50). The x_0 component gives

$$\begin{aligned}\left(b'\frac{d}{dt} - \frac{d\overline{G}}{dA}A_0\right)(\tilde{\mathbf{x}}_0\cdot\delta\boldsymbol{A}) + g'\frac{A_0}{n}\frac{d\delta\varphi_0}{dt} \\ = \frac{i_0\delta\varphi_0}{n^2N}\{(1 - G_0^{-1}g)\sin\varphi_0 + G_0^{-1}b\cos\varphi_0\} \\ \left(-g'\frac{d}{dt} - \frac{d\overline{B}}{dA}A_0\right)(\mathbf{x}_0\cdot\delta\boldsymbol{A}) + b'\frac{A_0}{n}\frac{d\delta\varphi_0}{dt} \\ = \frac{i_0\delta\varphi_0}{n^2N}\{(1 - G_0^{-1}g)\cos\varphi_0 - G_0^{-1}b\sin\varphi_0\}.\end{aligned}$$

From these equations two conditions necessary for stable locking are obtained. One is usually satisfied if the operating point is stable. The other gives

$$\cos(\varphi_0 + \psi) > 0.$$

This together with (54) uniquely determines the phase angle φ_0 relative to the injection signal. Thus, the locking behavior of the single-cavity multiple-device oscillator is identical to ordinary single-device oscillators.

X. Conclusion

The theory of the single-cavity multiple device oscillator has been developed using the eigenfunction approach. With the proposed circuit configuration it was found easy to get around the well-known moding problem of multiple-device oscillators. The noise behavior should be similar and the injection-locking behavior identical to those of ordinary single-device oscillators.

Acknowledgment

The author wishes to thank N. D. Kenyon, who reviewed the manuscript, and F. M. Magalhaes and J. Federico, who conducted experiments indispensable to the development of the theory.

References

[1] J. G. Josenhans, "Diamond as an insulating heat sink for a series combination of IMPATT diodes," *Proc. IEEE* (Lett.), vol. 56, Apr. 1968, pp. 762–763.

OTHER ARTICLES OF INTEREST

[2] D. L. Scharfetter, W. J. Evans, and R. L. Johnston. "Double-drift-region (p^+pnn^+) avalanche diode oscillation," *Proc. IEEE* (Lett.) vol. 58, July 1970, pp. 1131–1133.
[3] T. E. Seidel and D. L. Scharfetter, "High-power millimeter wave IMPATT oscillators with both hole and electron drift spaces made by ion implantation," *Proc. IEEE* (Lett.), vol. 58, July 1970, pp. 1135–1136.
[4] T. Misawa, "Microwave Si avalanche diodes with nearly-abrupt type junction," presented at 1966 Int. Electron Device Meeting, Washington, D. C.
[5] C. B. Swan, "The importance of providing a good heat sink for avalanching transit time oscillator diodes," *Proc. IEEE* (Lett.), vol. 55, Mar. 1967, pp. 451–452.
[6] ——, "Improved performance of silicon avalanche oscillators mounted on diamond heat sinks," *Proc. IEEE* (Lett.), vol. 55, Sept. 1967, pp. 1617–1618.
[7] H. Fukui, "Frequency locking and modulation of microwave silicon avalanche diode oscillators," *Proc. IEEE* (Lett.), vol. 54, Oct. 1966, pp. 1475–1477.
[8] J. F. Reintjes and G. T. Coate, *Principle of Radar.* New York: McGraw-Hill, 1952, pp. 712–716.
[9] C. T. Rucker, "A multiple-diode high-average-power avalanche-diode oscillator," *IEEE Trans. Microwave Theory Tech.* (Corresp.), vol. MTT-17, Dec. 1969, pp. 1156–1158.
[10] K. Kurokawa, "An analysis of Rucker's multidevice symmetrical oscillator," *IEEE Trans. Microwave Theory Tech.* (Corresp.), vol. MTT-18, Nov. 1970, pp. 967–969.
[11] K. Kurokawa and F. M. Magalhaes, "An *X*-band 10-watt multiple-IMPATT oscillator," *Proc. IEEE* (Lett.), vol. 59, Jan. 1971, pp. 102–103.
[12] G. Helmberg, *Introduction to Spectral Theory in Hilbert Space.* New York: Wiley-Interscience, 1969, p. 49.
[13] R. Courant and D. Hilbert, *Methods of Mathematical Physics,* vol. 1. New York: Wiley-Interscience, 1953.
[14] K. Kurokawa, *An Introduction to the Theory of Microwave Circuits.* New York: Academic Press, 1969, chs. 5, 6.
[15] ——, "Some basic characteristics of broadband negative resistance oscillator circuits," *Bell Syst. Tech. J.*, vol. 48, July–Aug. 1969, pp. 1937–1955.
[16] Y. Okabe and S. Okamura, "Analysis of stability and noise of oscillators in free-running, synchronized-running and parallel-running," *J. Inst. Elect. Commun. Eng.* (Tokyo), vol. 52-B, Dec. 1969, pp. 755–762.
[17] F. M. Magalhaes and K. Kurokawa, "A single-tuned oscillator for IMPATT characterizations," *Proc. IEEE* (Lett.), vol. 58, May 1970, pp. 831–832.

Microstrip IMPATT Oscillator with High Locking Figure of Merit

Abstract—Microstrip IMPATT oscillators have been built at *X*-band and *V*-band in order to obtain hybrid integrated power sources for locked amplifier applications. It is shown that using a microstrip circuit shorter than a half-wavelength gives the highest locking figure of merit and therefore the largest possible locking range for a given gain. A figure of merit of 0.10 has been obtained for a locking gain of 20 dB.

Microstrip IMPATT oscillators with a high locking figure of merit have been built in order to obtain power sources which can be phase-locked over a large frequency range for a given gain. The circuit[1,2] used for the *X*-band oscillator shown in Fig. 1 consists of a silicon avalanche diode in a V-package[3,4] which is mounted in a short microstrip cavity. The diode is connected between the ground plane and the center of a microstrip loop obtained by shorting both ends of a microstrip transmission line. The frequency of oscillation multiplied by the square root of the loop length is constant, as shown in Fig. 2. This result shows that the loop, which is much shorter than half a wavelength, acts to a first approximation as a lumped inductance in parallel with the diode. Because of the small amount of RF energy stored in the circuit, one expects to obtain the lowest circuit Q with the largest possible locking range for a given gain.

Solid-state RF sources require a dc bias circuit which does not affect the RF performance. A simple solution has been found for applying the dc bias to solid-state devices built into microstrip circuits. A thin dielectric sheet with a hole for the diode is mounted on a solid metal block which is the heat sink for the diode, as shown in Fig. 3. The dielectric sheet is metallized on both sides and provides an RF bypass as well as two terminals for applying the dc bias. This planar bypass configuration, which can be considered as a low impedance radial transmission line, has a much higher capacitance than the conventional coaxial RF bypass used in waveguide and coaxial circuits. Because of the low impedance RF bypass, spurious oscillations in the microstrip IMPATT oscillator are eliminated.

The RF coupling between the circuit and the coaxial transmission line required for measuring the output power is obtained with a microstrip line in series with a coupling capacitor. The coupling capacitance, the length of the stripline, and the impedance are adjusted for maximum output power. A similar circuit built at 30 GHz uses the same passive elements with a reduction of all physical dimensions. A transition with a tapered transformer from microstrip to RG 96/U waveguide is used, as shown in

Manuscript received July 25, 1969.

[1] S. Mao, H. L. Stover, G. L. Spence, and R. W. Dawson, "Integrated microwave transit-time oscillators," presented at the Internatl. Electron Devices Meeting, Washington, D. C., October 1967.

[2] R. J. Wagner, W. W. Gray, and P. V. Cooper, "*X*-band IMPATT microstrip power sources," *IEEE J. Solid-State Circuits*, vol. SC-3, pp. 221–225, September 1968.

[3] D. F. Ciccolella, "Microwave diode research," ASTIA Doc. AD-822016, October 1967.

[4] H. E. Elder, "Active solid-state devices," *Bell Sys. Tech. J.*, vol. 47, pp. 1323–1377, September 1968.

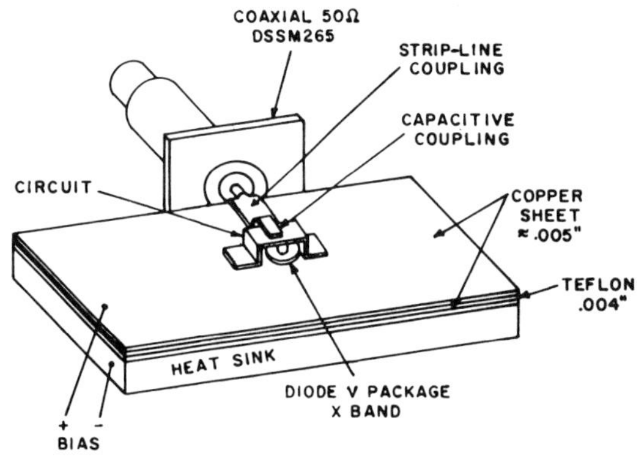

Fig. 1. Experimental assembly for *X*-band microstrip IMPATT oscillator.

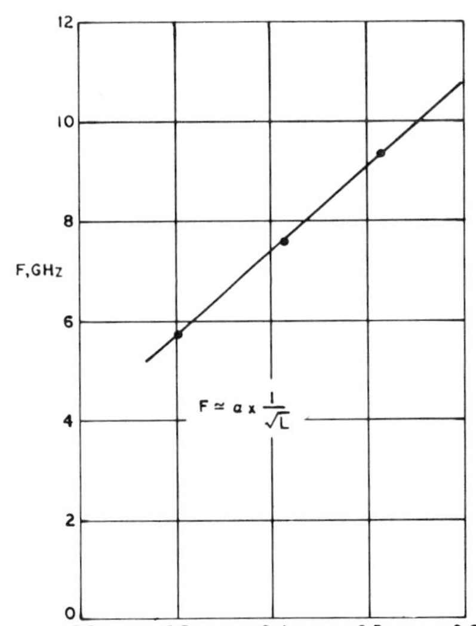

Fig. 2. Frequency variation versus loop length L in millimeters.

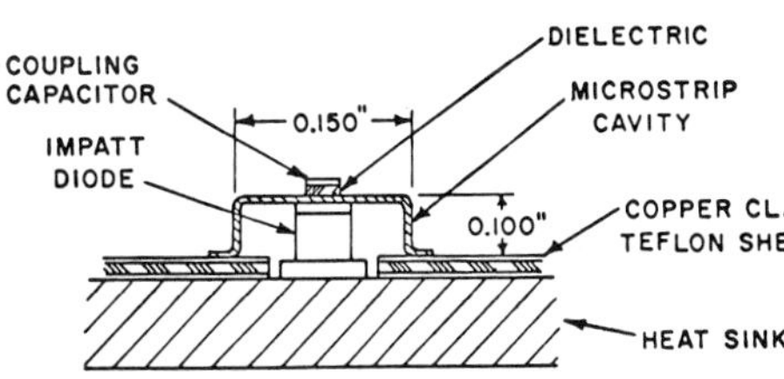

Fig. 3. Microstrip mount for *X*-band oscillator.

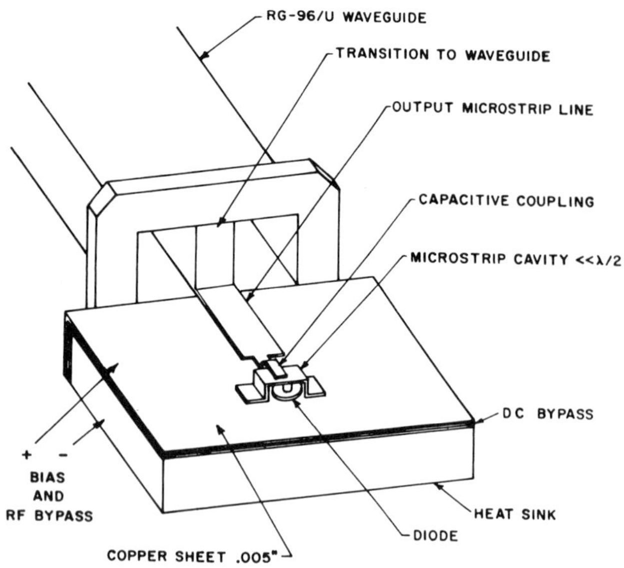

Fig. 4. 30-GHz microstrip IMPATT oscillator.

TABLE I

Circuit and Diode Parameters RF Output	*X*-Band Oscillator	*V*-Band Oscillator
Microstrip cavity dimensions (inch)	length l =0.150 width w=0.120 height h=0.100	l =0.070 w=0.070 h=0.020
Silicon IMPATT diode	*V*-type package[3,4]	Stud-mounted, epoxy-coated diode
Junction capacitance at V_B (pF)	0.44	0.22
Breakdown voltage V_B (volt)	80	24
Inductance across the diode (nH)	0.32	0.16
Coupling capacitance (pF)	0.1	0.01
Oscillator frequency (GHz)	9.6	28.6
RF output (mW)	225	135
Efficiency (percent)	2.3	2.2

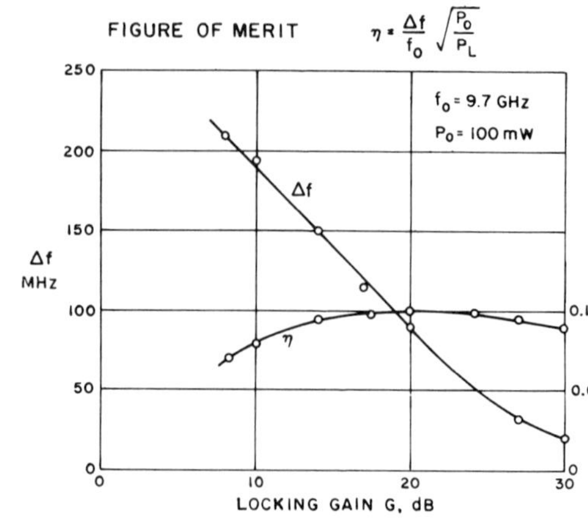

Fig. 5. Locking performance.

Fig. 4. The circuit parameters and the RF performance obtained with an *X*-band and a *V*-band microstrip oscillator are given in Table I. Locking measurements performed with the *X*-band oscillator are shown in Fig. 5. The locking figure of merit η is defined by

$$\eta = \frac{\Delta f}{f_0}\sqrt{\frac{P_0}{P_i}}$$

where Δf is the locking bandwidth, f_0 and P_0 the frequency and the power output of the free-running oscillator, and P_i the injected locking power.

The value of η is optimized by adjusting the impedance of the microstrip output line and the coupling capacitance. It is found that this adjustment does not affect the output power and the frequency of oscillation substantially.

ACKNOWLEDGMENT

The author wishes to thank B. C. DeLoach and D. F. Ciccolella for supplying the IMPATT diodes and for helpful discussions.

BERNARD GLANCE
Crawford Hill Laboratory
Bell Telephone Laboratories, Inc.
Holmdel, N. J. 07733

Microstrip High-Power High-Efficiency Avalanche-Diode Oscillator

SHING-GONG LIU, MEMBER, IEEE

Abstract—**A simple microstrip oscillator circuit has been designed and operated satisfactorily with high-power high-efficiency avalanche diodes. The power output obtained from a single diode chip is about 100 watts at 1 GHz with efficiencies of 25 to 30 percent. Mechanical tuning capability of a few hundred megahertz and a combined power output from series-connected diodes have been demonstrated using the circuit.**

An essential part of the circuit is a low-pass filter tuning section which enables the circuit to support high-order harmonics including the transit-time frequency and prevents them from getting to the load. Probe measurements of the electric field show strong second as well as third harmonics inside the circuit.

Introduction

THIS PAPER describes a simple microstrip oscillator circuit using avalanche diodes operating in the high-efficiency mode previously reported [1], [2]. Using this circuit, we have obtained from a single diode pulsed power of 100 watts at 1 GHz with efficiencies of 25 to 30 percent. Series-connected diodes with combined power output have also been achieved using the circuit. This result demonstrates the capability of integration of such high-efficiency diodes for system applications.

It will be recalled that the high-efficiency mode is initiated in avalanche diodes when the dc bias reaches a high current-density level. At lower current density, the diodes oscillate in the normal transit-time mode. The frequency of the anomalous-mode oscillation is lower than the transit-time oscillation, and the efficiency and power are many times higher than that from the transit-time mode.

Circuit Description and Impedance Calculation

Several proposed models as well as experimental evidence [2]–[5] indicate that in order to excite the high-efficiency mode of oscillation, the circuit must be capable of supporting higher order harmonics including the transit-time frequency.

The circuit described below allows the diode to be matched to the 50-ohm load at the fundamental frequency, i.e., the low high-efficiency mode frequency. At the same time, the diode sees a reactive termination at higher frequencies. As a result, the circuit supports high-order frequencies, yet they do not appear at the output terminal. In principle, the circuit is the same as the coaxial circuit reported by Levine and Liu [6].

Fig. 1 is a schematic diagram of two configurations of the microstrip circuit. Both circuits work well as will be described later. Basically, the circuit consists of a 50-ohm transmission line, with a short circuit at one end and a tuning section at the other end before the connection to the 50-ohm load. The diode chip is mounted directly onto the microstrip. The top side of the diode is connected to a pulser with an RF bypass capacitor to ground. The tuning section is essentially a low-pass filter. Fig. 1(a) has a π section and Fig. 1(b), a T section. The capacitance of the filter section is formed by a short open-circuit line segment. At the harmonic frequencies, the combined impedance presented at the position of the diode by the tuning section, the 50-ohm load, and the short circuit is nearly a pure reactance.

The circuit impedance presented at the position of the diode has been calculated up to the fifth harmonic using circuit parameters and frequency under optimum power-output conditions. The impedances for all harmonics are reactive except that a small resistive part of less than 2 ohms is presented at the second-harmonic frequency. The small resistive part is due to the nonsharp cutoff between the fundamental and second harmonic of the filter section. At the fundamental frequency the impedance seen at the diode looking toward the circuit is calculated to be about 10 ohms real and 10 ohms reactive. The calculation is based on transmission-line theory under the assumption that only the TEM mode is supported.

Fig. 2 shows an example of the result of such calculations on a Smith chart, normalized to 50 ohms. The dotted line with an arrow indicates the trend of change in impedance as one increases the value of capacitance in the low-pass filter tuning section. At the point where power is optimized, which is indicated by the solid dot on the dotted line, the impedances of harmonics are close to the outer circle of the Smith chart, or nearly purely reactive; the impedance at the fundamental frequency is about $14-j10$ ohms.

Experiments

Simple circuits were constructed for experimental purposes. First experimental circuits were made of conventional microstrip of metal on Teflon. The substrate has a dielectric constant of 2.3. The measured ratio of free space to microstrip wavelength of 1.4 for the microstrip line used checks well with published data. Fig. 3 shows one of the circuits with a π-type tuning section. The shorting position can be varied in steps using a fine screw threaded through the substrate to ground. The fine hole in the microstrip introduces a small inductive reactance which has a negligible effect on circuit performance. The two tunable capacitive elements are thin gold-plated steel plates. They can be moved along

Manuscript received May 16, 1969; revised July 16, 1969. This work was supported by the Air Force Avionics Laboratory, Wright-Patterson AFB, Ohio, under Contract F33615-68-C-1688. This paper was presented at the International Microwave Symposium, Dallas, Tex., May 5–7, 1969.

The author is with RCA Laboratories, Princeton, N. J

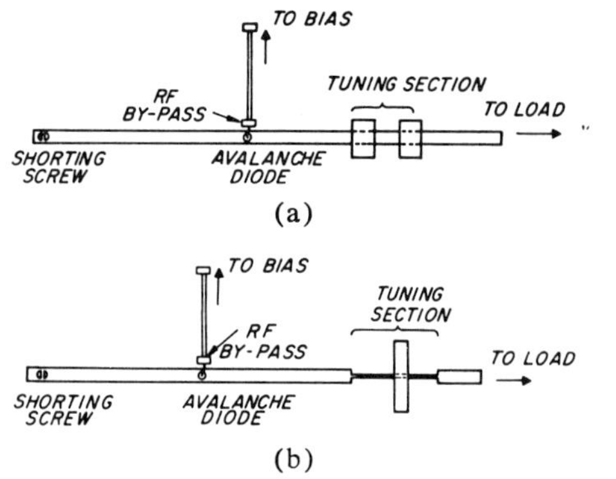

Fig. 1. Schematic diagram of microstrip high-efficiency avalanche-diode oscillator circuit with (a) π or (b) T tuning section.

Fig. 3. Experimental circuit with adjustable short-circuit and π tuning section. Tuning elements are held in position by magnetic bars underneath the substrate.

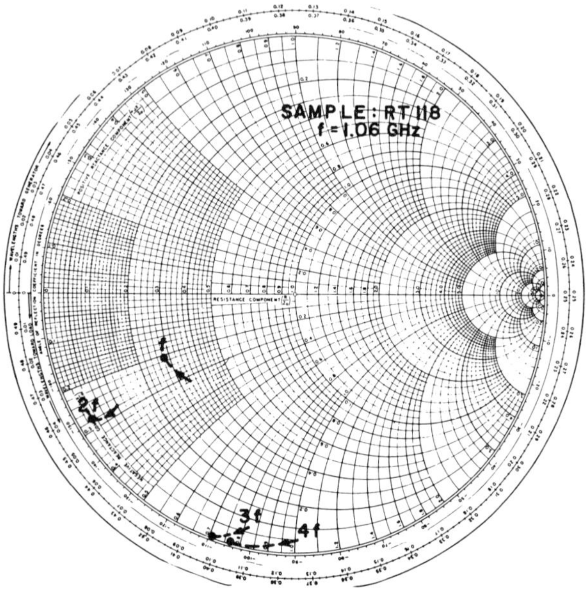

Fig. 2. Sample calculations of circuit impedance at fundamental and harmonic frequencies presented at the position of the diode. Arrows indicate change in impedance as the capacitance in the T tuning section increases.

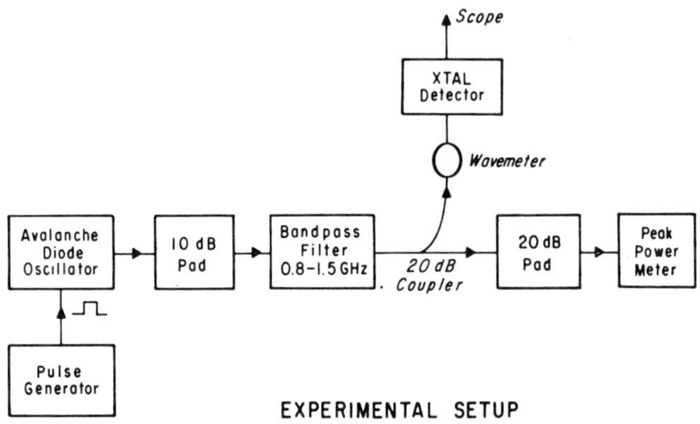

Fig. 4. Experimental setup for measurement of power and frequency.

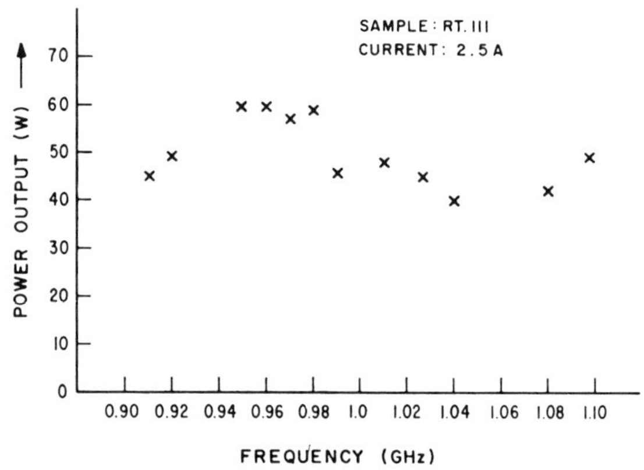

Fig. 5. Experimental results of power output versus frequency.

the line for tuning for optimum power output. Different values of capacitance can be obtained by changing plate size. The plates are held in position by magnet bars placed underneath the substrate. The bypass capacitor for the pulsed voltage is a chip capacitor connecting through the substrate to ground at a point next to the diode chip.

568

The diode is a p-n-n$^+$ silicon mesa structure. The diameter of the mesa is 0.018 to 0.023 inch. The junctions are formed by diffusion of boron into n-type silicon epitaxial wafers. The resistivity of the epitaxial layer is around 6 $\Omega \cdot$cm. The breakdown voltage of the diode is between 110 and 150 volts.

With the simple mechanically tunable experimental circuit just described, microwave oscillation in the high-efficiency mode at 1 GHz with a few hundred MHz tuning capability has been demonstrated. The experimental setup is shown in Fig. 4. Wide-band precision attenuators and peak power meters were used for power measurement. Bandpass filters are only used to ensure a single frequency in power measurement. Harmonic content in the output when filters are removed is very low, usually 15 to 20 dB below the fundamental.

The power output versus frequency is shown in Fig. 5. Here the frequency varies in steps over a range of about 25 percent, and the power variation is about 1.7 dB. The frequency is varied systematically by moving the position of the short circuit in steps, and correspondingly, the tuning segments. Frequency increases as the shorting position and the tuning segments move toward the diodes. In other words, the movement is only to compensate for the frequency change

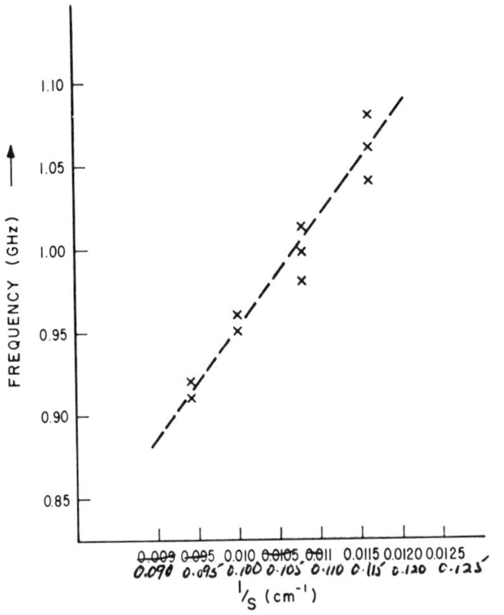

Fig. 6. Oscillation frequency as a function of distance between diode and the short circuit.

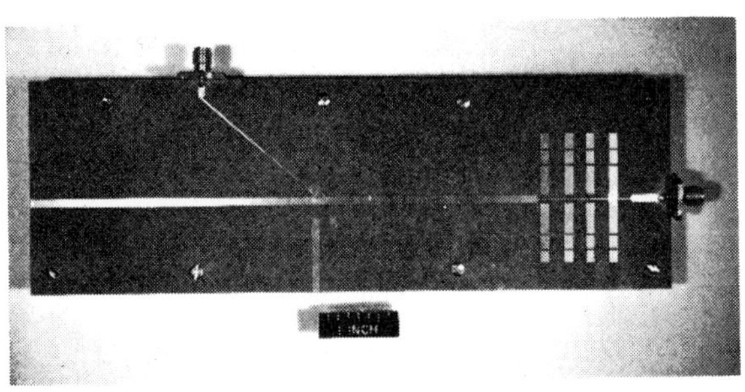

Fig. 7. Experimental circuit with T tuning section. One of the capacitive tuning elements is connected to the main line for optimum power output. The vertical line and dot near diode are not part of the circuit.

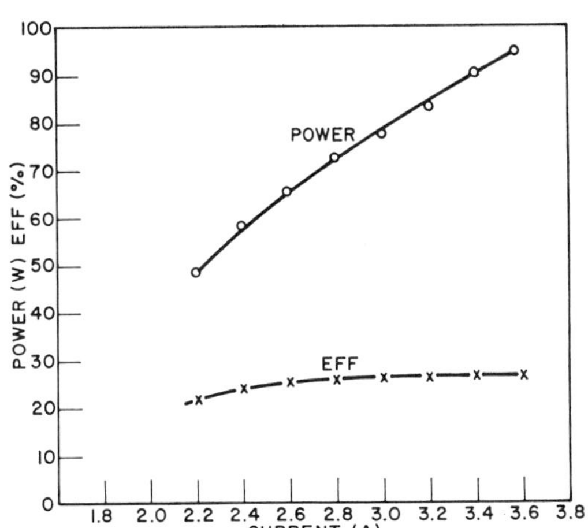

Fig. 8. Power output and efficiency versus pulsed current of an avalanche-diode oscillator operating at 1.06 GHz.

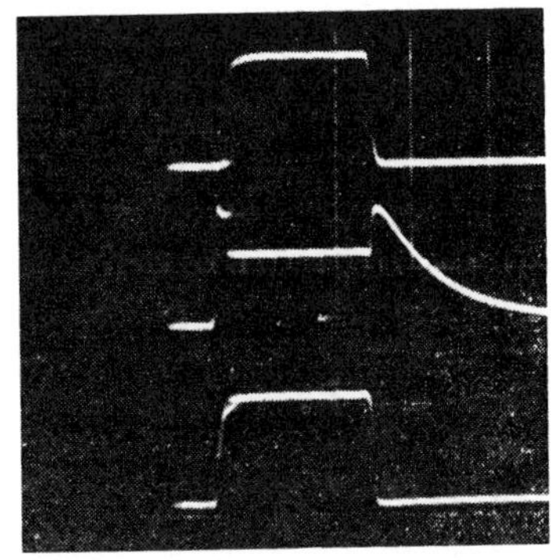

Fig. 9. Waveforms of detected RF envelope at 1.06 GHz (top), and average voltage (middle) and current (bottom) pulses at the diode. Scale: top—70 W/div; middle—100 V/div; bottom—2 A/div; time—0.5 μs/div.

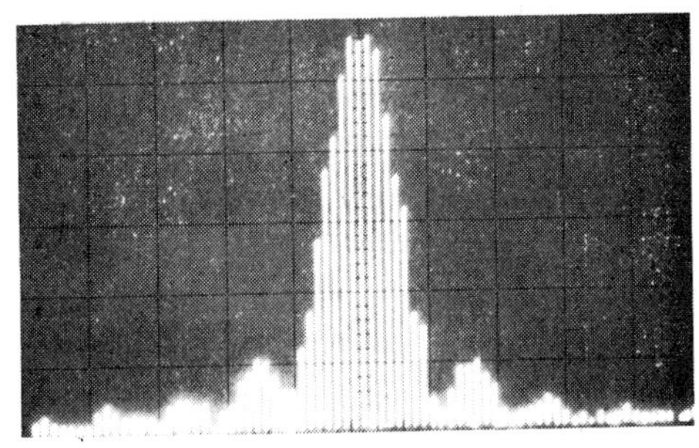

Fig. 10. Frequency spectrum of pulsed power output. Center frequency: 1.06 GHz. Horizontal scale: 3 MHz/div.

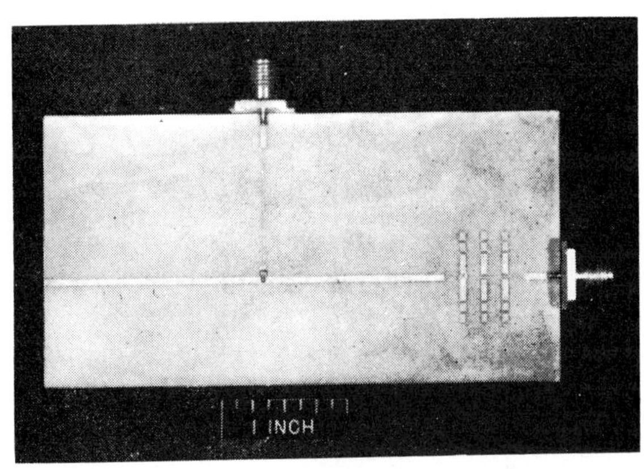

Fig. 11. High-power high-efficiency avalanche-diode oscillator using metal-on-alumina microstrip.

569

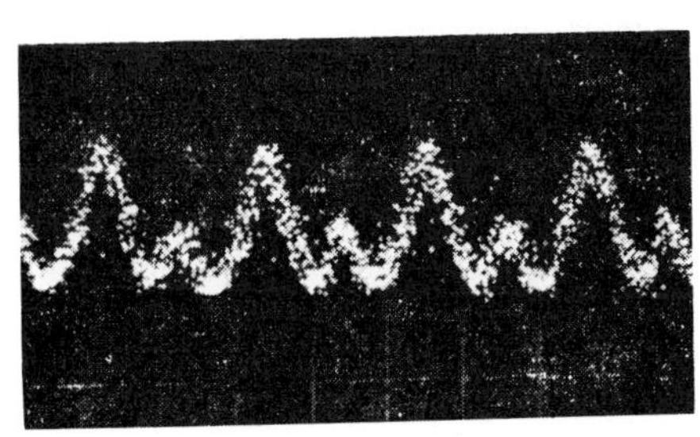

Fig. 12. Electric-field waveform corresponding to a probe location picking up a strong second-harmonic component. Scale: 0.5 ns/div.

OTHER ARTICLES OF INTEREST

while the impedance presented at the diode at different frequencies remains unchanged. A plot showing the relation between the distance from the short circuit to the diode and the oscillation frequency is shown in Fig. 6. The distance is approximately inversely proportional to the frequency of oscillation.

The above experimental results use circuits with a π-type tuning section. Experimental circuits gave also been constructed with the T-type tuning section. Such a circuit is shown in Fig. 7. A series of chemically etched capacitive elements is provided for convenient tuning for optimum power output. The elements are normally decoupled completely from the circuit; they can be connected independently to the high-impedance section of the main line. The high-efficiency mode of oscillation is started only when a capacitive element is added to the circuit at appropriate position.

The measured power output and efficiency versus bias current from the circuit shown in Fig. 7 are plotted in Fig. 8. The power output is 95 watts at 1.06 GHz, and the efficiency is 26.5 percent. Oscilloscope traces of the detected microwave envelope and the applied pulsed current and voltage waveforms across the diode are as shown in Fig. 9. The pulse width is 1 μs, and the repetition rate is 300 Hz. The average voltage across the diode drops from 150 to 100 volts during oscillation. This voltage drop is a characteristic of the high-efficiency mode operation. The current change is not noticeable since a high-impedance pulser was used in the experiment. The frequency spectrum of power output is shown in Fig. 10; the corresponding pulse width is 0.3 μs. The microstrip circuit has also been made with metal on alumina substrates, as shown in Fig. 11. The size is much reduced because of the high dielectric constant of the material.

One of the advantages of having the microstrip configuration is that fields along the circuit can be easily probed without seriously perturbing the oscillation. A small electric-field probe has been used to probe the harmonic content within the circuit. The probed signal was viewed on a panoramic receiver. It has been found that strong second as well as third harmonics are present in the circuit in addition to the fundamental frequency. Fourth and fifth harmonics are also present but with much less amplitude.

From the field distribution along the microstrip line, magnitudes of RF voltage at the diode were estimated. The magnitude of the second-harmonic voltage is comparable to that of the fundamental frequency, and the third harmonic, about one third of the fundamental. The probed signal has also been viewed on a sampling scope. A sampling oscilloscope waveform corresponding to a probe location picking up a strong second harmonic component is shown in Fig. 12.

Diodes have been successfully operated in series with combined power output on the microstrip circuit. Preliminary results of two series diodes produced 140 watts at 0.98 GHz with an efficiency of 23 percent. Diodes were first mounted in small pill boxes, and then connected in series between the main microstrip line and the bypass capacitor. A combined power output of 500 watts from four series-connected diodes has been achieved in coaxial circuitry [7]. Increased numbers of diodes on microstrip for higher power output are being investigated.[1]

CONCLUSIONS

In conclusion, a simple microstrip oscillator circuit has been designed and operated satisfactorily with high-efficiency high-power avalanche diodes. An essential part of the circuit is a low-pass filter tuning section which enables the circuit to support high-order harmonics and prevents them from getting to the load. Probe measurements of the electric field show strong second as well as third harmonics inside the circuit. The power output obtained from a single diode chip is about 100 watts at 1 GHz with efficiencies of 25 to 30 percent. Mechanical tuning capability of a few hundred megahertz and a combined power output from series-connected diodes have been demonstrated using the circuits.

ACKNOWLEDGMENT

The author wishes to thank J. J. Risko for his able technical assistance, and Dr. K. K. N. Chang and Dr. L. S. Nergaard for critical reading of the manuscript.

REFERENCES

[1] H. J. Prager, K. K. N. Chang, and S. Weisbrod, "High-power, high-efficiency silicon avalanche diodes at ultra high frequencies," *Proc. IEEE* (Letters), vol. 55, pp. 586–587, April 1967.

[2] S. G. Liu, "High-power punch-through avalanche diode microwave oscillators," *IEEE J. Solid-State Circuits*, vol. SC-3, pp. 213–217, September 1968.

[3] R. L. Johnston, D. L. Scharfetter, and D. J. Bartelink, "High-efficiency oscillations in germanium avalanche diodes below the transit-time frequency," *Proc. IEEE* (Letters), vol. 56, pp. 1611–1613, September 1968.

[4] C. P. Snapp, L. A. Stark, and B. Hoefflinger, "Performance and theory of avalanche resonance pumped IMPATT oscillators," presented at the International Electron Devices Meeting, Washington, D. C., October 1968.

[5] D. E. Iglesias and W. J. Evans, "High-efficiency CW IMPATT operation," *Proc. IEEE* (Letters), vol. 56, p. 1610, September 1968.

[6] P. A. Levine and S. G. Liu, "Tunable *L*-band high-power avalanche-diode oscillator," *1969 Internatl. Solid-State Circuit Conf., Digest of Technical Papers*, vol. 12, p. 84.

[7] S. G. Liu, "Stacked high-power avalanche-diode oscillators," *Proc. IEEE* (Letters), vol. 57, pp. 707–708, April 1969.

[1] *Note added in proof:* Three series-connected diodes made recently yielded 440 watts at 1.01 GHz with an efficiency of 23.4 percent.

Double-Drift-Region Ion-Implanted Millimeter-Wave IMPATT Diodes

THOMAS E. SEIDEL, MEMBER, IEEE, RONALD E. DAVIS, MEMBER, IEEE, AND DAVID E. IGLESIAS, MEMBER, IEEE

Abstract—A detailed experimental comparison between double-drift-region (DDR) and single-drift-region (SDR) millimeter-wave avalanche diodes is presented. For 50-GHz CW operation, DDR diodes have given a maximum of 1-W output power compared to 0.53 W for the SDR diodes, while maximum efficiencies of 14.2 percent for the DDR and 10.3 percent for the SDR diodes have been obtained. These results are in agreement with the theory of Scharfetter *et al.* [1] for DDR IMPATT diodes. Both the DDR and SDR diode measurements were made on room temperature, metal heat sinks. The DDR diodes were shown to operate at significantly lower junction temperatures for the same value of output power, indicating a potential reliability advantage.

Ion implantation was used to make the p drift region of the p^+p-n-n^+ 50-GHz DDR devices. Otherwise the fabrication (which includes diffusion and epitaxial technologies) and the microwave measurement methods were identical for both types of diodes. Capacitance measurements were compared with calculations to determine the desired doping concentrations for frequencies from 43 to 110 GHz. Experimental results for the higher frequency millimeter-wave region have been obtained on DDR structures with both p and n drift regions implanted. At 92 GHz an output power of 0.18 W and an efficiency of 7.4 percent have been obtained.

INTRODUCTION

THE double-drift-region (DDR) IMPATT diodes are a new class of IMPATT diodes in which nearly equal microwave power is obtained from the drift of holes and the drift of electrons [1]. Preliminary experimental results have been reported for DDR diodes [2]. In this paper these DDR diodes are compared in detail to conventional, single-drift-region (SDR) diodes where microwave power is obtained essentially from the drift of one type of carrier [3]. The millimeter-wave output power and efficiency are significantly higher for the DDR diodes than that obtained with the SDR diodes. At 50 GHz an efficiency of 14.2 percent and a power output of 1 W CW have been obtained with DDR diodes, which is the best efficiency and power–frequency product reported for any silicon IMPATT diodes.

The details of the theoretical advantages for the DDR diodes have been published [1]. In this paper we briefly review those theoretical advantages and then present our fabrication techniques, capacitance characterizations, and give comparative results for the microwave performance of DDR and SDR diodes at 50 GHz, together with new results for operation near 100 GHz.

THEORETICAL CONSIDERATIONS

Fig. 1 shows a schematic of the DDR and SDR structures with the doping profiles and the electric field distribution appropriate to the avalanche voltage condition. The single-drift diode has one drift region for electrons in the n part of the p^+n-n^+ structure. The double-drift diode has two drift regions, one for holes in the p part of the p^+p-n-n^+ structure and one for electrons in the n region. The added drift space increases the space-charge width and results in a smaller capacitance per unit area; hence the DDR diode may be made with a larger diameter to give the same capacitance as the SDR diode. This results in more output power at the same value of capacitive reactance, a larger working volume, and a lower thermal resistance for the DDR device.

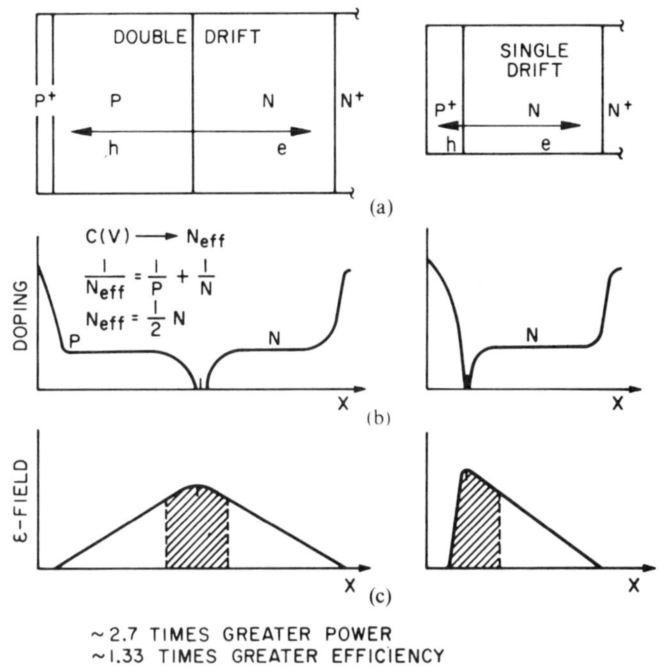

Fig. 1. Schematic comparison of double-drift-region (DDR) with single-drift-region (SDR) diodes. (a) Geometry. (b) Doping concentration. (c) Electric field distributions under avalanche bias conditions.

The schematic of the doping profile shows a nearly symmetric DDR device, where doping concentration and space-charge widths of n and p regions are nearly equal. These dopings and widths are, in turn, approximately the same as the doping and space-charge width of the SDR device. The relationship between the effective doping and the variation of capacitance with voltage is given by [4], [5]

$$N_{\text{eff}} = \frac{1}{q\varepsilon A^2} \frac{C^3}{(dC/dV)} \tag{1}$$

where

$$\frac{1}{N_{\text{eff}}} = \frac{1}{n} + \frac{1}{p}$$

q is the electron charge, ε is the dielectric constant, A is the diode area, and C and V are the capacitance and reverse voltages, respectively. The free carrier densities are n and p in the n and p drift layers. Hence a symmetric DDR device, where $n \approx p$, should have $N_{\text{eff}} \approx (1/2)$ n. It will be shown that this relationship is obeyed in experimental symmetric DDR diodes.

The electric field distributions are shown in Fig. 1(c). The "avalanche voltage" is $\sim 1/2$ the total applied voltage [6] for the

Manuscript received December 21, 1970.
The authors are with Bell Telephone Laboratories, Inc., Murray Hill, N. J. 07974.

SDR silicon diode, as represented by the hatched area. The DDR device has an added drift region so the avalanche region is shared by both p and n "drift" regions. In the DDR diode, only ~ 1/3 the total voltage is used for the avalanche zone, which results in an improved efficiency.

Using simple arguments [1], the output power is expected to be up to a factor of 4 higher for a DDR device than power output of an ideal-abrupt SDR device. The efficiency is expected to be a factor of 1.33 higher for a DDR device since the efficiency is proportional to V_D/V_D+V_A, where V_D and V_A are the voltages used for drift and avalanche zone, respectively [6]. The calculations of Scharfetter *et al.* [1] were made using small-signal analysis without thermal considerations. They imply that improved efficiency and power extends to large-signal CW operation, which is borne out by experimental results presented here for millimeter-wave operation.

The factor of 4 improvement in output power was obtained by the simple consideration that both voltage and current could be doubled. However, the real structure of the diodes results in an expected increase in output power closer to 2.7 than 4. This is due to a number of factors. First, the diffused SDR diodes have a small region where holes drift in the electric field. (See Fig. 1(a) for the SDR diode.) In addition, the electric field required for avalanche in the DDR diode is smaller than that required for avalanche in the SDR device. (See Fig. 1(c) for the DDR diode.) These factors lead to an operating voltage only ~1.3 times larger for the 50-GHz DDR diode than for the SDR diode. The current can be increased only by the increase in area and the thermal design limit of the diode. To obtain the same capacitance in a practical case for 50-GHz DDR and SDR diodes, the area is increased only by a factor of 1.7. The improved efficiency must be considered at the maximum burnout power. Using a value of 1.25 for an improved efficiency, the overall improvement in output power without considering the thermal design limits is $1.3 \times 1.7 \times 1.25 \cong 2.7$.

A "symmetric" DDR diode has been assumed above. The properties of holes and electrons are not the same in silicon; the avalanche coefficient for electrons is greater than for holes [7]. This results in a slight shift in the peak of the avalanche generated carriers towards the n^+ side of the diode [8]. The saturated drift velocities are the same, within experimental error [9], [10]. Hence, the electrical symmetry of the diode may be improved by making the hole drift region slightly shorter than the electron drift region. This effect is most important at higher frequencies (100 GHz).

Fabrication Techniques

The desired doping concentration for a symmetric 50-GHz DDR IMPATT is plotted as a bold line in Fig. 2. In principle, the p^+p-n-n^+ structure can be made using epitaxial techniques alone. However, the control of the doping concentrations and their distributions can be realized quickly [2], accurately, and with reproducibility using ion implantation techniques. The 50-GHz structure shown in Fig. 2 has been made using a combination of technologies: epitaxy to form the n and n^+ regions, ion implantation to form the p layer, and diffusion to form the p^+ layer.

First an n^+ substrate of 0.001-Ω · cm resistivity with an epitaxial layer of 1.2-μm thickness and doping concentration of 8×10^{16} donor/cm^3 was obtained. Then ^{11}B ions were implanted at 80 and 200 keV to compensate the donors and form the p region. The hatched regions in Fig. 2 are calculated distributions which were made using Gibbons' values [11]. The total dose ϕ was adjusted so that the concentrations of acceptor densities at the peaks of the implanted distributions were twice the donor concentration. In practice, this was done by implanting a variety of different doses between 6×10^{12} and 1.3×10^{13} boron ions/cm^2. After annealing at 850°C for 30 min [12], the sheet resistance of the implanted p regions was

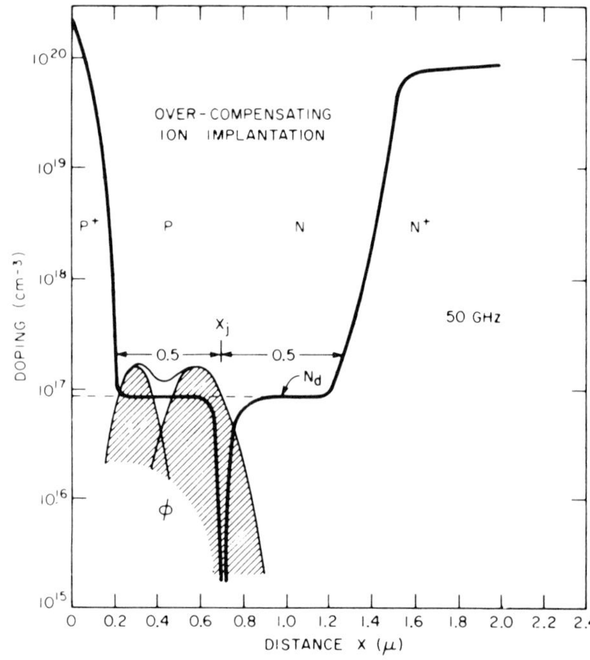

Fig. 2. Doping concentrations for a symmetric DDR 50-GHz IMPATT. The p region is made by implantation of 80- and 200-keV boron ions that overcompensate the donor concentration N_d in the epitaxial doping.

measured. From these sheet resistance measurements it was possible to select the dose value most suitable for use as a symmetric DDR diode [13].

This method required the plotting of sheet resistance against dose and fitting the resultant data to an equation of the form $\kappa/\phi - \phi_i$, where κ is a constant and ϕ_i is approximately the dose that is required to make the implanted region appear intrinsic. Data for sheet resistance are plotted in Fig. 3 for samples of two different epitaxial doping concentrations, 8×10^{16}/cm^3 and $\sim 4 \times 10^{16}$/cm^3. Higher doses were required to compensate and dope the p region to the same sheet resistance for the more heavily doped epitaxial layer, as expected. A dose value of $\phi = 1.15 \times 10^{13}$/cm$^2 \approx 2\phi_i$ for this case was used to produce the symmetric DDR 50-GHz IMPATT diodes whose microwave properties are given below. The dose was proportioned as 4.5×10^{12}/cm^2 to the 80-keV and 7×10^{12}/cm^2 to the 200-keV implantation.[1]

After implantation, the p^+ contact was diffused from a BBr_3 furnace system [16] at 875°C for 30 min, giving a diffusion depth of 0.2 μm. In principle, the p^+ contact could be implanted; however, it was more convenient to use diffusion. Before the p^+ diffusion was done, doping in the p region was assessed with the resistance plot before further processing steps were started.

A diffusion temperature of 930°C for 6 h was used to make the p^+ contacts for the SDR diodes, but the substrate and epitaxial layer used for the SDR diodes was the same as that used for the DDR diodes.

Using the methods described above, it is possible to make diodes for use at frequencies from 35 to above 110 GHz using the range of ion acceleration energies available between 300 and 30 keV. Alternatively, by utilizing the higher portion of this voltage range, one can implant both the drift spaces for operation above ~80 GHz by

[1] The resistance plot has also been used to choose implantation doses for *asymmetric* $p^+\pi$-n-n^+ structures. These devices have been used as "combination-mode" IMPATT–TRAPATT devices because they trigger into the CW TRAPATT oscillations very easily [14]. These diodes essentially have the geometry for DDR 50-GHz IMPATTs but give CW TRAPATT efficiencies of 20 percent at ~5 GHz [15]. CW TRAPATT operation at 5 GHz has not been obtained with conventional SDR diodes.

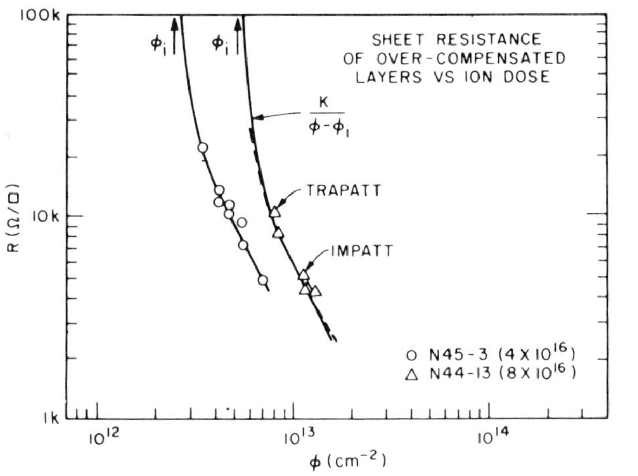

Fig. 3. Sheet resistance of overcompensated layers plotted against total boron dose ϕ for two epitaxial samples. N45-3 has a donor concentration of ~ $4 \times 10^{16}/\text{cm}^3$ and N44-13 about $8 \times 10^{16}/\text{cm}^3$. The data can be fitted to an equation of the form $K/(\phi - \phi_i)$, where ϕ_i is the dose that makes the epitaxial material approximately intrinsic. A dose of $2\phi_i$ is used for construction of a symmetric IMPATT.

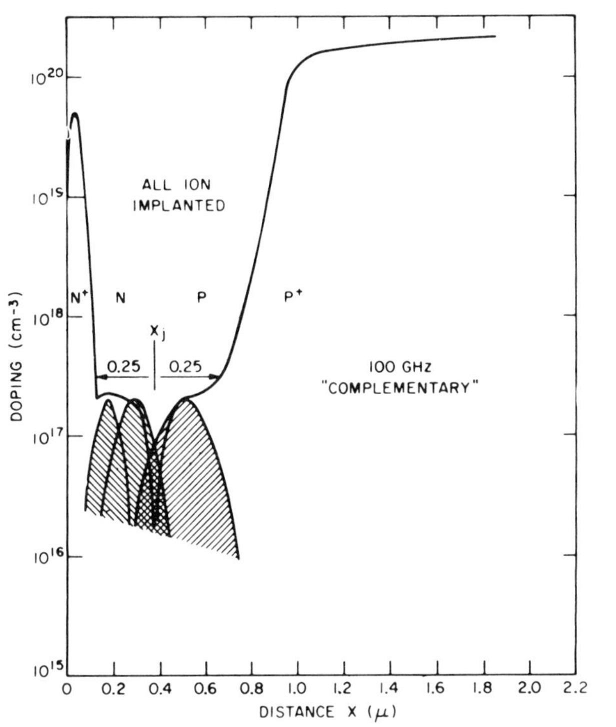

Fig. 4. Doping concentrations for a symmetric DDR 100-GHz IMPATT. The p region is made by implantation of 250-keV boron, multiple energy implantation of phosphorous at 300 and 150 keV for the n region, and a single implantation at 50-keV As for the n^+ region.

making n^+n-p-p^+ structures, which are complementary to the type shown in Fig. 2 for 50-GHz operation. The complementary structure is shown in Fig. 4 and makes use of the larger range of boron, as opposed to phosphorus, to form the deeper lying drift region. For random implants, the range of 300-keV boron is about 0.7 μm, which will allow this complementary structure to be made for frequencies above 80 GHz. The complementary diode is desirable because the control of doping is easier if implantations are done into both drift spaces, since no epitaxial doping and compensation are required. Diodes for 100-GHz operation were fabricated with the calculated doping concentrations shown in Fig. 4. The ions used, their energies, and doses are: As (50 keV, $5 \times 10^{14}/\text{cm}^2$) for the n^+ region, P(150 keV, $2.4 \times 10^{12}/\text{cm}^2$ and 240 keV, $3.5 \times 10^{12}/\text{cm}^2$) for

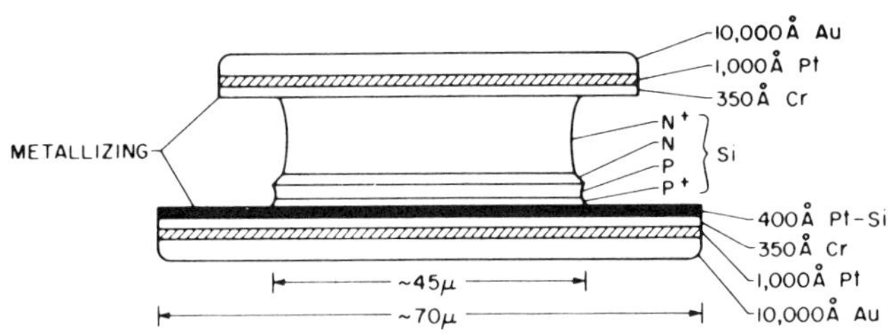

Fig. 5. Cross section of metallized 50-GHz DDR diode. PtSi is formed on the p^+ side; Cr-Au contacts are made to both sides with a Pt layer in between.

the n region, and B (200 keV, $4.5 \times 10^{12}/\text{cm}^2$) for the p region. An annealing temperature of 850°C was used for 30 min.

These processes define the internal structure of the diode. The diode slices were then thinned to 10–15 μm by removing the substrate by precision "Syton" polishing, as reported by Edwards *et al.* [17], using techniques like Mash *et al.* [18]. Slices were then metallized on both sides, given a photoresist pattern, and diodes separated by etching. A number of metallization methods have been used successfully; Fig. 5 shows a cross section of a completed diode featuring contact metallurgy similar to that used in beam-lead technology [19]. The Pt layer is used to prevent diffusion of Cr into Au at elevated operating temperatures. The PtSi is used on the p^+ side to give good ohmic contact (this is important for dc characterization and for microwave evaluations). The diodes used for the microwave comparisons of DDR and SDR operation were both prepared with a simpler metallization process that did not include the Pt layer between Cr and Au. These diodes were thermocompression bonded to 1/8-in-diameter gold-plated silver heat sinks with the junction side down. The diodes were then given a final "tailoring" etch to obtain the desired value of capacitance, e.g., 0.2 pF for use at 50 GHz. Larger area diodes were taken from the same wafer to evaluate the doping concentrations.

Evaluation of Doping Concentrations in DDR Diodes

Capacitance measurements were made on DDR diodes, SDR diodes, and on the epitaxial material before implantation. The effective doping concentration of 50- and 100-GHz DDR devices was calculated using (1) and differential capacitance measurements. The experimental points are plotted as effective density (doping) against the space-charge width ("distance") in Fig. 6. These solid curves drawn through the data have been calculated [5] using the distribution of net acceptors p and donors n, as shown by the dashed curves. The donor densities were measured before the implantation of the p region. The distance associated with the p and n concentration curves is the distance measured from just inside the p^+ contact.

The main point to notice is that the value of N_{eff} is ~1/2 the donor doping, which is characteristic of a symmetric DDR device. The doping concentration in the p region is constant within ±10 percent and its mean value is within 12 percent of the donor doping. The decrease in effective doping at small space-charge widths of ~0.2 μ is expected and is due to the straggling of boron ions in the first few tenths of a micron beyond the 200-keV boron range. The decrease in the effective doping at larger space-charge widths (0.8–0.9 μ) is unexpected and is reflected in the "dip" in the acceptor doping near the p^+-p interface. The dip is not very detrimental to DDR microwave operation, as indicated by the performance, but the influence of small variations in the doping such as this have not been completely investigated.

A similar capacitance analysis has been carried out for 100-GHz diodes. The effective doping is again one-half the actual doping on the p and n sides of the symmetric junction. A typical result is shown in the upper part of Fig. 6.

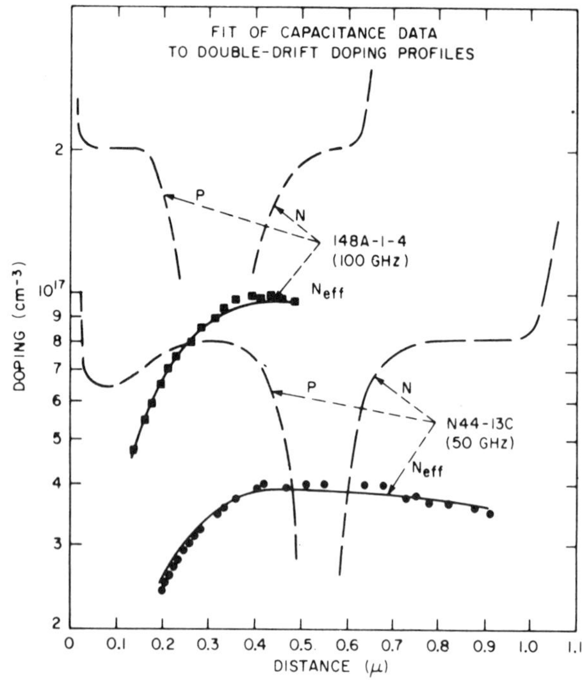

Fig. 6. Plot of doping against distance. Experimental points are "effective doping" calculated from differential capacitance and the distance is space-charge width obtained from capacitance. The dashed curves are the distribution of net acceptors and donors in the DDR devices. These curves are used to calculate the fit to the experimental points (solid curves).

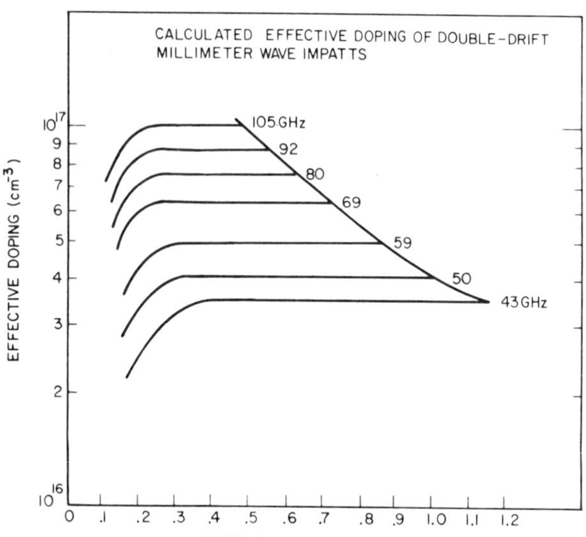

Fig. 7. Plot of effective doping versus space-charge width for DDR diodes to be used at various millimeter-wave frequencies. These curves can be used as design curves to check the doping characteristics of processed diodes.

In Fig. 7 the effective doping characteristics for various millimeter-wave frequencies are plotted. These results were calculated using the experimental results at 50 GHz and scaling the drift width linearly to other frequencies. The doping increases more than linearly with frequency, as is expected from the dependence of critical field for avalanche upon doping concentration and hence space-charge width. (The electric field at breakdown in an ideal-abrupt single-drift-region diode is $\sim 6.8 \times 10^5$ V/cm for a 100-GHz diode and is $\sim 6.0 \times 10^5$ V/cm for a 50-GHz diode.) In Table I, values of energies, doses, epitaxial layer thicknesses, and doping are given for the fabrication of p^+p-n-n^+ diodes using the "overcompensation" approach. Values for energies, doses, and epitaxial thicknesses are also given for fabrication of n^+n-p-p^+ diodes using the "all-ion implantation" method.

Uniformity of Breakdown Voltage

In order to gain some insight into the effect of ion implantation on the spread in diode parameters, the breakdown voltage V_B (at reverse current of 1 mA) was measured on 35 SDR diodes and 56 DDR diodes. Each group of diodes was fabricated from a different slice, but both slices were from the same lot of epitaxial silicon. Only the DDR diodes were implanted to form the p region, but both groups of diodes received a p^+ diffusion (deeper in the case of the SDR diodes). The comparison of the voltage distributions for the two groups is shown in Fig. 8 and the expected increase in V_B for the DDR diodes as a result of the extra p drift space is evident. The histograms also show that the added processing step of ion implantation does not result in any greater spread in V_B than was present in the original epitaxial slice, as seen by the SDR diode distribution. The mean-square deviation of DDR diode breakdown voltage is 0.4 percent, which demonstrates excellent uniformity of the implantation process.

Leakage Currents

It has been shown that diode oscillator efficiency decreases with increasing reverse saturation current [21]. This is due to an advance phase shift of the avalanche current. Since the p region of the DDR diodes was made by ion implantation, the damage produced by the implantation may play a role in limiting the efficiency. Reverse current densities were measured on unpassivated mesa DDR and SDR diodes. When the SDR diodes were prepared by diffusion at 930°C, giving the structure later shown in the insert of Fig. 12, the reverse currents of the SDR diodes were a factor of ~ 2 smaller than ion-implanted DDR diodes at comparable space-charge width. However, SDR diodes prepared at lower diffusion temperatures resulted in larger reverse currents than the DDR diode. The comparison of the reverse currents of DDR to SDR diodes depends upon the processing procedures of the SDR diode, but all currents were $2-8 \times 10^{-8}$ A/cm² at -5 V.

Based upon the microwave performance, we conclude that the annealing temperature of 850°C for 30 min is sufficient to reduce the damage and diode currents due to generation currents in the drift and avalanche regions to an acceptably small value. Not much improvement in efficiency may be expected by using higher annealing temperatures or implanting with channeling techniques since reverse currents in diodes implanted with 10^{15}/cm², 300-keV carbon ions can be substantially recovered after a 750°C anneal [22].

Microwave Measurements

The diode was mounted inside a waveguide cavity [3], [23] between the two broad sides of the guide, as shown in Fig. 9. Contact to the standoff structure of the diode was made by a cap that forms a radial-line coupling to the rectangular waveguide. Dc bias was applied through a 100-Ω resistor from a constant current power supply and through the center conductor of a cylindrical bypass capacitor. Tuning was done with a sliding short at one end of the cavity and an E-H tuner at the other. The same type of cavity was used for both 50- and 100-GHz operation with a WR-15 waveguide used for 50 GHz and a WR-10 for 100 GHz.

The geometry of the contacting cap was an important factor in determining oscillator performance. Testing was accomplished with the arrangement shown in Fig. 10. An isolator was placed immediately after the E-H tuner. Ten percent of the power is sampled with a 10-dB directional coupler to determine the frequency of oscillation and to check for parasitic oscillations. A switch was placed after the directional coupler which allowed connection to either a power meter or a calorimeter. A precision attenuator was used to reduce

TABLE I

DESIGN PARAMETERS FOR MILLIMETER-WAVE DDR IMPATTS

p^+p-n-n^+—Overcompensated Method

f (GHz)	Epi. Thickness (μm)	Epi. Dop. ($\times 10^{16}$/cm^3)	Boron Energies and Doses[a] (V, ϕ) (keV, $\times 10^{12}$/cm^2)			
43	1.36	6.8	(295 3.35)	(170 2.85)	(94 2.25)	(49 1.58)
50	1.20	8.0	(245 3.70)	(143 3.10)	(80 2.50)	(41 1.63)
59	1.05	10.0	(200 4.40)	(120 3.65)	(67 2.80)	(34 1.82)
69	0.93	12.5	(175 5.25)	(102 4.40)	(57 3.20)	(29 2.10)
80	0.83	15.0	(152 6.00)	(91 4.95)	(50 3.60)	
92	0.74	17.0	(140 6.60)	(82 5.30)	(44 3.70)	
105	0.67	20.0	(130 7.60)	(76 6.00)	(39 4.00)	

n^+n-p-p^+—All Ion Implantation Method

f (GHz)	Epi. Thickness (μm)	As, P, and B Energies and Doses[a] (V, ϕ) (keV, $\times 10^{12}$/cm^2)			
		As (n^+)	P(n)		B(p)
80	0.76	(50 4.5×10^2)	(315 4.5)	(210 3.15)	(265 4.8)
92	0.67	(50 4.5×10^2)	(275 4.0)	(175 2.75)	(232 4.65)
105	0.60		(240 3.5)	(150 2.4)	(200 4.50)

[a] These values are based upon most recent reliable measurements of range and straggling for ^{11}B in randomly oriented silicon. Phosphorus ranges and stragglings are from Gibbons [6] and recent measurements by Moline [20] are in substantial (±10 percent) agreement with Gibbons' calculations.

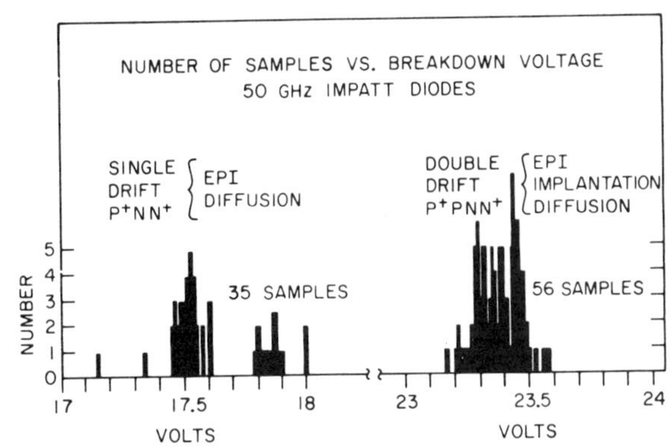

Fig. 8. Breakdown voltage distributions of DDR and SDR diodes.

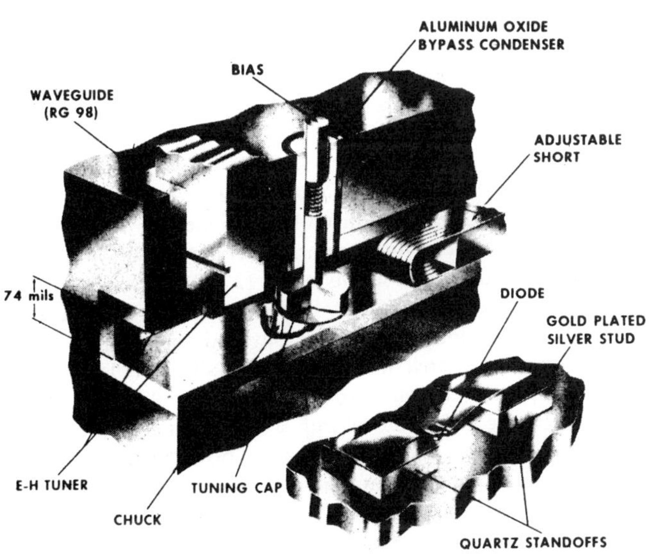

Fig. 9. Cutaway drawing of waveguide cavity of 40–60 GHz IMPATT diode oscillator. The insert shows the mounting of the diode with quartz standoffs.

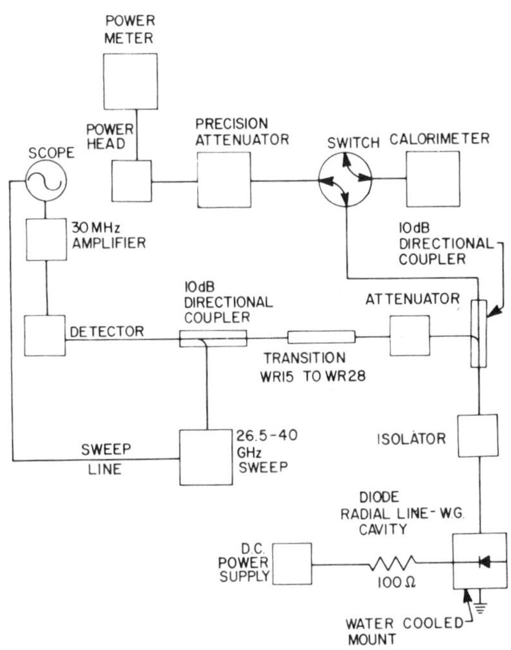

Fig. 10. Block diagram of testing components. Calibration for the power meter is based upon use of the calorimeter.

the signal to a power level within the range of the power meter. A calibration curve for the power meter was developed using the calorimeter as a standard. Power was measured at waveguide impedance, which is about 400 Ω for the $TE_{1,0}$ mode. The calorimeter could not be used as a power indicator while tuning the oscillator because of its slow response.

To observe the frequency of oscillation, a 26- to 40-GHz sweeper was used as a local oscillator. The sample signal from the directional coupler was fed to a second directional coupler for mixing with the sweep oscillator. A second harmonic of 52 to 80 GHz or a third harmonic of 79.5 to 120 GHz (for measurement of 100-GHz frequencies) was generated at the mixing diode. The difference signal was fed to an amplifier (30 MHz), detected, and displayed on the oscilloscope.

MICROWAVE RESULTS

50 GHz

The efficiency and output powers of DDR and SDR diodes were measured for the frequency range of 50 to 55 GHz and were compared. The efficiency was plotted against current density in Fig. 11

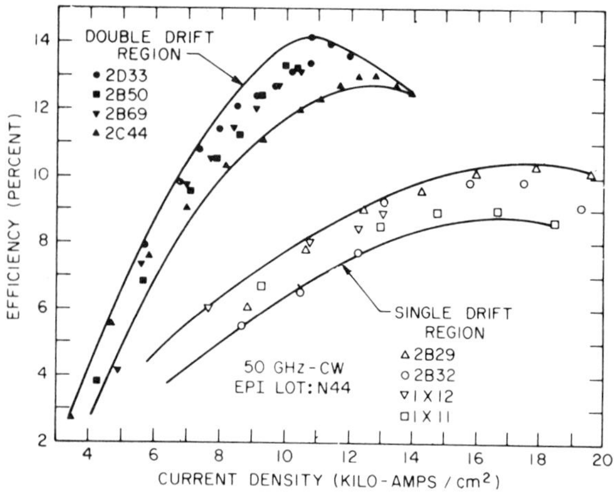

Fig. 11. Efficiency versus input current density for four DDR samples and four SDR samples.

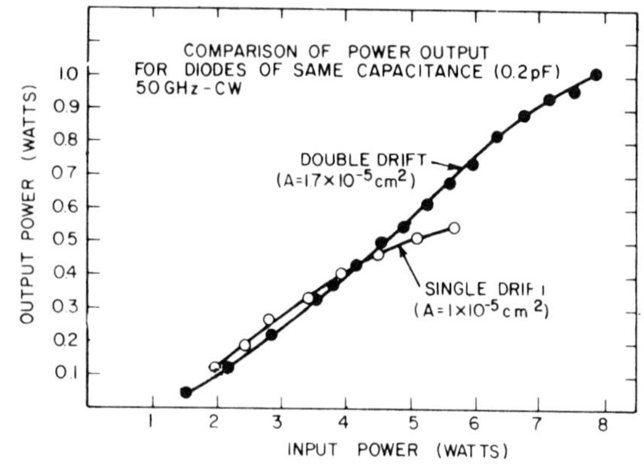

Fig. 13. Power output versus input power for DDR and SDR diodes.

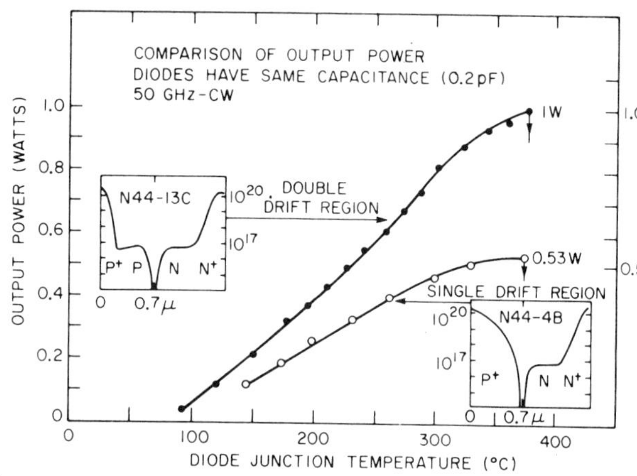

Fig. 12. Power output versus diode junction temperature for DDR and SDR samples with burnout powers of 1 and 0.53 W, respectively.

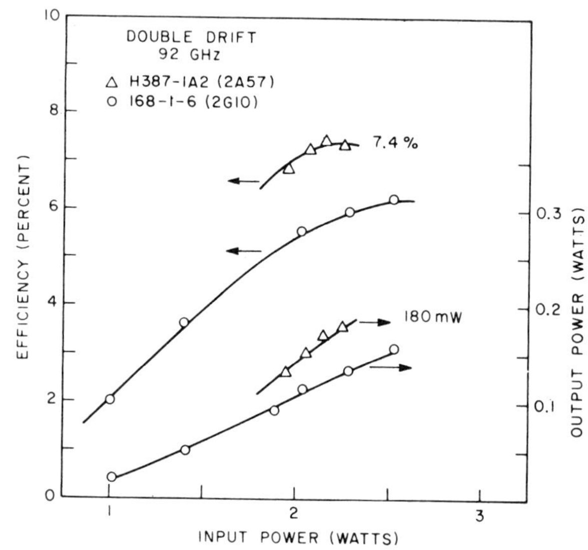

Fig. 14. Output power and efficiency versus input power for DDR diodes at 92 GHz.

for four different samples of the DDR design and four different samples of the SDR design. The four DDR samples were essentially the same, but had slightly different areas and mounting configuration, which could easily account for the spread in values. The double-drift samples have an area of $\sim 1.7 \times 10^{-5}\text{cm}^{+2}$ and the area of the single-drift samples is $\sim 1 \times 10^{-5}\text{cm}^{+2}$.

The maximum efficiency of the DDR samples was 14.2 percent, while the SDR samples gave a maximum efficiency of 10.3 percent. The higher efficiency is characteristic of double-drift operation and was reached at current densities of $\sim 10^4$ A/cm^2. The maximum current density that could be applied to the DDR devices (14 kA/cm^2) was smaller than the SDR devices, (20 kA/cm^2) because the higher voltage of the DDR device resulted in the same operating temperature at lower current densities. The maximum output powers of these four DDR samples were in the range 0.75 to 1.0 W, while the SDR samples ranged between 0.4 and 0.53 W. The operating temperatures were estimated from the product of the input power and the thermal resistance. These operating temperatures are shown in Fig. 12 for the two samples with the highest output powers obtained from DDR and SDR diodes.

The output power is plotted against average diode junction temperature. The thermal resistances were measured as $\sim$40°C/W and 62°C/W for the DDR and SDR diodes, respectively. The inserts in the figure show the doping profiles for the two devices as determined by capacitance techniques. The diodes were made with about the same capacitance at their breakdown voltages and the microwave measurements were made with the same cavity and circuit. The output powers at burnout are 1 W and 0.53 W, while the diode burnout temperature is $\sim$375°C, suggesting the possible formation of an Au–Si eutectic. This representation of power output data shows that the DDR diode operates at a lower temperature for the same output power. The lower temperature implies increased operating life and reliability from DDR diodes.

The plot of power output against power input is shown in Fig. 13 for the same two samples shown in Fig. 12. Higher efficiency is only obtained at higher input power. Efficiency is a function of the bias level (i.e., a function of current density and applied field), so the double-drift, larger area device required a sufficiently large current for the improved efficiency to become operative. We could have constructed the DDR diode with a smaller area and then the efficiency would be larger at a lower input power. This hypothetical device would burn out at lower input power and would not demonstrate the output power capability of the DDR device, but would demonstrate higher efficiency at a lower input power.

100 GHz

Diodes for use near 100 GHz were constructed using both the overcompensation technique and the all-ion implantation technique. Results obtained from the latter technique are shown in Fig. 14, where output power and efficiency are plotted against input power. The maximum efficiency is 7.4 percent and output power is 180 mW. The calorimeter at $\sim$100 GHz is not perfectly matched and the actual output power is believed to be conservatively estimated. These diodes have not yet burned out but are expected to burn out at about 3-W input power since the thermal resistance is ~125°C/W. These 92-GHz diodes give an output of 100 mW at operating temperatures of ~250°C. Somewhat better efficiencies are expected due to improved devices and circuits in the 100-GHz range, but a factor of ~2 improvement in output power and efficiency has already been obtained over SDR diodes. Parasitic resistive losses in the diode [24] and cavity are likely to be the most important sources of limited efficiency near 100 GHz.

Concluding Discussion

The principle of operation of the DDR diode has been discussed with the theoretical techniques for evaluating the doping profiles. The fabrication techniques are given in detail for 50- and 92-GHz diodes. The techniques use combinations of epitaxial, diffusion and ion-implantation technologies. The doping profiles of the DDR diodes have been evaluated using capacitance techniques and found to be in substantial agreement with the expected doping profiles. Plots of the effective doping as obtained from differential capacitance data are given for future design use for 43- to 110-GHz diodes.

Microwave results are given for both DDR and SDR 50-GHz devices that were made and measured in a carefully controlled fashion such that meaningful comparisons could be drawn. Both higher efficiencies and higher output powers have been obtained from DDR devices. The burnout power of 1 W at 50 GHz represents the highest figure of merit obtained for any CW, room temperature IMPATT microwave source, that is, the pf^2 (power–frequency squared) or the pf product is the highest. The pf^2 is more meaningful at the highest frequencies where the limit on output is electrical and not thermal [25]. The 14.2-percent efficiency is the highest CW efficiency of any silicon IMPATT at any frequency. At junction temperatures of ~250°C, output powers of 3/4 with ~11-percent efficiency at 50 GHz have been demonstrated. The 0.53-W and 10.3-percent efficiency results for the SDR diode are the best performance for single-drift 50-GHz IMPATTS.

For a given output power, the DDR design has a lower operating temperature and a larger diode size than SDR diodes, and these features are of considerable practical importance as they are expected to result in longer operating life and greater yield in fabrication.

Acknowledgment

The authors are grateful to R. Edwards for his suggestions and guidance and D. L. Scharfetter for suggesting the problem.

References

[1] D. L. Scharfetter, W. J. Evans, and R. L. Johnston, "Double-drift-region (p^+pnn^+) avalanche diode oscillators," *Proc. IEEE* (Lett.), vol. 58, July 1970, pp. 1131–1133.
[2] T. E. Seidel and D. L. Scharfetter, "High-power millimeter wave IMPATT oscillators with both hole and electron drift spaces made by ion implantation," *Proc. IEEE* (Lett.), vol. 58, July 1970, pp. 1135–1136.
[3] T. Misawa, "CW millimeter-wave IMPATT diodes with nearly abrupt junctions," *Proc. IEEE* (Lett.), vol. 56, Feb. 1968, pp. 234–235.
[4] J. L. Moll, *Physics of Semiconductors*. New York: McGraw-Hill, 1964, p. 125.
[5] D. L. Scharfetter, unpublished computer program used to calculate the effective doping of the DDR structure.
[6] D. L. Scharfetter and H. K. Gummel, "Large-signal analysis of a silicon Read diode oscillator," *IEEE Trans. Electron Devices*, vol. ED-16, Jan. 1969, pp. 64–77.
[7] C. A. Lee, R. A. Logan, R. L. Batdorf, J. J. Kleimack, and W. Wiegmann, "Ionization rates of holes and electrons in silicon," *Phys. Rev.*, vol. 134, 1964, p. A761.
[8] S. M. Sze, *Physics of Semiconductor Devices*. New York: Wiley, 1969, p. 205.
[9] C. B. Norris, Jr., and J. F. Gibbons, "Measurement of high-field carrier drift velocities in silicon by a time-of-flight technique," *IEEE Trans. Electron Devices*, vol. ED-14, Jan. 1967, pp. 38–43.
[10] T. E. Seidel and D. L. Scharfetter, "Dependence of hole velocity upon electric field and hole density for p-type silicon," *J. Phys. Chem. Solids*, vol. 28, 1967, p. 2563.
[11] J. F. Gibbons, "Ion implantation in semiconductors—Part I: Range distribution theory and experiments," *Proc. IEEE*, vol. 56, Mar. 1968, pp. 295–319.
[12] T. E. Seidel and A. U. MacRae, "The isothermal annealing of boron implanted silicon," *Radiat. Effects*, vol. 7, 1971, pp. 1–6.
[13] T. E. Seidel, presented at the 1970 Solid-State Device Res. Conf., Seattle, Wash.
[14] W. J. Evans, T. E. Seidel, and D. L. Scharfetter, "A novel TRAPATT oscillator design," *Proc. IEEE* (Lett.), vol. 58, Aug. 1970, pp. 1294–1295.
[15] W. J. Evans, "Computer experiments on TRAPATT diodes," *IEEE Trans. Microwave Theory Tech.*, vol. MTT-18, Nov. 1970, pp. 862–871.
[16] R. Schmidt, "High frequency transistor fabrication," presented at the 1966 IEEE Internat. Electron Devices Meeting, Washington, D. C.
[17] R. Edwards, D. F. Ciccollella, T. Misawa, D. E. Iglesias, and V. Decker, "Millimeter-wave silicon IMPATT diodes," presented at the IEEE Internat. Electron Devices Meeting, Washington, D. C., Oct. 1969, paper 2.3.
[18] D. H. Mash, G. D. Henshall, and B. A. Eates, "Thin silicon device technology," *J. Phys. D., Appl. Phys.*, vol. 3, Aug. 1970, pp. 1199–1204.
[19] M. P. Lepselter, "Beam lead technology," *Bell Syst. Tech. J.*, vol. 45, Feb. 1966, pp. 233–253.
[20] R. A. Moline, "Ion implanted phosphorus in silicon: Profiles using C-V analysis," *J. Appl. Phys.*, 1971.
[21] W. T. Read, "A proposed high-frequency negative resistance diode," *Bell Syst. Tech. J.*, vol. 37, Mar. 1958, pp. 401–446.
[22] K. A. Pickar and J. V. Dalton, "Lifetime effects in ion implanted silicon," *Radiat. Effects*, to be published.
[23] C. B. Swan, in *1967 NEREM Rec.*, pp. 22–23.
[24] B. C. DeLoach, Jr., "Thin skin IMPATTS," *IEEE Trans. Microwave Theory Tech.* (Corresp.), vol. MTT-18, Jan. 1970, pp. 72–74.
[25] D. L. Scharfetter, "Power-impedance frequency limitations of IMPATT oscillators calculated from a scaling approximation," *IEEE Trans. Electron Devices*, vol. ED-18, Aug. 1971, pp. 536–543.

OTHER ARTICLES OF INTEREST

B.S.T.J. BRIEF

A Low-Noise Metal-Semiconductor-Metal (MSM) Microwave Oscillator

By D. J. COLEMAN, JR., and S. M. SZE

(Manuscript received January 22, 1971)

I. INTRODUCTION

Low-noise microwave CW oscillations have been obtained from metal-semiconductor-metal (MSM) structures made from a 10-μm thin slice of silicon sandwiched between two PtSi Schottky barrier contacts. Microwave CW power up to 50 mW has been obtained at 5 GHz with efficiency up to 1.8 percent. The FM noise measure 1 MHz from the carrier is 22.8 dB which is considerably lower than that of a silicon avalanche oscillator. The mechanisms responsible for the microwave oscillation are (*i*) the exponential increase of the local carrier population due to injection of minority carriers at the forward-biased contact and (*ii*) the transit-time delay of injected carriers traversing the depletion region. By optimizing material and device parameters, it is believed that higher efficiency and higher power microwave oscillations can be obtained from the MSM and its related structures with the inherent low-noise characteristics.

II. DEVICE FABRICATION

Single-crystal n-type silicon wafers with 11Ω-cm resistivity (4×10^{14} cm^{-3} doping), ⟨111⟩ oriented, and with a dislocation density less than 100/cm^2 were Syton polished on both sides to a final thickness of 10 ± 2 μm. Platinum of 500 Å thickness was sputtered onto both sides of the wafer and was sintered to form approximately 1000 Å PtSi on both sides. Chromium of 300 Å was deposited on one side; this was followed by a 3000 Å layer of Au evaporation. The same depositions were then made on the other side. A standard photolithographic method was used to define circular patterns of gold dots with areas of 5×10^{-4} cm^2. The devices were separated by etching and were mounted onto V-type microwave packages.

III. DC CHARACTERISTICS

A schematic diagram of an MSM structure is shown in Fig. 1a. The band diagram at thermal equilibrium is shown in Fig. 1b for an n-type semiconductor where ϕ_{n1} and ϕ_{n2} are the barrier heights for the two metal-semiconductor contacts respectively. For the PtSi-Si-PtSi structure mentioned previously, $\phi_{n1} = \phi_{n2} = 0.85$ eV. Figure 1c shows the energy band diagram when a voltage is applied. We have electron current from the reverse-biased contact and hole current from the forward-biased contact.

The measured I-V characteristics at 300°K and 77°K of a representative device is shown in Fig. 2. The rapid increase in terminal current with applied voltage (above 30 volts) is caused by thermionic

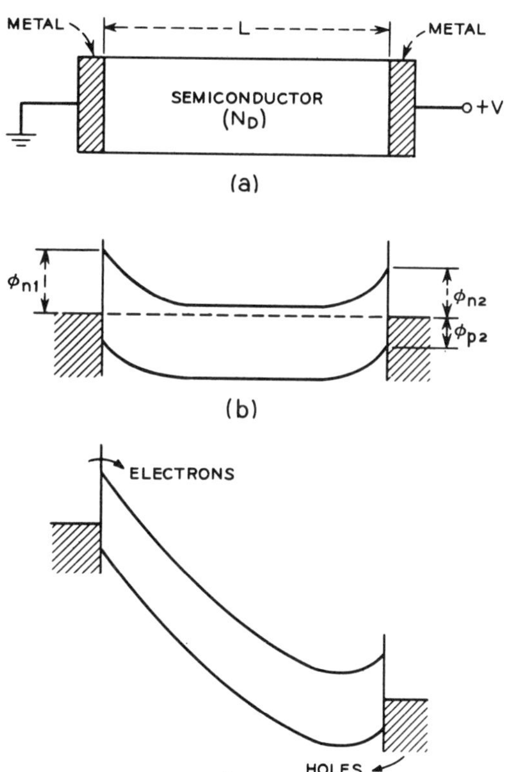

Fig. 1—(a) Schematic diagram of a metal-semiconductor-metal (MSM) structure. (b) Energy band diagram of an MSM structure in thermal equilibrium. (c) Energy band diagram of an MSM structure under biasing condition.

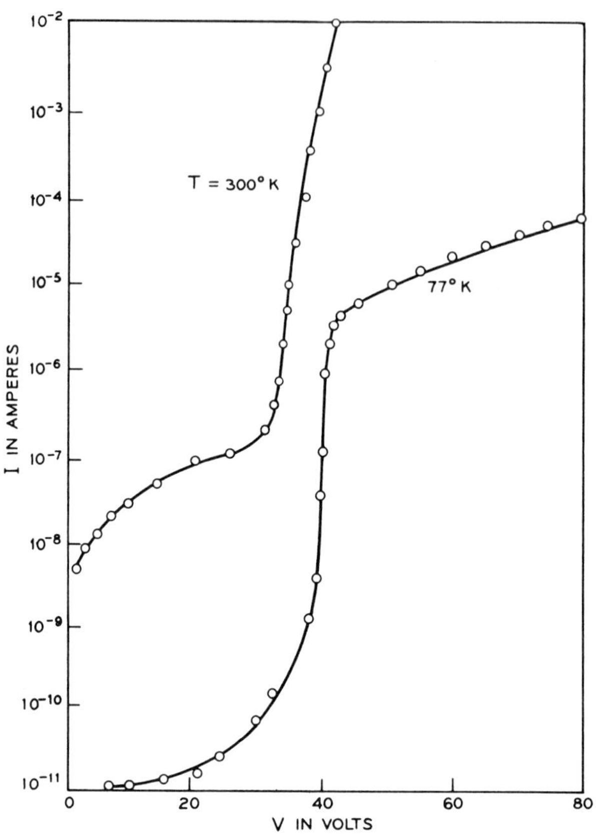

Fig. 2—Measured current vs voltage of a silicon MSM structure (PtSi-Si-PtSi) at two temperatures. The device parameters are $L = 10\ \mu m$, $N_D = 4 \times 10^{14}\ cm^{-3}$, $\phi_{n1} = \phi_{n2} = 0.85$ eV, and with an area of $5 \times 10^{-4}\ cm^2$.

hole injection into the semiconductor as the depletion layer of the reverse-biased contact reaches through the entire device thickness. This critical voltage is approximately given by $qNL^2/2\epsilon_s$, where N is the doping concentration, L the semiconductor thickness, and ϵ_s the dielectric permittivity.[1] The current increase is not due to avalanche multiplication as is apparent from the magnitude of the critical voltage and its negative temperature coefficient. At 77°K, the rapid increase is terminated at a current of about 10^{-5} amps. This saturated current is expected from the thermionic emission theory of hole injection[1] from the forward-biased contact with a hole barrier height (ϕ_{p2}) of about 0.15 eV.

IV. MICROWAVE PERFORMANCE

CW microwave performance of the MSM devices was measured in a coaxial Impatt circuit described by D. E. Iglesias.[2] Microwave power was obtainable over the entire C band of 4–8 GHz. The maximum power observed was 50 mW at 4.9 GHz. The maximum efficiency approached 1.8 percent. Figure 3 shows some of the measured microwave power versus current with frequency of operation indicated on each curve for three typical devices tested. The voltage indicated in parenthesis labeling each curve is the average bias voltage at the diode while oscillating. Because of the symmetry of the structure, it could be operated with either polarity of bias voltage, and similar results were obtained.

The highest-power unit was tested for FM noise when tuned to a frequency of 4.88 GHz. The FM single-sideband noise measure 1 MHz from the carrier frequency was found to be 22.8 dB at 7 mA bias

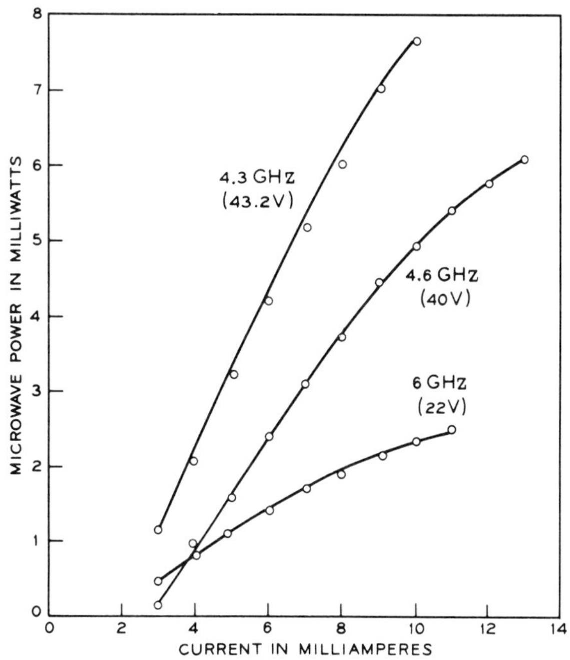

Fig. 3—CW microwave output vs input current for three Si MSM devices. Also indicated are the operating frequency and the average bias voltage while oscillating.

current. This noise measure is considerably lower than that of a silicon Impatt diode and is comparable to that of a GaAs transfer-electron oscillator.

The 6-GHz diode was used to build a stable negative conductance linear amplifier. A gain-bandwidth product of 200 MHz was obtained with 19 dB gain at 5 mA bias. The small signal noise measure was 15 ± 1 dB.

The mechanisms responsible for the microwave oscillations are believed to be (*i*) the rapid increase of carrier injection process caused by the decreasing potential barrier of the forward-biased metal-semiconductor contact and (*ii*) an apparent $3\pi/2$ transit angle of the injected carriers which traverse the semiconductor depletion region. For the 6-GHz diode, the thickness L is smaller. This results in higher frequency (since frequency is inversely proportional to L) and lower critical voltage (which is proportional to L^2). Since the main noise source for thermionic emission processes is the shot noise, one would expect a low noise measure. This is indeed observed experimentally.

If a large barrier height can be obtained for a p-type semiconductor, one can make a complementary MSM structure in the same way as described here. Since the reverse-biased metal-semiconductor contact serves mainly as a blocking contact until the reach-through voltage is obtained, it is conceivable that this contact can be replaced by a p-n junction such as p^+-n-metal structure. By optimizing the material parameters (such as doping profile, barrier heights, and semiconductor thickness) and device geometry and topology,[3] it is believed that higher efficiency and higher power microwave oscillations can be obtained from the MSM and its related structures with the inherent low-noise characteristics.

V. ACKNOWLEDGMENTS

We gratefully acknowledge the expert assistance of A. Loya who fabricated the devices. We wish also to thank A. R. Tretola for platinum deposition, J. Federico for FM noise measurements, J. M. Morris for linear amplifier noise measurements, and R. M. Ryder for many helpful discussions.

REFERENCES

1. Sze, S. M., Coleman, D. J., Jr., and Loya, A., "Current Transport in Metal-Semiconductor-Metal Structures," to be published in Solid State Electronics.
2. Iglesias, D. E., "Circuit for Testing High Efficiency Impatt Diodes," Proc. IEEE, *55* (1967), pp. 2065–67.
3. Coleman, D. J., Jr., and Sze, S. M., "A Negative Resistance Diode Circuit," U.S. Patent pending.